Marine Chemistry & Geoch

Subject Area Volumes from the Second Edition

Climate & Oceans edited by Karl K. Turekian
Elements of Physical Oceanography edited by Steve A. Thorpe
Marine Biology edited by John H. Steele
Marine Chemistry & Geochemistry edited by Karl K. Turekian
Marine Ecological Processes edited by John H. Steele
Marine Geology & Geophysics edited by Karl K. Turekian
Marine Policy & Economics guest edited by Porter Hoagland, Marine Policy Center, Woods Hole Oceanographic Institution, Woods Hole, Massachusetts
Measurement Techniques, Sensors & Platforms edited by Steve A. Thorpe
Ocean Currents edited by Steve A. Thorpe
The Coastal Ocean edited by Karl K. Turekian
The Upper Ocean edited by Steve A. Thorpe

MARINE CHEMISTRY & GEOCHEMISTRY

A DERIVATIVE OF ENCYCLOPEDIA OF OCEAN SCIENCES, 2ND EDITION

Editor

KARL K. TUREKIAN

AMSTERDAM • BOSTON • HEIDELBERG • LONDON • NEW YORK • OXFORD
PARIS • SAN DIEGO • SAN FRANCISCO • SINGAPORE • SYDNEY • TOKYO
Academic Press is an imprint of Elsevier

Academic Press is an imprint of Elsevier
32 Jamestown Road, London NW1 7BY, UK
30 Corporate Drive, Suite 400, Burlington, MA 01803, USA
525 B Street, Suite 1900, San Diego, CA 92101-4495, USA

Material in the work originally appeared in *Encyclopedia of Ocean Sciences* (Elsevier Ltd., 2001) and *Encyclopedia of Ocean Sciences*, 2nd Edition (Elsevier Ltd., 2009), edited by John H. Steele, Steve A. Thorpe and Karl K. Turekian.

British Library Cataloguing in Publication Data
A catalogue record for this book is available from the British Library

Library of Congress Control Number: 2009907116

ISBN: 978-0-08-096483-6

For information on all Elsevier publications
visit our website at www.elsevierdirect.com

Printed and bound by CPI Group (UK) Ltd, Croydon, CR0 4YY

Transferred to Digital Print 2012

CONTENTS

CHEMISTRY OF SEA WATER

TRACERS IN THE SEA

NATURAL RADIOACTIVE SPECIES

[†]Deceased.

POLLUTION OF THE OCEANS

MARINE DEPOSITS

COASTAL OCEAN

CYCLES OF THE NUCLIDES

AIR SEA EXCHANGES

APPENDICES

INDEX

INTRODUCTION: MARINE CHEMISTRY AND GEOCHEMISTRY

The field of marine chemistry and geochemistry has developed dramatically since the end of World War II. Mainly this spurt of refinement was the consequence of the development of novel techniques of measurement of both radioactive and radiogenic isotopes as well as the light stable isotopes. These new approaches were accompanied by the refinement of elemental analytical techniques for the determination of element concentrations, resulting in large part from the war effort.

The saltiness of the oceans was well known from ancient times and salt was used for domestic uses. But the major composition of sea water was yet to be determined. As early as 1820 William Wollaston, the discoverer of palladium and rhodium, predicted the presence of potassium in sea water and then proceeded to determine its concentration using gravimetric analytical techniques, based on the newly discovered platinum group elements and the formation of potassium-platinum compounds.

Both deep sea deposits and sea water were sampled first by the Challenger expedition (1872–1876). William Dittmar analyzed sea water from around the oceans collected by the Challenger expedition and showed the composition was fairly constant so far as the major components of the dissolved salts were concerned.

Over the past 60 years refined techniques of analysis have shown the patterns of distribution of the trace elements in ocean water profiles in all the oceans. These results are summarized by the late Yoshiyuki Nozaki in this volume. Similarly the understanding of the stable isotopes of oxygen, carbon and nitrogen in the oceans have played important roles in deciphering both the ancient temperature history of the oceans and the biological pathways of nutrient elements in the marine system.

The carbon system, tracked by radiocarbon has important consequences for both the study of ocean circulation and the ultimate fate of anthropogenically released carbon dioxide to the atmosphere.

Other forms of pollution are impacting the oceans resulting in enhanced anoxia especially in some coastal zones. The impact of trace metals from anthropogenic sources provide both tracers and in some cases health hazards for those relying on marine resources.

Some anthropogenically introduced radionuclides and synthetic compounds are useful as ocean circulation tracers. These include tritium, bomb radiocarbon and CFCs.

The role of hydrothermal vents in modifying the composition of sea water, as well as adding to deposits on the sea floor, has been explored since the 1970s. Especially the role of the high temperature reactions of sea water and hot basaltic rock have been shown to influence not only the chemistry of sea water but also the isotopic composition of oxygen as well as certain other isotope systems.

Extensive sampling of the oceans and the sediment piles provide information on nutrient cycling and the history of climate. The several organized efforts in these areas starting with GEOSECS (Geochemical Ocean Section Studies) and the Ocean Drilling Program have expanded our knowledge of how the ocean circulates and the reactivity of biological and other particles in the ocean column as well as the long-term history of the oceans. There are clearly secrets yet to be revealed using these methods of exploration.

Karl K. Turekian
Editor

CHEMISTRY OF SEA WATER

ORIGIN OF THE OCEANS

K. K. Turekian, Yale University, New Haven, CT, USA

Introduction

The oceans and the salts dissolved in them probably all formed early in the Earth's history. Although planetary degassing is occurring, as evidenced by studies of rare gases in rocks, the oceans, and the atmosphere, there is also a loss of water to the mantle via subduction. Supply from and loss to the mantle may be roughly equal at present.

Acquiring an Ocean

The oceans can be conceived of either as a feature established in the earliest history of the planet or as the result of a continuing supply of water and the constituents of sea water by degassing from the Earth's interior.

In 1951 W.W. Rubey, in his Presidential Address to the Geological Society of America, took the latter point of view. His arguments were colored by the knowledge available at that time of the age of the Earth and the age of the oldest rocks. Based on the analysis of lead isotopes in galenas, it was determined in the late 1940s that the Earth was about 3.2 billion years old. Some continental rocks, presumed to be relicts of ancient terrains, dated at about the same time also gave ages of about 3.2 billion years. On this basis it was assumed that the oldest rocks preserved a record of the dawn of Earth history. The contemporary oceans and ocean-derived sediments contain chemical species in quantities far in excess of those available from the weathering of crustal rocks (**Table 1**). These components were called 'excess volatiles' by Rubey, but Harold Urey suggested that they were really better characterized as 'excess solubles'. If these species all arrived with an early ocean, the early ocean would have had a radically different composition. It would dissolve rocks and also precipitate compounds different from those depositing from the present ocean.

If that were indeed the case, Rubey argued, the initial rocks should show the effects of a sudden supply of ocean water and hydrochloric and sulfuric acids, and carbon dioxide that ultimately dissolved rocks and formed the saline sea. The ancient rocks, however, do not look appreciably different from younger rocks; thus the absence of a difference in composition indicates that the oceans with their attendant excess anionic species have grown slowly with time. Indeed, it was argued that if a small fraction of the flux of water from fumaroles and hot springs were primary (from the Earth's interior) rather than meteoric or surface recycled water, then, over time, the oceans could be added to the surface from the interior so that the volume was increasing with time.

The discovery in 1955 that the Earth as a member of the solar system was really about 4.55 billion years old, and that the oldest rocks were considerably younger, ruled out having a record of the earliest days of the Earth's existence. In addition, from measurements of the hydrogen and oxygen isotopes of hot springs and in some cases tracking radioactive tritium from nuclear tests in hot springs, it was clear that all or most of the water in hot springs, and fumaroles was meteoric, and therefore determining a primary water flux was virtually impossible.

There is evidence, however, that there is planetary degassing, as revealed in the flux of radiogenic ^{40}Ar (**Figure 1**) and primordial ^{3}He (**Figure 2**) to the atmosphere. When these fluxes are used to model the flux of other gases (or their condensation products), two results are obtained. Gases that behave like ^{36}Ar (the nonradiogenic argon isotope) appear to have arrived at the Earth's surface in the earliest days of Earth history, while carbon dioxide and its condensation products, limestone and organic compounds, are being recycled via the processes associated with plate tectonics. One can assume that other chemically reactive analogues like water, behave in the same way. Like ^{36}Ar water may have been at the Earth's surface early in its history, and like carbon dioxide it is being recycled.

Table 1 Components of the oceans, atmosphere, and sedimentary rocks not derivable by weathering of primary silicate rocks

Chemical species	*Amount on Earth's surface not derived by weathering* ($\times 10^{20}$g)
Water	16 600
Total carbon as carbon dioxide	910
Chlorine	300
Nitrogen	42
Sulfur	22

After Rubey (1951).

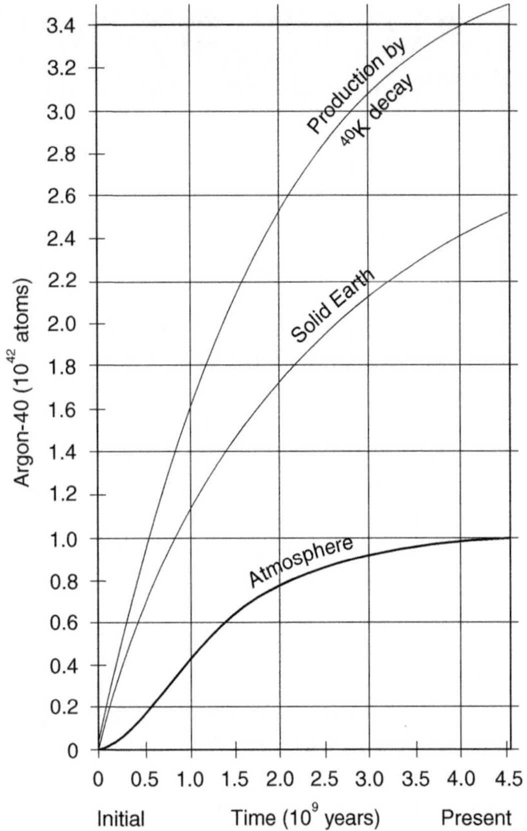

Figure 1 The degassing of radiogenic ^{40}Ar (from the radioactive decay of ^{40}K) from the solid Earth can be expressed by the equation $X_{SE}^{40}(t)/dt = \lambda K_0 exp(-\lambda t) - \propto exp(-\beta t)\ X_{SE}^{40}(t)$, where $X_{SE}^{40}(t)$ is ^{40}Ar in the solid Earth, λ is the radioactive decay constant for ^{40}K, α is the first-order degassing constant, and β is a measure of the exponential decrease in the efficiency of degassing over time as the result of the decrease in heat production from radioactive decay and loss of initial heat. K_0 is the amount of ^{40}K in the planet 4.55×10^9 years ago that will decay to ^{40}Ar. The lowest curve represents the growth of ^{40}Ar in the atmosphere over the past 4.5 billion years. The application of the values of α and β derived from the ^{40}Ar record when applied to the nonradiogenic isotope of argon (^{36}Ar) indicates that the nonradiogenic rare gases in the atmosphere largely had to be supplied to the atmosphere early in the Earth's history. $\alpha = 1.4 \times 10^{-9}, \beta = 1.2 \times 10^{-9}$, $K_0 = 3.85 \times 10^{42}$. Present-day flux of ^{40}Ar to atmosphere 1.5×10^{31} atoms per year.

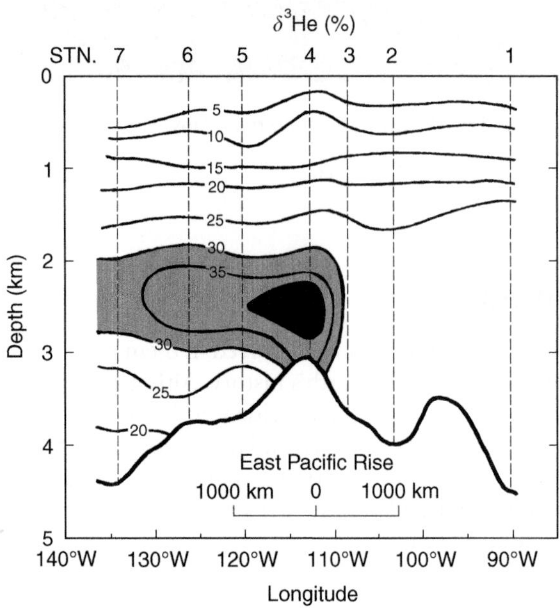

Figure 2 A section across the East Pacific Rise in the South Pacific showing ^3He in excess over that expected from the atmospheric ^3He dissolved in sea water when scaled to the more common ^4He. The difference between the dissolved isotopic ratio and atmospheric ratio is represented by δ^3He in percentage difference. ^3He is assumed to be primordial and indicates degassing of the mantle at the divergent plate boundaries featured as oceanic ridge systems. STNstation number of the sampling. (Reprinted with permission from Lupton and Craig 1981, copyright American Association for the Advancement of Science.)

We are thus left with the possibility that the oceans, including both the water and the soluble salts found therein, were present at the Earth's surface right at the beginning and subsequently have been subjected to recycling between crust and mantle via the plate tectonic cycle.

The Loss of Water from the Earth's Surface

If the oceans were effectively at the surface at the beginning, what processes could diminish the original inventory? There are two ways to lose water from the Earth's surface: (1) the photolytic dissociation of water vapor in the atmosphere with subsequent loss of hydrogen to space and the utilization of the released oxygen to oxidize reduced iron and sulfur compounds from the mantle; and (2) the loss of water, as well as of the oxidized components, to the mantle by way of hydrated minerals (such as clay, minerals, micas, and amphiboles) and minerals with oxidized iron and sulfur.

Although the first process is occurring today, it is not very efficient because of recombination with the abundant oxygen in the atmosphere as well as the fact that the supply of solar hydrogen to Earth may compensate for the loss of hydrogen. In the early history of the solar system, the flux of high-energy photons from the sun is thought to have been much greater than it is today, resulting in a very efficient conversion of water to hydrogen and oxygen. The stream of hydrogen from the atmosphere to outer space may have entrained heavier gases and caused the mass fractionation of the rare gases relative to each other and the fractionation of the isotopes of individual rare gases relative to solar composition.

We can estimate how much water was dissociated by considering the oxidation of the mantle. If the Earth started out with the oxidation state of chondrites (although not necessarily of total chondritic composition), it would take the oxygen from about one present-day ocean volume to reach the oxidation state of the mantle inferred from mantle-derived rocks. Alternatively, if we start with a more reduced ensemble, suggested by condensation calculations from a solar nebula, we will have primarily enstatite ($MgSiO_3$) and metallic iron as the original source to be oxidized. The oxidation reaction would then be as shown in eqn [1].

$$2\ MgSiO_3 + Fe + 2\ H_2O \rightarrow Mg_2SiO_4 + Fe_2Sio_4 + 2\ H_2 \qquad [1]$$

To oxidize the mass of the Earth with this initial make-up to the present oxidized level would require the dissociation of 60 present-day ocean volumes. Clearly, the results indicate hydrogen loss and planetary oxidation, although the exact starting material is not precisely defined.

In either case, the oxidation of the planet is the consequence of the photodissociation of water vapor and loss of hydrogen to space. In order to fractionate the rare gases by entrainment in the hydrogen streaming from Earth, this process must have occurred almost instantaneously in the early history of the Earth. Therefore, it is reasonable to assume that both water and rare gases were at the Earth's surface virtually at the time of formation of the Earth.

Sources of Water

Water reached Earth by one of two processes. The first is by way of the capture of water from the solar nebula as it cooled down. There is evidence from neon isotopes (**Figure 3**) that the initial imprinting of Earth with gases was likely to have been of solar composition. If this were the case, then the amount of water trapped by the accreting planet would have been large and provided the fundamental source of water. A second source is cometary impacts. Comets have often been described as dirty snowballs. They are rich in organic compounds and ice. Carbonaceous chondrites may be related to comets and, if so, their rare gas isotopic signatures are characteristically different from solar abundances and isotopic compositions owing to adsorption processes. Therefore, if water tracks the rare gases and if the neon isotope signature of mantle rocks indicates a solar origin for the terrestrially accumulating gases rather than a cometary origin, then most of the oceans owe their origin to direct capture of gases in the solar nebula cooled sufficiently to allow the production of water molecules.

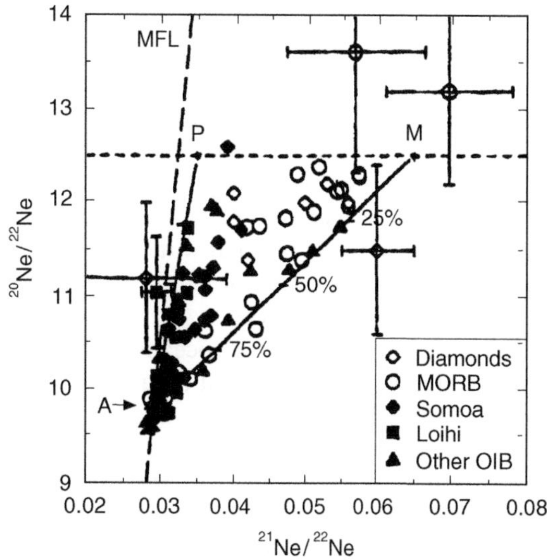

Figure 3 The distribution of neon isotopes in mantle-derived rocks, indicating the presence of an atmospheric component, a radiogenic component adding ^{21}Ne (produced by neutrons from uranium fission acting on oxygen and magnesium), and a solar component. It is this latter that indicates that gases in the mantle were derived from the capture of solar material in the early history of the Earth. M = MORB (midocean ridge basalts); P = plume or ocean island basalts (OIB); A = atmosphere. Solar neon is represented by the horizontal line at $^{20}Ne/^{22}Ne = 12.5$; MFL is the mass fractionation line. The presence of solar neon in ocean basalts was first identified by Craig and Lupton (Craig H and Lupton JE (1976) *Earth and Planetary Science Letters* 31: 369–385). (Reprinted with permission from Farley and Poreda (1993).

There have been claims that absorption spectra of sunlight indicate house-sized blocks of cometary ice entering our atmosphere. If the claim were substantiated the calculated rate of influx of these blocks of ice would be sufficient to provide the present volume of the oceans. However, these claims have not been independently substantiated and the consensus is that such blocks of ice are not responsible for the observations. Unless new measurements support this suggestion, we are constrained to accept the hypothesis of an initial 'watering' of the planet rather than a gradual accumulation.

The Composition of the Oceans

If indeed the oceans were present virtually from the day the Earth was formed, there is a further question: Was it salty like the present ocean? The salts dissolved in the ocean were probably dissolved from the original materials composing the Earth if low-temperature condensation compounds such as the chlorides, sulfates, and carbonates were present. Indeed, salts of these anions have been leached from primitive meteorites such as the carbonaceous chondrites as well as some other meteorites that have undergone some reheating. On the basis of theoretical calculations, similar compounds should have been present in the accreting Earth. With leaching of these compounds by the original water released from the Earth early in its history, it can be assumed that the oceans were always salty.

Indeed, the argument that oceans over several hundred million years ago might actually have been twice as saline as the contemporary ocean is based on the fact that there are many geological salt deposits and deep brines, often associated with oil fields. The assertion is based on a reasonable premise that salt deposits became important about 400 million years ago. The salt prior to that time had to be stored in the oceans, thus increasing its salinity by a factor of about two higher than the contemporary ocean.

The Future of the Oceans

If most of the water found on Earth was primarily in the oceans, with some dissolved in a molten mantle existing in the early days of the history of the Earth, then as we have seen, there was a decrease in the size of the original oceans. This decrease occurred because of photolysis of water vapor and subsequent loss of hydrogen from the atmosphere or the entrapment of water in hydrated minerals that were then subducted into the mantle.

The rate of photolytic loss must be considerably smaller at present compared to that on the early Earth because of the decrease in the extreme ultraviolet flux from the Sun. Also, the rate of subduction of the hydrated crust must be less now than it was early in the Earth's history because the driving forces for mantle convection and thus plate tectonics are gravitational heat from accumulation and fractionation and heat production by radioactive nuclides, both of which are waning with time. Therefore, the rate of supply of water to the mantle is now diminishing and there may actually be a release of the water stored in the mantle from previous times. As in the case of carbon dioxide, we may be in a steady-state of water supply from the mantle and return of water to the mantle, thereby maintaining the size of the oceans. At any rate, changes in the volume of water will probably not be large in future.

See also

Conservative Elements. Elemental Distribution: Overview. Hydrothermal Vent Fluids, Chemistry of. Mid-Ocean Ridge Geochemistry and Petrology. Volcanic Helium.

Further Reading

Craig H (1963) The isotopic geochemistry of water and carbon in geothermal areas. In: Tongiorgi E (ed.) *Nuclear Geology on Geothermal Areas, Proceedings of the First Spoleto Conference, Spoleto, Italy*, pp. 17–53. Pisa: V. Lischi Figli.

Farley KA and Poreda RJ (1993) Mantle neon and atmospheric contamination. *Earth and Planetary Science Letters* 114: 325–339.

Kump LR, Kasting JF, and Crane RC (1999) *The Earth System*. London: Prentice-Hall.

Lupton JE and Craig H (1981) A major helium-3 source at 15°S on the East Pacific Rise. *Science* 214: 13–18.

Lupton JE and Rubey WW (1951) Geologic history of sea water: an attempt to state the problem. *Geological Society of America* Bulletin 62: 1111–1147.

Turekian KK (1990) The parameters controlling planetary degassing based on ^{40}Ar systematics. In: Gopalan K, Gaur VK, Somayajulu BLK, and Macdougall JD (eds.) *From Mantle to Meteorites*, pp. 147–152. Delhi: Indian Academy of Science.

ELEMENTAL DISTRIBUTION: OVERVIEW

Y. Nozaki[†], University of Tokyo, Tokyo, Japan

Introduction

More than 97% of liquid water on the earth exists in the ocean. The ocean water contains approximately 3.5% by weight of dissolved salt. What is the elemental composition of the salts, how does it vary from place to place and with depth, and why? These are fundamental questions for which chemical oceanographers have sought answers. Despite more than a hundred years of intense investigation by modern chemical oceanography, the answers have not been fully elucidated. Nevertheless, we are now approaching complete understanding of the chemical composition of sea water and its variability in the ocean.

Historical Review

By the late nineteenth century it was well-established that the major components of sea water are extremely constant in their relative abundance, and comprise some ten constituents including Cl^-, Na^+, Mg^{2+}, SO_4^{2-} (*see* Conservative Elements). The analytical results reported by W. Dittmar in 1884 for waters collected during the British RMS Challenger Expedition (1872–1876) from the world's oceans were almost the same as today's values. The constancy of major chemical composition has led oceanographers to define 'salinity' as a fundamental property together with temperature to calculate the density of sea water. It was routine for classic physical oceanographers to titrate sea water for chloride (plus bromide) ion with silver nitrate standard solution, until the mid 1960s when salinity could be determined more practically by measurement of conductivity.

On the other hand, for minor elements, there has been little information gained since the establishment of major chemical composition of sea water. Measurements of trace constituents in sea water are difficult because of their very low abundance. There was a clear tendency for the reported concentrations of many trace elements to become lower and lower as time elapsed. This trend was, of course, not real but an artifact. It is a famous story that, to aid Germany's national deficit after World War I, the Nobel Prize winning chemist F. Haber attempted to recover gold from sea water which according to the current literature occurred at about $5 mg m^{-3}$. He completely failed however, but, after long and rigorous examination, he found that the concentration was ~1000 times less than that expected. Incidentally, Haber's value of gold concentration was two orders of magnitude higher compared to later reports (**Table 1**). Another good example may be found in the measurements of lead in sea water by Patterson and his associates (*see* Anthropogenic Trace Elements in the Ocean). The vertical profile of Pb in the North Pacific obtained by Schaule and Patterson in 1981 had concentrations about two orders of magnitude lower than those reported earlier (~1970) by the same workers although their 1981 values are believed to be accurate and real (**Table 1**).

Technical Challenge

It is now known that the most obvious reason for these trends is the continuous improvement in removing sources of contamination during sampling, handling, storage, and analysis. Significant efforts and advances in such field and laboratory techniques had been made until the GEOSECS (Geochemical Ocean Section Study) program started at around 1970. For example, polyvinyl chloride Niskin-bottle multisampling system together with CTD (conductivity–temperature–depth) sensors has routinely been employed in the hydrocasts, replacing the serial Nansen (metallic) bottle sampling method most widely used prior to that time. Yet, this was not enough for many trace metals except for barium, and an intercalibration exercise made in the early stage of the program did not produce any congruent results between laboratories. It was a significant and wise decision of the GEOSECS leaders that they focused on radionuclides and stable isotopes, which are almost free from contamination, and did not get involved in trace element geochemistry. Obviously, without having the real concentration data, any arguments that might be built upon them would be meaningless.

Obtaining clean (uncontaminated) water samples from various depths of the ocean is of prime importance in the study of trace metals. In this regard, various types of sampling bottles have been developed both domestically and commercially. They include the Cal-Tech Patterson sampler,

[†] Deceased.

Table 1 Estimated mean oceanic concentrations of the elements

Atomic number	*Element*	*Species*	*Type of distribution*[a]	*Oceanic mean concentration* ($ng\,kg^{-1}$)
1	Hydrogen	H_2O	—	—
2	Helium	Dissolved gas	c	7.6
3	Lithium	Li^+	c	180×10^3
4	Beryllium	$BeOH^+$	s + n	0.21
5	Boron	$B(OH)_3$	c	4.5×10^6
6	Carbon	Inorganic ΣCO_2	n	27.0×10^6
7	Nitrogen	Dissolved N_2	c	8.3×10^6
		NO_3^-	n	0.42×10^6
8	Oxygen	Dissolved O_2	inverse n	2.8×10^6
9	Fluorine	F^-	c	1.3×10^6
10	Neon	Dissolved gas	c	160
11	Sodium	Na^+	c	10.78×10^9
12	Magnesium	Mg^{2+}	c	1.28×10^9
13	Aluminum	$Al(OH)_3^0$	s	30
14	Silicon	$H_4SiO_4^0$	n	2.8×10^6
15	Phosphorus	$NaHPO_4^-$	n	62×10^3
16	Sulfur	SO_4^{2-}	c	898×10^6
17	Chlorine	Cl^-	c	19.35×10^9
18	Argon	Dissolved gas	c	0.62×10^6
19	Potassium	K^+	c	399×10^6
20	Calcium	Ca^{2+}	almost c	412×10^6
21	Scandium	$Sc(OH)_3^0$	(s + n)	0.7
22	Titanium	$Ti(OH)_4^0$	s + n	6.5
23	Vanadium	$NaHVO_4^-$	almost c	2.0×10^3
24	Chromium	CrO_4^{2-}(VI)	r + n	210
		$Cr(OH)_3^0$ (III)	r + s	2
25	Manganese	Mn^{2+}	s	20
26	Iron	$Fe(OH)_3^0$	s + n	30
27	Cobalt	$Co(OH)_2^0$?	s	1.2
28	Nickel	Ni^{2+}	n	480
29	Copper	$CuCO_3^0$	s + n	150
30	Zinc	Zn^{2+}	n	350
31	Callium	$Ga(OH)_4^-$	s + n	1.2
32	Germanium	$H_4GeO_4^0$	n	5.5
33	Arsenic	$HAsO_4^{2-}$(V)	r + n	1.2×10^3
		$As(OH)_3^0$(III)	r + s	5.2
34	Selenium	SeO_4^{2-}(VI)	r + n	100
		SeO_3^{2-}(IV)	r + n	55
35	Bromine	Br^-	c	67×10^6
36	Krypton	Dissolved gas	c	310
37	Rubidium	Rb^+	c	0.12×10^6
38	Strontium	Sr^{2+}	almost c	7.8×10^6
39	Yttrium	YCO_3^+	n	17
40	Zirconium	$Zr(OH)_5^-$	s + n	15
41	Niobium	$Nb(OH)_6^-$	almost c	0.37
42	Molybdenum	MoO_4^{2-}	c	10×10^3
43	Technetium	TcO_4^-	—	—
44	Ruthenium	RuO_4^-	?	< 0.005
45	Rhodium	$Rh(OH)_3^0$?	n	0.08
46	Palladium	$PdCl_4^{2-}$?	n	0.06
47	Silver	$AgCl_2^-$	n	2
48	Cadmium	$CdCl_2^0$	n	70
49	Indium	$In(OH)_3^0$	s	0.01
50	Tin	$SnO(OH)_3^-$	s	0.5
51	Antimony	$Sb(OH)_6^-$	s?	200
52	Tellurium	$Te(OH)_6^0$	r + s	0.05
		$TeO(OH)_3^-$	r + s	0.02
53	Iodine	IO_3^-	almost c	58×10^3
		I^-	(r + s)	4.4

(*Continued*)

Table 1 *Continued*

Atomic number	*Element*	*Species*	*Type of distribution*[a]	*Oceanic mean concentration* (ng kg^{-1})
54	Xenon	Dissolved gas	c	66
55	Cesium	Cs^+	c	306
56	Barium	Ba^{2+}	n	15×10^3
57	Lanthanum	$LaCO_3^+$	n	5.6
58	Cerium	$Ce(OH)_4^0$	s	0.7
59	Praseodymium	$PrCO_3^+$	n	0.7
60	Neodymium	$NdCO_3^+$	n	3.3
61	Promethium	—	—	—
62	Samarium	$SmCO_3^+$	n	0.57
63	Europium	$EuCO_3^+$	n	0.17
64	Gadolinium	$GdCO_3^+$	n	0.9
65	Terbium	$TbCO_3^+$	n	0.17
66	Dysprosium	$DyCO_3^+$	n	1.1
67	Holmium	$HoCO_3^+$	n	0.36
68	Erbium	$ErCO_3^+$	n	1.2
69	Thulium	$TmCO_3^+$	n	0.2
70	Ytterbium	$YbCO_3^+$	n	1.2
71	Lutetium	$LuCO_3^+$	n	0.23
72	Hafnium	$Hf(OH)_5^-$	s + n	0.07
73	Tantalum	$Ta(OH)_5^0$	s + n	0.03
74	Tungsten	WO_4^{2-}	c	10
75	Rhenium	ReO_4^-	c	7.8
76	Osmium	OsO_4^0	almost c	0.009
77	Iridium	$Ir(OH)_3^0$	s?	0.00013
78	Platinum	$PtCl_4^{2-}$	c	0.05
79	Gold	$AuOH(H_2O)^0$	c	0.02
80	Mercury	$HgCl_4^{2-}$	(s + n)	0.14
81	Thallium	Tl^+	c	13
82	Lead	$PbCO_3^0$	anth. + s	2.7
83	Bismuth	$Bi(OH)_3^0$	s	0.03
84	Polonium	$PoO(OH)_3^-$	s	—
85	Astatine	—	—	—
86	Radon	Dissolved gas	c	—
87	Francium	Fr^+	—	—
88	Radium	Ra^{2+}	n	0.00013
89	Actinium	$AcCO_3^+$	n	—
90	Thorium	$Th(OH)_4^0$	s	0.02
91	Protactinium	$PaO_2(OH)^0$	s	—
92	Uranium	$UO_2(CO_3)_3^{4-}$	c	3.2×10^3
93	Neptunium	NpO_2^+	—	—
94	Plutonium	$PuO_2(CO_3)(OH)^-$	(r + s)	—
95	Americium	$AmCO_3^+$	(s + n)	—

[a]c, conservative; n, nutrient-like; s, scavenged; r, redox sensitive; anth., anthropogenic.

modified Go-Flo bottles, and lever-action or X-type Niskin bottles. None of them are easy to keep clean and handle properly, and experience is needed in their operation depending on the type of bottle. Hydrowire is also important, since normal steel wire has rust and grease that can easily contaminate the water. To avoid this, some workers use plastic Kevler line and others use stainless-steel wire or a titanium armored cable.

With the rapid growth of semiconductor industries from the early 1970s, clean laboratory techniques also become more popular in the field of marine chemistry and helped considerably to reduce contamination from reagents, containers, and dust in the room atmosphere. Real oceanic concentrations of trace metals are so low that conventional analytical techniques prior to 1970 were not normally sensitive enough to detect them except in polluted or some coastal waters. Thus, significant efforts were also devoted to developing more sensitive and reliable methods using atomic absorption spectrophotometry, chemiluminesence detection, isotope dilution mass spectrometry, etc. As a result, in the late 1970s, data of some transition metals, like Cd, Cu, and Ni were obtained by the Massachusetts Institute of Technology group and soon after confirmed by others using different or modified methods. Their oceanic profiles were quite consistent with known biogeochemical cycling and scavenging processes in the ocean. Thus, these features have often been referred as 'oceanographically or geochemically consistent' distribution by subsequent workers. Since then, growing numbers of publications describing the oceanic distributions of trace elements in sea water based on modern technologies have appeared year by year.

Oceanic Profiles

It is now possible to compile, with reasonable confidence, the vertical profiles in the form of Periodic chart (**Figure 1**), where the data from the North Pacific have been chosen since physical processes that affect the distribution are relatively simple and well documented. **Figure 1** is an updated version of the original, including new data for Nb, Ta, Hf, Os, Ag, and rare earth elements. Now, there remains only one element, Ru on which no real data have been reported (*see* Platinum Group Elements and their Isotopes in the Ocean). However, confirmation is needed for many elements, including Sn, Hg, Rh, Pd, Au, Ir, Pt, etc., since they are based on a single study or on controversial results by different workers. Nevertheless, it is clear that the long-standing dream to establish the chemical composition of sea water is about to become a reality.

Trace elements follow one or more of the categories which are described below.

Conservative type Some of the trace elements such as U, W, and Re form stable ionic species, $UO_2(CO_3)_2^{2-}$, WO_4^{2-}, and ReO_4^- in sea water. Hence, their oceanic behavior is conservative (follow salinity) and their mean residence times in the ocean are generally long (e.g., $>10^5$ years). There is no significant variation in their concentration between different oceanic basins.

Recycled type (nutrient-like) Many others, e.g., Ni, Cd, Zn, Ge, and Ba, show a gradual increase in their concentration from the surface to deep water, much like nutrients (nitrate, phosphate, and silicate or alkalinity), suggesting their involvement in the biogeochemical cycle of biological uptake in the surface water and regeneration in deep waters. As a result of global ocean circulation, the deep-water concentrations of this type are higher in the Pacific than in the Atlantic.

Scavenged type Trace metals such as Al, Co, Ce, and Bi, show surface enrichment and depletion in deep waters, in contrast to the opposite trend in nutrient types. These elements are highly particle-reactive and are rapidly removed from the water column by sinking particulate matter and/or by scavenging at the sediment–water interface. Their mean oceanic residence times are short ($<10^2$–10^3 years). Interoceanic variations in their concentration can be large (e.g., Atlantic/Pacific concentration ratio $\sim$40 for Al) depending on kinetic balance between supply and removal for the specific basins.

Redox-controlled type Elements such as Cr, As, Se, and Te exist in sea water at more than one oxidation state. Their oceanic behavior is strongly dependent on the chemical form. Their reduced states are thermodynamically unstable in normal oxygenated waters but are probably formed through biological mediation. Reduced species can also be formed in anoxic basins, the Black Sea, Cariaco Trench, some fiords, and in organic-rich sediments.

Anthropogenic and transient type Finally, Pb and Pu are good examples of elements whose oceanic distributions are globally influenced by human activities (*see* Anthropogenic Trace Elements in the Ocean). Their oceanic distributions are changing with time. Although some others, such as Hg, Sn, Cd, and Ag, are deduced to be similarly influenced, their transient nature has not yet been proven through direct observation.

Particle Association and Speciation

One of the important features of **Figure 1** is that the concentration, even for trace elements, varies fairly smoothly and continuously with depth. This casts doubt on some erratic and highly discontinuous values unless there are obvious reasons for them, such as hydrothermal influence or difference in the water masses. The data shown in **Figure 1** are largely based on filtered samples and therefore, can be referred as 'dissolved concentration.' For conservative elements, it does not matter whether the water sample is filtered or not, since there is virtually no difference in the analytical results. For most nutrient-type elements, particle association in the open ocean is generally small ($<\sim5\%$) and therefore, the gross features of unfiltered samples remains the same as

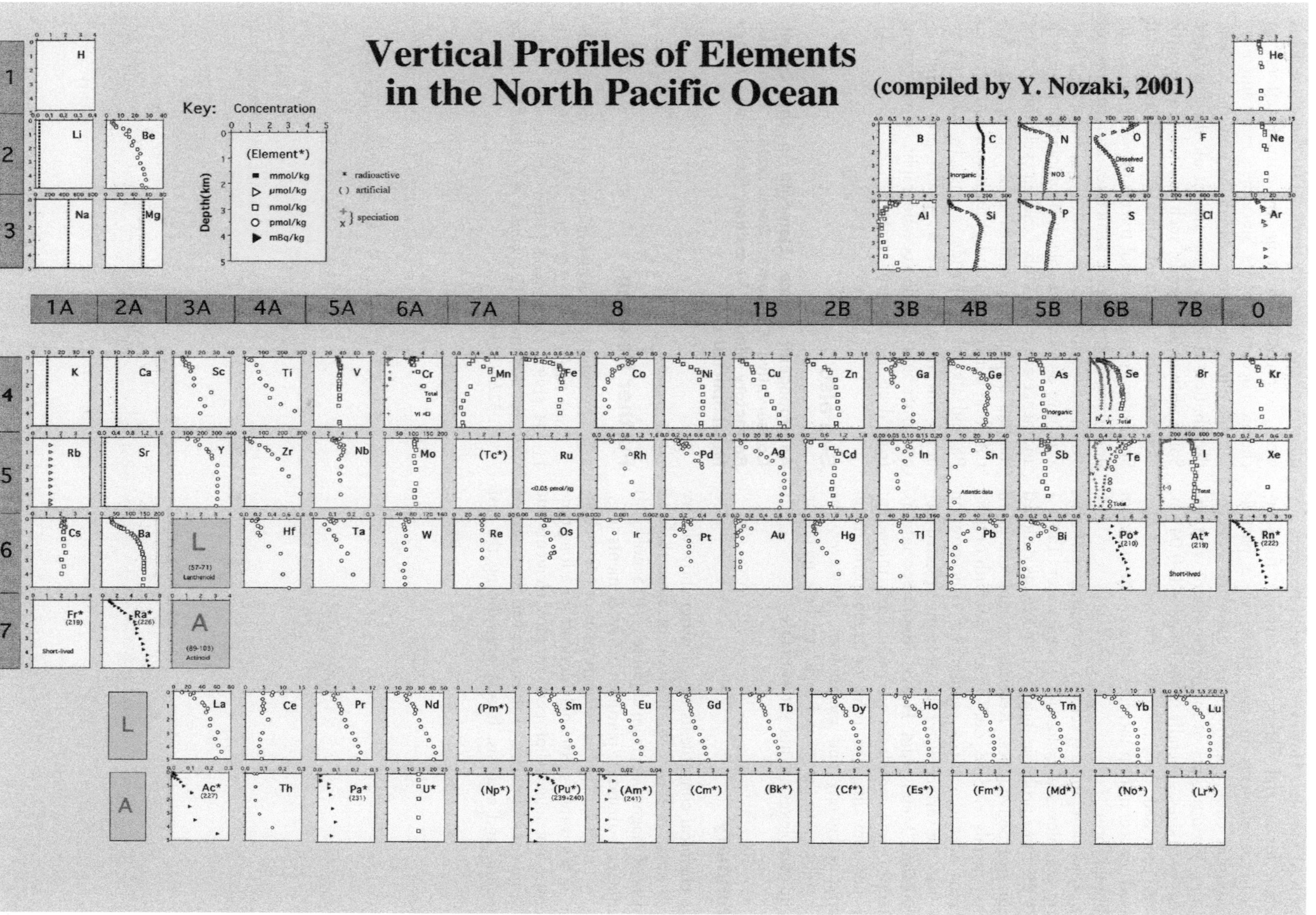

Figure 1 Vertical profiles of elements in the North Pacific Ocean.

dissolved samples. However, filtration becomes important for coastal waters, and certainly for scavenged-type elements in any place, since particle association could easily exceed dissolved concentration. Various types of membrane filters with different pore sizes, generally in the range $\sim 1\text{–}0.04\,\mu m$, have been used, and, therefore, the so-called 'dissolved' concentration includes different amounts of colloidal form.

Furthermore, ionic species of many trace elements form complexes with ligands in sea water. **Table 1** lists the most probable inorganic species as deduced from thermodynamics. Recently, complexation of heavy metals (e.g., Fe, Zn, Cu, and Co) with organic ligands which occur at nanomolar concentration levels in the surface water has been investigated (*see* Transition Metals and Heavy Metal Speciation). These organic complexes are particularly important for understanding the roles of trace elements in plankton biology and metabolism. However, the technologies to detect and separate these metal–organic complexes are not yet available.

Summary

The chemical constituents of sea water show a very wide range in their concentration of more than 15 orders of magnitude, from chlorine ($\sim 0.\ 5\,\text{mol kg}^{-1}$) to the least abundant platinum group elements (e.g., $\text{Ir} < \sim 1\,\text{fmol kg}^{-1}$). The measurement of trace elements in sea water is extremely difficult for a long time and was not achieved properly for most elements due to large contamination problems in various stages of sampling and analysis. It was only during the last quarter of the twentieth century that those problems were eventually overcome by current efforts of chemical oceanographers. Since then, more and more reliable data have accumulated and now the oceanic distribution is known for almost all the elements. Some of the trace metals, such as Al, and rare earth elements (*see* Transition Metals and Heavy Metal Speciation) serve as useful tracers of water masses in describing hydrographic structures and patterns of ocean circulation. Others, such as Fe, Si, and perhaps Zn, have an essential role in phytoplankton growth and hence affect the global carbon cycle through the 'biological pump.' Resemblance in the oceanic distribution between Cd and phosphorus, and between Ba and alkalinity, provides a basis by which those heavy metals can serve as useful proxies in reconstructing the paleo-oceanographic environment from deep-sea sediment strata.

See also

Anthropogenic Trace Elements in the Ocean. Conservative Elements. Conservative Elements. Refractory Metals. Transition Metals and Heavy Metal Speciation.

Further Reading

Nozaki Y (1997) A fresh look at element distribution in the North Pacific Ocean. *EOS Transactions, AGU* 78: 221.

Li YH (1991) Distribution patterns of elements in the ocean: a synthesis. *Geochimica et Cosmochimica Acta* 55: 3223–3240.

CONSERVATIVE ELEMENTS

D. W. Dyrssen, Gothenburg University, Göteborg, Sweden

Introduction

If 1 kg of sea water is evaporated and ignited according to a special procedure 35 g of solids are obtained. This is the normal (standard) salinity. Since the salinity is mainly changed by evaporation or by dilution with practically ion-free rain water the composition of the major ions in sea water is not changed by such processes. These constituents are considered to be conservative, and as a consequence their ratios are constant. Thus the concentration of a conservative constituent (element) at a salinity S is obtained by multiplying the values in **Table 1** by S/35.

Determinations

The salinity can be determined with five significant figures from conductivity measurements as well as by potentiometric titration of chloride + bromide in m g of sea water with v ml of t molar silver nitrate. Thereby the chlorinity is given by:

$$Cl = vt \cdot 107.87 \cdot 328.5233/1000m$$

where 107.87vt/1000 represents the mass in grams of pure silver that is necessary to precipitate the halogens in 328.5233 g of sea water. The relationship between salinity and chlorinity is:

$$S = 1.80655Cl$$

Sodium cannot be determined with four significant figures and the value in **Table 1** has been calculated from the ion balance

$$\sum n[X^{n+}] = \sum n[X^{n-}]$$

Potassium can be determined gravimetrically with a precision of 0.26% by precipitation with sodium tetraphenylborate.

Calcium (+strontium) and magnesium can be determined with four significant figures by titration procedures.

Strontium may be determined with three significant figures by various procedures.

Fluoride can be determined with three significant figures using a fluoride electrode. Otherwise, the recommended procedure is a spectrophotometric determination with lanthanum alizarin complexone.

Sulfate may be determined gravimetrically with a precision of 0.14% using precipitation of barium sulfate.

Boric acid, $B(OH)_3$, together with borate, $B(OH)_4^-$, may be determined with three significant figures by the spectrophotometric curcumin method.

Alkalinity is determined by titration with hydrochloric acid of the main basic constituent, HCO_3^-, together with minor basic components such as CO_3^{2-}, $B(OH)_4^-$, $SiO(OH)_3^-$, $H_2PO_4^-$, and HPO_4^{2-}. It can be determined with four significant figures.

When accurate methods are used for the determinations of the main constituents slight deviations from a conservative behavior may be detected. The deviations, which are due to some fundamental processes, will be discussed below.

Plankton Production

Plankton production involves the formation of hard parts (biogenic calcium carbonate and biogenic opal) in addition to soft material. The stoichiometry varies around:

$$(CH_2O)_{106}(NH_3)_{16}H_3PO_4(CaCO_3)_{20}(SiO_2)_{20}$$

With this stoichiometry the increase in alkalinity due to the uptake of nitrate

$$NO_3^- + H^+ + H_2O \Rightarrow NH_3(org) + 2O_2$$

is almost balanced by the biogenic formation of calcium carbonate

$$Ca^{2+} + HCO_3^- \Rightarrow CaCO_3(s) + H^+$$

However, when the production sinks below the euphotic zone the soft parts deteriorate

$$NH_3(org) + 2O_2 \Rightarrow H^+ + NO_3^- + H_2O$$

lowering the increase in alkalinity due to the dissolution of calcium carbonate

$$CaCO_3(s) + CO_2 + H_2O \Rightarrow Ca^{2+} + 2HCO_3^-$$

Table 1 The major constituents of average sea water with a salinity of 35

Constituent	*g kg^{-1} sea water*	*mol kg^{-1} sea water*
Sodium (Na^+)	10.76	0.4680
Potassium (K^+)	0.3992	0.01021
Magnesium (Mg^{2+})	1.292	0.05315
Calcium (Ca^{2+})	0.4128	0.01030
Strontium (Sr^{2+})	0.00815	0.000093
Fluoride (F^-)	0.00141	0.000074
Chloride (Cl^-)	19.344	0.54563
Bromide (Br^-)	0.06712	0.00084
Sulfate (SO_4^{2-})	2.712	0.02823
Alkalinity (A_t)	(0.143)[a]	0.00234[b]
Boron (B)	0.00445	0.000412

[a]Calculated from HCO_3^-, the principle base.
[b]Mol HCl kg^{-1} needed to titrate all bases to pH 4.5.

Table 2 World average major ions and silica drained into the oceans

Constituent	*Concentration (mmol l^{-1})*
Sodium (Na^+)	0.252
Potassium (K^+)	0.0486
Calcium (Ca^{2+})	0.353
Magnesium (Mg^{2+})	0.148
Chloride (Cl^-)	0.192
Sulfate (SO_4^{2-})	0.104
Hydrogen carbonate (HCO_3^-)	0.902
Silica ($Si(OH)_4$)	0.198

Ion balance: $\sum n[X^{n+}] = \sum n[X^{n-}] = 1.302$ meq l^{-1}

The CO_2 is supplied by microbial decomposition of carbohydrates

$$CH_2O(org) + O_2 \Rightarrow CO_2 + H_2O$$

In the past, the alkalinity (A_t) was considered to be a conservative property of sea water. With high precision titration techniques variations in $A_tS/35$ can be measured, as well as in $(Ca_t + Sr_t)S/35$.

The most important carbonate-secreting organisms in the oceans are foraminifera, coccolithophorides, and pteropods. The carbonate tests vary in size, appearance, crystal form (calcite or aragonite), and magnesium content. The solubility depends on the depth (pressure), temperature, and concentration of CO_2 besides the crystal form. For example, the pteropods which secrete shells of aragonite undergo dissolution at shallower depths than the coccolithophorides which secrete calcite shells.

River Inputs

The river inputs into the oceans vary between the oceans. Ten percent of the total inflow of 10^6 m^3 s^{-1} (1 Sverdrup, Sv) flows into the Arctic Ocean, whereby the normalized alkalinity ($A_tS/35$) is increased in the outflow along the east coast of Greenland. The average composition of the major ions in river water is presented in **Table 2**. The ratios are quite different from the ratios for the major elements in sea water that may be calculated from the concentrations in **Table 1**. For example, the Na/K ratio in river water is 5.2 while the ratio in sea water is 46. This is most likely due to ion exchange with the sediments:

$$NaR + K^+ \Leftrightarrow KR + Na^+$$

where R represents an aluminosilicate.

The ratio between magnesium and calcium is 0.42 in river water, but 5.2 in sea water. This may be explained by the fact that the organisms use only small amounts of magnesium, while the biogenic formation of calcium carbonate is a major process. The average ratios of SO_4^{2-}/Cl^- and HCO_3^-/Cl^- are 0.54 and 4.7 in river water; and 0.052 and 0.0043 in sea water. Only small amounts of sulfate are used by the organisms to produce essential sulfur-containing compounds, but when sea water reacts with hot basalt in rift zones sulfate is removed (see below). Hydrogen carbonate is removed upon the formation of biogenic calcium carbonate (see above).

The rivers also carry clay minerals into the ocean. The cation exchange capacity corresponds to $5.2 \cdot 10^{15}$ meq y^{-1}. This may be compared with the river input of cations of $41 \cdot 10^{15}$ meq y^{-1} (1.302 meq l^{-1} from **Table 2** in a flow of 10^6 m^3 s^{-1}). In the ocean the sodium, potassium, and magnesium displace calcium in the clay minerals by ion exchange.

Man is changing the composition of river waters. Besides pollutants and an increase of particulate matter, acid rain and increased concentration of carbon dioxide in the atmosphere causes the following reactions, especially with limestone in southern Europe:

$$CaCO_3(s) + H^+ \Rightarrow Ca^{2+} + HCO_3^-$$

$$CaCO_3(s) + CO_2(g) + H_2O \Rightarrow Ca^{2+} + 2HCO_3^-$$

$$CaCO_3(s) + CH_2O(org) + O_2 \Rightarrow Ca^{2+} + 2HCO_3^-$$

This has caused an increase in the normalized calcium concentration as well as the normalized alkalinity in the Baltic Sea. Weathering of silicate rocks is also dependent on the partial pressure of CO_2 in the atmosphere, for example:

$$CaAl_2Si_2O_8(s) + 6H_2O + 2CO_2(g) \Rightarrow$$
$$Ca^{2+} + 2HCO_3^- + 2AlOOH(s) + 2Si(OH)_4$$

Hot Vents

About 5000 $m^3\ s^{-1}$ of sea water reacts with fresh hot basalt in the rift zones. Hydrothermal reactions produce large changes in the composition of the sea salts. The reactions lead to a loss of magnesium, sulfate, and fluoride, but to an addition of calcium and potassium. The hydrothermal activity balances some of the river inputs and contributes to maintenance of the steady state and the conservation of the constant composition of sea salts.

If the hot vent lies in a depression such as the Atlantis II Deep in the Red Sea, the results of the hydrothermal reactions are very evident (see **Figure 1**). Water is removed together with magnesium and sulfate, and calcium is released from the basalt. Rock salt (NaCl), anhydrite ($CaSO_4$), and silica (SiO_2) are formed and the brine becomes saturated with these solids. Sulfide metals are coprecipitated with iron sulfide (FeS).

Brines are also formed upon the formation of sea ice in polar regions; thereby calcium carbonate and calcium sulfate are precipitated. These processes also occur upon evaporation of sea water, e.g. for the production of sodium chloride.

Marine Aerosols

Marine aerosols are formed by bursting of air bubbles at the sea's surface. Although some separations of the main constituents might occur by this process, the fallout of sea salts over the continents is in the order of 10^9 tons y^{-1}. Since the weight of the sea salts in the oceans is in the order of $13 \cdot 10^{17}$ tons the recycling time would be 1300 million years (My). The residence times calculated from the river inputs are 210 My for sodium and 100 My for chloride. This implies that weathering influences the concentrations of the major ions in the runoff from the continents.

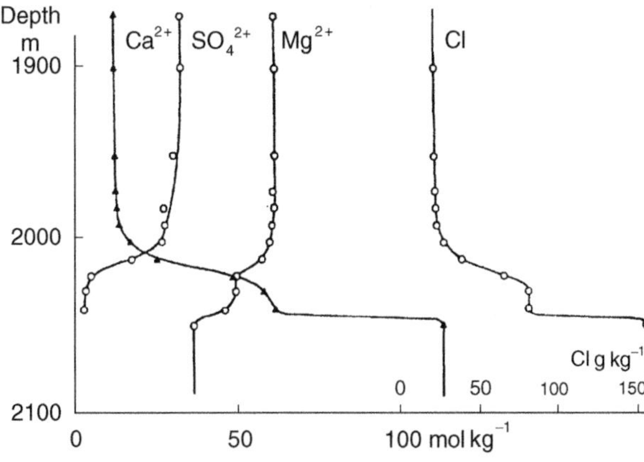

Figure 1 Depth profiles of chlorinity (Cl, g kg^{-1}), calcium, magnesium and sulfate (mol kg^{-1}) in the Red Sea Atlantis II deep on 22–23 March 1976.

Minor Constituents

Some minor constituents show a conservative type of distribution in the oceans. They are the alkali ions Li^+, Rb^+, and Cs^+, besides MoO_4^{2-} and Tl^+. Obviously, the organisms only use small amounts of molybdenum. However, in waters with anoxic basins (some fiords and the Black Sea), molybdate is depleted. In addition, WO_4^{2-}, ReO_4^{2-}, and U(VI) show conservative-type oceanic distributions.

Conclusions

The major constituents of sea water are present in practically constant proportions. The main ions include the alkali ions Na^+ and K^+ and the alkaline earth ions Mg^{2+} and Ca^{2+}, in addition to chloride (Cl^-) and sulfate (SO_4^{2-}). Their concentrations are also proportional to the salinity. In spite of the weathering and recycling processes and the hydrothermal reactions with hot basalt in rift zones there are only slight perturbations of the steady state. Deviations from a conservative behavior may be detected by accurate analytical methods and studies of ancient marine waters.

See also

Calcium Carbonates. Carbon Cycle. Hydrothermal Vent Deposits. Hydrothermal Vent Fluids, Chemistry of. Iron Fertilization. Nitrogen Cycle. River Inputs. Stable Carbon Isotope Variations in the Ocean.

Further Reading

Broecker WS and Peng T-H (1982) *Tracers in the Sea.* Palisades, New York: Lamont-Doherty Geological Observatory.

Chester R (1990) *Marine Geochemistry.* London: Unwin Hyman.

Degens ET and Ross DA (eds.) (1969) *Hot Brines and Recent Heavy Metal Deposits in the Red Sea.* New York: Springer-Verlag.

Dyrssen D and Jagner D (eds.) (1972) *The Changing Chemistry of the Oceans*. Stockholm: Almqvist Wiksell.

Dyrssen D (1993) The Baltic-Kattegat-Skagerrak estuarine system. *Estuaries* 16: 446–452.

Goldberg ED (ed.) (1974) *The Sea. Ideas and Observations on Progress in the Study of the Seas*. New York: John Wiley Sons.

Grasshoff K, Kremling K, and Ehrhardt M (eds.) (1999) *Methods of Seawater Analysis*, 3rd edn. Weinheim: Wiley-VCH.

Kremling K and Wilhelm G (1997) Recent increase of calcium concentrations in Baltic Sea waters. *Marine Pollution Bulletin* 34: 763–767.

Libes SM (1992) *Marine Biogeochemistry*. New York: John Wiley Sons.

Riley JP and Chester R (eds.) (1975) *Chemical Oceanography*, vol. 1, 2nd edn. London: Academic Press

Sillén LG (1963) How has the sea water got its present composition. *Svensk Kemisk Tidskrift* 75: 161–177.

TRACE ELEMENT NUTRIENTS

W. G. Sunda, National Ocean Service, NOAA, Beaufort, NC, USA

Published by Elsevier Ltd.

Introduction

Life in the sea is dependent on fixation of carbon and nitrogen by unicellular algae, ranging in size from <1 to over 100 μm in diameter. These so-called phytoplankton consist of eukaryotic algae, which photosyntheticly fix carbon into organic matter, and photosynthetic bacteria (cyanobacteria) that carry out both carbon and dinitrogen (N_2) fixation. Until recently, phytoplankton productivity in the ocean was thought to be primarily limited by available fixed nitrogen (nitrate, nitrite, ammonia, and various organic nitrogen compounds) and to a lesser extent phosphorus (orthophosphate and organic phosphorus compounds). However, in the past 20 years, enrichment experiments in bottles and in mesoscale patches of surface water have shown that iron regulates the productivity and species composition of planktonic communities in major regions of the world ocean, including the Southern Ocean, the equatorial and subarctic Pacific, and some coastal upwelling systems. In addition, it now appears that iron limits N_2 fixation by cyanobacteria in large regions of the subtropical and tropical ocean, and thus may control oceanic inventories of biologically available fixed nitrogen. Several other micronutrient metals (zinc, cobalt, manganese, and copper) have also been shown to stimulate phytoplankton growth in ocean waters, but their effect is usually much less than that of iron. However, these metals may play an important role in regulating the composition of phytoplankton communities because of large differences in trace metal requirements among species.

In this article interactions between trace element nutrients (iron, zinc, cobalt, manganese, copper, nickel, cadmium, molybdenum, and selenium) and phytoplankton in seawater are discussed. In these interactions, not only do the trace nutrients affect the growth and species composition of phytoplankton communities, but the phytoplankton and other biota (e.g., heterotrophic bacteria and zooplankton) have a profound influence on the distributions, chemistry, and biological availability of these elements (**Figure 1**). There are many aspects to consider, including (1) the sources, sinks, and cycling of trace element nutrients in the ocean; (2) the distribution of these elements in time and space, and their chemical speciation (or forms); (3) the interactions of these elements with phytoplankton at different levels of biological organization (molecular, cellular, population, community,

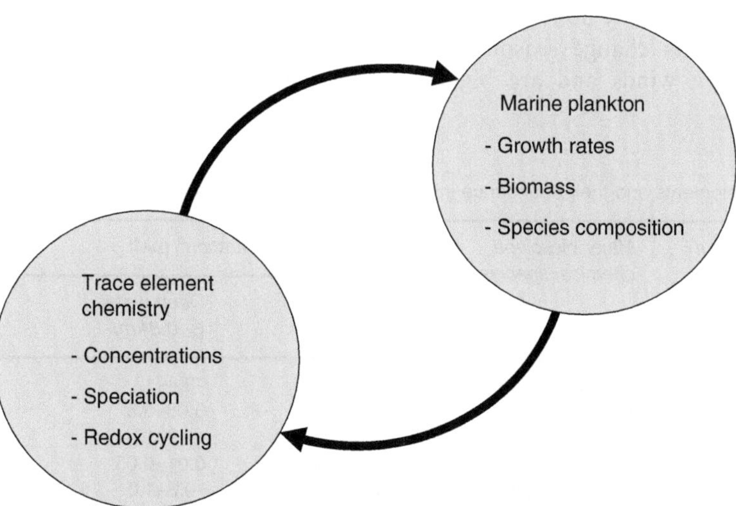

Figure 1 Conceptual diagram of the mutual interactions between trace element nutrients (Fe, Mn, Zn, Co, Cu, Cd, and Se) and phytoplankton in the sea. In these interactions, the chemistry of trace element nutrients, in terms of their concentrations, chemical speciation, and redox cycling, regulates the productivity and species composition of marine phytoplankton communities. These communities in turn regulate the chemistry and cycling of trace element nutrients through cellular uptake and assimilation, vertical transport of biogenic particles (intact cells and fecal pellets), grazer and bacterially mediated regeneration processes, production of organic chelators, and biological mediation of trace element redox transformations.

ecosystem); and (4) the role of planktonic communities in regulating the chemistry and cycling of these nutrient elements in seawater.

Distribution in Seawater

Knowledge of the distributions of trace element nutrients is essential to understanding the influence of these micronutrients on the productivity and species diversity of marine planktonic communities. Concentrations of filterable iron and zinc (that portion passing through a 0.2- or 0.4-μm-pore filter) typically are extremely low (only 0.02 to 0.1 nM) in surface waters of the open ocean. Cadmium, a nutrient analog for zinc, can reach values as low as 0.002 nM (**Table 1**). Concentrations of these and other trace element nutrients often increase by orders of magnitude in transects from the open ocean to coastal and estuarine waters due to inputs from continental sources, such as rivers, groundwater, eolian dust, and coastal sediments. Filterable iron can reach micromolar concentrations in estuaries and can approach 10–20 μM in rivers, 5 or 6 orders of magnitude higher than surface ocean values. This filterable iron occurs largely as colloidal particles (<0.4-μm diameter) consisting of iron oxides in association with organic matter. These are rapidly lost from estuarine and coastal waters via salt-induced coagulation and particulate settling. Because of this efficient removal, very little of the iron in rivers reaches the open sea, and most of the iron in ocean waters is derived from the deposition of mineral dust blown on the wind from arid regions of the continents. These eolian inputs change seasonally with variations in prevailing winds and are highest in waters downwind of arid regions such as North Africa and Central Asia. Areas far removed from these eolian sources, such as the South Pacific and the Southern Ocean, receive little atmospheric iron deposition and are among the most iron-limited regions of the oceans.

Because of the large gradients in trace metal concentrations between the open ocean and coastal waters, oceanic phytoplankton species have evolved the ability to grow at much lower available concentrations of iron, zinc, and manganese. In doing so they have been forced to rearrange their metabolic architecture (e.g., in the case of iron-rich protein complexes involved in photosynthesis) or to switch from scarce elements to more abundant ones in some critical metalloenzymes (e.g., Ni and Mn replacement of Fe in the antioxidant enzyme superoxide dismutase).

Concentrations of many trace element nutrients (zinc, cadmium, iron, copper, nickel, and selenium) increase with depth in the ocean, similar to increases observed for major nutrients (nitrate, phosphate, and silicic acid) (**Figures 2–4**). In the central North Pacific, filterable concentrations of zinc and cadmium increase by 80-fold and 400-fold, respectively, between the surface and 1000-m depth. The similarity between vertical distributions of these trace elements and major nutrients indicates that both sets of nutrients are subject to similar biological uptake and regeneration processes. In these processes, both major and trace element nutrients are efficiently removed from surface waters through uptake by phytoplankton. Much of these assimilated nutrients are recycled within the euphotic zone by the coupled processes of zooplankton grazing and excretion, viral lysis of cells, and bacterial degradation of organic

Table 1 Micronutrient elements and their abundance in ocean water and phytoplankton

Micronutrient element	*Major input source*	*Major dissolved chemical species*	*Dissolved concentration[a] (nM)*		*Element:carbon ratio in phytoplankton (μmol:mol)*
			Surface water	*Deep water (≥ 0.8 km)*	
Iron	Wind-born dust	Organic chelates	0.02–0.5	0.4–1	3–40
Manganese	Rivers	Mn^{2+}	0.1–5	0.08–0.5	2–30
Zinc	Wind-born dust	Organic chelates	0.05–0.2	2–10	1–40
Cobalt	Rivers	Organic chelates	0.007–0.03	0.01–0.05	0.1–3
Cadmium	Rivers	Organic chelates	0.002–0.3	0.3–1.0	0.2–8
Copper	Rivers	Organic chelates	0.5–1.4	1.5–5	2–6
Nickel	Rivers	Ni^{2+}	2–3	5–11	2–17
Molybdenum	Rivers	MoO_4^{2-}	100–110	100–110	0.05–0.8
Selenium	Rivers	Organic selenides, SeO_4^{2-}	0.5–1.0	1.5–2.3	1–2

[a]Dissolved is defined operationally as that passing through a 0.2- or 0.4-μm-pore filter.

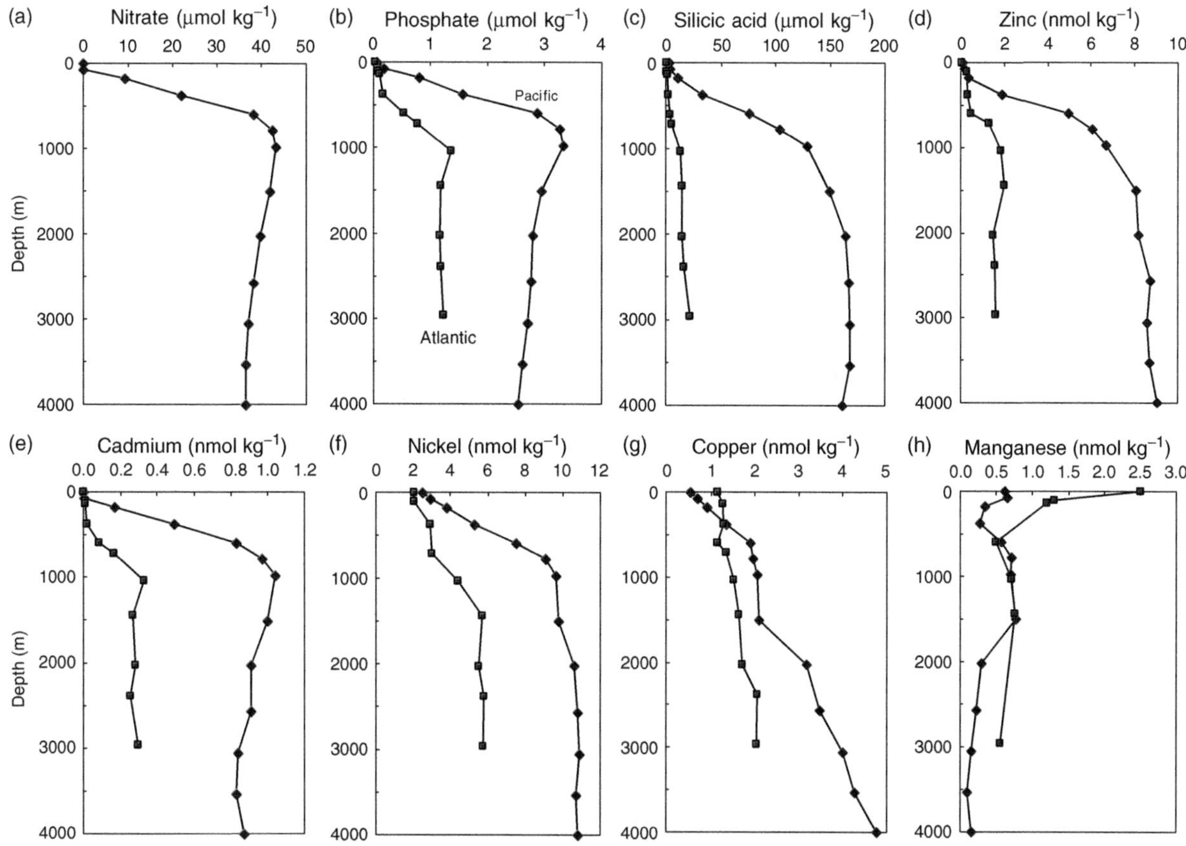

Figure 2 Depth profiles for major nutrients (nitrate (Pacific only), phosphate, and silicic acid) and filterable concentrations (that passing a 0.4-μm filter) of trace nutrient elements (zinc, cadmium, nickel, copper, and manganese) in the central North Pacific (diamonds, 32.7° N, 145.0° W, Sep. 1977) and North Atlantic (squares, 34.1° N, 66.1° W, Jul. 1979). Manganese concentrations in the Pacific were analyzed in acidified, unfiltered seawater samples. The units mol kg^{-1} are defined as moles per kilogram of seawater. Data from Bruland KW and Franks RP (1983) Mn, Ni, Cu, Zn and Cd in the western North Atlantic. In: Wong CS, Boyle E, Bruland KW, Burton JD, and Goldberg ED (eds.) *Trace Metals in Sea Water*, pp. 395–414. New York: Plenum.

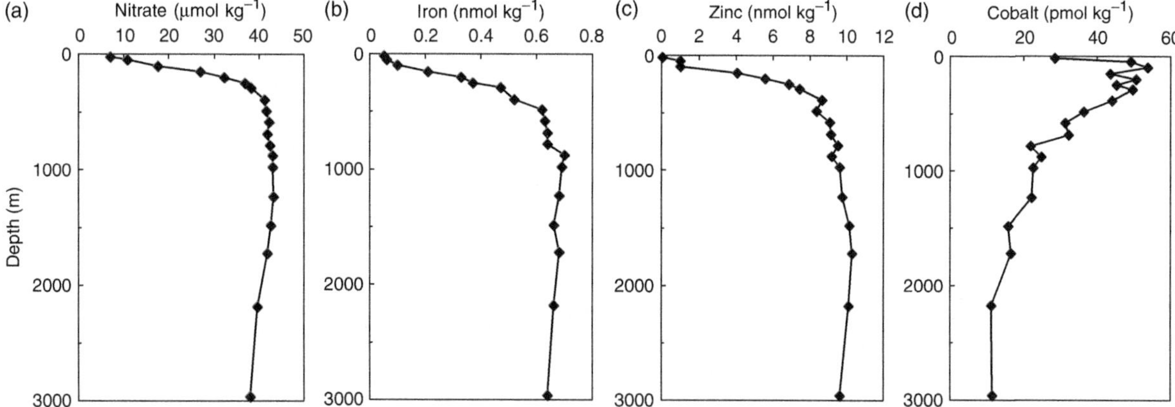

Figure 3 Depth profiles for nitrate and filterable concentrations of trace element nutrients (iron, zinc, and cobalt) in the subarctic North Pacific Ocean (ocean station *Papa*, 50.0° N, 145.0° W, Aug. 1987). Data from Martin JH, Gordon RM, Fitzwater S, and Broenkow WW (1989) VERTEX: Phytoplankton/iron studies in the Gulf of Alaska. *Deep-Sea Research* 36: 649–680.

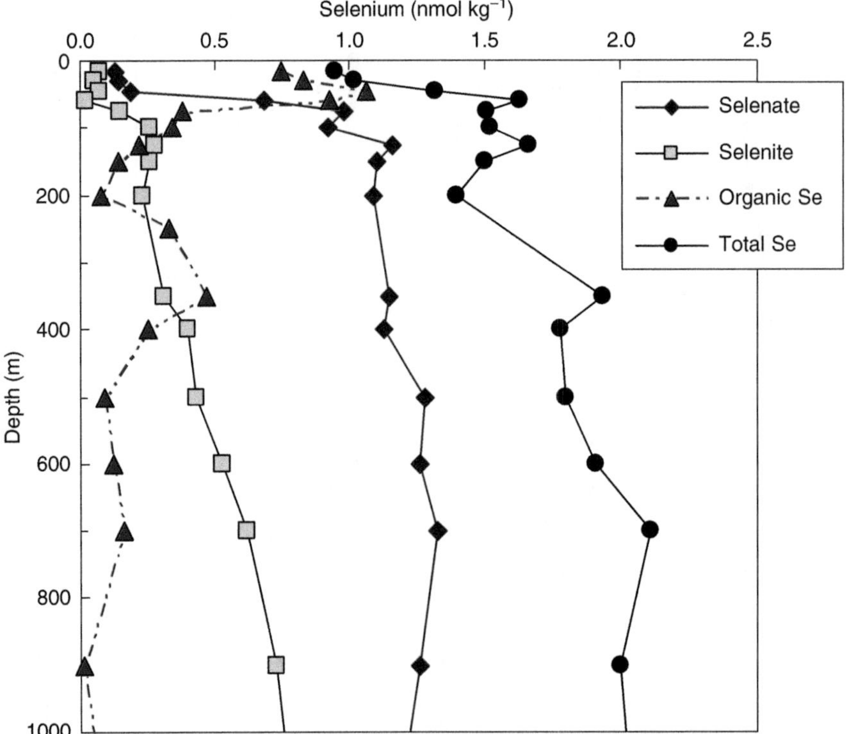

Figure 4 Depth profiles for concentrations of total selenium and different chemical forms of selenium (selenate, selenite, and organic selenium compounds) in filtered seawater samples from the eastern tropical North Pacific Ocean (18° N, 108° W; Oct.–Nov. 1981). Data are from Cutter GA and Bruland KW (1984) The marine biogeochemistry of selenium: A reevaluation. *Limnology and Oceanography* 29: 1179–1192.

materials. However, a portion of the assimilated nutrients is continuously lost from the euphotic zone by vertical settling of intact algal cells or zooplankton fecal pellets. The macro- and micronutrients contained within these settling biogenic particles are then returned to solution at depth in the ocean via bacterial degradation processes. Ultimately the uptake, settling, and regeneration processes deplete nutrients within the euphotic zone to low levels while concentrations at depth are increased. This process also transfers CO_2 to the deep sea and is often refered to as the biological CO_2 pump. The cycle is completed when the nutrient and CO_2 reservoirs at depth are returned to the surface via vertical advection (upwelling) and mixing processes.

The deep-water concentrations of both major nutrient elements (N, P, and Si) and many micronutrients (Zn, Cd, Ni, and Cu) are much higher in deep waters of the Pacific than the Atlantic (**Figure 2**) because of large-scale ocean circulation patterns, in which deep waters are formed via subduction at high latitudes in the North Atlantic and are returned to the surface via upwelling in the northern regions of the North Pacific and Indian Oceans. Because of these patterns, the deep North Pacific contains waters that have resided at the bottom for much longer (~1000 years) than the deep Atlantic waters and thus have had a much longer time to accumulate major nutrient and micronutrient elements from biological regeneration processes.

Several trace element nutrients (molybdenum, manganese, and cobalt) provide exceptions to the general trend of increasing concentrations with depth. Molybdenum occurs almost exclusively as soluble, nonreactive molybdate ions MoO_4^{2-}, which occur at a high concentration (~105 nM) relative to their biological demand (**Table 1**). Consequently, there is minimal biological removal of molybdenum from surface seawater and its concentration varies in proportion to salinity. By contrast, concentrations of manganese (**Figure 2(h)**) and cobalt (**Figure 3(d)**) are typically maximum near the surface and depleted at depth owing to deep-water scavenging processes.

Chemical Speciation

Trace element nutrients exist as a variety of chemical species in the sea, which strongly influences their chemical behavior and biological availability. All but Se and Mo occur as cationic metal ions that are complexed (bound) to varying degrees by inorganic and organic ligands (complexing agents) or are adsorbed onto or bound within particles. Many trace element nutrients (iron, copper, manganese, cobalt, and selenium) cycle between different oxidation states, which have quite different kinetic labilities (reaction rates), solubilities, binding strengths with organic ligands, and biological availabilities.

Nickel, zinc, and cadmium exist in normal oxygenated seawater as highly soluble divalent cations that are complexed to varying degrees by inorganic ligands (Cl^-, OH^-, and CO_3^{2-}) and organic chelators. Nickel is bound to only a small extent (0–30%) by organic ligands. By contrast, ~99% of the zinc ions and 70% of the cadmium are heavily complexed by unidentified strong organic ligands present at low concentrations in surface waters of the North Pacific. The strong chelation of zinc reduces the concentration of dissolved inorganic zinc to ~1 pM in surface seawater, sufficiently low to limit the growth of many algal species.

Manganese undergoes redox transformations, but is minimally bound to organic ligands. The stable redox species of manganese in oxygenated seawater, Mn(IV) and Mn(III) oxides, are insoluble, although Mn(III) can exist in some instances as soluble organic chelates. Mn(III) and Mn(IV) can be reduced chemically, photochemically, or biologically to dissolved Mn(II), which is fully soluble in seawater and is not appreciably bound by organic ligands. Although Mn(II) is unstable with respect to oxidation by molecular oxygen, the chemical kinetics of this reaction are exceedingly slow in seawater. However, Mn(II) oxidation is greatly accelerated by bacterial enzymes that catalyze Mn(II) oxidation to Mn(IV) oxides. The bacterial formation of Mn oxides, and subsequent removal via coagulation and settling of oxide particles, results in short residence times (20–40 years in the North Pacific) and low concentrations for manganese in deep-ocean waters (**Figure 2(h)**). Oxidation is absent or greatly diminished in the ocean's surface mixed layer due to photo-inhibition of the Mn-oxidizing bacteria. The absence of bacterially mediated oxidation of Mn(II) and minimal organic chelation often results in high concentrations of Mn^{2+} ions in surface seawater (**Table 1**; **Figure 2(h)**), enhancing the supply of Mn to phytoplankton.

Iron is the most biologically important trace metal nutrient, and its chemical behavior is perhaps the most complex. Its stable oxidation state in oxygen-containing waters is Fe(III), which forms sparingly soluble iron hydroxide and oxide precipitates. This oxide formation and the tendency of ferric ions to adsorb onto particle surfaces results in the scavenging of iron from seawater via particulate aggregation and settling processes. This removal results in short residence times for iron in deep-ocean waters (~50–100 years) and low concentrations (0.4–0.8 nM) despite the high crustal abundance of iron (it is the fourth most abundant element by weight). Most (>99%) of the dissolved ferric iron in seawater is bound to organic ligands which minimizes iron adsorption and precipitation, and thus reduces the removal of iron from seawater by particulate scavenging processes. Some of these organic ligands may be strong ferric chelators (siderophores) produced by bacteria to solubilize iron and facilitate intracellular iron uptake. Ferric iron can be reduced in seawater to highly soluble Fe(II) (ferrous iron) by a number of processes including photo-reduction of organic chelates in surface waters, biological reduction of iron at cell surfaces, and reduction by chemical reducing agents. Because ferrous iron binds much more weakly to organic chelators than ferric iron, the photo-reduction or biological reduction of iron in ferric chelates often results in the dissociation of iron from the chelates, which increases iron availability for biological uptake (see below). The released ferrous ions are unstable in oxygenated seawater, and are reoxidized to soluble ferric hydrolysis species, and recomplexed by organic ligands on timescales of minutes. Thus iron undergoes a dynamic redox cycling in surface seawater, which can greatly enhance its biological availability to phytoplankton.

Other micronutrient metals such as copper and cobalt also exist in multiple oxidation states and are heavily complexed by organic chelators. Copper can exist in seawater as thermodynamically stable copper(II), or as copper(I). Most (>99%) of the copper in near-surface seawater is heavily chelated by strong organic ligands present at low concentrations (2–3 nM in ocean waters). This chelation decreases free cupric (copper II) ion concentrations to very low levels (0.1–1 pM). Copper(II) can be reduced to Cu(I) by photochemical and biological processes or by reaction with chemical reducing agents, such as sulfur-containing organic ligands. The resultant Cu(I) can be reoxidized by reaction with molecular oxygen, but the effect of this redox cycling on the biological availability of copper is currently unknown.

The chemistry of cobalt is also highly complex. Cobalt exists in seawater as soluble cobalt(II) or as cobalt(III), which forms insoluble oxides at the pH of seawater. The formation of these oxides appears to be microbially mediated and is largely responsible for the removal of cobalt from deep-ocean waters and for the resultant low deep-ocean concentrations (**Figure 3(d)**). Much of the dissolved cobalt in seawater is strongly bound to organic ligands, and recent evidence suggests that this cobalt exists as kinetically inert cobalt(III) chelates. There is also evidence that these cobalt(III)-binding ligands are produced by marine cyanobacteria and that these ligands may facilitate microbial uptake of cobalt.

Selenium is a metalloid, which occurs immediately below sulfur in the periodic table. Consequently, its chemical behavior often mimics that of sulfur. Selenium exists in subsurface seawater primarily as soluble oxyanions selenate (SeO_4^{2-}; +6 oxidation state) and selenite (SeO_3^{2-}; +4 oxidation state). Phytoplankton preferentially take up selenite which depletes its concentration in surface seawater (**Figure 4**). Selenate is then taken up and depleted following the removal of selenite. The selenate and selenite ions taken up by phytoplankton are metabolically reduced to the selenide (−2 oxidation state) and used to synthesize selenomethionine and selenocysteine, chemical analogs of the sulfur-containing amino acids methionine and cysteine. In surface waters a majority of the selenium occurs as biologically regenerated organic selenide compounds of unknown chemical structure (**Figure 4**).

Biological Uptake

All trace elements are taken up intracellularly by specialized transport proteins (enzymes) on the outer membrane of plankton cells. Consequently, uptake rates generally follow Michaelis–Menten enzyme kinetics:

$$\text{Uptake rate} = V_{max}S/(K_s + S)$$

V_{max} is the maximum uptake rate, S is the concentration of the pool of chemical species that react with receptor sites on the transport protein, and K_s is concentration of the substrate pool at which half of the transport protein is bound, and the uptake rate is half of V_{max}. Virtually all of these proteins act as pumps and require energy for intracellular uptake. Each transport system reacts with a single chemical species or group of related chemical species and thus chemical speciation is extremely important in regulating cellular uptake. Uptake systems range from simple to highly complex depending on the chemical speciation of the nutrient element and its biological demand (requirement) relative to its external availability.

Uptake systems appear to be simplest for dissolved Mn(II), which is taken up in phytoplankton by a single high-affinity transport system that is under negative feedback regulation. In this negative feedback, as the concentration of dissolved Mn(II) decreases, the V_{max} of the transport system is increased to maintain relatively constant Mn uptake rates and intracellular concentrations.

Uptake systems for zinc, cadmium, cobalt(II), and copper(II) are somewhat more complex. The phytoplankton species examined to date have at least two separate zinc transport systems: a low-affinity system whose V_{max} is relatively constant, and an inducible high-affinity system. The low-affinity system has high V_{max} and high K_s values and transports zinc at high zinc ion concentrations. The high-affinity system is responsible for zinc uptake at low zinc ion concentrations, and has low K_s, and variable V_{max} values that are under negative feedback regulation. At sufficiently low concentrations of dissolved inorganic zinc species (~10 pM), the cellular uptake approaches limiting rates for the diffusion of labile inorganic zinc species to the cell surface. The existence of high- and low-affinity transport systems results in sigmoidal relationships between zinc uptake rates (and cellular Zn:C ratios) and concentrations of dissolved inorganic zinc species as seen in **Figure 5** for an oceanic diatom.

Cobalt and sometimes cadmium can metabolically substitute for zinc in many metalloenzymes. To facilitate this substitution, the uptake of these divalent metals is increased by over 100-fold in diatoms with decreasing dissolved inorganic zinc concentrations and resulting decreases in cellular zinc uptake rates (**Figure 5**). Uptake of Cd by this inducible transport system is repressed at high intracellular zinc levels, and under these conditions, cadmium leaks into the cell through the cell's Mn(II) transport system. Thus cellular uptake of cadmium in the ocean is regulated by complex interactions among dissolved inorganic concentrations of Cd, Zn, and Mn. Likewise, since cobalt uptake is repressed at high zinc ion concentrations, biological depletion of cobalt often does not occur until after zinc is depleted, as observed in the subarctic Pacific (**Figure 6**).

The binding and subsequent intracellular uptake of the above divalent metals (Zn^{2+}, Mn^{2+}, Cd^{2+}, Co^{2+}, and Cu^{2+}) by the various intracellular uptake systems are regulated by the concentration of dissolved inorganic metal species (free aquated ions and inorganic complexes with chloride ions, hydroxide ions, etc.). Organic complexation of these metals reduces their uptake by decreasing the concentration of dissolved inorganic metal species. This effect can be substantial in cases such as zinc, where up to 99% or more of the metal is bound to organic ligands in surface seawater.

Since iron is the most limiting of the trace element nutrients and its chemistry the most complex, it is perhaps not surprising that the transport systems for iron are the most varied and complex. Iron is highly

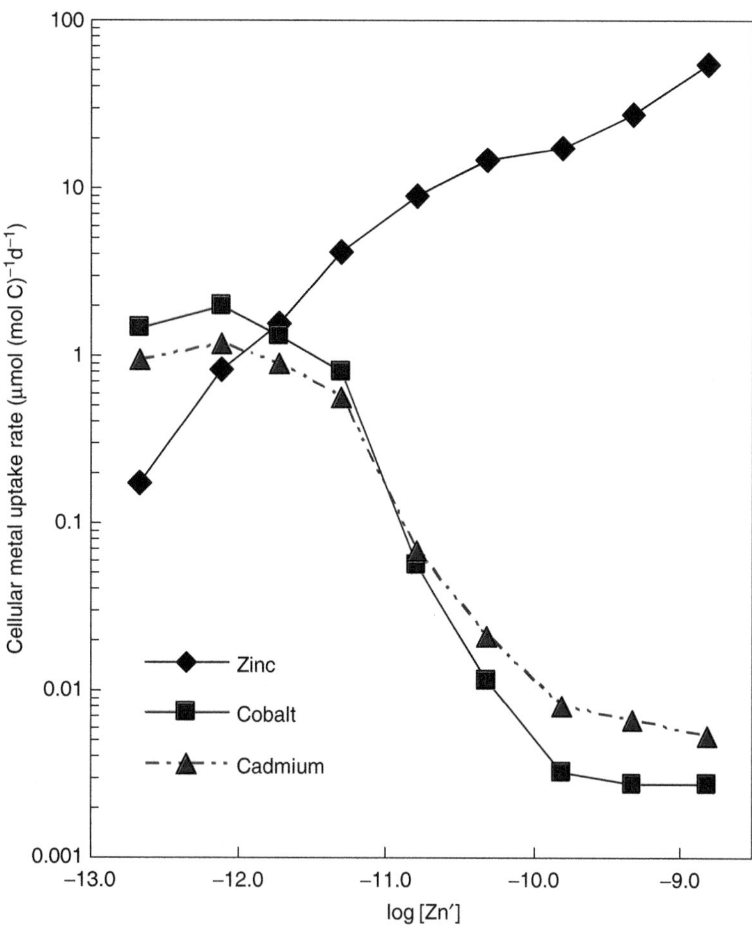

Figure 5 Cellular uptake rates for zinc, cobalt, and cadmium (normalized per mol of cell carbon) for the oceanic diatom *Thalassiosira oceanica* plotted as a function of the $\log_{10}$ of the molar concentration of dissolved inorganic zinc species (Zn′, aquated zinc ions plus inorganic zinc complexes). Dissolved inorganic cobalt and cadmium species in the seawater medium were held constant at concentrations of 1.5 and 2.7 pM (10^{-12} M), respectively. Uptake rates for cadmium and cobalt increase by at least 2 orders of magnitude when Zn′ concentrations decrease below 10^{-10} M. The large increase in uptake rates reflects the induction of high-affinity cellular transport systems for Cd and Co in response to declining intracellular Zn concentrations. Data are from Sunda WG and Huntsman SA (2000) Effect of Zn, Mn, and Fe on Cd accumulation in phytoplankton: Implications for oceanic Cd cycling. *Limnology and Oceanography* 45: 1501–1516.

bound as ferric oxides and organic chelates and prokaryotic and eukaryotic plankton cells have evolved different strategies to access these bound forms of iron. Prokaryotic cells (cyanobacteria and heterotrophic bacteria) have evolved high-affinity uptake systems that are induced under iron deficiency. These systems involve the biosynthesis and extracellular release of a variety of high-affinity iron chelators (siderophores) that strongly bind iron(III) in the surrounding seawater. The siderophore chelates are then actively taken up into the cells by transport proteins on the outer cell membrane. The siderophore chelates have different chemical structures, and different outer membrane siderophore transport proteins are needed to take up structurally distinct siderophores or groups of siderophores with similar chemical structures. Bacteria often take up not only their own siderophores, but those produced by other bacteria, resulting in complex ecological interactions among bacteria.

Eukaryotic phytoplankton do not appear to produce siderophores and there is little evidence for direct cellular uptake of ferric siderophore chelates. Instead there is mounting evidence for the utilization of a high-affinity transport system that accesses ferric complexes via their reduction at the cell surface and subsequent dissociation of the resulting ferrous-ligand complexes. The released ferrous ions bind to iron(II) receptors on iron transport proteins located on the outer cell membrane, which transport the iron into the cell. This intracellular transport involves the reoxidation of bound iron(II) to iron(III) by a copper protein, and thus copper is required for cellular iron uptake. The availability of iron to this transport

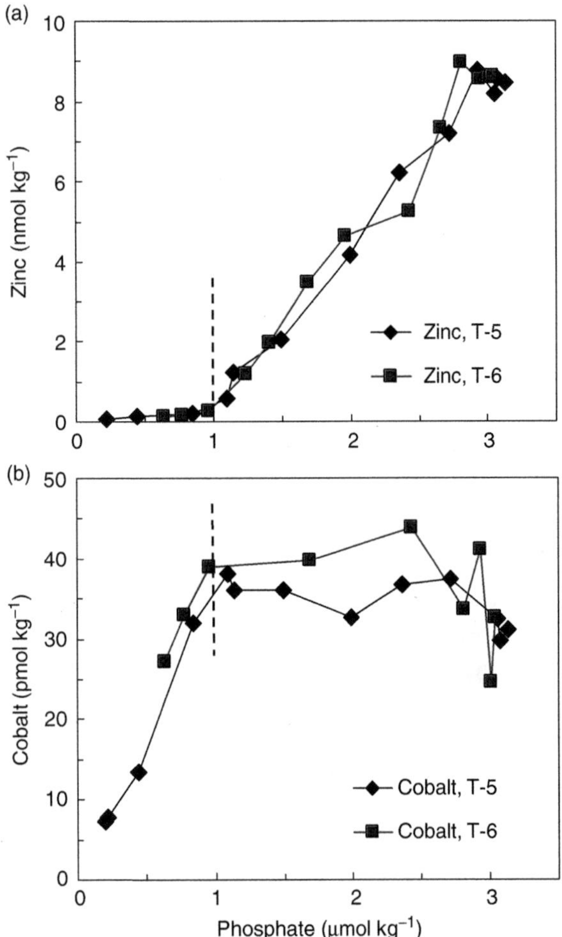

Figure 6 Plots of filterable zinc and cobalt concentrations vs. phosphate at two stations in the subarctic Pacific (Station T-5, 39.6° N, 140.8° W and Station T-6, 45.0° N, 142.9° W, Aug. 1987). The decrease in zinc with decreasing phosphate is caused by the simultaneous removal of both metals via cellular uptake and assimilation by phytoplankton. Cobalt becomes depleted by phytoplankton uptake only after zinc concentrations drop to very low levels ($<0.2\,\mathrm{nmol\,kg^{-1}}$). This pattern is consistent with metabolic replacement of cobalt for zinc, as observed in phytoplankton cultures (see **Figure 5**). Data plots after Sunda WG and Huntsman SA (1995) Cobalt and zinc interreplacement in marine phytoplankton: Biological and geochemical implications. *Limnology and Oceanography* 40: 1404–1417.

system is dependent on the reduction potential of the ferric complexes; consequently, readily reducible ferric species such as dissolved inorganic ferric hydroxide complexes are accessed much more readily by this system than are strongly bound ferric siderophore chelates. Thus, iron uptake by this system is highly dependent on the chemical speciation of iron in seawater. Photo-reductive dissociation of ferric chelates increases iron availability to this system, since the released ferrous ions can directly react with the membrane transport protein and the reoxidized ferric hydrolysis species are readily reduced and taken up.

Metabolic Requirements and their Relation to Other Limiting Resources

Trace element micronutrients are essential for the growth and metabolism of all marine algae and bacteria. They play critical roles in photosynthesis, respiration, and the assimilation and transformation of essential macronutrients (nitrogen, phosphorus, and silicic acid). Thus trace metal requirements can be influenced by the availability of light, CO_2, and major nutrients and the cycles of major nutrient elements are influenced by trace element nutrients. Of the micronutrient metals, iron is needed in the greatest amount and is the metal that most frequently limits algal growth. Iron serves essential metabolic functions in photosynthetic electron transport, respiration, nitrate assimilation, N_2 fixation, and detoxification of reactive oxygen species (e.g., superoxide radicals and hydrogen peroxide). Because of its heavy involvement in photosynthetic electron transport, cellular iron requirements increase with decreasing light intensity and photoperiod. Such effects can lead to iron–light co-limitation in low-light environments such as regions where the depth of the surface wind mixed layer greatly exceeds the depth of light penetration (as often occurs in the Southern Ocean and at high latitudes during the winter) or in the deep chlorophyll maximum at the bottom of the euphotic zone (the sunlit layer) in thermally stratified surface waters.

Iron also occurs in the enzymes (nitrate and nitrite reductases) involved in the reduction of nitrate to ammonium in phytoplankton and the enzyme complex (nitrogenase) that fixes nitrogen (reduces dinitrogen molecules to ammonia) in cyanobacteria. Both processes require cellular energy (in the form of ATP molecules) and reductant molecules (NADPH), and iron is also needed in high amounts for the photosynthetic production of the needed ATP and NADPH. Algal cells growing on nitrate need ~50% more iron to support a given growth rate than cells growing on ammonium. Consequently, iron can be especially limiting in oceanic upwelling systems (such as the equatorial and subarctic Pacific) where waters containing high nitrate concentrations, but low iron, are advected to the surface (see **Figures 3(a)** and **3(b)**). Even higher amounts of iron (up to 5 times as much) are needed for diazotrophic growth (growth on N_2) than for equivalent growth on ammonium due to high energetic (ATP) cost for nitrogen fixation and the large amount of iron in the nitrogenase enzyme complex. As a result, iron appears to limit N_2 fixation in large regions of the ocean and is thought to control oceanic inventories of fixed nitrogen. As a consequence, nitrogen is the primary limiting major nutrient in most ocean waters, while in lakes, where iron concentrations are much higher, phosphate is the primary limiting nutrient. Due to iron limitation of C fixation and N_2 fixation in major regions of the ocean, iron plays a significant role in regulating carbon and nitrogen cycles in the ocean. It thus helps regulate the biological CO_2 pump discussed earlier, which through transport of carbon to the deep ocean, controls the ocean/atmosphere CO_2 balance and CO_2-linked greenhouse warming. There is evidence that climatically driven variations in the input of iron-rich continental dust to the ocean has played an important role in regulating glacial–interglacial climate cycles.

Manganese occurs in the water-splitting complex of photosystem II, and thus is essential for photosynthesis. Consequently, like iron, it is needed in higher amounts for growth at low light. Manganese also occurs in superoxide dismutase, an antioxidant enzyme that removes toxic superoxide radicals, produced as byproducts of photosynthesis. Because it has fewer metabolic functions, its cellular growth requirement is less than that of iron. Manganese may limit algal growth in certain low-Mn environments such as the subarctic Pacific and Southern Ocean, where manganese additions have been observed to stimulate algal growth in bottle incubation experiments.

Zinc serves a variety of metabolic functions and has a cellular requirement similar to that for manganese. It occurs in carbonic anhydrase (CA), an enzyme critical to intracellular CO_2 transport and fixation. Higher amounts of this enzyme are needed at low CO_2 concentrations, leading to potential co-limitation by zinc and CO_2 in the ocean. However, the $\sim 35\%$ increase in CO_2 in the atmosphere and surface ocean waters from the burning of fossil fuels makes Zn–CO_2 co-limitation less likely in the modern ocean than in preindustrial times. Zinc also occurs in zinc finger proteins, involved in DNA transcription, and in alkaline phosphatase, needed to acquire phosphorus from organic phosphate esters, which dominate phosphate pools in low-phosphate ocean waters. Consequently, Zn and P may co-limit algal growth in regions where both nutrients occur at low concentrations such as the central gyre of the North Atlantic.

Cobalt, and sometimes cadmium, can substitute for zinc in many zinc enzymes such as CA, leading to complex interactions among the three metals in marine algae (**Figure 5**). The presence of cadmium in CA appears to explain its nutrient-like distribution in ocean waters (**Figure 2(e)**), and the identification of a unique Cd-CA enzyme in marine diatoms means that it functions as a micronutrient in these organisms. Cobalt also occurs in vitamin B_{12}, an essential vitamin required for growth of many eukaryotic algal species. This vitamin is synthesized only by bacteria, resulting in potential interactions between B_{12}-producing bacteria and B_{12}-requiring eukaryotic algae in the ocean. A specific requirement for cobalt not involving B_{12} is seen in marine cyanobacteria and bloom-forming prymnesiophytes (including *Emiliania huxleyi*), but the biochemical basis for this is not known. Both zinc and cobalt additions have been shown to stimulate phytoplankton growth in bottle incubation experiments in the subarctic Pacific and in some coastal upwelling regimes along the eastern margin of the Pacific, but the effects were modest relative to those for added iron. However, zinc addition had a large effect on algal species composition, and preferentially stimulated the growth of coccolithophores, an algal group largely responsible for calcium carbonate formation in the ocean. Biogenic $CaCO_3$ formation helps regulate the alkalinity (acid–base balance) of ocean water, which in turn affects oceanic CO_2 concentrations, and air–sea flux of this important greenhouse gas. By influencing the growth of coccolithophores, zinc could indirectly affect atmospheric CO_2 levels and global climate.

Copper occurs in cytochrome oxidase, a key protein in respiratory electron transport, and in plastocyanin, which substitutes for the iron protein cytochrome c_6 in photosynthetic electron transport in oceanic phytoplankton. It is also an essential component of the high-affinity iron transport system of many eukaryotic algae. Because copper is needed for iron uptake and can metabolically substitute for iron, co-limitations can occur for Cu and Fe, as observed in some diatoms.

Nickel and molybdenum, like iron, play important roles in nitrogen assimilation. Nickel occurs in the enzyme urease, and thus is required by phytoplankton grown on urea as a nitrogen source. It also occurs in Ni-superoxide dismutase found in many marine cyanobacteria, which, like the Mn and Fe forms of the enzyme, removes harmful superoxide radicals from cells. Little is currently known about the potential for nickel limitation in the ocean.

Molybdenum occurs with iron in the enzymes nitrate reductase and nitrite reductase and in nitrogenase, and consequently is utilized in nitrate assimilation and N_2 fixation. Along with the Fe–Mo enzyme, there are two other isoforms of nitrogenase, a primitive less-efficient form containing only iron in its active center, and another which contains iron and vanadium. Thus molybdenum is not absolutely essential for dinitrogen fixation, although the predominance of the more efficient Fe–Mo isoform in the modern ocean helps to minimize iron limitation of nitrogen fixation. Because of its high concentration in seawater (*c.* 105 nM), Mo does not appear to limit algal growth or N_2 fixation in the ocean.

The metalloid selenium is also essential for the growth of many marine phytoplankton. It occurs in glutathione peroxidase, an enzyme that degrades hydrogen peroxide, and thus is important in antioxidant protection. However, it is likely that selenium has other as-yet-unidentified metabolic functions. The potential for selenium limitation in the ocean is currently unknown.

Biological Feedback on Seawater Chemistry

Trace elements not only influence the productivity and species composition of planktonic communities, but the plankton have a profound effect on the chemistry and cycling of these elements on a variety of temporal and spatial scales (**Figure 1**). The most obvious example is the effect of algal uptake, particulate settling, and regeneration cycles on the vertical distribution and interocean transfer of trace element nutrients (Fe, Zn, Cd, Ni, Cu, and Se; **Figures 2–4**). In addition, bacteria largely mediate the removal of dissolved manganese and cobalt from subsurface seawater via the formation of Mn(IV) and Co(III) oxides.

There is evidence that the organic ligands that strongly bind iron, copper, zinc, and cobalt are produced either directly or indirectly by the biota. In the North Pacific, the organic ligands that strongly bind copper occur at highest levels at the depth of maximum productivity, and decrease below the euphotic zone. Ligands having the same copper-binding strength are produced by *Synechococcus*, an abundant group of oceanic cyanobacteria. There is evidence that these organisms produce the chelators to detoxify copper by decreasing free cupric ion concentrations. The organic ligands that strongly bind iron(III), cobalt, and zinc also have a beneficial effect. The iron ligands tightly bind ferric ions in soluble chelates and thereby minimize the abiotic removal of iron from seawater via the formation of insoluble ferric oxides or ferric ion adsorption onto particulate surfaces. Without such chelating ligands, iron concentrations would likely be much lower, and the productivity of the ocean would be greatly reduced. The Co(III)-binding ligands serve a similar function in limiting the formation of insoluble Co(III) oxides, a major mechanism for removal of cobalt from seawater. Recent culture experiments and seawater incubation experiments suggest that these ligands are produced by the cyanobacterial genus *Synechococcus*, whose growth may be limited by cobalt in some regions of the ocean.

Zinc chelators also serve a beneficial function, not only by minimizing abiotic scavenging of zinc in surface waters, but also by preventing the extremely efficient uptake systems of eukaryotic phytoplankton from completely depleting this essential micronutrient element from surface ocean waters.

Thus trace element nutrients and marine plankton comprise an interactive system in the ocean in which each exerts a controlling influence on the composition and dynamics of the other (**Figure 1**). On longer geological timescales, the feedback interactions between the biota and trace metal chemistry and availability have been profound. Currently, the air we breathe and virtually the entire ocean, with the exception of a few isolated anoxic basins (e.g., the Black Sea), contain free dioxygen molecules (O_2), generated over billions of years from its release as a byproduct of oxygenic photosynthesis. The presence of free O_2 sets the redox state of modern ocean toward oxidizing conditions, which as noted previously, limits the solubility of essential transition metals (Fe, Co, and Mn) whose stable oxidation states under these conditions are insoluble Co(III) and Mn(IV) oxides or sparingly soluble Fe(III) oxides. However, prior to the advent of oxygenic photosynthesis *c.* 3 billion years ago, the chemistry of the ocean was far different from that which exists today. There was no free oxygen and the entire ocean and Earth's surface was much more reducing. Under these conditions, the stable redox state of Fe, Mn, and Co was soluble Fe(II), Mn(II), and Co(II), and that of copper was Cu(I). Furthermore, the stable redox form of sulfur was sulfide (−2 oxidation state), rather sulfate (+6 oxidation state), which occurs in present-day seawater at a relatively high concentration (28 mM). The presence of moderate to high levels of sulfide greatly restricted the availability of zinc, copper, molybdenum, and cadmium, which form insoluble sulfide precipitates; but it had a much lesser impact on other metals (Mn^{2+}, Fe^{2+}, Co^{2+}, and Ni^{2+}) whose sulfides are much more soluble. Thus, early life in the ocean evolved in an environment of high availability of Fe, Mn, Co, and Ni and low availabilities of Zn, Mo, Cu, and Cd, contrasting the situation in the modern ocean. Given the utility of Fe as a redox catalyst and its relative abundance in the Earth's crust and ancient ocean, it is perhaps not surprising that this metal was utilized in the evolution of the major redox catylysts of life. It occurs in high amounts in the redox centers of nitrogenase responsible for dinitrogen fixation and in the various proteins and protein complexes involved in oxygenic photosynthesis (photosystem I, photosystem II, cytochorme b_6/f complex, ferredoxin, and cytochrome *c*). In addition, the abundant soluble manganese in the early ocean was utilized in the water-oxidizing centers of photosystem II. The combined action of these iron- and manganese-containing biological redox catalysts provided for efficient fixation of N_2 and CO_2 needed for production of proteins and other biological compounds. The concomitant release of O_2 from photosynthesis and sequestration of organic carbon in marine sediments and sedimentary rocks, slowly (over 1–2 billion years) oxidized ferrous iron to ferric oxides and sulfide species to soluble sulfate, ultimately resulting in the buildup of free O_2 first in the surface ocean and atmosphere, and gradually in the ocean as a whole. The precipitation of ferric oxides from the sea has resulted in the chronic Fe limitation of carbon fixation and N_2 fixation that we currently observe in the ocean. However, this negative effect is more than balanced by the usefulness of molecular oxygen in highly efficient oxygen-dependent respiration utilized by all present-day aerobic microbes, plants, and animals. Furthermore, the release of zinc, copper, molybdenum, and cadmium from insoluble sulfides allowed for the subsequent evolution of numerous new enzymes utilizing these metals, which appear to have evolved following the appearance of free O_2. Thus, evolution has involved a continuous feedback between biological systems and the surrounding chemical environment, with biological trace metal catalysts playing a central mediating role in this process.

Glossary

ATP Adenosine triphosphate; a high-energy compound produced in photosynthesis and respiration which is used as the main energy currency of cells.

Chemical speciation The different chemical forms of trace elements.

Chelate A strong complex between an organic ligand and a metal.

Chelation The reaction of a metal with an organic ligand to form a chelate.

Chelator An organic ligand that forms stable complexes with metal ions.

Cytochrome b_6/f complex An iron-rich protein complex involved proton pumping and ATP synthesis in photosynthesis.

Fe(II), Fe(III) Iron with oxidation states of +2 and +3, respectively, also referred to as ferrous and ferric iron.

Ferredoxin A soluble iron–sulfur protein involved in photosynthetic electron transport.

Ferric hydrolysis species Inorganic complexes of iron(III) with one to four hydroxide ions: $FeOH^{2+}$, $Fe(OH)_2^+$, $Fe(OH)_3$, and $Fe(OH)_4^-$.

Metalloenzyme An enzyme containing a metal as an essential functional component.

Mn(II), Mn(III), Mn(IV) Manganese with oxidation states of +2, +3, and +4, respectively.

NADPH The reduced form of nicotinamide adenine dinucleotide phosphate, which is produced in photosynthesis and serves as the primary reductant molecule in plant cells.

Nitrogenase An iron-containing enzyme complex responsible for nitrogen fixation.

Photosystem I and photosystem II The two photochemical reaction centers in photosynthesis.

Redox Chemical reduction and oxidation.

Siderophore A high-affinity organic ligand produced by bacteria to complex iron and facilitate its intracellular uptake.

Superoxide radical A free radical of chemical structure ($\bullet O_2^-$) formed from the single electron reduction of molecular oxygen.

See also

Carbon Cycle. Iron Fertilization. Nitrogen Cycle. Transition Metals and Heavy Metal Speciation.

Further Reading

Anbar AD and Knoll AH (2002) Proterozoic ocean chemistry and evolution: A bioinorganic bridge? *Science* 297: 1137–1142.

Barbeau K, Rue EL, Trick CG, Bruland KW, and Butler A (2003) Photochemical reactivity of siderophores produced by marine heterotrophic bacteria and cyanobacteria based on characteristic Fe(III) binding groups. *Limnology and Oceanography* 48: 1069–1078.

Boyle E, Edmond JM, and Sholkovitz ER (1977) The mechanism of iron removal in estuaries. *Geochimica Cosmochimica Acta* 41: 1313–1324.

Brand LE, Sunda WG, and Guillard RRL (1983) Limitation of marine phytoplankton reproductive rates by zinc, manganese and iron. *Limnology and Oceanography* 28: 1182–1198.

Bruland KW (1989) Complexation of zinc by natural organic ligands in the central North Pacific. *Limnology and Oceanography* 34: 269–285.

Bruland KW (1992) Complexation of cadmium by natural organic ligands in the central North Pacific. *Limnology and Oceanography* 37: 1008–1017.

Bruland KW and Franks RP (1983) Mn, Ni, Cu, Zn and Cd in the western North Atlantic. In: Wong CS, Boyle E, Bruland KW, Burton JD, and Goldberg ED (eds.)

Trace Metals in Sea Water, pp. 395–414. New York: Plenum.

Coale KH (1991) Effects of iron, manganese, copper, and zinc enrichments on productivity and biomass in the subarctic Pacific. *Limnology and Oceanography* 36: 1851–1864.

Coale KH, Johnson KS, Chavez FP, *et al.* (2004) Southern Ocean iron enrichment experiment, carbon cycling in high- and low-Si waters. *Science* 304: 408–414.

Crawford DW, Lipsen MS, Purdie DA, *et al.* (2003) Influence of zinc and iron enrichments on phytoplankton growth in the northeastern subarctic Pacific. *Limnology and Oceanography* 48: 1583–1600.

Cutter GA and Bruland KW (1984) The marine biogeochemistry of selenium: A reevaluation. *Limnology and Oceanography* 29: 1179–1192.

da Silva JJRF and Williams RJP (1991) *The Biological Chemistry of the Elements*. Oxford, UK: Clarendon.

Donat JR and Bruland KW (1995) Trace elements in the oceans. In: Salbu B and Steinnes E (eds.) *Trace Elements in Natural Waters*, pp. 247–281. Boca Raton, FL: CRC Press.

Duce RA and Tindale NW (1991) Atmospheric transport of iron and its deposition in the ocean. *Limnology and Oceanography* 36: 1715–1726.

Falkowski PG (1997) Evolution of the nitrogen cycle and its influence on the biological sequestration of CO_2 in the ocean. *Nature* 387: 272–275.

Ho T, Quigg A, Findel ZV, *et al.* (2003) The elemental composition of some marine phytoplankton. *Journal of Phycology* 39: 1145–1159.

Hutchins DA, Hare CE, and Weaver RS (2002) Phytoplankton iron limitation in the Humboldt Current and Peru Upwelling. *Limnology and Oceanography* 47: 997–1011.

Ito Y and Butler A (2005) Structure of synechobactins, new siderophores of the marine cyanobacterium *Synechococcus* sp. PCC 7002. *Limnology and Oceanography* 50: 1918–1923.

Johnson KS, Gordon RM, and Coale KH (1997) What controls dissolved iron concentrations in the world ocean? *Marine Chemistry* 57: 137–161.

Kustka AB, Sañudo-Wilhelmy S, Carpenter EJ, *et al.* (2003) Iron requirements for dinitrogen and ammonium supported growth in cultures of *Trichodesmium* (IMS 101): Comparison with nitrogen fixation rates and iron:carbon ratios of field populations. *Limnology and Oceanography* 48: 1869–1884.

Maldonado MT and Price NM (1996) Influence of N substrate on Fe requirements of marine centric diatoms. *Marine Ecology Progress Series* 141: 161–172.

Martin JH, Gordon RM, Fitzwater S, and Broenkow WW (1989) VERTEX: Phytoplankton/iron studies in the Gulf of Alaska. *Deep-Sea Research* 36: 649–680.

Morel FMM and Price NM (2003) Biogeochemical cycles of trace metals in the oceans. *Science* 300: 944–947.

Morel FMM, Reinfelder JR, Roberts SB, *et al.* (1994) Zinc and carbon co-limitation of marine phytoplankton. *Nature* 369: 740–742.

Rue EL and Bruland KW (1995) Complexation of iron(III) by natural organic ligands in the central North Pacific as determined by a new competitive ligand equilibration/adsorptive cathodic stripping voltammetric method. *Marine Chemistry* 50: 117–138.

Saito MA, Moffett JW, and Ditullio GR (2004) Cobalt and nickel in the Peru Upwelling region: A major flux of labile cobalt utilized as a micronutrient. *Global Biogeochemical Cycles* 18: GB4030.

Saito MA, Sigman DM, and Morel FMM (2003) The bioinorganic chemistry of the ancient ocean: The co-evolution of cyanobacterial metal requirements and biogeochemical cycles at the Archean–Proterozoic boundary? *Inorganica Chimica Acta* 356: 308–318.

Shaked Y, Kustka AB, and Morel FMM (2005) A general kinetic model for iron aquisition by eucaryotic phytoplankton. *Limnology and Oceanography* 50: 872–882.

Strzepek RF and Harrison PJ (2004) Photosynthetic architecture differs in coastal and oceanic diatoms. *Nature* 431: 689–692.

Sunda WG and Huntsman SA (1995) Cobalt and zinc interreplacement in marine phytoplankton: Biological and geochemical implications. *Limnology and Oceanography* 40: 1404–1417.

Sunda WG and Huntsman SA (1997) Interrelated influence of iron, light and cell size on marine phytoplankton growth. *Nature* 390: 389–392.

Sunda WG and Huntsman SA (2000) Effect of Zn, Mn, and Fe on Cd accumulation in phytoplankton: Implications for oceanic Cd cycling. *Limnology and Oceanography* 45: 1501–1516.

Turner DR and Hunter KA (eds.) (2001) *The Biogeochemistry of Iron in Seawater*. New York: Wiley.

PLATINUM GROUP ELEMENTS AND THEIR ISOTOPES IN THE OCEAN

G. E. Ravizza, Woods Hole Oceanographic Institution, Woods Hole, MA, USA

Introduction

The platinum group elements (PGEs) include three second-series transition metals, ruthenium (Ru), rhodium (Rh) palladium (Pd), and three third-series transition metals, osmium (Os), iridium (Ir) and platinum (Pt). The marine chemistry of this group of elements is the least understood and most poorly documented among the many elements in the periodic table. During the 1970s and 1980s when the attention of the marine chemistry community was focused on characterizing the distribution of all the elements in the ocean, direct measurement of the PGEs in sea water was, for the most part, beyond the reach of available analytical methods. Indeed, the most important attribute that links the marine chemistry of all the PGEs is their very low concentration in sea water. This group of metals accounts for 6 of the 10 least abundant elements in sea water.

This article has three objectives. First, to explain how the dissolved inventories of PGEs in the ocean are maintained at such low concentrations relative to most other elements. Second, to review the current status of our knowledge regarding the vertical distribution of these metals in the oceanic water column. Third, to present a brief overview of areas of marine PGE research that are the focus present research activity, and are likely to motivate future investigation. These areas are (1) sea water Os isotope geochemistry; (2) Ir and the other PGEs as tracers of extraterrestrial material in marine sediments; and (3) anthropogenic release of PGEs to the marine environment.

The Apparent Rarity of the PGEs in Sea Water

The underlying reason for the very low concentrations of all the PGEs in sea water has little to do with the aqueous chemistry of these elements. Rather it is the chemical partitioning of these elements within the deep earth that explains their relative scarcity in the sea water (**Table 1**). Comparison of the PGE content of meteoritic material, believed to represent the primordial material that constituted the undifferentiated earth, to ultramafic rocks, believed to be representative of the deep silicate earth, reveals a nearly uniform 100-fold depletion of the PGEs in the deep silicate earth. This depletion indicates that roughly 99% of the whole earth PGE inventory is sequestered within the earth's metallic core. Further comparison of the PGE concentrations of ultramafic rocks to estimated PGE concentrations of upper crustal rocks shows an additional more variable depletion of the PGEs in rocks exposed at the earth's surface relative to the deep silicate earth. This concentration contrast results from the fact that the PGEs are retained in the solid residue when the deep earth is melted to form the earth's crust. The net effect of the strong affinity of the PGEs for phases that reside in the deep earth is a strong depletion of the PGEs in the rocks typically exposed at the earth's

Table 1 Representative PGE concentrations in important earth reservoirs and their ratios

	Ru	*Rh*	*Pd*	*Os*	*Ir*	*Pt*
Chondrites (ppb)[a]	710	130	550	490	455	1010
Silicate earth (ppb)[b]	5	0.9	3.9	3.4	3.2	7.1
Upper crust (ppb)[b]	1.1	0.38	2	0.04	0.04	1.5
Sea water (pg kg^{-1})	2	100	60	10	0.1	50
Sea water (fmol kg^{-1})	20	100	550	50	0.5	260
Sea water/crust[c]	2.00×10^{-6}	0.00026	3.00×10^{-5}	0.00025	3.00×10^{-6}	3.30×10^{-5}
Log (Seawater/crust)	−5.7	−3.5	−4.5	−3.6	−5.5	−4.5

[a]Values from McDonough and Sun (1995) *Chem. Geol.* 120 p. 223.
[b]Values from Schmidt *et al.* (1997) Geochim. *Cosmochim. Act.* 61 p. 2977.
[c]Seawater/crust = (row 4)/(row 3) as a dimensionless ratio.

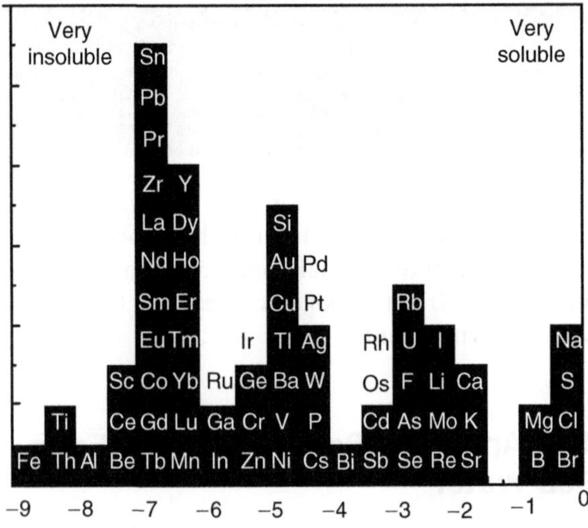

Figure 1 Histogram of seawater/upper crust partition coefficients. This parameter qualitatively represents gross patterns in elemental solubility across the periodic table. Note that the PGEs are neither particularly insoluble compared to other elements, nor particularly similar to one another. Data are derived from compilations by Taylor and McLennan (1985) *The continental crust, its composition and evolution.* Blackwell Scientific; Nozaki (1997) EOS v. 78, p. 221 and the PGE compilation in **Table 1**.

surface. In simplest terms the concentrations of the PGEs in sea water are low compared to other elements because there is a relatively smaller inventory of these elements in the earth's surficial environment.

In the context of the marine chemistry of the PGEs it is significant that the conceptual basis for considering these elements as a coherent group is more closely linked to their behavior in the deep earth than to their behavior in the ocean. Once the very low average crustal concentrations of the PGEs is taken into account it is clear that as a group these elements are not extremely insoluble in sea water (**Figure 1**). More importantly it also becomes apparent that there are obvious first order differences in the solubility of the PGEs. The striking contrast between the solid–solution partitioning of Os and Ir, two elements that exhibit very similar behavior in the deep earth, illustrates this clearly. The important general point here is that although the PGEs are often referred collectively, in a manner analogous to the rare earth elements, the PGEs do not exhibit systematic variations in charge or ionic radius that give rise to systematic similarities or differences in their marine chemistry.

Although the different PGEs do not share a coherent set of chemical affinities in the marine environment, the fact that all these elements occur in sea water at very low concentrations has two important implications that extend to all six elements in this group. The first and most obvious is that quantifying PGE distributions in sea water is analytically very challenging. Consequently, the cumulative set of published data constraining the water column distribution of these elements is quite small. Moreover, when different methodologies have been employed to measure the same element the results frequently disagree. The second implication of the low concentrations of the PGEs in sea water relates to understanding of the speciation of these metals in sea water. It is now generally accepted that many of the first series transition metals are strongly complexed in surface sea water by organic ligands that occur at nanomolar concentrations. Given that concentrations of the PGEs in sea water are 10^3–10^6 times lower than ligand concentrations it seems likely that the marine chemistry of the PGEs will also be strongly influenced by these ligands. Although this is largely a matter of speculation, the simple conceptual point is that a very complete description of all the more abundant species and their affinities for the various PGEs would be required to approach PGE speciation theoretically. Such a detailed description of sea water chemistry is unavailable, and as a result the true speciation of the PGEs in sea water is largely unconstrained. Finally it is noteworthy that the theoretical uncertainties regarding the speciation of the PGEs and the practical problems associated with the analysis of these elements in sea water are interrelated. A sound knowledge of the chemical form(s) of an element in sea water greatly facilitates the development of reliable methods for its separation and quantification.

Overview of Water Column PGE Data

Ruthenium (Ru)

There is little that can be said about the water column distribution of Ru because there are so few data available. The data that are available are limited to isolated analyses of surface sea water. The most recent work reports 20 fmol kg^{-1} for analysis of surface waters from the SIO (Scripps Institute of Oceanography) pier in southern California. Although this value seems reasonable given the relatively low concentration of Ru in upper crustal rocks (**Table 1**), there is no means of further evaluating the accuracy of this particular analysis, or of assessing how representative this single analysis is of the ocean in general. The likely valence of Ru in sea water is as Ru(IV) however, this assessment is based on very scant data for the stability constants for Ru in water. Ru enrichment in ferromanganese crusts has been interpreted as evidence that Ru is redox active in the marine environment, being subject to oxidation to an insoluble form that is coprecipitated with ferromanganese oxides. However, additional work is required to establish with any degree of certainty that this element is in fact redox active in the marine environment.

Rhodium (Rh)

The water column distribution of Rh has been investigated in only a single study. In analytical terms Rh is perhaps the most difficult of the PGEs to quantify because Rh has only one stable isotope, ^{103}Rh. Therefore, the efficiency of Rh preconcentration from sea water must be monitored using a short-lived radiotracer. This methodology was used to generate a full vertical profile from the eastern North Pacific (**Figure 2**). The significant features of this profile are the clear surface water depletion of Rh and the relatively large concentrations (approximately 100 fmol kg^{-1}) of Rh in deep waters.

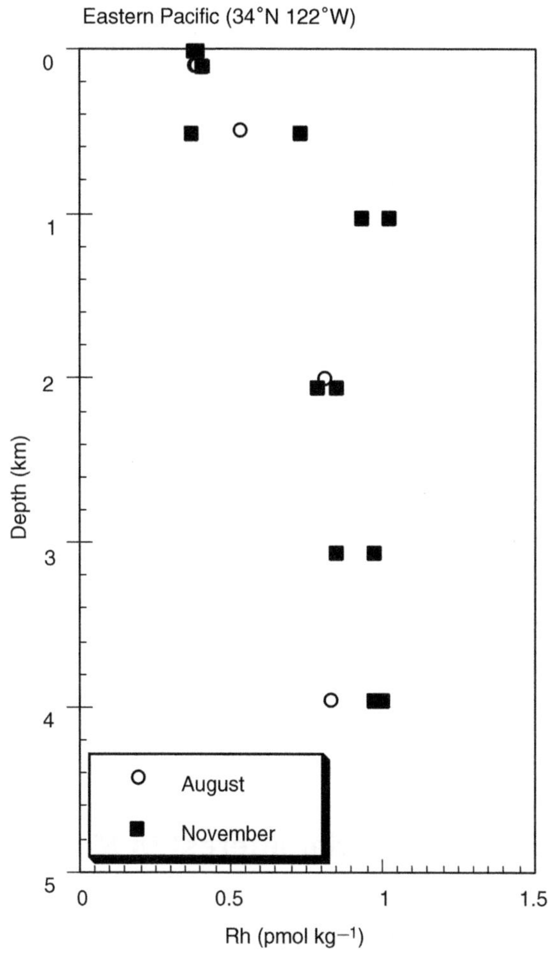

Figure 2 Profile of dissolved Rh in sea water. Data are from Bertine *et al.* (1993) *Marine Chemistry* 42: 199.

Distributions of this type are traditionally interpreted as the result of particulate scavenging in surface waters followed by remineralization at depth. Type of distribution contrasts strongly with that of Co, the first series transition metal which is located directly above Rh in the periodic table, illustrates that elements from the same group in the periodic table can exhibit very different chemical behavior. The contrasting behavior of Co and Rh is potentially related to the fact the Co has an active redox chemistry in the marine environment whereas Rh is believed to be stable only as Rh(III) complexes. It is unclear why the upper crustal partition coefficient calculated for Rh is so large (**Figure 1**); by analogy to other trivalent metals a much lower value would be expected.

Palladium (Pd)

Our knowledge of the distribution of Pd in the water column is based on a study by Lee in 1983, the first to report a full vertical profile of any PGE in sea water. Vertical profiles of filtered and unfiltered samples from two different stations in the Pacific both show a systematic increase in concentration with increasing depth in the water column. The pattern of depth variation closely mimics that of Ni in the same samples. This similarity in the vertical distribution of these two metals and their similar upper crustal partition co-efficients (**Figure 1**) have been rationalized in terms of similar outer electron configuration. Both metals are believed to be stable in their divalent form in sea water. Subsequent more detailed study of the marine chemistry of Ni demonstrates that organic complexation plays an important role in Ni speciation, and laboratory experiments show the same can be true for Pd. Thus it seems likely that complexation by organic ligands plays an important role in Pd speciation in sea water.

Osmium (Os)

Until very recently there were no data available reporting the concentration of Os in sea water. Since 1996, however, there have been several independent studies that focused on this problem making Os the PGE whose marine chemistry has been most extensively studied. Although the recent studies agree that deep water Os concentration is roughly 50–60 fmol kg^{-1} (**Table 2**), the vertical distribution of Os in the water column is still open to debate (**Figure 3**). Results from analyses of samples from the Indian Ocean led to the conclusion that Os behaves conservatively in sea water. A separate study in the Eastern Tropical North Pacific reported a 30% depletion in Os concentration within the core of the oxygen minimum

Table 2 Comparison of Os concentrations of fully oxic deep water from different studies

	Os (fmol kg^{-1})
Indian Ocean[a]	55–59
Eastern Pacific[a]	44–52
North Pacific[b]	53–55

[a]See Figure 3.
[b]Sharma *et al.* (2000) *Earth Planet. Sci. Lett.* 179, p. 139.

zone, and interpreted this as evidence that Os was subject to removal from sea water under reducing conditions.

Although it is widely believed that Os is redox active is sea water, as first indicated by the strong enrichment of Os in anoxic marine sediments, detailed knowledge of Os speciation in sea water does not exist. Inorganic speciation calculations considering the major ions in sea water indicate that in fully oxic sea water Os should be stable in it highest valence, Os(VIII), and exist as an oxyanion. As mentioned above in the case of Ru, the paucity of data constraining the stability constants for potential Os ligands in sea water precludes any rigorous assessment of the likely redox state of Os in sea water. Some working on separation of Os from sea water have suggested that Os is strongly complexed by organic ligands in sea water. The fact that Os(VIII) is highly reactive toward many organic compounds, and is subject to reduction by them in the laboratory, lends some credibility to this inference.

Iridium (Ir)

Among the stable elements that have been measured in sea water Ir is the least abundant, with concentrations on the order of 1 fmol kg^{-1}. Although a full vertical profile from the open ocean is not available, a vertical profile from the Baltic Sea has been reported (**Figure 4**). These data provide compelling evidence that Ir, unlike Os, is not subject to enhanced removal from solution under reducing conditions. Rather, these data are suggestive of Ir scavenging in Baltic surface waters, likely by Fe- and Mn-oxyhydroxides, and subsequent release in deeper anoxic waters. Ir(III) is likely to be the stable valence of Ir in sea water. Given that Ir and Rh are believed to exist in the $+3$ valence, have a d^6 and electron configuration, and reside in the same group in the periodic table, it is surprising that the apparent crustal partition coefficients for these two elements differ so dramatically (**Table 1**). This inconsistency suggests either that this simplistic view of the speciation of these metals is incorrect, or that there is a large systematic error in the available concentration data for Rh or Ir. The former seems more likely than the later, and Ir removal from sea water via oxidation to an insoluble form of Ir(IV) has been proposed in previous discussions of the marine chemistry of Ir.

Platinum (Pt)

Though relatively little work has been done on the water column distribution of Pt in recent years, several studies were conducted from the mid-1980s to the early 1990s. As is the case for Os, only a general consensus regarding deep-water concentrations was achieved, constraining values to fall between 1 and 0.3 pmol kg^{-1}. The depth variations reported in each of the three separate studies differed (**Figure 5**). Because each of these three studies employed different analytical methodologies, it is unclear to what extent the contrasting vertical profiles reflect true variability among the various ocean basins. Its seems unlikely that the strong near-surface Pt enrichment present in the Indian Ocean profile would be restricted to this ocean basin, or that deep water Pt concentrations would exhibit a fourfold difference between eastern and western Pacific. Though the differing vertical profiles suggest that some of the available sea water Pt data are subject to analytical artifact, it is uncertain which data are most reliable.

The uncertainties regarding the vertical distribution of Pt are mirrored in our understanding of the chemical form of Pt in sea water. There is agreement among different workers that the two relevant valances of Pt are Pt(II) and Pt(IV). Some workers argue that Pt(II), stabilized by strong chloro-complexes, is the primary form of Pt in sea water, whereas others argue that Pt(II) is only significant in surface waters and that Pt(IV) dominates in oxic deep water. This author believes these types of inferences must be regarded as largely speculative because the relevant complexing ligands are unknown and consequently appropriate redox potentials cannot be prescribed. Moreover marine chemistry is replete with examples of persistent disequilibrium and the slow kinetics of ligand exchange are a persistent theme in discussions of the aqueous chemistry of Pt. Consequently even if the required thermodynamic data were available for Pt and the other PGEs, they would not necessarily inform us of the true speciation of these metals in sea water.

Topics of Special Interest in Marine PGE Research

Os Isotope Geochemistry

The isotopic composition of Os in natural materials varies as a result of the decay of two long-lived

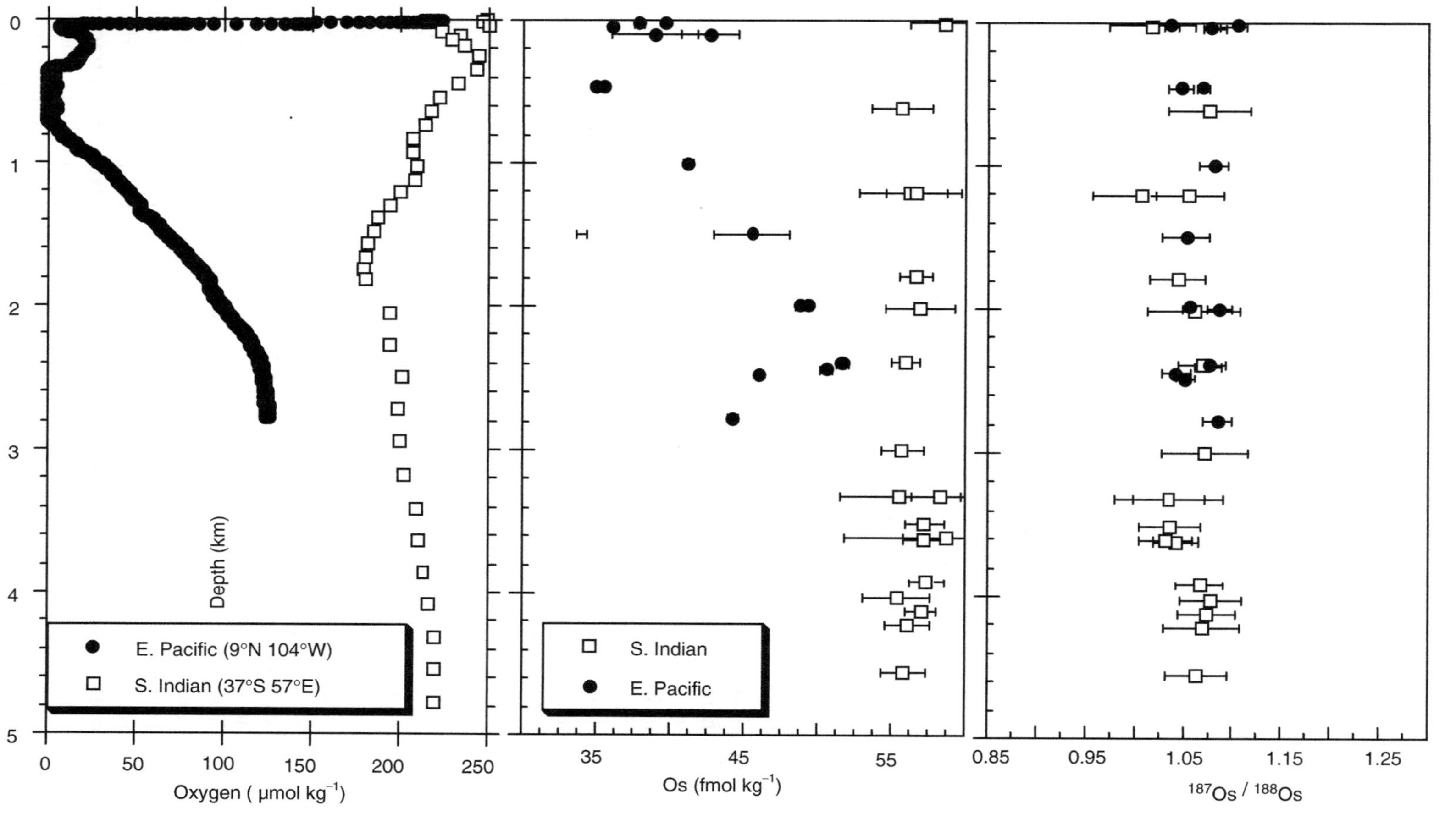

Figure 3 Vertical profiles of Os concentration and $^{187}Os/^{188}Os$ ratio in sea water from the Indian Ocean and the eastern tropical North Pacific. The Pacific data (Woodhouse *et al.* 1999 *EPSL* 173: 223) include analyses of both filtered and unfiltered samples; no systematic difference between the two is apparent. Note that the low Os concentrations in the Pacific profile coincide with the core of a very strong oxygen minimum zone. Indian Ocean data (Levassuer *et al.* 1998 *Science* 282: 272) indicate that Os behaves conservatively.

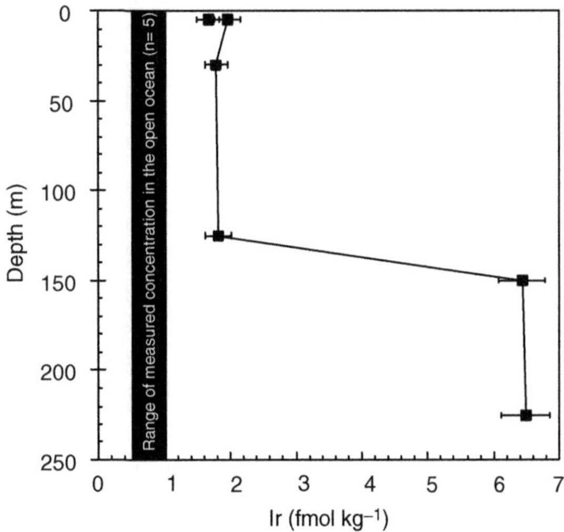

Figure 4 Dissolved Ir profile from the Baltic Sea plotted with the range of Ir concentrations reported for analyses of open ocean samples. Baltic Sea samples were filtered prior to acidification, open-ocean data were acidified and unfiltered. The abrupt increase in dissolved Ir at 150 m depth in the Baltic Sea profile coincides with complete depletion of dissolved oxygen. Anbar *et al.* (1996) *Science* 273: 1524.

naturally occurring radionuclides. The decay of ^{187}Re produces ^{187}Os and the decay of ^{190}Pt produces ^{186}Os; changes in Os isotopic composition that arise from these decay schemes are commonly reported as variations in $^{187}Os/^{188}Os$ and $^{186}Os/^{188}Os$, respectively (**Table 3**). The long half-life and low isotopic abundance of ^{190}Pt restricts the range of $^{186}Os/^{188}Os$ variations in most natural materials, making these isotopic analyses extremely challenging. Significant $^{186}Os/^{188}Os$ variability in marine deposits has yet to be documented. However, available Pt and Os concentration data from metalliferous sediments and marine manganese nodules demonstrate that these deposits have Pt/Os ratios among the highest measured in terrestrial materials. These data suggest that the Pt-Os decay scheme may be exploited in the future as a tool for dating these deposits, and provide the impetus for further investigating the geochemical processes that are responsible for producing the large Pt/Os ratio variation observed in marine deposits.

In contrast to the Pt-Os decay scheme, the Re-Os system gives rise to large variations in $^{187}Os/^{188}Os$ of marine deposits and is currently the subject of vigorous investigation. This work is motivated by two fundamentally important attributes of Os geochemistry; the relatively short marine residence time of Os, and the record of past variations in the $^{187}Os/^{188}Os$ of sea water preserved in marine sediments. Direct analyses of sea water do not yield evidence of any resolvable difference in the $^{187}Os/^{188}Os$ ratio between different ocean basins, but higher precision analyses of Mn crust surfaces do suggest that the $^{187}Os/^{188}Os$ of the Atlantic ocean may be slightly larger than in the Indian or Pacific basins. This isotopic contrast is extremely small compared to the large range in $^{187}Os/^{188}Os$ of sources of Os supplied to the ocean (**Table 4**). The nearly homogeneous character of modern sea water relative to oceanic inputs implies that the marine residence time of Os is poised close to the mixing time of the oceans. Spatial variations in $^{187}Os/^{188}Os$ of modern sea water are also small compared to the record of temporal variations in sea water $^{187}Os/^{188}Os$ preserved in marine sediments. Past variations in the $^{187}Os/^{188}Os$ of sea water provide a globally integrated record of Os input to ocean that can be exploited to make inferences about the geologic history of chemical weathering, and to identify extraterrestrial impacts in the sedimentary record. Detailed discussion of the marine Os isotope record is beyond the scope of this review.

The PGEs as Tracers of Extraterrestrial Material in Marine Sediments

Very large concentrations of the PGEs in extraterrestrial material relative to the average upper crustal material (**Table 1**) make the PGEs valuable indicators of the presence of particulate extraterrestrial material in marine sediments. For example, addition of 0.01% by weight chondritic material to a sediment with average crustal Ir and Os concentrations would roughly double the concentrations of these elements in the mixture relative to that of the starting material. Ir is more widely exploited than Os as a tracer of particulate extraterrestrial material because methods for low level Ir analysis were established earlier and are more widely available. The global Ir enrichment at the Cretaceous–Tertiary boundary, and the subsequent identification of a major extraterrestrial impact crater, provide the best known example of this type of research. Ir data have been applied in a similar manner to study numerous other event horizons in the geologic record. Other PGE analyses can be integrated into these studies to provide additional constraints on PGE source. For example, the Pt/Ir ratios typical of upper crustal material are roughly 10 times larger than in chondrites (**Table 1**). This type of contrast in element ratios can be used to help evaluate whether elevated Ir concentrations are truly related to an extraterrestrial PGE source, or are the result of natural enrichment of PGEs from the ambient environment. Many studies motivated by the controversy surrounding the interpretation of the Ir anomaly at the

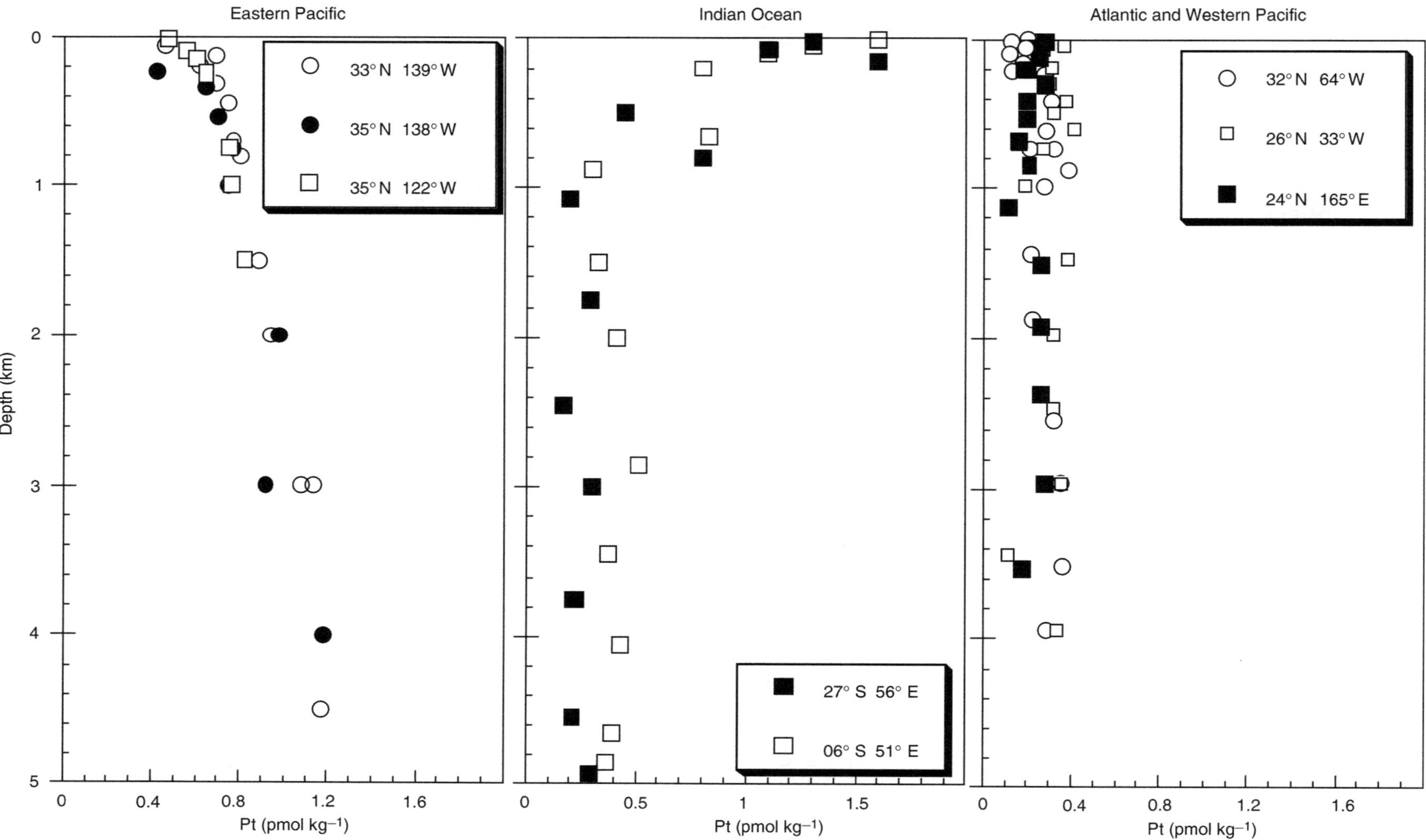

Figure 5 Dissolved Pt profiles from the Pacific, Indian and Atlantic Oceans. Although deep-water concentrations are similar to one another the vertical distribution of Pt differs dramatically among the various profiles. Data are from the following sources. Eastern Pacific: Hodge *et al.* (1986) *Analytical Chemistry* 58, p. 616; Indian: Jacinto and van den Berg (1992) *Nature* 338, p. 332; Atlantic and western Pacific: Colodner et al. (1993) *Analytical Chemistry* 65, p. 419 and Colodner (1991) The marine geochemistry of rhenium, platinum and iridium. Ph. D. thesis. MIT/WHOI Joint Program in Oceanography.

Table 3 Radioactive decay schemes that influence the isotopic composition of naturally occurring Os

Parent	Abundance	Half-life	Daughter
^{187}Re	62.6%	42 billion years	^{187}Os
^{190}Pt	0.0124%	449 billion years	^{186}Os

Compiled from walker *et al.* 1997 *Geochim. Cosmochim. Act.* 61, p. 4799.

Table 4 Comparison of $^{187}Os/^{188}Os$ ranges among ocean basins, sources of Os to sea water and Cenozoic sea water

	$^{187}Os/^{188}Os$
Atlantic Mn crust surfaces[a]	1.04–1.07
Indian Mn crust surfaces[a]	1.00–1.04
Pacific Mn crust surfaces[a]	1.00–1.04
Rivers[b]	0.64–2.94
Hydrothermal fluids[c]	0.11–0.39
Meteoritic material	0.12–0.14
Cenozoic sea water[d]	0.2– 1.06

[a]Burton *et al.* (1999) *Earth Planet. Sci. Lett.* 171, p. 185.
[b]Levasseur *et al.* (1999) *Earth Planet. Sci. Lett.* 174, p. 7.
[c]Sharma *et al.* (2000) *Earth Planet. Sci. Lett.* 179, p. 139.
[d]Pegram and Turekian (1999) *Geochim cosmochim. Act.* 63, p. 4053.

Cretaceous–Tertiary Boundary have demonstrated that a wide variety of natural processing can give rise to PGE enrichments unrelated to any extraterrestrial input of these elements. Therefore, it is important to emphasize that the PGEs are only one of several possible lines of evidence used to test impact hypotheses in the geologic record.

PGEs in marine sediments are important not only in the context of identifying specific extraterrestrial impact events, but also in quantifying the background flux of cosmic dust to the earth's surface. This flux is critically important to determining what level of Ir enrichment is likely to constitute evidence of an extraterrestrial impact. Slowly accumulating pelagic clays from the abyssal North Pacific are the best available records of the average long-term flux of extraterrestrial material to the earth's surface. This is because of their very slow accumulation rates, on the order of a few millimeters per thousand years. These slow accumulation rates reflect the fact that this region of the ocean is far removed from terrestrial sources of particulate material. Thus a few meters of sediment can provide a nearly continuous record of accumulation that spans several million years and maximizes the contribution of the background flux of extraterrestrial Ir relative to total Ir burial flux. The best example of such a record is the red clay sequence from LL44-GPC3, a core recovered from the North Pacific (**Figure 6**). However in such sediment records the influence of Ir that is scavenged from sea water and is not directly associated with extraterrestrial particles complicates interpretations. In the case of LL44-GPC3, lower than chondritic Os/Ir ratios and $^{187}Os/^{188}Os$ ratios much higher than those that characterize meteoritic material indicate that more than 50% of the average total Ir flux is derived from sea water. Determining the proportion of the seawater-derived Ir that originated from dissolution of cosmic dust and that which originated from terrestrial sources is very difficult. Analyses of dissolved Ir in rivers that accompany recent analyses of dissolved Ir in sea water suggest that riverine supply of Ir may account for more than half of the seawater-derived Ir that accumulates in deep-sea sediments. Uncertainties associated with these types of interpretations ultimately limit the precision and accuracy of estimates of the

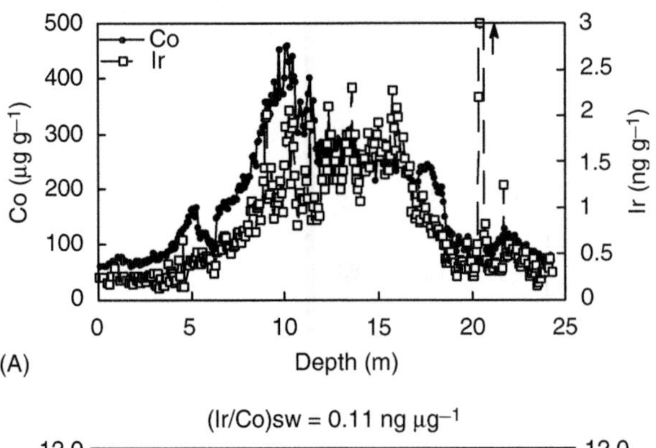

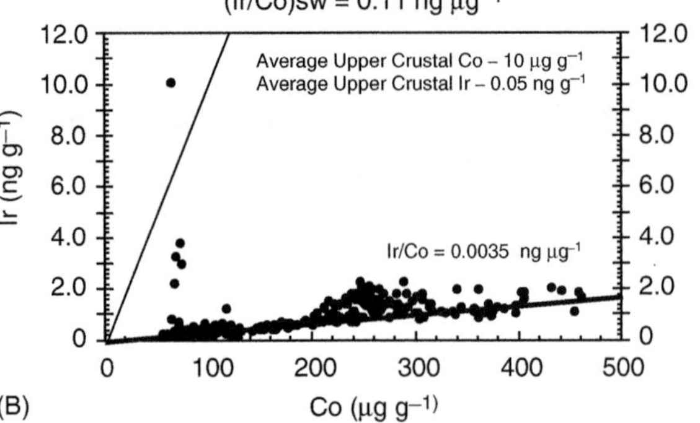

Figure 6 Concentration variations of Co and Ir vs. depth in LL44-GPC3 (A) and the same data plot as Ir vs. Co (B). The large Ir concentrations at 20 m depth (A) and 10 ng g^{-1} (B) correspond to the Cretaceous–Tertiary boundary Ir spike in this core. The slope of the Ir-Co trend represented by the bulk of the data is close to the Ir/Co ratio of average upper crust. This similarity suggests much of the Ir in this core may be derived from terrestrial rather than extraterrestrial sources. The steep line on the lower plot corresponds to the Ir/Co ratio of deep-water. In order for a significant fraction of the total Ir to occur as particulate extraterrestrial material Ir must be significantly more insoluble than Co, consistent with data from **Figure 1**. Data are from Kyte *et al.* (1993) *Geochimica et Cosmochimica Acta* 57: 1719.

long-term background flux of extraterrestrial Ir to the earth.

Anthropogenic Release of PGEs to the Marine Environment

Since the mid-1970s commercial demand for the PGEs, particularly Pt and Pd has been increasing rapidly (**Figure** 7). Although considerable effort is invested in recovering and recycling these metals, due in large part to their high cost, there is increasing evidence that release of these metals to the environment is giving rise to higher environmental concentrations in portions of the environment subject to anthropogenic perturbation. Among the many uses of PGEs, the utilization of Pt, Pd and more recently Rh, in automobile catalytic converters is the one pathway for anthropogenic PGE release to the environment that is best documented and most likely to lead to widespread dispersal of these metals. Although the most immediate impact of this mode of release is on land, anthropogenic PGEs also find their way to the marine environment. Direct release from storm sewers draining roadways, and indirect release from municipal sewage plants that treat road run-off with other wastewater streams are the two most likely modes of transport of autocatalyst PGEs to the marine environment. It is important to stress that the municipal waste streams may also carry PGEs associated with medical, dental, chemical and electronic applications.

Although marine chemists have been aware of the potential importance of anthropogenic PGE release for many years, the same analytical challenges that limit the amount of PGE data from pristine marine environments are also responsible for the paucity of information constraining the distribution and behavior of anthropogenic PGEs in the marine environment. The long-standing interest of marine chemists in anthropogenic PGEs is clearly illustrated by the fact that the report of dissolved Pd profiles in sea water (see above) in the mid 1980s was accompanied by data demonstrating elevated Pd concentrations in contaminated sediments from Japan. Release from autocatalysts was suggested as a possible source. In the intervening years there have been very few studies that have addressed this matter. Recent work in contaminated sediments from Boston Harbor in the north-eastern part of the USA shows that human activity has resulted in greater than fivefold increases in bulk sediment Pt and Pd concentrations, relative to background levels of approximately 1 ng g^{-1} (**Figure** 8). However, as enrichment of these metals predated the introduction of catalytic converters, there must be important sources of these metals to the marine environment other than autocatalysts. Temporal trends in the data show that although the concentrations of Ag and Pb in Boston Harbor sediments are decreasing, likely due to the cessation of sewage release, concentrations of Pt and Pd are either stable or increasing with time. This trend is consistent with a significant input of these metals from nonpoint sources such as

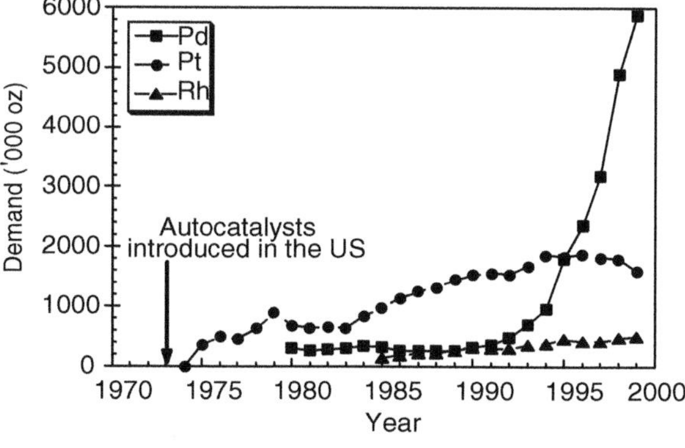

Figure 7 Representation of global demand for Pt, Pd and Rh by the automobile industry. Note that although no data are shown for Pd prior to 1980, Pd was used in some autocatalyst formulations prior to this time. These increasing demand trends suggest that PGEs release from autocatalysts is likely to become an increasingly important source of anthropogenic PGEs to the environment. Data are from the Johnson Matthey Platinum/2000 publication.

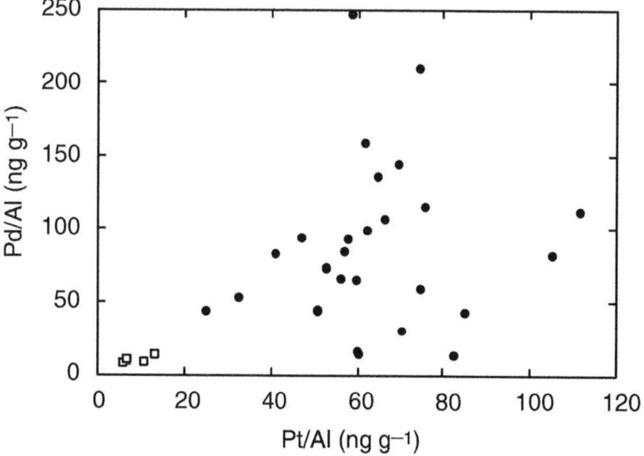

Figure 8 Plot of Pd/Al vs. Pt/Al in bulk sediment samples from Boston Harbor (●) and Massachusetts Bay (□) illustrating the influence of anthropogenic PGE release in this area. Pd and Pd concentrations are normalized to Al to eliminate grain size and dilution effects. All sediments from Massachusetts Bay are uninfluenced by human activity, based on depositional age estimates and Ag analyses. Contaminated sediments from Boston Harbor are enriched in Pt and Pd relative to pristine sediment. This is true for absolute Pt and Pd concentrations as well the Pt/Al and Pd/Al data shown above. Data are from Tuit *et al.* (2000) *Environmental Science Technology* 34: 927.

the release of untreated road run-off. The impact of human activity on levels of dissolved Pt and Pd in coastal waters is not well documented. Similarly, in the marine environment, the chemical form of anthropogenic PGEs, and the extent to which these metals are subject to biological uptake are also poorly known. These gaps in our knowledge of the marine chemistry of the PGEs will likely influence the future direction of marine PGE research.

Summary

The PGEs are among the least abundant elements in sea water. The low concentrations of these metals in sea water reflect their generally low concentration in earth surface material rather than uniformly low solubility. Although there is a general consensus regarding the approximate concentrations of these metals in sea water, their vertical distribution in the water column remains controversial and poorly documented. Improving our understanding of the marine chemistry of the PGEs both in the water column and in marine sediments is important to interpreting the marine Os isotope record, exploiting PGEs as tracers of extraterrestrial material in marine sediments, and understanding the consequences of anthropogenic release of PGEs to the marine environment.

Further Reading

Donat JR and Bruland KW (1995) Trace elements in the oceans. In: Steinnes E and Salbu B (eds.) *Trace Elements in Natural Waters*, ch. 11. Boca Raton: CRC Press.

Goldberg ED and Koide M (1990) Understanding the marine chemistries of the platinum group metals. *Marine Chemistry* 30: 249–257.

Helmers E and Kummerer K (eds) (1997) Platinum group elements in the environment – anthropogenic impact. *Environtal Science and Pollution Research* 4: 99.

Kyte FT (1988) The extraterrestrial component in marine sediments: description and interpretation. *Paleoceanography* 3: 235–247.

Lee DS (1983) Palladium and nickel in north-east Pacific waters. *Nature* 313: 782–785.

Peucker-Ehrenbrink B and Ravizza G (2001) The marine Os isotope record: a review. *Terra Nova*, in press.

Peucker-Ehrenbrink B (1996) Accretion of extraterrestrial matter during the last 80 million years and its effect on the marine osmium isotope record. *Geochimica et Cosmochimica Acta* 60: 3187–3196.

RARE EARTH ELEMENTS AND THEIR ISOTOPES IN THE OCEAN

Y. Nozaki[†], University of Tokyo, Tokyo, Japan

Introduction

The rare earth elements comprise fifteen lanthanide elements (atomic number, $Z = 57$–71) as well as yttrium ($Z = 39$) and scandium ($Z = 21$), although promethium ($Z = 61$) does not appear in nature due to its radioactive instability (**Figure 1**). They are an extremely coherent group in terms of chemical behavior and have recently been subjected to intense investigation in the field of marine geochemistry. All rare earth elements occur as a trivalent state with exception of Ce^{4+} and Eu^{2+}. In the lanthanide series, the progressive filling of electron in the shielded inner 4f shell with increasing atomic number leads to gradual decrease in the ionic radii from La^{3+} to Lu^{3+}, which is called 'the lanthanide contraction'. Consequently, small but systematic changes in chemical properties allow them to be used for unique tracers in studying fundamental processes that govern the cycling of rare earth elements in the ocean. For instance, the heavier lanthanides were predicted to be more strongly complexed in seawater and hence more resistant to removal by scavenging of particulate matter. Yttrium mimics the heavy lanthanides, particularly holmium, because of similarity in the ionic radii. However, Sc is a substantially smaller cation with a geochemical behavior that differs from other rare earths. In most literature, therefore, 'rare earth elements (REEs)' generally include only lanthanides and Y, and Sc is treated separately.

Two elements, Ce and Eu can take the other oxidation states. Although Ce is generally well accommodated within the strictly trivalent lanthanide series in igneous rocks, oxidation reaction of Ce^{3+} to Ce^{4+} proceeds in oxygenated aqueous systems. In seawater, the resulting Ce^{4+} hydrolyzes readily and tends to be removed by scavenging. For this reason, seawater is typically depleted in Ce relative to that expected from neighboring La and Pr, whereas Ce is often enriched in some authigenic minerals such as

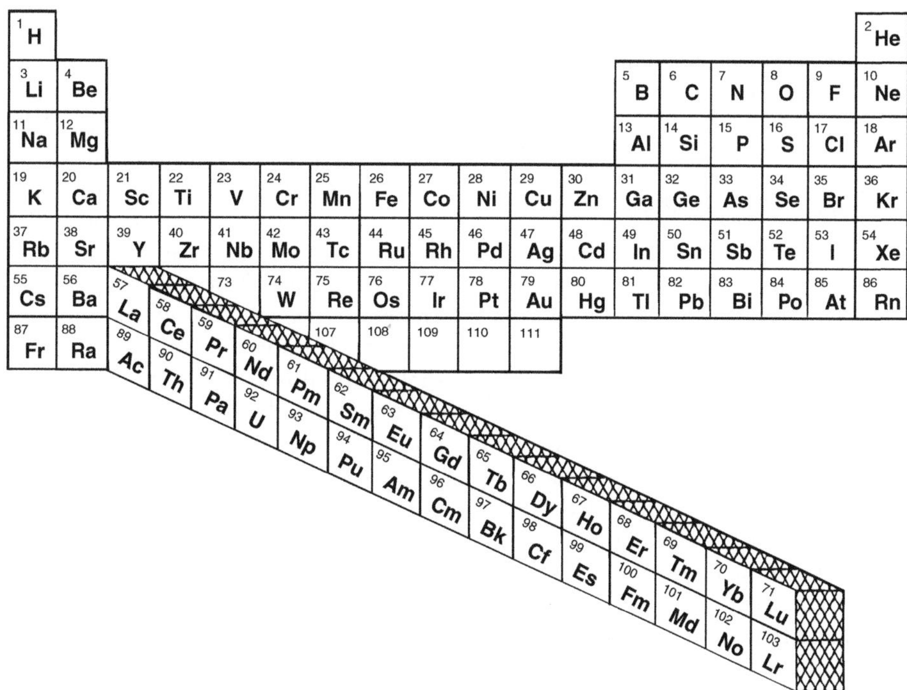

Figure 1 Periodic table showing the position of rare earth elements.

[†]Deceased.

manganese nodules and phosphorites. Europium can be reduced to Eu^{2+}, a larger cation that can be segregated from other REEs, during magmatic processes. Anomalous concentrations of Eu are not uncommon in various igneous and sedimentary rocks. However reduction of Eu does not normally take place within the ocean, although an Eu enrichment has often be encountered in hydrothermal fluids venting at midoceanic ridges. The anomalous behavior of Ce due to oxidation–reduction reactions can be best and quantitatively evaluated relative to the trivalent neighbors (La and Pr) in the lanthanides series without significant influence of the other processes affecting their oceanic distributions. This is a notable advantage of the element over the other transition metals, such as Mn and Fe, which behave individually affected by the oxidation states.

In addition, there are two geochemically important isotopes of the lanthanides, ^{143}Nd and ^{138}Ce. The ^{143}Nd is produced by decay of ^{147}Sm with a half-life of 10.6×10^{11} years. Natural variation of $^{143}Nd/^{144}Nd$ in terrestrial materials occurs depending on mantle/crust segregation of Sm and Nd, and the age of the rocks. Thus, the $^{143}Nd/^{144}Nd$ ratio may be used to constrain the sources of the REEs and mixing within the ocean. Likewise, the $^{138}Ce/^{142}Ce$ ratio may also be used to constrain homogenization of the element by oceanic mixing since the ^{138}Ce is produced by ^{138}La decay (half-life, 2.97×10^{11} years). However, the $^{138}Ce/^{142}Ce$ ratio has not been well exploited in marine geochemistry yet, because of its smaller natural variation as compared to that of Nd isotopes and analytical difficulty.

History and REE Normalization

The analysis of picomolar REE concentrations in seawater has been difficult due to lack of sensitivity in the conventional methods. Earlier attempts to measure REEs relied almost entirely on high-sensitivity instrumental neutron activation analysis. In 1963 reliable REE concentrations were reported in a few waters from the Eastern Pacific as well as those in a manganese nodule and a phosphorite and their significance was recognized. Basic features of the REEs in seawater were found: a progressive increase across the lanthanide series from the light Pr to the heaviest Lu when the seawater concentrations were divided by those of sediments, and that Ce is markedly depleted in the seawater but enriched in the manganese nodule relative to neighboring La and Pr, as expected from its 4+ valency state. Europium was normal relative to other trivalent REEs in all the sample. It was also noted that the concentrations of the heavy REEs (Ho, Yb and Lu) in the Pacific deep water were considerably higher than those in the surface water. Although these earlier findings had to be refined and confirmed by subsequent workers with more precise modern techniques, the fundamental aspects of the REE marine geochemistry were developed for that time. Prior to 1980, reliable data on the distribution of REEs in seawater were few. Since the early 1980s, a growing number of reports on the subject have become available. Now, several laboratories in the world are capable of determining REEs in seawater with precision between one and a few percent by use of isotope dilution thermal ionization mass spectrometry (ID-TIMS) or inductively coupled plasma mass spectrometry (ICPMS). Therefore, more detailed arguments are possible regarding geochemical processes controlling the concentration, distribution, fractionation and anomalous behaviors in the oceans.

When REE fractionation is discussed, it is common to normalize the data to the values in shale which are thought to be representative of the REEs in the upper continental crust. The shale-normalization not only helps to eliminate the well-known distinctive even–odd variation in natural abundance (the Oddo–Harkins effect) of REEs but also visualizes, to a first approximation, fractionation relative to the continental source. It should be noted, however, that different shale values in the literature have been employed for normalization, together with the ones of the Post-Archean Australian Sedimentary rocks (PAAS) adopted here (**Table 1**). Thus, caution must be paid on the choice of the shale values if one ought to interpret small anomalies at the strictly trivalent lanthanides such as Gd and Tb. Alternatively, for detailed arguments concerning fractionation between different water masses in the ocean, it has been recommended that the data are normalized relative to the REE values of a distinctive reference water mass, for example, the North Pacific Deep Water (NPDW, **Table 1**). The NPDW-normalization eliminates the common features of seawater that appeared in the shale-normalized REE pattern and can single out fractionation relative to the REEs in the dissolved end product in the route of the global ocean circulation.

The Oceanic Distributions

In 1982, the first 'oceanographically consistent' vertical profiles were reported for nine out of ten lanthanides that could be measured by the ID-TIMS method in the North Atlantic. Since then, a significant amount of data on the distribution of REEs have accumulated from various oceanic regions. For example, **Figure 2** shows the station locations where the REEs were measured in seawater, together with the

Table 1 Ionic radii and the average REE concentrations in shale and seawater used for normalization

Element	Atomic number Z	Ionic Radius (Å)[a]		PAAS[a] (μmol/kg)	NPDW[b] (pmol/kg)
		CN=6	CN=8		
Y	39	0.900	1.019	304	236.3
La	57	1.032	1.160	275	38.7
Ce	58	1.010	1.143	568	3.98
Pr	59	0.990	1.126	62.7	5.10
Nd	60	0.983	1.109	235	23.8
Sm	62	0.958	1.079	36.9	4.51
Eu	63	0.947	1.066	7.1	1.24
Gd	64	0.938	1.053	29.6	6.83
Tb	65	0.923	1.040	4.9	1.13
Dy	66	0.912	1.027	28.8	8.38
Ho	67	0.901	1.015	6.0	2.34
Er	68	0.890	1.004	17.0	7.94
Tm	69	0.880	0.994	2.4	1.23
Tb	70	0.868	0.985	16.3	8.37
Lu	71	0.861	0.977	2.5	1.46

[a]After *Taylor and McLennan (1985)* for trivalent cations; CN, coordination number.
[b]Dissolved (<0.04 μm) REE in the North Pacific Deep Water at 2500±100 m (after Alibo and Nozaki, 1999).

Nd concentrations in the surface water. Some of them are based on filtered samples, using generally a 0.4–1.0 μm membrane, of which results are called 'dissolved' concentrations. Many others, however, were measured on unfiltered and acidified seawaters, and hence can be ascribed to 'acid-soluble total' concentrations. The difference between the two is generally small, less than 5% for all trivalent REEs in the open oceans, even if the finer 0.04 μm membrane is used for filtration (**Table 2**), and gross features of their vertical profiles remain unchanged. Nevertheless, when fine structures of the REE patterns are discussed, filtration becomes critical in changing the pattern since the particulate fraction decreases from ~5% at the light and middle to less than 1% at the heavy REEs. Furthermore, there is an obvious exception for Ce of which more than 35% is associated with particles, being consistent with its 4+ oxidation state. Also, the REEs in unfiltered samples close to the bottom often show anomalously high concentrations due to resuspension of underlying sediments.

The vertical profiles of dissolved (<0.04 μm) REEs for different oceanic basins are shown in **Figure 3**. Except for Ce, all REEs show 'nutrient-like' gradual increase with depth. There are small but systematic differences in the profiles across the series, although the North Atlantic profiles show somewhat complex features dominated by horizontal advection of different water masses (from the above, the Sargasso Sea Surface Water, the North Atlantic Deep Water, and the Antarctic Bottom Water). For instance, in the Southern Ocean and the North Pacific, the light and middle REEs (e.g. Pr-Gd) almost linearly increase with depth, whereas the heavy REEs (e.g. Ho-Lu) and Y show convex features. The concentrations of the heavy REEs below 1500 m are in the order of North Atlantic<Southern Ocean<North Pacific much like those of nutrients, whereas those of the light REEs are Southern Ocean<North Atlantic<North Pacific, presumably reflected by scavenging intensities in those regions. Although the vertical profiles of the REEs are similar to those of nutrients, e.g. dissolved silica, there is a difference in that the REE concentrations never approach zero in the surface water like nutrients in the temperate oligotrophic zone. Analysis of interelement correlations indicate that the heavy REEs and Y better correlate with dissolved silica and alkalinity than with nitrate and phosphate. Within the REEs, the best correlations ($R^2>0.99$) can be found between neighboring trivalent REEs, and between Y and the heavy REEs. As a trivalent REE pair is apart in their atomic number, the correlation between the two becomes worse. Lanthanum often deviates in these general trends from the light REEs toward the heavy REEs.

The vertical profiles of Ce is unique among the REEs showing a decrease from the high concentrations in the surface waters to the low and nearly constant values in the deep waters (**Figure 3**). Such a distribution pattern can be seen for other least-soluble-elements, such as Al, Co, and Bi which also hydrolyze readily. The different profiles among the basins may be governed by the balance in the strength of external inputs (eolian+riverine) to the surface ocean and particle scavenging throughout the water column.

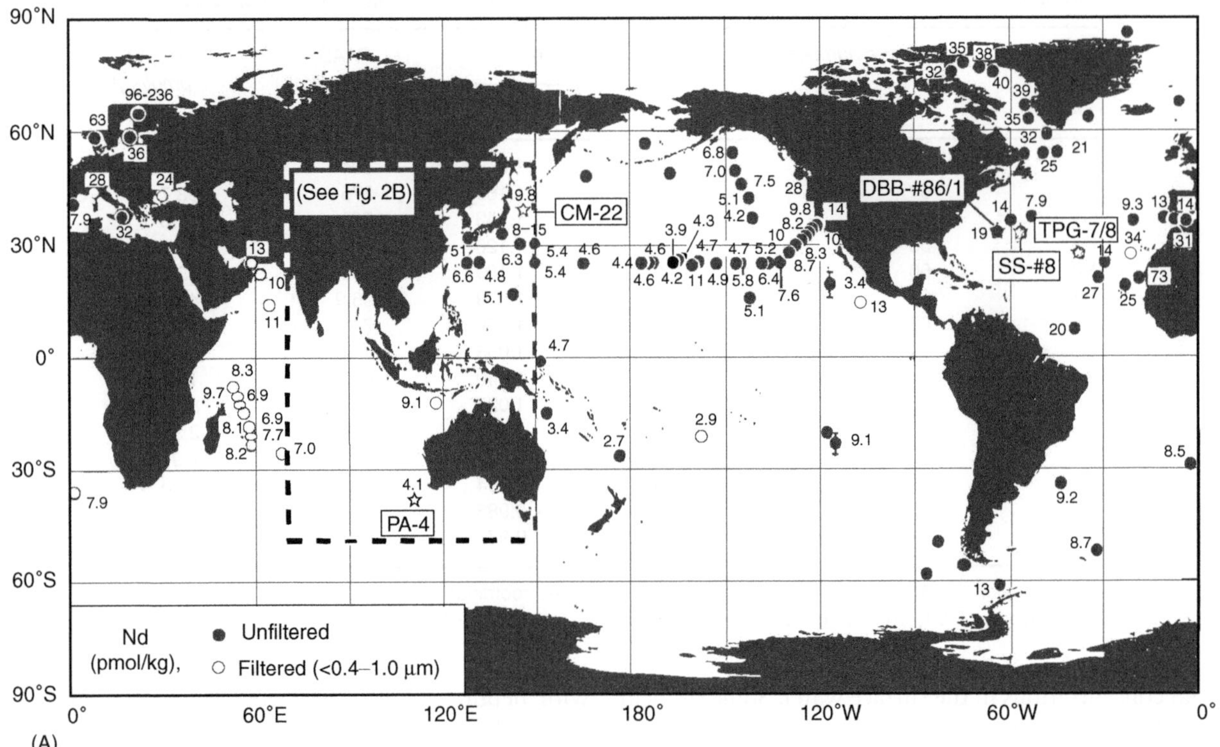

Figure 2 (A) World map for the REE data in the literature (Byrne and Sholkovitz, 1996) and Nd concentrations in the surface water (< 100 m in depth). The open and filled circles indicate the locations where filtered and unfiltered samples were analyzed. The stars indicate the station locations for which the profile data are shown in **Figure 3**.

REE Patterns

Shale-normalized dissolved REE patterns for the western North Pacific are shown in **Figure 4(A)**. Generalized features common to all seawaters are: a progressive heavier REE enrichment relative to the lighter ones and a pronounced depression at Ce. These features can be best understood by the conceptual model for interactions between REEs in solution and particles and subsequent removal of particulate matter by settling (**Figure 5**). The former is ascribed to a systematic increase in stability constant with atomic number of complexes of REE-ligands (mainly carbonate) in seawater. The latter is explained by preferential removal of Ce^{4+} species from seawater relative to trivalent REEs. It is also noted that La is always enriched compared to that expected by extrapolation from heavier Pr and Nd. Furthermore, the increasing trend from the light to the heavy REE is nonlinear and often has a marked break between Gd and Tb. These less-pronounced features have been discussed in terms of the basic physicochemical characteristics relating to absence for La and half filled for Gd of 4f electrons. However, the reasoning is somewhat controversial and has not yet been confirmed.

The NPDW-normalized REE patterns in the water column of the western North Pacific are shown in **Figure 4(B)**. In comparison with **Figure 4(A)**, it generally shows a flat pattern (no fractionation) including those between 400 and 600 m where the North Pacific Intermediate Water (NPIW) penetrates. The exception is that the surface samples (<200 m) indicate a middle REE enriched pattern being reflected by the sources and fractionation during scavenging of REEs. The different water masses have unique NPDW-normalized patterns (**Figure 6**). Thus, the REE patterns are useful as tracers in defining those water masses.

Redox Reaction of Ce, and Ce Anomalies

Cerium is oxidized in seawater according to the following equation,

$$Ce^{3+} + 2H_2O = CeO_2 + 4H^+ + e^-$$

where CeO_2 is highly insoluble species and rapidly removed by scavenging. This oxidation is considered to occur mainly in the surface water through

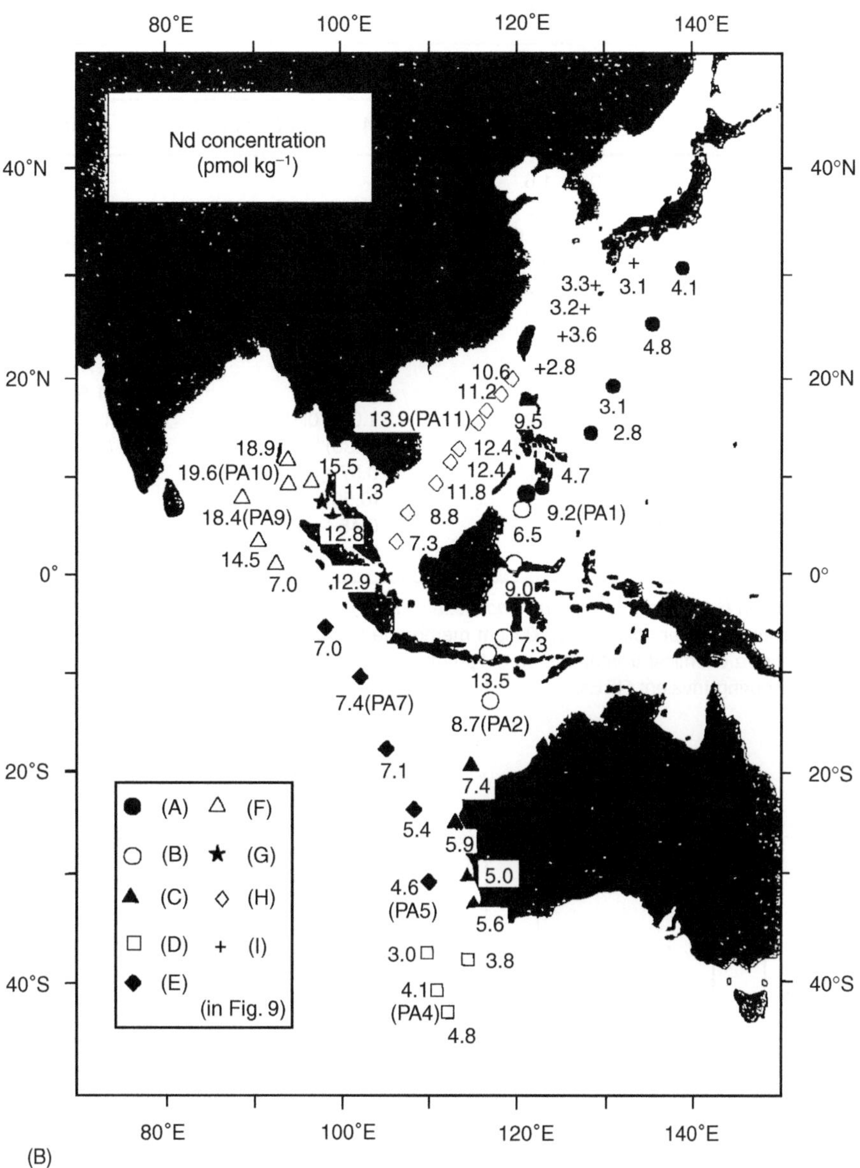

Figure 2 *Continued.* (B) The station locations occupied by R. V. Hakuho Maru during the 1996/97 Piscis Austrinus Expedition and the dissolved (<0.04 μm) Nd concentrations in the surface waters. The different symbols are used according to the different regions where NPDW-normalized REE patterns are grouped in **Figure 9**. The stations indicated by PA-numbers show the locations where the vertical profiles of dissolved REEs were determined.

bacterial mediation. There is no evidence that Ce oxidation continues in the deep sea. The deviation of Ce from other trivalent REEs in geochemical behavior is generally expressed as 'Ce-anomaly' which is defined by the following equation,

$$Ce/Ce^{*} = 2[Ce]/([La] + [Pr])$$

or

$$Ce/Ce^{*} = 3[Ce]/([La] + 2[Nd])$$

where [] indicates shale-normalized value of the REE concentration. The negative Ce-anomaly implies $Ce/Ce_{*} < 1$, vice versa. In the open oceans, the negative Ce-anomaly is developed with increasing depth (**Figure 7**) because of a decrease in Ce coinciding with increases in La and Pr (or Nd) concentrations.

In the aoxic basins, Ce^{4+} is reduced to Ce^{3+} and the Ce-anomaly tends to be smaller than those in oxygenated waters. The Black Sea provides a best example in which Ce/Ce_{*} sharply increases from less

Table 2 Rare earth elements associated with particulate matter in seawater

Atomic number Z	*Element*	*Particulate fraction (%)*			
		N.W. Pacific[a]	*W. Indian Ocean*[b]	*N. Atlantic*[c]	*Average*
39	Y	0.8±1.6	–	–	0.80
57	La	2.9±2.2	1.33±0.67	3.68±0.27	2.65
58	Ce	36.6±7.6	17.1±4.1	23.7±7.6	25.8
59	Pr	5.5±3.4	–	–	5.49
60	Nd	4.4±2.7	2.10±0.65	3.42±0.66	3.31
62	Sm	5.1±2.9	2.10±0.63	2.75±0.69	3.30
63	Eu	5.4±5.6	1.50±0.37	3.00±0.58	3.30
64	Gd	4.8±3.1	1.28±0.33	1.70±0.27	2.58
65	Tb	4.2±3.7	(1.2)[d]	(1.6)[d]	2.33
66	Dy	2.8±3.2	1.03±0.35	1.55±0.33	1.80
67	Ho	0.81±3.0	(0.80)[d]	(1.1)[d]	0.90
68	Er	0.45±3.3	0.57±0.19	0.63±0.45	0.55
69	Tm	–	(0.50)[d]	(0.62)[d]	0.56
70	Yb	–	0.48±0.20	0.61±0.33	0.55
71	Lu	–	0.44±0.21	0.48±0.34	0.46

[a]Acid-soluble particulate fraction (>0.04 μm), after *Alibo and Nozaki (1999).*
[b]Based on analysis on particulate matter collected on 0.4 μm membrane filters (Bertram and Elderfield, 1994).
[c]Sum of analyses on successive chemical treatments on 0.4 μm-filtered particles (Sholkovitz *et al.* 1994).
[d]Estimated value from neighboring trivalent REEs.

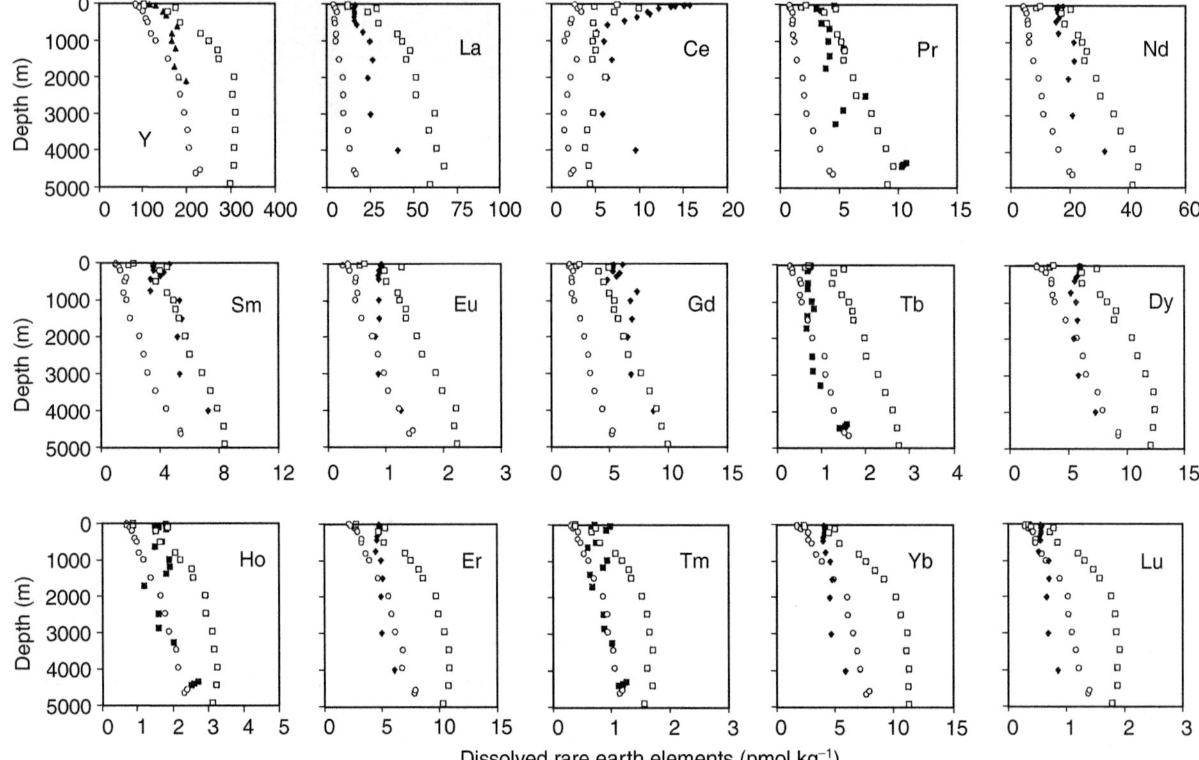

Figure 3 The vertical profiles of dissolved (< 0.04 μm) REEs in the western North Pacific (CM-22; □), and the Southern Ocean (PA-4; ○) from Alibo and Nozaki, unpublished. The North Atlantic profile data based on dissolved (< 0.4 μm) REEs (SS-#8; ♦) from Sholkovitz and Schneider (1991) and acid-soluble total concentrations for monoisotopic REEs (DBB-#86/1; ■) from De Baar *et al.* (1983) and yttrium (TPG-7/8; ▲) from Alibo *et al.* (1999).

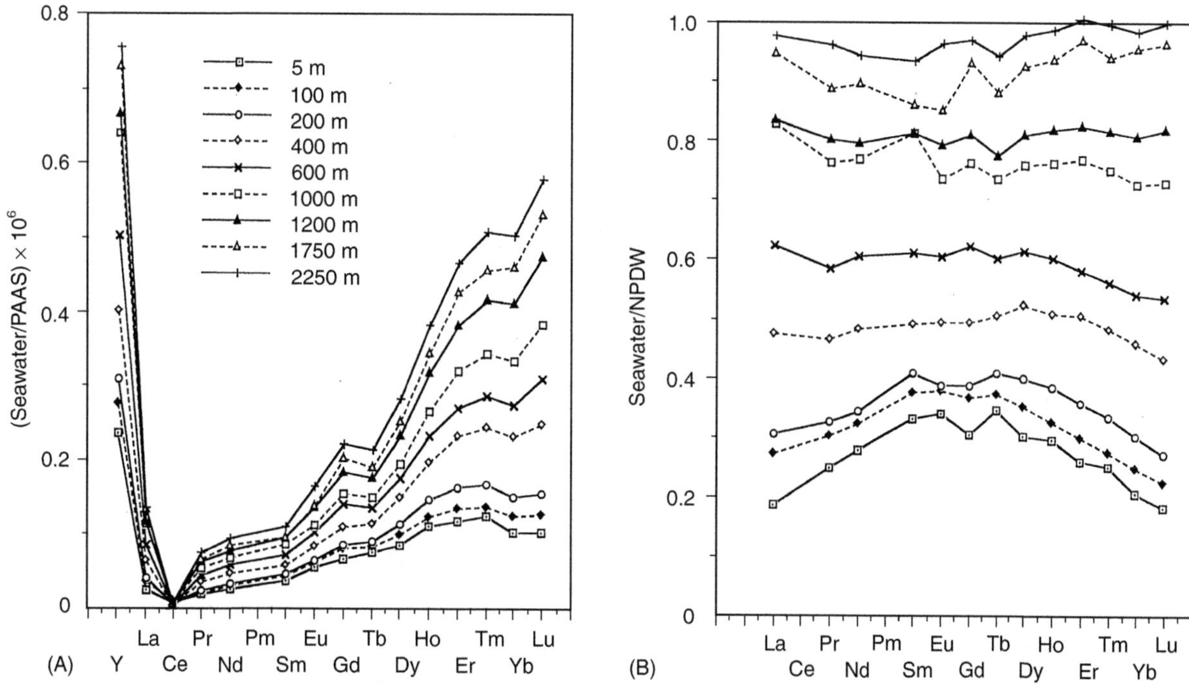

Figure 4 The PAAS-normalized (A) and the NPDW-normalized (B) patterns of dissolved REEs in the western North Pacific (Alibo and Nozaki, 1999).

than 0.1 at the oxic/aoxic boundary of ~100 m depth to ~1.0 (no anomaly) in reducing waters below 200 m (**Figure 8**). Even positive Ce anomaly can be found in the deep Cariaco Trench. Cerium reduction also takes place in coastal and hemipelagic sediment pore waters and can affect the distribution of Ce in the overlying waters.

Inputs of REEs to the Ocean

The most obvious source of REEs in the ocean is riverine input. The concentrations of dissolved REEs in river waters are significantly higher than those of seawaters. The behaviors of the dissolved REEs in rivers and estuaries have been intensively studied as a link of REE geochemistry between the crust and the ocean. Shale-normalized REE patterns in river and estuarine waters often show nonflat patterns indicating that fractionation takes place with respect to continental crust. One of the prominent features derived from those works is that large-scale removal of dissolved REEs (in particular, light REEs) occurs in the estuarine mixing zone. Planktonic uptake, coprecipitation with iron hydroxides, and salt-induced coagulation of colloids have been suggested as the removal mechanism. This reduces the effective fluvial flux of dissolved REEs to the ocean considerably and affects the budget calculation in the ocean. For example, the river flux of dissolved Nd is estimated to be 4.6×10^6 mol year^{-1} (**Table 3**), being lowered by a factor of 3–4 in the estuaries, and the mean oceanic residence time with respect to the river input is about 10^4 years. This Nd mean residence time appears too long in the light of heterogeneous distribution of its isotopic composition observed in the different oceanic basins.

The $^{143}Nd/^{144}Nd$ isotopic composition is generally expressed as,

$$\varepsilon_{Nd} = \left[\frac{\left(^{143}\mathrm{Nd}/^{144}\mathrm{Nd}\right)_{\mathrm{measured}}}{0.512638} - 1\right] \times 10^4$$

where the value of 0.512638 is referred to the $^{143}Nd/^{144}Nd$ ratio in the chondritic uniform reservoir (CHUR). Rivers entering into the Atlantic Ocean are influenced by old continental crust and have a relatively nonradiogenic ε_{Nd} value of -12, whereas values for rivers entering into the Pacific which is surrounded by young oceanic islands and island arcs are more radiogenic, -3 to -4. Rivers draining into the Indian Ocean have intermediate values around -9. The Nd isotopic compositions in the riverine input are almost faithfully reflected in seawater. The measured ε_{Nd} values for seawater also average -12 ± 2.5 for the Atlantic, -8 ± 2 for the Indian Ocean, and -3 ± 2 for the Pacific Ocean. Thus, the mean residence time of Nd must be short

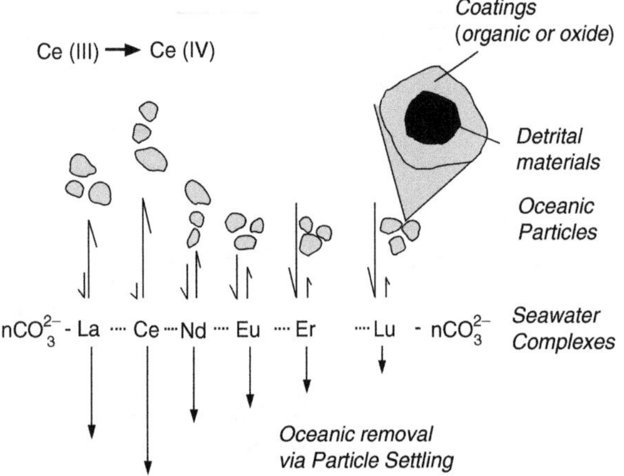

Figure 5 A conceptual model of REE fractionation between particles and seawater. Main features include (1) the systematic variation in the relative affinity of trivalent REEs for complexation to solution carbonates and binding to particles, (2) the enhanced formation of particulate Ce due to the oxidation of Ce(III) to Ce(IV), and presence of surface active coatings on detrital particles. These features lead to fractionation of REE between seawater and particles and to fractionation via the settling of large particles. After Sholkovitz *et al.* (1994).

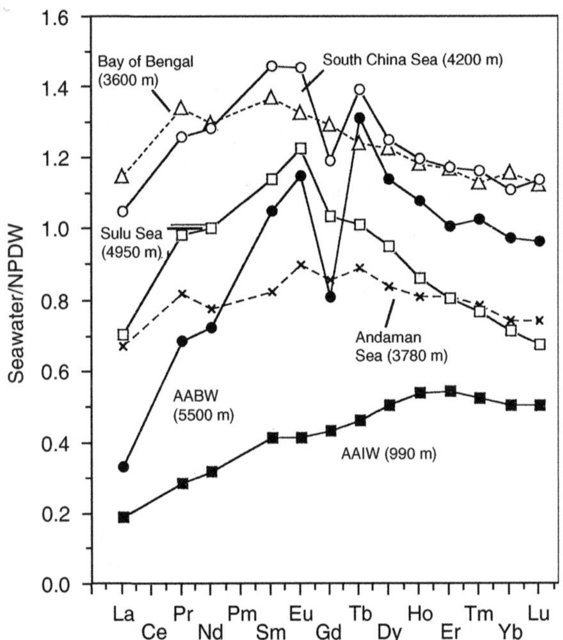

Figure 6 The characteristic features of NPDW-normalized patterns of dissolved REEs in different oceanic basins. Data from Alibo and Nozaki (unpublished) and Nozaki *et al.* (1999).

compared with the oceanic mixing time of $\sim 10^3$ years. This is in contrast to the Sr isotopes which show a constant $^{87}Sr/^{86}Sr$ ratio in seawater because of homogenization by oceanic mixing during its residence time of $\sim 10^6$ years. Some additional sources of REEs appear to exist in the ocean to maintain geochemical consistency for Nd isotopes.

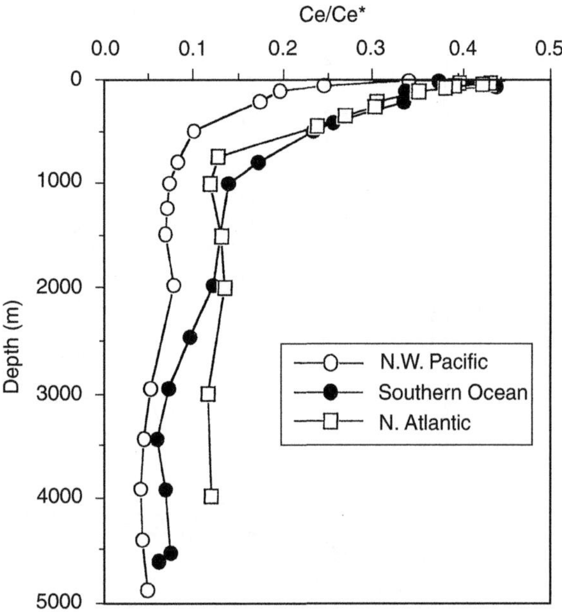

Figure 7 The vertical profiles of Ce-anomaly calculated from the profile data in **Figure 3**.

The concentrations of REEs in hydrothermal fluids venting from hot springs at mid oceanic ridges have recently been investigated. They are generally 1–2 orders of magnitude higher than those of ambient seawater with a distinctly positive Eu anomaly, but are also intensively removed by scavenging in the vicinity of hydrothermal vent fields. Consequently, the effective hydrothermal flux of REEs to the ocean is negligibly small compared to the fluvial flux.

Atmospheric input due to fallout of terrestrial aerosols and subsequent solubilization into seawater has been thought to be important, or even predominant in the open ocean for some reactive heavy metals such as iron and aluminum. However, the eolian fluxes are poorly quantified for the REEs as yet. Available estimates (**Table 3**), though highly uncertain, suggest that the eolian fluxes are 30–130% of the fluvial input of REEs. Terrestrial aerosols are transported through the atmosphere for relatively long distances from the sources by longitudinally prevailing winds such as the westerlies and the trade wind. This is in contrast to the fluvial input which enters the ocean across the land–sea interface. Thus, the relative importance of these REE sources may be reflected in the geographical distribution of REEs in the surface waters. The distribution of dissolved Nd in the surface waters (**Figure 2**) shows the higher concentrations in regions of strong coastal influence, such as Baffin Bay, the Bay of Bengal, and the South China Sea. In the open ocean, there is a general tendency of Atlantic > Indian Ocean > Pacific for the surface dissolved Nd. The surface waters of

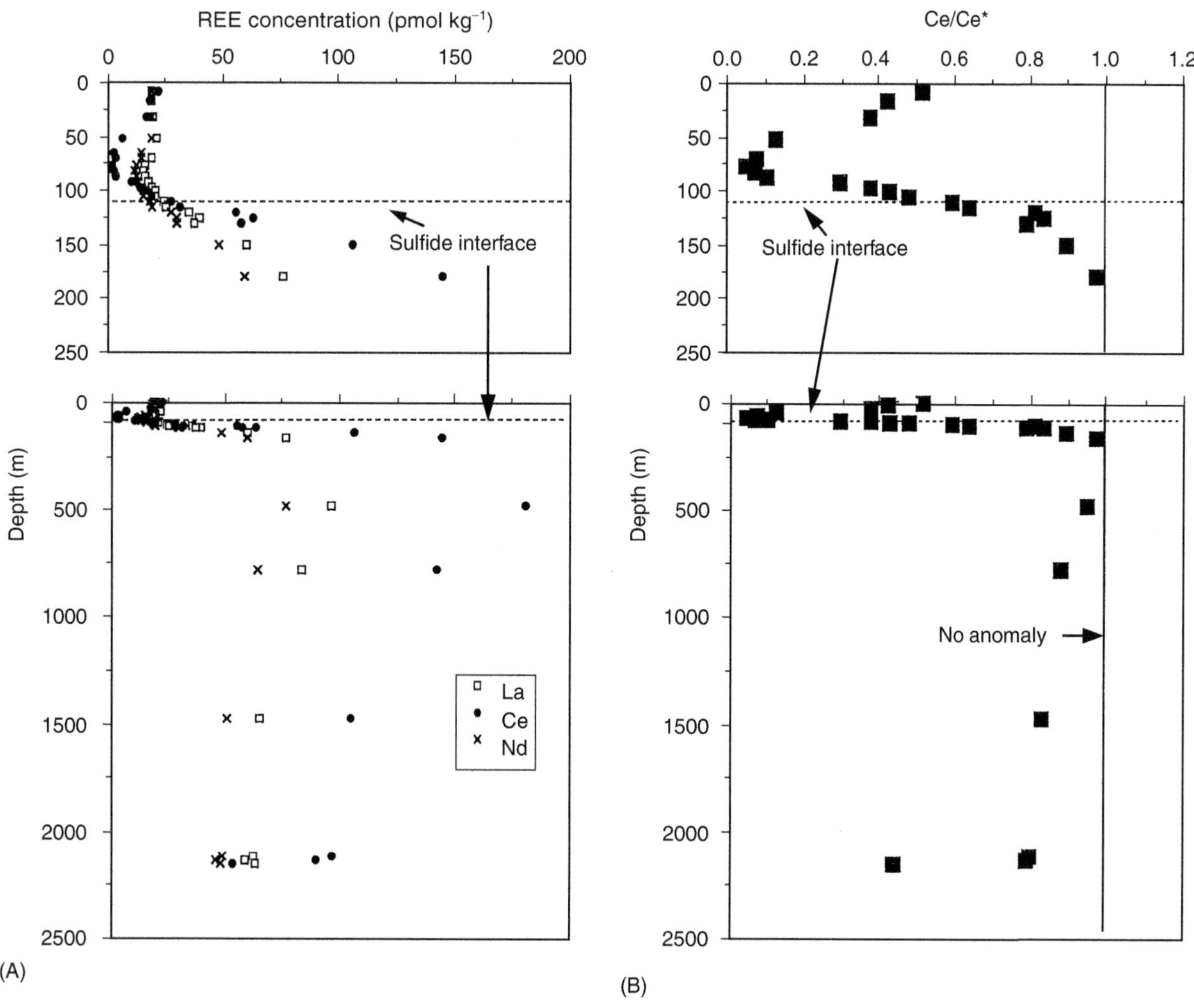

Figure 8 The vertical profiles of La, Ce, and Nd (A) and calculated Ce-anomaly (B) in the Black Sea. After German *et al.* (1991).

the different oceanic province have different NPDW-normalized REE patterns due to the input of REEs from local terrigneous sources (**Figure 9**). The Nd isotopic composition (**Figure 10**) that can serve as a 'fingerprint' of REE source also varies significantly with location. The geographical distributions of Nd concentration and its isotopic composition, and the NPDW-normalized REE pattern are hardly explained by the pattern of prevailing winds. All suggest that the REEs are supplied to the ocean from local sources and their mean residence times in the surface waters are relatively short as compared to homogenization by lateral mixing. The surface Nd distribution somewhat resembles that of fluvially and coastally derived ^{228}Ra with a half-life of 5.7 years which is predominantly supplied to the coastal and shelf waters from underlying sediments.

Since the direct riverine flux of dissolved Nd is not sufficient to yield its short mean residence time required for the oceanic Nd isotopic heterogeneity, some additional mechanism must be sought for the coastal and shelf sources of the REEs. Remineralization of REEs from river-transported sediments may occur near the coast and on the shelf. For instance, if only 1–2% of detrital Nd is released to the coastal water, then the mean oceanic residence time of Nd is shortened to as less than 1000 years. In reality, there is some evidence that the REEs are released from sediments to the overlying water in the outer high-salinity regions of the Amazon and Fly River estuaries. Fractionation can also take place in the sediment diagenesis, and hence, makes the REE patterns of the surface waters different from those of river waters, depending upon the oceanic province. The highly radiogenic ε_{Nd} values observed near the Indonesian Archipelago (**Figure 10**) also suggest that erosion of volcanic islands can be an important source of REEs to the ocean. Although quantification of these local sources of REEs is difficult at present, the coastal and shelf mechanism appears to be significant in the balance of dissolved REEs in the ocean (**Table 3**).

Table 3 Mean oceanic residence times derived from the particle reactivities and estimated remineralization fluxes of the REEs

Element	*Mean dissolved concentration (C_d)(pmol kg^{-1})*	*Atmospheric flux [a] (10^6mol y^{-1})*	*Riverine flux [a] (10^6mol y^{-1})*	*Remineralization flux [b] (10^6mol y^{-1})*	*Mean residence time (τ_{REE})[c] (y)*
Y	*220*	*8.3[d]*	*15[d]*	*155*	*1670*
La	30	4.1	5.0	72	500
Ce	4.5	8.5	6.3	107	50
Pr	3.8	1.1[e]	1.2[e]	19	240
Nd	20	4.4	4.6	59	400
Sm	4.0	0.91	1.1	11	400
Eu	1.0	0.20	0.3	2.8	410
Gd	4.7	0.85	1.4	10	520
Tb	1.1	0.13[e]	0.22[e]	2.3	570
Dy	6.5	0.72	1.2	9.9	740
Ho	2.0	0.15[e]	0.27[e]	1.1	1820
Er	6.5	0.45	0.8	2.4	2420
Tm	1.0	0.055[e]	0.14[e]	0.36	2430
Yb	6.5	0.32	1.1	2.2	2440
Lu	1.3	0.058	0.2	0.35	2890

[a] After *Greaves and Elderfield (1994)*. The values may be uncertain by a factor of 2–3.
[b] Remineralization flux = $(1.35 \times 10^9 \times C_d)/\tau_{REE}$ – Atmospheric flux – Riverine flux.
[c] Based on $\tau_{REE} = \tau_p/F_p$ for $\tau_p = 10$ years, $\gamma = 0.8 \pm 0.2$ and F_p values given in **Table 2**. The values may be uncertain by a factor of 2–3.
[d] Assumed to be 55 times of Ho flux from the crustal ratio.
[e] Estimated from neighboring trivalent REEs on the basis of shale (PAAS) composition.

Particle Reactivity and Mean Oceanic Residence Times of REEs

The major process that removes dissolved REEs from the water column is considered to be adsorptive scavenging by sinking particulate matter (**Figure 5**) which controls the distributions of not only REEs but also a number of reactive elements in the ocean. Some elements can also be removed by lateral transport along with currents and subsequent intensified particle scavenging and/or uptake at the sediment–water interface of the oceanic margins. This, 'boundary scavenging' is known to occur for ^{210}Pb and ^{231}Pa, but this mechanism is probably insignificant for dissolved REEs. The vertical profiles of dissolved REEs (**Figure 3**) strongly suggest that they are involved in the oceanic biogeochemical cycle. Marine particulate matter comprises biogenic organic matter, opaline silica, and carbonates as well as inorganic oxides and terrestrial detritus. The biogenic particles are produced in the surface water and transported by settling to the deep water and the seafloor where they are largely remineralized to the water. Despite a rough similarity in the vertical profile between the REEs and dissolved Si or alkalinity, however, active processes such as biological uptake and incorporation of REEs into calcareous and opaline skeletons do not seem to play any important role in this respect. For instance, the REE concentrations of shells are extremely low. Also, it is unlikely that the dissolved REEs are actively taken up into organic tissues.

Adsorptive scavenging of reactive metals has been clearly demonstrated, since some short-lived radionuclides such as ^{234}Th, ^{210}Po and ^{210}Pb are highly enriched in particulate matter collected by filtration and sediment traps. The scavenging processes of reactive metals may be governed by competitive reactions between binding affinity onto the particle surfaces and complex formation with ligands in seawater (**Figure 5**). Although marine particles are composed of various materials, their surfaces are thought to be coated with organic matter and oxides of manganese and iron. The fine suspended particles, that can be conventionally collected by filtration, predominate in the surface area onto which reactive metals are adsorbed and the standing stock of mass. In contrast, the large, fast sinking particles, that can be effectively collected by a sediment trap, virtually govern the vertical flux of materials. Those particles however, are frequently exchanged through aggregation and disaggregation during sinking through the water column. Since the vertical transport of particulate matter to the sedimentary sink is a process common to all associated elements, the overall rate of removal from the ocean for metals such as the

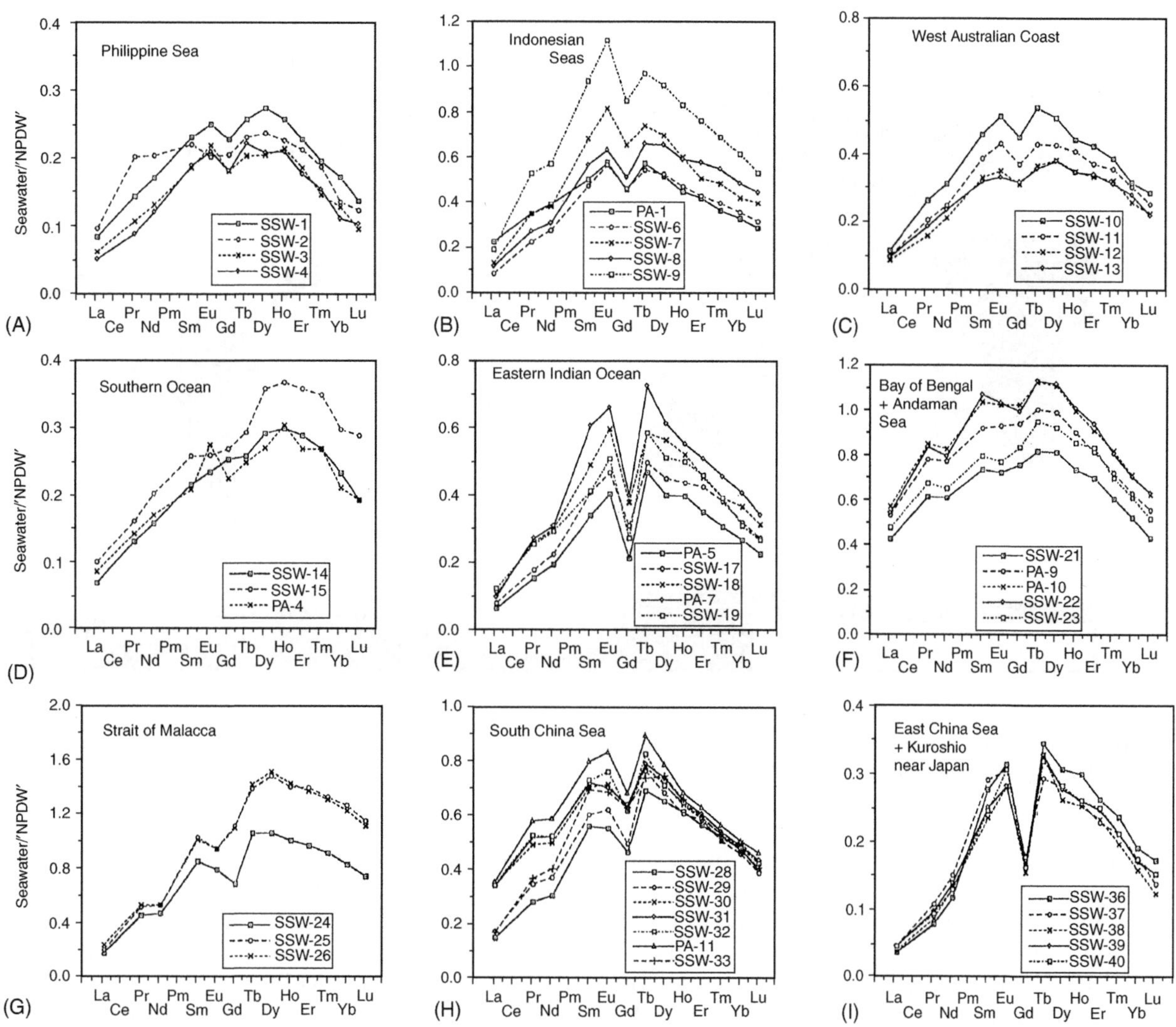

Figure 9 Comparison of NPDW-normalized REE patterns in the surface waters in the eastern Indian Ocean and its adjacent seas. See **Figure 2(B)** for the locations. Adapted from Amakawa *et al.* (2000).

REEs must depend on their partitioning between dissolved and particulate phases and a mean sinking velocity of the particles. Available data on the REEs associated with suspended particles are summarized in **Table 2**.

With the assumptions that a large portion of particulate REEs are exchangeable with dissolved REEs and that the suspended fine particles are incorporated into the large, fast-sinking particles through repeated aggregation and disaggregation and eventually removed from the ocean with a certain mean residence time (τ_p), the mean residence time of REEs (τ_{REE}) is given by the following equation.

$$\tau_{REE} = \tau_p/\gamma F_p$$

where F_p is the fraction of particulate REEs in seawater (**Table 2**) and γ is the portion that is effectively exchangeable with the dissolved REEs. Based on the studies of Th isotopic systematics in the water column, the fine particles, on which most reactive metals reside, settle with a mean speed of $\sim$1 m day^{-1} corresponding to the mean oceanic residence time of about 10 years. From the chemical leaching experiments of particulate matter and acidification treatment of both filtered and unfiltered seawaters to pH$\sim$1.5 to determine acid-soluble particulate fraction of REEs, the value of γ is deduced to be between 0.6 and 1.0. The mean residence times of REEs estimated using these values are given in **Table 3**. They range from 50 years for Ce to 2900 years for Lu. The short mean residence time of Ce is approximately equal to that of Th, being consistent with their predicted dominance of hydrolysis species in seawater. The 400 year residence time is well suited for Nd with its heterogeneous distribution of the isotopic ratios in the oceans.

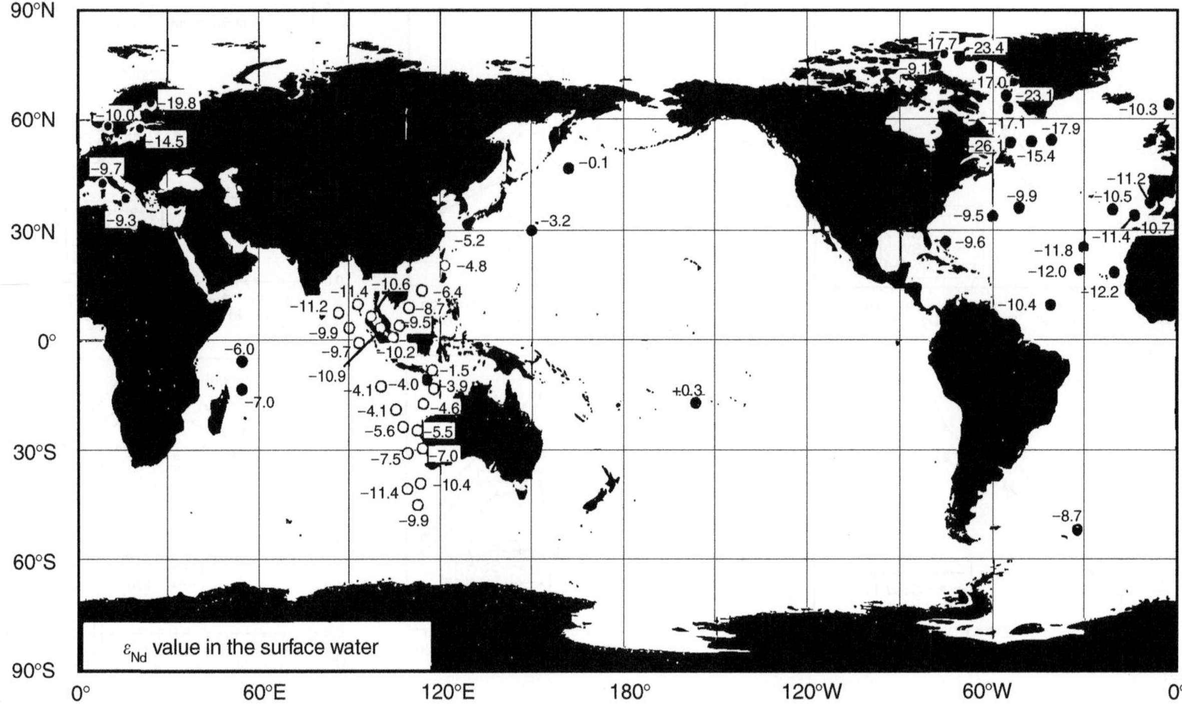

Figure 10 The distribution of $^{143}Nd/^{144}Nd$ isotopic composition (expressed as ε_{Nd} value) in the surface waters. Compiled from Byrne and Sholkovitz (1996) and Amakawa *et al.* (2000).

It is also clear that eolian and riverine sources alone are insufficient to yield such short residence times and there must be additional sources of REEs in the ocean. The most likely candidate of the potential REE source is remineralization of nearshore, coastal and shelf sediments as described earlier. The magnitude of this remineralization flux required to balance each REE in the ocean is given in **Table 3**. Those fluxes are quite large particularly for the light REEs, whereas the relative importance of remineralization decreases and becomes somewhat comparable with the sum of eolian and riverine inputs for the heavy REEs. This REE fractionation during remineralization on the shelf may contribute the different NPDW-normalized REE patterns observed in the surface waters (**Figure 9**), although the mechanism is not well understood.

Summary and Conclusion

Although the oceanic distributions of rare earth elements somewhat resemble those of nutrients, their behaviors are clearly different in that they are not actively taken up by photoplankton but passively scavenged by particles. Elemental reactivity with suspended particles, which controls, together with sinking of particle aggregates, the mean oceanic residence times of not only REEs but also a number of other heavy metals, is best understood by the competitive complexation reactions of dissolved REEs with ligands on the surface of particles and in solution. This yields the progressive enrichment from the light to the heavy REEs in the shale-normalized pattern as commonly observed in seawater. Remineralization of REEs from coastal and shelf sediments is thought to play a significant role in the global budget of REEs and the geochemical and isotopic consistency of Nd in the ocean. The REEs and Nd isotopes provide novel and unique oceanographic tracers or chemical problems in studying (1) particle scavenging processes, (2) redox sensitive geochemical processes with Ce, and (3) identification and modification of water masses. Applications to the wider problems are being explored.

See also

River Inputs. Tracers of Ocean Productivity.

Further Reading

Alibo DS and Nozaki Y (1999) Rare earth elements in seawater: particle association, shale-normalization, and Ce oxidation. *Geochimica et Cosmochimica Acta* 63: 363–372.

Alibo DS, Nozaki Y, and Jeandel C (1999) Indium and yttrium in North Atlantic and Mediterranean waters: comparison to the Pacific data. *Geochimica et Cosmochimica Acta* 63: 1991–1999.

Bertram C and Elderfield H (1993) The geochemical balance of the rare earth elements and neodymium isotopes in the oceans. *Geochimica et Cosmochimica Acta* 57: 1957–1986.

Byrne RH and Sholkovitz ER (1996) Marine chemistry and geochemistry of the lanthanides. In: Gschneider KA Jr and Eyring L (eds.) *Handbook on the Physics and Chemistry of Rare Earths*, vol. 23, pp. 497–593. Amsterdam: Elsevier Science.

DeBaar HJW, Bacon MP, and Brewer PG (1983) Rare-earth distributions with a positive Ce anomaly in the western North Atlantic Ocean. *Nature* 301: 324–327.

Elderfield H and Greaves MJ (1982) The rare earth elements in seawater. *Nature* 296: 214–219.

German CR, Holliday BP, and Elderfield H (1991) Redox cycling of rare earth elements in suboxic zone of the Black Sea. *Geochimica et Cosmochimica Acta* 55: 3553–3558.

Goldberg ED, Koide M, Schmitt RA, and Smith RH (1963) Rare-earth distributions in the marine environment. *Journal of Geophysical Research* 68: 4209–4217.

Greaves MJ, Statham PJ, and Elderfield H (1994) Rare earth element mobilization from marine atmospheric dust into seawater. *Marine Chemistry* 46: 255–260.

Lipin BR and McKay GH (eds.) (1989) *Geochemistry and mineralogy of rare earth elements. Reviews in Mineralogy* 21. Washington, DC: Mineralogical Society of America.

Nozaki Y (1997) A fresh look at elemental distribution in the North Pacific Ocean. *EOS Transactions* 78: 221.

Nozaki Y, Alibo DS, Amakawa H, Gamo T, and Hasumoto H (1999) Dissolved rare earth elements and hydrography in the Sulu Sea. *Geochimica et Cosmochimica Acta* 63: 2171–2181.

Sholkovitz ER, Landing WM, and Lewis BL (1994) Ocean particle chemistry: The fractionation of rare earth elements between suspended particles and seawater. *Geochimica et Cosmochimica Acta* 58: 1567–1579.

Sholkovitz ER and Schneider DL (1992) Cerium redox cycles and rare earth elements in the Sargasso Sea. *Geochimica et Cosmochimica Acta* 55: 2737–2743.

Taylor SR and McLennan SM (1985) *The Continental Crust: Its Composition and Evolution*. Oxford: Blackwell.

REFRACTORY METALS

K. J. Orians and C. L. Merrin,
The University of British Columbia,
Vancouver, BC, Canada

Introduction

The elements classified as 'refractory' here are those that are not readily dissolved in sea water. Their supply to the oceans is low relative to their abundance in the Earth's crust. In addition, they are rapidly removed from solution by interaction with the surfaces of sinking particles, a process referred to as 'scavenging.' This rapid removal means that they are in the oceans for only a short time before being removed to the seafloor. The average time they spend in the oceans, known as the oceanic residence time, ranges from a few tens to a few thousands of years. Both of these factors result in low concentrations in sea water relative to their abundance in the Earth's crust, a large range of oceanic concentrations, and distributions that typically reflect their sources.

The elements in this category exist as hydroxide species in sea water, mostly as $M(OH)_n^{X-n}$, where M is the metal, X is the oxidation state of the metal, and n is the number of hydroxide ligands in the complex. There is also the possibility that they may exist as organic complexes and/or in association with colloidal phases (particles $\leqslant \mu m$). Organic complexes have been shown to be important for iron, but not much is known about these other forms for many of these elements.

The most abundant of these elements, aluminum and iron (which comprise 8.23%, and 5.63%, respectively, of the Earth's crust, by weight) are also the most studied. The first reliable reports on dissolved aluminum in the oceans were made in the late 1970s. Since then there have been over 50 articles on the distribution of aluminum in the oceans. Reliable data on iron were not available until the late 1980s but, owing to the importance of iron in regulating primary production in some regions of the ocean, there has been a wealth of studies on this element in the past decade. Most of the other elements discussed here were not studied until the late 1980s or even the 1990s, and for many there are only a couple of articles on their oceanic distributions. There is still much we do not know about their distributions and the processes that control them in the oceans.

History

Advances in our understanding of trace metal distributions in the oceans began with the development of clean sampling methods in the late 1970s and have continued with the ongoing development of highly sensitive analytical methods. Clean sampling and handling methods are critical in the analysis of the more abundant refractory metals, aluminum and iron. Detection of the lower-abundance refractory metals, owing to their exceptionally low concentrations in sea water, has been limited by the sensitivity of available methods. Their analysis has greatly benefited from the increasing sensitivity of modern analytical instruments. The development of highly sensitive mass spectrometers that allow for aqueous sample introduction have revolutionized this field. Inductively coupled plasma mass spectrometers (ICP-MS) using quadrupole mass analyzers, made commercially available in the mid 1980s, and the magnetic and electric sector high-resolution ICP-MS instruments, available since the early 1990s, have allowed the detection of these elements without requirement for excessive sample processing and preconcentration steps.

Distributions

Rapid removal of the scavenged elements results in a low background concentration in the oceans and the potential for a large concentration range, depending on the variations in the magnitude of their sources. This is especially true for aluminum, where the concentrations vary by up to 2 orders of magnitude with depth at a given location, by 3 orders of magnitude from one major ocean to another, and by 4 orders of magnitude in extreme environments. This range reflects the large variations in dust sources to the oceans from place to place and rapid removal of aluminum from sea water away from these sources. The range of concentrations for all of the refractory elements, their oceanic average, and some typical concentrations found in surface (<100 m) and deep (<2000 m) waters of the central North Atlantic and North Pacific Oceans are presented in **Table 1**. Although the range of concentrations is largest for aluminum, this difference is most likely exaggerated in the data shown here, as aluminum has been studied in more diverse regions.

Table 1 Summary of the hydroxide speciation, concentrations, and residence times for the refractory elements in the oceans

Element (units)	*Probable main species in oxygenated sea water*	*Concentration*[a]					*Residence time (y)*[b]
		Range (Average)	*Pacific*		*Atlantic*		
			Surface	*Deep*	*Surface*	*Deep*	
Al ($nmol\,kg^{-1}$)	$Al(OH)_3^{\circ}$, $Al(OH)_4^-$	0.06–650 (1.5)	4	2	40	20	30–200
Ga ($pmol\,kg^{-1}$)	$Ga(OH)_4^-$	2–60 (17)	10	25	40	25	100–750
In ($pmol\,kg^{-1}$)	$In(OH)_3^{\circ}$	0.05–4.8 (0.1)	0.1	0.06	0.6	1.6	(similar to Al)
Sc ($pmol\,kg^{-1}$)	$Sc(OH)_3^{\circ}$	7–22 (16)	10	20	15	20	NA
Ti ($pmol\,kg^{-1}$)	$TiO(OH)_2^{\circ}$	4–250 (140)	5	200	50	200	100–700
Zr ($pmol\,kg^{-1}$)	$Zr(OH)_5^-$	12–360 (180)	20	270	80	150	800–5600
Hf ($pmol\,kg^{-1}$)	$Hf(OH)_5^-$	0.1–1.5 (0.7)	0.3	0.8	0.4	0.7	650–1300
Nb ($pmol\,kg^{-1}$)	$Nb(OH)_6^-$, $Nb(OH)_5^{\circ}$	2.8–4.2 (3.9)	2.8	3.8	NA	NA	5000+
Ta ($pmol\,kg^{-1}$)	$Ta(OH)_5^{\circ}$	0.08–0.21 (0.2)	0.09	0.21	NA	NA	5000+
Fe ($nmol\,kg^{-1}$)	Organic Fe, $Fe(OH)_3^{\circ}$, $Fe(OH)_2^+$	0.05–0.8 (0.5)	0.05	0.6	0.05	0.5	100–500
Bi ($fmol\,kg^{-1}$)	BiO^+, $Bi(OH)_2^+$, $Bi(OH)_3^{\circ}$	25–450 (140)	165	50	400	125	20
Th ($fmol\,kg^{-1}$)	$Th(OH)_4^{\circ}$ ($Th(OH)_5^-$?)	50–600 (80)	70	150	500	500	45

NA = data not available.
[a]Typical values for the central North Pacific and the central North Atlantic, where 'deep waters' are deeper then 2000 m.
[b]Residence times are estimated from deep water scavenging models for Al, Ga, Ti, Zr, Hf, Fe, and Th (radioactive disequilibria between ^{230}Th and ^{234}U, for Th); river input for Nb and Ta; and the total input to the surface ocean for Bi.

Table 2 The average abundance of the refractory elements in the Earth's crust, and their degree of enrichment, relative to aluminum, in the oceans

Element (M)	*Average crustal abundance*		*Average sea water concentration ($mol\,kg^{-1}$)*	$\frac{\text{Sea water}}{\text{crust}}$ ($\times 10^{15}$)	$(M/Al)_{SW}$	$\frac{(M/Al)_{sw}}{(M/Al)_{crust}}$
	By weight	*mol kg*$^{-1}$				
Al	8.23%	3.05×10^6	1.5×10^{-9}	0.5	1.00	(1.00)
Ga	15 ppm	215	1.7×10^{-11}	80	1.1×10^{-2}	160
In	0.1 ppm	0.87	1.0×10^{-13}	115	6.7×10^{-5}	230
Sc	22 ppm	489	1.6×10^{-11}	33	1.1×10^{-2}	65
Ti	0.57%	1.19×10^5	1.4×10^{-10}	1.2	9.3×10^{-2}	2.4
Zr	165 ppm	1810	1.8×10^{-10}	100	1.2×10^{-1}	200
Hf	3 ppm	16.8	7.1×10^{-13}	40	4.7×10^{-4}	90
Nb	20 ppm	215	3.9×10^{-12}	18	2.6×10^{-3}	40
Ta	2 ppm	11	2.0×10^{-13}	18	1.3×10^{-4}	40
Fe	5.63%	1.01×10^6	5.4×10^{-10}	0.5	3.6×10^{-1}	1
Bi	0.17 ppm	0.813	1.4×10^{-13}	170	9.3×10^{-5}	340
Th	9.6 ppm	41.4	8.0×10^{-14}	2	5.5×10^{-5}	4

Crustal abundances from Taylor (1964). Seawater concentrations from various references.

Also included in **Table 1** is an estimate of the average oceanic residence times for these elements. A shorter residence time indicates a more rapid removal from the oceans. Another way to estimate the relative reactivity of elements with crustal sources is to compare their sea water concentrations with their abundance in the Earth's crust. This is presented in **Table 2** (with a normalization to aluminum) and is shown graphically in **Figure 1**. If variations in the composition of continental materials and differences in the solubility of the elements from these materials are not too great, then to a first approximation the removal rate should be inversely related to the degree of enrichment in seawater. Relative to aluminum, iron is the only element considered here that is not enriched in the oceans. Titanium and thorium are very slightly enriched; indium, barium and yttrium are greatly enriched (by 2–3 orders of magnitude), and tantalum, niobium, gallium, scandium, zirconium, and hafnium show intermediate enrichment. The degree of enrichment does not correlate well with estimates of the residence times of these elements. These discrepancies will be discussed in the following sections.

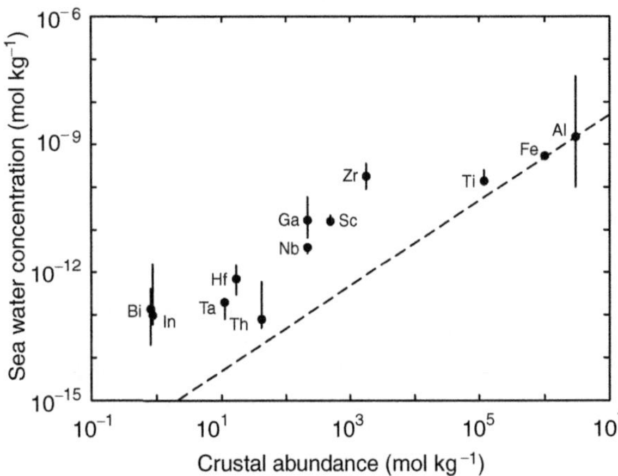

Figure 1 The average and range of concentrations of the elements in the deep ocean from 2000 m to the bottom plotted against their average abundance in the Earth's crust. (Crustal abundances from Taylor (1964) Seawater concentrations from various references.) The dotted line shows a 1:1 slope plotted through the average concentration of aluminum. Elements that plot above this line are enriched in sea water relative to aluminum and their abundance in the crust.

Distributions of scavenged elements in the oceans typically reflect their sources, as they are not in the oceans long enough to be homogenized by thermohaline circulation that mixes the oceans on a timescale of about a thousand years. Many of these elements have a surface source, primarily from continental dust that is partially dissolved in the surface waters. River sources are also possible, although much of the metals entering the ocean by this route is removed in coastal areas. Surface transects for aluminum and gallium (**Figure 2A** and **B**) show distribution expected with atmospheric dust as the dominant source. The lower values in coastal region result from increased removal in these highly productive waters, and restrict the transport of these elements to the open ocean. In contrast, zirconium and hafnium show a coastal source, possibly from rivers (**Figure 2C**). A bottom source is also evident in the vertical profiles of many refractory elements. The mechanism providing this source is not known, but could be from either dissolution of particles at the sediment surface (a process often referred to as 'remineralization') or a flux from the waters trapped within the sediments (an interstitial or pore water flux). Some of these elements (gallium, iron, and bismuth, for example) also show the effects of internal cycling within the ocean.

Aluminum

Aluminum (Al) is the most abundant metallic element in the Earth's crust. Aluminum is a trivalent metal with a strong tendency to hydrolyse in sea water to form the particle reactive species, $Al(OH)_3$ and $Al(OH)_4^-$. Both of these species are important in sea water, and there is a switch in the speciation within the pH and temperature range of sea water, from mostly $Al(OH)_4^-$ (75%) in surface waters to mostly $Al(OH)_3$ in deeper waters.

Dissolved aluminum has a large dynamic range in the oceans, from less than 0.06 nmol kg^{-1} in the mid-depth waters of the North Pacific to 650 nmol kg^{-1} in the surface microlayer of the Arabian Sea. Vertical distributions (**Figure 3**) typically show a surface maximum, due to eolian dust deposition, a mid-depth minimum, due to scavenging removal, and deep water concentrations that depend largely on the age of these waters. In the absence of recent deep water formation, the concentration is typically low in the bottom waters, with only a small increase (up to 2 nmol kg^{-1}) from sediment sources. The residence time of dissolved aluminum in the deep ocean, estimated using vertical advection diffusion (VAD) scavenging removal models, varies from 30 to 200 years, depending on the flux of particles from the overlying waters. The removal mechanism for dissolved aluminum is primarily via scavenging: a passive adsorption onto the surface of particles. There is some laboratory evidence for active uptake into biological soft tissues and/or silica frustules as well, but passive scavenging controls the distribution of aluminum in the oceans. The residence time in the surface ocean, estimated from soluble dust input, is about 3–4 years.

Surface water dissolved aluminum concentrations range from 0.3 to 10 nmol kg^{-1} in the Pacific Ocean, 0.2 to 85 nmol kg^{-1} in the Atlantic and Mediterranean, and 10 to 300 nmol kg^{-1} in the Arabian Sea. The surface distribution is tightly correlated with the magnitude of the dust fluxes in the region. The solubility of aluminum from eolian particles is about 5–10%. Fluvial input of aluminum, while significant, is rapidly removed in the estuaries and highly productive coastal regions (**Figure 2A**). Owing to the low background concentrations of aluminum in the oceans, and the large concentration of aluminum in crustal materials, aluminum is an excellent tracer of dust input to the oceans and of advective transport of water masses.

Deep water dissolved aluminum concentrations range from 0.5 to 2.0 nmol kg^{-1} in the Pacific Ocean, 8–30 in the Atlantic, and 135–170 in the Mediterranean. The western North Atlantic and the Mediterranean deep waters are high in dissolved aluminum from their recent contact with the surface (deep waters are 'young'). The western North Atlantic dissolved aluminum concentrations are 8–40

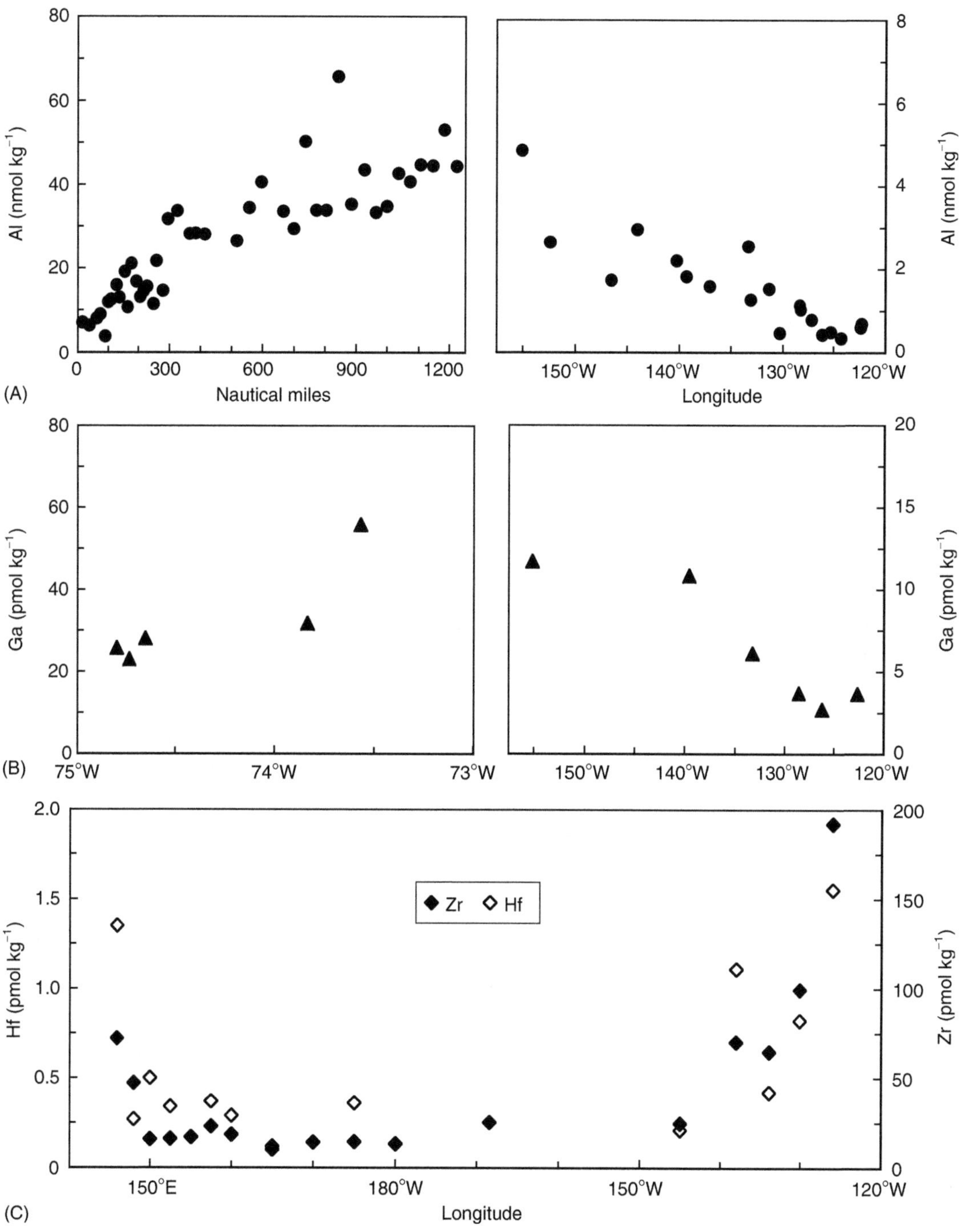

Figure 2 The surface distribution of (A) aluminum, (B) gallium, and (C) zirconium and hafnium from coastal to open ocean regions in the North Pacific and western North Atlantic. Note the scale changes between oceans. In all panels, the coastal locations are plotted at the outside edges, with more oceanic locations in towards the middle. The data for Al and Ga in the eastern North Pacific are from 28°N 155°W to 36°N 123°W (Orians and Bruland, 1986, 1988b). For the Atlantic, the Ga data are from 37°N 75°W to 36°N 73°W (Orians and Bruland, 1988b), and the Al data are from Rhode Island, across the Gulf Stream, into the Sargasso Sea, then down toward the Caribbean (Measures *et al*, 1984, stations #0-1206). For Zr and Hf the western North Pacific data are from 38°N 146°E to 16°N 169°W, and in the eastern North Pacific from 50°N 145°W to 49°N 126°W (McKelvey and Orians, 1993; McKelvey, 1994).

times higher than those in the North Pacific at similar depths – the largest interocean fractionation yet observed for any element (**Figure 3**). As the deep waters travel from the western North Atlantic to the eastern North Atlantic or on to the North Pacific, dissolved aluminum is continually removed from the water. Thus the lowest concentrations are found in the oldest waters of the North Pacific.

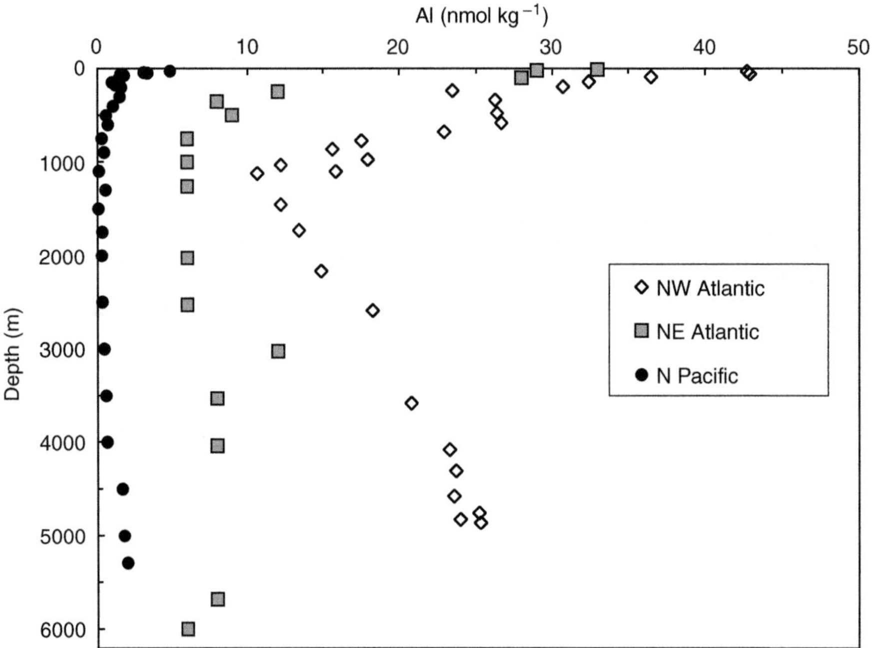

Figure 3 Depth profiles of dissolved aluminum in the central North Pacific (28°N 155°W; Orians and Bruland, 1986), in the eastern North Atlantic (31°W, 26°N; Hydes, 1983), and in the western North Atlantic (near Bermuda, EN120 Station 3; Measures *et al.*, 1986).

Gallium

Gallium (Ga) is a trivalent metal that exists as the strongly hydrolyzed $Ga(OH)_4^-$ ion and to a much lesser extent as $Ga(OH)_3$ (<2%) in sea water. Dissolved gallium ranges from 2 to 60 pmol l^{-1}, with the lowest values in the surface waters of the high-latitude North Pacific, and the highest values in the surface waters of the North Atlantic. Vertical distributions of dissolved gallium are complex. In the Pacific there is a surface minimum, a shallow subsurface increase then a mid-depth decrease, and concentrations increasing with depth below 1000 m to a maximum in the bottom waters (**Figure 4A**). Variable surface sources and advection of the mid-depth water masses complicate the distribution in the Atlantic Ocean. Profiles typically show concentrations increasing with depth without a sub-surface feature; at some stations there is also a surface maximum. The residence time of dissolved Ga in the deep ocean, estimated using VAD scavenging removal models, varies from 100 to 750 years.

Surface distributions (**Figure 2B**) show that eolian input is the dominant source of dissolved gallium to the surface ocean. The subsurface maximum seen in the Pacific is not an advective feature, but rather the result of internal cycles. The similarity in the ionic radii of gallium and iron, a known nutrient element, is thought to explain the subsurface maximum for gallium, which may be taken up along with iron in the surface waters and regenerated at the nutricline, with subsequent removal via scavenging in the deeper waters. There is a strong bottom source for gallium. The North Atlantic dissolved gallium concentrations are 2–6 times higher than those in the North Pacific at similar depths, showing a net scavenging with age in the deep waters. The residence times estimated for dissolved gallium are 3–5 times higher than those for aluminum at the same locations. A combination of the longer residence time and the likelihood that gallium is more soluble than aluminum from crustal materials, owing to its larger ionic radius, is thought to produce the observed Ga/Al enrichment in sea water of roughly two orders of magnitude (**Table 2, Figure 1**).

Indium

Indium (In) is a trivalent metal that exists as $In(OH)_3$ in sea water, with a minor contribution from $In(OH)_4^-$ (6%). Dissolved indium ranges from 0.05 to 4.7 pmol kg^{-1}, with the lowest values in the North Pacific and the highest values in the Mediterranean. Vertical profiles of dissolved indium in the Pacific show low concentrations (0.06–0.10 pmol kg^{-1}) that are relatively invariant with depth, with a slight suggestion of a surface maximum. In the Atlantic the concentrations increase gradually with depth from 0.6 to 1.7 pmol kg^{-1}, and in the Mediterranean the concentration is about 4 pmol kg^{-1} and relatively

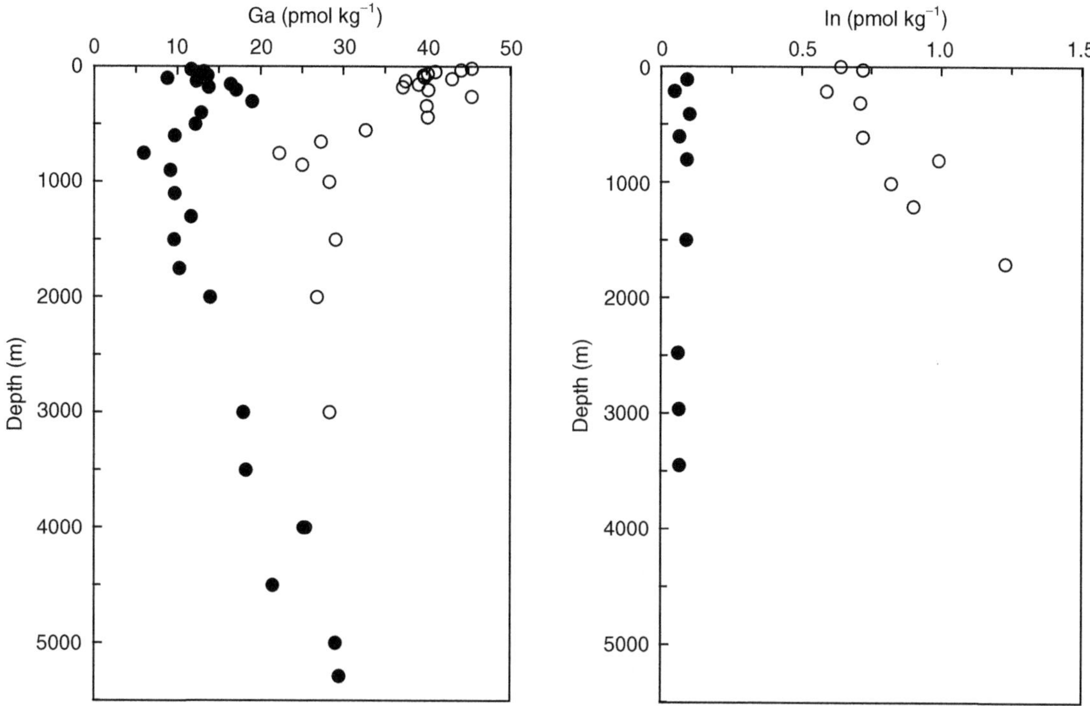

Figure 4 Depth profiles of (A) dissolved gallium in the central North Pacific (solid symbols; 28°N 155°W; Orians and Bruland, 1988) and in the western North Atlantic (open symbols; 32°N 64°W; Shiller, 1998), and (B) dissolved indium in the western North Pacific (solid symbols 34°N 142°E; Amakawa *et al.*, 1996) and in the eastern North Atlantic (open symbols; 26° N, 37°W; Alibo *et al.*, 1999).

invariant with depth. North Atlantic dissolved indium concentrations are 10–20 times higher than those in the North Pacific, showing a large degree of scavenging removal as deep waters age. The residence time of dissolved indium in the ocean is presumed to be similar to that of aluminum, owing to the similarity in their interocean fractionations and in their chemical speciation. The observed In/Al enrichment in sea water, as shown in **Table 2**, is not understood. It has been argued to be due to enrichment in the sources for indium, but it should be noted that increased dissolution of indium from dust can only account for a 10–20-fold enrichment, unless the solubility of aluminum from these sources (5–10%) has been grossly overestimated (there cannot be more than 100% dissolution!). The interocean differences in surface concentrations suggest that atmospheric sources are likely to dominate, as seen for gallium and aluminum.

Scandium

Scandium (Sc) exists in the +3 oxidation state as $Sc(OH)_3$ in sea water. Dissolved scandium has a nutrient-type vertical profile (**Figure 5A**), with low concentrations at the surface and enrichment at depth, yet no significant interocean fractionation is observed. Scandium was detected by neutron activation analysis in the early 1970s and there have not been any recent reports on scandium in the oceans. In the north-east Pacific, dissolved scandium is 7–11 pmol kg^{-1} in the surface water and 18–22 pmol kg^{-1} in deep waters. The Sargasso Sea has slightly higher scandium concentrations: 9–18 pmol kg^{-1} in the surface water and roughly 22 pmol kg^{-1} in the deep water. The residence time for scandium has not been determined.

Titanium

Titanium (Ti) exists in the +3 oxidation state as the neutral oxyhydroxide, $TiO(OH)_2$, in sea water. Dissolved titanium ranges from 4 to 300 pmol l^{-1}, with the lowest values in the surface waters of the North Pacific and the highest values in the deep waters. Vertical profiles of dissolved titanium in the Pacific show a minimum in the surface waters, with gradually increasing concentrations with depth to a maximum at the bottom. Dissolved titanium concentrations in surface waters range from 4 to 8 pmol l^{-1} in the North Pacific, from 50 to 100 pmol l^{-1} in the Atlantic, and from 100 to 150 pmol l^{-1} in the Mediterranean. Deep water data are scant in other regions, but the North Atlantic dissolved titanium concentrations may be slightly higher (up to 50%) than those in the North Pacific at comparable depths.

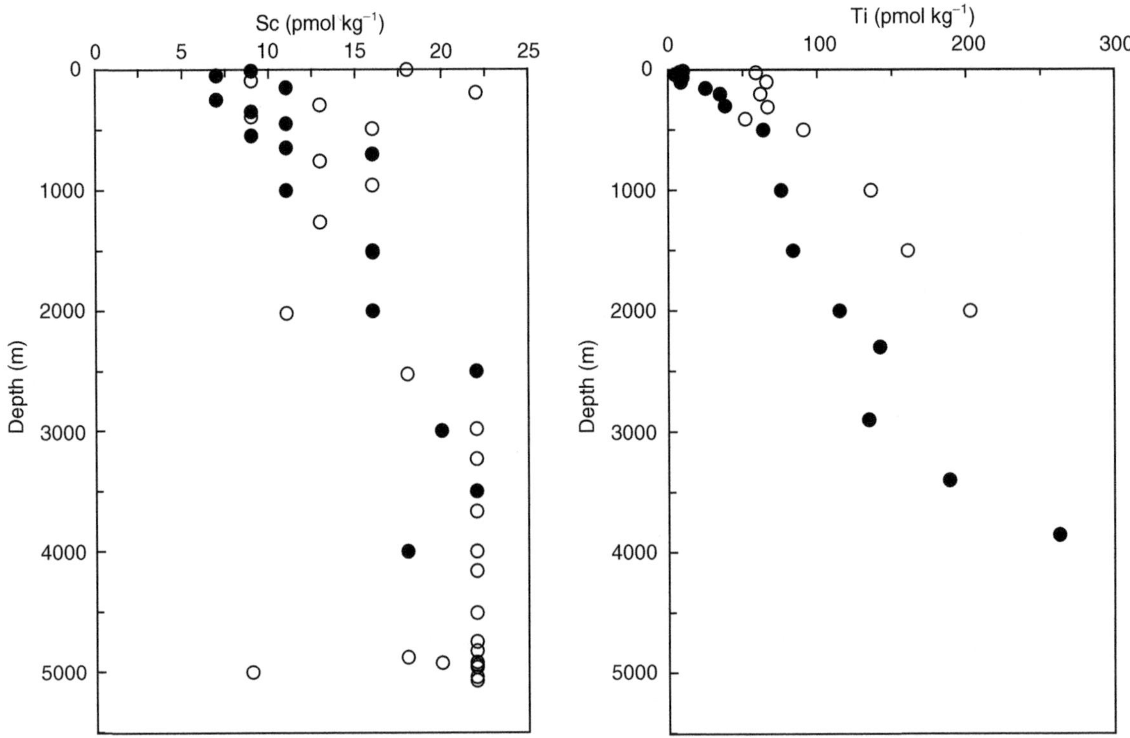

Figure 5 Depth profiles of (A) dissolved scandium in the central North Pacific (solid symbols; 28°N 122°W; Spencer *et al.*, 1970) and in the western North Atlantic (open symbols; 36°N 68°W; Brewer *et al.*, 1972), and (B) dissolved titanium in the North Pacific (solid symbols; 50°N 145°W; Orians *et al.*, 1990) and the western North Atlantic (32°N 64°W; Orians *et al.*, 1990).

The residence time of dissolved titanium in the high-latitude North Pacific is estimated to be 100–200 years, by VAD scavenging models. This is a region of high particle fluxes, where other elements are known to have a shorter than usual residence times. A comparison of titanium with aluminum and gallium at the same location shows that the residence time for titanium is about three times as long as for aluminum and 50% longer than for gallium. A global average is therefore expected to be higher – perhaps 500–700 years. The very small observed Ti/Al enrichment in sea water (**Table 2**) suggests that the residence time for titanium cannot be much longer than that for aluminum and that titanium is probably less soluble from continental materials. The interocean differences in surface concentrations suggest that atmospheric sources are likely to be important, as seen for aluminum and other metals in this group, but rivers have also been shown to be a significant source of titanium to the oceans, even after estuarine removal.

Zirconium and Hafnium

Zirconium (Zr) and hafnium (Hf) exist in the +4 oxidation state as $Zr(OH)_5^-$ and $Hf(OH)_5^-$ (with less than 2% in the neutral forms, $Zr(OH)_4$ or $Hf(OH)_4$). Owing to the lanthanide contraction, these two elements are very similar in their size and chemical properties. In most geological samples, their ratio remains nearly constant. In sea water, however, the ratio varies considerably. Both elements show a surface minimum, with concentrations gradually increasing to a maximum at the bottom (**Figure 6A, B**). The source to the bottom is similar to that seen for gallium and titanium. Surface distributions (**Figure 2C**) indicate that fluvial and/or reducing shelf sediments may be a significant source for zirconium and hafnium.

The concentration range for zirconium with depth is much larger than that for hafnium. This leads to a Zr/Hf atom ratio that increases from a near-crustal value (75–100) in the surface waters of the Pacific (even in the elevated costal waters shown in **Figure 2C**) to ~350 in the deep waters (**Figure 6C**). In the high-latitude North Atlantic, the ratio is higher in the surface (180–200), but not as high in the deep waters (~240, and quite variable from place to place). It appears that the Zr/Hf enrichment increases with the age of the water. Residence times estimated from river input suggest that zirconium has a longer residence time than hafnium (5600 versus 1300 y); estimates using VAD scavenging removal models, while shorter, lead to the same conclusion (see **Table 1**). The difference in residence time,

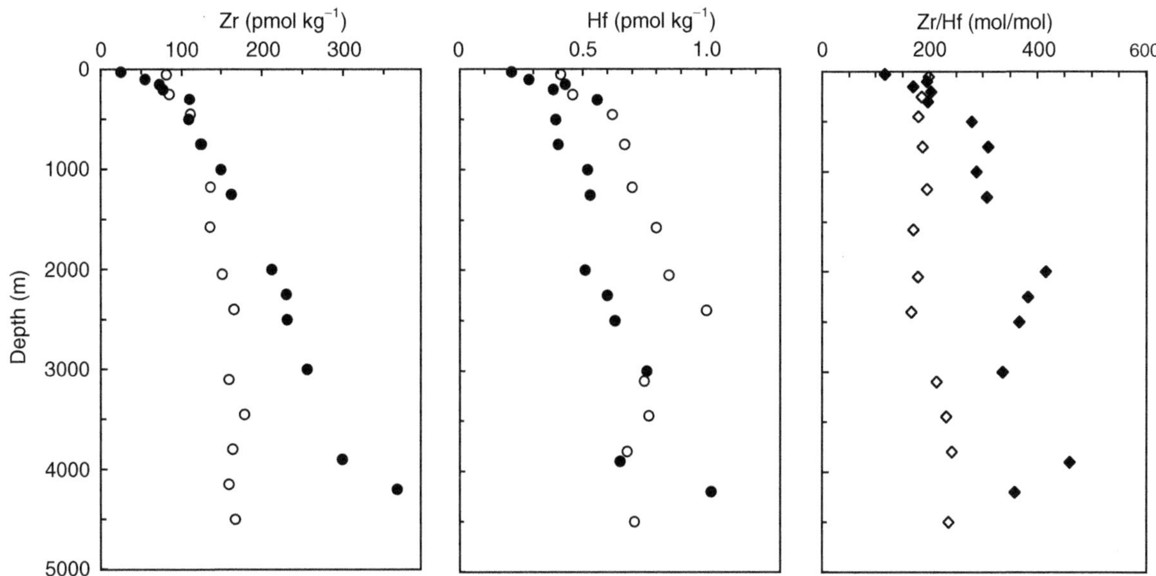

Figure 6 Depth profiles of (A) zirconium, (B) hafnium, and (C) the Zr/Hf atom ratio, in the North Pacific (solid symbols; 50°N 145°W; McKelvey and Orians, 1998) and in the North Atlantic (open symbols; 48°N 15°W; Godfrey *et al.*, 1996).

however, is not sufficient to explain the Zr/Hf enrichment found in deep waters. It is suspected that there must be a zirconium enrichment in the source to these bottom waters as well, although the details of this source are not known. Zirconium and hafnium are both enriched in sea water relative to aluminum, which is likely due to a combination of their longer residence times and the greater supply of these elements to the bottom and in coastal regions.

There appears to be little interocean fractionation for zirconium and hafnium; surface waters are higher in the Atlantic for both, and the deep waters are higher in the Pacific for zirconium (no change for hafnium). Atlantic data are limited to high latitudes, however, and may not be representative of the central gyre. The higher concentrations in the deep North Pacific imply a net input of dissolved zirconium to the deep waters as they age, consistent with the shape of the profile.

Niobium and Tantalum

Reports on the marine chemistry of niobium (Nb) and tantalum (Ta) are limited to one study in the Pacific Ocean. Niobium and tantalum are pentavalent metals that are predicted to exist in sea water either as hydroxides ($Nb(OH)_5$, $Nb(OH)_6^-$, $Ta(OH)_5$ and $Ta(OH)_6^-$) or possibly as oxyacids, similar to molybdenum and tungsten. Their distributions are not conservative like those of molybdenum and tungsten, but they may not be as particle-reactive as the other hydroxide-dominated species discussed. Dissolved niobium is low in the surface (3.0 pmol kg^{-1}) and increases to a nearly constant level from 400 m to the bottom (3.8 pmol kg^{-1}). Dissolved tantalum is low in the surface (0.08 pmol kg^{-1}) and gradually increases with depth to a maximum at the bottom (0.2–0.3 pmol kg^{-1}) (**Figure** 7). Residence times very crudely estimated from their predicted river sources range from 5000 to 60 000 years for both elements. The upper end of this range seems quite unlikely, given their distributions and relatively small enrichment in sea water relative to aluminum (**Table 2**). Residence times calculated from river sources are often overestimates for scavenged elements that are removed in estuaries and coastal environments, but niobium and tantalum may indeed be less reactive than the other refractory elements if they exist as oxyacids rather than hydroxides.

Iron

Iron (Fe) is the second most abundant metal in the Earth's crust. It is a group 8 element and its stable oxidation state in oxygenated seawater is Fe(III). Dissolved Fe(III) has a strong tendency to hydrolyze to form $Fe(OH)_3$ and $Fe(OH)_2^+$ in sea water. Iron is very insoluble with respect to precipitation of hydrous iron oxides and is expected to exist at extremely low concentrations in oxygenated seawater ($<$200 pmol kg^{-1}). Organic ligands that bind iron strongly, however, are found in both the Atlantic and Pacific Oceans at concentrations near 0.60 nmol kg^{-1}. These ligands may prevent loss of iron and allow higher concentrations of iron than would be expected from inorganic solubility alone. Under anaerobic conditions, Fe(II) is the thermodynamically

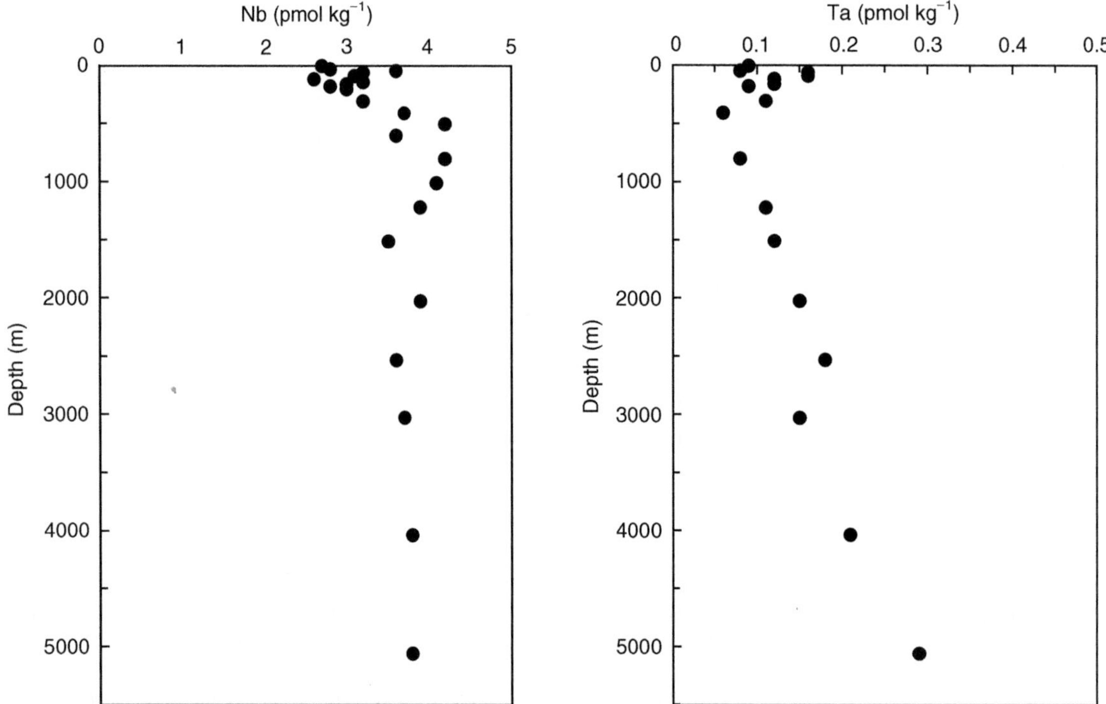

Figure 7 Depth profiles of (A) niobium and (B) tantalum in the western North Pacific (45°N 165°E; Sohrin *et al.*, 1998).

favored species. It exists primarily as the free ion with a minor contribution from the $FeCl^+$ complex. Oxidation kinetics for Fe(II) are rapid above pH 6 and, as a result, Fe(II) is rapidly oxidized to Fe(III) in oxygenated sea water.

Dissolved iron has a nutrient-type vertical profile, and is known to be a required element for phytoplankton growth. Surface water concentrations of dissolved iron in the Pacific, Atlantic, and Southern Oceans are typically $<0.2\,\text{nmol}\,\text{kg}^{-1}$ (average 0.07). A maximum is observed at around 500–700 m (a bit deeper than the maximum for nutrients, nitrate, or phosphate) with relatively uniform concentrations in deep waters. The average concentration below 500 m is $0.6\,\text{nmol}\,\text{kg}^{-1}$. Unlike other nutrient-type metals, there is no significant interocean fractionation for iron; the deep waters do not continue to accumulate iron as they travel from the Atlantic to the Pacific. This can be explained by a balance between regeneration of iron from biogenic matter and subsequent removal by scavenging; the residence time of iron is estimated to be quite short, on the order of 100–500 years (**Table 1**). An alternate hypothesis, that organic ligands control the solubility of iron and set the deep water concentration at 0.6 nmol kg^{-1} has also been proposed.

Dissolved iron enters the ocean via atmospheric, fluvial, hydrothermal, and sediment pathways. Rapid removal of iron from fluvial and hydrothermal sources limits the extent of their influence. Sediments can provide a more significant source; the flux of iron out of reducing sediments is large. Upwelling over the continental shelf brings elevated iron levels to the surface in many coastal regions. Atmospheric dust is the dominant source of iron to the surface of the open ocean. Owing to the low and nearly uniform levels of dissolved iron in surface waters, which do not follow variations in dust patterns, as dissolved aluminum does, the importance of this source has been questioned. The input of iron from atmospheric sources was observed in the central North Pacific by Bruland and colleagues. During a time of unusually strong stratification, when the upper half of the sunlit waters were cut off from the nutrient supply, the removal of iron by phytoplankton was limited to the lower half of the photic zone, thus allowing the build-up of iron from atmospheric dust in the upper waters (**Figure 8**).

Bismuth

Bismuth (Bi) exists in the +3 oxidation state in sea water, probably as the reactive cationic oxyhydroxide species, BiO^+ and $Bi(OH)_2^+$, with a minor contribution from $Bi(OH)_3$. Dissolved bismuth ranges from 25 to $450\,\text{fmol}\,\text{kg}^{-1}$, with the lowest concentrations in the deep North Pacific and the highest at mid-depth (600 m). Vertical profiles of bismuth are

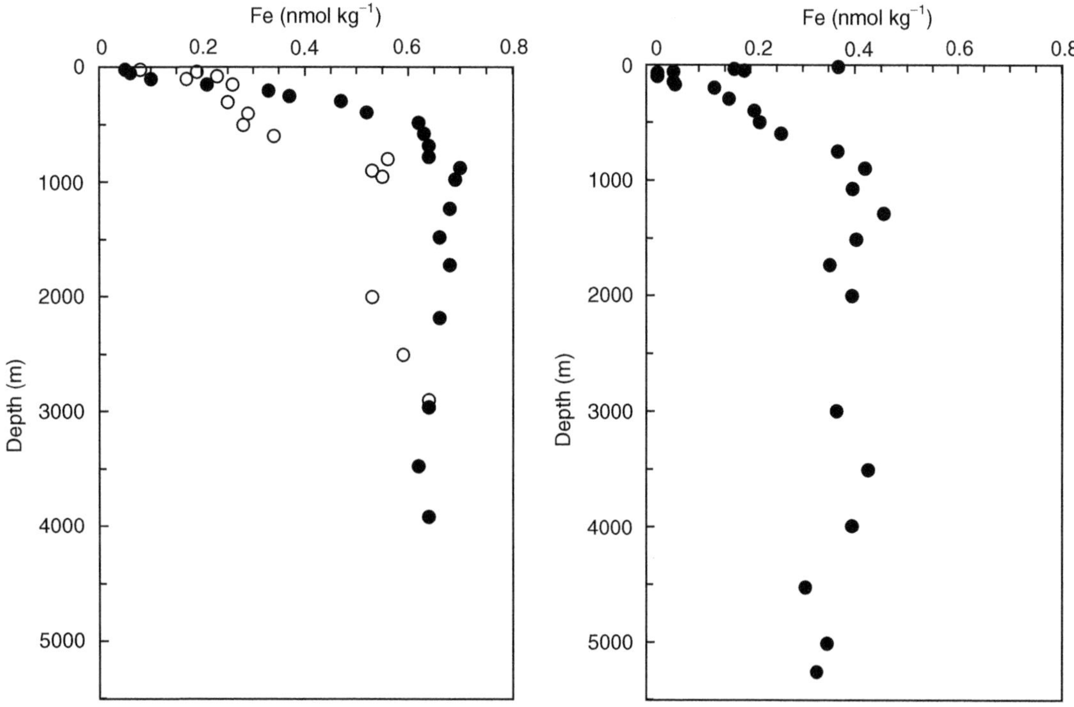

Figure 8 Depth profiles of dissolved iron in (A) the North Pacific (solid symbols; 50°N 145°W; Martin *et al.*, 1989) and the North Atlantic (open symbols; 47°N 20°W; Martin *et al.*, 1993), and (B) the central North Pacific (28°N, 155°W; Bruland *et al.*, 1994).

complex, as shown in **Figure 9**. They typically show high surface water concentrations, thought to reflect the dissolution of aerosol particles. Dissolved bismuth then decreases beneath the surface as a result of removal via scavenging onto particles, and increases again to a mid-depth maximum (500 fmol kg^{-1} in the Pacific; 370 fmol kg^{-1} in the Atlantic) at 600 m. Concentrations below 600 m decrease to their lowest values at the bottom, from scavenging in deep waters. The mid-depth maximum is roughly associated with the oxygen minimum and is possibly due to the dissolution of manganese phases, which may transport bismuth. In North Atlantic surface waters, bismuth varies between 200 and 400 fmol kg^{-1}, with a distribution consistent with a major atmospheric source. The residence time of bismuth, based on fluvial, atmospheric, and volcanic input to the upper ocean, is estimated to be very short – about 20 years. This estimate seems too short in light of the large enrichment seen for bismuth relative to aluminum in sea water (**Table 2, Figure 1**). The lower deep water concentrations in the Pacific are consistent with an increase in scavenging as water ages.

Thorium

Thorium (Th) is a naturally occurring radioactive element with four primary isotopes; ^{232}Th ($t_{1/2} = 14 \times 10^9$ y), ^{234}Th ($t_{1/2} = 24.1$ d), ^{230}Th ($t_{1/2} = 75\,200$ y), and ^{228}Th ($t_{1/2} = 1.91$ y), in order of abundance. The dominant chemical species of thorium is thought to be the neutral hydroxide, $Th(OH)_4$, but there are no data on the formation constant of $Th(OH)_5^-$. The average concentration of thorium in sea water is 80 fmol kg^{-1} and its residence time, estimated from the scavenging removal of ^{230}Th in the deep sea is 45 years. The only isotope with a primordial origin is the major isotope, ^{232}Th, the others are formed by *in situ* decay. The primary source for ^{232}Th to the oceans is believed to be dust deposition at the sea surface. The vertical profile for ^{232}Th shows a surface minimum (63 fmol kg^{-1}) and a gradual increase to the bottom (200 fmol kg^{-1}), indicating a bottom source as well. Data from the Atlantic are higher (400–600 fmol kg^{-1}) and show no discernible structure. The short residence time and small degree of enrichment in sea water relative to aluminum (**Table 1, Figure 2**) are both due to the high particle reactivity of thorium. It is perhaps surprising that there is any enrichment of Th/Al in sea water.

Discussion

The two indicators of metal reactivity in the oceans (their residence times and the degree of enrichment observed in sea water, relative to aluminum and their

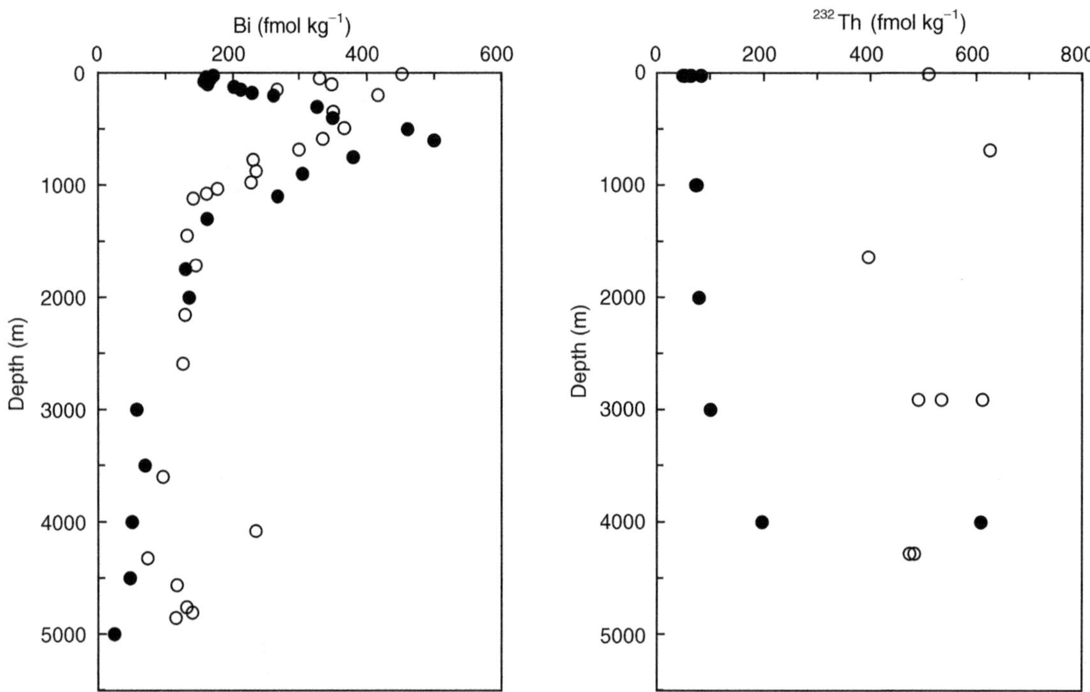

Figure 9 Depth profiles of (A) bismuth in the central North Pacific (solid symbols; 28°N 155°W; Lee *et al.*, 1985/86) and in the western North Atlantic (open symbols; near Bermuda; Lee *et al.*, 1985/86), and (B) ^{232}Th in the western North Pacific (solid symbols; 23°N 158°W; Roy-Barman *et al.*, 1996) and in the eastern North Atlantic (open symbols; 51°N, 43°W; Chen *et al.*, 1986).

abundance in the Earth's crust) do not agree particularly well for the elements discussed here. Short residence times and low sea water/crust ratios are observed for aluminum, iron, and thorium, while longer residence times and higher sea water/crust ratios are observed for gallium, zirconium, and hafnium. The rest are not in agreement, however. Bismuth and indium are both enriched and yet are thought to be rapidly removed. Titanium is not enriched but has an estimated residence time longer than gallium. Niobium and tantalum are not enriched as much as might be expected from their estimated residence times. It must be emphasized that we do not have enough information to place a high degree of confidence on the estimated residence times, or on the validity of using sea water enrichments to infer reduced removal intensity. Differences in the solubility of these elements from continental materials could be significant, but there are many other uncertainties as well.

There have been a number of theories regarding the geochemical basis for the relative reactivity of elements, and the factors that control their sea water concentrations. One theory is that the differential reactivity of strongly hydrolyzed elements may be related to the charge of the hydroxide species that dominates. Those that exist in an anionic form may be less particle-reactive (i.e., they may be less likely to adsorb on negatively charged particle surfaces) and may therefore have longer residence times and be relatively enriched in sea water. The elements with the shortest estimated residence times are bismuth, thorium, aluminum, and iron, all of which are predominantly neutral or cationic in seawater and, with the exception of bismuth, all of which have extremely low sea water concentrations, relative to their abundance in the Earth's crust. Some elements with intermediate residence times are gallium, hafnium, and zirconium, which are predominantly anionic and enriched in sea water. The other elements have poorly confined estimates of their residence times and are difficult to categorize, but do not appear to follow this trend.

Conclusion

The elements described here as refractory are not very soluble in water. They have low concentrations in sea water relative to their abundance in the Earth's crust, and short oceanic residence times. They can have extremely large concentration ranges in the oceans. The processes controlling the marine biogeochemistry of these elements are complex, and their distribution types vary considerably. The classic scavenged profile showing a surface maximum is observed only for aluminum, and even then only in some locations. Many of these elements increase gradually with depth, and some have subsurface maxima, mid-depth maxima, or nutrient-type profiles, showing the influence of internal cycles. Some are depleted as deep waters age, others are enriched. To date, there is no unifying theory to explain the distributions and the relative reactivities of hydroxide-dominated elements in the oceans. More information is needed before we are able to fully understand the processes that control these refractory elements in sea water.

See also

Anthropogenic Trace Elements in the Ocean. Metal Pollution.

Further Reading

Alibo DS, Nozaki Y, and Jeandel C (1999) Indium and yttrium in North Atlantic and Mediterranean waters: comparison to the Pacific data. *Geochimica et Cosmochimica Acta* 63: 1991–1999.

Amakawa H, Alibo DS, and Nozaki Y (1996) Indium concentration in Pacific seawater. *Geophysical Research Letters* 23: 2473–2476.

Brewer PG, Spencer DW, and Robertson DE (1972) Trace element profiles from the GEOSECS-II test station in the Sargasso Sea. *Earth and Planetary Science Letters* 16: 111–116.

Bruland KW, Orians KJ, and Cowen JP (1994) Reactive trace metals in the stratified central North Pacific. *Geochimica et Cosmochimica Acta* 58: 3171–3182.

Chen JH, Edwards RL, and Wasserburg GJ (1986) 238-U, 234-U and 232-Th in seawater. *Earth and Planetary Science Letters* 80: 241–251.

Godfrey LV, White WM, and Salters VJM (1996) Dissolved zirconium and hafnium distributions across a shelf break in the northeastern Atlantic Ocean. *Geochimica et Cosmochimica Acta* 60: 3995–4006.

Hydes DJ (1983) Distribution of aluminium in waters of the north east Atlantic 25° N to 35° N. *Geochimica et Cosmochimica Acta* 47: 967–973.

Johnson KS, Gordon RM, and Coale KH (1997) What controls dissolved iron in the world ocean? *Marine Chemistry* 57: 137–161.

Lee DS, Edmond JM, and Bruland KW (1985/86) Bismuth in the Atlantic and North Pacific: a natural analogue to plutonium and lead? *Earth and Planetary Science Letters* 76: 254–262.

Martin JH, Fitzwater SE, Gordon RM, Hunter CN, and Tanner SJ (1993) Iron, primary production, and carbon–nitrogen flux studies during the JGOFS North Atlantic bloom experiment. *Deep Sea Research* 40: 115–134.

Martin JH, Gordon RM, Fitzwater SE, and Broenkow WW (1989) VERTEX: phytoplankton /iron studies in the Gulf of Alaska. *Deep Sea Research* 36: 649–680.

McKelvey BA (1994) The Marine Geochemistry of Zirconium and Hafnium. PhD thesis, University of British Columbia.

McKelvey BA and Orians KJ (1993) Dissolved zirconium in the North Pacific Ocean. *Geochimica et Cosmochimica Acta* 57: 3801–3805.

McKelvey BA and Orians KJ (1998) The determination of dissolved zirconium and hafnium from seawater using isotope dilution inductively coupled plasma mass spectrometry. *Marine Chemistry* 60: 245–255.

Measures CI, Edmond JM, and Jickels TD (1986) Aluminum in the North West Atlantic. *Geochimica et Cosmo chimica Acta* 50: 1423–1429.

Measures CI, Grant B, Khadem M, Lee DS, and Edmond JM (1984) Distribution of Be, Al, Se, and Bi in the surface waters of the western North Atlantic and Caribbean. *Earth and Planetary Science Letters* 71: 1–12.

Orians KJ and Bruland KW (1986) The biogeochemistry of aluminum in the Pacific Ocean. *Earth and Planetary Science Letters* 78: 397–410.

Orians KJ and Bruland KW (1988a) Dissolved gallium in the open ocean. *Nature* 332: 717–719.

Orians KJ and Bruland KW (1988b) The marine geochemistry of dissolved gallium: a comparison with dissolved aluminum. *Geochimica et Cosmochimica Acta* 52: 2955–2962.

Orians KJ, Boyle EA, and Bruland KW (1990) Dissolved titanium in the open ocean. *Nature* 348: 322–325.

Roy-Barman M, Chen JH, and Wasserburg GJ (1996) 230Th–232Th systematics in the central Pacific Ocean: the sources and the fates of thorium. *Earth and Planetary Science Letters* 139: 351–363.

Shiller AM (1998) Dissolved gallium in the Atlantic Ocean. *Marine Chemistry* 61: 87–99.

Sohrin Y, Fujishima Y, Ueda K, *et al.* (1998) Dissolved niobium and tantalum in the North Pacific. *Geophysical Research Letters* 25: 999–1002.

Spencer DW, Robertson DE, Turekian KK, and Folsom TR (1970) Trace element calibrations and profiles at the GEOSECS test station in the Northeast Pacific Ocean. *Journal of Geophysical Research* 75: 7688.

Taylor SR (1964) Abundance of chemical elements in the continental crust: a new table. *Geochimica et Cosmochimica Acta* 28: 1273–1285.

Whitfield M and Turner DR (1987) The role of particles in regulating the composition of seawater. In: Stumm W (ed.) Aquatic Surface Chemistry, Ch. 17, pp. 457–493. New York: Wiley-Interscience.

METALLOIDS AND OXYANIONS

G. A. Cutter, Old Dominion University, Norfolk, VA, USA

Introduction

The concentrations and distributions of dissolved trace elements (typically called trace 'metals,' though not all trace elements are metals) in the world's oceans are due to a complex interaction between their purely chemical behavior (e.g., acid/base properties, oxidation state, solubility), the way in which they are delivered to the ocean (atmosphere, rivers, submarine hydrothermal vents), biological reactions, and water circulation (e.g., currents). To organize this somewhat chaotic and confusing situation, the kinds of trace element behavior are classified into four types: conservative, nutrient-like or recycled, scavenged, and hybrid or mixed. A conservative trace element behaves like the major dissolved elements that make up the bulk of the ocean's salinity (e.g., Na^+). These elements are only effected by the physical processes of mixing, or the addition (dilution) or removal (evaporation) of water. Since there are no chemical or biological reactions that affect these elements, they have rather uniform concentrations with ocean depth. In contrast, the nutrient-like trace element is taken up by phytoplankton in surface waters during photosynthesis (like the nutrient nitrate), and this organic matter-bound element begins to gravitationally settle into deep waters. However, organic matter is a precious commodity in the open and deep sea, so many levels of the food web (bacteria to zooplankton) consume this organic detritus, releasing some fraction of the bound trace element back into the water column. This recycling makes the nutrient-like trace element concentration lower at the surface and higher at depth, with the exact shape of the profile depending on the rate at which it is recycled. Many dissolved trace elements have high charge to atomic radius ratios, and electrostatically adsorb to particle surfaces; this process is loosely termed 'scavenging.' Thus, scavenged elements have distributions that depend on the number and type of particles (e.g., clay, phytoplankton) and the mode of introduction (e.g., atmosphere, hydrothermal vents). An excellent example of this interaction is given by lead, which is very particle-reactive and is introduced from atmosphere, with the resulting distribution showing a surface maximum and rapid decrease with depth. Finally, many trace elements display features of both scavenged and nutrient-like elements, with the distribution of the micronutrient iron being a good example (particle reactive, but also recycled).

Most of these classifications were developed for the metals that are cations (positively charged) in solution, but there are elements in periodic groups IVA, VA and B, VIA and B, VIIA, and VIII that actually form oxygen-containing anions. In general, these elements display their maximum potential oxidation state in sea water, and in aqueous solution undergo hydrolysis (e.g., $Mo^{+6} + 4H_2O \rightarrow MoO_4^{2-} + 8H^+$). The metalloid elements (antimony, Sb; arsenic, As; germanium, Ge; selenium, Se; tellurium, Te) all exist as oxyanions, as do the transition metals chromium (Cr), molybdenum (Mo), osmium (Os), rhenium (Re), tungsten (W), and vanadium (V). In addition to existing as anions, most of these trace elements can be found in multiple oxidation states (e.g., As(III) and As(V)), ensuring that the oxyanions probably have the most diverse behaviors of any trace elements in the ocean. Interestingly, the form in which an oxyanion exists in sea water, the 'chemical speciation,' strongly affects its biological and chemical reactivity.

In this review, the oceanic distributions of each element will be discussed in terms of its purely chemical properties, known biological behavior, and general geochemical considerations (inputs and outputs). The focus will be primarily on the dissolved ions rather than those associated with particles, since these are free to move with the water molecules and are available to the first trophic level in the ocean, phytoplankton. Because deeper waters in the Pacific Ocean are much older and have undergone more mixing than those in the Atlantic, most data will come from the Pacific to focus on the biological and chemical processes affecting the element, and not the mixing of different water masses. In addition, all concentrations will be expressed in fractions of a mole per liter rather than as mass per liter. This allows direct comparisons between elements (i.e., atom to atom) and is consistent with the principles of chemical and biological reactions (e.g., to make CO_2, it takes one atom of C and 2 atoms of O). As trace elements, the concentrations units will be nanomoles per liter (nmol l^{-1}; 10^{-9} mol l^{-1}), picomoles per liter (pmol l^{-1}; 10^{-12} mol l^{-1}), and femptomoles per liter (fmol l^{-1}; 10^{-15} mol l^{-1}). Nevertheless, low concentrations do not mean that the oxyanions are unimportant as either essential or toxic compounds, or as useful ocean tracers; these points will be highlighted below. To provide a logical order to this presentation, the periodic table will be followed from left to right.

The Elements

Vanadium

In oxygenated sea water with an average pH of 8, vanadium should be found in the +5 oxidation state, which undergoes hydrolysis to form vanadate, HVO_4^{2-}. In sea water with no oxygen ('anoxic') such as found in the Black Sea, V(V) can be reduced to V(IV) which is more reactive than V(V), and as a consequence, anoxic sediments have elevated concentrations of vanadium relative to sediments underlying oxic waters. Thus, sedimentary vanadium may act as a historical tracer of anoxic conditions. Vanadium has also been studied in sea water because it is enriched in fossil fuels and may be a potential pollutant. In this respect, vanadium does not have any established biological function, although the chemistries of phosphate and vanadate are similar, and therefore vanadate might be taken up into soft tissues (e.g., lipids) along with phosphate. The depth profile of dissolved vanadium in the North Pacific Ocean (**Figure 1A**) shows a surface concentration of ~32 nmol l^{-1} and an increase into deep waters to ~36 nmol l^{-1}. This slight surface depletion has also been observed at other locations in the Pacific and Atlantic Oceans and, based on their similarity to the depth profiles of phosphate, it appears that vanadium is taken up by phytoplankton in surface waters. This type of behavior is also found in estuaries where river and sea waters mix, and both phosphate and vanadium show removal. This means that processes at the ocean margins (e.g., in estuaries) reduce the amount of vanadium entering the oceans from rivers. In contrast, hydrothermal vents do not appear to be substantial sources or sinks of vanadium to the deep ocean.

Chromium

Owing to its use in many industrial processes and its high toxicity, considerable attention has been paid to chromium in group VIA. The two primary oxidation states of chromium are +6 and +3, which hydrolyze in water to form chromate, CrO_4^{2-}, and $Cr(OH)_3$, respectively.

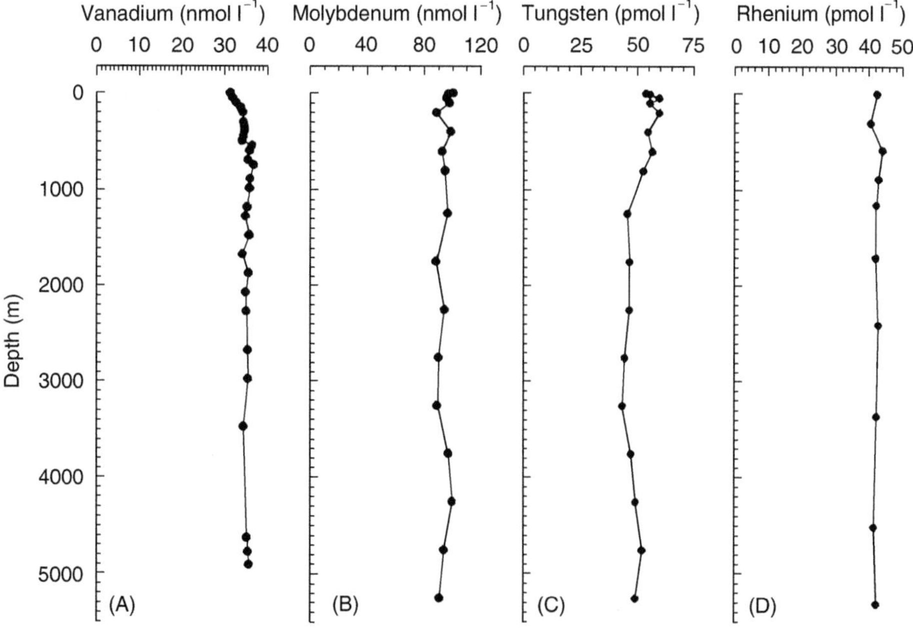

Figure 1 (A) Dissolved vanadium in the North Pacific Ocean, 11°N, 140°W. (Data from Collier RW (1979) Particulate and dissolved vanadium in the North Pacific Ocean. *Nature* 309: 441–444.) (B) Dissolved molybdenum in the North Pacific Ocean, 30°N, 159°50′W. (Data from Sohrin Y, Isshiki K and Kuwamoto T (1987) Tungsten in North Pacific waters. *Marine Chemistry* 22: 95–103.) (C) Dissolved tungsten in the North Pacific Ocean, 30°N, 159°50′W. (Data from Sohrin Y, Isshiki K and Kuwamoto T (1987) Tungsten in North Pacific waters. *Marine Chemistry* 22: 95–103.) (D) Dissolved rhenium in the North Pacific Ocean, 24°16′N, 169°32′W. (Data from Colodner D, Sachs J, Ravizza G, Turekian K, Edmond J and Boyle E (1993) The geochemical cycle of rhenium: a reconnaissance. *Earth and Planetary Science Letters* 117: 205–221.)

Thermodynamic calculations show that chromate is the expected form in oxygenated sea water, while the insoluble Cr(III) species would predominate in very low-oxygen (so called 'suboxic') or anoxic waters. However, it is important to note that thermodynamic calculations only predict elemental speciation at equilibrium (when the rates of formation and destruction are balanced), but they do not consider the rates of conversion themselves. For example, Cr(III) should not exist in oxygenated sea water, but its rate of oxidation to Cr(VI) is slow (days to months), meaning that Cr(III) can persist in oxic water ('kinetic stabilization'). In the eastern North Pacific Ocean, Cr(VI) displays a surface concentration of $\sim 3\,\text{nmol l}^{-1}$ (**Figure 2A**), but then decreases rapidly to a minimum of $1.7\,\text{nmol l}^{-1}$ at 300 m depth and increases below this to levels of $4–5\,\text{nmol l}^{-1}$ in the deeper waters. While chromate appears to display a mixture of scavenged and nutrient-like behavior, the Cr(VI) minimum occurs at the same depth as the widespread suboxic zone in the eastern Pacific. Indeed, at other sites in the North Pacific without a suboxic layer, Cr(VI) has only nutrient-like profiles. Thus, the data in **Figure 2A** suggest Cr(VI) to Cr(III) reduction, and correspondingly, Cr(III) shows a maximum at the same depth (**Figure 2B**), although the increase in Cr(III) ($\sim 0.6\,\text{nmol l}^{-1}$) is not as great as the Cr(VI) depletion ($\sim 1.3\,\text{nmol l}^{-1}$). This is likely due to the higher reactivity of Cr(III), which would be scavenged by particles, decreasing its concentration. While this all might seem in agreement with thermodynamic predictions, the existence of Cr(III) in fully oxygenated surface and deep waters (**Figure 2B**), argues that Cr(III) is kinetically stabilized (slow to oxidize).

Molybdenum

Because many trace elements such as iron function as essential nutrients in the ocean, considerable attention was paid to molybdenum, since it is a cofactor in the nitrogen-fixing enzyme nitrogenase,

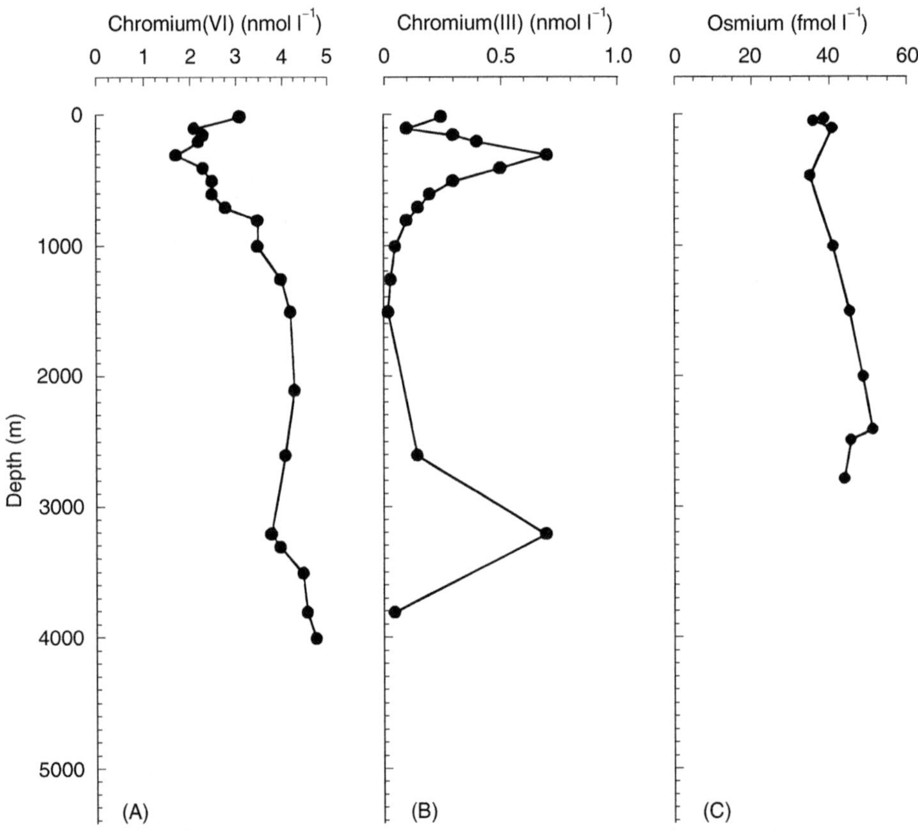

Figure 2 Dissolved Cr(VI) (A) and Cr(III) (B) in the eastern North Pacific Ocean, 23°N, 115°W. (Data from Murray JW, Spell B and Paul B (1983) The contrasting geochemistry of manganese and chromium in the eastern tropical Pacific Ocean. In: Wong CS *et al.* (eds) *Trace Metals in Seawater,* NATO Conference Services 4: *Marine Science* vol. 9, pp. 643–668. New York: Plenium Press.) (C) Dissolved rhenium in the eastern North Pacific Ocean, 9°46′N, 104°11′W. (Data from Woodhouse OB, Ravizza G, Falkner KK, Statham PJ and Peucker-Ehrenbrink B (1999) Osmium in seawater: vertical profiles of concentration and isotopic composition in the eastern Pacific Ocean. *Earth and Planetary Science Letters* 173: 223–233.)

and therefore essential for many algae and ecosystems. As a group VIA element like chromium, in oxic sea water dissolved molybdenum should be found in the +6 oxidation state as the hydrolysis product molybdate, MoO_4^{2-}, but in anoxic waters it is reduced to Mo(IV), which in the presence of hydrogen sulfide forms insoluble MoS_2. In fact, molybdenum is enriched in anoxic sediments by this mechanism and, like vanadium, can be used as a sediment tracer of past anoxia. In spite of its crucial biological role, molybdenum in the oxic ocean shows remarkably conservative behavior (**Figure 1B**), with no surface depletion and the highest concentration ($\sim 105\,nmol\,l^{-1}$) of any of the trace elements examined here. This does not mean that phytoplankton are not taking it up, but rather this uptake is trivial compared to its inputs. Hydrothermal vents also do not remove molybdenum, and the only waters where molybdenum removal is observed are in anoxic basins such as the Black Sea.

Tungsten

Completing the group VIA transition metal series, tungsten is chemically similar to molybdenum and is found in oxic sea water as W(VI) in the form of tungstate, WO_4^{2-}. Owing to difficulties in determining tungsten in sea water, there are few depth profiles for this dissolved element. In the North Pacific Ocean (**Figure 1C**), tungsten displays slightly higher concentrations in surface waters (average of $58\,pmol\,l^{-1}$) compared to deep waters (average of 49 $pmol\,l^{-1}$). Dissolved tungsten appears to have a slightly scavenged type of profile, with the surface enrichment likely due to the deposition of terrestrial dust (aerosols). In crustal rocks that are the source of most elements to the ocean by weathering, the abundance of tungsten is about one-third that of molybdenum, but its sea water concentration is over 1000 times less (compare **Figures 1B** and **C**). Since tungsten shows no strong removal in the open ocean, this observation suggests that tungsten must be removed in the coastal ocean. Indeed, profiles of tungsten in estuaries shows strong removal like that of iron; molybdenum shows no such strong removal in estuaries. Thus, while tungsten shows nearly conservative behavior in the open ocean, when the whole land–ocean system is considered, tungsten actually has a very active removal, which lowers its sea water concentration.

Rhenium

Interest in rhenium is for rather esoteric reasons, primarily because one of its radioactive isotopes, and that of its periodic table neighbor osmium, are useful for dating very ancient (>50 billion years) sediments (i.e., the $^{187}Re/^{187}Os$ ratio). In oxic sea water, Re(VII) is the stable oxidation state and after hydrolysis exists as the relatively unreactive perrhenate ion, ReO_4^-. In the North Pacific Ocean (**Figure 1D**), as well as in the Atlantic, the profile of dissolved rhenium is quite conservative, with an average of 44.3 ± 0.3 pmol l^{-1}. Using the same type of arguments used for tungsten, the very low concentration of crustal rhenium but relatively high sea water concentration (sea water rhenium and tungsten are nearly identical, but crustal tungsten is ~ 3000 times more abundant than rhenium) suggests that rhenium is very unreactive in the entire ocean system. Nevertheless, in anoxic systems such as the Black Sea, rhenium does show substantial decreases in concentration (nonconservative behavior) that have been attributed to removal at the surface of anoxic sediments. However, in the modern ocean these anoxic systems are too rare to substantially alter the distributions of rhenium in the water column.

Osmium

From the prior discussion of rhenium, it would seem logical to consider the oceanic behavior of the group VIII element osmium in terms of its use as a dating tool. In addition, the ratio of two of its isotopes ($^{187}Os/^{186}Os$) can trace inputs from extraterrestrial sources (e.g., meteors) and terrestrial sources (i.e., crustal weathering) to the oceans, which are recorded in marine sediments. There is some debate about the exact form of osmium in sea water, but thermodynamic calculations suggest that Os(VIII) as $H_3OsO_6^{2-}$ would be stable in oxic sea water. Determinations of dissolved osmium in sea water are very difficult, especially since osmium concentrations are over 1000 times lower than those of rhenium. Indeed, the profile of osmium in the North Pacific (**Figure 2C**) indicates not only that osmium is found at very low concentrations ($38\,fmol\,l^{-1}$) in surface waters, but also that its distribution is quite dynamic with depth (minimum at 460 m and rising to 51 fmol l^{-1} in deep waters). This profile is from the eastern North Pacific, which has the distinct suboxic layer, and bears a striking resemblance to that of Cr(VI) in the same region (**Figure 2A**). Thus, osmium may have nutrient-like behavior that is also affected by oxidation–reduction reactions (i.e., reduced to a more particle-reactive (but unidentified) form in the suboxic zone). In this respect, profiles in more oxygenated waters of the Atlantic and Indian Oceans show no such depletion in the upper water column. The nutrient-like distribution does not mean that it is used as a nutrient or nutrient substitute, but rather that it is only carried along in the organic matter cycle.

Germanium

Germanium is a group IVB metalloid, and therefore chemically quite different from the transition metals we have been considering so far. Being directly below silicon in the periodic table, germanium has quite similar chemistry and early studies of phytoplankton that make up their structural skeletons ('tests') of biogenic silica (e.g., diatoms), showed that they take up germanium along with silicon in a relatively constant atomic Ge : Si ratio of 10^{-6} : 1, a value that is nearly identical to the ratio in crustal rocks. Dissolved germanium exists as germanic (H_3GeO_4) in sea water just as silicon is found as silicic acid (H_3SiO_4). The depth profile of dissolved inorganic germanium in the North Pacific Ocean (**Figure 3A**) is undoubtedly nutrient-like, and not surprisingly exactly the same as that of silicon (i.e., uptake in the surface by siliceous phytoplankton; recycling at depth by the slow dissolution of biogenic silica). This

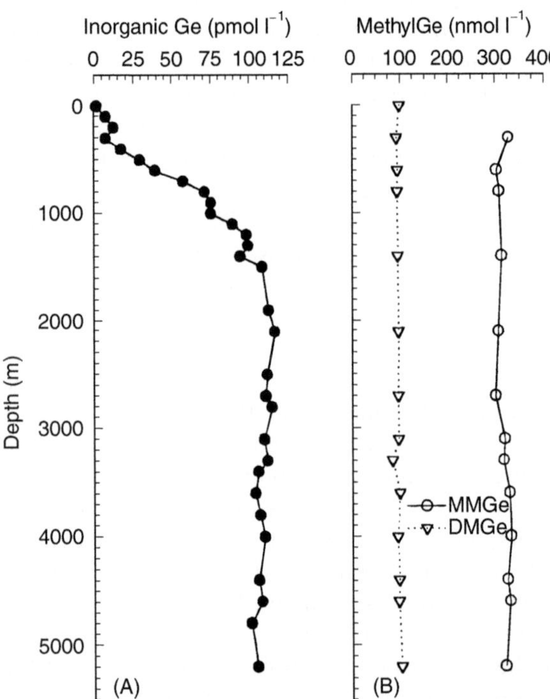

Figure 3 (A) Dissolved inorganic germanium in the North Pacific Ocean, 25°N, 175°E. (Data from Froehlich PN Jr and Andreae MO (1981) The marine geochemistry of germanium: ekasilicon. *Science* 213: 205–207.) (B) Dissolved methylgermanium compounds in the North Pacific Ocean, 25°N, 175°E. MMGe is monomethylgermanic acid and DMGe is dimethylgermanic acid. (Data from Lewis BL, Froelich PN and Andreae MO (1985) Methylgermanium in natural waters. *Nature* 313: 303–305.)

covariation is the primary reason why there is interest in marine germanium. The Ge : Si ratio derived from crustal weathering inputs to the ocean has a different value from that from hydrothermal vent fluids, and since it appears that the Ge : Si ratio in siliceous organisms records the water column value, the ancient record of crustal weathering versus hydrothermal inputs (i.e., plate spreading) to the oceans can be obtained. Unfortunately, it was discovered that there are methylated forms of germanium in sea water (monomethylgermanic acid, MMGe; dimethylgermanic acid, DMGe) which actually have higher concentrations than inorganic germanium, and for which there are no known silicon analogues. Thus, the two cycles seem to diverge, threatening the usefulness of the Ge : Si tracer. However, the distributions of these two methylated forms are very conservative in the open ocean (**Figure 3B**) and in estuaries. Indeed, methylgermanium is essentially inert and hence can build up to the observed 'high' concentrations. The source of these compounds still has not been found, although some production in anoxic, organic-rich waters has been documented. Nevertheless, the methylgermanium compounds do not really participate in the germanium cycle, and thus it seems that the Ge : Si ratio can still be used as a weathering versus hydrothermal input tracer.

Arsenic

The group VB element arsenic is usually linked with toxicity, and indeed most studies of this element are driven by such concerns. However, arsenic's toxicity is strongly affected by its chemical form and by the actual organisms being exposed. In oxygenated sea water, As(V) in the form of arsenate ($HAsO_4^{2-}$) is the stable form and, because of its nearly identical chemical properties to that of the nutrient phosphate, is highly toxic to phytoplankton. Interestingly, As(III), which can be found in anoxic waters as arsenite ($As(OH)_3$), is not toxic to phytoplankton but is highly toxic to higher organisms such as zooplankton and fish. While this might seem irrelevant (fish do not live in anoxic waters), many phytoplankton have a mechanism to detoxify arsenate by reducing it to arsenite and releasing it to the oxic water column. Other phytoplankton can methylate arsenate to form monomethyl- and dimethylarsenates (MMAs and DMAs, respectively), which are nontoxic. All of these processes make the marine arsenic cycle quite complicated, a feature that is common for all of the metalloid elements. In the North Pacific Ocean (**Figure 4A**) arsenate has nutrient-like behavior, with depletion in the surface and recycling at depth as for

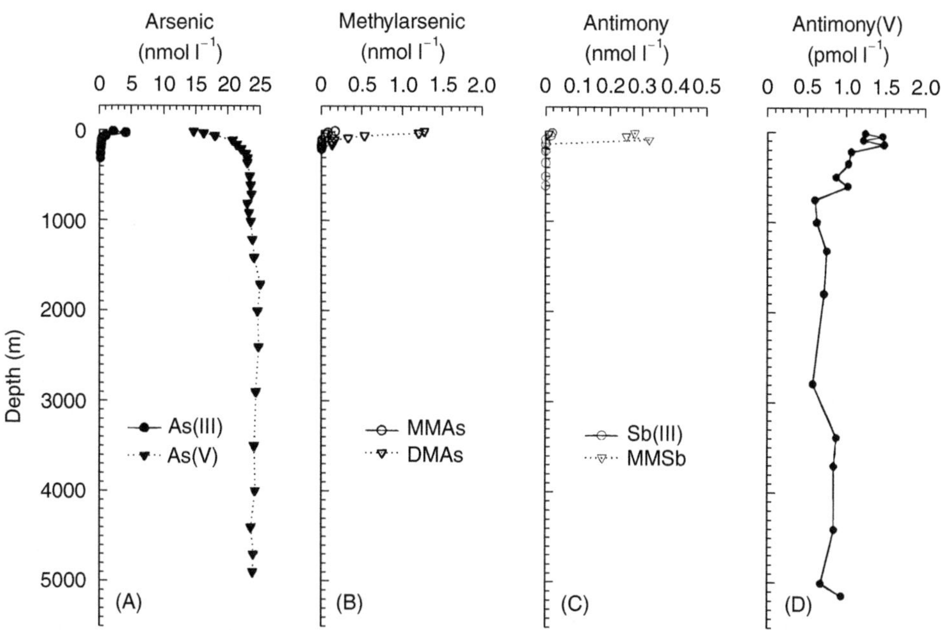

Figure 4 (A) Arsenate (As(V)) and arsenite (As(III)), and (B) monomethylarsenate (MMAs) and dimethylarsenate (DMAs) in the North Pacific Ocean, 30°46′N, 163°30′W. (Data from Andrese MO (1979) Arsenic speciation in seawater and interstitial waters: the influence of biological–chemical interactions on the chemistry of a trace element. *Limnology and Oceanography* 24: 440–452.) (C) Dissolved antimony(III) (Sb(III)) and monomethyl antimonate (MMSb), and (D) antimony(V) in the South Atlantic Ocean, 17°S, 25°W. (Data from Cutter GA, Cutter LS, Featherstone AM and Lohrenz SE (2001). Antimony and arsenic biogeochemistry in the western Atlantic Ocean. *Deep-Sea Research*, in press.)

phosphate; this is consistent with their chemistries and biochemistries. In surface waters, biologically produced arsenite and methylated arsenic compounds have their maximum concentrations (**Figures 4A** and **B**), and quickly decrease with depth. Again, these distributions are consistent with established biological processes, and therefore they are not due to scavenging (i.e., input from the atmosphere and then adsorption to particles). Moreover, data for methylated arsenic and As(III) at many sites in the world's oceans show that the amounts of these arsenic forms are roughly the inverse of the phosphate concentration – lower phosphate, higher methylarsenic and As(III). This is consistent with laboratory studies of phytoplankton where, under low-phosphate conditions, more arsenate is taken up and more methylated or reduced arsenic is produced in response to this stress. In addition, the fact that As(III) is found in oxygenated sea water at all (i.e., only stable in anoxic waters) demonstrates that its rate of oxidation is slow enough (half-life of months) that it can build up to almost 20% of the total dissolved arsenic in surface waters.

Antimony

Antimony is a group VB metalloid like arsenic but it has more metallic character and the chemistry of antimony is quite different than that of phosphorus or arsenic. Sb(V) is not as strong a Lewis acid as As(V) and in oxic sea water the stable form would be antimonate ($Sb(OH)_6^-$), while Sb(III), like As(III), would be $Sb(OH)_3$ in anoxic waters. Antimony also has methylated forms analogous to those of arsenic (i.e., MMSb, DMSb), although only MMSb has been found in the open ocean. Antimony is not as toxic as arsenic, and since it is used as a plasticizer and is enriched in fossil fuels, most of the interest in antimony has concerned its use as a pollution tracer (e.g. from the burning of plastics). Most of the data for dissolved antimony in the open ocean are from the Atlantic; the profiles in **Figure 4C** and **D** are typical for this element. The major form of dissolved antimony is antimonate, and it displays a profile consistent with mild scavenging (i.e., maximum of 1.5 nmol l^{-1} at the surface due to atmospheric or riverine input, lower concentrations below the surface layer via adsorption onto particles, some recycling near the sediment–water interface). Measurements of antimony in atmospheric particles (aerosols) and rain show that atmospheric input can explain the surface antimony maximum. Thus, the concentration and behavior of antimony are quite different from those of arsenic. However, MMSb and Sb(III) are found in the surface waters (although the concentration of SbIII is only 0.02 nmol l^{-1}), like the equivalent arsenic forms, but their concentrations do not negatively correlate with the concentration of phosphate (i.e., do not appear to be a result of detoxification reactions). Indeed, it is not clear what mechanisms are producing these forms of dissolved antimony, and bacterial production of MMSb (bacteria are good methylators) and the photochemical reduction of Sb(V) to Sb(III) cannot be ruled out.

Selenium

Selenium is a group VIB metalloid just below sulfur in the periodic table, and its chemistry and biochemistry is very similar to those of sulfur. The interest in this element is based on the fact that this trace element is both essential (e.g., a cofactor in antioxidant enzymes) and toxic, with the chemical form of the element strongly influencing its beneficial or toxic properties. As for sulfur, the most stable oxidation state in oxygenated sea water is Se(VI) as selenate (SeO_4^{2-}), while under suboxic conditions selenite ($HSeO_3^-$) would predominate. Selenium forms insoluble elemental Se(0) in anoxic waters, whereas sulfur exists as sulfide (S(II)). Nevertheless, there are numerous organic forms of selenide (Se(-II)) such as selenomethionine, that could be bound in soluble peptides (to be referred to as 'organic selenide'). There have also been recent measurements of the dissolved gas dimethylselenide ($(CH_3)_2Se$) in surface ocean, which results in a natural selenium input to the atmosphere. Laboratory and field studies have shown that selenite appears to be the most biologically preferred form of dissolved selenium by phytoplankton, while selenate is only taken up in the absence of selenite; data on the bioavailability of organic selenides suggest that these forms are the least available to marine phytoplankton. Depth profiles for dissolved selenium in the eastern North Pacific (**Figure 5A–C**) are from the same region where the chromium and osmium profiles (**Figure 2A–C**) were obtained, and the water column from ~200 to 800 m is suboxic. In surface waters, both selenite (**Figure 5A**) and selenate (**Figure 5B**) are very depleted, and then show nutrient-like profiles with increasing depth; this is consistent with biotic uptake of both forms, incorporation into organic matter (as organic selenides), and subsequent recycling. In contrast, organic selenide (**Figure 5C**) has a maximum at the surface and in the suboxic zone. Laboratory and field studies using organic matter show that selenium is primarily bound in proteins as organic selenide. When this organic matter degrades, dissolved organic selenide is released, which then sequentially oxidizes to selenite and then very slowly to selenate. This process explains how these unstable forms are introduced to the water column, with kinetic stabilization allowing them to persist. In the suboxic zone, organic selenide is stabilized and a maximum can develop.

Tellurium

Tellurium is a group VIB metalloid like selenium, but its lower position in the periodic table suggests that it has considerably more metallic character than selenium. Thus, Te(VI) exists as $Te(OH)_6$ but, unlike

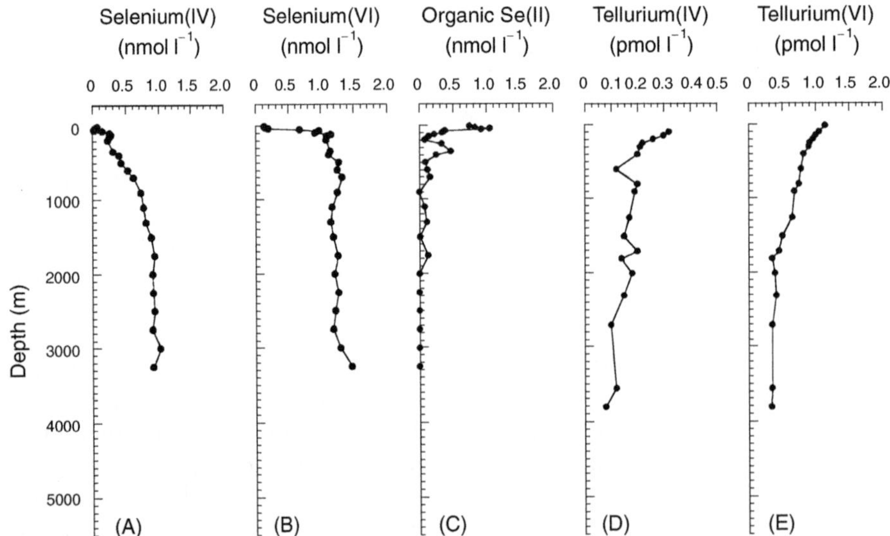

Figure 5 (A) Selenium(IV) (selenite), (B) selenium(VI) (selenate), and (C) organic selenide (organic Se(II)) in the eastern North Pacific Ocean, 18°N, 108°W. (Data from Cutter GA and Bruland KW (1984) The marine biogeochemistry of selenium: a re-evaluation. *Limnology and Oceanography* 29: 1179–1192.) (D) Tellurium(IV) and (E) tellurium(VI) in the eastern North Pacific Ocean, 7°N, 78°40′W. (Data from Lee DS and Edmond JM (1985) Tellurium species in seawater. *Nature* 313: 782–785.)

Se(VI), Te(IV) is the stable form as $Te(OH)_4$. There are few data for this element in the ocean, and profiles of tellurium in the eastern North Pacific (**Figures 5D** and **E**) show that its concentrations are ~1000 times less than those of selenium and both forms of tellurium show strongly scavenged behavior. It is interesting to note that the most abundant form of tellurium is Te(VI), but it is the least thermodynamically stable. Although there are numerous biological reasons to expect reduced species in oxic waters (i.e., as for arsenic and selenium), this observation is somewhat difficult to explain. The elevated concentrations of Te(VI) at the surface, relative to Te(IV), suggest that the atmospheric or riverine inputs of this element are enriched in this form, but virtually no atmospheric data are available to confirm this speculation.

Conclusions

Overall, the classic types of behavior described for trace elements are displayed by the oxyanions. Rhenium, molybdenum, and methylated germanium have conservative, salinity-like distributions with depth. This does not mean that they are truly unreactive, but rather that the reactions affecting them are minor. Vanadium, chromium, osmium, arsenic, and selenium show nutrient-like behavior, although this does not necessarily mean they are biologically required. Tungsten, antimony, and tellurium show scavenged behavior to some degree. Hybrid/mixed behavior was originally defined for trace elements with a single chemical form, but most of the oxyanions can exist in multiple chemical forms, creating a new type of hybrid distributions – oxidation–reduction behavior superimposed upon the other three categories. Finally, the types of behavior found for these elements are strongly affected by both chemical and biological processes, and thus it is appropriate to say that they are 'biogeochemically' cycled.

See also

Anthropogenic Trace Elements in the Ocean. Atmospheric Input of Pollutants. Carbon Cycle. Conservative Elements. Hydrothermal Vent Fluids, Chemistry of. Marine Silica Cycle. Metal Pollution. Nitrogen Cycle. Platinum Group Elements and their Isotopes in the Ocean. Refractory Metals. River Inputs. Transition Metals and Heavy Metal Speciation.

Further Reading

Bruland KW (1992) Trace elements in seawater. In: Riley JP Chester R (eds.) Chemical *Oceanography*, vol. 8, pp. 157–220. London: Academic Press.

Donat JR and Bruland KW (1995) Trace elements in the oceans. In: Salbu B and Steinnes E (eds.) *Trace Elements in Natural Waters*, pp. 247–281. Boca Raton, FL: CRC Press.

Libes SM (1992) *An Introduction to Marine Biogeochemistry.* New York: Wiley.

Millero FJ (1996) *Chemical Oceanography*, 2nd edn. Boca Raton, FL: CRC Press.

Pilson MEQ (1998) *An Introduction to the Chemistry of the Sea.* Reading, NJ: Prentice Hall.

Stumm W and Morgan JJ (1996) *Aquatic Chemistry: Chemical Equilibria and Rates in Natural Waters.* New York: Wiley.

TRANSITION METALS AND HEAVY METAL SPECIATION

J. Donat and C. Dryden,
Old Dominion University, Norfolk, VA, USA

Introduction

The transition metals and heavy metals (those with atomic weights greater than 20) enter the ocean via river runoff, wind-blown dust, diffusion from sediments, hydrothermal inputs resulting from reactions of sea water with newly formed ocean crust at mid-ocean seafloor spreading centers, and from anthropogenic activities. Some of these metals (e.g., manganese, iron, cobalt, nickel, copper and zinc) are extremely important micronutrients needed by phytoplankton for various metabolic functions. Several trace metals that are nonconservative with short oceanic residence time (e.g., manganese and aluminum, though the latter is not a heavy metal) are valuable as tracers for circulation and mixing in the ocean. Micronutrient metals, as well as metals like mercury, lead, and silver, which have no biochemical role, can be toxic very low concentrations.

Until recently, marine chemists and chemical oceanographers, using sample collection and analytical techniques of the time, could not accurately measure the naturally low concentrations of these metals in unpolluted sea water because of sample contamination problems and lack of instrumental sensitivity. Development of modern techniques for collection, storage, and analysis of uncontaminated samples, plus the development of highly sensitive analytical techniques and instrumentation, have only recently enabled marine trace metal chemists to determine accurate concentrations of these elements in sea water, furthering our understanding of their distributions and chemical behavior in the oceans. These procedural, analytical, and instrumental advancements led to the discoveries that the concentrations of many of these metals were orders of magnitude lower than previously believed, and that the depth distributions ('vertical profiles') of transition and heavy metal concentrations result from biological, physical, and geochemical processes in the oceans.

We now have a basic understanding of the concentrations and distributions of nearly all the naturally occurring elements in sea water. However, it has become increasingly clear that this information alone is insufficient for providing a complete understanding of the biological and geochemical interactions of these metals in the sea. Metals in sea water can exist in different physical forms (dissolved, colloidal, particulate) and chemical forms (ions, inorganic complexes, organic complexes, organometallic compounds) and in different oxidation states (collectively termed 'species') within a given chemical form. Knowing the distribution of a metal's total concentration among these various forms ('speciation') is critically important because the different forms can have very different biological and geochemical behaviors, and thus different fates and transport.

Before considering the speciation of the transition and heavy metals, we first present a brief overview of the concentrations and distributions of these elements.

Overview: Transition Metal and Heavy Metal Concentrations and Distributions

Concentrations of the transition metals and heavy metals vary both horizontally and vertically through the world's oceans. **Table 1** lists the ranges in the oceanic concentrations of the transition metals and heavy metals. For a representation of the North Pacific depth profiles of the elements in the periodic table, including the transition metals and heavy metals (*see* Elemental Distribution: Overview). The relative rates of supply and removal of the elements determine their horizontal and vertical distributions. These elements are supplied to the oceans primarily by riverine input, atmospheric precipitation, hydrothermal venting, and anthropogenic activities, and they are removed by adsorption onto sinking particles ('scavenging') or by incorporation into sinking biologically produced material by active uptake by phytoplankton. On the basis of their vertical profiles, these elements can be classified into one of the following categories: (1) conservative type, (2) scavenged type, (3) nutrient (recycled) type, and (4) mixed type. **Figure 1** shows the shapes of the vertical profiles for the conservative, scavenged, and nutrient (recycled) categories and lists the elements that display them.

Table 1 Oceanic concentrations of transition metals and heavy metals

Element	*Concentration units*[a]	*North Pacific*		*North Atlantic*	
		Surface	*Deep*	*Surface*	*Deep*
Sc	pmol l^{-1}	8	18	14	20
Ti	pmol l^{-1}	4–8	200–300	30–60	200
V	nmol l^{-1}	32	36	23	
Cr	nmol l^{-1}	3	5	3.5	4.5
Mn	nmol l^{-1}	0.5–3	0.08–0.5	1–3	0.25–0.5
Fe	nmol l^{-1}	0.02–0.5	0.5–1	1–3	0.25–0.5
Co	pmol l^{-1}	4–50	10–20	18–300	20–30
Ni	nmol l^{-1}	2	11–12	2	6
Cu	nmol l^{-1}	0.5–1.3	4.5	1.0–1.3	2
Zn	nmol l^{-1}	0.1–0.2	8.2	0.1–0.2	1.6
Ga	pmol l^{-1}	12	30	25–30	
Ge	pmol l^{-1}	5	100	1	20
Y	pmol l^{-1}	66–187	306–383		
Zr	pmol l^{-1}	12–95	275–325	100	
Nb	pmol l^{-1}	2.8	3.9		
Mo	nmol l^{-1}	93	105		
Rh	fmol l^{-1}	370	900		
Pd	pmol l^{-1}	0.18	0.66		
Ag	pmol l^{-1}	1–5	23	0.69–4.6	2.7–6.9
Cd	pmol l^{-1}	1–10	1000	1–10	350
In	pmol l^{-1}	0.09–1.8	0.07–0.09	2.7	0.9
Sn	pmol l^{-1}	4		10–20	8
Te	pmol l^{-1}	1.2	1	1–1.5	0.4–1
La	pmol l^{-1}	20	50–70	12–15	80–84
Hf	pmol l^{-1}	0.2–0.4	1–2	0.4	
Ta	pmol l^{-1}	0.09	0.3		
W	pmol l^{-1}	41	51		
Re	pmol l^{-1}	28–82		32–43	
Os	fmol l^{-1}		20	15	17
Ir	fmol l^{-1}	0.5	0.8		
Pt	pmol l^{-1}	0.4	0.3–1.2	0.2–0.4	0.2–0.4
Au	fmol l^{-1}	50–150		50–150	
Hg	pmol l^{-1}	0.5–10	2–10	1–7	1
Tl	pmol l^{-1}	60–80	80	60–70	60
Pb	pmol l^{-1}	14–50	3–6	100–150	20
Bi	pmol l^{-1}	0.2	0.02	0.25	

[a]1 nmol l^{-1} = 10^{-9} mol l^{-1}; 1 pmol l^{-1} = 10^{-12} mol l^{-1}; 1 fmol l^{-1} = 10^{-15} mol l^{-1}.

Conservative Type

Owing to their low reactivity, conservative type transition metals and heavy metals (V, Mo, W, Re, and Tl) are present in sea water at relatively high concentrations that are in constant proportion to salinity. Conservative metals have long mean oceanic residence times ($\gg 10^5$ y), their distributions are considerably homogeneous throughout the ocean due to the ocean's 1000-year circulation, and their concentrations are controlled by physical processes (e.g., advection and turbulent mixing).

Scavenged Type

Scavenged type transition metals and heavy metals (Mn, Co, Ga, In, Te, Pb, Bi, Ce) typically have strong interactions with particles, short mean oceanic residence times ($10^2 - 10^3$ y), and low concentrations. Their removal from sea water is dominated by adsorption onto the surfaces of particles and transport to the sediment via interactions with large, rapidly settling particles. Their depth profiles typically show enrichment in surface waters owing to sources from rivers and atmospheric dust, and rapid depletion to low concentrations at depth.

Nutrient (Recycled) Type

Metals having nutrient type distributions (Fe, Ni, Zn, Ge, Se, Y, Ag, Cd, Ba, La) are characterized by surface water depletion and enrichment at depth. Surface depletion is caused by biological uptake, and

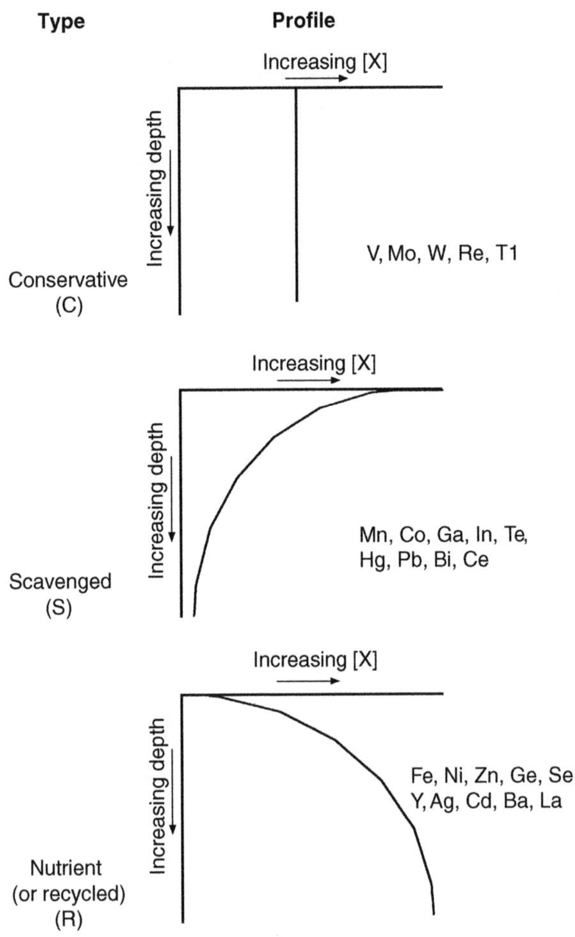

Figure 1 Oceanic profile classifications.

enrichment at depth is due to regeneration of the elements from particles back into solution by bacterial oxidation of the biological particulate matter. Deep waters of the North Pacific and Indian Ocean typically have higher concentrations of these elements than North Atlantic deep waters owing to biogeochemical cycles and ocean circulation.

Mixed Type

Some transition metals and heavy metals, such as Cu, Fe, Ga, Zr, Ti, La and other rare earths, have distributions that are influenced by both recycling and scavenging processes. For example, copper displays the characteristic surface depletion and deep-sea enrichment of the recycled element type; however, its concentration increases only gradually (almost linearly) with depth, indicating the effects of scavenging.

Modern Advances

Development of new analytical techniques, especially those that can be used at sea aboard ship, have enabled oceanographers to make large numbers of measurements of the concentration of a few transition metals across some ocean basins to construct two- and three-dimensional horizontal profiles, instead of just presenting an element's vertical profile. For example, two-dimensional ocean basin-scale distribution maps have been produced for aluminum and iron. These two-dimensional distribution maps can help identify the input and distribution mechanisms of an element and can be useful as tracers of water mass movements.

Although such detailed information has been obtained for a few transition metals and heavy metals, initial measurements of the oceanic concentrations and distributions need to be made for elements such as Ti, Ga, Ru, Pd, Ir, Pt, Au, Re Te, Zr, and Hf in many ocean basins before simple vertical and horizontal profiles can be constructed. Using newly developed analytical techniques, researchers have begun to obtain initial data on these metals. For example, the first concentration data on iridium in sea water (North Pacific) have been reported. Iridium concentrations ranged from 0.5×10^{-15} mol l^{-1} in North Pacific surface waters and increased with depth to a maximum of 0.8×10^{-15} mol l^{-1} near the bottom.

Speciation

Introduction

Knowing the oceanic concentrations and distributions is only part of the picture in understanding the biological and geochemical interactions of transition metals and heavy metals. Dissolved metals can exist in different oxidation states and chemical forms ('species'). These forms include free solvated ions, organometallic compounds, organic complexes (e.g., metals bound to proteins or humic substances), and inorganic complexes (e.g., metals bound to Cl^-, OH^-, CO_3^{2-}, SO_4^{2-}, etc.). Knowledge of the concentrations of these various species of a transition metal or a heavy metal, in conjunction with its distribution and concentration, is critical to understanding how the various chemical species interact biologically and geochemically. For example, the nutrient availability and toxicity of several transition metals have been found to be proportional to the concentrations of their free metal ions and not their total concentrations. Complexation of a metal by an organic ligand will decrease the concentration of the free ion form of the metal, thereby decreasing its toxicity or bioavailability. Organic complexation may also decrease or increase adsorption of metals onto metal oxide particles. These examples illustrate the importance of speciation information for fully understanding a metal's oceanic biogeochemical cycle.

Inorganic Speciation

Inorganic forms of the transition metals and heavy metals in sea water include hydrated metal ions, complexes with inorganic ligands, and species with different oxidation states. Transition metals and heavy metals with different oxidation states can exist in sea water when the potential required to change valence states falls within the range of the sea water's oxidizing/reducing potentials. Examples of transition metals and heavy metals having multiple oxidation states in sea water include Fe(II)/Fe(III), Mn(II)/Mn(IV), Cr(III)/Cr(VI), and Cu(I)/Cu(II). In oxygenated sea water, the thermodynamically stable form is usually the higher of the two oxidation states. However, species whose existence is thermodynamically unfavorable (i.e., usually the lower oxidation states) can be produced biochemically (e.g., by photosynthesis) and/or chemically (e.g., by photochemistry), as a result of the input of solar energy.

Calculational estimates of the inorganic speciation of many of the transition metals and heavy metals in sea water have been given in two landmark papers by Turner *et al.* and Byrne *et al.* (see Further Reading). The extent to which a metal is complexed by inorganic ligands is expressed by the inorganic side-reaction coefficient, α. This, in turn, is calculated from eqn[1] where β is the overall conditional stability constant for the inorganic complex MX_i of the transition or heavy metal M with the inorganic ligand X_i, and $X[_i']$ is the concentration of uncomplexed X_i.

$$\alpha = 1 + \sum_i \beta_{MX_i}[X_i'] \qquad [1]$$

The inorganic side-reaction coefficient, α, is also equal to the ratio of the sum of the concentrations of all inorganic species of the metal $M([M'])$ to the concentration of its free hydrated cation M $[M^{n+}]$ (eqn [2]).

$$\alpha = \frac{[M']}{[M^{n+}]} \qquad [2]$$

For zinc and the first transition series metals manganese, iron, cobalt, and nickel, the free hydrated divalent cation form dominates the dissolved inorganic speciation. The trivalent metal cations Al^{3+}, Ga^{3+}, Tl^{3+}, Fe^{3+}, and Bi^{3+} are strongly hydrolyzed (i.e., they form strong complexes with OH^-). With respect to complexation by OH^-, the inorganic side-reaction coefficients of the strongly hydrolyzed metals range from $10^{5.76}$ for Al^{3+} to $10^{20.4}$ for Tl^{3+}, and their inorganic speciation is strongly influenced by pH and temperature. For example, at a pH of 7.6, α for Al^{3+} increases 300-fold as the temperature is increased from 5 to 25°C; and at a temperature of 5°C, α for Al^{3+} increases 4000-fold as the pH increases from 7.6 to 8.2 (**Table 2**).

Table 2 Influence of pH and temperature on the α of Al^{3+}

pH	*Temperature (°C)*	α
7.6	5	$10^{5.76}$
7.6	25	$10^{7.23}$
8.2	5	$10^{9.39}$

Source: Byrne *et al.* (1988).

Other important inorganic species are the chloride and carbonate complexes. Chloride complexes are important in the inorganic speciation of Ag^+, Cd^{2+}, and Hg^{2+}. Unlike the strongly hydrolyzed metals, chloride dominated metals are only moderately affected by temperature and pH. Of this group, Hg^{2+} is complexed by chloride to the greatest extent. The side reaction coefficient of Hg^{2+} with respect to chloride is $10^{15.10}$ at 5°C. Carbonate complexes dominate the inorganic speciation of the lanthanides and some actinides (e.g., U(VI) and La(III)). These carbonate complexes are considerably influenced by temperature and pH, although less than the strongly hydrolyzed metal cations.

Organic Speciation

Organic forms of the transition metals and heavy metals in sea water include complexes with organic ligands (e.g., metals bound to proteins or humic substances) and organometallic compounds in which the metal is covalently bound to carbon (e.g., methyl forms of As, Ge, Hg, Sb, Se, Sn, and Te; ethyl-Pb forms; butyl-Sn forms). A most interesting discovery is that 90% of the germanium in open-ocean sea water exists in methylated forms so stable to degradation that they have been called the 'Teflon of the sea.' Methyl forms of metals are generally highly toxic because these compounds are soluble in cell walls and accumulate in cells. This accumulation is one example of how a nonessential metal can become biologically available.

The organically complexed fraction of certain transition metals and heavy metals in sea water has been reliably estimated only relatively recently, and attempts have been made to characterize the nature of these complexes. Early studies of metal

Table 3 Techniques used to determine the speciation of copper in natural waters

Technique	*Limitations/considerations*	*References*[a]
Fixed-potential amperometry (FPA)	Applicable to high [Cl^-] solutions only and low organic ligand concentrations ($\leq$1000 mol l^{-1})	Waite and Morel (1983); Hering *et al.* (1987)
Copper ion-selective electrode (ISE)	Limited sensitivity and chloride interferences	Belli and Zirino (1993); Zirino *et al.*, 1998
Biological assays	Assumes only free metal ion activity causes biological inhibition	Sunda and Ferguson (1983); Hering *et al.* (1987)
Solid-phase extraction (SPE)	May underestimate the extent of organically complexed copper in oceanic surface waters	Mills and Quinn (1981); Hanson and Quinn (1983); Donat *et al.* (1986)
Competitive equilibration with MnO_2	Assumes only Cu^{2+} adsorbs to MnO_2	van den Berg (1982)
Differential pulse anodic stripping voltammetry (DPASV)	Assumes only inorganic copper is detected and that natural copper complexes dissociate too slowly to be detected	Coale and Bruland (1988); Donat *et al.* (1994)
Competitive ligand equilibration/ adsorptive cathodic stripping voltammetry (CLE/CSV)	Assumes that samples at equilibrium during measurement and that natural copper complexes are not detected (i.e., not electroactive)	van den Berg (1985); Donat and Bruland (1990)

[a]See Further Reading list.

complexation showed little agreement between values for ligand concentrations, conditional stability constants, and the extent to which copper was organically complexed, which ranged from 0 to 100%. Organic speciation work on copper, zinc, and iron shows that the organically complexed fraction dominates the dissolved speciation of these metals in oceanic surface waters and is critically important in controlling the free metal ion concentrations of these metals. Although the chemical nature and complete chemical characteristics of the complexing ligands remains unknown, preliminary investigations have shown that the ligands are generally hydrophillic and of low molecular weight.

Methods for determining the speciation of transition metals and heavy metals in natural waters include fixed-potential amperometry (FPA), ion-selective electrodes (ISE), biological assays, solid-phase extraction (SPE), competitive equilibration with $MnO_2(s)$, differential pulse anodic stripping voltammetry (DPASV), and competitive ligand equilibration with adsorptive cathodic stripping voltammetric detection (CLE/CSV). **Table 3** lists the methods utilized for copper speciation with pertinent limitations and considerations. These techniques involve physical isolation or detection of one of the metal's species, or of a metal species not originally present in the sample but created for the speciation determination by introduction of a competing ligand. The speciation methods must operate under some general constraints: (1) samples must be at equilibrium, and (2) the technique must detect only the species intended.

Copper The fraction of organically complexed copper in sea water has been determined throughout many of the world's oceans including the Pacific, Atlantic, and Indian. The percentage of organic copper found in these oceans ranges from 89% to 99.9%. In the surface waters of the North Pacific (i.e., the upper 200 m), more than 99.7% of total dissolved Cu(II) is organically complexed (**Figure 2A**). The organic complexation is dominated by two copper-complexing ligands (or classes of ligands), L_1 and L_2. The stronger L_1 ligand class has an average concentration of $\sim$1.8 nmol l^{-1} in the upper 100 m and from the surface down to 200 m and its concentration exceeds that of dissolved copper (**Figure 2B**). The great strength of the L_1 class and its excess concentration relative to dissolved copper causes the inorganic copper fraction to account for less than 0.3% of total dissolved copper, and causes the free hydrated Cu^{2+} to account for only about 0.012% of total dissolved copper. A comparison of **Figure 2C** with **Figure 2B** shows that while dissolved copper ranges only from 0.3 to 1.5 nmol l^{-1} (a factor of 5), the Cu^{2+} concentration ranges from 10^{-13} to 10^{-10} (a thousand-fold)!

Measurements made in the Sargasso Sea revealed concentrations of the stronger L_1 copper-complexing ligand class to be equal to or less than the dissolved copper concentration, causing the weaker L_2 ligand class to dominate organic copper speciation, with a concomitant increase in the inorganic copper fraction and free Cu^{2+} concentration. Some evidence exists that the ligand concentrations and extent of organic complexation can vary seasonally.

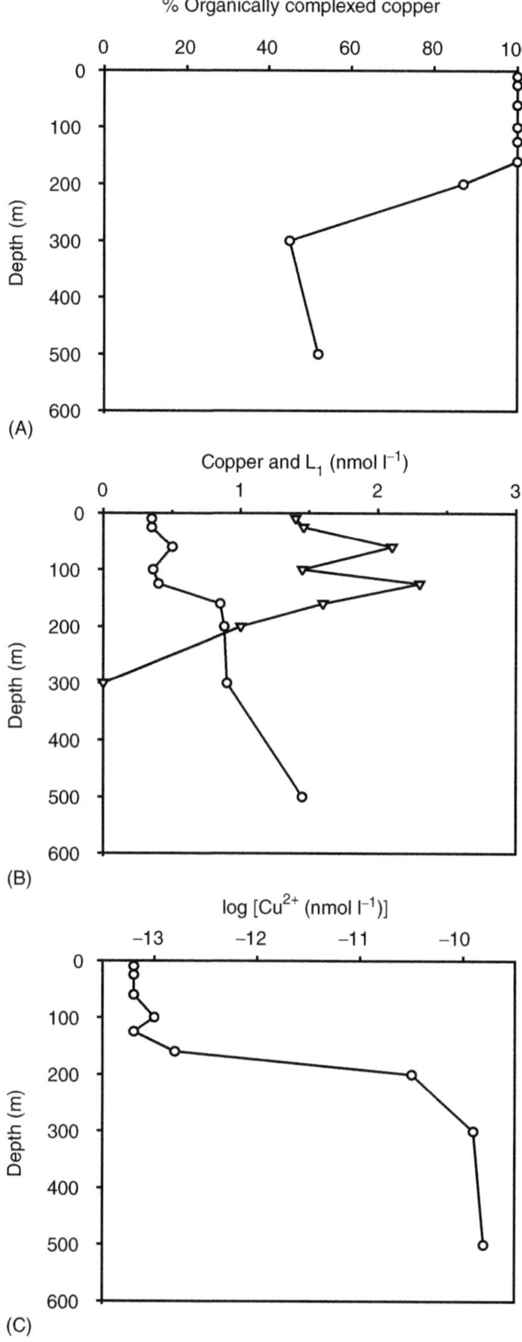

Figure 2 North Pacific surface waters dissolved Cu(II) speciation: (A) depth profile of L_1, the stronger copper-complexing organic ligand; (B) dissolved Cu(II) depth profile; (C) depth profile of free Cu^{2+} ion as logarithmic concentration values.

Zinc The fraction of organically complexed zinc found in North Pacific waters averages 98.7% (**Figure 3A**). As with copper, organic complexation of zinc is dominated by a relatively zinc-specific organic ligand (or ligand class) in surface waters shallower than 200 m (**Figure 3B**). In this upper 200 m, the zinc-complexing ligand averages 1.2 nmol l^{-1} and exceeds the concentration of dissolved zinc at depths above 300 m (**Figure 3B**). The high degree of organic complexation of zinc in the upper 300 m is caused by the excess in ligand relative to that of dissolved zinc and the strength of its zinc complexes. Organic complexation of zinc reduces the concentration of inorganic zinc species to 2×10^{-12} mol l^{-1}. Concentrations of free Zn^{2+} vary with depth from $\sim 10^{-11.8}$ mol l^{-1}, at depths less than 200 m, increasing to $\sim 10^{-8.6}$ mol l^{-1} at a depth of 600 m.

Iron Fe^{3+} forms complexes with natural organic ligands (like humic substances) that help keep this very insoluble cation in solution at elevated levels in estuarine and coastal waters. In the North Pacific and in the North Sea, researchers have determined that more than 99% of dissolved Fe(II) is bound with an extremely strong ligand class whose concentration ranges from 1 to 5 nmol l^{-1} and is in excess of the ambient dissolved iron concentration. These ligands have conditional stability constants consistent with low molecular weight organic substances called siderophores, which are produced by bacteria to specifically obtain iron. The availability of iron to aquatic primary producers has become the focus of many research projects since experiments have shown that in certain areas of the world's oceans iron availability is very low and may regulate productivity and perhaps influence atmospheric levels of carbon dioxide.

Other metals Organic complexation of other dissolved transition metals and heavy metals (i.e., Cd, Pb, Co, Ni, and Fe) has been investigated only much more recently and the information on these metals is not as defined or as extensive as for copper, iron and zinc. Recent measurements of dissolved cadmium in the North Pacific revealed that 70% was bound by cadmium-specific organic ligands found only at depths less than 175 m. Inorganic cadmium concentrations varied from 0.7×10^{-12} mol l^{-1} in surface waters to 800×10^{-12} mol l^{-1} at 600 m. The free Cd^{2+} concentration ranged from 20×10^{-15} mol l^{-1} in the surface, where organic complexation dominates the speciation, to 22×10^{-12} mol l^{-1} at 600 m where chloro complexes appear to dominate the inorganic speciation.

In the North Pacific, measurements of dissolved lead in the surface waters revealed that 50% was organically complexed by one class of strong organic ligands found to have concentrations between 0.2 and 0.5 nmol l^{-1}. The free Pb^{2+} surface water concentration as a result of inorganic and organic complexation was $\sim 0.4 \times 10^{-12}$ mol l^{-1}.

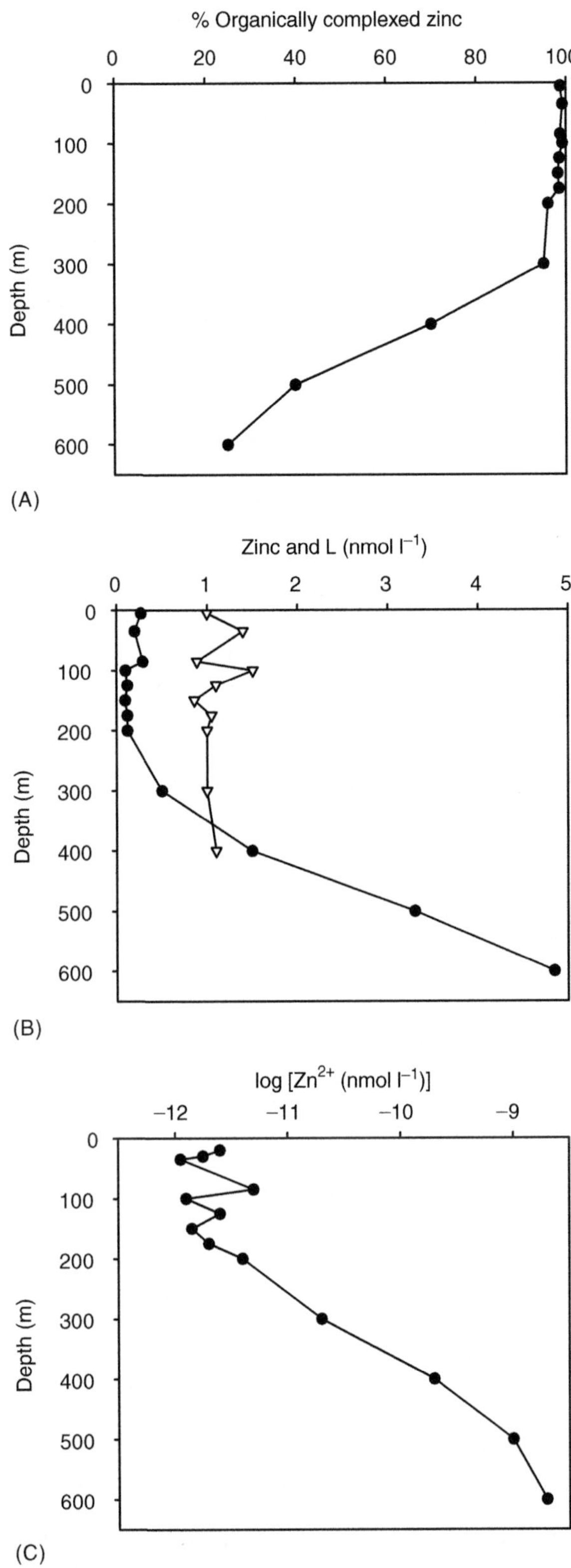

Figure 3 North Pacific zinc speciation: (A) depth profile of zinc-complexing organic ligand presented as percentage of organically complexed zinc; (B) dissovled zinc depth profile; (C) Zn^{2+} ion depth profile as logarithmic concentration values.

Organic complexation of dissolved cobalt and nickel in the open ocean has not been reported; however, organically complexed cobalt and nickel in estuarine and coastal samples have been found. The fraction of organic complexation is highly variable from estuary to coastal ocean. About 50% of the dissolved cobalt in coastal sea water was found to be organically complexed. In UK coastal waters and south San Francisco Bay, 30–50% of the nickel was bound in extremely strong organic complexes.

The information presented in this section demonstrates the importance of organic complexation of several transition metals and heavy metals. These organic ligands exist at low concentrations and form very strong complexes (i.e., they have high conditional stability constants). Although the actual chemical structures of these complexing organic ligands are still unknown, new analytical techniques may soon uncover their structure.

How Speciation Relates to Biology

Early researchers suggested that some organic compounds present in sea water in trace quantities may influence the primary production of marine communities by reducing toxic free metal concentrations (especially Cu^{2+}) to nontoxic levels. Data show that maximum levels of organically complexed copper occur in the surface euphotic zone at depths near the productivity maximum, and decrease dramatically below the vernal mixed layer in the North Pacific. The speciation of dissolved zinc is dominated by organic complexes and it may suggest a biological influence, as discussed for copper. Yet, the reasons for organic zinc speciation are not completely understood and only speculations exist.

Laboratory evidence exists for production of a strong copper-binding ligand by four marine phytoplankton (three species of eukaryotes and one prokaryote). The ligand that was produced has identical copper-complexing strength (i.e., similar conditional stability constants) to that of the stronger ligand observed in surface waters of the North Pacific and Sargasso Sea. The production of this L_1-like ligand may demonstrate a detoxification mechanism used by phytoplankton to lower the free Cu^{2+} concentration. Laboratory studies of the sensitivity of phytoplankton to varying Cu^{2+} concentrations revealed the following trend: cyanobacteria were the most sensitive; diatoms were the least sensitive; and coccolithophores and dinoflagellates showed intermediate sensitivity. Using this laboratory work, researchers are theorizing how cyanobacteria might produce strong L_1 ligands to lower the free Cu^{2+}

concentration in oceanic surface waters to levels at which their growth would not be impacted ($<10^{-12}$ mol l^{-1}). During an upwelling event, cyanobacterial production of the L_1 ligand might not exceed the newly upwelled Cu^{2+}, therefore cyanobacteria abundance would decline. Actual field evidence is supporting the speculation that species composition and seasonal species successions of phytoplankton are influenced by Cu^{2+} concentrations, especially in high-nutrient–low-chlorophyll areas.

Growth limitation experiments, like those for copper, have also been performed for iron, zinc, and manganese. These experiments showed that sufficiently low free ion activities of these nutrient metals could result in species shifts in phytoplankton communities. Iron is perhaps the most important nutrient transition metal to phytoplankton and its speciation is extremely complex and is not known with any reliability. Forms of iron that are speculated to have biological importance are organic Fe(III) complexes, Fe(III) oxides, and Fe(III)–siderophore complexes. Unlike Cu^{2+} which acts as a toxin, increased free Zn^{2+} concentrations in upwelled water could enhance reproduction of phytoplankton communities. Manganese in sea water, which shows no evidence of any organic complexation, appears to be maintained by photochemical reduction processes and photoinhibition of microbial oxidation of Mn^{2+}. Low manganese concentrations could potentially limit oceanic productivity if not supplied in sufficient quantities by atmosphere or horizontal mixing. Therefore, the distributions of Zn^{2+}, Mn^{2+}, and dissolved iron have important consequences for species composition and species succession of a phytoplankton community.

Oceanic concentrations of dissolved cadmium may be outside the range causing cadmium toxicity. However, in estuarine and riverine areas, anthropogenic sources could supply excessive cadmium inputs, leading to cadmium toxicity in aquatic phytoplankton. On the other hand, some researchers have shown that cadmium can promote growth of zinc-limited oceanic phytoplankton by substituting for zinc in certain macromolecules, thereby causing growth at lower than expected free Zn^{2+} concentrations. It has been speculated that this biochemical substitution of cadmium for zinc by phytoplankton could account for the nutrient-type oceanic distribution of cadmium.

Summary

Major advances in procedural, analytical, and instrumental techniques have advanced our knowledge of the concentrations, distributions, and speciation of the transition metals and heavy metals in the oceans, and therefore our understanding of their biogeochemical cycling. For most of the transition metals and heavy metals we have a first-order understanding of their oceanic distributions, and now with more data and better sea-going analytical techniques, basin-wide cross-sections of the distributions of some metals (e.g., aluminum, manganese, and iron) are becoming available. These basin-wide distributions allow more interpretation of sources and fates of these metals. Mediation by light and microorganisms dominates the biogeochemical cycling of certain metals such as copper, iron, and manganese. Organic complexation has come into the forefront of metal speciation research. Not only has the evidence for the existence of organic complexation been overwhelming, but organic ligands dominate the speciation of copper, zinc, and iron in oceanic surface waters. Organic complexation of certain metals in the oceans has important biological implications (i.e., controlling availability of metals as nutrients and toxicants) for phytoplankton.

See also

Carbon Cycle. Metal Pollution. Tracers of Ocean Productivity.

Further Reading

Belli SL and Zirino A (1993) Behavior and calibration of the copper(II) ion-selective electrode in high chloride media and marine waters. *Analytical Chemistry* 65: 2583–2589.

Brand LE, Sunda WG, and Guillard RRL (1986) Reduction of marine phytoplankton reproduction rates by copper and cadmium. *Journal of Experimental Marine Biology and Ecology* 96: 225–250.

Broecker WS and Peng TH (1982) *Tracers in the Sea*. New York: Eldigio Press.

Bruland KW (1983) Trace elements in sea-water. In: Riley JP and Chester R (eds.) *Chemical Oceanography*, vol. 8, pp. 157–220. London: Academic Press.

Bruland KW, Donat JR, and Hutchings DA (1991) Interactive influences of bioactive trace metals on biological production in oceanic waters. *Limnology and Oceanography* 36: 1555–1577.

Bruno J (1990) The influence of dissolved carbon dioxide on trace metal speciation in seawater. *Marine Chemistry* 30: 231–240.

Burton JD and Statham PJ (1988) Trace metals as tracers in the ocean. *Philosophical Transactions of the Royal Society of London Series A* 325: 127–145.

Byrne RH, Kump LR, and Cantrell KJ (1988) The influence of temperature and pH on trace metal speciation in seawater. *Marine Chemistry* 25: 163–181.

Coale KH and Bruland KW (1990) Spatial and temporal variability in copper complexation in the North Pacific. *Deep-Sea Research* 37: 317–336.

Donat JR and Bruland KW (1990) A comparison of two voltammetric techniques for determining zinc speciation in Northeast Pacific Ocean waters. *Marine Chemistry* 28: 301–323.

Donat JR and Bruland KW (1995) Trace elements in the oceans. In: Steinnes E and Salbu B (eds.) *Trace Elements in Natural Waters*, pp. 247–281. Boca Raton, FL: CRC Press.

Donat JR, Lao KA, and Bruland KW (1994) Speciation of dissolved copper and nickel in South San Francisco Bay: a multi-method approach. *Analytica Chimica Acta* 284: 547–571.

Donat JR, Statham PJ, and Bruland KW (1986) An evaluation of a C-18 solid phase extraction technique for isolating metal–organic complexes from central North Pacific Ocean waters. *Marine Chemistry* 18: 85–99.

Hanson AKJ and Quinn JG (1983) The distribution of organically complexed copper and nickel in the mid-Atlantic Bight. *Canadian Journal of Fisheries and Aquatic Sciences* 20: 151–161.

Hering JG, Sunda WG, Ferguson RL, and Morel FMM (1987) A field comparison of two methods for the determination of copper complexation: bacterial bioassay and fixed-potential amperometry. *Marine Chemistry* 20: 299–312.

Li YH (1991) Distribution patterns of the elements in the ocean. *Geochimica et Cosmochimica Acta* 55: 3223–3240.

Millero FJ (1992) Stability constants for the formation of rare earth inorganic complexes as a function of ionic strength. *Geochimica et Cosmochimica Acta* 56: 3123–3132.

Mills GL and Quinn JG (1981) Isolation of dissolved organic matter and copper–organic complexes from estuarine waters using reverse-phase liquid chromatography. *Marine Chemistry* 10: 93–102.

Nozaki Y (1997) A fresh look at element distribution in the North Pacific. *Eos, Transactions of the AGU* 78: 221.

Quinby-Hunt MS and Turekian KK (1983) Distribution of elements in sea water. *Eos, Transactions of the AGU* 64: 130–131.

Rainbow PS and Furness RW (eds.) (1990) *Heavy Metals in the Marine Environment*. Boca Raton, FL: CRC Press.

Sunda WG and Ferguson RL (1983) Sensitivity of natural bacterial communities to additions of copper and to cupric ion activity: a bioassay of copper complexation in seawater. In: *Trace Metals in Sea Water*, NATO Conference Series 4, Marine Science, Vol. 9, pp. 871–890. New York: Plenum Press.

Turner DR, Whitfield M, and Dickson AG (1981) The equilibrium speciation of dissolved components in freshwater and seawater at 25°C and 1 atm pressure. *Geochimica et Cosmochimica Acta* 45: 855–881.

van den Berg CMG (1982) Determination of copper complexation with natural organic ligands in seawater by equilibration with MnO_2. II. Experimental procedures and application to surface seawater. *Marine Chemistry* 11: 323–342.

van den Berg CMG (1985) Determination of the zinc complexing capacity in seawater by cathodic stripping voltammetry of zinc–APDC complex ions. *Marine Chemistry* 16: 121–130.

Waite TD and Morel FMM (1983) Characterization of complexing agents in natural waters by copper(II)/copper(I) amperometry. *Analytical Chemistry* 55: 1268–1274.

Wong CS, Boyle E, Bruland KW, Burton JD, and Goldberg ED (eds.) (1983) *Trace Metals in Seawater*. New York: Plenum Press.

Whitfield M and Turner DR (1987) The role of particles in regulating the composition of seawater. In: Stumm W (ed.) *Aquatic Surface Chemistry*, pp. 457–493. New York: Wiley.

Zirino A, DeMarco DJ, VanderWeele DA, and Belli SL (1998) Direct measurement of copper(II) (aq) in seawater at pH 8 with the jalpaite ion-selective electrode. *Marine Chemistry* 61: 173–184.

HYDROTHERMAL VENT FLUIDS, CHEMISTRY OF

K. L. Von Damm, University of New Hampshire, Durham, NH, USA

Introduction

It was not until 1977 that we knew that fluids exit from the seafloor along the global midocean ridge system. With the first discovery of hydrothermal venting at the Galapagos Spreading Center, our ideas on how elements cycle through the oceans and lithosphere, and even how and where life on our planet may have originated were fundamentally and irrevocably changed. Although these fluids were only a few tens of degrees hotter than ambient sea water ($<30°C$ vs. $2°C$), on the basis of their chemical compositions it was immediately clear that these fluids were derived from reactions at much higher temperatures between sea water and the oceanic crust. Less than two years later, spectacular jets of hot ($\geqslant 350°C$) and black water were discovered several thousand kilometers away on the northern East Pacific Rise, and 'black smokers' and 'chimneys' (**Figure 1**) entered the oceanographic lexicon.

From a chemical oceanography perspective, hydrothermal venting and hydrothermal vent fluids provide both new source and sink mechanisms for elemental cycling in the ocean, and therefore possible resolutions to a number of the outstanding chemical flux imbalances. They therefore play a fundamental role in regulating the chemistry of the oceans through geological time. From a biological oceanography perspective, hydrothermal vents provide us with new ecosystems based on chemosynthetic, rather than photosynthetic energy. Of the >500 new species discovered at these sites, the archea and other microbiological components are attracting increasing interest for both biotechnological applications and 'origin of life' questions on our own and other planets. From a physical oceanographic perspective, hydrothermal vents provide an input of both heat and materials into the oceanic mid-depth circulation. From a geological oceanographic perspective they provide an efficient means of removing heat from newly formed oceanic crust, as well as a means of altering the elements recycled into the mantle when the oceanic crust formed at spreading centers is later subducted back into the Earth's interior.

Where are Hydrothermal Vents Found?

Hydrothermal vents are now known to exist at approximately 30 locations on the global midocean ridge system (**Figure 2**). Initially some people had speculated that venting would only be found on intermediate- or faster-spreading ridges (i.e., ridges with full spreading rates of at least $60\,mm\,yr^{-1}$); we now know of numerous locations on slow-spreading ridges (e.g., Mid-Atlantic Ridge) where they occur. Known sites occur at depths from 800 to $>3600\,m$, with spreading rates from <20 to $>150\,mm\,y^{-1}$ (full rate), on both bare basalt and sedimented-covered ridges, as well as on seafloor where ultramafic rock types are known to outcrop, and at temperatures up to 405°C. If one looks at the global distribution of known vent sites, it is obvious that many of the sites are in relatively close proximity to nations that operate submersibles, and are clustered disproportionately in the north Atlantic and eastern Pacific. Although not strictly part of the midocean ridge system, venting associated with back arc spreading centers is also known from a number of sites in the western Pacific. No sites have yet been discovered in the Indian Ocean, although cruises to this ocean are now planned. Similarly, no sites are known in the south Atlantic, or at high latitudes. This is an exploration issue, not a lack of their existence in these areas.

While initially vents were thought to occur at the mid-point of ridge segments, this was a largely self-fulfilling prophesy, as this is where exploration for them was focused. There is increasing evidence that more venting occurs on the magmatically robust portions of the ridge, rather than on those areas that are deemed to be magma-starved, on the basis of their morphological characteristics.

How are they Found?

Vent fields have been discovered in numerous ways, but surveys of the overlying water column and camera tows are the most common systematic approaches employed today. The venting of hot/warm water often forms a plume with unique temperature, salinity, reduced light transmittance, and other specific chemical signals several hundred meters above the ridge. The presence of such a plume is often the first indication that a given section of ridge is hydrothermally active. Cameras, or other

Figure 1 Pictures of black smokers. (A)–(C) Brandon Vent a black smoker with 405°C fluid temperatures on the Southern East Pacific Rise at 21°33′S at a depth of 2834 m; in (A) and (B), shown instrumented with a Hobo recording temperature probe and in (C) with one of *Alvin*'s manipulators. (D) Nadir vent at 17°26′S at a depth of 2562 m and fluid temperatures of 343°C shown prior to sampling in 1998. In the foreground is *Alvin*'s basket with water sampling equipment. (E) Water sampling of Nadir vent fluids.

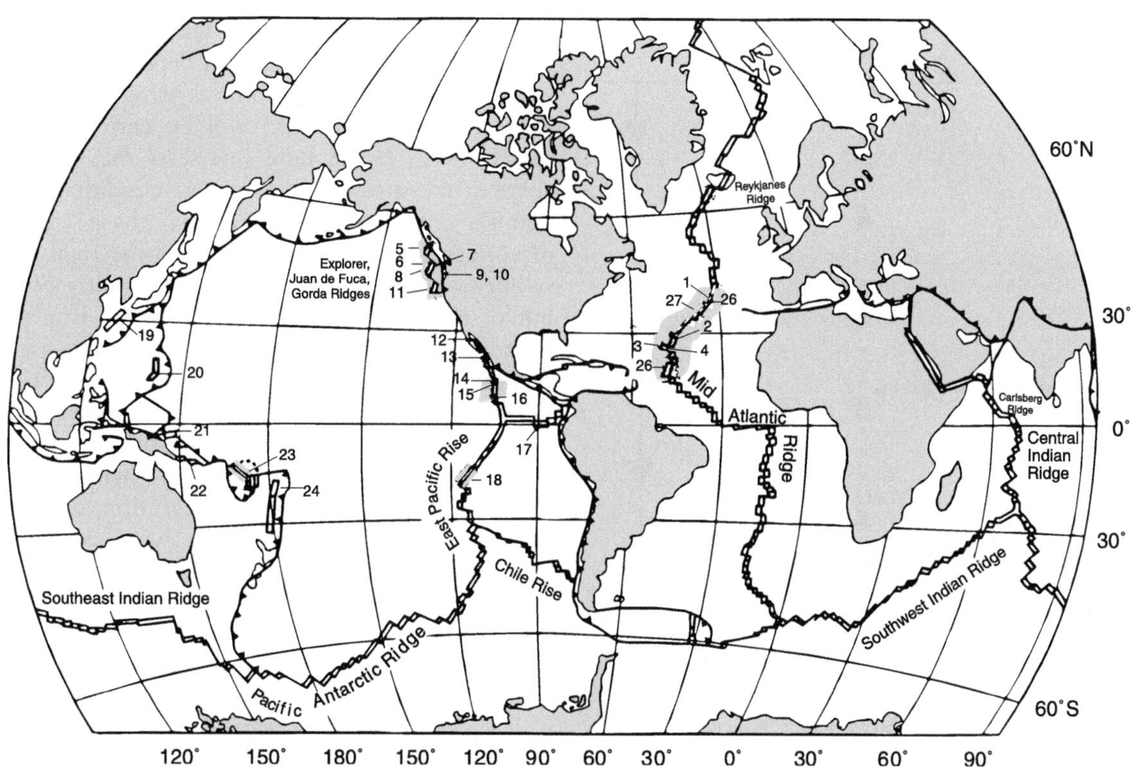

Figure 2 Global distribution of known (sampled) hydrothermal systems. (Modified from Baker *et al.* (1995).)

1	Lucky Strike (37°17′N)	15	EPR (11°N)
2	Broken Spur (29°N)	16	EPR (10°N)
3	TAG (26°N)	17	Galapagos (86°W)
4	Snakepit/MARK (23°N)	18	EPR (17°–19°S)
5	Explorer Ridge (49°45′N	19	Okinawa Trough (27°30′N)
6	Middle Valley	20	Mariana Trough
7	Endeavour (48°)	21	Manus Basin
8	Axial Volcano (46°N)	22	Woodlark Basin
9	Cleft—North (45°)	23	N. Fiji Basin (17°S)
10	Cleft—South (44°40′N)	24	Lau/Valu Fa (22°S)
11	Escanaba Trough	25	Logatchev (14°45′N)
12	Guaymas Basin (27°N)	26	Menez Gwen (37°50′N)
13	EPR (21°N)	27	Rainbow (36°16′N)
14	EPR (13°N)		

deep-towed survey vehicles may also find visual evidence for communities of vent-specific organisms, sulfide structures (both active and extinct), and murky water. These surface ship surveys may then lead to the use of an ROV (remotely operated vehicle) or submersible, in what has been called a 'nested survey strategy.' Sites have also been found by the fortuitous dredging up of both pieces of hydrothermal chimneys and vent-specific animals.

Controls on Fluid Compositions

Hydrothermal vent fluids are primarily, if not entirely, sea water that has been altered due to reaction within the oceanic crust at high temperatures (at least 350°C, and more likely >400°C) and elevated pressures (at least 80 bar, and more typically at least 250 bar). The two primary controls on vent fluid compositions are phase separation and water–rock reaction (**Figure 3**). Unlike pure water, sea water is a two-component system, containing both water and salt (primarily NaCl, especially for hydrothermal fluids from which the magnesium and sulfate have been essentially quantitatively removed). The critical point for sea water is higher than that for distilled water – 407°C and 298 bar rather than 374°C and 220 bar – and most importantly the two-phase curve therefore continues beyond the critical point

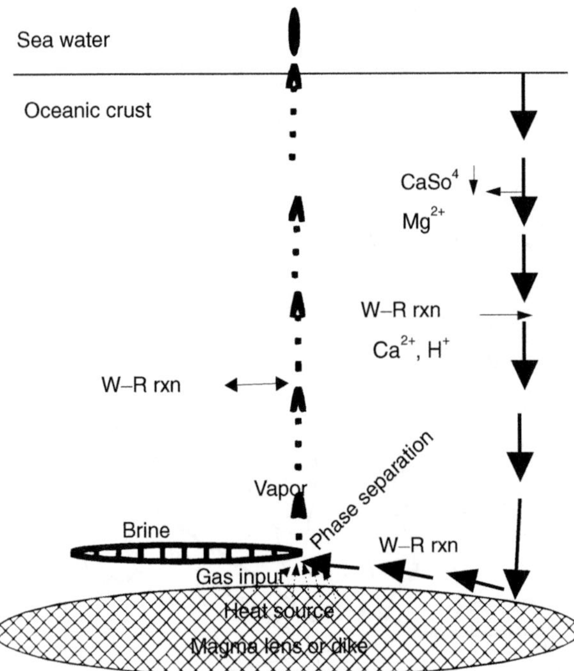

Figure 3 Schematic of a hydrothermal flow cell. $W\text{–}R_{rxn}$ = water–rock reaction. (Von Damm 1995 in Humphris *et al.*)

(**Figure 4**). If fluids intersect the two-phase curve at pressure and temperature conditions less than the critical point, they will undergo subcritical phase separation or boiling. The vapor phase formed by this process will contain some salt, the exact amount and composition of which will depend on the pressure and temperature conditions at which the phase separation occurred, and will be enriched in dissolved gases. If the fluid intersects the two phase curve at pressure and temperature conditions higher than the critical point, supercritical phase separation or condensation will occur, and, rather than a low-salinity vapor being formed, a high-salinity brine or liquid phase will condense. In most, if not all, hydrothermal fluids that have been sampled, phase separation has occurred as evidenced by the chlorinities of these fluids, which can be from ~6% to 200% of those in sea water. While the change in the absolute chlorinity of seafloor hydrothermal fluids is primarily the result of phase separation, changes in the other dissolved ions in sea water also reflect substantial reaction of the fluids with the rock, mostly basalt, but also ultramafics and sometimes sediments. Seafloor hydrothermal fluids therefore do not show the 'constancy of composition' and elemental ratios known to characterize the major chemical species in sea water. The two primary mechanisms for determining the composition of hydrothermal fluids are therefore phase separation and water–rock interaction. This does not mean that other mechanisms may not be important. Two that have been identified but whose global importance is not yet known are magmatic degassing and biological uptake/removal. On the East Pacific Rise at 9°50′N latitude unusually high gas concentrations and gas/heat ratios have been observed in the hydrothermal fluids that suggest that we are seeing the degassing of a recently resupplied magma chamber. This has not yet been observed at any other sites, although it has been observed for ~7 years at this site. The potential biological effects on fluid compositions have long been speculated on, but there are few quantitative data at this time that can directly address this question. As the 'black smoker' fluids are at temperatures well beyond the known bounds of life on this planet, if biological effects are present they are most likely to be found in fluids with temperatures less than ~110°C.

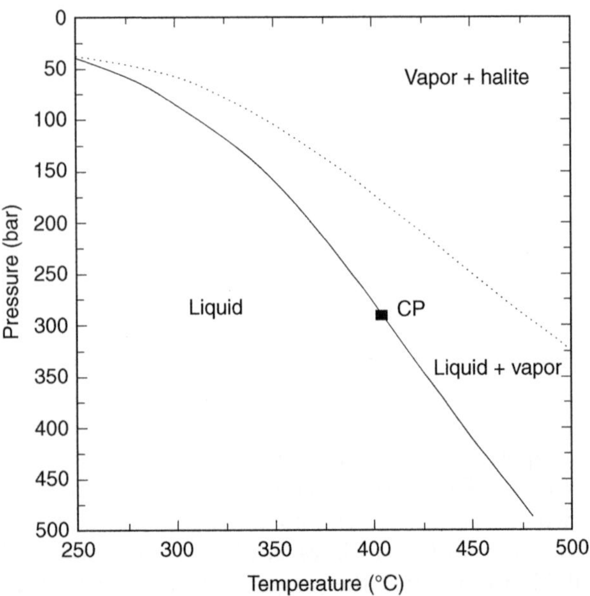

Figure 4 Two-phase curve for sea water, showing the location of the critical point (CP) for sea water (407°C, 298 bar), and the relative locations of the liquid, liquid + vapor, and vapor + halite stability fields. Note that, unlike for pure water, the two-phase curve continues beyond the critical point.

The Division of Fluids by Temperature and Styles of Venting

Hydrothermal fluids are often subdivided into two categories: high-temperature or focused flow, and low-temperature or diffuse flow. These terms are often poorly defined in the contexts in which they are used, and may mean different things to different authors. High-temperature fluids, generally > 200–250°C, are usually focused jets of water that are exiting from constructional features that either are precipitated from the hydrothermal fluids themselves when they encounter cold, alkaline sea water or are formed by precipitation due to the mixing of hydrothermal fluids with sea water, or, and more usually, both of these processes. Although these chimneys comprise metal sulfides and sulfates, other minerals are also present, usually in lesser quantities, although on sediment-covered ridges carbonates, for example, may also be abundant in these chimney or mound structures. Low-temperature fluids, generally $\leqslant 35^{\circ}C$, but sometimes approaching $100^{\circ}C$, have usually deposited (if they ever contained) much of their metal load below the seafloor and hence usually neither smoke nor have specific constructional features associated with their fluid flow. Because of this lack of specific structure, the fluid flow is often less organized, and hence is referred to as 'diffuse.' Some authors use this term to refer to fluids exiting directly from the basaltic substrate. These sites are usually well-colonized by various types of vent megafauna, making it difficult, if not impossible, to identify a specific orifice. Other authors use this same term to refer to fluids that are oozing out of large sulfide structures. At some sites, only high-temperature 'black smoker' vents are found, while at other sites only the lower-temperature diffuse venting is found (as at the Galapagos Spreading Center where venting was first discovered). At other sites, both focused and diffuse flow may occur right next to one another, and, at least at one site, we have now observed low-temperature diffuse flow evolving into high-temperature focused flow over several years. For all types of 'low temperature and/or diffuse flow,' all of the fluids that have been sampled to date appear to contain some fluid that has reacted at significantly greater temperatures than are measured directly in these fluids. Hence one could argue that, at least at the ridge axes, all fluids are ultimately high-temperature fluids and that some of them have undergone significant mixing with sea water at some depth, likely relatively shallow, within the oceanic crust. With the continued inability to drill these systems, our knowledge of their subsurface hydrology is rudimentary at best. One might note that fluids with temperatures in the ~ 100–$200^{\circ}C$ range have been left out of the above classification. This is because there are few fluids that have been sampled in this temperature range, which may reflect a true lack of abundance of fluids at such temperatures, or simply a sampling gap.

The Influence of Volcanic Events

In the late 1980s there was a real dichotomy in our understanding of the controls on seafloor hydrothermal systems. We knew of individual vents that were chemically stable for years at a time, yet each individual vent had a distinctive chemistry, and rapid changes in the hydrothermal plumes above these sites had been noted. In 1991 some of this puzzle was resolved with the first opportunity to sample a mid-ocean ridge hydrothermal site immediately following a volcanic eruption and to study the evolution of this site on timescales now approaching decadal. The discovery of an eruption on the East Pacific Rise at 9°45–52′N was fortuitous, but it could be sampled and documented with the DSV *Alvin* within weeks of the volcanic event. Since that time using the US Navy's SOSUS system, it has been possible to determine real-time when some of these events are occurring on the Juan de Fuca and Gorda Ridges, but immediate response to these events have been limited to surface ship observations due to logistical constraints. The events studied so far have been characterized by a volcanic eruption at the seafloor, the intrusion of a dike, or both. The knowledge gained from responding to these events has revolutionized our understanding of these systems, especially their pronounced temporal variability on very short timescales (much less than days to weeks). Presumably we will be able to study more of these events and to gain a sense of their frequency, perhaps as a function of spreading rate, as we begin to instrument more of the ridge crest with hydrophones to detect the T-phase signals associated with these events.

The Influence of Tectonic/Cracking Events

Magmatic events have provided new insights into the processes that drive hydrothermal systems and their fluid compositions on intermediate- to fast-spreading ridges, but presumably tectonic events are more important (or at least more frequent than magmatic events) on slow-spreading ridges. As we have not yet been able to observe one of these events, we cannot assess their importance. In a relatively small cracking event on a fast-spreading ridge observed with a seismic array, changes in fluid temperatures were marked, and changes in fluid compositions were profound, leading to major changes in the biological communities existing at this site. Presumably the observation of one or more tectonic events on slow-spreading ridges will also provide critical new insights into how hydrothermal systems on these types of ridges function and evolve.

On-axis versus Off-axis

The discussion above has focused on the axial component of midocean ridge hydrothermal systems. Many debates have focused on the importance of these axial systems compared to hydrothermal systems on the ridge flanks. The net amount of heat transported and the total geographical extent of the flanks is much larger than that of the axial parts of the ridges, the division often being drawn at 1 My, resulting in a partitioning of 70%:30% relative heat loss. Hence, the relative chemical anomaly and volume of fluids expelled on the axis versus the flanks of the ridge will determine which of these parts of the midocean ridge hydrothermal system is most important for the cycles of individual chemical elements and species. It has been difficult to acquire quantitative data on the fluids exiting from the flanks, one of the reasons being that sites of flank fluid venting are difficult to identify because water column plumes are not present, nor are the distinctive animal communities (distinctive both for their animal types as well as their white color on the black basalt) found at the axial sites of venting. In the last several years drilling by the Ocean Drilling Program on the flanks of the Juan de Fuca Ridge has provided important new information on the chemistry and hydrology of these systems, as well as demonstrating a viable method for further approaching the flank question.

Observed Fluid Compositions

Compositions of hydrothermal fluids not only vary widely but also are almost always very different from those of sea water (**Table 1**). While some of the low temperature diffusely venting fluids may be close to sea water in their major element compositions, they will often have very different compositions of dissolved gases (e.g., H_2S, CO_2, CH_4, H_2, and He) and will usually be highly enriched in iron and manganese compared to local ambient sea water. Compared to sea water, hydrothermal fluids have lost essentially all of their magnesium and sulfate, and are highly enriched in H_2S, CH_4, H_2, He, Si, Li, Fe, and Mn. As hydrothermal fluids are very acid, they also have no alkalinity, and with the loss of sulfate, chloride becomes the major, and almost only anion (bromide is present in much lower concentrations). The behavior of the cations is more variable. As the amount of chloride present is a result of the phase separation history of the fluids, and the fluids must maintain electroneutrality, to determine whether a particular cation has been added to or removed from the fluid,

Table 1 Range of physical parameters and chemical compositions for hydrothermal vent fluids[a]

Parameter	*Units*[b]	*Vents*	*Seawater*
Temperature	°C	> 2–405	~1–5 (bottom water)
Depth (pressure)	m (bars)	800–3600 (80–360)	
pH 25°C, 1 atm		2.5–7.8	7.8
$Alkalinity_{total}$	meq kg^{-1}	−2.7–10.6	2.3
Cl	mmol kg^{-1}	31–1245	545
SO_4^{2-}	mmol kg^{-1}	0	28
H_2S	mmol kg^{-1}	0–110	0
Si	mmol kg^{-1}	2.7[b]–24.0	0.032–0.180
Li	mmol kg^{-1}	<0.012–2.35	0.026
Na	mmol kg^{-1}	<15–924	465
K	mmol kg^{-1}	<1–58.7	10
Ca	mmol kg^{-1}	<0.2–109	10
Mg	mmol kg^{-1}	0	53
Sr	μmol kg^{-1}	<1–348	87
Fe	mmol kg^{-1}	0–18.7	<0.001
Mn	mmol kg^{-1}	0–4.48	<0.001
Cu	μmol kg^{-1}	0–310	0.007
Zn	μmol kg^{-1}	0–900	0.01
Cd	nmol kg^{-1}	<10–1000	1
Co	nmol kg^{-1}	<5–2570	0.03
Pb	nmol kg^{-1}	50–2200	0.01
B	μmol kg^{-1}	416–1630	416
Al	μmol kg^{-1}	0–20.0	0.02
Br	μmol kg^{-1}	29–1880	840
CO_2	mmol kg^{-1}	2.3–375	2.3
CH_4	μmol kg^{-1}	0.0003–6800	0.0003
H_2	mmol kg^{-1}	<0.001–38	0.0000003

[a]Only includes mid-ocean ridge systems, both bare rock and sediment-covered.
[b]From high-temperature vents.

not only the absolute concentration but also its ratio to chloride must be compared to the local ambient sea water value. Hydrothermal fluids are also very reducing, often containing large amounts of H_2S, H_2, and CH_4. In most, but not all hydrothermal fluids (fluids that are formed immediately after a volcanic eruption being a major exception), Li, K, Rb, Cs, Ca, Sr, Si, the transition metals in reduced forms, including Fe, Mn, Cu, and Zn, are enriched in hydrothermal fluids, the cause being water–rock interaction. Sodium may be either enriched or depleted with respect to the chloride content of the fluids, the loss being due to albitization, with a concomitant gain in the calcium-content of the fluids. It is the loss of magnesium to form magnesiumhydroxy silicates that, along with other aluminosilicate reactions, generates and then maintains the acidity of the fluids. In the case of sulfate, some is lost as anhydrite ($CaSO_4$) in the downflow zone, while some is reduced to sulfide (as H_2S).

The compositions of fluids exiting from a single hydrothermal vent may vary widely over time. The cases in which this has been observed are increasing, and are usually associated with vents where a known magmatic event has occurred. The variation in the composition of a single vent can vary from vapor to brine, and may encompass almost the entire range of known compositions. In contrast, some sites of venting are known where the fluid compositions have been stable during the time interval over which they have been sampled. None of these vents with constant compositions have known 'magmatic' or 'tectonic' events associated with them, although several of these sites have now been sampled over times of ~ 15 years. Presumably, these vents are in a period of steady-state venting, although we do not have adequate constraints to determine how long after an eruptive event, or at what spreading rates, this may occur. While our data has increased on the temperature and chemical characteristics of vents during their early histories, few data exist for their waning stage(s). Presumably all this variability – or lack thereof – and the timescale(s) on which it occurs can ultimately be tied to the nature of the heat source at a given site. Aside from the most general characteristics related to the presence or absence of a seismic low-velocity zone, and the depth at which it occurs, little is known about the specifics of the heat sources at sites that are hydrothermally active, especially in contrast to those that are not.

The composition of a hydrothermal fluid cannot be correlated to, or predicted by, such known physical parameters as the depth of the seafloor on which it occurs, the spreading rate of the ridge on which it occurs, and so on. As the fluid compositions in most cases are probably due to the achievement of steady-state, if not true thermodynamic equilibrium, of the fluids with the rock substrate, some of the measured compositions can be tied to either the measured exit temperature or the presumed *in situ* conditions within the hydrothermal system itself. While there has been some success, especially recently, with understanding the chemical controls on these systems using thermodynamic modeling, a major limitation in many cases remains the proximity to the critical point of both pure water and sea water.

The Flux Question

One of the driving questions for the study of seafloor hydrothermal systems is to understand their net flux to the ocean in terms of energy (thermal and chemical) and mass. The thermal, or heat, energy they carry is believed to be relatively well constrained, as various independent ways of estimating this flux provide similar values. The mass of chemicals they add and/or remove remains problematic. In some cases whether hydrothermal activity is a net source or sink for particular elements remains unresolved as well. In addition to absolute concentrations, or concentrations normalized to the chloride content, the isotopic signature of various species can also be used to constrain the source and sink terms, as well as helping to identify the important processes occurring within the hydrothermal circulation cell.

Summary and Conclusions

Hydrothermal venting along the global midocean ridge system is a process that is widespread throughout the ocean basins and impacts all of the oceanographic disciplines. Our studies of these systems remain in their infancy, however, and we do not yet completely understand the controls on the chemistry of these systems, the controls on the locations of individual vent sites, their overall importance to ocean chemistry, productivity, and circulation, and their net effects on the structure and composition of the oceanic crust.

See also

Hydrothermal Vent Deposits. Mid-Ocean Ridge Geochemistry and Petrology.

Further Reading

Cowen JP and Baker ET (1998) Topical studies in oceanography: detection of and response to mid-ocean ridge magmatic events. *Deep-Sea Research II* 45(12): 2503–2766.

Elderfield H and Schultz A (1996) Mid-ocean ridge hydrothermal fluxes and the chemical composition of the ocean. *Annual Review of Earth and Planetary Science* 24: 191–224.

Humphris SE, Zierenberg RA, Mullineaux LS, and Thomson RE (eds.) (1995) *Seafloor Hydrothermal Systems: Physical, Chemical, Biological, and Geological Interactions*. AGU Monograph 91. Washington, DC: American Geophysical Union.

Parson LM, Walker CL, and Dixon DR (eds.) (1995) *Hydrothermal Vents and Processes*. London: Geological Society: Geological Society Special Publication 87.

Seyfried WE Jr (1987) Experimental and theoretical constraints on hydrothermal alteration processes at mid-ocean ridges. *Annual Review of Earth and Planetary Science* 15: 317–335.

Von Damm KL (1990) Seafloor hydrothermal activity: black smoker chemistry and chimneys. *Annual Review of Earth and Planetary Science* 18: 173–204.

PHOTOCHEMICAL PROCESSES

N. V. Blough, University of Maryland, College Park, MD, USA

Introduction

Life on Earth is critically dependent on the spectral quality and quantity of radiation received from the sun. The absorption of visible light (wavelengths from 400 to 700 nm) by pigments within terrestrial and marine plants initiates a series of reactions that ultimately transforms the light energy to chemical energy, which is stored as reduced forms of carbon. This complex photochemical process, known as photosynthesis, not only provides all of the chemical energy required for life on Earth's surface, but also acts to decrease the level of a major greenhouse gas, CO_2, in the atmosphere. By contrast, the absorption of ultraviolet light in the UV-B (wavelengths from 280 to 320 nm) and UV-A (wavelengths from 320 to 400 nm) by plants (as well as other organisms) can produce seriously deleterious effects (e.g. photoinhibition), leading to a decrease in the efficiency of photosynthesis and direct DNA damage (UV-B), as well as impairing or destroying other important physiological processes. The level of UV-B radiation received at the Earth's surface depends on the concentration of ozone (O_3) in the stratosphere where it is formed photochemically. The destruction of O_3 in polar regions, leading to increased levels of surface UV-B radiation in these locales, has been enhanced by the release of man-made chlorofluorocarbons (CFCs), but may also be influenced in part by the natural production of halogenated compounds by biota.

These biotic photoprocesses have long been recognized as critical components of marine ecosystems and air–sea gas exchange, and have been studied extensively. However, only within the last decade or so has the impact of abiotic photoreactions on the chemistry and biology of marine waters and their possible coupling with atmospheric processes been fully appreciated. Light is absorbed in the oceans not only by phytoplankton and water, but also by colored dissolved organic matter (CDOM), particulate detrital matter (PDM), and other numerous trace light-absorbing species. Light absorption by these constituents, primarily the CDOM, can have a number of important chemical and biological consequences including: (1) reduction of potentially harmful UV-B and UV-A radiation within the water column; (2) photo-oxidative degradation of organic matter through the photochemical production of reactive oxygen species (ROS) such as superoxide (O_2^-), hydrogen peroxide (H_2O_2), the hydroxyl radical (OH) and peroxy radicals (RO_2); (3) changes in metal ion speciation through reactions with the ROS or through direct photochemistry, resulting in the altered biological availability of some metals; (4) photochemical production of a number of trace gases of importance in the atmosphere such as CO_2, CO, and carbonyl sulfide (COS), and the destruction of others such as dimethyl sulfide (DMS); (5) the photochemical production of biologically available low molecular weight (LMW) organic compounds and the release of available forms of nitrogen, thus potentially fueling the growth of microorganisms from a biologically resistant source material (the CDOM). These processes provide the focus of this article.

Optical Properties of the Abiotic Constituents of Sea Waters

CDOM is a chemically complex material produced by the decay of plants and algae. This material, commonly referred to as gelbstoff, yellow substance, gilvin or humic substances, can be transported from land to the oceans by rivers or be formed directly in marine waters by as yet poorly understood processes. CDOM is the principal light-absorbing component of the dissolved organic matter (DOM) pool in sea waters, far exceeding the contributions of discrete dissolved organic or inorganic light-absorbing compounds. CDOM absorption spectra are broad and unstructured, and typically increase with decreasing wavelength in an approximately exponential fashion (**Figure 1**). Spectra have thus been parameterized using the expression [1].

$$a(\lambda) = a(\lambda_0) \cdot e^{-S(\lambda - \lambda_0)} \qquad [1]$$

$a(\lambda)$ and $a(\lambda_0)$ are the absorption coefficients at wavelength λ and reference wavelength λ_0, respectively, and S defines how rapidly the absorption increases with decreasing wavelength. Absorption coefficients are calculated from relation [2], where A is the absorbance measured across pathlength, r.

$$a(\lambda) = \frac{2.303 \cdot A(\lambda)}{r} \qquad [2]$$

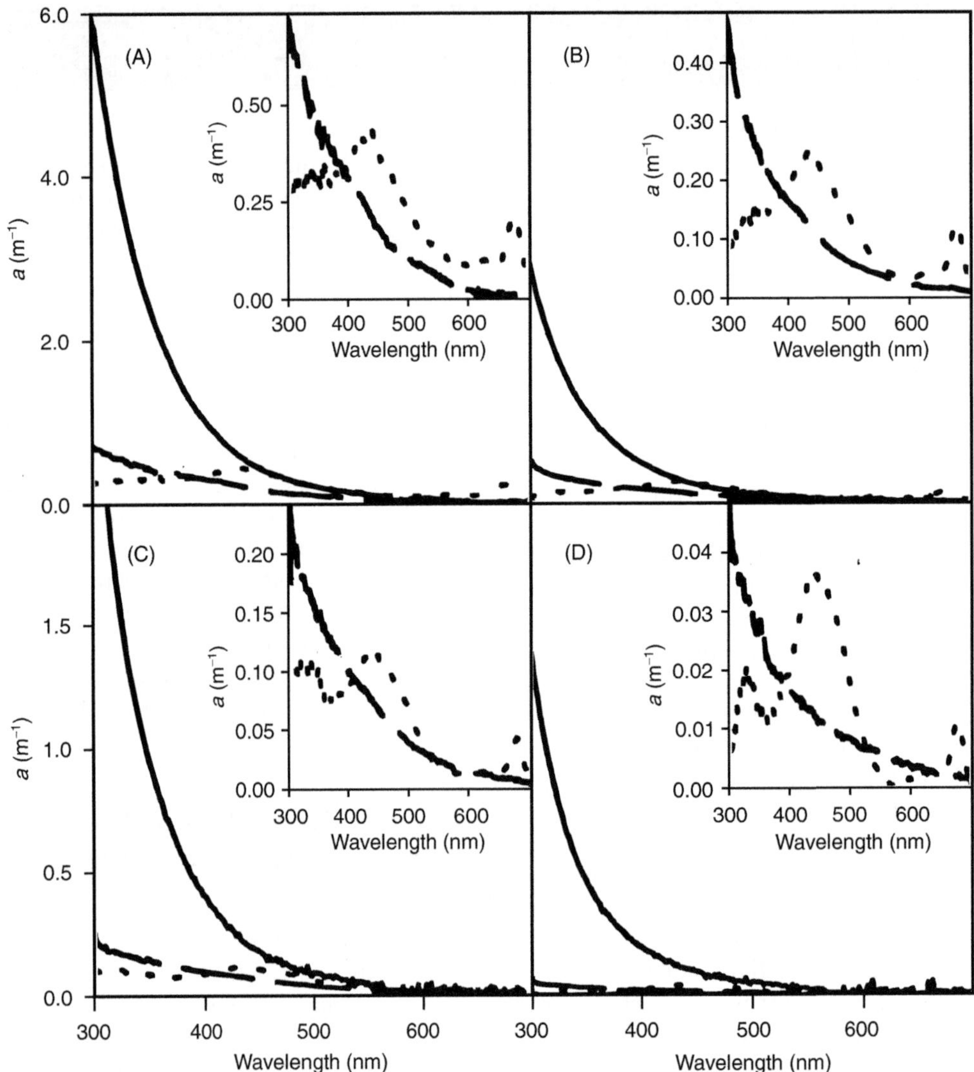

Figure 1 Absorption spectra of CDOM (—), PDM (– – –) and phytoplankton (– – –) from surface waters in the Delaware Bay and Middle Atlantic Bight off the east coast of the USA in July 1998: (A) mid-Delaware Bay at 39° 9.07′ N, 75° 14.29′ W; (B) Mouth of the Delaware Bay at 38° 48.61′ N, 75° 5.07′ W; (C) Mid-shelf at 38° 45.55′ N, 74° 46.60′ W; (D) Outer shelf at 38° 5.89′ N, 74° 9.07′ W.

Due to the exponential increase of $a(\lambda)$ with decreasing λ, CDOM absorbs light strongly in the UV-A and UV-B, and thus is usually the principal constituent within marine waters that controls the penetration depth of radiation potentially harmful to organisms (**Figure 1**). Moreover, for estuarine waters and for coastal waters strongly influenced by river inputs, light absorption by CDOM can extend well into the visible wavelength regime, often dominating the absorption by phytoplankton in the blue portion of the visible spectrum. In this situation, the amount and quality of the photosynthetically active radiation available to phytoplankton is reduced, thus decreasing primary productivity and potentially affecting ecosystem structure. High levels of absorption by CDOM in these regions can also seriously compromise the determination of phytoplankton biomass through satellite ocean color measurements.

As described below, the absorption of sunlight by CDOM also initiates the formation of a variety of photochemical intermediates and products. The photochemical reactions producing these species ultimately lead to the degradation of the CDOM and the loss, or bleaching, of its absorption. This process can act as a feedback to alter the aquatic light field.

Particulate detrital material (PDM), operationally defined as that light-absorbing material retained on a GFF filter and not extractable with methanol, is a composite of suspended plant degradation products and sediment that also exhibits an exponentially rising absorption with decreasing wavelength (**Figure 1**); eqn [2] has thus been used to parameterize this material as well. However, the values of S acquired for this material are usually smaller than those of the CDOM. In estuarine and near-shore waters, and in shallow coastal waters subject to resuspension of bottom sediments, PDM can contribute substantially to the total water column absorption. However, in most marine waters, the PDM is a rather minor constituent. Little is known about its photochemical reactivity.

Other light-absorbing trace organic compounds such as flavins, as well as inorganic compounds such as nitrate, nitrite, and metal complexes, do not contribute significantly to the total water column absorption. However, many of these compounds are quite photoreactive and will undergo rapid transformation under appropriate light fields.

Photochemical Production of Reactive Oxygen Species

CDOM is the principal abiotic photoreactive constituent in marine waters. Available evidence suggests that the photochemistry of this material is dominated by reactions with dioxygen (O_2) in a process known as photo-oxidation. In this process, O_2 can act to accept electrons from excited states, radicals (highly reactive species containing an unpaired electron) or radical ions generated within the CDOM by the absorption of light. This leads to the production of a variety of partially reduced oxygen species such as superoxide (O_2^-, the one-electron reduction product of O_2), hydrogen peroxide (H_2O_2, the two-electron reduction product of O_2), peroxy radicals (RO_2, formed by addition of O_2 to carbon-centered radicals, R) and organic peroxides (RO_2H), along with the concomitant oxidation of the CDOM (**Figure 2**). Many of these reduced oxygen species as well as the hydroxyl radical (OH), which is generated by other photochemical reactions, are also quite reactive. These reactive oxygen species or ROS can undergo additional secondary reactions with themselves or with other organic and inorganic seawater constituents. The net result of this complex series of reactions is the light-induced oxidative degradation of organic matter by dioxygen (**Figure 2**). This process leads to the consumption of O_2, the production of oxidized carbon gases (CO_2, CO, COS), the

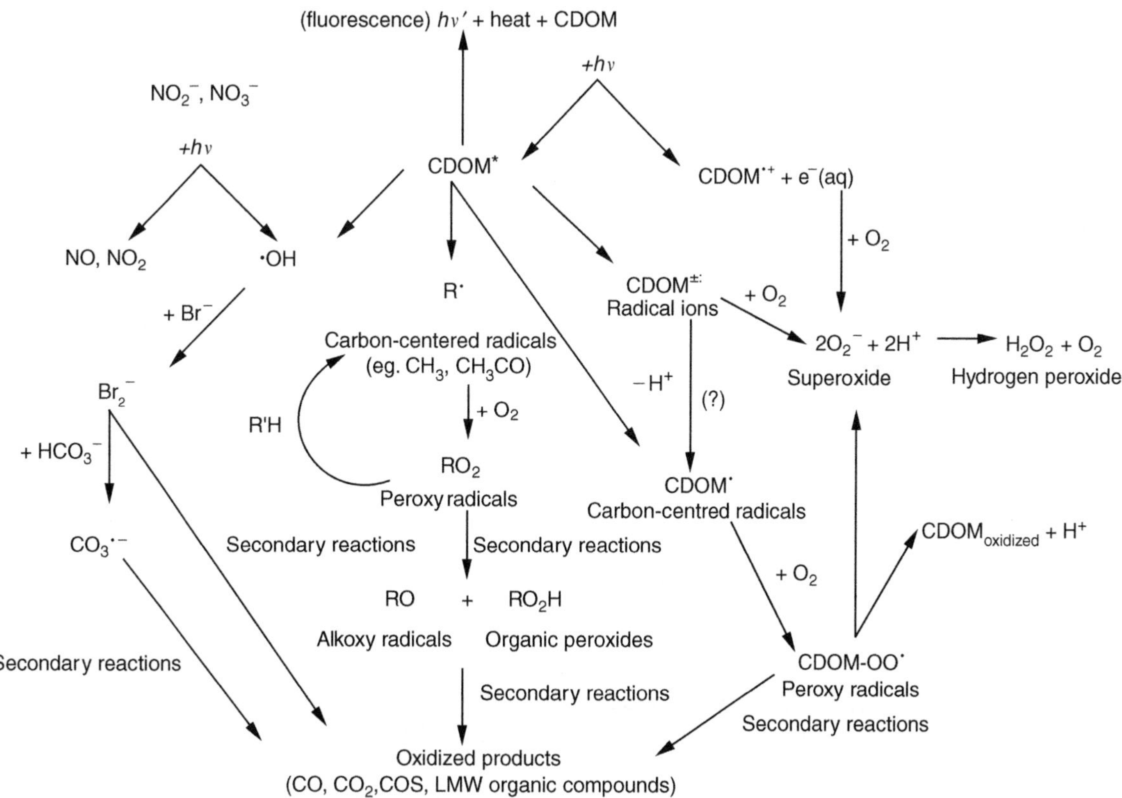

Figure 2 Schematic representation of the photochemical and secondary reactions known or thought to occur following light absorption by CDOM. For a more detailed description of these reactions see the text, Blough and Zepp (1995), and Blough (1997). Not shown in this diagram are primary and secondary reactions of metal species; for a description of these processes, see Helz *et al.* (1994) and Blough and Zepp (1995).

formation of a variety of LMW organic compounds, the release of biologically available forms of nitrogen, and the loss of CDOM absorption. Through direct photochemical reactions and reactions with the ROS, the speciation of metal ions is also affected.

These photochemical intermediates and products are produced at relatively low efficiencies. About 98–99% of the photons absorbed by CDOM are released as heat, while another ~1% are re-emitted as fluorescence. These percentages (or fractions) of absorbed photons giving rise to particular photoresponses are known as quantum yields (Φ). The Φ for the production of H_2O_2 and O_2^- (the two reduced oxygen species produced with highest efficiency), are approximately one to two orders of magnitude smaller than those for fluorescence, ranging from ~0.1% at 300 nm to ~0.01% at 400 nm. The Φ for other intermediates and products range even lower, from ~0.01% to 0.000 0001% (see below). The Φ for most of the intermediates and products created from the CDOM are highest in the UV-B and UV-A, and fall off rapidly with increasing wavelength; yields at visible wavelengths are usually negligible (see for example, **Figure 3**).

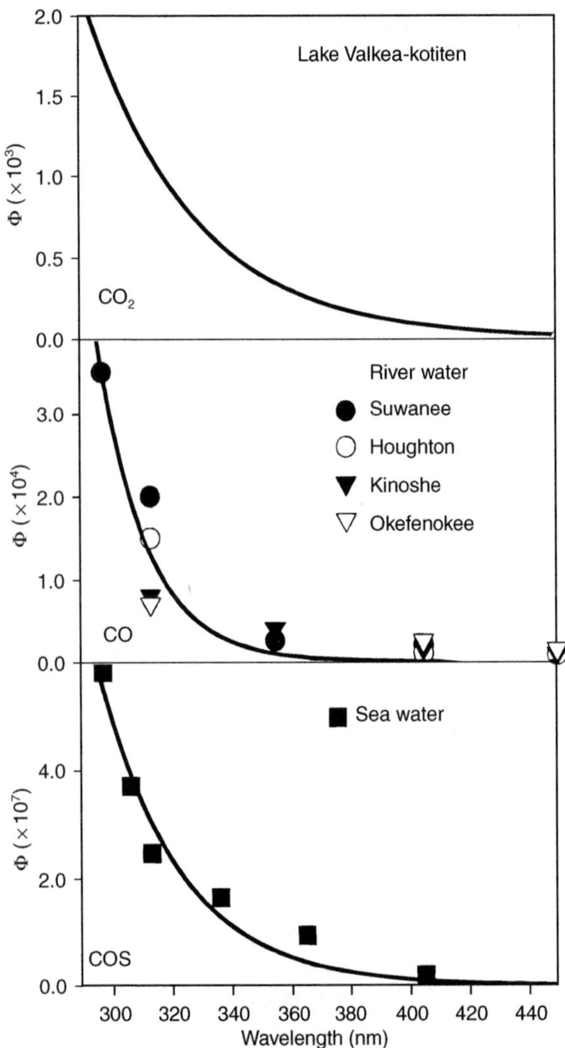

Figure 3 Wavelength dependence of the quantum yields (Φ) for the photochemical production of CO_2, CO, and COS. Data have been replotted from those dependencies originally reported in Vähätalo *et al.* (2000), Valentine and Zepp (1993), and Weiss *et al.* (1995) for CO_2, CO, and COS, respectively.

The hydroxyl radical, a very powerful oxidant, can be produced by the direct photolysis of nitrate and nitrite (eqns [I]–[III]).

$$NO_3^- + h\nu \rightarrow O^- + NO_2 \qquad [I]$$

$$NO_2^- + h\nu \rightarrow O^- + NO \qquad [II]$$

$$O^- + H_2O \rightarrow OH + OH^- \qquad [III]$$

The Φ values for these reactions are relatively high, about 7% for nitrite and about 1–2% for nitrate. However, because of the relatively low concentrations of these compounds in most marine surface waters, as well as their low molar absorptivities in the ultraviolet, the fraction of light absorbed is generally small and thus fluxes of OH from these sources also tend to be small. Recent evidence suggests that OH, or a species exhibiting very similar reactivity, is produced through a direct photoreaction of the CDOM; quinoid moieties within the CDOM may be responsible for this production. Quantum yields are low, ~0.01%, and restricted primarily to the ultraviolet. In estuarine and near-shore waters containing higher levels of iron, the production of OH may also occur through the direct photolysis of iron–hydroxy complexes or through the Fenton reaction (eqn [IV]).

$$Fe^{2+} + H_2O_2 \rightarrow Fe^{3+} + OH + OH^- \qquad [IV]$$

Compounds that do not absorb light within the surface solar spectrum are also subject to photochemical modification through indirect or 'sensitized' photoreactions. In this case, the ROS or intermediates produced by direct photoreactions of a light-absorbing constituent such as CDOM can react secondarily with the nonabsorbing compounds. DMS and COS, two trace gases of some importance to the atmosphere, are thought to be destroyed and created, respectively, by sensitized photoreactions in marine surface waters.

Photochemical Production and Consumption of LMW Organic Compounds and Trace Gases

The photolysis of CDOM produces a suite of LMW organic compounds and a number of trace gases. The production of these species presumably occurs through radical and fragmentation reactions arising from the net oxidative flow of electrons from CDOM to O_2 (see above), although the exact mechanism(s) have yet to be established. Most LMW organic compounds produced contain three or fewer carbon atoms and include such species as acetaldehyde, acetate, acetone, formaldehyde, formate, glyoxal, glyoxalate, methylglyoxal, propanal, and pyruvate. The Φ values for the production of individual compounds are low, ~0.001–0.0001%, with wavelengths in the UV-B the most effective; efficiencies decrease rapidly with increasing wavelength. Available evidence indicates that the Φ for O_2^- and H_2O_2 production are about one to two orders of magnitude larger than those for the LMW organic compounds, so it appears that the sum of the production rates for the known LMW organic compounds is small with respect to the flux of photochemical equivalents from CDOM to O_2.

Most, if not all, of these products are rapidly taken up and respired by bacteria to CO_2. Numerous investigators have presented evidence supporting enhanced microbial activity in waters exposed to sunlight, with bacterial activities increasing from 1.5- to almost 6-fold depending presumably on the length and type of light exposure, and the concentration and source of the CDOM. Recently, biologically labile nitrogen-containing compounds such as ammonia and amino acids have also been reported to be produced photochemically from CDOM. Because CDOM is normally considered to be biologically refractive, this recent work highlights the important role that abiotic photochemistry plays in the degradation of CDOM, not only through direct photoreactions, but also through the formation of biologically available products that can be respired to CO_2 or used as nutrients by biota. A recent estimate suggests that the utilization of biologically labile photoproducts could account for as much as 21% of the bacterial production in some near-surface waters.

Carbon dioxide and carbon monoxide are major products of the direct photolysis of CDOM (**Figure 3**). Quantum yields for CO production are about an order of magnitude smaller than those for O_2^- and H_2O_2 production, ranging from ~0.01% at 300 nm to ~0.001% at 400 nm. Available data indicate that the Φ for CO_2 range even higher, perhaps as much as 15–20-fold. The yields for CO_2 production must thus approach, if not exceed, those for O_2^- and H_2O_2. This result is somewhat surprising, since it implies that about one CO_2 is produced for each electron transferred from the CDOM to O_2, further implying a high average redox state for CDOM. Although CDOM (i.e. humic substances) is known to contain significant numbers of carboxyl moieties that could serve as the source of the CO_2, the yield for CO_2 production, relative to O_2^- and H_2O_2, would be expected to fall rapidly as these groups were removed photochemically; available evidence suggests that this does not occur. An alternative explanation is that other species, perhaps the CDOM itself, is acting as an electron acceptor. Regardless of mechanism, existing information indicates that CO_2 is the dominant product of CDOM photolysis (**Figure 3**).

A recent estimate suggests that the annual global photoproduction of CO in the oceans could be as high as 0.82×10^{15} g C. Assuming that CO_2 photoproduction is 15–20 times higher than that for CO, values for CO_2 formation could reach from 12 to 16×10^{15} g C y^{-1}. To place these numbers in perspective, the estimated annual input of terrestrial dissolved organic carbon to the oceans (0.2×10^{15} g C y^{-1}) is only 1.3–1.7% of the calculated annual CO_2 photoproduction, which is itself about 2–3% of the oceanic dissolved organic carbon pool. These calculated CO_2 (and CO) photoproduction rates may be high due to a number of assumptions, including (1) the complete absorption of UV radiation by the CDOM throughout the oceans, (2) constant quantum yields (or action spectra) for production independent of locale or light history, and (3) neglecting mass transfer limitations associated with physical mixing. Nevertheless, these estimates clearly highlight the potential impact of abiotic photochemistry on the oceanic carbon cycle. Moreover, the products of this photochemistry are generated in near-surface waters where exchange with the atmosphere can take place readily.

Like the LMW organic compounds, bacteria can oxidize CO to CO_2; this consumption takes place in competition with the release of CO to the atmosphere. Due to its photochemical production, CO exists at supersaturated concentrations in the surface waters of most of the Earth's oceans. Recent estimates indicate that global oceanic CO emissions could range from 0.013×10^{15} g y^{-1}–1.2×10^{15} g y^{-1} (see above). The upper estimate is based on calculated photochemical fluxes (see below) and the assumption that all CO produced is emitted to the atmosphere. The lower estimate was calculated using air–sea gas exchange equations and extensive measurements of CO concentrations in the surface waters and atmosphere of the Pacific Ocean. The source of the significant discrepancy between these two estimates has yet to be resolved. Depending on the answer, CO emitted to the atmosphere from the oceans could play a significant role in controlling OH levels in the marine troposphere.

Carbonyl sulfide (COS) is produced primarily in coastal/shelf waters, apparently by the CDOM-photosensitized oxidation of organosulfur compounds. UV-B light is the most effective in its formation, with Φ decreasing rapidly from $\sim 6 \times 10^{-7}$ at 300 nm to $\sim 1 \times 10^{-8}$ by 400 nm (**Figure 3**). The principal sinks of seawater COS are release to the atmosphere and hydrolysis to CO_2 and H_2S. Accounting for perhaps as much as one-third of the total source strength, the photochemical production of COS in the oceans is probably the single largest source of COS to the atmosphere, although more recent work has revised this estimate downward. Smaller amounts of carbon disulfide (CS_2) are also generated photochemically in surface waters through CDOM sensitized reaction(s); Φ values decrease from $\sim 1 \times 10^{-7}$ at 313 nm to 5×10^{-9} at 366 nm. The CS_2 emitted to the atmosphere can react with OH to form additional COS in the troposphere. Although it was previously thought that the oxidation of COS in the stratosphere to form sulfate aerosol could be important in determining Earth's radiation budget and perhaps in regulating stratospheric ozone concentrations, more recent work suggests that other sources contribute more significantly to the background sulfate in the stratosphere.

Dimethyl sulfide (DMS), through its oxidation to sulfate in the troposphere, acts as a source of cloud condensation nuclei, thus potentially influencing the radiative balance of the atmosphere. DMS is formed in sea water through the microbial decomposition of dimethyl sulfonioproprionate (DMSP), a compound believed to act as an osmolyte in certain species of marine phytoplankton. The flux of DMS to the atmosphere is controlled by its concentration in surface sea waters, which is controlled in turn by the rate of its decomposition. Estimates indicate that 7–40% of the total turnover of DMS in the surface waters of the Pacific Ocean is due to the photosensitized destruction of this compound, illustrating the potential importance of this pathway in controlling the flux of DMS to the atmosphere.

In addition to these compounds, the photochemical production of small amounts of nonmethane hydrocarbons (NMHC) such as ethene, propene, ethane, and propane has also been reported. Production of these compounds appears to result from the photolysis of the CDOM, with Φ values of the order of 10^{-7}–10^{-9}. The overall emission rates of these compounds to the atmosphere via this source are negligible with respect to global volatile organic carbon emissions, although this production may play some role in certain restricted locales exhibiting stronger source strengths, or in the marine environment remote from the dominant terrestrial sources.

The photolysis of nitrate and nitrite in sea water produces nitrogen dioxide (NO_2) and nitric oxide (NO), respectively (eqns [I] and [II]). Previous work indicated that the photolysis of nitrite could act as a small net source of NO to the marine atmosphere under some conditions. However, this conclusion seems to be at odds with estimates of the steady-state concentrations of superoxide and the now known rate constant for the reaction of superoxide with nitric oxide ($6.7 \times 10^9\,M^{-1}\,s^{-1}$) to form peroxynitrite in aqueous phases (eqn [V]).

$$O_2^- + NO \rightarrow {}^-OONO \qquad [V]$$

The peroxynitrite subsequently rearranges in part to form nitrate (eqn [VI]).

$${}^-OONO \rightarrow NO_3{}^- \qquad [VI]$$

Even assuming a steady-state concentration of O_2^- (10^{-12} M) that is about two orders of magnitude lower than that expected for surface sea waters ($\sim 10^{-10}$ M), the lifetime of NO in surface sea waters would be only ~ 150 s, a timescale too short for significant exchange with the atmosphere except for a thin surface layer. Moreover, even in this situation, the atmospheric deposition of additional HO_2 radicals to this surface layer (to form O_2^-) would be expected to act as an additional sink of the NO (flux capping). It appears that most if not all water bodies exhibiting significant steady-state levels of O_2^-, produced either photochemically or thermally, should act as a net sink of atmospheric NO and probably of NO_2 as well. Further, although less is known about the steady-state levels of peroxy radicals in sea waters due largely to their unknown decomposition routes, their high rate constants for reaction with NO (1–$3 \times 10^9\,M^{-1}\,s^{-1}$) indicate that they should also act as a sink of NO. In fact, methyl nitrate, a trace species found in sea waters, may in part be produced through the aqueous phase reactions (eqns [VII] and [VIII]) with the methylperoxy radical (CH_3OO) generated through a known photochemical reaction of CDOM (or through atmospheric deposition) and the NO arising from the photolysis of nitrite (or through atmospheric deposition).

$$CH_3OO + NO \rightarrow CH_3OONO \qquad [VII]$$

$$CH_3OONO \rightarrow CH_3ONO_2 \qquad [VIII]$$

The concentrations of NO and NO_2 in the troposphere are important because of the involvement of these gases in the formation of ozone.

The atmospheric deposition of ozone to the sea surface can cause the release of volatile iodine compounds to the atmosphere. There is also evidence that methyl iodide can be produced (as well as destroyed) by photochemical processes in surface sea waters. The release of these volatile iodine species from the sea surface or from atmospheric aqueous phases (aerosols) by these processes may act as a control on the level of ozone in the marine troposphere via iodine-catalyzed ozone destruction.

Trace Metal Photochemistry

A lack of available iron is now thought to limit primary productivity in certain ocean waters containing high nutrient, but low chlorophyll concentrations (the HNLC regions). This idea has spurred interest in the transport and photochemical reactions of iron in both seawaters and atmospheric aerosols. Very little soluble Fe(II) is expected to be available at the pH and dioxygen concentration of surface seawaters due to the high stability of the colloidal iron (hydr)oxides. The photoreductive dissolution of colloidal iron oxides by CDOM is known to occur at low pH; this process is also thought to occur in seawaters at high pH, but the reduced iron appears to be oxidized more rapidly than its detachment from the oxide surface. However, some workers have found that CDOM-driven cycles of reduction followed by oxidation increases the chemical availability, which was strongly correlated with the growth rate of phytoplankton. Significant levels of Fe(II) are also known to be produced photochemically in atmospheric aqueous phases (at lower pH) and could serve as a source of biologically available iron upon deposition to the sea surface.

Manganese oxides are also subject to reductive dissolution by light in surface seawaters. This process produces Mn(II), which is kinetically stable to oxidation in the absence of bacteria that are subject to photoinhibition. These two effects lead to the formation of a surface maximum in soluble Mn(II), in contrast to most metals which are depleted in surface waters due to biological removal processes. Other examples of the impact of photochemical reactions on trace metal chemistry are provided in Further Reading.

Photochemical Calculations

Global and regional estimates for the direct photochemical production (or consumption) of a particular photoproduct (or photoreactant) can be acquired with knowledge of the temporal and spatial variation of the solar irradiance reaching the Earth's surface combined with a simple photochemical model (eqn [3]).

$$F(\lambda, z) = E_D(\lambda, z) \cdot \Phi_i(\lambda) \cdot a_{Di}(\lambda) \qquad [3]$$

Here $F(\lambda, z)$ is the photochemical production (or consumption) rate; $E_D(\lambda, z)$ is the downwelling irradiance at wavelength, λ, and depth, z, within the water column; a_{Di} is the diffuse absorption coefficient for photoreactive constituent i; $\Phi_i(\lambda)$ is the quantum yield of this ith constituent. $E_D(\lambda, z)$ is well approximated by eqn [4].

$$E_D(\lambda, z) = E_{D0}(\lambda) \cdot e^{-K_d(\lambda) \cdot z} \qquad [4]$$

$E_{D0}(\lambda)$ is the downwelling irradiance just below the sea surface and $K_d(\lambda)$ is the vertical diffuse attenuation coefficient of downwelling irradiance. $K_d(\lambda)$ can be approximated by eqn [5].

$$K_d(\lambda) \approx \frac{\sum a_i(\lambda) + \sum b_{bi}(\lambda)}{\mu_D} \qquad [5]$$

where $\sum a_i(\lambda)$ and $\sum b_{bi}(\lambda)$ are the total absorption and backscattering coefficients, respectively, of all absorbing and scattering constituents within the water column, and μ_D is the average cosine of the angular distribution of the downwelling light. This factor accounts for the average pathlength of light in the water column, and for direct solar light is approximately equal to cos θ, where θ is the solar zenith angle (e.g. $\mu_D \sim 1$ when the sun is directly overhead). The diffuse absorption coefficient, a_{Di}, is given by eqn [6].

$$a_{Di} = \frac{a_i}{\mu_D} \qquad [6]$$

This model assumes that the water column is homogeneous, that $K_d(\lambda)$ is constant with depth, and that upwelling irradiance is negligible relative to $E_D(\lambda, z)$. Combining eqns [3], [4] and [6] gives eqn [7].

$$F(\lambda, z) = \frac{E_{D0}(\lambda).e^{-K_d(\lambda) \cdot z} \cdot \Phi_i(\lambda) \cdot a_i(\lambda)}{\mu_D} \qquad [7]$$

This equation allows calculation of the spectral dependence of the production (consumption) rate as a function of depth in the water column, assuming knowledge of $E_{D0}(\lambda)$, $K_d(\lambda)$, $a_i(\lambda)$ and $\Phi_i(\lambda)$, all of which can be measured or estimated (**Figure 4**).

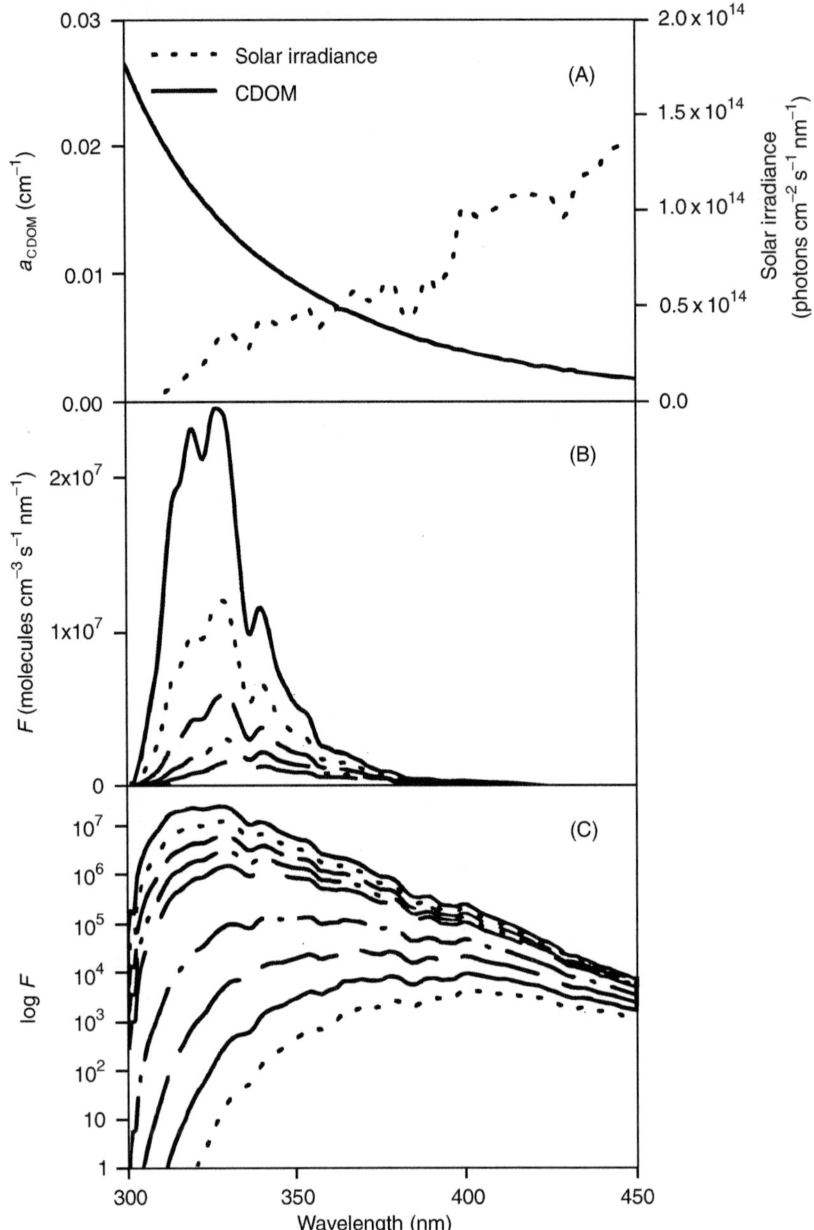

Figure 4 Spectral dependence of CO photoproduction rates with depth, plotted on a linear (B) and logarithmic (C) scale. Depths in (B) are (from top to bottom): surface, 0.5, 1, 1.5, and 2 m. Depths in (C) are (from top to bottom): surface, 0.5, 1, 1.5, 2, 4, 6, 8, and 10 m. These spectral dependencies were calculated using eqn [7], the wavelength dependence of the quantum yield for CO shown in **Figure 3**, and the CDOM absorption spectrum and surface solar irradiance shown in (A). The attenuation of irradiance down the water column in this spectral region was assumed to be only due to CDOM absorption, a reasonable assumption for coastal waters (see **Figure 1**). Note the rapid attenuation in production rates with depth in the UV-B, due to the greater light absorption by CDOM in this spectral region.

Integrating over wavelength provides the total production (consumption) rate at each depth.

Integration of eqn [7] from the surface to depth z provides the spectral dependence of the photochemical flux (Y) over this interval

$$Y(\lambda, z) = \frac{E_{D0}(\lambda).(1 - e^{-K_d(\lambda)\cdot z}) \cdot \Phi_i(\lambda) \cdot a_i(\lambda)/K_d(\lambda)}{\mu_D} \quad [8]$$

which upon substitution of eqn [4] becomes,

$$Y(\lambda, z) = E_{D0}(\lambda).\left(1 - e^{-K_d(\lambda)\cdot z}\right) \cdot \Phi_i(\lambda) \cdot \frac{a_i(\lambda)}{\sum a_i(\lambda) + \sum b_{bi}(\lambda)} \quad [9]$$

In most, but not all seawaters, the total absorption will be much greater than the total backscatter,

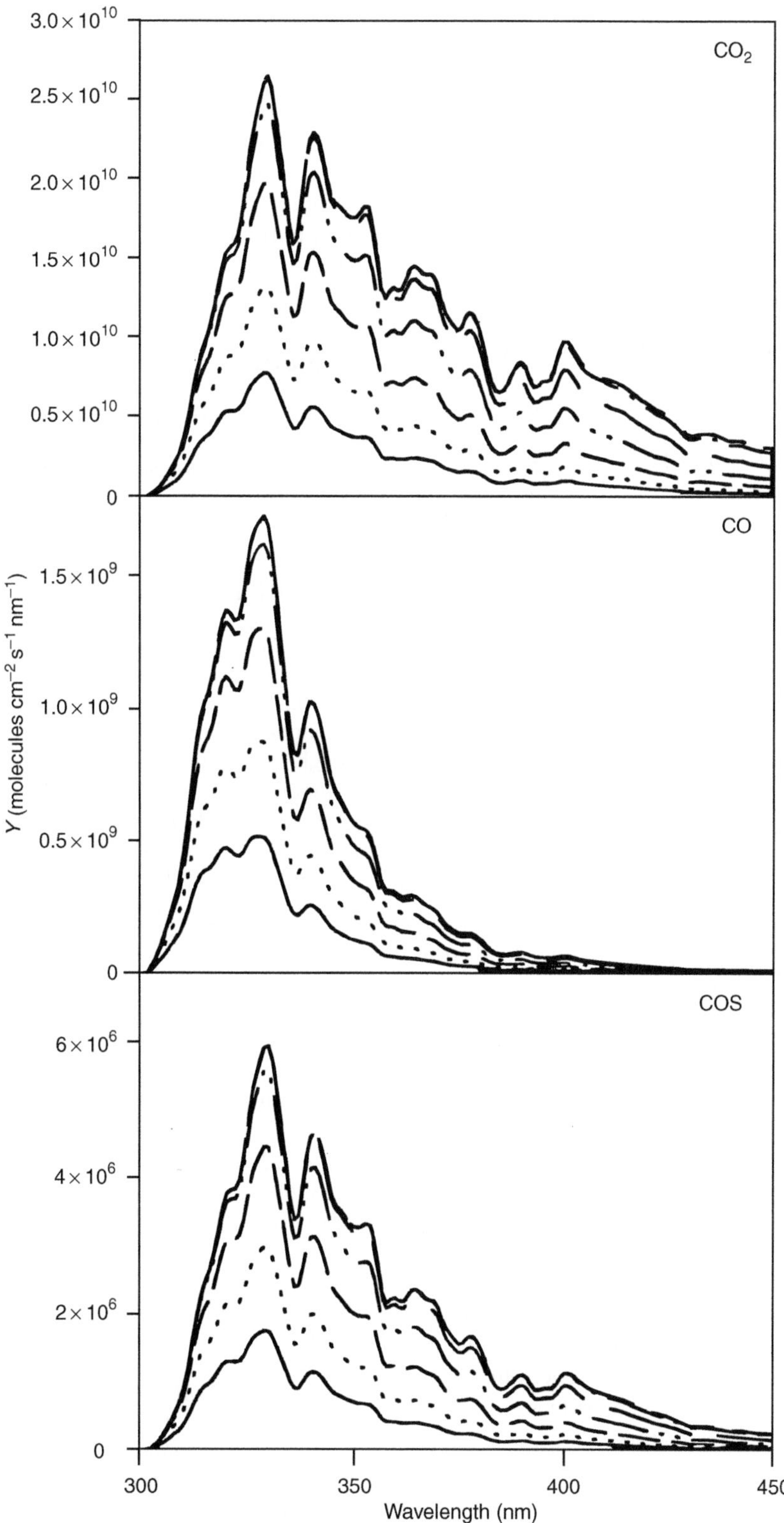

Figure 5 Spectral dependence of the photochemical flux with depth for CO_2, CO, and COS. Fluxes with depth are from the surface to 0.25, 0.5, 1.0, 2.0, and 4 m, respectively (bottom spectrum to top spectrum). Below 4 m, increases in the flux are nominal. These spectral dependencies were calculated using eqn [10], the wavelength dependence of the quantum yields for CO_2, CO and COS shown in **Figure 3**, and the surface solar irradiance shown in **Figure 4A**. CDOM is assumed to absorb all photons in this spectral region (see **Figures 1** and **4**).

$\sum a_i(\lambda) \gg \sum b_{bi}(\lambda)$, and thus the backscatter can be ignored; this approximation is not valid for most estuarine waters and some coastal waters, where a more sophisticated treatment would have to be applied. This approximation leads to the final expression for the variation of the spectral dependence of the flux with depth (**Figure 5**),

$$F(\lambda, z) = E_{D0}(\lambda).\left(1 - e^{-K_d(\lambda)\cdot z}\right) \cdot \Phi_i(\lambda) \cdot \frac{a_i(\lambda)}{\sum a_i(\lambda)} \quad [10]$$

The spectral dependence of the total water column flux ($z \to \infty$) is then given by,

$$F(\lambda) = E_{D0}(\lambda) \cdot \Phi_i(\lambda) \cdot \frac{a_i(\lambda)}{\sum a_i(\lambda)} \quad [11]$$

with the total flux obtained by integrating over wavelength,

$$F \int_{\lambda} E_{D0}(\lambda).\Phi_i(\lambda) \cdot \frac{a_i(\lambda)}{\sum a_i(\lambda)} d\lambda \quad [12]$$

To obtain global estimates of photochemical fluxes, many investigators assume that the absorption due to CDOM, a_{CDOM}, dominates the absorption of all other seawater constituents in the ultraviolet, and thus that $a_{CDOM}(\lambda)/\sum a_i(\lambda) \approx 1$.While this approximation is reasonable for many coastal waters, it is not clear that this approximation is valid for all oligotrophic waters. This approximation leads to the final expression for flux,

$$\mathrm{Y} \int_{\lambda} \mathrm{E}_{D0}(\lambda).\Phi_i(\lambda) d\lambda \quad [13]$$

which relies only on the surface downwelling irradiance and the wavelength dependence of the quantum yield for the photoreaction of interest. Uncertainties in the use of this equation for estimating global photochemical fluxes include (1) the (usual) assumption that $\Phi(\lambda)$ acquired for a limited number of samples is representative of all ocean waters, independent of locale or light history, and (2) differences in the spatially and temporally averaged values of $E_{D0}(\lambda)$ utilized by different investigators.

Conclusions

The absorption of solar radiation by abiotic sea water constituents initiates a cascade of reactions leading to the photo-oxidative degradation of organic matter and the concomitant production (or consumption) of a variety of trace gases and LMW organic compounds (**Figure 2**), as well as affecting trace metal speciation. The magnitude and impact of these processes on upper ocean biogeochemical cycles and their coupling with atmospheric processes are just beginning to be fully quantified and understood. There remains the need to examine possible couplings between atmospheric gas phase reactions and photochemical reactions in atmospheric aqueous phases.

See also

Air–Sea Transfer: Dimethyl Sulfide, COS, CS_2, NH_4, Non-Methane Hydrocarbons, Organo-Halogens. Air–Sea Transfer: N_2O, NO, CH_4, CO.

Further Reading

Blough NV (1997) Photochemistry in the sea-surface microlayer. In: Liss PS and Duce R (eds.) *The Sea Surface and Global Change*, pp. 383–424. Cambridge: Cambrige University Press.

Blough NV and Green SA (1995) Spectroscopic characterization and remote sensing of non-living organic matter. In: Zepp RG and Sonntag C (eds.) *The role of Non-living Organic Matter in the Earth's Carbon Cycle*, pp. 23–45. New York: John Wiley.

Blough NV and Zepp RG (1995) Reactive oxygen species in natural waters. In: Foote CS, Valentine JS, Greenberg A, and Liebman JF (eds.) *Reactive Oxygen Species in Chemistry*, pp. 280–333. New York: Chapman & Hall.

de Mora S, Demers S, and Vernet M (eds.) (2000) *The Effects of UV Radiation in the Marine Environment.* Cambridge: Cambridge University Press.

Häder D-P, Kumar HD, Smith RC, and Worrest RC (1998) Effects of UV-B radiation on aquatic ecosystems. *Journal of Photochemistry and Photobiology B* 46: 53–68.

Helz GR, Zepp RG, and Crosby DG (eds.) (1994) *Aquatic and Surface Photochemistry.* Ann Arbor, MI: Lewis Publishers.

Huie RE (1995) Free radical chemistry of the atmospheric aqueous phase. In: Barker JR (ed.) *Progress and Problems in Atmospheric Chemistry*, pp. 374–419. Singapore: World Scientific Publishing Co.

Kirk JTO (1994) *Light and Photosynthesis in Aquatic Ecosystems*. Cambridge: Cambridge University Press.

Moran MA and Zepp RG (1997) Role of photoreactions in the formation of biologically labile compounds from dissolved organic matter. *Limnology and Oceanography* 42: 1307–1316.

Thompson AM and Zafiriou OC (1983) Air–sea fluxes of transient atmospheric species. *Journal of Geophysical Research* 88: 6696–6708.

Vähätalo AV, Salkinoja-Salonen M, Taalas P, and Salonen K (2000) Spectrum of the quantum yield for photochemical mineralization of dissolved organic carbon in a humic lake. *Limnology and Oceanography* 45: 664–676.

Valentine RL and Zepp RG (1993) Formation of carbon monoxide from the photodegradation of terrestrial dissolved organic carbon in natural waters. *Environmental Science Technology* 27: 409–412.

Weiss EW, Andrews SS, Johnson JE, and Zafiriou OC (1995) Photoproduction of carbonyl sulfide in south Pacific Ocean waters as a function of irradiation wavelength. *Geophysical Research Letters* 22: 215–218.

Zafiriou OC, Blough NV, Micinski E, *et al.* (1990) Molecular probe systems for reactive transients in natural waters. *Marine Chemistry* 30: 45–70.

Zepp RG, Callaghan TV, and Erickson DJ (1998) Effects of enhanced solar ultraviolet radiation on biogeochemical cycles. *Journal of Photochemistry and Photobiology B* 46: 69–82.

IRON FERTILIZATION

K. H. Coale, Moss Landing Marine Laboratories, CA, USA

Introduction

The trace element iron has been shown to play a critical role in nutrient utilization and phytoplankton growth and therefore in the uptake of carbon dioxide from the surface waters of the global ocean. Carbon fixation in the surface waters, via phytoplankton growth, shifts the ocean–atmosphere exchange equilibrium for carbon dioxide. As a result, levels of atmospheric carbon dioxide (a greenhouse gas) and iron flux to the oceans have been linked to climate change (glacial to interglacial transitions). These recent findings have led some to suggest that large-scale iron fertilization of the world's oceans might therefore be a feasible strategy for controlling climate. Others speculate that such a strategy could deleteriously alter the ocean ecosystem, and still others have calculated that such a strategy would be ineffective in removing sufficient carbon dioxide to produce a sizable and rapid result. This article focuses on carbon and the major plant nutrients, nitrate, phosphate, and silicate, and describes how our recent discovery of the role of iron in the oceans has increased our understanding of phytoplankton growth, nutrient cycling, and the flux of carbon from the atmosphere to the deep sea.

Major Nutrients

Phytoplankton growth in the oceans requires many physical, chemical, and biological factors that are distributed inhomogenously in space and time. Because carbon, primarily in the form of the bicarbonate ion, and sulfur, as sulfate, are abundant throughout the water column, the major plant nutrients in the ocean commonly thought to be critical for phytoplankton growth are those that exist at the micromolar level such as nitrate, phosphate, and silicate. These, together with carbon and sulfur, form the major building blocks for biomass in the sea. As fundamental cellular constituents, they are generally thought to be taken up and remineralized in constant ratio to one another. This is known as the Redfield ratio (Redfield, 1934, 1958) and can be expressed on a molar basis relative to carbon as 106C : 16N : 1P.

Significant local variations in this uptake/regeneration relationship can be found and are a function of the phytoplankton community and growth conditions, yet this ratio can serve as a conceptual model for nutrient uptake and export.

The vertical distribution of the major nutrients typically shows surface water depletion and increasing concentrations with depth. The schematic profile in **Figure 1** reflects the processes of phytoplankton uptake within the euphotic zone and remineralization of sinking planktonic debris via microbial degradation, leading to increased concentrations in the deep sea. Given favorable growth conditions, the nutrients at the surface may be depleted to zero. The rate of phytoplankton production of new biomass, and therefore the rate of carbon uptake, is controlled by the resupply of nutrients to the surface waters, usually via the upwelling of deep waters. Upwelling occurs over the entire ocean basin at the rate of approximately 4 m per year but increases in coastal and

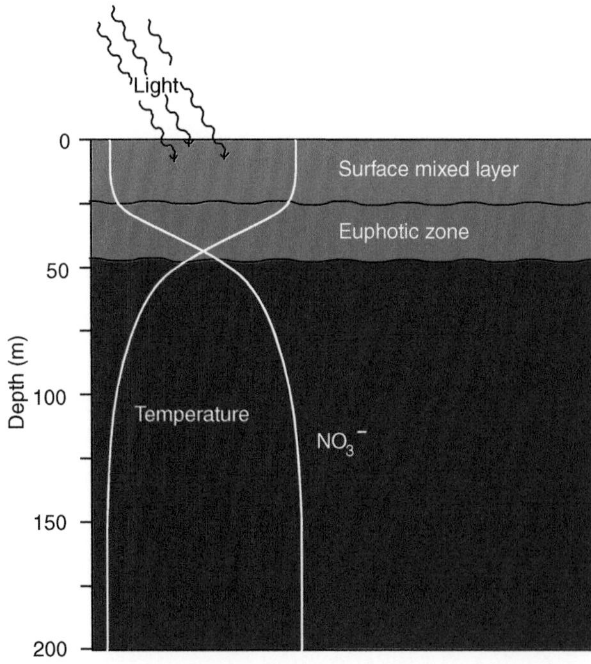

Figure 1 A schematic profile indicating the regions of the upper water column where phytoplankton grow. The surface mixed layer is that region that is actively mixed by wind and wave energy, which is typically depleted in major nutrients. Below this mixed layer temperatures decrease and nutrients increase as material sinking from the mixed layer is regenerated by microbial decomposition.

regions of divergent surface water flow, reaching average values of 15 to 30 or greater. Thus, those regions of high nutrient supply or persistent high nutrient concentrations are thought to be most important in terms of carbon removal.

Nitrogen versus Phosphorus Limitation

Although both nitrogen and phosphorus are required at nearly constant ratios characteristic of deep water, nitrogen has generally been thought to be the limiting nutrient in sea water rather than phosphorus. This idea has been based on two observations: selective enrichment experiments and surface water distributions. When ammonia and phosphate are added to sea water in grow-out experiments, phytoplankton growth increases with the ammonia addition and not with the phosphate addition, thus indicating that reduced nitrogen and not phosphorus is limiting. Also, when surface water concentration of nitrate and phosphate are plotted together (**Figure 2**), it appears that there is still residual phosphate after the nitrate has gone to zero.

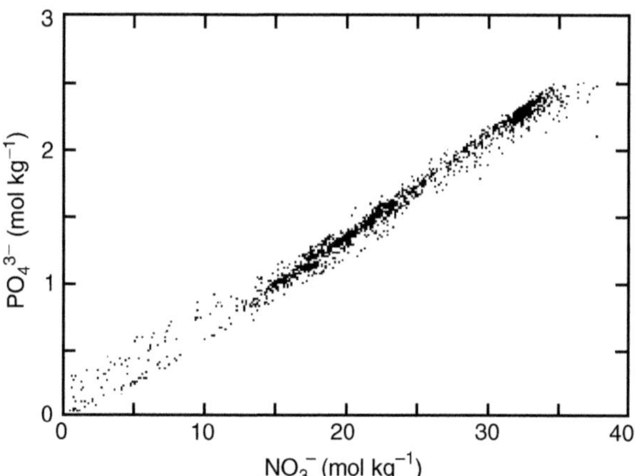

Figure 2 A plot of the global surface water concentrations of phosphate versus nitrate indicating a general positive intercept for phosphorus when nitrate has gone to zero. This is one of the imperical observations favoring the notion of nitrate limitation over phosphate limitation.

The notion of nitrogen limitation seems counterintuitive when one considers the abundant supply of dinitrogen (N_2) in the atmosphere. Yet this nitrogen gas is kinetically unavailable to most phytoplankton because of the large amount of energy required to break the triple bond that binds the dinitrogen molecule. Only those organisms capable of nitrogen fixation can take advantage of this form of nitrogen and reduce atmospheric N_2 to biologically available nitrogen in the form of urea and ammonia. This is, energetically, a very expensive process requiring specialized enzymes (nitrogenase), an anaerobic microenvironment, and large amounts of reducing power in the form of electrons generated by photosynthesis. Although there is currently the suggestion that nitrogen fixation may have been underestimated as an important geochemical process, the major mode of nitrogen assimilation, giving rise to new plant production in surface waters, is thought to be nitrate uptake.

The uptake of nitrate and subsequent conversion to reduced nitrogen in cells requires a change of five in the oxidation state and proceeds in a stepwise fashion. The initial reduction takes place via the nitrate/nitrite reductase enzyme present in phytoplankton and requires large amounts of the reduced nicotinamide–adenine dinucleotide phosphate (NADPH) and of adenosine triphosphate (ATP) and thus of harvested light energy from photosystem II. Both the nitrogenase enzyme and the nitrate reductase enzyme require iron as a cofactor and are thus sensitive to iron availability.

Ocean Regions

From a nutrient and biotic perspective, the oceans can be generally divided into biogeochemical provinces that reflect differences in the abundance of macronutrients and the standing stocks of phytoplankton. These are the high-nitrate, high-chlorophyll (HNHC); high-nitrate, low-chlorophyll (HNLC); low-nitrate, high-chlorophyll (LNHC); and low-nitrate, low-chlorophyll (LNLC) regimes (**Table 1**). Only the HNLC and LNLC regimes are relatively stable, because the high phytoplankton

Table 1 The relationship between biomass and nitrate as a function of biogeochemical province and the approximate ocean area represented by these regimes

	High-chlorophyll	*Low-chlorophyll*
High-nitrate	Unstable/coastal (5%)	Stable/Subarctic/Antarctic/ equatorial Pacific (20%)
Low-nitrate	Unstable/coastal (5%)	Oligotrophic gyres (70%)

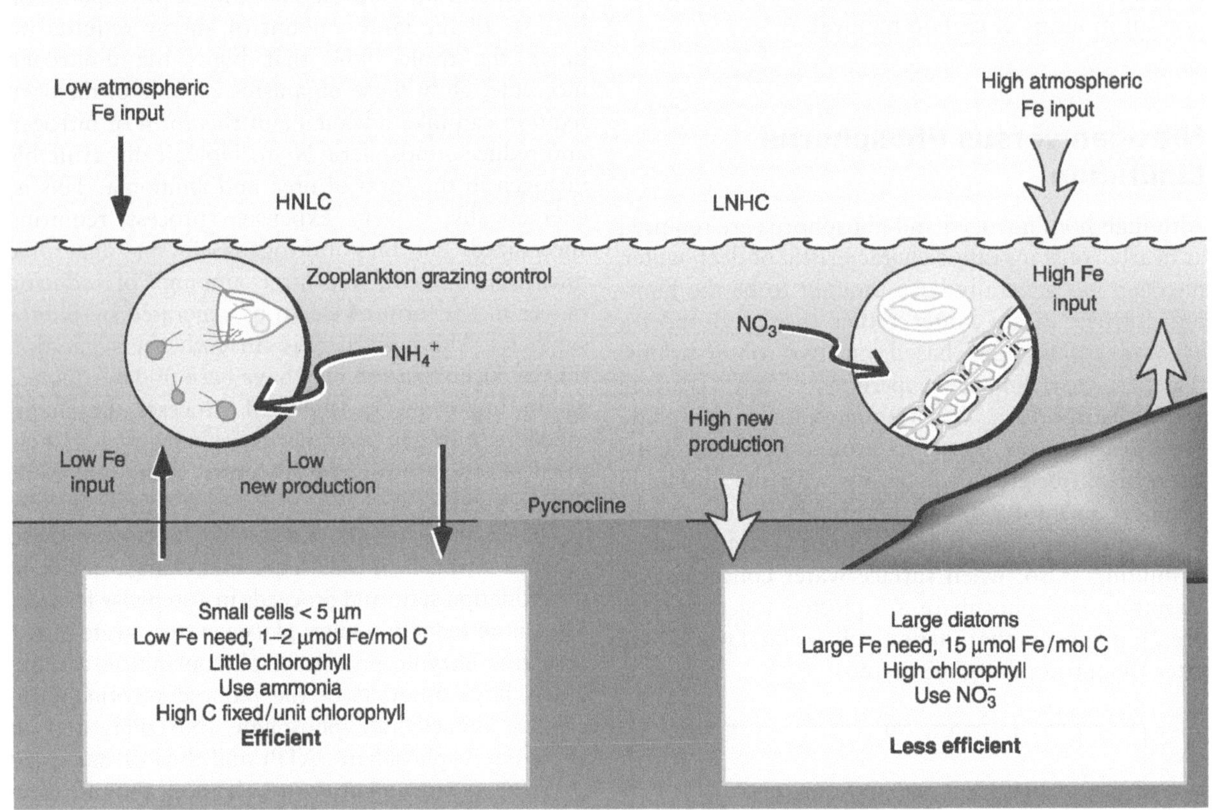

Figure 3 A schematic representation of the 'iron theory' as it functions in offshore HNLC regions and coastal transient LNHC regions. It has been suggested that iron added to the HNLC regions would induce them to function as LNHC regions and promote carbon export.

growth rates in the other two systems will deplete any residual nitrate and sink out of the system. The processes that give rise to these regimes have been the subject of some debate over the last few years and are of fundamental importance relative to carbon export (**Figure** 3).

High-nitrate, Low-chlorophyll Regions

The HNLC regions are thought to represent about 20% of the areal extent of the world's oceans. These are generally regions characterized by more than $2\,\mu\text{mol}\ l^{-1}$ nitrate and less than $0.5\,\mu\text{g}\ l^{-1}$ chlorophyll-*a*, a proxy for plant biomass. The major HNLC regions are shown in **Figure 4** and represent the Subarctic Pacific, large regions of the eastern equatorial Pacific and the Southern Ocean. These HNLC regions persist in areas that have high macronutrient concentrations, adequate light, and physical characteristics required for phytoplankton growth but have very low plant biomass. Two explanations have been given to describe the persistence of this condition. (1) The rates of zooplankton grazing of the phytoplankton community may balance or exceed phytoplankton growth rates in these areas, thus cropping plant biomass to very low levels and recycling reduced nitrogen from the plant community, thereby decreasing the uptake of nitrate. (2) Some other micronutrient (possibly iron) physiologically limits the rate of phytoplankton growth. These are known as top-down and bottom-up control, respectively.

Several studies of zooplankton grazing and phytoplankton growth in these HNLC regions, particularly the Subarctic Pacific, confirm the hypothesis that grazers control production in these waters. Recent physiological studies, however, indicate that phytoplankton growth rates in these regions are suboptimal, as is the efficiency with which phytoplankton harvest light energy. These observations indicate that phytoplankton growth may be limited by something other than (or in addition to) grazing. Specifically, these studies implicate the lack of

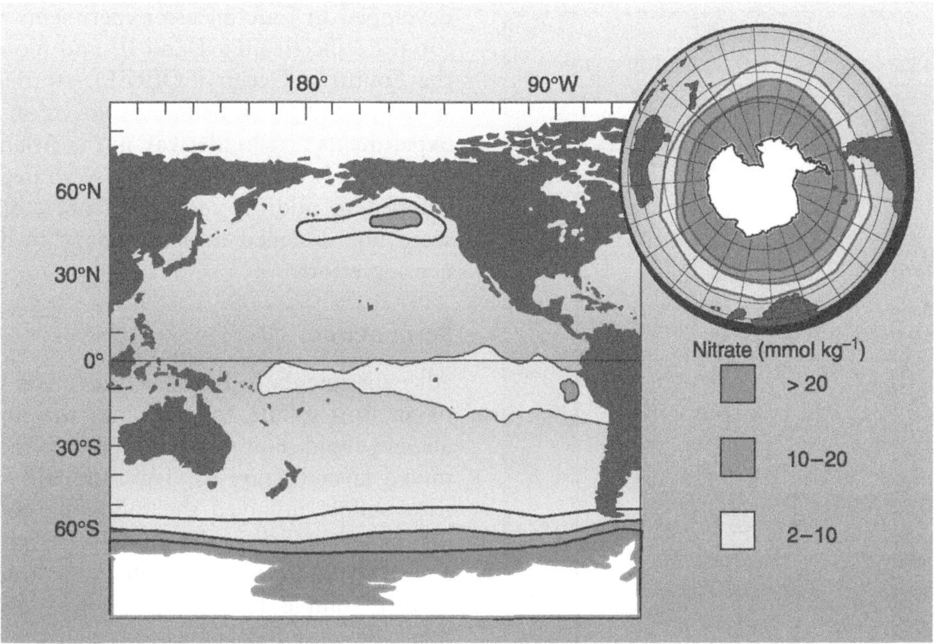

Figure 4 Current HNLC regions of the world's oceans covering an extimated 20% of the ocean surface. These regions include the Subarctic Pacific, equatorial Pacific and Southern Ocean.

sufficient electron transport proteins and the cell's ability to transfer reducing power from the photocenter. These have been shown to be symptomatic of iron deficiency.

The Role of Iron

Iron is a required micronutrient for all living systems. Because of its d-electron configuration, iron readily undergoes redox transitions between Fe(II) and Fe(III) at physiological redox potentials. For this reason, iron is particularly well suited to many enzyme and electron carrier proteins. The genetic sequences coding for many iron-containing electron carriers and enzymes are highly conserved, indicating iron and iron-containing proteins were key features of early biosynthesis. When life evolved, the atmosphere and waters of the planet were reducing and iron was abundant in the form of soluble Fe(II). Readily available and at high concentration, iron was not likely to have been limiting in the primordial biosphere. As photosynthesis evolved, oxygen was produced as a by-product. As the biosphere became more oxidizing, iron precipitated from aquatic systems in vast quantities, leaving phytoplankton and other aquatic life forms in a vastly changed and newly deficient chemical milieu. Evidence of this mass Fe(III) precipitation event is captured in the ancient banded iron formations in many parts of the world. Many primitive aquatic and terrestrial organisms have subsequently evolved the ability to sequester iron through the elaboration of specific Fe(II)-binding ligands, known as siderophores. Evidence for siderophore production has been found in several marine dinoflagellates and bacteria and some researchers have detected similar compounds in sea water.

Today, iron exists in sea water at vanishingly small concentrations. Owing to both inorganic precipitation and biological uptake, typical surface water values are on the order of 20 pmol l^{-1}, perhaps a billion times less than during the prehistoric past. Iron concentrations in the oceans increase with depth, in much the same manner as the major plant nutrients (**Figure 5**).

The discovery that iron concentrations in surface waters is so low and shows a nutrient-like profile led some to speculate that iron availability limits plant growth in the oceans. This notion has been tested in bottle enrichment experiments throughout the major HNLC regions of the world's oceans. These experiments have demonstrated dramatic phytoplankton growth and nutrient uptake upon the addition of iron relative to control experiments in which no iron was added.

Criticism that such small-scale, enclosed experiments may not accurately reflect the response of the HNLC system at the level of the community has led to several large-scale iron fertilization experiments in the equatorial Pacific and Southern Ocean. These have been some of the most dramatic

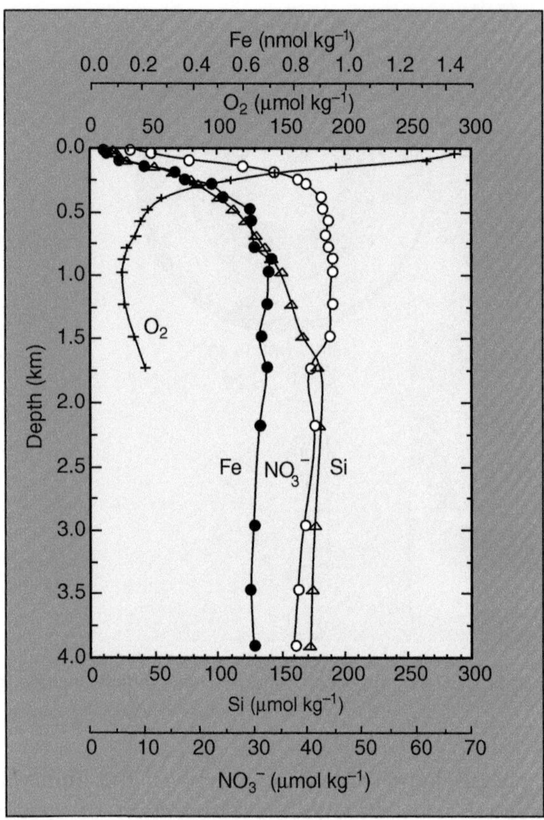

Figure 5 The vertical distributions of iron, nitrate, silicate, and oxygen in sea water. This figure shows how iron is depleted to picomolar levels in surface waters and has a profile that mimics other plant nutrients.

oceanographic experiments of our times and have led to a profound and new understanding of ocean systems.

Open Ocean Iron Enrichment

The question of iron limitation was brought into sharp scientific focus with a series of public lectures, reports by the US National Research Council, papers, special publications, and popular articles between 1988 and 1991. What was resolved was the need to perform an open ocean enrichment experiment in order to definitively test the hypothesis that iron limits phytoplankton growth and nutrient and carbon dioxide uptake in HNLC regions. Such an experiment posed severe logistical challenges and had never been conducted.

Experimental Strategy

The mechanics of producing an iron-enriched experimental patch and following it over time was developed in four release experiments in the equatorial Pacific (IronEx I and II) and more recently in the Southern Ocean (SOIREE). At this writing, a similar strategy is being employed in the Caruso experiments now underway in the Atlantic sector of the Southern Ocean. All of these strategies were developed to address certain scientific questions and were not designed as preliminary to any geoengineering effort.

Form of Iron

All experiments to date have involved the injection of an iron sulfate solution into the ship's wake to achieve rapid dilution and dispersion throughout the mixed layer (**Figure 6**). The rationale for using ferrous sulfate involved the following considerations: (1) ferrous sulfate is the most likely form of iron to enter the oceans via atmospheric deposition; (2) it is readily soluble (initially); (3) it is available in a relatively pure form so as to reduce the introduction of other potentially bioactive trace metals; and (4) its counterion (sulfate) is ubiquitous in sea water and not likely to produce confounding effects. Although mixing models indicate that Fe(II) carbonate may reach insoluble levels in the ship's wake, rapid dilution reduces this possibility.

New forms of iron are now being considered by those who would seek to reduce the need for subsequent infusions. Such forms could include iron lignosite, which would increase the solubility and residence time of iron in the surface waters. Since this is a chelated form of iron, problems of rapid precipitation are reduced. In addition, iron lignosulfonate is about 15% Fe by weight, making it a space-efficient form of iron to transport. As yet untested is the extent to which such a compound would reduce the need for re-infusion.

Although solid forms of iron have been proposed (slow-release iron pellets; finely milled magnetite or iron ores), the ability to trace the enriched area with an inert tracer has required that the form of iron added and the tracer both be in the dissolved form.

Inert Tracer

Concurrent with the injection of iron is the injection of the inert chemical tracer sulfur hexafluoride (SF_6). By presaturating a tank of sea water with SF_6 and employing an expandable displacement bladder, a constant molar injection ratio of Fe : SF_6 can be achieved (**Figure 6**). In this way, both conservative and nonconservative removal of iron can be quantified. Sulfur hexafluoride traces the physical properties of the enriched patch; the relatively rapid shipboard detection of SF_6 can be used to track and

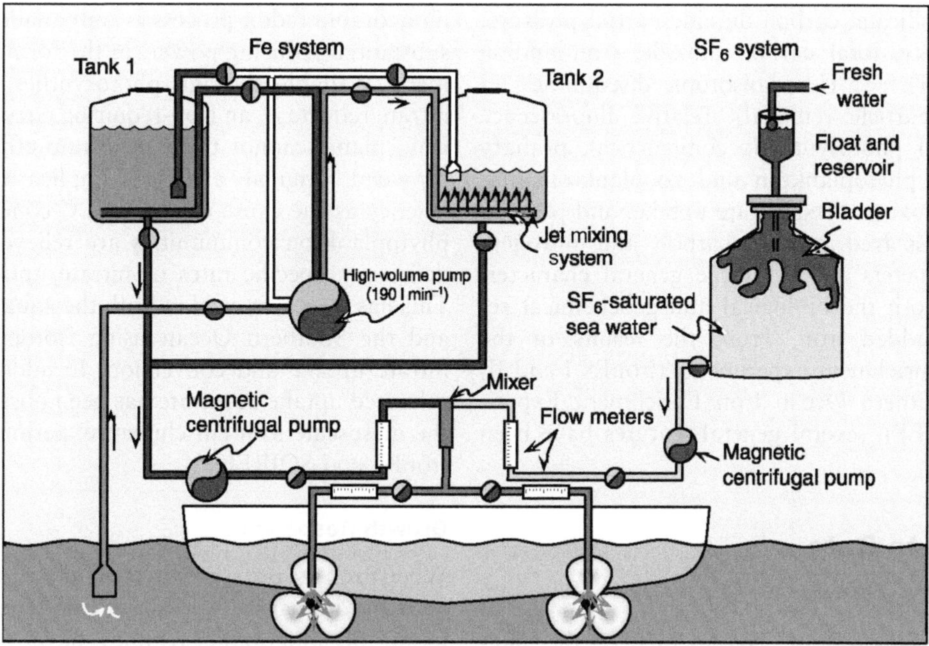

Figure 6 The iron injection system used during the IronEx experiments utilized two polyethylene tanks that could be sequentially filled with sea water and iron sulfate solution while the other was being injected behind the ship's propellers. A steel tank of sea water saturated with 40 g of sulfur hexafluoride (SF_6) was simultaneously mixed with the iron sulfate solution to provide a conservative tracer of mixing.

map the enriched area. The addition of helium-3 to the injected tracer can provide useful information regarding gas transfer.

Fluorometry

The biophysical response of the phytoplankton is rapid and readily detectable. Thus shipboard measurement of relative fluorescence (F_v/F_m) using fast repetition rate fluorometry has been shown to be a useful tactical tool and gives nearly instantaneous mapping and tracking feedback.

Shipboard Iron Analysis

Because iron is rapidly lost from the system (at least initially), the shipboard determination of iron is necessary to determine the timing and amount of subsequent infusions. Several shipboard methods, using both chemiluminescent and catalytic colorimetric detection have proven useful in this regard.

Lagrangian Drifters

A Lagrangian point of reference has proven to be very useful in every experiment to date. Depending upon the advective regime, this is the only practical way to achieve rapid and precise navigation and mapping about the enriched area.

Remote Sensing

A variety of airborne and satellite-borne active and passive optical packages provide rapid, large-scale mapping and tracking of the enriched area. Although SeaWiffs was not operational during IronEx I and II, AVHRR was able to detect the IronEx II bloom and airborne optical LIDAR was very useful during IronEx I. SOIREE has made very good use of the more recent SeaWiffs images, which have markedly extended the observational period and led to new hypotheses regarding iron cycling in polar systems.

Experimental Measurements

In addition to the tactical measurements and remote sensing techniques required to track and ascertain the development of the physical dynamics of the enriched patch, a number of measurements have been made to track the biogeochemical development of the experiment. These have typically involved a series of underway measurements made using the ship's flowing sea water system or towed fish. In addition, discrete measurements are made in the vertical dimension at every station occupied both inside and outside of the fertilized area. These measurements include temperature salinity, fluorescence (a measure of plant biomass), transmissivity (a measure of suspended particles), oxygen, nitrate, phosphate, silicate, carbon dioxide partial pressure, pH, alkalinity, total carbon dioxide, iron-binding ligands, ^{234}Th : ^{238}U radioisotopic disequilibria (a proxy for particle removal), relative fluorescence (indicator of photosynthetic competence), primary production, phytoplankton and zooplankton enumeration, grazing rates, nitrate uptake, and particulate and dissolved organic carbon and nitrogen. These parameters allow for the general characterization of both the biological and geochemical response to added iron. From the results of the equatorial enrichment experiments (IronEx I and II) and the Southern Ocean Iron Enrichment Experiment (SOIREE), several general features have been identified.

Findings to Date

Biophysical Response

The experiments to date have focused on the high-nitrate, low-chlorophyll (HNLC) areas of the world's oceans, primarily in the Subarctic, equatorial Pacific and Southern Ocean. In general, when light is abundant many researchers find that HNLC systems are iron-limited. The nature of this limitation is similar between regions but manifests itself at different levels of the trophic structure in some characteristic ways. In general, all members of the HNLC photosynthetic community are physiologically limited by iron availability. This observation is based primarily on the examination of the efficiency of photosystem II, the light-harvesting reaction centers. At ambient levels of iron, light harvesting proceeds at suboptimal rates. This has been attributed to the lack of iron-dependent electron carrier proteins at low iron concentrations. When iron concentrations are increased by subnanomolar amounts, the efficiency of light harvesting rapidly increases to maximum levels. Using fast repetition rate fluorometry and non-heme iron proteins, researchers have described these observations in detail. What is notable about these results is that iron limitation seems to affect the photosynthetic energy conversion efficiency of even the smallest of phytoplankton. This has been a unique finding that stands in contrast to the hypothesis that, because of diffusion, smaller cells are not iron limited but larger cells are.

Nitrate Uptake

As discussed above, iron is also required for the reduction (assimilation) of nitrate. In fact, a change of oxidation state of five is required between nitrate and the reduced forms of nitrogen found in amino acids and proteins. Such a large and energetically unfavorable redox process is only made possible by substantial reducing power (in the form of NADPH) made available through photosynthesis and active nitrate reductase, an iron-requiring enzyme. Without iron, plants cannot take up nitrate efficiently. This provided original evidence implicating iron deficiency as the cause of the HNLC condition. When phytoplankton communities are relieved from iron deficiency, specific rates of nitrate uptake increase. This has been observed in both the equatorial Pacific and the Southern Ocean using isotopic tracers of nitrate uptake and conversion. In addition, the accelerated uptake of nitrate has been observed in both the mesoscale iron enrichment experiments to date, IronEx and SOIREE.

Growth Response

When iron is present, phytoplankton growth rates increase dramatically. Experiments over widely differing oceanographic regimes have demonstrated that, when light and temperature are favorable, phytoplankton growth rates in HNLC environments increase to their maximum at dissolved iron concentrations generally below 0.5 nmol l^{-1}. This observation is significant in that it indicates that phytoplankton are adapted to very low levels of iron and they do not grow faster if given iron at more than 0.5 nmol l^{-1}. Given that there is still some disagreement within the scientific community about the validity of some iron measurements, this phytoplankton response provides a natural, environmental, and biogeochemical benchmark against which to compare results.

The iron-induced transient imbalance between phytoplankton growth and grazing in the equatorial Pacific during IronEx II resulted in a 30-fold increase in plant biomass (**Figure** 7). Similarly, a 6-fold increase was observed during the SOIREE experiment in the Southern Ocean. These are perhaps the most dramatic demonstrations of iron limitation of nutrient cycling, and phytoplankton growth to date and has fortified the notion that iron fertilization may be a useful strategy to sequester carbon in the oceans.

Heterotrophic Community

As the primary trophic levels increase in biomass, growth in the small microflagellate and heterotrophic bacterial communities increase in kind. It appears that these consumers of recently fixed carbon (both particulate and dissolved) respond to the food source and not necessarily the iron (although some have been found to be iron-limited). Because their division rates are fast, heterotrophic bacteria, ciliates, and flagellates can rapidly divide and

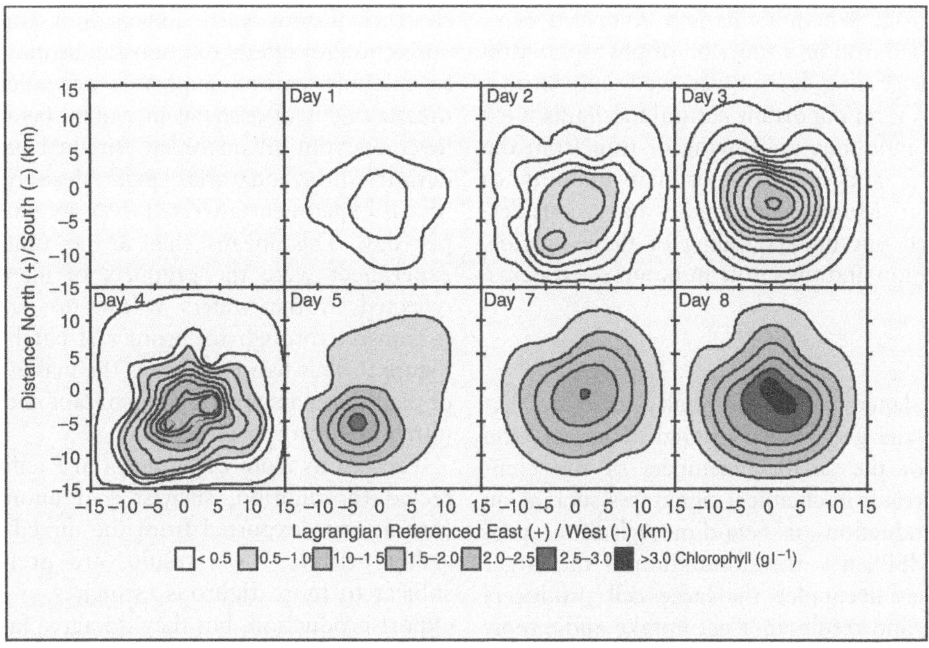

Figure 7 Chlorophyll concentrations during IronEx II were mapped daily. This figure shows the progression of the phytoplankton bloom that reached over 30 times the background concentrations.

respond to increasing food availability to the point where the growth rates of the smaller phytoplankton can be overwhelmed by grazing. Thus there is a much more rapid turnover of fixed carbon and nitrogen in iron replete systems. M. Landry and co-workers have documented this in dilution experiments conducted during IronEx II. These results appear to be consistent with the recent SOIREE experiments as well.

Nutrient Uptake Ratios

An imbalance in production and consumption, however, can arise at the larger trophic levels. Because the reproduction rates of the larger micro- and mesozooplankton are long with respect to diatom division rates, iron-replete diatoms can escape the pressures of grazing on short timescales (weeks). This is thought to be the reason why, in every iron enrichment experiment, diatoms ultimately dominate in biomass. This result is important for a variety of reasons. It suggests that transient additions of iron would be most effective in producing net carbon uptake and it implicates an important role of silicate in carbon flux. The role of iron in silicate uptake has been studied extensively by Franck and colleagues. The results, together with those of Takeda and co-workers, show that iron alters the uptake ratio of nitrate and silicate at very low levels (**Figure 8**). This is thought to be brought about by the increase in nitrate uptake rates relative to silica.

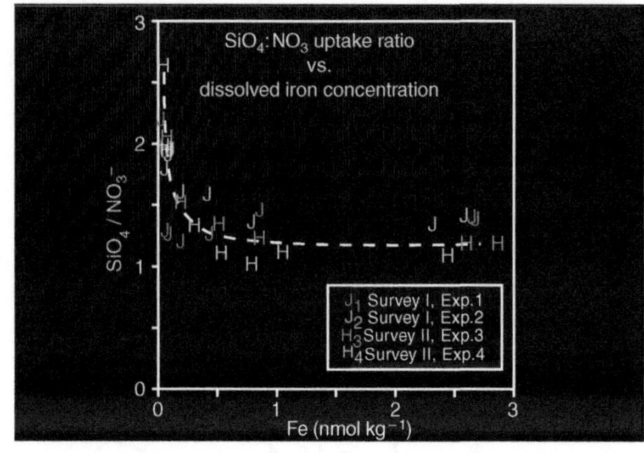

Figure 8 Bottle enrichment experiments show that the silicate : nitrate uptake ratio changes as a function of the iron added. This is thought to be due to the increased rate of iron uptake relative to silicate in these experimental treatments.

Organic Ligands

Consistent with the role of iron as a limiting nutrient in HNLC systems is the notion that organisms may have evolved competitive mechanisms to increase iron solubility and uptake. In terrestrial systems this is accomplished using extracellularly excreted or membrane-bound siderophores. Similar compounds have been shown to exist in sea water where the competition for iron may be as fierce as it is on land. In open ocean systems where it has been measured, iron-binding ligand production increases with the addition of iron. Whether this is a competitive response to added iron or a function of phytoplankton biomass and grazing is not yet well understood. However, this is an important natural mechanism for reducing the inorganic scavenging of iron from the surface waters and increasing iron availability to phytoplankton. More recent studies have considerably advanced our understanding of these ligands, their distribution and their role in ocean ecosystems.

Carbon Flux

It is the imbalance in the community structure that gives rise to the geochemical signal. Whereas iron stimulation of the smaller members of the community may result in chemical signatures such as an increased production of beta-dimethylsulfoniopropionate (DMSP), it is the stimulation of the larger producers that decouples the large cell producers from grazing and results in a net uptake and export of nitrate, carbon dioxide, and silicate.

The extent to which this imbalance results in carbon flux, however, has yet to be adequately described. The inability to quantify carbon export has primarily been a problem of experimental scale. Even though mesoscale experiments have, for the first time, given us the ability to address the effect of iron on communities, the products of surface water processes and the effects on the midwater column have been difficult to track. For instance, in the IronEx II experiment, a time-series of the enriched patch was diluted by 40% per day. The dilution was primarily in a lateral (horizontal/isopycnal) dimension. Although some correction for lateral dilution can be made, our ability to quantify carbon export is dependent upon the measurement of a signal in waters below the mixed layer or from an uneroded enriched patch. Current data from the equatorial Pacific showed that the IronEx II experiment advected over six patch diameters per day. This means that at no time during the experiment were the products of increased export reflected in the waters below the enriched area. A transect through the IronEx II patch is shown in **Figure 9**. This figure indicates the massive production of plant biomass with a concomitant decrease in both nitrate and carbon dioxide.

The results from the equatorial Pacific, when corrected for dilution, suggest that about 2500 t of carbon were exported from the mixed layer over a 7-day period. These results are preliminary and subject to more rigorous estimates of dilution and export production, but they do agree favorably with estimates based upon both carbon and nitrogen budgets. Similarly, thorium export was observed in this experiment, confirming some particle removal.

The results of the SOIREE experiment were similar in many ways but were not as definitive with respect to carbon flux. In this experiment biomass increased 6-fold, nitrate was depleted by 2 μmol l^{-1} and carbon dioxide by 35–40 microatmospheres (3.5–4.0 Pa). This was a greatly attenuated signal relative to IronEx II. Colder water temperatures likely led to slower rates of production and bloom evolution and there was no observable carbon flux.

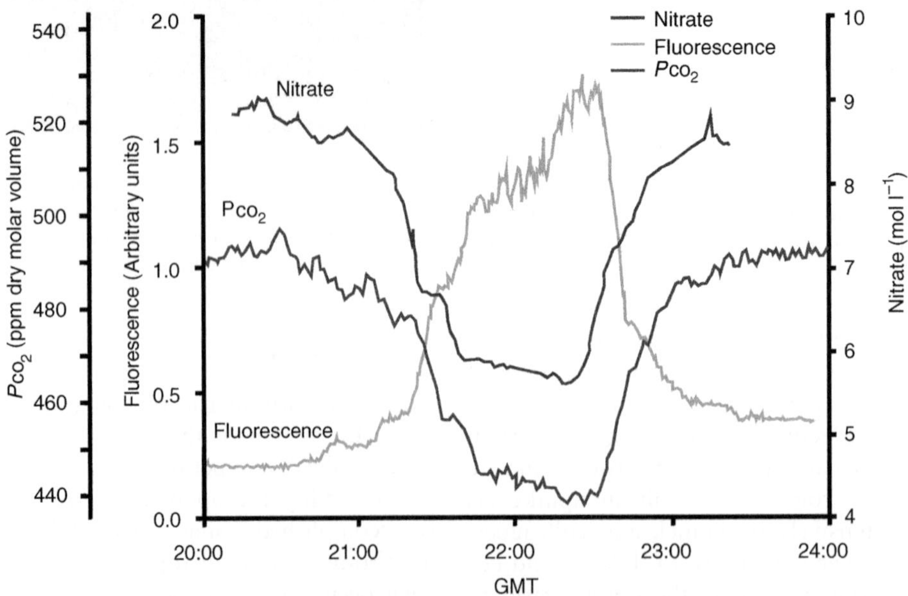

Figure 9 A transect through the IronEx II patch. The *x*-axis shows GMT as the ship steams from east to west through the center of the patch. Simultaneously plotted are the iron-induced production of chlorophyll, the drawdown of carbon dioxide, and the uptake of nitrate in this bloom.

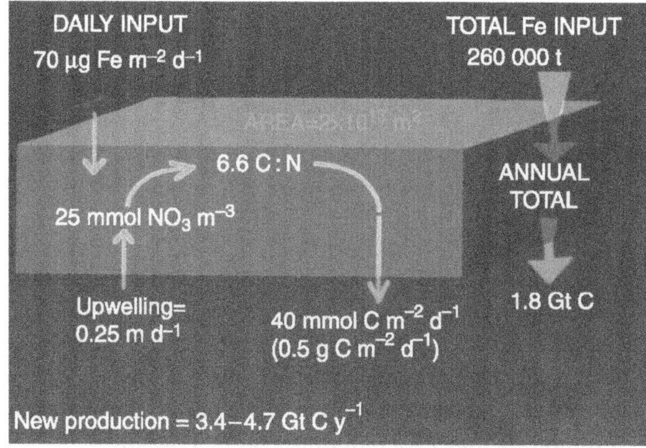

Figure 10 Simple calculations of the potential for carbon export for the Southern Ocean. These calculations are based on the necessary amount of iron required to efficiently utilize the annual upwelled nitrate and the subsequent incorporation into sinking organic matter. An estimated 1.8×10^9 t (Gt) of carbon export could be realized in this simple model.

Original estimates of carbon export in the Southern Ocean based on the iron-induced efficient utilization of nitrate suggest that as much as 1.8×10^9 t of carbon could be removed annually (**Figure 10**). These estimates of carbon sequestration have been challenged by some modelers yet all models lack important experimental parameters which will be measured in upcoming experiments.

Remaining Questions

A multitude of questions remain regarding the role of iron in shaping the nature of the pelagic community. The most pressing question is whether iron enrichment accelerates the downward transport of carbon from the surface waters to the deep sea? More specifically, how does iron affect the cycling of carbon in HNLC, LNLC, and coastal systems? Recent studies indicate that coastal systems may be iron-limited and the iron requirement for nitrogenase activity is quite large, suggesting that iron may limit nitrogen fixation, but there have been limited studies to test the former and none to test the latter. If iron does stimulate carbon uptake, what are the spatial scales over which this fixed carbon may be remineralized? This is crucial to predicting whether fertilization is an effective carbon sequestration mechanism.

Given these considerations, the most feasible way to understand and quantify carbon export from an enriched water mass is to increase the scale of the experiment such that both lateral dilution and sub-mixed-layer relative advection are small with respect to the size of the enriched patch. For areas such as the equatorial Pacific, this would be very large (hundreds of kilometers on a side). For other areas, it could be much smaller.

The focus of the IronEx and SOIREE experiments has been from the scientific perspective, but this focus is shifting toward the application of iron enrichment as a carbon sequestration strategy. We have come about rapidly from the perspective of trying to understand how the world works to one of trying to make the world work for us. Several basic questions remain regarding the role of natural or anthropogenic iron fertilization on carbon export. Some of the most pressing questions are: What are the best proxies for carbon export? How can carbon export best be verified? What are the long-term ecological consequences of iron enrichment on surface water community structure, midwater processes, and benthic processes? Even with answers to these, there are others that need to be addressed prior to any serious consideration of iron fertilization as an ocean carbon sequestration option.

Simple technology is sufficient to produce a massive bloom. The technology required either for a large-scale enrichment experiment or for purposeful attempts to sequester carbon is readily available. Ships, aircraft (tankers and research platforms), tracer technology, a broad range of new Autonomous Underwater Vehicles (AUVs) and instrument packages, Lagrangian buoy tracking systems, together with aircraft and satellite remote sensing systems and a new suite of chemical sensors/*in situ* detection technologies are all available, or are being developed. Industrial bulk handling equipment is available for large-scale implementation. The big questions, however, are larger than the technology.

With a slow start, the notion of both scientific experimentation through manipulative experiments, as well as the use of iron to purposefully sequester carbon, is gaining momentum. There are now national, international, industrial, and scientific concerns willing to support larger-scale experiments. The materials required for such an experiment are inexpensive and readily available, even as industrial by-products (of paper, mining, and steel processing).

Given the concern over climate change and the rapid modernization of large developing countries such as China and India, there is a pressing need to address the increased emission of greenhouse gases. Through the implementation of the Kyoto accords or other international agreements to curb emissions (Rio), financial incentives will reach into the multi-billion dollar level annually. Certainly there will soon be an overwhelming fiscal incentive to investigate, if not implement, purposeful open ocean carbon sequestration trials.

A Societal Challenge

The question is not whether we have the capability of embarking upon such an engineering strategy but whether we have the collective wisdom to responsibly negotiate such a course of action. Posing the question another way: If we do not have the social, political and economic tools or motivation to control our own population and greenhouse gas emissions, what gives us the confidence that we have the wisdom and ability to responsibly manipulate and control large ocean ecosystems without propagating yet another massive environmental calamity? Have we as an international community first tackled the difficult but obvious problem of overpopulation and implemented alternative energy technologies for transportation, industry, and domestic use?

Other social questions arise as well. Is it appropriate to use the ocean commons for such a purpose? What individuals, companies, or countries would derive monetary compensation for such an effort and how would this be decided?

It is clear that there are major scientific investigations and findings that can only benefit from large-scale open ocean enrichment experiments, but certainly a large-scale carbon sequestration effort should not proceed without a clear understanding of both the science and the answers to the questions above.

Glossary

ATP	Adenosine triphosphate
AVHRR	Advanced Very High Resolution Radiometer
HNHC	High-nitrate high-chlorophyll
HNLC	High-nitrate low-chlorophyll
IronEx	Iron Enrichment Experiment
LIDAR	Light detection and ranging
LNHC	Low-nitrate high-chlorophyll
LNLC	Low-nitrate low-chlorophyll
NADPH	Reduced form of nicotinamide–adenine dinucleotide phosphate
SOIREE	Southern Ocean Iron Enrichment Experiment

See also

Nitrogen Cycle. Phosphorus Cycle.

Further Reading

Abraham ER, Law CS, Boyd PW, *et al.* (2000) Importance of stirring in the development of an iron-fertilized phytoplankton bloom. *Nature* 407: 727–730.

Barbeau K, Moffett JW, Caron DA, Croot PL, and Erdner DL (1996) Role of protozoan grazing in relieving iron limitation of phytoplankton. *Nature* 380: 61–64.

Behrenfeld MJ, Bale AJ, Kobler ZS, Aiken J, and Falkowski PG (1996) Confirmation of iron limitation of phytoplankton photosynthesis in Equatorial Pacific Ocean. *Nature* 383: 508–511.

Boyd PW, Watson AJ, Law CS, *et al.* (2000) A mesoscale phytoplankton bloom in the polar Southern Ocean stimulated by iron fertilization. *Nature* 407: 695–702.

Cavender-Bares KK, Mann EL, Chishom SW, Ondrusek ME, and Bidigare RR (1999) Differential response of equatorial phytoplankton to iron fertilization. *Limnology and Oceanography* 44: 237–246.

Coale KH, Johnson KS, Fitzwater SE, *et al.* (1996) A massive phytoplankton bloom induced by an ecosystem-scale iron fertilization experiment in the equatorial Pacific Ocean. *Nature* 383: 495–501.

Coale KH, Johnson KS, Fitzwater SE, *et al.* (1998) IronEx-I, an *in situ* iron-enrichment experiment: experimental design, implementation and results. *Deep-Sea Research Part II* 45: 919–945.

Elrod VA, Johnson KS, and Coale KH (1991) Determination of subnanomolar levels of iron (II) and total dissolved iron in seawater by flow injection analysis with chemiluminescence dection. *Analytical Chemistry* 63: 893–898.

Fitzwater SE, Coale KH, Gordon RM, Johnson KS, and Ondrusek ME (1996) Iron deficiency and phytoplankton growth in the equatorial Pacific. *Deep-Sea Research Part II* 43: 995–1015.

Greene RM, Geider RJ, and Falkowski PG (1991) Effect of iron lititation on photosynthesis in a marine diatom. *Limnology Oceanogrography* 36: 1772–1782.

Hoge EF, Wright CW, Swift RN, *et al.* (1998) Fluorescence signatures of an iron-enriched phytoplankton community in the eastern equatorial Pacific Ocean. *Deep-Sea Research Part II* 45: 1073–1082.

Johnson KS, Coale KH, Elrod VA, and Tinsdale NW (1994) Iron photochemistry in seawater from the Equatorial Pacific. *Marine Chemistry* 46: 319–334.

Kolber ZS, Barber RT, Coale KH, *et al.* (1994) Iron limitation of phytoplankton photosynthesis in the Equatorial Pacific Ocean. *Nature* 371: 145–149.

Landry MR, Ondrusek ME, Tanner SJ, *et al.* (2000) Biological response to iron fertilization in the eastern equtorial Pacific (Ironex II). I. Microplankton community abundances and biomass. *Marine Ecology Progress Series* 201: 27–42.

LaRoche J, Boyd PW, McKay RML, and Geider RJ (1996) Flavodoxin as an *in situ* marker for iron stress in phytoplankton. *Nature* 382: 802–805.

Law CS, Watson AJ, Liddicoat MI, and Stanton T (1998) Sulfer hexafloride as a tracer of biogeochemical and physical processes in an open-ocean iron fertilization experiment. *Deep-Sea Research Part II* 45: 977–994.

Martin JH, Coale KH, Johnson KS, *et al.* (1994) Testing the iron hypothesis in ecosystems of the equatorial Pacific Ocean. *Nature* 371: 123–129.

Nightingale PD, Liss PS, and Schlosser P (2000) Measurements of air–gas transfer during an open ocean algal bloom. *Geophysical Research Letters* 27: 2117–2121.

Obata H, Karatani H, and Nakayama E (1993) Automated determination of iron in seawater by chelating resin concentration and chemiluminescence detection. *Analytical Chemistry* 65: 1524–1528.

Redfield AC (1934) On the proportions of organic derivatives in sea water and their relation to the composition of plankton. *James Johnstone Memorial Volume*, pp. 177–192. Liverpool: Liverpool University Press.

Redfield AC (1958) The biological control of chemical factors in the environment. *American Journal of Science* 46: 205–221.

Rue EL and Bruland KW (1997) The role of organic complexation on ambient iron chemistry in the equatorial Pacific Ocean and the response of a mesoscale iron addition experiment. *Limnology and Oceanography* 42: 901–910.

Smith SV (1984) Phosphorus versus nitrogen limitation in the marine environment. *Limnology and Oceanography* 29: 1149–1160.

Stanton TP, Law CS, and Watson AJ (1998) Physical evolutation of the IronEx I open ocean tracer patch. *Deep-Sea Research Part II* 45: 947–975.

Takeda S and Obata H (1995) Response of equatorial phytoplankton to subnanomolar Fe enrichment. *Marine Chemistry* 50: 219–227.

Trick CG and Wilhelm SW (1995) Physiological changes in coastal marine cyanobacterium *Synechococcus* sp. PCC 7002 exposed to low ferric ion levels. *Marine Chemistry* 50: 207–217.

Turner SM, Nightingale PD, Spokes LJ, Liddicoat MI, and Liss PS (1996) Increased dimethyl sulfide concentrations in seawater from *in situ* iron enrichment. *Nature* 383: 513–517.

Upstill-Goddard RC, Watson AJ, Wood J, and Liddicoat MI (1991) Sulfur hexafloride and helium-3 as sea-water tracers: deployment techniques and continuous underway analysis for sulphur hexafloride. *Analytica Chimica Acta* 249: 555–562.

Van den Berg CMG (1995) Evidence for organic complesation of iron in seawater. *Marine Chemistry* 50: 139–157.

Watson AJ, Liss PS, and Duce R (1991) Design of a small-scale *in situ* iron fertilization experiment. *Limnology and Oceanography* 36: 1960–1965.

MARINE CHEMICAL AND MEDICINE RESOURCES

S. Ali and C. Llewellyn, Plymouth Marine Laboratory, Plymouth, UK

Introduction

The marine environment consists of several defined habitats ranging from the sea surface microlayer which encompasses the first few microns of the water column, through the bulk water column itself, down to the ocean floor and the subsurface sediments underneath which can be found hydrothermal vents, cold seeps, hydrocarbon seeps, and saturated brines, as well as a wide range of mineral and geological variation. It has become increasingly apparent that within all these oceanic layers there is a diversity of micro- and macroorganisms capable of generating a plethora of previously undescribed molecules through novel metabolic pathways which could be of value to both industry and the clinic. The biological diversity in some marine ecosystems may exceed that of the tropical rain forests and this is supported by the presence of 34 out of the 36 phyla of life. This biodiversity stems from the wide range of environmental conditions to which marine organisms have adapted for survival, including extremes of pH (acid and alkali), temperature (high and low), salinity, pressure, and chemical toxicity (complex polycyclic hydrocarbons, heavy metals).

Marine organisms currently being exploited for biotechnology include sponges, tunicates, bryozoans, mollusks, bacteria, cyanobacteria, macroalgae (seaweeds), and microalgae. These organisms have produced compounds with good activities for a range of infectious and noninfectious disease with high specificity for the target molecule (usually an enzyme). Targets of marine natural products which may be clinically relevant include ion channels and G-protein-coupled receptors, protein serine-threonine kinases, protein tyrosine kinases, phospholipase A_2, microtubule-interfering agents (of which the largest number identified are of marine origin), and DNA-interactive compounds. In addition to small organic molecules, marine organisms are increasingly being recognized as a potential source of novel enzymes which could be of industrial and pharmaceutical importance. More than 30 000 diseases have been clinically described, yet less than one-third of these can be treated based on symptoms and only a small number can be cured. Thus, the potential market for novel marine compounds for clinical development is enormous. In addition to providing new molecules for direct clinical intervention, the marine environment is also rich in compounds which are finding uses as natural additives in foods, as nutritional supplements including color additives and antioxidants, and as vitamins, oils, and cofactors which enhance general well-being. Marine organisms are also increasingly providing new solutions to developments in such diverse fields as bioremediation, biocatalysis and chemistry, materials science, nanotechnology, and energy. Some of the potential uses of marine products are summarized in **Figure 1**.

The oceans have long been a source of nutrients, additives, and medicines derived from marine mammals and fish; however, this article focuses on some of the potential which is harbored in predominantly microscopic organisms which are now being increasingly studied for novel bioactive compounds and chemicals and may provide a sustainable alternative source for new compounds and processes.

Novel Metabolites and Drug Discovery

Marine organisms have long been recognized as a source of novel metabolites with applications in human disease therapy. Particular emphasis has been placed on the invertebrates such as sponges, mollusks, tunicates, and bryozoans, but more recently advances in genetics and microbial culture have led to a growing interest in cyanobacteria and marine bacteria. For example, a number of anticancer drugs have been derived from marine sources such as sponges which have proven difficult to cultivate and their metabolites display a structural complexity which often precludes total chemical synthesis as an option for potential drug candidates. In recent years, studies have suggested that many of these complex molecules may in fact be the product of microbes which live in a symbiotic relationship with the sponge and that some of these molecules may be the final product of reactions carried out by different organisms. A major challenge within marine biotechnology will be to ascertain the nature of the organisms present in the symbiotic relationship and to identify the pathways involved in metabolite production. A recent advance in molecular biology with the development of metagenomics has opened up the possibility of organism-independent cultivation of genetic material and subsequent screening and characterization of that

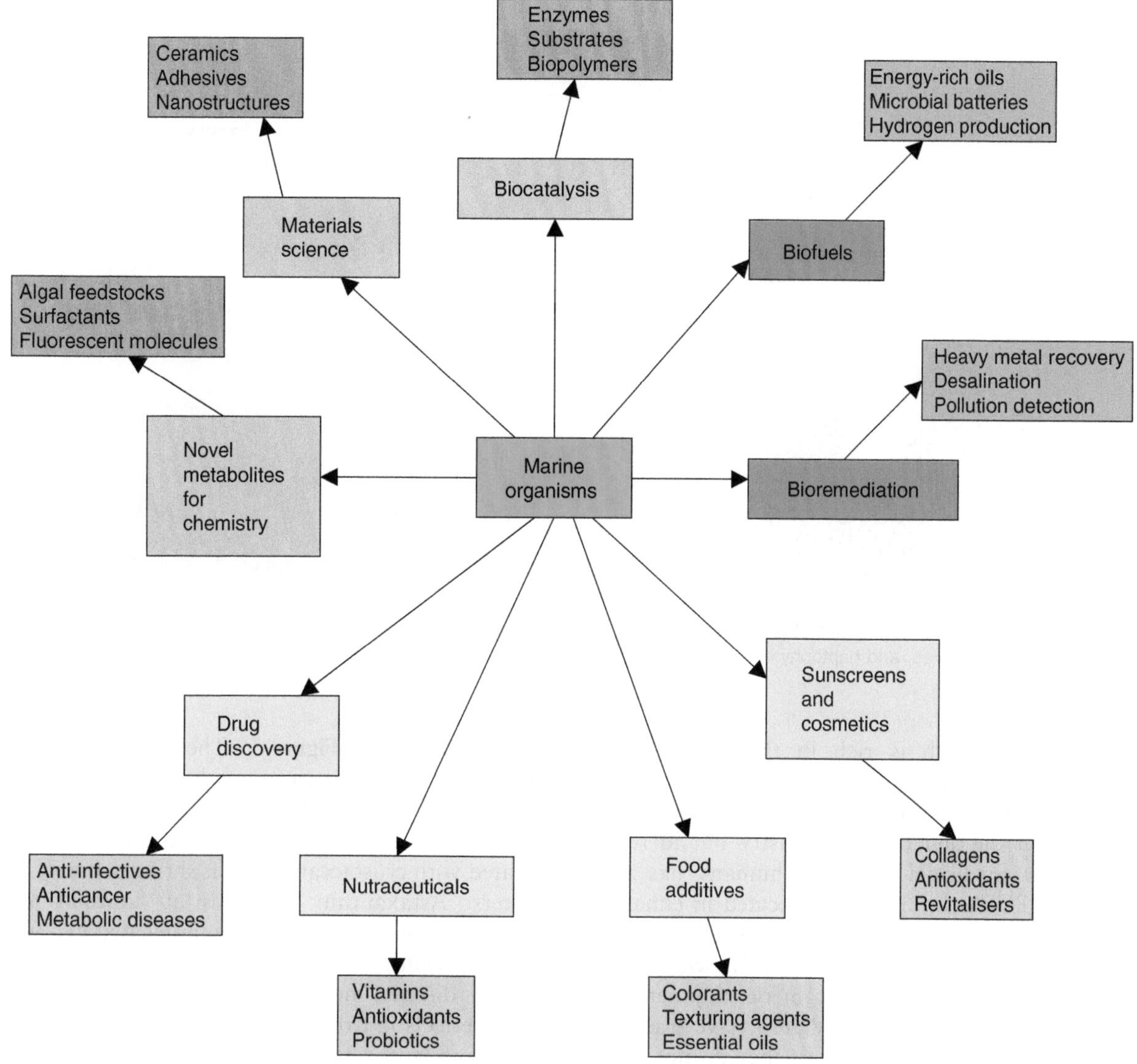

Figure 1 Uses for marine organisms and their products in the chemical, pharmaceutical, energy, and environmental industries.

DNA for novel metabolic pathways and enzymes. This provides a powerful tool for accessing difficult-to-culture microbes which exist in complex symbiotic relationships.

Recent advances in the isolation and culture of marine bacteria using both flow cytometry and microencapsulation-based methods have yielded a vast array of previously unknown bacteria (**Figure 2**). Increasingly, these bacteria are being tested for the presence of bioactive compounds with activities against a diverse range of human and infectious diseases including cancer, HIV, hepatitis C, malaria, and those caused by the increasingly drug-resistant common bacterial pathogens (e.g., *Staphylococcus aureus*, *Enterococcus faecalis*, *Mycobacterium tuberculosis*). For example, the antibiotic abyssomicin C has been isolated from an actinomycete which was cultured from marine sediment collected in the Sea of Japan. Abyssomicin has been shown to interfere with the synthesis of the essential cofactor folic acid in bacteria and is active against the methicillin-resistant *Staphylococcus aureus* (MRSA) pathogen. Thus, marine bacteria represent a vast untapped source for novel compounds with the potential for development as novel drugs.

Marine-derived Nutraceuticals and Food Additives

In recent years, there has been an upsurge in the consumption of nutritional supplements such as vitamins and cofactors which are essential to cellular function. One traditional supplement has been

Figure 2 Examples of marine microbial cultures. (a) Bacteria isolated from the English Channel. (b) Microalgal cultures of chlorophytes, cryptophytes, and haptophytes.

cod-liver oil, which is rich in the omega-3 and omega-6 long-chain polyunsaturated fatty acids (PUFAs). Fatty acids have long been used as supplements in the aquaculture industry but in recent years their medicinal value to humans has been proposed. PUFAs have been implicated in enhanced blood circulation and brain development, particularly docosahexaenoic acid (omega-3), which plays an important role in early brain development and neurite outgrowth. Although PUFAs have traditionally been extracted from fish oils they do not naturally occur there but rather accumulate from the diet of the fish. The primary source of PUFAs are marine microbes, and in recent years the isolation and characterization of microbes which produce these fatty acids have allowed their production in other organisms. At present considerable effort is being expended on obtaining high-level production of PUFAs in plants using genes from microalgae with the aim of eventually producing PUFAs in plants which have traditionally been a source of natural oils (e.g., linseed oil, rapeseed oil). The technology to grow and extract oils from such plants in large scale already exists and the development of genetically modified plants which can produce PUFAs in a readily accessible and sustainable form for an increasing market is highly desirable.

Consumer-led demand for naturally occurring food colorings and antioxidants has resulted in increased interest in photosynthetic pigments and in particular the carotenoids which occur within marine microalgae (**Figure 3**). There is, for example, widespread use of the carotenoid astaxanthin; this is a pigmented antioxidant produced by many microalgae and is responsible for the red color often associated with crustaceans such as shrimps, crabs, and lobsters. Astaxanthin possesses an unusual antioxidant activity which has been implicated in a wide range of health benefits such as preventing cardiovascular disease, modulating the immune system as well as effects in cancer, diabetes, and ocular health. Its antioxidant activities may also have a neuroprotective effect. It has been used extensively in the feed of farmed fish as a nutritional supplement and is partly responsible for the strong coloration often observed in farmed salmon, a fish which naturally accumulates astaxanthin in the wild resulting in the pink hue of its flesh. Other carotenoids such as beta-carotene and lutein are used widely as food coloring and antioxidants. Inclusion, for example, in the diets of chickens leads to a darkening of the egg yolk resulting in a rich yellow color.

Another group of pigments of commercial importance is that of the phycobiliproteins; these are used as colorings, in cosmetics, and as fluorescent dyes for flow cytometry and in immunological assays. More recent research suggests that phycobiliproteins have anticancer and anti-inflammatory properties. A more unusual and unique pigment which is being used increasingly in personal care products and may also possess anticancer and anti-HIV activity is 'marennine', a blue-green pigment

(a)

(b)

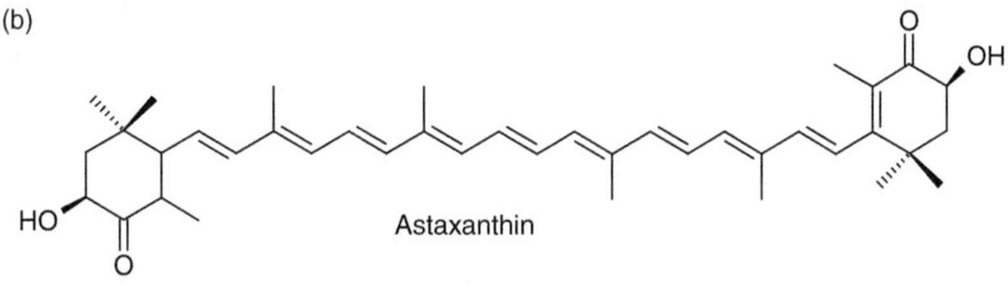

OH

HO

Lutein

B,B-carotene

Figure 3 (a) Any one species of microalgae contains an array of carotenoid pigments, as shown here with the fractionated isolates obtained from the chromatographic analysis of an *Emiliania huxleyi* extract. (b) Chemical structures of cartenoids widely used as color additives and antioxidants.

produced by *Haslea ostrearia*. Several species of microalgae and, in particular, *Haematococcus*, *Dunaliella*, and *Spirulina* are now grown on large commercial scale to accommodate the growing demand for natural pigments.

In addition to being rich in phycobiliproteins, *Spirulina*, a filamentous cyanobacterium, contains a wide variety of nutrients including potentially beneficial proteins, lipids, vitamins, and antioxidants. *Spirulina* is also reported to have various beneficial effects including antiviral activity, immunomodulatory effects, and a role in modulating metabolic function in humans which could be of value in managing diseases involving lipids and carbohydrates such as diabetes. Furthermore, studies indicate that pretreatment with *Spirulina* may reduce the toxic side effects observed with some drugs on mammalian organs such as the heart and kidneys.

The marine environment is a rich source for naturally occurring antioxidants and pigments with a diverse range of microorganisms producing a unique and valuable resource. The potential for the discovery of new pigments and other additives which can be used to replace some of the existing artificial additives currently being used in the food industry is significant.

Sunscreens

The continuous exposure of marine organisms to strong sunlight has resulted in some, primarily the macro- and microalgae, evolving compounds which provide a very good screen against ultraviolet (UV) light. These organisms have the ability to synthesize small organic molecules, called mycosporine-like amino acids (MAAs), which are capable of absorbing UV light very efficiently and thus prevent DNA damage. Over 20 different MAAs occur in nature with a wide range of marine organisms utilizing them, including corals, anemones, limpets, shrimp, sea urchins, and some vertebrates including fish and fish eggs. MAAs are widely distributed across the marine environment; however, they can only be synthesized by certain types of bacteria or algae. For example, red seaweeds and some bloom-forming phytoplankton species are a particularly rich source of MAAs. Studies have revealed that in addition to their screening ability, some MAAs, such as mycosporine-glycine, have antioxidant properties. The ability of naturally occurring compounds such as MAAs to act as effective sunscreens has resulted in some interest from the commercial sector as to the value of these compounds in creams and cosmetics.

Biocatalysis

The ability of enzymes to synthesize complex chiral molecules with high efficiency and precision is of considerable interest within the pharmaceutical and chemical industries and marine bacteria present a new source for novel enzymes with not only unusual synthetic properties but also potentially valuable catalytic and structural properties. This arises from the ability of marine bacteria to grow under extreme conditions such as high and low temperature, high pressure (extreme depth), high salinity, and extremes of pH. This has opened up the potential to isolate naturally occurring small molecules which are difficult to synthesize in the laboratory but which could be of value in synthetic organic chemistry as intermediates. The existence of a large number of potentially novel enzymes in these same organisms which are capable of performing diverse chemical modifications not readily amenable by standard chemical synthesis also opens the route to novel chemical modification of synthetic molecules using biocatalysis and biotransformation.

Growth of microbes under these extreme conditions has led to proteins which possess different temperature optima and improved stability, which has been exploited in the development of new processes and methods such as the polymerase chain reaction, a method for amplifying specific fragments of DNA, and which depends on a thermostable DNA polymerase isolated from a thermophilic microorganism. It has been suggested that enzymes which display high salt tolerance may be of value in the development of enzyme reactions to be performed in organic solvents as they appear to be less prone to denaturing under dehydrating conditions.

Marine bacteria make up the largest potential single source of novelty in the world's oceans and of these the major component are the actinobacteria which includes the actinomycetes. Actinomycetes are readily isolated from the marine environment and consequently are the best studied of the actinobacteria but the other more difficult to culture members are now being identified using advanced culturing and molecular techniques. The actinomycetes in particular hold the promise of tremendous diversity and to date have been underexploited. Terrestrial actinomycetes are responsible for about half of the known bioactive molecules isolated from natural sources to date and include antibiotics, antitumor compounds, immunosuppressants, and novel enzymes. Consequently, the isolation of new organisms from the environment and their analysis for novel metabolites has been a cornerstone in drug discovery. In recent years, however, terrestrial organisms have divulged less novelty than before, and advances in microbiology and genetics have now made the exploitation of marine-derived actinomycetes more attractive. The recognition that the world's oceans are rich in biological diversity and that extreme environmental conditions (e.g., high pressure and temperature at deep-sea hydrothermal vents) have not repressed the development of organisms to form distinct ecological niches suggests that these habitats will be a rich source of chemical novelty.

Although much emphasis has been placed on isolating organisms from extreme environments in the search for novel biocatalysts, the general marine environment should not be ignored. Both micro- and macroalgae have been demonstrated to produce novel enzymes with possible applications in biocatalysis such as the haloperoxidases, enzymes capable of introducing halogen atoms into metabolites. For example, two species of tropical red macroalgae produce halogenated compounds as a defense against predators and such compounds are being tested for medical applications. The availability of haloperoxidases with different catalytic functions would be of use in generating new types of halogenated molecules for the chemical and pharmaceutical industries.

The realization that viruses are the most abundant biological agents in the marine environment and the discovery of highly diverse, ancient, giant viruses with genomes comparable in size to the smallest microbes opens up new sources of genetic diversity. Current indications are that the oceans contain a wide variety of both DNA and RNA viruses with survival strategies which mimic those of terrestrial viruses yet these marine viruses encode a great many proteins of unknown function. Most marine viruses are assumed to be bacteriophages because virus particles are most commonly detected in the vicinity of bacteria, and bacteria are the most abundant organisms in the oceans.

Recent studies have revealed that marine viruses encode unexpected and novel proteins which would not be expected to occur within a virus genome. For example, the giant algal viruses have been shown to encode novel glycosylases, potassium pumps, and a pathway for the synthesis of complex sphingolipids. This biochemical diversity indicates that marine viruses could be a rich source for exploitation in the future for new types of carbohydrate and lipid as well as new proteins and enzymes.

Bioremediation

Pollution of the marine environment is a growing concern particularly with the continuous discharge of both industrial and domestic waste into rivers and estuaries leading to concerns about the impact such pollution could have on long-term human health. The discovery of marine microorganisms capable of detoxifying heavy metals and utilizing complex hydrocarbons as an energy source has provided a new impetus to develop natural solutions to the problems of environmental pollution. However, it should be remembered that toxic substances are not the only causes of marine distress and that the utilization of fertilizers and the disposal of sewage can also result in an imbalance in the marine ecology, resulting in the formation of large, often toxic, algal blooms which although not always a direct threat to human health do lead to widespread ecological damage. Thus, the discovery of microbes capable of growing in the presence of high concentrations of ammonia could be of value in the treatment of wastewater, and an understanding of the anaerobic oxidation of ammonia could lead to the development of new chemical processes. The same organisms also possess unusual metabolic intermediates such as hydrazine and produce unusual lipids which could also be of value in the search for new chemical intermediates.

Heterocyclic molecules containing sulfur, nitrogen, and oxygen are among the most potent pollutants and inevitably contaminate the marine environment to a considerable extent. The use of microbes to degrade and detoxify such compounds is gaining considerable interest as a process which is environmentally friendly and would represent a long-term solution to removing heterocyclic contaminants. A particularly rich source of such organisms is the marine environment where growth in close proximity to sulfur-rich hydrothermal vents or adjacent to hydrocarbon (oil) seeps on the ocean floor has produced a plethora of microorganisms with metabolisms adapted to the utilization of a wide variety of carbon-, nitrogen-, and sulfur-based chemistries. Again, these organisms are also a very rich source of enzymes with previously unknown characteristics such as unusual substrate specificity, which could be of great value to the chemicals industry where they could be utilized in the production of new or difficult-to-synthesize compounds because they can perform reactions which are difficult to duplicate using traditional synthetic chemistry methods.

Microbial Fuel Cells and Biofuels

The use of marine organisms to produce fuels has also been proposed. The generation of electricity through the degradation of organic matter has recently been demonstrated to occur in marine sediments and may be mediated by complex communities of marine microorganisms. These organisms degrade complex organic matter such as carbohydrates and proteins to simpler molecules such as acetate which are then used by electricity-generating bacteria to reduce metals such as iron and manganese. By replacing the naturally occurring metals with an anode these bacteria, under anoxic conditions, will supply the electrons needed to produce an electric current to a cathode linked to the anode by wires and exposed to the oxygen in the water column. It has been suggested that this type of system could be used to supply the electricity needed to operate equipment in regions where access is difficult and so eliminate the need to replace batteries. Microbial fuel cells would be self-sustaining, would not require the preprocessing of fuels to function efficiently, and would not contribute net CO_2 to the atmosphere nor produce toxic waste as with conventional batteries.

Another area of intense study is the development of renewable biofuels with much focus being given to developing terrestrial plant species to produce the precursors to biodiesel. Microalgae present a potential alternative source of hydrocarbons for the generation of biofuels as some species naturally produce significantly more oil (per year per unit area of land) than terrestrial oil seed crops. Several marine species such as *Porphyridiuim*, *Chlorella*, and *Tetraselmis* are currently under investigation as sources rich in hydrocarbons suitable for biofuel production.

Biomaterials

Another area of interest is the development of novel biomaterials inspired by marine organisms. Areas of particular interest are the mechanism of calcium- and silica-based structure formation which is found in many phytoplankton, formation of hard chitinous shells in many larger marine organisms such as oysters and crabs, as well as the very powerful bioadhesives produced by mussels and barnacles. Proteins form the basis of a number of naturally occurring adhesive molecules which display a number of attractive features, particularly for clinical and other specialist applications. These features include the ability to adhere strongly to both smooth and uneven surfaces with a high degree of bonding strength and the ability to form and maintain bonds in very humid and wet conditions. This bonding ability seems to be strongly linked to the presence of hydroxylated tyrosine residues (L-dopa; L-3,4-dihydroxyphenylalanine) in such proteins and it is thought that adhesion involves interactions between the hydroxyl groups and the target surface. The development of powerful adhesives which can cure rapidly under wet conditions and are nontoxic would be of particular value in the clinic. At present there is considerable interest in using bioadhesives in the field of ophthalmology where the use of alternatives to sutures in, for example, corneal grafts is desired in order to reduce the risks of irritation and scarring to the eye following surgery. Another area where the use of bioadhesives is being actively researched is in drug delivery where the ability to attach naturally occurring polymers which can slowly release a drug over time would be useful. This is particularly relevant for poorly soluble biological drugs based on antibodies and other large proteins which can be difficult to administer. The potential contribution that marine-derived biomaterials and bioadhesives could make to such fields is enormous.

Chitin is the second most abundant natural polysaccharide after cellulose and is found in the exoskeletons of crustaceans such as crabs and shrimp as well as in the cell walls of fungi and cuticles of insects. The deacetylation of chitin produces chitosan, a biopolymer with great potential in medicine. Chitosan and its derivatives possess numerous applications due to their properties which include reactive functional groups, gel-forming capability, low toxicity, and high adsorption capacity, as well as complete biodegradability and antibacterial and antifungal activities. These properties make chitosan particularly attractive in areas of research such as drug delivery and tissue engineering where a nontoxic, biodegradable scaffold with antimicrobial activity would be particularly attractive. Both chitin and chitosan can influence the immune system and are being studied extensively as biomaterials in the development of supports for accelerated wound healing. These chitosan-based materials are also being modified to improve adhesion to wound sites and for the incorporation of antimicrobial agents to minimize the risk of infection. Chitosan and its derivatives are also being used to develop scaffolds for applications in tissue engineering to grow cells to form complex structures which could ultimately be used to replace damaged tissues and organs.

The elaborate silica-based structures (frustules) which are exhibited by many diatoms have been of interest to materials scientists for many years and recent studies have begun to reveal some of the characteristics that are present in these silica shells, including an understanding of the proteins and other molecules involved in structure formation (**Figure 4**). The highly precise nature of the structures has led to suggestions that the silica structures can be used directly as either templates for microfabrication or as materials for use in microprocesses such as filters in microfluidics. By understanding and manipulating the growth environment of any given diatom it may be possible to modify the precise geometry of the natural silica shells it produces and the resultant frustules could then be modified using standard microengineering techniques to create new nanostructures with potential applications in the development of medical devices. The glass-like properties of diatom frustules, the remains of which form diatomite (diatomaceous earth), have over 300 recorded commercial applications. The fine pores present in the frustules make them especially useful in filtration processes and the bulk of diatomite is used for this purpose. It has also been suggested that frustules might have applications in the development of new optical devices.

Figure 4 Scanning electron microscope images of (a) the diatom *Thalassiosira*, (b) the dinoflagellate *Gonyualax*, and (c) the dinoflagellate *Peridinium* showing the intricacies of cell walls as inspirations to novel biomaterials.

Future Prospects

The marine environment, representing 70% of the Earth's surface, is a vast untapped resource for new chemicals and enzymes, often with characteristics which are considerably different to anything discovered in the terrestrial environment. A major hurdle to the study and exploitation of this resource has been the inaccessibility of the oceans but advances in science and technology are now providing new approaches to isolating and characterizing the organisms present. This should lead to a considerable increase in the number of novel marine-derived chemicals and medicines available to the market in the near future.

Glossary

Actinomycetes Major group of bacteria commonly isolated from low-nutrient environments. Rich source of unusual small molecules (e.g., antibiotics) and enzymes.

Cyanobacteria Aquatic photosynthetic bacteria widely found throughout nature.

Macroalga(e) Seaweed(s).

Microalga(e) Microscopic single-cell plants.

Nutraceuticals Natural chemicals, usually contained in foods, with potential benefits to human health.

See also

Global Marine Pollution. Inverse Modeling of Tracers and Nutrients. Metal Pollution. Nuclear Fuel Reprocessing and Related Discharges. Radioactive Wastes.

Further Reading

Gullo VP, McAlpine J, Lam KS, Baker D, and Petersen F (2006) Drug discovery from natural products. *Journal of Industrial Microbiology and Biotechnology* 33: 523–531.

Higuera-Ciapara I, Félix-Valenzuela L, and Goycoolea FM (2006) Astaxanthin: A review of its chemistry and applications. *Critical Reviews in Food Science and Nutrition* 46: 185–196.

Hussein G, Sankawa U, Goto H, Matsumoto K, and Watanabe H (2006) Astaxanthin, a carotenoid with potential in human health and nutrition. *Journal of Natural Products* 69: 443–449.

Khan Z, Bhadouria P, and Bisen PS (2005) Nutritional and therapeutic potential of *Spirulina*. *Current Pharmaceutical Biotechnology* 6: 373–379.

König GM, Kehraus S, Seibert SF, Abdel-Lateff A, and Müller D (2005) Natural products from marine organisms and their associated microbes. *ChemBioChem* 7: 229–238.

Lovley DR (2006) Microbial fuel cells: Novel microbial physiologies and engineering approaches. *Current Opinion in Biotechnology* 17: 327–332.

Marszalek JR and Lodish HF (2005) Docosahexaenoic acid, fatty acid-interacting proteins, and neuronal function: Breastmilk and fish are good for you. *Annual Review of Cell and Developmental Biology* 21: 633–657.

Napier JA and Sayanova O (2005) The production of very-long-chain PUFA biosynthesis in transgenic plants: Towards a sustainable source of fish oils. *Proceedings of the Nutritional Society* 64: 387–393.

Newman DJ and Hill RT (2005) New drugs from marine microbes: The tide is turning. *Journal of Industrial Microbiology and Biotechnology* 33: 539–544.

Shi C, Zhu Y, Ran X, Wang M, Su Y, and Cheng T (2006) Therapeutic potential of chitosan and its derivatives in regenerative medicine. *Journal of Surgical Research* 133: 185–192.

Shick JM and Dunlap WC (2002) Mycosporine-like amino acids and related gadusols: Biosynthesis, accumulation, and UV-protective functions in aquatic organisms. *Annual Review of Physiology* 64: 223–262.

Suttle CA (2005) Viruses in the sea. *Nature* 437: 356–361.

Wilt FH (2005) Developmental biology meets materials science: Morphogenesis of biomineralized structures. *Developmental Biology* 280: 15–25.

TRACERS IN THE SEA

LONG-TERM TRACER CHANGES

F. von Blanckenburg, Universität Bern, Bern, Switzerland

Introduction

Ocean tracers that record long-term changes preserve certain water column information within the sediment. This information comprises (1) the tracers' fluxes in the past, such as erosional input from the continents, hydrothermal activity at mid-ocean ridges, input of extraterrestrial material, or carbonate recycling; (2) the distribution of water masses in the past and the state of the past global thermohaline circulation. Inorganic isotope tracers whose isotope ratios are modified by radioactive decay in their source materials are ideally suited for these studies. Their original water column values can be measured in materials such as biogenic carbonates, ferromanganese crusts and nodules, and the authigenic phase of deep-sea sediments.

Studies of tracer fluxes in the past are favored by those tracers whose residence time in the ocean (τ, defined below) is long relative to the turnover time of the thermohaline circulation (1500 y), such as Sr and Os. Tracers of which τ is of the order of, or shorter than the oceans turnover time (Nd, Hf, Pb, Be) offer the ability to label water masses isotopically. In this case, long-term isotope changes of these intermediate-τ tracers are potentially caused by variations of the thermohaline circulation. However, secular variations of these isotope tracers can also be caused by regional variations in these tracers' fluxes, mostly resulting from changes in weathering. It is not always straightforward to distinguish between these two causes of tracer variations.

Certainly the globally uniform seawater isotope evolution of Sr, Os, and potentially also Be, offer excellent tools for isotope stratigraphy on long (My) timescales.

Definitions and Concepts

Long-term tracers are those elements whose isotopic compositions provide information on the physical and chemical state of the oceans on timescales of several thousands of years to millions of years (My). For example, paleo-oceanographers aim to reconstruct past water mass distributions and the mode of the thermohaline circulation. For this purpose it would be desirable to reconstruct past oceanographic water mass characteristics such as salinity, temperature, silica, or phosphorus content from the sedimentary record. Similarly, the reconstruction of the past land–sea transfer of certain tracers is desirable in order to reconstruct changes in the weathering history of the continents. However, these present-day tracers are usually not conserved in the sedimentary record. Even if they were precipitated chemically and stored in sediments, their changes in concentration as measured in a sedimentary column back through time cannot be directly related to past water mass properties. This is because the tracers' concentrations in sediments depend on factors such as sedimentation rate, diagenesis, partitioning into a certain phase, uptake by organisms, and additions of the same element by detrital hemipelagic or aeolian material. Therefore ocean chemists make use of proxy tracers which are not routinely analyzed in surveys of present-day water masses, because their measurement presents a considerable effort compared with tracers such as salinity and temperature. However, their characteristics can be directly related to those well-known oceanographic seawater tracers.

The conditions that need to be met for an element to be of use as a proxy tracer are that (1) it conserves a characteristic chemical or isotopic property when transferred from the water column into the sediment; (2) the elements or their isotopic composition can be extracted from the sediment; (3) the age of the sediment is known so that changes of the tracer over time can be reconstructed.

One such proxy makes use of element ratios. For example, the ratio of Cd to Ca in foraminiferal tests is a proxy for the PO_4 content of the past water mass in which the foraminifera formed and therefore provides information on the past thermohaline circulation. This tracer is explained in detail in the article on trace elements in foraminiferal tests. Similarly, the ratio of the intermediate uranium decay products ^{231}Pa and ^{230}Th, measured in bulk sediment, may under certain conditions provide information on the advection of water masses in the overlying water column in the past (*see* Cosmogenic Isotopes and Uranium-Thorium Series Isotopes in Ocean Profiles). Isotope ratios are ideally suited as long-term proxy tracers. Some of these isotope ratios are characteristic of certain seawater properties and can be measured in sediments, regardless of the actual tracer partitioning, concentration, or location of precipitation.

A property describing the behavior of a tracer in sea water is the residence time τ. If the tracer's fluxes in and out of an ocean basin are invariant with time, the tracer is at steady state and the residence time can be calculated from the tracer's ocean inventory:

$$\tau = \frac{\text{Inventory}}{\text{Flux}_{\text{in}}} = \frac{\text{Inventory}}{\text{Flux}_{\text{out}}}$$

A tracer suitable as a water mass tracer has a short global residence time (τ) relative to the ocean's mixing time. This ensures that isotope 'fingerprints' characteristic of a certain water mass are prevented from being completely dispersed by the global thermohaline circulation. While the global ocean mixing time is difficult to assess, a meaningful quantity is the time it takes for one turnover of the global deep water circulation, which is c. 1500 y. Tracers with τ of this order have the potential to preserve distinct water mass labels. It can be assumed that a conservative (i.e. nonreactive) tracer would be almost perfectly homogenized within 10 000–20 000 years. Tracers with τ in excess of this period will only record changes in the global flux of this tracer, regardless of the water mass, the location of the input, or the location of the samples taken.

Isotope Tracers Used

Much use is made of the stable isotopes of carbon as a paleo-water mass isotope 'fingerprint'. The $^{13}C/^{12}C$ ratio in the tests of foraminifera depends on the relative position of the overlying water mass within the thermohaline circulation system. However, these isotope ratios are modified during the incorporation into organisms, depend on availability of nutrients, and like the isotopes of oxygen, also depend on seawater temperature. (These tracers are dealt with in the relevant articles; please refer to the See also section.)

Isotope ratios of inorganic trace metals which are the topic of this chapter are not modified when incorporated into the sediment (note that some minor isotope fractionation might occur on incorporation into the sediment, but usually such shifts are either smaller than analytical precision or they are removed by the internal correction procedures of the techniques used). The variation in isotope ratios only varies

Table 1 Long-term isotope tracers currently in use

Tracer	*Isotopes*	*Sources*	*Average deep-water concentration*	*Global deep-water residence time*
Strontium (Sr)	^{87}Sr (stable) ← ^{87}Rb ($T_{1/2}$ = 48.8 Gy) ^{86}Sr (stable, primordial)	Mostly chemical weathering of the continental crust and carbonates Hydrothermal solutions from mid-ocean ridges Dissolution of marine carbonates	7.6 μg g^{-1}	2–4 My
Osmium (Os)	^{187}Os (stable) ← ^{187}Re ($T_{1/2}$ = 43 Gy) ^{188}Os (stable, primordial)	Erosion of the continental crust (chemical weathering important) Leaching of abyssal peridotites Cosmic dust and spherules	10 fg g^{-1}	8000–40 000 y
Neodymium (Nd)	^{143}Nd (stable) ← ^{147}Sm ($T_{1/2}$ = 106 Gy) ^{144}Nd (stable, primordial)	Erosion of the continental crust	4 pg g^{-1}	~1000–2000 y
Hafnium (Hf)	^{176}Hf (stable) ← ^{177}Lu ($T_{1/2}$ = 37.3 Gy) ^{177}Hf (stable, primordial)	Erosion of the continental crust Hydrothermal solutions at mid-ocean ridges	0.18 pg g^{-1}	~1000–2000 y?
Lead (Pb)	^{208}Pb (stable) ← ^{232}Th ($T_{1/2}$ = 14.0 Gy) ^{207}Pb (stable) ← ^{235}U ($T_{1/2}$ = 0.704 Gy) ^{206}Pb (stable) ← ^{238}U ($T_{1/2}$ = 4.47 Gy) ^{204}Pb (stable, primordial)	Erosion of the continental crust Hydrothermal solutions at mid-ocean ridges (minor) Today: industrial Pb	1 pg g^{-1}	40 y (Atlantic) 80–200 y (Pacific)
Be	^{10}Be (cosmogenic, $T_{1/2}$ = 1.5 My)	^{10}Be: atmospheric precipitation by rain	1000 atoms/g	250 y (Atlantic)
Beryllium	^{9}Be (stable, primordial)	^{9}Be: erosion of the continental crust	0.25 pg g^{-1}	600 y (Pacific)

by radioactive decay of the parent isotope of at least one isotope in the tracer's sources or cosmogenic production. The elements currently in use for paleo-oceanography are given in **Table 1**.

As apparent from the properties listed in **Table 1**, ocean chemists have a variety of tracers at hand, covering a range of residence times and chemical behaviors. Those tracers varying due to radioactive decay have distinct isotopic compositions in their various source materials (**Table 2**). This makes them particularly useful both as water mass tracers, and to reconstruct the flux from these various sources into the oceans. It may be surprising to find the cosmogenic nuclide ^{10}Be in this list of otherwise radiogenic tracers. The reason is that Be behaves very similarly to the other tracers in that the ratio $^{10}Be/^{9}Be$ is distinct in different water masses. Given that ^{10}Be is the only tracer of which the flux into the oceans is known, τ can be calculated precisely from its water column concentration. Further, the continent-derived isotope ^{9}Be is the only tracer of which the flux into the oceans can be calculated from the $^{10}Be/^{9}Be$ ratio.

Examples of the isotopes of Nd and Be as water mass labels are shown in **Figure 1a** and **b**. The isotope variations of Nd are so small that the $^{143}Nd/^{144}Nd$ ratio is reported normalized to a ratio typically found in chondritic meteorites ('CHUR'):

$$\varepsilon_{Nd} = \left(\frac{^{143}\mathrm{Nd}/^{144}\mathrm{Nd}_{\mathrm{sample}}}{^{143}\mathrm{Nd}/^{144}\mathrm{Nd}_{\mathrm{CHUR}}} - 1 \right) \times 10^4$$

The salinity contours in **Figure 1** define water masses, such as North Atlantic Deep Water (NADW), Antarctic Intermediate Water (AAIW), and Antarctic Bottom Water (AABW). Note that ε_{Nd} is −13.5 in NADW, and $^{10}Be/^{9}Be$ is *c.* 0.5×10^{-7}. In the southern circumpolar water ε_{Nd} is −9, and $^{10}Be/^{9}Be$ is 1×10^{-7}. $^{10}Be/^{9}Be$, and in particular ε_{Nd}, mimic the shape of salinity. Incorporation of these tracers into the sediment at a given location potentially provides information on the distribution and mixing of water masses at this location back through time.

Table 2 Isotope ratios of source materials

Isotope ratio	*Pacific mid-ocean ridges*	*Average upper continental crust*	*Cosmic dust*
$^{87}Sr/^{86}Sr$	0.7028	0.72	N/A
$^{187}Os/^{188}Os$	0.125 (abyssal peridotites)	1.26	0.126
$^{143}Nd/^{144}Nd$	0.5132	0.5121	
ε_{Nd}	+10	−11.4	N/A
$^{176}Hf/^{177}Hf$	+20	−10	
ε_{Nd}	0.2834	0.2825	N/A
$^{206}Pb/^{204}Pb$	18.5	19.3	
$^{207}Pb/^{204}Pb$	15.5	15.7	N/A
$^{208}Pb/^{204}Pb$	38.0	39.1	

ε_{Nd} and ε_{Hf} are $^{143}Nd/^{144}Nd$ and $^{176}Hf/^{177}Hf$ ratios, respectively, normalized to a chondritic value CHUR. $^{143}Nd/^{144}ND_{CHUR}$ = 0.512638; $^{176}Hf/^{177}Hf_{CHUR}$ = 0.282772; (N/A)L: Not Available.

$$\varepsilon_{Nd} = \left(\frac{^{143}\mathrm{Nd}/^{144}\mathrm{Nd}_{\mathrm{sample}}}{^{143}\mathrm{Nd}/^{144}\mathrm{Nd}_{\mathrm{CHUR}}} - 1 \right) * 10^4 \quad \varepsilon_{Hf} = \left(\frac{^{176}\mathrm{Hf}/^{177}\mathrm{Hf}_{\mathrm{sample}}}{^{176}\mathrm{Hf}/^{177}\mathrm{Hf}_{\mathrm{CHUR}}} - 1 \right) * 10^4$$

The schematic global distribution of deep-water isotope ratios of all tracers discussed here is shown in **Figure 4A–F**. Note that the variability decreases with increasing residence time. $^{87}Sr/^{86}Sr$ is perfectly homogenized (**Figure 2A**). The only location worldwide at which a different Sr isotope ratio has been measured in sea water is the restricted Baltic Sea, where riverine dilution halves the open-ocean salinity and leads to a distinct $^{87}Sr/^{86}Sr$ only just detectable by modern analytical methods. $^{187}Os/^{188}Os$, with an estimated τ of 8000–40 000 y, shows only a minute difference between the Atlantic and the other oceans (**Figure 2B**) show clear gradients between Atlantic and Pacific deep water. This is because the Atlantic receives the highest flux of continental erosion products (aeolian dust, river particulate matter, river dissolved matter) per unit open-ocean area. Furthermore, all this material is derived from old continental crust with an isotope composition distinct from younger rocks (**Table 2**). Labrador Sea water, for example, receives erosion products from Archean cratons with a unique isotope composition. In contrast, the Pacific receives most of its tracer input from the surrounding volcanic arcs, which have isotope compositions different from the continental crust surrounding the Atlantic. The Indian Ocean has ratios intermediate between the Atlantic and the Pacific for all of these tracers. Whether this is due to mixing of Atlantic water masses (advected through the circumpolar current) and Pacific water (advected via the Indonesian throughflow), or due to internal sources unique to the Indian Ocean is currently not known. $^{10}Be/^{9}Be$ ratios are lower in the Atlantic because the North Atlantic receives a higher flux of terrigenous ^{9}Be. This keeps the ^{9}Be concentration uniform worldwide, whereas ^{10}Be increases along the advective flow path as expected from a nutrient-type tracer.

Materials and Methods used in Long-term Tracer Studies

It is important that sedimentary materials chosen for long-term tracer studies are true chemical or biogenic precipitates formed in the water column. Contamination by terrestrial detrital material (fine clays from

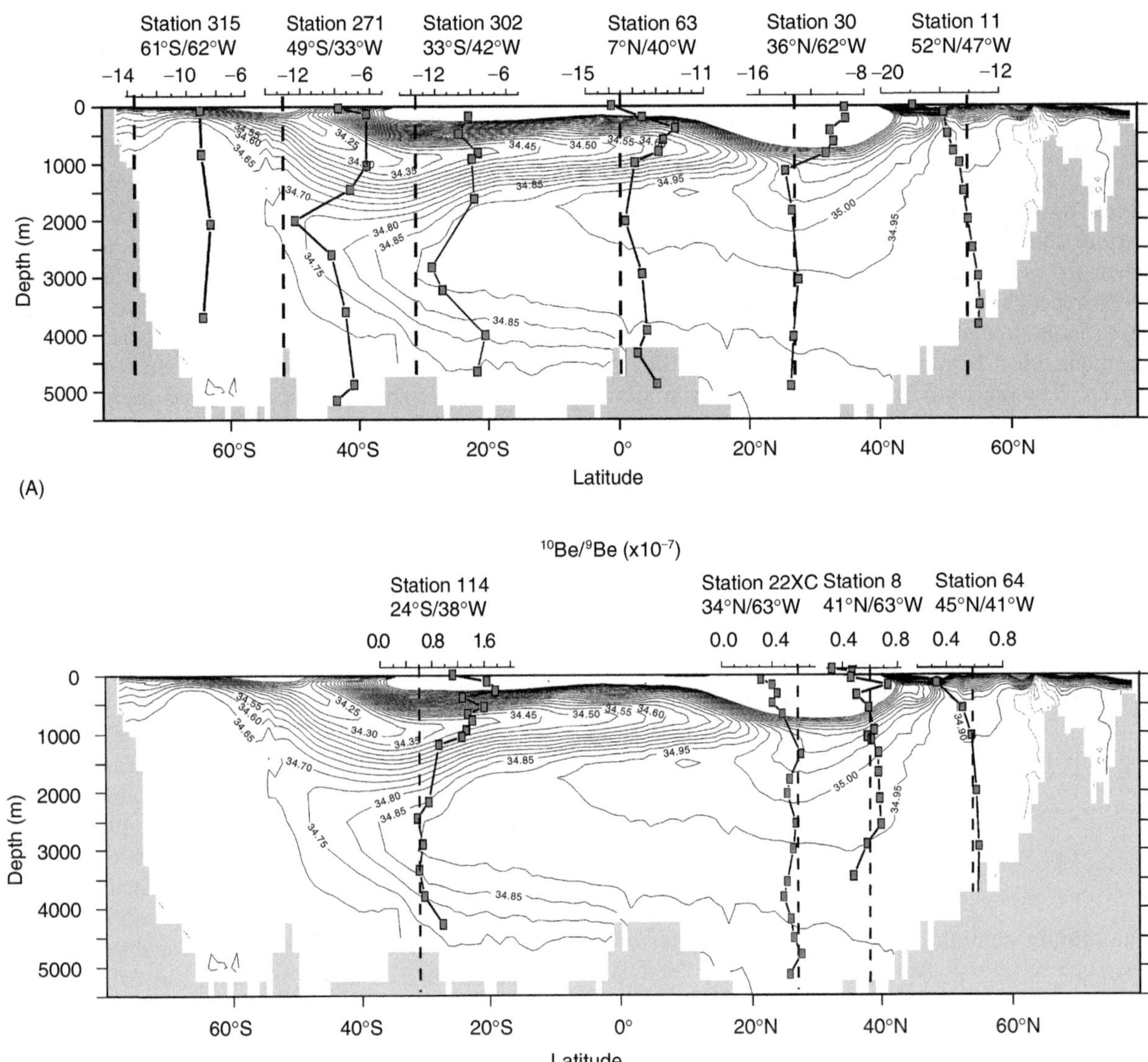

Figure 1 (A) Salinity contours of Atlantic sea water with superimposed dissolved Nd isotope compositions. Stippled line gives the typical composition of NADW ($\varepsilon_{Nd} = -13$). Note the pronounced tongue of NADW with intermediate salinities and ε_{Nd} of -13 spreading south, and the tongues of AABW and AAIW with lower salinities and ε_{Nd} of -9 spreading north. (Reprinted with permission from von Blanckenburg F (1999) Tracing past ocean circulation? *Science* 286: 1862–1863. Copyright 1999 American Association for the Advancement of Science.) (B) Salinity contours of Atlantic sea water with superimposed dissolved Be isotope compositions. Stippled line gives the typical composition of NADW ($^{10}Be/^{9}Be = 0.6 \times 10^{-7}$). Note the pronounced tongue of AAIW with lower salinities and $^{10}Be/^{9}Be$ of 1×10^{-7} spreading north. (Data reproduced with permission from Ku *et al.*, 1990; Xu, 1994; Measures *et al.*, 1996).

aeolian or hemipelagic sources) and material affected by chemical alteration through diagenetic processes has to be avoided. The isotopes of Sr are usually measured on carbonates or barite, while those of Be, Nd, or Hf are extracted from chemical sediments, such as ferromanganese (Fe-Mn) crusts, manganese nodules, the authigenic phase of deep-sea sediments, or marine phosphorites, argillites, and glauconites.

Results

Sr

Sr isotopes represent the best-studied long-term tracer, as well-dated carbonate sequences are readily available, the extraction and analysis is simple, and the long residence time ensures worldwide homogenization. Therefore, a single isotope evolution curve has

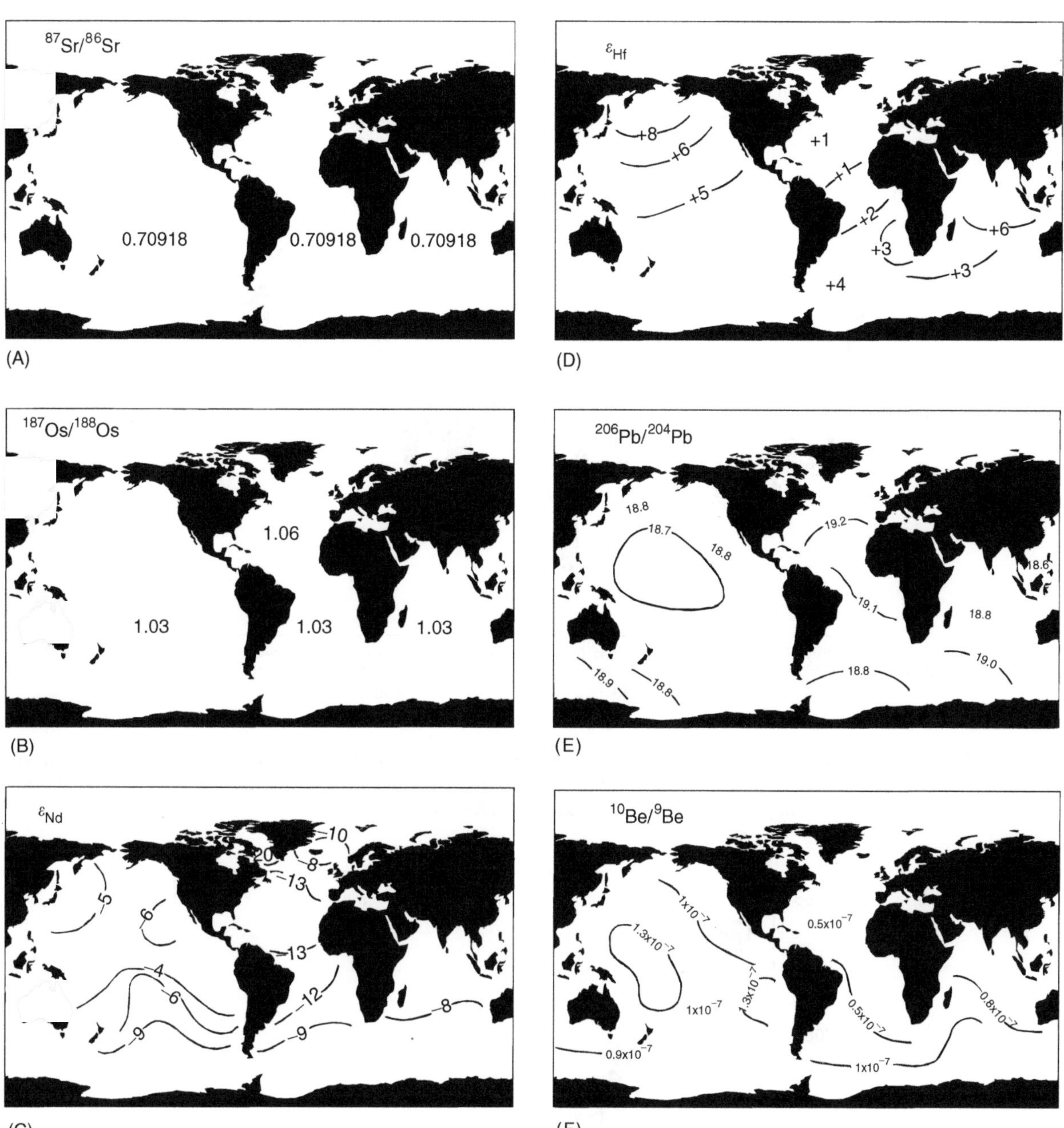

Figure 2 (A) Map of modern $^{87}Sr/^{86}Sr$ ratios in sea water (relative to 0.710248 for the Sr isotope standard SRM 987 (McArthur, 1994). Note that the long τ of Sr (several My) allows for perfect homogenization and uniform isotope ratios in all basins. (B) $^{187}Os/^{188}Os$ isotope ratios in modern sea water as measured on the surface of hydrogenous Fe-Mn crusts (Burton *et al.*, 1999b). τ of a few tens of thousands of years just allows for a small difference between the North Atlantic (with its rich input of old continental material) and the other oceans. (C) Dissolved Nd isotope compositions in modern deep sea water as measured in Mn nodules (reproduced with permission from Albarède and Goldstein, 1992), adjusted for more recent measurements of Nd in deep sea water (see references in von Blanckenburg, 1999). A τ of c. 1000 y allows for distinct gradients between basins, depending on the age and Sm/Nd ratio of their surrounding continental erosion sources. (D) ε_{Hf} measured in the surface layer of hydrogenetic Fe-Mn crusts and nodules (Albarède *et al.*, 1998). Similar to Nd, distinct gradients exist between basins. (E) Pre-anthropogenic $^{206}Pb/^{204}Pb$ ratios measured in the surface layer of hydrogenetic Fe-Mn crusts and nodules. (Abouchami and Goldstein, 1995. Reprinted from *Geochimica et Cosmochimica Acta*, 60, von Blanckenburg F, O'Nions RK, Hein JR, Distribution and sources of pre-anthropogenic lead isotopes in deep ocean water as derived from Fe-Mn crust, 4957–4963, Copyright (1996), with permission from Elsevier Science.). Pre-anthropogenic Pb cannot be measured in modern sea water which is dominated by industrial Pb. The short τ (40–200 y.) results in distinct signatures between and within basins. (F) $^{10}Be/^{9}Be$ ratios in modern sea water from both the dissolved phase and the surface layer of hydrogenetic Fe-Mn crusts. (Reprinted from *Earth and Planetary Letters*, 141, von Blanckenburg F, O'Nions RK, Belshaw NS, Gibb A, Hein JR, Global distribution of Beryllium isotopes in deep ocean water as derived from Fe-Mn crusts, 213–226, Copyright (1996) with permission from Elsevier Science.) ^{10}Be is a cosmogenic nuclide that enters the ocean by precipitation from the atmosphere, whereas ^{9}Be is stable and enters the oceans by erosion. The global deep-water τ of ^{10}Be is 600 y, allowing for preservation of distinct gradients between the basins.

emerged dating back through the entire Phanerozoic (**Figure 3**), and at a much higher resolution for the Cenozoic (**Figure 4**). The curve was time-calibrated using biostratigraphy and, in some cases, geochronology of intercalated ash layers. This makes Sr isotopes a very useful tool for stratigraphy. Ages of high confidence can be determined for those periods where $^{87}Sr/^{86}Sr$ underwent strong changes. For some periods

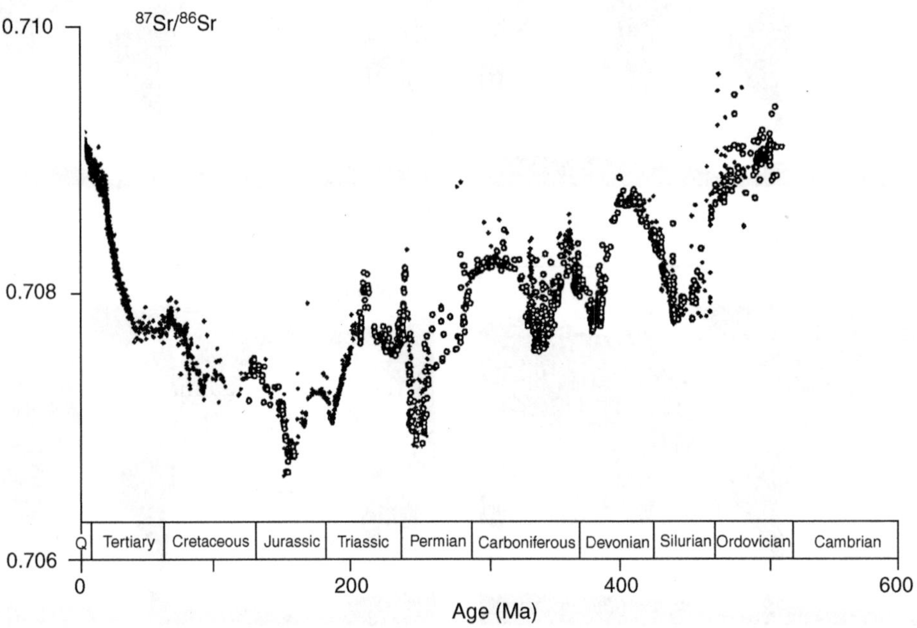

Figure 3 $^{87}Sr/^{86}Sr$ variations for the Phanerozoicum based on samples of brachiopods, belemnites, conodonts, foraminifera, and samples of micritic matrix. (Adapted from *Chemical Geology*, 161, Veizer J, Ala D, Azmy K *et al.*, $^{87}Sr/^{86}Sr$, $\delta^{13}C$, $\delta^{13}O$ evolution of Phanerozoic sea water, 59–88, Copyright (1999), with permission from Elsevier Science.)

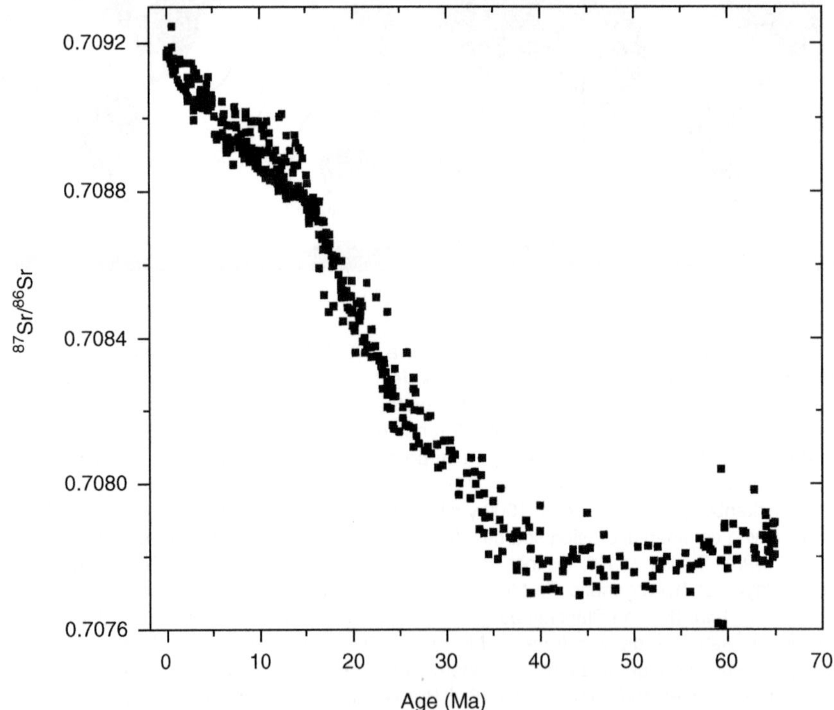

Figure 4 Cenozoic evolution of marine $^{87}Sr/^{86}Sr$, based primarily on analyses of foraminifera from all oceans. Data sources are as in McArthur (1994).

unique ages cannot be assigned. For example, the slope in $^{87}Sr/^{86}Sr$ in the Early Tertiary period is too low to allow for high-resolution stratigraphy.

The changes in the $^{87}Sr/^{86}Sr$ ratio are controlled by several processes. These are (1) the mid-ocean ridge flux, which is in turn controlled by the spreading rates of the seafloor; (2) the rate of chemical weathering, in particular that of feldspar and calcite; (3) the areal extent of the continents above sea level; (4) changes in the carbonate compensation depth (CCD). Seawater Sr isotope ratios vary within a small range over time. This is because of the long and efficient mixing of Sr, and also because dissolution of continental carbonate buffers seawater Sr isotope compositions within a narrow range.

Some of the trends visible in **Figure 3** are a relatively slow decrease in $^{87}Sr/^{86}Sr$ ratios from the Cambrian to the Jurassic period, upon which are superimposed a number of relatively large fluctuations. The sharp decline and following rise at the Permian/Triassic boundary are spectacular, and are thought to reflect either an extreme climate and weathering change, or the sudden mixing of a previously stratified ocean. A second main trend is a relatively rapid increase in $^{87}Sr/^{86}Sr$ ratios from the Jurassic to the present, upon which a number of relatively small fluctuations are superimposed.

Much discussion has been stimulated by the strong Cenozoic increase in $^{87}Sr/^{86}Sr$ (**Figure 4**) that has been linked by some workers to the uplift of the Himalayas and the ensuing delivery of high $^{87}Sr/^{86}Sr$ by Himalayan rivers. This view was challenged by the recent observation that $^{187}Os/^{188}Os$ (**Figure 5**), showing a similar and simultaneous increase, cannot be attributed to the dissolved flux draining the rising Himalayas. Therefore there must be a different cause for the rise in Sr too, possibly a worldwide increase in weathering rate. Similar attention was focused on the pronounced rise over the past 2.5 My. One possibility is that the latter can be explained simply by changes in sea level during the glaciations. However, calculations have shown that sea level variations of 200–300 m would be required – far in excess of the *c.* 100–150 m of change believed to have taken place during the Quarternary period. Therefore a much more plausible explanation is a change in the

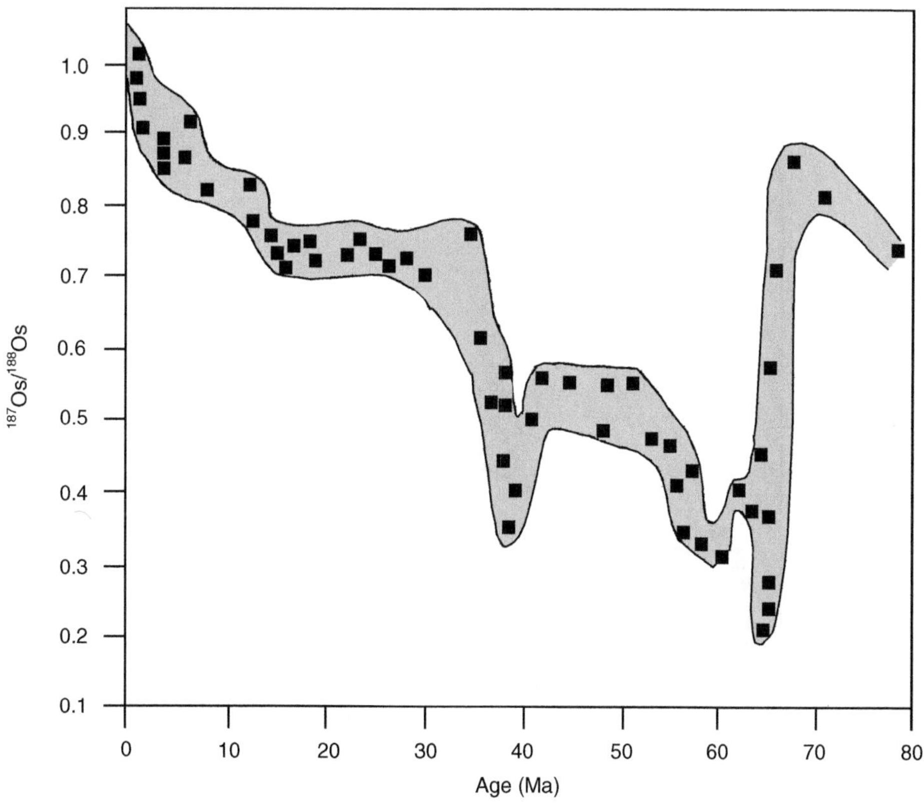

Figure 5 Marine $^{187}Os/^{188}Os$ record for the past 80 My in all oceans from H_2O_2-leached metalliferous and hydrogenetic sediments. Note that the pronounced excursion to low ratios at the K/T boundary (65 My) is explained by a meteorite impact. (Reprinted from *Geochimica et Cosmochimica Acta*, 63, Pegram WJ, Turekian KK. The osmium isotopic composition change of Cenozoic sea water as inferred from a deep-sea core corrected for meteoritic contributions, 4053–4088, Copyright (1999), with permission from Elsevier Science.)

terrigenous Sr input supplied by rivers to the oceans. Either a change in the dissolved Sr flux may have occurred or a change in the isotopic composition of rivers, or a combination of both. The increased erosion and availability of weatherable mineral surfaces resulting from the build up of glaciers during the northern hemisphere glaciation might have provided these changes.

Os

Because of the exceedingly small differences in $^{187}Os/^{188}Os$ between the different ocean basins, Os has a potential similar to Sr to serve as an isotope stratigraphic tool. It is particularly valuable for carbonate-poor pelagic clays and metalliferous sediments, from which Os is extracted by leaching techniques. The Os isotope curve (**Figure 5**) bears many similarities to the Sr isotope curve (**Figure 4**), with the exception of three excursions to low $^{187}Os/^{188}Os$ ratios. The spectacular drop at the Cretaceous/Tertiary boundary is probably of meteorite impact origin. The second, mid-Paleocene decrease and also the slow recovery following the impact can be explained by exposure of coastal sediments imprinted by the meteoritic Os from the K/T boundary. The Eocene-Oligocene excursion ($\sim$33 Ma) has been explained by an increased supply of

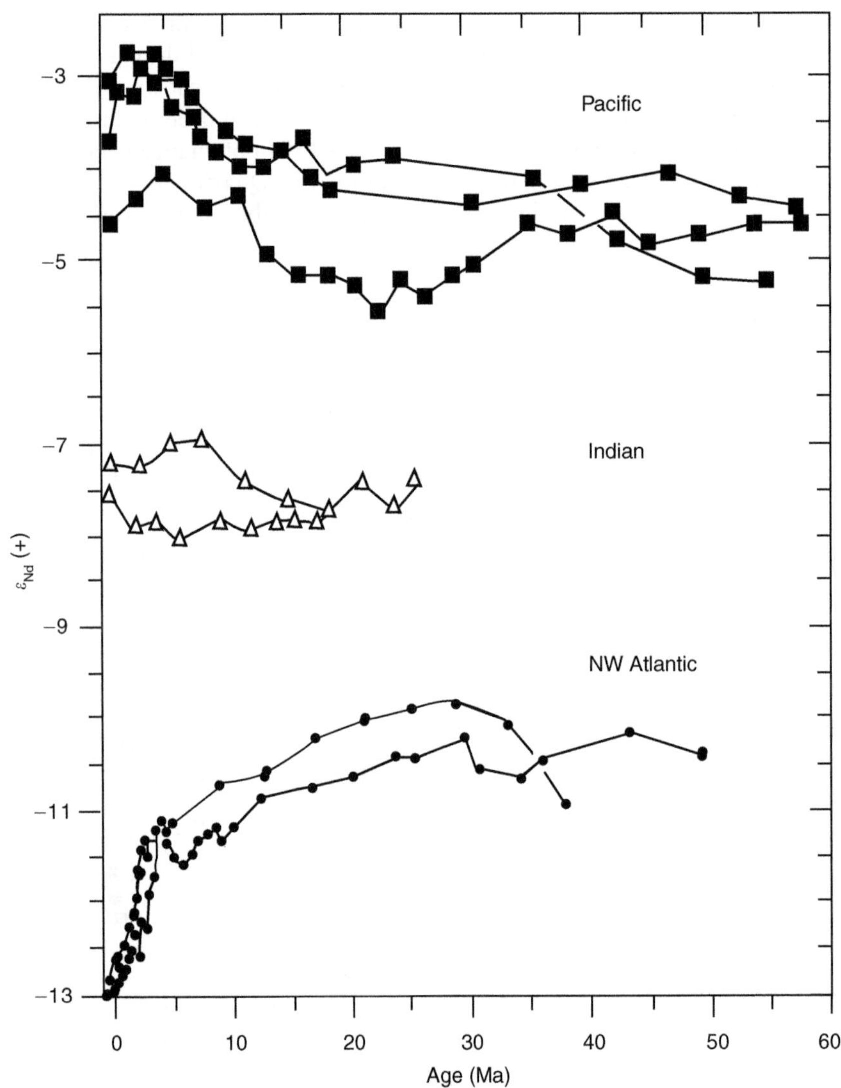

Figure 6 Nd isotope variations in Cenozoic sea water based on the analyses of hydrogenetic Fe-Mn crusts. Note that despite the assumed τ of more than 1000 y the oceans have kept their Nd isotope provinciality observed today throughout the last 50 My. (Reprinted from *Earth and Planetary Science Letters*, 155, O'Nions RK, Frank M, von Blanckenburg F, Ling HF. Secular variations of Nd and Pb isotopes in ferromanganese crusts from the Atlantic, Indian, and Pacific Oceans, 15–28, Copyright (1998) with permission from Elsevier Science.)

nonradiogenic Os from peridotite weathering (**Table 2**).

As is the case for $^{87}Sr/^{86}Sr$, there is a strong increase of $^{187}Os/^{188}Os$ towards more radiogenic values over the past ~14 My. As stated above, this has been linked to the rise of the Himalayas, but recent analyses of Himalayan river waters for Os isotope compositions do not support this view. A more likely possibility is the weathering of ancient crystalline terranes exposed by physical erosion, or the weathering of black shales. These organic-rich sediments have a high Re/Os ratio. Therefore old black shales have the potential to supply Os with a very high $^{187}Os/^{188}Os$ ratio to sea water.

Nd

Nd isotopes, analyzed with low-time resolution in Fe-Mn crusts and given as ε_{Nd} units, are presented in **Figure 6**. The most outstanding feature is that the provinciality, observed in ε_{Nd} of the modern oceans (**Figure 2C**), has been a feature prevailing as far back as 55 Ma. It is thought that despite the large isotopic variability in source materials, the τ of Nd is sufficiently long to allow for efficient intra-basin homogenization. This produces the basins' characteristic Nd isotope blend. Significant variations in ε_{Nd} are mainly observed for the past 5 My. In the Pacific a decrease in ε_{Nd} over the past 5 My might be due to an increased flow of AABW (with low ε_{Nd}, **Figure 2C**), a rearrangement of the thermohaline circulation following the opening of the Indonesian throughways for exchange of thermocline waters, or an increase in dust input. The strong decrease in north-west Atlantic ε_{Nd} over the past 3–4 My has been linked to a strengthening of NADW production following closure of the Panama gateway (suppressing northward flow of AABW and AAIW high in ε_{Nd}). However, a pronounced decrease of ε_{Nd} in a shallow ferromanganese crust off Florida has ocurred as early as 8–5 Ma. This has been ascribed to a decreasing inflow of Pacific water through the narrowing Panama gateway. Thus, a change in the amount and style of weathering associated with the onset of northern hemisphere glaciation at 3 Ma is a more likely explanation for the Pleistocene decrease in ε_{Nd}. In particular the Labrador Sea, a major source of NADW, is surrounded by ancient rocks with ε_{Nd} as low as -40 and is supplying deep water with ε_{Nd} of -20 to NADW (**Figure 2C**). An increase in weathering of this component has the potential to drive the Nd in NADW towards lower compositions.

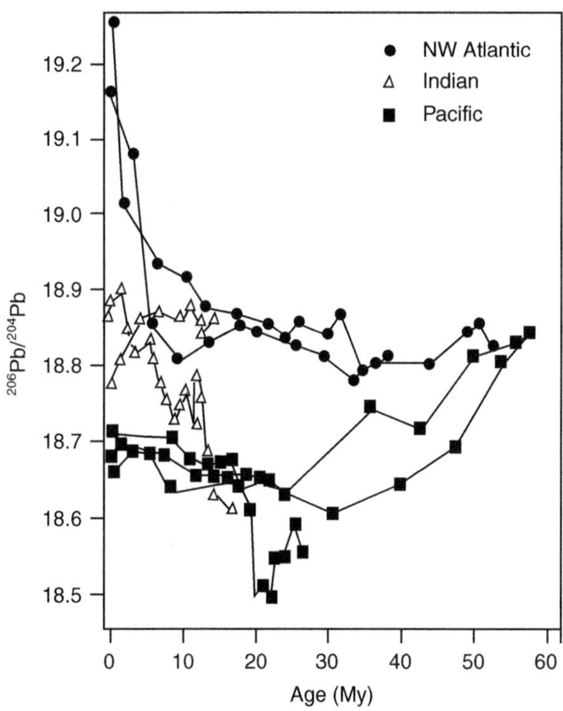

Figure 7 Pb isotope variations in Cenozoic sea water based on the analyses of hydrogenetic Fe-Mn crusts. Because of the short τ of Pb the oceans have maintained distinct isotope signals throughout the past 50 My. There is more intra-basin variability with time because the short τ does not allow such efficient lateral homogenization within basins as is the case for Nd. Therefore, local changes in erosion are much more visible in Pb isotope variations. (Reprinted from *Geochimica et Cosmochimica Acta*, 63, Frank M, O'Nions RK, Hein JR, Banaker VK, 1689–1708, Copyright (1999) with permission from Elsevier Science.)

Pb

No information can be obtained on natural Pb from modern sea water, because of the strong contamination by industrial Pb. Therefore, the pre-anthropogenic Pb distribution has to be obtained from chemical sediments. $^{206}Pb/^{204}Pb$ time-series, analyzed in Fe-Mn crusts (**Figure 7**), show patterns of changes that are less clear than those of Nd. Relative differences even within ocean basins are much larger than those observed for Nd. This may be expected from the short residence time of Pb (**Table 1**), which does not allow for lateral within-basin homogenization to the same degree as Nd. Therefore, local sources dominate the natural Pb budget, and their changes in flux introduce strong isotope variability. For example, in the Indian Ocean a crust located close to the circumpolar current shows a distinctly different history from the more northerly one, experiencing strong changes. The pronounced increase in north-west Atlantic $^{206}Pb/^{204}Pb$ can be attributed, as ε_{Nd}, to a change in NADW production, but is more likely due to a change in weathering of the

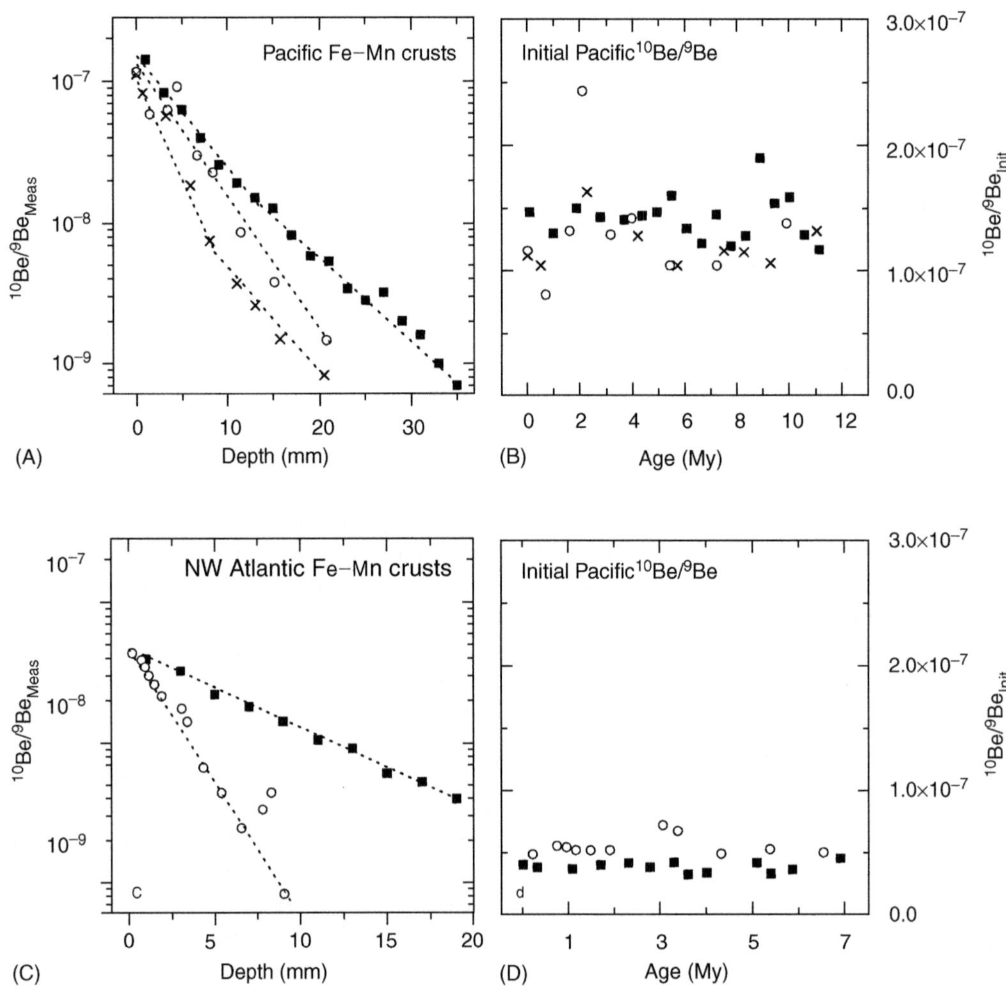

Figure 8 $^{10}Be/^{9}Be$ ratios in Pacific (A, B) and Atlantic (C, D) Fe-Mn crusts (Ling et al., 1997; von Blanckenburg and O'Nions, 1999). The smooth exponential decrease with depth usually observed in the measured data (as fitted following the stippled lines in A, C, where some individual outliers are attributed to alteration) is compatible with radioactive decay of ^{10}Be only ($T_{1/2} = 1.5$ My). A kink in the fits to the data is due to changes in growth rates. Changes in the initial ratio would be visible in the form of offsets in the data. At the time resolution of the samples taken (several hundred thousand years) these are not observed. Therefore, the derived growth rates can be used to time-correct the $^{10}Be/^{9}Be$ ratios for radioactive decay and to calculate initial ratios. The results shown in B and D indicate that the $^{10}Be/^{9}Be$ ratios in both the Pacific and the Atlantic have been within the range of modern sea water, for the last 7–10 My, despite presumably widely varying erosional input into the ocean.

glaciated areas or a change in the provenance of the erosional products.

Be

$^{10}Be/^{9}Be$ has been analyzed in Fe-Mn crusts as chronometer for the past 10 My. The smooth logarithmic decrease in $^{10}Be/^{9}Be$ (**Figure 8A** and C) is compatible with radioactive decay of ^{10}Be. This allows for calculation of growth rates and reconstruction of the initial $^{10}Be/^{9}Be$ ratios. Quite unexpectedly these initial $^{10}Be/^{9}Be$ ratios have been within the present-day range of the respective ocean basins through the past 7–10 My. Since the modern ocean basins do display strongly different $^{10}Be/^{9}Be$ ratios (**Figure 2F**), and since a τ of *c.* 600 y favors exchange of Be between basins, the relative constancy of these ratios with time suggests that large-scale changes in deep-water circulation have not taken place within the measured period.

Discussion and Conclusion

Changes in the style of weathering have the potential to change the isotope composition of tracers released to the sea. Leaching experiments on fresh mechanically (glacially) weathered rocks have shown a strongly incongruent release (meaning the release of tracers with varying isotope compositions depending

Table 3 Incongruent release of isotopes from strongly mechanically weathered continental rocks

Isotope ratio	*Isotope change*	*Original fresh material*	*Experimental evidence*
$^{87}Sr/^{86}Sr$	0.725→0.795	Young granitic moraine, Wind River Range, Wyoming, USA	Ammonium acetate leach River water composition HCl leach (Blum and Erel, 1995)
$^{187}Os/^{188}Os$	1.5→9.5	Young granitic moraine, Wind River Range, Wyoming, USA	HCl leach Young terrestrial Fe-Mn coatings Peucker-Ehrenbrink and Blum (1998)
ε_{Nd}	−26→−42	Greenland river bedload Baffin Bay deep-sea sediment	HCl leach (own work)
$^{206}Pb/^{204}Pb$	15.2→22.0	Greenland river bedload Baffin Bay deep-sea sediment	HCl and HBr leach (own work)

on the mode of liberation from rocks) of radiogenic Sr, Os, and Pb, and nonradiogenic Nd (up to 16 ε_{Nd} units, **Table 3**). Strongly chemically weathered rocks do not show such an incongruent release. Therefore, some of the long-term changes seen in the isotope evolution of seawater Sr, Os, Nd, and Pb might be attributable to these effects. Certainly the evolution of Sr and Os is only controlled by weathering, and the relative contribution of various sources, such as continental rocks, MORB (Mid Ocean Ridge Basalt), carbonate recycling, peridotite weathering, and cosmic dust. The uniformity between basins makes Sr and Os isotopes reliable stratigraphic tools.

$^{10}Be/^{9}Be$ appears to be a robust dating tool for the past ~10 My, due to its relatively constant observed initial ratio. It remains to be demonstrated whether Nd, Hf, and Pb have real value as tracers of past ocean circulation, as expected from their distinct water mass compositions, or whether their changes back through time merely record local changes in weathering.

Clearly the time resolution achievable from Fe-Mn crust studies is not sufficient to answer these questions. Much more insight will be obtained when these radiogenic tracers and Be isotopes are applied to sediments allowing tracer change studies at the resolution of a few thousand years. This will allow more reliable studies on the relationship between climate change, ocean circulation, and continental weathering.

Suggested Reading

The topic of radiogenic seawater tracers is too novel to be covered by a single monograph. All information is spread between numerous publications in international journals. Faure (1986) gives a general introduction into radiogenic isotope techniques. Analytical methods are summarized in a monograph by Potts (1987). Broecker and Peng (1982) provide a much-cited introduction into the topic of tracers in the sea. Ferromanganese crusts have recently been summarised by Hein *et al.* (1999). McArthur (1994) has reviewed the material suitable for Sr isotope analysis in carbonates, covering all ages of deposits from recent to the Precambrian. A summary of the suitability of marine clay minerals for isotope analyses is given by Stille *et al.* (1992). A brief summary on radiogenic seawater tracers, containing useful cross-references, has been published by the author (von Blanckenburg, 1999).

See also

Authigenic Deposits. Carbon Cycle. Cenozoic Climate – Oxygen Isotope Evidence. Cenozoic Oceans – Carbon Cycle Models. Cosmogenic Isotopes. Mid-Ocean Ridge Geochemistry and Petrology. Rare Earth Elements and their Isotopes in the Ocean. River Inputs. Stable Carbon Isotope Variations in the Ocean. Uranium-Thorium Series Isotopes in Ocean Profiles.

Further Reading

Broecker WS and Peng TH (1982) *Tracers in the Sea.* Palisades: Lamont-Doherty Geological Observatory.

Faure G (1986) *Principles of Isotope Geology.* John Wiley & Sons.

Hein JR, Koschinsky A, Bau M, Manheim FT, Kang JK, and Roberts L (1999) Cobalt-rich ferromanganese crusts in the Pacific. In: Cronan DS (ed.) *Handbook of Marine Mineral Deposits*, pp. 239–279. Boca Raton: CRC Press.

McArthur JM (1994) Recent trends in strontium isotope stratigraphy. *Terra Nova* 6: 331–358.

Potts PJ (1987) *A Handbook of Silicate Rock Analysis*. Blackie.

Stille P, Chaudhuri S, Kharaka YK, and Clauer N (1992) Neodymium, strontium, oxygen and hydrogen isotope compositions of waters in present and past oceans: a review. In: Clauer N and Chaudhuri S (eds.) *Isotopic Signatures and Sedimentary Rocks*, p. 555. Berlin: Springer Verlag.

von Blanckenburg F (1999) Tracing past ocean circulation? *Science* 286: 1862–1863.

NOBLE GASES AND THE CRYOSPHERE

M. Hood, Intergovernmental Oceanographic Commission, Paris, France

Introduction

Ice formation and melting strongly influence a wide range of water properties and processes, such as dissolved gas concentrations, exchange of gases between the atmosphere and the ocean, and dense water formation. As water freezes, salt and gases dissolved in the water are expelled from the growing ice lattice and become concentrated in the residual water. As a result of the increased salt content, this residual water becomes more dense than underlying waters and sinks to a level of neutral buoyancy, carrying with it the dissolved gas load. Dense water formation is one of the primary mechanisms by which atmospheric and surface water properties are transported into the interior and deep ocean, and observation of the effects of this process can answer fundamental questions about ocean circulation and the ocean–atmosphere cycling of biogeochemically important gases such as oxygen and carbon dioxide. Because it is not possible to determine exactly when and where dense water formation will occur, it is not an easy process to observe directly, and thus information about the rates of dense water formation and circulation is obtained largely through the observation of tracers. However, when dense water formation is triggered by ice formation, interaction of surface water properties with the ice and the lack of full equilibration between the atmosphere and the water beneath the growing ice can significantly modify the concentrations of the tracers in ways that are not yet fully understood. Consequently, the information provided by tracers in these ice formation areas is often ambiguous.

A suite of three noble gases, helium, neon, and argon, have the potential to be excellent tracers in the marine cryosphere, providing new information about the interactions of dissolved gases and ice, the cycling of gases between the atmosphere and ocean, and mixing and circulation pathways in high latitude regions of the world's oceans and marginal seas. The physical chemistry properties of these three gases span a wide range of values, and these differences cause them to respond to varying degrees to physical processes such as ice formation and melting or the transfer of gas between the water and air. By observing the changes of the three tracers as they respond to these processes, it is possible to quantify the effect the process has on the gases as a function of the physical chemistry of the gases. Subsequently, this 'template' of behavior can be used to determine the physical response of any gas to the process, using known information about the physical chemistry of the gas. Although this tracer technique is still being developed, results from laboratory experiments and field programs have demonstrated the exciting potential of the nobel gases to provide unique, quantitative information on a range of processes that it is not possible to obtain using conventional tracers.

Noble Gases in the Marine Environment

The noble gases are naturally occurring gases found in the atmosphere. **Table 1** shows the abundance of the noble gases in the atmosphere as a percentage of the total air composition, and the concentrations of the gases in surface sea water when in equilibrium with the atmosphere.

Other sources of these gases in sea water include the radioactive decay of uranium and thorium to helium-4 (^{4}He), and the radioactive decay of potassium (^{40}K) to argon (^{40}Ar). For most areas of the surface ocean, these radiogenic sources of the noble gases are negligible, and thus the only significant source for these gases is the atmosphere.

The noble gases are biogeochemically inert and are not altered through chemical or biological reactions, making them considerably easier to trace and quantify as they move through a system than other gases whose concentrations are modified through reactions. The behavior of the noble gases is largely determined by the size of the molecule of each gas and the natural affinity of each gas to reside in a

Table 1 Noble gases in the atmosphere and sea water

Gas	*Abundance in the atmosphere (%)*	*Concentration in seawater ($cm^3\,g^{-1}$)*
Helium	0.0005	3.75×10^{-8}
Neon	0.002	1.53×10^{-7}
Argon	0.9	2.49×10^{-4}

gaseous or liquid state. The main physical chemistry parameters of interest are the solubility of the gas in liquid, the temperature dependence of this solubility, and the molecular diffusivity of the gas. The suite of noble gases have a broad range of these properties, and the behavior of the noble gases determined by these properties, can serve as a model for the behavior of most other gases.

One unique characteristic of the noble gases that makes them ideally suited as tracers of the interactions between gases and ice is that helium and neon are soluble in ice as well as in liquids. It has been recognized since the mid-1960s that helium and neon, and possibly hydrogen, should be soluble in ice because of the small size of the molecules, whereas gases having larger atomic radii are unable to reside in the ice lattice. These findings, however, were based on theoretical treatises and carefully controlled laboratory studies in idealized conditions. It was not until the mid-1980s that this process was shown to occur on observable scales in nature, when anomalies in the concentrations of helium and neon were observed in the Arctic.

The solubility of gases in ice can be described by the same principles governing solubility of gases in liquids. Solubility of gases in liquids or ice occurs to establish equilibrium, where the affinities of the gas to reside in the gaseous, liquid, and solid state are balanced. The solubility process can be described by two principle mechanisms:

1. creation of a cavity in the solvent large enough to accommodate a solute molecule;
2. introduction of the solute molecule into the liquid or solid surface through the cavity.

In applying this approach to the solubility of gases in ice, it follows that if the atomic radius of the solute gas molecule is smaller than the cavities naturally present in the lattice structure of ice, then the energy required to make a cavity in the solvent is zero, and the energy required for the solubility process is then only a function of the energy required to introduce the solute molecule into the cavity. For this reason, the solubility of a gas molecule capable of fitting in the ice lattice is greater than its solubility in a liquid. The solubilities of helium and neon in ice have been determined in two separate laboratory studies, and although the values agree for the solubility of helium in ice, the values for neon disagree. The size of neon is very similar to the size of a cavity in the ice lattice, and the discrepancies between the two reported values for the solubility of neon in ice may result from small differences in the experimental procedure.

During ice formation, most gases partition between the water and air phases to try to establish equilibrium under the changing conditions, whereas helium and neon additionally partition into the ice phase. As water freezes, salt and gases are rejected from the growing ice lattice, increasing the concentrations of salt and gas in the residual water. Helium and neon partition between the water and ice reservoirs according to their solubility in water and ice. The concentrations of the gases in the residual water that have been expelled from the ice lattice, predominantly oxygen and nitrogen, can become so elevated through this process that the pressure of the dissolved gases in the water exceeds the *in situ* hydrostatic pressure and gas bubbles form. The gases then partition between the water, the gas bubble, and the ice according to the solubilities of the gases in each phase. This three-phase partitioning process can occur either at the edge of the growing ice sheet at the ice–water interface, or in small liquid water pockets, called 'brine pockets' in salt water systems, entrained in the ice during rapid ice formation.

Table 2 Noble gas partitioning in three phases

Partition phases	*Helium*	*Neon*	*Argon*
Bubble to water	106.8	81.0	18.7
Bubble to ice	56.9	90.0, 56.3	∞
Ice to water	1.9	0.9, 1.4	0

Table 2 quantitatively describes how the noble gases partition between the three phases when a system containing these three phases is in equilibrium in fresh water at 0°C. The numbers represent the amount of the gas found in one phase relative to the other. For example, the first row describes the amount of each gas that would reside in the gaseous bubble phase relative to the liquid phase; thus for helium, there would be 106.8 times more helium present in the bubble than in the water. This illustrates the small solubility of helium in water and its strong affinity for the gas phase. Because helium is 1.9 times more soluble in ice than in water, helium partitions less strongly between the bubble and ice phases compared to the partition between the bubble and water phases. The two numbers shown for neon represent the two different estimates for the solubility of neon in the ice phase. One estimate suggests that neon is less soluble in ice than in water, whereas the other suggests that it is more soluble in ice.

Application of the Noble Gases as Tracers

The noble gases have been used as tracers of air–sea gas exchange processes for more than 20 years.

Typically, the noble gases are observed over time at a single location in the ocean along with other meteorological and hydrodynamic parameters to characterize and quantify the behavior of each of the gases in response to the driving forces of gas exchange such as water temperature, wind speed, wave characteristics, and bubbles injected from breaking waves. Because both the amount and rate of a gas transferred between the atmosphere and ocean depend on the solubility and diffusivities of the gas, the noble gases have long been recognized as ideal tracers for these processes. In addition, argon and oxygen have very similar molecular diffusivities and solubilities, making argon an excellent tracer of the physical behavior of oxygen. By comparing the relative concentration changes of argon and oxygen over time, it is possible to account for the relative contributions of physical and biological processes (such as photosynthesis by phytoplankton in the surface ocean) to the overall concentrations, thus constraining the biological signal and allowing for estimates of the biological productivity of the surface ocean.

The observations of anomalous helium and neon concentrations in ice formation areas and the suggestion that these anomalies could be the result of solubility of these gases in the ice were made in 1983, and since that time, a number of laboratory and field studies have been conducted to characterize and quantify these interactions. The partitioning of the noble gases among the three phases of gas, water, and ice creates a very distinctive 'signature' of the noble gas concentrations left behind in the residual water. Noble gas concentrations are typically expressed in terms of 'saturation', which is the concentration of a gas dissolved in the water relative to its equilibrium with the atmosphere at a given temperature. For example, a parcel of water at standard temperature and pressure containing the concentrations of noble gases shown in column 2 of **Table 1** would be said to have a saturation of 100%. Saturations that deviate from this 100% can arise when equilibration with the atmosphere is incomplete, either because the equilibration process is slow relative to some other dynamic process acting on the system (for example, rapid heating or cooling, or injection of bubbles from breaking waves), or because full equilibration between the water and atmosphere is prevented, as in the case of ice formation.

Typical saturations for the noble gases in the surface ocean range from 100 to 110% of atmospheric equilibrium, due mostly to the influx of gas from bubbles. Ice formation, however, can lead to quite striking saturations of -70 to -60% for helium and neon and $+230$% for argon in the relatively undiluted residual water. Ice melting can also lead to large anomalous saturations of the noble gases, showing the reverse of the freezing pattern for the gas saturations, where helium and neon are supersaturated while argon is undersaturated with respect to the atmosphere.

The interactions of noble gases and ice have been well-documented and quantified in relatively simple freshwater systems. Observations of large noble gas anomalies in a permanently ice-covered antarctic lake were quantitatively explained using the current understanding of the solubility of helium and neon in ice and the partitioning of the gases in a three-phase system. Characteristics of ice formed from salt water are more complex than ice formed from fresh water, and the modeling of the system more complex. Using a set of equations developed in 1983 and measurements of the ice temperature, salinity, and density, it is possible to calculate the volume of the brine pockets in the ice and the volume of bubbles in the ice. With this type of information, a model of the ice and the dissolved gas balance in the various phases in the ice and residual water can be constructed. Such an ice model was developed during a field study of gas–ice interactions in a seasonally ice-covered lagoon, and the model predicted the amount of argon, nitrogen, and oxygen measured in bubbles in similar types of sea ice. No measurements are available for the amount of helium and neon in the bubbles of sea ice to verify the results for these gases. It is also possible to predict the relative saturations of the noble gases in the undiluted residual water at the ice–water interface, and this unique fingerprint of the noble gases can then serve as a tracer of the mixing and circulation of this water parcel as it leaves the surface and enters the interior and deep ocean. In this manner, the supersaturations of helium from meltwater have been successfully used as a tracer of water mass mixing and circulation in the Antarctic, and the estimated sensitivity of helium as a tracer for these processes is similar to the use of the conventional tracer, salinity, for these processes.

As an illustration of the ways in which the noble gases can be used to distinguish between the effects of ice formation, melting, injection of air bubbles from breaking waves, or temperature changes on dissolved gases, **Figure 1** shows a vector diagram of the characteristic changes of helium compared to argon resulting from each of these processes.

From a starting point of equilibrium with the atmosphere (100% saturation), both helium and argon saturations increase as a result of bubbles injected from breaking waves. Ice formation increases the saturation of argon and decreases the saturation of helium, whereas ice melting has the opposite effect.

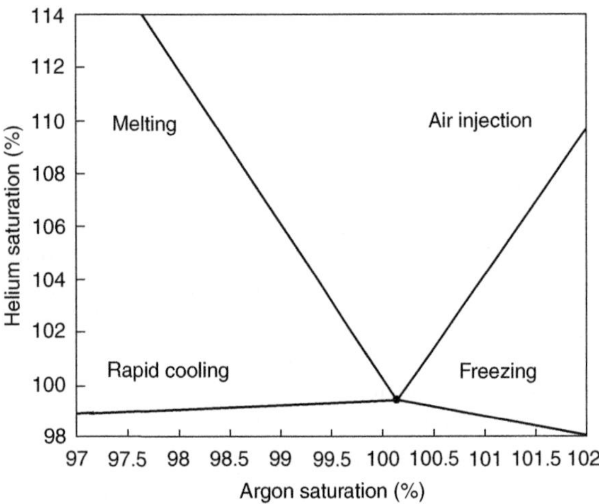

Figure 1 Vector diagram of helium and argon saturation changes in response to upper ocean processes.

Changes in temperature with no gas exchange with the atmosphere to balance this change can lead to modest changes in the saturations of the gases, where the gas saturations decrease with decreasing temperature and increase with increasing temperature. The trends presented here are largely qualitative indicators, since quantitative assessment of the changes depend on the exact nature of the system being studied. However, this diagram does illustrate the general magnitude of the changes that these processes have on the noble gases and conversely, the ability of the noble gases to differentiate between these effects.

Conclusions

The use of the noble gases as tracers in the marine cryosphere is in its infancy. Our understanding of the interactions of the noble gases and ice have progressed from controlled, idealized laboratory conditions to natural freshwater systems and simple salt water systems, and the initial results from these studies are extremely encouraging. This technique is currently being developed more fully to provide quantitative information about the interactions of dissolved gases and ice, and to utilize the resulting effects of these interactions to trace water mass mixing and circulation in the range of dynamic ice formation environments. Water masses in the interior and deep ocean originating in ice formation and melting areas have been shown to have distinct noble gas ratios, which are largely imparted to the water mass at the time of its formation in the surface ocean. By understanding and quantifying the processes responsible for these distinct ratios, we will be able to learn much about where and how the water mass was formed and the transformations it has experience since leaving the surface ocean. These issues are important for our understanding of the global cycling of gases between the atmosphere and the ocean and for revealing the circulation pathways of water in the Arctic, Antartic, and high latitude marginal seas. The noble gases could represent a significant addition to the set of tracers typically used to study these processes.

See also

CFCs in the Ocean. Long-Term Tracer Changes. Oxygen Isotopes in the Ocean. Stable Carbon Isotope Variations in the Ocean. Tritium–Helium Dating.

Further Reading

Bieri RH (1971) Dissolved noble gases in marine waters. *Earth and Planetary Science Letters* 10: 329–333.

Cox GFN and Weeks WF (1982) Equations for determining the gas and brine volumes in sea ice samples, *USA Cold Regions Research and Engineering Laboratory Report* 82-30, Hanover, New Hampshire.

Craig H and Hayward T (1987) Oxygen supersaturations in the ocean: biological vs. physical contributions. *Science* 235: 199–202.

Hood EM, Howes BL, and Jenkins WJ (1998) Dissolved gas dynamics in perennially ice-covered Lake Fryxell, Antarctica. *Limnology and Oceanography* 43(2): 265–272.

Hood EM (1998) *Characterization of Air–sea Gas Exchange Processes and Dissolved Gas/ice Interactions Using Noble Gases*. PhD thesis, MIT/WHOI, 98–101.

Kahane A, Klinger J, and Philippe M (1969) Dopage selectif de la glace monocristalline avec de l'helium et du neon. *Solid State Communications* 7: 1055–1056.

Namoit A and Bukhgalter EB (1965) Clathrates formed by gases in ice. *Journal of Structural Chemistry* 6: 911–912.

Schlosser P (1986) Helium: a new tracer in Antarctic oceanography. *Nature* 321: 233–235.

Schlosser P, Bayer R, Flodvik A, *et al.* (1990) Oxygen-18 and helium as tracers of ice shelf water and water/ice interaction in the Weddell Sea. *Journal of Geophysical Research* 95: 3253–3263.

Top Z, Martin S, and Becker P (1988) A laboratory study of dissolved noble gas anomaly due to ice formation. *Geophysical Research Letters* 15: 796–799.

Top Z, Clarke WB, and Moore RM (1983) Anomalous neon–helium ratios in the Arctic Ocean. *Geophysical Research Letters* 10: 1168–1171.

TRITIUM–HELIUM DATING

W. J. Jenkins, University of Southampton, Southampton, UK

Introduction: Tritium in the Oceans

Tritium (^3H) is the heaviest isotope of hydrogen. Its nucleus consists of one proton (making it hydrogen) and two neutrons. Inasmuch as it is chemically hydrogen, tritium exists within the global environment primarily as part of the water molecule. Thus it is a potentially useful tracer of the hydrologic cycle, and an ideal tracer of water motions within the ocean. Tritium is radioactive, decaying with a half-life of 12.45 years to the stable, inert daughter isotope ^3He. Because of its geologically short half-life, there is very little natural tritium in the environment. Small quantities are created by cosmic ray spallation (i.e. the smashing of atomic nuclei into small fragments by high-energy cosmic rays) in the upper atmosphere. The balance between production and radioactive decay leads to a global natural tritium inventory of approximately 4 kg.

This natural inventory was dwarfed by the production of tritium by the atmospheric testing of nuclear fusion weapons during the 1950s and early 1960s. During this period, several hundred kilograms of tritium were released, largely late in the test series, and primarily in the Northern Hemisphere. The detonations generally injected the tritium into the stratosphere, where it was quickly oxidized to form water vapor. Over a period of a few years, the tritiated water vapor was transferred, largely at mid-latitudes, to the troposphere, where it was rapidly 'rained out' to the earth's surface. The delivery of bomb tritium to the earth's surface was monitored by a number of WMO/IAEA (World Meteorological Organization (UN)/International Atomic Energy Authority) precipitation sampling stations. The pattern and timing of this delivery has been shown to consist of two primary components: a dominant northern, spike-like component, and a weaker southern component. Due to the geographic nature of the coupling between the stratosphere and the troposphere, tritium concentrations were elevated in both components toward higher latitudes, and weaker near the equator (**Figure 1**). Tritium levels in precipitation over land also tended to increase with altitude.

The northern component reflects the more immediate injection of bomb tritium into the northern hemispheric hydrologic system because virtually all of the major detonations occurred in the Northern Hemisphere. Prior to the bomb tests, the concentration of natural tritium in rainfall was of the order of 5–10 tritium units (1 TU = 1 tritium atom per 10^{18} normal hydrogen atoms). During the mid-1960s, tritium concentrations of more than several thousand TU were recorded in higher latitude, mid-continental locales such as Chicago, USA or Ottawa, Canada. The southern component, on the other hand, is much weaker in amplitude and more smeared out in time

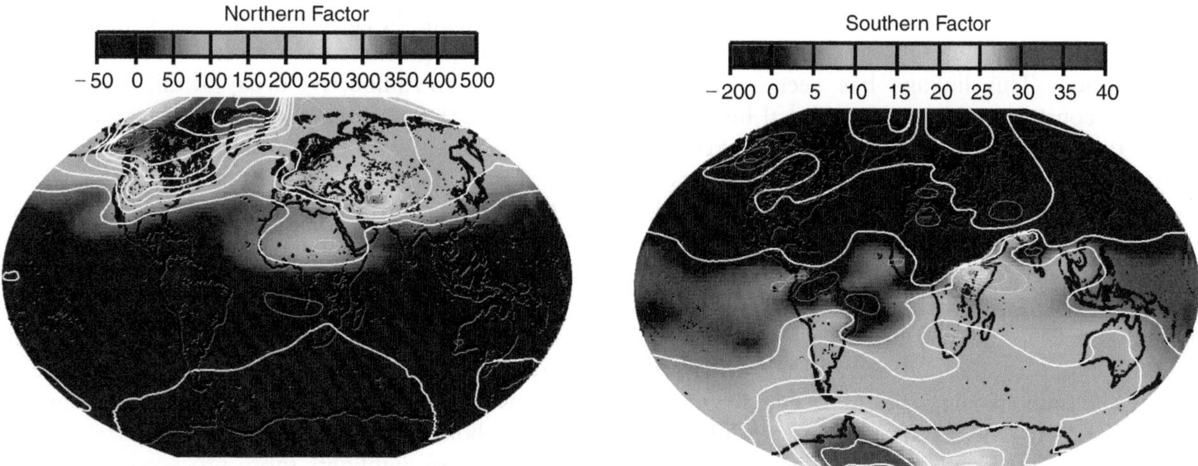

Figure 1 Spatial pattern of the two dominant principal components of bomb tritium in precipitation. These were derived from a statistical analysis of the time variation of bomb tritium in precipitation by S. Doney.

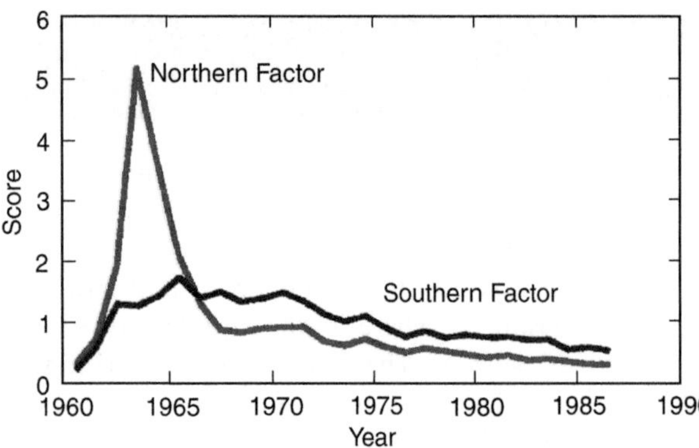

Figure 2 Time history of the two dominant principal components of tritium in precipitation. Note that the northern component (red curve) is more spike-like and the southern component (blue curve) is more 'smeared out' in time.

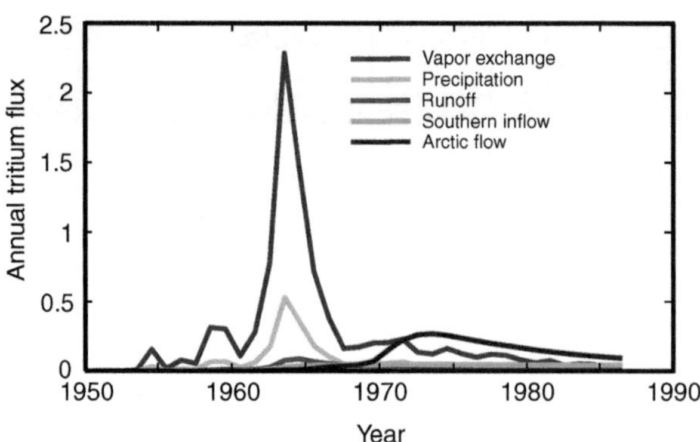

Figure 3 The various modes of tritium deposition to the North Atlantic over time. Note that during the peak of bomb-testing, vapor deposition was dominant, but that after the early 1970s, the influx of fresh water from the Arctic plays a prominent role.

since it results primarily from the cross-equatorial leakage of northern hemispheric tritium with few local sources (**Figure 2**).

Providing the production of bomb tritium is well known, the patterns and time variations of tritium concentrations in rain, environmental, and ocean waters provide useful insights into the hydrologic cycle and ocean circulation. Unfortunately, direct observation of environmental tritium levels was limited because the development of analytical techniques lagged events. Efforts are ongoing to reconstruct tritium records in precipitation by analysis of this isotope in tree rings. This has been made possible by the relatively recent development and improvement of high-sensitivity techniques of tritium measurement by ^{3}He regrowth.

The deposition of tritium to the oceans occurs both by direct precipitation and by vapor exchange. Vapor exchange is a two-way process, and in general dominates over the direct precipitation. There are relatively few direct measures of tritium concentration in atmospheric water vapor, but studies indicate that it is closely related to levels in precipitation. This linkage has been exploited in order to construct tritium depositional histories for ocean basins from tritium in precipitation records.

Another pathway whereby tritium enters the ocean is through continental runoff and river flow. Tritium deposited to the continents ultimately flows to the oceans via lakes, rivers and groundwater flow, but is retained within the continental hydrosphere for time-scales of many years, thereby introducing a delayed input to the oceans. Further, when computing the time-evolving tritium inventory within an ocean basin, it is necessary to consider inflow and outflow across the basin's boundaries.

The relative importance of the various inputs to the ocean varied with time. An analysis of the tritium budget for the North Atlantic Ocean, for example, shows that water vapor exchange (the magenta curve in **Figure 3**) and direct precipitation (the cyan curve in **Figure 3**) were the dominant inputs of tritium during the mid-1960s when the tritium 'spike' occurred. By the 1970s, however, the major input became the inflow of low salinity water from the Arctic (the dark blue curve in **Figure 3**). A substantial inventory of bomb tritium had been delivered to and held up within the Arctic fresh-water system, to be released more gradually to the subpolar oceans, and subsequently to the North Atlantic.

This input can be seen in the distribution of tritium in surface waters as observed during the early 1980s (**Figure 4**). **Figure 4** shows the intrusion of tritium-labeled waters along the east coast of Greenland and the Labrador Sea (red areas). This is superimposed on a general southward-decreasing trend. In response to the deposition of tritium, North Atlantic surface water concentrations rose rapidly, reaching values approaching 18 TU, or about 40 times greater than natural, prebomb, surface ocean levels. After peaking in 1964, surface water concentrations decreased, in part due to radioactive decay of this isotope, but also due to the dilution of surface waters with older, lower tritium waters from below, and lower concentration Southern Hemisphere waters. Consequently, the surface water decrease observed is significantly faster than the radioactive decay timescale.

The penetration of tritium into the oceans provides us with a direct visualization of the large-scale ventilation of the oceans. As a time-dependent dye, it stains water that has been in contact with the surface since the bomb tests in the 1960s. The time evolution of this picture highlights those processes that occur on decade time-scales that are important for climate change. **Figure 5** is a north–south section taken through the western North Atlantic in the early 1980s. The section shows how far the dye has penetrated along the pathway of the planetary-scale overturning circulation ('the global conveyor') and is an important quantitative measure of the rate of this overturning on decade timescales.

The boundary between high- and low-tritium waters at depths of 4–5 km corresponds to the transition region between the subtropical and subpolar gyres. In the subpolar gyre, deep convection injected tritium into deep and intermediate waters. In the subtropical gyre, subsurface penetration occurs from the north, primarily along deep western boundary currents. Otherwise, bomb tritium is restricted to the upper 1 km, tracing the bowl-like structure of the main thermocline, which it penetrates by subduction of fluid by a combination of wind stress convergence (a process called 'Ekman pumping', i.e. convergence of surface waters due to wind forcing effectively pushes water downward) and southward penetration under lighter, warmer waters.

A time series of tritium in the Sargasso Sea near Bermuda shows the penetration of this bomb tritium into the subtropical North Atlantic (**Figure 6**). To compensate for predictable radioactive decay, the concentration of tritium has been decay-corrected to one point in time (arbitrarily chosen here to be 1981, the approximate mid-point of the series). Two relatively sudden increases in tritium concentrations occur in the deep waters. The first appears at a depth of about 1500 m in the late 1970s, whereas the deeper one arrives in the late 1980s. These increases signal the arrival of waters that had been 'ventilated' or exposed to the surface since the bomb tests. The delayed arrival provides a measure of the transit time of properties southward from the outcropping regions, important knowledge for ocean climate models.

The time series, however, is dominated in the upper waters by the downward penetration of bomb

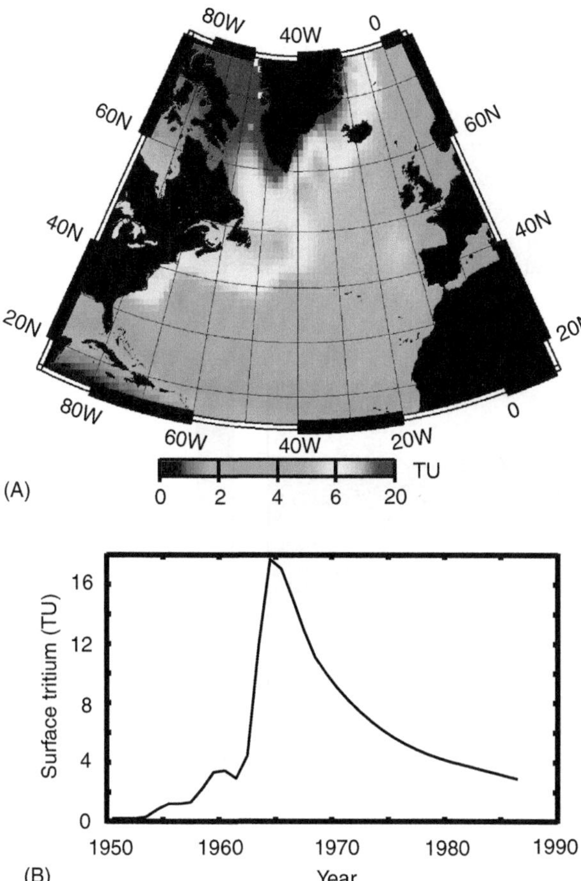

Figure 4 North Atlantic surface water tritium concentrations: (A) geographical distribution in the early 1980s; (B) variation with time in the subtropics.

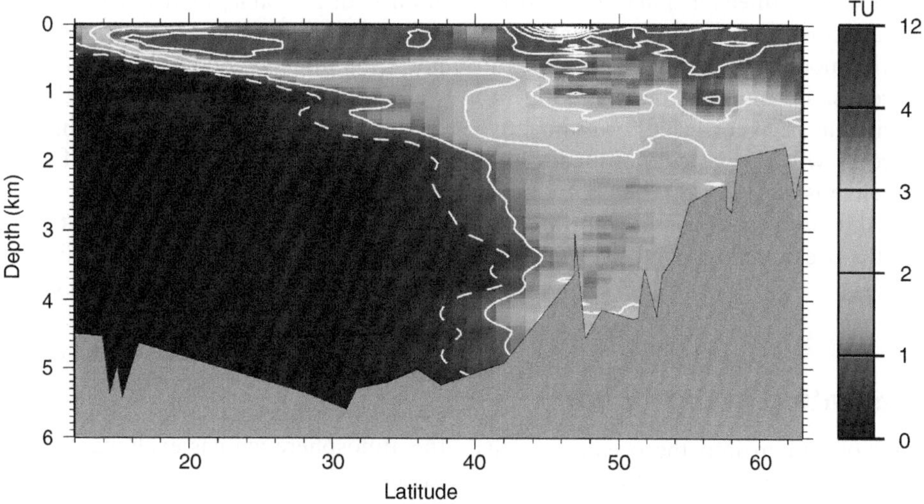

Figure 5 A North Atlantic tritium meridional section taken in the early 1980s.

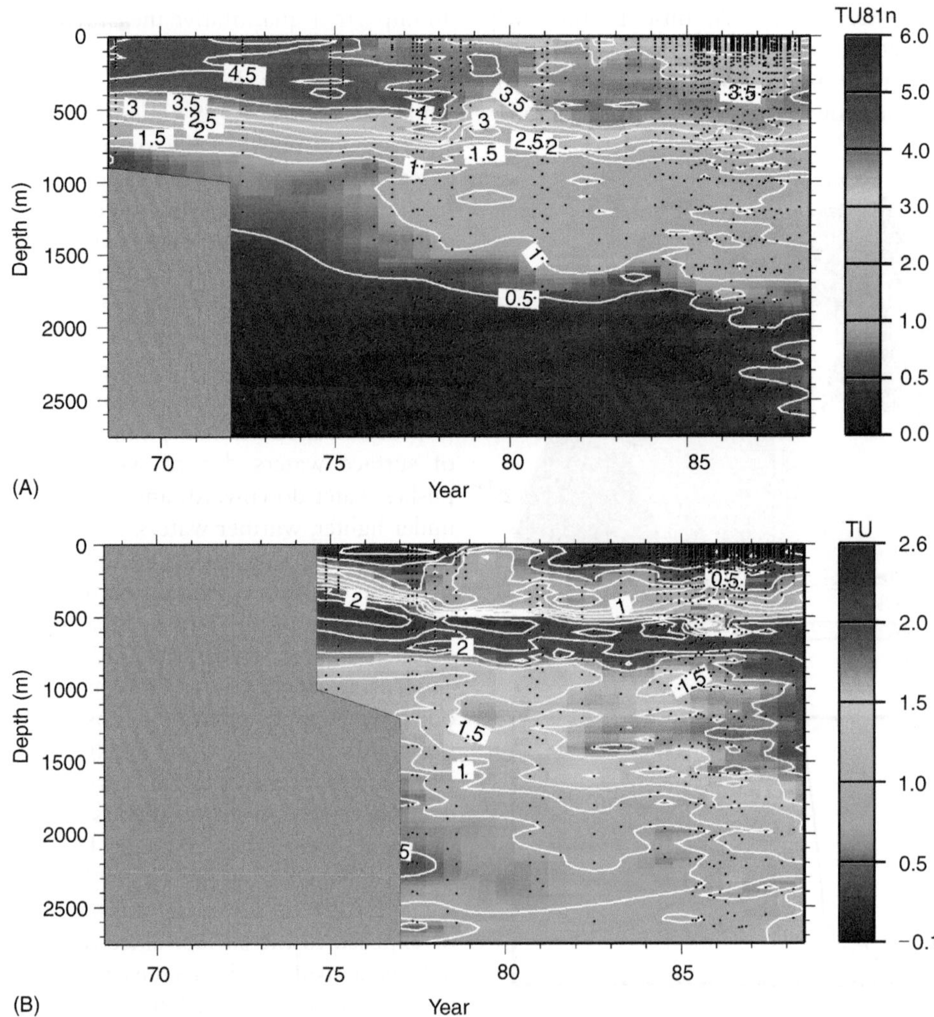

Figure 6 A Bermuda time series of (A) tritium and (B) ^{3}He. The tritium concentrations have been decay-corrected (that is, corrected for the effects of radioactive decay) to a fixed point in time (1981). This allows the effects of dilution and fluid motions to be seen.

tritium into the main thermocline. The tritium 'spike' first appears as a surface-intensified maximum at the beginning of the record, but then subsequently descends into the thermocline at a rate of about 20 m y^{-1}. As it descends, its intensity decreases due to dilution (the series has been corrected for radioactive decay). The rate at which this maximum descends into the thermocline is vital information for climate modeling; i.e., this information is important for predicting how the ocean will respond to changes on decade timescales.

Tritium-^{3}He Dating in the Ocean

!The penetration of tritium into the oceans, and its subsequent evolution, provides us with valuable information on ocean ventilation and large-scale circulation on multiyear and multidecade timescales. However, it is possible to use this tracer in combination with its stable, inert daughter ^{3}He to extend its utility to much shorter timescales, and provide a powerful measure of circulation and ventilation, as well as the rates of biological and chemical processing in the oceans. The manner in which this is accomplished can be seen in the following thought experiment.

Imagine a parcel of water at the sea surface (**Figure 7**). Tritium within this fluid parcel is decaying, producing its daughter product ^{3}He. (Half of the tritium decays to ^{3}He in 12.45 years, while in 24.9 years, one-quarter would be left, and in 37.4 years, only one-eighth would remain, etc.) However, because it is at the sea surface, this ^{3}He will be lost to the atmosphere via gas exchange. Thus no excess or 'tritiugenic' ^{3}He would accumulate. However, should this water parcel sink below the surface and lose contact with the atmosphere, tritiugenic ^{3}He would

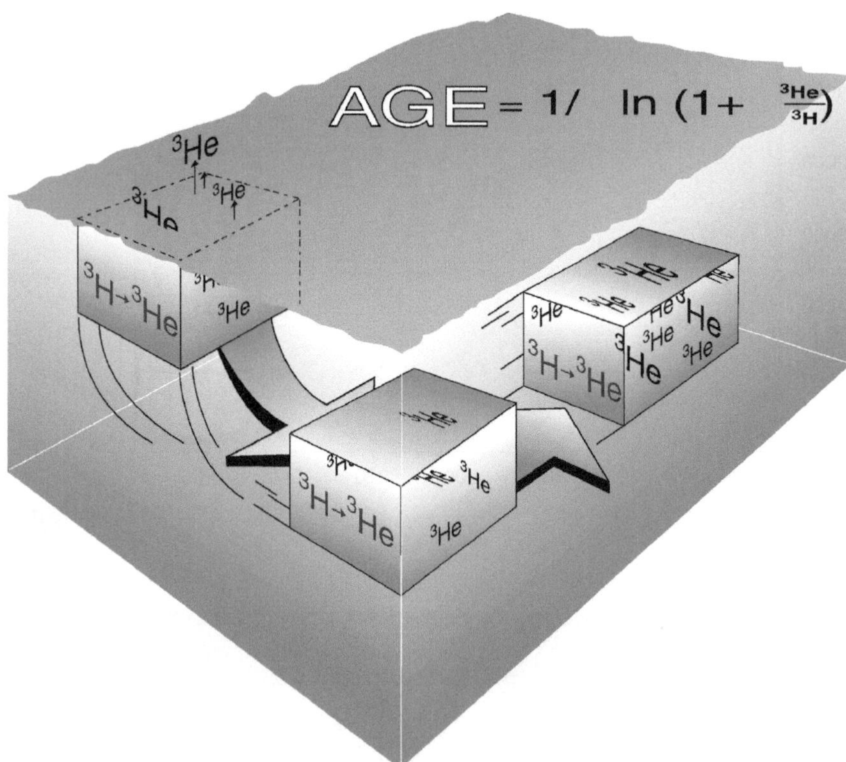

Figure 7 The tritium–helium dating concept.

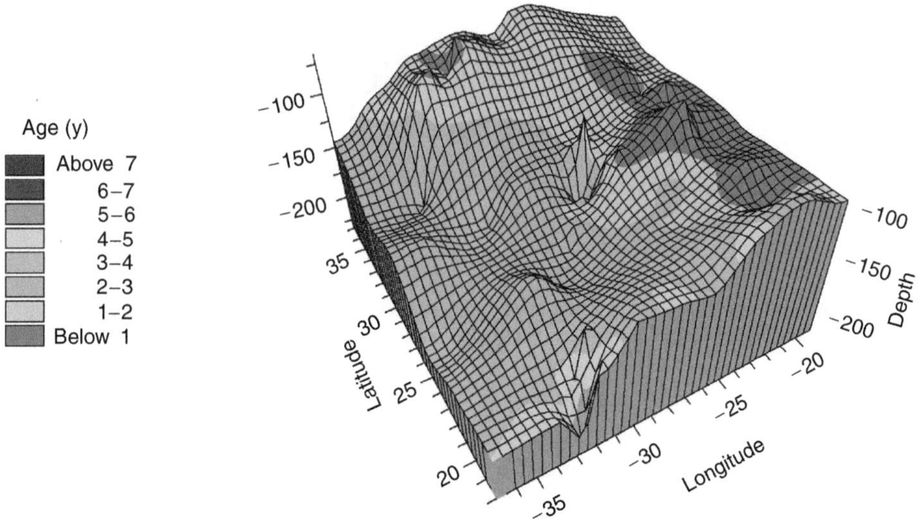

Figure 8 The distribution of tritium–helium age on a constant density surface (26.4 kg m^{-3}) in the subtropical North Atlantic.

accumulate at a predictable rate. By measuring both the tritium concentration and the accumulated ^{3}He in the fluid parcel, the time that has elapsed since the fluid was last in contact with the surface can be determined according to the equation:

$$\tau = 17.96 \ln\left(1 + \frac{[^3\mathrm{He}]}{[^3\mathrm{H}]}\right)$$

where τ is the tritium–^{3}He age in years, and [^{3}He] and [^{3}H] are the concentrations of ^{3}He and tritium in the water, respectively. For typical surface water concentrations of a few tritium units, elapsed times as short as a month or two can be detected, and the upper limit to the dating technique is of the order of 10–20 years (see discussion below). This range of timescales is ideal for studying shallow-ocean circulation, ventilation, and biogeochemical processing.

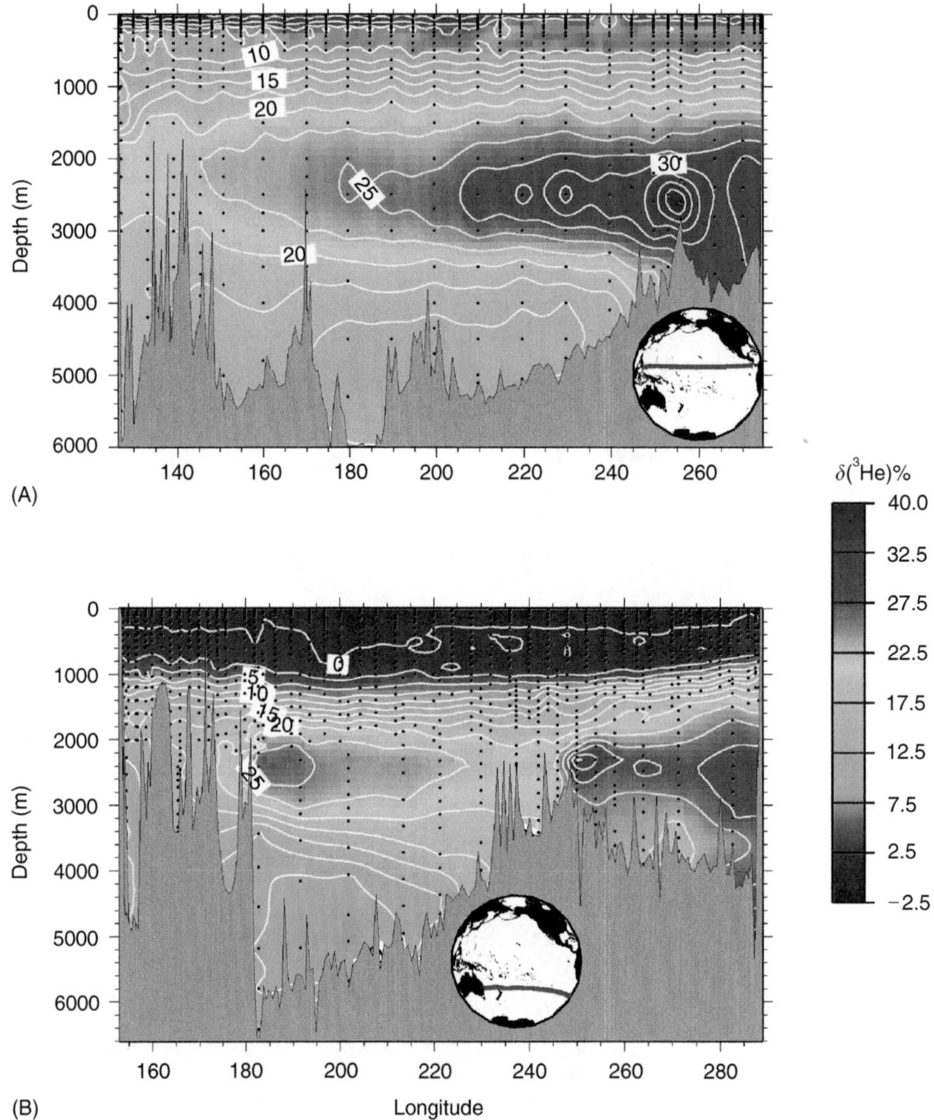

Figure 9 Two deep Pacific zonal sections of ^{3}He. Data are presented as isotope ratio anomaly (%), relative to atmospheric helium. Samples were processed during the WOCE Pacific hydrographic expeditions from 1989 to 1994.

In the subtropics, where wind-stress is convergent, water tends to be forced downward from the surface ocean into the thermocline. This downwelling is an important process for ventilation of the thermocline, and for driving the shallow gyre circulation. **Figure 8** shows the measured tritium–helium age as measured on a constant density surface (1024 kg m^{-3}) in the eastern subtropical North Atlantic in the early 1990s. Water is youngest in the north-east, where the horizon rises toward the ocean surface. In fact, this horizon intercepts the base of the wintertime mixing layer, and the tritium–helium age of the water is less than one year, indicating that it was in active contact with the previous winter's surface mixed layer. The age of the water increases monotonically as the layer deepens to the southwest, consistent with a south-westward flow associated with the large-scale circulation of the gyre. The next logical step would be to use the observed age-gradients to compute fluid velocities.

Before applying this technique quantitatively, however, there are two significant concerns that need to be considered. The first is the possible release of volcanic helium from submarine hydrothermal activity. This injection occurs at active volcanic centers, predominantly along midocean ridges, and to a lesser extent at near-axial seamounts. This helium is a mixture of primordial helium inherited during the earth's formation from the presolar nebula and radiogenic helium produced by the decay of long-lived radioactive U and Th isotopes in the deep earth. The injection of this helium is visible on a very large scale in the deep Pacific Ocean, where plumes of this helium extend across the basin (**Figure 9**). These plumes provide compelling evidence of ocean–crust interaction, terrestrial degassing, and trace deep ocean circulation (*see* Volcanic Helium).

As this helium tends to be enriched in ^{3}He compared to atmospheric helium, it may be confused with tritiugenic ^{3}He. Such injections tend to occur in deeper waters, away from the surface where one would tend to use tritium–^{3}He dating. Moreover, calculations indicate that despite the impressive signature in abyssal waters, the actual flux of volcanic ^{3}He is smaller than the tritiugenic production rate due to bomb tests. Clearly, however, caution should be exercised in areas where the two signals can interfere. The shallow North Atlantic, in particular, is well suited to tritium–^{3}He dating, partly because water masses tend to be younger there, and partly because seafloor spreading rates (and hence the rate of injection of volcanic ^{3}He into the deep water) are low. (One would expect, on average, that volcanic activity would be related to seafloor spreading rates.)

A second concern arises from the behavior of the tritium–^{3}He age in response to mixing. Returning to the model concept discussed earlier, it must be recognized that water does not circulate in discrete 'parcels' but is subjected to mixing. In general, this manifests itself in a 'nonlinear' response in the tritium–^{3}He age. For example, consider two fluid parcels that undergo mixing in equal proportions (**Figure 10**).

We consider, for simplicity, the case where the two are mixed in equal proportions, but the arguments apply equally well for an arbitrary mixture. In general, the tritium–^{3}He age of the mixture would be calculated from its tritium and ^{3}He concentrations, and will be different from the average of the component ages. That is, the age of the mixture is not equivalent to the mixture of the ages. The results for three example cases are shown in **Table 1**.

In the first case, the average age of the two water masses should be slightly more than 22 years, but the tritium concentration of the mixture is dominated by water mass A, which is the younger water mass. In the second case, the mixture is significantly older than 'average age', again because it is dominated by the higher tritium component. Only when the two components are of equal tritium concentration (case 3) does the mixture age more closely match the average of the components. Even here, there is a deviation due to the logarithmic nature of the age dependence.

Consideration of the scenarios presented in **Table 1** reveals that when water masses mix, the tritium–^{3}He age of the resultant mixture is weighted in favor of the water mass component with the greater tritium concentration. The implication of this is that a small admixture of a young, relatively tritium-rich water mass will depress the tritium–^{3}He age disproportionately. Therefore, there will be a tendency for the tritium–^{3}He age to be an underestimate of the true age in the presence of mixing. Although it seems a serious concern, consideration of real-world oceanographic situations indicates that this is not a significant problem for timescales of less than a decade.

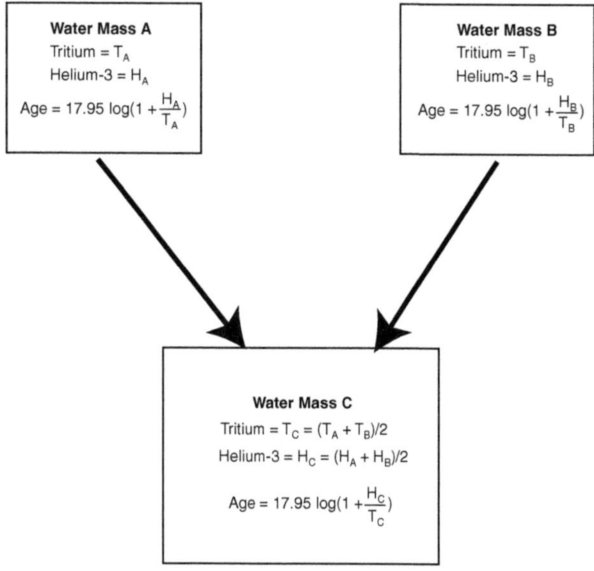

Figure 10 The effect of mixing on the tritium–helium age.

Table 1 Examples of water mass mixing effects on the tritium–helium age

		[^{3}H]	*[^{3}He]*	*Age (y)*
Case 1	Watermass A	10	1	1.71
	Watermass B	1	10	43.04
	50 : 50 Mixture	5.5	5.5	12.45
Case 2	Watermass A	10	100	43.04
	Watermass B	1	0.1	1.71
	50 : 50 Mixture	5.5	50.05	41.51
Case 3	Watermass A	10	10	17.95
	Watermass B	10	1	1.71
	50 : 50 Mixture	5	5.5	13.32

The effects of mixing on the tritium–^{3}He age have been quantified by the development of an advection–diffusion equation for the age. This is accomplished by combining the definition of the tritium–^{3}He age (τ) with the advection–diffusion equations for tritium and ^{3}He.

$$\frac{\partial \tau}{\partial t} + \vec{u} \cdot \nabla \tau = \nabla(k \nabla \tau) + 1 - k\left(\frac{\nabla [^3\mathrm{He}]}{[^3\mathrm{He}]} + 2\frac{[^3\mathrm{H}]}{[^3\mathrm{H}]}\right) . \nabla \tau$$

where κ is the turbulent diffusivity and u is the fluid velocity. The equation appears similar to that of an ideal age tracer (A), governed by

$$\vec{u}\cdot\nabla A = \nabla(k\nabla A) + 1$$

except for the presence of the unsteady (time derivative) term and the last term on the right. The unsteady term arises from the fact that the parent distributions are changing with time, and the age distribution is adjusting accordingly. The last term appears more as a pseudovelocity that is a direct manifestation of the nonlinear mixing behavior exemplified in the two-water-mass thought experiment described earlier. Although the equation appears complex, the key point is that all the terms are observable. That is, given field observations of the tracers, the terms can be computed to within a value of κ. The effects on the shallowest surfaces are small. Analysis of actual distributions within the shallow North Atlantic, for example, shows that deviations from 'ideal' behavior are negligibly small. Moreover, combining the age distributions with other tracers, for example salinity, and with geostrophic constraints, permits the determination of absolute velocities within the main thermocline to a resolution of order 0.1 cm s^{-1}.

See also

Elemental Distribution: Overview.

Further Reading

Clarke WB, Jenkins WJ, and Top Z (1976) Determination of tritium by mass spectrometric measurement of ^{3}He. *International Journal of Applied Radioisotopes* 27: 515.

Doney SC, Glover DM, and Jenkins WJ (1992) A model function of the global bomb tritium distribution in precipitation, 1960–1986. *Journal of Geophysical Research* 97: 5481–5492.

Doney SC, Jenkins WJ, and Östlund HG (1993) A tritium budget for the North Atlantic. *Journal of Geophysical Research* 98(C10): 18069–18081.

Jenkins WJ (1978) Helium isotopes from the solid earth. *Oceanus* 21: 13.

Jenkins WJ (1992) Tracers in oceanography. *Oceanus* 35: 47–56.

Jenkins WJ (1998) Studying subtropical thermocline ventilation and circulation using tritium and ^{3}He. *Journal of Geophysical Research* 103: 15817–15831.

Jenkins WJ and Smethie WM (1996) Transient tracers track ocean climate signals. *Oceanus* 39: 29–32.

VOLCANIC HELIUM

J. E. Lupton, Hatfield Marine Science Center, Newport, OR, USA

Introduction

Volcanic activity along the global mid-ocean ridge system and at active seamounts introduces a helium-rich signal into the ocean basins that can be used to trace patterns of ocean circulation and mixing. Helium is extracted from oceanic volcanic rocks by circulating sea water and then injected into the ocean as helium dissolved in submarine hydrothermal vent fluids. Hydrothermal venting produces plumes in the ocean that are highly enriched in a variety of tracers, including heat, helium, manganese, iron, methane, and hydrogen. Among these, volcanic helium is a particularly useful tracer because it has such a high concentration in hydrothermal fluids relative to the background values ofhelium in sea water, and because it is stable and conservative, i.e., helium does not decay radioactively and is not affected by any chemical or biological processes. By making careful measurements of the relative abundance of heliumisotopes, it is possible to trace hydrothermal helium plumes for thousands of kilometers from the source regions.

There are two stable isotopes of helium, ^{3}He and ^{4}He, which vary in their ratio by over three orders of magnitude in terrestrial samples. The Earth's atmosphere is well mixed with respect to helium and contains helium with a uniform isotopic composition of $^3He/^4He = 1.39 \times 10^{-6}$. Atmospheric helium is a convenient standard for helium isotope determinations, and terrestrial $^3He/^4He$ ratios are usually normalized to the air ratio and expressed in units of R/R_A, where $R = {}^3He/^4He$ and $R_A = ({}^3He/^4He)_{air}$. In contrast to atmospheric helium ($R/R_A = 1$), the radiogenic helium produced by α-decay of U and Th series isotopes has a much lower ratioof $R/R_A 0.1$, while the volcanic helium that is derived from the Earth's mantle is highly enriched in ^{3}He ($R/R_A = 5-30$). Thus volcanic helium has an isotopic composition distinct from other sources such as atmospheric helium or the helium produced by radioactive decay. This ^{3}He-rich mantle helium is sometimes called 'primordial' helium, since it is thought to be the remnant of a primitive component trapped in the Earth's interior since the time of its formation. This trapped component probably had $^3He/^4He = 1 \times 10^{-4}$or 100 R_A, similar to the helium found trapped in meteorites or in the solar wind, but has been modified to $R = 30R_A$ by dilution with radiogenic helium since the time the Earth was formed. Although there is a wide variety of volcanic sources in the oceans, including subduction zone volcanoes and hot spot volcanoes, most of the oceanic volcanic helium is derived from activity along the global mid-oceanridge system. While the $^3He/^4He$ ratio of mantle helium shows a wide range of variation, the helium from mid-ocean ridgesfalls in a much narrower range of $R/R_A = 7-9$.

In order of decreasing importance, the most abundant forms of helium in sea water are dissolved atmospheric helium, volcanic helium, and to a lesser degree radiogenic helium from sediments. Thereis also an input of pure ^{3}He into the oceans from tritium(^{3}H), the radioactive isotope of hydrogen, which decays to^{3}He with a half-life of 12.4 years. Because tritium isgenerally found only in the upper ocean, ^{3}He from tritium decay(tritiogenic helium) is only significant at depths less than about 1000 m.

Although there are only two isotopes of helium, it is still possible to clearly distinguish submarine volcanic helium from the other components because of its high $^3He/^4He$ ratio and because volcanic helium is introduced at mid-depth rather thanat the ocean surface or on the abyssal plain.

Units

For samples highly enriched in helium such as volcanic rocks and hydrothermal vent fluids, the helium isotope ratio is usually expressed in the R/R_A notation described above. However, for the relatively small variations observed in sea water samples, the $^3He/^4He$ variations are usually expressed as $\delta(^3He)$, which is the percentage deviation from the ratio in air, defined as in eqn [1].

$$\delta(^3He) = 100[(R/R_A) - 1] \qquad [1]$$

Here again $R = {}^3He/^4He$ and $R_A = ({}^3He/^4He)_{air}$. Thus $R/R_A = 1.50$ is equivalent to $\delta(^3He) = 50\%$.

History and Background

The first attempt to detect nonatmospheric helium in the oceans was made by Suess and Wänke in 1965, who predicted that the deep oceans should contain excess ^{4}He due to U and Th decay in sediments and in the ocean crust. Although they were correct about the existence of radiogenic helium in the oceans, their measurements were of insufficient precision to detect any ^{4}He enrichment above the dissolved air component. It is now known that the input of^{3}He-rich volcanic helium has a greater effect on both the^{3}He/^{4}He ratio and the ^{4}He concentration insea water than does the input of radiogenic helium.

Mantle or volcanic helium was first detectedon the Earth as an excess in the ^{3}He/^{4}He ratio indeep Pacific waters. Although this oceanic ^{3}He excess is derived from the helium residing in oceanic volcanic rocks, it was not until about five years later that mantle helium was directly measured in the volcanic rocks themselves. Clarke *et al.* in 1969 reported a 21% excess in the ^{3}He concentration at mid-depth above that expected forair-saturated water, and correctly attributed this excess to a flux ofprimordial helium leaking from the Earth's interior into the oceans and inturn into the atmosphere (see **Figure 1**). Using a box model for oceanic helium, they were able to estimate the global ^{3}He flux from the oceans into the atmosphere at 2 atoms ^{3}He cm^{-2}, a number that isstill in reasonable agreement with more recent flux estimates of 4–5atoms ^{3}He cm^{-2}.

The discovery of excess ^{3}He inthe oceans from localized sources distributed along the global mid-ocean ridge system led immediately to the use of this tracer for oceanographic studies. The Geochemical Ocean Sections Study (GEOSECS), which began in 1972, provided the first maps of the global distribution of helium in the oceans. Since then, several other oceanographic programs, including the World Ocean Circulation Experiment (WOCE), have added to our knowledge of the global helium distribution.

To illustrate the presence of volcanic helium in the oceans, a typical helium profile in the north Pacific Ocean is shown in **Figure 2**. The figure shows the-vertical variation in the ^{3}He/^{4}He ratio expressed as $\delta(^3\text{He})$ in%, and the ^{4}He concentration in nmol kg^{-1}. The values expected for air-saturated water (dashed lines) are shown for comparison. For the calculation of air-saturated values it is assumed that each water parcel equilibrated with the atmosphere at the potential temperature of the sample. This profile exhibits a broad maximum in the deep water, reaching avalue of $\delta(^3\text{He}) = 25.0\%$ at ~1850 m depth. Although this station is located at a distance of over 1500 km from the nearest active spreading center, the profile still exhibits a clear excess in ^{3}He/^{4}He in the 1500–3500 m depth range due to input of volcanic helium from the mid-ocean ridge system. The secondary maximum in the $\delta(^3He)$ profile at ~350 m depth is due to excess ^{3}He produced by tritium decay. That this peak is tritiogen-ichelium is evident because the peak in $\delta(^3\text{He})$ at 350 m depth is absent from the ^{4}He profile, indicating input of pure ^{3}He as would be expected for tritium decay. At the ocean surface $\delta(^3\text{He}) = -1.4\%$, which is very close to the expected value of $\delta(^3\text{He}) = -1.35\%$ for water in equilibriumwith air (^{3}He is slightly less soluble in water than^{4}He).

The absolute ^{4}He concentration (**Figure 2B**) also increases with depth, but not as dramatically as the ^{3}He/^{4}He ratio. Part of the ^{4}He increase is due to the

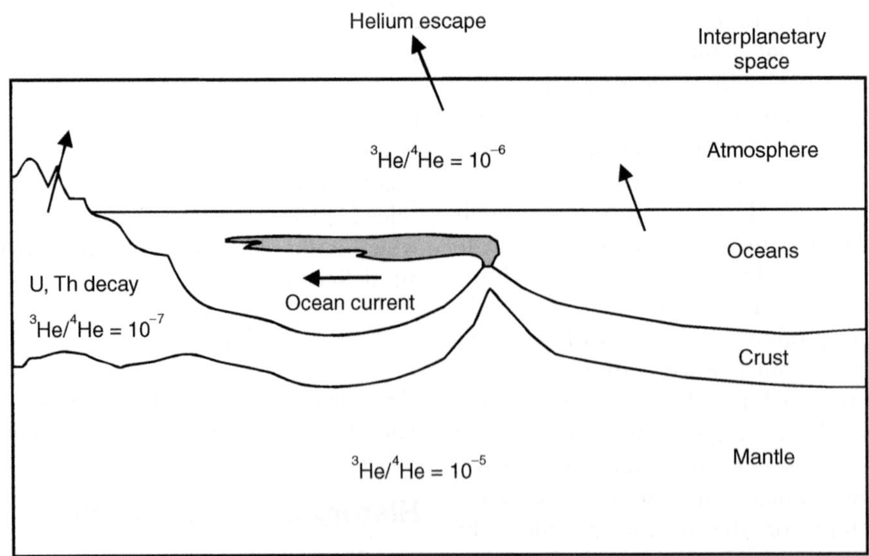

Figure 1 A schematic of the terrestrial helium budget, indicating the flux of helium from the Earth's mantle into the oceans, and in turn into the atmosphere.

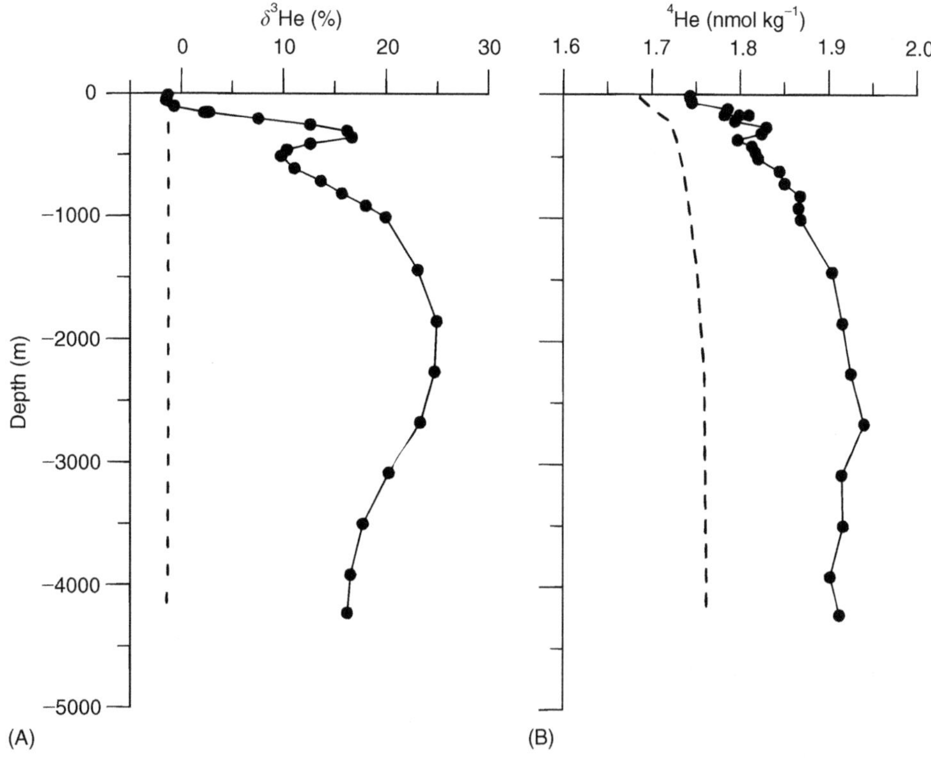

Figure 2 A typical helium profile collected at 28.5°N, 121.6°W in the north Pacific Ocean. (A) The ^{3}He/^{4}Heratio expressed as $\delta(^3\mathrm{He})\%$ plotted versus depth. The sharppeak at ~350 m depth is due to tritium decay, while the broad maximum centered at ~2000 m depth is due to volcanic helium introduced along themid-ocean ridge system. The dashed line represents the $\delta(^3\mathrm{He})$ for sea water in equilibrium with air. (B) The ^{4}He concentration plotted versus depth for the same samples. The dashed line represents the ^{4}He concentration expected for sea water in equilibrium with air.

higher solubility of heliumin the colder deep waters, as shown by the expected solubility values for air-saturated water (dashed line). However, much of the^{4}He excess above solubility equilibrium is due to the finite amountof ^{4}He present in the volcanic helium signal. At ~2500 m depth, the profile has $^4\mathrm{He} = 1.92\,\mathrm{nmol\,kg^{-1}}$, about 10%higher than the value of $1.75\,\mathrm{nmol\,kg^{-1}}$ for air-saturated water at those conditions.

The distinct isotopic signature of oceanic volcanic helium can be seen by plotting the ^{3}He concentration versus the ^{4}He concentration as shown in **Figure 3**. In this plot the slope of the trends corresponds to the isotopic ratio of the end-member helium that has been added to the water samples. The thin solid line corresponds to the atmospheric ratio ($^3\mathrm{He}/^4\mathrm{He} = 1.39 \times 10^{-6}$ or $R/R_A = 1$), and addition of air would cause the values to migrate along this line. As expected, the range of equilibrium solubility values falls directly on the atmospheric line. Although the measured samples (filled circles) near the ocean surface also fall on this line, the deeper samples fall off the atmospheric trend, defining a much steeper slope. This steeper slope is direct evidence that the helium that has been added tothe deep ocean has a higher ^{3}He/^{4}He ratio than air.

Mid-ocean Ridge Helium

The input of volcanic helium has affected the helium content of all the major ocean basins, although the magnitude of this effect varies greatly. To a large degree, the amount of the excess volcanic helium in each of the ocean basins is controlled by the relative strength of the hydrothermal input, which is in turn roughly proportional to the spreading rate of the ridges. In the Pacific Ocean, where the fastest ridge-crest spreading rates are found, the ^{3}He/^{4}He values at mid-depth average $\delta(^3\mathrm{He}) = 20\%$ for the entire Pacific basin (**Figure 4**). The Indian Ocean, which has ridges spreading at intermediate rates, has $\delta(^3He)$ values averaging about 10–15%. Finally, the Atlantic Ocean, which is bisected by the slow-spreading Mid-Atlantic Ridge, has the lowest ^{3}He enrichments, averaging $\delta(^3\mathrm{He}) = 0-5\%$ (**Figure 4**).

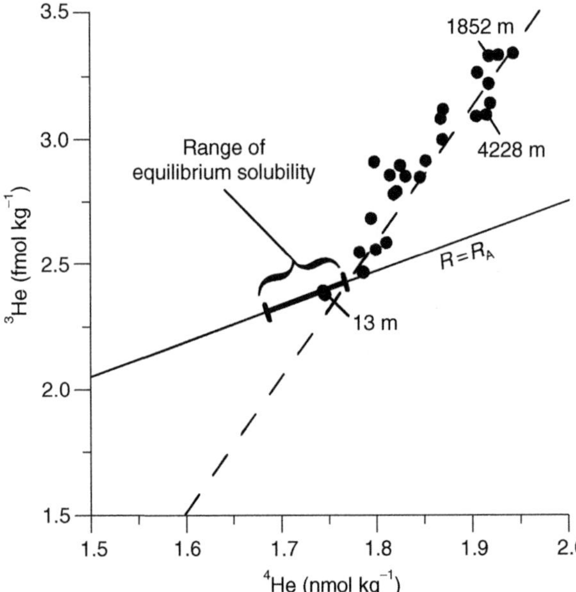

Figure 3 The ^{3}He concentration (in fmol kg^{-1} or 10^{-15} mol kg^{-1}) plotted versus ^{4}He concentration (in nmol kg^{-1}or 10^{-9} mol kg^{-1}) for the samples shown in **Figure 2**. In this plot the slope of any trend corresponds to the isotopic ratio of the end-member helium that has been added to the water samples. The depths in meters of three representative samples are indicated. The thick solid line represents the range of equilibrium solubility values expected for air saturated water (Weiss, 1970; 1971). As expected, the equilibrium solubility values fall on the thin solid line, which is the mixing relation expected for air helium ($R = R_A$). The steep slope of the dashed line, which isa best fit to the sea water samples, indicates that helium with an elevated ^{3}He/^{4}He ratio ($R > R_A$) has been added.

It has been recognized for several decades that the distribution of mantle ^{3}He has great potential for delineating the patterns of circulation and mixing of deep and intermediate water masses. This potential is probably greatest in the Pacific Ocean, becauseof the strong ^{3}He signal in that ocean. The helium field atmid-depth in the Pacific has been mapped in considerable detail (**Figure 5**). This work has identified several distinct helium plumes emanating from active hydrothermal systems distributed along the mid-ocean ridges. In the eastern equatorial Pacific, two jets of helium-rich water originate at latitude 10°N and at 14°S on the crest of the East Pacific Rise (EPR)and protrude westward into the interior of the basin. Between these two helium jets there is a minimum in the ^{3}He signal on the Equator. This distinct pattern in the helium distribution requires westward transport atmid-depth in the core of these helium plumes, and suggests eastward transport on the Equator (see dashed arrows in **Figure 5**). A separate helium plume is present in the far northeast Pacific produced by input on the Juan de Fuca and Gorda Ridges(JdFR). Although this helium signal is weaker than the helium plumes from the EPR, the JdFR helium is still traceable as a distinct plume that trends south-west into the interior of the north Pacific basin. Farther south at ~20°N, a low-^{3}He tongue penetrates from the west, implying eastward transport at this latitude. Thus the helium field defines a cyclonic (clockwise) circulation pattern at ~2000 mdepth in the northeast Pacific.

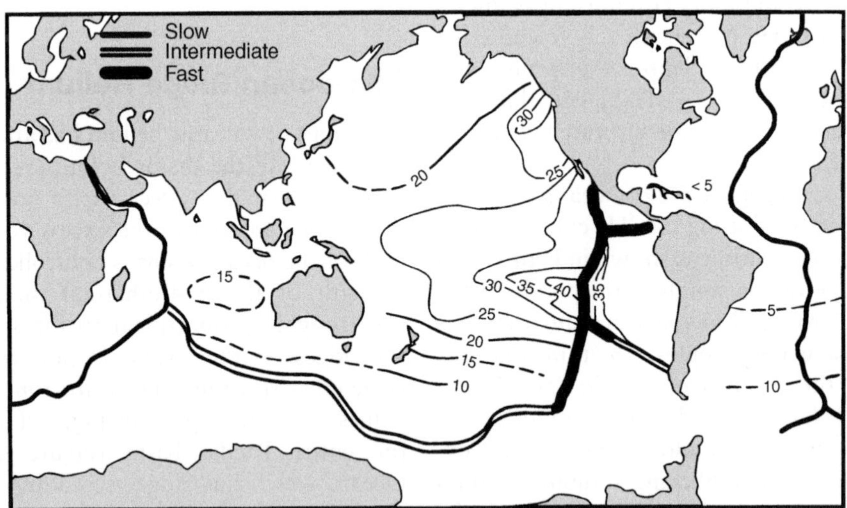

Figure 4 Map of $\delta(^3\text{He})\%$ at mid-depth in the world ocean. The location of the mid-ocean ridges is shown, and the relative spreading rate is indicated by the width of the lines.

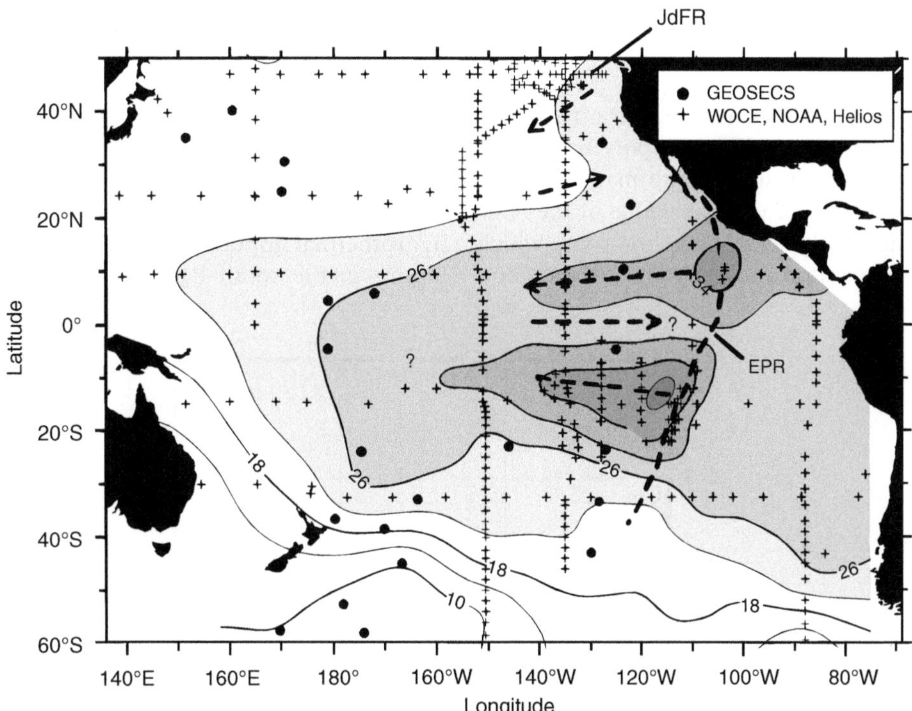

Figure 5 Map of $\delta(^3He)\%$ contoured on a surface at 2500 m depth in the Pacific. Contour interval is 4%. The major helium sources lie along the East Pacific Rise (EPR) and the Juan de Fuca Ridge (JdFR) systems. The dashed arrows indicate areas where the helium plumes define regional circulation patterns. Data along WOCE lines P4 and P6 are from Jenkins (unpublished data). All other data are from Lupton (1998).

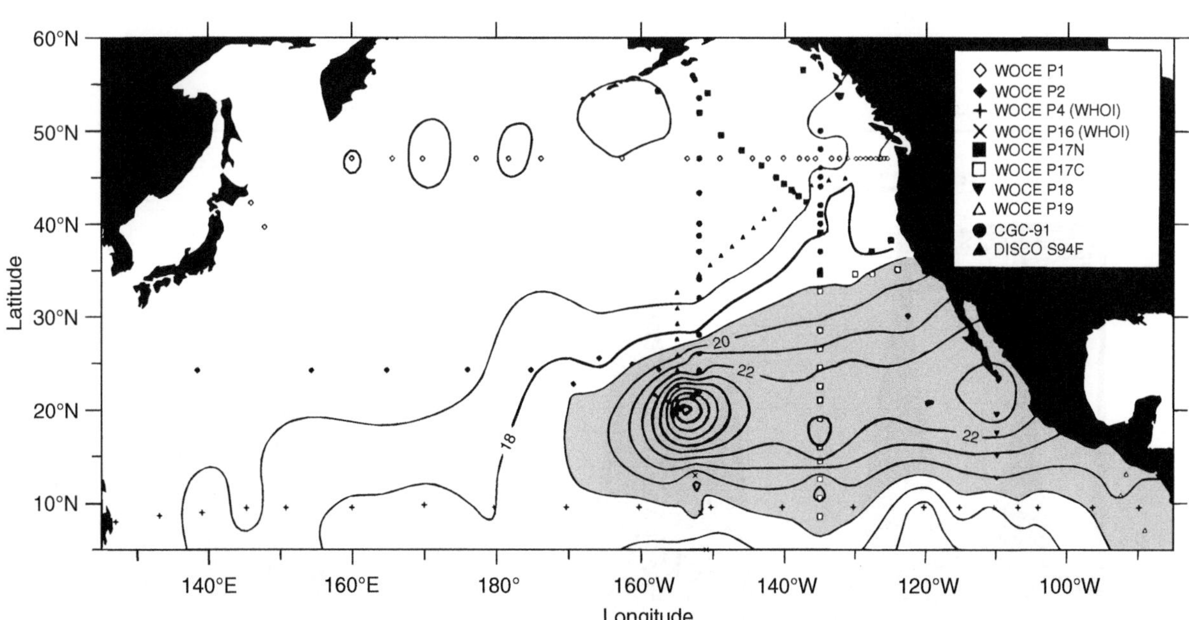

Figure 6 Map of $\delta(^3He)\%$ contoured on a surface at 1100 m depth in the north Pacific, showing the broad lateral extent of a helium plume emanating from Loihi Seamount on the south-eastern flank of the Island of Hawaii. As indicated in the key, data are from several different expeditions.

Hot Spot Helium

In addition to the volcanism along the global mid-ocean ridge system, the oceans are also affected by hot spot volcanoes. Over 100 hot spots have been identified on the Earth's surface,and many of them are located within the ocean basins. One of the best-known examples is the Hawaiian hot spot, which over time has generated the Hawaiian Islands and the Hawaiian-Emperor seamount chain. Unlike mid-ocean ridges, which are submarine, many hot spot volcanoes are subaerial and do not necessarily have direct input into the oceans. Furthermore, hot spot volcanoes have not been explored extensively for their volcanic and hydrothermal activity. Nevertheless, there are several known examples of submarine hydrothermal input at hot spots. Macdonald Seamount in the south Pacific has active vents on its

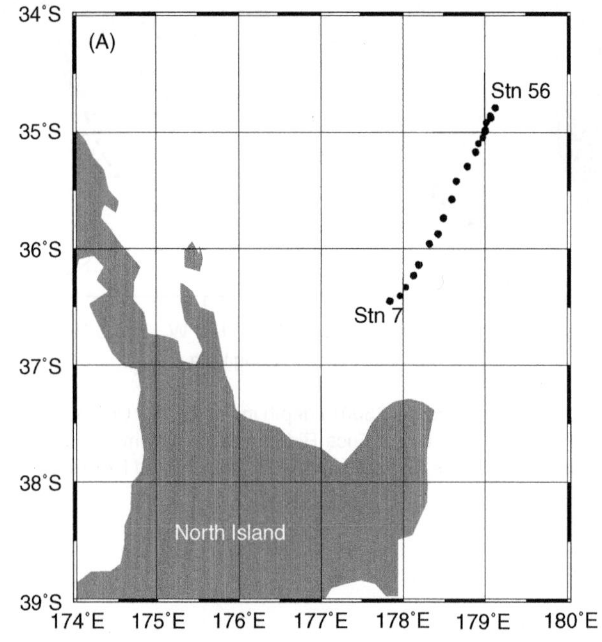

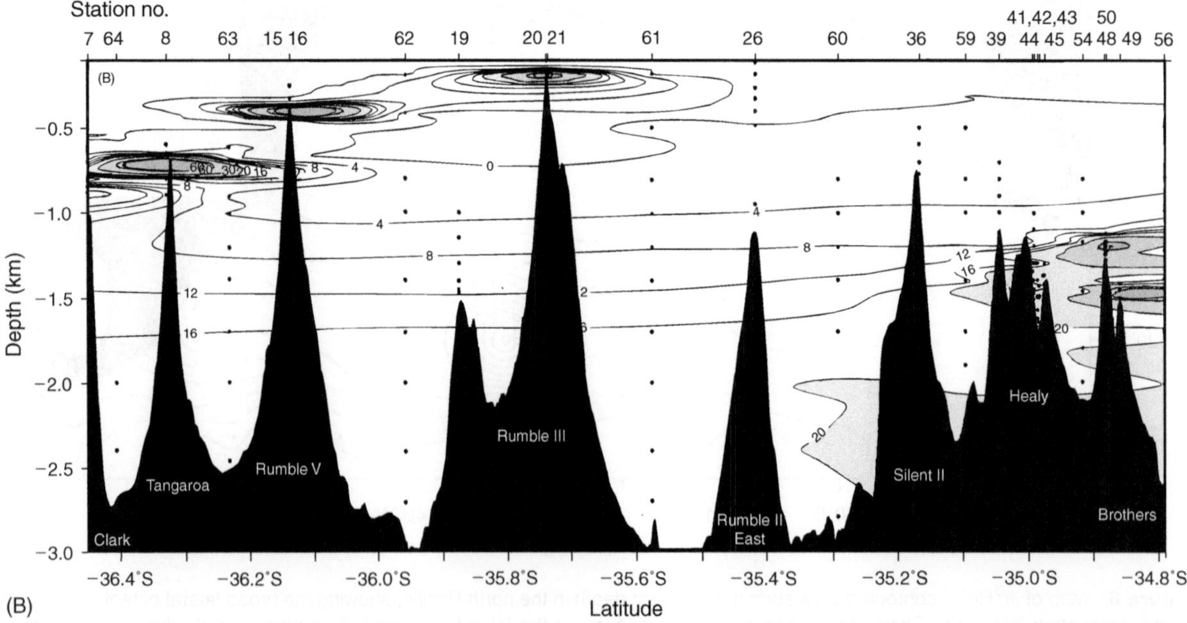

Figure 7 (A) Map showing location of hydrographic stations along the southern end of the Kermadec Arc northeast of New Zealand. (B) $\delta(^3He)\%$ contoured in section view along the southern end of the Kermadec Arc, showing ^3He-rich water-column plumes emanating from several of these subduction zone volcanoes. From de Ronde et al. (2000).

summit that inject volatiles into the overlying water column at a depth of ~130 m. Loihi Seamount, situated on the south-eastern flank of the island of Hawaii, also has active ventsnear its summit at a depth of ~1000 m. Loihi is of considerable importance because it is thought to be the current locus of the Hawaiian hot spot. Loihi lavas and hydrothermal fluids contain helium with a very primitive signature of$R/R_A = 25-30$, indicating a deep mantle origin.

It has been known for some time that hydrothermal venting on Loihi Seamount produces water column plumes that can bedetected with tracers such as temperature, manganese, iron, and methane. However, these tracers are not useful for far-field studies of the Loihi plume because they are either rapidly removed from the water column or arepresent in low concentrations. Because helium is a stable, conservative tracerthat is highly enriched in Loihi vent fluids, the helium signal from Loihi is detectable at considerable distances from the Hawaiian Islands. As shown in **Figure 6**, a map of $\delta(^3He)$ on a surface at 1100 m depth reveals a ^{3}He-rich plume that extends eastward from the Hawaiian Islands for several thousand kilometers, reaching the coast of Mexico at its greatest extent. This far-field plume produced by the Hawaiian hot spot clearly defines an eastward transport at ~1000 m depth in this region ofthe north Pacific. Furthermore, because the end-member helium introduced at Loihi has a ^{3}He/^{4}He ratio three times higher than mid-ocean ridge helium, it should be possible to distinguish the Loihihelium from mid-ocean ridge helium with accurate measurements of^{3}He and ^{4}He concentrations. The ability to distinguish hot spot helium from mid-ocean ridge helium has been demonstrated for the Loihi helium plume near Hawaii but not yet in the far-field.

Subduction Zone Helium

Submarine volcanism also occurs alongconvergent margins, particularly in regions where two oceanic plates are converging. However, very little is known about the incidence of submarine hydrothermal activity associated with this type of volcanism. Studies of subaerial volcanoes at convergent margins have shown that these volcanoes emit mantle helium with an isotopic ratio of $R/R_A = 3-7$, lower than in mid-ocean ridges. Thus the volcanic helium from subduction zones represents a third type of mantle helium that isisotopically distinct from mid-ocean ridge and hot spot helium.

One clear example of oceanic helium plumesfrom subduction zone volcanism is shown in **Figure 7**, which shows the results of a survey along the southern end of the Kermadec Arc northeast of New Zealand. The Kermadec Arc consists of a series of discrete volcanoes generated by the subduction of the Pacific plate beneath the Australian plate. At the southern end of the arcthese volcanoes are submarine, while farther north some of them are subaerial volcanoes, including Curtis Island, Macauley Island, and Raoul Island. The survey shown in **Figure 7** consisted of a series of hydrographic casts along the arc, and many of the casts were lowered directly over the summits of these arc volcanoes. The section shown in **Figure 7B** shows a series of ^{3}He-rich found at a variety of depths between 150 m for Rumble III volcano down to 1400 m for Brothers volcano. A plot of ^{3}He versus ^{4}He concentration for these samples (notshown), indicated an average ^{3}He/^{4}He ratio of $R/R_A = 6$, in agreement with previous studies of helium from subaerial subduction zone volcanoes. Although the lateral extent of the helium plumes from the Kermadec Arc is not known, this survey confirms that subduction zone volcanoes do produce helium plumes that can be used to trace ocean currents. Furthermore, these subduction zone plumes are potentially quite valuable for tracer studies, since they occur at a wide variety of depths and are generally much shallower than plumes produced at mid-ocean ridges (**Figure 7B**).

See also

Hydrothermal Vent Deposits. Hydrothermal Vent Fluids, Chemistry of. Mid-Ocean Ridge Geochemistry and Petrology. Noble Gases and the Cryosphere. Tritium–Helium Dating.

Further Reading

Clarke WB, Beg MA, and Craig H (1969) Excess ^{3}He in the sea: evidence for terrestrial primordial helium. *Earth and Planetary Science Letters* 6: 213–220.

Craig H, Clarke WB, and Beg MA (1975) Excess ^{3}He in deep water on the East Pacific Rise. *Earth and Planetary Science Letters* 26: 125–132.

Craig H and Lupton JE (1981) Helium-3 and mantle volatiles in the ocean and the oceanic crust. In: Emiliani C (ed.) The Sea, vol. 7, pp. 391–428. New York: Wiley.

Krylov A Ya, Mamyrin BA, Khabarin L, Maxina TI, and Silin Yu I (1974) Helium isotopes in ocean floor bedrock. *Geokhimiya* 8: 1220–1225.

Lupton JE (1983) Terrestrial inert gases: isotope tracer studies and clues to primordial components in the mantle. *Annual Review of Earth and Planetary Science* 11: 371–414.

Lupton JE (1995) Hydrothermal plumes: near and far field. In: Humphris S et al. (eds.) *Seafloor Hydrothermal Systems, Physical, Chemical, Biological, and Geological Interactions*, Geophysical Monograph Series, vol. 91, pp. 317–346. Washington, DC: American Geophysical Union.

Lupton JE (1998) Hydrothermal helium plumes in the Pacific Ocean. *Journal of Geophysical Research* 103: 15855–15868.

Lupton JE and Craig H (1975) Excess ^{3}He in oceanic basalts: evidence for terrestrial primordial helium. *Earth and Planetary Science Letters* 26: 133–139.

Suess HE and Wänke H (1965) On the possibility of a helium flux through the ocean floor. *Progress in Oceanography* 3: 347–353.

Weiss RF (1970) Helium isotope effect in solution in water and seawater. *Science* 168: 247–248.

Weiss RF (1971) Solubility of helium and neon in water and seawater. *Journal of Chemical Engineering Data* 16: 235–241.

CFCS IN THE OCEAN

R. A. Fine, University of Miami, Miami, FL, USA

Introduction

The oceans, atmosphere, continents, and cryosphere are part of the tightly connected climate system. The ocean's role in the climate system involves the transport, sequestration, and exchange of heat, fresh water, and carbon dioxide (CO_2) between the other components of the climate system. When waters descend below the ocean surface they carry with them atmospheric constituents. Some of these are gases such as carbon dioxide and chlorofluorocarbons (CFCs). The CFCs can serve as a physical analog for CO_2 because they are biologically and chemically inert in oceans. In the oceans the distribution of CFCs provides information on which waters have been in contact with the atmosphere in the past few decades. The CFCs also give information on the ocean's circulation and its variability on timescales of months to decades. The timescale information is needed to understand and to assess the ocean's role in climate change, and its capacity to take up anthropogenic constituents from the atmosphere. Thus, the advantage of using tracers like CFCs for ocean circulation studies is the added dimension of time; their time history is fairly well known, they are an integrating quantity and an analog for oceanic anthropogenic CO_2 uptake, and they provide an independent test for time integration of models.

Tracers serve as a 'dye' with which to follow the circulation of ocean waters. There are conventional ocean tracers such as temperature, salinity, oxygen, and nutrients. There are stable isotope tracers such as oxygen-18, carbon-13, and there are radioactive tracers both naturally occurring (such as the uranium/thorium series, and radium), and those produced both naturally and by the bomb tests (such as tritium and carbon-14). The bomb contributions from the latter two are called transient tracers, as are the CFCs, because they have been in the atmosphere for a short time. This implies an anthropogenic source and a nonsteady input function.

Atmospheric Source

The chlorofluorocarbons, CFCs, are synthetic halogenated methanes. Their chemical structures are as follows: CFC-11 is CCl_3F, CFC-12 is CCl_2F_2, and CFC-113 is CCl_2FCClF_2. For completeness the compound carbon tetrachloride, CCl_4, is also included in this article as its atmospheric source, measurement, and oceanic distribution are similar to those of the CFCs. The CFCs have received considerable attention because they are a double-edged environmental sword. They are a threat to the ozone layer, and a greenhouse gas. The CFCs are used as coolants in refrigerators and air conditioners, as propellants in aerosol spray cans, and as foaming agents. These chemicals were developed over 50 years ago when no one realized that they might cause environmental problems. When released CFCs are gases that have two sinks, the predominant one being the atmosphere, and to a lesser extent the oceans. Most of the CFCs go up into the troposphere, where they remain for decades. In the oceans and in the troposphere the CFCs pose no problem. However, some escape into the stratosphere where they are a threat to the ozone layer. Due to their role in UV absorption they have been correlated with the increased incidence of skin cancers. Since the recognition of the CFCs as an environmental problem in the 1970s and the signing of the Montreal Protocol in 1987, the use of CFCs has been phased out. The atmospheric concentrations have just started to decrease. This is an important international step toward correcting the dangerous trend of stratospheric ozone depletion.

The atmospheric CFC concentrations became significant after the 1940s. The concentrations increased exponentially until the mid-1970s, and then increased linearly until the 1990s at a rate of about 5% per year. The production and release data for CFCs tabulated by the Chemical Manufacturers Association (CMA) were used (**Figure 1**) to reconstruct the atmospheric time histories for the Northern and Southern Hemispheres. Since 1979 the atmospheric concentrations have been based on actual measurements at various sampling stations around the globe, and these are checked against the CMA production and release estimates. The curves in **Figure 1** show all CFCs including CCl_4 increasing with time, with CFC-11 leveling off and actually decreasing in the late 1990s. The atmospheric increase of all the CFCs slowed markedly after the Montreal Protocol. The

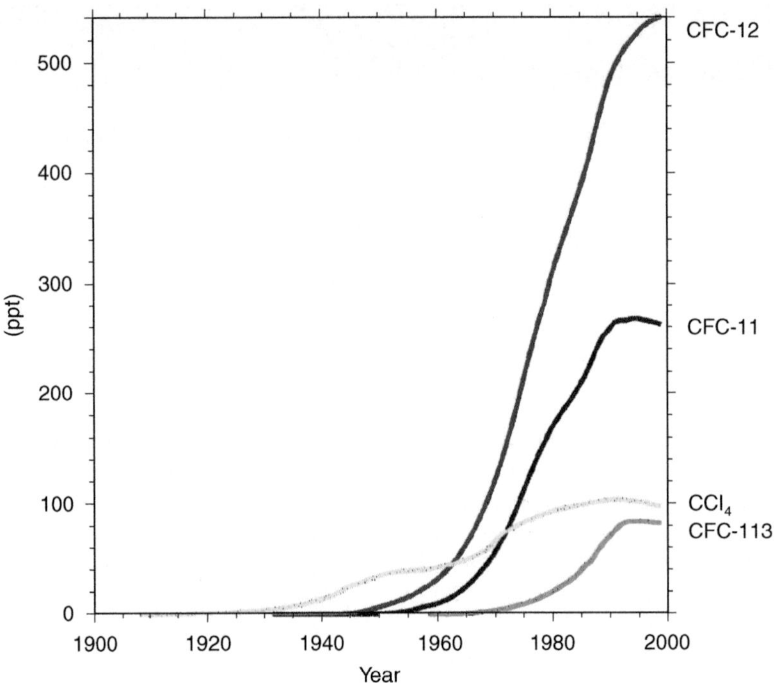

Figure 1 Northern Hemisphere atmospheric time histories. (Atmospheric data from Walker *et al.*, 2000.)

uncertainties in the reconstructed pre-1979 atmospheric histories depend on the atmospheric lifetimes of the compounds. The range of lifetimes for the compounds are 29–76 years for CFC-11, 77–185 years for CFC-12, 54–143 years for CFC-113, and 21–43 years for CCl_4. The uncertainties are a few percent, and they are highest for the early period. The continuous direct atmospheric measurements, which began in the late 1970s, are uncertain to within 1–2%. It is important to evaluate the uncertainties in the atmospheric source function, because they translate into uncertainties when used to put timescales on ocean processes. Because of their long atmospheric residence times, the CFCs are homogeneously distributed in the Northern and Southern Hemisphere, with the Northern Hemisphere about 8% higher than the Southern Hemisphere.

CFCs in the Oceans

Analytical Techniques

Water samples collected from the ocean are measured for CFCs and CCl_4 using an electron capture detection gas chromatography system. Analysis of water samples is done onboard ship, usually within hours of collection. The unit of measure is pmol kg^{-1} or 10^{-12} moles kg^{-1}. These are extremely low level concentrations that are easily susceptible to contamination from shipboard refrigerants, solvents, lubricants, etc.

Chemical Stability

Under oxygenated oceanic conditions both CFC-11 and CFC-12 are believed to be chemically stable. CFC-11 has been shown to be unstable in anoxic marine waters; both CFC-11 and CFC-12 have been shown to be unstable in anoxic sediments. The compound CCl_4 undergoes temperature-dependent hydrolysis, which limits its usefulness in the ocean when sea surface temperatures exceed $\sim 18°C$. CFC-113 also has some stability problems at higher temperatures.

Gas Flux and Solubility

The CFCs are gases, and like other gases they get into the ocean via air–sea exchange. There is a direct correlation between gas exchange rate and wind speed, and the direction of the gas flux between the air and ocean is from high to low concentration. For CFCs the atmospheric concentrations generally exceed those in the ocean. The concentration of CFCs dissolved in the surface layer of the oceans is dependent upon the solubility, atmospheric concentration, and other physical factors affecting the gas saturation including upwelling, entrainment due to mixing, ice cover, etc. The solubility of CFCs and CCl_4 has been measured in the laboratory. The accuracy of the measurements is about 1.5% and precision is about 0.7%. The solubility increases with decreasing temperature, at a rate of about 4% for 1°C. Therefore, the colder the water the higher the CFC concentration. At a constant salinity the temperature effect is about two times greater for CFCs than for oxygen. The solubility is only slightly dependent on the salinity, and it decreases with increasing salinity.

Surface Saturation

The approach to equilibrium condition or the saturation state is dependent on the mixed layer depth and air–sea transfer rate. It takes from days up to a few weeks after a change in temperature or salinity for 'normal' (not very deep) oceanic surface layers to come to equilibrium with the present atmosphere. While the surface waters of the world's oceans are close to equilibrium with the present day atmospheric concentration of CFCs, there are exceptions. At times of rapid warming, such as in the spring, the surface waters will tend to be a few percent supersaturated with the gas due to lack of time to equilibrate with the atmosphere. Likewise at times of rapid cooling the surface waters will be a few percent undersaturated with the gas. Typically there are undersaturations within a few degrees of the equator due to upwelling of deeper less saturated waters. In high latitudes, where there are deep convective mixed layers that do not readily equilibrate with the atmosphere, there are likely to be undersaturations of as much as 60%. These have been observed in the Labrador Sea. The undersaturations in the high latitude water mass source regions need to be taken into account when using the CFCs to put timescales on oceanic processes.

Oceanic Distribution

The compounds CFC-11 and CFC-12 were first measured in the oceans in the late 1970s. The first systematic and intensive survey was carried out in the tropical North and South Atlantic oceans starting in the early 1980s. Since then CFCs have been part of the measurements made during physical oceanography field work. A global survey was conducted as part of the World Ocean Circulation Experiment during the 1990s. Typical vertical profiles versus pressure for stations in the North Atlantic and North Pacific oceans are presented in **Figure 2** along with other properties. Although CFC-12 has higher concentrations in the atmosphere, CFC-11 is more soluble in sea water, so its concentrations are about twice that of CFC-12. Note that there are measurable concentrations of CFCs in the western North Atlantic that reach to the ocean bottom, while they reach to only 1000 m in the North Pacific. The difference between the CFC concentrations of the North Atlantic as compared with the North Pacific, reflects the formation of deep waters in the North Atlantic and the absence in the North Pacific. Concentrations generally decrease as the ocean depth increases. However, there may be subsurface concentration maxima due to the lateral intrusion of water that has been in more recent contact with the atmosphere (see applications below). The concentrations of CFCs and oxygen should behave similarly except where the biological effects on the oxygen distribution cause the differences, for example, the oxygen minimum at mid-depth.

Combining a series of vertical profiles, as in **Figure 2**, will give a slice or section through the ocean. Sections through the eastern Pacific and Atlantic are shown in **Figure 3**. The absence of CFCs in the deep waters of the Pacific Ocean shows the relative isolation of the deep Pacific from contact with the atmosphere on timescales of decades. In contrast, the North Atlantic north of 35°N has CFCs in deep and bottom waters, because these waters form in the high latitudes of the North Atlantic and easily spread equatorward on timescales of 10–20 years. As part of the density-driven, thermohaline circulation some of these waters will eventually be transported into the Pacific, but it will take hundreds of years. The upper waters of both oceans are in contact with the atmosphere on much shorter timescales. These upper waters are part of the wind-driven circulation.

CFC Ages in the Ocean

Age Calculations

One of the main advantages of using CFCs as tracers of ocean circulation is that the time-dependent source function permits the calculation of timescales for these processes. A tracer age is the elapsed time since a water parcel was last exposed to the atmosphere. The tracer-derived age is the elapsed time since a subsurface water mass was last in contact with the atmosphere. Two estimates of 'age' can be calculated, one from the CFC-11/CFC-12 ratio and one from the partial pressure of either dissolved CFC. In both cases, the atmospheric value of either the ratio or partial pressure with which the water had equilibrated is compared to the atmospheric source function to determine the corresponding date.

To normalize the concentrations for the effects on the solubility of temperature and salinity CFCs are expressed in terms of their partial pressures, pCFC, where the pCFC is the concentration divided by the

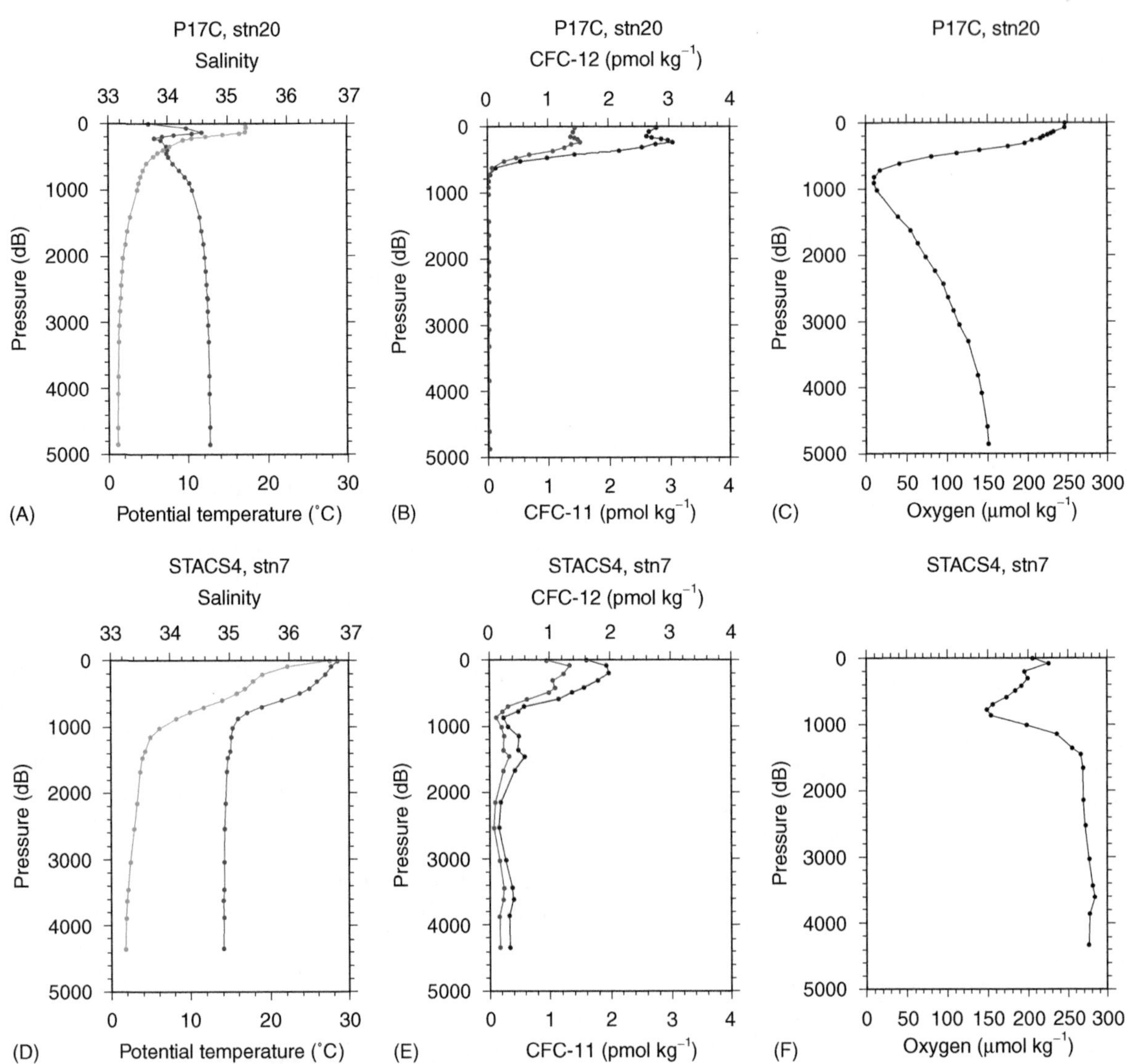

Figure 2 Vertical profiles of oceanographic data. (A) North Pacific salinity and potential temperature, (B) North Pacific CFC-11 and CFC-12, (C) North Pacific oxygen, (D) North Atlantic salinity and potential temperature, (E) North Atlantic CFC-11 and CFC-12, (F) North Atlantic oxygen. North Pacific World Ocean Circulation Experiment cruise P17C station 20, 33°N, 135°W, June 1991; North Atlantic Subtropical Atlantic Climate Studies cruise station 7, 26.5°N, 76°W, June 1990. (North Atlantic data from Johns *et al.* (1997) *Journal of Physical Oceanography* 27: 2187–2208; Pacific data from Fine *et al.* (2001) *Journal of Geophysical Research.*)

solubility of the gas. This value is then adjusted for what the percent surface saturation is thought to be based on the measured temperature and salinity, then matched to the atmospheric time histories, and a corresponding year is assigned to the water mass. This age is an average of the water parcel. The pCFC is used to calculate the age of upper ocean waters, because at low concentrations the effects of dilution will bias the age toward the older components of a mixture.

The age can also be calculated using the ratio of two CFCs; instead of using one pCFC the ratio of two pCFCs are used. In this case no assumptions are needed about surface equilibrium saturation at the time of water mass formation. Since the atmospheric changes in the ratio of CFC-11/CFC-12 have remained unchanged since the mid-1970s, this restricts the application of the ratio age for CFC-11 and CFC-12 to waters dating back further than 1975. However, either CFC-11 or CFC-12 can be combined with CFC-113 to extend age estimates to the present. Similarly they can be combined with CCl_4 to extend age estimates further into the past. Unlike the pCFC age, the ratio ages are actually the ages of the CFC-bearing components. **Figure 4** shows sections of CFC ratio ages from the eastern North Atlantic and North Pacific oceans. Note that the intermediate and deep waters of the eastern North Atlantic (between 2000

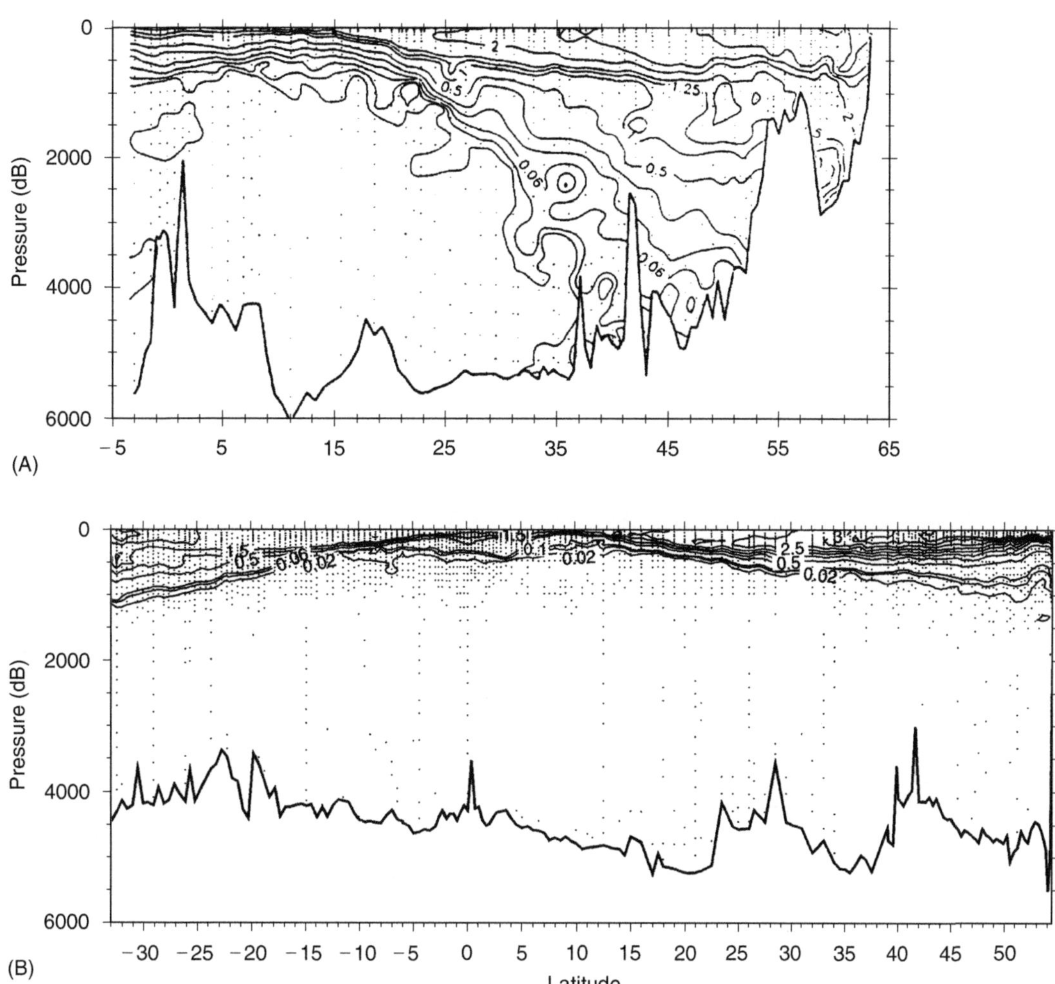

Figure 3 (A) Sections versus pressure of CFC-11 concentrations (pmol kg^{-1}) in the eastern Atlantic (latitude 65°N–5°S) along 20°W in summer 1988. (B) Sections of CFC-11 concentrations (pmol kg^{-1}) in the eastern Pacific (latitude 54°N–32°S) mostly along 135°W in summer 1991. (North Atlantic data from Doney SC and Bullister JB (1992) *Deep-Sea Research* 39: 1857–1883; Pacific data from Fine *et al.* (2001) *Journal of Geophysical Research.*)

and 4000 m) have CFCs younger than 30 years north of 45°N, because of their proximity to the formation regions, whereas this is not the case in the North Pacific. In the Pacific below 2000 m the water column has been isolated from interaction with the atmosphere on similar timescales (except for the far western South Pacific).

Caveats for Using CFC Ages

There are several caveats to the use of CFC ages. Both ages – partial pressure and ratio – may be subject to biases when there is mixing of more than one water mass component. Because of nonlinearities in the source functions and solubilities, neither age mixes linearly in multicomponent systems over the entire concentration range observed in the ocean. The atmospheric source function is nonlinear for much of the input history; however, it can be approximated as being linear between the late 1960s and 1990. The solubilities are nonlinear functions of temperature, but they are approximately linear over ranges of a few degrees. Thus, for some regions of the ocean, these nonlinearities are not significant.

The different types of ages are appropriate for putting timescales on different processes. For thermocline ventilation, where equilibrated water is subducted and mixed isopycnally along extensively outcropping density surfaces, the water subducted within a given year mixes with water subducted in previous years. In this situation, a water parcel is a mixture of water that has left the surface over a period of several years. The average age of this water parcel can be represented by the pCFC age if the change of CFC concentration in the source region is constant with respect to time. This has been

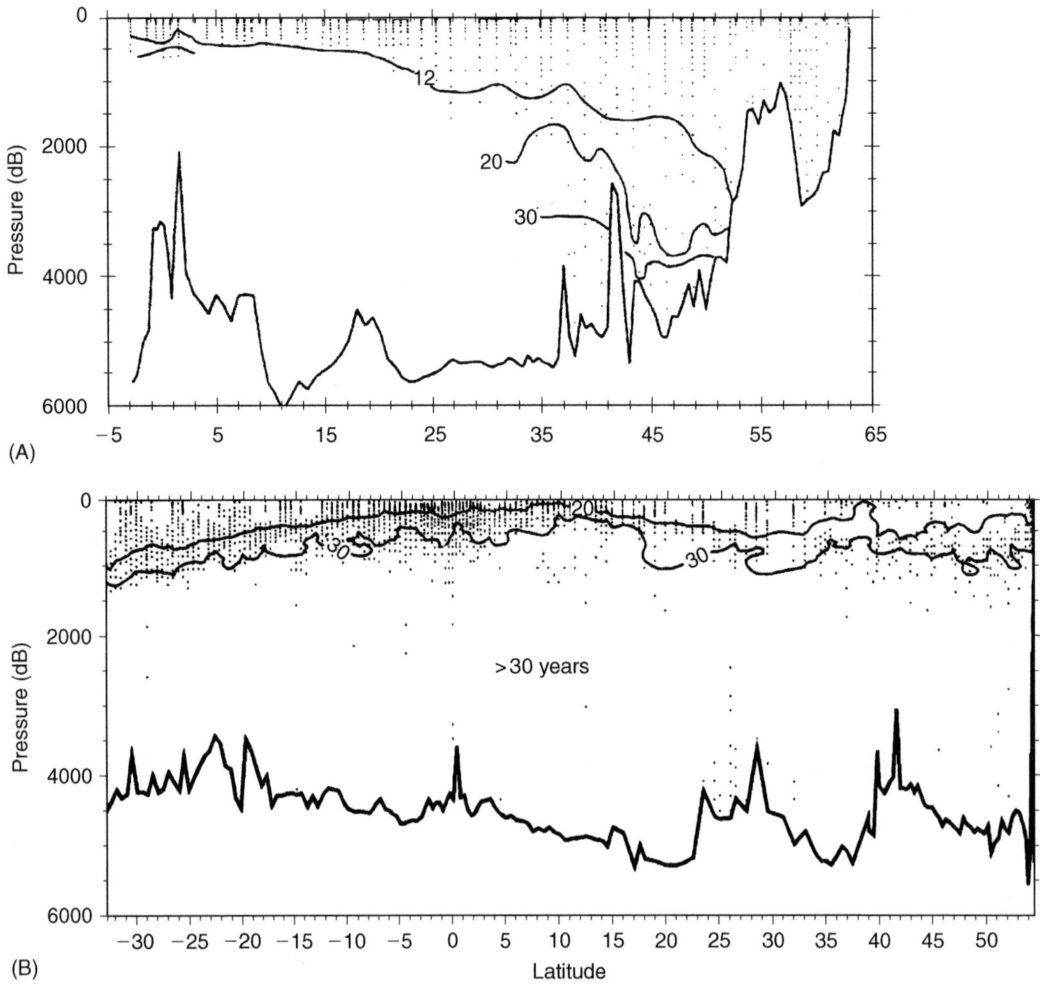

Figure 4 (A) Sections of CFC-11/CFC-12 ratio ages (years) in the eastern Atlantic (latitude 65°N–5°S) along 20°W in summer 1988. (B) Sections of CFC-11/CFC-12 ratio ages (years) in the eastern Pacific (latitude 54°N–32°S) mostly along 135°W in summer 1991. (North Atlantic data from Doney SC and Bullister JB (1992) *Deep-Sea Research* 39: 1857–1883; Pacific data from Fine *et al.* (2001) *Journal of Geophysical Research.*)

confirmed for the North Atlantic thermocline in the eastern basin by comparing pCFC ages to tritium/He-3 ages.

In regions where surface waters are converted to deep and bottom waters which then spread into a background of low-tracer water, the high CFC concentrations of the cold surface water are diluted by entrainment and mixing. The resulting pCFC age is much too young for the average age of the mixture and much too old for the CFC-bearing component. However, a tracer ratio is conserved in this situation, and the corresponding ratio age represents that of the youngest component of the mixture, not the average age of the water parcel. Thus, there are different estimates of ages that can be derived from CFC-11 and CFC-12, and the associated timescales can be expanded in regions where CFC-113 and CCl_4 data are available.

In most high latitude intermediate and deep-water source regions the age clock is not reset to zero due to lack of time to equilibrate deep mixed layers with the atmosphere. Thus, water masses will start out with an age of a few years (rather than zero), that is, they are not completely renewed during formation. This additional age is called a relic age which can be estimated from observations of the tracers at the water mass formation regions. The relic age can then be subtracted from the tracer ages calculated downstream from the water mass formation regions.

Applications of CFCs to Ocean Processes

Examples of the application of CFCs to understanding oceanographic processes are divided into four subjects: the thermohaline circulation, upper ocean circulation, model constraints, and biogeochemical processes.

Thermohaline Circulation

There is a close coupling of the surface waters in high latitudes to the deep ocean through the density-driven thermohaline circulation. During the process of deep-water formation, atmospheric constituents such as CFCs are introduced into the newly formed water. In recent years, major advances in our knowledge of the thermohaline circulation can be attributed to information derived from transient tracer data, particularly for two reasons. First, the development of analytical techniques so that oceanographers can easily produce large quantities of high quality data. Tracer oceanographers have benefited from multi-investigator programs like the World Ocean Circulation Experiment. The following highlights some of the advances that have come about in our understanding of the thermohaline circulation to which observations of CFCs have contributed:

- Discovery of a new water mass component of North Atlantic Deep Water (NADW), called Upper Labrador Sea Water, location of its formation region and contributing processes, and timescales of eastward spreading along the equator.
- Identification of Denmark Straits Overflow Water as the primary source of bottom water of the western subpolar basin.
- Confirmation of the structure and continuity of the Deep Western Boundary Current throughout the western North Atlantic Ocean, and extension into the South Atlantic.
- Extension of the CFCs well into the interior of the western North Atlantic show the importance of deep recirculation gyres in ventilating the interior basins, and in slowing the equatorward transport to timescales of <30 years with effective spreading rates of 1–2 cm s^{-1}.
- Contribution to quantifying formation rates and decadal climate variability in the Arctic, Greenland and Labrador Seas.
- Estimates for the formation rates of Weddell Sea Deep and Bottom Waters, production rate of Antarctic Bottom Water and pathways and timescales for spreading into the North Atlantic.

Upper Ocean Circulation

The use of CFCs for upper ocean processes has involved the application of concentrations to deduce sources and circulation pathways, and application of pCFC ages. The following highlights some of the advances that have come about in our understanding of the upper ocean circulation to which observations of CFCs have contributed:

- Identification of the Sea of Okhotsk and Alaskan Gyre as important location for the ventilation of North Pacific Intermediate Water, these waters then spread into the subtropics on a timescale of <20 years.
- Quantification of the flux of water from the mixed layer into thermocline and intermediate layers of the North and South Pacific.
- Contribution to the description of sources and pathways of water masses transported from the Pacific through the Indonesian Seas into the Indian Ocean.
- Quantification of the sources of northern and southern water and the processes needed to ventilate the tropical Pacific and Atlantic, including advection, diapycnal and vertical mixing.
- Observation that pathways of the most recently ventilated Antarctic Intermediate Waters are into the eastern South Indian Ocean, while at that level there appears to be flow of older waters from the South Pacific into the western Indian Ocean.
- Quantification of subduction and formation rates for subtropical underwaters and in the North Atlantic its interannual variability that is negatively correlated with intermediate waters of the eastern subpolar gyre.

Model Constraints

In general CFC concentrations and inventories have been used in comparison with model simulated concentrations and inventories. The time-dependent nature of the CFCs provides a stringent test of a model's ability to integrate property distributions over time. The following highlights some of the advances that have come about in our ability to put constraints on models from the use of CFCs in models:

- Dilution of CFCs transported by the Deep Western Boundary Current and effect on tracer ages.
- Testing the sensitivity of a model for correct simulation of formation rates, pathways, and spreading rates.
- Testing the sensitivity of a model for correct simulation of ocean model velocity fields.
- Determining the model sensitivity to subgrid scale mixing for purposes of estimating ventilation rates.
- The importance of considering seasonal variations in the upper oceans as part of the tracer boundary

conditions when trying to simulate subduction processes.

- Demonstration in a model simulation that eddy transport is required to transport South Indian subtropical gyre waters across the equator along the western boundary.
- Use of CFCs to validate model parameterizations of gas fluxes.

Biogeochemical Processes

The tracers provide a method for calculating rates of biogeochemical fluxes that is independent of direct biological measurements. Again the age information from the CFCs is used to calculate rates for these processes. (*see* Nitrogen Isotopes in the Ocean). The following highlights some of the advances that have come about in our understanding of biogeochemical processes to which observations of CFCs have contributed:

- Apparent oxygen utilization rates from the central Arctic that are so high, they need to be balanced by transport of high production water from over the continental shelves.
- Quantification of moderate biological consumption and initially low oxygen concentrations in the Arabian Sea are needed to maintain the low oxygen layer.
- Calculation of denitrification rates for the Arabian Sea and Bay of Bengal.

Conclusions

The advantage of oceanic tracers like CFCs is that they can be used to provide timescale information for oceanographic processes. Direct application of the timescale information from the CFCs is used to calculate fluxes of atmospheric constituents, such as CO_2. The oceans have taken up a considerable portion of the anthropogenic CO_2 released to the atmosphere. A large part of the uptake involves water mass formation in high latitudes. The rate at which these waters are transported into the interior will have an effect on the rate at which anthropogenic CO_2 is taken up.

See also

Carbon Dioxide (CO_2) Cycle. Nitrogen Isotopes in the Ocean. Tritium–Helium Dating.

Further Reading

Broecker WS and Peng T-H (1982) *Tracers in the Sea.* Palisades, NY: Lamont-Doherty Geological Observatory, Columbia University.

Fine RA (1995) Tracers, time scales and the thermohaline circulation: the lower limb in the North Atlantic Ocean. *Reviews of Geophysics* 33: 1353–1365.

Rowland FS and Molina MJ (1994) Ozone depletion: 20 years after the alarm. *Chemical & Engineering News* 72: 8–13.

Schlosser P and Smethie WS (1994) *Transient Tracers as a Tool to Study Variability of Ocean Circulation. Natural Climate Variability on Decadal-to-Century Time Scales*, pp. 274–288. Washington, DC: National Academic Press.

Smethie WS, Fine RA, Putzka A, and Jones EP (2000) Reaching the flow of North Atlantic Deep Water using chlorofluorocarbons. *Journal of Geophysical Research* 105: 14 297–14 323.

Walker SJ, Weiss RF, and Salameth PK (2000) Reconstructed histories of the annual mean atmospheric mole fractions for the halocarbons CFC-11, CFC-12, CFC-113 and carbon tetrachloride. *Journal of Geophysical Research* 105: 14 285–14 296.

CHLORINATED HYDROCARBONS

J. W. Farrington, Woods Hole Oceanographic Institution, MA, USA

Introduction

These chemicals are considered in a pollution category because both deliberate and accidental release to the environment of several of these types of compounds, for example the industrial chemicals such as PCBs (polychlorinated biphenyls) and the chlorinated pesticides *p,p'*DDT (dichlorodiphenyl-trichloroethane; formal chemical name 1,1′-(2,2,2-trichloroethylidene)-bis (4-chlorobenzene)), have had unintended adverse environmental effects on diverse plants and animals and on people. Initially, chemicals such as PCBs and DDT were beneficial to human civilization: PCBs as industrial chemicals allowing economical, safe delivery of electricity, and DDT as a pesticide eradicating vector pests of human health concern and agricultural crop pests. Only after these chemicals had entered widespread use did it become apparent that there were environmental problems, although in hindsight there was evidence of potential problems early in the history of their manufacture and use.

Chlorinated hydrocarbons are chemicals made up of the elements carbon (C) and hydrogen (H) at the combine for the 'hydrocarbon' part of the molecule, and chlorine atoms (Cl) substituted for hydrogen where a hydrogen atom was normally bonded to a carbon atom. Examples of structures of chlorinated hydrocarbons are given in **Figure 1**.

Chlorinated hydrocarbons have a wide range of molecular weights (related to size), and complexity, i.e., there are various distinct configurations or arrangements of constituent atoms. For example, there are 209 individual chlorobiphenyls (known collectively as congeners) making up the family of chemicals known as PCBs. Not all of these are present in the commercial chemical mixtures of PCBs, but there are usually 20–50 chlorobiphenyl congeners in a given commercial mixture.

Smaller molecules among the class of chlorinated hydrocarbons, such as tetrachloroethylene, trichloroethylene, and carbon tetrachloride are used for activities such as degreasing of machinery and dry cleaning. Presently, these compounds are of environmental concern for the oceans only in the near-shore coastal areas where their presence in sewage effluents and contaminated or polluted ground water interfacing with coastal sea water results in elevated concentrations in coastal waters near sources of input.

Environmental and human health concerns associated with chlorinated pesticides and PCBs have evolved over the past several decades into wider concerns with organochlorine compounds of all types, ranging from those found in plastic trash bags to the chemicals of Agent Orange defoliant used by the United States during the war in Vietnam. Among the chemicals of greatest concern on a per unit amount basis are tetrachlorodibenzodioxins; often the name is shortened in general public use to 'dioxins'. Assessments of risks to human health and wildlife for the various chlorinated hydrocarbon pesticides and industrial chemicals are often expressed relative to tetrachlorodibenzodioxin risks.

There are an estimated 10 000 to 11 000 organochlorines in commercial production and many thousands more may be present, but are as yet unidentified, as by-products of the production. In addition, processes such as chlorination of sewage effluent to kill bacteria, result in active forms of chlorine which react with natural organic chemicals in the sewage to produce a myriad of organochlorines; perhaps hundreds to thousands depending on the effluent and the chlorination conditions. Small amounts of organochlorines are also reported to result from various combustion processes, both natural fires and volcanic eruptions, and human-controlled processes such as incineration of wastes.

Analytical chemical, biochemical and molecular biological methods can detect very low concentrations of these compounds in environmental samples, including marine organisms ($ng\,g^{-1}$, or about one unit mass of chlorinated hydrocarbon molecule per billion unit masses of tissue molecules), sea water (ng per 1000 liters or kg, $ng\,kg^{-1}$), and marine sediments (nanograms per gram of sediment, $ng\,g^{-1}$).

Given the widespread occurrence of these compounds, and the known or suspected adverse environmental and human health effects at elevated concentrations, there is a challenge, as with most chemicals of environmental concern, in establishing a 'safe' concentration in environmental samples. This has stimulated intense debate among environmental activists, the chemical industry, researchers, government officials and the public about the adverse environmental effects and human health risks

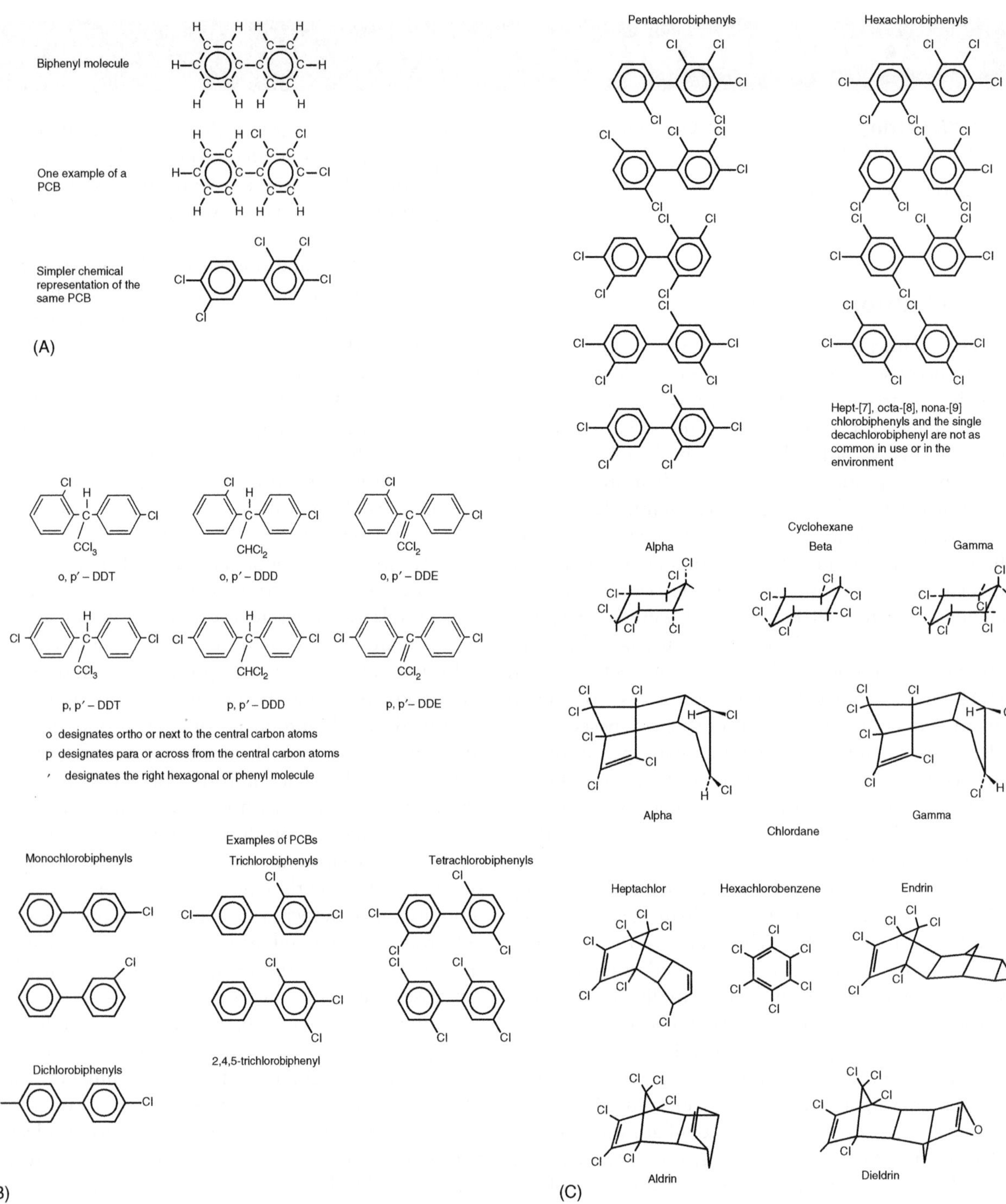

Figure 1 Chemical structures of (A) polychlorinated biphenyls (PCBs); (B) dichlorodiphenyltrichloroethane (DDT) and the metabolites dichlorophenyldichloroethane (DDD) and DDE; (C) other chlorinated pesticides. C, carbon atoms; H, hydrogen atom; Cl, chlorine atom.

associated with low amounts of organochlorine compounds, including chlorinated hydrocarbons.

Lessons learned about the environmental behavior and adverse effects of chlorinated hydrocarbons provide guidance about what to expect for the more general class of organochlorine compounds. Much information is available about chlorinated pesticides such as the DDTs (in this article DDT includes *p,p′*- and *o,p*-DDT and the immediate biodegradation and metabolism products DDE and DDD, see **Figure 1**) and PCBs. Thus, DDT and PCBs are used herein as examples for the larger class of chlorinated hydrocarbons. However, the approach of using examples is pursued with the caveat that movement of each of the chlorinated hydrocarbons through the environment, and biological effects associated with each chemical, are specific in important details for each chemical. Although not discussed further here, it is important to note that these chemicals can and have been used as tracers of general processes acting on naturally occurring organic chemicals in marine ecosystems.

History

DDT

DDT (dichlorodiphenyltrichloroethane) is not a naturally occurring compound. It was first synthesized in 1874 and its insecticidal properties were discovered in 1939. Initial large use of DDT as an insecticide began in 1944 and continued into the late 1960s. It was used with success against vectors of human diseases such as malaria and with dramatic effect in agriculture around the world in controlling insect pests. The environmental problems associated with DDT in terms of adverse effects on nontarget organisms such as birds were brought to popular attention in the highly influential book '*Silent Spring*' by Rachael Carson in 1962. Further studies of DDT in the mid 1960s to early 1970s documented the presence of DDT and several other chlorinated pesticides in marine organisms at all major sectors in the marine food web. Analyses of samples from organisms dwelling in the deep part of the oceans, for example 4000–5000 m depth in the North Atlantic, and from Arctic and Antarctic marine ecosystems contained measurable concentrations of DDT. Evidence of adverse effects on nontarget terrestrial, freshwater and marine organisms, especially birds, resulted in curtailed use or bans on the use of DDT in several developed countries in the early 1970s. The legacy of past releases to the environment is present in marine ecosystems in the form of measurable concentrations of DDT compounds. In addition, the use of DDT continues in a few countries.

PCBs

PCBs (polychlorinated biphenyls) have been used industrially since 1929. Industrial mixtures of PCBs are known by commercial names, e.g., Aroclors (United States), Kaneclor (Japan), Chlophen (Germany), Sovol (former USSR), Fenchlor (France). PCBs were widely used in insulating fluids in transformers and capacitors, as well as hydraulic systems, surface coatings, flame retardants, inks and other minor uses.

Concerns about human health effects associated with halogenated aromatic compounds such as PCBs and halowax (polychlorinated naphthalenes) date to the 1930s and 1940 with the reports of rashes and liver abnormalities for workers in manufacturing plants and electricians. The identification of polychlorinated biphenyls (PCBs) as chemicals of environmental concern dates to the late 1960s when they were reported nearly simultaneously by three different research groups to be present in seabirds and seabird eggs in three different coastal ecosystems.

Subsequent research in the late 1960s and early 1970s confirmed the widespread presence of PCBs in numerous ecosystems, their relative persistence in the environment, and several instances of known or suspected adverse effects associated with various organisms exposed to and incorporating PCBs into their tissues, e.g., mink and chickens fed on fish or fish meal. In 1968, contamination of rice oil used in food preparation at a location in western Japan by PCBs from a leaking transformer caused human health effects for people who consumed the polluted food; this was designated the Yusho incident. A similar incident occurred in 1979–81 in Taiwan: the Yuncheng incident. Detailed studies of the PCB oil involved in these incidents suggested that some or all of the observed adverse effects may have resulted from the presence of small amounts of chlorinated bibenzonfurans or chlorinated dibenzodioxins.

By 1971, the concerns about human health and environmental impacts led Monsanto, the producer of PCBs in the United States, to a voluntary ban on sales of PCBs except for closed systems use. Monsanto ceased all production in 1977 and there was no large-scale increase in imports. PCBs were banned from production and further use in the United States in 1978. Equipment that already contained PCBs, e.g., transformers, were allowed to remain in use but restrictions were placed on the disposal of PCBs when the equipment was decommissioned. Delegates from 122 countries completed a draft treaty on persistent organic pollutants (POPs) in December 2000. The POPs that were initially addressed and banned from further use include chlordane, DDT, dieldrin, endrin, heptachlor, mirex, toxaphene, PCBs, hexachlorbenzene, chlorinated dibenzofurans, and chlorinated dibenzodioxins. Limited selective use of DDT for human disease vector control is allowed in some countries.

Distribution in the Marine Environment

Early 1970s

Analyses during the early 1970s of various species of marine biota for DDT family chemicals and PCBs established widespread distribution of these chemicals in many areas of the world's oceans from equatorial to polar regions and to depths of 4000–5000 m in the Atlantic Ocean. Biota with significant lipid (fat) in their body tissues tended to accumulate DDT family compounds and PCBs much more than biota with lower lipid content. Marine mammals and birds accumulated higher concentrations of DDT compounds and PCBs, presumably due to being near the top of the food web (biomagnification of contaminants), and having a high body lipid content (especially for marine mammals and birds).

Analyses of surface seawater samples (both dissolved and particulate) and samples of the atmosphere over the oceans established the presence of low concentrations (ranges of 0.01–1 ng kg^{-1} water or 0.001 ng m^{-3} of air) of DDT family compounds and PCBs. Very low concentrations and difficulties in avoiding contamination from the sampling ship and the sampling gear made measurement of deep seawater samples problematic. The few deep-water samples analyzed documented that the concentrations of DDT and PCBs were not higher than about 0.001 ng kg^{-1} of water. Confirmatory measurements were not made for years thereafter because of intense debate among chemical oceanographers about how to make reliable measurements for these compounds at very low concentrations in sea water. In contrast, the underlying sediments had accumulated sufficient concentrations, because of sorption on particles and deposition, to allow undisputed measurements of both DDT and PCB in deep ocean surface sediments.

Several surveys and research programs documented much higher concentrations of DDT and PCBs in coastal waters and coastal ecosystems compared to open ocean ecosystems; especially near urban areas for DDT and PCBs, and in coastal regions near agricultural drainage areas for DDT, as might be expected given patterns of use for these compounds and probable release to the environment.

1980s to the Present Day

Open Ocean Recent measurements of PCBs and DDT family compounds in surface sea water and air samples over the ocean on a regional oceanic scale are few and are exemplified in **Figure 2(A–D)**. Despite the paucity of data, some important findings are evident. There are higher concentrations of DDT in surface sea waters and air overlying the oceans near south-east Asia in comparison to the other areas sampled. This is consistent with environmental concerns associated with continued use of DDT in the Asian continent, South Asia subcontinent, and Oceania areas beyond the years when DDT use was curtailed or eliminated in the developed countries of North America, Europe, Japan, and Australia. The PCB concentrations in the surface sea water are in the range of 1–60 pg kg^{-1} and in the overlying atmosphere 3–600 pg m^{-3}. The distribution of concentrations is more even across the areas sampled compared to DDT (**Figure 2**). This is consistent with the continued presence of PCBs cycling in the environment as a result of past releases, leakage from landfills and products containing PCBs still in use, and perhaps continued new uses even though PCB manufacture has been eliminated or severely reduced in many countries.

Recent progress with measuring low concentrations of PCBs in deep ocean waters has enabled a few measurements of deep ocean waters. A depth profile of the sum of concentrations of several chlorobiphenyl congeners for a station in the eastern North Atlantic (**Figure 3**) documents higher concentrations in surface waters with decreasing concentrations with depth as expected due to the greater contact of the surface waters with the atmosphere and contemporary environment. However, all the concentrations are very low in comparison to concentrations found in near-shore waters, lakes, and rivers. Concentrations in the deepest waters are below or at detection limits of the analytical methods used. PCBs in mid-depth and deeper waters are most likely a result of the flux of particles from the surface water carrying sorbed PCBs in and on particles to deeper waters. There has been considerable progress during the 1980s and 1990s in understanding the role and details of particles as conveyers of chemicals from the surface ocean to deep waters and sediments. As the particles sink through the deep waters, desorption and disaggregation of particles and subsequent desorption releases PCBs. Therefore, it is likely, though not proved due to lack of a series of data over time, that current deep water concentrations of PCBs reflect inputs from particle fluxes over the total time of PCB use and release to the environment; for example inputs from peak use in North America and Europe during the 1950s to 1960s.

Deep-water sediments contain low concentrations of DDT and PCBs in the range of 1–100 ng g^{-1} dry weight or parts per billion. Relatively few deep-ocean benthic (bottom dwelling) animals have been analyzed, but those that have been analyzed contain detectable concentrations of DDT and PCBs in the range of 0.001–1 μg g^{-1} dry weight. Analyses of a few samples of mid-water fish in the deep ocean document the presence of PCBs in a pattern that reflects metabolism of the PCBs after uptake. Although small in numbers of samples analyzed,

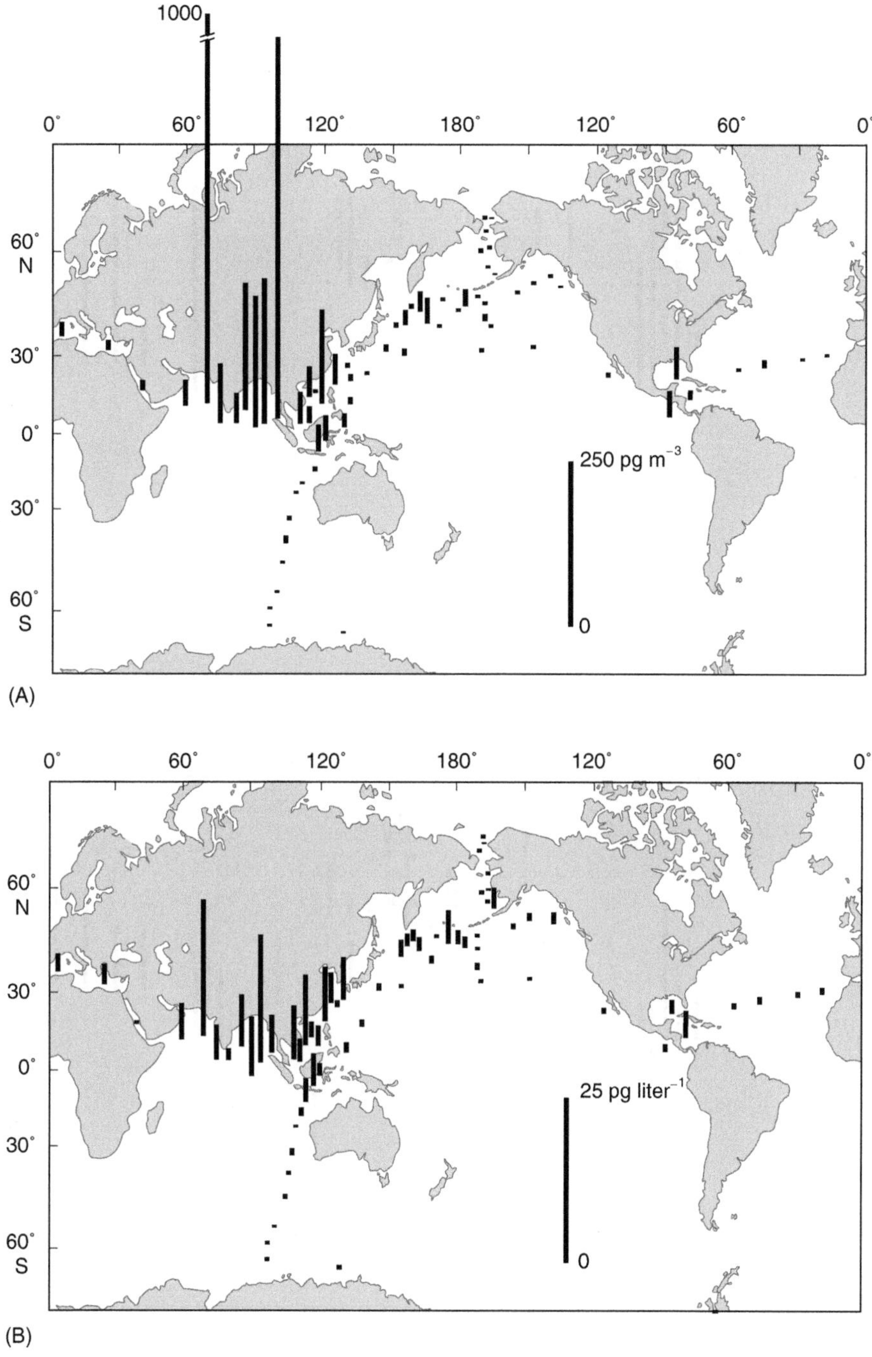

Figure 2 Concentrations of DDT (A,B) and PCBs (C,D) in the atmosphere and surface sea water (B,D) 1989–90. (Adapted with permission from Iwata *et al.*, 1993.)

assessment of fish liver enzymes specific to metabolism of compounds such as PCBs indicate activation of these enzymes and that the fish have responded biologically to the presence of PCBs.

Recent data and an assessment (1993–94) of the flux of PCBs to the North Atlantic open ocean have estimated, as an example, that 14 000 mol year^{-1} of a tetrachlorobiphenyl congener was sequestered in the north-western Atlantic Ocean with 75% of the flux being in the deep-ocean pelagic sector and only 25% in the coastal–continental shelf area. This same assessment provided good reasoning that the flux of PCBs from the atmosphere to the North Atlantic has decreased over the past two decades.

Coastal Ocean Aquatic toxicological studies during the 1960s to the present have established the range of sensitivities of several different species and

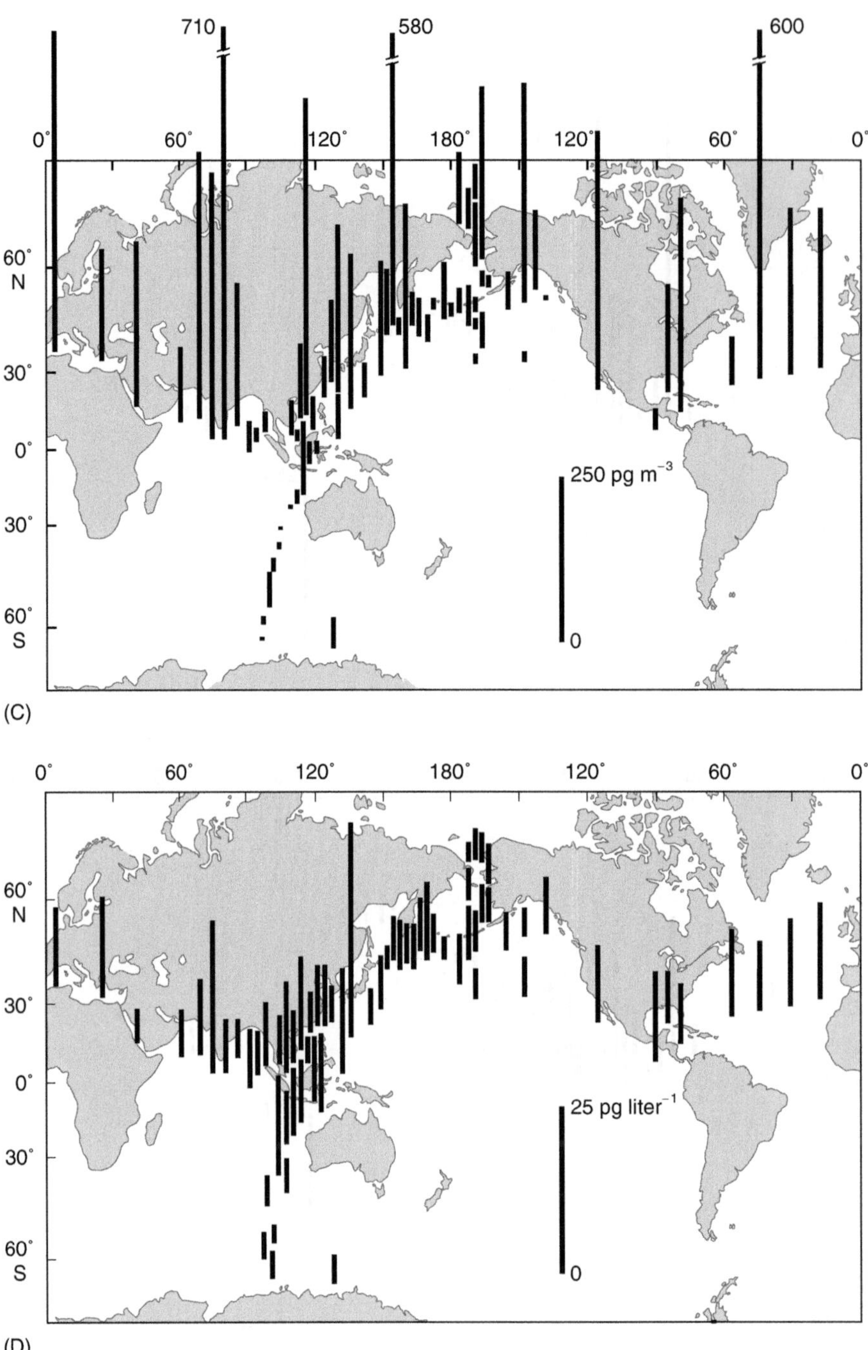

Figure 2 *Continued*

life stages of marine organisms to chlorinated hydrocarbon pesticides and industrial chemicals such as PCBs. These studies established lethal tolerance limits for these chemicals in aquatic habitats, usually in the microgram per liter range. Such concentrations are rarely encountered today in estuarine or coastal areas, except in the immediate vicinity of exceptional discharges of the chemical. Most current risk assessments for chlorinated hydrocarbons are related to longer term, subacute effects on human health, fish and wildlife feeding on marine organisms. There are various sublethal effects on various organisms that are beyond the scope of this article. They are similar for many types of natural and nonnatural stress on marine organisms and ecosystems. These are summarized in generic form in **Table 1**. However, the caveats below apply, as do the cautions in **Table 1**. Depending on the chemical and the species involved, concentrations in water, sediment, or in

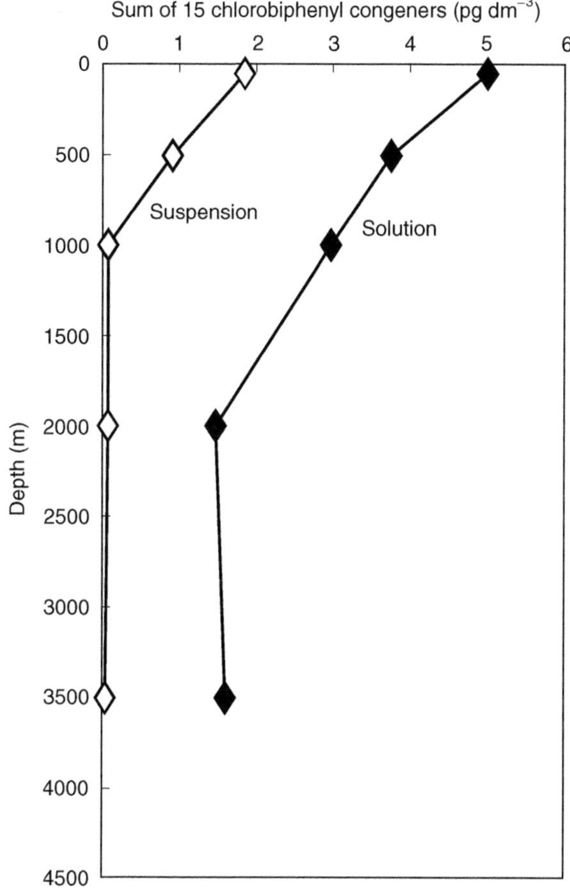

Figure 3 Depth profile of PCB concentrations in sea water, May, 1992. at 47°N, 20°W. (Adapted from Petrick *et al.*, 1996.)

Table 1 General types of responses to PCB contamination for marine organisms. (PCB concentrations in tissues or habitat types eliciting a given intensity of response varies with species and ecosystem)

Level of biological organization	*Types of response*[a]
Biochemical-cellular	Toxication Metabolic impairment Cellular damage Detoxication
Organismal	Physiological change Behavioral change Susceptibility to disease Reproductive effort Larval viability Immune responses
Population	Age, size structure Recruitment Mortality Biomass Adjustments in reproductivity and other demographic characteristics
Community	Species abundance Species distribution Biomass Trophic interactions

[a]Responses are mostly adverse effects, but some are beneficial in offering protection against adverse effects.

Adapted with permission from Farrington JW and McDowell JE (1994) Toxic chemicals in Buzzards Bay: Sources, fates, and effects. In: Costa JE, Gibson V and Pedersen JM (eds) *A Synthesis of Pollutant Inputs to Buzzards Bay*. Buzzards Bay Project Technical Report Series BBP94-30, 18 October 1994. Marion, MA, USA.

the organism's food and tissues eliciting a given effect can range over many orders of magnitude from parts per million to parts per trillion.

The near-shore and estuarine waters of the coastal ocean contain elevated concentrations of DDT and PCBs in comparison to the open ocean. Therefore, attention has been focused on obtaining more data for the coastal ocean. The data sets are more numerous and provide better geographic and temporal coverage for coastal areas of developed countries but much less so for most of the developing countries. Sufficient data have been collected in several areas and sufficient laboratory experiments have been completed to provide a reasonable general understanding of the inputs, fates and effects of DDT and PCBs in coastal ecosystems. **Figure 4** shows a general depiction of the cycling of PCBs in a coastal ecosystem. One key aspect of this biogeochemical cycle is the uptake by animals of DDT and PCBs both from food sources and from water across membrane surfaces such as gills. Exceptions are air-breathing organisms such as birds and marine mammals for which the predominant source is food. Another key aspect of the biogeochemical cycle is sorption of DDT and PCBs onto particles and deposition to sediments. During inadvertent or deliberate discharges or releases to the environment, a portion of these compounds move through coastal ecosystems with portions lost to the atmosphere and transported elsewhere and to be deposited by dust or aerosols, and by rain and snow.

Even though the chlorinated hydrocarbons are among the chemicals more resistant to chemical or biological alteration in the environment, there are physical–chemical (e.g., sorption–desorption, transfer from water to air), microbial transformation and degradation, and animal enzyme modifications or transformations, that change the mixture of compounds as the chemicals move through the environment. For example, the mixture of chlorobiphenyl congeners found in a lobster were dramatically different from the original mixture discharged in a

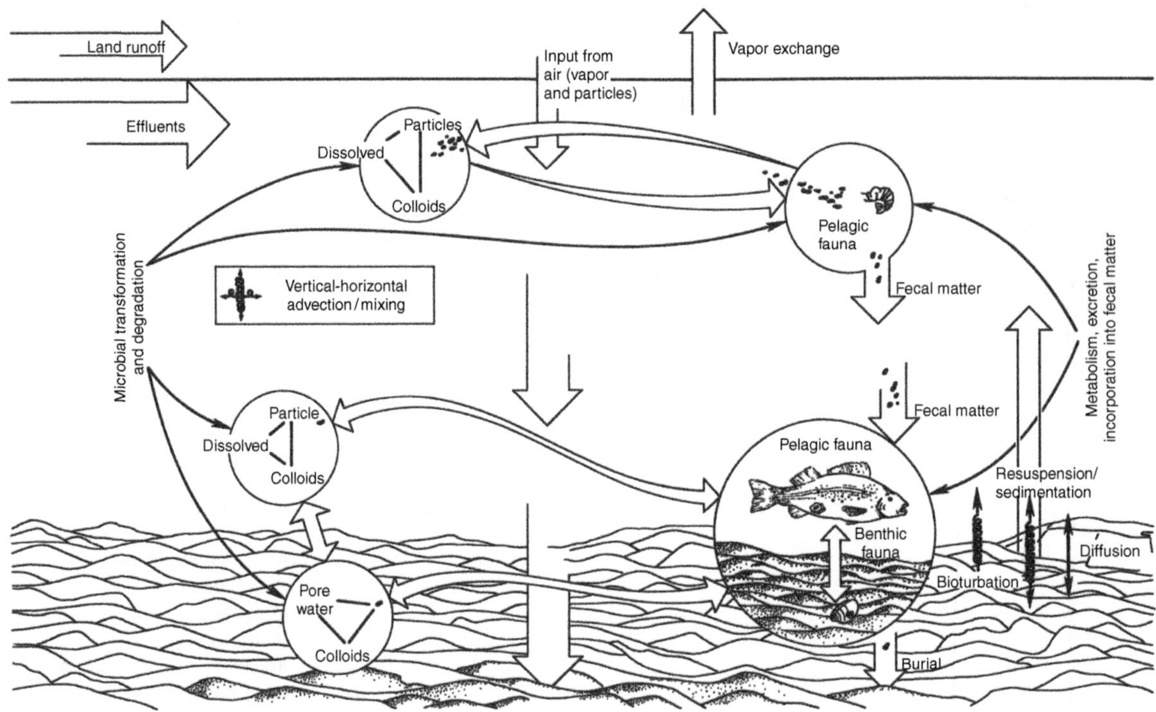

Figure 4 Biogeochemical cycle for chlorinated hydrocarbon pesticides and PCBs in coastal ecosystems.

nearby effluent from a capacitor manufacturing facility. Those chlorobiphenyl congener mixtures were also different in composition compared to the PCB congener mixture in flounder caught in the same area. The sediment congener mixture for the habitat of both the lobster and flounder had yet another composition. Biological effects of chlorinated hydrocarbons can be specific to each individual chemical in terms of mode of action and potency of action. Thus, the presence of complex and diverse mixtures of these chemicals in various compartments of an ecosystem introduces significant complications to the task of providing an assessment of ecological and human health risks associated with a given site of chlorinated hydrocarbon contamination. In addition, the presence of other chemicals of environmental concern in many of the same areas, means that the present knowledge base is insufficient to provide a high degree of accuracy to a quantitative risk assessment for ecological and human health concerns.

In several coastal areas near urban harbors or major industrial production or use areas, sediments accumulated high concentrations and substantive amounts of DDT or PCBs. Once discharges were reduced or eliminated with curtailment of production and use of DDT and PCBs, the accumulations of these compounds in sediments continued to be of concern as a source of contamination for coastal ecosystems. The DDT and PCB contaminated sediments can leak REB and DDT or PCBs to the

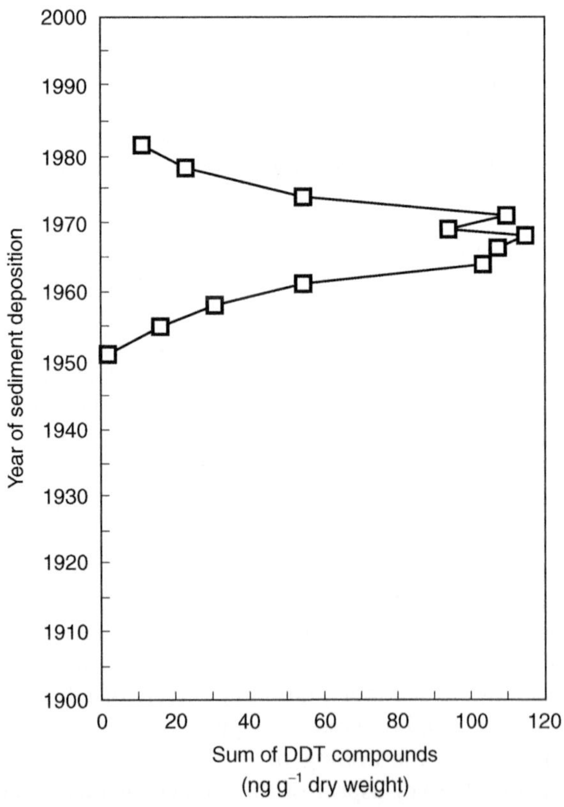

Figure 5 Depth profile of DDT concentrations in San Pedro Shelf sediments, Southern California, USA, documenting the historical input of DDT to the area. (Adapted from Eganhouse and Kaplan, 1988.)

overlying water column or DDJ can be taken up from sediments and water in the spaces or pores between sediment particles (pore waters) by animals living in or on the sediments (**Figure 4**).

Much of the present scientific effort related to DDT and PCBs is focused towards three broad issues: (1) in support of remediation and clean-up of areas of high levels of concentrations as a result of past practices; (2) preventing or limiting mistakes made in developed countries from occurring in developing countries; and (3) tracking the spatial and temporal trends of concentrations of these compounds in marine ecosystems, especially coastal ecosystems.

There are two principal approaches available to track the trend in concentrations over time. One approach is to find areas where coastal sediments are accumulating at a steady and sufficient rate, and are reasonably undisturbed by activities such as mixing of the upper layers by organisms or storm turbulence. Sediments deposited in waters with very low or no oxygen content have limited or no mixing by organisms and usually meet the criteria. Cores of sediments can be carefully sliced or sectioned at fine intervals and analyzed to provide a historical record, layer by layer, of DDT and PCB concentrations. This has been accomplished with success in several coastal

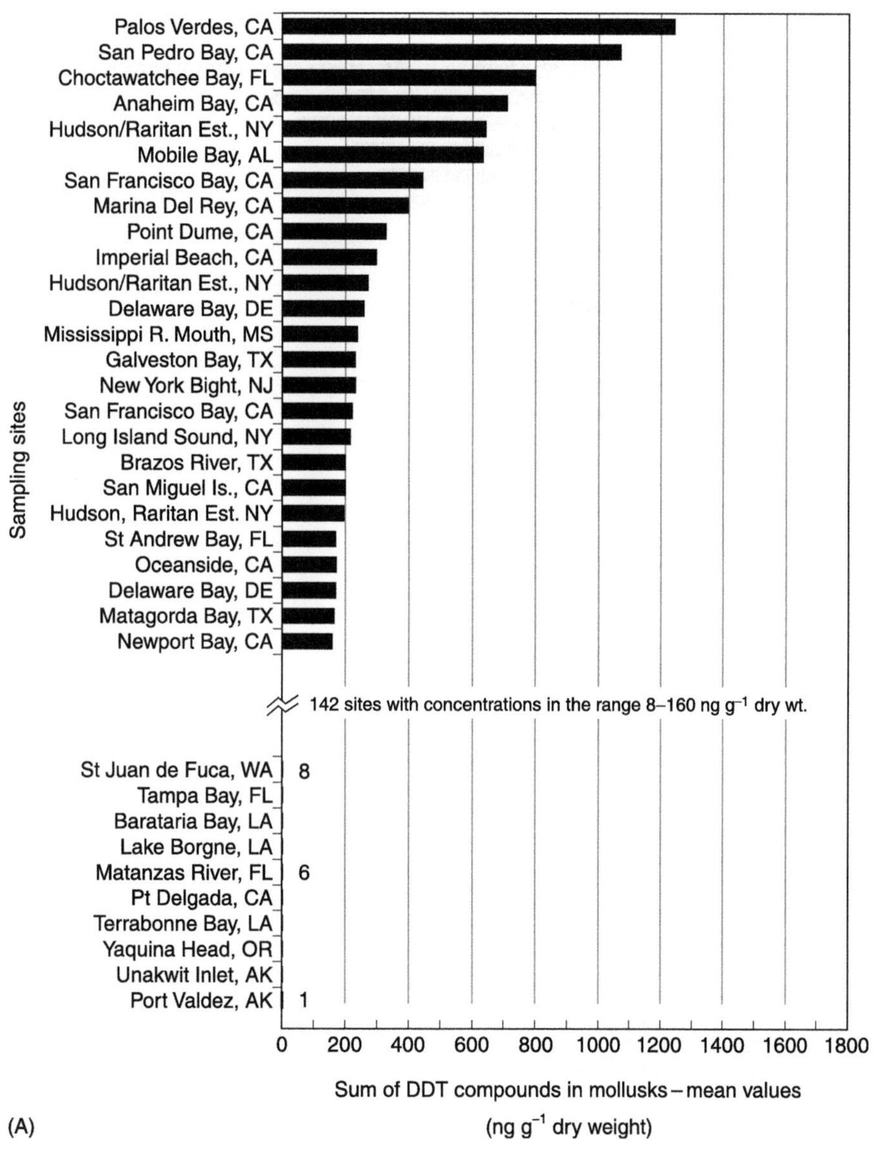

Figure 6 Concentrations of (A) DDT and (B) PCBs in Mussel Watch Stations US NOAA Status and Trends Program (Mussels and Oysters) 1986. (Adapted from *A Summary of Data on Tissue Contamination from the First Three Years (1986–1988) of the Mussel Watch Project.* NOAA Technical Memorandum NOS OMA 49. NOAA Office of Oceanography and Marine Assessment, Ocean Assessments Division, Rockville, MD US, 1989.)

areas, one example being the San Pedro Shelf, Southern California, US coastal area (**Figure 5**).

The other approach that incorporates geographic or spatial assessments with the time series measurements has been incorporated into a monitoring strategy for assessing and monitoring concentrations of several chemicals of environmental concern. This involves the use of bivalves, mainly mussels and oysters, as sentinels of biologically available contaminants such as DDT and PCBs: the 'Mussel Watch' approach. Prototypes of such a program were evaluated in the 1970s in the US, Canada, Europe, and Australia and there are currently several operational programs such as the Mussel Watch component of the US National Oceanographic and Atmospheric Administration (NOAA) National Status and Trends Program.

DDT and PCB concentrations from samples of mussels and oysters obtained between 1986 and 1988 at over 150 stations located around the US coast are summarized in **Figure 6**. The higher concentrations of both DDT and PCBs correspond to known or suspected sources of inputs from industrial facilities making or using these chemicals, or are near urban areas. Generally it is accepted that concentrations of DDT began to decrease dramatically in portions of ecosystems for some areas of the US coast as a result of curtailed input. Concentrations of PCBs also decreased in the few areas measured during the 1970s and early 1980s as a result of curtailed

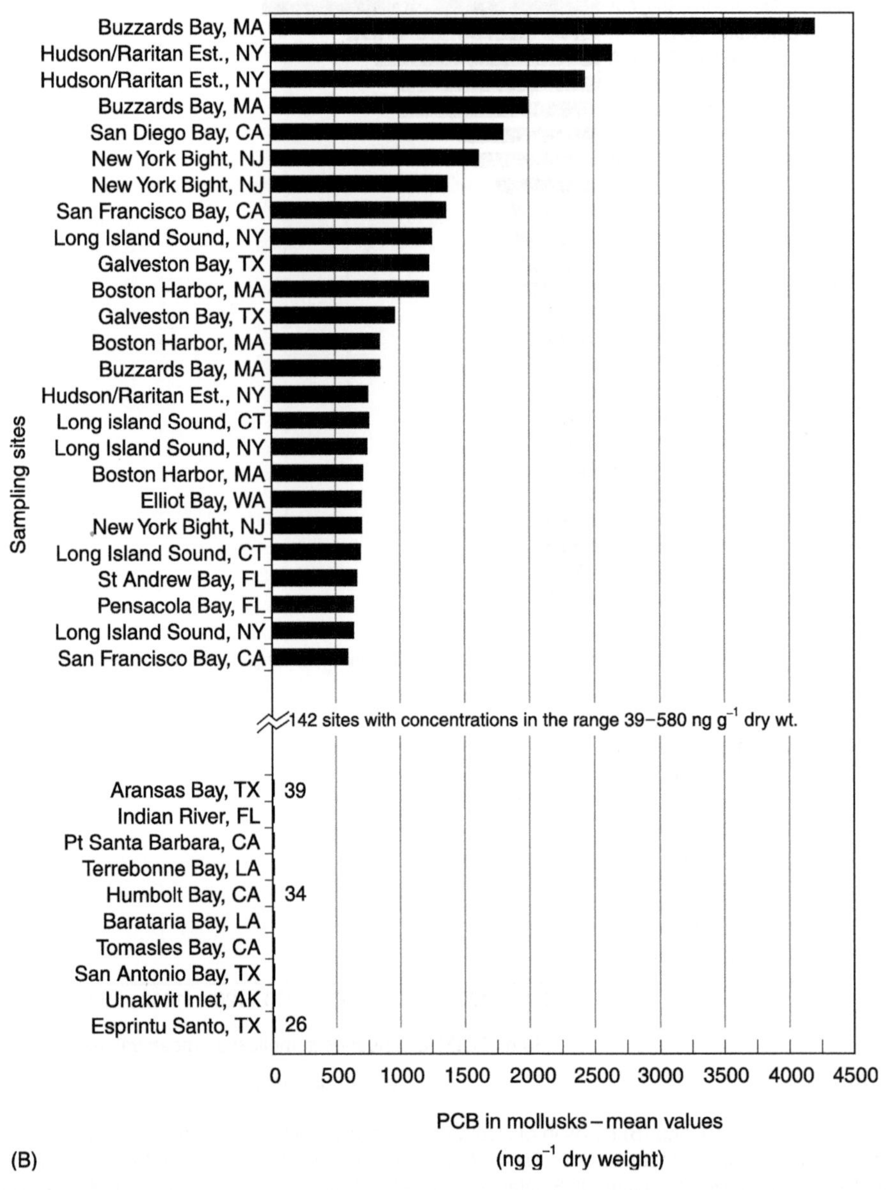

Figure 6 *Continued*

manufacture and use. The exceptions where concentrations remained elevated were generally in ecosystems with a significant burden of DDT or PCBs in surface sediments as a result of past inputs.

The trends in DDT and PCB concentrations, and two other chlorinated hydrocarbon pesticides, chlordane and dieldrin, in bivalve tissues at locations in the US coastal area for 1986 to 1995 are summarized in **Table 2**. The decrease noted from limited sampling for a few areas in the 1970s and early 1980s continues for some locations. For many other locations, examination of the data indicates that the concentrations are so low that general global and regional biogeochemical cycles are causing only a slow further decrease. A few stations continue to maintain elevated concentrations and for most this can be attributed to continuing contamination of the bivalves from nearby surface sediments containing high concentrations of the compounds.

Similar types of 'Mussel or Oyster Watch' data have been collected in some European countries (e.g., France) with similar results. Prototypes of this approach have been carried out in the 1990s for developing countries of Central and South America and South-east Asia under the auspices of UNESCO-IOC and UNEP. Other time trends of DDT and PCB concentrations have been assessed such as concentrations in cod liver oil collected from samples in the southern Baltic Sea from 1971 to 1989 (**Figure 7**). Consistent with the preceding discussion, DDT concentrations decrease by a factor of three to four comparing 1971–1974 with 1987–1989 and PCB concentrations decrease at a slower rate.

The following summary of one aspect of the PCB and chlorinated pesticide saga illustrates the importance of understanding the global, regional, and local biogeochemical cycles of these compounds and their relationship to environmental and human health risks. PCBs and several chlorinated pesticides released to the environment in developed countries of the Northern Hemisphere enter the atmosphere from land and from surface ocean waters in the tropics, subtropics, and temperate zones. Subsequently these compounds are transported by atmospheric circulation patterns to Arctic regions, and enter Arctic ecosystems by precipitation and dry deposition. There may be several cycles of precipitation and volatilization back to the atmosphere before these compounds reach the Arctic. Contamination of the Arctic aquatic ecosystems results in the transfer of these compounds through the food web and biomagnification in marine mammals. Inuits, a native Arctic region or Northern peoples, hunt several of these marine mammals and eat their tissues. The resulting contamination of mother's breast milk transfers these chemicals to infants. There are good reasons to be concerned that subsequent normal development of the children is impaired or slowed. This is the net result of actions of human civilization and complex environmental processes operating over decades and distances of thousands of kilometers.

Table 2 Trends in concentrations of selected chlorinated hydrocarbons 1986–95 in bivalves (mussels and oysters), US coastal areas[a]

Chemical	*Number of sampling locations*		
	Increased	*Decreased*	*No trend*
Chlordane	1	81	104
DDT	1	38	147
Dieldrin	1	32	153
PCBs	1	37	148

[a]Sites of several elevated concentrations are indicated in Figure 6. Data obtained from the US Department of Commerce, National Oceanic and Atmospheric Administration World Wide Web Site, October, 2000 http://state-of-coast.noaa.gov/bulletins. Data compiled by Dr Thomas P. O'Connor, US NOAA.

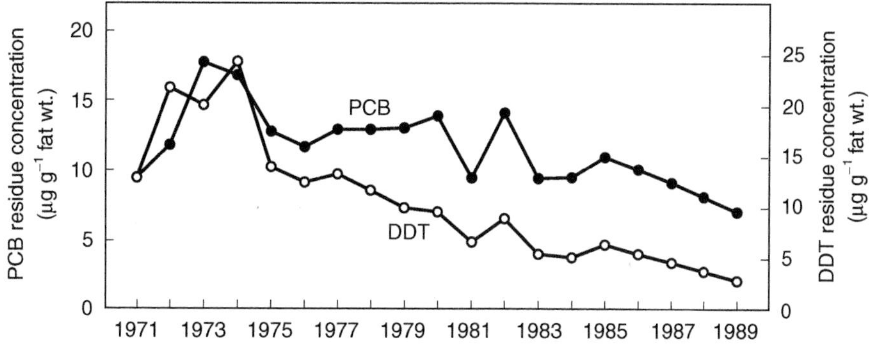

Figure 7 Time trends of DDT (○) and PCB (•) concentrations in cod liver oil from the southern Baltic, 1971–89. (Adapted from Kannan *et al.*, 1992.)

Conclusions

Humanity was fortunate to learn valuable lessons from early results that indicated widespread regional and global transport of chlorinated pesticides and PCBs accompanied by environmental and human health problems. Much knowledge has been gained about the biogeochemical cycles and about ecological and biological effects of these chemicals in the oceans. This knowledge has been used to guide policy and management actions in many instances. Otherwise, the deplorable situation faced by Inuits might be much more severe and widespread and natural resource populations, including oceanic species, might have been more severely impacted. Despite policy and management actions in many developed countries limiting or eliminating production and release of many of these compounds, there are still concerns about the legacy of past releases to the environment present in coastal ocean surface sediments in several locations. There are serious coastal environmental and human health concerns associated with continued uses of several of these chlorinated hydrocarbons in developing countries.

Further Reading

Dawe CJ and Stegeman JJ (eds.) (1991) Symposium on Chemically Contaminated Aquatic Food Resources and Human Cancer Risk. Environmental Health Perspectives 90: 3–149.

Eganhouse RP and Kaplan IR (1988) Depositional history of recent sediments from San Pedro Shelf, California: reconstruction using elemental, isotopic composition, and molecular markers. *Marine Chemistry* 24: 163–191.

Erickson MD (1997) *Analytical Chemistry of PCBs*, 2nd edn. New York: Lewis Publishers.

Farrington JW (1991) Biogeochemical processes governing exposure and uptake of organic pollutant compounds in aquatic organisms. *Environmental Health Perspectives* 90: 75–84.

Fowler S (1990) Critical review of selected heavy metal and chlorinated hydrocarbon concentrations in the marine environment. *Marine Environmental Research* 29: 1–64.

Giesy J and Kannan K (1998) Dioxin-like and non dioxin-like toxic effects of polychlorinated biphenyls (PCBs): implications for risk assessment. *Critical Reviews in Toxicology* 28(6): 511–569.

Goldberg ED (1991) Halogenated hydrocarbons, past, present and near-future problems. *Science of the Total Environment* 100: 17–28.

Gustafsson O, Gschwend PM, and Buesseler KO (1997) Settling removal rates of PCBs into the northwestern Atlantic derived from ^{238}U-^{234}Th disequilibria. *Environmental Science and Technology* 31: 3544–3550.

Iwata H, Tanabe S, Sakai N, and Tatsukawa R (1993) Distribution of persistent organochlorines in the oceanic air and surface seawater and the role of the ocean on their global transport and fate. *Environmental Science and Technology* 27: 1080–1098.

Kannan K, Falandysz J, Yamashita N, Tanabe S, and Tatsakawa R (1992) Temporal trends of organochlorine concentrations in cod-liver oil from the southern Baltic proper, 1971–1989. *Marine Pollution Bulletin* 24: 358–363.

O'Connor TP (1991) Concentrations of organic contaminants in mollusks and sediments at NOAA National Status and Trends sites in the coastal and estuarine United States. *Environmental Health Perspectives* 90: 69–73.

Petrick G, Schulz-Bull DE, Martens V, Scholz K, and Duinker JC (1996) An in-situ filtration/extraction system for the recovery of trace organics in solution and on particles tested in deep ocean water. *Marine Chemistry* 54: 97–103.

Schwarzenbach RP, Gschwend PM, and Imboden DM (1993) *Environmental Organic Chemistry*. New York: John Wiley.

Thornton J (2000) *Pandora's Poison. Chlorine, Health and a New Environmental Strategy*. Cambridge, MA: MIT Press.

Waid JS (ed.) (1986) *PCBs and the Environment*, vol. 1–3. Boca Raton: CRC Press.

TRACER RELEASE EXPERIMENTS

A. J. Watson, University of East Anglia, Norwich, UK
J. R. Ledwell, Woods Hole Oceanographic Institution, Woods Hole, MA, USA

Introduction

Since the mid 1980s, analytical and engineering techniques have been developed to enable the compound sulfur hexafluoride (SF_6) to be used as a tracer for oceanographic experiments. SF_6 is a stable and inert substance with an exceptionally low level of detection, and its use enables large bodies of water to be unambiguously marked, allowing the investigator to keep track of a particular parcel of water. Three kinds of experiment have thus far made use of this technique: (1) measurement of mixing and transport integrated over large regions; (2) estimates of gas transfer velocities at the surface of the sea; (3) open ocean iron enrichment experiments. This article briefly describes the techniques used, and the major results from each of these types of process study.

The Tracer

Sulfur hexafluoride is an inert perfluorine, routinely detectable in sea water at $\sim 0.01\,\text{fmol}\,\text{kg}^{-1}$ by electron-capture gas chromatography ($1\,\text{fmol} = 10^{-15}$ mol). At room temperature and pressure SF_6 is a gas, but it forms a dense ($\rho = 1880\,\text{kg}\,\text{m}^{-3}$) liquid at pressures exceeding 20 bar. It is extremely stable in the environment and, other than being an asphyxiant, the pure compound has no known toxic effects. It is produced commercially largely (~80%) for use as a gaseous insulator in high-voltage installations. Much of this industrial production eventually finds its way into the atmosphere. The atmospheric mixing ratio was about 4×10^{-12} in 1999, and is growing at about 7% per year. Its solubility is very low, so that the surface concentrations in equilibrium with the atmospheric concentration are on the order of $1\,\text{fmol}\,\text{kg}^{-1}$. The combination of very low detection limit, nontoxicity, low marine background concentration, ease of analysis and inertness make SF_6 a nearly ideal tracer.

SF_6 is included in the Kyoto Protocol because, molecule-for-molecule, it is a powerful greenhouse gas with a long (>1000 years) lifetime in the atmosphere. The signatory nations are thus committed to controlling the rate of its production. However, for any realistic future emission scenario, SF_6 will remain insignificant (<1%) as a contributor to the anthropogenic greenhouse effect for the foreseeable future.

Mixing Experiments in the Deep Ocean

To measure diapycnal mixing (i.e. mixing across-density surfaces) by tracer release, the tracer is released, as near as possible, onto a single, well-defined density surface, and its subsequent spread onto neighboring surfaces is monitored. If the mixing occurs in accordance with Fick's law, the square of the mean width of the concentration distribution increases linearly with time, the rate of increase being a direct measure of the diffusivity. The advantage of this strategy compared to the documentation of temperature or velocity microstructure, is that it gives an unambiguous measurement integrated over a substantial time and space scale. In practice, in the open ocean these scales are of order months or years, and hundreds or thousands of kilometers – hence also the method's main disadvantage, which is that it must be done on a large scale.

At the time of writing, five experiments of this kind have been initiated in the open ocean. The first two, relatively small-scale releases, were made in the ocean-floor basins off the coast of Southern California and the remaining three in the thermocline of the North Atlantic, the deep Brazil Basin and the central Greenland Sea. Below we describe the release method used in most of these experiments, and the results of the North Atlantic experiment in more detail. Mixing rates from all five experiments are then compared.

Release Method

Sulfur hexafluoride is very insoluble; for small-scale experiments it can be dissolved by presaturating drums or tanks of water with the gas. However, the practical limit for the amount which can be injected in this way is a few moles, sufficient for tracer experiments on the 10–100 km scale only. For large open ocean releases, we designed an injection package which releases liquid SF_6 into water by pumping it through fine orifices at high pressure, so that it breaks into an emulsion of fine droplets on contact with the sea. These droplets are sufficiently small that they dissolve before they have settled an appreciable distance. The apparatus is designed to allow the accurate delivery of SF_6 at rates of up to $3\,kg\,h^{-1}$ onto a given 'target' density surface at any depth greater than 200 m in the ocean, when towed behind a ship on a conducting cable.

In use, the injector was controlled by a computer in the laboratory of the ship. The output of the CTD was used to calculate in real time the density of the water at the package, and compare it to the 'target' density. The computer issued commands to the automated winch to haul in wire if the density was higher than the target, or pay out if it was significantly lower. During the North Atlantic Tracer Release Experiment (NATRE) this system was able to deliver tracer with an overall RMS accuracy of ± 2 m from the target surface. With such an injection system, it is practical to initiate experiments using several hundred kilograms of tracer, sufficient to enable investigations at the ocean-basin scale.

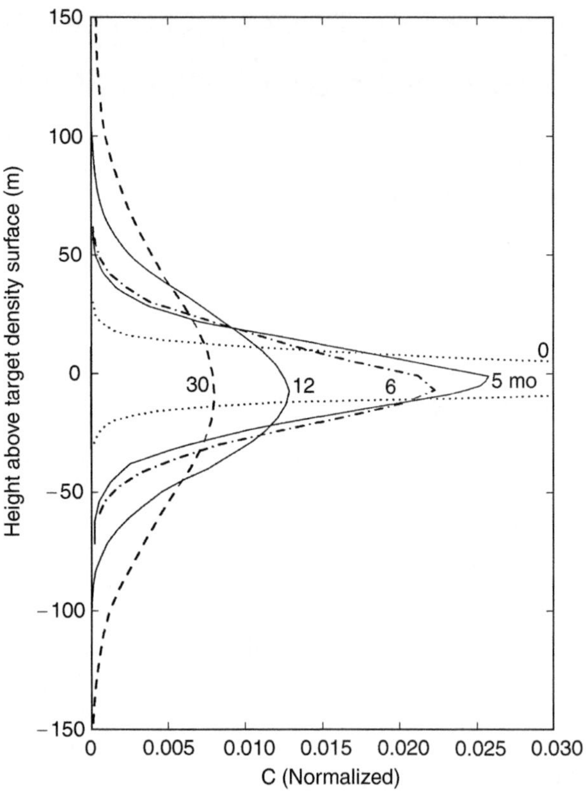

Figure 1 Mean vertical profiles from NATRE at 0, 5, 6, 12 and 30 months after the initial survey. The SF_6 concentration has been averaged on isopycnal surfaces, approximately, and plotted versus height above the target isopycnal surface using the mean relation between depth and density for the 12-month survey. The profiles are normalized to have equal areas. The initial profile (… .) is allowed to run off the graph so that the others are clear.

NATRE: Overview of Results

The tracer results from NATRE have been reported in detail. Major findings were that the diapycnal diffusivity was $0.12\,cm^2\,s^{-1}$ for the first 6 months, and then $0.17\,cm^2\,s^{-1}$ for the subsequent 24 months. The mean vertical profile for each survey was nearly Gaussian, and as a set they illustrate an approximate solution of the diffusion equation in one dimension (**Figure 1**).

The result that the diapycnal diffusivity in the pycnocline is of order $0.1\,cm^2\,s^{-1}$ confirms estimates based on internal wave dynamics and on measurements of turbulent dissipation rates. Some analyses of the penetration of transient tracers into the deep pycnocline also have implied diffusivities on the order of $0.1\,cm^2\,s^{-1}$. Values of diffusivity of $1\,cm^2\,s^{-1}$ were inferred by Munk's classic 'abyssal recipes' analysis, but this was for depths between 1000 and 4000 m and included boundary processes as well as interior processes. It is now clear that $1\,cm^2\,s^{-1}$ is an overestimate for the interior pycnocline.

The lateral dispersion of the tracer revealed surprisingly efficient mechanisms of stirring at scales from 0.1 to 30 km. The lateral diffusivity setting the width of tracer streaks at 6 months was found to be about $2\,m^2\,s^{-1}$. The mechanism is not well understood, but may be due to shear dispersion by vortices generated during the adjustment to diapycnal mixing events. The experiment did confirm the predictions of C. Garrett, that a tracer patch remains streaky only for a year or so, after which time the exponential growth of the area actually tainted by the tracer streaks catches up with a power-law growth of the overall area occupied by the tracer patch.

It is important in a tracer study of mixing in the ocean to measure the hydrodynamic forcing, and also to measure hydrodynamic parameters that are believed to be useful for estimating diffusivities, so that existing theories can be tested. Several groups were involved in profiling fine structure and microstructure during NATRE. Dissipation of turbulent kinetic energy and temperature variance measured by profiling instruments gave estimates of diapycnal diffusivity which agreed closely with the tracer results. Measurements of the fine structure have helped reveal the roles of shear and double diffusive gradients in driving the mixing.

Dependency of Diapycnal Mixing and Buoyancy Frequency from Tracer Release Experiments

Figure 2 shows diapycnal diffusivities as a function of buoyancy period for the deep ocean tracer release experiments so far published. Except for the recent

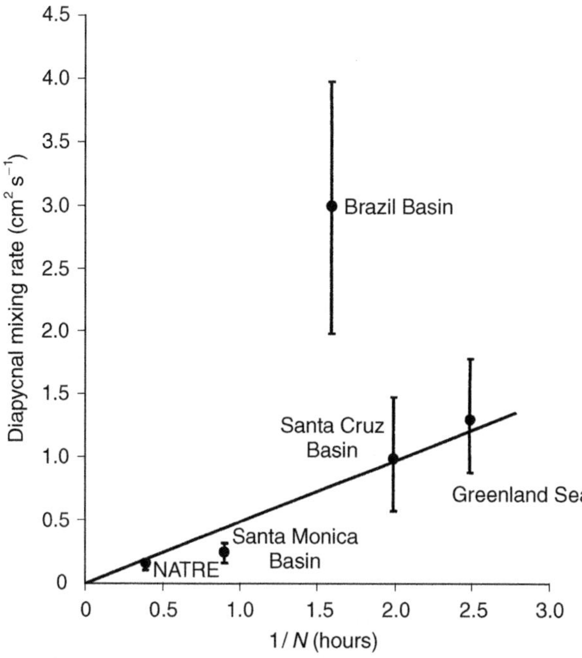

Figure 2 Vertical mixing coefficients for five tracer release experiments in the open ocean, plotted as a function of 1/*N* where *N* is the buoyancy frequency. For discussion see text.

Brazil Basin experiment, the data indicate correlation between the two, as would be expected if the forcing were in some sense held constant. However, the reader should beware of such relationships, as the Brazil Basin result shows. There is evidence from internal wave phenomenology and energy dissipation measurements that the diffusivity in the interior of the ocean, when driven only by the background internal wave field, is independent of the buoyancy period with a value of approximately $0.05\,cm^2\,s^{-1}$. The only one of our experiments that has been conducted in the interior of the ocean, well away from boundaries, was NATRE, and that was probably influenced by salt fingering. The measurements shown in **Figure 2** are those made before the tracer-containing water had time to contact the boundaries; mixing increased dramatically in the California basin experiments once such contact occurred. Nevertheless, the energy input for all but the NATRE site may have been enhanced by the proximity of the boundaries. If this is the reason why most of these experiments show elevated values, it is evident that in many situations of interest, the diffusivity, and presumably the energy flux through the internal wave field, must be enhanced even at considerable distance from boundaries.

Gas Exchange Experiments

The rate of air–sea gas transfer is a parameter which is needed in a wide range of biogeochemical studies. Gas exchange is dependent on environmental conditions that affect the near-surface turbulence in the sea and which are not easily reproduced in laboratory facilities, such as wind speed, sea state, and the chemical state of the air–sea interface. In laboratory wind-wave facilities for example, a strong dependence on wind speed is observed, but the functional form depends on the experimental set-up. As a consequence, though substantial theoretical understanding has been gained from experiments in laboratory facilities, there has also been a need to assemble a body of gas transfer measurements made at sea.

The first aqueous use of SF_6 as a tracer was the measurement of gas exchange in lakes by R. Wanninkhof, in 1985. Lake experiments are comparatively easy to set up and perform, and give absolute estimates of gas exchange. The basis of the technique is to keep track of the total amount of gas in the lake. The results of the first experiment gave unambiguous evidence in a field situation, for a strong dependence of gas exchange on wind speed, and the data form the calibration for the 'Liss–Merlivat' formulation of gas exchange. However, the gas exchange rates found in that experiment, when scaled and applied to carbon dioxide, are lower by about a factor of two than might be expected from an analysis of the global ^{14}C budget of the ocean. This uncertainty in marine gas exchange rates remains unresolved up to the present. In recent years, many investigators who need to parameterize gas exchange as a function of wind speed, have bracketed the uncertainty by applying both the Liss–Merlivat relation (scaled to agree with the lake SF_6 experiment), and a relation due to Wanninkhof that is scaled to agree with global ^{14}C values.

The Dual Tracer Technique

This long-standing uncertainty in marine gas exchange rates provided a good reason to adapt the lake SF_6 technique to the measurement of gas exchange at sea. However, whereas in a lake it was easy to determine the total amount of tracer present and the area over which it is spread, in the open ocean the tracer release is unenclosed and dilutes into a constantly larger volume of water. A means must be found to account for this dilution. Theoretically, this could be accomplished by releasing a nonvolatile conservative tracer with the gaseous one, and then use the change in ratio between the two to define gas exchange rates. In practice, no such ideal conservative nonvolatile tracer is available, so instead SF_6 and ^3He were released, two volatile tracers having very different molecular diffusivities. When the water column is well mixed and of constant depth H, the ratio $r = c_2/c_1$ of the concentrations of the tracers (in excess of any concentration in equilibrium with the atmosphere) evolves according to the equation:

$$\frac{1}{r}\frac{\mathrm{d}r}{\mathrm{d}t} = -\frac{1}{H}(k_2 - k_1)$$

where k_1 and k_2 are the gas transfer velocities appropriate to each tracer. This suggested that in the right environment, that is a shallow sea, well-mixed and with a constant depth, measurement of the tracer ratio could be used to define the difference between the two gas transfer rates. If another relation between the gas transfer rates could be defined, the 'dual tracer' technique would enable absolute values for k_1 and k_2 to be derived.

For this second relation, dual tracer experimenters have used a power law dependence of gas transfer velocities on Schmidt number (the ratio of kinematic viscosity of water to the diffusivity of the gas):

$$\frac{k_1}{k_2} = \left(\frac{Sc_1}{Sc_2}\right)^n$$

For most conditions in which bubbles and spray are not affecting gas exchange, $n = 0.5$. This result is derived from models and supported by measurements, in the laboratory and on lakes. At very low wind speeds when the sea is glassy smooth, this relation does not hold and $n = 0.67$ is the theoretical result, but this condition is very rarely met at sea. In rough seas where substantial bubble-mediated gas transfer may occur, the theory is more complex and different assumptions have been made to derive absolute values under these conditions. Recent theoretical work suggests that the square-root assumption is reasonably accurate even in the presence of bubble-mediated transfer, though care is needed in scaling the results obtained using these insoluble tracers to more soluble gases such as carbon dioxide. In one experiment, a third tracer, bacterial spores specially treated to be suitable for this purpose, were used as a nonvolatile tracer, and these results also support the use of the square-root law.

Figure 3 shows a compilation of results from dual-tracer experiments at sea. The dual-tracer results confirm the strong dependence of gas exchange on wind speed. They generally lie between the Liss–Merlivat and Wanninkhof parameterizations. In the light of recent results, concerning the effect of ubiquitous natural organic films, we can hypothesize that the trends in these data are due to the decreasing effect of organics as one moves away from coastally influenced sites out into the open ocean. The Wanninkhof parameterization, being tuned to global ^{14}C exchange rate, is most affected by the open ocean and the Liss–Merlivat formulation, originally calibrated from the result of lake experiments, the most affected by organics. The two data sets lie in between these. Georges Bank might be expected to be less coastally influenced than the North Sea, and the trend in the results is consistent with that expectation.

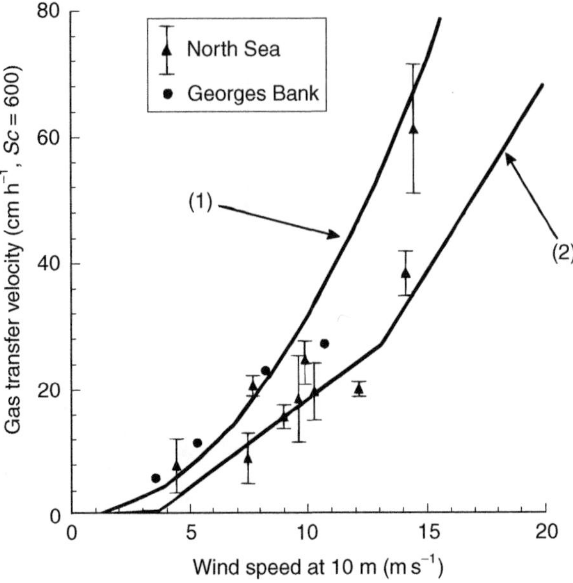

Figure 3 Compilation of dual tracer gas exchange measurements. The North Sea results include some previously published data for which revised wind speeds have been estimated using the procedures detailed by P. D. Nightingale. Data from (1) Wanninkhof (1992) and (2) Liss and Merlivat (1986).

Small-scale Surface Patch Experiments for Biogeochemical Studies

A practical problem in carrying out open-sea dual-tracer gas exchange experiments was the difficulty of keeping track of the released tracer patch. To overcome this, in the late 1980s instrumentation was built which took advantage of the uniquely fast gas chromatographic analysis for SF_6. Gas chromatography is normally a slow, batch process, but for SF_6 using a molecular sieve column, the actual separation takes only 30 seconds and the entire analysis can be completed in three minutes. Thus it was possible to build an instrument which continually measured the concentration of SF_6 in a supply of water, and use this to 'chase' the tracer patch from a ship. This opened the possibility of using the tracer to guide experiments to investigate the biology or chemistry of an accurately marked patch of surface water, over a period of days to weeks. Such 'lagrangian' experiments have frequently been performed in the past using drogued drifting buoys to mark movement of water. However, an early observation from the trial tracer experiments made in the English Channel was that such buoys do not normally stay co-located with a patch of water marked by a tracer release. Surface buoys are subject to windage and tend to slip downwind of the marked water.

In-situ Iron Enrichment Experiments

The first use of the tracer technique to guide biogeochemical studies was in the IRONEX experiments in the equatorial Pacific. At about the time the tracer-release technique had been developed for gas-exchange experiments, the idea was suggested of testing the 'Iron hypothesis' of phytoplankton limitation by releasing a large amount of iron in the surface waters of, for example, the equatorial Pacific. A difficulty was that if the experiment was too small, then the iron-enriched patch would be easily lost, whereas if it were large enough to be easily found (probably ~100 km in scale) then it would be logistically difficult and expensive. The use of the tracer release to guide a 10-km scale experiment was an obvious next step, and the design for such a study was published in 1991.

The first two unenclosed iron-enrichment studies were carried out in 1993 and 1995 in the Equatorial Pacific. In both, nanomolar concentrations of iron were induced in the surface layer by release of iron sulfate, the patches being labeled by SF_6 addition. The SF_6(<1 mol in total) was added in a constant ratio to the initial addition of iron, the tracer component was then used as a guide to keep track of the affected patch of ocean. Sampling could be reliably categorized as 'in' or 'out' the patch, even after all the measurable iron had disappeared from solution. In the second study, the main experiment included reseeding the patch with iron, but not tracer, twice after the initial release. Important secondary aims of the tracer component of the experiments have been the study of mixing rates both horizontally in the mixed layer, and vertically across the thermocline.

Figure 4 shows a summary of the results for the effect of the iron releases on surface water fugacity of carbon dioxide (fCO_2) from Ironex I and II. fCO_2 is plotted against SF_6 measured in the water on paired samples, for various times following the initiation of the experiment. Such a plot shows whether the fCO_2 (or any other variable of interest) develops a

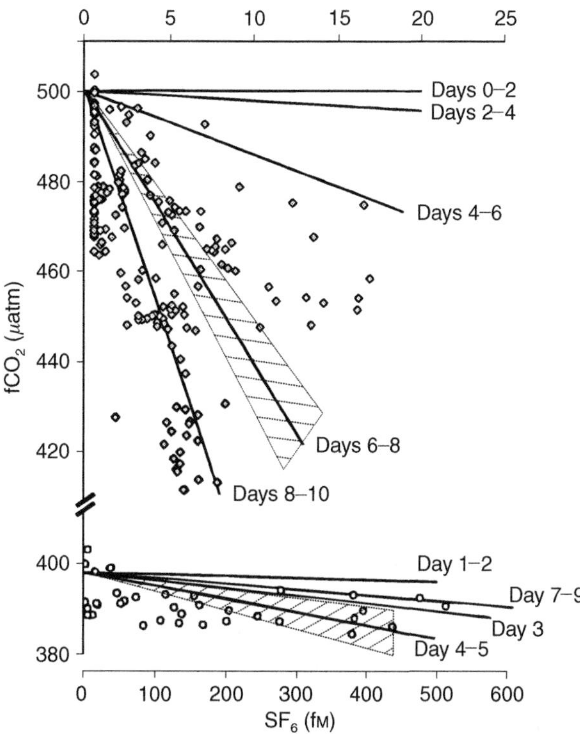

Figure 4 Linear regressions of sea surface values of fugacity of CO_2 (fCO_2) with SF_6 concentration, for specified periods after the start of the Ironex I (lower) and Ironex II (upper). Representative data points are shown, for the 'day 4–5' period during Ironex I (○), and the 'day 6–8' period during the Ironex II (◇). The hashed region around the two regression lines which correspond to these data shows the confidence limit (3-σ) on the slope of the line. (Data from Cooper *et al.*, 1996; Law *et al.*, 1998; Watson *et al.*, 1994.)

relationship with the tracer concentration over time. It is a useful summary of the effects observed even if the evolution of the patch shape is complex and not readily mapped in space. Data at low and background SF_6 show the 'control' condition, outside the patch, while data at high SF_6 show the evolution of the center of the marked water.

Figure 4 shows the contrasting results of the experiments. Ironex II produced an intense bloom of diatoms which fixed substantial carbon, resulting in a drawdown of carbon dioxide in the surface water which at its peak amounted to 70–80 μatm below the starting 'outside patch' condition. The drawdown continued to build up beyond the first week of the experiment, and a substantial signal was left in the water even after the bloom began to fade. By contrast, during Ironex I (shown on the same fCO_2 scale) the effect on the carbon concentration of the surface water was small, only 10–20% of that seen on the Ironex II, and it was already fading by the end of the first week. As measured by carbon uptake, the response of the two experiments in the first 3–4 days is similar. The divergence between the two time histories after that time is probably attributable to the fact that there were further additions of iron to the patch water on days 3 and 7 of Ironex II, but only the single initial iron enrichment during Ironex I, after which the disappearance of the added Fe, presumably by sedimentation, occurred very quickly. The simplest possible interpretation of the Ironex results is therefore that iron supply, when increased in the equatorial Pacific, allows diatoms to bloom and the chemistry of the water to change, providing that the iron concentration is elevated for several days at least.

Conclusion

Several further applications of the tracer technique are presently under way. Two 'large scale' experiments in the open ocean are being actively monitored, in the Greenland Sea and the Brazil Basin. Numerous useful subsurface experiments can be imagined. However, because of the conflict between such subsurface release experiments and the use of SF_6 as a transient tracer, there is a need to establish a forum by which the wider oceanographic community can have input into the planning of prospective release experiments.

Small-scale releases in surface waters should not normally compromise the transient tracer signal. One obvious application now under way is that of iron fertilization experiments to examine the extent to which 'high nutrient low chlorophyll' regions other than the equatorial Pacific are limited by iron availability. The recent Southern Ocean Iron Enrichment Experiment (SOIREE) has shown unequivocal evidence that iron supply does affect the biology of that region. This experiment was carried out during sometimes stormy weather, confirming that the patch-tracking technique works well in the open ocean under storm conditions.

To summarize, experiments using SF_6 tracer in the open ocean are now reduced to practice, if not routine. Three experiments at the 1000-km scale have so far been initiated, to measure ocean mixing on these scales. There have been more than twenty smaller scale experiments, of increasing sophistication, since they were first begun in 1986. For topics to which they are suited, such as iron limitation, biogeochemical budgets, gas exchange and diapycnal mixing rates, these experiments have enabled something of the precision of the land-based laboratory investigation to be brought to bear in at-sea oceanography.

See also

Long-Term Tracer Changes. Tracers of Ocean Productivity.

Further Reading

Cooper DJ, Watson AJ, and Nightingale PD (1996) Large decrease in ocean-surface CO_2 fugacity in response to in-situ iron fertilization. *Nature* 383: 511–513.

Law CS, Watson AJ, Liddicoat MI, and Stanton T (1998) Sulphur hexaflouride as a tracer of biogeochemical and physical processes in an open-ocean iron fertilisation experiment. *Deep-Sea Research II* 45: 977–994.

Ledwell JR, Montgomery ET, Polzin KL, *et al.* (2000) Evidence for enhanced mixing over rough topography in the abyssal ocean. *Nature* 403: 179–182.

Ledwell JR, Watson AJ, and Law CS (1998) Mixing of a tracer in the pycnocline. *Journal of Geophysical Research* 103: 21499–21529.

Watson AJ, Law CS, Van Scoy K, *et al.* (1994) Minimal effect of iron fertilization on sea-surface carbon dioxide concentrations. *Nature* 371: 143–145.

Watson AJ, Messias M-J, Fogelqvist E, *et al.* (1999) Mixing and convection in the Greenland sea from a tracer release. *Nature* 401: 902–904.

Watson AJ, Upstill-Goddard RC, and Liss PS (1991) Air–sea gas exchange in rough and stormy seas measured by a dual-tracer technique. *Nature* 349: 145–147.

TRACERS OF OCEAN PRODUCTIVITY

W. J. Jenkins, University of Southampton, Southampton, UK

Introduction

Primary production is the process whereby inorganic carbon is fixed in the sunlit (euphotic) zone of the upper ocean, and forms the base of the marine food pyramid. It occurs when marine phytoplankton use sunlight energy and dissolved nutrients to convert inorganic carbon to organic material, thereby releasing oxygen. The total amount of carbon fixed during photosynthesis is called gross production, whereas the amount of carbon fixed in excess of internal metabolic costs is referred to as net production. It is understood that a significant fraction of the carbon fixed in this manner is rapidly recycled by a combination of grazing by zooplankton and *in situ* bacterial oxidation of organic material. New production is that portion of net production that is supported by the introduction of new nutrients into the euphotic zone. Traditionally, this has been regarded as production fueled by nitrate as opposed to more reduced forms of nitrogen, such as ammonia and urea. Some portion of the fixed carbon sinks out of the euphotic zone in particulate form, or is subducted or advected away as dissolved organic material from the surface layers by physical processes. This flux is regarded collectively as export production. The ratio of new (export) to net production, referred to as the f-ratio (e-ratio) can vary between 0 and 1, and is believed to be low in oligotrophic ('blue water'), low productivity regions, and higher in eutrophic, high productivity regions. Finally, net community production is the total productivity in excess of net community metabolic cost. On sufficiently long space- and time-scales, it can be argued that new, net community, and export production should be equivalent in magnitude.

Net production has been measured 'directly' by radiocarbon incubation experiments, whereby water samples are 'spiked' with radiocarbon-labeled bicarbonate, and the net rate of transfer of the radioisotope into organic matter phases determined by comparison of light versus dark incubations. Global maps of net productivity have been constructed on the basis of such measurements, and current estimates indicate a global fixation rate of order 50 GT C a^{-1} (1 GT $= 10^{15}$ g). Rates of export, new, and net community production are more difficult to determine directly, yet are of equal importance as determinants of biogeochemically important fluxes on annual through centennial timescales.

Geochemical tracer techniques have been used to make such estimates, and offer significant advantages in that they are fundamentally nonperturbative, and integrate over relatively large space-scales and long time-scales. Conversely, such determinations must be viewed from the perspective that they are indirect measures of biogeochemical processes, and have characteristic implicit space- and time-scales, as well as boundary conditions, and sometimes ambiguities and model dependence. Further, the specific tracer or physical system used to obtain production estimates determines the type of productivity measured. Thus any treatment of geochemical tracer estimates must include a discussion of these attributes.

Measuring Oceanic Productivity with Tracers

Just a few approaches will be discussed here. Other techniques have been used with some success, particularly with relation to particle interceptor traps, but this section will concentrate on basic mass budgeting approaches using water column distributions or seasonal cycling of tracers. There are three basic, yet fundamentally independent approaches that can be used.

1. Aphotic zone oxygen consumption rates that, when vertically integrated, provide a net water column oxygen demand that can then be related stoichiometrically to a carbon export flux.
2. Seasonal timescale euphotic zone mass budgets, particularly of oxygen, carbon, and carbon isotopes, which lead to estimates of net community production.
3. Tracer flux-gauge measurements of physical mechanisms of nutrient supply to the surface ocean, which place lower bounds on rates of new production.

These techniques, summarized in **Figure 1**, yield estimates of subtly different facets of biological production. On annual timescales, however, these different modes of production should be very close to equivalent, and hence the results of these various

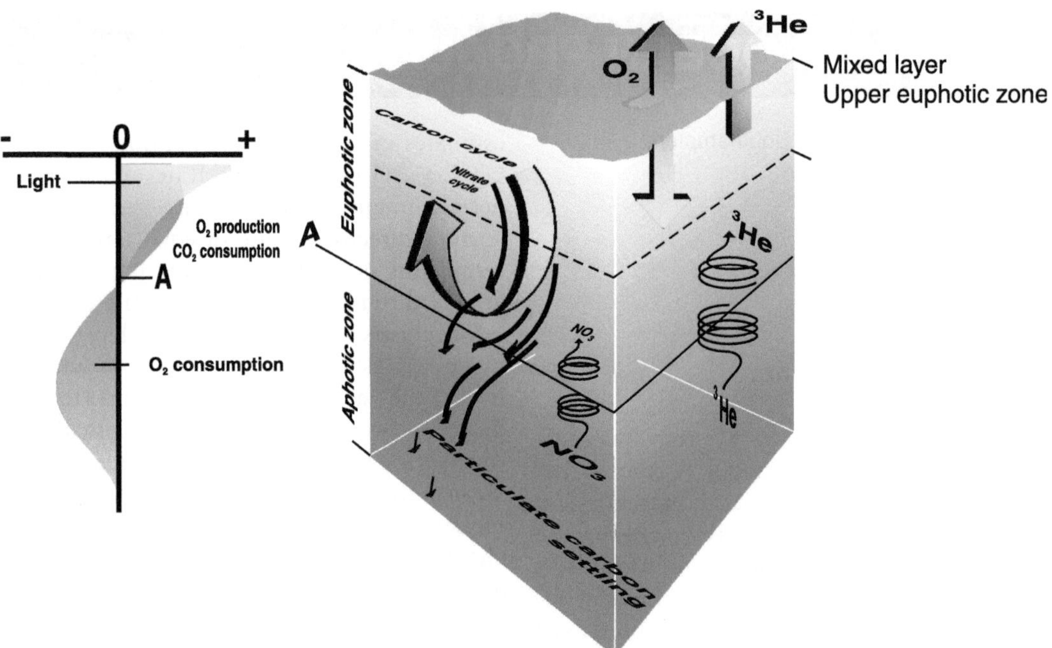

Figure 1 A schematic of the upper ocean, showing material fluxes and various tracer constraints on primary production.

measurement approaches should be comparable. As shown below, their quantitative agreement coupled with their essential independence lends an inductive support to the validity of their results.

Aphotic Zone Oxygen Consumption Rates

In the surface ocean air–sea gas exchange controls the composition of dissolved gases and phytoplankton release oxygen. Below, in the aphotic (nonsunlit) zone, oxygen is generally undersaturated, because bacterially mediated oxidation of sinking organic material consumes oxygen. Credible estimates of aphotic zone oxygen consumption rates have been made since the 1950s. However, the earliest quantitative linkage to primary production was in 1982. The principle behind it is dating water masses and dividing the age of the water mass into the observed oxygen deficit. Another approach involves correlating water mass age along streamlines with oxygen concentration (older water has less oxygen). This dating can be achieved by a technique such as tritium-^{3}He dating, which uses the ingrowth of the stable, inert noble gas isotope ^{3}He from the decay of the radioactive heavy isotope of hydrogen (tritium), according to:

$$^3\mathrm{H} \overset{12.45\mathrm{y}}{\rightarrow} 3\mathrm{He}$$

If surface waters are in good gas exchange contact with the atmosphere, then very little ^{3}He will accumulate due to tritium decay. Once isolated from the surface, this ^{3}He can accumulate. From the measurement of both isotopes in a fluid parcel, a tritium-^{3}He age can be computed according to:

$$\tau = \lambda^{-1}\ln\left(1 + \frac{[^3\mathrm{He}]}{[^3\mathrm{H}]}\right)$$

where λ is the decay probability for tritium, and τ is the tritium-^{3}He age (usually given in years). Under typical Northern Hemispheric conditions with current technology, times ranging from a few months to a few decades can be determined.

Although a conceptually simple approach, under normal circumstances mixing must be accounted for because it can affect the apparent tritium-^{3}He age in a nonlinear fashion. Furthermore, in regions of horizontal oxygen gradients, lateral mixing may significantly affect apparent oxygen consumption rates. For example, following a fluid parcel as it moves down a streamline, mixing of oxygen out of the parcel due to large-scale gradients will masquerade as an augmentation of oxygen consumption rates. These issues can be accounted for by determining the three-dimensional distributions of these properties, and applying the appropriate conservation equations. With additional constraints provided by geostrophic velocity calculations, these effects can be separated and absolute oxygen

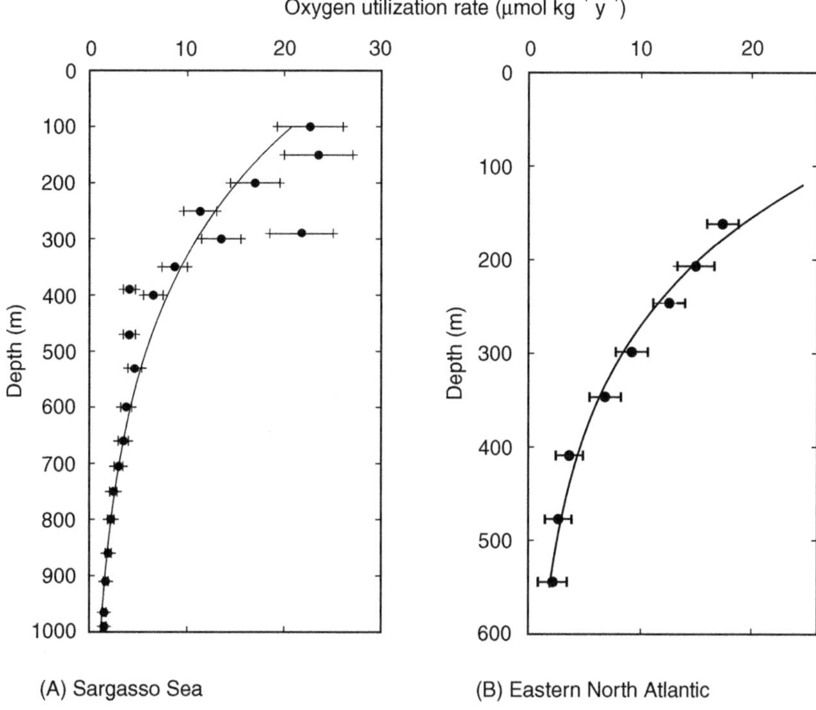

Figure 2 Aphotic zone oxygen consumption rates as a function of depth for two locales in the subtropical North Atlantic. These consumption rates are based on tritium-^{3}He dating and other tracer techniques.

consumption rates can be computed as a function of depth. **Figure 2** shows profiles of oxygen consumption rates as functions of depth for two locales in the subtropical North Atlantic. Integration of these curves as a function of depth gives net water column oxygen demands of 6.5 ± 1.0 mol m^{-2} a^{-1} for the Sargasso Sea and 4.7 ± 0.5 mol m^{-2} a^{-1} in the eastern subtropical North Atlantic. Using the molar ratio of oxygen consumed to carbon oxidized for organic material (170 : 117), the flux of carbon from the euphotic zone above required to support such an oxygen demand can be calculated for the two regions (4.5 ± 0.7 and 3.2 ± 0.4 mol C m^{-2} a^{-1}).

The character of these estimates bears some consideration. Firstly, according to the definitions of primary production types described earlier, this represents a determination of export productivity. Secondly, the determinations represent an average over timescales ranging from several years to a decade or more. This is the range of ages of the water masses for which the oxygen utilization rate has been determined. Thirdly, the corresponding space-scales are of order 1000 km, for this is the region over which the age gradients were determined. Fourthly, although the calculation was done assuming that the required carbon flux was particulate material, it cannot distinguish between the destruction of a particulate rain of carbon and the *in situ* degradation of dissolved organic material advected along with the water mass from a different locale. These characteristics must be borne in mind when comparing this with other estimates.

Seasonal Euphotic Zone Mass Budgets

There have been three basically independent approaches to estimating net community production based on observation of the seasonal cycles of oxygen and carbon in the upper ocean. Photosynthesis in the euphotic zone results in the removal of inorganic carbon from the water column, and releases oxygen (**Figure 3**). Recycling of organic material via respiration and oxidation consumes oxygen and produces CO_2 in essentially the same ratios. It is only that carbon fixation that occurs in excess of these processes, i.e., processes that result in an export of organic material from the euphotic zone, or a net biomass increase, that leaves behind an oxygen or total CO_2 (ΣCO_2) signature. Estimates of productivity based on euphotic zone oxygen or carbon budgets are consequently estimates of net community production. Such productivity estimates are characterized by seasonal to annual timescales, and space-scales of order of a few hundred kilometers.

In subtropical waters, excess oxygen appears within the euphotic zone just after the onset of

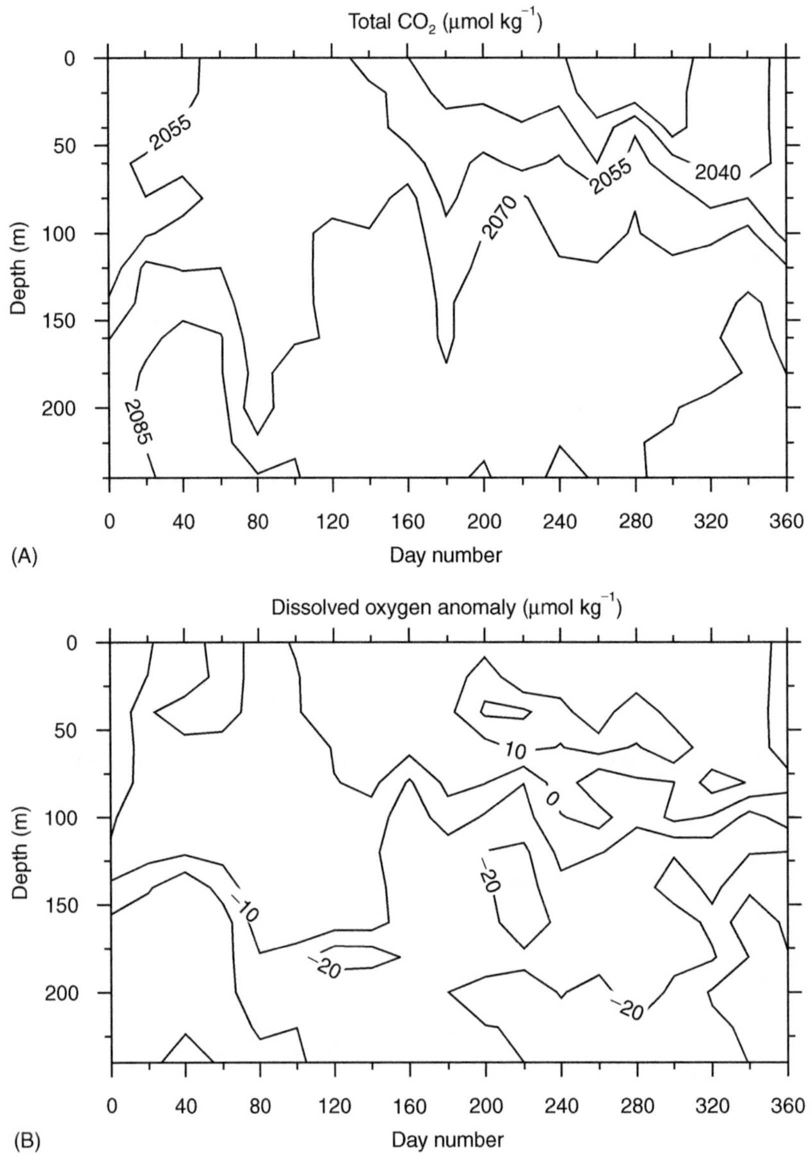

Figure 3 Euphotic zone seasonal cycles of total inorganic carbon (A) and oxygen (B) near Bermuda. Note the build-up of oxygen anomaly and reduction of total CO_2 in the euphotic zone during the summer months due to photosynthetic activity.

stratification, and continues to build up throughout summer months. Use of the seasonal accumulation of photosynthetic oxygen in the upper ocean to estimate primary production is complicated by the fact that it tends to be lost to the atmosphere by gas exchange at the surface. Furthermore, temperature changes due to seasonal heating and cooling will change the solubility of the gas, further driving fluxes of oxygen across the air–sea interface. In addition, bubble trapping by surface waves can create small supersaturations. While such processes conspire to complicate the resultant picture, it is possible to use observations of noble gases (which do not undergo biological and chemical processing) and upper ocean physical models to interpret the seasonal cycle of oxygen. These calculations have been successfully carried out at a variety of locations, including the subtropical North Atlantic and the North Pacific. In the Sargasso Sea, estimates of oxygen productivity range from 4.3 to 4.7 mol m^{-2} a^{-1}. Using the molar ratio of oxygen released to carbon fixed in photosynthesis of 1.4 : 1, the carbon fixation rate is estimated to be 3.2 ± 0.4 mol m^{-2} a^{-1}.

There is also a net seasonal decrease in ΣCO_2 attributable to photosynthesis at these locations. Such decreases are simpler to use in productivity estimates, principally because air–sea interaction has a much weaker influence on ΣCO_2. On the other hand, precise measurements are required because the photosynthetically driven changes are much smaller compared with the background ΣCO_2 levels. Because of these differences, estimates based on ΣCO_2 seasonal cycles offer an independent measure of euphotic zone mass budgets.

Finally, differences in the carbon isotopic ratio between organic and inorganic carbon, as well as atmospheric CO_2, allow the construction of yet a third mass budget for the euphotic zone. There is a clear carbon isotope signature that can be modeled as a function of primary production, air–sea exchange, and mixing with deeper waters.

Tracer Flux-gauge Determinations

The third tracer constraint that may be used to determine primary production involves the use of 'tracer flux gauges' to estimate the flux of nutrients to

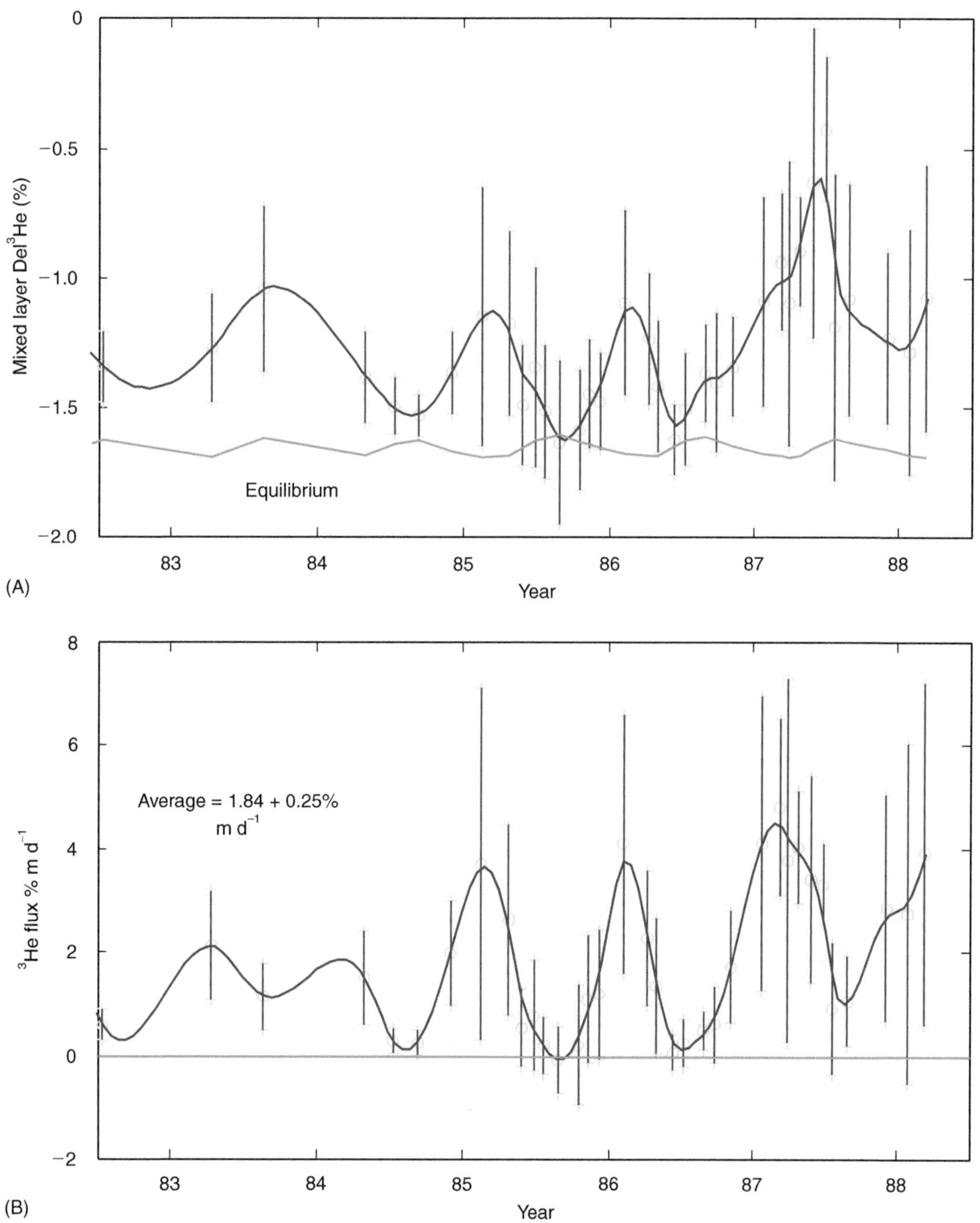

Figure 4 An approximately 6 year history of surface water ^{3}He isotope ratio anomalies (A) and computed flux to the atmosphere near Bermuda (B).

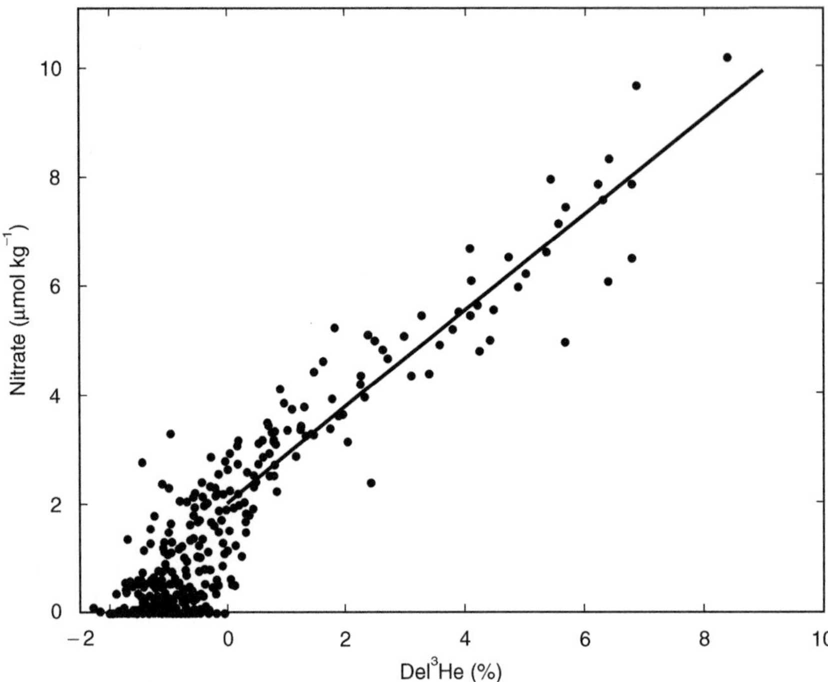

Figure 5 The correlation of ^{3}He isotope ratio anomaly (in %) and nitrate (in $\mu molkg^{-1}$) in the upper ocean near Bermuda for the period 1985–88.

the euphotic zone. This approach relies on the premise that the physical mechanisms that serve to transport nutrients to the euphotic zone from the nutrient-rich waters below also carry other tracers in fixed proportion. If the rate at which these other tracers are transported can be determined, and the nutrient to tracer ratio at the 'source' is known, then the corresponding nutrient flux may be inferred; that is:

$$F_{\text{Nutrient}} = \left[\frac{\text{Nutrient}}{\text{Tracer}}\right]_{\text{Source}} \times F_{\text{Tracer}}$$

Inasmuch as there may be alternate, biologically mediated pathways (such as zooplankton migration), such a calculation would serve as an underestimate to the total nutrient flux.

Measurements of the rare, inert isotope ^{3}He in the mixed layer of the Sargasso Sea near Bermuda reveal a persistent excess of this isotope over solubility equilibrium with the atmosphere (**Figure** 4). The existence of this excess implies a flux of this isotope to the atmosphere, which can be calculated using the estimated gas exchange rate. Although ^{3}He is produced in the water by the *in situ* decay of tritium, it can be shown that only about 10% of the observed flux can be explained by tritium decay within the euphotic zone. The greater portion of this ^{3}He flux arises from the upward 'exhalation' of old tritium-produced ^{3}He from the waters below. That is, the ^{3}He flux observed leaving the surface ocean is largely the loss of this isotope from the main thermocline.

The ocean–atmosphere flux of ^{3}He shows a pronounced seasonal variation, with the greatest fluxes in the winter months. The winter maximum is due to high rates of gas exchange (more vigorous winter winds lead to higher gas exchange rates) and deeper winter convection. This is the time history of the ^{3}He flux out of the upper ocean. The time history of the ^{3}He flux to the upper ocean may be different. However, the annual mean fluxes must be the same, since the winter mixed layer penetrates below the bottom of the euphotic zone. The annual average ^{3}He flux from the ocean surface near Bermuda is 1.84 ± 0.25%-m d^{-1}. To estimate the flux of ^{3}He entering the euphotic zone from below, this flux must be corrected for the *in situ* production of ^{3}He by the decay of tritium within the euphotic zone, which produces a ^{3}He flux of 0.20 ± 0.02%-m d^{-1}. The resultant flux is thus 1.64 ± 0.25%-m d^{-1}.

Insofar as there is a strong correlation between the concentrations of this isotope and nutrients within the waters below the euphotic zone (older waters are richer in both ^{3}He and nutrients), the ratio of ^{3}He to nutrient can be employed to compute nutrient flux. **Figure** 5 is a composite plot of ^{3}He versus nitrate in the upper 600 m over a 3 year period. The slope of the relationship is 0.87 ± 0.05 μmol kg^{-1}‰$^{-1}$.

Applying the flux equation presented above, a nitrate flux of $0.56 \pm 0.16\,\mathrm{mol\,m^{-2}\,a^{-1}}$ is computed. Using the average biological C : N ratio of 6.6, this leads to a carbon fixation rate of $3.7 \pm 1.0\,\mathrm{mol\,m^{-2}\,a^{-1}}$. The estimate thus obtained is a local, annual-scale measure of new production.

A similar calculation can be made by observing the long-term (decade timescale) trends in thermocline ^{3}He inventories. The long-term evolution of ^{3}He inventory in the thermocline must respond to the opposing processes of production by tritium decay and 'exhalation' upward to the euphotic zone. Knowing the former gives the latter. Using nutrient-^{3}He ratios, a gyre-scale, decadal average estimate of the nutrient flux to the euphotic zone can be obtained. A detailed analysis of the long-term trends of tritium and ^{3}He in the upper 1000 m of the Sargasso Sea, coupled with the observed nitrate : ^{3}He ratios, yields an estimate of $0.70 \pm 0.20\,\mathrm{mol\,m^{-2}\,a^{-1}}$. This leads to a somewhat higher carbon fixation rate of $4.6 \pm 1.3\,\mathrm{mol\,m^{-2}\,a^{-1}}$. This estimate differs from the surface layer flux calculation in that it is a much longer-term average, since it depends on the very long-term evolution of isotopes in the thermocline. Moreover, it represents a very large-scale gyre-scale determination, rather than a local measure: horizons within the thermocline probably connect to regions of higher productivity further north.

Comparing Tracer-derived Estimates

Although the various techniques described here are based on differing assumptions, and measure different types of production, they should be mutually consistent on annual or greater timescales. **Table 1** is a comparison between the various estimates near Bermuda in the Sargasso Sea. A weighted average of these determinations gives a productivity of $3.6 \pm 0.5\,\mathrm{mol\,(C)\,m^{-2}\,a^{-1}}$ for the Sargasso Sea near Bermuda. The determinations are within uncertainties of each other, although they utilize different tracer systems, are reliant on different assumptions, and are virtually independent of each other. This agreement provides some confidence as to their accuracy.

Table 1 Comparison of tracer-derived estimates near Bermuda in the Sargasso Sea

Type of determination	*Type of production*	*Technique used*	*Carbon flux ($\mathrm{mol\,m^{-2}a^{-1}}$)*
Aphotic zone oxygen consumption rates	Export production	Tritium-^{3}He dating	4.5 ± 0.7
Euphotic zone cycling	Net community	Oxygen cycling	3.2 ± 0.4
		Carbon isotopes	3.8 ± 1.3
Tracer flux-gauge	New production	Mixed layer ^{3}He	3.7 ± 1.0
		Thermocline budgets	4.6 ± 1.3

See also

Air–Sea Transfer: N_2O, NO, CH_4, CO. Carbon Cycle. Tritium–Helium Dating.

Further Reading

Falkowski PG and Woodhead AD (1992) *Primary Productivity and Biogeochemical Cycles in the Sea.* New York: Plenum Press.

Jenkins WJ (1995) Tracer based inferences of new and export primary productivity in the oceans. IUGG, Quadrennial Report 1263–1269.

Williams PJ and le B (1993) On the definition of plankton production terms. *ICES Marine Science Symposium* 197: 9–19.

INVERSE MODELING OF TRACERS AND NUTRIENTS

R. Schlitzer, Alfred Wegener Institute for Polar and Marine Research, Bremerhaven, Germany

Introduction

Seawater contains a large variety of dissolved and particulate constituents, commonly referred to as tracers. Many of these tracers (such as the dissolved nutrients, phosphate, nitrate, and silicate, or the radioactive carbon isotope ^{14}C (radiocarbon)) are natural and are part of the ocean system since geological times. Other tracers, such as the chlorofluorocarbons (CFCs) and various radioactive isotopes from nuclear bomb testing, are of anthropogenic origin, and have been introduced into the ocean only during the last decades. The distributions of anthropogenic tracers in the ocean change markedly on decadal or annual timescales. Naturally occurring tracers, like the nutrients, are believed to be near steady state. However, recent data indicate decadal and inter-annual changes in ocean oxygen and nutrient distributions, likely caused by changes in circulation and biogeochemical processes associated with global climate change.

Some ocean tracers, such as radiocarbon and the CFCs, act as passive tracers. They are chemically inert and are being transported and redistributed in the ocean by currents after entering the ocean from the atmosphere. When dense surface waters sink and spread in the ocean interior, they carry the tracer signal with them and produce tongues of high tracer concentrations along their spreading paths. An example of such tracer tongues and concentration maxima in the ocean interior can be seen in the zonal CFC section in **Figure 1**. Note the pronounced CFC maxima at about 1500- and 3500-m depth at the western boundary, revealing the two cores of the deep western boundary current (DWBC) carrying tracer laden waters southward at the western side of the basin. The spatial extent of the tongues and the concentration gradients along the tongues can, in principle, be used to determine the pathways and velocities of the currents. Such inferences require advanced mathematical methods, some of which are described below.

Other tracers (like the macronutrients phosphate, nitrate, and silicate, or the micronutrients Fe and Zn) are chemically and biologically active and are a prerequisite for biological production in the ocean's surface layer. Surface nutrient concentrations are very low in most regions of the world ocean due to

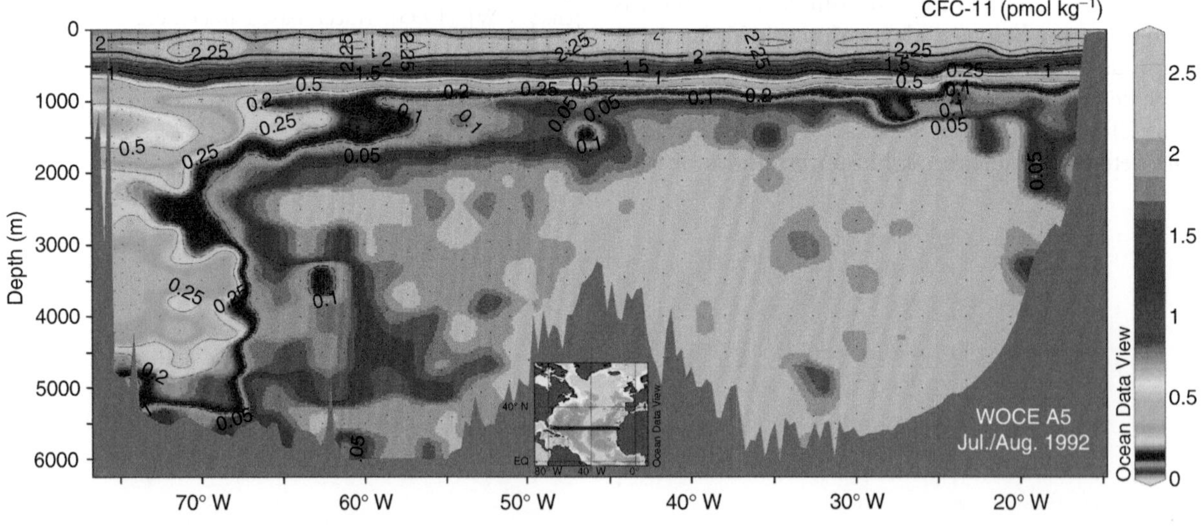

Figure 1 Zonal section of chlorofluorocarbon 11 at about 24° N in the Atlantic, clearly showing the traces of the upper and lower branches of the deep western boundary current (DWBC). Chlorofluorocarbons (CFCs) are man-made gases that invade the ocean from the atmosphere in increasing amounts since the 1940s. Before that time the ocean was CFC-free. High CFC concentrations in the ocean interior are clear indications of vigorous subsurface currents.

biological utilization. Biologically produced organic particles as well as calcium carbonate and opal shells sink and dissolve in deeper parts of the water column, thereby returning nutrients to the dissolved pool, while utilizing and drawing down dissolved oxygen. Overall, the biogeochemical processes act as a vertical nutrient and carbon pump creating significant vertical gradients and pronounced subsurface nutrient maxima at the depths where particle remineralization occurs. Subsurface nutrient maxima are often correlated with oxygen minima, clearly indicating the remineralization of organic material. The signatures of these processes in the nutrient and oxygen distributions are large and easily detected (**Figure 2**) and reveal information about the underlying processes. The strength of the biological

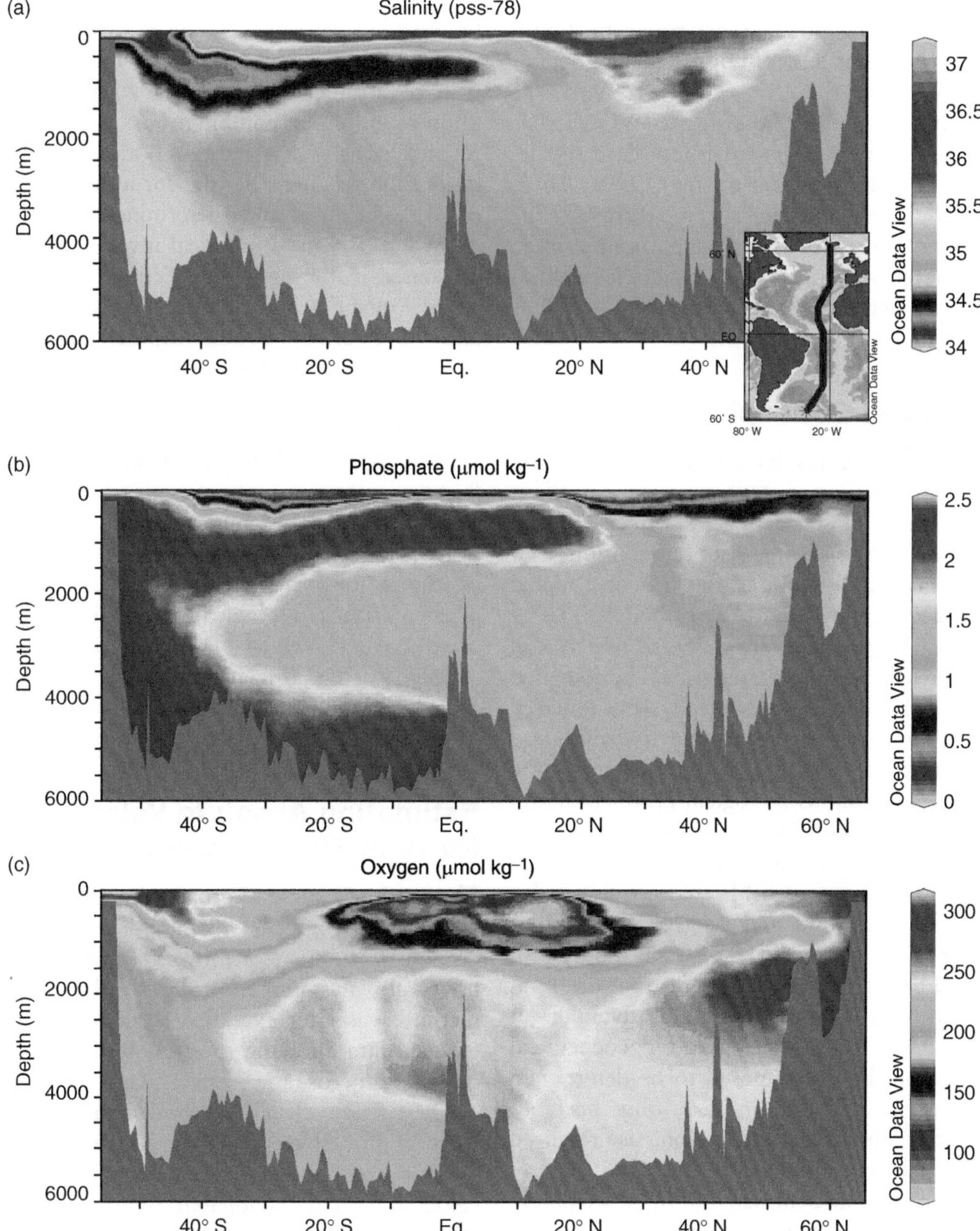

Figure 2 Meridional sections of (a) salinity, (b) phosphate, and (c) oxygen along WOCE section A16 in the Atlantic. In addition to ocean circulation, the phosphate and oxygen distributions are also affected by biological nutrient utilization near the surface and by subsurface remineralization of sinking organic material. Organic matter remineralization releases dissolved nutrients while utilizing dissolved oxygen, and can be clearly identified between 200- and 1500-m depth in the tropical and subtropical areas.

production, for instance, can be inferred from the observed strength of the vertical nutrient gradient, and the depth range of particle remineralization can be determined from the observed vertical position and spatial extent of the nutrient maxima and oxygen minima. Differences in the vertical structure of different nutrients (e.g., phosphate, silicate, carbon, and alkalinity) reveal differences in the remineralization depths of organic material, $CaCO_3$, and opal.

Basin-wide observations of a variety of oceanic tracers have been conducted since the 1950s. The first coordinated and global tracer program, GEOSECS, produced tracer data of unprecedented quality and coverage during the 1970s. More recently, the World Ocean Circulation Experiment (WOCE) has provided an even more detailed tracer data set describing the distributions during the 1990s. These data are publicly available in electronic form over the Internet or as colored distribution plots in printed atlases. Availability of original tracer data is essential for the inverse methods described below.

Inverse Model Concepts

Deriving quantitative results about the underlying biogeochemical processes from oceanic nutrient and tracer data, and separating the effects of biogeochemistry from the effects of circulation is a challenging task, and requires the use of coupled physical/biogeochemical numerical models. There are a variety of possible approaches, which can broadly be divided into two categories. The so-called 'forward models' assume rates of physical and biogeochemical processes to be known *a priori*, and require ocean currents (or the physical forcing at the ocean surface), as well as biological production and particle remineralization rates to be specified as input. Oceanic tracer concentrations are treated as unknowns, and the tracer fields that would evolve under the assumed flows and biogeochemical parameters are then simulated. Forward models, in general, lead to mathematical systems that are relatively easy to solve. The conceptual disadvantage of forward models is that physical and biogeochemical rate information that is supposed to be determined from the tracer data and, in fact, only becomes available after the data evaluation, is required *a priori* to enable the model run.

'Inverse models', in contrast, follow a seemingly more intuitive approach, treating the measured tracer concentrations and other auxiliary knowledge formally as knowns, whereas the physical and biogeochemical parameters, to be determined on the basis of the tracer data, are treated as unknowns. As described in more detail below, the inverse approach generally leads to underdetermined mathematical systems that are much harder to solve than the systems encountered in forward models. Error and resolution analysis are two issues of particular importance when solving underdetermined inverse problems. First, the solved-for physical and biogeochemical parameters depend directly on the tracer data, and errors in the data propagate into errors in the solution. Second, owing to the incompleteness of information in underdetermined systems, the unknowns are usually not fully resolved. Instead, only specific linear combinations of unknowns may be well constrained by the data, while individual unknowns or other combinations of unknowns may remain poorly determined. Both, error and resolution analysis are essential for a quality assessment of the solution of underdetermined systems.

Two widely used practical inverse approaches are presented in the next two sections ('Estimating absolute velocities and nutrient fluxes across sections' and 'Estimating carbon export fluxes with the adjoint method'). These serve as examples to describe details of the mathematical methods and to list achievable results. The first method, the section inverse approach, infers nutrient, carbon, and tracer fluxes across sets of sections, based on hydrographic, tracer, and nutrient data along these sections. The second example describes an application of the adjoint method for the determination of ocean currents, biological productivity, and downward particle fluxes. This method is specifically adapted for the utilization of many different tracers and can handle problems with heterogeneous and sparse data coverage.

Estimating Absolute Velocities and Nutrient Fluxes across Sections

The section inverse approach developed by Wunsch exploits hydrographic data along sets of oceanographic sections, and allows estimation of absolute flow velocities perpendicular to the sections. The method was later extended to include nutrient and oxygen data allowing to estimate nutrient and oxygen fluxes across the sections.

The section inverse method is based on the geostrophic principle, which for a given pair of hydrographic stations allows calculating the geostrophic velocity $v_g(z)$ perpendicular to the connecting line between the two stations as function of depth z. A reference velocity b has to be added to the geostrophic velocity to obtain the absolute flow velocity

$$v(z) = v_g(z) + b \quad [1]$$

at a given station pair. The reference velocity b is left unknown by the geostrophic calculation. Determining these unknowns was a major challenge in physical oceanography for many decades. There is one unknown reference velocity b_i for every station pair in the sections considered. Thus, for global networks consisting of hundreds of sections, as in the case of the WOCE survey, the number of unknown velocities b_i may amount to several thousands.

Steady-state mass and tracer conservation equations for all subdomains formed by the intersecting hydrographic sections (see **Figure 3**) are written as constraint equations for the unknown b_i. For a given domain the conservation equation of tracer C is expressed as

$$\sum_i (\bar{c}\bar{v}_{g_i} + \bar{c}b_i) \cdot d_i \cdot \Delta x_i = 0 \qquad [2]$$

where the summation is over all station pairs along the boundary of the domain, $\bar{c}\bar{v}_{g_i}$ is the vertically averaged tracer flux density arising from the geostrophic flow, $\bar{c}$ is the vertically averaged tracer concentration, d_i is the mean water depth of pair i, and Δx_i is the distance between the two stations of the pair. Mixing terms are ignored for the sake of simplicity. Mass budget equations are obtained by replacing c with density ρ. It is important to note that the budget equation [2] is linear in the unknown b_i, and that the known components of the fluxes involving the geostrophic velocities can be moved to the right-hand side. Formulation of conservation equations for all domains (whole water column and suitably defined layers) and for all tracers considered results in a set of linear equations

$$Ab = -\Gamma \qquad [3]$$

where b is the vector of m unknown reference velocities (m is the total number of station pairs), A is an $n \times m$ matrix containing the coefficients of the n budget equations, and Γ is an n vector with the known right-hand sides $\sum_i \bar{c}\bar{v}_{gi} \cdot d_i \cdot \Delta x_i$ of the budget

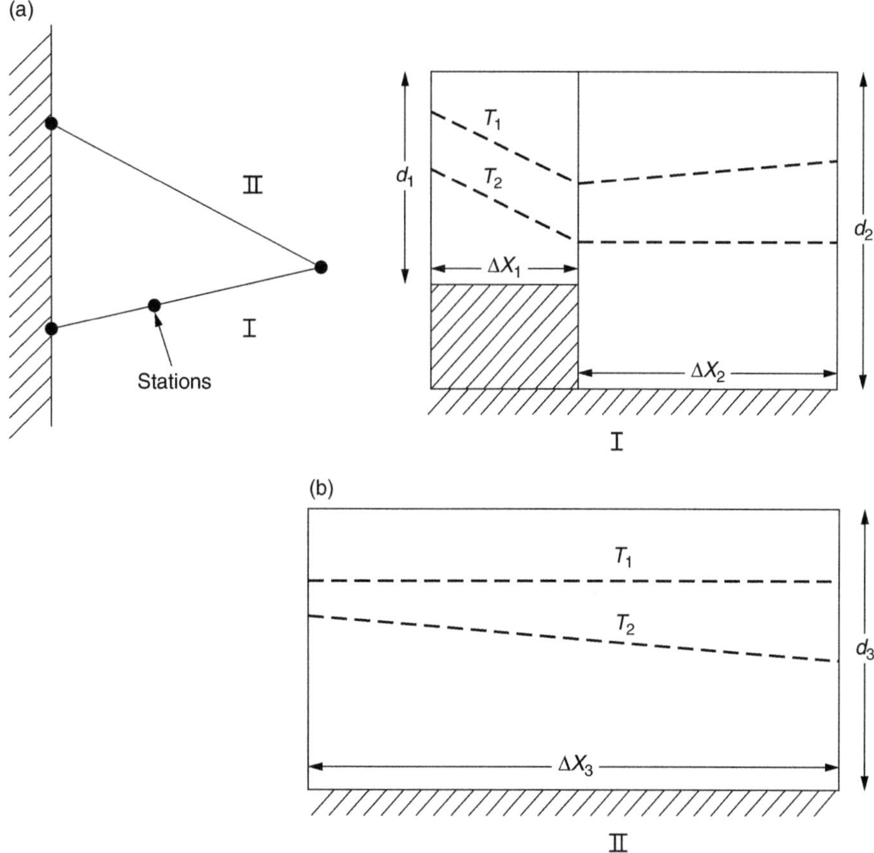

Figure 3 Simple examples of (a) a three-station pair defining sections I and II, and (b) the station configuration and bottom depths d_i in sections I and II, where dashed lines are supposed layer interfaces. Reproduced from Wunsch C (1978) The North Atlantic general circulation west of 50° W determined by inverse methods. *Reviews of Geophysics* 16: 583–620, with permission from the American Geophysical Union.

equations. The objective is to solve the system of linear equations for b. Once this is achieved, the absolute flow velocities can be calculated according to [1], and the fluxes of any tracer across a given cross section can be obtained by integrating $c \cdot v$ over the interface.

Finding the solution b of [3] is complicated for two reasons: (1) the number of equations n is typically much smaller than the number of unknowns m, leaving [3] underdetermined, and (2) the right-hand side Γ is based on hydrographic measurements and therefore contains errors. Because of the errors in Γ, there is no exact solution to [3], in general, and one has to resort to the least-squares solution that minimizes the residuals $r = Ab + \Gamma$. It should be noted that row and column weighting of [3] is usually applied before calculating solutions. The reader should refer to the literature for a discussion of the weighting process and a description of the various error contributions to be considered for Γ.

Solving [3] relies on the singular value decomposition of the coefficient matrix A:

$$A = U \cdot S \cdot V^{\mathrm{T}} \qquad [4]$$

where U and V are orthogonal matrices of dimension $n \times n$ and $m \times m$, respectively, and S is a diagonal matrix with the same dimension $n \times m$ as A. The superscript T indicates transposition. The diagonal of S contains the nonnegative singular values s_i ($i = 1, \ldots, n$) sorted in decreasing size. The number of singular values greater than zero, p, is equal to the rank of A and reveals the actual number of independent equations in [3]. All subsequent singular values $s_{p+1}, \ldots, s_n$ are zero, and A can be rewritten as

$$A = U_p \cdot S_p \cdot V_p^{\mathrm{T}} \qquad [5]$$

where U_p and V_p are trimmed versions of U and V of dimension $n \times p$ and $m \times p$, respectively, containing only the first p columns of U and V. S_p is the $p \times p$ submatrix of S consisting of the first p rows and columns only.

With these quantities, the smallest least-squares solution $\hat{b}$ of [3], the covariance matrix of the solution $\mathrm{cov}(\hat{b})$, and the resolution matrix R are given by

$$\hat{b} = -V_p \cdot S_p^{-1} \cdot U_p^{\mathrm{T}} \cdot \Gamma \qquad [6]$$

$$\mathrm{cov}(\hat{b}) = V_p \cdot S_p^{-2} \cdot V_p^{\mathrm{T}} \qquad [7]$$

$$R = V_p \cdot V_p^{\mathrm{T}} \qquad [8]$$

In these expressions, the superscript '-1' indicates the inverse of the matrix, and superscript '-2' indicates the product of the inverse matrix with itself. The resolution matrix R describes the relationship between the optimal solution $\hat{b}$ obtained from the underdetermined system [3] with the 'true' solution b that one would find if [3] contained sufficient information and was full rank:

$$\hat{b} = R \cdot b \qquad [9]$$

Every component $\hat{b}_j$ of the solution of the underdetermined system [3] can thus be represented as a linear combination of the 'true' unknowns $\hat{b}_j = \sum_k r_{jk} \cdot b_k$ involving coefficients from row j of R. The analysis of the resolution matrix R and the deciphering of the real significance of the calculated $\hat{b}_j$ are important steps for a meaningful interpretation of the results of underdetermined systems. Only if the resolution matrix R is diagonal (this is the case if [3] is full rank) will the calculated unknowns represent the 'true' unknowns.

The section inverse approach described above has been applied by Ganachaud and Wunsch using hydrographic, nutrient, and oxygen data from 20 WOCE sections worldwide. Results for section integrated top-to-bottom net nitrate fluxes and divergences are shown in **Figure 4**. Net nitrate transports are found to be southward throughout the Atlantic and Indian Oceans and northward in the South Pacific. However, for many sections the uncertainties are of similar magnitude as the flux values themselves, leaving the net transports essentially indistinguishable from zero. Nitrate divergences in many regions of the world ocean indicate nutrient sinks in the upper part of the water column and nutrient sources below, consistent with the concept of biological production in surface waters and subsequent particle remineralization and release of nutrients below.

Estimating Carbon Export Fluxes with the Adjoint Method

The section inverse approach described above requires hydrographic, nutrient, and tracer data along all interfaces of the domains for which budget equations are formulated. It is assumed that tracer fields and flows are in steady state, precluding the use of time-dependent tracers, like the CFCs. The budget equations could be generalized to include the time rate of change of tracer inventory inside the domain. However, one would need repeated tracer measurements inside the domain and along all its boundary for all times considered in the model to correctly describe the temporal inventory changes and the

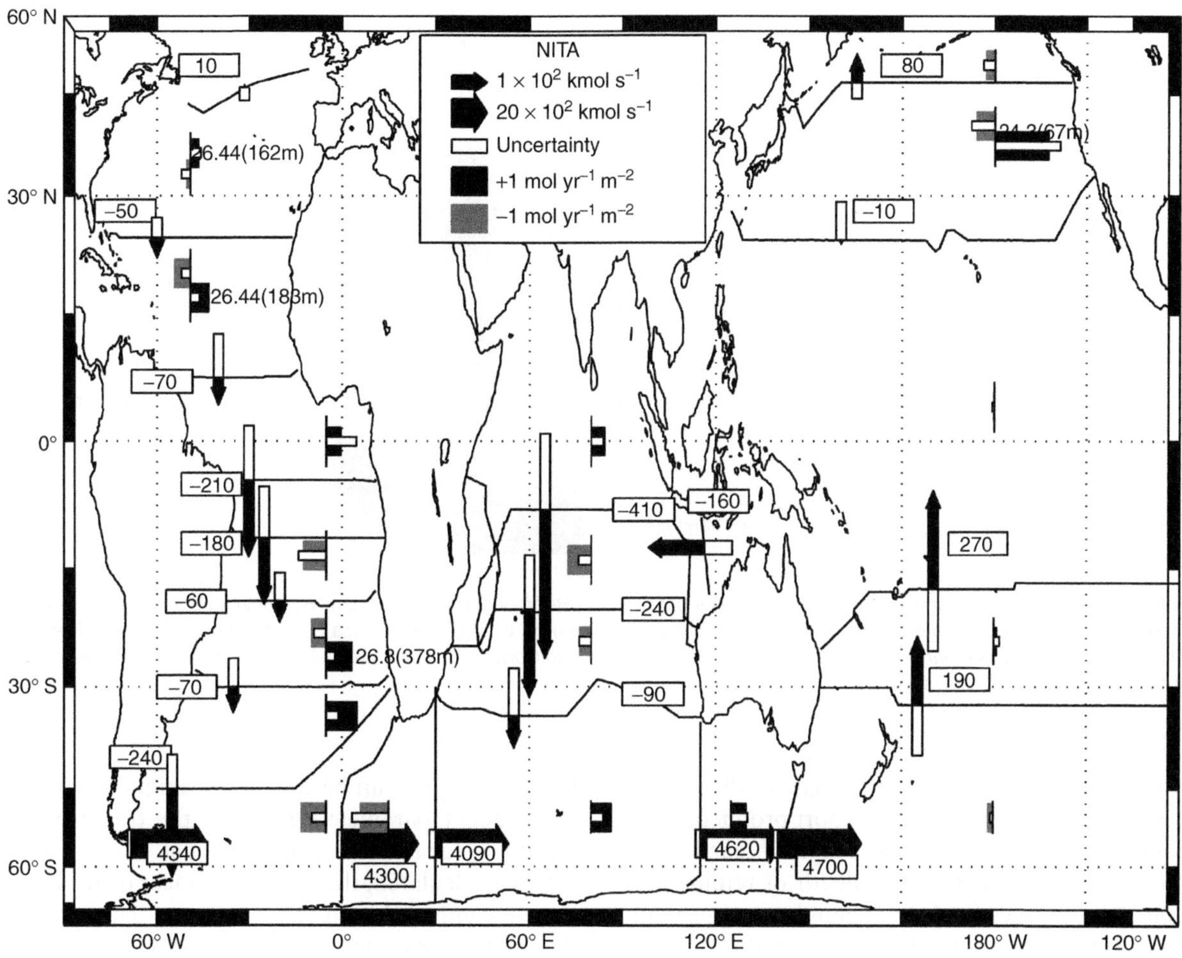

Figure 4 Global dissolved nitrate transports and divergences. The length of each arrow corresponds to the nitrate transports between continents. The open boxes behind each arrow indicate the uncertainty (one standard deviation). Between sections, nitrate divergences are indicated by the solid boxes, either top-to-bottom (single box) or surface/deep (double box). Adapted from Ganachaud A and Wunsch C (2002) Oceanic nutrient and oxygen transport and bounds on export production during the World Ocean Circulation Experiment. *Global Biogeochemical Cycles* 16 (doi:10.1029/2000GB001333), with permission from the American Geophysical Union.

tracer fluxes across the boundaries. These massive data requirements cannot be met for any known transient tracer, and it seems necessary to adapt the inverse methodology for inclusion of sparse time-dependent and steady-state tracers.

A hybrid model consisting of forward and inverse steps and utilizing the Lagrange multiplier method of constrained variational optimization for fitting the model to tracer data has been developed for this purpose. The model exploits data for many tracers, including nutrients, radiocarbons, and CFCs. The objective of the model is to find optimal three-dimensional (3-D) global ocean flows, biological production rates, and depth-dependent downward particle fluxes that explain the observed tracer, nutrient, and oxygen distributions best. The optimization is done iteratively, varying the flows as well as the biogeochemical parameters systematically until the agreement between model simulated tracer fields and observations is optimal.

The particular model has a rectangular grid (**Figure 5**), where grid cell boundaries are not required to match lines of available data. The layout of the grid is decoupled from the that of the available data, and individual grid cells (boxes) may be void of any data. Model tracer values are defined at the center of the boxes, whereas flows are defined on the interfaces. Biological production of particulate material occurs in the top model layers representing the euphotic zone. Particle fluxes below the euphotic zone are assumed to decrease with depth following a functional relationship from the literature

$$j_P(z) = a \cdot (z/z_{EZ})^{-b} \qquad [10]$$

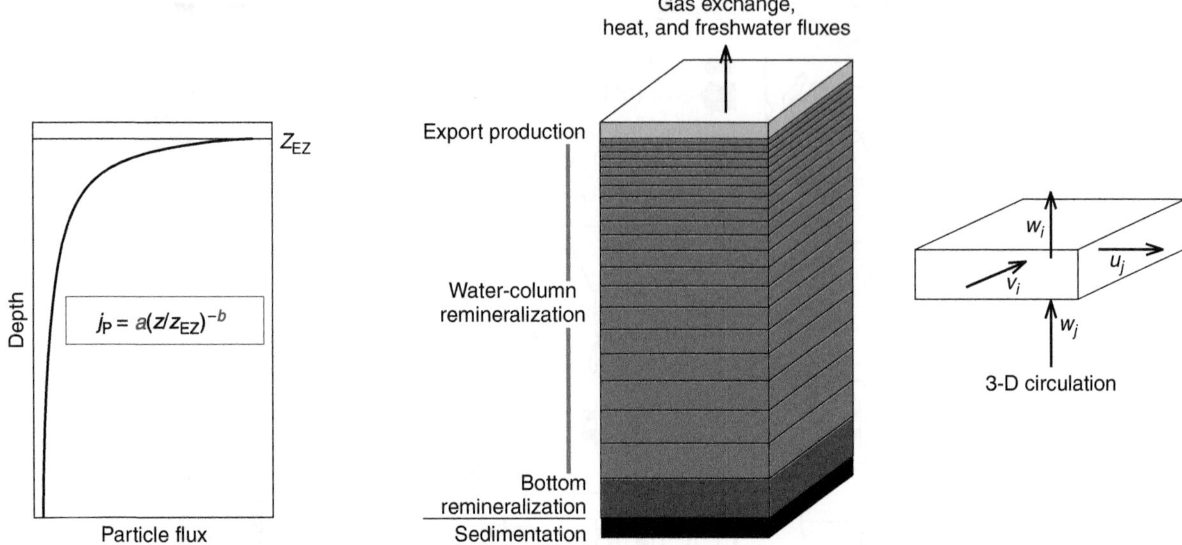

Figure 5 Vertical model grid and definition of model parameters. Reproduced from Schlitzer R, Usbeck R, and Fiscjer H (2004) Inverse modeling of particulate organic carbon fluxes in the South Atlantic. In: Wefer G, Mulitza S, and Ratmeyer V (eds.) *The South Atlantic in the Late Quaternary – Reconstruction of Material Budget and Current Systems*, pp. 1–19. Berlin: Springer, with permission of Springer Science Business Media.

where a is the particle flux at the base of the euphotic zone, z_{EZ}, and represents the export production. The parameter b determines the shape of particle flux profile and thus controls the depth of remineralization. Large values for b correspond to steep particle flux decreases and thus large remineralization rates just below the euphotic zone, whereas values for b close to zero result in almost constant particle fluxes with depth with little remineralization in the water column and most of the particle export reaching the ocean floor.

Export production a and remineralization parameter b vary geographically, and goal of the model runs is to infer optimal values for a and b (in addition to flow velocities v) based on the available nutrient and tracer data.

Figure 6 shows a schematic overview of the computational strategy. All quantities listed under model parameters are to be determined by the model. These parameters are combined in the vector of independent parameters p^*. They have to be initialized to start the calculation (see the literature for initialization strategies), and they will be varied in a systematic way in the course of the calculations. Using the initial independent parameters, the model simulates the distributions of all steady-state (temperature, salinity, oxygen, phosphate, nitrate, total inorganic carbon, alkalinity, and radiocarbon) and transient (CFCs) tracers, as in normal forward models. The simulated tracer concentrations m_i are combined in a vector of dependent parameters $\tilde{p}$. All dependent parameters in $\tilde{p}$ can be calculated uniquely from the independent parameters p^* using the tracer budget equations for the boxes of the model.

Once calculated, the simulated tracer distributions can be compared with measurements. Traditionally, this model/data comparison is done subjectively by analyzing the misfit fields and hypothesizing possible causes for the misfits. Model flows or biogeochemical parameters are then modified, hoping that new simulations with the modified parameters lead to more realistic tracer fields. This manual tuning has been used successfully with small box models; however, for problems with many thousands of parameters, such as the one described above, it is impractical and not successful in most cases.

The Lagrange method of constrained optimization (termed 'adjoint method' in meteorology and oceanography) offers an alternative to manual parameter tuning. This method varies and optimizes the independent parameters p^* automatically, while satisfying the set of model equations consisting of all tracer budgets for all boxes exactly. The model equations are usually represented in homogeneous form $E_j = 0$. The adjoint method can be applied to very large problems with hundred thousands of parameters or more. Here, the evaluation of the model performance is done objectively using a suitably defined cost function F. A cost function typically contains terms for all unwanted features of the model, most importantly terms that penalize deviations between model simulated tracer values m_i and

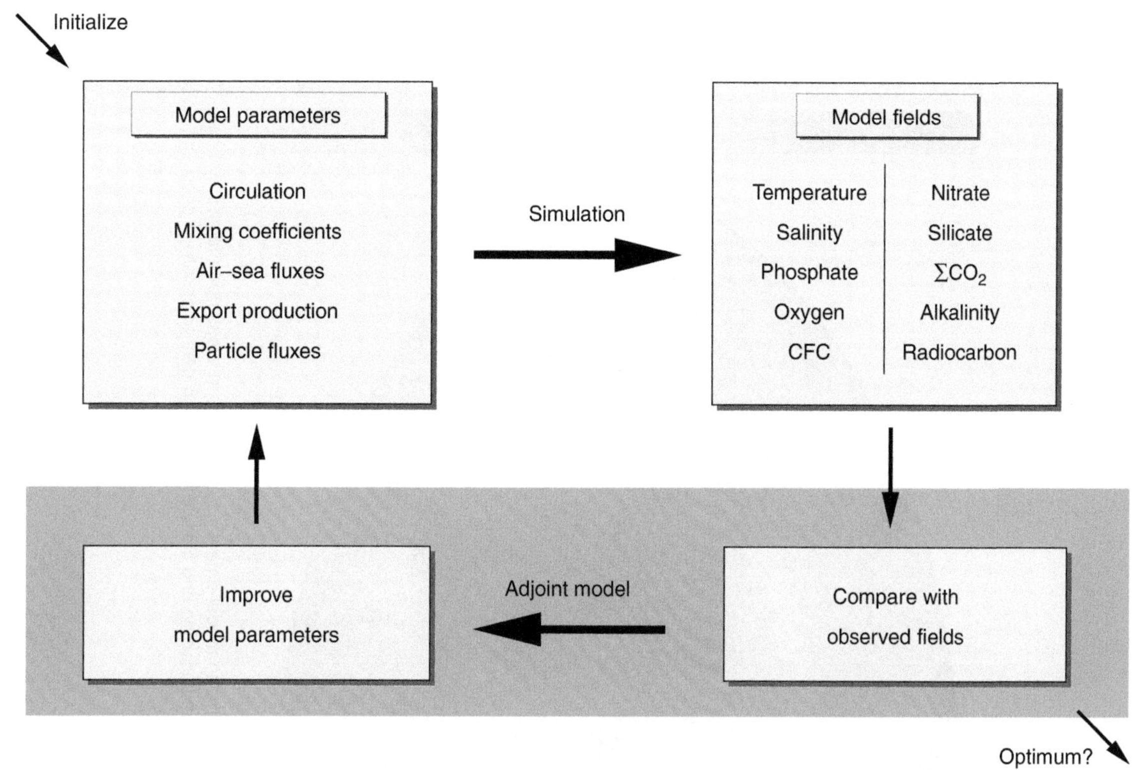

Figure 6 Schematic overview of model calculations performed for every iteration of the optimization process.

observations $d_i \pm \sigma_i$:

$$F(p^*, \tilde{p}) = \cdots + \left[\frac{m_i - d_i}{\sigma_i}\right]^2 + \cdots \qquad [11]$$

A large value of F indicates that the simulated tracer concentrations differ much from the observations. Bringing the model close to the data, therefore, is equivalent to minimizing F, subject to satisfying the model equations (tracer budget equations) $E_j = 0$ exactly.

Details on the Lagrange multiplier method used for the constrained minimization of F can be found in textbooks and the literature on data assimilation or constrained data fitting. Overall, this method allows calculating the direction in parameter space $-\nabla F_{p^*}$ of steepest decrease of F (the negative gradient of F with respect to p^*), which can then be used by a descent algorithm to arrive at a new, modified vector of independent parameters p^*. It is guaranteed that new simulations using the modified p^* will lead to more realistic tracer simulations and a smaller value of the cost function F. This procedure is repeated until no significant further decrease in F can be achieved or a limit on the number of iterations is reached (see **Figure 6**). The adjoint step in the calculations replaces the subjective evaluation of misfits mentioned above and provides an automatic 'learning' step of the model based on the current model/data misfits. The computational cost of the adjoint run is comparable to the cost of the simulation. The number of iterations required to reach the minimum of F depends on the initial p^*, but is usually large. It is therefore important to use efficient implementations of the simulation and adjoint steps to keep the computation time of a single iteration as short as possible.

Figure 7 shows the observed and simulated radiocarbon values in the bottom waters of the world ocean after optimization. Following common practice, ^{14}C concentrations are given in Δ-notation expressing the per mille ^{14}C concentration difference of a given sample from the ^{14}C standard (wood grown in 1890, decay corrected to 1950). A sample with $\Delta^{14}C = -100‰$, for instance, has a ^{14}C concentration 100‰ (or 10%) lower than that of the standard. In agreement with the data and consistent with the general concept of the global thermohaline circulation, the model yields highest $\Delta^{14}C$ values in the North Atlantic (−70‰ to −80‰) and lowest radiocarbon concentrations in the northeast Pacific (−235‰). $\Delta^{14}C$ values in the Southern and Indian Oceans are intermediate and range between −150 and −165‰. There are clear signs of tongues with elevated $\Delta^{14}C$ values along the major deep and bottom water spreading paths. This includes the

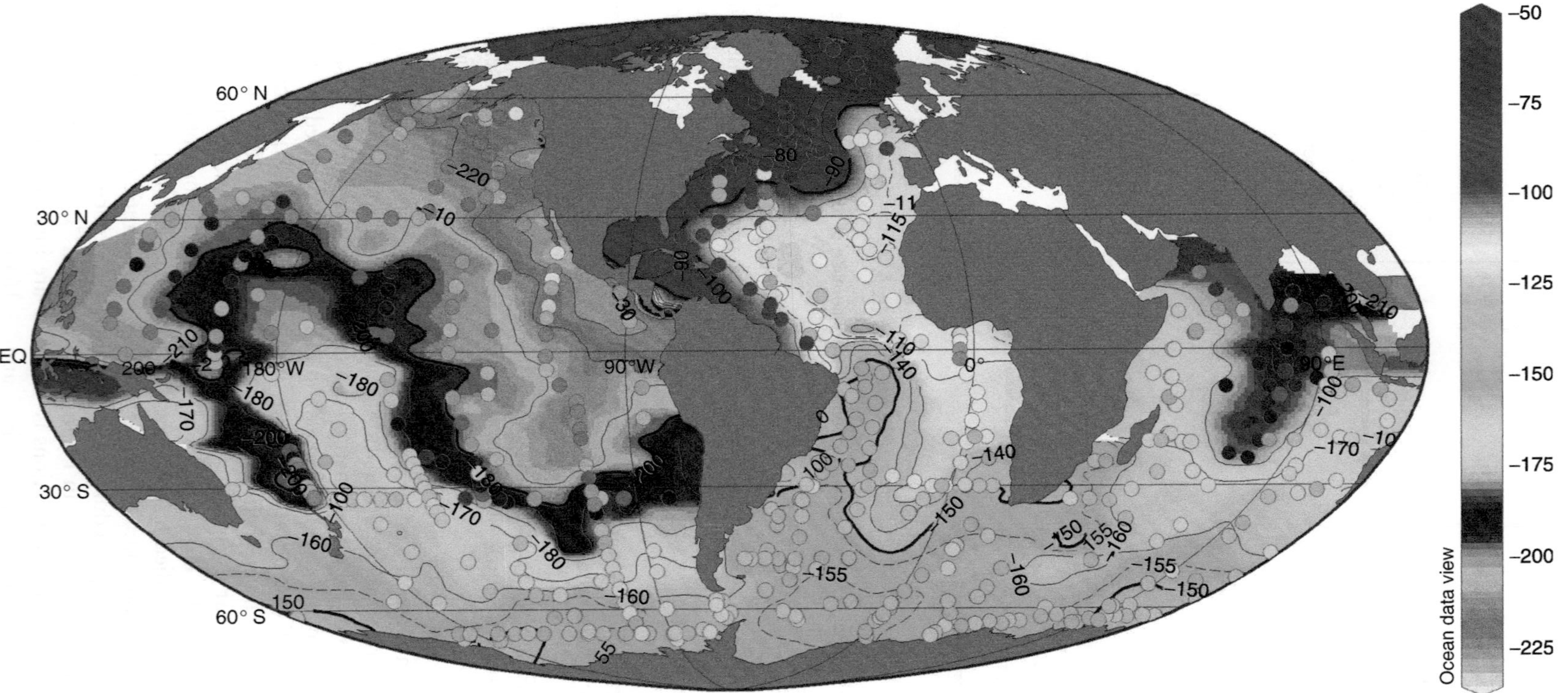

Figure 7 Bottom water $\Delta^{14}C$ simulated by the model (color-shaded field) and from data (colored dots). The mean model-data $\Delta^{14}C$ difference in areas not affected by bomb-^{14}C is only +1.3‰, and the root-mean-square difference amounts to only 5.2‰. From Schlitzer R (2006) Assimilation of radiocarbon and chlorofluorocarbon data to constrain deep and bottom water transports in the world ocean. *Journal of Physical Oceanography* 37: 259–276.

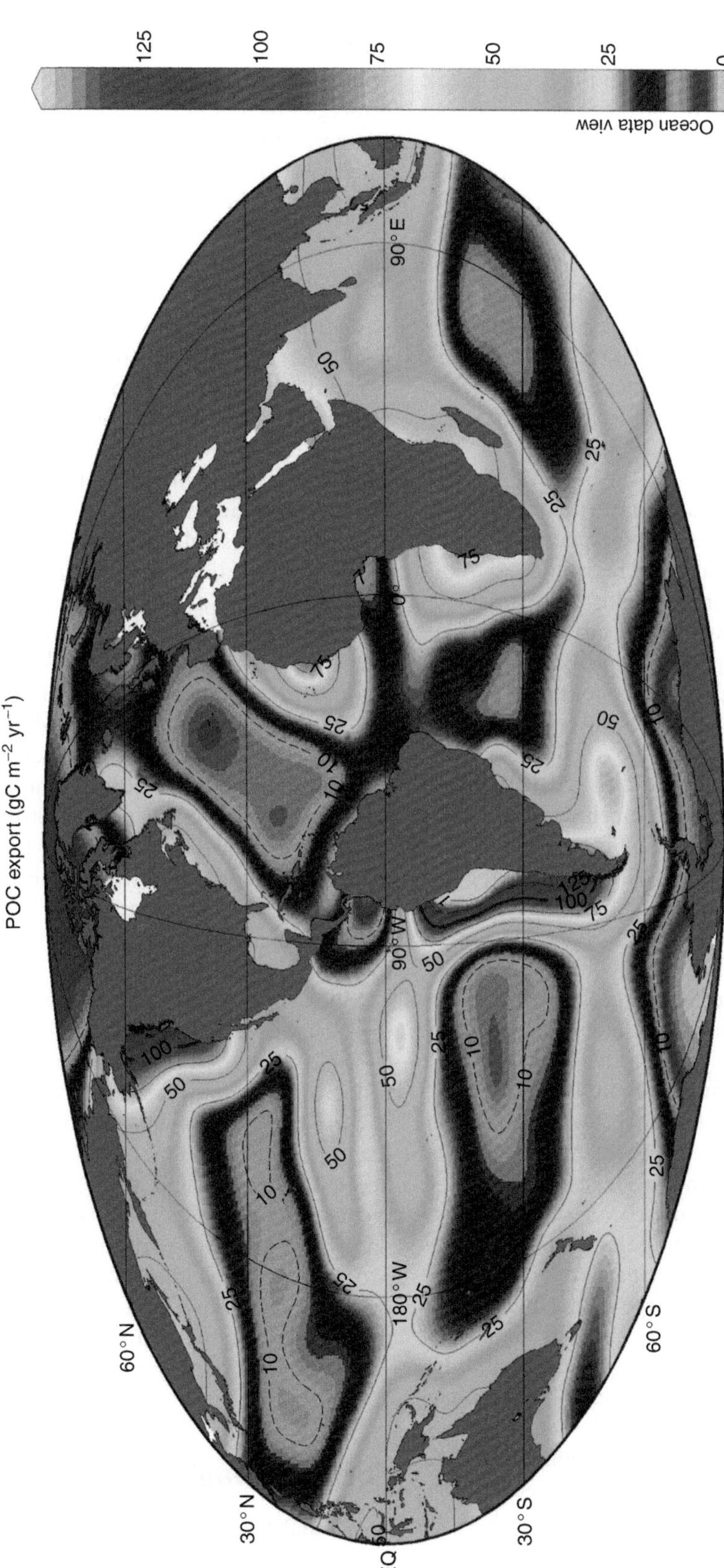

Figure 8 Export production of particulate organic carbon (POC) from the inverse model. Redrawn from Schlitzer R (2000) Applying the adjoint method for global biogeochemical modeling. In: Kasibhatla P, Heimann M, Rayner P, Mahowald N, Prinn RG, and Hartley DE (eds.) *AGU Geophysical Monograph Series, Vol. 114: Inverse Methods in Global Biogeochemical Cycles*, pp. 107–124. Washington, DC: American Geophysical Union.

southward flow of North Atlantic Deep Water (NADW) in the North Atlantic and the northward spreading of Antarctic Bottom Water (AABW) in the western South Atlantic as well as in the South Pacific and the Indian Oceans. The quantitative agreement between simulated and measured $\Delta^{14}C$ values is excellent, and the model-data difference is only 1.3‰ on average. The root-mean-square (rms) difference amounts to only 5.2‰ and is of the same order of magnitude as the uncertainties of the radiocarbon data. In the deep North Atlantic and parts of the Southern Ocean, the measurements are systematically higher than the model simulations, owing to the contribution of bomb-^{14}C in these waters. Bomb-radiocarbon is not included in the model, and contaminated data values are not assimilated in the model.

The model fluxes of particulate organic carbon (POC) at the base of the euphotic zone (carbon export production) necessary to reproduce the observed oxygen and nutrient fields are shown in **Figure 8**. The spatial patterns of carbon export resemble the patterns of primary production in published maps showing high fluxes in coastal upwelling areas off West Africa, along the West American coast, in the Arabian Sea and Bay of Bengal, in the northwest Atlantic and north and tropical Pacific, in the area of the Indonesian archipelago, and in the Southern Ocean. Highest values in the productive areas are on the order of 5–10 $\mathrm{mol\,C\,m^{-2}\,yr^{-1}}$; in the oligotrophic, open-ocean regions they amount to between 0.5 and 2 $\mathrm{mol\,C\,m^{-2}\,yr^{-1}}$. The export of POC in the model predominantly occurs at mid-latitudes and in the Southern Ocean, and the contribution of the Northern Hemisphere is comparatively small. Globally integrated, the POC export in the model amounts to about 10 $\mathrm{Gt\,C\,yr^{-1}}$.

Error analysis of the solution vector p^* is possible but computationally very expensive. An eigenvector/eigenvalue analysis of the inverse Hessian matrix H^{-1} (the Hessian is the square matrix of second partial derivatives of F with respect to parameters p^*) reveals directions in parameter space that are well determined (eigenvectors associated with large eigenvalues; values of F increase rapidly when moving away from optimal point along the direction of the eigenvector) and directions that are only poorly determined by the model (eigenvectors associated with smallest eigenvalues; values of F increase slowly when moving away from optimal point along the direction of the eigenvector). The ratio of largest and smallest eigenvalues of H^{-1} is a measure of the anisotropy of F around the optimal solution p^*. For relatively small problems with a few hundred parameters in p^*, the Hessian matrix, its inverse, and the associated eigenvalues and eigenvectors can actually be calculated and a rigorous error analysis of the solution is possible. For large problems with hundred thousands of parameters, this will remain impossible for the foreseeable future. Strategies for obtaining sensitivity information at least for some directions in parameter space or for finding the largest eigenvalues and associated eigenvectors are described in the literature.

Conclusion

Recent progress in applications of inverse models to problems in physical and biogeochemical oceanography clearly shows that inverse methodology is well advanced and being used by a growing number of researchers. Mathematical techniques exist that exploit the available data better than before and produce new and important scientific results. These methods successfully cope with problems, such as sparseness of data and incompleteness of information. The advances were possible because of the tremendous technological progress in computer hardware, combined with breakthroughs in the development of efficient and innovative algorithms. These algorithms finally allowed the numerical application of mathematical principles, such as the Lagrange multiplier method, whose theory was established for centuries already. Still, the widespread use of inverse methods would not have been possible, if there had not been at the same time an increased awareness among scientists for the need of integrated, global databases and an increased willingness to contribute to these data sets.

Much more data will become available in the future, enabling inverse-type studies that are still impossible today. While ship observations will continue to be important, the new autonomous floats, gliders, or profiling instruments on moorings will provide data in near real time, even from remote and inaccessible regions. New satellite sensors will complement the water column data with high-resolution data from the ocean surface, and new geochemical programs, such as GEOTRACES, will produce high-quality data of the ocean's trace elements and isotopes, including micronutrients such as Fe and Zn, in unprecedented quality and coverage.

Further Reading

Ganachaud A and Wunsch C (2002) Oceanic nutrient and oxygen transport and bounds on export production during the World Ocean Circulation Experiment. *Global Biogeochemical Cycles* 16 (doi:10.1029/2000GB 001333).

Gill PE, Murray W, and Wright MH (1981) *Practical Optimization*. London: Academic Press.

Hestenes MR (1975) *Optimization Theory*. New York: Wiley.

Kasibhatla P, Heimann M, Hartley D, Mahowald N, Prinn R, and Rayner P (eds.) (1998) *AGU Geophysical Monograph Series, Vol. 114: Inverse Methods in Global Biogeochemical Cycles*. Washington, DC: American Geophysical Union.

Menke W (1984) *Geophysical Data Analysis: Discrete Inverse Theory*. San Diego, CA: Academic Press.

Rintoul S and Wunsch C (1991) Mass, heat, oxygen and nutrient fluxes and budgets in the north Atlantic Ocean. *Deep Sea Research* 38(supplement): S355–S377.

Schlitzer R (2000) Applying the adjoint method for global biogeochemical modelling. In: Kasibhatla P, Heimann M, Rayner P, Mahowald N, Prinn RG, and Hartley DE (eds.) *AGU Geophysical Monograph Series, Vol. 114: Inverse Methods in Global Biogeochemical Cycles*, pp. 107–124. Washington, DC: American Geophysical Union.

Schlitzer R (2002) Carbon export fluxes in the Southern Ocean: Results from inverse modeling and comparison with satellite based estimates. *Deep Sea Research II* 49: 1623–1644.

Schlitzer R (2004) Export production in the Equatorial and North Pacific derived from dissolved oxygen, nutrient and carbon data. *Journal of Oceanography* 60: 53–62.

Schlitzer R (2006) Assimilation of radiocarbon and chlorofluorocarbon data to constrain deep and bottom water transports in the world ocean. *Journal of Physical Oceanography* 37: 259–276.

Schlitzer R, Usbeck R, and Fiscjer H (2004) Inverse modeling of particulate organic carbon fluxes in the South Atlantic. In: Wefer G, Mulitza S, and Ratmeyer V (eds.) *The South Atlantic in the Late Quaternary – Reconstruction of Material Budget and Current Systems*, pp. 1–19. Berlin: Springer.

Tarantola A (1983) *Inverse Problem Theory: Methods for Data Fitting and Model Parameter Estimation*. New York: Elsevier.

Thacker WC and Long RB (1988) Fitting dynamics to data. *Journal of Geophysical Research* 93: 1227–1240.

Wunsch C (1978) The North Atlantic general circulation west of 50° W determined by inverse methods. *Reviews of Geophysics* 16: 583–620.

Wunsch C (1996) *The Ocean Circulation Inverse Problem*. Cambridge, UK: Cambridge University Press.

Relevant Websites

http://cdiac.esd.ornl.gov
– Carbon Dioxide Information Analysis Center.

http://whpo.ucsd.edu
– Clivar and Carbon Hydrographic Data Office.

http://www.coriolis.eu.org
– Coriolis.

http://www.ewoce.org
– eWOCE: Electronic Atlas of WOCE Data.

http://www.nodc.noaa.gov
– National Oceanographic Data Center, NOAA.

http://odv.awi.de
– Ocean Data View.

NATURAL RADIOACTIVE SPECIES

URANIUM-THORIUM DECAY SERIES IN THE OCEANS: OVERVIEW

M. M. R. van der Loeff, Alfred-Wegener-Institut für Polar und Meereforschung Bremerhaven, Germany

Introduction

Natural radioactivity provides tracers in a wide range of characteristic timescales and reactivities, which can be used as tools to study the rate of reaction and transport processes in the ocean. Apart from cosmogenic nuclides and the long-lived radioisotope K-40, the natural radioactivity in the ocean is primarily derived from the decay series of three radionuclides that were produced in the period of nucleosynthesis preceding the birth of our solar system: Uranium-238, Thorium-232, and Uranium-235 (a fourth series, including Uranium-233, has already decayed away). The remaining activity of these so-called primordial nuclides in the Earth's crust, and the range of half-lives and reactivities of the elements in their decay schemes, control the present distribution of U-series nuclides in the ocean.

The Distribution of Radionuclides of the Uranium Thorium Series in the Ocean

Distribution of ^{238}U, ^{235}U, ^{234}U, and ^{232}Th (see Uranium-Thorium Series Isotopes in Ocean Profiles)

Uranium is supplied to the ocean by rivers. In sea water it is stabilized by a strong complexation as uranyl carbonate $UO_2(CO_3)_3^{4-}$, causing its long residence time in the ocean. U follows closely the distribution of salinity with ^{238}U (in dpm l^{-1}) = 0.0704* salinity. (Note: dpm = disintegrations per minute. The SI Unit Bq, 60 dpm = 1 Bq, is not used in the literature on natural radioactivity in the ocean.) Under anoxic conditions, U is reduced from the soluble (VI) to the insoluble (IV) oxidation state and rapidly removed from sea water. Reductive removal occurs especially in sediments underlying high productivity or low-oxygen bottom waters. Locally this may influence the U–salinity relationship. Salinity-corrected U contents have a variation of 3.8% in the world ocean and are about 1% higher in the Pacific than in the Atlantic Ocean. At lower salinities in estuaries, salinity-corrected U contents are much more variable as a result of removal and release processes and of interaction with organic complexants and colloids.

^{235}U is chemically equivalent to ^{238}U and occurs with a ^{235}U/^{238}U activity ratio of 0.046. As a result of the preferential mobilization of ^{234}U during chemical weathering, the river supply of ^{234}U activity exceeds the supply of ^{238}U, causing a ^{234}U/^{238}U ratio in the ocean greater than unity. The isotopic composition of uranium in sea water with salinity 35 is shown in **Table 1**.

Like U, ^{232}Th is a component of the Earth's crust and is present in the lithogenic fraction of every marine sediment. As a result of its high particle reactivity, Th is rapidly removed from the water column. The ^{232}Th activity in the ocean is very low (around 3×10^{-5} dpm l^{-1} or 0.1 ng/kg) and its distribution can be compared to that of other particle-reactive elements like Al or Fe.

Distribution of Isotopes from the Three Decay Series

In all three decay series, isotopes of relatively soluble elements like U, Ra, and Rn, decay to isotopes of highly particle-reactive elements (Th, Pa, Po, Pb), and vice versa (**Figure 1**), resulting in widely different distributions in the water column (**Table 2**) (*see* Uranium-Thorium Series Isotopes in Ocean Profiles).

In a closed system, given enough time, all nuclides in a decay series reach secular equilibrium. This means that growth is balanced by decay, and that all intermediate nuclides have the same activity. In a natural open system, however, reaction and transport

Table 1 Average isotopic uranium composition of sea water with salinity 35

Parameter	*Value*
^{235}U + ^{238}U concentration	3.238 ng g^{-1}
^{235}U/^{238}U activity ratio	0.0460
^{234}U/^{238}U activity ratio	1.144 ± 0.002
Isotope activity	
^{238}U	2.46 dpm l^{-1}
^{234}U	2.82 dpm l^{-1}
^{235}U	0.113 dpm l^{-1}

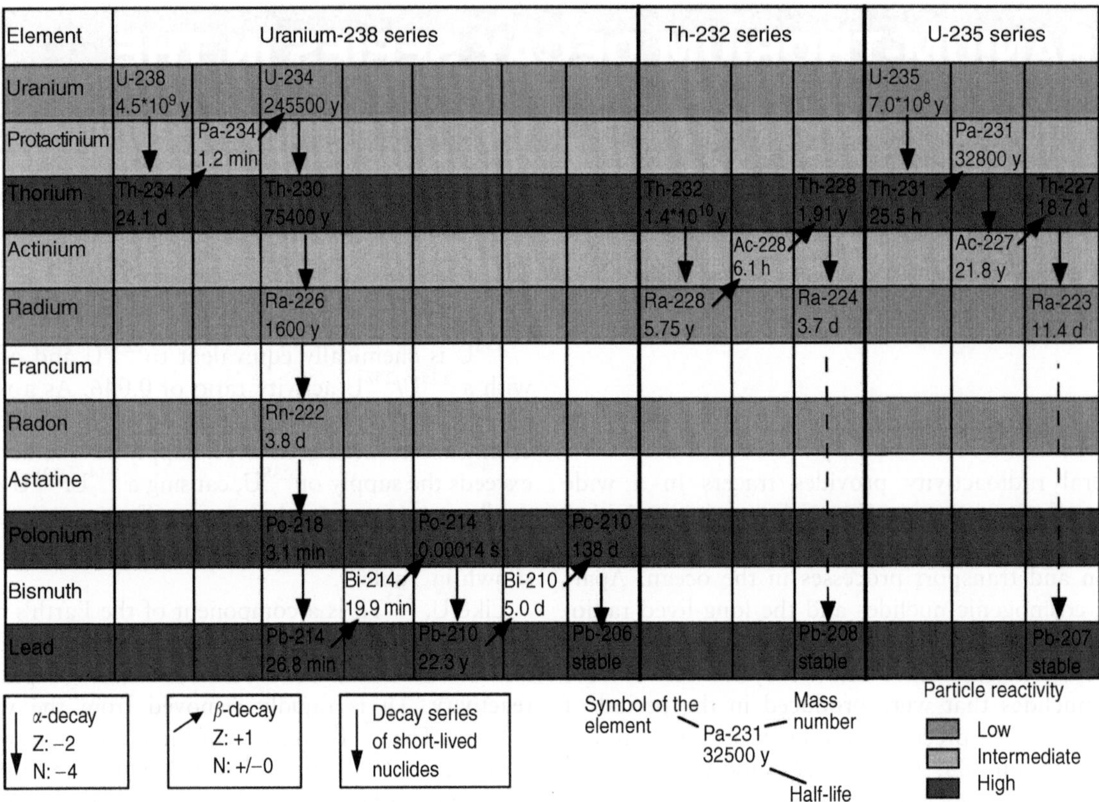

Figure 1 The natural uranium-thorium decay series, colored according to particle reactivity. The arrows represent decay with the changes in atomic number (Z) and number of nucleons (N) indicated. All three series end with a stable lead isotope.

Table 2 List of the elements (with isotopes of half-life $t_{1/2} > 1$ day) in the U decay series with their scavenging residence time in deep and surface ocean and their estimated particle-water partition coefficient K_d, showing the relative mobility of U, Ra, and Rn

Element	*Scavenging residence time (years)*		K_d (cm^3 g^{-1})
	Deep sea	*Surface ocean*	
U	450 000		500
Pa	130	<1	1×10^6
Th	30	<1	1×10^7
Ac	decays (>30)		$0.4–2 \times 10^5$
Ra	1000		$0.2–3 \times 10^4$
Rn	decays	gas exchange with atmosphere	0
Pb	50–100	2	1×10^7
Po	decays (>2)	0.6	2×10^7

cause a separation between parent and daughter nuclides. The resulting disequilibria between parent and daughter nuclide can be used to calculate the rate of the responsible processes.

Disequilibrium: The Basis for Flux Calculations (Figure 2)

Mobile Parent with Particle-reactive Daughter (Table 3)

Tracers in this group are produced in the water column and removed on sinking particles, a process called scavenging. They allow us to determine particle transport rates in the ocean.

In a simple box model the total daughter activity (A_D^t) is determined by decay (decay constant $\lambda = \ln(2)/t_{1/2}$), ingrowth from the parent nuclide (activity A_p, production rate of daughter nuclide $P_D = \lambda A_p$) and removal on sinking particles J (**Figure 3**):

$$\frac{dA_D^t}{dt} = P_D - \lambda A_D^t - J = \lambda(A_P - A_D^t) - J \qquad [1]$$

In steady-state the flux is directly related to the depletion of the daughter with respect to the parent:

$$J = \lambda(A_P - A_D^t) \qquad [2]$$

and the residence time of the daughter nuclide with respect to scavenging, τ_{sc}, is given by the quotient of

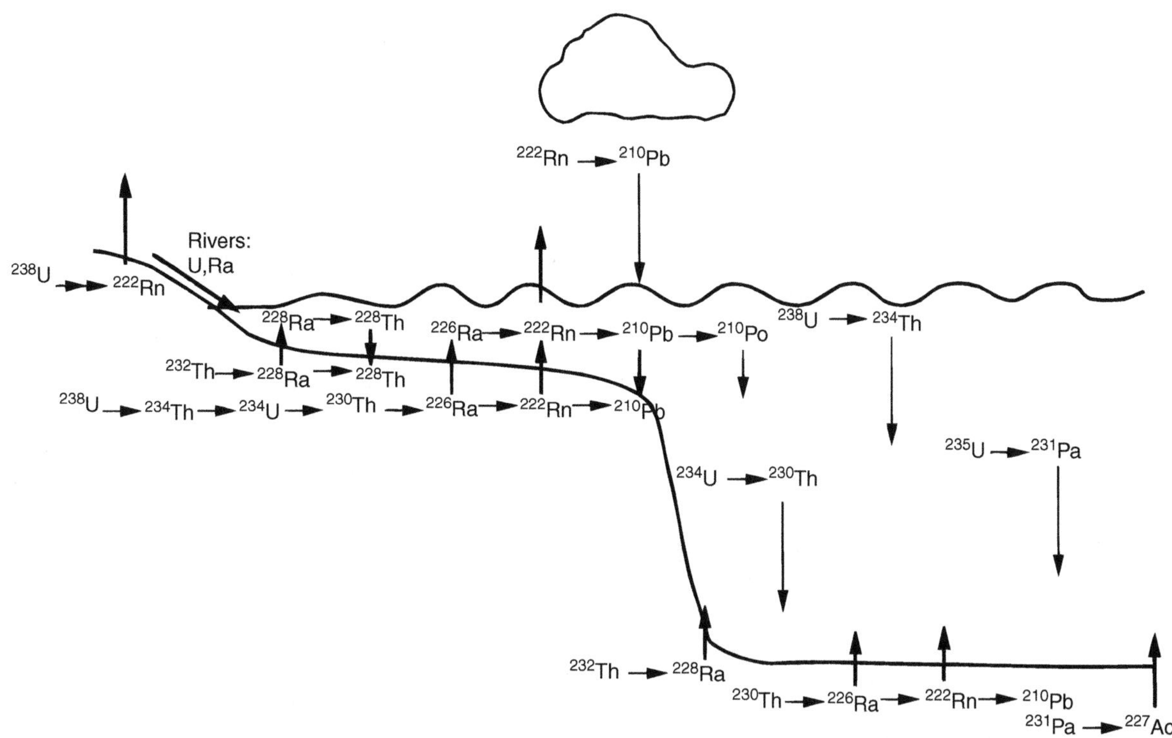

Figure 2 Schematic diagram of radioactive decay (horizontal arrows) and typical transport processes in ocean and atmosphere that can be traced by the nuclides described here (vertical arrows). (Adapted with permission from Ernst WG and Morin JG (eds.) 1980 *The Environment of the Deep Sea*. Englewood Cliffs, NJ: Prentice Hall.)

Table 3 Isotope pairs with mobile parents and particle-reactive daughters

Mother	*Daughter*	*Half-life*	*Source*	*Oceanographic applications*
^{234}U	^{230}Th	75200 y	water column	sediment trap calibration, reconstruction of past vertical rain, sediment focusing
^{238}U	^{234}Th	24.1 d	water column	export production, calibration of shallow sediment trap resuspension budgets, bioturbation
^{228}Ra	^{228}Th	1.9 y	water column, in deep sea and continental shelf	scavenging in coastal waters, bioturbation
^{235}U	^{231}Pa	32500 y	water column	boundary scavenging, paleoproductivity, refined sediment trap calibration
^{226}Ra	^{210}Pb	22.3 y	water column, atmosphere	boundary scavenging, bioturbation
^{210}Pb	^{210}Po	138 d	water column	scavenging in surface ocean

activity and removal rate:

$$\tau_{sc} = \frac{A_D^t}{J} = \frac{A_D^t}{\lambda(A_P - A_D^t)} \qquad [3]$$

Elements in this group are described below.

Thorium ^{230}Th is produced at a known rate from ^{234}U in sea water. The highly reactive element is rapidly adsorbed onto particles and transported down in the water column when these particles sink out. As the adsorption is reversible, a steady-state distribution is achieved, in which both particulate and dissolved activities increase linearly with depth (*see* Uranium-Thorium Series Isotopes in Ocean Profiles). At any depth, disregarding horizontal advection, the vertical flux of $^{230}Th_{XS}$ ('xs' meaning in excess of the activity supported by the parent nuclide, in this case the small amount of ^{234}U on the sinking particles) must equal its production from

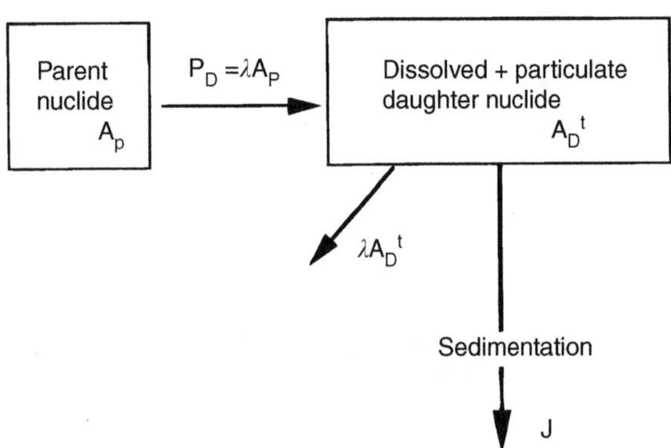

Figure 3 Schematic diagram of the scavenging of a particle-reactive daughter nuclide (decay constant λ) produced in the water column from a soluble parent.

^{234}U in the overlying water column (depth z in meters), which amounts to:

$$P_{230} = \lambda_{230} A_{234} z = 9.19 \times 10^{-6} (y^{-1}) \; 2820 \; (\text{dpm m}^{-3}) z(\text{m}) = 0.0259 z \; (\text{dpm m}^{-2} y^{-1}) \quad [4]$$

This known, constant ^{230}Th flux, depending only on water depth, is a powerful tool to quantify errors in the determination of rain rates of other components of the particle flux, either by sediment traps or through the accumulation rate of a marine sediment. The collection efficiency of sediment traps, known to be highly variable and dependent on trap design, turbulence, and flow rates, can be derived from a comparison of the intercepted ^{230}Th flux F_{230} with the theoretical flux P_{230} (see below for a refinement of this procedure using ^{231}Pa). The vertical rain rate R_i of any component i of the particle flux can be derived from the ratio of the concentration C_i to the ^{230}Th activity in the particles A_{230}, using:

$$R_i = \frac{C_i}{A_{230}} P_{230} \quad [5]$$

In a similar way, the past flux of $^{230}Th_{xs}$ to the sea floor, $^0F_{230}$, derived from decay-corrected ^{230}Th activities ($^0A_{230}$) in dated sediment core sections, can be compared to the theoretical rain rate. The ratio $\Psi = {}^0F_{230}/P_{230}$, the focusing factor, is used to determine to what extent the sediment core location has been subject to focusing or winnowing during certain geological periods. The preserved vertical rain rate of sediment components corrected for such redistribution effects follows in analogy to eqn [5]:

$$R_i = \frac{C_i}{{}^0A_{230}} P_{230} \quad [6]$$

^{234}Th is produced from the decay of ^{238}U in sea water. In the deep ocean, approximately 3% of its activity is on particles and removal is so slow compared with its half-life (24.1 days) that total (dissolved + particulate) ^{234}Th is in secular equilibrium with ^{238}U. In coastal and productive surface waters, however, scavenging (**Figure 3**) causes a strong depletion of ^{234}Th (**Figure 4**). Following eqn [2], the depth-integrated depletion in the surface water yields the export flux of ^{234}Th. If required, the calculation can be refined to include advection and nonsteady-state situations. The resulting flux of ^{234}Th out of the surface layer of the ocean is the most suitable way to calibrate shallow sediment traps. The export flux of other constituents, like organic carbon or biogenic silica, can be derived from the export flux of ^{234}Th if the ratio of these constituents to particulate ^{234}Th in the vertical flux is known. This ratio is variable and depends, for example, on particle size, and the uncertainty in the determination of this ratio limits the quality of ^{234}Th-based estimates of export production from the upper ocean.

A very similar situation exists near the seafloor, where resuspended sediment particles scavenge ^{234}Th from the bottom water. The resulting depletion of ^{234}Th in the benthic nepheloid layer (BNL) is a measure of the intensity of the resuspension-sedimentation cycle on a timescale of weeks. The tracer thus shows whether a nepheloid layer is advected over large distances or sustained by local resuspension.

Mass balance requires that the activity removed from surface waters and from the BNL is balanced by excess activities below (i.e. activities in excess of the activities supported by ^{238}U). Excess activities have sometimes been observed in mineralization horizons in the water column below the euphotic zone and are common in the surface sediment. The distribution of excess ^{234}Th in the sediment is used to calculate bioturbation rates on short timescales.

The half-life of 1.9 years makes ^{228}Th useful as a tracer for particle flux on a seasonal or interannual timescale. However, due to the highly inhomogeneous distribution of its parent ^{228}Ra, the interpretation is much more complicated than in the case of ^{234}Th, for example.

As regards multiple Th isotopes as an in situ coagulometer, it has been shown that Th isotopes in the ocean are in reversible exchange between the particulate and dissolved form (**Figure 5**) and in steady-state, including radioactive decay we have:

$$\frac{A^{\text{part}}}{A^{\text{diss}}} = \frac{k_1}{\lambda + k_{-1}} \quad [7]$$

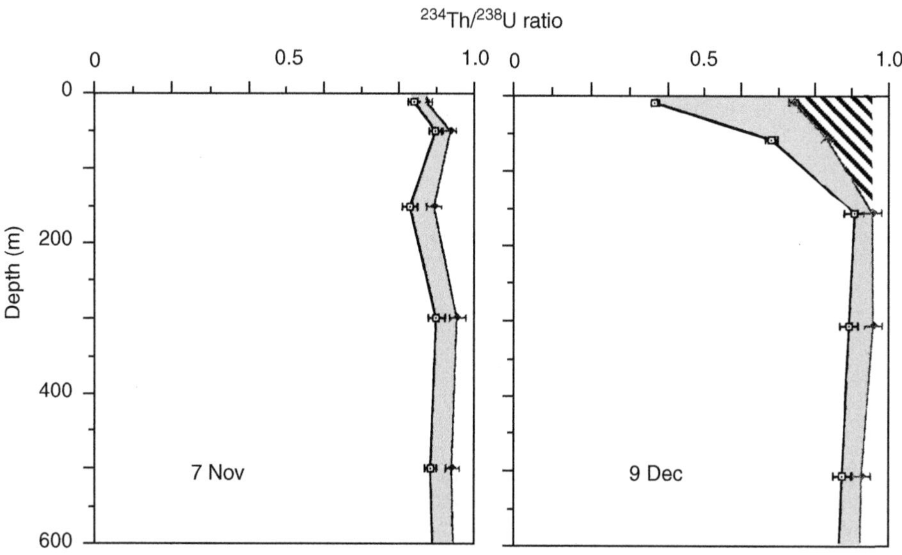

Figure 4 ^{234}Th : ^{238}U ratio before (left) and during (right) a plankton bloom in the Bransfield Strait, Antarctic Peninsula. The left profile in each diagram represents dissolved, the right profile total ^{234}Th activities. More ^{234}Th was adsorbed to particles (shaded) in the bloom. Total ^{234}Th was probably in equilibrium with ^{238}U in November, but became depleted in the surface water in December (hatched) due to particle export. (Adapted from Scavenging and particle flux: seasonal and regional variations in the Southern Ocean (Atlandic sector). *Marine Chemistry* 35, Rutgers van der Loeff and Berger, 553–567, Copyright (1991) with permission from Elsevier.)

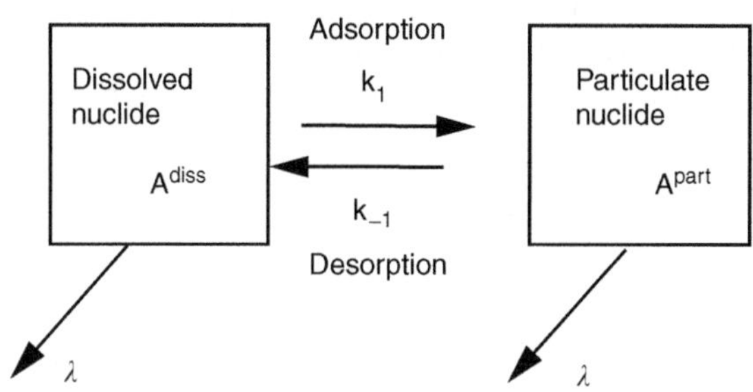

Figure 5 Box model of the reversible exchange between the dissolved and particulate form of a nuclide with adsorption and desorption rate constants k_1 and k_{-1}.

where A^{part} and A^{diss} are the particulate and dissolved activities, k_1 is the desorption rate constants. If the distribution of two or more isotopes (usually ^{234}Th and ^{230}Th or ^{228}Th and ^{234}Th) between dissolved and particulate forms is known, k_1 and k_{-1} can be calculated. Values for $1/k_1$ derived for thorium are on the order of a month in bloom situations and >1 year in clear deep water, much longer than expected from adsorption theory. This is explained when thorium adsorbs to colloidal-sized particules and the rate limiting steps, which determine the distribution of the tracers over dissolved and filterable form, are the coagulation and disaggregation with rate constants k_2 and k_{-2} respectively (**Figure 6**). Thus, when aggregation is clearly slower than adsorption $((k_2 \ll k_1)$ thorium isotopes provide a way to derive particle aggregation rates *in situ*.

Protactinium ^{231}Pa is produced from the decay of ^{235}U in sea water. The behavior of ^{231} Pa is very similar to that of ^{230}Th, and these two uranium daughters are produced throughout the water column in a constant activity ratio, given by the production rate of ^{231}Pa divided by the production rate of ^{230}Th or $A_{235}\ \lambda_{231}/A_{234}\lambda_{230} = 0.093$. The major application of ^{231}Pa lies in the combined use of these two tracers, whose exact production ratio is known. The approximately 10 times lower reactivity of ^{231}Pa allows it to be transported laterally over larger distances than ^{230}Th before being scavenged. The resulting basin-wide fractionation between ^{231}Pa and ^{230}Th is the basis for the use of the $^{231}Pa/^{230}Th$ ratio as a tracer of productivity. In areas of high particle flux the particles have a $^{231}Pa/^{230}Th$ ratio >0.093, whereas particles sinking in low productivity gyres have a ratio <0.093. The $^{231}Pa/^{230}Th$ ratio stored in the sediment, after proper correction for decay since deposition, is a powerful tool for the reconstruction of paleoproductivity. The fractionation between Th and Pa depends on particle composition and has been found to be much lower when opal is abundant. The tracer loses much of its value in a diatom-dominated system like the Southern Ocean.

A related application of the $^{231}Pa/^{230}Th$ ratio is a correction to the ^{230}Th-based calibration of sediment trap efficiency. The removal of both nuclides from sea water can be divided into a vertically scavenged component (V_{230}; V_{231}) and a component

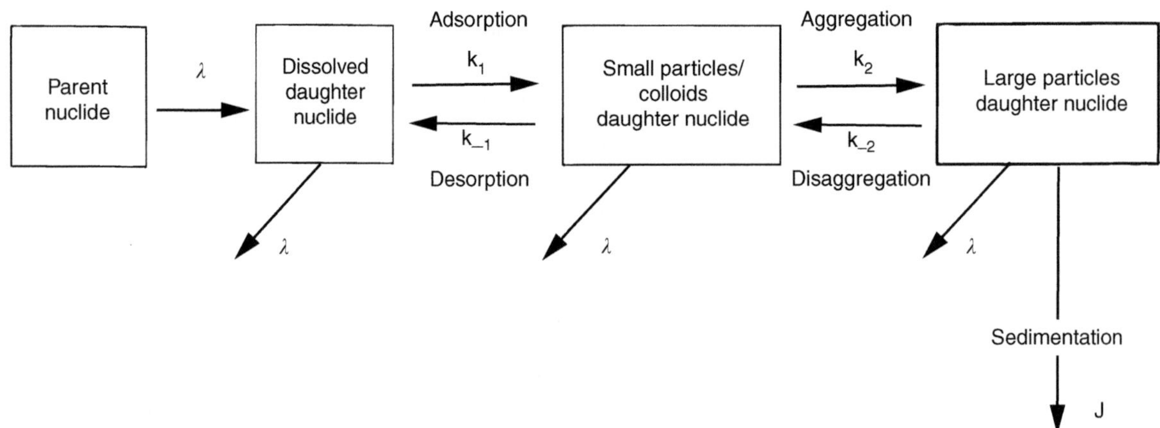

Figure 6 Conceptual model, including the models depicted in **Figure 3** and **Figure 5**, of the processes thought to control scavenging of radionuclides. (Adapted from Seasonality in the flux of natural radionuclides and plutonium in the deep Sargasso Sea. *Deep-Sea Research* 32, Bacon MP *et al.*, 273–286, Copyright 1985 with permission from Elsevier Science.)

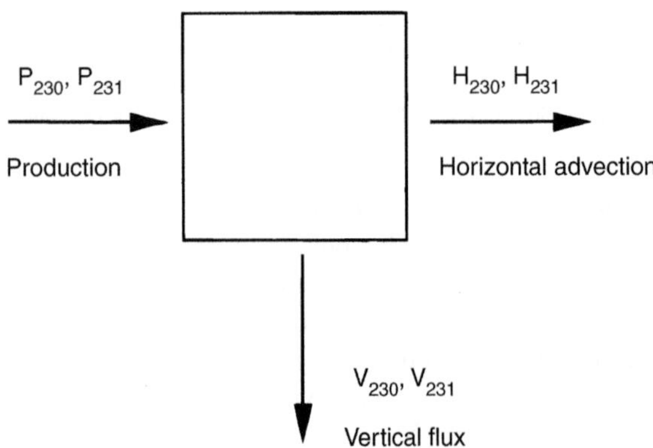

Figure 7 Box model used to derive the vertical flux of ^{230}Th.

transported horizontally by eddy mixing or advection (H_{230}; H_{231}) (**Figure 7**).

$$P_{230} = V_{230} + H_{230} \quad [8]$$

$$P_{231} = V_{231} + H_{231} \quad [9]$$

The calibration of sediment traps is based on the comparison of the intercepted ^{230}Th flux F_{230} with the predicted vertical flux V_{230}. In the original ^{230}Th-based calibration procedure (eqn [4]), H_{230} is neglected and F_{230} is compared directly to the production rate P_{230}. Since P_{230} (eqn [4]) and P_{231} are known and the V_{230}/V_{231} ratio can be measured as the $^{230}Th_{XS}/^{231}Pa_{XS}$ ratio in the sediment trap material, it is sufficient to estimate the H_{230}/H_{231} ratio from water column distributions to solve eqns [8] and [9] for V_{230} and obtain a refined estimate of trapping efficiency.

Lead ^{210}Pb (half-life 22.3 years) is produced from ^{222}Rn, the immediate daughter of ^{226}Ra. ^{222}Rn emanation from land is the major source of ^{210}Pb deposition from the atmosphere (**Figure 2**). ^{222}Rn emanation from surface sea water accounts for only 2% of ^{222}Rn in the atmosphere, but is a significant source in remote areas like the Antarctic Ocean. Below the surface water, seawater ^{226}Ra becomes the most important source.

The high particle reactivity makes ^{210}Pb a tracer for particle flux. This is shown most clearly by the good correlation between the fluxes of ^{210}Pb and of biogenic material in sediment traps. Thus, low ^{210}Pb activities (or ^{210}Pb/^{226}Ra ratios) in surface and deep water, high ^{210}Pb fluxes in traps, and high inventories in the sediment all point to high particle fluxes and consequently high productivity. (Note, however, that in hemipelagic sediments in productive ocean areas the redox cycling of Mn can cause additional near-bottom scavenging of Pb.) Due to this removal on biogenic particles, ^{210}Pb shows strong boundary scavenging similar to ^{231}Pa, with accumulation rates in the sediments of productive (especially eastern) ocean boundaries that are far above local production and atmospheric deposition, whereas the flux to deep-sea sediments in oligotrophic central gyre regions can be very low. Consequently, the flux of ^{210}Pb into and its inventory in surface sediments is highly variable in space. But as long as the (yearly averaged) scavenging conditions do not change with time, the ^{210}Pb flux to the sediment at a certain location can be considered constant, a prerequisite for the interpretation of ^{210}Pb profiles to derive sedimentation and bioturbation rates.

Due to the relatively well-known production and input rates of ^{210}Pb, the scavenging residence time τ_{sc}

can be derived from the distribution of ^{210}Pb in the ocean (compare eqn[3]). τ_{sc} was found to increase from about 2 years in the surface ocean to about 35 years in the deep Atlantic and 150 years in the deep Pacific, a result that is used to understand the behavior of stable lead. This illustrates how ^{210}Pb is a useful analog for stable lead, the study of which is complicated by the extreme risk of contamination (*see* Anthropogenic Trace Elements in the Ocean).

Polonium ^{210}Po, the immediate daughter of ^{210}Pb, is highly particle-reactive. The 138-day half-life of ^{210}Po makes the $^{210}Po/^{210}Pb$ tracer pair a suitable extension to ^{234}Th as tracer for seasonal particle flux from the surface ocean. The non-homogeneous distribution and reactivity of the parent ^{210}Pb implies that ^{210}Po can only be used if concurrent accurate measurements are made of ^{210}Pb.

As a result of the strong affinity for organic material and cytoplasm, ^{210}Po accumulates in the food chain and $^{210}Po/^{210}Pb$ activity ratios from around 3 in phytoplankton to around 12 in zooplankton have been reported. A high excess ^{210}Po activity is therefore indicative of a pathway including zooplankton. The preference of Po for organic material in comparison with Pb and Th, which may adsorb on any surface, can be exploited to distinguish between the fluxes of organic carbon and other components of the particle flux.

Reactive Parent with Mobile Daughter (Table 4)

This type of tracer is used to quantify diffusion, advection, and mixing rates of water masses, for example, the distribution of ^{222}Rn near the seafloor. The parent, ^{226}Ra, has a far higher activity in marine sediments (^{222}Rn emanation rate A^s_{226} of order 100 dpm l^{-1} wet sediment) than in the bottom water (A^w_{226} of order 0.2 dpm l^{-1}). This gradient causes a diffusion of the daughter ^{222}Rn from the sediment into the water column, and a typical vertical distribution as shown in **Figure 8**.

The distribution of ^{222}Rn, A_{222}, can be described by the diffusion-reaction equation:

$$\frac{dA_{222}}{dt} = \lambda(A_{226} - A_{222}) + D\frac{d^2A_{222}}{dz^2} \quad [12]$$

where D is the diffusion coefficient. This yields in steady state:

$$A_{222} = A_{226} - (A^o_{222} - A_{226})e^{-\sqrt{(\lambda/D)}z} \quad [13]$$

A solution valid for the sediment and the water column (if z is defined positive as the distance to the interface), where A^o_{222} signifies the ^{222}Rn activity at the interface (**Figure 8**). In the sediment, this corresponds to an integrated depletion of:

$$I_s = (A^s_{226} - A^0_{222})\sqrt{\frac{D}{\lambda}} \quad [14]$$

maintained by a ^{222}Rn release rate of:

$$F_s = \lambda I_s \quad [15]$$

In the water column, this flux causes an excess activity which is transported upwards by turbulent diffusion (coefficient K). The integrated ^{222}Rn excess in the bottom water is given by:

$$I_w = (A^0_{222} - A^w_{226})\sqrt{\frac{K}{\lambda}} \quad [16]$$

maintained by a supply from the sediment

$$F_w = \lambda I_w \quad [17]$$

Note that mass balance requires that $F_s = F_w$ and that the depletion in the sediment equals the excess in the water column ($I_s = I_w$). The example shows how the

Table 4 Isotope pairs with a particle-reactive parent and a mobile daughter

Mother	*Daughter*	*Half-life*	*Source*	*Oceanographic application*
^{231}Pa	^{227}Ac	22 y	deep-sea sediments	ocean circulation, upwelling
^{232}Th	^{228}Ra	5.8 y	all terrigenous sediments	tracing of shelf water sources, mixing in deep-sea and surface water
^{230}Th	^{226}Ra	1600 y	deep-sea sediments	ocean circulation, ground-water inputs
^{228}Th	^{224}Ra	3.6 d	^{232}Th (sediment) ^{228}Ra (sediment + water column)	mixing in shelf waters and estuaries
^{227}Th	^{223}Ra	11.4 d	^{235}U (sediment) ^{231}Pa (sediment + water column)	mixing in shelf waters and estuaries
^{226}Ra	^{222}Rn	3.8 d	(deep-sea) sediments	mixing in bottom water, air–sea gas exchange, ground-water inputs

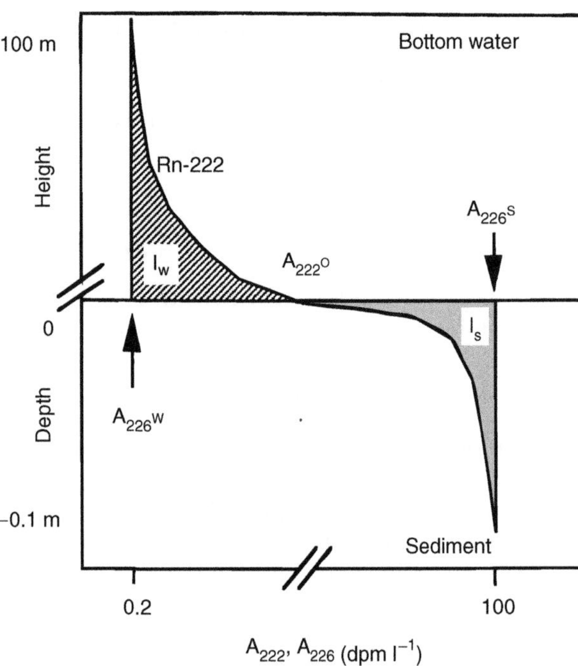

Figure 8 Generalized distribution of ^{226}Ra and ^{222}Rn in surface sediments and bottom water (note change in horizontal and vertical scales). The cumulative ^{222}Rn depletion in the sediment (I_s, shaded) is balanced by the ^{222}Rn excess in the bottom water (I_w, hatched). The vertical extent of the disequilibrium is 3 orders of magnitude larger in the water column than in the sediment, corresponding to the 6 orders of magnitude difference in diffusion coefficient (on the order of $10\,cm^2\,s^{-1}$ in the bottom water as opposed to $10^{-5}\,cm^2\,s^{-1}$ in the sediment).

diffusion coefficient in the sediment and the vertical eddy diffusion coefficient in the bottom water can be derived from measurements of the vertical distribution of this tracer using eqn [13].

Elements in this group are described below.

Actinium ^{227}Ac is produced by the decay of ^{231}Pa. Over 99% of ^{231}Pa produced in the water column resides in the sediment, with highest specific activities in slowly accumulating deep-sea sediments. As actinium is relatively mobile, it is released to the pore water and from there to the overlying water, very similar to the behavior of ^{226}Ra and ^{228}Ra. This results in a strong signal from the deep seafloor on top of a background concentration, which is given by the distribution of ^{231}Pa in the ocean. The nuclide is therefore a potential tracer for vertical mixing and advection (e.g. upwelling) on a decennium timescale.

Radium Radium is relatively mobile and the major source of the isotope ^{226}Ra is the production from the ^{230}Th in the upper layer of sediments. Just like ^{227}Ac, this source is strongest over deep-sea sediments with a slow accumulation rate. The intermediate reactivity of radium (**Table 2**) and its half-life (1600 years) in the order of the ocean mixing time (around 1000 years) explain its distribution as a 'biointermediate' element: ^{226}Ra activities are low in surface waters but never become depleted. They increase with depth and with the age of water masses in the conveyor-belt circulation to reach highest values in the deep north Pacific around 340 dpm m^{-3}. Extensive attempts in the GEOSECS program to use the isotope as a tracer of ocean circulation and water mass age proved unsuccessful as a result of the diffuse nature of the source. Even a normalization with barium, an element that can to a certain extent be regarded as a stable analog of radium, could not sufficiently account for this variation.

Ground waters sometimes have high ^{226}Ra activities. The isotope can then be used to trace groundwater inputs to the coastal ocean.

^{228}Ra is also produced in marine sediments, but in contrast to ^{226}Ra and ^{227}Ac, its parent ^{232}Th is present in the terrigenous fraction of all sediments irrespective of water depth. In combination with the relatively short half-life (5.8 years), this results in a distribution in the open ocean with enhanced concentrations near the seafloor of the deep ocean and near the continental slope, while the activities can accumulate to highest values over extensive continental shelf areas. The vertical distribution in the deep sea (**Figure 9**) resembles the exponential decay that would be expected in a one-dimensional (1-D) model with the source in the seafloor, vertical mixing, and radioactive decay (eqn [13]). This would allow the tracer to be used to derive the vertical mixing rate in the deep ocean. However, it has been shown that even in a large ocean basin like the north-east Atlantic, horizontal mixing is so strong that the vertical distribution is influenced by inputs from slope sediments, making the 1-D model inadequate.

The inputs of shelf waters to the open ocean cause the high activities in the surface waters, illustrated by a typical profile in **Figure 9**. This surface water signal has a strong gradient from the continental shelf to the inner ocean, which has been used to derive horizontal eddy diffusion coefficients in a way analogous to eqn [13]. As the distribution of ^{228}Ra has been shown to vary with time, a steady-state distribution can usually not be assumed, and a repeated sampling is required. Moreover, the horizontal distribution is affected by advection and vertical diffusion, making the interpretation rather complicated. The combination of various radium isotopes (see below), can alleviate some of these problems.

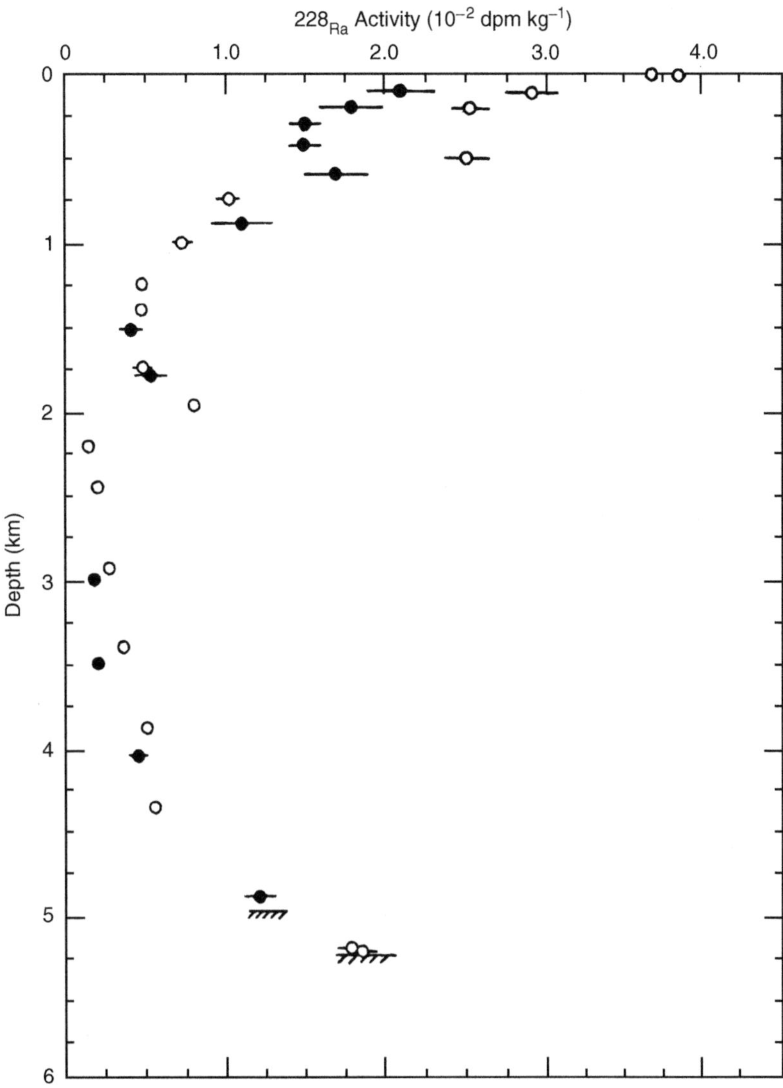

Figure 9 Typical ^{228}Ra activity profiles in the water column in the North Atlantic, showing the enrichment near the seafloor and in the surface water. (Adapted with permission from Ivanovich M and Harmon RS (eds) 1992 *Uranium-series Disequilibrium*, 2nd edn).

In surface current systems away from the continents, ^{228}Ra becomes a powerful tracer for waters that have been in contact with the continental shelf. The ^{228}Ra enrichment in surface waters in the equatorial Pacific point to shelf sources off New Guinea, from where the isotope is carried eastward in the North Equatorial Counter Current. In this plume, the vertical distribution of the isotope has been used to derive vertical mixing rates. A very high accumulation of ^{228}Ra is observed in the transpolar drift in the central Arctic Ocean, a signal derived from the extensive Siberian shelves.

Due to their short half lives, ^{224}Ra (3.4 days) and ^{223}Ra (11.4 days) are interesting only in the immediate vicinity of their sources. In the open ocean they are close to secular equilibrium with their parents ^{228}Th and ^{227}Ac, but in coastal waters these tracers are being developed to study mixing rates. Their distribution is controlled here by sources in the estuary and on the shelf, mixing and decay. Horizontal mixing rates have been obtained from the distribution of ^{223}Ra and ^{224}Ra across the shelf using eqn [13]. As with ^{228}Ra, this procedure is limited to cases where the mixing can be considered to be one-dimensional, but the steady-state requirement is more easily met at these short timescales. The ^{223}Ra/^{224}Ra activity ratio, which decays with a half-life of 5.4 days, yields the age of a water mass since its contact with the source, irrespective of the nature of the mixing process with offshore waters.

Table 5 Summary of the processes that can be investigated using the natural uranium-thorium decay series

Processes	*Tracers*
Particle fluxes	
Boundary scavenging	$^{231}Pa/^{230}Th$, ^{210}Pb
(Paleo) productivity	$^{231}Pa/^{230}Th$, ^{210}Pb
Export production	^{234}Th
Scavenging, trace metal behavior	^{234}Th, ^{230}Th, ^{210}Pb, ^{210}Po
Sediment trap efficiencies	^{234}Th, ^{230}Th, ^{231}Pa
Aggregation rates of particles and colloids	Joint Th isotopes
Sediment redistribution in bottom water	^{230}Th
Resuspension near seafloor	^{234}Th, ^{210}Pb
Water masses	
Shelf interaction/horizontal mixing rates	^{228}Ra, ^{224}Ra, ^{223}Ra
Vertical mixing rates	^{222}Rn, ^{228}Ra, ^{227}Ac
Upwelling	^{227}Ac
Ground-water inputs	^{226}Ra, ^{222}Rn
Gas exchange	
Exchange with atmosphere	^{222}Rn

Radon With its half-life of 3.8 days, the readily soluble gas ^{222}Rn is in secular equilibrium with its parent ^{226}Ra in the interior ocean. At the boundaries of the ocean, however, inputs from sediments and release to the atmosphere create concentration gradients carrying useful kinetic information. The distribution of excess ^{222}Rn near the seafloor is used to quantify vertical diffusion (see above, **Figure** 8) and ground-water inputs; the depletion of ^{222}Rn in surface waters has been used to quantify the air–sea gas exchange rate.

Summary

The accurate clocks provided by the uranium-thorium decay series enable us to extract rate information from the measurement of radioactive disequilibria in the ocean. Among the wide spectrum of available tracers, a mother–daughter pair with appropriate reactivities and half-lives can be found for a multitude of processes related to particle transport, water mass transport and mixing, and gas exchange (**Table** 5).

Nomenclature

A^{diss} dissolved activity
A_P parent activity
A^{part} particulate activity
A_{222} ^{222}Rn activity
A^o_{222} ^{222}Rn activity at sediment–water interface
A_{226} ^{226}Ra activity
A^w_{226} ^{226}Ra activity in the bottom water
A^s_{226} radon emanation rate in sediment
A_{230} ^{230}Th activity in the particles
A_{234} activity of ^{234}U
$^0A_{230}$ decay-corrected ^{230}Th activities
A_{235} ^{235}U activity
A^t_D total daughter activity
C_i concentration of component i
D diffusion coefficient
F_s ^{222}Rn release rate
F_w ^{222}Rn input rate
F_{230} intercepted ^{230}Th flux
$^0F_{230}$ past flux of $^{230}Th_{xs}$ to the seafloor
H_{230} horizontal flux of ^{230}Th
H_{231} horizontal flux of ^{231}Pa
I_s ^{222}Rn depletion in the sediment
I_w ^{222}Rn excess in the bottom water
J sedimentaion rate
K turbulent diffusion coefficient
K_d particle-water partition coefficient
λ decay constant
k_1 adsorption rate constant
k_{-1} desorption rate constant
k_2 coagulation rate constant
k_{-2} disaggregation rate constant
N number of nucleons
P_D production rate
P_{230} production rate of ^{230}Th
$^{231}Pa_{xs}$ excess activity of ^{231}Pa
P_{231} production rate of ^{231}Pa
t time
$t_{1/2}$ half-life
$^{230}Th_{xs}$ excess activity of ^{230}Th
R_i rain rate of component i
V_{230} vertical flux of ^{230}Th
V_{231} vertical flux of ^{231}Pa
z depth
Z atomic number
λ_{230} decay constant of ^{230}Th
λ_{231} decay constant of ^{231}Pa
τ_{sc} scavenging residence time
Ψ focusing factor

See also

Anthropogenic Trace Elements in the Ocean. Hydrothermal Vent Fluids, Chemistry of. Tracers of Ocean Productivity. Uranium-Thorium Series Isotopes in Ocean Profiles.

Further Reading

Bacon MP and Anderson RF (1982) Distribution of thorium isotopes between dissolved and particulate forms in the deep sea. *Journal of Geophysical Research* 87: 2045–2056.

Broecker WS and Peng T-H (1982) *Tracers in the Sea.* Columbia University, New York: Lamont-Doherty Geological Observatory. Eldigio Press.

Cochran JK (1992) The oceanic chemistry of the Uranium and Thorium-series nuclides. In: Ivanovich M and Harmon RS (eds.) *Uranium-series Disequilibrium: Applications to Earth, Marine, and Environmental Sciences,* 2nd edn. pp. 334–395. Oxford: Clarendon Press.

Firestone RB (1998) *Table of Isotopes*, 8th edn. In: Baglin CM (ed) and Chu SYF (CD-ROM ed) New York: Wiley.

Grasshoff K, Kremling K, and Ehrhardt M (1999) *Methods of Seawater Analysis, 3rd edn, pp. 365–397.* Weinheim: Wiley-VCH.

Santschi PH and Honeyman BD (1991) Radioisotopes as tracers for the interactions between trace elements, colloids and particles in natural waters. In: Vernet J-P (ed.) *Trace Metals in the Environment 1. Heavy Metals in the Environment*, pp. 229–246. Amsterdam: Elsevier..

URANIUM-THORIUM SERIES ISOTOPES IN OCEAN PROFILES

S. Krishnaswami, Physical Research Laboratory, Ahmedabad, India

Introduction

Natural radioactivity in the environment originates from two sources. First, primordial radionuclides which were incorporated into the Earth at the time of its formation are still present in it because of their long half-lives. ^{238}U, ^{235}U, ^{232}Th and their decay series (**Figure 1**), ^{40}K, ^{87}Rb and ^{187}Re are examples of this category. Second, cosmic ray-produced isotopes which are generated continuously in the atmosphere and earth's crust through interactions of cosmic rays with their constituents. ^{3}H, ^{14}C and ^{10}Be are some of the isotopes belonging to this group. The distribution of all these isotopes in the oceans is governed by their supply, radioactive decay, water mixing and their biogeochemical reactivity (the tendency to participate in biological and chemical processes) in sea water. Water circulation plays a dominant role in the dispersion of isotopes which are biogeochemically 'passive' (e.g. ^{3}H, Rn), whereas biological uptake and release, solute–particle interactions and chemical scavenging exert major control in the distribution of biogeochemically 'active' elements (e.g. C, Si, Th, Pb, Po). Systematic study of the isotopes of these two groups in the sea can yield important information on the physical and biogeochemical processes occurring in sea water.

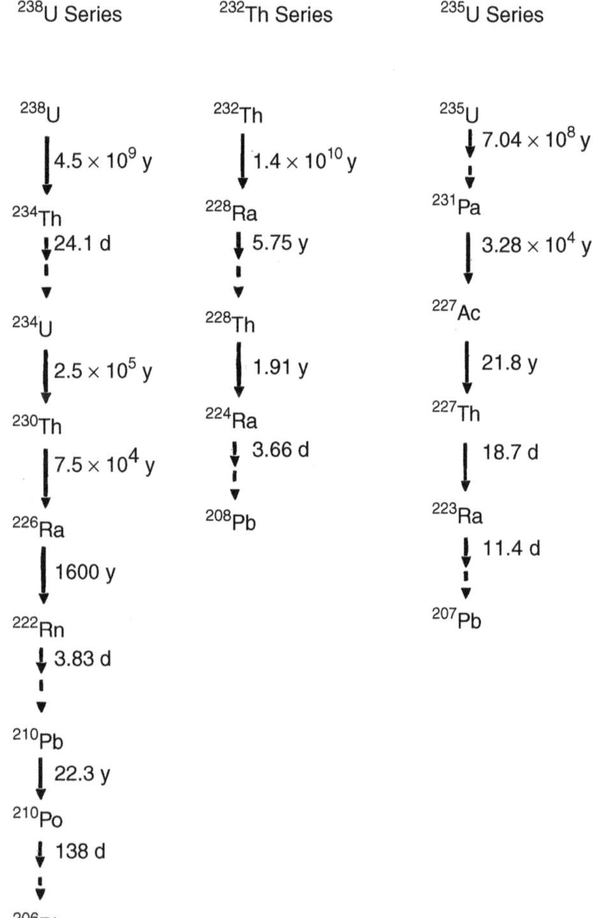

Figure 1 ^{238}U, ^{232}Th and ^{235}U decay series: Only the isotopes of interest in water column process studies are shown.

Supply of U/Th Isotopes to the Sea

These nuclides enter the oceans through three principal pathways.

Fluvial Transport

This is the main supply route for ^{238}U, ^{235}U, ^{234}U and ^{232}Th to the sea. These isotopes are transported both in soluble and suspended phases. Their dissolved concentrations in rivers depend on water chemistry and their geochemical behavior. In rivers, uranium is quite soluble and is transported mainly as uranyl carbonate, $UO_2(CO_3)_3^{-4}$, complex. The dissolved uranium concentration in rivers is generally in the range of 0.1–1.0 μg l^{-1}. During chemical weathering ^{235}U is also released to rivers in the same ^{235}U/^{238}U ratio as their natural abundance (1/137.8). This is unlike that of ^{234}U, a progeny of ^{238}U (**Figure 1**) which is released preferentially to solution due to α-recoil effects. As a result, the ^{234}U/^{238}U activity ratios of river waters are generally in excess of that in the host rock and the secular equilibrium value of 1.0 and often fall in the range of 1.1–1.5.

The concentration of dissolved ^{232}Th in rivers, $\sim$0.01 μg l^{-1} is significantly lower than that of ^{238}U, although their abundances in the upper continental crust are comparable. This is because ^{232}Th (and other Th isotopes) is more resistant to weathering and is highly particle-reactive (the property to be associated with particles) in natural waters and hence is rapidly adsorbed from solution to particles. It is likely that even the reported dissolved ^{232}Th concentrations are upper limits, as recent results, based on smaller volume samples and high sensitivity mass-spectrometric measurements seem to show that dissolved ^{232}Th in rivers is associated with smaller particles ($<0.45\,\mu$m size). Similar to ^{232}Th, the bulk of ^{230}Th and ^{210}Pb is also associated with particles in rivers and hence is transported mainly in particulate form from continents.

^{226}Ra and ^{228}Ra are two other members of the U-Th series (**Figure 1**) for which dissolved concentration data are available for several rivers, these show that they are present at levels of $\sim$0.1 d.p.m. l^{-1}. The available data show that there are significant differences between the abundances of U, Ra isotopes and ^{232}Th in the host rocks and in river waters. The various physicochemical processes occurring during the mobilization and transport of these nuclides contribute to these differences.

Rivers also transport U/Th series nuclides in particulate phase to the sea. These nuclides exist in two forms in the particulate phase, one as a part of their lattice structure and the other as surface coating resulting from their adsorption from solution. Analysis of suspended particulate matter from rivers shows the existence of radioactive disequilibria among the members of the same radioactive decay chain. In general, particulate phases are characterized by ^{234}U/^{238}U, ^{226}Ra/^{230}Th activity ratios <1 and ^{230}Th/^{234}U and ^{210}Pb/^{226}Ra>1, caused by preferential mobilization of U and Ra over Th and Pb isotopes.

Soluble and suspended materials from rivers enter the open ocean through estuaries. The interactions of sea water with the riverine materials can modify the dissolved concentrations of many nuclides and hence their fluxes to the open sea. Studies of U/Th series isotopes in estuaries show that in many cases their distribution is governed by processes in addition to simple mixing of river and sea water. For example, in the case of U there is evidence for both its addition and removal during transit through estuaries. Similarly, many estuaries have ^{226}Ra concentration higher than that expected from water mixing considerations resulting from its desorption from riverine particles and/or its diffusion from estuarine sediments. Estuaries also seem to act as a filter for riverine ^{232}Th.

The behavior of radionuclides in estuaries could be influenced by their association with colloids. Recent studies of uranium in Kalix River show that a significant part is bound to colloids which is removed in the estuaries through flocculation. Similarly, colloids seem to have a significant control on the ^{230}Th–^{232}Th distribution in estuarine waters.

In situ Production

Radioactive decay of dissolved radionuclides in the water column is an important supply mechanism for several U/Th series nuclides. This is the dominant mode of supply for ^{234}Th, ^{228}Th, ^{230}Th, ^{210}Po, ^{210}Pb, and ^{231}Pa. The supply rates of these nuclides to sea water can be precisely determined by measuring the concentrations of their parents. This is unlike the case of nuclides supplied via rivers whose fluxes are relatively more difficult to ascertain because of large spatial and temporal variations in their riverine concentrations and their modifications in estuaries.

Supply at Air–sea and Sediment–water Interfaces

A few of the U/Th nuclides are supplied to the sea via atmospheric deposition and diffusion through sediment pore waters. Decay of ^{222}Rn in the atmosphere to ^{210}Pb and its subsequent removal by wet and dry deposition is an important source of dissolved ^{210}Pb to the sea. As the bulk of the ^{222}Rn in the atmosphere is of continental origin, the flux of ^{210}Pb via this route depends on factors such as distance from land and aerosol residence times. ^{210}Po is also deposited on the sea surface through this source, but its flux is $<10\%$ of that of ^{210}Pb. Leaching of atmospheric dust by sea water can also contribute to nuclide fluxes near the air–sea interface, this mechanism has been suggested as a source for dissolved ^{232}Th.

Diffusion out of sediments forms a significant input for Ra isotopes, ^{227}Ac and ^{222}Rn into overlying water. All these nuclides are produced in sediments through α-decay (**Figure 1**). The recoil associated with their production enhances their mobility from sediments to pore waters from where they diffuse to overlying sea water. Their diffusive fluxes depend on the nature of sediments, their accumulation rates, and the parent concentrations in them. ^{234}U is another isotope for which supply through diffusion from sediments may be important for its oceanic budget.

In addition to diffusion out of sediments, ^{226}Ra and ^{222}Rn are also introduced into bottom waters through vent waters associated with hydrothermal circulation along the spreading ridges. The flux of ^{226}Ra from this source though is comparable to that from rivers; its contribution to the overall ^{226}Ra budget of the oceans is small. This flux, however, can overwhelm ^{226}Ra diffusing out of sediments along the ridges on a local scale.

Distribution in the Oceans

Uranium

^{238}U and ^{235}U are progenitors of a number of particle-reactive nuclides in sea water which find applications in the study of several water column and sedimentary processes. The study of uranium distribution in the sea is therefore essential to a better understanding of the radioactive disequilibrium between ^{238}U–^{234}Th, ^{234}U–^{230}Th, ^{238}U–^{234}U, and ^{235}U–^{231}Pa in sea water. Uranium in sea water is almost entirely in solution as $UO_2(CO_3)_3^{-4}$. Considerable data on its concentration and $^{234}U/^{238}U$ activity ratios are available in the literature, most of which are based on α-spectrometry. These results show that uranium concentration in salinity normalized open ocean sea water (35‰) are the same within experimental uncertainties, $3.3 \pm 0.2\,\mu g\ l^{-1}$. Measurements with highly sensitive mass-spectrometric techniques also yield quite similar values, but with a much better precision (~0.2%) and narrower range, 3.162–3.282 ng g^{-1} 35‰ salinity water (**Figure 2**). The ~3.8% spread even in the recent data is intriguing and is difficult to account for as uranium is expected to be uniformly distributed in the oceans because of its long residence time, ~(2–4) × 10^5 years. More controlled sampling and analysis of uranium in sea water are needed to address this issue better. The mass-spectrometric measurements of uranium have also provided data showing that the $^{238}U/^{235}U$ atomic ratio in sea water is 137.17–138.60, identical within errors to the natural abundance ratio of 137.88.

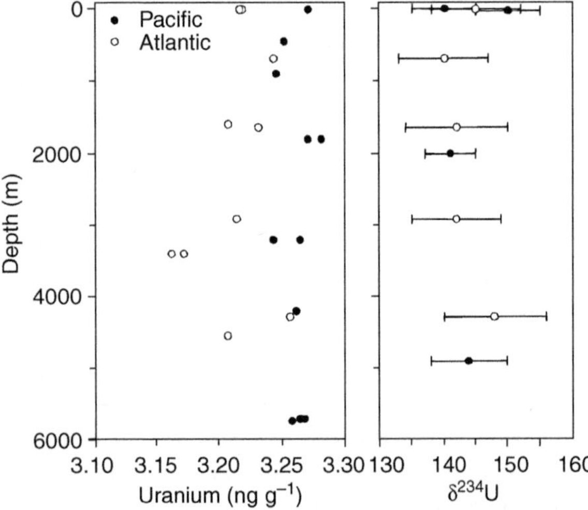

Figure 2 ^{238}U concentration (ng g^{-1} 35‰ salinity water) and $\delta(^{234}U)$ in the Pacific (●) and the Atlantic (○) waters. Data from Chen *et al.* (1986).

Studies of uranium distribution in anoxic marine basins (e.g., the Black Sea and the Saanich Inlet) have been a topic of interest as sediments of such basins are known to be depositories for authigenic uranium. These measurements show that even in these basins, where H_2S is abundant, uranium exists predominantly in +6 state and its scavenging removal from the water column forms only a minor component of its depositional flux in sediments.

The preferential mobilization of ^{234}U during weathering and its supply by diffusion from deep-sea sediments causes its activity in sea water to be in excess of that of ^{238}U. The $^{234}U/^{238}U$ activity ratio of sea water, determined by α-spectrometry, indicates that it is quite homogenous in open ocean waters with a mean value of 1.14 ± 0.02. Mass-spectrometric measurements have confirmed the above observations of ^{234}U excess with a much better precision and have also led to the use of 'δ notation' to describe ^{234}U–^{238}U radioactive disequilibrium.

$$\delta(^{234}U) = [(R_s/R_e) - 1] \times 10^3 \qquad [1]$$

where R_s and R_e are $^{234}U/^{238}U$ atomic ratios in sample and at radioactive equilibrium respectively. The $\delta(^{234}U)$ in the major oceans (**Figure 2**) are same within analytical precision and average 144 ± 2. Coralline $CaCO_3$ and ferromanganese deposits forming from sea water incorporate $^{234}U/^{238}U$ in the ratio of 1.144, the same as that in seawater. The decay of excess ^{234}U in these deposits has been used as a chronometer to determine their ages and growth rates.

Th Isotopes

Among the U/Th series nuclides, the Th isotopes (^{232}Th, ^{230}Th, ^{228}Th, and ^{234}Th), because of their property to attach themselves to particles, are the most extensively used nuclides to investigate particle cycling and deposition in the oceans, processes which have direct relevance to carbon export, solute-particle interactions and particle dynamics. ^{232}Th, ^{230}Th and ^{228}Th are generally measured by α-spectrometry and ^{234}Th by β or γ counting. Highly sensitive mass-spectrometric techniques have now become available for precise measurements of ^{232}Th and ^{230}Th in sea water.

Dissolved ^{232}Th concentration in sea water centers around a few tens of picograms per liter. It is uncertain if the measured ^{232}Th is truly dissolved or is associated with small particles/colloids. Some ^{232}Th profiles show a surface maximum which has been attributed to its release from atmospheric dust.

^{234}Th is continuously produced in sea water from the decay of ^{238}U at a nearly uniform rate of ~2.4 atoms l^{-1} min^{-1}. It has been observed that ^{234}Th activity in the surface ~200 m is generally deficient relative to its parent ^{238}U suggesting its removal by particles, the mechanism of how this is accomplished, however, is not well understood. This result has been attested by several studies (**Figure 3**). The residence time of Th in the upper layers of the ocean is determined based on ^{234}Th–^{238}U disequilibrium and the relation;

$$\tau = \left[\frac{R}{(1-R)}\right]\tau_\lambda \qquad [2]$$

where R is the ^{234}Th/^{238}U activity ratio and τ_λ is the radioactive mean life of ^{234}Th (36.8 days). More complex models considering reversible Th exchange, particle remineralization, aggregation and breakup have also been used to treat the ^{234}Th data which allow better understanding of processes regulating both particle and Th cycling. All these studies demonstrate that Th removal by particle scavenging is ubiquitous in surface water and occurs very rapidly, on timescales of a few days to a few months. Much of this variability in the residence time of Th appears to be dictated by particle concentration, short residence times are typical of coastal and biologically productive areas where particles are generally more abundant. These observations have prompted the use of the ^{234}Th–^{238}U pair as a survey tool to determine the export fluxes of carbon from the euphotic zone. The results, though encouraging, suggest the need for a more rigorous validation of the assumptions and parameters used.

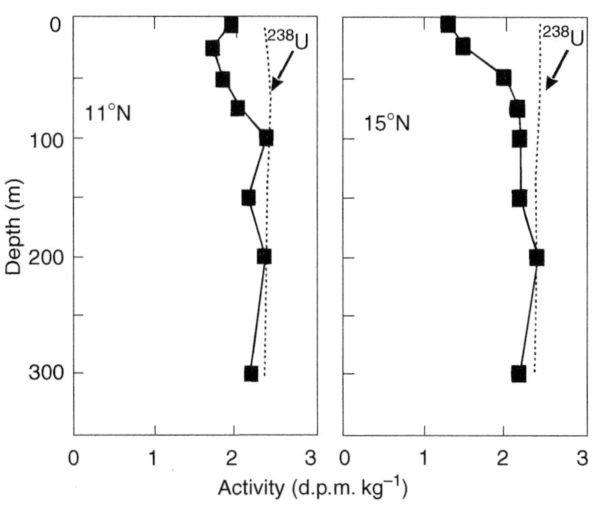

Figure 3 ^{234}Th –^{238}U profiles from the Arabian Sea. Note the clear deficiency of ^{234}Th in the upper layers relative to ^{238}U. (Modified from Sarin *et al.*, 1996.)

^{228}Th activity in the sea exhibits significant lateral and depth variations with higher concentration in the surface and bottom waters and low values in the ocean interior (**Figure 4**). This pattern is governed by the distribution of its parent ^{228}Ra, which determines its production (see section on Ra isotopes). Analogous to ^{234}Th, the distribution of ^{228}Th in the upper layers of the sea is also determined by particle scavenging which causes the ^{228}Th/^{228}Ra activity ratio to be <1, the disequilibrium being more pronounced near coasts where particles are more abundant. The residence time of Th in surface waters calculated from ^{234}Th–^{228}U and ^{228}Th–^{228}Ra pairs yields similar values. Profiles of ^{228}Th activity in bottom waters show a decreasing trend with height above the sediment–water interface. In many of these profiles ^{228}Th is in radioactive equilibrium with ^{228}Ra and in a few others it is deficient. Some of these profile data have been used as a proxy for ^{228}Ra to derive eddy diffusion rates in bottom waters.

Systematic measurements of ^{230}Th activity–depth profiles in soluble and suspended phases of sea water have become available only during the past two decades. ^{230}Th is produced from ^{234}U at a nearly uniform rate of ~2.7 atoms l^{-1} min^{-1}. The dissolved ^{230}Th activity in deep waters of the North Atlantic is ~(5–10) × 10^{-4} d.p.m. l^{-1} and in the North Pacific it is ~2 times higher. In comparison, the particle ^{230}Th concentrations are about an order of magnitude lower (**Figure 5**). These values are far less than would be expected if ^{230}Th were in radioactive

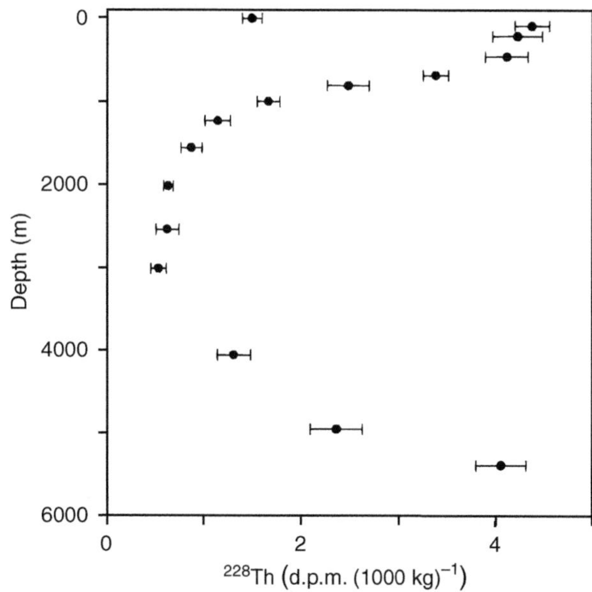

Figure 4 ^{228}Th distribution in the Pacific. The higher activity levels of ^{228}Th in near-surface and near-bottom waters reflect that of its parent ^{228}Ra. Data from Nozaki *et al.* (1981).

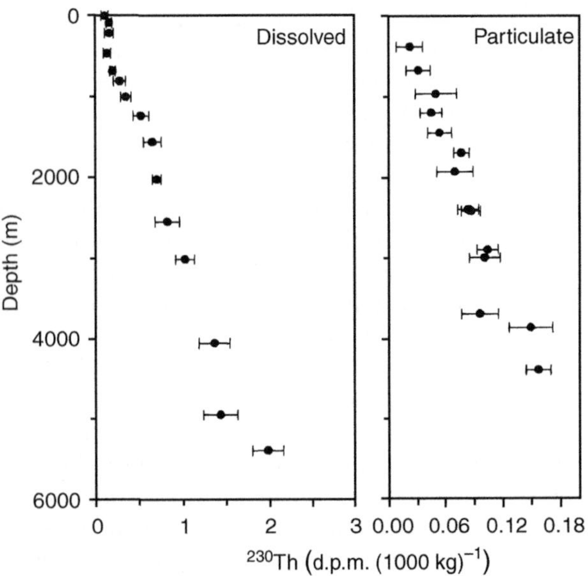

Figure 5 Water-column distributions of dissolved and particulate ^{230}Th. Dissolved ^{230}Th data from the North Pacific (Nozaki *et al.* 1981) and particulate ^{230}Th from the Indian Ocean (Krishnaswami *et al.* 1981). The steady increase in the ^{230}Th activities in both the phases is evident.

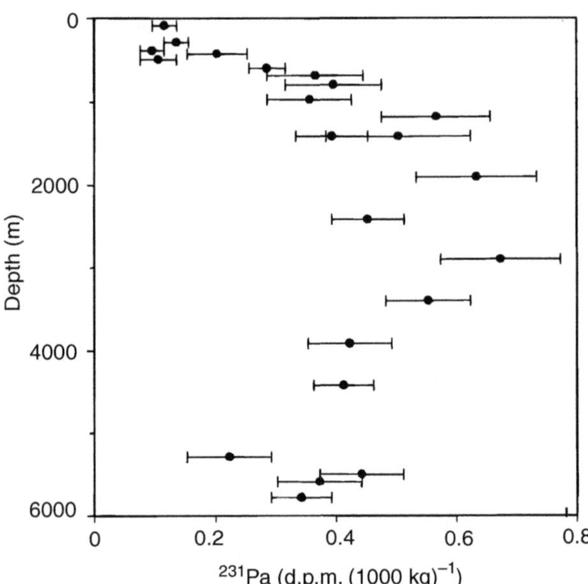

Figure 6 ^{231}Pa distribution in the north-west Pacific. Data from Nozaki and Nakanishi (1985).

equilibrium with ^{234}U, $\sim$2.7 d.p.m. l^{-1}, reinforcing the intense particle-reactive nature of Th isotopes and the occurrence of particle scavenging throughout the seawater column. More importantly, these studies showed that both the soluble and particulate ^{230}Th activities increase steadily with depth (**Figure 5**), an observation which led to the hypothesis of reversible exchange of Th between soluble and suspended pools to explain its distribution. In this model the equations governing the distribution of Th in the two phases are:

Suspended Th:

$$k_1 C = (\lambda + k_2)\bar{C} \quad [3]$$

$$S\frac{d\bar{C}}{dz} + k_1 C - (\lambda + k_2)\bar{C} = 0 \quad [4]$$

Soluble Th:

$$P + k_2\bar{C} = (\lambda + k_1)C \quad [5]$$

where P is the production rate of ^{230}Th, C and $\bar{C}$ are the ^{230}Th concentrations in soluble and suspended phases, k_1 and k_2 are the first order adsorption and desorption rate constants, respectively, and S is the settling velocity of particles. Analysis of Th isotope data using this model suggests that adsorption of Th occurs on timescales of a year or so, whereas its release from particles to solution is much faster, i.e. a few months, and that the particles in sea are at equilibrium with Th in solution. Modified versions of the above model include processes such as particle aggregation and breakup, remineralization and release of Th to solution. The timescales of some of these processes also have been derived from the Th isotope data.

^{231}Pa, ^{210}Po, and ^{210}Pb

These three isotopes share a property with Th, in that all of them are particle reactive. ^{231}Pa is a member of the ^{235}U series (**Figure 1**) and is produced in sea water at a rate of $\sim$0.11 atoms l^{-1} min^{-1}. Analogous to ^{230}Th, ^{231}Pa is also removed from sea water by adsorption onto particles, causing its activity to be quite low and deficient relative to ^{235}U (**Figure 6**). The ^{231}Pa/^{235}U activity ratio in deep waters of the western Pacific is $\sim 5 \times 10^{-3}$. Measurements of ^{230}Th/^{231}Pa ratios in dissolved, suspended, and settling particles have led to a better understanding of the role of their scavenging by vertically settling particles in the open ocean in relation to their removal on continental margins. The dissolved ^{230}Th/^{231}Pa in sea water is $\sim$5, less than the production ratio of $\sim$10.8 and those in suspended and settling particles of $\sim$20, indicating that ^{230}Th is preferentially sequestered onto settling particles. This, coupled with the longer residence time of ^{231}Pa (a few hundred years) compared to ^{230}Th (a few tens of years), has led to the suggestion that ^{231}Pa is laterally transported from open ocean areas to more intense scavenging regimes such as the continental margins, where it is removed. The measurements of settling fluxes of ^{230}Th and ^{231}Pa using sediment traps and $^{230}Th/^{231}Pa$ ratios in sediments from various oceanic regions support this connection.

^{210}Po is supplied to sea almost entirely through its *in situ* production from the decay of ^{210}Pb (**Figure 1**), a minor contribution comes from its atmospheric deposition at the air–sea interface. ^{210}Po is deficient relative to ^{210}Pb in surface waters ($^{210}Po/^{210}Pb$ ~ 0.5, **Figure 7**), the deficiency being more pronounced in biologically productive regimes. The residence time of ^{210}Po in surface waters of the world oceans is in the range of 1 ± 0.5 years. The $^{210}Po/^{210}Pb$ ratio at the base of the euphotic zone falls between 1.0 and 2.0 and often exceeds the secular equilibrium value of unity (**Figure 7**), below ~ 200 m ^{210}Po and ^{210}Pb are in equilibrium. The ^{210}Po profiles in the upper thermocline have been modeled to obtain eddy diffusion coefficients and derive fluxes of nutrients into the euphotic zone from its base. The nature of ^{210}Po profiles in the thermocline and the observation that it is enriched in phyto- and zooplankton indicates that it is a 'nutrient like' element in its behavior and organic matter cycling significantly influences its distribution in the sea. The strong dependence of ^{210}Po removal rate on chlorophyll *a* abundance in various oceans (**Figure 8**) is another proof for the coupling between ^{210}Po and biological activity. In deep and bottom waters, ^{210}Po and ^{210}Pb are generally in equilibrium except in areas of hydrothermal activity where Fe/Mn oxides cause preferential removal of ^{210}Po resulting in $^{210}Po/^{210}Pb$ activity ratio <1.

The studies of ^{210}Pb–^{226}Ra systematics in the oceans have considerably enhanced our understanding of scavenging processes, particularly in the deep sea and the marine geochemistries of lead and its chemical homologues. ^{210}Pb occurs in excess over ^{226}Ra in surface water (**Figure 9**) resulting from its supply from the atmosphere. This excess, however, is less than that would be expected from the known

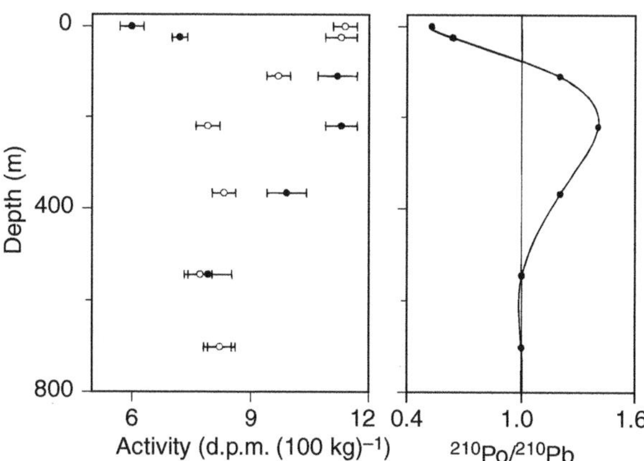

Figure 7 ^{210}Po–^{210}Pb disequilibrium in the Indian Ocean. ^{210}Po (●) is deficient relative to ^{210}Pb (○) near the surface and is in excess at 100–200 m. Data from Cochran *et al.* (1983).

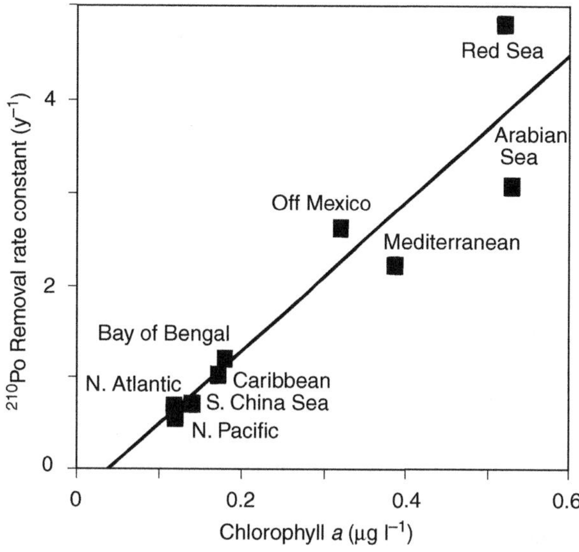

Figure 8 Interrelation between ^{210}Po scavenging rate and chlorophyll *a* concentrations in various oceanic regions. (Modified from Nozaki *et al.*, 1998.)

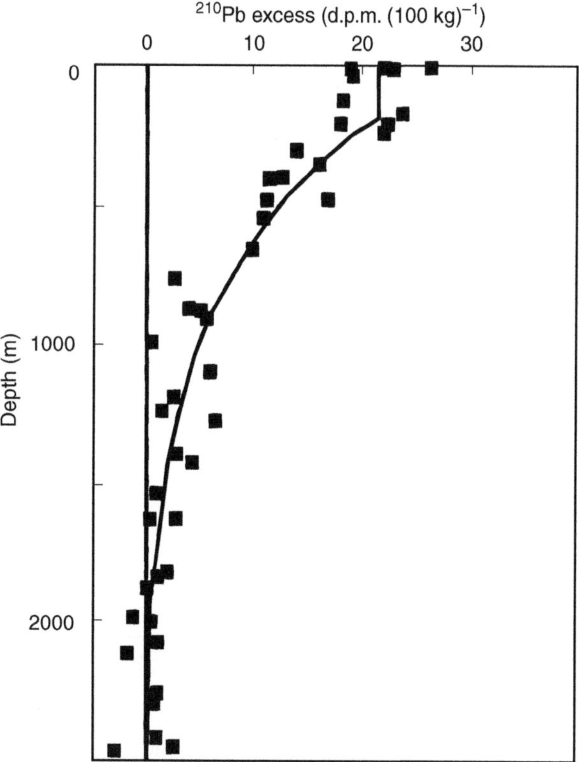

Figure 9 ^{210}Pb excess over ^{226}Ra in the upper thermocline from several stations of the Pacific. This excess results from its atmospheric deposition. (Modified from Nozaki *et al.*, 1980.)

supply rate of ^{210}Pb from the atmosphere if it is removed only through its radioactive decay. This led to the proposal that ^{210}Pb is scavenged from surface to deep waters on timescales of a few years. In many profiles, excess ^{210}Pb shows exponential decrease with depth (**Figure 9**), which has been modeled to derive apparent eddy diffusion coefficients. Measurements of ^{210}Pb–^{226}Ra in the deep sea produced a surprise result in that ^{210}Pb was found to be deficient relative to ^{226}Ra with ^{210}Pb/^{226}Ra of $\sim$0.5 (**Figure 10**). This was unexpected from the available estimates of the residence time of lead in the deep sea, i.e., a few thousands of years, orders of magnitude more than ^{210}Pb mean-life. Numerous subsequent studies have confirmed this deficiency of ^{210}Pb, though with significant variability in its extent and has led to the conclusion that ^{210}Pb is rapidly and continuously removed from the deep sea on timescales of $\sim$50–200 years. The residence time is much shorter, $\sim$2–5 years, in anoxic basins such as the Cariaco Trench and the Black Sea. Two other important findings of these studies are that the extent of ^{210}Pb–^{226}Ra disequilibrium increases from open ocean regimes to continental margins and topographic highs and that there is a significant concentration gradient in ^{210}Pb activity from ocean interior to ocean margins. These results coupled with ^{210}Pb data in suspended and settling particles form the basis for the proposal that ^{210}Pb is removed from deep sea both by vertically settling particles and by lateral transport to margins and subsequent uptake at the sediment–water interface. Processes contributing to enhanced uptake in continental margins are still being debated; adsorption on Fe/Mn oxides formed due to their redox cycling in sediments and the effect of higher particle fluxes, both biogenic and continental, have been suggested. It is the ^{210}Pb studies which brought to light the role of continental margins in sequestering particle-reactive species from the sea, a sink which is now known to be important for other nuclides such as ^{231}Pa and ^{10}Be.

^{222}Rn

The decay of ^{226}Ra in water generates the noble gas ^{222}Rn; both these are in equilibrium in the water column, except near the air–sea and sea–sediment interfaces. ^{222}Rn escapes from sea water to the atmosphere near the air–sea boundary, causing it to be deficient relative to ^{226}Ra, whereas close to the sediment–water interface ^{222}Rn is in excess over ^{226}Ra due to its diffusion out of bottom sediments (**Figure 11**). These disequilibria serve as tracers for mixing rate studies in these boundary layers. In addition, the surface water data have been used to derive ^{222}Rn emanation rates and parameters pertaining to air–sea gas exchange.

^{222}Rn excess in bottom waters decreases with height above the interface, however, the ^{222}Rn activity profiles show distinct variations. Commonly

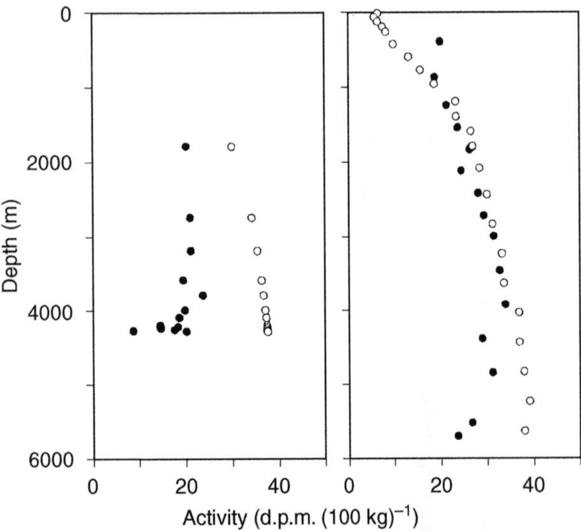

Figure 10 ^{210}Pb (●)–^{226}Ra (○) disequilibrium in sea water. The deficiency of ^{210}Pb in the ocean interior is attributed to its removal by vertically settling particles and at the ocean margins. Data from Craig *et al.* (1973), Chung and Craig (1980) and Nozaki *et al.* (1980).

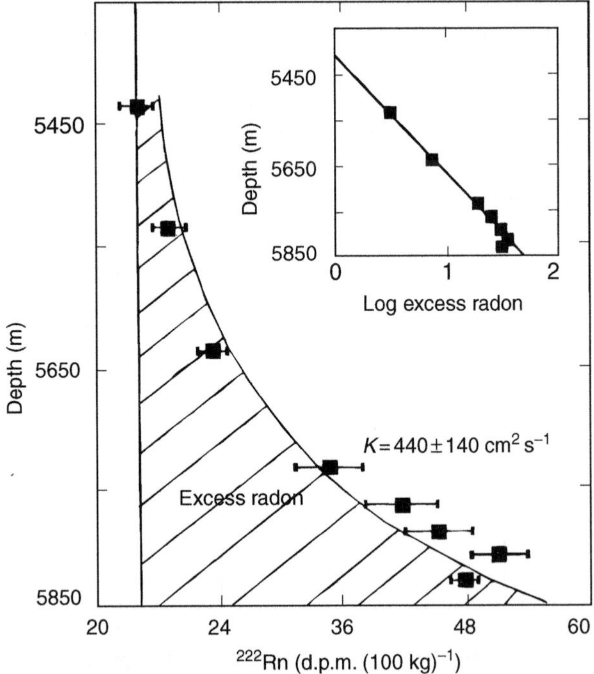

Figure 11 Example of bottom water ^{222}Rn profile in the Atlantic. The calculated vertical eddy diffusion coefficient is also given. (Modified from Sarmiento *et al.*, 1976.)

the ^{222}Rn activity decreases exponentially with height above bottom (**Figure 11**) which allows the determination of eddy diffusion coefficient in these waters. In these cases the ^{222}Rn distribution is assumed to be governed by the equation:

$$K\frac{d^2C}{dz^2} - \lambda C = 0 \qquad [6]$$

where K is the eddy diffusion coefficient and z height above bottom with ^{222}Rn activity C. The values of K calculated from the ^{222}Rn data span about two orders of magnitude, 1–100 cm^2 s^{-1}. Other types of ^{222}Rn profiles include those with a two-layer structure and those without specific trend suggesting that its transport via advection and eddy diffusion along isopycnals and non-steady-state condition also need to be considered while describing its distribution. These studies also demonstrated a strong dependence between ^{222}Rn-based eddy diffusion and the stability of bottom water column.

Ra Isotopes

Ra isotopes, particularly, ^{226}Ra and ^{228}Ra have found extensive applications in water circulation studies. All the Ra isotopes, ^{224}Ra, ^{223}Ra, ^{228}Ra, and ^{226}Ra enter the oceans mainly through diffusion from sediments and by desorption from river particulates and are commonly measured by α and γ counting techniques. ^{224}Ra and ^{223}Ra, because of their very short half-lives (**Figure 1**), are useful for studying mixing processes occurring on timescales of a few days to a few weeks which restricts their utility to regions close to their point of injection such as coastal and estuarine waters (**Figure 12**). The half-life of ^{228}Ra is also short, 5.7 years, and hence its concentration decreases with increasing distance from its source, the sediment–water interface, e.g., from coast to open sea (**Figure 13**) surface waters to ocean interior and height above the ocean floor (**Figure 14**). These distributions have been modeled, by treating them as a balance between eddy diffusion and radioactive decay (eqn [6]), to determine the rates of lateral and vertical mixing occurring on timescales of 1–30 years in the thermocline and near bottom waters.

^{226}Ra is the longest lived among the Ra isotopes, with a half-life comparable to that of deep ocean mixing times. The potential of ^{226}Ra as a tracer to study large-scale ocean mixing was exploited using a one-dimensional vertical advection–diffusion model to describe its distribution in the water column. Subsequent studies brought to light the importance of biological uptake and cycling in influencing ^{226}Ra distribution, processes which were later included in the ^{226}Ra model.

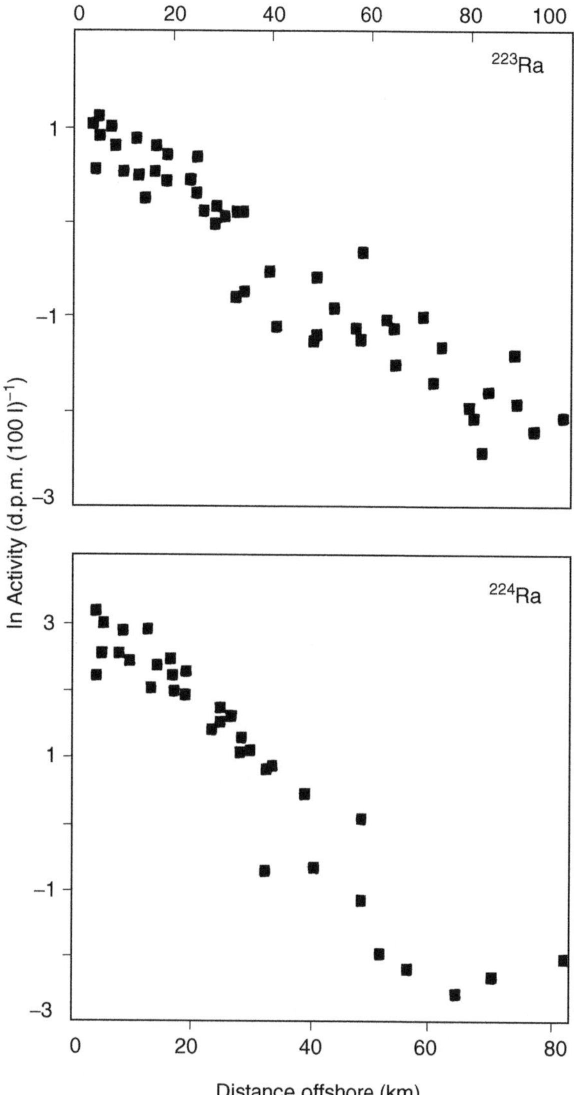

Figure 12 Distributions of ^{223}Ra and ^{224}Ra activities as a function of distance off-shore from Winyah Bay off Carolina Coast, USA. These profiles have been modeled to yield horizontal eddy diffusion coefficients. (Modified from Moore, 1999.)

Figure 15 shows typical profiles of ^{226}Ra in the oceans. Its concentration in surface waters falls in the range of 0.07 ± 0.01 d.p.m. l^{-1} which steadily increases with depth such that its abundance in the deep waters of the Pacific>Indian>Atlantic (**Figure 15**). ^{226}Ra concentration in the North Pacific bottom water is $\sim$0.4 d.p.m. l^{-1}, some of the highest in the world's oceans.

^{226}Ra distribution in the ocean has been modeled to derive eddy diffusivities and advection rates taking into consideration its input by diffusion from sediments, loss by radioactive decay, and dispersion

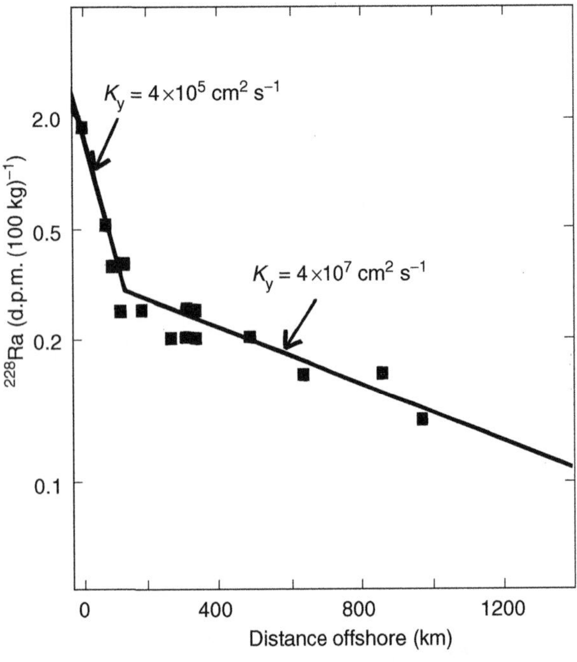

Figure 13 ^{228}Ra distribution as a function of distance from the coast off California. Values of horizontal eddy diffusion coefficient can be derived from these profiles. Note that ^{228}Ra mixes farther into the open sea than ^{223}Ra and ^{224}Ra (**Figure 12**) because of its longer half-life. (Modified from Cochran, 1992.)

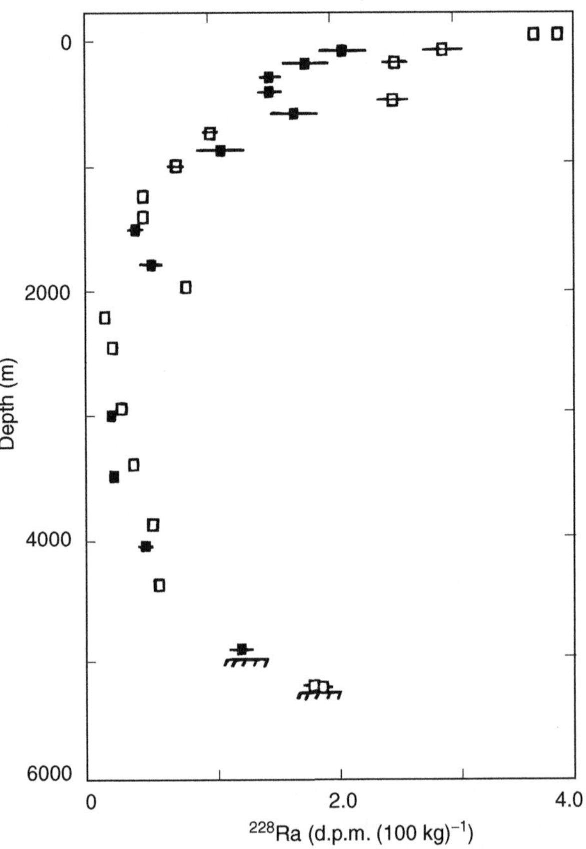

Figure 14 Example of ^{228}Ra depth profile in the North Atlantic. The high concentrations near the surface and near the sediment–water interface is due to its supply by diffusion from sediments. Lateral transport also plays an important role in determining surface water concentrations. (Modified from Cochran, 1992.)

through water mixing, particulate scavenging and regeneration. It has been shown that particulate scavenging and regeneration plays a crucial role in contributing to the progressive increase in ^{226}Ra deep water concentration from the Atlantic to the Pacific. Attempts to learn more about particulate transport processes in influencing ^{226}Ra distribution using Ba as its stable analogue and Ra–Ba and Ra–Si correlations have met with limited success and have clearly brought out the presence of more ^{226}Ra in deep waters than expected from their Ba content (**Figure 16**). This 'excess' is the nascent ^{226}Ra diffusing out of deep sea sediments and which is yet to take part in particulate scavenging and recycling. Such excesses are quite significant and are easily discernible in the bottom waters of the eastern Pacific.

^{227}Ac

The first measurement of ^{227}Ac in sea water was only reported in the mid-1980s. These results showed that its concentration increases steadily from surface to bottom water (**Figure 17**) and that its activity in ocean interior and deep waters is considerably in excess of its parent ^{231}Pa (**Figure 17**). The diffusion of ^{227}Ac out of bottom sediments is the source of its

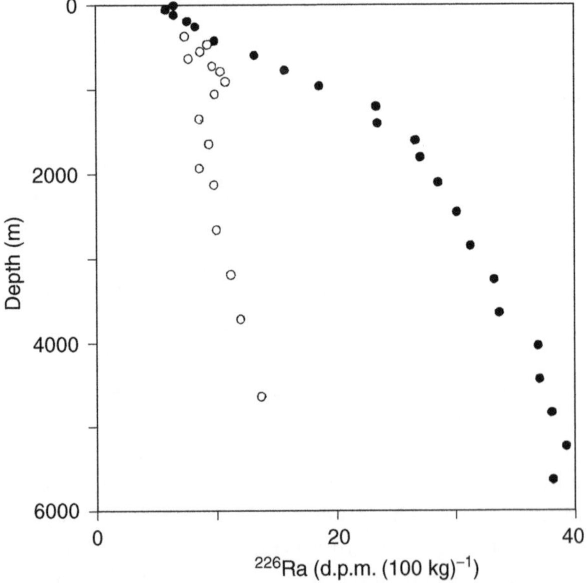

Figure 15 Typical distributions of ^{226}Ra in the water column of the Pacific (●) and Atlantic (○) oceans. Data from Broecker *et al.* (1976) and Chung and Craig (1980).

excess in bottom waters, analogous to those of Ra isotopes. Measurements of ^{227}Ac in pore waters have confirmed this hypothesis. ^{227}Ac distribution can serve as an additional tracer in studies of water mixing processes occurring on decadal timescales, thus complementing the ^{228}Ra applications.

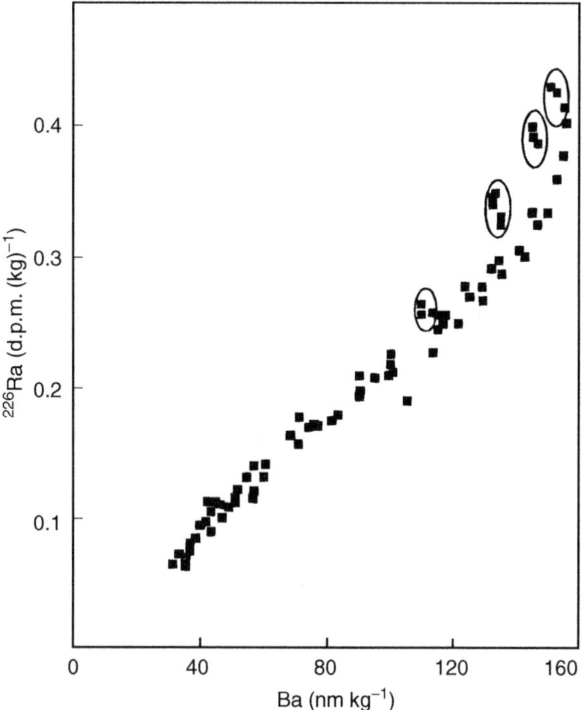

Figure 16 Ra–Ba correlation in the north-east Pacific. The presence of 'excess Ra' (enclosed in ellipses) is clearly discernible in bottom waters. (Modified from Ku *et al.*, 1980.)

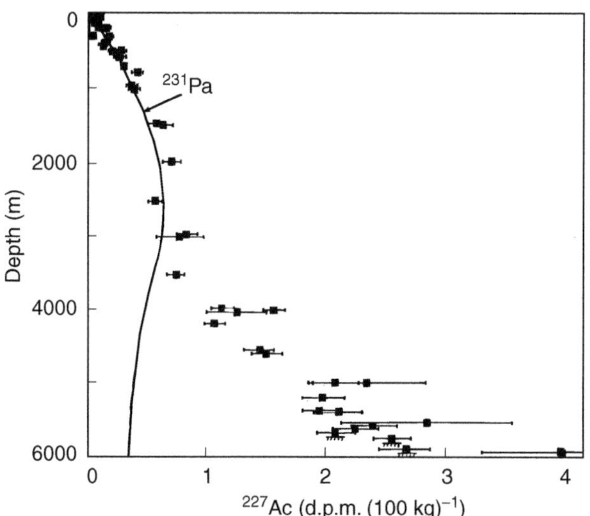

Figure 17 ^{227}Ac profile in the Pacific Ocean. Its large excess over ^{231}Pa is due to its diffusion out of sediments. (Modified from Nozaki, 1984.)

Summary

The distribution of U/Th series nuclides in the sea is regulated by physical and biogeochemical processes occurring in the water column and at the air–sea and sea–sediment interfaces. These processes often create radioactive disequilibria among the members of the U/Th decay chains. These disequilibria serve as powerful 'tools' to examine and quantify several processes in the sea, such as water circulation on various timescales (days to thousands of years), particle-scavenging, solute–particle interactions, particle dynamics and transformation and air–sea gas exchange. The understanding of these processes and elucidation of their timescales have direct relevance to studies such as dispersal of chemical species in the sea, contaminant transport and sites of their removal and particulate carbon fluxes through the water column. Recent advances in sampling and measurements of U/Th series nuclides have considerably enhanced the scope of their application in the study of water column processes.

See also

River Inputs. Uranium-Thorium Decay Series in the Oceans Overview.

Further Reading

Anderson RF, Bacon MP, and Brewer PG (1983) Removal of Th-230 and Pa-231 from the open ocean. *Earth and Planetary Science Letters* 62: 7–23.

Anderson PS, Wasserburg GJ, Chen JH, Papanastassiou DA, and Ingri J (1995) ^{238}U–^{234}U and ^{232}Th–^{230}Th in the Baltic sea and in river water. *Earth and Planetary Science Letters* 130: 218–234.

Bacon MP and Anderson RF (1982) Distribution of thorium isotopes between dissolved and particulate forms in the deep sea. *Journal of Geophysical Research* 87: 2045–2056.

Bhat SG, Krishnaswami S, Lal D, and Rama and Moore WS (1969) Thorium-234/Uranium-238 ratios in the ocean. *Earth and Planetary Science Letters* 5: 483–491.

Broecker WS, Goddard J, and Sarmiento J (1976) The distribution of ^{226}Ra in the Atlantic Ocean. *Earth and Planetary Science Letters* 32: 220–235.

Broecker WS and Peng JH (1982) *Tracers in the Sea*. New York: Eldigio Press, Lamont-Doherty Geological Observatory.

Chen JH, Edwards RL, and Wesserburg GJ (1986) ^{238}U, ^{234}U and ^{232}Th in sea water. *Earth and Planetary Science Letters* 80: 241–251.

Chen JH, Edwards RL, and Wasserburg GJ (1992) Mass spectrometry and application to uranium series disequilibrium. In: Ivanovich M and Harmon RS (eds.)

Uranium Series Disequilibrium: Applications to Earth, Marine and Environmental Sciences, 2nd edn pp. 174–206. Oxford: Clarenden Press.

Chung Y and Craig H (1980) ^{226}Ra in the Pacific Ocean. *Earth and Planetary Science Letters* 49: 267–292.

Coale KH and Bruland KW (1985) Th-234 : U-238 disequilibria within the California Current. *Limnology and Oceanography* 30: 22–33.

Cochran JK (1992) The oceanic chemistry of the uranium and thorium series nuclides. In: Ivanovich M and Harmon RS (eds.) *Uranium Series Disequilibrium Applications to Earth, Marine and Environmental Sciences,* 2nd edn, pp. 334–395. Oxford: Clarenden Press.

Cochran JK, Bacon MP, Krishnaswami S, and Turekian KK (1983) ^{210}Po and ^{210}Pb distribution in the central and eastern Indian Ocean. *Earth and Planetary Science Letters* 65: 433–452.

Craig H, Krishnaswami S, and Somayajulu BLK (1973) ^{210}Pb–^{226}Ra radioactive disequilibrium in the deep sea. *Earth and Planetary Science Letters* 17: 295.

Dunne JP, Murray JW, Young J, Balistrieri LS, and Bishop J (1997) ^{234}Th and particle cycling in the central equatorial Pacific. *Deep Sea Research II* 44: 2049–2083.

Krishnaswami S (1999) Thorium: element and geochemistry. In: Marshall CP and Fairbridge RW (eds.) *Encyclopedia of Geochemistry,* pp. 630–635. Dordrecht: Kluwer Academic.

Krishnaswami S, Sarin MM, and Somayajulu BLK (1981) Chemical and radiochemical investigations of surface and deep particles of the Indian ocean. *Earth and Planetary Science Letters* 54: 81–96.

Krishnaswami S and Turekian KK (1982) U-238, Ra-226 and Pb-210 in some vent waters of the Galapagos spreading center. *Geophysical Research Letters* 9: 827–830.

Ku TL, Huh CA, and Chen PS (1980) Meridional distribution of ^{226}Ra in the eastern Pacific along GEOSECS cruise track. *Earth and Planetary Science Letters* 49: 293–308.

Ku TL, Knauss KG, and Mathieu GG (1977) Uranium in open ocean: concentration and isotopic composition. *Deep Sea Research* 24: 1005–1017.

Moore WS (1992) Radionuclides of the uranium and thorium decay series in the estuarine environment, In: Ivanovich M and Harmon RS (eds.) *Uranium Series Disequilibrium. Applications to Earth, Marine and Environmental Sciences,* 2nd edn. pp. 334–395. Oxford: Clarenden Press.

Moore WS (1999) Application of ^{226}Ra, ^{228}Ra, ^{223}Ra and ^{224}Ra in coastal waters to assessing coastal mixing rates and ground water discharge to the oceans. *Proceedings of the Indian Academy of Sciences (Earth and Planetary Sciences)* 107: 109–116.

Nozaki Y (1984) Excess, Ac-227 in deep ocean water. *Nature* 310: 486–488.

Nozaki Y, Dobashi F, Kato Y, and Yamamoto Y (1998) Distribution of Ra isotopes and the ^{210}Pb and ^{210}Po balance in surface sea waters of the mid-northern hemisphere. *Deep Sea Research I* 45: 1263–1284.

Nozaki Y and Nakanishi T (1985) ^{231}Pa and ^{230}Th profiles in the open ocean water column. *Deep Sea Research* 32: 1209–1220.

Nozaki Y, Turekian KK, and Von Damm K (1980) ^{210}Pb in GEOSECS water profiles from the north Pacific. *Earth and Planetary Sciences Letters* 49: 393–400.

Nozaki Y, Horibe Y, and Tsubota H (1981) The water column distributions of thorium isotopes in the western north Pacific. *Earth and Planetary Sciences Letters* 54: 203–216.

Roy-Barman M, Chen JH, and Wasserburg GJ (1996) ^{230}Th–^{232}Th systematics in the central Pacific Ocean: the sources and fate of thorium. *Earth and Planetary Science Letters* 139: 351–363.

Sarin MM, Rengarajan R, and Ramaswamy V (1996) ^{234}Th scavenging and particle export fluxes from upper 100 m of the Arabian Sea. *Current Science* 71: 888–893.

Sarmiento JL, Feely HW, Moore WS, Bainbridge AE, and Broecker WS (1976) The relationship between vertical eddy diffusion and buoyancy gradient in the deep sea. *Earth and Planetary Letters* 32: 357–370.

COSMOGENIC ISOTOPES

D. Lal, Scripps Institute of Oceanography, University of California San Diego, La Jolla, CA, USA

Introduction

In different settings, spanning from the extraterrestrial to the terrestrial, naturally occurring nuclides offer unique possibilities for being deployed as tracers for studying a great variety of physical, chemical, and biological processes, occurring over a wide range of timescales. This article discusses the continuous production of several stable and radioactive isotopes as a result of nuclear reactions of cosmic ray particles in the Earth's atmosphere and the hydrosphere, and their potential usefulness as tracers for studying oceanic processes. The great merit of cosmic ray produced (cosmogenic) isotopes as tracers lies in the fact that their source functions in the different geospheres can be determined, and that several nuclides with a wide range of half-lives and chemical properties are available.

Cosmic radiation, which consists of energetic H, He and heavier nuclei, with kinetic energies much greater than tens of mega electronvolts (MeV), (with particles of energies much beyond 10^{10} MeV), produce a great variety of nuclides by their interactions with target nuclei in the atmosphere, hydrosphere and the lithosphere. The predominant cosmic ray interaction is fragmentation of the target nuclei by primary and secondary particles of the cosmic radiation. Some nuclides are produced following the capture of thermal (very slowly moving) neutrons by target nuclei, which are abundant in the secondary cosmic radiation as a result of slowing down of fast neutrons emitted in energetic cosmic ray-produced nuclear reactions.

Radiocarbon, ^{14}C, was the first cosmic ray-produced isotope to be discovered in 1947 in sewage methane. Soon thereafter it was applied for archaeological/anthropological dating. This discovery was a milestone in the use of cosmic ray-produced (cosmogenic) isotopic changes as a tool for learning about planetary sciences. It laid the foundations of the field of cosmic ray geophysics/geochemistry. Subsequently, in the early 1960s, about 25 cosmogenic radionuclides produced in the earth's atmosphere, with half-lives ranging from ~1 h to millions of years were detected. The driving force for the studies of cosmic ray-produced nuclides was the realization that if they could be detected in different dynamic reservoirs of the geospheres, they could be used as tracers to obtain important information about the timescales involved in the transport of materials through the atmosphere to the hydrosphere, oceans, and the cryosphere, and that in some cases they could be used as clocks to introduce timescales into the diverse proxy records of earth's climate. Oceans are central to the dynamic interplay between the dynamic reservoirs, and therefore considerable emphasis has been placed on understanding the nature of the principal mixing/transfer processes, of the marine biogeochemical cycles, and of the large-scale ocean circulation. All oceanic investigations, in one way or the other, are linked to the central question of what processes control the earth's climate. Geochemical tracers serve as tools to understand these processes and their rates.

As mentioned above, the field of cosmic ray produced (cosmogenic) tracers caught roots in 1947, with the discovery of ^{14}C. It grew rapidly thereafter in the late 1950s/early 1960s, and to date it is still one of the frontier areas in modern geochemistry. There are two reasons for this sustained hold and value of the cosmogenic tracers: continued development of new and powerful techniques for measuring their distribution in natural settings at very low concentrations, and the emergence of new biogeochemical questions which crop up as our understanding of the terrestrial climate system improves. However, there are often no other (suitable) tracers available to study the short- and long-term behavior of oceans on large space scales. This article considers the essentials of the cosmogenic tracers, their potentials, and how new advances continue to keep this field growing.

Terrestrial Cosmogenic Isotopes and their Production Rates

Most of the cosmic ray energy (>98%) is dissipated in the earth's atmosphere in the nuclear reactions they produce. The atmospheric column represents about 13 mean free paths for nuclear interactions of fast protons and neutrons. After traversal through the atmosphere, the secondary particles of the cosmic radiation continue to produce nuclear reactions with the surficial terrestrial reservoirs: the hydrosphere, the cryosphere, and the lithosphere, but at a much lower rate. The techniques of the 1950s for studying the cosmogenic nuclides were barely adequate to study the isotopes produced in the atmosphere. However, because of their value in understanding geophysical and geochemical processes, they were studied fairly extensively until the 1970s. A number of technical developments in the 1970s have made it easier to study the 'atmospheric' cosmogenic isotopes in the ocean environs, and even the cosmogenic isotopes produced *in situ* in terrestrial materials, including the hydrosphere and the cryosphere.

The cosmogenic tracer-based information obtained from studies of the lithosphere and the cryosphere is of great value in interpreting the oceanic records. **Table 1** lists a suite of isotopes, which serve (or should serve) as useful tracers in geophysical and geochemical studies. This list includes nominally potentially useful 20 nuclides, having half-lives exceeding two weeks, which are produced in the atmosphere, hydrosphere and in the lithosphere. The target elements from which they are produced in the earth's atmosphere, and from principal elements in surficial matter are also listed in **Table 1**.

This article is concerned primarily with nuclides that are useful as tracers in oceanography, and therefore shorter-lived nuclides have been excluded from **Table 1**. Since cosmic ray intensity is appreciably reduced at sea level due to nuclear interactions in the atmosphere isotope production rates (per gram target element per second), in surficial materials, are appreciably smaller than in the atmosphere. Therefore, nuclides, which can be produced in nuclear interactions with the major elements present in the atmosphere, N and O, have their principal source in the atmosphere. The next most abundant element, Ar, in the atmosphere occurs at an abundance of only 0.93% (v/v) in the atmosphere. Nuclear interactions with surficial materials can therefore be an important source of some of the nuclides produced from Ar, and for those isotopes which have mass numbers greater than 40, since ^{40}Ar is the most abundant nuclear isotope in the atmosphere. Permanent constituent gases heavier than Ar have very low abundances in the atmosphere. The abundances of the next heavier gases, Kr and Xe, are ~1 and 0.1 ppm (v/v), respectively.

The production rates of several cosmogenic isotopes in the earth's atmosphere are given in **Table 2**, along with their estimated global inventories. Some of the cosmogenic isotopes are also produced directly, *in situ*, in the upper layers of the oceans. The source strengths of cosmic ray-produced nuclei in the oceans, due to their production in the atmosphere, and direct production in the ocean water, are given in **Table 3**. Atmospheric production is the dominant source of all nuclides in **Table 3**, except for ^{36}Cl, where its *in situ* oceanic production exceeds the atmospheric production by about 50%. In the case of ^{37}Ar the two source strengths are comparable, and in the cases of ^{32}P and ^{33}P *in situ* production in the oceans is an order of magnitude lower.

Table 1 Potentially useful cosmogenic nuclides (arranged in order of mass numbers) with half-lives exceeding 2 weeks, produced in the Earth's atmosphere and in surficial materials

Nuclide(s)	*Half-life*	*Main targets*[a]
^{3}H	12.3 y	O, Mg, Si, Fe (N, O)
^{3}He, ^{4}He	S	O, Mg, Si, Fe (N, O)
^{7}Be	53 d	O, Mg, Si, Fe (N, O)
^{10}Be	1.5×10^{6} y	O, Mg, Si, Fe (N, O)
^{14}C	5730 y	O, Mg, Si, Fe (N)
^{20}Ne, ^{21}Ne, ^{22}Ne	S	Mg, Al, Si, Fe
^{22}Na	2.6 y	Mg, Al, Si, Fe (Ar)
^{26}Al	7.1×10^{5} y	Si, Al, Fe (Ar)
^{32}Si	~150 y	(Ar)
^{32}P	14.3 d	(Ar)
^{33}P	25.3 d	(Ar)
^{35}S	87 d	Fe, Ca, K, Cl (Ar)
^{36}Cl	3.0×10^{5} y	Fe, Ca, K, Cl (Ar)
^{37}Ar	35 d	Fe, Ca, K (Ar)
^{39}Ar	268 y	Fe, Ca, K (Ar)
^{41}Ca	1.0×10^{5} y	Ca, Fe (Kr)
^{53}Mn	3.7×10^{6} y	Fe (Kr)
^{54}Mn	312 d	Fe (Kr)
^{59}Ni	7.6×10^{4} y	Ni, Fe (Kr)
^{60}Fe	1.5×10^{6} y	Ni (Kr)
^{81}Kr	2.3×10^{5} y	Rb, Sr, Zr (Kr)
^{129}I	1.6×10^{7} y	Te, Ba, La, Ce (Xe)

[a]Elements from which most production occurs; those in parentheses give the main targets from which they are produced in the Earth's atmosphere.
s, stable.

Isotopes produced in the earth's atmosphere (**Table 2**), are introduced in the upper ocean:

1. in wet precipitation, in the case of isotopes which are removed directly (^{3}H), or as attached to aerosols (^{7}Be, ^{10}Be, ^{22}Na, ^{26}Al, ^{32,33}P, ^{32}Si, ^{35}S, ^{36}Cl);
2. by air–sea exchange of ^{14}C (as $^{14}CO_2$), and of isotopes of rare gases (^{3}He, ^{37}Ar, ^{39}Ar and ^{81}Kr).

Besides direct *in situ* production of isotopes in the ocean waters (**Table 2**), some isotopes are also introduced to the oceans with river runoff as a result of weathering of the crustal materials in which they are produced, e.g., ^{10}Be, ^{26}Al, ^{41}Ca, and ^{53}Mn (**Table 2**). To date, those introduced by weathering of crustal materials have not been either studied or identified as important, and estimates of the strength of this source, are not presented here.

Table 2 Production rates of several isotopes in the Earth's atmosphere; arranged in order of decreasing half-lives[a]

Isotope	*Half-life*	*Production rate (atoms cm^{-2}s^{-1})*		*Global inventory*
		Troposphere	*Total atmosphere*	
^{3}He	Stable	6.7×10^{-2}	0.2	3.2×10^{3} tons[b]
^{10}Be	1.5×10^{6} y	1.5×10^{-2}	4.5×10^{-2}	260 tons
^{26}Al	7.1×10^{5} y	3.8×10^{-5}	1.4×10^{-4}	1.1 tons
^{81}Kr[c]	2.3×10^{5} y	5.2×10^{-7}	1.2×10^{-6}	8.5 kg
^{36}Cl	3.0×10^{5} y	4×10^{-4}	1.1×10^{-3}	15 tons[e]
^{14}C	5730 y	1.1	2.5	75 tons
^{39}Ar[d]	268 y	4.5×10^{-3}	1.3×10^{-2}	52 kg
^{32}Si	~150 y	5.4×10^{-5}	1.6×10^{-4}	0.3 kg
^{3}H	12.3 y	8.4×10^{-2}	0.25	3.5 kg
^{22}Na	2.6 y	2.4×10^{-5}	8.6×10^{-5}	1.9 g
^{35}S	87 d	4.9×10^{-4}	1.4×10^{-3}	4.5 g
^{7}Be	53 d	2.7×10^{-2}	8.1×10^{-2}	3.2 g
^{37}Ar	35 d	2.8×10^{-4}	8.3×10^{-4}	1.1 g
^{33}P	25.3 d	2.2×10^{-4}	6.8×10^{-4}	0.6 g
^{32}P	14.3 d	2.7×10^{-4}	8.1×10^{-4}	0.4 g

[a]Based on Lal and Peters (1967).
[b]The inventory of this stable nuclide is based on its atmospheric inventory, which includes an appreciable contribution from crustal degassing of ^{3}He.
[c]Based on atmospheric ^{81}Kr/Kr ratio of $(5.2\pm0.4)\times10^{-13}$.
[d]Based on atmospheric ^{39}Ar/Ar ratio of (0.107 ± 0.004) d.p.m. l^{-1} Ar (STP).
[e]Includes a rough estimate of ^{36}Cl produced by the capture of neutrons at the earth's surface.

Table 3 Source functions of cosmogenic nuclides in the oceans, for nuclides of half-lives >10 days, arranged in order of increasing half-lives

Nuclide	*Half-life*	*Principal target element (s)*		*Global average surface injection rate*[a]	*Integrated* in situ *oceanic production rate*
		Atmosphere	*In ocean water*	*(atoms cm^{-2}min^{-1})*	*(atoms cm^{-2}min^{-1})*
^{32}P	14.3 d	Ar	Cl, S, K	5.82×10^{-3}	7.6×10^{-4}
^{33}P	25.3 d	Ar	Cl, S, K	6.93×10^{-3}	2.9×10^{-4}
^{37}Ar	35.0 d	Ar	K, Ca	9.10×10^{-6}	8.1×10^{-6}
^{7}Be	53.3 d	N,O	O	1.27	6.0×10^{-3}
^{35}S	87.4 d	Ar	Cl, Ca, K	2.84×10^{-2}	5.1×10^{-4}
^{22}Na	2.6 y	Ar	Na	3.75×10^{-3}	3.9×10^{-4}
^{3}H	12.3 y	N,O	O, ^{2}H	1.39×10^{1}	1.2×10^{-2}
^{32}Si	~150 y	Ar	S, Ca	9.60×10^{-3}	2.5×10^{-5}
^{39}Ar	269 y	Ar	K, Ca	2.00×10^{-1}	1.2×10^{-5}
^{14}C	5,730 y	N,O	O	1.20×10^{2}	9.0×10^{-3}
^{41}Ca[a]	1.0×10^{5} y	—	Ca	—	2.4×10^{-5} *(n)*[a]
^{81}Kr	2.1×10^{5} y	Kr	Sr	2.30×10^{-5}	1.9×10^{-8} *(n)*[c]
^{36}Cl[b]	3.0×10^{5} y	Ar	Cl	6.60×10^{-2}	1.06×10^{-1} *(n)*[c]
^{26}Al	7.2×10^{5} y	Ar	S, K, Ca	8.40×10^{-3}	6.8×10^{-6}
^{10}Be	1.6×10^{6} y	N,O	O	2.70	1.8×10^{-3}

[a]Flux to oceans from rivers should be included to take into account production in rocks and soil by ^{40}Ca(n,γ) ^{41}Ca reaction; this estimate is not given here because of large uncertainties in these calculations.
[b]As above, due to ^{35}Cl(n,γ) ^{36}Cl reaction.
[c]*(n)*, The *in situ* production of ^{41}Ca, ^{81}Kr and ^{36}Cl in the oceans is primarily due to the relevant thermal neutron capture reaction. Note the 50% greater *in situ* production of ^{36}Cl in the oceans compared to its atmospheric source.

Table 4 Approximate steady-state fractional inventories of cosmic ray produced radioisotopes in exchange reservoirs[a]

Exchange reservoir	*Radioisotope*													
	^{10}Be	^{26}Al	^{36}Cl	^{81}Kr	^{14}C	^{32}Si	^{39}Ar	^{3}H	^{22}Na	^{35}S	^{7}Be	^{37}Ar	^{33}P	^{32}P
Atmosphere	2.3×10^{-3}	1.4×10^{-6}	1.1×10^{-6}	0.96	1.9×10^{-2}	2.0×10^{-3}	0.99	7.2×10^{-2}	0.27	0.65	0.71	0.99	0.80	0.84
Land surface	0.29[b]	0.29[b]	0.29[b]	0	4×10^{-2}	0.29[b]	0	0.27	0.21	0.1	0.08	0	5.6×10^{-2}	4.7×10^{-2}
Mixed oceanic layer	5.7×10^{-6}	1.4×10^{-5}	1.4×10^{-2}	6×10^{-4}	2.2×10^{-2}	3.5×10^{-3}	0	0.35	0.44	0.24	0.20	0	0.13	0.11
Deep oceanic layer	10^{-4}	7×10^{-5}	0.69	3.5×10^{-2}	0.92	0.68	0.01	0.3	8×10^{-2}	4×10^{-3}	2×10^{-3}	0	7×10^{-4}	10^{-4}
Oceanic sediments	0.71	0.71	0	0	4×10^{-3}	2.8×10^{-2}	0	0	0	0	0	0	0	0
Half-life (y)	1.5×10^{6}	7.1×10^{5}	3.0×10^{5}	2.3×10^{5}	5730	~150	268	12.3	2.6	87d	53d	35d	25.3d	14.3d

[a]Approximate calculations based on Lal and Peters (1967). Values given as zero imply very small fractional inventories.
[b]Part of the inventory may in fact be carried as silt or dust to the oceans before decay.

Pathways of Isotopes to the Oceans and their Approximate Inventories/Concentrations in the Atmosphere, Hydrosphere, and Sediments

The applications of cosmogenic isotopes as tracers depend on three principal factors: (1) their source function; (2) their half-lives; and (3) their chemical properties. These considerations decide how the fractional inventories of different tracers are distributed on the earth in the atmosphere, hydrosphere, and the sediments. The work of Lal and Peters, using simplified models for the pathways of the isotopes considering six mixing/exchange reservoirs is still quite instructive. These estimates of fractional inventories of 14 isotopes amongst these reservoirs are shown in **Table 4**.

Table 4 shows that most of the global inventory of the long-lived isotopes, ^{10}Be and ^{26}Al is in the oceanic sediments whereas that of another long-lived isotope, ^{36}Cl is in the oceans; this is a manifestation of their chemical properties. Analogous to ^{36}Cl, the chemical behavior determines the large fractional inventories of ^{14}C and ^{32}Si in the oceans. The inventories of the long-lived ^{81}Kr, and also of ^{39}Ar (~270 y half-life), are primarily in the atmosphere primarily because of the low abundances of Kr and Ar in the atmosphere. It should be noted that generally the applications of an isotope are favored in the reservoir where its inventory is the largest. However, this is not always true. For example, in the cases of ^{39}Ar and ^{33,32}P, in spite of their low inventories in the oceans, they have valuable applications in studies of oceanic processes.

Approximate theoretical estimates of isotope concentrations in the oceans (disintegrations per minute (d.p.m.) per tonne of sea water) are presented in **Table 5**. The values are in the range of 10^{-5}–250 d.p.m. t^{-1}. The corresponding specific elemental concentrations are very low, with isotope/element ratios lying in range of 10^{-19}–10^{-10}.

The concentrations of a large number of naturally occurring radioactive and stable nuclides, those produced in nuclear reactions caused by cosmic radiation, and those produced by energetic particles in radioactive disintegrations and in nuclear decays of naturally occurring long-lived nuclides, have been measured in the past five decades in the marine environment. Dramatic improvements in the radiometric techniques in the past two decades have allowed their measurements to be done fairly reliably. The database on the distribution of the cosmogenic and other tracers in the oceans is therefore growing steadily. Their

Table 5 Approximate specific radioactivities of cosmic ray produced isotopes in the ocean

Radioisotope	*Half-life (y)*	*Average specific radio activity in oceans*	
		d.p.m. per tonne water	*d.p.m. per g element*
^{10}Be	1.5×10^6	10^{-3}	1.6×10^3
^{26}Al	7.1×10^5	1.2×10^{-5}	1.2×10^{-3}
$^{81}Kr^a$	2.3×10^5	7×10^{-6}	2.1×10^{-2}
^{36}Cl	3.0×10^5	0.55	3×10^{-5}
^{14}C	5730	260	10
^{32}Si	~150	1.5×10^{-2}	2.3×10^{-2}
$^{39}Ar^b$	268	2.9×10^{-3}	5×10^{-3}
^{3}H	12.3	36	3.3×10^{-4}

^{a}Based on atmospheric $^{81}Kr/Kr$ ratio of $(5.2 \pm 0.4) \times 10^{-13}$.
^{b}Based on atmospheric $^{39}Ar/Ar$ ratio of (0.107 ± 0.004) d.p.m. per liter Ar.
(Based on Lal and Peters (1967)).

measurements to date are in good agreement with the theoretically predicted values of their distribution in the oceans (cf. **Tables 3** and **4**).

In many cases these nuclides serve as tracers for the study of physical, chemical, and biological processes in the oceans. Several radiotracers successfully provide chronology of sediments, corals, and manganese nodules, but learning about large-scale ocean circulation is another matter. The ability merely to make measurements of a tracer in the marine environment is not sufficient to use it as an effective tracer for delineating important oceanic variables. Tracer data must be examined in terms of ocean models. Constructing ocean models is an iterative process between data acquisition and model building, forcing model outputs, to become compatible with the observations. The oceanic processes are very complex, exhibiting significant spatial and temporal variability on a wide range of scales. For the tracer data to be useful in developing meaningful coupled atmosphere–ocean circulation models, which may be considered as the goal of tracer studies, one would require three-dimensional tracer data with sufficient resolution in the horizontal direction. The latter are not available, except for ^{14}C, where a considerable database is growing as a result of recent WOCE (World Ocean Circulation Experiment) expeditions.

Tracers in Oceanography: Why We Need Them and What We Learn From Them

The oceans represent a large mass of water endowed with a large amount of diverse substances and heat. The dissolved and particulate oceanic 'complex' is in continuous exchange with the land surface and the atmosphere. An appreciable part of the dissolved phases is recycled within the oceans through biogeochemical cycles, which are maintained by the large-scale oceanic circulation. The latter is a manifestation of the continuous exchange of heat between the atmosphere and the ocean. Large-scale oceanic circulation replenishes nutrients in the surface waters, which are rapidly removed by biological productivity. Biological recycling changes the chemical makeup of ocean waters at all depths.

Thus there is a complex cause–effect relationship with significant feedbacks between oceanic circulation, biogeochemical cycling within the oceans, and composition of sea water. Understanding these processes is essential for understanding oceanographic processes, earth's climate, terrestrial biogeochemistry, and the proxy records contained in the oceanic sediments. Success in achieving this goal requires sensitive multidisciplinary techniques in which tracers play an important part.

Chemical and isotopic tracers have been used successfully for the past five decades. Oceanic water masses are conventionally characterized by their chemical and isotopic composition, and temperature. A central problem in oceanography is to understand the origins and the processes which determine the evolution of different water masses. Radioactive isotope tracers provide additional information on timescales, specifically on the rate constants of different processes. The most attractive feature of radioisotopes is that they provide time integrals of evolution of water masses through space, influenced by exchange/mixing processes, and radioactive decay of the tracer, which introduces the element of time in the model(s). In steady-state situations, all losses and gains balance out. By combining with information on stable isotopes, one can then determine effective time required for the water mass to reach equilibrium between gain and loss terms, i.e. get an estimate of the effective equilibration time of the water mass as it evolves.

Tracers fall into two broad categories:

1. Transient tracers which are introduced sporadically in a system, e.g., radionuclides introduced by testing of nuclear weapons, and from discharges from nuclear reactors.
2. Steady tracers which are introduced continuously in a system, e.g., those produced by nuclear interactions of cosmic rays on the earth, and by radioactive decay of dissolved uranium in the oceans.

It is convenient to further designate the tracers according to their chemical behavior in the system:

1. conservative tracers, which follow the motion of the 'fluid' in the system;
2. nonconservative tracers, which do not follow the motion of the 'fluid' in the system.

The first naturally occurring radiotracer to be used in oceanography was ^{226}Ra. The first successful tracer measurements of the cosmogenic ^{14}C with a view to understanding timescales in large-scale water circulation were made in 1960 and demonstrated the great value of this tracer in oceanography. The discovery of cosmogenic ^{32}Si in marine siliceous sponges opened up the possibility of using this as a tracer for studying biogenic silica fluxes to the deep sea, and the nutrient cycle of silicon. As the techniques for the measurement of weak activities of the nuclides became available, additional nuclides were measured in the oceans. To date 12 cosmogenic nuclides have been studied in the oceans, some during the 1960s, several during the 1980s. It is important to realize that all tracers are important because of their particular unique attributes (cf. **Table 4**).

Table 6 lists tracers, which are studied in oceanographic research, together with ^{36}Cl (which is included for its potential usefulness for determining the average source strength of cosmic ray neutrons in the past 0.5–0.7 My (million years)). The usefulness of cosmogenic tracers depends on their half-lives, chemical properties, and their source functions, which can be appreciated from their expected distribution in the geospheres (**Table 4**), and by the ease with which they can be measured. (**Table 4** does not include ^{4}He, ^{20}Ne, ^{21}Ne, ^{22}Ne, ^{22}Na, ^{35}S, ^{36}Cl, ^{37}Ar, ^{41}Ca, ^{53}Mn, ^{52}Ni, ^{60}Fe and ^{81}Kr, which are either not useful as oceanic tracers because of their short half-lives, or very long half-lives, or have very low cosmogenic production rates. However, with technical developments these nuclides may eventually become useful.) The long-lived cosmogenic radionuclide, ^{129}I is not included in **Tables 3** and **4** because (1) its half-life is rather long (15.7 My) to be useful for studying oceanic processes, (2) it is continuously produced in the oceans, and in ocean sediments in the spontaneous fission of ^{238}U, and (3) it has been added to the oceans in appreciable amounts in the last five decades by human activities; such as nuclear weapons' testing and processing of nuclear plants (which have raised the prenuclear age inventory of ^{129}I in the oceans of about 100 kg by more than an order of magnitude.

Several cosmogenic tracers also qualify as transient tracers at the present time, because of an appreciable contribution from anthropogenic sources (**Table 3**). Thus, the nuclides ^{3}H, ^{14}C (produced in appreciable amounts in nuclear weapons testing), tritugenic ^{3}He and ^{129}I (which has also been produced in large amounts by nuclear weapons tests and operation of nuclear power plants), serve as (useful) transient tracers in some geophysical reservoirs.

Table 6 Important characteristics and principal applications of selected cosmogenic tracers

Isotope	*Half-life*	*Principal applications*
Isotopes which do not form compounds		
^{3}He	Stable	Air–sea exchange; escape of helium from the atmosphere
^{37}Ar	35 d	Air–sea exchange; tropospheric circulation
^{39}Ar	268 y	Air–sea exchange; vertical mixing in oceans
^{81}Kr	2.3×10^{5} y	Ground water ages, and constancy of cosmic radiation
Isotopes which label constituent molecules in the atmosphere and the hydrosphere		
^{3}H (H_2O)	12.3 y	Characterizing water molecules in the atmosphere, hydrosphere and cryosphere
^{14}C (CO_2, CO_3, HCO_3)	5730 y	Characterization of the carbon cycle reservoirs
^{32}Si ($HSiO_3$, SiO_2)	~150 y	Biogeochemical cycle of silicon
^{32}P, ^{33}P (DIP, DOP)	14.3, 25.3 d	Biogeochemical cycle of phosphorus
Isotopes which attach to aerosols/particles		
^{7}Be	53 d	Atmospheric circulation, vertical mixing in surface ocean waters
^{10}Be	1.5 10^{6} y	Role of particle scavenging in the coastal and open oceans; dating of sediments and accretions
^{26}Al	7.1×10^{5} y	Role of particle scavenging in the coastal and open oceans; dating of marine sediments and accretions
^{32}Si ($HSiO_3$, SiO_2)	~150 y	Labeling the dissolved oceanic silicon pool; atmospheric circulation
^{32}P, ^{33}P	14.3, 25.3 d	Labeling the dissolved oceanic phosphorus pool; tropospheric circulation

Note: Not included here are ^{36}Cl and ^{129}I for reasons discussed in the text.

An important consideration in the use of a transient tracer is knowledge of its source strength. Anthropogenic sources are generally not well defined; however, in the case of ^{14}C, an important advantage is that its contribution to the atmospheric CO_2 reservoir is well known (i.e., the excess ^{14}C amount relative to ^{12}C), and this precisely defines its source function. In the case of anthropogenic ^{3}H, this is not the case, but a great advantage is that one can measure both the ^{3}H and tritugenic ^{3}He in a water sample and defines an 'age' of the water mass. These measurements are by no means easy, however, but a large database of information has been produced which has yielded very useful insights into large-scale ocean circulation in the upper ocean.

There are also two important nonnuclear transient tracers, chlorofluorocarbons CFC-11 and CFC-12, which have also proven very useful in view of their known (changing) relative concentrations in the atmosphere. These behave essentially as conservative tracers; any CFC losses would not be expected to change their ratio in the oceanic water mass. The use of new tracers (F113, CCl_4) has extended the timescales of CFC in both directions.

In practice, one has to work with tracers of different properties, and each of its properties can be taken advantage of, as its special attribute. Even nonconservative tracers, e.g., ^{14}C and ^{32}Si, have their own significance and merit. In fact, in the oceans the only conservative tracers of natural origin are ^{3}H (half-life 12.3 y) and ^{3}He (stable).

New Techniques for Measurements of Tracers in the Oceans in the 1980s and 1990s

By the end of the 1970s, the field of cosmogenic tracers had clearly recognized the usefulness of most of the cosmogenic tracers, with sufficient measurements at hand in each case. After isolated studies of a few tracers, e.g., ^{14}C, ^{10}Be, in individual water samples, it became apparent that oceans can yield their secrets only with multiple tracer attack. In early multiple tracer investigations, detailed information regarding the nature and rate of processes responsible for the formation of the Antarctic Bottom water was obtained by including the tracers ^{3}He and ^{14}C. The field was expanded in the 1980s and 1990s with larger-scale exploitation of several tracers for answering specific questions. This came about due to a fruitful combination of events and discoveries, which gave a tremendous fillip to both chemical oceanography and tracer studies including nuclides belonging to U-Th series. Foremost was the decision to study oceans in a systematic manner, along geochemical sections (GEOSECS), using a suite of tracers. GEOSECS expeditions were successfully carried out to the principal oceans in 1972–78 and resulted in fairly accurate tracer data. The GEOSECS concept was very successful; it rested on the necessity for making more precise measurements of several tracers and ocean properties in addition this integrated study resulted in information about temporal changes in the property profiles at the same stations after an elapse of one to two decades since the site was occupied in the GEOSECS expedition. and finally, it was an artful and timely combination of theory and experiment, which gave a tremendous boost to the field of learning about oceanographic processes.

The 1980s also marked an era of dramatic advance in the techniques of measurements of long-lived cosmogenic radionuclides ^{14}C, ^{10}Be, and ^{26}Al in the oceans, in sediments, and in manganese nodules using AMS (accelerator mass spectrometry). This opened up new windows for observing in detail a host of physical, chemical, and biological processes. The ease with which these nuclides can be measured allowed long series of measurements in space and time to be obtained. Examples of this development are the direct measurements of ^{10}Be and ^{26}Al concentration profiles in sea water in the principal oceans; and profiles of ^{10}Be concentrations in marine sediments and in manganese nodules which opened up a new field of investigation in marine beryllium geochemistry.

A new field, the study of P-biodynamics in surface waters using cosmogenic ^{32}P and ^{33}P surfaced in the late 1980s. This tracer application was not held up for want of a technique. In this case, it was not realized that these short-lived nuclides (half-lives, 14.3 and 25.3 days) in fact had about the appropriate half-lives for studying timescales of exchange of phosphorus between dissolved inorganic P, organic P and plankton. Concurrently, technical advances were also being made to measure short-lived radionuclides in ocean waters, where the AMS technique does not offer any gain in detection sensitivity, e.g. ^{32}P (half-life, 14.3 days), ^{33}P (half-life, 25.3 days) and ^{32}Si (half-life, 150 years). By using a standard liquid scintillation counting system to simultaneously measure both ^{32}P and ^{33}P activities, much higher sensitivity is attainable than by using low-level counters, especially for samples of low specific radioactivity.

Examples of Oceanic Data Derived Using Cosmogenic Tracers

With this foreground what has been learnt about the oceanic processes using cosmogenic tracers is now briefly examined. We are obviously interested in learning about cycles of principal nutrient elements, and processes by which the ocean chemistry evolves (as regulated by aeolian and fluvial fluxes from the land, biogeochemical cycles within the oceans, and large-scale oceanic circulation). This problem can be approached in bits and pieces only, and then the interconnections and feedbacks examined. A comprehensive mosaic of all the interactions and controls may or may not be achieved. The records of present day ocean biogeochemical processes are recorded in the sediments. It is therefore important to study the chronology and the makeup of ocean sediments to get a comprehensive picture of the temporal evolution of ocean chemistry and climate through aeons. The suite of tracers listed in **Table 3** has provided sufficient information on oceanic processes in four broad fields.

1. Biogeochemical cycling of nutrients and trace elements
2. Chronology of marine sediments and manganese nodules
3. Principal features of large-scale oceanic circulation
4. Biogeochemical and ocean circulation controls on climate.

In each of these studies, the task is complemented by the availability of radiotracers belonging to the U-Th series. It should be stressed that tracers each have some particular unique features for studying critical problems in oceanography. All tracers are not created equal: some are more equal than others. This social expression also finds a rightful place in the realm of oceans. This can be illustrated by citing unique features of two of the cosmogenic nuclides, ^{14}C (half-lifem, 5730 years) and ^{10}Be (half-life, 1.5 My), which have a special status among all natural tracers.

The great 'virtues' or attributes of ^{14}C are that (1) it is a carbon isotope, and is introduced in to the carbon cycle reservoirs as carbon dioxide in the earth's atmosphere, and (2) its half-life is well suited to study late Quaternary events and processes, including dating of sediments and timing of deep and bottom water formation.

In the oceans, ^{14}C does not behave as a conservative tracer, since carbon (and its compounds) is not distributed uniformly in the oceans. But this does not present any problems; rather its studies allow determination of carbon fluxes within the ocean. In fact, if ^{14}C was a conservative tracer, it could not have been used to date marine sediments. It has been added in significant amounts to the atmosphere as a result of nuclear weapons tests during the 1950s and 1960s. Consequently, ^{14}C can be used both as a 'steady-state' and as a 'transient' tracer.

The second most attractive ocean tracer is the cosmogenic ^{10}Be, which serves to delineate pathways of particle active elements through the water column, and is useful for dating sediments and in manganese nodules to about 10 My BP. The particle active nature of ^{10}Be leads to its preferential deposition in the coastal regions of the oceans. Recent studies have demonstrated that using special chemical techniques, the activity of cosmogenic ^{26}Al can be measured in the oceanic environment, and it has been suggested that it should be a useful tracer for studying changes in the past biological productivity of the oceans. This application arises from the higher chemical reactivity of ^{26}Al, compared to ^{10}Be. If this suggestion is borne out from future studies, ^{26}Al would constitute an invaluable tracer for studying temporal and spatial variations in biological productivity. Its studies would complement the information obtained using the cosmogenic ^{10}Be.

Recent measurements of cosmogenic ^{32}P (half-life, 14.3 days) and ^{33}P (half-life, 25.3 days) in surface ocean waters have opened up new possibilities of quantifying P-biodynamics with complementary information on eddy diffusivity in the waters, based on the cosmogenic ^{7}Be (**Table 3**). A wealth of new ^{32}P, ^{33}P data have been added on the distribution of cosmogenic ^{32}P and ^{33}P in the surface ocean waters, and in plankton.

Epilogue

A large number of cosmogenic tracers are available for oceanic studies, and the use of these tracers has steadily increased to date. But of course, tracers are not the complete answer to the mysteries of the ocean. It is necessary to learn how to use tracers, how to model them, how to combine them with other tracers, singly and multiply, etc. The usefulness and application of a tracer cannot be discussed on an absolute basis, because such an approach would result in a largely academic discussion. A tracer may have the appropriate physical and chemical attributes, but its source strength may be too weak, or its source function may not be known at the present time. Tracer suitability has therefore to be evaluated periodically as frontiers of knowledge expand. With the freedom in thinking about what type of tracer measurements can be made, a great deal of valuable information will probably be derived from it; however, one has to think about applying it within practical constraints. An important constraint is the number of measurements, which can be made within a reasonable period, and within available financial resources. Today, the principal constraint is the inadequacy of spatial and temporal coverage of the tracer data. With more synoptic data, tracer modeling could also be refined. Any shortcomings in these foil the goal of these studies. We are now in the mode of understanding details of ocean circulation and chemistry; not just the integrated overall features. In the earliest applications of cosmogenic tracers just a few measurements of radiocarbon in the oceans were sufficient to show that the upper ocean had a turnover time of about 600 years. In the case of the atmosphere, the mean time for removal of aerosols was similarly learned very quickly from the observed fallout of nuclear weapons tests-injected radioactivities (^{90}Sr, ^{3}H, etc.) and from the fallout of cosmogenic ^{22}Na, ^{7}Be and ^{32}P. With continued applications of these tracers, it has been shown that oceanic processes are indeed complex, especially at the interfaces (air–sea, mixed layer–intermediate waters–deep waters, and water–sediment), and in the polar oceans, which determine formation of bottom–deep waters. Thus, our hasty perception of these tracers being a panacea quickly changed, even back in the 1970s.

What is the basic nature of large-scale oceanic circulation? The radiochemists used the *Kw* model in its simple one-dimensional form; *K* denotes the eddy diffusivity, and *w* the advection velocity. At this time a stimulating discussion of essential mathematical approaches for treating the tracer data was presented. Simple material balance calculations by Lal showed that an appreciable amount of carbon was added to the dissolved carbon inventory (J-flux) at depths by sinking biogenic particles. This important aspect of resetting the ^{14}C clock in the deep sea by J-flux is now being examined on a global basis under WOCE experiments, and constitutes a critical parameter in climatic feedback processes. Subsequently, tracer data showed that an important transfer of oceanic properties occurred across pycnoclines. Modeling of tracer data in fact reveals model inadequacies and fosters development of more appropriate physical models. There are several very basic issues that are recognized, but not well understand. For example, what are the roles of tides and internal waves in large-scale ocean circulation? These questions have been asked several times, but not yet attacked properly due to our present limitations. Today, we are far from a synthesis of comprehensive models which are capable of providing an interactive atmosphere–ocean coupled model which can respond to changes in climate, or predict climatic changes as the model is run.

There are several academic and technical issues which we are confronting today:

1. complexity of the ocean system; variable response at different time and space scales;
2. lack of three-dimensional tracer data;
3. lack of information on temporal and spatial changes in tracer distribution;
4. lack of understanding of physical, chemical, and biological processes.

However, there are proven methods based on ^{14}C, ^{10}Be, ^{26}Al, ^{39}Ar, ^{32}Si, ^{7}Be, ^{33}P and ^{32}P. Improved tracer modeling will emerge only with further advances in techniques for their measurement, and with a better understanding of the atmospheric and oceanic circulation, mixing, and biogeochemical processes. On the question of ease of measurement, the radionuclide ^{39}Ar is an important case in point here. It is a conservative tracer, ideally suited for studying vertical mixing in the oceans, but to date very few measurements have been made, since they are very time consuming. Another example is that of ^{32}Si. Although it has been measured at several stations in the Atlantic, Pacific and Indian oceans, these measurements are not currently precise enough to make detailed mixing and transport models to define the silica cycle in the oceans. They are, however, useful to determine vertical J-fluxes, one-diemnsional *K*/ω ratios, and the latitudinal inventories of ^{32}Si in the oceans.

Physical oceanography provides the theoretical basis for oceanic mixing and circulation, but the experimental data necessary to understand the nature of this circulation must be based on present day and proxy observations of chemical composition of sea water in space and time. Directed global scale coordination between scientists to study important oceanic processes, such as physical and biological controls on biological production, and export of carbon, are rapidly providing new insights and accelerated developments of realistic models. An example is the coordinated US Joint Global Ocean Flux Studies (JGOFS) in the Equatorial Pacific Ocean in 1992, during a four-month period which coincided with the maximum intensity of the warm El Niño event, and another three-month period during well-developed cool surface-water conditions. The combined physical/chemical and biological data produced, which included ^{234}Th concentrations of sinking particulates, led to new insights about the roles of dissolved organic carbon, microzooplankton grazing, nutrient and CO_2 fluxes, and highlighted the importance of physical, in contrast to biological, processes in this region, where net carbon fluxes out of the system are very small as a result of highly efficient biological cycling.

This discussion would not be complete without mention of the 'purposeful tracer experiment' carried out as a component of WOCE, using SF_6 released in the open ocean. The results confirmed the earlier estimates of very low diapycnal diffusivity, of the order of $0.1\,cm^2\,s^{-1}$, implying that heat, salt, and tracers must penetrate the thermocline primary by transport along, rather than across, density surfaces.

Clearly, understanding of the large-scale ocean circulation will come from multiple approaches, with directed research to simultaneously understand transfer and mixing of 'properties' and 'substances' and their relationship to climatic changes. It has taken five decades to develop techniques to make relevant oceanographic measurements, and one should expect rapid advances in our knowledge of oceanic processes as a result of innovative research, and international observational programs such as Tropical Oceans and the Global Atmosphere (TOGA) and WOCE.

It must be realized that a state must be reached where experiments and theory go hand in hand, leading to the development of better (more realistic) models, and acquisition of critical tracer data. In the absence of a knowledge of the processes involved, models employed often yield very erroneous results. Thus, whereas even a few tracer data are quite informative (since a few data points can be treated only with zero order models), any attempts to understand oceanic processes in detail pose a serious challenge. A few examples are considered here, where tracer data have contributed to the development of realistic models. As mentioned earlier, simple one-dimensional models were developed earlier on using two parameters K and w, to consider vertical transfer of tracers through an oceanic column. Even today these are used, in the absence of better alternatives, and in reality, because of a lack of tracer data in the three-dimensional space. The result is that as yet the general validity of the K-w models in space is not known or their dependence on climate. The latter arises because there are experimental tracer data for ocean waters only during the Holocene.

The recent significant developments in oceanic general circulation are a result of transient tracer experiments, and order of magnitude improvements in a number of fields, including orbit dynamics, gravity field estimation and atmospheric variability. High accuracy data on ocean surface elevation by satellite altimetry is leading to hopes of complete theee-dimensional time-evolving estimates of ocean circulation, which would also improve estimates of oceanic heat, and several property fluxes.

There has been prolific growth in the field of tracer oceanography within the last two decades, but there are acute limitations in providing a consistent picture of the interrelationships between physical, chemical, and biological processes, which are needed to develop a coupled atmosphere–ocean model that responds to climate in an interactive manner. It is not clear how this will be achieved in the near future.

Finally, it is gratifying to see that the cosmogenic radiotracer field has evolved highly from an academic curiosity in the 1950s and 1960s to its present-day form, wherein it aims to become an integral part of realistic atmosphere–ocean global atmosphere and ocean circulation models.

See also

Carbon Cycle. Carbon Dioxide (CO_2) Cycle. Phosphorus Cycle. Radiocarbon. Stable Carbon Isotope Variations in the Ocean.

Further Reading

Broecker WS (1981) Geochemical tracers and ocean circulation. In: Warren BA and Wunsch C (eds.) *Evolution of Physical Oceanography*, pp. 434–460. Cambridge, MA: MIT Press.

Broecker WS and Peng TH (1982) *Tracers in the Sea*. New York: Lamont-Doherty Geological Observatory.

Jenkins WJ (1992) Tracer based inferences of new primary production in the sea. In: Falkowski PG and Woodhead AD (eds.) *Primary Productivity and Biogeochemical Cycles in the Sea*. New York: Plenum Press.

Lal D (1962) Cosmic ray produced radionuclides in the sea. *J. Ocean. Soc.* Japan: 20th Anniv. Vol. 600–614.

Lal D and Peters B (1967) Cosmic ray produced radioactivity on the earth. *Handbuch der Physik* 46(2): 551–612.

Lal D (1999) An overview of five decades of studies of cosmic ray produced nuclides in the oceans. *Science of the Total Environment* 237/238: 3–13.

Ledwell JR, Watson AJ, and Law CS (1993) Evidence for slow mixing across the pycnocline from an open ocean tracer-release experiment. *Nature* 364: 701–703.

Libby WF, Anderson EC, and Arnold JR (1949) Age determination by radiocarbon content: world-wide assay of natural radiocarbon. *Science* 109: 227–228.

Measures CI and Edmond JM (1982) Beryllium in the water column of the Central Pacific. *Nature* 297: 51–53.

Murray JW, Barber RT, Roman MR, Bacon MP, and Feely RA (1994) Physical and biological controls on carbon cycling in the equatorial Pacific. *Science* 266: 58–65.

Raisbeck GM and Yiou F (1999) ^{129}I in the oceans: origins and applications. *Science of the Total Environment* 237/238: 31–41.

Schlosser P and Smethie WM Jr (1995) Transient tracers as a tool to study variability of ocean circulation. In: *Natural Climate Variability on Decade-to-century Time Scales*, pp. 274–289. Washington: National Research Council.

RADIOCARBON

R. M. Key, Princeton University, Princeton, NJ, USA

Introduction

In 1934 F.N.D. Kurie at Yale University obtained the first evidence for existence of radiocarbon (carbon-14, ^{14}C). Over the next 20 years most of the details for measuring ^{14}C and for its application to dating were worked out by W.F. Libby and co-workers. Libby received the 1960 Nobel Prize in chemistry for this research.

The primary application of ^{14}C is to date objects or to determine various environmental process rates. The ^{14}C method is based on the assumption of a constant atmospheric formation rate. Once produced, atmospheric ^{14}C reacts to form $^{14}CO_2$, which participates in the global carbon cycle processes of photosynthesis and respiration as well as the physical processes of dissolution, particulate deposition, evaporation, precipitation, transport, etc. Atmospheric radiocarbon is transferred to the ocean primarily by air–sea gas exchange of $^{14}CO_2$. Once in the ocean, $^{14}CO_2$ is subject to the same physical, chemical, and biological processes that affect CO_2. While alive, biota establish an equilibrium concentration of radiocarbon with their surroundings; that is, ^{14}C lost by decay is replaced by uptake from the environment. Once the tissue dies or is removed from an environment that contains ^{14}C, the decay is no longer compensated. The loss of ^{14}C by decay can then be used to determine the time of death or removal from the original ^{14}C source. After death or removal of the organism, it is generally assumed that no exchange occurs between the tissue and its surroundings; that is, the system is assumed to be closed. As a result of the ^{14}C decay rate, the various reservoir sizes involved in the carbon cycle, and exchange rates between the reservoirs, the ocean contains approximately 50 times as much natural radiocarbon as does the atmosphere.

Carbon-14 is one of three naturally occurring carbon isotopes; ^{14}C is radioactive, has a half-life of 5730 years and decays by emitting a β-particle with an energy of about 156 keV. On the surface of the earth, the abundance of natural ^{14}C relative to the two stable naturally occurring carbon isotopes is $^{12}C:{}^{13}C:{}^{14}C = 98.9\% : 1.1\% : 1.2\times10^{-10}\ \%$. Natural radiocarbon is produced in the atmosphere, primarily by the collision of cosmic ray produced neutrons with nitrogen according to the reaction [I].

$$^{1}_{0}n + ^{14}_{7}N \Rightarrow ^{14}_{6}C + ^{1}_{1}H \qquad [I]$$

where n is a neutron and H is the proton emitted by the product nucleus. Similarly, the decay of ^{14}C takes place by emission of a β-particle and leads to stable nitrogen according to reaction (II),

$$^{14}_{6}C \Rightarrow ^{14}_{7}N + \beta^{-} + \bar{\upsilon} + Q \qquad [II]$$

where $\bar{\upsilon}$ is an antineutrino and Q is the decay energy.

The atmospheric production rate varies somewhat and is influenced by changes in the solar wind and in the earth's geomagnetic field intensity. A mean of 1.57 atom $cm^{-2}\,s^{-1}$ is estimated based on the long-term record preserved in tree rings and a carbon reservoir model. This long-term production rate yields a global natural ^{14}C inventory of approximately 50 t ($1t = 10^6$ g). Production estimates based on the more recent record of neutron flux measurements tend to be higher, with values approaching 2 atom $cm^{-2}\,s^{-1}$. **Figure 1** shows the atmospheric history of ^{14}C from AD 1511 to AD 1954 measured by Minze Stuiver (University of Washington) using tree growth rings. The strong decrease that occurs after about AD 1880 is due to dilution by anthropogenic addition of CO_2 during the industrial revolution by the burning of fossil fuels (coal, gas, oil). This dilution has come to be known as the Suess effect (after Hans E. Suess).

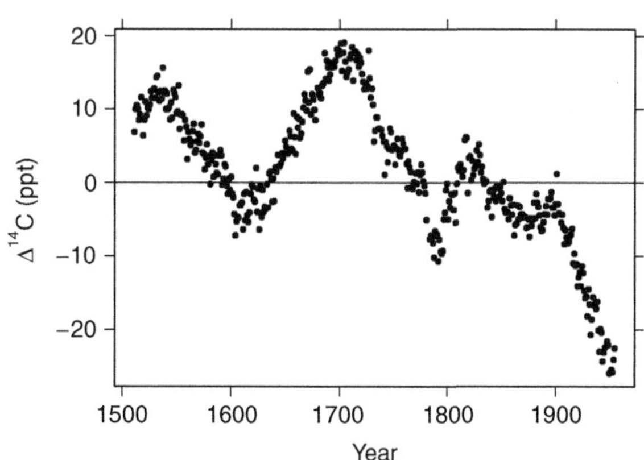

Figure 1 Atmospheric history of $\Delta^{14}C$ measured by M. Stuiver in tree rings covering AD 1511 to AC 1954. Most of the decrease over the last hundred years is due to the addition of anthropogenic CO_2 to the atmosphere during the industrial revolution by the burning of fossil fuels.

Prior to 16 July 1945 all radiocarbon on the surface of the earth was produced naturally. On that date, US scientists carried out the first atmospheric atomic bomb test, known as the Trinity Test. Between 1945 and 1963, when the Partial Test Ban Treaty was signed and atmospheric nuclear testing was banned, approximately 500 atmospheric nuclear explosions were carried out by the United States (215), the former Soviet Union (219), the United Kingdom (21) and France (50). After the signing, a few additional atmospheric tests were carried out by China (23) and other countries not participating in the treaty. The net effect of the testing was to significantly increase ^{14}C levels in the atmosphere and subsequently in the ocean. Anthropogenic ^{14}C has also been added to the environment from some nuclear power plants, but this input is generally only detectable near the reactor.

It is unusual to think of any type of atmospheric contamination – especially by a radioactive species – as beneficial; however, bomb-produced radiocarbon (and tritium) has proven to be extremely valuable to oceanographers. The majority of the atmospheric testing, in terms of number of tests and ^{14}C production, occurred over a short time interval, between 1958 and 1963, relative to many ocean circulation processes. This time history, coupled with the level of contamination and the fact that ^{14}C becomes intimately involved in the oceanic carbon cycle, allows bomb-produced radiocarbon to be valuable as a tracer for several ocean processes including biological activity, air–sea gas exchange, thermocline ventilation, upper ocean circulation, and upwelling.

Oceanographic radiocarbon results are generally reported as $\Delta^{14}C$, the activity ratio relative to a standard (NBS oxalic acid, 13.56 dpm per g of carbon) with a correction applied for dilution of the radiocarbon by anthropogenic CO_2 with age corrections of the standard material to AD1950. $\Delta^{14}C$ is defined by eqn [1].

$$\Delta^{14}C = \delta^{14}C - 2(\delta^{13}C + 25)\left(1 + \frac{\delta^{14}C}{1000}\right) \qquad [1]$$

$\delta^{14}C$ is given by eqn [2] and the definition of $\delta^{13}C$ is analogous to that for $\delta^{14}C$.

$$\delta^{14}C = \left[\frac{{}^{14}C/C|_{smp} - {}^{14}C/C|_{std}}{{}^{14}C/C|_{std}}\right] \times 1000 \qquad [2]$$

The first part of the second term in the right side of eqn [1] $2(\delta^{13}C + 25)$, corrects for fractionation effects. The factor of 2 accounts for the fact that ^{14}C fractionation is expected to be twice as much as for ^{13}C and the additive constant 25 is a normalization factor conventionally applied to all samples and based on the mean value of terrestrial wood. The details of ^{14}C calculations can be significantly more involved than expressed in the above equations; however, there is a general consensus that the calculations and reporting of results be done as described by Minze Stuiver and Henry Polach in a paper specifically written to eliminate differences that existed previously. $\Delta^{14}C$ has units of parts per thousand (ppt). That is, 1 ppt means that $^{14}C/^{12}C$ for the sample is greater than $^{14}C/^{12}C$ for the standard by 0.001. In these units the radioactive decay rate of ^{14}C is approximately 1 ppt per 8.1 years.

The number of surface ocean measurements made before any bomb-derived contamination are insufficient to provide the global distribution before input from explosions. It is now possible to measure $\Delta^{14}C$ values in the annual growth rings of corals. By establishment of the exact year associated with each ring, reconstruction of the surface ocean $\Delta^{14}C$ history is possible. Applying the same procedure to long-lived mollusk shells extends the method to higher latitudes than is possible with corals. Whether corals or shells are used, it must be demonstrated that the coral or shell incorporates ^{14}C in the same ratio as the water in which it grew or at least that the fractionation is known. This method works only over the depth range at which the animal lived. **Figure 2** shows the $\Delta^{14}C$ record from two Pacific coral reefs measured by Ellen Druffel. Vertical lines indicate the period of atmospheric nuclear tests (1945–1963). The relatively small variability over the first ~300 years of the record includes variations due to weather events, climate change, ocean circulation, atmospheric production, etc. The last 50 years of the sequence records the

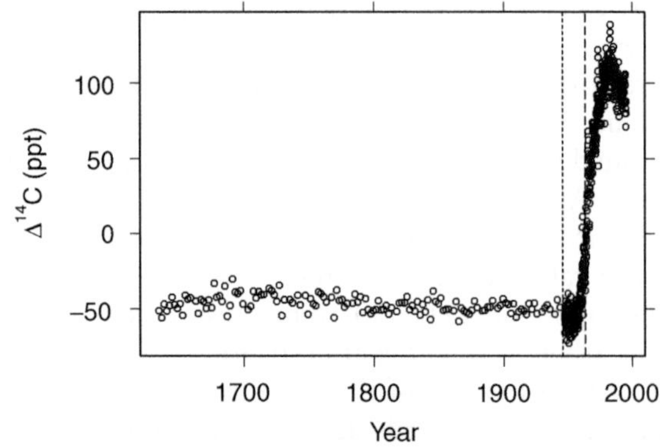

Figure 2 Long-term history of $\Delta^{14}C$ in the surface Pacific Ocean measured by E. Druffel in two coral reefs. The vertical lines surround the period of atmospheric nuclear weapons testing. The oceanic response to bomb contamination is delayed relative to the atmosphere because of the relatively long equilibration time between the ocean and atmosphere for $^{14}CO_2$.

invasion of bomb-produced $\Delta^{14}C$. Worth noting is the fact that the coral record of the bomb signal is lagged. That is, the coral values did not start to increase immediately testing began nor did they cease to increase when atmospheric testing ended. The lag is due to the time for the northern and southern hemisphere atmospheres to mix ($\sim$1 year) and to the relatively long time required for the surface ocean to equilibrate with the atmosphere with respect to $\Delta^{14}C$ ($\sim$10 years). Because of the slow equilibration, the surface ocean is frequently not at equilibrium with the atmosphere. This disequilibrium is one of the reasons why pre-bomb surface ocean results, when expressed as ages rather than ppt units, are generally 'old' rather than 'zero' as might be expected.

Figure 3A shows measured atmospheric $\Delta^{14}C$ levels from 1955 to the present in New Zealand (data from T.A. Rafter, M.A. Manning, and co-workers) and Germany (data from K.O. Munnich and co-workers) as well as older estimates based on tree ring measurements (data from M. Stuiver). The beginning of the significant increase in the mid-1950s marks the atmospheric testing of hydrogen bombs. Atmospheric levels increased rapidly from that point until the mid-1960s. Soon after the ban on atmospheric testing, levels began a decrease that continues up to the present. The rate of decrease in the atmosphere is about 0.055 y^{-1}. Also clearly evident in the figure is that the German measurements were significantly higher than those from New Zealand between approximately 1962 and 1970. The difference reflects the facts that most of the atmospheric tests were carried out in the Northern Hemisphere and that approximately 1 year is required for atmospheric mixing across the Equator. During that interval some of the atmospheric $^{14}CO_2$ is removed. Once atmospheric testing ceased, the two hemispheres equilibrated to the same radiocarbon level.

Figure 3B shows detailed Pacific Ocean $\Delta^{14}C$ coral ring data (J.R. Toggwelier and E. Druffel). This surface ocean record shows an increase during the 1960s; however, the peak occurs somewhat later than in the atmosphere and is significantly less pronounced. Careful investigation of coral data also demonstrates the north–south difference evidenced in the atmospheric record.

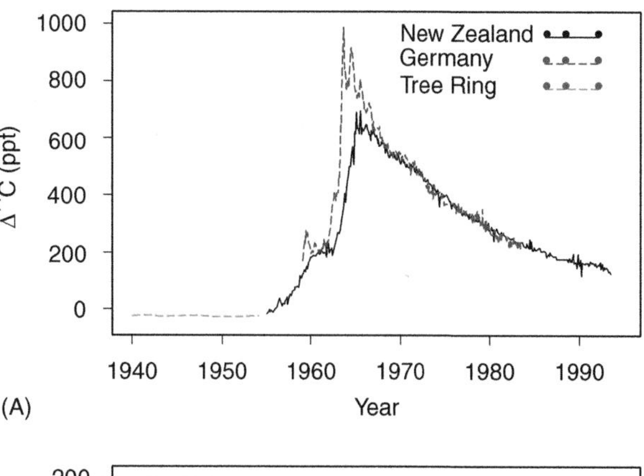

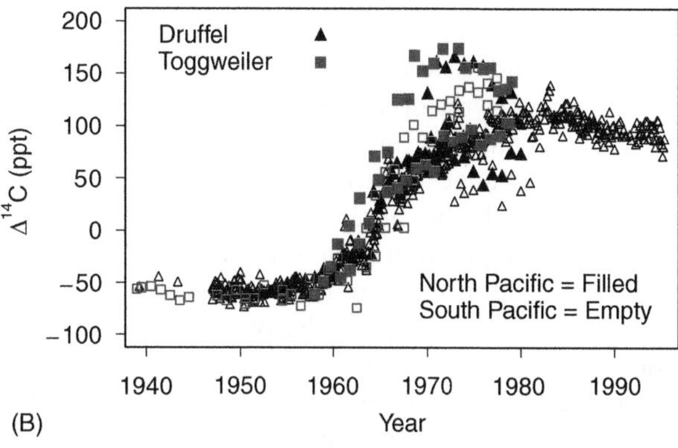

Figure 3 (A) Detailed atmospheric $\Delta^{14}C$ history as recorded in tree rings for times prior to 1955 and in atmospheric gas samples from both New Zealand and Germany subsequently. The large increase in the late 1950s and 1960s was due to the atmospheric testing of nuclear weapons (primarily fusion devices). The hemispheric difference during the 1960s is because most atmospheric bomb tests were carried out north of the Equator and there is a resistance to atmospheric mixing across the equator. Atmospheric levels began to decline shortly after the ban on atmospheric bomb testing. (B) $\Delta^{14}C$ in the surface Pacific Ocean as recorded in the annual growth rings of corals. The same general trend seen in the atmosphere is present. The bomb contamination peak is broadened and time-lagged relative to the atmosphere due to both mixing and to the time required for transfer from the atmosphere to the ocean.

Sampling and Measurement Techniques

The radiocarbon measurement technique has existed for only 50 years. The first ^{14}C measurement was made in W.F. Libby's Chicago laboratory in 1949 and the first list of ages was published in 1951. A necessary prerequisite to the age determination was accurate measurement of the radiocarbon half-life. This was done in 1949 in Antonia Engelkeimer's laboratory at the Argonne National Laboratory. Between 1952 and 1955 several additional radiocarbon dating laboratories opened. By the early 1960s several important advances had occurred including the following.

- Significantly improved counting efficiency and lower counting backgrounds, resulting in much greater measurement precision and longer time-scale over which the technique was applicable.
- Development of the extraction and concentration technique for sea water samples.

- More precise determination of the half-life by three different laboratories.
- Recognition by Hans Suess, while at the USGS and Scripps Institution of Oceanography, that radiocarbon in modern samples (since the beginning of the industrial revolution) was being diluted by anthropogenic CO_2 addition to the atmosphere and biosphere.
- Recognition that atmospheric and oceanic $\Delta^{14}C$ levels were increasing as a result of atmospheric testing of nuclear weapons.

During the 1970s and 1980s incremental changes in technique and equipment further increased the precision and lowered the counting background. With respect to the ocean, this was a period of sample collection, analysis, and interpretation. The next significant change occurred during the 1990s with application of the accelerator mass spectrometry (AMS) technique to oceanic samples. This technique counts ^{14}C atoms rather than detecting the energy released when a ^{14}C atom decays. The AMS technique allowed reduction of the sample size required for oceanic $\Delta^{14}C$ determination from approximately 250 liters of water to 250 milliliters! By 1995 the AMS technique was yielding results that were as good as the best prior techniques using large samples and decay counting. This size reduction and concurrent automation procedures had a profound effect on sea water $\Delta^{14}C$ determination. Many of the AMS techniques were developed and most of the oceanographic AMS $\Delta^{14}C$ measurements have been made at the National Ocean Sciences AMS facility in Woods Hole, Massachusetts, by Ann McNichol, Robert Schneider, and Karl von Reden under the initial direction of Glenn Jones and more recently John Hayes.

The natural concentration of ^{14}C in sea water is extremely low ($\sim 1 \times > 10^9$ atoms kg^{-1}). Prior to AMS, the only available technique to measure this low concentration was radioactive counting using either gas proportional or liquid scintillation detectors. Large sample were needed to obtain high precision and to keep counting times reasonable. Between about 1960 and 1995 most subsurface open-ocean radiocarbon water samples were collected using a Gerard–Ewing sampler commonly known as a Gerard barrel. The final design of the Gerard barrel consisted of a stainless steel cylinder with a volume of approximately 270 liters. An external scoop and an internal divider running the length of the cylinder resulted in efficient flushing while the barrel was lowered through the water on wire rope. When the barrel was returned to the ship deck, the water was transferred to a gas-tight container and acidified to convert carbonate species to CO_2. The CO_2 was swept from the water with a stream of inert gas and absorbed in a solution of sodium hydroxide. The solution was returned to shore where the CO_2 was extracted, purified, and counted. When carefully executed, the procedure produced results which were accurate to 2–4 ppt based on counting errors alone. Because of the expense, time, and difficulty, samples for replicate analyses were almost never collected.

With the AMS technique only 0.25 liter of sea water is required. Generally a 0.5 liter water sample is collected at sea and poisoned with $HgCl_2$ to halt all biological activity. The water is returned to the laboratory and acidified, and the CO_2 is extracted and purified. An aliquot of the CO_2 is analyzed to determine $\delta^{13}C$ and the remainder is converted to carbide and counted by AMS. Counting error for the AMS technique can be <2 ppt, however, replicate analysis shows the total sample error to be approximately 4.5 ppt.

Sampling History

Soon after the radiocarbon dating method was developed, it was applied to oceanic and atmospheric samples. During the 1950s and 1960s most of the oceanographic samples were limited to the shallow waters owing to the difficulty of deep water sampling combined with the limited analytical precision. The majority of the early samples were collected in the Atlantic Ocean and the South-west Pacific Ocean. Early sample coverage was insufficient to give a good description of the global surface ocean radiocarbon content prior to the onset of atmospheric testing of thermonuclear weapons; however, repeated sampling at the same location was sufficient to record the surface water increase due to bomb-produced fallout. A very good history of radiocarbon activity, including the increase due to bomb tests and subsequent decrease, exists primarily as a result of the work of R. Nydal and co-workers (Trondheim) and K. Munnich and co-workers (Heidelberg).

The primary application of early radiocarbon results was to estimate the flux of CO_2 between the atmosphere and ocean and the average residence time in the ocean. Sufficient subsurface ocean measurements were made, primarily by W. Broecker (Lamont–Doherty Earth Observatory LDEO) and H. Craig (Scripps Institution of Oceanography SIO), to recognize that radiocarbon had the potential to be an important tracer of deep ocean circulation and mixing rates.

During the 1970s the Geochemical Ocean Sections (GEOSECS) program provided the first full water column global survey of the oceanic radiocarbon distribution. The GEOSECS cruise tracks were approximately meridional through the center of the major ocean basins. Radiocarbon was sampled with a station spacing of approximately 500 km and an average of 20 samples per station. All of the GEOSECS $\Delta^{14}C$ measurements were made by G. Östlund (University of Miami) and M. Stuiver (University of Washington) using traditional β counting of large-volume water samples with a counting accuracy of ~4 ppt. GEOSECS results revolutionized what was known about the oceanic $\Delta^{14}C$ distribution and the applications for which radiocarbon is used.

During the early 1980s the Atlantic Ocean was again surveyed for radiocarbon as part of the Transient Tracers in the Ocean (TTO) North Atlantic Study (NAS) and Tropical Atlantic Study (TAS) programs and the South Atlantic Ventilation Experiment (SAVE). Sampling for these programs was designed to enable mapping of property distributions on constant pressure or density surfaces with reasonable gridding uncertainty. The radiocarbon

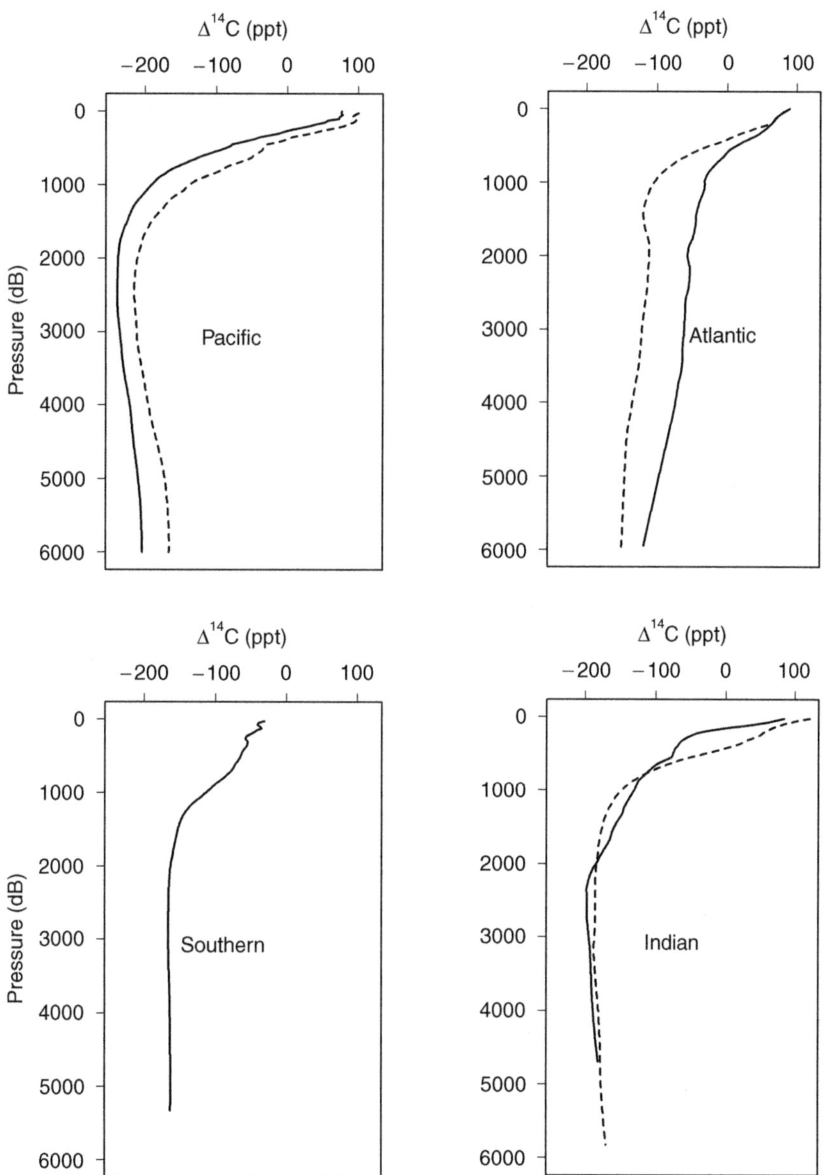

Figure 4 Average vertical $\Delta^{14}C$ profiles for the major ocean basins. Except for the Southern Ocean the dotted line is for the Southern Hemisphere and the solid line for the Northern Hemisphere. The Pacific and Southern Ocean profiles were compiled from WOCE data; the Atlantic profiles from TTO and SAVE data; and the Indian Ocean profiles from GEOSECS data. In approximately the upper 1000 m(= 1000 dB) of each profile, the natural $\Delta^{14}C$ is contaminated with bomb-produced radiocarbon.

portion of these programs was directed by W. Broecker. Östlund made the $\Delta^{14}C$ measurements with $\delta^{13}C$ provided by Stuiver using the GEOSECS procedures. Comparison of TTO results to GEOSECS gave the first clear evidence of the penetration of the bomb-produced radiocarbon signal into the subsurface North Atlantic waters. The French carried out a smaller scale (INDIGO) ^{14}C program in the Indian Ocean during this time with Östlund and P. Quay (University of Washington) collaborating. These data also quantified upper ocean changes since GEOSECS and relied on the same techniques.

The most recent oceanic survey was carried out during the 1990s as part of the World Ocean Circulation Experiment (WOCE). This program was a multinational effort. The US ^{14}C sampling effort was heavily focused on the Pacific (1991–1993) and Indian oceans (1995–1996) since TTO and SAVE had provided reasonable Atlantic coverage. R. Key (Princeton University) directed the US radiocarbon effort with collaboration from P. Schlosser (LDEO) and Quay. In the deep Pacific where gradients were known to be small, most radiocarbon sampling was by the proven large-volume β technique. The Pacific thermocline, however, was sampled using the AMS technique. Shifting techniques allowed thermocline waters to be sampled at approximately 2–3 times the horizontal density used for large volume sampling. Östlund and Stuiver again measured the large-volume samples while the AMS samples were measured at the National Ocean Sciences AMS facility (NOSAMS) at Woods Hole Oceanographic Institution. By 1994 the analytical precision at NOSAMS had improved to the point that all US Indian Ocean WOCE ^{14}C sampling used this technique. WOCE sampling increased the total number of ^{14}C results for the Pacific and Indian Oceans by approximately an order of magnitude. Analysis of the Pacific Ocean samples was completed in 1998. US WOCE ^{14}C sampling in the Atlantic was restricted to two zonal sections in the north-west basin using the AMS technique. Analysis of the Atlantic and Indian Ocean samples is expected to be finished during 2000–2001.

$\Delta^{14}C$ Distribution and Implications for Large-scale Circulation

The distribution of radiocarbon in the ocean is controlled by the production rate in the atmosphere, the spatial variability and magnitude of $^{14}CO_2$ flux across the air–sea interface, oceanic circulation and mixing, and the carbon cycle in the ocean. **Figure 4** shows average vertical radiocarbon profiles for the Pacific, Atlantic, Southern, and Indian oceans with the dotted line being southern basin and solid line northern basin. All of the profiles have higher $\Delta^{14}C$ in shallow waters, reflecting proximity to the atmospheric source. The different collection times combined with the penetration of the bomb-produced signal into the upper thermocline negate the possibility of detailed comparison for the upper 600–800 dB (deeper for the North Atlantic). Detailed comparison is justified for deeper levels. The strongest signal in deep and bottom waters is that the North Atlantic is significantly younger (higher $\Delta^{14}C$) than the South Atlantic, while the opposite holds for the Pacific. Second, the average age of deep water increases ($\Delta^{14}C$ decreases) from Atlantic to Indian to Pacific. Third, the Southern Ocean $\Delta^{14}C$ is very uniform below approximately 1800 dB at a level (~ -160 ppt). This is similar to the near bottom water values for all three southern ocean basins. All three differences are directly attributable to the large-scale thermohaline circulation.

Figure 5 shows meridional sections for the Atlantic, Indian and Pacific oceans using subsets of the data from **Figure 4**. As with **Figure 4**, the $\Delta^{14}C$ values in the upper water column have been increased by invasion of the bomb signal. The pattern of these contours, however, is generally representative of the natural $\Delta^{14}C$ signal. The $\Delta^{14}C = -100‰$ contour can be taken as the approximate demarcation between the bomb-contaminated waters and those having only natural radiocarbon.

Comparison of the major features in each section shows that the meridional $\Delta^{14}C$ distributions in the Pacific and Indian Oceans are quite similar. The greatest difference between these two is that the Indian Ocean deep water (1500–3500 m) is significantly younger than Pacific deep waters. In both oceans:

- The near bottom water has higher $\Delta^{14}C$ than the overlying deep water.
- The deep and bottom waters have higher $\Delta^{14}C$ at the south than the north.
- The lowest $\Delta^{14}C$ values are found as a tongue extending southward from the north end of the section at a depth of ~ 2500 m.
- Deep and bottom water at the south end of each section is relatively uniform with $\Delta^{14}C \sim -160$ ppt.
- The $\Delta^{14}C$ gradient with latitude from south to north is approximately the same for both deep waters and for bottom waters.
- The $\Delta^{14}C$ contours in the thermocline shoal both at the equator and high latitudes. (This feature is suppressed in the North Indian Ocean owing to the limited geographic extent and the influence of flows through the Indonesian Seas region and from the Arabian Sea.)

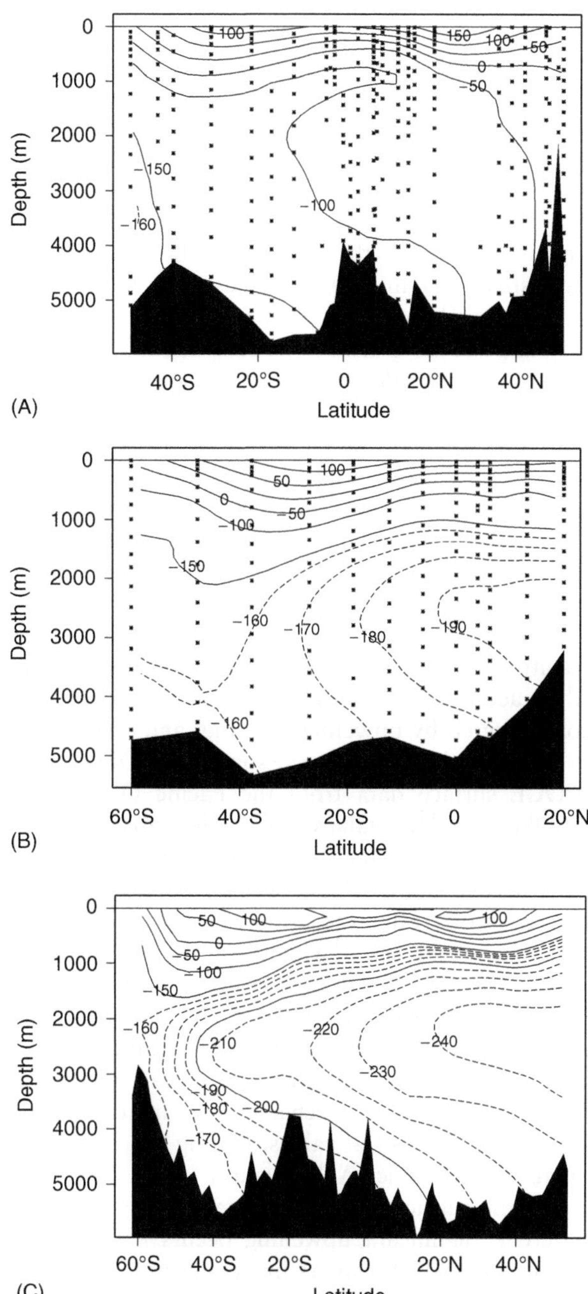

Figure 5 Typical meridional sections for each ocean compiled from a subset of the data used for **Figure 4**. The deep water contour patterns are primarily due to the large-scale thermohaline circulation. The highest deep water $\Delta^{14}C$ values are found in the North Atlantic and the lowest in the North Pacific. The natural $\Delta^{14}C$ in the upper ocean is contaminated by the influx of bomb-produced radiocarbon.

In the Atlantic Ocean the pattern in the shallow water down through the upper thermocline is similar to that in the other oceans. The $\Delta^{14}C$ distribution in the deep and bottom waters of the Atlantic is, however, radically different. The only similarities to the other oceans are (1) the $\Delta^{14}C$ value for deep and bottom water at the southern end of the section, (2) a southward-pointing tongue in deep water, and (3) the apparent northward flow indicated by the near-bottom tongue-shaped contour. Atlantic deep water has higher $\Delta^{14}C$ than the bottom water, and the deep and bottom waters at the north end of the section have higher rather that lower $\Delta^{14}C$ as found in the Indian and Pacific. Additionally, the far North Atlantic deep and bottom waters have relatively uniform values rather than a strong vertical gradient.

The reversal of the Atlantic deep and bottom water $\Delta^{14}C$ gradients with latitude relative to those in the Indian and Pacific is due to the fact that only the Atlantic has the conditions of temperature and salinity at the surface (in the Greenland–Norwegian Sea and Labrador Sea areas) that allow formation of a deep water mass (commonly referred to as North Atlantic Deep Water, NADW). Newly formed NADW flows down slope from the formation region until it reaches a level of neutral buoyancy. Flow is then southward, primarily as a deep western boundary current constrained by the topography of the North American slope. In its southward journey, NADW encounters and overrides northward-flowing denser waters of circumpolar origin. This general circulation pattern can be very clearly demonstrated by comparing the invasion of the bomb-produced tritium and radiocarbon signals obtained during GEOSECS to those from the TTO programs. This large circulation pattern leads to the observed $\Delta^{14}C$ distribution in the deep Atlantic.

Since neither the Pacific nor the Indian Ocean has a northern hemisphere source of deep water, the large-scale circulation is simpler. The densest Pacific waters originate in the Southern Ocean and flow northward along the sea floor (Circumpolar Deep Water, CDW). In the Southern Ocean, CDW is partially ventilated, either by direct contact with the atmosphere or by mixing with waters that have contacted the atmosphere, resulting in somewhat elevated $\Delta^{14}C$. As CDW flows northward, it ages, warms, mixes with overlying water, and slowly upwells. This upwelling, combined with mixing with overlying lower thermocline waters, results in the water mass commonly known as Pacific Deep Water (PDW). PDW has the lowest $\Delta^{14}C$ values found anywhere in the oceans. The long-term mean flow pattern for PDW is somewhat controversial; however, the radiocarbon distribution supports a southward flow with the core of the flow centered around 2500 m. WOCE results further imply that if there is a mean southward flow of PDW, it may be concentrated toward the eastward and westward boundaries rather than uniformly distributed zonally. **Figure 6** shows a zonal Pacific WOCE $\Delta^{14}C$ section at 32°S contoured at the same intervals as the previous sections. PDW is identified by the minimum layer between 2000 and 3000 m. The PDW core appears segregated into two channels, one against the South American slope and the other over the Kermadec Trench. The actual minimum values in the latter were found at ~170°W, essentially abutting the western wall of the trench. The northward-flowing CDW is also clearly indicated in this section by the relatively high $\Delta^{14}C$ values near the bottom between 140°W and the Date Line.

Little has been said about the natural $\Delta^{14}C$ values found in the upper ocean where bomb-produced radiocarbon is prevalent. GEOSECS samples were collected only ~10 years after the maximum in atmospheric $\Delta^{14}C$. GEOSECS surface water measurements almost always had the highest $\Delta^{14}C$ values. Twenty years later during WOCE, the maximum $\Delta^{14}C$ was generally below the surface.

Broecker and Peng (1982, p. 415, Figures 8–19) assembled the few surface ocean $\Delta^{14}C$ measurements made prior to bomb contamination for comparison to the GEOSECS surface ocean data. For the Atlantic and Pacific Oceans, their plot of $\Delta^{14}C$ versus latitude shows a characteristic 'M' shape with maximum $\Delta^{14}C$ values of approximately −50 ppt centered in the main ocean gyres between latitudes 20° and 40°. Each ocean had a relative minimum $\Delta^{14}C$ value of approximately −70 ppt in the equatorial latitudes, 20° S to 20° N and minima at high latitudes ranging from −70 ppt for the far North Atlantic to −150 ppt for the other high latitudes. Pre-bomb measurements in the Indian Ocean are extremely sparse; however, the few data that exist imply a similar distribution. The GEOSECS surface ocean data had the same 'M' shape; however, all of the values were significantly elevated owing to bomb-derived contamination and the pattern was slightly asymmetric about the equator with the Northern Hemisphere having higher values since most of the atmospheric bomb tests were carried out there. The 'M' shape of $\Delta^{14}C$ with latitude is due to circulation patterns, the residence time of surface water in an ocean region, and air–sea gas exchange rates. At mid-latitudes the water column is relatively stable and surface waters reside sufficiently long to absorb a significant amount of ^{14}C from the atmosphere. In the equatorial zone, upwelling of deeper (and therefore lower $\Delta^{14}C$) waters lowers the surface ocean value. At high latitudes, particularly in the Southern Ocean, the near-surface water is relatively unstable, resulting in a short residence time. In these regions $\Delta^{14}C$ acquired from the atmosphere is more than compensated by upwelling, mixing, and convection.

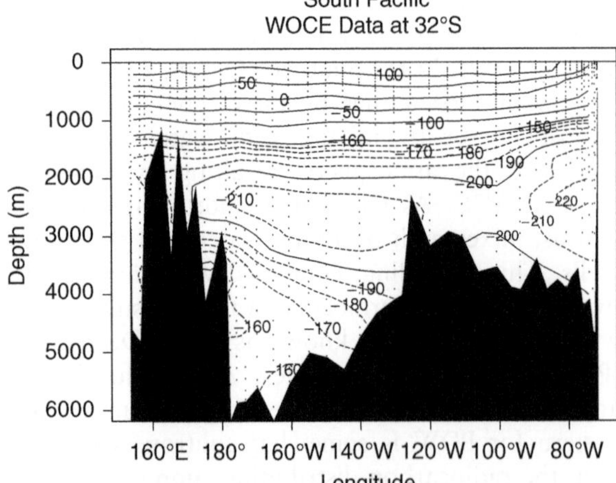

Figure 6 Zonal section of $\Delta^{14}C$ in the South Pacific collected during the WOCE program. The two minima at 2000–2500 m depth are thought to be the core of southward-flowing North Pacific Deep Water. Northward-flowing Circumpolar Deep Water is identified by the relatively high values in the Kermadec Trench area at the bottom between 140°W and the Date Line.

Figure 7 shows a comparison for GEOSECS and WOCE surface data from the Pacific Ocean. The GEOSECS $\Delta^{14}C$ values are higher than WOCE everywhere except for the Equator. The difference is due to two factors. First, GEOSECS sampling occurred shortly after the atmospheric maximum. At that time the air–sea $\Delta^{14}C$ gradient was large and the surface ocean $\Delta^{14}C$ values were dominated by air–sea gas exchange processes. Second, by the 1990s, atmospheric $\Delta^{14}C$ levels had declined significantly and sufficient time had occurred for ocean mixing to compete with air–sea exchange in terms of controlling the surface ocean values. During the 1990s, the maximum oceanic $\Delta^{14}C$ values were frequently below the surface. Near the Equator the situation is different. Significant upwelling occurs in this zone. During GEOSECS, waters upwelling at low latitude in the Pacific were not yet contaminated with bomb radiocarbon. Twenty years later, the upwelling waters had acquired a bomb radiocarbon component.

While surface ocean $\Delta^{14}C$ generally decreased between GEOSECS and WOCE, values throughout the upper kilometer of the water column generally increased as mixing and advection carried bomb-produced radiocarbon into the upper thermocline. The result of these processes on the bomb-produced $\Delta^{14}C$ signal can be visualized by comparing GEOSECS and WOCE depth distributions. **Figure 8** shows such a comparison. To produce this figure the WOCE data from section P16 (152°W) were gridded (center panel). GEOSECS data collected east of the data line were then gridded to the same grid (top

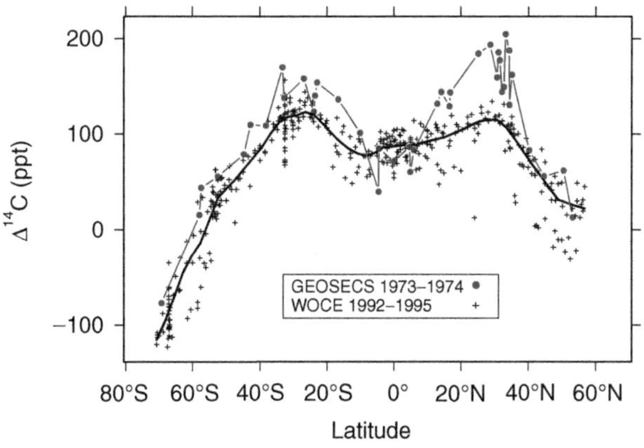

Figure 7 Distribution of Δ^{14}C in the surface Pacific Ocean as recorded by the GEOSECS program in the early 1970s and the WOCE program in the early 1990s. From **Figure 3B** it follows that GEOSECS recorded the maximum bomb-contamination. Over the 20 years separating the programs, mixing and advection dispersed the signal. By the time of WOCE the maximum contamination level was found below the surface at many locations. The asymmetry about the Equator in the GEOSECS data is a result of most atmospheric bomb tests being executed in the Northern Hemisphere.

panel). Once prepared, the two sections were simply subtracted grid box by grid box (bottom panel). One feature of **Figure 8** is the asymmetry about the Equator. The difference at the surface in **Figure 8** reflects the same information (and data) as in **Figure 7**. The greatest increase (up to 60 ppt) along the section is in the Southern Hemisphere mid-latitude thermocline at a depth of 300–800 m. This concentration change decreases in both depth and magnitude toward the Equator. All of the potential density isolines that pass through this region of significant increase (dashed lines in the bottom panel) outcrop in the Southern Ocean. These outcrops (especially during austral winter) provide the primary pathway by which radiocarbon is entering the South Pacific thermocline. In the North Pacific the surface ocean decrease extends as a blob well into the water column (>200 m). This large change is due to the extremely high surface concentrations measured during GEOSECS and to subsurface mixing and ventilation processes that have diluted or dispersed the peak signal. The values contoured in the bottom panel represent the change in Δ^{14}C between the two surveys, not the total bomb Δ^{14}C.

WOCE results from the Indian Ocean are not yet available. Once they are, changes since GEOSECS in the South Indian Ocean should be quite similar to those in the South Pacific because the circulation and ventilation pathways are similar. Changes in the North Indian Ocean are difficult to predict owing to water inputs from the Red Sea and the Indonesian

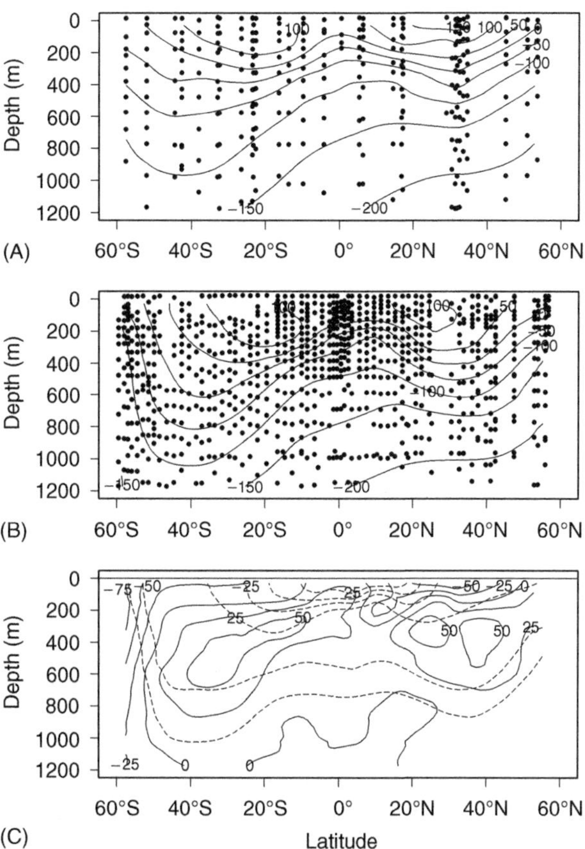

Figure 8 Panel (C) shows the change in the meridional eastern Pacific thermocline distribution of Δ^{14}C between the GEOSECS (1973–1974, (A)) and WOCE (1991–1994, (B)) surveys. The change was computed by gridding each section then finding the difference. The dashed lines in (C) indicate constant potential density surfaces. Negative near-surface values indicate maximum concentration surfaces moving down into the thermocline after GEOSECS. The region of greatest increase in the southern hemisphere is ventilated in the Southern Ocean.

throughflow region and to the changing monsoonal circulation patterns.

Göte Östlund and Claes Rooth described radiocarbon changes in the North Atlantic Ocean using data from GEOSECS (1972) and the TTO North Atlantic Study (1981–1983). The pattern of change they noted is different from that in the Pacific because of the difference in thermohaline circulation mentioned previously. Prior to sinking, the formation waters for NADW are at the ocean surface long enough to pick up significant amounts of bomb radiocarbon from the atmosphere. The circulation pattern coupled with the timing of GEOSECS and TTO sampling resulted in increased Δ^{14}C levels during the latter program. The significant changes were mostly limited to the deep water region north of 40°N latitude. When the WOCE Atlantic samples are analyzed, we expect to see changes extending farther southward.

Separating the Natural and Bomb Components

Up to this point the discussion has been limited to changes in radiocarbon distribution due to oceanic uptake of bomb-produced radiocarbon. Many radiocarbon applications, however, require not the change but the distribution of either bomb or natural radiocarbon. Ocean water measurements give the total of natural plus bomb-produced $\Delta^{14}C$. Since these two are chemically and physically identical, no analytical procedure can differentiate one from the other. Far too few $\Delta^{14}C$ measurements were made in the upper ocean prior to contamination by the bomb component for us to know what the upper ocean natural $\Delta^{14}C$ distribution was.

One separation approach derived by Broecker and co-workers at LDEO uses the fact that $\Delta^{14}C$ is linearly anticorrelated with silicate in waters below the depth of bomb-^{14}C penetration. By assuming the same correlation extends to shallow waters, the natural $\Delta^{14}C$ can be estimated for upper thermocline and near surface water. Pre-bomb values for the ocean surface were approximated from the few pre-bomb surface ocean measurements. The silicate method is limited to temperate and low-latitude waters since the correlation fails at high latitudes, especially for waters of high silicate concentration. More recent work by S. Rubin and R. Key indicates that potential alkalinity (alkalinity + nitrate normalized to salinity of 35) may be a better co-variable than silicate and can be used at all latitudes. **Figure 9** illustrates the silicate and PALK correlations using the GEOSECS data set. Regardless of the co-variable, the correlation is used to estimate pre-bomb $\Delta^{14}C$ in contaminated regions. The difference between the measured and estimated natural $\Delta^{14}C$ is the bomb-produced $\Delta^{14}C$.

In **Figure 10** the silicate and potential alkalinity (PALK) methods are illustrated and compared. The upper panel (A) shows the measured $\Delta^{14}C$ and estimates of the natural $\Delta^{14}C$ using both methods. The bomb $\Delta^{14}C$ is then just the difference between the measured value and the estimate of the natural value (B). For this example, taken from the mid-latitude Pacific, the two estimates are quite close; however this is not always true.

In **Figure 11A** the upper 1000 m of the Pacific WOCE $\Delta^{14}C$ section shown in **Figure 5C** is reproduced. **Figure 11B** shows the estimated natural $\Delta^{14}C$ using the potential alkalinity method. The shape of the two contour sets is quite similar; however, the contour values and vertical gradients are very different, illustrating the strong influence of bomb-produced radiocarbon on the upper ocean. The integrated difference between these two sections would yield an estimate of the bomb-produced $\Delta^{14}C$ inventory for the section.

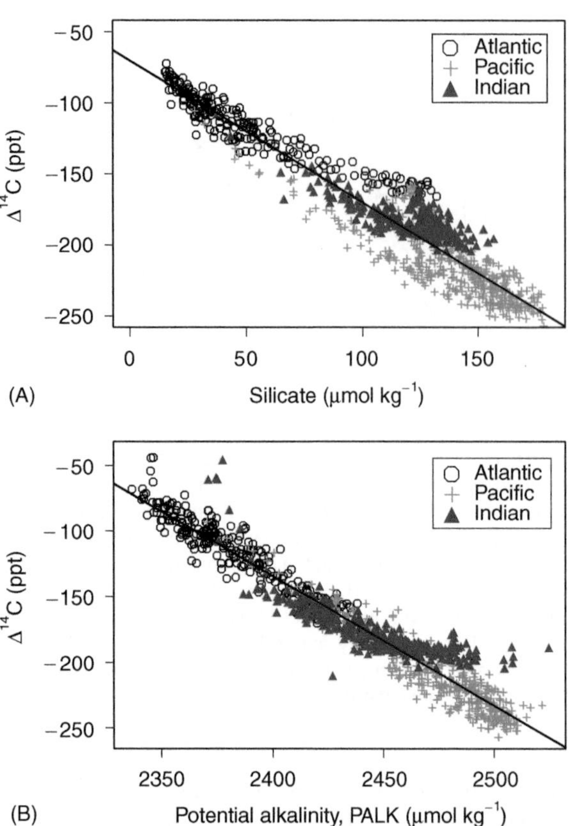

Figure 9 Comparison of the correlation of natural $\Delta^{14}C$ with silicate (A) and potential alkalinity (PALK = [alkalinity + nitrate] × 35/salinity) (B) using the GEOSECS global data. Samples from high southern latitudes are excluded from the silicate relation. The presence of tritium was used to surmise the presence of bomb-$\Delta^{14}C$. The somewhat anomalous high PALK values from the Indian Ocean are from upwelling–high productivity zones and may be influenced by nitrogen fixation and/or particle flux.

Oceanographic Applications

As illustrated, the $\Delta^{14}C$ distribution can be used to infer general large-scale circulation patterns. The most valuable applications for radiocarbon derive from the fact that it is radioactive and has a half-life appropriate to the study of deep ocean processes and that the bomb component is transient and is useful as a tracer for upper ocean processes. A few of the more common uses are described below.

Deep Ocean Mixing and Ventilation Rate, and Residence Time

Since the first subsurface measurements of radiocarbon, one of the primary applications has been the

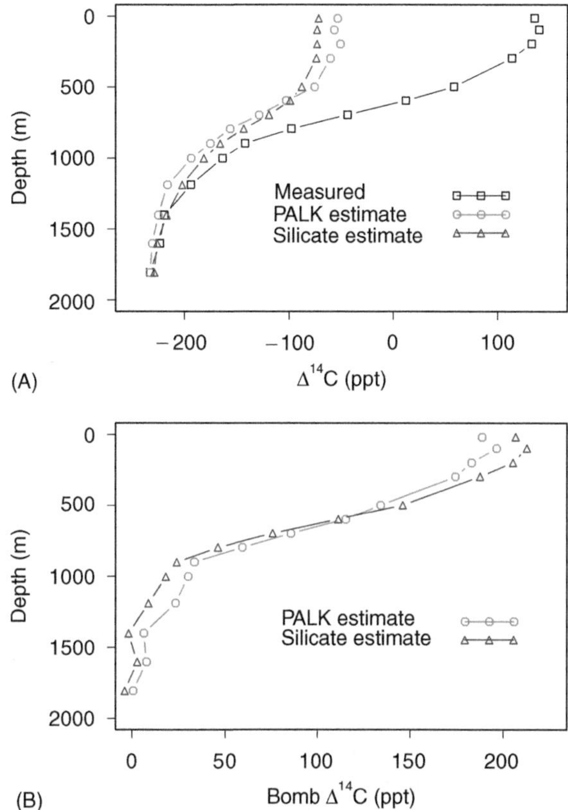

Figure 10 Panel (A) compares measured $\Delta^{14}C$ from a mid-latitude Pacific WOCE station with natural $\Delta^{14}C$ estimated using the silicate and potential alkalinity methods. Bomb-$\Delta^{14}C$, the difference between measured and natural $\Delta^{14}C$, estimated with both methods is compared in (B). Integration of estimated bomb-$\Delta^{14}C$ from the surface down to the depth where the estimate approaches zero yields an estimate of the bomb-$\Delta^{14}C$ inventory. Inventory is generally expressed in units of atoms per unit area.

determination of deep ocean ventilation rates. Most of these calculations have used a box model to approximate the ocean system. The first such estimates yielded mean residence times for the various deep and abyssal ocean basins of 350–900 years. Solution of these models generally assumes a steady-state circulation, identifiable source water regions with known $\Delta^{14}C$, no mixing between water masses, and no significant biological sources or sinks. Another early approach assumed that the vertical distribution of radiocarbon in the deep and abyssal ocean could be described by a vertical advection–diffusion equation. This type of calculation leads to estimates of the effect of biological particle flux and dissolution and to the vertical upwelling and diffusion rates. The 1D vertical advection–diffusion approach has been abandoned for 2D and 3D calculations as the available data and our knowledge of oceanic processes have increased.

When the GEOSECS data became available, box models were again used to estimate residence times and mass fluxes for the abyssal ocean. In this case the model had only four boxes, one for the deep region (>1500 m) of each ocean. New bottom water formation (NADW and Antarctic Bottom Water, AABW) were included as inputs to the Atlantic and Circumpolar boxes. Upwelling was allowed in the Atlantic, Pacific, and Indian boxes and exchange was considered between the Circumpolar box and each of the other three ocean boxes. Results from this calculation gave mean replacement times of 510, 250, 275, and 85 years for the deep Pacific, Indian, Atlantic, and Southern Ocean, respectively, and 500 years for the deep waters of the entire world. Upwelling rates were estimated at 4–5 m y^{-1} and mass transports generally agreed with contemporary geostrophic calculations. Applying the same model to more recent data sets would yield the same results.

Oxygen Utilization Rate

Radiocarbon can be used to determine the rate of biological or geochemical processes such as the rate at which oxygen is consumed in deep ocean water. The simplest example of this would be the case of a water mass moving away from a source region at a steady rate, undergoing constant biological oxygen uptake and not subject to mixing. In such a situation the oxygen utilization rate could be obtained from the slope of oxygen versus ^{14}C in appropriate units. The closest approximation to this situation is the northward transport of CDW in the abyssal Pacific, although the mixing requirement is only approximate. **Figure 12** shows such a plot for WOCE Pacific Ocean samples from deeper than 4000 m and north of 40°S. In this case, apparent oxygen utilization (saturated oxygen concentration at equilibration temperature − measured oxygen concentration) rather than oxygen concentration is plotted, to remove the effect of temperature on oxygen solubility. The least-squares slope of 0.83 μmol kg^{-1} per ppt converts to 0.1 μmol kg^{-1} y^{-1} for an oxygen utilization rate. Generally, mixing with other water masses must be accounted for prior to evaluating the gradient. With varied or additional approximations, very similar calculations have been used to estimate the mean formation rates of various deep water masses.

Ocean General Circulation Model Calibration

Oceanographic data are seldom of value for prediction. Additionally, the effect of a changing

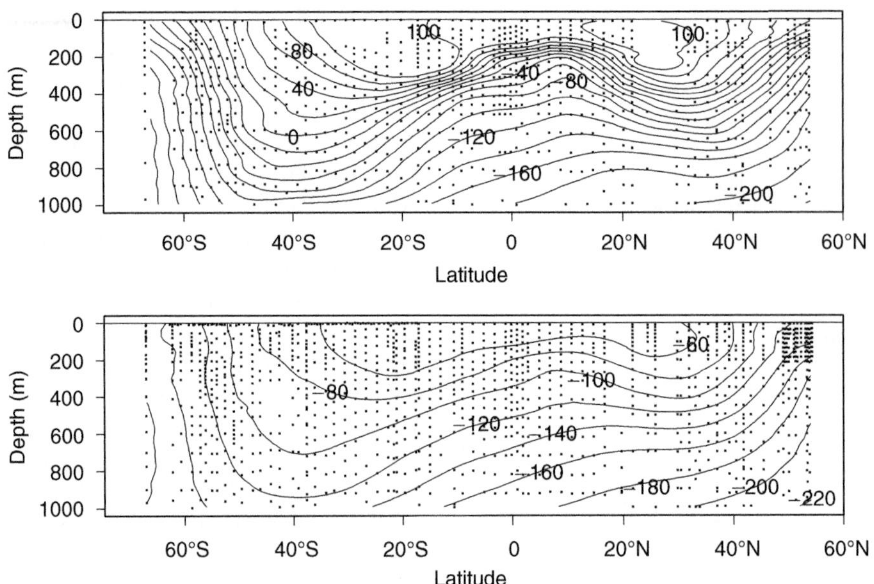

Figure 11 Upper thermocline meridional sections along 152° W in the central Pacific. (A) The same measured data as in **Figure 5C** (B) An estimate of thermocline $\Delta^{14}C$ values prior to the invasion of bomb-produced radiocarbon.

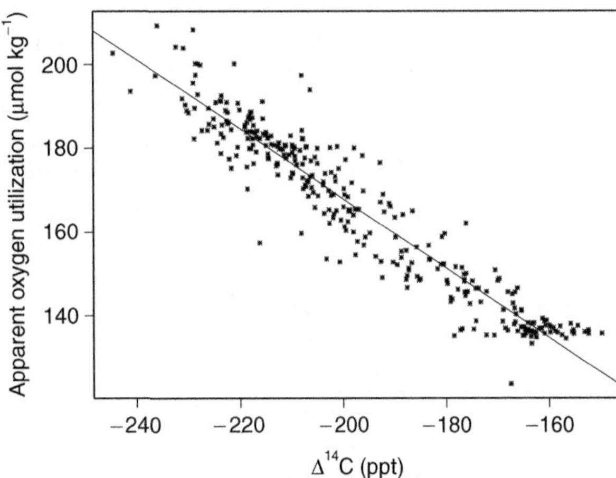

Figure 12 Apparent oxygen utilization plotted against measured $\Delta^{14}C$ for WOCE Pacific Ocean samples taken at depths greater than 4000 m and north of 40°S. The slope of the line (-0.831 ± 0.015) can be used to estimate an approximate oxygen utilization rate of 0.1 μmol kg^{-1} y^{-1} if steady state and no mixing with other water masses is assumed.

oceanographic parameter on another parameter can be difficult to discern directly from data. These research questions are better investigated with numerical ocean models. Before an ocean model result can be taken seriously, however, the model must demonstrate reasonable ability to simulate current conditions. This generally requires that various model inputs or variables be 'tuned' or calibrated to match measured distributions and rates. Radiocarbon is the only common measurement that can be used to calibrate the various rates of abyssal processes in general circulation models. M. Fiadeiro carried out the first numerical simulation for the abyssal Pacific and used the GEOSECS ^{14}C data to calibrate the model. J.R. Toggweiler extended this study using a global model.

Both the Fiadeiro and Toggweiler models, and all subsequent models that include the deep water $\Delta^{14}C$, are of coarse resolution owing to current computer limitations. As the much larger WOCE ^{14}C data set becomes available, the failure of these models, especially in detail, becomes more evident. Toggweiler's model, for example, has advective mixing in the Southern Ocean that is significantly greater than supported by data. Additionally, the coarse resolution of the model prevents the formation of, or at least retards the importance of, deep western boundary currents. Significant model deficiencies appear when the bomb-^{14}C distribution and integrals at the time of GEOSECS and WOCE are compared with data.

During the last 10 years the number and variety of numerical ocean models has expanded greatly, in large part because of the availability and speed of modern computers. The Ocean Carbon Model Intercomparison Project (OCMIP) brought ocean modelers together with data experts in the first organized effort to compare model results with data, with the long-term goals of understanding the processes that cause model differences and of improving the prediction capabilities of the models. The unique aspect of this study was that each participating group

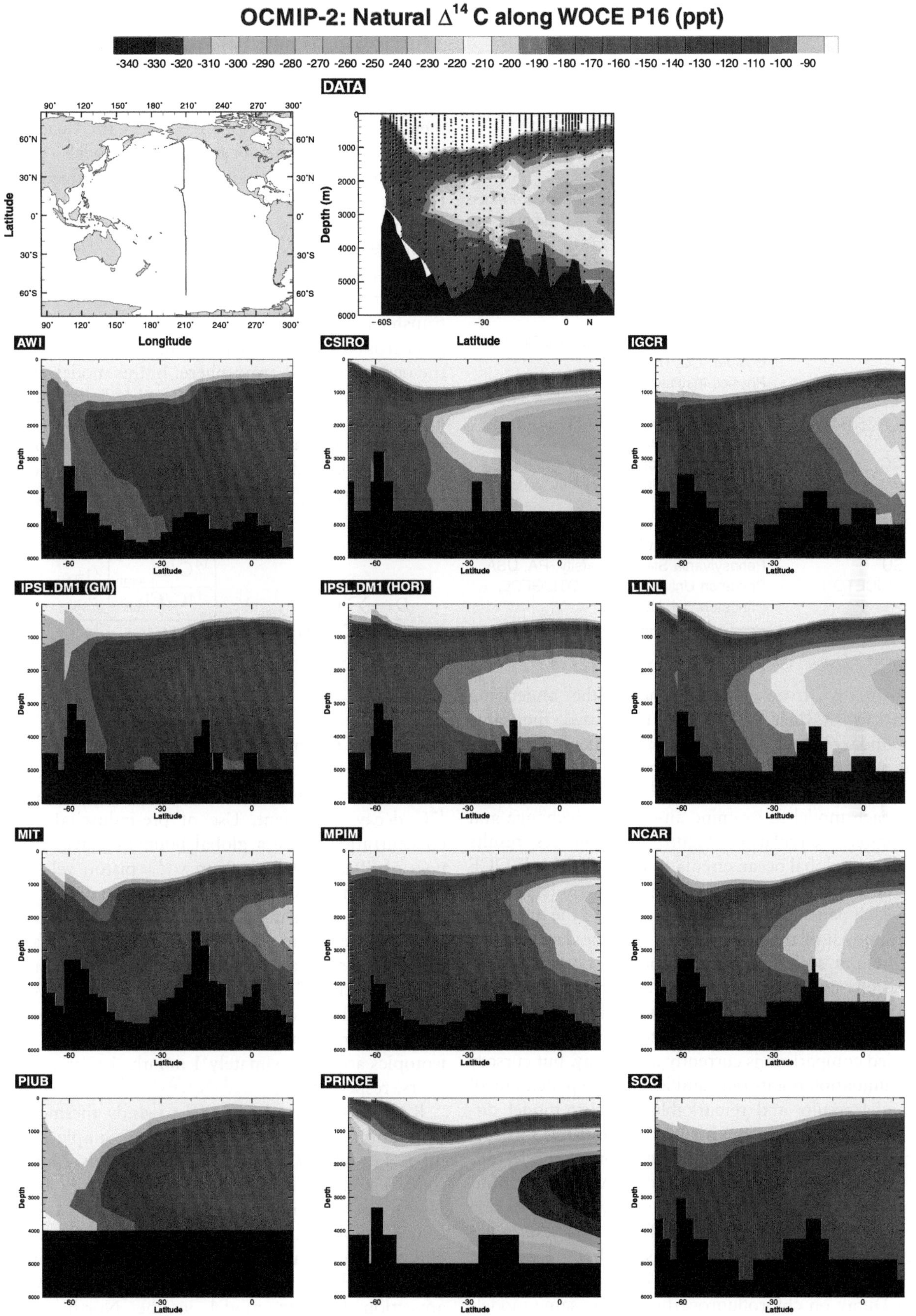
OCMIP-2: Natural Δ14 C along WOCE P16 (ppt)
-340 -330 -320 -310 -300 -290 -280 -270 -260 -250 -240 -230 -220 -210 -200 -190 -180 -170 -160 -150 -140 -130 -120 -110 -100 -90
DATA
Latitude
Longitude
Depth (m)
Latitude
AWI
CSIRO
IGCR
IPSL.DM1 (GM)
IPSL.DM1 (HOR)
LLNL
MIT
MPIM
NCAR
PIUB
PRINCE
SOC

Figure 13 (Right) Global ocean circulation model results from 12 different coarse-resolution models participating in OCMIP-2 compared to WOCE data for natural $\Delta^{14}C$ on a meridional Pacific section. The model groups are identified in each subpanel and in **Table 1** All of the models used the same chemistry and boundary conditions.

Table 1 OCMIP-2 participants

	Model groups
AWI	Alfred Wegener Institute for Polar and Marine Research, Bremerhaven, Germany
CSIRO	Commonwealth Science and Industrial Research Organization, Hobart, Australia
IGCR/CCSR	Institute for Global Change Research, Tokyo, Japan
IPSL	Institut Pierre Simon Laplace, Paris, France
LLNL	Lawrence Livermore National Laboratory, Livermore, CA, USA
MIT	Massachusetts Institute of Technology, Cambridge, MA, USA
MPIM	Max Planck Institut fur Meteorologie, Hamburg, Germany
NCAR	National Center for Atmospheric Research, Boulder, CO, USA
PIUB	Physics Institute, University of Bern, Switzerland
PRINCETON	Princeton University AOS, OTL/GFDL, Princeton, NJ, USA
SOC	Southampton Oceanography Centre/ SUDO/Hadley Center, UK Met. Office
	Data groups
PMEL	Pacific Marine Environmental Laboratory, NOAA, Seattle, WA, USA
PSU	Pennsylvania State University, PA, USA
PRINCETON	Princeton University AOS, OTL/GFDL, Princeton, NJ, USA

essentially 'froze' development of the underlying physics in their model and then used the same boundary conditions and forcing in order to eliminate as many potential variables as possible. Radiocarbon, both bomb-derived and natural, were used as tracers in each model to examine air–sea gas exchange and long-term circulation. **Figure 13** compares results from 12 global ocean circulation models with WOCE data from section P16. The tag in the top left corner of each panel identifies the institution of the modeling group. All of the model results and the data are colored and scaled identically and the portion of the section containing bomb radiocarbon has been masked. While all of the models get the general shape of the contours, the concentrations vary widely. Detailed comparison is currently under way, but cursory examination points out significant discrepancies in all model results and remarkable model-to-model differences. Similar comparisons can be made focusing on the bomb component. Discussion of model differences is beyond the scope of this work. For information, see publications by the various groups having results in **Figure 13** (listed in **Table 1**). These radiocarbon results are not yet published, but an overview of the OCMIP-2 program can be found in the work of Dutay on chlorofluorocarbon in the same models (see Further Reading).

Air–Sea Gas Exchange and Thermocline Ventilation Rate

Radiocarbon has been used to estimate air–sea gas exchange rates for almost as long as it has been measured in the atmosphere and ocean. Generally, these calculations are based on box models, which have both included and excluded the influence of bomb contamination. W. Broecker and T.-H. Peng summarized efforts to estimate air–sea transfer rates up to 1974 and gave examples based on GEOSECS results using both natural and bomb-^{14}C and a stagnant film model. In this, the rate-limiting step for transfer is assumed to be molecular diffusion of the gas across a thin layer separating the mixed layer of the ocean from the atmosphere. In this model, if one assumes steady state for the ^{14}C and ^{12}C distribution and uniform $^{14}C/^{12}C$ for the atmosphere and surface ocean then the amount of ^{14}C entering the ocean must be balanced by decay. For this model the solution is given by eqn [3].

$$\frac{D}{z} = \frac{\sum[CO_2]|_{ocean}}{\sum[CO_2]|_{mix}} \frac{V}{A} \frac{\left[\frac{^{14}C/C|_{ocean}}{^{14}C/C|_{mix}}\right]\frac{\alpha_{^{14}CO_2}}{\alpha_{CO_2}}}{1 - \left[\frac{^{14}C/C|_{mix}}{^{14}C/C|_{atm}}\right]\frac{\alpha_{^{14}CO_2}}{\alpha_{CO_2}}}\lambda \quad [3]$$

Here D is the molecular diffusivity of CO_2, z is the film thickness, α_i is the solubility of i, V and A are the volume and surface area of the ocean, and λ is the ^{14}C decay coefficient. Use of pre-industrial mean concentrations gave a global boundary layer thickness of 30 μm ($D/z \sim 1800\,m\ y^{-1}$ = piston velocity). The film thickness is then used to estimate gas residence times either in the atmosphere or in the mixed layer of the ocean. For CO_2 special consideration must be made for the chemical speciation in the ocean, and for $^{14}CO_2$ further modification is necessary for isotopic effects. The equilibration times for CO_2 with respect to gas exchange, chemistry, and isotopics are approximately 1 month, 1 year, and 10 years, respectively.

Radiocarbon has been used to study thermocline ventilation using tools ranging from simple 3-box models to full 3D ocean circulation models. Many of the 1D and 2D models are based on work by W. Jenkins using tritium in the North Atlantic. In a recent example, R. Sonnerup and co-workers at the University of Washington used chlorofluorocarbon data to calibrate a 1D (meridional) along-isopycnal advection–diffusion model in the North Pacific with WOCE data. [4] is the basic equation for the

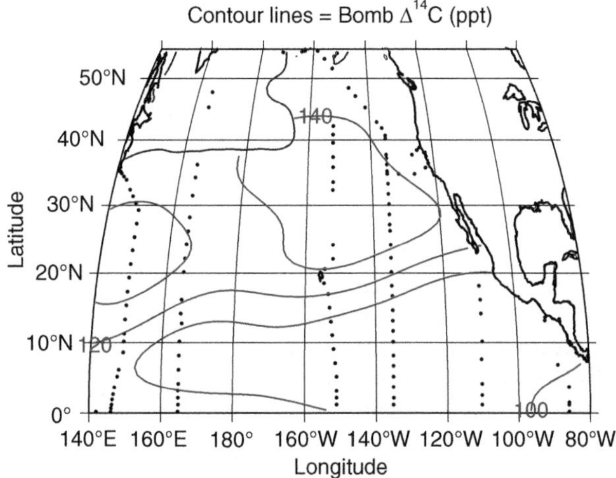

Figure 14 Bomb-Δ^{14}C on the potential density surface $\sigma_\theta = 26.1$ in the North Pacific. The blue line is the wintertime outcrop of the surface based on long-term climatology. The Sea of Okhotsk is a known region of thermocline ventilation for the North Pacific.

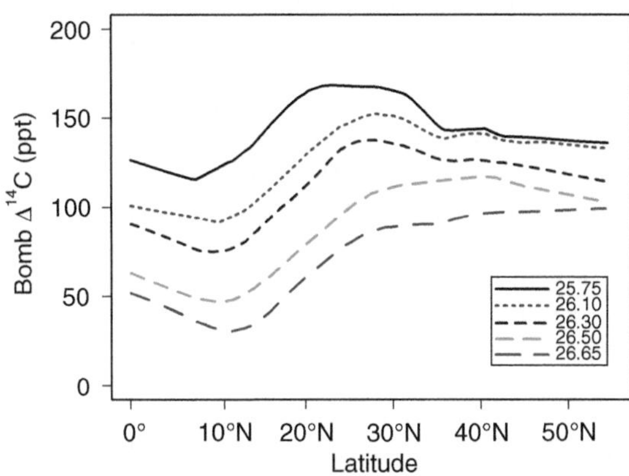

Figure 15 Meridional distribution of bomb-Δ^{14}C on potential density surfaces in the North Pacific thermocline.

model.

$$\frac{dC}{dt} = -v\frac{dC}{dx} + K\frac{d^2C}{dx^2} \qquad [4]$$

In eqn [4] C is concentration, K is along-isopycnal eddy diffusivity, $-v$ is the southward component of along isopycnal velocity, t is time, and x is the meridional distance. Upper-level isopycnal surfaces outcrop at the surface. Once the model is calibrated, the resulting values are used to investigate the distribution of other parameters. The original work and the references cited there should be read for details, but **Figure 14** shows an objective map of the bomb-^{14}C distribution on the potential density surface 26.1 for the North Pacific and **Figure 15** summarizes the bomb-^{14}C distribution as a function of latitude. These figures illustrate the type of data that would be input considerations to an investigation of thermocline ventilation.

Conclusions

Since the very earliest measurements, radiocarbon has proven to be an extremely powerful tracer, and sometimes the only available tracer, for the study of many oceanographic processes. Perhaps the most important of these today are large-scale deep ocean mixing and ventilation processes and the calibration of numerical ocean models. The first global survey of the radiocarbon distribution collected on the GEOSECS program resulted in radical changes in the way the abyssal ocean is viewed. The newer and much denser WOCE survey will certainly add significant detail and precision to what is known and will probably result in other, if not so many, totally new discoveries. Progress with this tracer today is due largely to the decrease in required sample size from ~250 liters to ~250 milliliters and to the availability and application of fast, inexpensive computers.

Glossary

dpm Disintegrations per minute: a measure of the activity of a radioactivesubstance frequently used rather than concentration.

$t_{1/2}$ Half-life: time required for one half of the atoms of a radio active species to decay.

λ Decay constant for a radioactive species = ln(2)/$t_{1/2}$

Mean life, λ^{-1} Average time expected for a given radioactive atom to decay.

Abyssal Very deep ocean, often near bottom.

Steady state Unchanging situation over long time interval relative to the process under consideration; frequently assumed state for the deep and abyssal ocean with respect to many parameters.

See also

Atmospheric Input of Pollutants. Carbon Cycle. Carbon Dioxide (CO_2) Cycle. Elemental Distribution: Overview. Long-Term Tracer Changes. Marine Silica Cycle. Ocean Carbon System, Modeling of. Radioactive Wastes. Stable Carbon Isotope Variations in the Ocean. Tritium–Helium Dating.

Further Reading

Broecker WS, Gerard R, Ewing M, and Heezen BC (1960) Natural radiocarbon in the Atlantic Ocean. *Journal of Geophysical Research* 65(a): 2903–2931.

Broecker WS and Peng T-H (1974) Gas exchange rates between air and sea. *Tellus* 26: 21–34.

Broecker WS and Peng T-H (1982) *Tracers in the Sea.* Columbia University, Palisades, NY: Lamont-Doherty Geological Observatory.

Broecker WS, Sutherland S, Smethie W, Peng T-H, and Östlund G (1995) Oceanic radiocarbon: separation of natural and bomb components. *Global Biogeochemical Cycles* 9(2): 263–288.

Craig H (1969) Abyssal carbon radiocarbon in the Pacific. *Journal of Geophysical Research* 74(23): 5491–5506.

Druffel ERM and Griffin S (1999) Variability of surface ocean radiocarbon and stable isotopes in the southwestern Pacific. *Journal of Geophysical Research* 104(C10): 23607–23614.

Dutay JC, Bullister JL, and Doney SC (2001) Evaluation of ocean model ventilation with CFC-11: comparison of 13 global ocean models. *Global Biogeochemical Cycles.* (in press).

Fiadeiro ME (1982) Three-dimensional modeling of tracers in the deep Pacific Ocean, II. Radiocarbon and circulation. *Journal of Marine Research* 40: 537–550.

Key RM, Quay PD, Jones GA, *et al.* (1996) WOCE AMS radiocarbon I: Pacific Ocean results (P6, P16 and P17). *Radiocarbon* 38(3): 425–518.

Libby WF (1955) *Radiocarbon Dating.* Chicago: University of Chicago Press.

Östlund HG and Rooth CGH (1990) The North Atlantic tritium and radiocarbon transients 1972–1983. *Journal of Geophysical Research* 95(C11): 20147–20165.

Schlosser P, Bönisch G, and Kromer B (1995) Mid-1980s distribution of tritium, ^{3}He, ^{14}C and ^{39}Ar in the Greenland/Norwegian Seas and the Nansen Basin of the Arctic Ocean. *Progress in Oceanography* 35: 1–28.

Sonnerup RE, Quay PD, and Bullister JL (1999) Thermocline ventilation and oxygen utilization rates in the subtropical North Pacific based on CFC distributions during WOCE. *Deep-Sea Research I* 46: 777–805.

Stuiver M and Polach HA (1977) Discussion: Reporting of ^{14}C data. *Radiocarbon* 19(3): 355–363.

Stuiver M and Quay P (1983) Abyssal water carbon-14 distribution and the age of the World Ocean. *Science* 219: 849–851.

Taylor RE, Long A, and Kra RS (eds.) (1992) *Radiocarbon After Four Decades, An Interdisciplinary Perspective.* New York: Springer.

Toggweiler JR, Dixon K, and Bryan K (1989) Simulations of radiocarbon in a coarse-resolution World Ocean model. 1. Steady state prebomb distributions. *Journal of Geophysical Research* 94(C6): 8217–8242.

SINGLE COMPOUND RADIOCARBON MEASUREMENTS

T. I. Eglinton and A. Pearson, Woods Hole Oceanographic Institution, Woods Hole, MA, USA

Introduction

Many areas of scientific research use radiocarbon (carbon-14, ^{14}C) measurements to determine the age of carbon-containing materials. Radiocarbon's ~5700-year half-life means that this naturally occurring radioisotope can provide information over decadal to millennial timescales. Radiocarbon is uniquely suited to biogeochemical studies, where much research is focused on carbon cycling at various spatial and temporal scales. In oceanography, investigators use the ^{14}C concentration of dissolved inorganic carbon (DIC) to monitor the movement of water masses throughout the global ocean. In marine sediment geochemistry, a major application is the dating of total organic carbon (TOC) in order to calculate sediment accumulation rates. Such chronologies frequently rely on the premise that most of the TOC derives from marine biomass production in the overlying water column.

However, the ^{14}C content of TOC in sediments, as well as other organic pools in the ocean (dissolved and particulate organic matter in the water column) often does not reflect a single input source. Multiple components with different respective ages can contribute to these pools and can be deposited concurrently in marine sediments (**Figure 1**). This is particularly true on the continental margins, where fresh vascular plant debris, soil organic matter, and fossil carbon eroded from sedimentary rocks can contribute a significant or even the dominant fraction of the TOC. This material dilutes the marine input and obscures the true age of the sediment. Although such contributions from multiple organic carbon

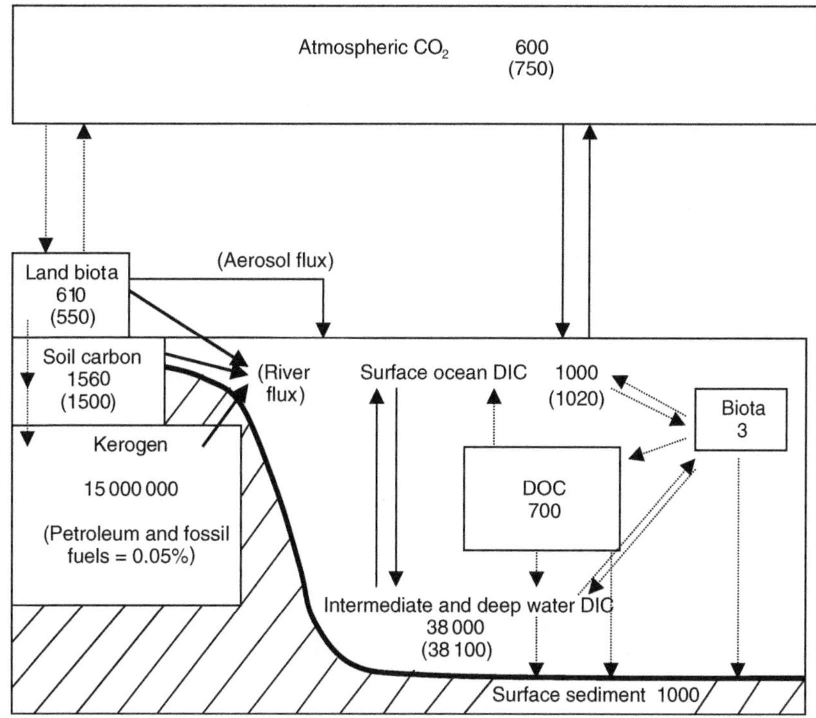

Figure 1 Major global reservoirs involved in active production, exchange and cycling of organic carbon. Reservoir sizes are shown in Gt carbon (1 GtC = 10^{15} g C). Numbers in parentheses are based on 1980s values; numbers without parentheses are estimates of the pre-anthropogenic values. Fluxes primarily mediated by biological reactions are shown with dashed arrows; physical transport processes are shown with solid arrows. (Modified after Siegenthaler and Sarmiento (1993) and Hedges and Oades (1997).)

sources can complicate the development of TOC-based sediment chronologies, these sediment records hold much important information concerning the cycling of organic carbon both within and between terrestrial and marine systems. The challenge, then, is to decipher these different inputs by resolving them into their individual parts.

Most of the allochthonous, or foreign, sources represent carbon with lower ^{14}C concentrations ('older' $\Delta^{14}C$; radiocarbon ages) than the fraction of TOC originating from phytoplanktonic production. The only exception is the rapid transport and sedimentation of recently synthesized terrestrial plant material, which is in equilibrium with the ^{14}C concentration of atmospheric CO_2. Other sources of nonmarine carbon typically are of intermediate (10^3–10^4 years) or 'infinite' $\Delta^{14}C$; (beyond the detection limit of 50–60 000 years) radiocarbon age, depending on the amount of time spent in other reservoirs such as soils, fluvial deposits, or carbon-rich rocks.

It is only at the molecular level that the full extent of this isotopic heterogeneity resulting from these diverse organic carbon inputs is expressed. Isotopic analysis of individual biomarker compounds was employed originally to study the stable carbon isotope (^{13}C) distribution in lipids of geological samples. It proved to be a useful tool to describe the diversity of carbon sources and metabolic pathways as well as to link specific compounds with their biological origins. Recently, this approach was expanded into a second isotopic dimension by the development of a practical method to achieve compound-specific ^{14}C analysis. Not only do these new ^{14}C analyses of individual biomarker molecules provide a tool for dating sediments, but they are another source of fundamental information about biogeochemical processes in the marine environment.

Carbon Isotopes

Carbon in the geosphere is composed of the stable isotopes ^{12}C (98.9%) and ^{13}C (1.1%), and the cosmogenic radionucleotide, ^{14}C (radiocarbon). Upon production, ^{14}C is incorporated quickly into atmospheric CO_2, where it occurs as approximately 10^{-10}% of the total atmospheric abundance of CO_2. The distribution of the minor isotopes relative to ^{12}C is governed by thermodynamic and kinetic fractionation processes[1], in addition to the radioactive decay associated with ^{14}C.

[1] This article assumes the reader is familiar with the conventions used for reporting stable carbon isotopic ratios, i.e., $\delta^{13}C(ppt) = 1000[R/R_{PDB}) - 1]$ where $R \equiv {}^{13}C/{}^{12}C$. For further explanation, see the additional readings listed at the end of this article.

^{14}C Systematics

Today, most radiocarbon data are obtained through the use of accelerator mass spectrometry (AMS) rather than by counting individual decay events. In particular, the advantage of AMS is its small carbon requirement (micrograms to milligrams); this ability to analyze small samples is critical to the compound-specific ^{14}C approach, where sample sizes typically range from tens to hundreds of micrograms. These sample sizes are dictated by natural concentrations of the analytes in geochemical samples (often $<1\ \mu g\ g^{-1}$ dry sediment), and by the capacity of the techniques used to isolate the individual compounds in high purity.

Raw AMS data are reported initially as fraction modern (f_m) carbon (eqn [1]).

$$f_m = \frac{R_{sn}^{14/12}}{R_{std}^{14/12}} \qquad [1]$$

$R^{14/12} \equiv {}^{14}C/{}^{12}C$ (some laboratories use $R = {}^{14}C/{}^{13}C$), sn indicates the sample has been normalized to a constant ^{13}C fractionation equivalent to $\delta^{13}C = -25$ ppt, and std is the oxalic acid I (HOxI) or II (HOxII) modern-age standard, again normalized with respect to ^{13}C.

For geochemical applications, data often are reported as $\Delta^{14}C$ values (eqn [2]).

$$\Delta^{14}C = \left[f_m\left(\frac{e^{\lambda(y-x)}}{e^{\lambda(y-1950)}}\right) - 1\right] \times 1000 \qquad [2]$$

Here, $\lambda = 1/8267(y^{-1})(= t_{1/2}/\ln 2)$, y equals the year of measurement, and x equals the year of sample formation or deposition (applied only when known by independent dating methods, for example, by the use of ^{210}Pb). This equation standardizes all $\Delta^{14}C$ values relative to the year AD 1950. In oceanography, $\Delta^{14}C$ is a convenient parameter because it is linear and can be used in isotopic mass balance calculations of the type shown in eqn [3].

$$\Delta^{14}C_{total} = \sum_i (\chi_i \Delta^{14}C_i) \sum_i \chi_i = 1 \qquad [3]$$

The 'radiocarbon age' $\Delta^{14}C$; of a sample is defined strictly as the age calculated using the Libby half-life of 5568 years (eqn [4]).

$$\text{Age} = -8033\, ln(f_m) \qquad [4]$$

For applications in which a calendar date is required, the calculated ages subsequently are converted are using calibration curves that account for past natural variations in the rate of formation of ^{14}C. However, the true half-life of ^{14}C is 5730 years, and this true value should be used when making decay-related corrections in geochemical systems.

^{14}C Distribution in the Geosphere

Natural Processes

Atmospheric $^{14}CO_2$ is distributed rapidly throughout the terrestrial biosphere, and living plants and their heterotrophic consumers (animals) are in equilibrium with the $\Delta^{14}C$ value of the atmosphere. Thus, in radiocarbon dating, the ^{14}C concentration of a sample is strictly an indicator of the amount of time that has passed since the death of the terrestrial primary producer. When an organism assimilates a fraction of pre-aged carbon, an appropriate 'reservoir age' $\Delta^{14}C$; must be subtracted to correct for the deviation of this material from the age of the atmosphere. Therefore, reservoir time must be considered when interpreting the ^{14}C 'ages' $\Delta^{14}C$; of all of the global organic carbon pools other than the land biota.

For example, continuous vertical mixing of the ocean provides the surface waters with some abyssal DIC that has been removed from contact with atmospheric CO_2 for up to 1500 years. This process gives the ocean an average surface water reservoir age of about 400 years ($\Delta^{14}C = -50$ ppt). A constant correction factor of 400 years often is subtracted from the radiocarbon dates of marine materials (both organic and inorganic). There are regional differences, however, and in upwelling areas the true deviation can approach 1300 years.

An example of the actual range of $\Delta^{14}C$ values found in the natural environment is shown in **Figure 2A**. This figure shows the distribution of ^{14}C in and around Santa Monica Basin, California, USA, prior to significant human influence. The basin sediments are the final burial location for organic matter derived from many of these sources, and the TOC $\Delta^{14}C$ value of -160 ppt represents a weighted average of the total organic carbon flux to the sediment surface.

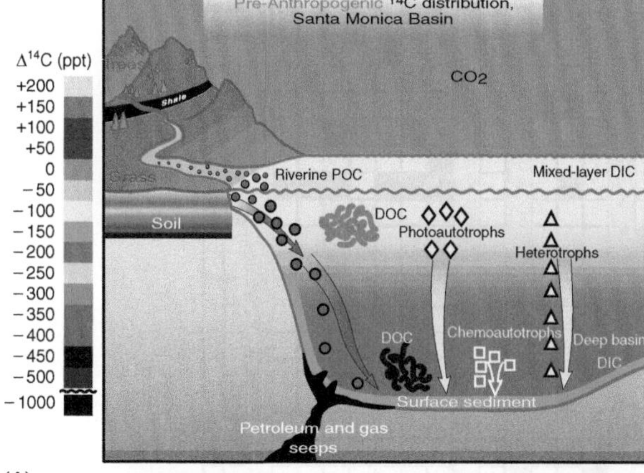

(A)

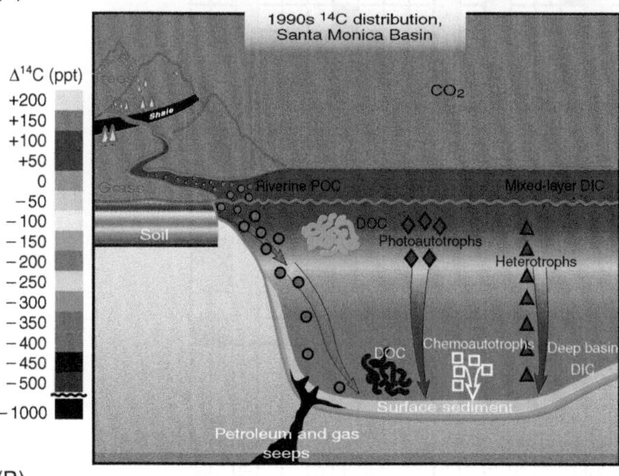

(B)

Figure 2 $\Delta^{14}C$ values for bulk carbon reservoirs in the region of Santa Monica Basin, California, USA: (A) prior to human influence ('pre-bomb' $\Delta^{14}C$;), and (B) contemporary values ('post-bomb' $\Delta^{14}C$;). Modified after Pearson (2000).

Anthropogenic Perturbation

In addition to natural variations in atmospheric levels of ^{14}C, anthropogenic activity has resulted in significant fluctuations in ^{14}C content. The utilization of fossil fuels since the late nineteenth century has introduced ^{14}C- (and ^{13}C-) depleted CO_2 into the atmosphere (the 'Suess effect' $\Delta^{14}C$;). In sharp contrast to this gradual change, nuclear weapons testing in the 1950s and 1960s resulted in the rapid injection of an additional source of ^{14}C into the environment. The amount of ^{14}C in the atmosphere nearly doubled, and the $\Delta^{14}C$ of tropospheric CO_2 increased to greater than $+900$ ppt in the early 1960s. Following the above-ground weapons test ban treaty of 1962, this value has been decreasing as the excess ^{14}C is taken up by oceanic and terrestrial sinks for CO_2. This anthropogenically derived $^{14}CO_2$ 'spike' $\Delta^{14}C$; serves as a useful tracer for the rate at which carbon moves through its global cycle. Any carbon reservoir currently having a $\Delta^{14}C$ value >0 ppt has taken up some of this 'bomb-^{14}C' $\Delta^{14}C$;. Carbon pools that exhibit no increase in $\Delta^{14}C$ over their 'pre-bomb' $\Delta^{14}C$; values have therefore been isolated from exchange with atmospheric CO_2 during the last 50 years. This contrast between 'pre-bomb' $\Delta^{14}C$; and 'post-bomb' $\Delta^{14}C$; $\Delta^{14}C$ values can serve as an excellent tracer of biogeochemical processes over short timescales. These changes in ^{14}C concentrations can be seen in the updated picture of the Santa Monica Basin regional environment shown in **Figure 2B**, where bomb-^{14}C has invaded everywhere except for the deep basin waters and older sedimentary deposits.

In general, the global distribution of organic ^{14}C is complicated by these interreservoir mixing and exchange processes. The more end-member sources contributing organic carbon to a sample, the more complicated it is to interpret a measured $\Delta^{14}C$ value, especially when trying to translate that value to chronological time. Source-specific ^{14}C dating is needed, and this requires isotopic measurements at the molecular level.

Compound-specific ^{14}C Analysis: Methods

The ability to perform natural-abundance ^{14}C measurements on individual compounds has only recently been achieved. This capability arose from refinements in the measurement of ^{14}C by AMS that allow increasingly small samples to be measured, and from methods that resolve the complex mixtures encountered in geochemical samples into their individual components. Here we describe the methods that are currently used for this purpose.

Selection of Compounds for ^{14}C Analysis

The organic matter in marine sediments consists of recognizable biochemical constituents of organisms (carbohydrates, proteins, lipids, and nucleic acids) as well as of more complex polymeric materials and nonextractable components (humic substances, kerogen). Among the recognizable biochemicals, the lipids have a diversity of structures, are comparatively easy to analyze by gas chromatographic and mass spectrometric techniques, and are resistant to degradation over time. These characteristics have resulted in a long history of organic geochemical

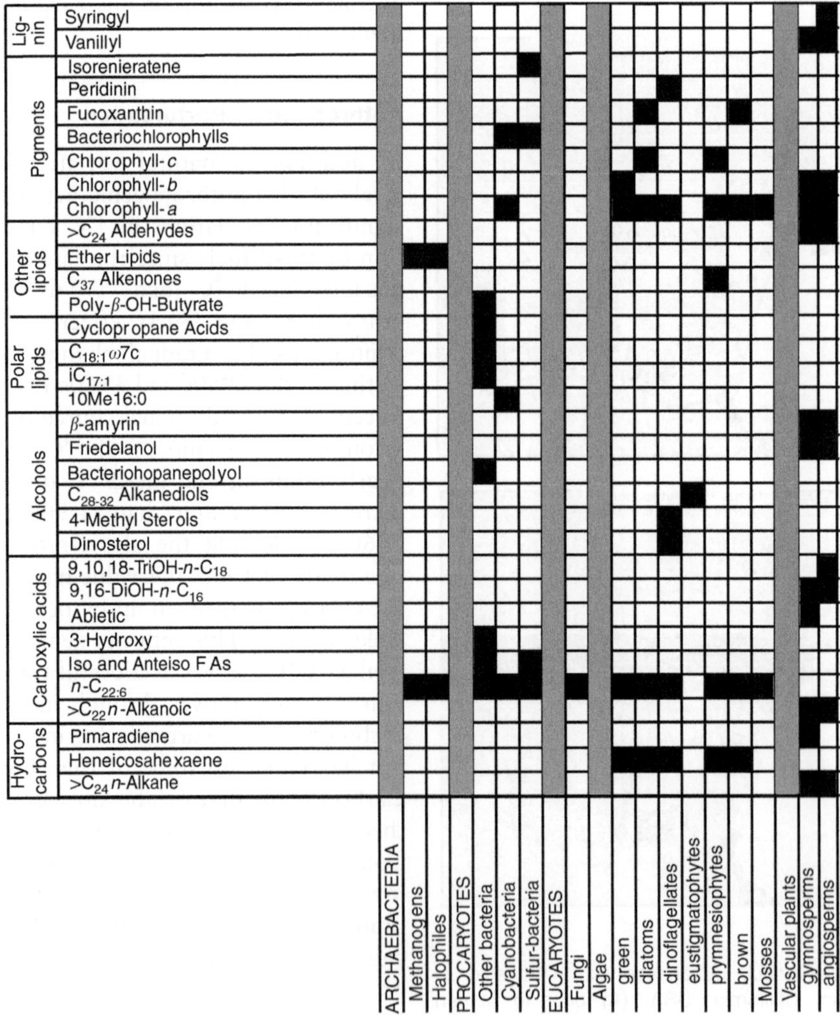

Figure 3 Common source assignments of lipid biomarkers (Modified from Hedges and Oades (1997).)

studies aimed at identifying and understanding the origins of 'source-specific' $\Delta^{14}C$; lipid 'biomarker' $\Delta^{14}C$; compounds. Frequently, lipids from several organic compound classes are studied within the same sample (**Figures 3** and **4**).

Although many of the most diagnostic compounds are polar lipids that are susceptible to modification during sediment diagenesis (e.g., removal of functional groups, saturation of double bonds), several retain their marker properties through the preservation of the carbon skeleton (**Figure 5**). Thus sterols (e.g., cholesterol) are transformed to sterenes and ultimately steranes. The isotopic integrity of the compound is also preserved in this way.

It is sometimes the case that families of compounds can also be characteristic of a particular source. For example, plant waxes comprise homologous series of *n*-alkanes, *n*-alkanols, and *n*-alkanoic acids (**Figure 3**). As a result, ^{14}C measurements of a compound class can yield information with similar specificity to single compound ^{14}C analysis, with the benefits of greater total analyte abundance and, potentially, simpler isolation schemes.

Compound Separation and Isolation

Procedures for single-compound ^{14}C analysis are quite involved, requiring extraction, purification, modification and isolation of the target analytes (**Figure 5**). For lipid analyses, the samples are processed by extracting whole sediment with solvents such as methylene chloride, chloroform, or methanol to obtain a total lipid extract (TLE). The TLE is then separated into compound classes using solid–liquid chromatography. The compound classes elute on the basis of polarity differences, from least polar (hydrocarbons) to most polar (free fatty acids) under normal-phase chromatographic conditions. Individual compounds for ^{14}C analysis are then isolated from these polarity fractions. Additional chromatographic steps or chemical manipulations may be included to reduce the number of components in each fraction prior to single compound isolation, or to render the compounds amenable to isolation by the method chosen. These steps may include silver nitrate-impregnated silica gel chromatography (separation of saturated from unsaturated compounds), 'molecular sieving' $\Delta^{14}C$; (e.g., urea adduction, for separation of branched/cyclic compounds from straight-chain compounds), and derivatization (for protection of functional groups, such as carboxyl or hydroxyl groups, prior to gas chromatographic separation).

For ^{14}C analysis by AMS, tens to hundreds of micrograms of each individual compound must be isolated from the sample of interest. Isolation of individual biomarkers from geochemical samples such as marine sediments and water column particulate matter requires separation techniques with high resolving power. To date, this has been most effectively achieved through the use of automated preparative capillary gas chromatography (PCGC; **Figure 6**).

A PCGC system consists of a commercial capillary gas chromatograph that is modified for work on a semipreparative, rather than analytical, scale. Modifications include a large-volume injection system; high-capacity, low-bleed 'megabore' $\Delta^{14}C$; (e.g., 60 m length $\times$ 0.53 mm inner diameter $\times$ 0.5 μm stationary phase film thickness) capillary columns; an effluent splitter; and a preparative trapping device in which isolated compounds are collected in a series of cooled U-tube traps. Approximately 1% of the effluent passes to a flame ionization detector (FID) and the remaining

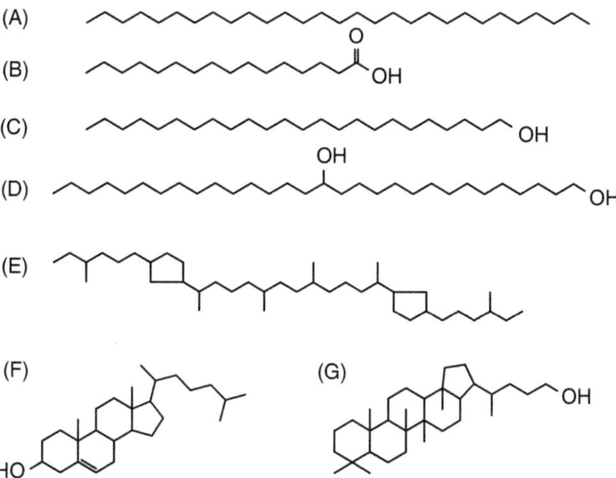

Figure 4 Selected example structures (carbon skeletons and functional groups) for biomarkers shown in Table 1: (A) *n*-C_{29} alkane; (B) *n*-$C_{16:0}$ alkanoic acid; (C) *n*-C_{24} alkanol; (D) C_{30} alkanediol; (E) $C_{40:2cy}$ isoprenoid; (F) C_{27} Δ^5-sterol (cholesterol); (G) C_{32} hopanol.

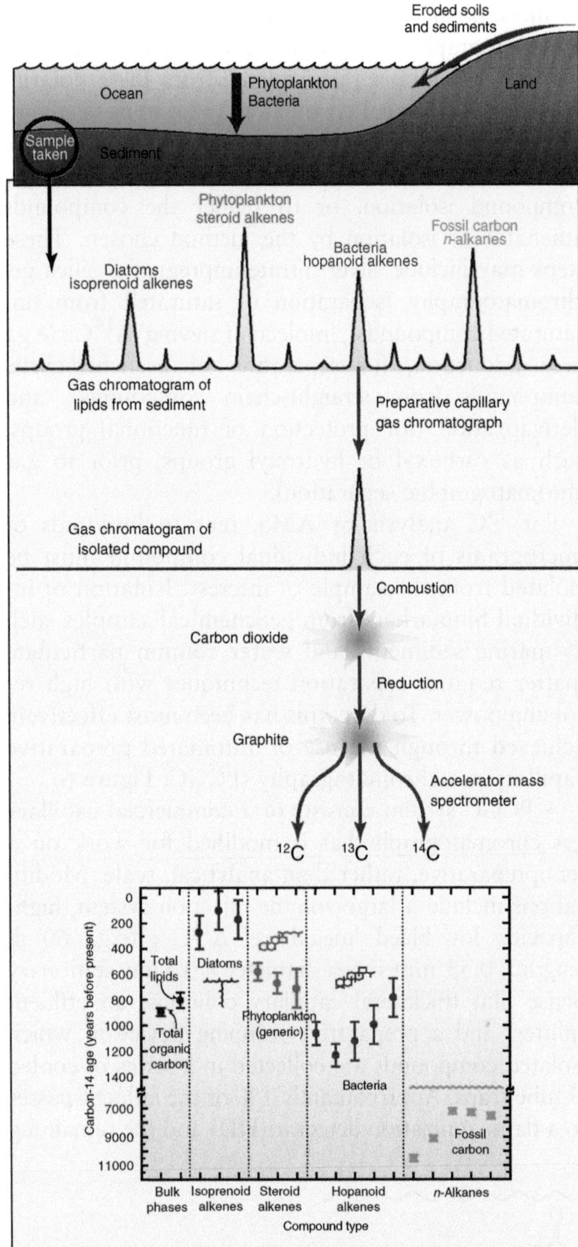

Figure 5 Schematic diagram showing steps for the isolation and ^{14}C analysis of individual sedimentary lipids.

99% is diverted to the collection system. The traps are programmed to receive compounds of interest on the basis of chromatographic retention time windows determined from the FID trace. Computerized synchronization of the trapping times permits collection of multiple identical runs (often > 100 consecutive injections). Using PCGC, baseline resolution of peaks can be achieved at concentrations > 100-fold higher than typical analytical GC conditions, allowing up to 5 μg of carbon per chromatographic peak, per injection, to be separated (to achieve greater resolution, typical loadings are usually about 1 μg of carbon per peak). An example of a typical PCGC separation is shown in **Figure** 7, where ~40–130 μg of individual sterols (as their acetate derivatives) were resolved and isolated from a total sterol fraction obtained from Santa Monica Basin surface sediment.

Another practical means of isolating individual components from compound mixtures is high-performance liquid chromatography (HPLC). While the resolving power of HPLC is lower, this technique is particularly suited to polar, nonvolatile, or thermally unstable analytes that are difficult to separate by GC. It also offers higher loading capacity than capillary GC.

In addition to chromatographic resolution and capacity, two additional aspects that require consideration are the potential for contamination of the analytes during the isolation procedure, and corrections for carbon associated with any derivative groups that have been appended to the molecule of interest. Regarding the former, entrainment 'bleed' $\Delta^{14}C$; of chromatographic stationary phase can result in significant carbon contamination of the isolated compound, unless steps are taken to avoid this problem (e.g., use of ultra-low bleed GC columns, removal of contaminants after the chromatographic isolation). This problem is likely to be most acute in HPLC when reversed-phase chromatographic phases are used. Comparison of yields and the $\Delta^{13}C$ compositions of the isolated compound and the CO_2 resulting from its combustion are effective means of assessing potential contamination problems.

AMS Measurement of ^{14}C

The purified compounds are sealed in evacuated quartz tubes with CuO as an oxidant. The material is combusted to CO_2, purified, and then reduced to graphite over cobalt or iron catalyst. The mixture of graphite and catalyst is loaded into a cesium sputter ion source. ^{14}C-AMS analysis is performed using special methods necessary for the accurate determination of $\Delta^{14}C$ in samples containing only micrograms, rather than milligrams, of carbon. AMS targets containing < 150 μg of carbon are prone to machine-induced isotopic fractionation, which appears to be directly related to the lower levels of carbon ion current generated by these samples. Therefore, small samples are analyzed with identically prepared, size-matched small standards to compensate for these effects. The f_m values that are calculated relative to these standards no longer show a size-dependent fractionation.

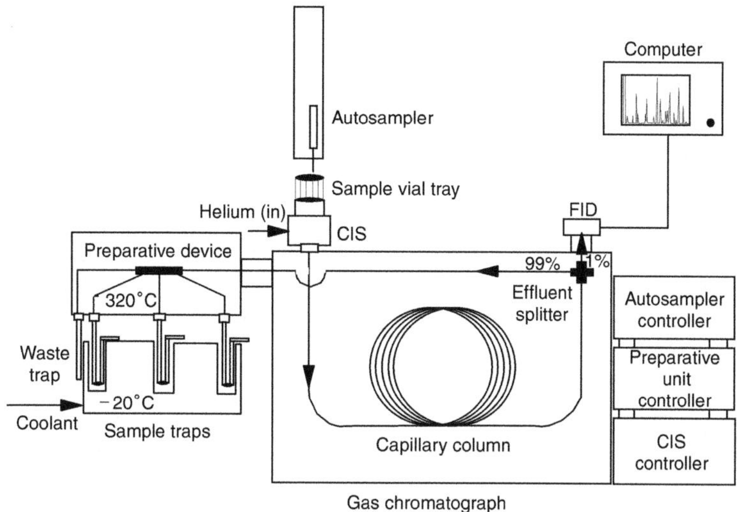

Figure 6 Diagrammatic representation of a preparative capillary gas chromatograph (PCGC) system.

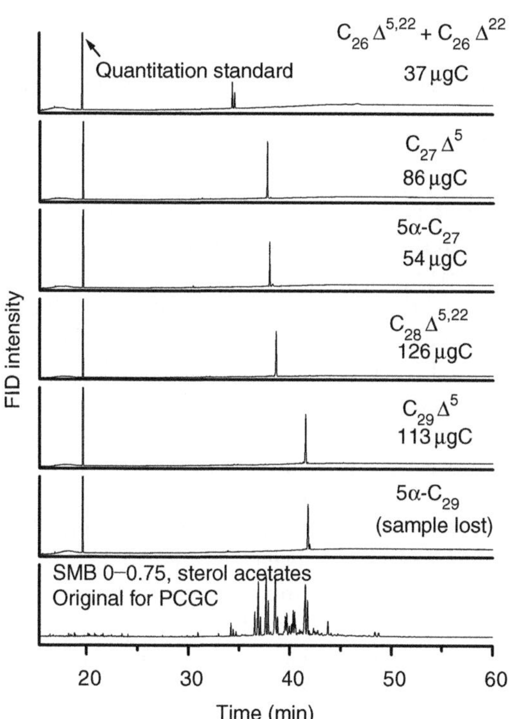

Figure 7 An example PCGC series, showing the total original mixture and the six individual, trapped compounds. In this case the analytes are sterols (as their acetate derivatives). (From Pearson (2000).)

Examples of Applications

Lipid Biomarkers in Santa Monica Basin Sediments

As one example of the application of single-compound radiocarbon analysis, we show a detailed data set for a range of lipid biomarkers extracted from marine sediments. This work focused on the upper few centimeters of a core from Santa Monica Basin. The basin has a high sedimentation rate, and its suboxic bottom waters inhibit bioturbation. As a result, laminated cores recovered from the basin depocenter allow decadal resolution of recent changes in the ^{14}C record. On the timescale of radiocarbon decay, these samples are contemporary and have no in situ ^{14}C decay. However, the $\Delta^{14}C$ values of the end-member carbon sources have changed (**Figure 2A, B**). 'Bomb-^{14}C' $\Delta^{14}C$; has invaded the modern surface ocean phytoplankton and the terrestrial biota, and through subsequent sedimentation of their organic detritus, this bomb-^{14}C is carried to the underlying sediments. The contrast between 'pre-bomb' $\Delta^{14}C$; and 'post-bomb' $\Delta^{14}C$; $\Delta^{14}C$ values, or the relative rate of bomb-^{14}C uptake, therefore is a useful tracer property. It can help distinguish biogeochemical processes that transfer carbon within years or decades (source-specific lipids that now contain bomb-^{14}C) from biogeochemical processes that do not exchange with atmospheric CO_2 on a short timescale (lipids that remain free from bomb-^{14}C).

Compound-specific $\Delta^{14}C$ values for 31 different lipid biomarker molecules are shown in **Figure 8** for sedimentary horizons corresponding to pre-bomb (before AD 1950) and post-bomb (1950–1996) eras. These organic compounds represent phytoplanktonic, zooplanktonic, bacterial, archaeal, terrestrial higher plant, and fossil carbon sources. The lipid classes include long-chain *n*-alkanes, alkanoic (fatty) acids, *n*-alcohols, C_{30} mid-chain ketols and diols, sterols, hopanols, and C_{40} isoprenoid side chains of the ether-linked glycerols of the *Archaea*.

The data show that the carbon source for the majority of the analyzed biomarkers is marine euphotic

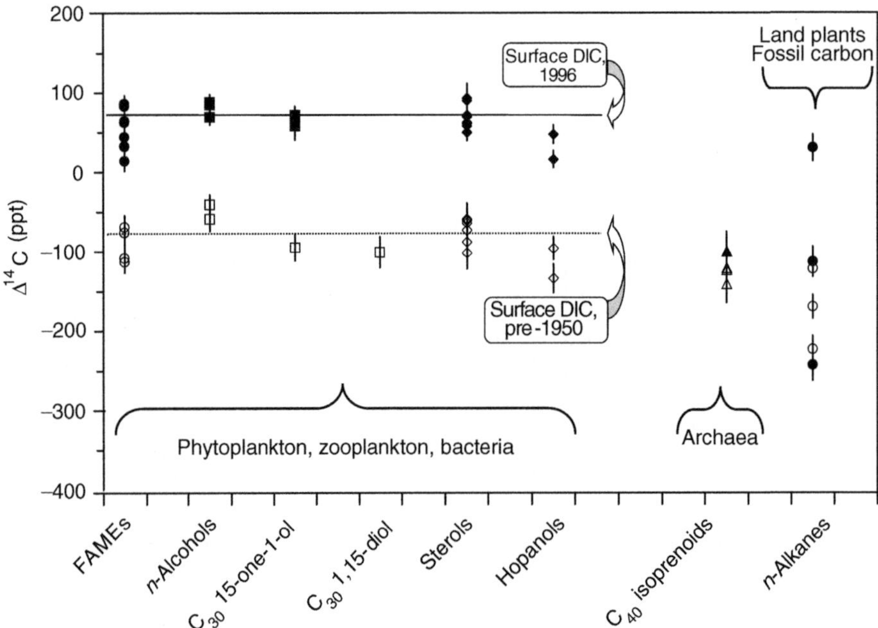

Figure 8 $\Delta^{14}C$ data for individual lipids extracted from Santa Monica Basin sediments. The solid symbols represent compounds extracted from the post-bomb sedimentary horizon (AD 1950–1996). The hollow symbols represent compounds extracted from the pre-bomb sedimentary horizon (deposited prior to AD 1950). (Modified after Pearson (2000).)

zone production. Most of the lipids from 'pre-bomb' $\Delta^{14}C$; sediments have $\Delta^{14}C$ values equal to the $\Delta^{14}C$ of surface water DIC at this time (dotted line), while most of the lipids from 'post-bomb' $\Delta^{14}C$; sediments have $\Delta^{14}C$ values equal to the $\Delta^{14}C$ of present-day surface water DIC (solid line).

However, it is clear that two of the lipid classes do not reflect carbon originally fixed by marine photoautotrophs. These are the *n*-alkanes, for which the $\Delta^{14}C$ data are consistent with mixed fossil and contemporary terrestrial higher plant sources, and the archaeal isoprenoids, for which the $\Delta^{14}C$ data are consistent with chemoautotrophic growth below the euphotic zone. This is just one example of the way in which compound-specific ^{14}C analysis can distinguish carbon sources and biogeochemical processes simultaneously. The large number of compounds that appear to record the $\Delta^{14}C$ of surface water DIC, and therefore marine primary production, points to the potential for numerous tracers of marine biomass; these are the target compounds of interest when developing refined sediment chronologies. In particular, the sterols appear to be particularly effective tracers of surface ocean DIC, and hence suitable for this purpose.

Monosaccharides in Oceanic High-molecular-weight Dissolved Organic Matter

The second example illustrates the utility of single compound, as well as compound class, ^{14}C measurements as ocean process tracers. In this case, the process of interest is the cycling of dissolved organic matter (DOM) in the ocean. Much progress has been made in characterizing this large carbon pool. A significant fraction of the DOM pool is composed of high-molecular-weight (HMW) compounds (>1 kDa), and a substantial fraction of this HMW DOM is known to be comprised of complex polysaccharides. Evidence suggests that these polysaccharides are produced in the surface ocean as a result of primary productivity, and/or attendant heterotrophic activity, and should therefore carry a bomb-influenced ^{14}C signature. Similar polysaccharides have been detected in HMW DOM well below the surface mixed layer, implying that these compounds are transported to the deep ocean. Two possible mechanisms can explain these observations: (1) advection of DOM associated with ocean circulation, and/or (2) aggregation and vertical transport followed by disaggregation/dissolution at depth. Because the timescales of aggregation and sinking processes are short relative to deep water formation and advective transport, ^{14}C measurements on polysaccharides in HMW DOM provide means of determining which mechanism is dominant.

Figure 9 shows vertical ^{14}C profiles for DIC and DOM as well as ^{14}C results for selected samples of sinking and suspended particulate organic matter (POM), HMW DOM, and monosaccharides isolated from selected depths at a station in the North-east Pacific Ocean. Individual monosaccharides were obtained by hydrolysis of HMW DOM, and purified and

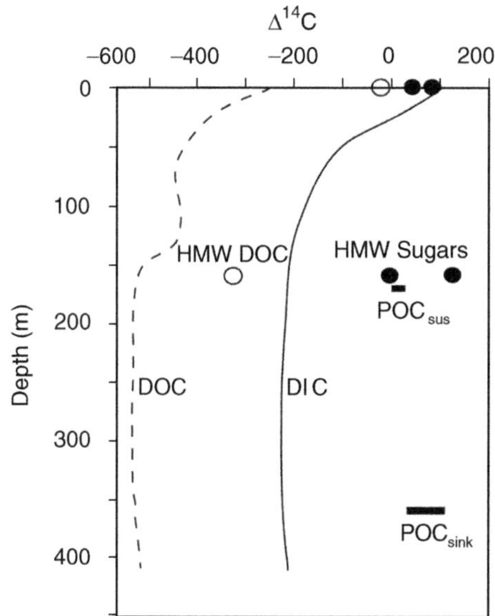

Figure 9 $\Delta^{14}C$ values (ppt) for different fractions of carbon in the North-East Pacific: solid bars labeled POC_{sus} and POC_{sink} correspond to suspended and sinking POC, respectively; solid and dashed lines show depth profiles for DIC and DOC, respectively; open circles are total HMW DOC and closed circles are individual sugars (monosaccharides) isolated from the HMW DOC fractions. (Modified from Aluwihare (1999).)

isolated by HPLC. The similarity of $\Delta^{14}C$ values of individual monosaccharides implies that they derive from a common polysaccharide source. Such similarities in ^{14}C lend support to the utility of ^{14}C measurements at the compound class level. Furthermore, the similarity between $\Delta^{14}C$ values of these compounds and surface ocean DIC indicates that they are either directly or indirectly the products of marine photoautotrophy. The deep ocean (1600 m) data shows the presence of bomb-radiocarbon in the monosaccharides. Their enrichment in ^{14}C relative to DIC, and similarity to suspended POC at the same depth, suggests that this component of HMW DOM is injected into the deep ocean by vertical transport as particles.

Summary

The ability to perform single-compound ^{14}C measurements has only recently been realized, and as a consequence its application as a tracer in ocean sciences remains in its infancy. The above examples highlight potential applications of single-compound ^{14}C measurements as tools for understanding the biogeochemical cycling of organic matter in the ocean. There are several other areas of study where this approach holds great promise. For example, ^{14}C measurements of vascular plant biomarkers (e.g., plant waxes, lignin-derived phenols) in continental shelf sediments provide constraints on the timescales over which terrestrial organic matter is delivered to the ocean. The 'infinite ^{14}C age' $\Delta^{14}C$; signature that polycyclic aromatic hydrocarbons and other fossil fuel-derived contaminants carry provides an effective means of tracing their inputs to the coastal ocean relative to contributions from natural processes (e.g., biomass burning). As methods are streamlined, it is anticipated that single compound ^{14}C measurements will find increasing application in marine biogeochemistry.

See also

Ocean Carbon System, Modeling of. Radiocarbon. Stable Carbon Isotope Variations in the Ocean.

Further Reading

Aluwihare LI (1999) *High Molecular Weight (HMW) Dissolved Organic Matter (DOM) in Seawater: Chemical Structure, Sources and Cycling.* PhD thesis, Massachusetts Institute of Technology/Woods Hole Oceanographic Institution.

Eglinton TI, Aluwihare LI, Bauer JE, Druffel ERM, and McNichol AP (1996) Gas chromatographic isolation of individual compounds from complex matrices for radiocarbon dating. *Analytical Chemistry* 68: 904–912.

Eglinton TI, Benitez-Nelson BC, Pearson A, *et al.* (1997) Variability in radiocarbon ages of individual organic compounds from marine sediments. *Science* 277: 796–799.

Faure G (1986) *Principles of Isotope Geology.* New York: Wiley.

Hayes JM (1993) Factors controlling the ^{13}C content of sedimentary organic compounds: principles and evidence. *Marine Geology* 113: 111–125.

Hedges JI (1992) Global biogeochemical cycles: progress and problems. *Marine Chemistry* 39: 67–93.

Hedges JI and Oades JM (1997) Comparative organic geochemistries of soils and marine sediments. *Organic Geochemistry* 27: 319–361.

Hoefs J (1980) *Stable Isotope Geochemistry.* New York: Springer-Verlag.

Pearson A (2000) *Biogeochemical Applications of Compound-Specific Radiocarbon Analysis.* PhD thesis, Massachusetts Institute of Technology/Woods Hole Oceanographic Institution.

Siegenthaler U and Sarmiento JL (1993) Atmospheric carbon dioxide and the ocean. *Nature* 365: 119–125.

Tuniz C, Bird JR, Fink D, and Herzog GF (1998) *Accelerator Mass Spectrometry: Ultrasensitive Analysis for Global Science.* Boca Raton, FL: CRC Press.

Volkman JK, Barrett SM, Blackburn SI et al. Microalgal biomarkers: a review of recent research developments. Organic Geochemistry 29: 1163–1179.

POLLUTION OF THE OCEANS

GLOBAL MARINE POLLUTION

A. D. McIntyre, University of Aberdeen, Aberdeen, UK

This aritcle is designed to provide an entry to the historical and global context of issues related to pollution of the oceans.

Until recently, the size and mobility of the oceans encouraged the view that they could not be significantly affected by human activities. Freshwater lakes and rivers had been degraded for centuries by effluents, particularly sewage, but, although from the 1920s coastal oil pollution from shipping discharges was widespread, it was felt that in general the open sea could safely dilute and disperse anything added to it. Erosion of this view began in the 1950s, when fallout from the testing of nuclear weapons in the atmosphere resulted in enhanced levels of artificial radionuclides throughout the world's oceans. At about the same time, the effluent from a factory at Minamata in Japan caused illness and deaths in the human population from consumption of mercury-contaminated fish, focusing global attention on the potential dangers of toxic metals. In the early 1960s, buildup in the marine environment of residues from synthetic organic pesticides poisoned top predators such as fish-eating birds, and in 1967 the first wreck of a supertanker, the *Torrey Canyon*, highlighted the threat of oil from shipping accidents, as distinct from operational discharges.

It might therefore be said that the decades of the 1950s and 1960s saw the beginnings of marine pollution as a serious concern, and one that demanded widespread control. It attracted the efforts of national and international agencies, not least those of the United Nations. The fear of effects of radioactivity focused early attention, and initiated the establishment of the International Commission on Radiological Protection (ICRP), which produced a set of radiation protection standards, applicable not just to fallout from weapons testing but also to the increasingly more relevant issues of operational discharges from nuclear reactors and reprocessing plants, from disposal of low-level radioactive material from a variety of sources including research and medicine, and from accidents in industrial installations and nuclear-powered ships. Following Minamata, other metals, in particular cadmium and lead, joined mercury on the list of concerns. Since this was seen, like radioactive wastes, as a public health problem, immediate action was taken. Metals in seafoods were monitored and import regulations were put in place. As a result, metal toxicity in seafoods is no longer a major issue, and since most marine organisms are resilient to metals, this form of pollution affects ecosystems only when metals are in very high concentrations, such as where mine tailings reach the sea.

Synthetic organics, either as pesticides and antifoulants (notably tributyl tin (TBT)) or as industrial chemicals, are present in seawater, biota, and sediments, and affect the whole spectrum of marine life, from primary producers to mammals and birds. The more persistent and toxic compounds are now banned or restricted, but since many are resistant to degradation and tend to attach to particles, the seabed sediment acts as a sink, from which they may be recirculated into the water column. Other synthetic compounds include plastics, and the increasing use of these has brought new problems to wildlife and amenities.

Oil contaminates the marine environment mainly from shipping and offshore oil production activities. Major incidents can release large quantities of oil over short periods, causing immense local damage; but in the longer term, more oil reaches the sea via operational discharges from ships. These and other threats to the ocean are controlled by the International Convention for the Prevention of Pollution from Ships, administered by the International Maritime Organisation of the UN. Pollution is generated by human activities, and the most ubiquitous item is sewage, which is derived from a variety of sources: as a direct discharge; as a component of urban wastewater; or as sludge to be disposed of after treatment. Sewage in coastal waters is primarily a public health problem, exposing recreational users to pathogens from the local population. The dangers are widely recognized, and many countries have introduced protective legislation. At the global level, the London Dumping Convention controls, among other things, the disposal of sewage sludge. As well as introducing pathogens, sewage also contributes carbon and nutrients to the sea, adding to the substantial quantities of these substances reaching the marine environment from agricultural runoff and industrial effluents. The resulting eutrophication is causing major ecosystem impacts around the world, resulting in excessive, and sometimes harmful, algal blooms.

The need for a global approach to ocean affairs was formally brought to the attention of the United Nations in 1967, and over the next 15 years, while sectoral treaties and agreements were being introduced, negotiations for a comprehensive regime led in 1982 to the adoption of the UN Convention on the Law of the Sea (UNCLOS). This provided a framework for the protection and management of marine resources. In the ensuing years, concepts such as sustainable development and the precautionary approach came to the fore and were endorsed in the Declaration of the Rio Summit, while proposals for Integrated Coastal Zone Management (ICZM) are being widely explored nationally and advanced through the Intergovernmental Oceanographic Commission.

Early ideas of pollution were focused on chemical inputs, but following the Group of Experts on the Scientific Aspects of Marine Pollution (GESAMP) definition pollution is now seen in a much wider context, encompassing any human activity that damages habitats and amenities and interferes with legitimate users of the sea. Thus, manipulation of terrestrial hydrological cycles, and other hinterland activities including alterations in agriculture, or afforestation, can profoundly influence estuarine regimes, and are seen in the context of pollution. In particular, it is now recognized that excessive fishing can do more widespread damage to marine ecosystems than most chemical pollution, and the need for ecosystem-based fisheries management is widely recognized.

Over the years, assessments of pollution effects have altered the priority of concerns, which today are very different from those of the 1950s. Thanks to the rigorous control of radioactivity and metals, these are not now major worries. Also, decades of experience with oil spills have shown that, after the initial damage, oil degrades and the resilience of natural communities leads to their recovery. Today, while the effects of sewage, eutrophication, and harmful algal blooms top the list of pollution concerns, along with the physical destruction of habitats by coastal construction, another item has been added: aquatic invasive species. The particular focus is on the transfer of harmful organisms in ships' ballast water and sediments, which can cause disruption of fisheries, fouling of coastal industry, and reduction of human amenity. In conclusion, most of the impacts referred to above are on the shallow waters and the shelf, associated with continental inputs and activities. The open ocean, although subject to contamination from the atmosphere and from vessels in shipping lanes, is relatively less polluted United Nations Environment Programme (UNEP).

See also

Anthropogenic Trace Elements in the Ocean. Atmospheric Input of Pollutants. Chlorinated Hydrocarbons. Metal Pollution. Radioactive Wastes.

Further Reading

Brackley P (ed.) (1990) *World Guide to Environmental Issues and Organisations*. Harlow: Longman.

Brune D, Chapman DV, and Gwynne DW (1997) Eutrophication. In: *The Global Environment*, ch. 30. Weinheim: VCH.

Coe JM and Rogers DB (1997) *Marine Debris*. Berlin: Springer.

de Mora SJ (ed.) (1996) *Tributyltin Case Study of an Environmental Contaminant*. Cambridge, UK: Cambridge University Press.

GESAMP (1982) *Reports and Studies No. 15: The Review of the Health of the Oceans*. Paris: UNESCO.

Grubb M, Koch M, and Thomson K (1993) *The 'Earth Summit' Agreements: A Guide and Assessment*. London: Earth Scan Publications.

HMSO (1981) *Eighth Report of the Royal Commission on Environmental Pollution, Oil Pollution of the Sea, Cmnd 8358*. London: HMSO.

Hollingworth C (2000) Ecosystem effects of fishing. *ICES Journal of Marine Science* 57(3): 465–465(1).

International Maritime Organization (1991) *IMO MARPOL 73/78 Consolidated Edition MARPOL 73/78*. London: International Maritime Organization.

International Oceanographic Commission (1998) *Annual Report of the International Oceanographic Commission*. Paris: UNESCO.

Kutsuna M (ed.) (1986) *Minamata Disease*. Kunamoto: Kunammoyo University.

Matheickal J and Raaymakers S (eds.) (2004) Second International Ballast Water Treatment R & D Symposium, 21–23 July 2003: Proceedings. *GloBallast Monograph Series No. 15*. London: International Maritime Organization.

National Research Council (1985) *Oil in the Sea*. Washington, DC: National Academies Press.

Park PK, Kester DR, and Duedall IW (eds.) (1983) *Radioactive Wastes and the Ocean*. New York: Wiley.

Pravdic V (1981) *GESAMP the First Dozen Years*. Nairobi: UNEP.

Pritchard SZ (1987) *Oil Pollution Control*. Wolfeboro, NH: Croom Helm.

Sinclair M and Valdimarsson G (eds.) (2003) *Responsible Fisheries in the Marine Ecosystem*. Oxford, UK: CABI.

Tolba MK, El-Kholy OA, and El-Hinnawi E (1992) *The World Environment 1972–1992: Two Decades of Challenge*. London: Chapman and Hall.

UNEP (1990) The State of the Marine Environment. *UNEP Regional Seas Reports and Studies No. 115*. Nairobi: UNEP.

UNO (1983) *The Law of the Sea: Official Text of UNCLOS*. New York: United Nations.

World Commission on Environment and Development (1987) *Our Common Future*. Oxford, UK: Oxford University Press.

METAL POLLUTION

G. E. Millward and A. Turner,
University of Plymouth, Plymouth, UK

Introduction

Marine pollution has been defined as 'the introduction by man, directly or indirectly, of substances or energy to the marine environment resulting in deleterious effects such as hazards to human health; hindrance of marine activities, including fishing; impairment of the quality for the use of sea water; and reduction in amenities' (GESAMP, 1990). Approximately 45% of people on Earth live within 150 km of the coast and marine pollution occurs as a consequence of increases in population density and industrialization. The problems of marine pollution are generally limited to nearshore waters rather than the open ocean, with the main impacted areas being estuaries, fjords, rias, and their adjoining shelf seas (**Figure 1**).

In the marine environment, metals such as iron, vanadium, copper, and zinc are essential for certain biochemical reactions in organisms, but even in moderately contaminated estuaries these metals contribute to stress in marine biota. By virtue of their toxic and bioaccumulative properties both cadmium and mercury are regarded as 'Black List' substances, while lead is on the 'Grey List'. These elements have little or no biochemical function and, while tolerable in minute quantities, exhibit toxic effects above critical concentrations. Mercury has a complex marine chemistry and exists in various forms, such as inorganic mercury, organically complexed mercury (with natural dissolved organic carbon), as a dissolved gas, Hg^0, and as the methylated species monomethyl mercury (MMHg) and dimethyl mercury (DMHg). Both MMHg and DMHg are present in the water column in sediments and in the tissues of

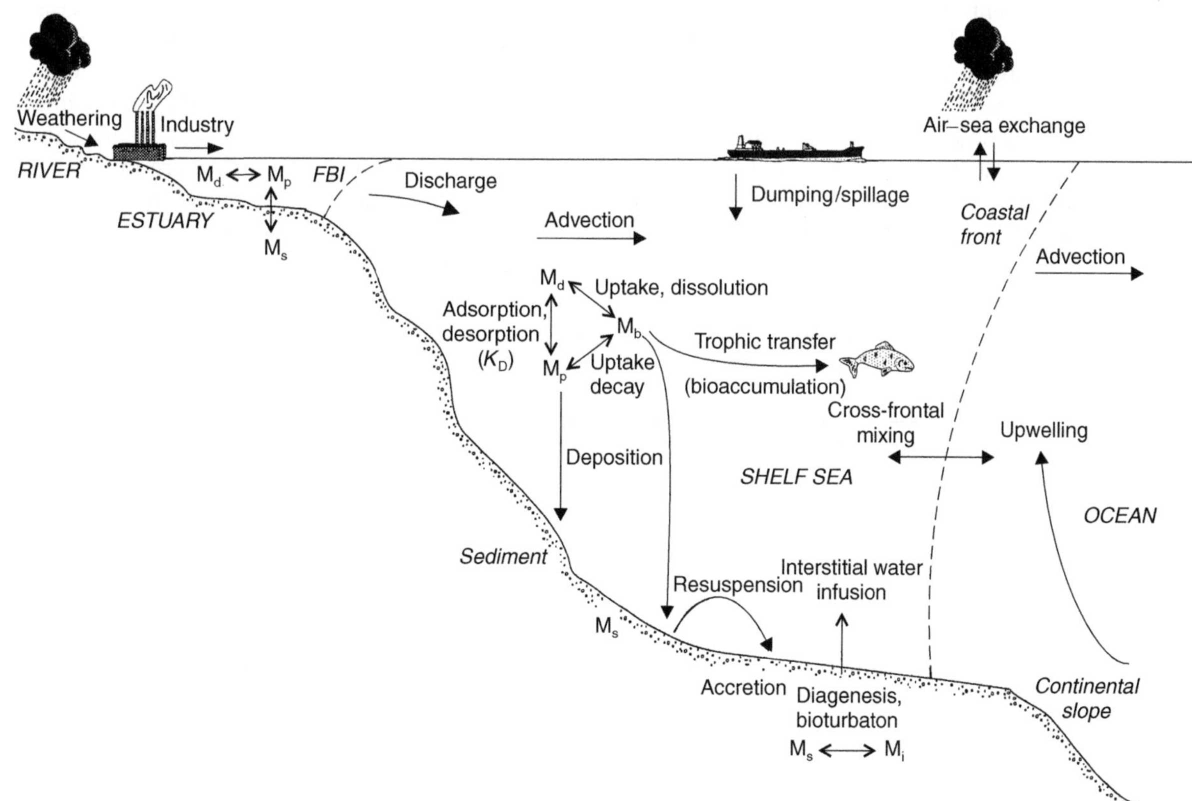

Figure 1 Processes affecting the transport and biogeochemistry of metal pollutants in estuaries and shelf seas. FBI = fresh water–brackish water interface. Metal compartments are designated. M_d, dissolved; M_p, suspended particulate; M_s, sediment; M_i, interstitial water; M_b, biogenic particulate.

marine organisms. Thus, depending on their physicochemical state or bioavailability, metals will impact upon different parts of the marine food web and in some cases bioaccumulation and/or biomagnification occurs, which may, ultimately, expose humans to a potential health hazard.

When attempting to assess the biogeochemical pathways and health impact of metals it is crucial to determine the total concentration accurately and where possible to identify and quantify the physical and chemical forms, or species. The analytical determination of metals in sea water has had a difficult history and many measurements reported in the literature prior to about 1985 should be treated with caution. Major strides have been made in the minimization of contamination during sample collection, storage, and preparation and in the application of sensitive analytical techniques, sometimes coupled with methods for the separation of metal species. The concentrations of dissolved metals have been revised downwards in recent years as a consequence of the introduction of these advances, together with improvements in analytical quality assurance, including appropriate use of certified reference materials.

The sources and pathways of metals through the coastal environment are complex (see **Figure 1**). Interfacial processes play a key role in their passage from the land to the sea. In estuaries the composition of river water may be modified by physicochemical processes at the fresh water–brackish water interface (FBI), where strong gradients of salinity, temperature, concentration and type of suspended particulate matter (SPM), pH and dissolved oxygen exist. Metal exchanges, between the dissolved and particulate phases, take place under the influence of these gradients and this process is quantified by the partition coefficient (K_D, $1\ kg^{-1}$) (eqn [1]).

$$K_D = \frac{[M_p]}{[M_d]} \qquad [1]$$

Here $[M_p]$ is the metal concentration of SPM in nmol kg^{-1} (or $\mu g\,kg^{-1}$), and $[M_d]$ is the dissolved metal concentration in $nmol\,l^{-1}$ (or $\mu g\,l^{-1}$). Coastal sediments can contain elevated concentrations of dissolved metals in their interstitial waters which may be exchanged across the sediment–water interface via molecular diffusion or by resuspension and in soft sediments by enhanced diffusion due to bioturbation from burrowing organisms (**Figure 1**). The mercury cycle is complicated by the fact that microbial activity, in sediments and the water column, can produce DMHg and Hg^0, both of which are volatile and can exchange across the air–sea interface.

Anthropogenic and Natural Inputs

Dissolved and particulate metals in rivers and estuaries are derived from natural weathering process in the catchment area, and reflect the geological composition of the watershed (see **Table 1** for the crustal abundance of selected metals) and the local climatic conditions. Natural concentrations of metals can be augmented in catchment areas that are mineralized, and there may be a significant anthropogenic perturbation downstream because of mineral extraction

Table 1 Fluxes of metals to the atmosphere from natural and anthropogenic sources. The interference factor is the ratio of the anthropogenic flux to the natural flux. The generic term 'combustion' refers to various combinations of coal, oil, and wood combustion and refuse incineration

Metal	*Crustal abundance ($nmol\ g^{-1}$)*	*Atmospheric emission rate ($t\ y^{-1}$)*			*Major uses of metals and their compounds*
		Natural	*Anthropogenic*	*Interference factor*	
Cadmium	2	1.4	7.6	5.3	Nonferrous metal production; cement/fertilizer manufacture; combustion
Copper	510	28	35	1.2	Nonferrous metal production; biocides; combustion
Mercury	0.4	2.5	3.6	1.5	Chlorine cells; gold mining operations; combustion
Lead	80	12	332	27	Petroleum additive; nonferrous metal production; combustion
Zinc	2000	45	131	3.0	Nonferrous metal production; steel/iron manufacturing; cement production

and processing. In densely populated regions, metals originate from a wide range of industrial, domestic and agricultural uses, and their inputs into river systems have increased significantly over the past two centuries. Regulated dredging and dumping of metal pollutants at sea, inadvertent spills, and illegal discharge all add to the complexity of anthropogenic inputs to the aquatic environment. Thus, our ability to unravel natural versus anthropogenic inputs is often complicated by the significant and uncontrolled human perturbation of catchments and their river systems. Only where metal compounds are entirely of anthropogenic origin, such as tributyl tin, can the human impact be evaluated. In the case of lead, however, it has been possible to identify man-made inputs via the application of inductively coupled plasma mass spectrometry (ICP-MS) to the determination of lead isotopic ratios (e.g., ^{206}Pb to ^{207}Pb) which have distinct signature in leaded gasoline.

Because a significant proportion of the marine environment has been altered by anthropogenic activities, natural concentrations values for dissolved and particulate metals are difficult to obtain unambiguously. Baseline values are often assumed from analyses of samples from remote systems that are considered to be 'pristine' or from metal analyses of sediment horizons dated as being prior to the industrial revolution. Another approach to assessing man's impact on the global ocean is to compare the rates of metal emission to the atmosphere from natural and anthropogenic sources. In **Table 1** the 'interference factor' is >1 for all metals, with relatively high values for lead and cadmium, suggesting that there is a significant anthropogenic alteration of their natural cycles.

Macrotidal. estuaries have strong internal cycles and particles may be retained within the system for years, and in large estuaries for decades. Thus, estuaries are a significant repository for metals, although no systematic inventories of the sediment metal burden have been made. Suspended particles advecting from estuaries into shelf seas are trapped in the coastal margin and estimates show that $\sim$90% of the fluvial suspended load (and associated metals) of the Mississippi, St. Lawrence, Rhône, and rivers in the south east of the United States is deposited in the coastal margin. Early diagenesis of deposited material may result in release of metals into sediment pore waters and, since the dissolved phase is generally considered to be more bioavailable, the composition of interstitial waters could be more important in the overall toxicity of the sediments than is their total metal content. Accurate quantification of interstitial water composition in estuaries and shelf seas is hindered because of the heterogeneous distribution of sediment texture and because a satisfactory method has not yet been developed for application in shallow waters that are highly dynamic.

Particulate metals deposited in the coastal margin are slowly advected onto the continental slope by seabed currents and wave action. Sediment diagenesis and diffusion releases dissolved metals into the oceanic water column, where they may be involved in upwelling processes at the shelf break (**Figure 1**). Comparisons of the relative magnitudes of the combined river and atmospheric fluxes with the upwelling flux suggests that the latter is greater by a factor 2 for copper, of 2–7 for zinc, and of about 10 for cadmium. In contrast for inorganic mercury the upwelling flux to shelf seas is half the magnitude of the combined river and atmospheric input, while for methylated mercury the main source to shelf seas is the upwelling flux.

Distributions

The temporal and spatial distributions of dissolved metal pollutants are highly dependent on two important processes in the coastal boundary zone.

Local hydrodynamics. Water is dispersed in estuaries and coastal waters according to the local hydrodynamics, which can be characterized by the flushing time. The dispersion and dilution of metal pollutants from point and diffuse sources, and therefore their range of concentrations, will be affected by the flushing time. A flushing time of 0.5 y for coastal waters is typical for the North Sea and Irish Sea, a value of 2 y is representative of the waters around Bermuda, while a value of 5 y is representative of a semi-enclosed sea such as the Baltic. Waters with longer flushing times will register a slower response to changes in metal inputs, whereas changes in metal concentrations will be detected earlier in waters with shorter flushing times.

Particle–water interactions. Metal partitioning between the dissolved and particulate phases is a crucial factor because solutes are transported in a different way to particles. The latter experiences gravitational settling and aggregation, as well as advection and mixing. The concentrations of SPM and the types of SPM play a significant role on the fraction of metal carried in the particulate phase. Lead has a relatively high K_D, largely owing to its tendency to complex with carboxyl and phonolic groups that dominate the surfaces of natural particles. In contrast, the relatively low K_D for cadmium is the result of its ability to complex with chloride

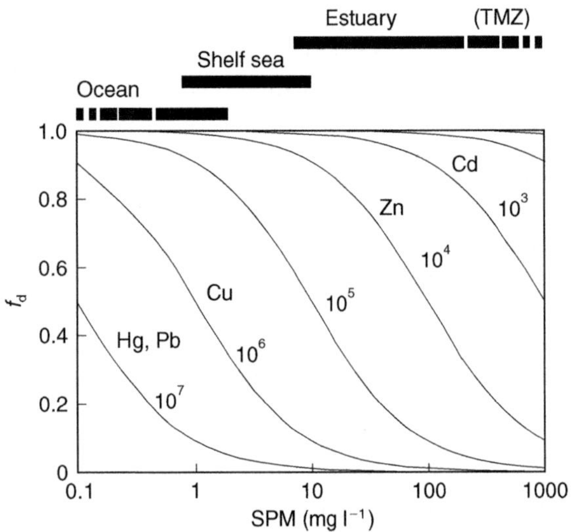

Figure 2 The fraction of metal in the dissolved phase (f_d) as a function of the concentration of suspended particulate matter (SPM). The bands at the top of the diagram represent the general ranges of SPM concentrations from the open ocean through to the estuarine turbidity maximum zone (TMZ). The numbers next to the lines are values of the partition coefficients, K_D, used to estimate the value of f_d. Metals are associated with their typical K_D values and are shown next to the appropriate line.

Table 2 Distributions of concentrations of dissolved cadmium, mercury, and lead (pmol l^{-1}) in rivers, estuaries, the English Channel and the North Atlantic

Location	*Cadmium*	*Mercury*	*Lead*
River background	90	2–20	1 000
Scheldt, Belgium	180–800	3.4–14	240–810
Seine, France	900–1 800	2.5–40	<2 300–16 000
English Channel	100–130	1.5–2.5	115–150
N. Atlantic Surface	1–10	1–7	100–150
N. Atlantic Deep	350	1	20

ions in sea water. **Figure 2** shows the fraction of metals in the dissolved phase (f_d) calculated as a function of SPM concentration, using representative K_D values for sea water. In estuaries and coastal waters, significant fractions of the metals are associated with particulate matter, but, as the SPM concentration declines through the coastal margin and into the ocean, the dissolved phase assumes more importance Because of their reactivity with particles many metals have short residence times in the coastal ocean, in the range 50–1000 y.

Distributions in Estuaries and Coasts

Table 2 illustrates the trends in the concentrations of dissolved metals from river to ocean. The Scheldt and Seine estuaries have important anthropogenic sources of the metals compared to the riverine inputs. Estuarine chemistry in the Scheldt is complicated by the discharge of a high organic load that renders the waters of the upper estuary anoxic during most of the year. Thus, the concentrations of dissolved cadmium, mercury and lead in the water column can be relatively low as a result of the formation of sparingly soluble metal sulfides. However, the sediment interstitial waters of the Scheldt can contain up to 25 000 pmol l^{-1} of dissolved cadmium. In the Scheldt and Seine estuaries, analyses of mercury speciation have shown the presence of the Hg^0 form and microbial mediation appears to have transformed inorganic mercury into MMHg and DMHg.

In assessing the distributions of dissolved metals in the coastal margin, their concentrations should be normalized with respect to salinity because a higher concentration at one location may be due to lower salinity (or greater fluvial influence). Distributions of dissolved metals in estuarine waters can vary linearly with salinity, the slope of which is dependent on the relative dissolved metal concentrations in the river and the sea, i.e., the 'end-member' concentrations. An example of conservative behavior, in which the concentration of dissolved cadmium varies in proportion to the amount of mixing between river water and sea water, is shown in **Figure 3A** for the Humber estuary plume. The observed behavior for cadmium is due to its affinity for the dissolved phase, even in turbid waters. The temporal change in the slope also shows that the distribution of the metal is highly responsive to changing inputs between different flow regimes. Total dissolved mercury (**Figure 3B**) displays non-conservation behaviour with a maximum concentration as a result of inputs from point sources along the banks of the Humber Estuary. Dissolved lead also behaves nonconservatively (**Figure 3C**) and the scatter of data arises because of diffuse atmospheric inputs. The dissolved lead is maintained at relatively low concentrations by its propensity to react with particles (**Figure 2**).

Distribution in the North Atlantic Ocean

Dissolved cadmium has a higher concentration in the deep waters of the North Atlantic Ocean owing to its uptake by phytoplankton in surface waters and recycling at depth; it exhibits nutrient-like behavior. Dissolved mercury has almost no gradient through the water column, because of a significant loss of Hg^0 to the atmosphere. However, in the deep waters of the North Atlantic, higher concentrations of MMHg and DMHg have been detected, possibly as a

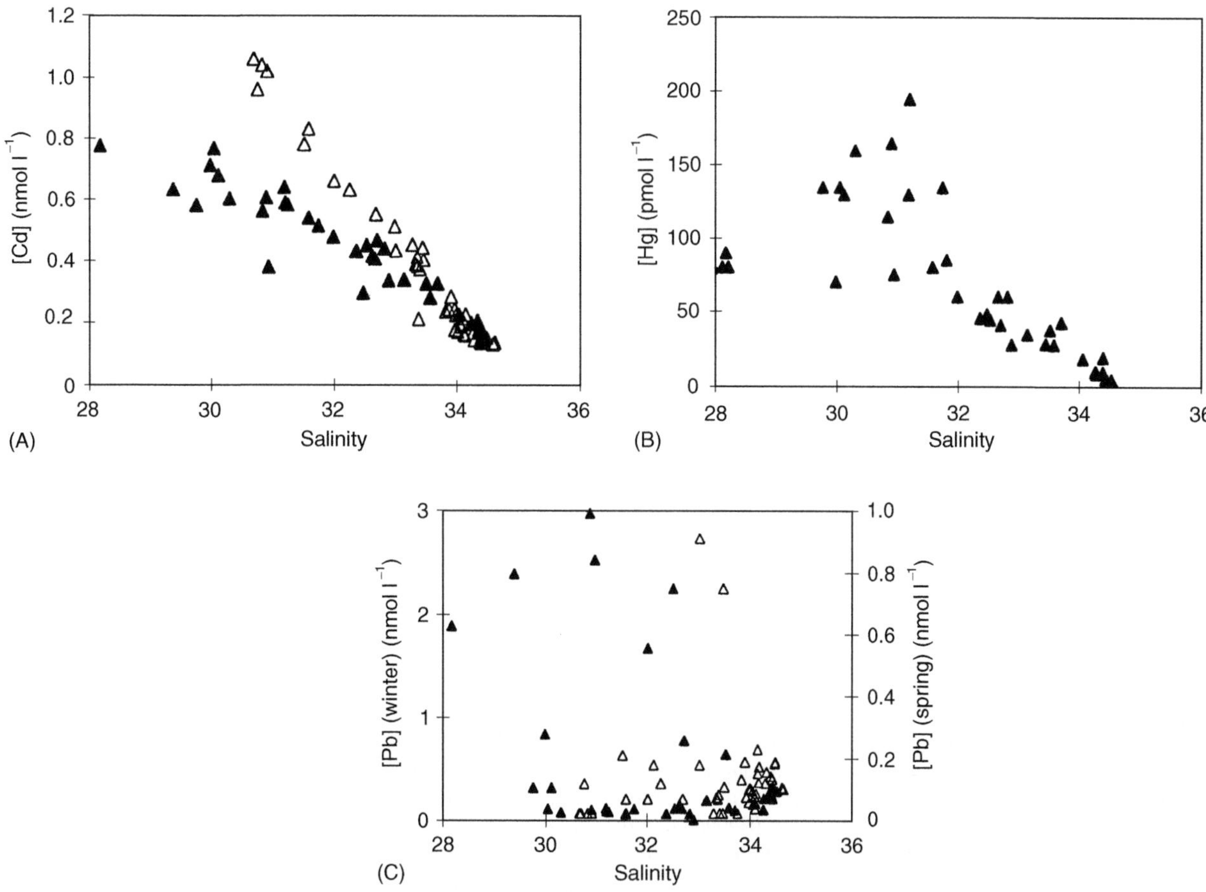

Figure 3 Concentrations of dissolved metals in the Humber Estuary as a function of salinity in winter with high fluvial input (solid symbols) and spring with reduced fluvial input (open symbols): (A) cadmium; (B) total mercury; (C) lead.

consequence of remineralization of phytoplankton. In contrast, lead has higher concentrations in the surface waters of the Atlantic, reflecting an atmospheric input. The strength of the atmospheric lead source appears to be declining as a result of a decrease in the use of leaded petrol. The monitoring of dissolved lead in the waters around Bermuda over a 15-year period shows that lead concentrations have decreased significantly. Since the SPM concentrations in these waters are low, the changes in dissolved lead concentrations are controlled almost exclusively by the flushing time of the water and the changing lead inputs. Following the decline in the atmospheric input, these waters have relaxed to near-background concentrations of dissolved lead (**Figure 4**).

Environmental Impact

The sediments of coastal regions reflect the long-term accumulation of metal contamination. Assessments of the anthropogenic component of metals in sediments require that the grain size must be accounted for and the corrected metal concentration compared with an uncontaminated reference material. For sediments with similar grain sizes, normalization is achieved with respect to a major element that is unaffected by anthropogenic inputs, such as aluminum, lithium, or rubidium. The enrichment factor (*EF*) is then defined as in eqn [2].

$$EF = \frac{[M_p]/[Al_p]}{[M_r]/[Al_r]} \qquad [2]$$

Here $[M_p]$ and $[M_r]$ are the metal concentrations in particulate matter and in crustal rock, respectively, and $[Al_p]$ and $[Al_r]$ are the concentrations of aluminum (or any suitable reference element) in particulate matter and crustal rock, respectively. **Table 3** lists *EF*s for SPM or fine sediment in contrasting estuaries. Enrichment factors are close to unity for the baseline sediment and in the 'pristine' Lena Estuary, while the greatest *EF* values are encountered for cadmium in the Rhine (impacted by the production of phosphate fertilizers) and the Scheldt, and for copper in Restronguet Creek (impacted by historical mining activity). The general sequence of *EF*s

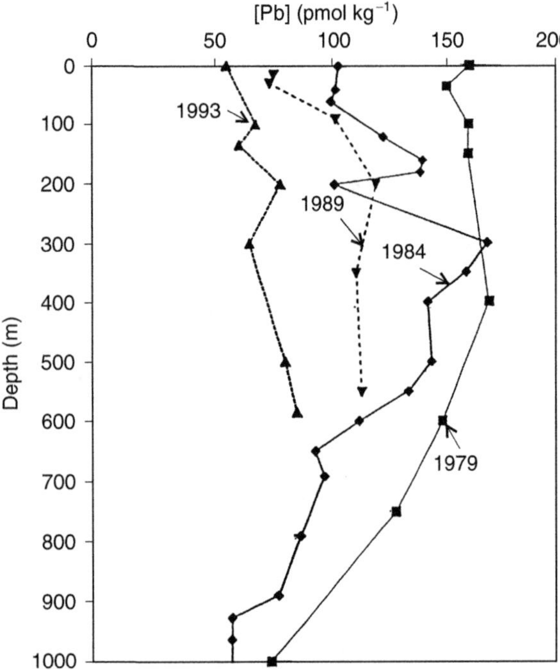

Figure 4 Vertical profiles of dissolved lead in the north Atlantic near Bermuda. The low point at 200 m in 1984 could not be accounted for by the authors, even though the analysis had been replicated and checked, and was thought to be a residual from the deep mixed layer from the previous winter. (Reprinted from Wu and Boyle (1997), copyright 1997, with permission from Elsevier Science.)

Table 3 Enrichment factors calculated according to eqn [2] for metals in suspended particulate matter or fine sediment from estuarine and coastal environments

Location	*Cadmium*	*Copper*	*Lead*	*Zinc*
Rhine Estuary, The Netherlands	310	21	84	23
Scheldt Estuary, Belgium	38	3.1	8.9	4.9
Seine Estuary, France	18	3.1	9.6	3.3
Humber Estuary, UK	N/A	2.1	7.4	2.4
Restronguet Creek, UK	N/A	88	28	28
Lena Estuary, Russia	N/A	0.8	1.4	1.1
Baseline (Norwegian Coastal Sediments)	0.6	0.2	2.2	1.9

N/A, not available.

is Cd > Pb > Cu, Zn; reflecting the relative significance of anthropogenic inputs to the estuarine environment and modification by different particle–water reactivities. The Scheldt Estuary has a history of metal pollution; and in comparison with other estuaries, Baeyens (see Further Reading) classifies the Scheldt as 'moderately polluted for all metals in the dissolved phase and fairly highly polluted in the particulate phase, especially for cadmium.' Efforts are being made to reduce the concentrations of cadmium in SPM in the Scheldt, and in the low-salinity region concentrations have declined from 400 nmol g^{-1} in 1978 to 79 nmol g^{-1} in 1995.

Table 4 Mean concentration (nmol g^{-1}) of metals in *Fucus* spp. in estuaries and coastal waters

Location	*Cadmium*	*Copper*	*Lead*	*Zinc*
Hardangefjord, Norway	63	330	220	21 000
Humber Estuary, UK	46	680	39	8 500
Restronguet Creek, UK	7.3	14 000	140	26 000
Baseline (Western Irish Sea)	4.4	60	6	1 100

Metal contamination can also be registered in indicator organisms (that is, organisms that are able to accumulate metals rather than regulate them), affording a measure of the contamination of the marine food chain. Across a broad range of phyla (from macroalgae to dophins), copper, zinc, and possibly cadmium exhibit a relatively low spread of concentrations, indicating efficient regulation of these metals. For lead, the concentrations are lower in vertebrates, which may be due to effective regulation or reduced bioavailability of the metal. Mercury is exceptional because of its biomagnification along the food chain as a result of its being present mainly as methyl mercury which is eliminated slowly from the organism. Mercury is retained by long-lived species, such as seals and dolphins, owing to their biochemical ability to isolate mercury as mercuric selenide granules. *Fucus* is a representative indicator species and baseline concentrations of selected metals occur in samples from an uncontaminated area of the Western Irish Sea (**Table 4**). Concentrations of copper, lead, and zinc are generally highest in industrialized estuaries (e.g., Humber) and fiords (e.g., Hardangefjord) and those that drain mineralized catchments and old mine workings (e.g., Restronguet Creek). Elevated concentrations of metals impact the growth, respiration, reproduction, recruitment and species diversity of marine organism, for example in Restronguet Creek the absence of bivalves has been ascribed to high levels of copper and zinc which prevent the settlement of juvenile bivalves.

Human Health

The toxicity of metals depends on their rate of excretion from an organism and their chemical form.

The adverse effects of metals on human health were recognized in the 1950s and 1960s following catastrophic events involving mercury. Inorganic mercury is normally excreted by humans and poses little hazard to the general public. However, organic mercury compounds, such as methylated forms, are not readily excreted. Methylated mercury compounds can pass to all tissues in the body after absorption from the gastrointestinal tract. They can cross diffusion barriers and penetrate membranes, such as the blood–brain barrier (causing irreversible brain damage) and the placenta (rendering methylated mercury concentrations in fetal blood higher than those in the mother). The most significant outbreak of neurological, and often fatal illnesses occurred among the residents of Minamata in Japan. Chemical companies had discharged or dumped tonnes of mercury compounds into Minamata Bay for decades and these accumulated in the tissues of shellfish and fish. Consequently, large doses were passed onto the local, fish-eating population. Eventually the Bay was sealed off with nets to prevent organisms contaminated with mercury from escaping and affecting other areas. Over several decades biogeochemical processes and hydrodynamic flushing of mercury from the Bay have resulted in concentrations of mercury falling below government standards. Presently, the World Health Organization (WHO) regards a tolerable daily intake of total mercury (inorganic + organic) to be 50 $\mu g\ d^{-1}$ for an adult of 70 kg.

Human exposure to high concentrations of cadmium are rare and current concern centers around the chronic toxicity caused by long-term exposure to low levels of the metal. Bone disorders are one manifestation of chronic cadmium exposure. Cadmium is present in all tissues of adults, with the most significant amounts found in the liver and kidney, and the concentrations tend to increase with age. The WHO regards a tolerable daily intake of cadmium to be 70 $\mu g\ d^{-1}$ for an adult of 70 kg.

Conclusions

Despite coastal waters in the vicinity of urban and industrial regions being contaminated with metals, there exists no evidence of significant pollution that poses a threat to human health, except on a local scale (and usually in shellfish) or where control has been poor. Since the dominant temporary or ultimate sink for metal contaminants is the sediment, an important goal of current research is to understand the mechanisms and extent to which contaminants are extracted by organisms (i.e., contaminant bioavailability) and transferred within the marine food chain.

Glossary

Advection Horizontal water motion.

Bioavailable metals Dissolved and particulate metals that are accessible to organisms during normal metabolic activity.

Bioaccumulative metals Metals that can be regulated and reside in the organism and are added to over its life.

Biomagnified metals Metals that are not regulated by organisms that can acquire an even larger body burden of metals.

Bioturbation Reworking of bottom sediment by burrowing marine organisms.

Diagenesis Release of particulate metals into the dissolved phase under suboxic conditions.

Flushing time The time required for an existing body of water to be exchanged with surrounding water.

Upwelling Vertical, upward movement of water at the shelf break, often tidally induced.

See also

Anthropogenic Trace Elements in the Ocean. Atmospheric Input of Pollutants. Land–Sea Global Transfers. Metalloids and Oxyanions. Pore Water Chemistry. Refractory Metals. River Inputs. Transition Metals and Heavy Metal Speciation.

Further Reading

Baeyens W (1998) Evolution of trace metal concentrations in the Scheldt estuary (1978–1995). A comparison with estuarine and ocean levels. *Hydrobiologia* 366: 157–167.

Clark RB (1998) *Marine Pollution* 4th edn. Oxford: Clarendon Press.

Ebinghaus R (ed.) (1997) *Regional and Global Cycles of Mercury: Source, Fluxes and Mass Balances*. NATO Series, Amsterdam: Kluwer Press.

GESAMP: Group of Experts on Scientific Aspects of Marine Pollution (1990) *The State of the Marine Environment*. Nairobi, Kenya: United Nations Environment Programme.

Langston WJ and Bebianno MJ (eds.) (1998) *Metal Metabolism in Aquatic Environments*. London: Chapman and Hall.

Lowry R, Cramer RN, and Rickards LJ (1992) *North Sea Project CD ROM and Users Guide*. Swindon: British Oceanographic Data Centre, Natural Environment Research Council of the United Kingdom.

Mantoura RFC, Martin J-M, and Wollast R (1991) *Ocean Margin Processes in Global Change*. Dahlem Workshop Reports No. 9. Chichester: Wiley.

Oslo and Paris Commissions (1993) *North Sea Quality Status Report* – 1993. Fredensborg, Denmark: Olsen & Olsen.

Salbu B and Steinnes E (eds.) (1995) *Trace Elements in Natural Waters*. Boca Raton, FL: CRC Press.

Sindermann CJ Ocean Pollution: *Effects on Living Resources and Humans*. Boca Raton, FL: CRC Press.

Wu J and Boyle EA (1997) Lead in the western North Atlantic Ocean: completed response to leaded gasoline phaseout. *Geochimica et Cosmochimica Acta* 61: 3279–3283.

ANTHROPOGENIC TRACE ELEMENTS IN THE OCEAN

E. A. Boyle, Massachusetts Institute of Technology, Cambridge, MA, USA

Introduction

Human activities have increased the fluxes of several chemical elements into the ocean above natural levels. Despite convincing evidence for this enhancement of elemental fluxes (see Further Reading section for references relevant to the discussion in this article), there is only one element – lead (Pb) – where abundant evidence proves that open-ocean seawater concentrations are substantially higher than they were in preindustrial times. For a few other elements – e.g. cadmium (Cd) and mercury (Hg) – there is some evidence suggesting a detectable anthropogenic impact (or models indicating that an anthropogenic enhancement must exist even if it has not been observed). For most other elements, the size of the oceanic reservoir of these elements overwhelms relatively large anthropogenic fluxes, and it may require centuries of further inputs before the human impact can be discerned.

Anthropogenic Lead in the Ocean

Sampling and analysis for Pb have been difficult because of low concentrations and abundant contamination sources: lead paint, lead weights, and gasoline exhausts. The latter source is now substantially reduced, so perhaps the current Pb contamination problem is less serious than it was in the 1980s.

Patterson and co-workers were the first to call attention to the overwhelming anthropogenic augmentation of Pb fluxes into the environment, particularly that resulting from alkyl leaded gas utilization. Patterson's evidence began with the demonstration that Pb deposition in remote Greenland snows had increased by two orders of magnitude. In the late 1970s/early 1980s, his laboratory obtained the first valid data (uncontaminated and properly analyzed) for the vertical distribution of Pb in water sample profiles from the North Atlantic, North Pacific, and South Tropical Pacific (**Figure 1**). These data demonstrated that the highest concentrations of Pb occurred in the surface ocean and that concentrations decreased with increasing depth in the water column. At that time, the highest Pb concentrations were found in the North Atlantic Ocean (160×10^{-12} moles kg^{-1} at the surface decreasing to 26×10^{-12} moles kg^{-1} at 3000 m water depth). The high Pb concentrations in this basin are emitted from the major industrial nations surrounding the basin. Lower Pb concentrations were seen in North Pacific surface waters (60×10^{-12} moles kg^{-1} at the surface decreasing to 5×10^{-12} moles kg^{-1} at 3500 m depth), and the lowest concentrations in the south tropical Pacific (20×10^{-12} moles kg^{-1} at the surface decreasing to 4×10^{-12} moles kg^{-1} at 4000 m). In Patterson's view, this evidence proved the anthropogenic origin of Pb in the modern ocean.

The lead industry attempted to discredit Patterson's evidence on environmental Pb pollution by many spurious arguments. Although they were not clever enough, they might have attempted to discredit Patterson's interpretation of his oceanic Pb data by pointing out that similar comparative concentration variations occur for aluminum (Al) in the ocean, even though oceanic Al is entirely of natural origin. Al is released from terrestrial dusts blown into the ocean and 'scavenged' onto sinking biologically produced particles that remove it from the deep ocean. Therefore Al is high in the surface ocean, highest downwind of major dust sources such as north-west Africa, and lowest in the South Pacific because of low dust inputs to the surface and cumulative scavenging in the deep waters.

Lead was phased out of gasoline in the USA (followed soon by Canada and Japan, and somewhat later by western Europe and a few other countries) beginning in 1970 when the US Environmental Protection Agency mandated emissions controls on gasoline exhausts. Originally, controls on lead emissions were not the goal of the regulations; the regulations were formulated to minimize emissions of hydrocarbons, nitrogen oxides, and carbon monoxide. As it turned out, the technological fix for those problems was to use catalytic converters on the exhaust stream, and the activity of the catalysts was destroyed by lead exhausts. Hence leaded gasoline could not be used with catalytic converters, and regulations mandated the elimination of leaded gasoline. Later, regulations specifically directed at minimizing lead emissions into the environment were also enacted.

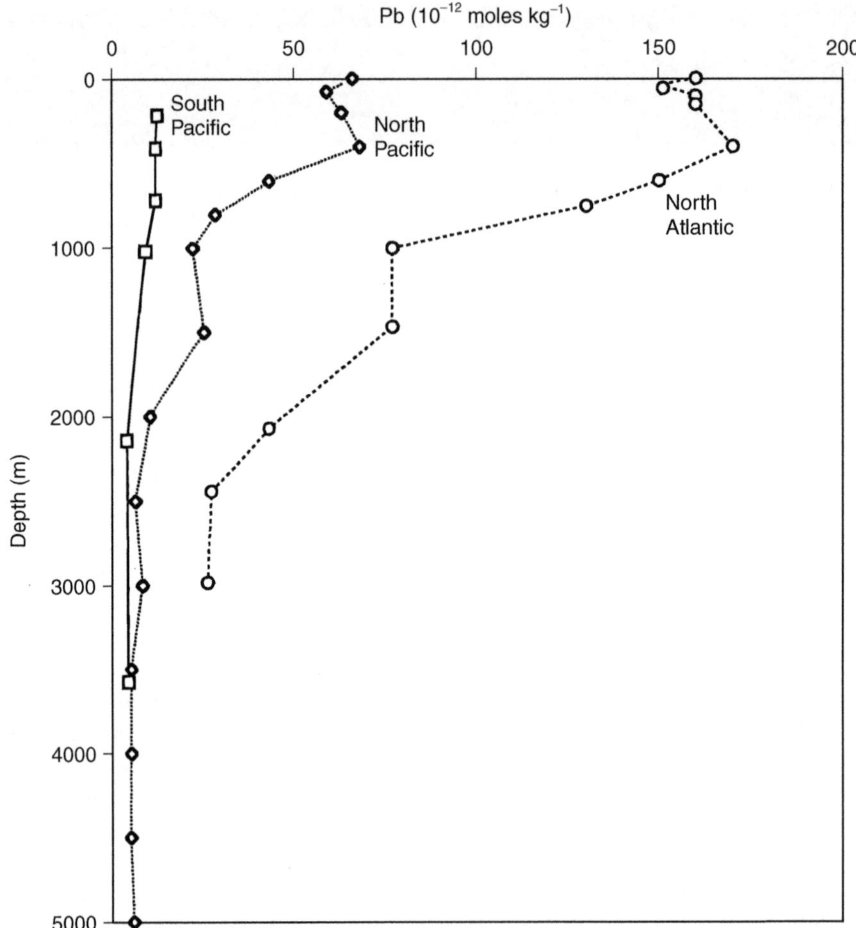

Figure 1 Oceanic Pb profiles obtained by Patterson and co-workers on samples collected between 1976 and 1981.

Changing leaded gasoline utilization patterns offer an opportunity to establish that Pb decreases in the ocean in response to the decreasing Pb emissions from leaded gasoline. Pb gas utilization increased from its introduction in the 1920s until the 1970s, then decreased rapidly – resulting in nearly complete elimination from the USA by 1990 and from western Europe by the turn of the century (**Figure 2**).

Surface waters of the ocean should respond within a few years to changes in fluxes from the atmosphere, and the upper layers of the ocean should respond on decadal timescales. These expectations are based upon the penetration of the nuclear bomb tritium into the ocean and natural radioisotope ^{210}Pb (half-life 22.3 years). Studies of tritium penetration show that surface waters ventilate the upper thermocline of the ocean on a timescale of several years to several decades. ^{210}Pb is supplied to the ocean from wind-blown aerosols which have acquired ^{210}Pb from the decay of the radioactive noble gas ^{222}Rn. ^{222}Rn is released into the atmosphere as a decay product of naturally occurring crystal ^{238}U which decays in several stages to the immediate parent of ^{222}Rn, ^{226}Ra. The flux of ^{210}Pb out of the atmosphere has been measured at numerous sites, therefore the flux of ^{210}Pb into the ocean can be estimated reasonably well. In the surface ocean, ^{210}Pb is rapidly scavenged from sea water by newly formed biological material whose residues eventually sink out of the surface waters carrying ^{210}Pb. The steady-state concentration of ^{210}Pb in the mixed layer of the ocean is thus determined by a balance between ^{210}Pb supply from the atmosphere and ^{210}Pb removal by sinking particles. The time constant for this process is estimated by dividing the measured surface water ^{210}Pb reservoir by the flux of ^{210}Pb from the atmosphere. The rate of removal varies with the rate of biological activity, but in typical open-ocean 'deserts' such as the Sargasso Sea, ^{210}Pb is removed from surface waters every 2 years. Thus as the flux of anthropogenic lead from the atmosphere decreases, the concentration of lead in surface waters should follow this decrease with a 2 year lag.

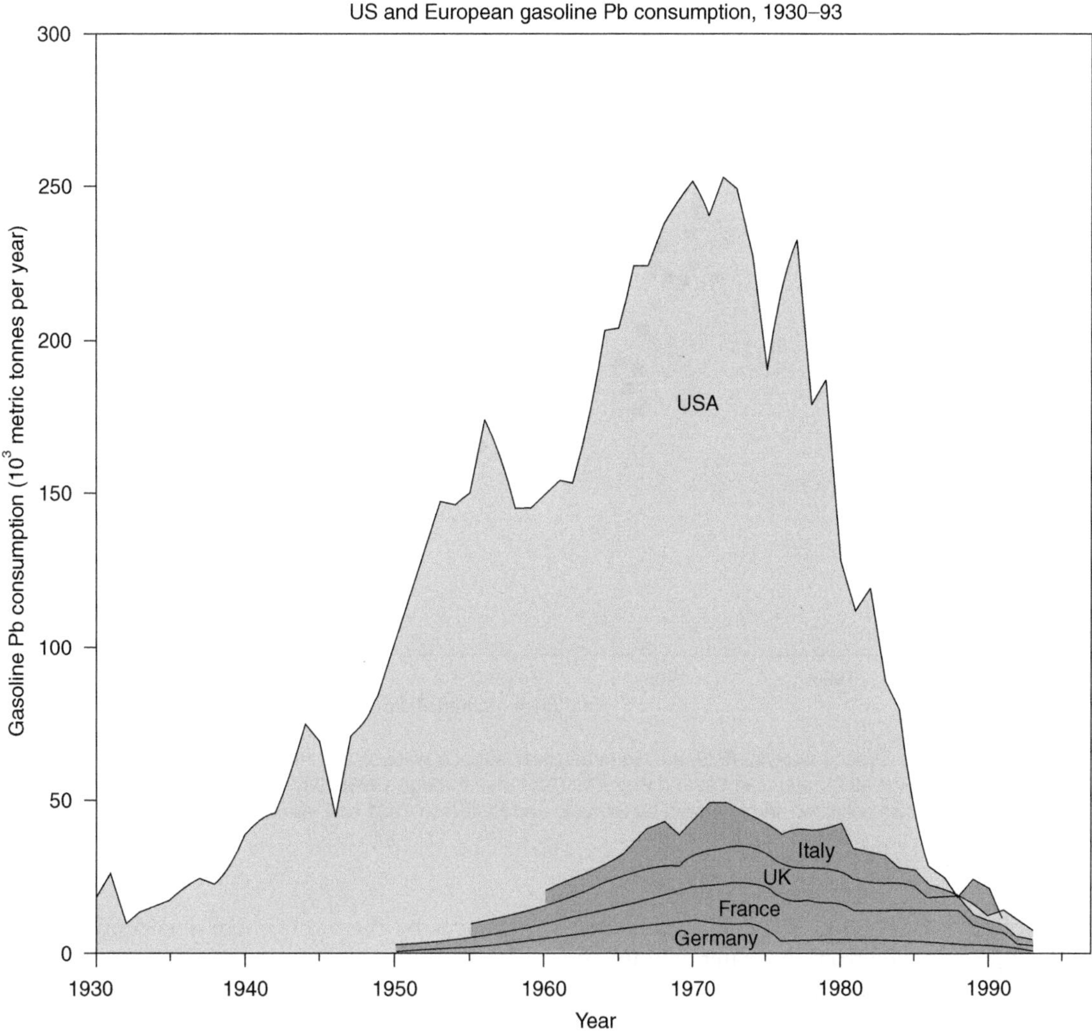

Figure 2 Alkyl Pb gas consumption, 1930–93. Four European countries (accounting for 70% of EC gasoline consumption) stacked to compare with the much higher US consumption.

Decreasing concentrations of Pb in surface waters of the Atlantic Ocean have been documented by observations from 1979 until the present (**Figure 3**). The decreasing concentration of Pb in the upper layers of the ocean is also observed (**Figure 4**), with a slower response in the deeper waters that are replaced on decadal timescales. These decreases are seen in other parts of the Atlantic and North Pacific as well. This evidence amply demonstrates that the phasing out of leaded gasoline in the USA has been closely followed by decreases of the Pb concentration of the ocean on the appropriate timescale.

In the period before direct observations of Pb in the ocean, the response of oceanic Pb to earlier changes in emissions are documented by the Pb content of reef-building corals. Corals precipitate calcium carbonate skeletons with annual variations in their density that can be counted in the same way as tree rings, hence providing a chronology of the time of deposition. In corals from near Bermuda, concentrations of Pb increased from very low levels in the 1880s to higher levels in the 1920s, in pace with the emissions of Pb from high temperature industrial activities (smelting, coal combustion, etc.) (**Figure 5**). With the introduction of leaded gasoline, Pb increased far more until the mid-1970s, when coralline Pb began to decrease with the phasing out of leaded gasoline, confirming the picture provided by direct observations in this period.

Other information on the sources and dispersion of anthropogenic lead can be derived from the stable isotope ratios of lead (e.g. $^{206}Pb/^{207}Pb$) which vary from one mining source to another because of differing Pb/U and Pb/Th for geological sources of lead. In particular, US leaded gasoline for many years had a $^{206}Pb/^{207}Pb$ ratio of >1.20, whereas European

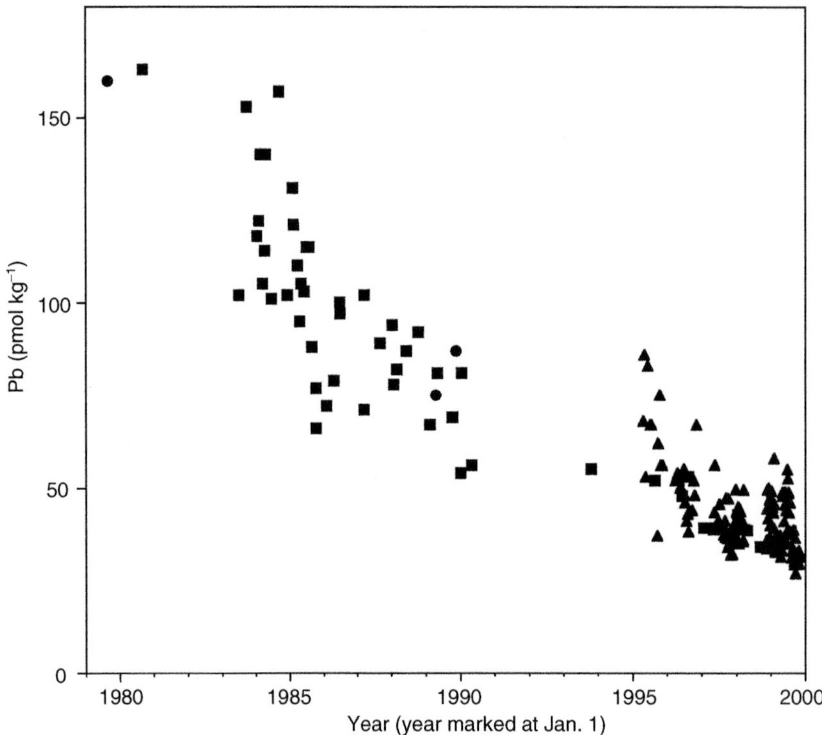

Figure 3 Pb in Bermuda surface waters, 1979–99. Squares (near surface waters) and triangles (40–50 m depth) are samples collected and analyzed by the MIT group (see Wu and Boyle 1997 for data through 1996; 1997–2000 from Boyle *et al.*, unpublished); circular symbols are samples collected and analyzed by Schaule and Patterson 1981 and Veron *et al* 1993.

leaded gasoline was <1.15. Eventually anthropogenic Pb will be removed from the water column and reside in marine sediments, and there is already evidence from surficial sediments of the North Atlantic Ocean for significant quantities of anthropogenic Pb.

Anthropogenic Mercury in the Ocean

Volatile mercury (Hg) and organomercury compounds are emitted from the land into the atmosphere naturally from wildfires, volcanoes, and microbial activity. Mercury is emitted into the atmosphere by humans as a result of high temperature processes (e.g. smelting, coal combustion, incineration) combined with commercial uses of elemental mercury (e.g. thermometers, batteries), as well as disposal of mercury-laden wastes (e.g. from gold mining operations) that are then converted into volatile forms in the environment. Estimates indicate that the anthropogenic mercury emissions from the land into the atmosphere exceeded natural sources by approximately a factor of three during the past century, not nearly so large as the Pb emission enhancement, but nonetheless substantial. A large fraction of this Hg is carried over great distances by the atmosphere and deposited into the ocean.

As for Pb, contamination is a major problem in Hg sampling and analysis (perhaps even more so), because Hg^0 is volatile and broken mercury thermometer residues exist in most chemical laboratories.

Evidence for anthropogenic perturbation of the oceanic Hg reservoir is less straightforward than it is for Pb. Observations in the modern Atlantic marine atmosphere show that the Northern Hemisphere has two- to three-fold higher Hg concentrations than the Southern Hemisphere, as might be expected for an anthropogenic source. However, land area is also higher in the Northern Hemisphere, and it is likely that even natural emissions are higher in the Northern Hemisphere, so observation of an inter-hemisphere difference does not by itself prove anthropogenic perturbation. Measurements of Hg in sea water have attempted to determine elemental (Hg^0), organometallic (e.g. CH_3Hg, $(CH_3)_2Hg$), and 'reactive' forms (ionic and inorganically complexed Hg, organically complexed Hg, and labile organic and particulate Hg). Reported Hg concentrations range from 10^{-11} to 10^{-9} moles kg^{-1}. There are no direct seawater measurements or indirect seawater

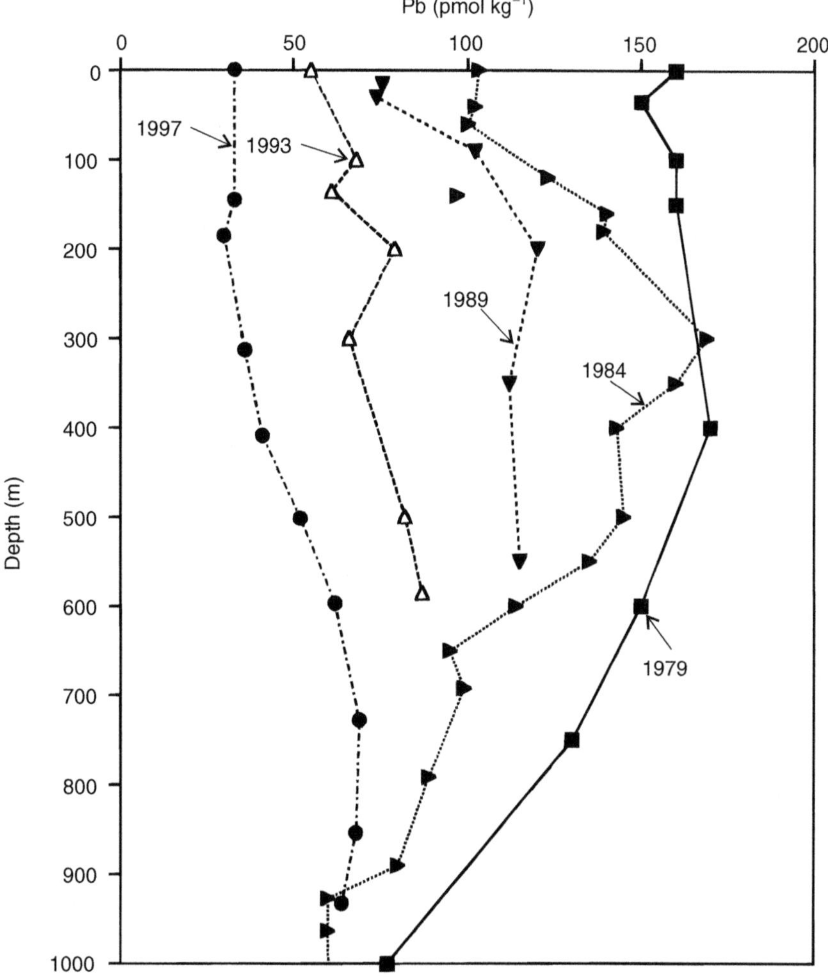

Figure 4 Pb in ocean profiles in the Sargasso Sea near Bermuda: 1979 data from Schaule and Patterson 1983; 1984 data from Boyle *et al.* (1986); data from Veron *et al* 1993; 1993 data from Wu and Boyle 1997; 1997 data (Wu and Boyle, unpublished).

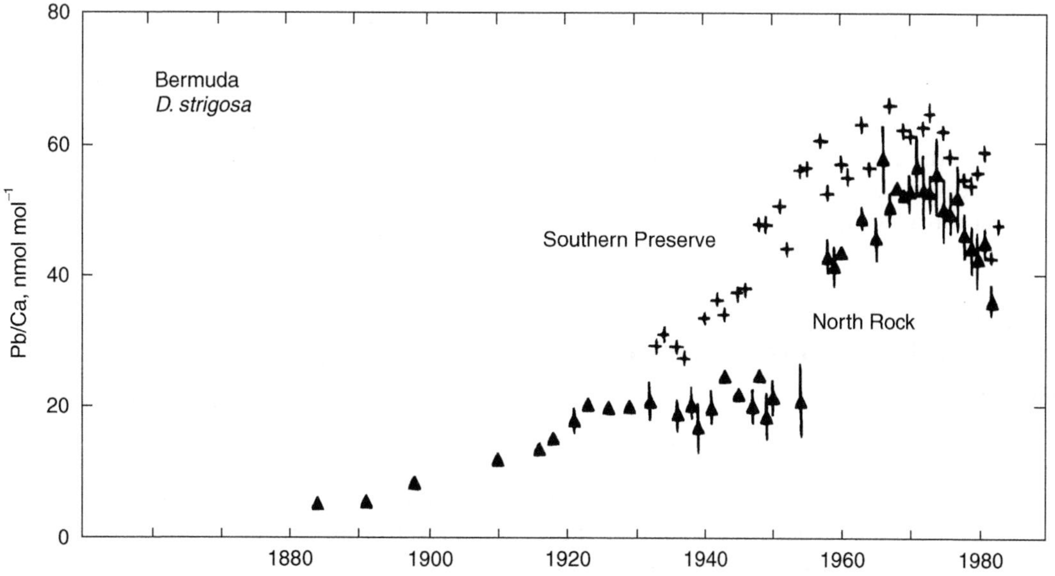

Figure 5 Pb in corals near Bermuda in annually laminated layers deposited between 1880 and 1997 (Shen and Boyle, 1987).

indicators such as corals which have yet documented an anthropogenic enhancement, although it has been estimated that Hg concentrations in the surface ocean have increased from 0.5 to 1.5 pmol kg^{-1} during the past century.

Other Anthropogenic Elements in the Ocean

Several other elements have had significant anthropogenic enhancements of global fluxes, but in most cases, the size of the oceanic reservoir of the elements is too large for an observable shift in open-ocean concentrations. A two- to three-fold increase in the concentration of cadmium (Cd) has been observed in corals near Bermuda during the past century. However, the concentrations of Cd in surface waters near Bermuda are low even today (certainly $< 10^{-11}$ moles kg^{-1} and sometimes as low as 10^{-12} moles kg^{-1}; because concentrations in deep waters are two or three orders of magnitude higher, the oceanic reservoir is large compared with the anthropogenic enhancement of Cd in surface waters. Although concentrations of Zn in corals were not measured, a similar anthropogenic enhancement of emissions to the atmosphere may well have caused slight enhancements to the Zn concentrations of the surface ocean, but evidence of such an increase has not been reported.

Because of the use of tributyl tin as a stabilizer in PVC plastic (hence an incineration source for inorganic Sn) and as an antifouling agent in marine paints, there has been some interest in the fate of anthropogenic Sn in the oceanic environment. Inorganic Sn is very low in the surface waters of the Sargasso Sea ($\sim 3 \times 10^{-12}$ moles kg^{-1}). Tributyl tin has been observed in enclosed harbors, but not in the open ocean.

On the whole, however, the size of the ocean precludes major enhancements of oceanic trace metal concentrations. Even where the enhancement is clear, such as for Pb, the concentrations are quite low and there is no reason to suppose that these enhancements pose a threat to marine biota.

Anthropogenic Trace Metal Enhancements in Coastal Waters and Embayments

As opposed to minimal evidence for anthropogenic trace metal contamination of open-ocean waters, it is clear that major impacts on the trace metal chemistry of coastal sea water have occurred and that these have sometimes had serious consequences. The discharge of untreated industrial and human wastes into coastal waters, and the release of toxic metals and metal compounds from antifouling paints have clearly increased metal concentrations in shallow ocean waters in many coastal regions.

One of the most notorious examples of coastal pollution was in Minimata Bay, Japan where from the 1930s until the 1960s a chemical company dumped tonnes of mercury, which contaminated the bay sediments. This Hg pollution entered the marine food chain. Thousands of people developed severe neurological symptoms as a result of eating Hg-contaminated fish. This is perhaps the most extreme example where pollution in the marine environment has demonstrably harmed humans directly.

More recently, it has been demonstrated that high Cd, Zn, Cu, and Pb concentrations in Atlantic coastal waters of southern Spain have risen by many orders of magnitude due to acid mine drainage from the Rio Tinto massive sulfides. As yet, little is known about any human health effects from this trace element pollution, although the region has an active coastal fishing industry.

There is too much detail in the many local examples to discuss here, but there is little doubt that many coastal embayments proximate to big cities have elevated metal levels, and that an understanding of the consequences of this pollution is still being developed. Unfortunately, proper measurement capabilities did not exist until the past two decades, so there is usually no direct evidence for the degree of enhancement of dissolved metals relative to pre-industrial times. One exception to this situation is in San Francisco Bay, where the anthropogenic enhancement of dissolved cadmium has been documented by measuring the Cd content of benthic foraminifera.

An anthropogenic impact on metals in coastal marine sediments has been demonstrated repeatedly. Lead has been shown to be enriched in coastal sediments. Even more exotic elements such as silver (photography) and osmium (electron microscopy) are enriched in the sewage of some cities and can be traced from their source.

Conclusions

The anthropogenic dominance of lead in the open-ocean environment has been conclusively demonstrated. Although there is evidence that the fluxes of other elements to the ocean have been enhanced to a significant degree, there is less direct evidence documenting their enhancements in the open-ocean environment. Coastal waters and sediments near major cities show large enhancements in several trace metals.

See also

Metal Pollution. Radioactive Wastes. Refractory Metals. Transition Metals and Heavy Metal Speciation.

Further Reading

Bruland KW and Franks RP (1983) Mn, Ni, Cu, Zn, and Cd in the western North Atlantic. In: Wong EBCS, Bruland KW, Burton JD, and Goldberg ED (eds.) *Trace Metals in Seawater*, pp. 395–414. New York: Plenum.

Byrd JT and Andreae MO (1982) Tin and methyltin species in seawater; concentrations and fluxes. *Science* 218: 565–569.

Chow TJ, Bruland KW, Bertine K *et al.* (1973) Lead pollution: records in Southern California coastal sediments. *Science* 181: 551–552.

Fitzgerald WF, Engstrom DR, Mason RP, and Nater EA (1997) The case for atmospheric mercury contamination in remote areas. *Environmental Science and Technology* 32: 1–7.

Flegal AR and Patterson CC (1983) Vertical concentration profiles of lead in the Central Pacific at 15N and 20S. *Earth and Planetary Science Letters* 64: 19–32.

Hamelin B, Grousset F, and Sholkovitz ER (1990) Pb isotopes in surficial pelagic sediments from the North Atlantic. *Geochimica et Cosmochimica Acta* 54: 37–47.

Hamelin B, Ferrand JL, Alleman L, and Nicolas E (1997) Isotopic evidence of pollutant lead transport from North America to the subtropical North Atlantic gyre. *Geochimica et Cosmochimica Acta* 61: 4423.

Harrison RM and Laxen DPH (1981) *Lead Pollution Causes and Control.* London: Chapman and Hall.

Helmers E and van bulleted Loeff MMR (1993) Lead and aluminum in Atlantic surface waters (50°N to 50°S) reflecting anthropogenic and natural sources in the eolian transport. *Journal of Geophysical Research* 98: 20261–20273.

Mason RP, Fitzgerald WF, and Morel FMM (1994) The biogeochemical cycling of elemental mercury: anthropogenic influences. *Geochimica et Cosmochimica Acta* 58: 3191–3198.

Jenkins WJ (1980) Tritium and He-3 in the Sargasso Sea. *Journal of Marine Research* 38: 533–569.

Measures CI and Edmond JM (1988) Aluminum as a tracer of the deep outflow from the Mediterranean. *Journal of Geophysical Research* 93: 591–595.

Measures CI and Edmond JM (1990) Aluminium in the south Atlantic: steady state distribution of a short residence time element. *Journal of Geophysical Research* 95: 5331–5340.

Measures CI, Edmond JM, and Jickells T (1986) Aluminum in the northwest Atlantic. *Geochimica et Cosmochimica Acta* 50: 1423–1429.

Orians KJ and Bruland KW (1985) Dissolved aluminum in the central North Pacific. *Nature* 316: 427–429.

Orians KJ and Bruland KW (1986) The biogeochemistry of aluminum in the Pacific Ocean. *Earth and Planetary Science Letters* 78: 397–410.

Murozumi M, Chow TJ, and Patterson C (1969) Chemical concentrations of pollutant lead aerosols, terrestrial dusts and sea salts in Greenland and Antarctic snow strata. *Geochimica et Cosmochimica Acta* 33: 1247–1294.

Nriagu JO (1989) A global assessment of natural sources of atmospheric trace metals. *Nature* 338: 47–49.

Ravizza GE and Bothner MH (1996) Osmium isotopes and silver as tracers of anthropogenic metals in sediments from Massachusetts and Cape Cod bays. *Geochimica et Cosmochimica Acta* 60: 2753–2763.

Schaule B and Patterson CC (1983) Perturbations of the natural lead depth profile in the Sargasso Sea by industrial lead. In: Wong CS, Boyle EA, Bruland KW, Burton JD, and Goldberg ED (eds.) *Trace Metals in Seawater*, pp. 487–504. New York: Plenum.

Schaule BK and Patterson CC (1981) Lead concentrations in the northeast Pacific: evidence for global anthropogenic perturbations. *Earth and Planetary Science Letters* 54: 97–116.

Shen GT and Boyle EA (1987) Lead in corals: reconstruction of historical industrial fluxes to the surface ocean. *Earth and Planetary Science Letters* 82: 289–304.

Shen GT and Boyle EA (1988) Thermocline ventilation of anthropogenic lead in the western North Atlantic. *Journal of Geophysical Research* 93: 15715–15732.

Shen GT, Boyle EA, and Lea DW (1987) Cadmium in corals as a tracer of historical upwelling and industrial fallout. *Nature* 328: 794–796.

Sherrell RM, Boyle EA, and Hamelin B (1992) Isotopic equilibration between dissolved and suspended particulate lead in the Atlantic Ocean: evidence from Pb-210 and stable Pb isotopes. *Journal of Geophysical Research* 97: 11257–11268.

Slemr F and Langer E (1992) Increase in global atmospheric concentrations of mercury inferred from measurements over the Atlantic Ocean. *Nature* 355: 434–437.

Turekian KK, Benninger LK, and Dion EP (1983) Be-7 and Pb-210 total deposition fluxes at New Haven, Connecticut and at Bermuda. *Journal of Geophysical Research* 88: 5411–5415.

van Geen A and Luoma SN (1999) A record of estuarine water contamination from the Cd content of foraminiferal tests in San Francisco Bay, California. *Marine Chemistry* 64: 57.

van Geen A, Adkins JF, Boyle EA, Nelson CH, and Palanques A (1997) A 120 year record of metal contamination on an unprecedented scale from mining of the Iberian Pyrite Belt. *Geology* 25: 291–294.

Veron AJ, Church TM, Flegal AR, Patterson CC, and Erel Y (1993) Response of lead cycling in the surface

Sargasso Sea to changes in tropospheric input. *Journal of Geophysical Research* 98: 18269–18276.

Veron AJ, Church TM, Rivera-Duarte I, and Flegal AR (1999) Stable lead isotope ratios trace thermohaline circulation in the subarctic North Atlantic. *Deep-Sea Research II* 46: 919–935.

Wolff EW and Peel DA (1985) The record of global pollution in polar snow and ice. *Nature* 313: 535–540.

Wu JF and Boyle EA (1997) Lead in the western North Atlantic Ocean: completed response to leaded gasoline phaseout. *Geochimica et Cosmochimica Acta* 61: 3279–3283.

ATMOSPHERIC INPUT OF POLLUTANTS

R. A. Duce, Texas A&M University, College Station, TX, USA

Introduction

For about a century oceanographers have tried to understand the budgets and processes associated with both natural and human-derived substances entering the ocean. Much of the early work focused on the most obvious inputs – those carried by rivers and streams. Later studies investigated sewage outfalls, dumping, and other direct input pathways for pollutants. Over the past decade or two, however, it has become apparent that the atmosphere is also not only a significant, but in some cases dominant, pathway by which both natural materials and contaminants are transported from the continents to both the coastal and open oceans. These substances include mineral dust and plant residues, metals, nitrogen compounds from combustion processes and fertilizers, and pesticides and a wide range of other synthetic organic compounds from industrial and domestic sources. Some of these substances carried into the ocean by the atmosphere, such as lead and some chlorinated hydrocarbons, are potentially harmful to marine biological systems. Other substances, such as nitrogen compounds, phosphorus, and iron, are nutrients and may enhance marine productivity. For some substances, such as aluminum and some rare earth elements, the atmospheric input has an important impact on their natural chemical cycle in the sea.

In subsequent sections there will be discussions of the input of specific chemicals via the atmosphere to estuarine and coastal waters. This will be followed by considerations of the atmospheric input to open ocean regions and its potential importance. The atmospheric estimates will be compared with the input via other pathways when possible. Note that there are still very large uncertainties in all of the fluxes presented, both those from the atmosphere and those from other sources. Unless otherwise indicated, it should be assumed that the atmospheric input rates have uncertainties ranging from a factor of 2 to 4, sometimes even larger.

Estimating Atmospheric Contaminant Deposition

Contaminants present as gases in the atmosphere can exchange directly across the air/sea boundary or they may be scavenged by rain and snow. Pollutants present on particles (aerosols) may deposit on the ocean either by direct (dry) deposition or they may also be scavenged by precipitation. The removal of gases and/or particles by rain and snow is termed wet deposition.

Direct Deposition of Gases

Actual measurement of the fluxes of gases to a water surface is possible for only a very few chemicals at the present time, although extensive research is underway in this area, and analytical capabilities for fast response measurements of some trace gases are becoming available. Modeling the flux of gaseous compounds to the sea surface or to rain droplets requires a knowledge of the Henry's law constants and air/sea exchange coefficients as well as atmospheric and oceanic concentrations of the chemicals of interest. For many chemicals this information is not available. Discussions of the details of these processes of air/sea gas exchange can be found in other articles in this volume.

Particle Dry Deposition

Reliable methods do not currently exist to measure directly the dry deposition of the full size range of aerosol particles to a water surface. Thus, dry deposition of aerosols is often estimated using the dry deposition velocity, v_d. For dry deposition, the flux is then given by:

$$F_d = v_d \cdot C_a \quad [1]$$

where F_d is the dry deposition flux (e.g., in $g\,m^{-2}\,s^{-1}$), v_d is the dry deposition velocity (e.g., in $m\,s^{-1}$), and C_a is the concentration of the substance on the aerosol particles in the atmosphere (e.g., in $g\,m^{-3}$). In this formulation v_d incorporates all the processes of dry deposition, including diffusion, impaction, and gravitational settling of the particles to a water surface. It is very difficult to parameterize accurately the dry deposition velocity since each of these processes is acting on a particle population, and they are each dependent upon a number of factors, including wind speed, particle size, relative humidity, etc. The following are dry deposition velocities that have been used in some studies of atmospheric deposition of particles to the ocean:

- Submicrometer aerosol particles, $0.001\,\mathrm{m\,s^{-1}} \pm$ a factor of three
- Supermicrometer crustal particles not associated with sea salt, $0.01\,\mathrm{m\,s^{-1}} \pm$ a factor of three
- Giant sea-salt particles and materials carried by them, $0.03\,\mathrm{m\,s^{-1}} \pm$ a factor of two

Proper use of eqn [1] requires that information be available on the size distribution of the aerosol particles and the material present in them.

Particle and Gas Wet Deposition

The direct measurement of contaminants in precipitation samples is certainly the best approach for determining wet deposition, but problems with rain sampling, contamination, and the natural variability of the concentration of trace substances in precipitation often make representative flux estimates difficult using this approach. Studies have shown that the concentration of a substance in rain is related to the concentration of that substance in the atmosphere. This relationship can be expressed in terms of a scavenging ratio, S:

$$S = C_r \cdot \rho \cdot C_{a/g}{}^{-1} \qquad [2]$$

where C_r is the concentration of the substance in rain (e.g., in $\mathrm{g\,kg^{-1}}$), ρ is the density of air ($\sim 1.2\,\mathrm{kg\,m^{-3}}$), $C_{a/g}$ is the aerosol or gas phase concentration in the atmosphere (e.g., in $\mathrm{g\,m^{-3}}$), and S is dimensionless. Values of S for substances present in aerosol particles range from a few hundred to a few thousand, which roughly means that 1 g (or 1 ml) of rain scavenges $\leq 1\,\mathrm{m^3}$ of air. For aerosols, S is dependent upon such factors as particle size and chemical composition. For gases, S can vary over many orders of magnitude depending on the specific gas, its Henry's law constant, and its gas/water exchange coefficient. For both aerosols and gases, S is also dependent upon the vertical concentration distribution and vertical extent of the precipitating cloud, so the use of scavenging ratios requires great care, and the results have significant uncertainties. However, if the concentration of an atmospheric substance and its scavenging ratio are known, the scavenging ratio approach can be used to estimate wet deposition fluxes as follows:

$$F_r = P \cdot C_r = P \cdot S \cdot C_{a/g} \cdot \rho^{-1} \qquad [3]$$

where F_r is the wet deposition flux (e.g., in $\mathrm{g\,m^{-2}\,year^{-1}}$) and P is the precipitation rate (e.g., in $\mathrm{m\,year^{-1}}$), with appropriate conversion factors to translate rainfall depth to mass of water per unit area. Note that $P \cdot S \cdot \rho^{-1}$ is equivalent to a wet deposition velocity.

Atmospheric Deposition to Estuaries and the Coastal Ocean

Metals

The atmospheric deposition of certain metals to coastal and estuarine regions has been studied more than that for any other chemicals. These metals are generally present on particles in the atmosphere. Chesapeake Bay is among the most thoroughly studied regions in North America in this regard. **Table 1** provides a comparison of the atmospheric and riverine deposition of a number of metals to Chesapeake Bay. The atmospheric numbers represent a combination of wet plus dry deposition directly onto the Bay surface. Note that the atmospheric input ranges from as low as 1% of the total input for manganese to as high as 82% for aluminum. With the exception of Al and Fe, which are largely derived from natural weathering processes (e.g., mineral matter or soil), most of the input of the other metals is from human-derived sources. For metals with anthropogenic sources the atmosphere is most important for lead (32%).

There have also been a number of investigations of the input of metals to the North Sea, Baltic Sea, and Mediterranean Sea. Some modeling studies of the North Sea considered not only the direct input pathway represented by the figures in **Table 1**, but also considered Baltic Sea inflow, Atlantic Ocean inflow and outflow, and exchange of metals with the

Table 1 Estimates of the riverine and atmospheric input of some metals to Chesapeake Bay

Metal	*Riverine input (10^6 g year^{-1})*	*Atmospheric input (10^6 g year^{-1})*	*% Atmospheric input*
Aluminum	160	700	81
Iron	600	400	40
Manganese	1300	13	1
Zinc	50	18	26
Copper	59	3.5	6
Nickel	100	4	4
Lead	15	7	32
Chromium	15	1.5	10
Arsenic	5	0.8	14
Cadmium	2.6	0.4	13

Data reproduced with permission from Scudlark JR, Conko KM and Church TM (1994) Atmospheric wet desposition of trace elements to Chesapezke Bay: (CBAD) study year 1 results. *Atmospheric Environment* 28: 1487–1498.

sediments, as well as the atmospheric contribution to all of these inputs. **Figure 1** shows schematically some modeling results for lead, copper, and cadmium. Note that for copper, atmospheric input is relatively unimportant in this larger context, while atmospheric input is somewhat more important for cadmium, and it is quite important for lead, being approximately equal to the inflow from the Atlantic Ocean, although still less than that entering the North Sea from dumping. As regards lead, note that approximately 20% of the inflow from the Atlantic to the North Sea is also derived from the atmosphere. This type of approach gives perhaps the most accurate and in-depth analysis of the importance of atmospheric input relative to all other sources of a chemical in a water mass.

Nitrogen Species

The input of nitrogen species from the atmosphere is of particular interest because nitrogen is a necessary nutrient for biological production and growth in the ocean. There has been an increasing number of studies of the atmospheric input of nitrogen to estuaries and the coastal ocean. Perhaps the area most intensively studied is once again Chesapeake Bay. **Table 2** shows that approximately 40% of all the nitrogen contributed by human activity to Chesapeake Bay enters via

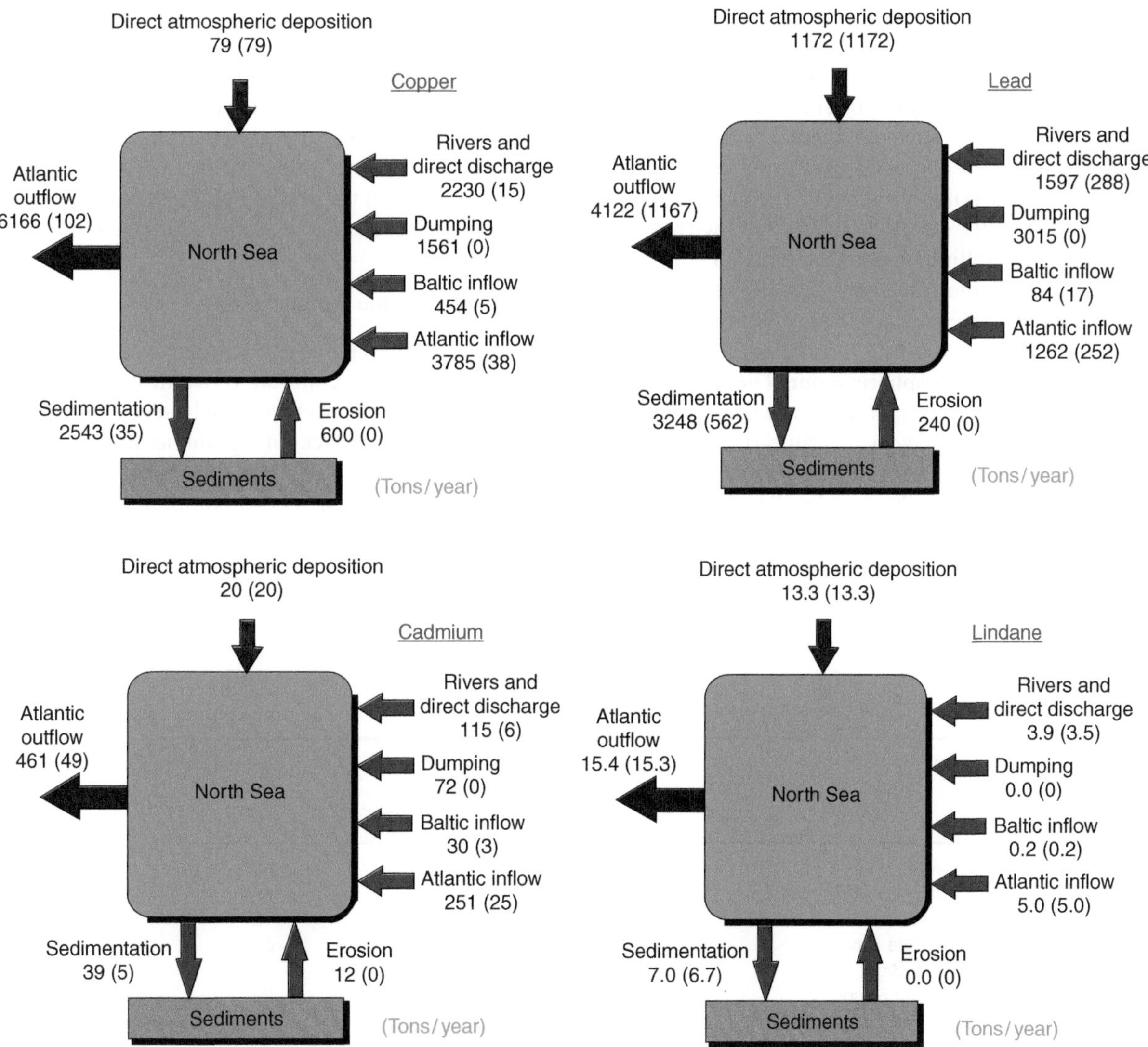

Figure 1 Input of copper, lead, cadmium, and lindane to the North Sea. Values in parentheses denote atmospheric contribution. For example, for copper the atmospheric contribution to rivers and direct discharges is 15 tons per year. (Figure reproduced with permission from Duce, 1998. Data adapted with permission from van den Hout, 1994.)

Table 2 Estimates of the input of nitrogen to Chesapeake Bay

Source	*Total input (10^9g year^{-1})*	*Areal input rate (g m^{-2}year^{-1})*	*% of the total*
Animal waste	5	0.4	3
Fertilizers	48	4.2	34
Point sources	33	2.9	24
Atmospheric precipitation			
nitrate	35	3.1	25
ammonium	19	1.7	14
Total	140	12.3	100

Data reproduced with permission from Fisher D, Ceroso T, Mathew T and Oppenheimer M (1988) *Polluted Coastal Waters: The Role of Acid Rain.* New York: Environmental Defense Fund.

precipitation falling directly on the Bay or its watershed. These studies were different from most earlier studies because the atmospheric contributions were considered not only to be direct deposition on the water surface, but also to include that coming in via the atmosphere but falling on the watershed and then entering the Bay. Note from **Table 2** that atmospheric input of nitrogen exceeded that from animal waste, fertilizers, and point sources. In the case of nitrate, about 23% falls directly on the Bay, with the remaining 77% falling on the watershed. These results suggest that studies that consider only the direct deposition on a water surface (e.g., the results shown in **Table 1**) may significantly underestimate the true contribution of atmospheric input. The total nitrogen fertilizer applied to croplands in the Chesapeake Bay region is $\sim 5.4\,\mathrm{g\,m^{-2}\,year^{-1}}$, while the atmospheric nitrate and ammonium nitrogen entering the Bay is $\sim 4.8\,\mathrm{g\,m^{-2}\,year^{-1}}$. Chesapeake Bay is almost as heavily fertilized from atmospheric nitrogen, largely anthropogenic, as the croplands are by fertilizer in that watershed!

Results from studies investigating nitrogen input to some other estuarine and coastal regions are summarized in **Table 3**. In this table atmospheric sources for nitrogen are compared with all other sources, where possible. The atmospheric input ranges from 10% to almost 70% of the total. Note that some estimates compare only direct atmospheric deposition with all other sources and some include as part of the atmospheric input the portion of the deposition to the watershed that reaches the estuary or coast.

Synthetic Organic Compounds

Concern is growing about the input of a wide range of synthetic organic compounds to the coastal ocean. To date there have been relatively few estimates of the atmospheric fluxes of synthetic organic compounds to the ocean, and these estimates have significant uncertainties. These compounds are often both persistent and toxic pollutants, and many have relatively high molecular weights. The calculation of the atmospheric input of these compounds to the

Table 3 Estimates of the input of nitrogen to some coastal areas[a]

Region	*Total atmospheric input[b] (10^9g year^{-1})*	*Total input all sources (10^9g year^{-1})*	*% Atmospheric input*
North Sea	400[c]	1500	27[c]
Western Mediterranean Sea	400[c]	577[d]	69[c]
Baltic Sea	500	~1200	42
Chesapeake Bay	54	140	39
New York Bight	–	–	13[c]
Long Island Sound	11	49	22
Neuse River Estuary, NC	1.7	7.5	23

[a] Data from several sources in the literature.
[b] Total from direct atmospheric deposition and runoff of atmospheric material from the watershed.
[c] Direct atmospheric deposition to the water only.
[d] Total from atmospheric and riverine input only.

Table 4 Estimates of the input of synthetic organic compounds to the North Sea

Organic compound	*Atmospheric input (10^6 g $year^{-1}$)*	*Input from other sources (10^6 g $year^{-1}$)*	*% Atmospheric input*
PCB	40	3	93
Lindane	36	3	92
Polycyclic aromatic hydrocarbons	80	90	47
Benzene	400	500	44
Trichloroethene	300	80	80
Trichloroethane	90	60	94
Tetrachloroethene	100	10	91
Carbon tetrachloride	6	40	13

Data reproduced with permission from Warmerhoven JP, Duiser JA, de Leu LT and Veldt C (1989) *The Contribution of the Input from the Atmosphere to the Contamination of the North Sea and the Dutch Wadden Sea.* Delft, The Netherlands: TNO Institute of Environmental Sciences.

coastal ocean is complicated by the fact that many of them are found primarily in the gas phase in the atmosphere, and most of the deposition is related to the wet and dry removal of that phase. The atmospheric residence times of most of these compounds are long compared with those of metals and nitrogen species. Thus the potential source regions for these compounds entering coastal waters can be distant and widely dispersed.

Figure 1 shows the input of the pesticide lindane to the North Sea. Note that the atmospheric input of lindane dominates that from all other sources. **Table 4** compares the atmospheric input to the North Sea with that of other transport paths for a number of other synthetic organic compounds. In almost every case atmospheric input dominates the other sources combined.

Atmospheric Deposition to the Open Ocean

Studies of the atmospheric input of chemicals to the open ocean have also been increasing lately. For many substances a relatively small fraction of the material delivered to estuaries and the coastal zone by rivers and streams makes its way through the near shore environment to open ocean regions. Most of this material is lost via flocculation and sedimentation to the sediments as it passes from the freshwater environment to open sea water. Since aerosol particles in the size range of a few micrometers or less have atmospheric residence times of one to several days, depending upon their size distribution and local precipitation patterns, and most substances of interest in the gas phase have similar or even longer atmospheric residence times, there is ample opportunity for these atmospheric materials to be carried hundreds to thousands of kilometers before being deposited on the ocean surface.

Metals

Table 5 presents estimates of the natural and anthropogenic emission of several metals to the global atmosphere. Note that ranges of estimates and the best estimate are given. It appears from **Table 5** that anthropogenic sources dominate for lead, cadmium, and zinc, with essentially equal contributions for copper, nickel, and arsenic. Clearly a significant fraction of the input of these metals from the atmosphere to the ocean could be derived largely from anthropogenic sources.

Table 6 provides an estimate of the global input of several metals from the atmosphere to the ocean and compares these fluxes with those from rivers. Estimates are given for both the dissolved and particulate forms of the metals. These estimates suggest that rivers are generally the primary source of particulate

Table 5 Emissions of some metals to the global atmosphere

Metal	*Anthropogenic emissions (10^9 g $year^{-1}$)*		*Natural emissions (10^9g $year^{-1}$)*	
	Range	*Best estimate*	*Range*	*Best estimate*
Lead	289–376	332	1–23	12
Cadmium	3.1–12	7.6	0.15–2.6	1.3
Zinc	70–194	132	4–86	45
Copper	20–51	35	2.3–54	28
Arsenic	12–26	18	0.9–23	12
Nickel	24–87	56	3–57	30

Data reproduced with permission from Duce *et al.*, 1991.

Table 6 Estimates of the input of some metals to the global ocean

Metal	*Atmospheric input*		*Riverine input*	
	Dissolved (10^9 g year^{-1})	*Particulate (10^9 g year^{-1})*	*Dissolved (10^9 g year^{-1})*	*Particulate (10^9 g year^{-1})*
Iron	1600–4800	14 000–42 000	1100	110 000
Copper	14–45	2–7	10	1 500
Nickel	8–11	14–17	11	1 400
Zinc	33–170	11–55	6	3 900
Arsenic	2.3–5	1.3–3	10	80
Cadmium	1.9–3.3	0.4–0.7	0.3	15
Lead	50–100	6–12	2	1 600

Data reproduced with permission from Duce *et al.*, 1991.

metals in the ocean, although again a significant fraction of this material may not get past the coastal zone. For the dissolved phase atmospheric and riverine inputs are roughly equal for metals such as iron, copper, and nickel; while for zinc, cadmium, and particularly lead atmospheric inputs appear to dominate. These estimates were made based on data collected in the mid-1980s. Extensive efforts to control the release of atmospheric lead, which has been primarily from the combustion of leaded gasoline, are now resulting in considerably lower concentrations of lead in many areas of the open ocean. For example, **Figure 2** shows that the concentration of dissolved lead in surface sea water near Bermuda has been decreasing regularly over the past 15–20 years, as has the atmospheric lead concentration in that region. This indicates clearly that at least for very particle-reactive metals such as lead, which has a short lifetime in the ocean (several years), even the open ocean can recover rather rapidly when the anthropogenic input of such metals is reduced or ended. Unfortunately, many of the other metals of most concern have much longer residence times in the ocean (thousands to tens of thousands of years).

Figure 3 presents the calculated fluxes of several metals from the atmosphere to the ocean surface and from the ocean to the seafloor in the 1980s in the tropical central North Pacific. Note that for most metals the two fluxes are quite similar, suggesting the potential importance of atmospheric input to the

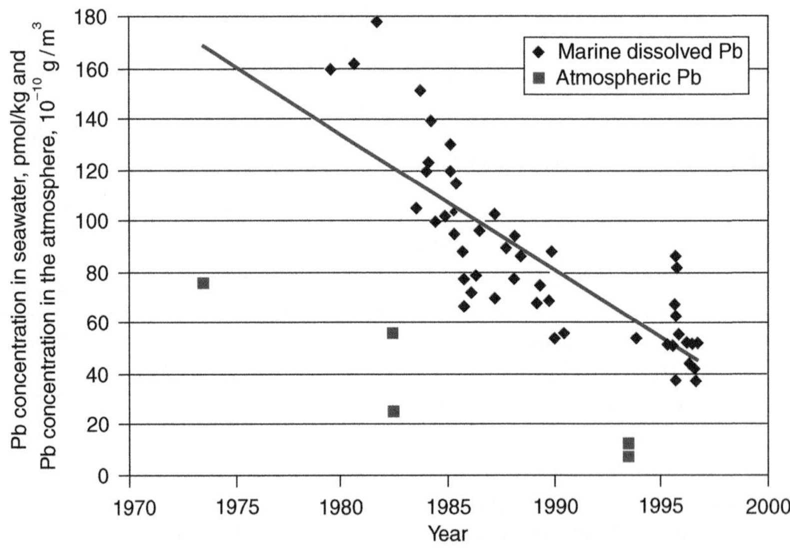

Figure 2 Changes in concentration of atmospheric lead at Bermuda and dissolved surface oceanic lead near Bermuda from the mid-1970s to the mid-1990s. (Data reproduced with permission from Wu J and Boyle EA (1997) Lead in the western North Atlantic Ocean: Completed response to leaded gasoline phaseout. *Geochimica et Cosmochimica Acta* 61: 3279–3283; and from Huang S, Arimoto R and Rahn KA (1996) Changes in atmospheric lead and other pollution elements at Bermuda. *Journal of Geophysical Research* 101: 21 033–21 040.)

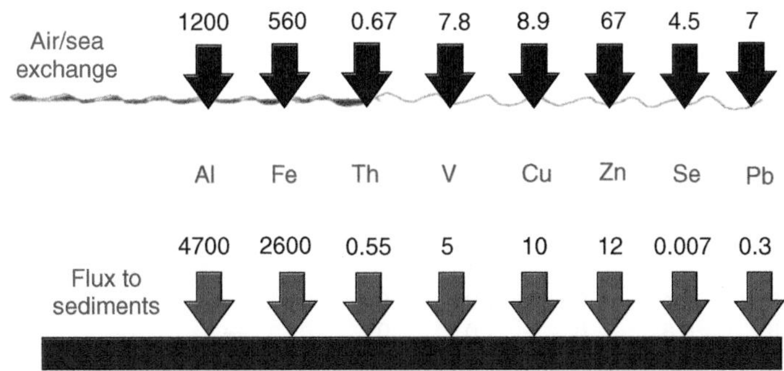

Figure 3 A comparison of the calculated fluxes of aluminum (Al), iron (Fe), thorium (Th), vanadium (V), copper (Cu), zinc (Zn), selenium (Se), and lead (Pb) (in 10^{-9} g cm^{-2} $year^{-1}$) from the atmosphere to the ocean and from the ocean to the sediments in the central tropical North Pacific. For each metal note the relative similarity in the two fluxes, except for lead and selenium. (Reproduced with permission from Duce, 1998.)

marine sedimentation of these metals in this region. Lead and selenium are exceptions, however, as the atmospheric flux is much greater than the flux to the seafloor. The fluxes to the seafloor represent average fluxes over the past several thousand years, whereas the atmospheric fluxes are roughly for the present time. The atmospheric lead flux is apparently much larger than the flux of lead to the sediments, primarily because of the high flux of anthropogenic lead from the atmosphere to the ocean since the introduction of tetraethyllead in gasoline in the 1920s. (The atmospheric flux is much lower now than in the 1980s, as discussed above.) However, in the case of selenium the apparently higher atmospheric flux is an artifact, because most of the flux of selenium from the atmosphere to the ocean is simply marine-derived selenium that has been emitted from the ocean to the atmosphere as gases, such as dimethyl selenide (DMSe). DMSe is oxidized in the atmosphere and returned to the ocean, i.e., the selenium input is simply a recycled marine flux. Thus, care must be taken when making comparisons of this type.

Nitrogen Species

There is growing concern about the input of anthropogenic nitrogen species to the global ocean. This issue is of particular importance in regions where nitrogen is the limiting nutrient, e.g., the oligotrophic waters of the central oceanic gyres. Estimates to date suggest that in such regions atmospheric nitrogen will in general account for only a few percent of the total 'new' nitrogen delivered to the photic zone, with most of the 'new' nutrient nitrogen derived from the upwelling of nutrient-rich deeper waters and from nitrogen fixation in the sea. It is recognized, however, that the atmospheric input is highly episodic, and at times it may play a much more important role as a source for nitrogen in surface waters. **Table 7** presents a recent estimate of the current input of fixed nitrogen to the global ocean from rivers, the atmosphere, and nitrogen fixation. From the numbers given it is apparent that all three sources are likely important, and within the uncertainties of the estimates they are roughly equal. In the case of rivers, about half of the nitrogen input is anthropogenic for atmospheric input perhaps the most important information in **Table 7** is that the organic nitrogen flux appears to be equal to or perhaps significantly greater than the inorganic (i.e., ammonium and nitrate) nitrogen flux. The source of the organic nitrogen is not known, but there are indications that a large fraction of it is anthropogenic in origin. This is a form of atmospheric nitrogen input to the ocean that had not been considered until very recently, as there had been few measurements of organic nitrogen input to the ocean before the mid-1990s. The chemical forms of this organic nitrogen are still largely unknown.

Of particular concern are potential changes to the input of atmospheric nitrogen to the open ocean in

Table 7 Estimates of the current input of reactive nitrogen to the global ocean

Source	*Nitrogen input (10^{12} g $year^{-1}$)*
From the atmosphere	
Dissolved inorganic nitrogen	28–70
Dissolved organic nitrogen	28–84
From rivers (dissolved inorganic + organic nitrogen)	
Natural	14–35
Anthropogenic	7–35
From nitrogen fixation within the ocean	14–42

Data reproduced with permission from Cornell S, Rendell A and Jickells T (1995) Atmospheric inputs of dissolved organic nitrogen to the oceans. *Nature* 376: 243–246.

Table 8 Estimates of anthropogenic reactive nitrogen production, 1990 and 2020

Region	*Energy (NO_x)*					*Fertilizer*				
	1990 (10^{12}g N year^{-1})	*2020*	Δ	*Factor*	*% of total increase*	*1990 (10^{12}g N year^{-1})*	*2020*	Δ	*Factor*	*% of total increase*
USA/Canada	7.6	10.1	2.5	1.3	10	13.3	14.2	0.9	1.1	1.6
Europe	4.9	5.2	0.3	1.1	1	15.4	15.4	0	1.0	0
Australia	0.3	0.4	0.1	1.3	0.4	—	—	—	—	—
Japan	0.8	0.8	0	1.0	0	—	—	—	—	—
Asia	3.5	13.2	9.7	3.8	39	36	85	49	2.4	88
Central/South America	1.5	5.9	4.4	3.9	18	1.8	4.5	2.7	2.5	5
Africa	0.7	4.2	3.5	6.0	15	2.1	5.2	3.1	2.5	6
Former Soviet Union	2.2	5.7	3.5	2.5	15	10	10	0	1.0	0
Total	21	45	24	2.1	100	79	134	55	1.7	100

Data adapted with permission from Galloway *et al.*, 1995.

the future as a result of increasing human activities. The amount of nitrogen fixation (formation of reactive nitrogen) produced from energy sources (primarily as NO_x, nitrogen oxides), fertilizers, and legumes in 1990 and in 2020 as a result of human activities as well as the current and predicted future geographic distribution of the atmospheric deposition of reactive nitrogen to the continents and ocean have been evaluated recently. **Table 8** presents estimates of the formation of fixed nitrogen from energy use and production and from fertilizers, the two processes which would lead to the most important fluxes of reactive nitrogen to the atmosphere. Note that the most highly developed regions in the world, represented by the first four regions in the table, are predicted to show relatively little increase in the formation of fixed nitrogen, with none of these areas having a predicted increase by 2020 of more than a factor of 1.3 nor a contribution to the overall global increase in reactive nitrogen exceeding 10%. However, the regions in the lower part of **Table 8** will probably contribute very significantly to increased anthropogenic reactive nitrogen formation in 2020. For example, it is predicted that the production of reactive nitrogen in Asia from energy sources will increase $\leq$ fourfold, and that Asia will account for almost 40% of the global increase, while Africa will have a sixfold increase and will account for 15% of the global increase in energy-derived fixed nitrogen. It is predicted that production of reactive nitrogen from the use of fertilizers in Asia will increase by a factor of 2.4, and Asia will account for ~88% of the global increase from this source. Since both energy sources (NO_x, and ultimately nitrate) and fertilizer (ammonia and nitrate) result in the extensive release of reactive nitrogen to the atmosphere, the predictions above indicate that there should be very significant increases in the atmospheric deposition to the ocean of nutrient nitrogen species downwind of such regions as Asia, Central and South America, Africa, and the former Soviet Union.

This prediction has been supported by numerical modeling studies. These studies have resulted in the generation of maps of the 1980 and expected 2020 annual deposition of reactive nitrogen to the global ocean. **Figure 4** shows the expected significant increase in reactive nitrogen deposition from fossil fuel combustion to the ocean to the east of all of Asia, from Southeast Asia to the Asian portion of the former Soviet Union; to the east of South Africa, northeast Africa and the Mideast and Central America and southern South America; and to the west of northwest Africa. This increased reactive nitrogen transport and deposition to the ocean will provide new sources of nutrient nitrogen to some regions of the ocean where biological production is currently nitrogen-limited. There is thus the possibility of significant impacts on regional biological primary production, at least episodically, in these regions of the open ocean.

Synthetic Organic Compounds

The atmospheric residence times of many synthetic organic compounds are relatively long compared with those of the metals and nitrogen species, as mentioned previously. Many of these substances are found primarily in the gas phase in the atmosphere, and they are thus very effectively mobilized into the atmosphere during their production and use. Their long atmospheric residence times of weeks to months leads to atmospheric transport that can often be hemispheric or near hemispheric in scale. Thus

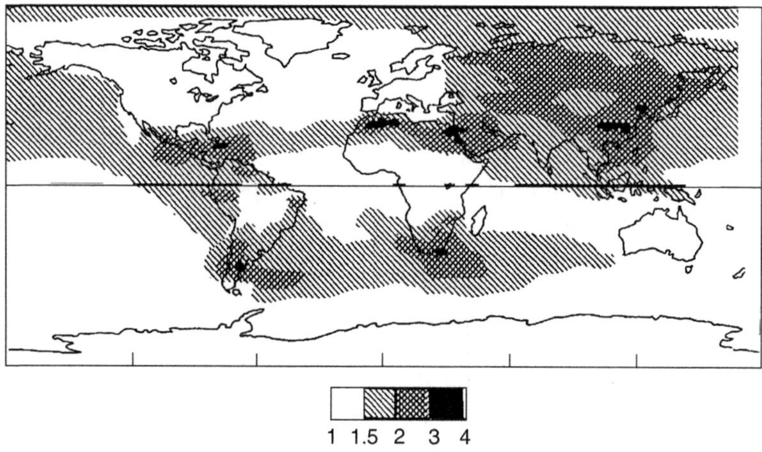

Figure 4 The ratio of the estimated deposition of reactive nitrogen to ocean and land surfaces in 2020 relative to 1980. (Figure reproduced with permission from Watson AJ (1997) Surface Ocean–Lower Atmosphere Study (SOLAS). *Global Change Newsletter, IGBP* no. 31, 9–12; data in figure adapted with permission from Galloway JN, Levy H and Kasibhatla PS (1994) Year 2020: Consequences of population growth and development on deposition of oxidized nitrogen. *Ambio* 23: 120–123.).

Table 9 Estimates of the atmospheric input of organochlorine compounds to the global ocean

Ocean	*ΣHCH (10^6g year^{-1})*	*ΣDDT (10^6g year^{-1})*	*ΣPCB (10^6g year^{-1})*	*HCB (10^6g year^{-1})*	*Dieldrin (10^6g year^{-1})*	*Chlordane (10^6g year^{-1})*
North Atlantic	850	16	100	17	17	8.7
South Atlantic	97	14	14	10	2.0	1.0
North Pacific	2600	66	36	20	8.9	8.3
South Pacific	470	26	29	19	9.5	1.9
Indian	700	43	52	11	6.0	2.4
Global input via the atmosphere	~4700	~170	~230	~80	~40	~22
Global input via rivers	~60	~4	~60	~4	~4	~4
% Atmospheric input	~99%	~98%	~80%	~95%	~91%	~85%

(Data reproduced with permission from Duce *et al.*, 1991.)

atmospheric transport and deposition in general dominates all other sources for these chemicals in sea water in open ocean regions.

Table 9 compares the atmospheric and riverine inputs to the world ocean for a number of synthetic organochlorine compounds. Note that the atmosphere in most cases accounts for 90% or more of the input of these compounds to the ocean. **Table 9** also presents estimates of the input of these same organochlorine compounds to the major ocean basins. Since most of these synthetic organic compounds are produced and used in the northern hemisphere, it is not surprising that the flux into the northern hemisphere ocean is greater than that to southern hemisphere marine regions. There are some differences for specific compounds in different ocean basins. For example, HCH (hexachlorocyclohexane) and DDT have a higher input rate to the North Pacific than the North Atlantic, largely because of the greater use of these compounds in Asia than in North America or Europe. On the other hand, the input of PCBs (polychlorinated biphenyls) and dieldrin is higher to the North Atlantic than the North Pacific, primarily because of their greater use in the continental regions adjacent to the North Atlantic.

Conclusions

The atmosphere transports materials to the ocean that are both harmful to marine life and that are essential for marine biological productivity. It is now apparent that atmospheric transport and deposition of some metals, nitrogen species, and synthetic organic compounds can be a significant and in some cases dominant pathway for these substances entering both estuarine and coastal waters as well as some open ocean regions. Atmospheric input clearly must be considered in any evaluation of material fluxes to marine ecosystems. However, the uncertainties in the atmospheric fluxes of these materials to the ocean are large. The primary reasons for these large uncertainties are:

- The lack of atmospheric concentration data over vast regions of the coastal and open ocean, particularly over extended periods of time and under varying meteorological conditions;
- The episodic nature of the atmospheric deposition to the ocean;
- The lack of accurate models of air/sea exchange, particularly for gases;
- The inability to measure accurately the dry deposition of particles; and
- The inability to measure accurately the air/sea exchange of gases.

See also

Chlorinated Hydrocarbons. Metal Pollution. Refractory Metals. Transition Metals and Heavy Metal Speciation.

Further Reading

Duce RA (1998) *Atmospheric Input of Pollution to the Oceans*, pp. 9–26. Proceedings of the Commission for Marine Meteorology Technical Conference on Marine Pollution, World Meteorological Organization TD-No. 890, Geneva, Switzerland.

Duce RA, Liss PS, Merrill JT, *et al.* (1991) The atmospheric input of trace species to the world ocean. *Global Biogeochemical Cycles* 5: 193–259.

Galloway JN, Schlesinger WH, Levy H, Michaels A, and Schnoor JL (1995) Nitrogen fixation: anthropogenic enhancement – environmental reponse. *Global Biogeochemical Cycles* 9: 235–252.

Jickells TD (1995) Atmospheric inputs of metals and nutrients to the oceans: their magnitude and effects. *Marine Chemistry* 48: 199–214.

Liss PS and Duce RA (1997) *The Sea Surface and Global Change*. Cambridge: Cambridge University Press.

Paerl HW and Whitall DR (1999) Anthropogenically-derived atmospheric nitrogen deposition, marine eutrophication and harmful algal bloom expansion: Is there a link? *Ambio* 28: 307–311.

Prospero JM, Barrett K, Church T, *et al.* (1996) Nitrogen dynamics of the North Atlantic Ocean – Atmospheric deposition of nutrients to the North Atlantic Ocean. *Biogeochemistry* 35: 27–73.

van den Hout KD (ed.) (1994) *The Impact of Atmospheric Deposition of Non-Acidifying Pollutants on the Quality of European Forest Soils and the North Sea.* Report of the ESQUAD Project, IMW-TNO Report No. R 93/329.

NUCLEAR FUEL REPROCESSING AND RELATED DISCHARGES

H. N. Edmonds, University of Texas at Austin, Port Aransas, TX, USA

Introduction

Oceanographers have for some decades made use of human perturbations to the environment as tracers of ocean circulation and of the behavior of similar substances in the oceans. The study of the releases of anthropogenic radionuclides – by-products of the nuclear industry – by nuclear fuel reprocessing plants is an example of such an application. Other examples of the use of anthropogenic substances as oceanographic tracers addressed in this Encyclopedia include tritium and radiocarbon, largely resulting from atmospheric nuclear weapons testing in the 1950s and 1960s, and chlorofluorocarbons (CFCs), a family of gases which have been used in a variety of applications such as refrigeration and polymer manufacture since the 1920s. As outlined in this article, the source function of nuclear fuel reprocessing discharges, which is very different from that of either weapons test fallout or CFCs, is both a blessing and a curse, defining the unique applicability of these tracers but also requiring much additional and careful quantification.

The two nuclear fuel reprocessing plants of most interest to oceanographers are located at Sellafield, in the UK, which has been discharging wastes into the Irish Sea since 1952, and at Cap de la Hague, in north-western France, which has been discharging wastes into the English Channel since 1966. These releases have been well documented and monitored in most cases. It should be noted, however, that the releases of some isotopes of interest to oceanographers have not been as well monitored throughout the history of the plants because they were not of radiological concern, were difficult to measure, or both. This is particularly true for ^{99}Tc and ^{129}I, which are the isotopes currently of greatest interest to the oceanographic community but for which specific, official release data are only available for the last two decades or less. The radionuclides most widely studied by ocean scientists have been ^{137}Cs, ^{134}Cs, ^{90}Sr, ^{125}Sb, ^{99}Tc, ^{129}I, and Pu isotopes. The Sellafield plant has dominated the releases of most of these, with the exception of ^{125}Sb and, particularly in the 1990s, ^{129}I. In general, the Sellafield releases have been particularly well documented and are the easiest for interested scientists and other parties to obtain. The figures of releases presented in this article are based primarily on the Sellafield data, except where sufficient data are available for Cap de la Hague.

The location of the Sellafield and Cap de la Hague plants is important to the oceanographic application of their releases. As discussed below, the releases enter the surface circulation of the high latitude North Atlantic and are transported northwards into the Norwegian and Greenland Seas and the Arctic Ocean. They have been very useful as tracers of ventilation and deep water formation in these regions, which are of great importance to global thermohaline circulation and climate. In addition, the study of the dispersion of radionuclides from the reprocessing plants at Sellafield and Cap de la Hague serves as an analog for understanding the ultimate distribution and fate of other wastes arising from industrial activities in north-western Europe.

Much of the literature on reprocessing releases in the oceans has focused on, or been driven by, concerns relating to contamination and radiological effects, and they have not found the same broad or general oceanographic application as some other anthropogenic tracers despite their great utility and the excellent quality of published work. This may be attributable to a variety of factors, including perhaps (1) the contaminant-based focus of much of the work and its publication outside the mainstream oceanographic literature, (2) the complications of the source function as compared with some other tracers, and (3) the comparative difficulty of measuring many of these tracers and the large volumes of water that have typically been required. This article addresses the historic and future oceanographic applications of these releases.

A Note on Units

In most of the literature concerning nuclear fuel reprocessing releases, amounts (both releases and ensuing environmental concentrations) are expressed in activity units. The SI unit for activity is the becquerel (Bq), which replaced the previously common curie

(Ci), a unit which was based on the activity of one gram of radium. These two units are related as follows:

$$1\,\text{Bq} = 1\,\text{disintegration per second}$$
$$1\,\text{Ci} = 3.7 \times 10^{10}\,\text{Bq}$$

The conversion from activity to mass or molar units is dependent on the half-life of the isotope in question. Activity (A) is the product of the number of atoms of the isotope present (N) and its decay constant (λ): $A = \lambda N$. The decay constant, λ, is related to the half-life by $t_{1/2} = (ln2)/\lambda$. Using this equation, it is simple to convert from Bq to mol, given the decay constants in units of s^{-1}, and Avogadro's number of 6.023×10^{23} atoms mol^{-1}.

It is also worth noting that the total activity of naturally occurring radionuclides in sea water is approximately 12 kBq m^{-3}. Most of this activity is from the long-lived naturally occuring isotope, $^{40}K(t_{1/2} = 1.25 \times 10^{9}$ years). Activities contributed from anthropogenic radionuclides greatly exceed this (millions of Bq m^{-3}) in the immediate vicinity of the Sellafield and La Hague outfall pipes. However, most oceanographic studies of these releases involve measurements of much lower activities: up to 10s of Bq m^{-3} in the case of ^{137}Cs, and generally even lower for other radionuclides, for instance mBq m^{-3} or less for plutonium isotopes and ^{129}I.

Nuclear Fuel Reprocessing and Resulting Tracer Releases

Origin and Description of Reprocessing Tracers

Nuclear fuel reprocessing involves the recovery of fissile material (plutonium and enriched uranium) and the separation of waste products from 'spent' (used) fuel rods from nuclear reactors. In the process, fuel rods, which have been stored for a time to allow short-lived radionuclides to decay, are dissolved and the resulting solution is chemically purified and separated into wastes of different composition and activity. Routine releases from the plants to the environment occur under controlled conditions and are limited to discharge totals dictated by the overseeing authorities of each country. The discharges have varied over the years as a function of the amount and type of fuel processed and changes in the reprocessing technology. The discharge limits themselves have changed, in response to monitoring efforts and also spurred by technological advances in waste-treatment capabilities. As a general rule, these changes have resulted in decreasing releases for most nuclides (most notably cesium and the actinide elements), but there are exceptions. For instance, the Enhanced Actinide Removal Plant (EARP) was constructed at the Sellafield site in the 1990s in order to enable the additional treatment and subsequent discharge of a backlog of previously stored wastes. Although the new technology enabled the removal of actinide elements from these wastes, it is not effective at removing ^{99}Tc. Therefore an allowance was made for increased discharge of ^{99}Tc, up to 200 TBq per year. The resulting pulse of increased ^{99}Tc discharges from Sellafield beginning in 1994 is currently being followed with great interest, as is discussed in more detail below.

The end result of these processes is the availability of a suite of oceanographic tracers with different discharge histories (e.g., in terms of the timing and magnitude of spikes) and a range of half-lives, and thus with a range of utility and applicability to studies of oceanographic processes at a variety of spatial and temporal scales. A brief summary of the reprocessing radionuclides most widely applied in oceanography is shown in **Table 1**, and examples of discharge histories are given in **Figure 1**. It has also been particularly useful in some cases to measure the ratio of a pair of tracers, for instance ^{134}Cs/^{137}Cs, ^{137}Cs/^{90}Sr, or potentially, ^{99}Tc/^{129}I. The use of isotope ratios can (1) provide temporal information and aid the estimation of rates of circulation, (2) mitigate the effects of mixing which will often alter individual concentrations more than ratios, and (3) aid in distinguishing the relative contributions of different sources of the nuclides in question. Finally, differences in the chemical behavior of the different elements released can be exploited to study different processes. Although most of the widely applied tracers are largely conservative in sea water and therefore serve as tracers of water movement, the actinides, particularly Pu, have a high affinity for particulate material and accumulate in the sediments. These tracers are then useful for studying sedimentary processes.

The Reprocessing Tracer Source Function

The primary difference between the north-western European reprocessing releases and other anthropogenic tracers used in oceanography is the nature of their introduction to the oceans. Reprocessing releases enter the oceans essentially at a point source, rather than in a more globally distributed fashion as is the case for weapons test fallout or the chlorofluorocarbons. Thus, reprocessing tracers are excellent, specific tracers for waters originating from north-western Europe. Because these waters are transported to the north into the Nordic Seas and

Table 1 Summary of the major reprocessing tracers and their applications

Isotope	*Half-life (years)*	*Sources*	*Applications*
^{137}Cs	30	Weapons testing, reprocessing (mostly Sellafield), Chernobyl	The 'signature' reprocessing tracer, used in the earliest studies of Sellafield releases. Has been applied in European coastal waters, the Nordic Seas, Arctic Ocean, and deep North Atlantic
^{90}Sr	28	Weapons testing, reprocessing (mostly Sellafield)	In combination with ^{137}Cs, early tracer for Sellafield discharges. The $^{137}Cs/^{90}Sr$ ratio was used to distinguish reprocessing and weapons sources
^{134}Cs	2.06	Reprocessing and Chernobyl	In combination with ^{137}Cs, this short-lived isotope has been used in the estimation of transit times, e.g., from Sellafield to the northern exit of the Irish Sea
^{125}Sb	2.7	Reprocessing, mostly Cap de la Hague	Circulation in the English Channel and North Sea, into the Skaggerak and Norwegian coastal waters
^{99}Tc	213 000	Reproducing – large pulse from Sellafield beginning in 1994. Some weapons testing	Surface circulation of European coastal waters, the Nordic Seas, Arctic Ocean, and East and West Greenland Currents
^{129}I	15.7×10^6	Reprocessing, mostly Cap de la Hague since 1990, some weapons testing	Deep water circulation in the Nordic Seas, Arctic Ocean, and Deep Western Boundary Current of the Atlantic Ocean. It has also been measured in European coastal waters and the Gulf of Mexico
^{241}Pu ^{239}Pu ^{240}Pu ^{238}Pu	14 24 000 6000 88	Weapons testing, reprocessing (mostly Sellafield)	Not widely used as circulation tracers but often measured in conjunction with other reprocessing radionuclides. Other transuranic elements that have been measured include ^{241}Am and ^{237}Np

Adapted in part from Dahlgaard (1995).

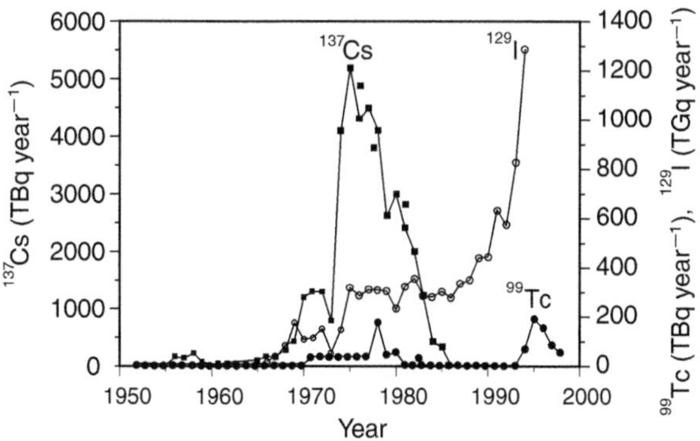

Figure 1 Examples of the source functions of reprocessing tracers to the oceans, illustrating some of the differences in magnitude and timing of the releases. ^{137}Cs data, from Gray *et al.* (1995) and ^{99}Tc data, courtesy of Peter Kershaw, are for liquid discharges from Sellafield. ^{129}I release information is courtesy of G. Raisbeck and F. Yiou, and combines available information for Sellafield from 1966 and Cap de la Hague from 1975. Note the different scales used for the releases of the three isotopes.

Arctic Ocean, reprocessing releases are particularly sensitive tracers of the climatically important deep water formation processes that occur in these regions.

There are several complications associated with this point source tracer introduction, however. Comparison with CFCs indicates some of these difficulties. Within each hemisphere (northern or southern), CFCs are well mixed throughout the troposphere, and the time history of their concentrations is well known. Their entry into the oceans occurs by equilibration with surface waters, and the details of their solubilities as a function of temperature and salinity have been well characterized. The primary complication is that in some areas equilibrium saturation is not reached, and so assumptions must be made about the degree of equilibration. Where they have been necessary, these assumptions appear to be fairly robust. In the case of reprocessing tracers, the discharge amounts have not always been as well known, although this situation has improved continuously. One major difficulty arises in translating a discharge amount, in kg or Bq per year, or per month, to a concentration in sea water some time later. In order to do this, the surface circulation of the coastal regions must be very well known. This circulation is highly variable, on daily, seasonal, interannual and decadal timescales, further complicating the problem. Also, other sources of the radionuclides under consideration must often be taken into account. Depending on the tracer, these sources may be important and/or numerous, and due to the secrecy involved in many aspects of the operation of nuclear installations the necessary information may not always be available.

Other Sources of Anthropogenic Radionuclides

Other sources of radionuclides to the oceans complicate the interpretation of Sellafield and Cap de la Hague tracers to varying extents, depending on the isotope under consideration, the location, and the time. First there is the need to understand the mixing of signals from these two plants. Other sources may include fallout from nuclear weapons tests, uncontrolled releases due to nuclear accidents, dumping on the seabed, other reprocessing plants, atmospheric releases from Sellafield and Cap de la Hague, and unknown sources. Many of these other sources are small compared to the Sellafield and Cap de la Hague releases. The primary complicating source for many isotopes is nuclear weapons test fallout, which peaked in the 1950s and again, more strongly, in 1962–63. This source has been particularly important for ^{137}Cs and ^{90}Sr. The Chernobyl accident in 1986 also released significant amounts of radioactive materials which have themselves found use as oceanographic tracers. In terms of comparison to releases from Sellafield, the Chernobyl accident has been most important with respect to ^{134}Cs and ^{137}Cs.

In addition, it has recently been noted that in the case of some nuclides, particularly ^{137}Cs and Pu isotopes, the sediments of the Irish Sea have become a significant ongoing source of tracers to the North Atlantic. Although generally considered a conservative tracer, ^{137}Cs exhibits some affinity for particulate material and a significant amount has accumulated in the sediments around Sellafield. With the continuing reduction of ^{137}Cs activities in liquid effluents from Sellafield, release from the sediments, either by resuspension of the sediments or by reequilibration with the reduced seawater concentrations, has become a relatively large (though still small in an absolute sense compared to the liquid discharges of the 1970s) contributor to the current flux out of the Irish Sea, and will continue to be such for years to come.

A further complication of the use of some tracers derived from Sellafield and Cap de la Hague, but one that cannot be lamented, is the continuing reduction of the releases, as well as the radioactive decay of those with the shorter half-lives, such as ^{137}Cs which was released in large quantities over 20 years ago. With respect to many of these complications, and for several other reasons, the reprocessing radionuclides attracting the most attention in the oceanographic community today are ^{99}Tc and ^{129}I. Both are long lived and have fairly small relative contributions from weapons testing and other sources. Unlike most radionuclides, their releases increased in the 1990s, and the recent releases of each are largely dominated by a single source: ^{99}Tc by Sellafield and ^{129}I by Cap de la Hague. Significant advances have been and are being made in the measurement of these isotopes, allowing their measurement on smaller sample volumes. The activity of ^{129}I is so low that it is measured by accelerator mass spectrometry. This technique allows measurement of ^{129}I on 1 liter seawater samples. Advances in ^{99}Tc measurement, using both radiochemical and mass-spectrometric (ICP-MS) techniques, are continuing. The primary limitation of ^{99}Tc studies continues to be the comparative difficulty of its measurement. This is also true to some extent for ^{129}I, for although the sample sizes have been greatly reduced the measurement requires highly specialized technology. A further complication for ^{129}I is its volatility, and the fact that as much as 10% of the reprocessing discharges have been released directly to the atmosphere. Nevertheless, the promise of both tracers is such that these difficulties are likely to be overcome.

Regional Setting and Circulation of Reprocessing Discharges

A summary of the regional circulation into which the liquid effluents from Sellafield and Cap de la Hague are released is presented in **Figure 2**. It is particularly important to note that studies of the reprocessing discharges have contributed greatly to the development of this detailed picture of the regional circulation. Briefly, from the Sellafield site the waste stream is carried north out of the Irish Sea, around the coast of Scotland, through the North Sea, and into the northward-flowing Norwegian Coastal Current (NCC). Transport across the North Sea occurs at various latitudes: some fraction of the reprocessing releases 'short-circuits' across the northern part of the North Sea, while some flows farther south along the eastern coast of the UK before turning east and north. Recent studies of the EARP ^{99}Tc pulse from Sellafield have suggested that the rate and preferred transport path of Sellafield releases across the North Sea into the NCC may vary in relation to climatic conditions in the North Atlantic such as the North Atlantic Oscillation (NAO).

Radionuclides discharged from the Cap de la Hague reprocessing plant flow north-east through the English Channel and into the North Sea,

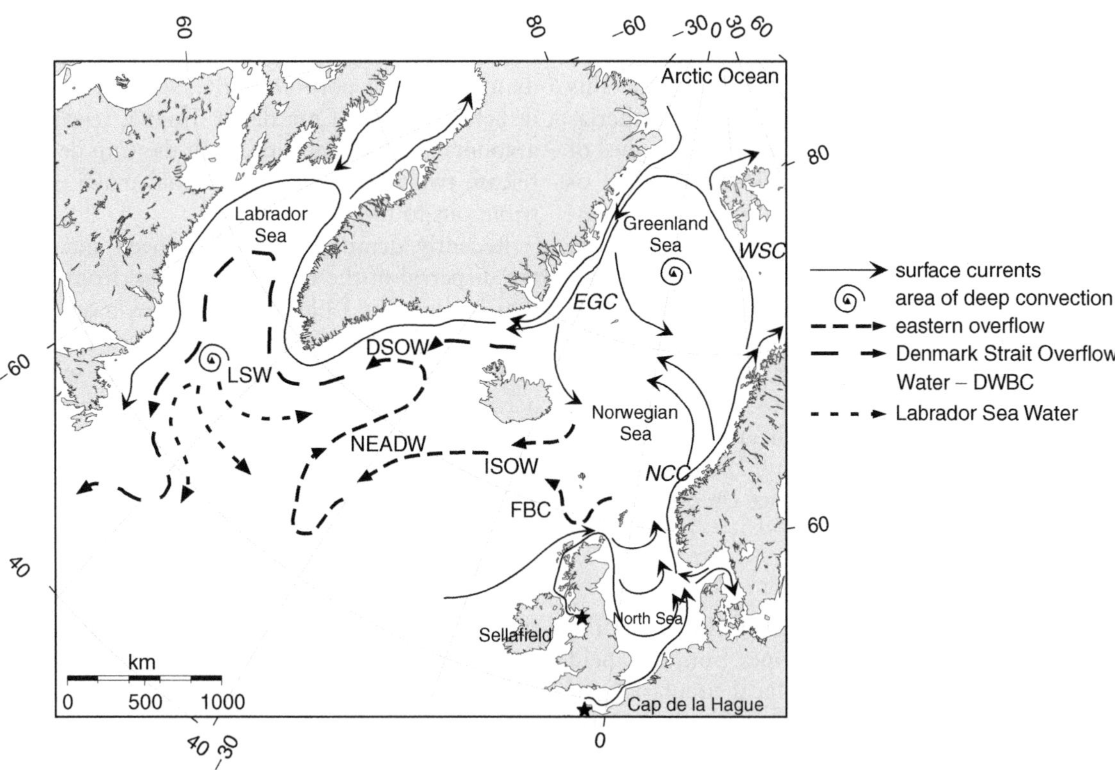

Figure 2 Map of the northern North Atlantic and Nordic Seas, indicating the locations of the Sellafield and Cap de la Hague reprocessing plants (stars), and the major circulation pathways relevant to the discussion of the dispersal of reprocessing wastes. Surface currents are indicated with thin solid lines, and deep waters, (deep overflows from the Nordic Seas, the resulting Deep Western Boundary Current (DWBC) and Labrador Sea Water (LSW)), with heavy dashed lines. Curled arrows indicate the two major areas where deep waters are formed by convective processes. Additional abbreviations: NCC, Norwegian Coastal Current; WSC, West Spitsbergen Current; EGC, East Greenland Current; FBC, Faroe Banks Channel; ISOW, Iceland Scotland Overflow Water; NEADW, Northeast Atlantic Deep Water; DSOW, Denmark Strait Overflow Water.

following the coast and joining the Sellafield releases in the NCC. A small amount of the Sellafield releases and some of the Cap de la Hague releases, which flow closer to the coast, flow east through the Skaggerak and Kattegat to enter the Baltic Sea.

The NCC is formed of a mixture of coastal waters (containing the reprocessing tracers) and warm, saline North Atlantic surface waters. Dilution of the reprocessing signal with Atlantic water continues along the northward flow path of the NCC. There is evidence from reprocessing tracers that turbulent eddies between the NCC and Atlantic water result in episodic transport westward into the surface waters of the Norwegian Sea, in addition to a fairly well-defined westward advective transport towards Jan Mayen Island. The NCC branches north of Norway, with one branch, the North Cape (or Nordkap) Current, flowing eastward through the Barents Sea and thence into the Kara Sea and Arctic Ocean, and the remainder flowing north and west as part of the West Spitsbergen Current (WSC). This latter flow branches in the Fram Strait west of Spitsbergen, with some recirculation to the west and south joining the southward flowing East Greenland Current (EGC), and the remainder entering the Arctic Ocean. The majority of surface outflow from the Arctic is through the Fram Strait into the EGC, thus the bulk of the reprocessing nuclides entering the Arctic Ocean will eventually exit to the Nordic (Greenland, Iceland, and Norwegian) Seas and the North Atlantic. The presence of high concentrations of radionuclides derived from reprocessing in the surface waters of the Barents and Nordic Seas is also an indication of their utility as tracers of deep-water ventilation and formation in these regions, as discussed below.

The Use of Reprocessing Releases in Oceanography

Historical Background

The first papers reporting measurements of Sellafield-derived ^{137}Cs in coastal waters appeared in the early 1970s. Much of this early work arose from monitoring efforts by the Division of Fisheries

Research of the UK's Ministry of Agriculture, Fisheries, and Food, which holds a joint oversight role on the Sellafield discharges and which continues to this day, now as the Centre for Environment, Fisheries and Aquaculture Science, to be a leader in studies of the distribution and oceanographic application of reprocessing radionuclides. In the late 1970s scientists at the Woods Hole Oceanographic Institution published the first of a number of papers detailing the unique utility of the reprocessing tracers and demonstrating their application. Leading studies of reprocessing tracers in the oceans have been undertaken by researchers in numerous other countries affected by the radiological implications of the releases, including France, Germany, Denmark, Canada, Norway, and the former Soviet Union.

Coastal and Surface Circulation

A great deal of work has been published using the documented releases of radioisotopes from Sellafield and Cap de la Hague to study the local circulations of the Irish Sea, North Sea, and English Channel. Early studies of the Sellafield releases examined a variety of isotopes, including ^{134}Cs, ^{137}Cs, ^{90}Sr, and Pu isotopes. Most attention focused on ^{137}Cs and ^{90}Sr, and their activity ratio, because: (1) the releases of these two isotopes were well documented, (2) they had been studied extensively since the 1950s and 1960s in weapons test fallout; and (3) the ^{137}Cs/^{90}Sr ratio in reprocessing releases (particularly those from Sellafield) was significantly higher than in global fallout and thus could be used to distinguish the sources of these isotopes in a given water sample. The use of the ^{134}Cs/^{137}Cs ratio enabled estimates of transit times, assuming the initial ratio in the releases was constant and making use of the short half-life of ^{134}Cs. The short-lived isotope ^{125}Sb has been used as a specific tracer of the circulation of Cap de la Hague discharges through the English Channel and North Sea, into the Baltic and the Norwegian Coastal Current.

Summaries of transit times and dilution factors for the transport of Sellafield and Cap de la Hague discharges to points throughout the North Sea, Norwegian Coastal Current, Barents and Kara Seas, Greenland Sea, and East and West Greenland Currents have been published in recent reviews. Numbers are not included in this article because they are currently under revision. In terms of transport to the NCC, where the two waste streams are generally considered to merge, the consensus has been that the transit time from Sellafield to about 60°N is three to four years, and that from Cap de la Hague to the same area is one to two years. Compilations of 'transfer factors,' which relate observed concentrations to the discharge amounts, and factor in the transit times, suggest that the two reprocessing waste streams meet in approximately equal proportions in the NCC. In other words, if Sellafield and Cap de la Hague released equal amounts of a radionuclide, with the Cap de la Hague release two years later, they would make equal contributions to the NCC.

Recently, detailed studies have been undertaken of the dispersal of the EARP ^{99}Tc pulse from Sellafield, which began in 1994. These studies have suggested substantially shorter circulation times for Sellafield releases to northern Scottish coastal waters and across the North Sea to the NCC than previously accepted. For instance, the ^{99}Tc pulse reached the NCC within 2.5 years, rather than three to four. It has been suggested that this is a real difference between sampling periods, resulting from climatically induced circulation changes in the North Atlantic. Some of the difference may also be related to the fact that much more detailed data, both in terms of seawater sampling and regarding the releases, are available on this recent event. The continuing passage of the EARP ^{99}Tc signal promises to be very useful in the Nordic Seas and Arctic Ocean as well.

Deep-water Formation in the North Atlantic and Arctic Oceans

In addition to surface water flows, deep-water formation processes within the Arctic Ocean and Nordic Seas have been elucidated through the study of reprocessing releases. In a classic presentation of reprocessing tracer data, Livingston showed that the surface water distribution of ^{137}Cs in the Nordic Seas in the early 1980s was marked by high concentrations at the margins of the seas, highlighting the delivery of the isotope in the northward-flowing NCC and WSC to the west and the return flow in the southward-flowing EGC to the east. The opposite distribution was found in the deep waters, with higher concentrations in the center of the Greenland Basin than at the margins, as a result of the ventilation of the Greenland Sea Deep Water by deep convective processes in the center of the gyre.

In the Arctic Ocean, elevated ^{137}Cs and ^{90}Sr concentrations and ^{137}Cs/^{90}Sr ratios at 1500 m at the LOREX ice station near the North Pole in 1979 indicated that deep layers of the Arctic Ocean were ventilated from the shelves. Similar observations in deep water north of Fram Strait provided early evidence suggesting a contribution of dense brines from the Barents Sea shelf to the bottom waters of the Nansen Basin. Reprocessing tracers, particularly those, like ^{129}I and ^{99}Tc, which have only a small contribution from other sources such as weapons fallout, hold great promise for illuminating eastern Arctic Ocean deep water ventilation processes from the Barents and Kara Sea shelves. The NCC delivers reprocessing tracers to the Barents and Kara relatively rapidly ($\sim$ 5 years) and more importantly in high concentration. With high tracer concentrations in the area of interest as a source water, the reprocessing tracers may be particularly sensitive tracers of a contribution of dense shelf waters to the deep Arctic Ocean.

In addition to the deep waters formed through convection in the polar regions (most notably in the Greenland Sea) which fill the deep basins of the Nordic Seas, intermediate waters are formed which subsequently overflow the sills between Greenland, Iceland, and Scotland and ventilate the deep North Atlantic. The presence of ^{137}Cs and ^{90}Sr from Sellafield was reported in the overflow waters immediately south of the Denmark Straits sampled during the Transient Tracers in the Ocean (TTO) program in 1981. Later, it was demonstrated that reprocessing cesium and strontium could be distinguished in the deep waters as far as TTO Station 214, off the Grand Banks of Newfoundland. No samples were taken for reprocessing radionuclides further south as part of that study, but it was clear in retrospect that the reprocessing signal had traveled even further in the Deep Western Boundary Current (DWBC). Recent work on ^{129}I has highlighted the utility of reprocessing radionuclides as tracers of northern source water masses and the DWBC of the Atlantic. Profiles in stations south of the overflows show much clearer tracer signals for ^{129}I than for the CFCs, and ^{129}I has been detected in the DWBC as far south as Cape Hatteras. As with the cesium studies of the 1980s, it is likely that sampling further south will reveal the tracer there as well.

Conclusions

Releases of radionuclides from the nuclear fuel-reprocessing plants at Sellafield and Cap de la Hague have provided tracers for detailed studies of the circulations of the local environment into which they are released, namely the Irish Sea, English Channel, and North Sea, and for the larger-scale circulation processes of the North Atlantic and Arctic Oceans. These tracers have very different source functions for their introduction into the oceans compared to other widely used anthropogenic tracers, and in some cases compared to each other. The fact that they are released at point sources makes them highly specific tracers of several interesting processes in ocean circulation, but the nature of the releases has complicated their quantitative interpretation to some extent.

Recent advances have been made in the measurement of ^{129}I and ^{99}Tc, long-lived tracers whose releases have increased in recent years and which experience little complication from other sources. These two tracers hold great promise for elucidating deep water formation and ventilation processes in the North Atlantic, Nordic Seas, and Arctic Ocean in the years to come.

See also

CFCs in the Ocean. Radioactive Wastes.

Further Reading

Aarkrog A, Dahlgaard H, Hallstadius L, Hansen H, and Hohm E (1983) Radiocaesium from Sellafield effluents in Greenland waters. *Nature* 304: 49–51.

Dahlgaard H (1995) Transfer of European coastal pollution to the Arctic: radioactive tracers. *Marine Pollution Bulletin* 31: 3–7.

Gray J, Jones SR, and Smith AD (1995) Discharges to the environment from the Sellafield Site, 1951–1992. *Journal of Radiological Protection* 15: 99–131.

Jefferies DF, Preston A, and Steele AK (1973) Distribution of caesium-137 in British coastal waters. *Marine Pollution Bulletin* 4: 118–122.

Kershaw P and Baxter A (1995) The transfer of reprocessing wastes from north-west Europe to the Arctic. *Deep-Sea Research II* 42: 1413–1448.

Kershaw PJ, McCubbin D, and Leonard KS (1999) Continuing contamination of north Atlantic and Arctic waters by Sellafield radionuclides. *Science of Total Environment* 237/238: 119–132.

Livingston HD (1988) The use of Cs and Sr isotopes as tracers in the Arctic Mediterranean Seas. *Philosophical Transactions of the Royal Society of London* A 325: 161–176.

Livingston HD, Bowen VT, and Kupferman SL (1982) Radionuclides from Windscale discharges I: non-equilibrium tracer experiments in high-latitude oceanography. *Journal of Marine Research* 40: 253–272.

Livingston HD, Bowen VT, and Kupferman SL (1982) Radionuclides from Windscale discharges II: their dispersion in Scottish and Norwegian coastal circulation. *Journal of Marine Research* 40: 1227–1258.

Livingston HD, Swift JH, and Östlund HG (1985) Artificial radionuclide tracer supply to the Denmark Strait Overflow between 1972 and 1981. *Journal of Geophysical Research* 90: 6971–6982.

RADIOACTIVE WASTES

L. Føyn, Institute of Marine Research, Bergen, Norway

Introduction

The discovery and the history of radioactivity is closely connected to that of modern science. In 1896 Antoine Henri Becquerel observed and described the spontaneous emission of radiation by uranium and its compounds. Two years later, in 1898, the chemical research of Marie and Pierre Curie led to the discovery of polonium and radium.

In 1934 Frédéric Joliot and Irène Curie discovered artificial radioactivity. This discovery was soon followed by the discovery of fission and the enormous amounts of energy released by this process. However, few in the then limited community of scientists working with radioactivity believed that it would be possible within a fairly near future to establish enough resources to develop the fission process for commercial production of energy or even think about the development of mass-destruction weapons.

World War II made a dramatic change to this. The race that began in order to be the first to develop mass-destruction weapons based on nuclear energy is well known. Following this came the development of nuclear reactors for commercial production of electricity. From the rapidly growing nuclear industry, both military and commercial, radioactive waste was produced and became a problem. As with many other waste problems, discharges to the sea or ocean dumping were looked upon as the simplest and thereby the best and final solution.

The Sources

Anthropogenic radioactive contamination of the marine environment has several sources: disposal at sea, discharges to the sea, accidental releases and fallout from nuclear weapon tests and nuclear accidents. In addition, discharge of naturally occurring radioactive materials (NORM) from offshore oil and gas production is a considerable source for contamination.

The marine environment receives in addition various forms of radioactive components from medical, scientific and industrial use. These contributions are mostly short-lived radionuclides and enter the local marine environment through diffuse outlets like muncipal sewage systems and rivers.

Disposal at Sea

The first ocean dumping of radioactive waste was conducted by the USA in 1946 some 80 km off the coast of California. The International Atomic Energy Agency (IAEA) published in August 1999 an 'Inventory of radioactive waste disposal at sea' according to which the disposal areas and the radioactivity can be listed as shown in **Table 1**.

Figure 1 shows the worldwide distribution of disposal-points for radioactive waste.

The majority of the waste disposed consists of solid waste in various forms and origin, only 1.44% of the total activity is contributed by low-level liquid waste. The disposal areas in the north-east Atlantic and the Arctic contain about 95% of the total radioactive waste disposed at sea.

Most disposal of radioactive waste was performed in accordance with national or international regulations. Since 1967 the disposals in the north-east Atlantic were for the most part conducted in accordance with a consultative mechanism of the Organization for Economic Co-operation and Development/Nuclear Energy Agency (OECD/NEA).

The majority of the north-east Atlantic disposals were of low-level solid waste at depths of 1500–5000 m, but the Arctic Sea disposals consist of various types of waste from reactors with spent fuel to containers with low-level solid waste dumped in fairly shallow waters ranging from about 300 m depth in the Kara Sea to less than 20 m depth in some fiords on the east coast of Novaya Zemlya.

Most of the disposals in the Arctic were carried out by the former Soviet Union and were not reported internationally. An inventory of the USSR disposals was presented by the Russian government in 1993. Already before this, the good collaboration between Russian and Norwegian authorities had led

Table 1 Worldwide disposal at sea of radioactive waste[a]

Area	Activity
North-west Atlantic Ocean	2.94 PBq
North-west Atlantic Ocean	2.94 PBq
Arctic Ocean	38.37 PBq
North-east Pacific Ocean	0.55 PBq
West Pacific Ocean	0.89 PBq

[a]PBq (petaBq) = 10^{15} Bq (1 Bq1 = disintegration s^{-1}). The old unit for radioactivity was Curie (Ci); 1 Ci3.7 × 10^{10} Bq.

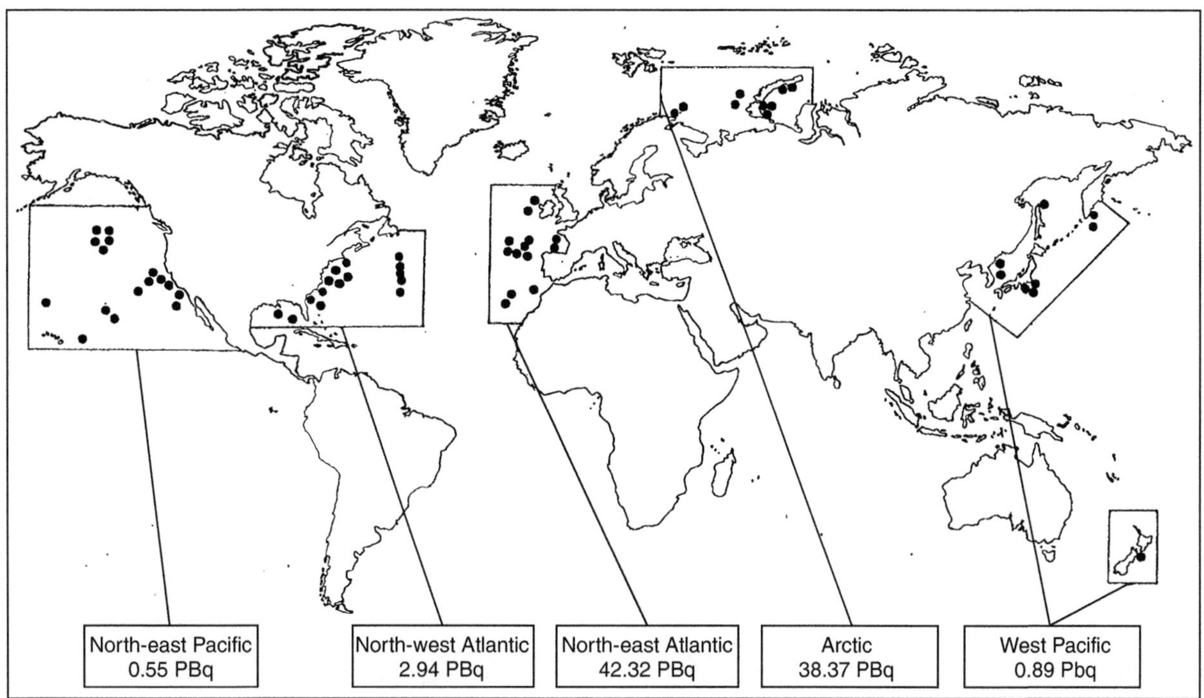

Figure 1 The worldwide location for disposal of radioactive waste at sea.

to a joint Norwegian–Russian expedition to the Kara Sea in 1992 followed by two more joint expeditions, in 1993 and 1994. The main purpose of these expeditions was to locate and inspect the most important dumped objects and to collect samples for assessing the present environmental impact and to assess the possibility for potential future leakage and environmental impacts.

These Arctic disposals differ significantly from the rest of the reported sea disposals in other parts of the world oceans as most of the dumped objects are found in shallow waters and some must be characterized as high-level radioactive waste, i.e. nuclear reactors with fuel. Possible releases from these sources may be expected to enter the surface circulation of the Kara Sea and from there be transported to important fisheries areas in the Barents and Norwegian Seas. **Figures 2** and **3** give examples of some of the radioactive waste dumped in the Arctic. Pictures were taken with a video camera mounted on a ROV (remote operated vehicle). The ROV was also equipped with a NaI-detector for gamma-radiation measurements and a device for sediment sampling close to the actual objects.

The Global Convention on the Prevention of Marine Pollution by Dumping of Wastes and Other Matter was adopted by an Intergovernmental Conference in London in 1972. The convention named the London Convention 1972, formerly the London Dumping Convention (LDC), addressed from the very beginning the problem of radioactive waste. But it was not until 20 February 1994 that a total prohibition on radioactive waste disposal at sea came into force.

Discharges to the Sea

Of the total world production of electricity about 16% is produced in nuclear power plants. In some countries nuclear energy counts for the majority of the electricity produced, France 75% and Lithuania 77%, and in the USA with the largest production of nuclear energy of more than 96 000 MWh this accounts for about 18% of the total energy production.

Routine operations of nuclear reactors and other installations in the nuclear fuel cycle release small amounts of radioactive material to the air and as liquid effluents. However, the estimated total releases of ^{90}Sr, ^{131}I and ^{137}Cs over the entire periods of operation are negligible compared to the amounts released to the environment due to nuclear weapon tests.

Some of the first reactors that were constructed used a single-pass cooling system. The eight reactors constructed for plutonium production at Hanford, USA, between 1943 and 1956, pumped water from Columbia River through the reactor cores then delayed it in cooling ponds before returning it to the river. The river water and its contents of particles and components were thereby exposed to a great neutron

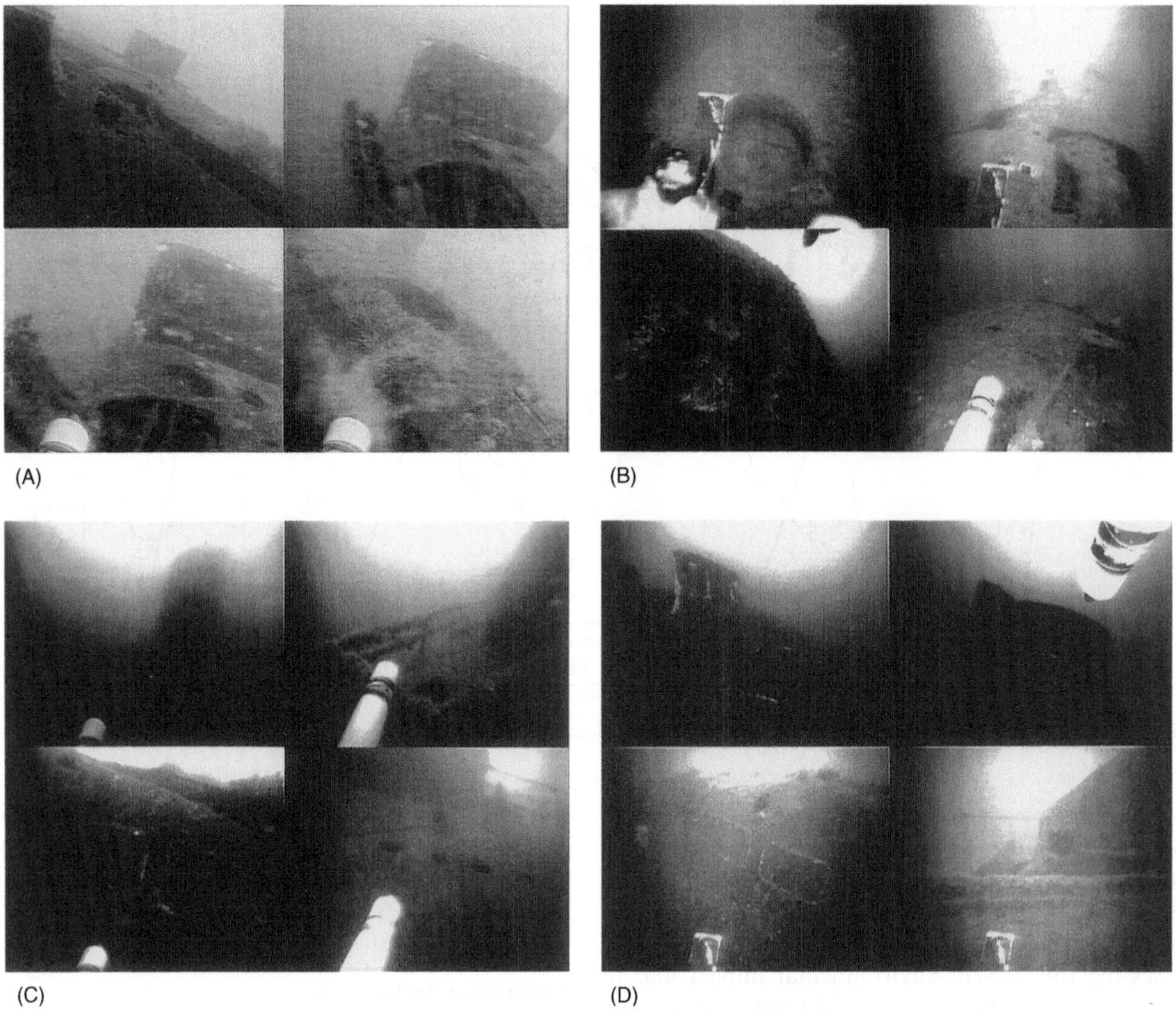

Figure 2 (A)–(D) Pictures of a disposed submarine at a depth of *c.* 30 m in the Stepovogo Fiord, east coast of Novaya Zemyla. The submarine contains a sodium-cooled reactor with spent fuel. Some of the hatches of the submarine are open which allows for 'free' circulation of water inside the vessel.

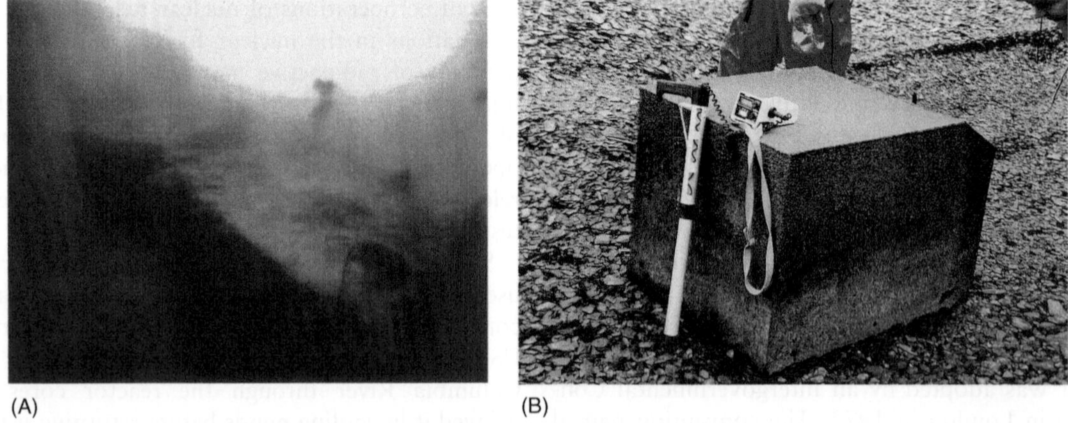

Figure 3 (A) Containers of low level solid waste at the bottom of the Abrosimov Fiord at a depth of *c.* 15 m and (B) a similar container found washed ashore.

flux and various radioactive isotopes were created. In addition corrosion of neutron-activated metal within the reactor structure contributed to the radioactive contamination of the cooling water. Only a limited number of these radionuclides reached the river mouth and only ^{32}P, ^{51}Cr, ^{54}Mn and ^{65}Zn were detected regularly in water, sediments and marine organisms in the near-shore coastal waters of the US Pacific Northwest.

Reactors operating today all have closed primary cooling systems that do not allow for this type of contamination. Therefore, under normal conditions production of electricity from nuclear reactors does not create significant amounts of operational discharges of radionuclides. However, the 434 energy-producing nuclear plants of the world in 1998 created radioactive waste in the form of utilized fuel. Utilized fuel is either stored or reprocessed.

Only 4–5% of the utilized nuclear fuel worldwide is reprocessed. Commercial, nonmilitary, reprocessing of nuclear fuel takes place in France, Japan, India and the United Kingdom. Other reprocessing plants defined as defense-related are in operation and producing waste but without discharges. For example in the USA, at the Savannah River Plant and the Hanford complex, about 83 000 m^3 and 190 000 m^3, respectively, of high-level liquid waste was in storage in 1985.

Reprocessing plants and the nuclear industry in the former Soviet Union have discharged to the Ob and Yenisey river systems ending up in the Arctic ocean. In 1950–51 about 77×10^6 m^3 liquid waste of 100 PBq was discharged to the River Techa. The Techa River is connected to the River Ob as is the Tomsk River where the Tomsk-7, a major production site for nuclear weapons plutonium, is situated. Other nuclear plants, such as the Krasnoyarsk industrial complex, have discharged to the Yenisey river. Large amounts of radioactive waste are also stored at the sites.

Radioactive waste stored close to rivers has the potential of contaminating the oceans should an accident happen to the various storage facilities.

The commercial reprocessing plants in France at Cap de la Hague and in the UK at Sellafield have for many years, and still do, contributed to the radioactive contamination of the marine environment. They both discharge low-level liquid radioactive effluents to the sea. Most important, however, these discharges and their behavior in the marine environment have been and are still thoroughly studied and the results are published in the open literature. The importance of these discharges is extensive as radionuclides from Sellafield and la Hague are traced throughout the whole North Atlantic. Most important is Sellafield; **Table 2** summarizes the reported discharge of some important radionuclides.

Table 2 Total discharges of some radionuclides from Sellafield 1952–92

^{3}H	39 PBq
^{90}Sr	6.3 PBq
^{134}Cs	5.8 PBq
^{137}Cs	41.2 PBq
^{238}Pu	0.12 PBq
^{239}Pu	0.6 PBq
^{241}Pu	21.5 PBq
^{241}Am	0.5 PBq

In addition a range of other radionuclides have been discharged from Sellafield, but prior to 1978 the determination of radionuclides was, for many components, not specific. Technetium (^{99}Tc) for instance was included in the 'total beta' determinations with an estimated annual discharge from 1952 to 1970 below 5 TBq and from 1970 to 1977 below 50 TBq. Specific determination of ^{99}Tc in the effluents became part of the routine in 1978 when about 180 TBq was discharged followed by about 50 TBq in 1979 and 1980 and then an almost negligible amount until 1994.

The reason for mentioning ^{99}Tc is that this radionuclide, in an oceanographic context, represents an almost ideal tracer in the oceans. Technetium is most likely to be present as pertechnetate, TcO_4^-, totally dissolved in seawater; it acts conservatively and moves as a part of the water masses. In addition the main discharges of technetium originate from point sources with good documentation of time for and amount of the release. The discharges from Sellafield are a good example of this. From 1994 the UK authorities have allowed for a yearly ^{99}Tc discharge of up to 200 TBq.

Based on surveys before and after the discharges in 1994, 30 TBq (March–April) and 32 TBq (September–October), the transit time for technetium from the Irish Sea to the North Sea was calculated to be considerably faster than previous estimations of transit times for released radionuclides. This faster transport is demonstrated by measurements indicating that the first discharge plume of ^{99}Tc had reached the south coast of Norway before November 1996 in about 2.5 years compared to the previously estimated transit time of 3–4 years.

Other reprocessing plants may have discharges to the sea, but without a particular impact in the world oceans. The reprocessing plant at Trombay, India, may, for example, be a source for marine contamination.

Accidental Releases

Accidents resulting in direct radioactive releases to the sea are not well known as most of them are connected to wreckage of submarines. Eight nuclear submarines with nuclear weapons have been reported lost at sea, two US and six former USSR. The last known USSR wreck was the submarine *Komsomolets* which sank in the Norwegian Sea southwest of Bear Island, on 7 April 1989. The activity content in the wreck is estimated by Russian authorities to be 1.55–2.8 PBq ^{90}Sr and 2.03–3 PBq ^{137}Cs and the two nuclear warheads on board may contain about 16 TBq ^{239,240}Pu equivalent to 6–7 kg plutonium. Other estimates indicate that each warhead may contain 10 kg of highly enriched uranium or 4–5 kg plutonium.

On August 12th 2000, the Russian nuclear submarine *Kursk* sank at a depth of 108 meters in the Barents Sea north of the Kola peninsula. Vigorous explosions in the submarine's torpedo-chambers caused the wreckage where 118 crew-members were entrapped and lost their lives. *Kursk*, and Oscar II attack submarine, was commissioned in 1995 and was powered by two pressurized water reactors. *Kursk* had no nuclear weapons on board. Measurements close to the wreck in the weeks after the wreckage showed no radioactive contamination indicating that the primary cooling-systems were not damaged in the accident. A rough inventory calculation estimates that the reactors at present contain about 56 000 TBq. Russian authorities are planning for a salvage operation where the submarine or part of the submarine will be lifted from the water and transported to land. Both a possible salvage operation or to leave the wreck where it is will be create a demand for monitoring as the location of the wreck is within important fishing grounds.

The wreckage of *Komsomolets* in 1989 and the attempts to raise money for an internationally financed Russian led salvage operation became very public. The Russian explanation for the intensive attempts of financing the salvage was said to be the potential for radioactive pollution. The wreck of the submarine is, however, located at a depth of 1658 m and possible leaching of radionuclides from the wreck will, due to the hydrography of the area, hardly have any vertical migration and radioactive components will spread along the isopycnic surfaces gradually dispersing the released radioactivity in the deep water masses of the Nordic Seas. An explanation for the extensive work laid down for a salvage operation and for what became the final solution, coverage of the torpedo-part of the hull, may be that this submarine was said to be able to fire its torpedo missiles with nuclear warheads from a depth of 1000 m.

In 1990, the Institute of Marine Research, Bergen, Norway, started regular sampling of sediments and water close to the wreck of *Komsomolets*. Values of ^{137}Cs were in the range 1–10 Bq per kg dry weight sediment and 1–30 Bq per m^3 water. No trends were found in the contamination as the variation between samples taken at the same date were equal to the variation observed from year to year. Detectable amounts of ^{134}Cs in the sediment samples indicate that there is some leaching of radioactivity from the reactor.

Accidents with submarines and their possible impact on the marine environment are seldom noticed in the open literature and there is therefore little common knowledge available. An accident, however, that is well known is the crash of a US B-52 aircraft, carrying four nuclear bombs, on the ice off Thule air base on the northwest coast of Greenland in January 1968. Approximately 0.4 kg plutonium ended up on the sea floor at a depth of 100–300 m. The marine environment became contaminated by about 1 TBq ^{239,240}Pu which led to enhanced levels of plutonium in benthic animals, such as bivalves, sea-stars and shrimps after the accident. This contamination has decreased rapidly to the present level of one order of magnitude below the initial levels.

Fallout from Nuclear Weapon Tests and Nuclear Accidents

Nuclear weapon tests in the atmosphere from 1945 to 1980 have caused the greatest man-made release of radioactive material to the environment. The most intensive nuclear weapon tests took place before 1963 when a test-ban treaty signed by the UK, USA and USSR came into force. France and China did not sign the treaty and continued some atmospheric tests, but after 1980 no atmospheric tests have taken place.

It is estimated that 60% of the total fallout has initially entered the oceans, i.e. 370 PBq ^{90}Sr, 600 PBq ^{137}Cs and 12 PBq ^{239,240}Pu. Runoff from land will slightly increase this number. As the majority of the weapon tests took place in the northern hemisphere the deposition there was about three times as high as in the southern hemisphere.

Results from the GEOSECS expeditions, 1972–74, show a considerable discrepancy between the measured inventories in the ocean of 900 PBq ^{137}Cs, 600 PBq ^{90}Sr and 16 PBq ^{239,240}Pu and the estimated input from fallout. The measured values are far higher than would be expected from the assumed fallout data. Thus the exact input of anthropogenic radionuclides may be partly unknown or the geographical coverage of the measurements in the oceans were for some areas not dense enough for accurate calculations.

Another known accident contributing to marine contamination was the burn-up of a US satellite (SNAP 9A) above the Mozambique channel in 1964 which released 0.63 PBq ^{238}Pu and 0.48 TBq ^{239}Pu; 73% was eventually deposited in the southern hemisphere.

The Chernobyl accident in 1986 in the former USSR is the latest major event creating fallout to the oceans. Two-thirds of the c.100 PBq ^{137}Cs released was deposited outside the Soviet Union. The total input to the world oceans of ^{137}Cs from Chernobyl is estimated to be from 15–20 Pbq, i.e. 4.5 PBq in the Baltic Sea; 3–5 PBq in the Mediterranean Sea, 1.2 PBq in the North Sea and about 5 PBq in the northeast Atlantic.

Natural Occurring Radioactive Material

Oil and gas production mobilize naturally occurring radioactive material (NORM) from the deep underground reservoir rock. The radionuclides are primarily ^{226}Ra, ^{228}Ra and ^{210}Pb and appear in sludge and scales and in the produced water. Scales and sludge containing NORM represent an increasing amount of waste. There are different national regulations for handling this type of waste. In Norway, for example, waste containing radioactivity above 10 Bq g^{-1} is stored on land in a place specially designed for this purpose. However, there are reasons to believe that a major part of radioactive contaminated scales and sludge from the worldwide offshore oil and gas production are discharged to the sea.

Reported NORM values in scales are in the ranges of 0.6–57.2 Bq g^{-1} ^{226}Ra + ^{228}Ra (Norway), 0.4–3700 Bq g^{-1} (USA) and 1–1000 Bq g^{-1} ^{226}Ra (UK).

Scales are an operational hindrance in oil and gas production. Frequent use of scale-inhibitors reduce the scaling process but radioactive components are released to the production water adding to its already elevated radioactivity. More than 90% of the radioactivity in produced water is due to ^{226}Ra and ^{228}Ra having a concentration 100–1000 times higher than normal for seawater.

The discharge of produced water is a continuous process and the amount of water discharged is considerable and increases with the age of the production wells. As an example, the estimated amount of produced water discharged to the North Sea in 1998 was 340 million m^3 and multiplying by an average value of 5 Bq^{-1} of ^{226}Ra in produced water, the total input of ^{226}Ra to the North Sea in 1998 was 1.7 TBq.

Discussion

The total input of anthropogenic radioactivity to the world's oceans is not known exactly, but a very rough estimate gives the following amounts: 85 PBq dumped, 100 PBq discharged from reprocessing and 1500 PBq from fallout. Some of the radionuclides have very long half-lives and will persist in the ocean, for example ^{99}Tc has a half-life of 2.1×10^5 years, ^{239,240}Pu, 2.4×10^4 years and ^{226}Ra, 1600 years. ^{137}Cs, ^{90}Sr and ^{228}Ra with half-lives of 30 years, 29 years and 5.75 years, respectively, will slowly decrease depending on the amount of new releases.

In an oceanographic context it is worth mentioning the differences in denomination between radioactivity and other elements in the ocean. The old denomination for radioactivity was named after Curie (Ci) and 1 g radium was defined to have a radioactivity of 1 Ci; 1 Ci 3.7×10^{10} Bq and 1 PBq 27 000 Ci. Therefore released radioactivity of 1 PBq can be compared to the radioactivity of 27 kg radium.

The common denominations for major and minor elements in seawater are given in weight per volume. For comparison if 1 PBq or 27 kg radium were diluted in 1 km^3 of seawater, this would give a radium concentration of 0.027 μg l^{-1} or 1000 Bq l^{-1}. Calculations like this clearly visualize the sensitivity of the analytical methods used for measuring radioactivity. In the Atlantic Ocean for example radium (^{228}Ra) has a concentration of 0.017–3.40 mBq l^{-1}, whereas ^{99}Tc measured in surface waters off the southwest coast of Norway is in the range of 0.9–6.5 mBq l^{-1}.

Measured in weight the total amount of radionuclides do not represent a huge amount compared to the presence of nonradioactive components in seawater. The radioisotopes of cesium and strontium are both important in a radioecological context since they have chemical behavior resembling potassium and calcium, respectively. Cesium follows potassium in and out of the soft tissue cells whereas strontium follows calcium into bone cells and stays. Since uptake and release in organisms is due to the chemical characteristics and rarely if the element is radioactive or not, radionuclides such as ^{137}Cs and ^{90}Sr have to compete with the nonradioactive isotopes of cesium and strontium.

Oceanic water has a cesium content of about 0.5lμg^{-1} and a strontium content of about 8000 μg l^{-1}. Uptake in a marine organism is most likely to be in proportion to the abundance of the radioactive and the nonradioactive isotopes of the actual element. This can be illustrated by the following example. The sunken nuclear submarine *Komsomolets* contained an estimated (lowest) amount of 1.55 PBq ^{90}Sr (about 300 g) and 2.03 PBq

^{137}Cs (about 630 g). If all this was released at once and diluted in the immediate surrounding 1 km^3 of water the radioactive concentration would have been 1550 Bq l^{-1} for ^{90}Sr and 2030 Bq l^{-1} for ^{137}Cs, the concentration in weight per volume would have been 0.000 3 μg l^{-1} ^{90}Sr and 0.000 63 μg l^{-1} ^{137}Cs. This means that even if the radioactive material was kept in the extremely small volume of 1 km^3, compared to the volume of the deep water of the Norwegian Sea available for a primary dilution, the proportion of radioactive to nonradioactive isotopes of strontium and cesium, available for uptake in marine organisms, would have been about 2.7×10^6 and 7.9×10^{-5}, respectively.

From the examples above it can be seen that if uptake, and thereby impact, in marine organisms follows regular chemical–physiological rules there is a 'competition' in seawater in favor of the nonradioactive isotopes for elements normally present in seawater. Measurable amount of radionuclides of cesium and strontium are detected in marine organisms but at levels far below the concentrations in freshwater fish. Average concentrations of ^{137}Cs in fish from the Barents Sea during the period with the most intensive nuclear weapon tests in that area, 1962–63, never exceeded 90 Bq kg^{-1} fresh weight, whereas fallout from Chernobyl resulted in concentrations in freshwater fish in some mountain lakes in Norway far exceeding 10 000 Bq kg^{-1}.

For radionuclides like technetium and plutonium, which will persist in the marine environment, uptake will be based only on the actual concentrations in seawater of radionuclide. The levels of ^{99}Tc, for example, increased in seaweed (*Fucus vesiculosus*) from 70 Bq per kg dry weight (December 1997) to 124 Bq kg^{-1} in January 1998 in northern Norway which reflected the increased concentration in the water as the peak of the technetium plume from Sellafield reached this area.

Previously the effects of anthropogenic radioactivity have been based on the possible dose effect to humans. Most of the modeling work has been concentrated on assessing the dose to critical population groups eating fish and other marine organisms. But even if the radiation from anthropogenic radionuclides to marine organisms is small compared to natural radiation from radionuclides like potassium, ^{40}K, the presence of additional radiation may give a chronic exposure with possible effects, at least on individual marine organisms.

The input of radioactivity, NORM, from the offshore oil and gas production may also give reason for concern. The input will increase as it is a continuous part of the production. Even if radium as the main radionuclide is not likely to be taken up by marine organisms the use of chemicals like scale inhibitors may change this making radium more available for marine organisms.

Conclusion

The sea began receiving radioactive waste from anthropogenic sources in 1946, in a rather unregulated way in the first decades. Both national and international regulations controlling disposals have now slowly come into force. Considerable amounts are still discharged regularly from nuclear industries and the practice of using the sea as a suitable wastebasket is likely to continue for ever. In 1994 an international total prohibition on radioactive waste disposal at sea came into force, but the approximately 85 PBq of solid radioactive waste that has already been dumped will sooner or later be gradually released to the water masses.

Compared to other wastes disposed of at sea the amount of radioactive waste by weight is rather diminutive. However, contrary to most of the 'ordinary' wastes in the sea, detectable amounts of anthropogenic radioactivity are found in all parts of the world oceans and will continue to contaminate the sea for many thousands of years to come. This means that anthropogenic radioactive material has become an extra chronic radiation burden for marine organisms. In addition, the release of natural occurring radionuclides from offshore oil and gas production will gradually increase the levels of radium, in particular, with a possible, at present unknown, effect.

However, marine food is not, and probably never will be, contaminated at a level that represents any danger to consumers. The ocean has always received debris from human activities and has a potential for receiving much more and thereby help to solve the waste disposal problems of humans. But as soon as a waste product is released and diluted in the sea it is almost impossible to retrieve. Therefore, in principal, no waste should be disposed of in the sea without clear documentation that it will never create any damage to the marine environment and its living resources. This means that with present knowledge no radioactive wastes should be allowed to be released into the sea.

See also

Nuclear Fuel Reprocessing and Related Discharges. Single Compound Radiocarbon Measurements. Uranium-Thorium Decay Series in the Oceans Overview.

Further Reading

Guary JC, Gueguenia P, and Pentreath RJ (eds.) (1988) *Radionuclides: A Tool for Oceanography*. London, New York: Elsevier Applied Science.

Hunt GJ, Kershaw PJ, and Swift DJ (eds.) (1998) Radionuclides in the oceans (RADOC 96–97). Distribution, Models and Impacts. *Radiation Protection Dosimetry* 75: 1–4.

IAEA (1995) *Environmental impact of radioactive releases*; Proceedings of an International Symposium on Environmental Impact of Radioactive Releases. Vienna: International Atomic Energy Agency.

IAEA (1999) *Inventory of Radioactive Waste Disposals at Sea*. IAEA-TECDOC-1105 Vienna: International Atomic Energy Agency. pp. 24 A.1–A.22.

HYPOXIA

N. N. Rabalais, Louisiana Universities Marine Consortium, Chauvin, LA, USA

Introduction: Definitions

Hypoxic (low-oxygen) and anoxic (no-oxygen) waters have existed throughout geologic time. Presently, hypoxia occurs in many of the ocean's deeper environs, open-ocean oxygen-minimum zones (OMZs), enclosed seas and basins, below western boundary current upwelling zones, and in fjord. Hypoxia also occurs in shallow coastal seas and estuaries, where their occurrence and severity appear to be increasing, most likely accelerated by human activities (**Figure 1**). A familiar term used in the popular press and literature, 'dead zone', used for coastal and estuarine hypoxia, refers to the fish and shellfish killed by the suffocating conditions or the failure to catch these animals in bottom waters when the oxygen concentration in the water covering the seabed is below a critical level.

Based on laboratory or field observations or both, the level of oxygen stress and related responses of invertebrate and fish faunas vary. The units are often determined by oxygen conditions that are physiologically stressful, but these levels also differ depending on the organisms considered, and the pressure, temperature, and salinity of the ambient waters. The numerical definition of hypoxia varies as do the units used, but hypoxia has mostly been defined as dissolved oxygen levels lower than a range of 3–2 $\mathrm{ml\,l^{-1}}$, with the consensus being in favor of 1.4 $\mathrm{ml\,l^{-1}}$ (= 2 $\mathrm{mg\,l^{-1}}$ or ppm). This value is approximately equivalent to 30% oxygen saturation at 25 °C and 35 salinity (psu). Below this concentration, bottom-dragging trawl nets fail to capture fish, shrimp, and swimming crabs. Other fishes, such as rays and sharks, are affected by oxygen levels below 3 $\mathrm{mg\,l^{-1}}$, which prompts a behavioral response to evacuate the area, up into the water column and shoreward. Water-quality standards in the coastal waters of Long Island Sound, New York, and Connecticut, USA, consider that dissolved oxygen conditions below 5 $\mathrm{mg\,l^{-1}}$ result in behavioral effects in marine organisms and fail to support living resources at sustainable levels.

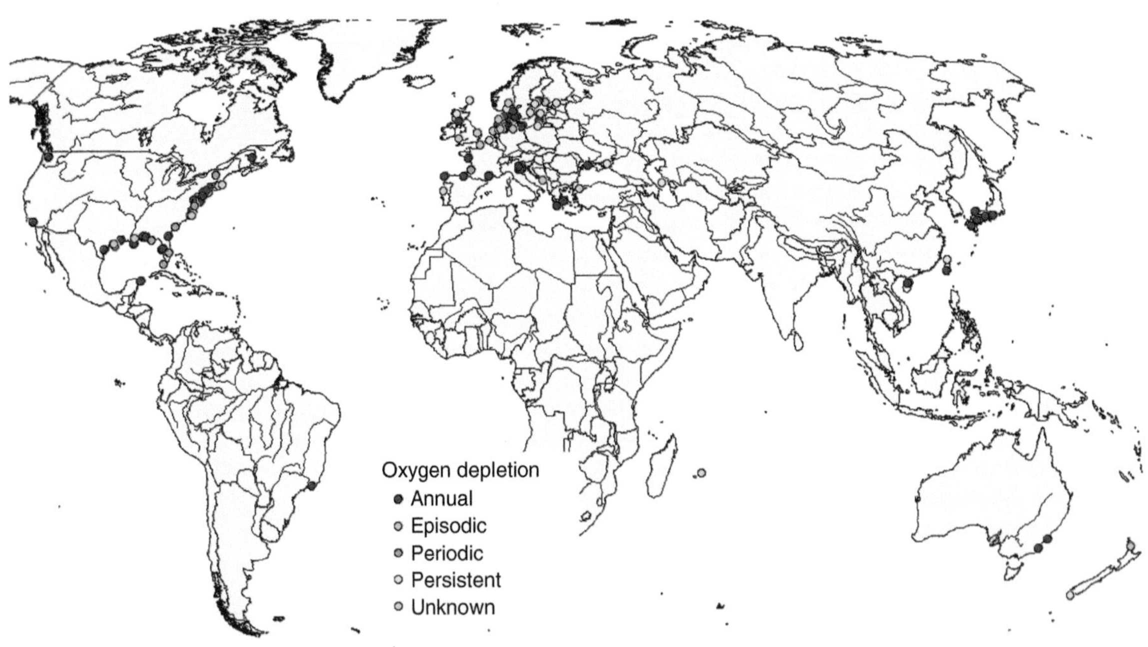

Figure 1 Distribution of coastal ocean hypoxic areas; excludes naturally occurring hypoxia, such as upwelling zones and OMZs. Reproduced from Díaz RJ, Nestlerode J, and Díaz ML (2004) A global perspective on the effects of eutrophication and hypoxia on aquatic biota. In: Rupp GL and White MD (eds.) *Proceedings of the 7th International Symposium on Fish Physiology, Toxicology and Water Quality, Tallinn, Estonia, May 12-15, 2003. EPA 600/R-04/049*, pp. 1–33. Athens, GA: Ecosystems Research Division, US Environmental Protection Agency, with permission from Robert J. Díaz.

The most commonly used definition for oceanic waters is dissolved oxygen content less than 1 ml l^{-1} (or 0.7 mg l^{-1}). Disoxyic or disaerobic refers to oxygen levels between 0.1 and 1.0 ml l^{-1}. OMZs are usually defined as waters less than 0.5 ml l^{-1} dissolved oxygen.

Causes

Hypoxia occurs where the consumption of oxygen through respiratory or chemical processes exceeds the rate of supply from oxygen production via photosynthesis, diffusion through the water column, advection, or mixing. The biological and physical water-column characteristics that support the development and maintenance of hypoxia include (1) the production, flux, and accumulation of organic-rich matter from the upper water column; and (2) water-column stability resulting from stratification or long residence time. Dead and senescent algae, zooplankton fecal pellets, and marine aggregates contribute significant amounts of organic detritus to the lower water column and seabed. Aerobic bacteria consume oxygen during the decay of the carbon and deplete the oxygen, particularly when stratification prevents diffusion of oxygen. Stratification is the division of the water column into layers with different densities caused by differences in temperature or salinity or both. Hypoxia will persist as long as oxygen consumption rates exceed those of supply. Oxygen depletion occurs more frequently in estuaries or coastal areas with longer water residence times, with higher nutrient loads and with stratified water columns.

Hypoxia is a natural feature of many oceanic waters, such as OMZs and enclosed seas, or forms in coastal waters as a result of the decomposition of high carbon loading stimulated by upwelled nutrient-rich waters. Hypoxia in many coastal and estuarine waters, however, is but one of the symptoms of eutrophication, an increase in the rate of production and accumulation of carbon in aquatic systems. Eutrophication very often results from an increase in nutrient loading, particularly by forms of nitrogen and phosphorus. Nutrient over-enrichment from anthropogenic sources is one of the major stressors impacting estuarine and coastal ecosystems, and there is increasing concern in many areas around the world that an oversupply of nutrients is having pervasive ecological effects on shallow coastal waters. These effects include reduced light penetration, increased abundance of nuisance macroalgae, loss of aquatic habitat such as seagrass or macroalgal beds, noxious and toxic algal blooms, hypoxia and anoxia, shifts in trophic interactions and food webs, and impacts on living resources.

Hypoxic Systems

Oxygen-minimum Zones

Persistent hypoxia is evident in mid-water OMZs, which are widespread in the world oceans where the oxygen concentrations are less than 0.5 ml l^{-1} (or about 7.5% oxygen saturation, $<22\,\mu M$). They occur at different depths from the continental shelf to upper bathyal zones (down to 1300 m). Many of the OMZs form as a result of high primary production associated with coastal upwelled nutrient-rich waters. Their formation also requires stagnant circulation, long residence times, and the presence of oxygen-depleted source waters. The extensive OMZ development in the eastern Pacific Ocean is attributed to the fact that intermediate depth waters of the region are older and have overall oxygen concentrations lower than other water masses. The largest OMZs are at bathyal depths in the eastern Pacific Ocean, the Arabian Sea, the Bay of Bengal, and off southwest Africa. The upper boundary of an OMZ may come to within 10 or 50 m of the sea surface off Central America, Peru, and Chile. The OMZ is more than 1000-m thick off Mexico and in the Arabian Sea, but off Chile, the OMZ is <400-m thick. Along continental margins, minimum oxygen concentrations occur typically between 200 and 700 m. The area of the ocean floor where oceanic waters permanently less than 0.5 ml l^{-1} impinge on continental margins covers 10^6 km^2 of shelf and bathyal seafloor, with over half occurring in the northern Indian Ocean. These permanently hypoxic waters account for 2.3% of the ocean's continental margin. These hypoxic areas are not related to eutrophication, but longer-term shifts in meteorological conditions and ocean currents may increase their prevalence in the future with global climate change. Shifts in ocean currents have been implicated in the increased frequency of continental shelf hypoxia along the northwestern US Pacific coast of Oregon.

Deep Basins, Enclosed Seas, and Fjord

Many of the existing permanent or periodic anoxic ocean environments occur in enclosed or semi-enclosed waters where a mass of deep water is bathymetrically isolated from main shelf or oceanic water masses by surrounding landmasses or one or more shallow sills. In conjunction with a pycnocline, the bottom water volume is restricted from exchange with deep open water. Examples of hypoxic and anoxic basins include anoxic deep water fjord, such as Saanich Inlet, the deeper basins of the Baltic, the basin of the Black Sea, the Japanese Seto Inland Sea, deep waters of the Sea of Cortez, Baja California, and Santa Barbara Basin in the southern California borderland.

Coastal Seas and Estuaries

Periodic hypoxia or anoxia also occurs on open continental shelves, for example, the northern Gulf of Mexico and the Namibian and Peruvian shelves where upwelling occurs. More enclosed shelves such as the northern Adriatic and the northwestern shelf of the Black Sea also have periodic hypoxia or anoxia. In these instances, there is minimal exchange of shelf-slope water and/or high oxygen demand on the shallow shelf. Estuaries, embayments, and lagoons are susceptible to the formation of hypoxia and anoxia if the water residence time is sufficiently long, especially where the water column is stratified. Light conditions are also important in these coastal habitats as a limiting factor on phytoplankton growth, which, if excessive, contributes to high organic loading within the confined waters.

Coastal ecosystems that have been substantially changed as a result of eutrophication exhibit a series of identifiable symptoms, such as reduced water clarity, excessive, noxious, and, sometimes, harmful algal blooms, loss of critical macroalgal or seagrass habitat, development or intensification of oxygen depletion in the form of hypoxia or anoxia, and, in some cases, loss of fishery resources. More subtle responses of coastal ecosystems to eutrophication include shifts in phytoplankton and zooplankton communities, shifts in the food webs that they support, loss of biodiversity, changes in trophic interactions, and changes in ecosystem functions and biogeochemical processes.

In a review of anthropogenic hypoxic zones in 1995, Díaz and Rosenberg noted that no other environmental variable of such ecological importance to estuarine and coastal marine ecosystems around the world has changed so drastically, in such a short period of time, as dissolved oxygen. For those reviewed, there was a consistent trend of increasing severity (either in duration, intensity, or size) where hypoxia occurred historically, or hypoxia existed presently when it did not occur before. While hypoxic environments have existed through geologic time and are common features of the deep ocean or adjacent to areas of upwelling, their occurrence in estuarine and coastal areas is increasing, and the trend is consistent with the increase in human activities that result in nutrient over-enrichment.

The largest human-caused hypoxic zone is in the aggregated coastal areas of the Baltic Sea, reaching 84 000 km^2. Hypoxia existed on the northwestern Black Sea shelf historically, but anoxic events became more frequent and widespread in the 1970s and 1980s, reaching over areas of the seafloor up to 40 000 km^2 in depths of 8–40 m. There is also evidence that the suboxic zone of the open Black Sea enlarged toward the surface by about 10 m since 1970. The condition of the northwestern shelf of the Black Sea, in which hypoxia covered up to 40 000 km^2, improved over the period 1990–2000 when nutrient loads from the Danube River decreased, but may be experiencing a worsening of hypoxic conditions more recently.

Similar declines in bottom water dissolved oxygen have occurred elsewhere as a result of increasing nutrient loads and cultural eutrophication, for example, the northern Adriatic Sea, the Kattegat and Skaggerak, Chesapeake Bay, Albemarle-Pamlico Sound, Tampa Bay, Long Island Sound, New York Bight, the German Bight, and the North Sea. In the United States, over half of the estuaries experience hypoxia at some time over an annual period and many experience hypoxia over extensive areas for extended periods on a perennial basis. The number of estuaries with hypoxia or anoxia continues to rise.

Historic data on Secchi disk depth in the northern Adriatic Sea in 1911 through the present, with few interruptions of data collection, provide a measure of water transparency that could be interpreted to depict surface water productivity. These data coupled with surface and bottom water dissolved oxygen content determined by Winkler titrations and nutrient loads outline the sequence of eutrophication in the northern Adriatic Sea. Similar historical data from other coastal areas around the world demonstrate a decrease in water clarity due to phytoplankton production in response to increased nutrient loads that are paralleled by declines in water column oxygen levels.

There are strong relationships between river flow and nutrient flux into the Chesapeake Bay and northern Gulf of Mexico and phytoplankton production and biomass and the subsequent fate of that production in spring deposition of chlorophyll *a*. Further there is a strong relationship between the deposited chlorophyll *a* and the seasonal decline of deep-water dissolved oxygen. Excess nutrients in many watersheds are driven by agricultural activities and atmospheric deposition from burning of fossil fuels. The link with excess nutrients in more urban areas, such as Long Island Sound, is with the flux of nutrients associated from numerous wastewater outfalls.

Swift currents that move materials away from a river delta and that do not permit the development of stratification are not conducive to the accumulation of biomass or depletion of oxygen, for example in the Amazon and Orinoco plumes. Similar processes off the Changjiang (Yantze River) and high turbidity in the plume of the Huanghe (Yellow River) were once thought to be reasons why hypoxia did not develop in those coastal systems. Incipient indications of the beginning of symptoms of cultural eutrophication were becoming evident at the terminus of both these systems as nutrient loads increased. The severely reduced, almost minimal, flow of the Huanghe has prevented the formation of hypoxia, but other coastal ecosystem problems remain. There is, however, now a hypoxic area off the Changjiang Estuary and harmful algal blooms are more frequent in the East China Sea. The likelihood that more and more coastal systems, especially in developing countries, where the physical conditions are appropriate will become eutrophic with accompanying hypoxia is worrisome.

Northern Gulf of Mexico

The hypoxic zone on the continental shelf of the northern Gulf of Mexico is one of the largest hypoxic zones in the world's coastal oceans, and is representative of hypoxia resulting from anthropogenic activities over the last half of the twentieth century (**Figure 2**). Every spring, the dissolved oxygen levels in the coastal waters of the northern Gulf of Mexico decline and result in a vast region of oxygen-starved water that stretches from the Mississippi River westward along the Louisiana shore and onto the Texas coast. The area of bottom covered by hypoxic water can reach 22 000 km^2, and the volume of hypoxic waters may be as much as 10^{11} m^3. Hypoxia in the Gulf of Mexico results from a combination of natural and human-influenced factors. The Mississippi River, one of the 10 largest in the world, drains 41% of the land area of the lower 48 states of the US and delivers fresh water, sediments, and nutrients to the Gulf of Mexico. The fresh water, when it enters the Gulf, floats over the denser saltier water, resulting in stratification, or a two-layered system. The stratification, driven primarily by salinity, begins in the spring, intensifies in the summer as surface waters warm and winds that normally mix the water subside, and dissipates in the fall with tropical storms or cold fronts.

Hypoxic waters are found at shallow depths near the shore (4–5 m) to as deep as 60 m. The more typical depth distribution is between 5 and 35 m. The hypoxic water is not just located near the seabed, but may rise well up into the water column, often occupying the lower half of a 20-m water column (**Figure 3**). The inshore/offshore distribution of hypoxia on the Louisiana shelf is dictated by winds and currents. During typical winds from the southeast, downwelling favorable conditions force the hypoxic bottom waters farther offshore. When the wind comes from the north, an upwelling favorable current regime promotes the movement of the hypoxic bottom waters close to shore. When the hypoxic waters move onto the shore, fish, shrimp, and crabs are trapped along the beach, resulting sometimes in a 'jubilee' when the stunned animals are easily harvested by beachgoers. A more negative result is a massive fish kill of all the sea life trapped without sufficient oxygen.

Hypoxia occurs on the Louisiana coast west of the Mississippi River delta from February through November, and nearly continuously from mid-May through mid-September. In March and April, hypoxic water masses are patchy and ephemeral. The hypoxic

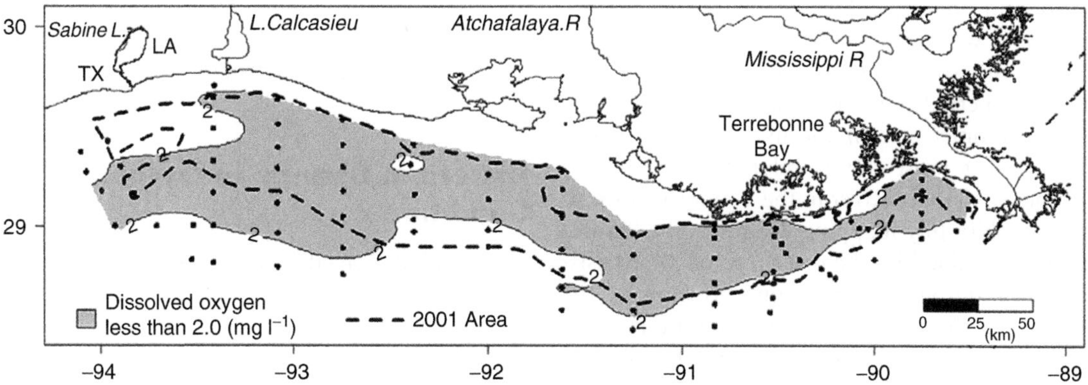

Figure 2 Similar size and expanse of bottom water hypoxia in mid-July 2002 (shaded area) and in mid-July 2001 (outlined with dashed line). Data source: N. N. Rabalais, Louisiana Universities Marine Consortium.

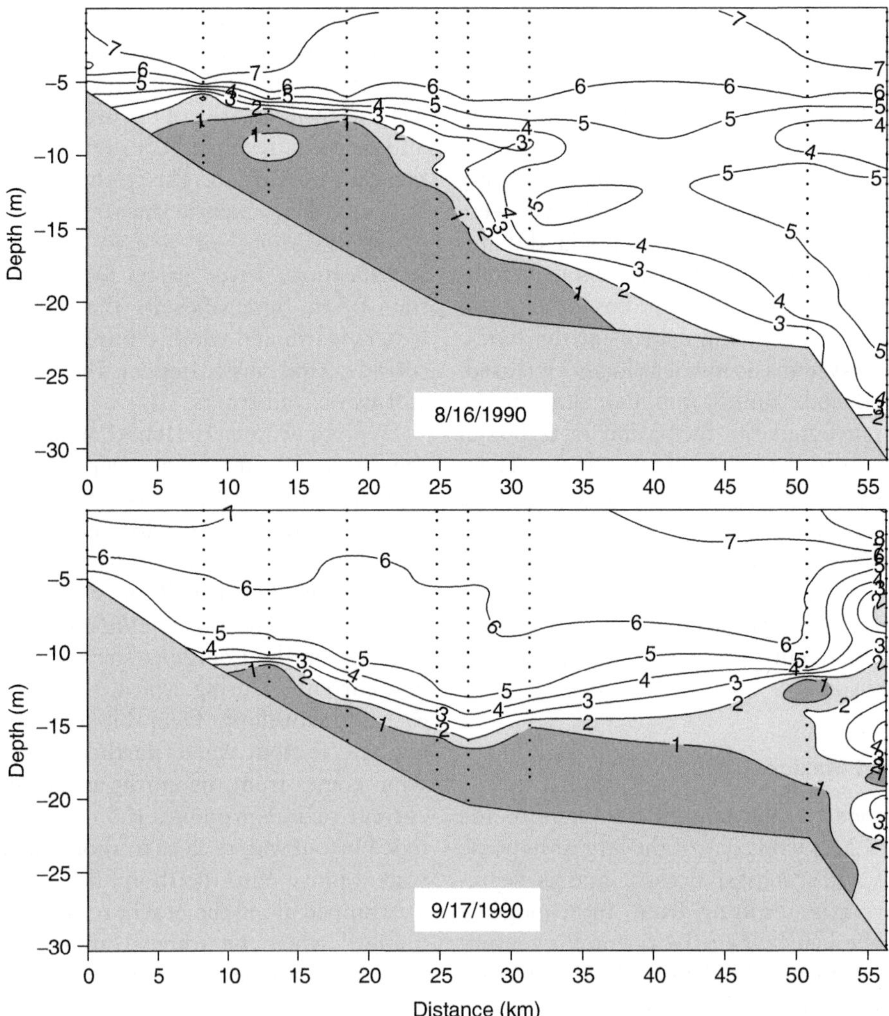

Figure 3 Contours of dissolved oxygen (mg l^{-1}) across the continental shelf of Louisiana approximately 200 km west of the Mississippi River delta in summer. The distribution across the shelf in August is a response to an upwelling favorable oceanographic regime and that of September to a downwelling favorable oceanographic regime. These contours also illustrate the height above the seabed that hypoxia can reach, i.e., over half the water column. Data source: N. N. Rabalais, Louisiana Universities Marine Consortium.

zone is most widespread, persistent, and severe in June, July, and August, and often well into September, depending on whether tropical storm activity begins to disrupt the stratification and hypoxia. Anoxic waters occur periodically in midsummer.

The midsummer size of the hypoxic zone varies annually, and is most closely related to the nitrate load of the Mississippi River in the 2 months prior to the typically late-July mapping exercise. The load of nitrate is determined by the discharge of the Mississippi River multiplied by the concentration of the nitrate, so that the amount of water coming into the Gulf of Mexico is also a factor. The relationship of the size of hypoxia, however, is stronger with the load of nitrate than with the total river water discharge or any other nutrient or combination of nutrients. Changes in the severity of hypoxia over time are related mostly to the change in nitrate concentration in the Mississippi River (80%), the remainder to changes in increased discharge (20%).

Historical Change in Oxygen Conditions

Historical dissolved oxygen data such as those for the northern Adriatic Sea beginning in 1911 are not commonly available. A solution is to turn to the sediment record for paleoindicators of long-term transitions related to eutrophication and oxygen deficiency. Biological, mineral, or chemical indicators of plant communities, level of productivity, or conditions of hypoxia preserved in sediments, where sediments accumulate, provide clues to prior hydrographic and biological conditions.

Data from sediment cores taken from the Louisiana Bight adjacent to the Mississippi River where sediments accumulate with their biological and chemical indicators document increased recent eutrophication and increased organic sedimentation in bottom waters, with the changes being more apparent in areas of chronic hypoxia and coincident with the increasing nitrogen loads from the Mississippi River system beginning in the 1950s. This evidence comes as an increased accumulation of diatom remains and marine-origin carbon accumulation in the sediments.

Benthic microfauna and chemical conditions provide several surrogates for oxygen conditions. The mineral glauconite forms under reducing conditions in sediments, and its abundance is an indication of low-oxygen conditions. (Note that glauconite also forms in reducing sediments whose overlying waters are $>2\,\mathrm{mg\,l^{-1}}$ dissolved oxygen.) The average glauconite abundance in the coarse fraction of sediments in the Louisiana Bight was $\sim$5.8% from 1900 to a transition period between 1940 and 1950, when it increased to $\sim$13.4%, suggesting that hypoxia 'may' have existed at some level before the 1940–50 time period, but that it worsened since then.

Benthic foraminiferans and ostracods are also useful indicators of reduced oxygen levels because oxygen stress decreases their overall diversity as measured by the Shannon–Wiener diversity index (SWDI) and shifts community composition. Foraminiferan and ostracod diversity decreased since the 1940s and early 1950s, respectively. While present-day foraminiferan diversity is generally low in the Louisiana Bight, comparisons among assemblages from areas of different oxygen depletion indicate that the dominance of *Ammonia parkinsoniana* over *Elphidium* spp. (A–E index) was much more pronounced in oxygen-depleted compared to well-oxygenated waters. The A–E index has also proven to be a strong, consistent oxygen-stress signal in other coastal areas, for example, Chesapeake Bay and Long Island Sound. The A–E index from sediment cores increased significantly after the 1950s, suggesting increased oxygen stress (in intensity or duration) in the last half century. *Buliminella morgani*, a hypoxia-tolerant species, known only from the Gulf of Mexico, dominates the present-day population (>50%) within areas of chronic seasonal hypoxia, and has also increased markedly in recent decades. *Quinqueloculina* sp., a hypoxia-intolerant foraminiferan, was a conspicuous member of the fauna from 1700 to 1900, indicating that oxygen stress was not a problem prior to 1900, but this species is no longer present on northern Gulf of Mexico shelf in the Louisiana Bight.

Multiple lines of evidence from sediment cores indicate an overall increase in phytoplankton productivity and continental shelf oxygen stress (in intensity or duration) in the northern Gulf of Mexico adjacent to the plume of the Mississippi River, especially in the last half of the twentieth century. The changes in these indicators are consistent with the increases in river nitrate-N loading during that same period.

OMZ intensity and distribution vary over geological timescales as a result of shifts in productivity or circulation over a few thousands to 10 ky. These changes affect expansions and contractions of the oxygen-depleted waters both vertically and horizontally. Paleoindicators, including foraminiferans, organic carbon preservation, carbonate dissolution, nitrogen isotopes, and Cd:Ca ratios that reflect productivity maxima and shallow winter mixing of the water column, are used to trace longer-term changes in OMZs, similar to studies of continental shelf sediment indicators.

Consequences

Direct Effects

The obvious effects of hypoxia/anoxia are displacement of pelagic organisms and selective loss of demersal and benthic organisms. These impacts may be aperiodic so that recovery occurs; may occur on a seasonal basis with differing rates of recovery; or may be permanent so that a shift occurs in long-term ecosystem structure and function. As the oxygen concentration falls from saturated or optimal levels toward depletion, a variety of behavioral and physiological impairments affect the animals that reside in the water column or in the sediments or that are attached to hard substrates (**Figure 4**). Hypoxia acts as an endocrine disruptor with adverse effects on reproductive performance of fishes, and loss of secondary production may therefore be a widespread environmental consequence of hypoxia. Mobile animals, such as shrimp, fish, and some crabs, flee waters where the oxygen concentration falls below $3\text{–}2\,\mathrm{mg\,l^{-1}}$.

As dissolved oxygen concentrations continue to fall, less mobile organisms become stressed and move up out of the sediments, attempt to leave the seabed, and often die (**Figure 5**). As oxygen levels fall from 0.5 toward $0\,\mathrm{mg\,l^{-1}}$, there is a fairly linear decrease in benthic infaunal diversity, abundance, and biomass. Losses of entire higher taxa are features of the

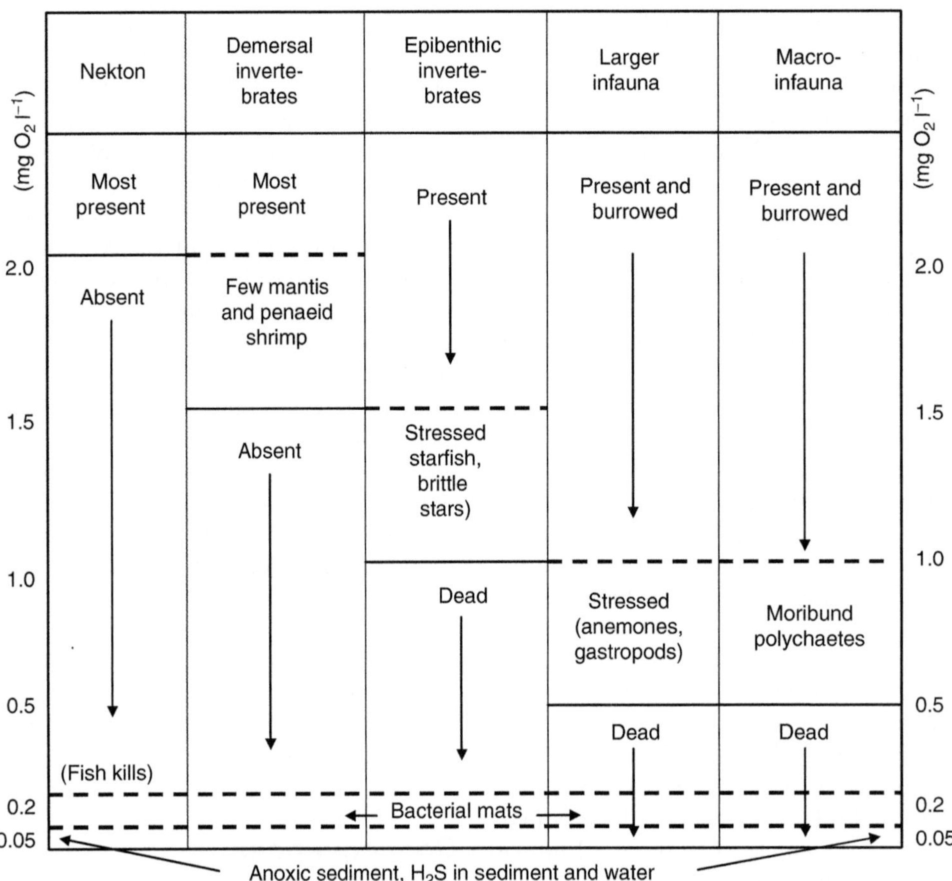

Figure 4 Progressive changes in fish and invertebrate fauna as oxygen concentration decreases from 2 mg l^{-1} to anoxia. From Rabalais NN, Harper DE, Jr., and Tuner RE (2001) Responses of nekton and demersal and benthic fauna to decreasing oxygen concentrations. In: Rabalais NN and Turner RE (eds.) *Coastal and Estuarine Studies 58: Coastal Hypoxia: Consequences for Living Resources and Ecosystems*, pp. 115–128. Washington, DC: American Geophysical Union.

depauperate benthic fauna in the severely stressed seasonal hypoxic/anoxic zone of the Louisiana inner shelf in the northern Gulf of Mexico. Larger, longer-lived burrowing infauna are replaced by short-lived, smaller surface deposit-feeding polychaetes, and certain typical marine invertebrates are absent from the fauna, for example, pericaridean crustaceans, bivalves, gastropods, and ophiuroids. Long-term trends for the Skagerrak coast of western Sweden in semi-enclosed fiordic areas experiencing increased oxygen stress showed declines in the total abundance and biomass of macroinfauna, abundance and biomass of mollusks, and abundance of suspension feeders and carnivores. These changes in benthic communities result in an impoverished diet for bottom-feeding fish and crustaceans and contribute, along with low dissolved oxygen, to altered sediment biogeochemical cycles. In waters of Scandinavia and the Baltic, there was a reduction of 3 million t in benthic macrofaunal biomass during the worst years of hypoxia occurrence. This loss, however, may be partly compensated by the biomass increase that occurred in well-flushed organically enriched coastal areas not subject to hypoxia.

Where oxygen minimum zones impinge on continental margins or sea mounts, they have considerable effects on benthic assemblages. The benthic fauna of OMZs consist mainly of smaller-sized protozoan and meiofaunal organisms, with few or no macrofauna or megafauna. The few eukaryotic organisms are nematodes and foraminiferans. Meiofauna appear to be more broadly tolerant of oxygen depletion than are macrofauna. The numbers of metazoan meiofaunal organisms, primarily nematodes, are not reduced in OMZs, presumably due to abundant particulate food and reduced predation pressure. In hypoxic waters of the northern Gulf of Mexico, harpacticoid copepod meiofauna are reduced at low oxygen levels, but the nematodes maintain their densities. Benthic macrofauna are found in all hypoxic sediments of the northern Gulf of Mexico, although the density is severely reduced

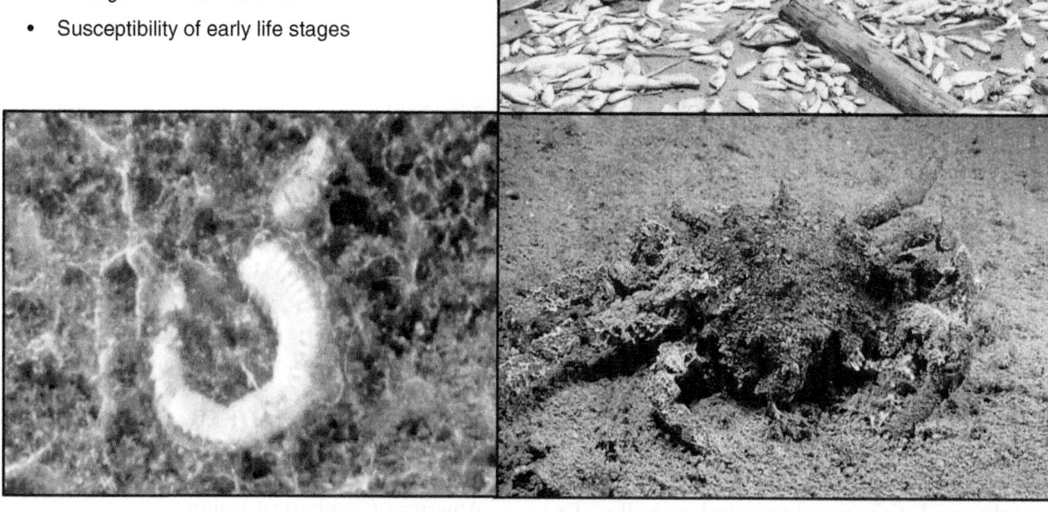

Figure 5 Effects of hypoxia on fishery resources and the benthic communities that support them. Upper right: Dead demersal and bottom-dwelling fishes killed by the encroachment of near-anoxic waters onto a Grand Isle, Louisiana, beach in August 1990. Photo provided by K. M. St. Pé. Lower right: dead spider crab (family Majidae) at sediment surface. Photo provided by Franklin Viola. Lower left: dead polychaete (family Spionidae) and filamentous sulfur bacteria. Photo provided by Franklin Viola.

below $0.5\,\mathrm{mg\,l^{-1}}$, and the few remaining organisms are polychaetes of the families Ampharetidae and Magelonidae and some sipunculans.

While permanent deep-water hypoxia that impinges on 2.3% of the ocean's continental margin may be inhospitable to most commercially valuable marine resources, they support the largest, most continuous reducing ecosystems in the world oceans. Large filamentous sulfur bacteria, *Thioploca* and *Beggiatoa*, thrive in hypoxic conditions of $0.1\,\mathrm{ml\,l^{-1}}$. OMZ sediments characteristically support large bacteria, both filamentous sulfur bacteria and giant spherical sulfur bacteria with diameters of 100–300 μm. The filamentous sulfur bacteria are also characteristic of severely oxygen depleted waters in the northern Gulf of Mexico.

Secondary Production

An increase in nutrient availability results in an increase of fisheries yield to a maximal point; then there are declines in various compartments of the fishery as further increases in nutrients lead to seasonal hypoxia and permanent anoxia in semi-enclosed seas. Documenting loss of fisheries related to the secondary effects of eutrophication, such as the loss of seabed vegetation and extensive bottom water oxygen depletion, is complicated by poor fisheries data, inadequate economic indicators, increase in overharvesting that occurred at the time that habitat degradation progressed, natural variability of fish populations, shifts in harvestable populations, and climatic variability.

Eutrophication often leads to the loss of habitat (rooted vegetation or macroalgae) or low dissolved oxygen, both of which may lead to loss of fisheries production. In the deepest bottoms of the Baltic proper, animals have long been scarce or absent because of low oxygen availability. This area was $20\,000\,\mathrm{km^2}$ until the 1940s. Since then, about a third of the Baltic bottom area has intermittent oxygen depletion. Lowered oxygen concentrations and increased sedimentation have changed the benthic fauna in the deeper parts of the Baltic, resulting in an impoverished diet for bottom fish. Above the halocline in areas not influenced by local pollution, benthic biomass has increased due mostly to an increase in mollusks. On the other hand, many reports document instances where local pollution resulting in severely depressed oxygen levels has greatly impoverished or even annihilated the soft-bottom macrofauna.

Eutrophication of surface waters accompanied by oxygen-deficient bottom waters can lead to a shift in dominance from demersal fishes to pelagic fishes. In the Baltic Sea and Kattegatt where eutrophication-related ecological changes occurred mainly after World War II, changes in fish stocks have been both positive (due to increased food supply; e.g., pike perch in Baltic archipelagos) and negative (e.g., oxygen deficiency reducing Baltic cod recruitment and eventual harvest). Similar shifts are inferred with limited data on the Mississippi River-influenced shelf with the increase in two pelagic species in bycatch from shrimp trawls and a decrease in some demersal species. Commercial fisheries in the Black Sea declined as eutrophication led to the loss of macroalgal habitat and oxygen deficiency, amid the possibility of overfishing. After the mid-1970s, benthic fish populations (e.g., turbot) collapsed, and pelagic fish populations (small pelagic fish, such as anchovy and sprat) started to increase. The commercial fisheries diversity declined from about 25 fished species to about five in 20 years (1960s to 1980s), while anchovy stocks and fisheries increased rapidly. The point on the continuum of increasing nutrients versus fishery yields remains vague as to where benefits are subsumed by environmental problems that lead to decreased landings or reduced quality of production and biomass.

Future Expectations

The continued and accelerated export of nitrogen and phosphorus to the world's coastal ocean is the trajectory to be expected unless societal intervention takes place (in the form of controls or changes in culture). The largest increases are predicted for southern and eastern Asia, associated with predicted large increases in population, increased fertilizer use to grow food to meet the dietary demands of that population, and increased industrialization. The implications for coastal eutrophication and subsequent ecosystem changes such as worsening conditions of oxygen depletion are significant.

Further Reading

Díaz RJ, Nestlerode J, and Díaz ML (2004) A global perspective on the effects of eutrophication and hypoxia on aquatic biota. In Rupp GL and White MD (eds.) *Proceedings of* the 7th International *Symposium on Fish Physiology, Toxicology and Water Quality*, EPA 600/R-04/049, pp. 1–33. Tallinn, Estonia, 12–15 May 2003. Athens, GA: Ecosystems Research Division, US EPA.

Díaz RJ and Rosenberg R (1995) Marine benthic hypoxia: A review of its ecological effects and the behavioural responses of benthic macrofauna. *Oceanography and Marine Biology Annual Review* 33: 245–303.

Gray JS, Wu RS, and Or YY (2002) Review. Effects of hypoxia and organic enrichment on the coastal marine environment. *Marine Ecology Progress Series* 238: 249–279.

Hagy JD, Boynton WR, and Keefe CW (2004) Hypoxia in Chesapeake Bay, 1950–2001: Long-term change in relation to nutrient loading and river flow. *Estuaries* 27: 634–658.

Helly J and Levin LA (2004) Global distributions of naturally occurring marine hypoxia on continental margins. *Deep-Sea Research* 51: 1159–1168.

Mee LD, Friedrich JJ, and Gomoiu MT (2005) Restoring the Black Sea in times of uncertainty. *Oceanography* 18: 100–111.

Rabalais NN and Turner RE (eds.) (2001) *Coastal and Estuarine Studies 58: Coastal Hypoxia – Consequences for Living Resources and Ecosystems*. Washington, DC: American Geophysical Union.

Rabalais NN, Turner RE, and Scavia D (2002) Beyond science into policy: Gulf of Mexico hypoxia and the Mississippi River. *BioScience* 52: 129–142.

Rabalais NN, Turner RE, Sen Gupta BK, Boesch DF, Chapman P, and Murrell MC (2007) Characterization and long-term trends of hypoxia in the northern Gulf of Mexico: Does the science support the Action Plan? *Estuaries and Coasts* 30(supplement 5): 753–772.

Turner RE, Rabalais NN, and Justić D (2006) Predicting summer hypoxia in the northern Gulf of Mexico: Riverine N, P and Si loading. *Marine Pollution Bulletin* 52: 139–148.

Tyson RV and Pearson TH (eds.) *Geological Society Special Publication No. 58: Modern and Ancient Continental Shelf Anoxia*, 470pp. London: The Geological Society.

Relevant Websites

http://www.gulfhypoxia.net
– Hypoxia in the Northern Gulf of Mexico.

CARBON SEQUESTRATION VIA DIRECT INJECTION INTO THE OCEAN

E. E. Adams, Massachusetts Institute of Technology, Cambridge, MA, USA
K. Caldeira, Stanford University, Stanford, CA, USA

Introduction

Global climate change, triggered by a buildup of greenhouse gases, is emerging as perhaps the most serious environmental challenge in the twenty-first century. The primary greenhouse gas is CO_2, whose concentration in the atmosphere has climbed from its preindustrial level of *c.* 280 to >380 ppm. Stabilization at no more than 500–550 ppm is a target frequently discussed to avoid major climatic impact.

The primary source of CO_2 is the burning of fossil fuels – specifically gas, oil, and coal – so stabilization of atmospheric CO_2 concentration will clearly require substantial reductions in CO_2 emissions from these sources. For example, one commonly discussed scenario to stabilize at 500 ppm by the mid-twenty-first century suggests that about 640 Gt CO_2 (*c.* 175 Gt C) would need to be avoided over 50 years, with further emission reductions beyond 50 years. As references, a 1000 MW pulverized coal plant produces 6–8 Mt CO_2 (*c.* 2 Mt C) per year, while an oil-fired single-cycle plant produces about two-thirds this amount and a natural gas combined cycle plant produces about half this amount. Thus the above scenario would require that the atmospheric emissions from the equivalent of 2000–4000 large power plants be avoided by approximately the year 2050.

Such changes will require a dramatic reduction in our current dependence on fossil fuels through increased conservation and improved efficiency, as well as the introduction of nonfossil energy sources like solar, wind, and nuclear. While these strategies will slow the buildup of atmospheric CO_2, it is probable that they will not reduce emissions to the required level. In other words, fossil fuels, which currently supply over 85% of the world's energy needs, are likely to remain our primary energy source for the foreseeable future. This has led to increased interest in a new strategy termed carbon capture and storage, or sequestration. The importance of this option for mitigating climate change is highlighted by the recent *Special Report on Carbon Dioxide Capture and Storage* published by the Intergovernmental Panel on Climate Change, to which the reader is referred for more information.

Carbon sequestration is often associated with the planting of trees. As they mature, the trees remove carbon from the atmosphere. As long as the forest remains in place, the carbon is effectively sequestered. Another type of sequestration involves capturing CO_2 from large, stationary sources, such as a power plant or chemical factory, and storing the CO_2 in underground reservoirs or the deep ocean, the latter being the focus of this article. There has been much attention paid recently to underground storage with several large-scale field sites in operation or being planned. Conversely, while there have been many studies regarding use of the deep ocean as a sink for atmospheric carbon, there have been only a few small-scale field studies.

Why is the ocean of interest as a sink for anthropogenic CO_2? The ocean already contains an estimated 40 000 Gt C compared with about 800 Gt C in the atmosphere and 2200 Gt C in the land biosphere. As a result, the amount of carbon that would cause a doubling of the atmospheric concentration would only change the ocean concentration by about 2%. In addition, natural chemical equilibration between the atmosphere and ocean would result in about 80% of present-day emissions ultimately residing in the ocean. Discharging CO_2 directly to the ocean would accelerate this slow, natural process, thus reducing both peak atmospheric CO_2 concentrations and their rate of increase. It is noted that a related strategy for sequestration – not discussed here – would be to enhance the biological sink using nutrients such as iron to fertilize portions of the world's oceans, thus stimulating phytoplankton growth. The phytoplankton would increase the rate of biological uptake of CO_2, and a portion of the CO_2 would be transported to ocean depths when the plankton die.

The indirect flux of CO_2 to the ocean from the atmosphere is already quite apparent: since preindustrial times, the pH of the surface ocean has been reduced by about 0.1 units, from an initial surface pH of about 8.2. **Figure 1** illustrates what could happen to ocean pH under conditions of continued atmospheric release of CO_2. Under the conditions simulated, the pH of the surface would drop by over 0.7 units. Conversely, by injecting some of the CO_2

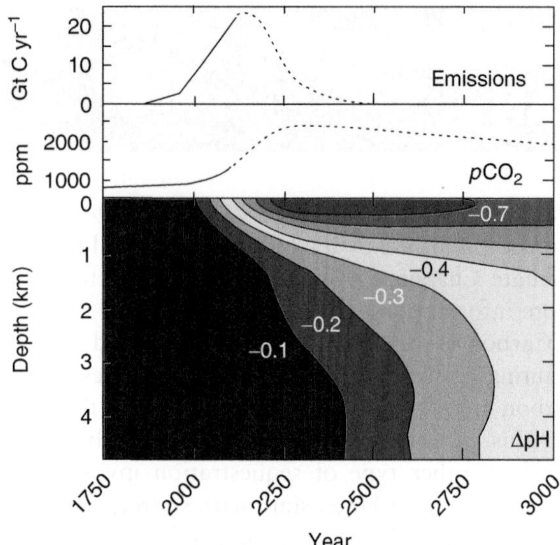

Figure 1 Model simulations of long-term ocean pH changes, averaged horizontally, as a result of atmospheric CO_2 emissions shown in the top panel. Reprinted from Caldeira K and Wickett ME (2003) Anthropogenic carbon and ocean pH. *Nature* 425: 365.

to the deep ocean, the change in pH could be more uniformly distributed.

Ocean sequestration of CO_2 by direct injection assumes that a relatively pure CO_2 stream has been generated at a power plant or chemical factory and transported to an injection point. To better understand the role the ocean can play, we address the capacity of the ocean to sequester CO_2, its effectiveness at reducing atmospheric CO_2 levels, how to inject the CO_2, and possible environmental consequences and issues of public perception.

Capacity

How much carbon can the ocean sequester? At over 70% of the Earth's surface and an average depth of 3800 m, the ocean has enormous storage capacity; based on physical chemistry, the amount of CO_2 that could be dissolved in the deep ocean far exceeds the estimated available fossil energy resources of 5000–10 000 Gt C. However, a more realistic criterion needs to be based on an understanding of ocean biogeochemistry and expected environmental impact.

CO_2 exists in seawater in various forms as part of the carbonate system:

$$CO_2(aq) + H_2O \leftrightarrow H_2CO_3(aq) \leftrightarrow H^+ + HCO_3^- \leftrightarrow 2H^+ + CO_3^{2-} \quad [1]$$

Dissolving additional CO_2 increases the hydrogen ion concentration (lowering the pH), but the change is buffered by the fact that total alkalinity is conserved, which results in carbonate ion being converted into bicarbonate. Thus, the principal reactions occurring when CO_2 is dissolved in seawater are

$$CO_2 + H_2O + CO_3^{2-} \rightarrow 2HCO_3^- \quad [2]$$

$$CO_2 + H_2O + H^+ \rightarrow HCO_3^- \quad [3]$$

which result in a decrease in pH and carbonate ion, and an increase in bicarbonate ion.

Reduced pH is one of the principal environmental impacts threatening marine organisms, the other being the concentration of CO_2 itself. At short travel times from the injection point, the changes in pH and CO_2 concentration will be greatest, which suggests that injection schemes should achieve the maximum dilution possible to minimize potential acute impacts in the vicinity of injection. See further discussion below.

At longer travel times, injected carbon would be distributed widely in the oceans and any far-field impact of the injected CO_2 on the oceans would be similar to the impact of anthropogenic CO_2 absorbed from the atmosphere. As indicated above, such changes are already taking place within the surface ocean, where the pH has been reduced by about 0.1 unit. Adding about 2000 Gt CO_2 to the ocean would reduce the average ocean pH by about 0.1 unit, while adding about 5600 Gt CO_2 (about 200 years of current emissions) would decrease the average ocean pH by about 0.3 units. (It should be noted that with stabilization of atmospheric CO_2 at 550 ppm, natural chemical equilibration between the atmosphere and ocean will result in eventual storage of over 6000 Gt CO_2 in the ocean.)

The impacts of such changes are poorly understood. The deep-ocean environment has been relatively stable and it is unknown to what extent changes in dissolved carbon or pH would affect these ecosystems. However, one can examine measured spatial and temporal variation in ocean pH to understand how much change might be tolerated. The spatial variability within given zoogeographic regions and bathymetric ranges (where similar ecosystems might be expected), and the temporal variability at a particular site, have both been found to vary by about 0.1 pH unit. If it is assumed that a change of 0.1 unit is a threshold tolerance, and that CO_2 should be stored in the bottom half of the ocean's volume (to maximize retention), nearly 1000 Gt CO_2 might be stored, which exceeds the 640 Gt CO_2 over 50 years estimated above. It is important to recognize that the long-term changes in ocean pH would ultimately be much the same whether the CO_2 is released into the atmosphere or the deep ocean. However, in the shorter term, releasing the CO_2 in the deep ocean will diminish the pH change in the near-surface ocean, where marine biota are most plentiful. Thus, direct injection of CO_2 into the deep ocean could reduce adverse impacts presently occurring in the surface ocean. In the long run, however, a sustainable solution to the problem of climate change must ultimately entail a drastic reduction of total CO_2 emissions.

Effectiveness

Carbon dioxide is constantly exchanged between the ocean and atmosphere. Each year the ocean and atmosphere exchange about 350 Gt CO_2, with a net ocean uptake currently of about 8 Gt CO_2. Because of this exchange, questions arise as to how effective ocean sequestration will be at keeping the CO_2 out of the atmosphere. Specifically, is the sequestration permanent, and if not, how fast does the CO_2 leak back to the atmosphere. Because there has been no long-term CO_2 direct-injection experiment in the ocean, the long-term effectiveness of direct CO_2 injection must be predicted based on observations of other oceanic tracers (e.g., radiocarbon) and on computer models of ocean circulation and chemistry.

As implied earlier, because the atmosphere and ocean are currently out of equilibrium, most CO_2 emitted to either media will ultimately enter the ocean. The percentage that is permanently sequestered depends on the atmospheric CO_2 concentration, through the effect of atmospheric CO_2 on surface ocean chemistry (see **Table 1**). At today's concentration of *c.* 380 ppm, nearly 80% of any carbon emitted to either the atmosphere or the ocean would be permanent, while at a concentration of 550 ppm, 74% would be permanent. Of course, even at equilibrium, CO_2 would continue to be exchanged between the atmosphere and oceans, so the carbon that is currently being injected is not exactly the same carbon that will reside in equilibrium.

Table 1 Percent of injected CO_2 permanently sequestered from the atmosphere as a function of atmospheric CO_2 stabilization concentration

Atmospheric carbon dioxide concentration (ppm)	*Percentage of carbon dioxide permanently sequestered*
350	80
450	77
550	74
650	72
750	70
1000	66

Based on data in IPCC (2005) *Special Report on Carbon Dioxide Capture and Storage.* Prepared by Working Group III of the Intergovernmental Panel on Climate Change. Cambridge, UK: Cambridge University Press. http://arch.rivm.nl/env/int/ipcc/pages_media/SRCCS-final/IPCCSpecialReportonCarbondioxideCaptureandStorage.htm (accessed Mar. 2008) and references therein.

For CO_2 injected to the ocean today, the net quantity retained in the ocean ranges from 100% (now) to about 80% as equilibrium between the atmosphere and oceans is approached. (A somewhat greater percentage will ultimately be retained as CO_2 reacts with ocean sediments over a timescale of thousands of years.) The nomenclature surrounding ocean carbon storage can be somewhat confusing. The percentage retained in the ocean shown in **Figure 2** is the fraction of injected CO_2 that has never interacted with the atmosphere. **Table 1** shows the fraction of CO_2 that contacts the atmosphere that remains permanently in the ocean. So, for example, for a 550 ppm atmosphere, even as the 'retained fraction' approaches zero (**Figure 2**), the amount permanently stored in the ocean approaches 74% (see **Table 1**). The exact time course depends on the location and depth of the injection.

Several computer modeling studies have studied the issue of retention. The most comprehensive summary is the Global Ocean Storage of Anthropogenic Carbon (GOSAC) intercomparison study of several ocean general circulation models (OGCMs). In this study a number of OGCMs simulated the fate of CO_2, injected over a period of 100 years at seven locations and three depths, for a period of 500 years. The CO_2 retained as a function of time, averaged over the seven sites, is shown in **Figure 2**. While there is variability among models, they all show that retention increases with injection depth, with most simulations predicting over 70% retention after 500 years for an injection depth of 3000 m.

The time required for injected carbon to mix from the deep ocean to the atmosphere is roughly equal to the time required for carbon to mix from the atmosphere to the deep ocean. This can be estimated through observations of radiocarbon (carbon-14) in the ocean. Correcting for mixing of ocean waters from different sources, the age of North Pacific deep water is in the range of 700–1000 years, while other basins, such as the North Atlantic, have overturning times of 300 years or more. These estimates are consistent with output from OGCMs and, collectively, suggest that outgassing of the 20% of injected carbon would occur on a timescale of 300–1000 years.

It is important to stress that leakage to the atmosphere would take place gradually and over large areas of the ocean. Thus, unlike geological sequestration, it

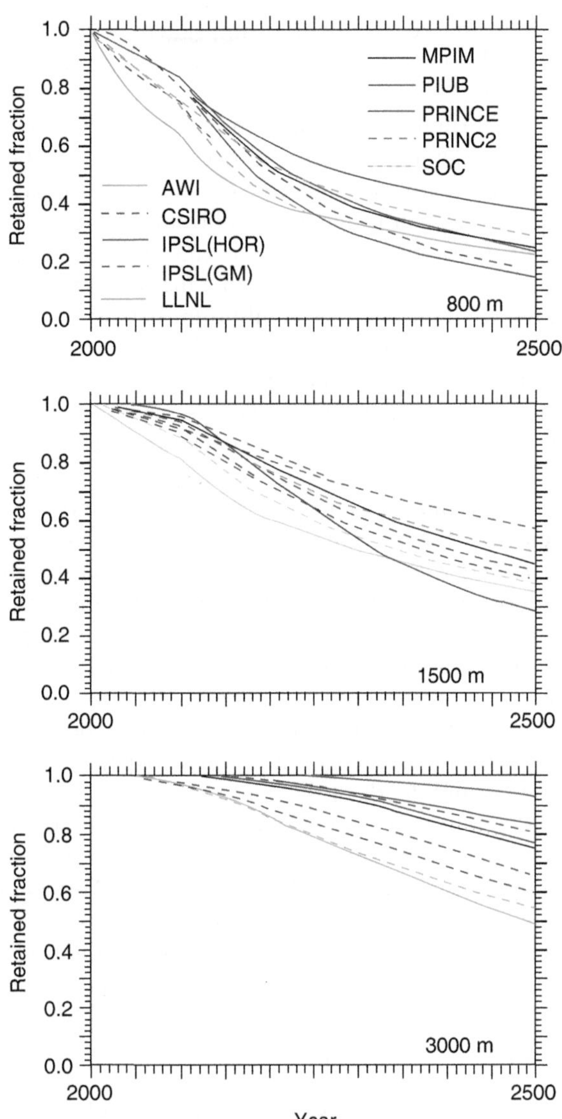

Figure 2 Model-intercomparison study reported by Orr in 2004 showing fraction of CO_2, injected from 2000 through 2100, that remains isolated from the atmosphere as a function of time and injection depth. Results are averaged over seven injection locations. Most of the CO_2 that does interact with the atmosphere remains in the ocean (see **Table 1**), so the amount of CO_2 remaining in the ocean is much greater than shown here. Reprinted with permission from IEA Greenhouse Gas R&D Programme.

would not be possible to produce a sudden release that could lead to harmful CO_2 concentrations at the ocean or land surface.

Injection Methods

The first injection concept was proposed by the Italian physicist Cesare Marchetti, who thought to introduce CO_2 into the outflow of the Mediterranean Sea, where the relatively dense seawater would cause the CO_2 to sink as it entered the Atlantic Ocean. As illustrated in **Figure 3**, a number of options have been considered since then.

Understanding these methods requires some background information on the CO_2–seawater system. Referring to **Figure 4**, at typical ocean pressures and temperatures, pure CO_2 would be a gas above a depth of 400–500 m and a liquid below that depth. Liquid CO_2 is more compressible than seawater, and would be positively buoyant (i.e., it will rise) down to about 3000 m, but negatively buoyant (i.e., it will sink) below that depth. At about 3700 m, the liquid becomes negatively buoyant compared to seawater saturated with CO_2. In seawater–CO_2 systems, CO_2 hydrate ($CO_2 \cdot nH_2O$, $n \sim 5.75$) can form below *c.* 400 m depth depending on the relative compositions of CO_2 and H_2O. CO_2 hydrate is a solid with a density about 10% greater than that of seawater.

The rising droplet plume has been the most studied and is probably the easiest scheme to implement. It would rely on commercially available technology to inject the CO_2 as a stream of buoyant droplets from a bottom manifold. Effective sequestration can be achieved by locating the manifold below the thermocline, and dilution can be increased by increasing the manifold length. Even better dilution can be achieved by releasing the CO_2 droplets from a moving ship whose motion provides additional dispersal. Although the means of delivery are different, the plumes resulting from these two options would be similar, each creating a vertical band of CO_2-enriched seawater over a prescribed horizontal region.

Another promising option is to inject liquid CO_2 into a reactor where it can react at a controlled rate with seawater to form hydrates. While it is difficult to achieve 100% reaction efficiency, laboratory and field experiments indicate that negative buoyancy, and hence sinking, can be achieved with as little as about 25% reaction efficiency. The hydrate reactor could be towed from a moving ship to encourage dilution, or attached to a fixed platform, where the large concentration of dense particles, and the increased seawater density caused by hydrate dissolution, would induce a negatively buoyant plume.

The concept of a CO_2 lake is based on a desire to minimize leakage to the atmosphere and exposure to biota. This would require more advanced technology and perhaps higher costs, as the depth of the lake should be at least 3000 m, which exceeds the depths at which the offshore oil industry currently works. The CO_2 in the lake would be partly in the form of solid hydrates. This would limit the CO_2 dissolution into the water column, further slowing leakage to the atmosphere from that shown in **Figure 2**, which

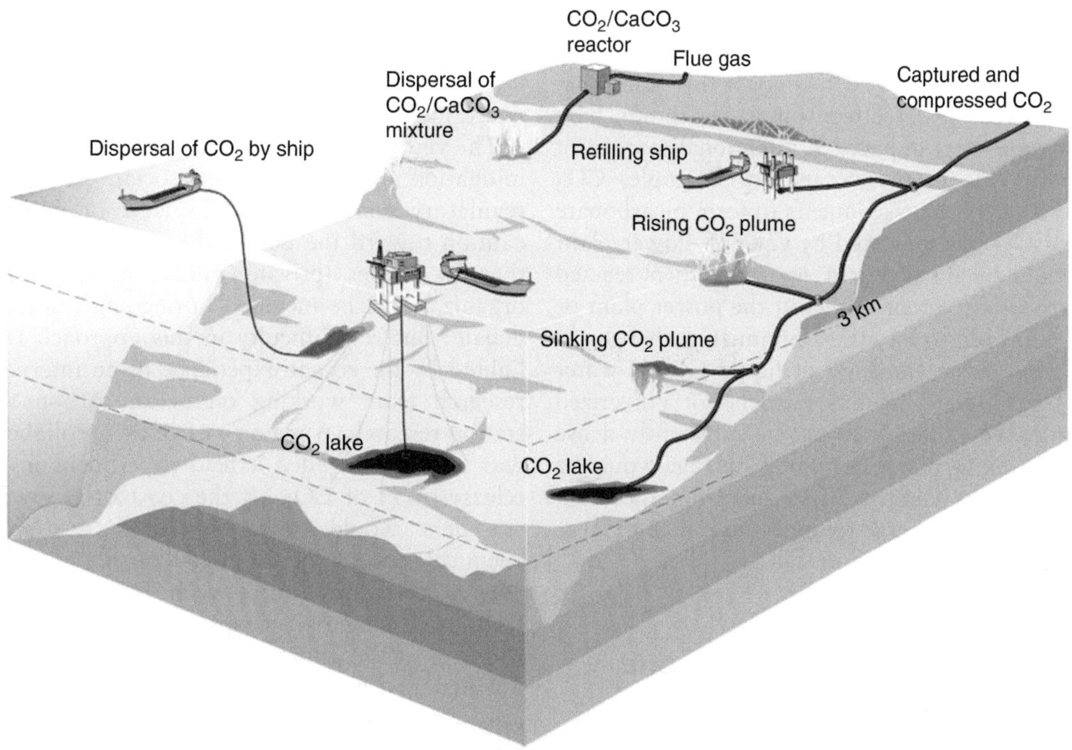

Figure 3 Different strategies for ocean carbon sequestration. Reprinted with permission from IPCC (2005) *Special Report on Carbon Dioxide Capture and Storage*, figure TS-9 (printed as *Special Report on Safeguarding the Ozone Layer and the Global Climate System*, Figure 6.1). Prepared by Working Group III of the Intergovernmental Panel on Climate Change. Cambridge, UK: Cambridge University Press. http://arch.rivm.nl/env/int/ipcc/pages_media/SRCCS-final/IPCCSpecialReportonCarbondioxideCaptureandStorage.htm, with permission from the Intergovernmental Panel on Climate Change.

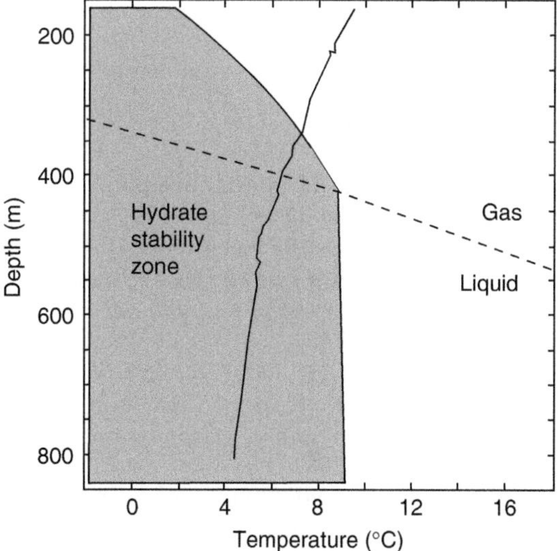

Figure 4 Phase diagram for CO_2 including typical ocean temperature profile (solid line). Reprinted from Brewer PG, Peltzer E, Aya I, *et al.* (2004) Small scale field study of an ocean CO_2 plume. *Journal of Oceanography* 60(4): 751.

assumes that CO_2 is injected into the water column. It is also possible that various approaches could be engineered to physically contain CO_2 on the seafloor and isolate the CO_2 from the overlying water column (and perhaps the sediments); however, this would entail an additional cost.

Another method that has received attention is injecting a dense CO_2–seawater mixture at a depth of 500–1000 m, forming a sinking bottom gravity current. CO_2-enriched seawater is less than 1% heavier than seawater, but this is sufficient to promote a sinking density current, especially if the current were formed along a submarine canyon. However, the environmental impacts would be greater with this option due to the concentrated nature of the plume, and its contact with the seafloor.

As discussed earlier, the deep ocean equilibrates with the surface ocean on the scale of 300–1000 years, and by injecting anthropogenic CO_2 into the deep ocean, the surface-to-deep mixing timescale is effectively bypassed. Anthropogenic CO_2 also equilibrates with carbonate sediments, but over a much longer time, about 6000 years. Technical means could also be used to bypass this timescale, thereby increasing the effectiveness and diminishing the environmental impacts of intentional storage of carbon dioxide in the ocean. For example, CO_2 reacts with carbonate sediments to form bicarbonate ions (HCO_3^-) as indicated by eqn [2]. Power plant CO_2 could be dissolved in seawater, then reacted with crushed limestone, either at the power plant or at the point of release, thus minimizing changes in plume pH. Or an emulsion of liquid CO_2-in-water could be stabilized by fine particles of pulverized limestone; the emulsion would be sufficiently dense to form a sinking plume, whose pH change would be buffered by the limestone. Drawbacks of these approaches include the cost to mine and transport large quantities of carbonate minerals.

Local Environmental Impacts and Public Perception

Environmental impacts may be the most significant factor determining the acceptability of ocean storage, since the strategy is predicated on the notion that impacts to the ocean will be significantly less than the avoided impacts of continued emission to the atmosphere. Earlier, environmental impacts were discussed from the global viewpoint. Here, we examine the environmental impacts near the injection point.

A number of studies have summarized potential impacts to different types of organisms, including adult fish, developmental fish, zooplankton, and benthic fauna. While earlier studies focused mainly on lethal impacts to coastal fauna exposed to strong acids, recent data have focused on deep-water organisms exposed to CO_2, and have included sublethal effects. Impacts include respiratory stress (reduced pH limits oxygen binding and transport of respiratory proteins), acidosis (reduced pH disrupts an organism's acid/basis balance), and metabolic depression (elevated CO_2 causes some animals to reach a state of torpor).

Data generally show that CO_2 causes greater stress than an equivalent change in pH caused by a different acid, that there are strong differences in tolerance among different species and among different life stages of the same species, and that the duration of stress, as well as the level of stress, are important. While some studies imply that deep organisms would be less tolerant than surface organisms, other studies have found the opposite. Likewise, some animals are able to avoid regions of high CO_2 concentration, while others appear less able to. Results generally suggest that lethal effects can be avoided by achieving high near-field dilution. However, more research is needed to resolve impacts, especially at the community level (e.g., reduced lifespan and reproduction effects).

The viability of ocean storage as a greenhouse gas mitigation option hinges on social, political, and regulatory considerations. In view of public precaution toward the ocean, the strategy will require that all parties (private, public, nongovernmental organizations) be included in ongoing research and debate. But the difficulty in this approach is highlighted by the recent experience of an international research team working on ocean carbon sequestration research. A major part of their collaboration was to have included a field experiment involving release of 5 t of CO_2 off the coast of Norway. Researchers would have monitored the physical, chemical, and biological effects of the injected CO_2 over a period of about a week. However, lobbying from environmental groups caused the Norwegian Minister of Environment to rescind the group's permit that had previously been granted. Such actions are unfortunate, because field experiments of this type are what is needed to produce data that would help policymakers decide if full-scale implementation would be prudent.

See also

Air-Sea Transfer: Dimethyl Sulfide, COS, CS_2, NH_4, Non-Methane Hydrocarbons, Organo-Halogens. Air-Sea Transfer: N_2O, NO, CH_4, CO. Carbon Cycle. Carbon Dioxide (CO_2) Cycle. Ocean Carbon System, Modeling of.

Further Reading

Alendal G and Drange H (2001) Two-phase, near field modeling of purposefully released CO_2 in the ocean. *Journal of Geophysical Research* 106(C1): 1085–1096.

Brewer PG, Peltzer E, Aya I, *et al.* (2004) Small scale field study of an ocean CO_2 plume. *Journal of Oceanography* 60(4): 751–758.

Caldeira K and Rau GH (2000) Accelerating carbonate dissolution to sequester carbon dioxide in the ocean: Geochemical implications. *Geophysical Research Letters* 27(2): 225–228.

Caldeira K and Wickett ME (2003) Anthropogenic carbon and ocean pH. *Nature* 425: 365.

Giles J (2002) Norway sinks ocean carbon study. *Nature* 419: 6.

Golomb D, Pennell S, Ryan D, Barry E, and Swett P (2007) Ocean sequestration of carbon dioxide: Modeling the deep ocean release of a dense emulsion of liquid

CO_2-in-water stabilized by pulverized limestone particles. *Environmental Science and Technology* 41(13): 4698–4704.

Haugan H and Drange H (1992) Sequestration of CO_2 in the deep ocean by shallow injection. *Nature* 357(28): 1065–1072.

IPCC (2005) *Special Report on Carbon Dioxide Capture and Storage*. Prepared by Working Group III of the Intergovernmental Panel on Climate Change. Cambridge, UK: Cambridge University Press. http://arch.rivm.nl/env/int/ipcc/pages_media/SRCCS-final/IPCCSpecialReporton CarbondioxideCapture-andStorage.htm (accessed Mar. 2008).

Ishimatsu A, Kikkawa T, Hayashi M, and Lee KS (2004) Effects of CO_2 on marine fish: Larvae and adults. *Journal of Oceanography* 60: 731–741.

Israelsson P and Adams E (2007) Evaluation of the Acute Biological Impacts of Ocean Carbon Sequestration. *Final Report for US Dept. of Energy, under grant DE-FG26-98FT40334*. Cambridge, MA: Massachusetts Institute of Technology.

Kikkawa T, Ishimatsu A, and Kita J (2003) Acute CO_2 tolerance during the early developmental stages of four marine teleosts. *Environmental Toxicology* 18(6): 375–382.

Ohsumi T (1995) CO_2 storage options in the deep-sea. *Marine Technology Society Journal* 29(3): 58–66.

Orr JC (2004) *Modeling of Ocean Storage of CO_2 – The GOSAC Study*, Report PH4/37, 96pp. Paris: Greenhouse Gas R&D Programme, International Energy Agency.

Ozaki M, Minamiura J, Kitajima Y, Mizokami S, Takeuchi K, and Hatakenka K (2001) CO_2 ocean sequestration by moving ships. *Journal of Marine Science and Technology* 6: 51–58.

Pörtner HO, Reipschläger A, and Heisler N (2004) Biological impact of elevated ocean CO_2 concentrations: Lessons from animal physiology and Earth history. *Journal of Oceanography* 60(4): 705–718.

Riestenberg D, Tsouris C, Brewer P, *et al.* (2005) Field studies on the formation of sinking CO_2 particles for ocean carbon sequestration: Effects of injector geometry on particle density and dissolution rate and model simulation of plume behavior. *Environmental Science and Technology* 39: 7287–7293.

Sato T and Sato K (2002) Numerical prediction of the dilution process and its biological impacts in CO_2 ocean sequestration. *Journal of Marine Science and Technology* 6(4): 169–180.

Vetter EW and Smith CR (2005) Insights into the ecological effects of deep-ocean CO_2 enrichment: The impacts of natural CO_2 venting at Loihi seamount on deep sea scavengers. *Journal of Geophysical Research* 110: C09S13 (doi:10.1029/2004JC002617).

Wannamaker E and Adams E (2006) Modeling descending carbon dioxide injections in the ocean. *Journal of Hydraulic Research* 44(3): 324–337.

Watanabe Y, Yamaguchi A, Ishida H, *et al.* (2006) Lethality of increasing CO_2 levels on deep-sea copepods in the western North Pacific. *Journal of Oceanography* 62: 185–196.

MARINE DEPOSITS

AUTHIGENIC DEPOSITS

G. M. McMurtry, University of Hawaii at Manoa, Honolulu, HI, USA

Introduction

There are many kinds and forms of authigenic mineral deposits in the deep-sea. Here we concentrate on the more abundant and, in some cases, potentially economic types, namely ferromanganese deposits, phosphorites, marine barite, and the authigenic silicate minerals in deep-sea sediments. Hydrothermal deposits and a class of sedimentary iron and manganese oxides associated with seafloor hydrothermal activity – the metalliferous sediments found as a basal sequence in the deep-sea sedimentary column and in deep basins adjacent to active spreading centers – are described elsewhere in the encyclopedia. Ferromanganese deposits are here divided into nodules and crusts on the basis of morphology and environment of formation. The sources of the major and minor metal enrichments in these deposits are contrasted between continental weathering and submarine volcanism, with important enrichment influences from the marine biosphere. Marine barite has a similar mixed origin with strong biological influence. Knowledge of the origins of submarine phosphorites has been greatly aided by studies of recent seafloor deposits. There are a variety of authigenic silicate minerals in deep-sea sediments. The more abundant zeolites and clays such as smectite form mainly from diagenetic alteration of metastable volcanic glasses, but many other reaction paths are known and suggested.

Ferromanganese Deposits

Marine ferromanganese deposits are complex assemblages of authigenic minerals and detrital components. They are broadly characterized by shape and by environment of formation as deep-sea nodules and crusts. Ferromanganese nodules are commonly found on the abyssal seafloor at and near the top of the underlying pelagic sediment cover. Crusts are usually found on the steep slopes of islands and seamounts but are also found with the pelagic sediment on the deep-sea floor where sedimentation is extremely low. Occasionally, nodules may grow to coalesce into a pavement or crust of ferromanganese deposits. Conversely, crusts can grow around small fragments of the usually lithified sediment or basalt substrate of a seamount or island slope to create 'pseudo' or 'seamount nodules' that are otherwise indistinguishable from crustal pavements.

Ferromanganese Nodules

Deep-sea ferromanganese nodules are classified as three major types on the basis of their origin: hydrogenetic, early diagenetic, and mixed hydrogenetic–early diagenetic. In this same scheme, crusts are described as primarily hydrogenetic, with comparatively minor diagenetic and hydrothermal occurrences (**Figure 1**). Hydrogenetic nodules obtain their ferromanganese oxides as precipitates from sea water. However, they are rarely if ever pure and also contain a variety of detrital minerals and biogenic debris from the pelagic sediment and fallout from the overlying water column and atmosphere. Early diagenetic nodules obtain their ferromanganese oxides from metals dissolved from the sediment and transported within the pore waters of the upper, peneliquid sediment layer (**Figure 1**). Because manganese and minor metals such as nickel and copper are more readily mobile in pore waters under suboxic seafloor conditions, diagenetic nodules are enriched in these metals relative to concentrations of iron and cobalt. The hybrid or mixed type nodule has both hydrogenetic and diagenetic components. These nodules are usually found at the sediment–water interface where their outer layers record their most recent position. The upper, relatively smooth hydrogenetic top is composed of iron- and cobalt-rich ferromanganese oxides from the bottom water and the lower, relatively bumpy or 'botroyidal' diagenetic bottom is composed of manganese-, nickel-, and copper-rich ferromanganese oxides from the sediment.

Attempts have been made to classify nodules on the basis of morphology and to relate their occurrence to the composition of the underlying sediment. Such field classifications were made to predict metal enrichment and genesis, but reliance upon morphological data for other than abundance information from either seafloor photography or remote acoustic imaging has not yet been widely practiced in exploration. Nodule abundance appears to correlate more strongly with siliceous ooze than with red clays, and both siliceous ooze and red clays have higher nodule abundance than does carbonate ooze.

Ferromanganese nodules and crusts are mainly composed of three types of manganese oxide minerals: vernadite or δ-MnO_2; birnessite; and todorokite. These

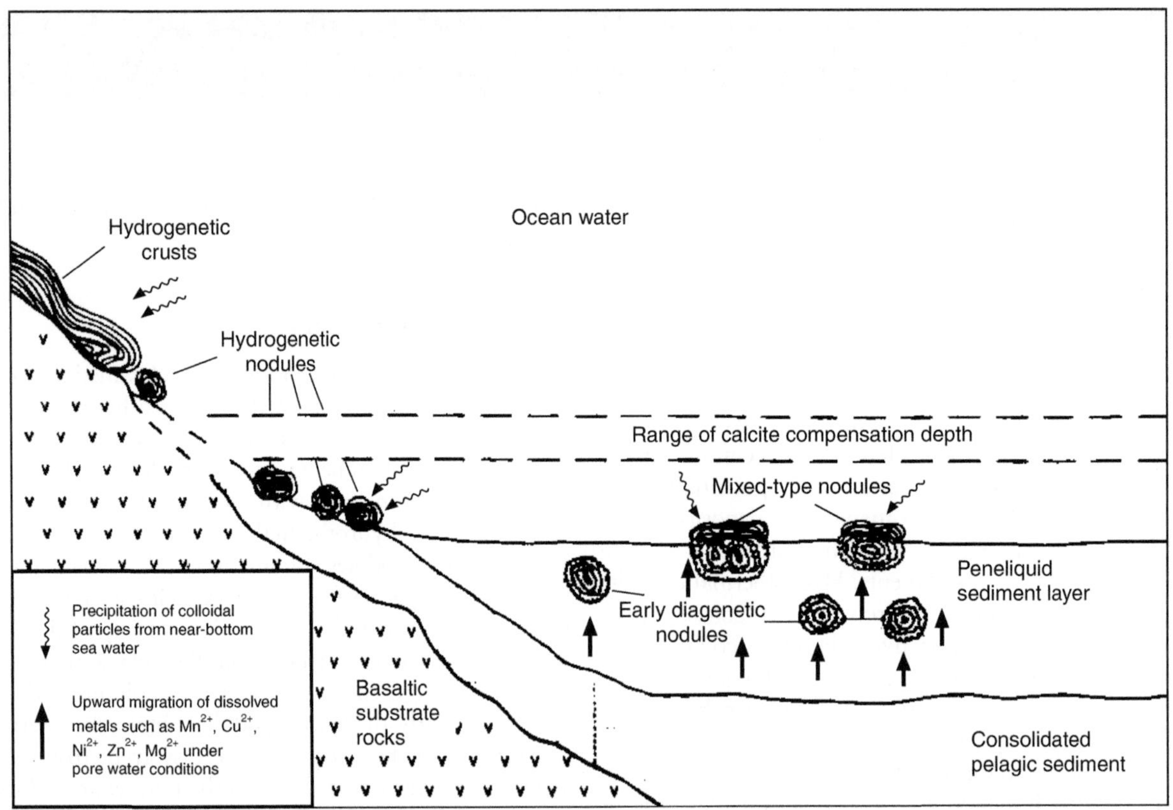

Figure 1 Schematic figure of the principal types of ferromanganese nodules (diameters vary from 1 to 15 cm) and crusts (thicknesses vary from <1 to 10 cm) distributed in pelagic sediments and on seamounts and island slopes. (After Halbach P (1986) Processes controlling the heavy metal distribution in Pacific ferromanganese nodules and crusts. *Geologische Rundschau* 75: 235–247.)

minerals are listed in order of decreasing E_h or oxidation potential of the formation environment, and increasing crystal order. In vernadite, all manganese occurs as Mn^{4+}, which allows for crystal lattice substitution of metals such as Co^{3+} and Pb^{4+}. Todorokite contains some manganese as Mn^{2+}, which allows for crystal lattice substitution of metals such as Cu^{2+} and Ni^{2+}. Birnessite is an intermediate form. Iron oxides are primarily found as closely intermingled X-ray-amorphous FeOOH and occasionally as goethite or lepidocrocite. Other minerals include detrital quartz, plagioclase feldspar, clay minerals such as smectites, authigenic zeolites, apatite, barite, and calcite.

The global distribution of ferromanganese nodules on the seafloor (**Figure 2**) largely reflects the inverse distribution of thickest sediment cover. Nodules, which accumulate at rates of several millimeters per million years, cannot grow where sediments are accumulating at rates that are greater than 10^3 times faster (several centimeters per thousand years). Otherwise, rapid burial and dissolution would be their fate. Paradoxically, nodules are found to lie atop sediments that are accumulating at rates up to 10^3 times faster; this phenomenon has been explained by bottom current winnowing and gentle bumping or tilting of nodules during the burrowing activities of benthic organisms. Nodules are truly pelagic deposits and are rarely found near the continents or near islands where rapid hemipelagic sedimentation occurs. Exceptions are found on the Blake Plateau off eastern Florida, within the Drake Passage off South America, and near the Cape of Good Hope off Southern Africa, where bottom currents are strong and sweep sediments from these areas (**Figure 2**). Highly concentrated nodule fields occur between the Clarion and Clipperton fracture zones just north of the equatorial high productivity zone in the Eastern Pacific, just south of this zone in the Indian Ocean, and near the Antarctic Convergence in the Southeast Pacific.

Source of major metals The sources of the major metals in ferromanganese nodules and crusts – iron and manganese – are commonly described as products of the weathering of the continents, seafloor hydrothermal activity, and sediment diagenesis. Other than a small but distinct cosmogenic component (see Ferromanganese Crusts below), the iron and manganese must ultimately come from the Earth's crust and mantle. The

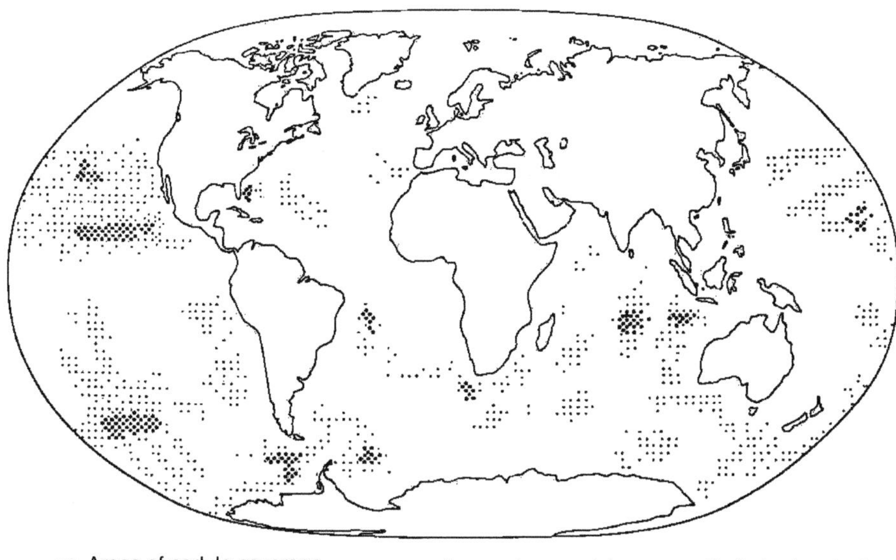

Figure 2 Distribution of ferromanganese nodule abundance on the world ocean floor. (After Cronan, 1980.)

fraction coming from the continents is more geologically recycled than that from the ocean crust, and diagenesis in this regard is clearly not an ultimate source but a process of further recycling of these metals within the sediment column. The relative importance of continental weathering versus seafloor hydrothermal activity as a source of iron and manganese to pelagic sediments and marine ferromanganese deposits has been debated since the 1890s. Early work by able geochemists on both sides of this issue was hampered by a lack of convincing evidence. For example, on the basis of early distribution maps of manganese and other trace metals in surface Atlantic sediments and evidence for iron and manganese enrichments in atmospheric dust, it was suggested that continental weathering and atmospheric transport was an important pathway of these metals to the ocean. However, the manganese distribution in Atlantic sediments can more readily be explained by hydrothermal enrichments from hot springs along the Mid-Atlantic Ridge, unknown at that time but well-known today. Likewise, early arguments for extensive hydrothermal iron and manganese enrichments from volcanic hot springs located along the mid-ocean ridges were compelling, but were not confirmed until the discovery of hot springs there in 1977. Water column trace-metal analytical techniques have improved to the point that a very convincing argument can be made for continental weathering and atmospheric transport as a path for at least some fraction of the iron and manganese in marine ferromanganese deposits. Another approach, which additionally answers the old question of the relative contributions of these two sources to ferromanganese deposits at any given location of the seafloor, was made for the Pacific (**Figure 3**) using a 'Co-chronometer'. This dating method is based upon the inverse relationship to growth rate of some minor metals' concentrations in hydrogenetic ferromanganese crust deposits, more painstakingly measured by radiometric techniques such as U-series (excess ^{230}Th) or ^{10}Be dating. Using the relatively easy Co-chronometer, hundreds of ferromanganese deposits were measured to construct the detailed map of ferromanganese crust growth rates for the Pacific in **Figure 3**. The highest growth rates are associated with active spreading along the East Pacific Rise, Juan de Fuca Ridge, and Galapagos Rift, and with submarine arc volcanism in the Mariana Island arc. The deposits in these areas therefore receive most of their iron and manganese from hydrothermal sources. The slowest crust growth rates and highest cobalt enrichments are located near the Mid-Pacific Mountains seamount province west of Hawaii, far from active hot springs and the continents.

Source of minor metals Relative to sea water and the Earth's crust, ferromanganese nodules are enriched in nickel, copper, cobalt, and a host of other minor metals such as the platinum-group elements (PGE) and the rare-earth elements (REE). Concentrations of nickel and copper reach up to 2 wt% in 'high-grade' nodules from fields between the Clarion and Clipperton fracture zones, with cobalt concentrations approaching 0.5 wt%. Broadly, nodules from the Pacific appear to be more enriched

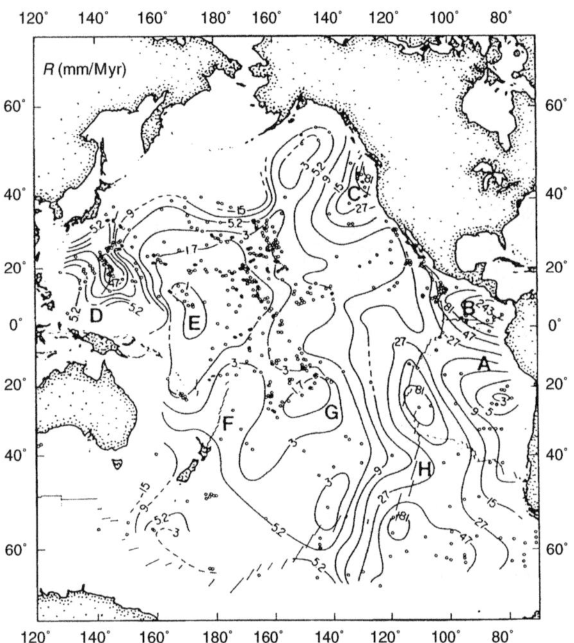

Figure 3 Distribution of ferromanganese crust accumulation rates in the Pacific Ocean. (After Manheim FT and Lane-Bostwick CM (1988) Cobalt in ferromanganese crusts as a monitor of hydrothermal discharge on the Pacific sea floor. *Nature* 335, 59–62.)

in these metals than those from the Indian and Atlantic Oceans, but metal concentrations can vary widely within basins and nodule fields. For at least the better-studied Pacific, concentrations of nickel and copper show good correlation to the nodule Mn/Fe ratio, suggesting that manganese-rich, todorokite-containing diagenetic nodules are the major depositories of these metals (**Figure 4A, B**). There is also a regional trend of nickel and copper enrichments and Mn/Fe ratios in the Pacific that is highest in the equatorial NE Pacific, the Peru Basin, and the SE Pacific near the Antarctic Convergence. The distribution of nodule cobalt concentrations shows little correlation to the Mn/Fe ratio, however, being enriched instead near the Mid-Pacific Mountains and in the South Central Pacific (**Figure 4C**).

The ultimate source for the minor metals in ferromanganese nodules is again the Earth's crust and mantle, but with minor metals the case for continental weathering versus seafloor hydrothermal activity is stronger. First, hydrothermal ferromanganese crusts, like many rapidly accumulating diagenetic crusts, are known to have very low concentrations of minor metals. Second, studies of copper and nickel distributions in pelagic surface sediments suggest that the highest concentrations are nearest to regions of high surface productivity, implying that plankton are involved in the enrichment process. There is an inverse relationship of depth of copper and nickel concentration in Pacific nodules with the sedimentary concentration of calcium carbonate tests from planktonic organisms, with deeper sediment deposits containing lesser $CaCO_3$ because of increased dissolution. Comparative studies of copper, nickel, and cobalt in plankton and the labile or easily mobilized fraction of pelagic sediment and associated deep-sea nodules has also suggested that diagenesis of organic matter in the sediments leads to an enrichment of these metals in the nodules. All told, scavenging by organisms in the surface waters enriches the particle rain in minor metals that originally enter the world's oceans by winds and rivers, and further concentration of these metals occurs within the surface sediments during suboxic diagenesis, with ferromanganese nodules as the final metals depository.

Ferromanganese Crusts

Crust deposits differ from nodules in form, occurrence, and composition. Hydrogenetic crusts are principally composed of vernadite with iron oxides and minor detrital mineral and carbonate fluoroapatite (CFA) contaminates. They grow at extremely slow rates from less than 1 mm to tens of millimeters per million years. Because the cobalt flux is for the most part invariant throughout the world's oceans, cobalt concentrations approach 2 wt% on comparatively shallow seamount slopes between 800 and 2000 m water depth within the Western Equatorial Pacific (**Figure 3** – compare with **Figure 4C**) where crust growth rates are slowest. Above approximately 800 m dilution and coverage of crustal pavements by principally carbonate-rich sediments limits their growth, and below 2000 m incorporation of increasing fluxes of seamount and wind-blown detritus and ferromanganese oxides causes crust growth rates to increase, effectively decreasing the cobalt concentration. The oxygen minimum zone (OMZ) presently intersects seamounts and island slopes in the Central Pacific at between 500 and 1500 m water depth. Although not presently anoxic enough to inhibit manganese oxide precipitation in this area, the OMZ was probably more intense in the past, thereby providing a mechanism for both transport of dissolved Mn^{2+} and inhibition of crust growth. Cobalt concentrations also show a general trend of decreasing values within older crustal layers, suggesting that past seafloor conditions favored increased manganese fluxes and growth.

Relative to nodules, crusts are on average enriched in Fe, Ca, P, Ti, Pb, Ce, As, and Pt, as well as cobalt, and are depleted in Si, Al, Ni, Cu, and Zn. Manganese concentrations are similar. The relative enrichments of calcium and phosphorus in crusts reflect the more widespread incorporation of CFA in seamount crusts

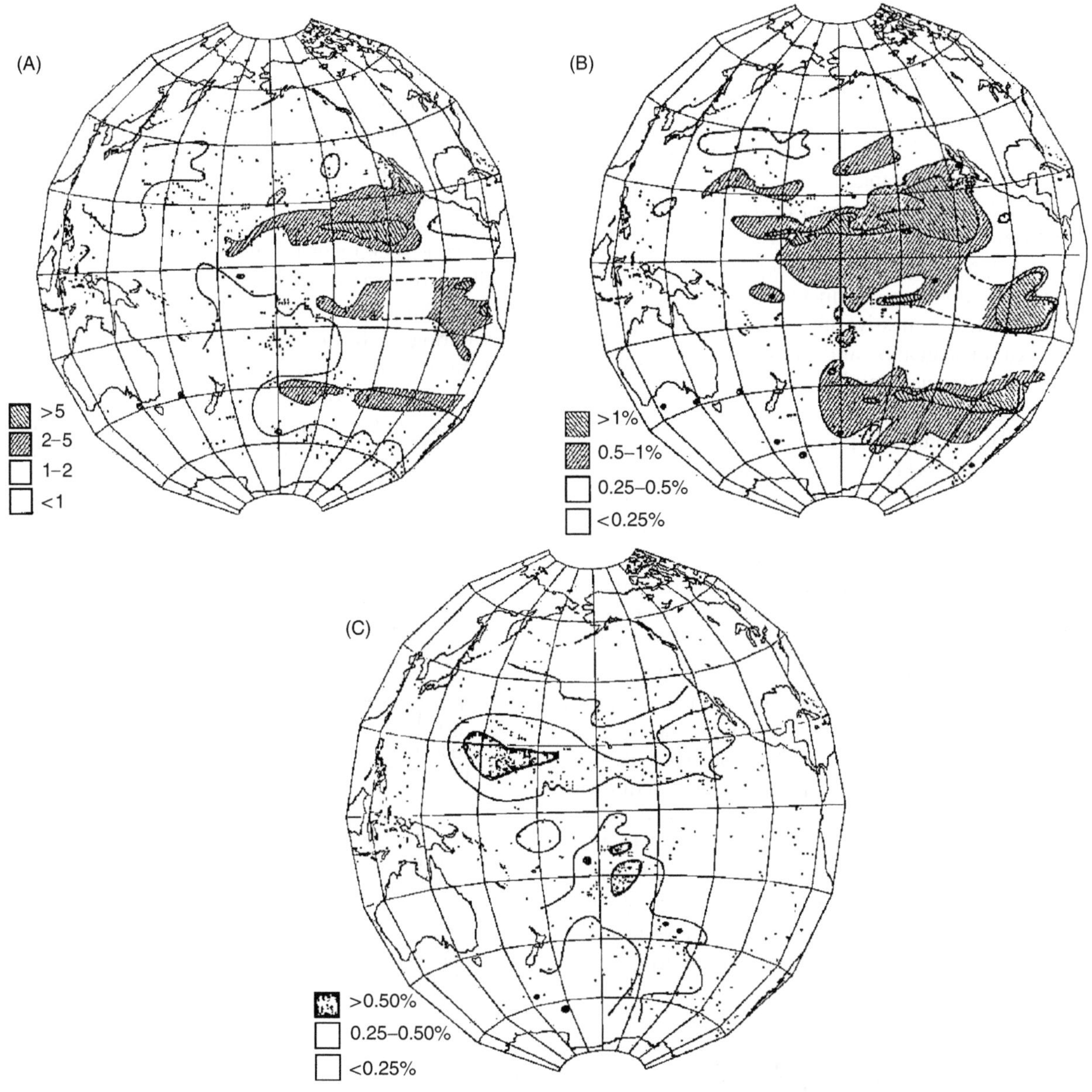

Figure 4 (A) Distribution of Mn/Fe ratios in ferromanganese nodules from the Pacific. (B) Distribution of nickel concentrations in ferromanganese nodules from the Pacific. Copper shows a similar distribution. (C) Distribution of cobalt concentrations in ferromanganese nodules from the Pacific. (After Calvert SE (1978) In: *Sea Floor Development: Moving into Deeper Water*. London: The Royal Society.)

(see Phosphorites below), whereas the higher silicon and aluminum contents of nodules reflect their origin within loose, aluminosilicate-rich sediments of the abyssal seafloor. The mechanism for cobalt enrichment is oxidative scavenging, whereby dissolved Co^{2+} is oxidized to Co^{3+} on the manganese oxide surface. This mechanism also explains the relative enrichments of cerium, and possibly some of the lead and titanium enrichments in crusts. Lower concentrations of iron (and arsenic, an oxyanion in sea water scavenged by positively charged FeOOH; MnO_2 surfaces are negatively charged at sea water pH) in nodules reflect a greater diagenetic component that favors mobilization of manganese over iron. Because of their extremely slow growth in areas relatively free of detrital input, crusts also accumulate large amounts of cosmogenic debris. Enrichments of platinum and PGE metals can be partially explained by this source, but several lines of geochemical evidence indicate that the majority of platinum and the other PGE (Ir, Os, Pd, Rh, Ru) are scavenged from sea water. Oxidation and reduction mechanisms have been proposed for both incorporation and postdepositional remobilization of the PGE in crusts and nodules.

Phosphorites

Submarine phosphorites are principally composed of CFA or francolite with usually large admixtures of detrital impurities from the sediments in which they form or replace during authigenesis. Concentrations of P_2O_5 commonly range from 5 to 28 wt%, reaching values of 35 wt%, with minor element enrichments of strontium, yttrium, and the REE. There are three principal types of phosphorite deposits in the world's oceans: offshore or continental shelf-slope nodules and concretions; massive insular deposits originating from sea bird guano; and 'seamount' phosphorites that often associate with ferromanganese crusts as layers, veins, impregnations, and substrate material (**Figure 5**). Submarine phosphorites were originally thought to be fossil deposits no younger than Miocene age, like those marine phosphorites mined on land, but were subsequently found to be actively forming in shelf-slope areas off Western South America, South-west Africa, and Australia. Studies of the deposits off Peru and Chile have demonstrated their contemporaneous origin near anoxic sediments where the OMZ impacts the continental shelf-slope there at between 200 and 400 m (**Figure 6**). Upwelling currents bring waters rich in dissolved PO_4 onto the continental slope and shelf, increasing surface productivity and enhancing the reduction potential in sediments underlying the OMZ. Bacterial consumption of organic matter within the sediment releases $HPO_4{}^{2-}$ that can combine with dissolved Ca^{2+}, F^- and $CO_3{}^{2-}$ in pore waters to form authigenic precipitates of CFA. This process is thought to occur in sediments near the boundaries of the OMZ where the hydrogen ion concentrations (low pH) produced during bacterial consumption and in CFA precipitation are diluted by relatively alkaline (high pH) sea waters. Associated sediment diagenetic reactions involving dolomite formation and Mg^{2+} exchange in clays reduce the concentration of apatite precipitation inhibiting Mg^{2+} in pore waters, aiding CFA formation.

The distribution of phosphorites throughout the Phanerozoic rock record is highly irregular, correlating broadly with periods of warm climate and higher sea level. Off South America and South-west Africa, U-series dating of the deposits has demonstrated their correlation with warm, interglacial periods back to 150 000 years before present. Increased chemical weathering of the continents during warm periods may provide enhanced PO_4 fixation into hemipelagic

Figure 5 Photograph of a ferromanganese crust on a phosphorite substrate recovered from the summit of Schumann seamount, north of Kauai, Hawaii. Note the inclusions of altered basalt clasts from the seamount within the phosphorite, which has likely replaced limestone. (From G. McMurtry and D. L. VonderHaar, unpublished data.)

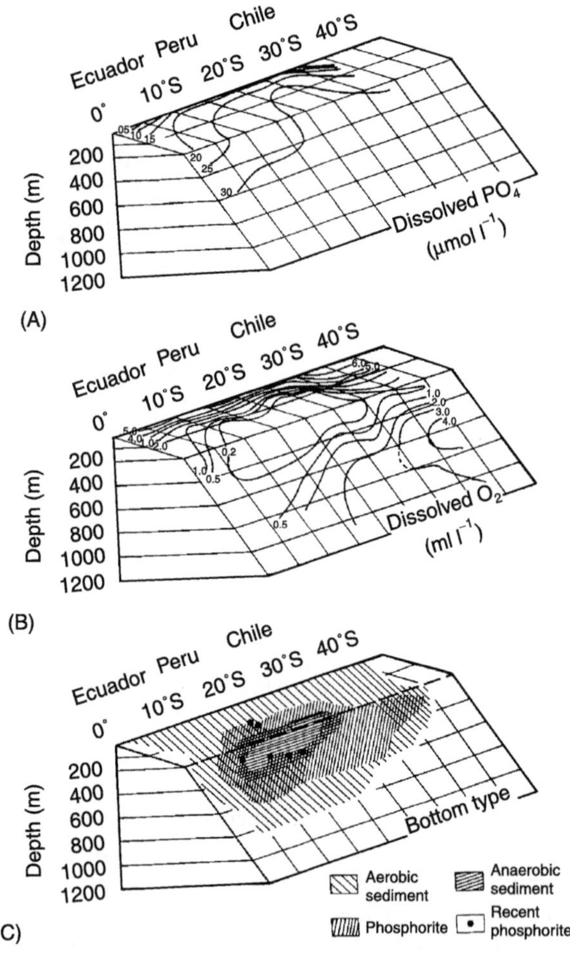

Figure 6 Schematic figure of the western South American shelf and slope, showing contours of dissolved phosphate and oxygen in the bottom waters and the locations of aerobic and anaerobic sediments, phosphorite and Recent phosphorite deposits. (After Burnett WC, Veeh HH and Soutar A (1980) U-series, oceanographic and sedimentary evidence in support of recent formation of phosphate nodules off Peru. In: *Marine Phosphorites*, SEPM Special Publ. no. 29, pp. 61–72.)

sediments. Concurrent encroachment of the OMZ onto the continental shelf-slope during marine transgressions increases the extent of organic matter preservation. When combined, these effects are thought to promote initial phosphorite formation. Subsequent marine regressions may enhance the initial enrichment by winnowing detritus from older phosphatic sediments, producing deposits of economic value.

Insular phosphorites are relatively easily understood deposits from guano in the nesting areas of ancient sea birds that created the ore deposit by transporting and further biologically concentrating scarce dissolved phosphate in surface sea waters within the flight radius of the island. The island of Nauru in the central Equatorial Pacific is a prime example. Although some are presently submerged, these deposits have all formed subaerially and are no older than Tertiary.

Comparatively less is known about the formation of seamount phosphorites. Substrate samples show cryptocrystalline CFA matrix replacement of preexisting limestones that often contain altered basalt breccia and other evidence of high-energy deposition upon the summits and upper slopes of Cretaceous seamounts in the Pacific (**Figure** 5). Within ferromanganese crusts, these deposits show the effects of lowered redox conditions (e.g., platinum remobilization) with impregnation and replacement of older ferromanganese oxides. Strontium and oxygen isotope-derived ages of CFA formation appear to center on the Eocene–Oligocene (36 Ma) and Oligocene–Miocene (24 Ma) boundaries, with some evidence for a minor Middle Miocene event at about 15 Ma. Paleotracking of the Pacific Cretaceous seamounts shows that many were close to the equatorial high productivity belt during the late Cretaceous and Paleocene–Eocene periods, where their then-shallower summits and slopes could have intersected the OMZ in a region of equatorial upwelling. Later depositional episodes would require a greatly expanded and more intense OMZ than the present one.

Marine Barite and Authigenic Silicates

Marine Barite

Barite ($BaSO_4$) is a widespread mineral in deep-sea sediments, varying between 1 and 10 wt% on a carbonate-free basis. It is the predominant barium phase in the ocean. $BaSO_4$ is known to compose a solid solution series with $SrSO_4$ as celestobarite in the skeletal portions of some marine organisms (i.e., the *Xenophyophoria*) and is often found in association with marine organic matter, such as in suspended particles and fecal pellets. Barite is also a well-known gangue mineral in hydrothermal deposits on land and has more recently been found as a principal component of moderate-temperature hydrothermal chimneys on mid-ocean ridges and volcanically active seamounts. Early work on the distribution of marine barite in deep-sea sediments had difficulty distinguishing the relative importance of hydrothermal sources from continental weathering as cycled through the marine biosphere. Distribution maps of marine barite in Pacific deep-sea sediments and the more quantitative barium accumulation rate (**Figure** 7) show both an association with the equatorial high productivity zone and the East Pacific Rise–Bauer Basin areas that are heavily influenced by metalliferous hydrothermal deposition. The key question is how much of the relatively dispersed, fine-grained (usually $< 2\,\mu m$) marine barite particles in deep-sea sediments result from hydrothermal plume fallout and bottom current redistribution of ridge-crest metalliferous sediments versus those carried to the seafloor in the particle rain from surface productivity.

The association of marine barite with organic matter complicates the interpretation of occurrence of increased barite deposition found along midocean ridges because greater preservation of organic matter also occurs with increased carbonate sedimentation above the calcite compensation depth. Additionally, basin-scale dispersal of hydrothermal particles appears limited, especially for the relatively dense barite. Studies of marine barite saturation show that barite is below saturation in the water column but rapidly approaches saturation in the pore waters of deep-sea sediments. Discrete barite particles are nevertheless found in the microenvironments of suspended

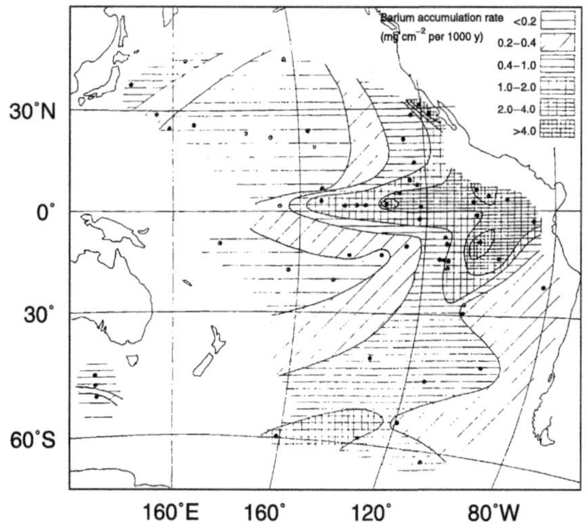

Figure 7 Distribution of barium accumulation rates (units of mg cm^{-2} per 1000 years) in the Pacific sediments. (After Boström K, Joensuu O, Moore C *et al.* (1973), Geochemistry of barium in pelagic sediments. *Lithos 6*: 159–174.)

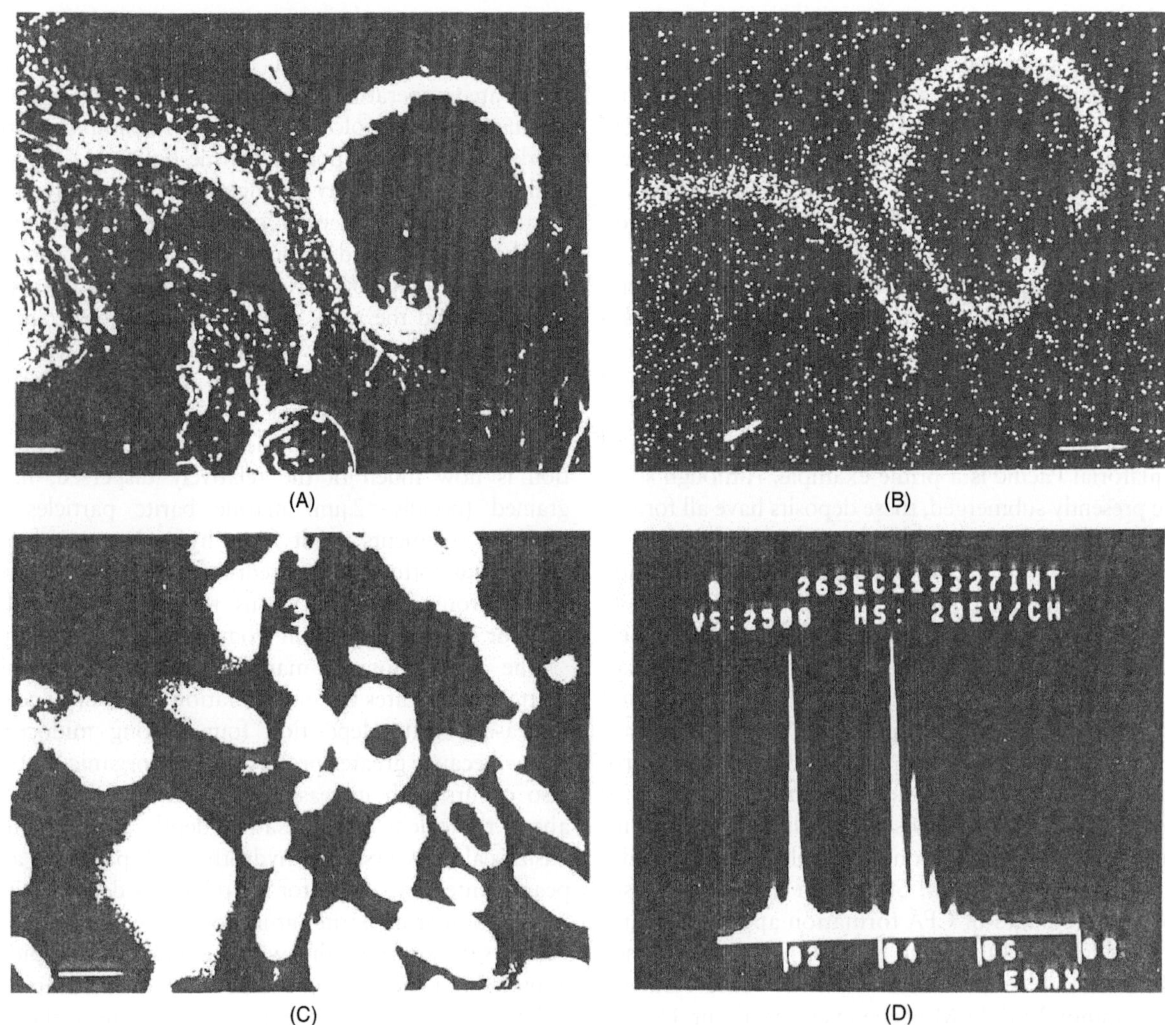

Figure 8 Photograph of barite crystals inside a ferromanganese nodule. (A) Scanning electron micrograph of foraminifera shell walls that have been replaced by barite. (B) Barium elemental X-ray map produced by EMP. (C) Magnification of (A) showing individual barite crystals. (D) Elemental spectrum of the barite crystals: S Kα at 2300 eV; Ba Lα, β, γ at 4460 eV. (After Lalou C, Brichet E, Poupeau G, Romany P and Jehanno C (1979) Growth rates and possible age of a North Pacific manganese nodule. In Bischoff JL and Piper DZ (eds) *Marine Geology and Oceanography of the Central Pacific Manganese Nodule Province*, pp. 815–834. New York and London: Plenum Press.)

biogenic matter and in fecal pellets that rapidly fall to the seafloor; their dissolution on the seafloor and in transit is likely the dominant control on the barium concentration in the deep ocean. Within the sediment and in ferromanganese deposits, barite can form authigenically as discrete particles and diagenetically as replacement of biogenic skeletal remains (**Figure 8**).

Authigenic Silicates

Zeolites Low-temperature alteration of basalt and metastable volcanic glasses on pillow rims, in layers, and dispersed throughout deep-sea sediments can produce a variety of diagenetic alteration phases. (Low temperature is defined as ranging from modern ambient seafloor conditions, or near 0°C to up to 150°C, the low-temperature metamorphism limit.) These include X-ray-amorphous palagonite, the zeolites phillipsite, clinoptilolite, analcite (plus several others of rarer occurrence), smectites, and authigenic K-feldspar (**Table 1**). Phillipsite is a hydrated potassium- and sodium-rich aluminosilicate that forms elongated and sometimes twinned crystals from 8 to 250 μm length with numerous inclusions that indicate rapid growth. Oxygen isotope values of 34‰ for phillipsite indicate formation in the marine environment at modern seafloor temperatures. Phillipsite appears metastable on geological timescales, becoming rarer beyond Cenozoic age. The mineral is found globally at the sediment–sea water interface and continues to grow within the sediment column until it dissolves and disappears in the deepest

Table 1 Summary of observed and suggested diagenetic reactions in deep-sea sediments

Opal-A → opal-CT → quartz
Opal-A → quartz
Basaltic glass → palagonite
Volcanic glass → smectite minerals
Volcanic glass → zeolites
Basalt → smectite minerals
Basalt + H_4SiO_4 → smectite minerals
Basaltic glass + H_4SiO_4 → phillipsite
Poorly cyrstalline smectite → phillipsite
Andesitic and rhyolitic glass + H_4SiO_4 → clinoptilolite
Andesitic and rhyolitic glass → smectite minerals
Phillipsite + H_4SiO_4 → clinoptilolite
Phillipsite + smectite + H_4SiO_4 → clinoptilolite + palygorskite
Opal-A + $Al(OH)_4^-$ + K^+ → clinoptilolite
Opal-A + $Al(OH)_4^-$ + K^+ → opal-CT + clinoptilolite
Plagioclase + K^+ + H_4SiO_4 → K-feldspar + Na^+ + Ca^{2+}
→ K-feldspar
? → albite
Clinoptilolite + Na^+ → analcite + K^+ + quartz
Clinoptilolite → K-feldspar + quartz
Clinoptilolite + Na^+ → analcite + K-feldspar + quartz
H_4SiO_4 + Mg^{2+} + $Al(OH)_4^-$ → palygorskite
H_4SiO_4 + Mg^{2+} → sepiolite
Volcanic glass + Mg^{2+} + H_4SiO_4 → palygorskite
Smectite + Mg^{2+} + H_4SiO_4 → palygorskite
Clinoptilolite + palygorskite + calcite → K-feldspar + dolomite + quartz
Clinoptilolite + sepiolite + calcite → K-feldspar + dolomite + quartz
Amorphous hydroxides (mainly Fe) + H_4SiO_4 + Mg^{2+} → Fe-montmorillonite
Nontronite + Mg^{2+} + reduced sulfur → saponite + FeS_2
Dissolved silica adsorption by clay minerals
Amorphous aluminosilicate reconstitution?
Smectite → mixed-layer illite–smectite?

After Kastner (1981).

drill holes (usually below 500 m depth). Marine clinoptilolite forms finer grained (<45 μm), platy crystals of relatively silicon-rich hydrated potassium- and sodium-rich aluminosilicate (high Si/Al ratio). Clinoptilolite is more frequently encountered at depths >100 m in the sediment column and persists to great depths and geological age, suggesting that it is thermodynamically stable in the deep-sea. Analcite is a rarer, sodium-rich zeolite in deep-sea sediments that displays a general trend of increasing abundance with geological age which parallels that of clinoptilolite. Both clinoptilolite and analcite can form directly from alteration of more siliceous volcanic glass (andesite, rhyolite, from active island arcs and explosive continental volcanism), by diagenetic reaction of phillipsite and dissolved silica (clinoptilolite) or Na/K exchange with clinoptilolite (analcite).

Palygorskite and sepiolite Authigenic clay minerals (<2 μm particle size) in the deep-sea include the fibrous minerals palygorskite and sepiolite and the smectite family of expandable phyllosilicates. Both palygorskite and sepiolite are rare in recent marine sediments, occuring more often in Eocene and older sediments. Palygorskite is a hydrous silicate containing Mg, Al and Fe^{3+}, whereas sepiolite is almost a pure hydrous magnesium silicate. Pore water solutions of alkaline pH with high concentrations of dissolved silica and magnesium favor the formation of both minerals. Fine fibrous textures and overgrowths of siliceous tests and opal-CT attest to their authigenesis. Reactions range from the diagenetic alteration of (mainly silicic and intermediate) volcanic ash, either directly or indirectly via smectite with biogenic silica, formation upon magnesium release after conversion of biogenic opal-A to opal-CT, and reaction of biogenic silica tests with marine pore waters, including hypersaline brines enriched in magnesium (**Table 1**).

Smectites Most clay minerals in deep-sea sediments are detrital phases from continental weathering. These minerals compose the bulk of the nonbiogenic sediments in the <2 μm fraction and include clay-sized quartz and the phyllosilicates kaolinite, illite, chlorite, and smectite. The smectite

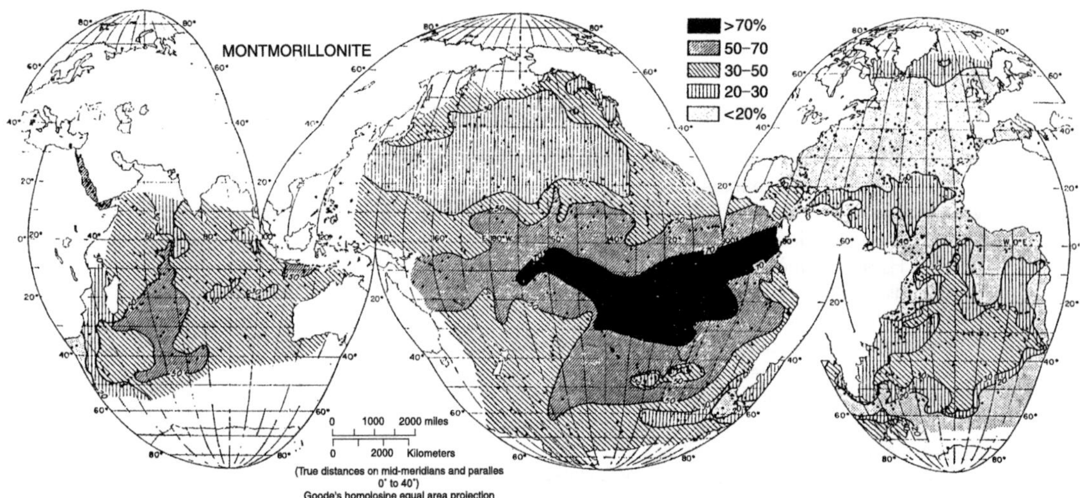

Figure 9 Distribution of montmorillonite (smectite) abundance in the $<2\,\mu m$ fraction of sediments in the world ocean, carbonate-free basis. (After Griffin JJ, Windom H and Goldberg ED (1968) The distribution of clay minerals in the world ocean. *Deep-Sea Research* 15: 433–459.)

group of expandable clay minerals includes these common end-members: magnesium-rich saponite, found mostly as a diagenetic or low-temperature metamorphic product of basalt; iron-rich nontronite, often found as a low-temperature ($<100°C$) hydrothermal deposit; and aluminum-rich beidellite predominantly derived from volcanic ash alteration on land. The surface distribution of marine smectite in the world's oceans shows broad areas, such as in the south-eastern Pacific where $>70\%$ of the mineral's abundance is far from land and blankets the active East Pacific Rise spreading center (**Figure 9**). Most of this smectite is iron-rich montmorillonite. Up to 50% of this mineral has been described as authigenically formed from the seafloor temperature alteration of volcanic glass, with the remaining 50% from detrital sources. Early oxygen isotope studies of deep-sea montmorillonite suggested that much of it formed pedogenically on land, but iron-rich montmorillonite is not typically found in windblown dust. More recent oxygen isotope work has suggested either a low-temperature submarine hydrothermal origin for the iron-montmorillonite or formation from low-

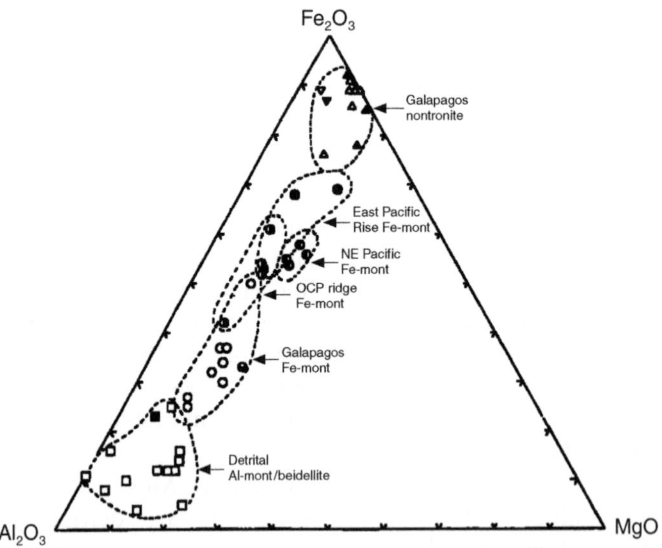

Figure 10 Triangle plot of octahedral Mg–Al–Fe composition for marine smectites. (After McMurtry GM, Wang CH and Yeh HW (1983) Chemical and isotopic investigations into the origin of clay minerals from the Galapagos hydrothermal mounds field. *Geochimica et Cosmochimica Acta* 47: 475–489.)

temperature hydrothermal alteration of ash within subaerial volcanoes that is periodically deposited in the ocean by explosive volcanism. Nontronite and iron-montmorillinite have also been suggested to form at seafloor temperatures by reaction of hydrothermally derived iron oxides with biogenic silica. One of the difficulties in assigning a precise origin to this mineral class is the possibility of physical admixtures of truly authigenic iron-rich smectites with more aluminum-rich smectites derived from the continents (**Figure 10**).

Glossary

Aluminosilicate A silicate containing aluminum in coordination with oxyhydroxides and/or in substitution for silicon in SiO_4 tetrahedra.

Authigenesis New origin. Process of formation of new minerals in place.

Detrital Formed from detritus, usually from rocks, minerals, or sediments from elsewhere than the depositional site.

Diagenesis Changed origin. Recombination or rearrangement of a mineral that results in a new mineral, usually postdepositionally.

Hemipelagic Deep-sea sediment that accumulates near the continental margin, so that the sediment contains abundant continentally derived material and rates of sedimentation are high.

Hypersaline Excessive salinity, much greater than the normal salinity of sea water.

Metalliferous Metal bearing, usually enriched toward economic extraction of the metals.

Metastable State of a phase that is stable toward small disturbance, but is capable of reaction if sufficiently disturbed.

Pedogenesis Soil origin. Mineral formation within the soil.

Pelagic Open ocean environment. A marine sediment with that fraction derived from the continents indicating deposition from a dilute suspension distributed throughout deep-sea water.

Phyllosilicate Layered or sheet silicate mineral, formed by sharing three of the four oxygens in neighboring silicon tetrahedra.

Plankton Aquatic organisms that drift, or swim weakly. Can be either plants (phytoplankton) or animals (zooplankton).

Redox Abbreviation for reduction–oxidation, usually expressed as a potential.

Seamount Underwater mountain, 1000 m or higher elevation from seafloor base. Morphology may be peaked or flat-topped, with the latter called guyot.

Suboxic Condition lacking free oxygen, but not extremely reducing.

Zeolite Any of the minerals of the zeolite group. Aluminosilicate minerals with an open framework structure that allows for easily reversible hydration, gas adsorption, and either cation or anion exchange.

See also

Clay Mineralogy. Hydrothermal Vent Deposits. Platinum Group Elements and their Isotopes in the Ocean. Pore Water Chemistry. Rare Earth Elements and their Isotopes in the Ocean. River Inputs. Tracers of Ocean Productivity. Transition Metals and Heavy Metal Speciation. Uranium-Thorium Decay Series in the Oceans Overview.

Further Reading

Bentor YK (ed.) (1980) *Marine Phosphorites; a Symposium*. Oklahoma: SEPM Special Publication no. 29.

Burns RG and Burns VM (1981) Authigenic oxides. *The Sea*, vol. 7, pp. 875–914. New York: Wiley.

Chamley H (ed.) (1989) *Clay Sedimentology*. Berlin: Springer-Verlag.

Cronan DS (1974) Authigenic minerals in deep-sea sediments. In: Goldberg ED (ed.) *The Sea*, vol. 5, pp. 491–525. New York: Wiley.

Cronan DS (ed.) (1980) *Underwater Minerals*. London: Academic Press.

Cronan DS (ed.) (2000) *Handbook of Marine Mineral Deposits*. Boca Raton, FL: CRC Press.

Glasby GP (ed.) (1977) *Marine Manganese Deposits. Elsevier Oceanography Series*. Amsterdam: Elsevier.

Glenn CR, Prévot-Lucas L, and Lucas J (eds.) (2000) *Marine Authigenesis: from Global to Microbial*. Oklahoma: SEPM Special publication no. 66.

Halbach P, Friedrich G, and von Stackelberg U (eds.) (1988) *The Manganese Nodule Belt of the Pacific Ocean: Geological Environment, Nodule Formation, and Mining Aspects*. Stuttgart: F. Enke Verlag.

Kastner M (1981) Authigenic silicates in deep-sea sediments: formation and diagenesis. In: Emiliani C (ed.) *The Sea*, vol. 7, pp. 915–980. New York: Wiley.

Manheim FT (1986) Marine cobalt resources. *Science* 232: 600–608.

Margolis SV and Burns RG (1976) Pacific deep-sea manganese nodules: their distribution, composition, and origin. *Annual Review of Earth and Planetary Science* 4: 229–263.

CALCIUM CARBONATES

L. C. Peterson, University of Miami, Miami, FL, USA

Introduction

The ocean receives a continual input of calcium from riverine and groundwater sources and from the hydrothermal alteration of oceanic crust at mid-ocean ridge spreading centers. Balancing this input is the biological precipitation of calcium carbonate ($CaCO_3$) by shell-and skeleton-building organisms in both shallow marine and open-ocean environments. In the deep sea, the primary contributors to the carbonate budget of open-ocean sediments are the skeletal remains of calcareous plankton that have settled down from the surface after death. Seafloor sediments consisting of more than 30% by weight calcium carbonate are traditionally referred to as calcareous or carbonate ooze; such oozes accumulate at the rate of 1–4 cm per 1000 years and cover roughly half of the ocean bottom. Carbonate oozes are the most widespread biogenous sediments in the ocean.

While the biological production of calcium carbonate in oversaturated surface waters determines the input of carbonate to the deep sea, it is the dissolution of carbonate in undersaturated deep waters that has the dominant control on calcium carbonate accumulation in the open ocean. Since carbonate production rates in the surface ocean today greatly exceed the rate of supply of calcium, this 'compensation' through dissolution must occur in order to keep the system in steady-state. Increased dissolution at depth is largely a function of the effect of increasing hydrostatic pressure on the solubility of carbonate. However, superimposed on this bathymetric effect are regional preservation patterns related to differences in carbonate input and the carbonate chemistry of deep water masses. Carbonate oozes in the deep sea serve as a major reservoir of calcium and carbon dioxide on the Earth's surface. Their spatial and temporal accumulation patterns in the marine stratigraphic record are thus a primary source of data about the carbonate chemistry and circulation of past oceans, as well as of the global geochemical cycle of CO_2.

Carbonate Producers

The most important carbonate producers in the open ocean are planktonic coccolithophorids and foraminifera, unicellular phytoplankton and zooplankton respectively, which inhabit the upper few hundred meters of the water column (**Figure 1**). Coccolithophorids are the dominant carbonate-precipitating organisms on Earth. During part of their life cycle, they produce a skeletal structure (the coccosphere) consisting of loosely interlocking plates, often button-like in appearance, known as coccoliths. Deep-sea carbonates generally contain only the individual coccoliths, as the intact coccospheres are rarely preserved. Foraminifera produce a calcareous shell, or 'test', a few hundred microns in size that sinks after death or reproduction to the sea floor. Both coccolithophorids and the planktonic foraminifera construct their skeletal elements out of the mineral calcite, the more stable polymorph of $CaCO_3$. Calcareous sediments dominated by one or the other component are termed coccolith oozes or foraminiferal oozes, although in reality most carbonate-rich sediments are a mixture of both.

Coccolithophorids made their first appearance in the geological record in the earliest Jurassic, while planktonic foraminifers evolved somewhat later in the middle Jurassic. The appearance of these two dominant pelagic carbonate producers, and their rapid diversification in the Cretaceous, would have had major effects upon the carbonate geochemistry of the oceans. Before this, most carbonate was deposited in shallow seas, accounting for the high proportion of limestones among older rocks on the continents. Since the Mesozoic, deep-ocean basins have become enormous sinks for carbonate deposition.

Smaller contributions to the deep-sea carbonate budget come from a variety of other sources. Pteropods, free-swimming pelagic gastropods, construct a relatively large (several millimeters) but delicate shell out of the metastable form of $CaCO_3$ known as aragonite. However, while pteropods can be unusually abundant in certain environments, the increased solubility of aragonite leads to very restricted preservation of the shells and pteropod oozes are relatively rare in the ocean. In the vicinity of shallow, tropical carbonate platforms such as the Bahamas or Seychelles Bank, shedding of aragonitic bank-top sediments derived from algal and coral production can lead to aragonite-rich 'periplatform oozes' in deep waters around the perimeters of the platform.

(A)

(B)

Figure 1 Carbonate oozes in the deep sea are dominated by the skeletal remains of (A) planktonic foraminifera (× 50 magnification) and (B) coccolithophorids (× 6000 magnification). Specimens shown here were isolated from a Caribbean sediment core.

In general, contributions from bottom-dwelling organisms (e.g. benthonic foraminifera, ostracods, micromollusks) are negligible in deep-sea sediments.

Carbonate Distribution and Dissolution

The distribution of carbonate sediments in the ocean basins is far from uniform (**Figure 2**). If it were possible to drain away all of the ocean's water, carbonate oozes would be found draped like snow over the topographic highs of the seafloor and to be largely absent in the deep basins. The lack of carbonate-rich sediments in the deepest parts of the world's oceans has been recognized since the earliest investigations. Although surface productivity and dilution by noncarbonate sediment sources can locally influence the concentration of carbonate in

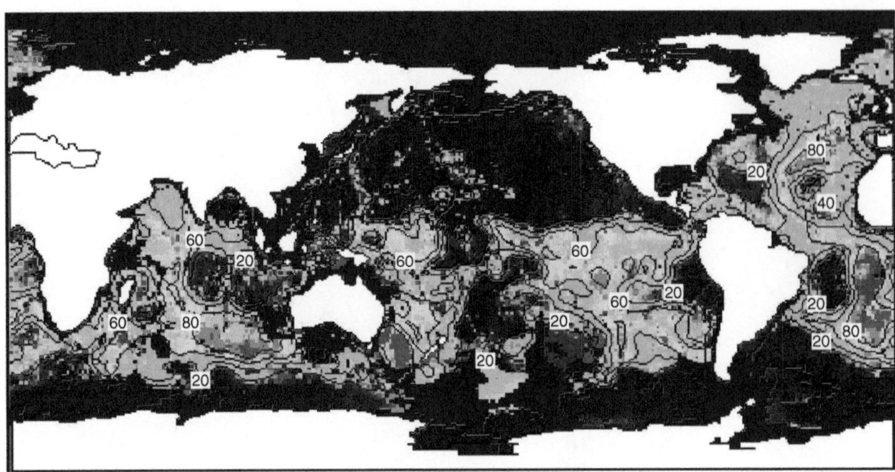

Figure 2 Global distribution of calcium carbonate (weight-% $CaCO_3$) in surface sediments of the ocean. Data compilation from Archer (1996); reproduced with permission from the American Geophysical Union.

deep-sea sediments, the clear-cut relationship between calcium carbonate content and water depth indicates that carbonate dissolution plays the major role in governing carbonate distribution patterns. To a first approximation, the dissolution of carbonate on the seafloor is a function of the corrosiveness or saturation state of the overlying bottom waters.

The amount of calcium carbonate that will dissolve in sea water if thermodynamic equilibrium is reached is governed by the following reaction:

$$CaCO_3(s) \leftrightarrow Ca^{2+}(aq) + CO_3{}^{2-}(aq)$$

At equilibrium, the rate of carbonate dissolution is equal to the rate of its precipitation and the sea water is said to be saturated with respect to the carbonate phase. In the deep sea, the degree of calcium carbonate saturation (D) can be expressed as:

$$D = \frac{[Ca^{2+}]_{seawater} \times [CO_3{}^{2-}]_{seawater}}{[Ca^{2+}]_{saturation} \times [CO_3{}^{2-}]_{saturation}}$$

where $[Ca^{2+}]_{seawater}$ and $[CO_3{}^{2-}]_{seawater}$ are the *in situ* concentrations in the water mass of interest and $[Ca^{2+}]_{saturation}$ and $[CO_3{}^{2-}]_{saturation}$ are the concentrations of these ions at equilibrium, or saturation, at the same conditions of pressure and temperature. Since shell formation and dissolution cause the concentration of $[Ca^{2+}]$ to vary by less than 1% in the ocean, the degree of calcium carbonate saturation (D) can be simplified and expressed in terms of the concentration of the carbonate ions only:

$$D = \frac{[CO_3{}^{2-}]_{seawater}}{[CO_3{}^{2-}]_{saturation}}$$

D is thus a measure of the degree to which a seawater sample is saturated with respect to calcite or aragonite, and so provides a measure of the strength of the driving force for dissolution. Values of $D > 1$ indicate oversaturation while values of $D < 1$ indicate undersaturation and a tendency for calcium carbonate to dissolve. Since the saturation carbonate ion concentration increases with increasing pressure and decreasing temperature, calcium carbonate is more soluble in the deep sea than at the surface. At the depth in the water column where $D = 1$, the transition from oversaturated to undersaturated conditions is reached. This depth is known as the saturation horizon (**Figure 3**). Aragonite is always more soluble than calcite, and its respective saturation horizon is shallower, because the saturation carbonate ion concentration for aragonite is always higher for the same conditions of pressure and temperature.

Observations from studies of surface sediments have allowed definition of regionally varying levels in the ocean at which pronounced changes in the presence or preservation of calcium carbonate result from the depth-dependent increase of dissolution on the seafloor. The first such level to be identified was simply the depth boundary in the ocean separating carbonate-rich sediments above from carbonate-free sediments below. This level is termed the calcite (or carbonate) compensation depth (CCD) and represents the depth at which the rate of carbonate dissolution on the seafloor exactly balances the rate of carbonate supply from the overlying surface waters. Because the supply and dissolution rates of carbonate differ from place to place in the ocean, the depth of the CCD is variable. In the Pacific, the CCD is typically found at depths between about 3500 and 4500 m. In the North Atlantic and parts of the South Atlantic, it is found

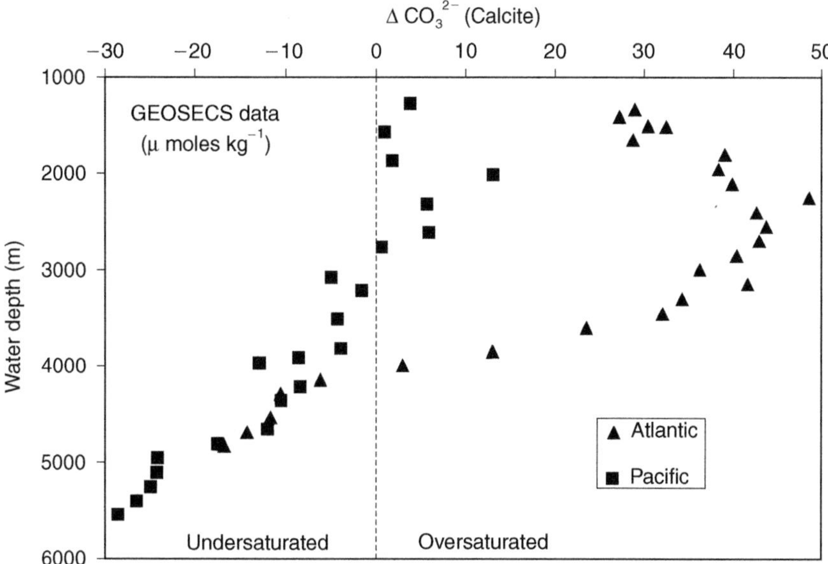

Figure 3 Bathymetric profiles of calcium carbonate (calcite) saturation for hydrographic stations in the Atlantic and Pacific Oceans (data from Takahashi *et al.* 1980). Carbonate saturation here is expressed as ΔCO_3^{2-}, defined as the difference between the *in situ* carbonate ion concentration and the saturation carbonate ion concentration at each depth $\Delta CO_3^{2-} = [CO_3^{2-}]_{seawater} - [CO_3^{2-}]_{saturation}$. The saturation horizon corresponds to the transition from waters oversaturated to waters undersaturated with respect to calcite ($\Delta CO_3^{2-} = 0$). This level is deeper in the Atlantic than in the Pacific because Pacific waters are CO_2-enriched and $[CO_3^{2-}]$-depleted as a result of thermohaline circulation patterns and their longer isolation from the surface. The Atlantic data are from GEOSECS Station 59 (30°12′S, 39°18′W); Pacific data come from GEOSECS Station 235 (16°45′N,161°23′W).

closer to a depth of about 5000 m. Close to continental margins the CCD tends to shoal, although much of this apparent rise can be attributed to carbonate dilution by terrigenous input from the continents. Rarely does carbonate ooze accumulate on seafloor that is deeper than about 5 km.

In practice, the CCD is identified by the depth transition from carbonate ooze to red clay or siliceous ooze that effectively defines the upper limit of the zone of no net $CaCO_3$ accumulation on the seafloor. Given the practical difficulty (e.g. analytical precision, redeposition) of determining the depth level at which the carbonate content of sediment goes to zero, some investigators choose instead to recognize a carbonate critical depth (CCrD), defined as the depth level at which carbonate contents drop to $<10\%$ of the bulk sediment composition. The CCrD lies systematically and only slightly shallower than the CCD. A similar boundary to the CCD can be recognized marking the lower depth limit of aragonite-bearing sediment in the ocean, the aragonite compensation depth or ACD. Because of the greater solubility of aragonite as compared with calcite, the ACD is always much shallower than the CCD.

Above the CCD, the level at which significant dissolution of carbonate first becomes apparent is called the lysocline. As originally defined, the term lysocline was used to describe the depth level where a pronounced decrease in the preservation of foraminiferal assemblages is observed. It thus marks a facies boundary separating well-preserved from poorly preserved assemblages on the seafloor. This level is now more specifically referred to as the foraminiferal lysocline to differentiate it from the coccolith lysocline and pteropod lysocline, which may differ in depth because of varying resistance to dissolution or differences in solubility (in the case of the aragonitic pteropods). In addition, it is customary to recognize a sedimentary or carbonate lysocline as the depth at which a noticeable decrease in the carbonate content of the sediment begins to occur.

In theory, the lysocline records the sedimentary expression of the saturation horizon, that is the depth-dependent transition from waters oversaturated to waters undersaturated with respect to carbonate solubility (**Figure** 4). The lysocline thus marks the top of a depth zone, bounded at the bottom by the CCD, over which the bulk of carbonate dissolution in the ocean is expected to occur in response to saturation state-driven chemistry. The thickness of this sublysocline zone, as indicated by the vertical separation between the lysocline and CCD, is variable and is governed by the rate of carbonate supply, the actual dissolution gradient, and

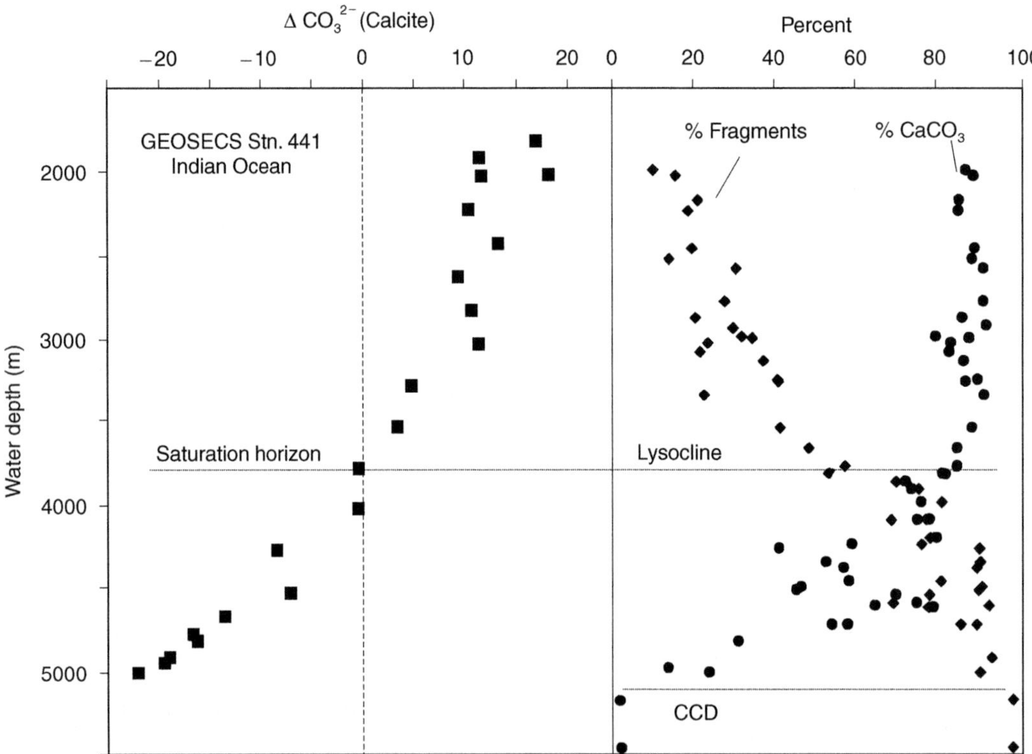

Figure 4 Comparison of carbonate saturation profile for the eastern equatorial Indian Ocean with measurements of foraminiferal fragmentation and carbonate content (weight-%) from depth-distributed modern sediment samples in this region. The saturation horizon with respect to calcite ($\Delta CO_3^{2-} = 0$) occurs locally in the water column at a depth of 3800 m. This level corresponds with both the foraminiferal lysocline and carbonate lysocline as recognized in the sediments. The carbonate compensation depth (CCD) in this region is found at a depth of approximately 5000 m. Increased foraminiferal fragmentation and decreases in sedimentary carbonate content are the result of dissolution and carbonate loss below the lysocline. Carbonate saturation data are from GEOSECS Station 441 (5°2′S, 91°47′E; Takahashi *et al.* 1980); modern sediment data are from Peterson and Prell (1985).

potentially by noncarbonate dilution in certain regions of the ocean.

While the term lysocline was originally used to define a preservational boundary, it has also been used in a fundamentally different sense to denote the depth at which dissolution rates of carbonate on the seafloor greatly accelerate. Whether these levels may or may not coincide, and the nature of their relationship to the saturation horizon or 'chemical lysocline', has been the subject of much discussion and debate. One of the reasons for uncertainty in this regard is the fact that both the carbonate content (%) of a sediment sample and the preservation of the calcareous microfossil assemblages there in can be surprisingly poor indicators of the extent to which dissolution has occurred. For example, the loss of carbonate (L) from sediment, expressed as a weight percentage of the total sediment, is given by:

$$L = 100(1 - R_o/R)$$

where R_o and R are the initial and final values of the noncarbonate (or residual) material. Thus, for a sample initially containing 95% carbonate and a R_o value of 5%, 50% of the carbonate in the sample must be dissolved in order to double the noncarbonate fraction and reduce the carbonate content to 90%. Since the carbonate fraction of the pelagic rain in the open ocean often approaches 95%, this inherent insensitivity means that significant loss of carbonate can occur before detectable changes in the carbonate content are observed. As a consequence, the carbonate lysocline, traditionally defined as the level where the carbonate content of sediments begins to sharply decrease with water depth, may lie deeper than the depth at which significant loss of carbonate to dissolution actually begins to occur.

Dissolution leads to an increase in surface area during the etching of carbonate skeletal material. Etching produces roughness and widens pores, leading to weakening and ultimately to breakage. Because of their larger size, planktonic foraminifera have usually been the subject of dissolution studies that focus on the preservation state of the microfossils themselves. Planktonic foraminifera have a wide range of morphologic characteristics that enhance their abilities to remain suspended in the upper water column while alive. These same characteristics largely dictate their resistance to dissolution after death. Taxa living in warm, tropical surface waters, where density is generally low, tend to be open-structured with thin shells and porous walls. Taxa that live deeper in cooler, denser subsurface waters, or in colder surface waters at high latitudes, tend to be more heavily calcified with thicker shells and small or closed up pores. On the seafloor, the thin-shelled, more fragile species tend to dissolve more readily than the robust taxa. In effect, this means that individual species each have their own 'lysocline', which can be offset shallower or deeper from the foraminiferal lysocline determined from the total assemblage. There are additional consequences of this selective preservation of taxa that must be considered in paleo-oceanographic or paleoclimatic studies. For example, the selective preservation of more heavily calcified taxa tends to impart a generally 'cooler' appearance to the overall microfossil population and can bias attempts to derive paleotemperature information from seafloor assemblages, as well as other population properties such as diversity.

For carbonate particles produced in the upper ocean, settling rates play an important role in their distribution and preservation. Smaller planktonic foraminifers settle at about 150–250 $\mathrm{m\,d^{-1}}$, while larger ($>250\,\mu\mathrm{m}$) foraminifers may settle as much as 2000 $\mathrm{m\,d^{-1}}$. These rates are rapid enough that little dissolution is thought to occur in the water column. Solitary coccoliths, on the other hand, sink at rates of 0.3 to $\sim 10\,\mathrm{m\,d^{-1}}$, slow enough that dissolution within the water column should theoretically prevent their ever reaching the ocean bottom. However, sediment trap studies have shown that transport by fecal pellets is the dominant process by which small phytoplankton skeletons are transferred to the seafloor. Protection offered by the organic fecal pellet covering may also protect the coccoliths after deposition and account for the fact that the coccolith lysocline is generally observed to lie somewhat deeper than the foraminiferal lysocline.

While the seafloor depths of the lysocline and CCD can be readily identified from sedimentary criteria, this information is of limited use without realistic knowledge of the rates at which calcium carbonate is lost from the sediments to dissolution. In practice, it is much easier to determine carbonate accumulation in the deep sea than it is to estimate carbonate loss. Yet the latter information is clearly needed in order to close sediment budgets and to reconstruct changes in the carbonate system.

Carbonate-rich sediments deposited above the saturation horizon should experience little in the way of saturation-driven dissolution because they lie in contact with waters oversaturated with respect to calcite. Nevertheless, evidence for significant supra-lysoclinal dissolution has been found in a number of studies. Much of this dissolution at shallower water depths is thought to be driven by chemical reactions associated with the degradation of organic carbon in the sediments. Organic carbon arriving at the seafloor is generally respired as CO_2 or remineralized to other organic compounds by benthic organisms. The metabolic CO_2 generated by organisms that live within the sediment can contribute to the dissolution of calcite even above the lysocline by increasing the chemical corrosivity of the pore waters. Studies of organic matter diagenesis in deep-sea sediments suggest that rates of supralysoclinal dissolution vary greatly with location, ranging from minimal loss to $>40\%$ calcite loss by weight. Temporal and spatial changes in the rain rate of organic carbon relative to carbonate can affect this process.

Whether above or below the lysocline, carbonate dissolution is mostly confined to the bioturbated surface sediment layer (typically ≤ 10 cm in the deep sea). As carbonate is depleted from this bioturbated layer, older 'relict' carbonate is entrained from the sediments below. This results in 'chemical erosion' and can produce substantial hiatuses or gaps in the record. Dissolution, and hence erosion, eventually stops when nonreactive materials fill up the mixed layer and isolate the underlying sediment from the overlying water. Many clay layers interbedded within carbonate-rich sequences are likely produced by this mechanism; the resulting lithologic contrasts often show up as subsurface seismic horizons which can be traced for long distances and tell a story of changing dissolution gradients and carbonate chemistry in the past.

Basin-to-basin Fractionation in the Modern Ocean

Superimposed on the general depth-dependent decrease of carbonate accumulation observed everywhere in the deep sea are preservation patterns that differ between the major ocean basins. Today, carbonate-rich sediments tend to accumulate in the Atlantic Ocean, while more carbonate-poor sediments are generally found at comparable water depths in the Indian and Pacific Oceans. This modern pattern is largely the product of the ocean's thermohaline circulation and has been termed 'basin-to-basin fractionation'. In the Atlantic, deep and bottom waters tend to be produced at high latitudes because cold temperatures and high sea surface salinities lead to the formation of dense water masses that sink and spread at depth. These young, relatively well oxygenated and [CO_3^{2-}]-enriched waters tend to depress the depth of the saturation horizon and allow carbonate to accumulate over much of the Atlantic basin, as manifested by a deep lysocline and CCD. In contrast, neither the Indian nor Pacific Oceans today experience surface conditions that allow deep or bottom waters to form; water masses at depth in these basins largely originate in the Atlantic sector as part of what is sometimes described as the ocean's conveyor belt circulation, with a general upwelling of waters from depth balancing the formation and sinking of deep waters in the Atlantic source areas. Since deep and bottom waters in the Indian and Pacific Oceans are further removed from their modern source areas in the Atlantic, they tend to be CO_2-enriched and [CO_3^{2-}]-depleted because of their greater age and the cumulative effects of organic matter remineralization along their flow path. In particular, the *in situ* decrease in [CO_3^{2-}] concentration leads to an increase in undersaturation of the water masses and a progressive shoaling of the saturation horizon (**Figure 3**). Thus, Indian and Pacific deep waters are generally more corrosive to the biogenic carbonate phases than Atlantic waters at comparable depth, the lysocline and CCD are shallower, and a smaller area of the seafloor experiences conditions suitable for carbonate preservation and accumulation. This pronounced modern pattern of basin-to-basin fractionation is illustrated by the fact that roughly 65% of the present Atlantic seafloor is covered by carbonate ooze, while only 54% of the Indian Ocean floor and 36% of the Pacific Ocean floor share that distinction. Naturally, if thermohaline circulation patterns have changed in the past, then carbonate preservation and accumulation patterns will change accordingly. The mapping and reconstruction of such trends has emerged as a powerful paleoceanographic tool.

Temporal Changes in Carbonate Accumulation and Preservation

The patterns of carbonate accumulation and preservation in the deep sea contain important information about the chemistry and fertility of ancient oceans. Numerous studies have now shown that variations in the carbonate system have occurred on a variety of timescales, both within and between ocean basins. On a local or even regional scale, such variations can often be used as a correlation tool. This has come to be known as 'preservation stratigraphy'.

A number of criteria have commonly been used as indicators of the intensity of carbonate dissolution in deep-sea sediments. Variations in the measured carbonate content of sediments are commonly used to correlate between cores in a region, but are difficult to interpret strictly in terms of dissolution and changing deep-water chemistry. This is because the weight percent carbonate content of a sample can also be affected by changing carbonate input (i.e. surface production) and by dilution from noncarbonate sources. More useful are indices based on some direct measure of preservation state, such as the percentage of foraminiferal fragments in a sample relative to whole shells (**Figure 5**). However, while clearly recording dissolution, preservation-based indices can also be affected by other factors, including ecologic changes that may introduce variable proportions of solution-susceptible species into a region over time.

Because carbonate dissolution is a depth-dependent process, it is best studied where existing seafloor topography allows for sampling of sediments over a broad depth range. Given this sampling strategy, one way to circumvent the problems of using measured carbonate content and other relative dissolution indices (e.g. fragmentation) is to calculate carbonate accumulation histories for the individual sampling locations and examine depth-dependent differences in accumulation rates and patterns. To do so requires an accurate knowledge of sedimentation rates (e.g. cm per thousand years) and measurements of sediment bulk density (in $g\,cm^{-3}$), in addition to the data on carbonate content. The product of these three measures yields a mass accumulation rate for the carbonate component expressed in g per cm^2 per thousand years. Differences in accumulation between depth-distributed sites can provide insights into dissolution gradients and carbonate loss.

As the relative importance of calcium supply from weathering and carbonate production vary through time, the depth of the CCD must adjust to control dissolution and to keep calcium levels in balance. Studies of CCD behavior during the Cenozoic (**Figure 6**) have generally shown that CCD fluctuations were similar in the various ocean basins and were likely to have been driven by a global mechanism, such as a change in sea level and/or hypsometry of the ocean basins or a change in supply of calcium to the oceans. There are, however, clear ocean-to-ocean differences in this general pattern that are likely to have been the result of changes in regional productivity and the interbasinal exchange of deep and surface waters. By examining such differences, estimates of past circulation and of the relative differences in carbonate productivity in different regions can be determined from regional offsets in the depth of the CCD.

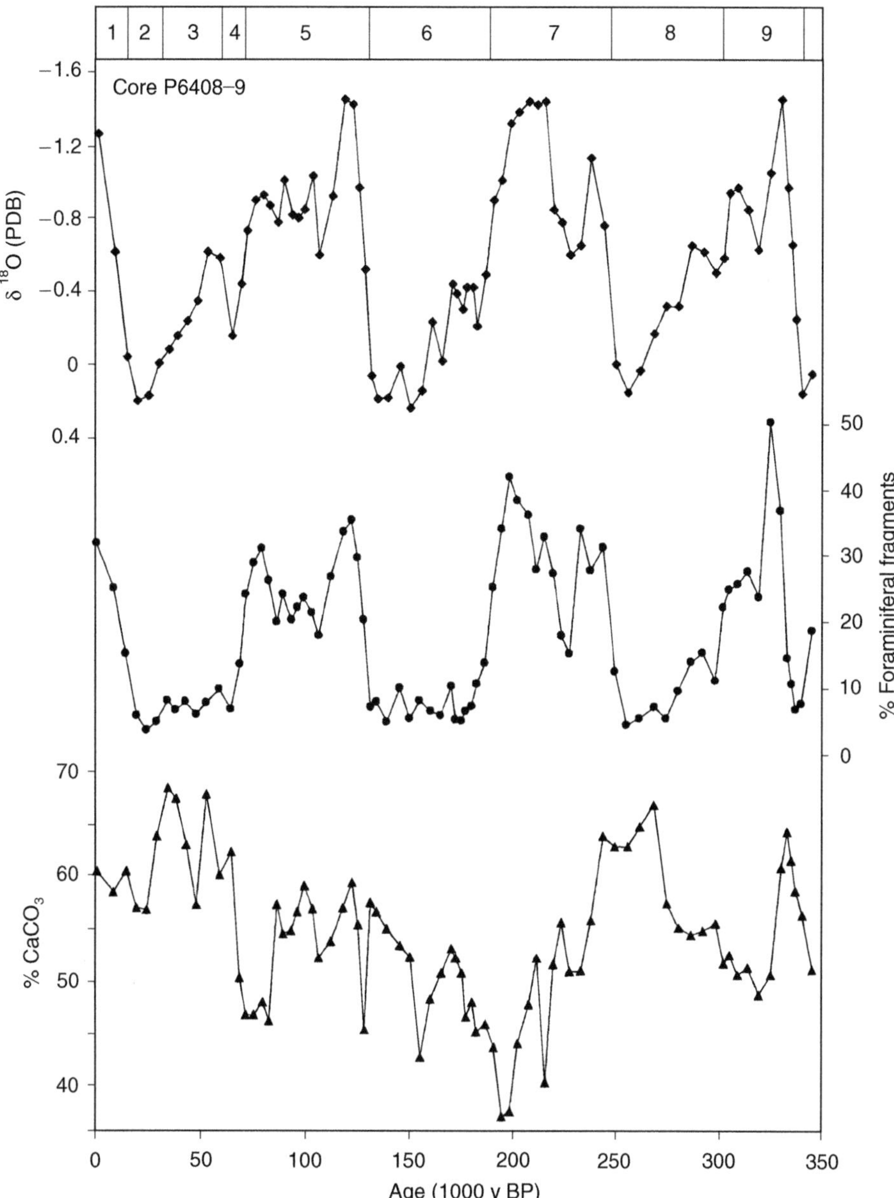

Figure 5 Measurements of foraminiferal fragmentation and calcium carbonate content (weight-%) spanning the last 350 000 years in Caribbean sediment core P6408-9. Stratigraphy and age control come from the oxygen isotope ($\delta^{18}O$) record shown at the top of the figure; odd-numbered stages are warm, interglacial intervals and even-numbered stages indicate cold, glacial climates with greatly expanded Northern Hemisphere ice cover. Variations in the ratio of foraminiferal fragments to whole shells can be directly related to the intensity of carbonate dissolution on the seafloor. Greatly increased preservation (i.e. decreased numbers of fragments)during cold, glacial stages indicates reductions in the chemical corrosivity of deep Caribbean waters in response to climate and ocean circulation changes. Note that variations in carbonate content at this location are not as clearly linked to climate-induced changes in deep-water chemistry as the fragmentation record. This is because the carbonate content of the sediments can also be affected by carbonate productivity at the surface and by dilution on the seafloor by noncarbonate sediment types. (Unpublished data from L. Peterson.)

Seafloor Diagenesis

With time and burial, carbonate oozes undergo a progressive sequence of diagenesis and are transformed first to chalk and then to limestone through a combination of gravitational compaction, dissolution, reprecipitation, and recrystallization. Porosity is reduced from about 70% in typical unconsolidated carbonate oozes to roughly 10% in cemented limestones, while overall volume decreases by about one-third. Drilling results have shown that the transformation from ooze to chalk typically

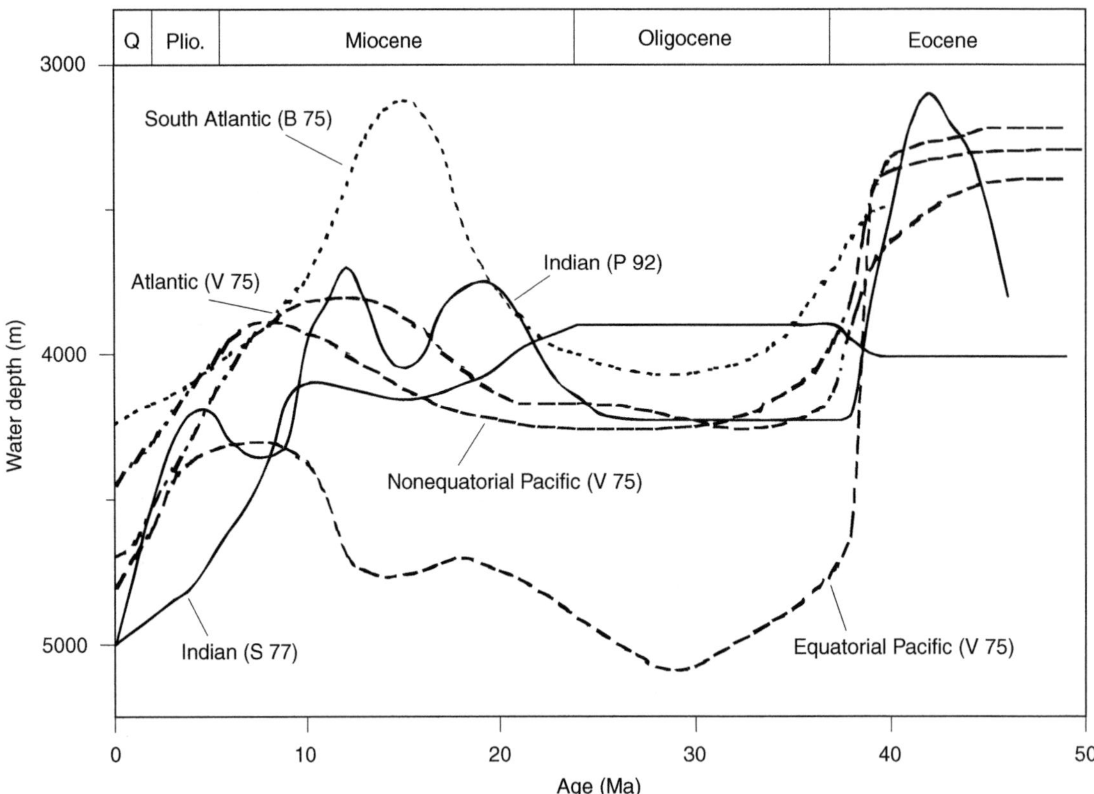

Figure 6 Compilation of reconstructed variations in the depth of the CCD from selected studies covering the last 50 million years for different oceanic regions. The overall similarity of the CCD behavior between regions suggests a common forcing mechanism, such as global sea level or a long-term change in the supply of calcium to the ocean. Variations between the oceans are probably the result of differences in regional surface productivity and deep circulation patterns. Cited CCD studies include: V75, van Andel (1975); B75, Berger and Roth (1975); S77, Sclater *et al.* (1977); P92, Peterson *et al.* (1992).

occurs within a few hundred meters of burial, while limestones are produced by further cementation under about 1 km of burial. Although the transformation of ooze to chalk to limestone is the expected diagenetic sequence, smaller scale reversals in lithification are often observed. Such reversals in pattern have led to the concept of diagenetic potential, which simply states that different sediments will take different lengths of time to reach equal stages of lithification depending upon the original character of the deposited sediment. Such factors as the original proportions of coccoliths to foraminifera (affecting grain size), the amount of dissolution experienced before burial, sedimentation rates, and numerous other subtle factors can influence the diagenetic potential of a carbonate sediment. To the extent that these factors reflect original oceanographic conditions, the sub-bottom acoustic reflectors that result from changing lithification state and diagenetic potential preserve a history of paleooceanographic events that can often be traced across large regions within ocean basins.

See also

Carbon Dioxide (CO_2) Cycle. Cenozoic Climate – Oxygen Isotope Evidence. Cenozoic Oceans – Carbon Cycle Models. Ocean Carbon System, Modeling of. Pore Water Chemistry.

Further Reading

Archer DE (1996) An atlas of the distribution of calcium carbonate in sediments of the deep sea. *Global Biogeochemical Cycles* 10: 159–174.

Arrhenius G (1988) Rate of production, dissolution and accumulation of biogenic solids in the ocean. *Palaeogeography, Palaeoclimatology and Palaeoecology* 67: 1119–1146.

Berger WH (1976) Biogenous deep sea sediments: production, preservation and interpretation. In: Riley JP and Chester R (eds.) *Chemical Oceanography*, vol. 5, pp. 266–388. London: Academic Press.

Berger WH and Roth PH (1975) Oceanic micropaleontology: progress and prospect. *Reviews of Geophysics and Space Physics* 13: 561–585.

Broecker WS and Peng T-H (1982) *Tracers in the Sea.* Palisades, NY: Lamont-Doherty Geological Observatory Press.

Emerson S and Archer DE (1992) Glacial carbonate dissolution cycles and atmospheric pCO_2: a view from the ocean bottom. *Paleoceanography* 7: 319–331.

Jahnke RA, Craven DB, and Gaillard J-F (1994) The influence of organic matter diagenesis on $CaCO_3$ dissolution at the deep-sea floor. *Geochimica Cosmochimica Acta* 58: 2799–2809.

Milliman JD (1993) Production and accumulation of calcium carbonate in the ocean: budget of a nonsteady state. *Global Biogeochemical Cycles* 7: 927–957.

Peterson LC and Prell WL (1985) Carbonate dissolution in recent sediments of the eastern equatorial Indian Ocean: Preservation patterns and carbonate loss above the lysocline. *Marine Geology* 64: 259–290.

Peterson LC, Murray DW, Ehrmann WU, and Hempel P (1992) Cenozoic carbonate accumulation and compensation depth changes in the Indian Ocean. In: Duncan RA, Rea DK, Kidd RB, von Rad U, and Weissel JK (eds.) *Synthesis of Results from Scientific Drilling in the Indian Ocean*, Geophysical Monograph 70, pp. 311–333. Washington, DC: American Geophysical Union.

Schlanger SO and Douglas RG (1974) The pelagic ooze-chalk-limestone transition and its implications for marine stratigraphy. In: Hsü KJ and Jenkyns HC (eds.) *Pelagic Sediments on Land and Under the Sea*, Special Publication of the International Association of Sedimentologists, 1, pp. 117–148. Oxford: Blackwell.

Sclater JG, Abbott D, and Thiede J (1977) Paleobathymetry and sediments of the Indian Ocean. In: Heirtzler JR, Bolli HM, Davies TA, Saunders JB, and Sclater JG (eds.) *Indian Ocean Geology and Biostratigraphy*, pp. 25–60. Washington, DC: American Geophysical Union.

Takahashi T, Broecker WS, Bainbridge AE, and Weiss RF (1980) *Carbonate Chemistry of the Atlantic, Pacific and Indian Oceans: The Results of the GEOSECS Expeditions, 1972–1978*, Lamont-Doherty Geological Observatory Technical Report 1, CU-1-80.

van Andel TH (1975) Mesozoic-Cenozoic calcite compensation depth and the global distribution of calcareous sediments. *Earth and Planetary Science Letters* 26: 187–194.

CLAY MINERALOGY

H. Chamley, Université de Lille 1, Villeneuve d'Ascq, France

Introduction

Clay constitutes the most abundant and ubiquitous component of the main types of marine sediments deposited from outer shelf to deep sea environments. The clay minerals are conventionally comprised of the $<2\,\mu m$ fraction, are sheet- or fiber-shaped, and adsorb various proportions of water. This determines a high buoyancy and the ability for clay to be widely dispersed by marine currents, despite its propensity for forming aggregates and flocs. Clay minerals in the marine environments are dominated by illite, smectite, and kaolinite, three families whose chemical composition and crystalline status are highly variable. The marine clay associations may include various amounts and types of other species, namely chlorite and random mixed layers, but also vermiculite, palygorskite, sepiolite, talc, pyrophyllite, etc. The clay mineralogy of marine sediments is therefore very diverse according to depositional environments, from both qualitative and quantitative points of view.

As clay minerals are considered to be dependent on chemically concentrated environments, and as they commonly form in surficial conditions on land especially through weathering and soil-forming processes, their detrital versus authigenic origin in marine sediments has been widely debated. The transition from continental fresh to marine saline water, marked by a rapid increase of dissolved chemical elements, was the central point of discussion and arose from both American and European examples. In fact the mineralogical changes recorded at the land-to-sea transition are either important or insignificant, are characterized in estuarine sediments by various, sometimes opposite trends impeding consistent geochemical explanations, and often vanish in open marine sediments. The changes observed at the fresh-to-saline water transition in the clay mineral composition essentially proceed from differential settling processes or from mixing between different sources, and not from chemical exchanges affecting the crystalline network. Such a historical debate underlines the interest in investigating the sensitive clay mineral associations for understanding and reconstructing environmental conditions. This article will consider the general distribution and significance of clay minerals in recent sediments, some depositional and genetic environments, and a few examples of the use of clay assemblages to reconstruct paleoclimatic and other paleoenvironmental changes.

General Distribution and Significance

As a result of extensive reviews made by both American and Russian research teams the general characters of the clay mineral distribution in deep sea sediments have been known since the late 1970s. The maps published by various authors demonstrate the dominant control of terrigenous sources, which comprise either soils and paleosoils or rocks. The impact of soils on the marine clay sedimentation is largely dependent on weathering intensity developing on land, and therefore on the climate. For instance, kaolinite mostly forms under intense warm, humid conditions characterizing the intertropical regions, and prevails in the clay fraction of corresponding marine sediments. By contrast chlorite and illite chiefly derive from physical weathering of crystalline and diagenetic sedimentary rocks outcropping widely in cold regions, and therefore occur abundantly in high latitude oceans. The kaolinite/chlorite ratio in marine sediments constitutes a reliable indicator of chemical hydrolysis versus physical processes in continental weathering profiles and therefore of climatic variations occurring on the land masses.

Other clay minerals are also able to bear a clear climatic message, as for instance the amount of random mixed layers and altered smectite in temperate regions, the crystalline status of illite in temperate to warm regions, and the abundance of soil-forming Al-Fe smectite in subarid regions. Detailed measurements on X-ray diffraction diagrams, electron microscope observations and geochemical analyses allow precise characterization of the different continental climatic environments from data obtained on detrital sedimentary clays.

Some terrigenous clay minerals in recent sediments reflect both climatic and non-climatic influences. For instance, the distribution of illite (**Figure 1**), a mineral that primarily derives from the erosion of mica-bearing rocks, shows increased percentages in high latitude oceans due to predominant physical

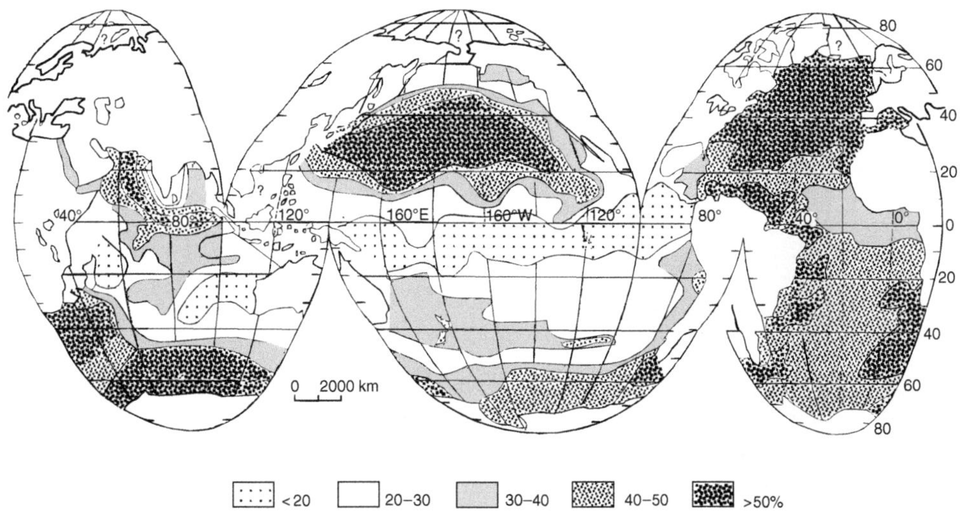

Figure 1 Worldwide distribution of illite in the clay fraction of surface sediments in the ocean. (After Windom, 1976. Reproduced with permission from Chamley, 1989.)

weathering, but also in a few low latitude regions depending on active erosion of tectonically rejuvenated, high altitude domains (e.g., supply by Indus and Ganges river drainage systems of Himalayan material to the northern Indian Ocean). The abundance of illite in the Atlantic Ocean, especially in its high latitude and northern parts, is due to several converging causes: cold to temperate climate, extensive outcrops of crystalline and metamorphic rocks, active erosion and river input, relative narrowness of the ocean favoring the ubiquitous transportation of the mineral particles, etc. Abundant illite percentages centered on the 30° parallel of latitude in the North Pacific result from aeolian supply by high altitude jet streams blowing from eastern Asia, and subsequent rainfall above the ocean. The general distribution of illite in marine sediments therefore proceeds from direct and indirect climatic control, meteorological conditions, petrographic and tectonic characteristics, physiography, river influx, etc.

All clay minerals may potentially be reworked from continental outcrops and transported over long distances until they settle on the ocean bottom. This is the case for nearly all geochemical types of smectite minerals (except perhaps for some very unstable ferriferous varieties formed in dense saline brines of the Red Sea), and also of palygorskite and sepiolite, two fibrous species wrongly suspected to not undergo significant transport. For instance, palygorskite and sepiolite are widely transported by wind and or water and deposited as detrital aggregates around the Tertiary basins bordering Africa and Arabia, where they initially formed under arid and evaporative conditions.

The clay mineral family whose distribution is the most complex and dependent on various detrital and autochthonous processes is the smectite group. Moderately crystalline smectites of diverse chemical types form pedogenically by chemical weathering under temperate conditions (essentially by degradation of illite and chlorite), and are supplied by erosion to sediments of mid-latitude regions where they are associated with various types and amounts of random mixed layers. Climate is also the dominant factor in warm, subarid regions where Al-Fe smectite forms in vertisolic soils and is reworked towards the ocean. Fairly high percentages of Fe-smectite characterize the low latitude eastern Pacific basins, where clay minerals in the clay-sized fraction are accessory relative to Fe and Mn oxides, and result from *in situ* hydrogenous genesis. In addition, smectites of Fe, Mg, and even Al types may form by alteration of volcanic rocks, a process which is more intense in well drained, subaerial conditions (hydrolysis) than in submarine environments (halmyrolysis).

The diversity of the factors controlling the distribution of clay minerals in modern deep sea sediments is widely used to trace the influence of continental climate, geological and petrographic sources, tectonics, morphological barriers, etc., and also to identify the nature, direction and intensity of transportation agents. As an example, the distribution of smectite and illite in the western Indian Ocean depends on different source provinces as well as on land geology, climate, volcanism, aeolian and marine currents (**Figure 2**). The terrigenous sources and climatic conditions relieved by north-to-south or south-to-north surface to deep currents are responsible for

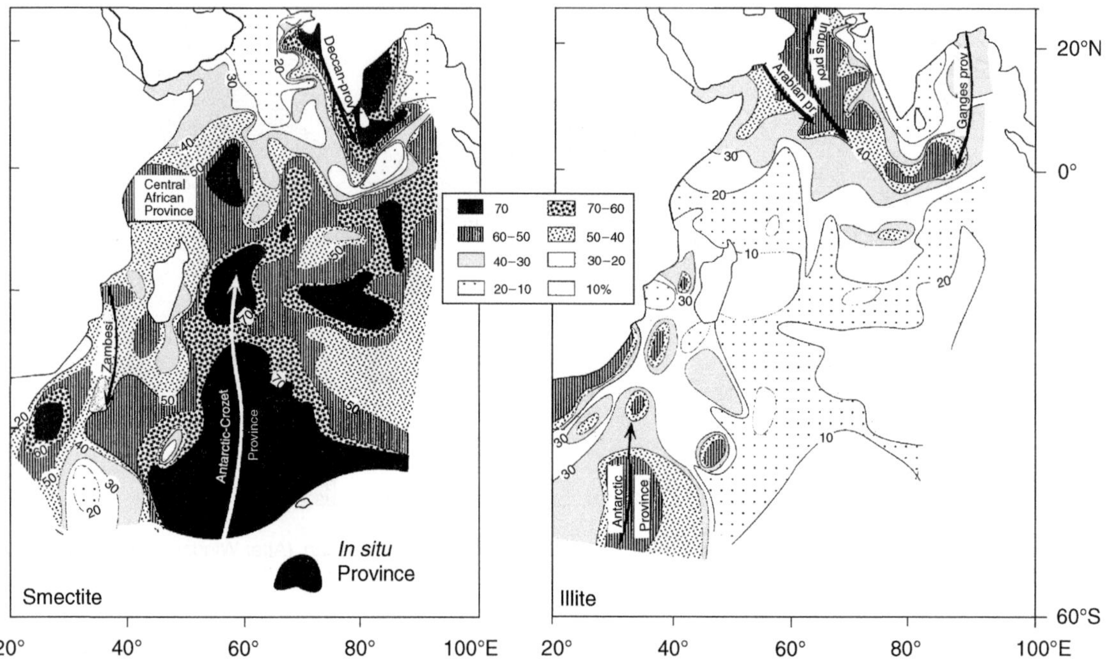

Figure 2 Distribution of smectite and illite in the western Indian Ocean, and related source provinces. (After Kolla *et al.*, 1976. Reproduced with permission from Chamley, 1989.)

long-distance transportation of Antarctic-derived smectite in the Crozet and Madagascar basins, of abundant volcanogenic smectite derived from Deccan traps erosion off the Indian coasts, of Himalayan illite in the Indus and Ganges deep sea fans, of illite associated with up to 30% palygorskite off Arabian and especially on submarine ridges (i.e., aeolian supply), and of illite associated with soil-derived kaolinite off Southeastern Africa. Both illite and smectite are dominantly inherited from various terrestrial rocks and soils, including Antarctic outcrops responsible for illite dominance to the west of the Indian Ocean (35°C) and for smectite dominance to the east (45–75°E). An *in situ* smectite-rich province located in the southern ocean around 55°S and 70°E is attributed to the submarine alteration of volcanic rocks. Volcanic contributions are also suspected in the Central Indian basin and in the vicinity of Indonesia. Of course such investigations constitute very useful guidelines for reconstructing past climatic, oceanographic, and physiographic conditions.

Marine Autochthonous Processes

From Volcanic to Hydrothermal and Hydrogenous Environments

Until the 1970s, the submarine weathering of **volcanic** material (basalt, glass, ash) was often considered to be responsible for important *in situ* formation of clay minerals, especially of smectite, in deep sea sediments. Effectively basalt altered by surficial oxidation and hydration may give way to Mg-smectite, sometimes Fe-smectite, frequently associated with celadonite (a glauconite-like Fe-Al micaceous species), phillipsite (a Na-rich zeolite), calcium carbonates, Fe-Mn oxyhydroxides, etc. The more amorphous, the smaller sized and the more porous the volcanic material (e.g., pumiceous ashes), the more intense the submarine formation of clay. In fact the clay minerals resulting from halmyrolysis of volcanic material are quantitatively limited and essentially located at close vicinity to this material (e.g., altered volcaniclastites or basalts); they are unable to participate in a large way in the formation of the huge amounts of clay incorporated in deep-sea sediments. Additional arguments contradicting the importance of volcanic contribution to deep-sea clay consist of the frequent absence of correlation between the presence of volcanic remains and that of smectite, and in the non-volcanogenic chemistry of most marine smectites (e.g., aluminum content, rare earth elements, strontium isotopes). The shape of smectite particles observed by electron microscopy is typical of volcanic influence only in restricted regions marked by high volcanic activity, especially explosive activity. Notice that local overgrowths of lath systems oriented at 60° from each other may characterize marine clay particles and especially smectites, but they are neither specifically related to volcanic environments nor associated with noticeable increase of smectite proportion or specific change in the clay chemical or isotopic composition. The intrusion of basalt sills in soft marine sediments may determine some metamorphic effects and the very local formation of ordered mixed layers (corrensite), chlorite, and associated non-clay minerals.

The **hydrothermal** impact on deep-sea sedimentation is fundamentally characterized by *in situ* precipitation of Fe-Mn oxyhydroxides relatively depleted in accessory transition elements (Co, Cu, Ni), and locally by the deposition of massive sulfides near the vents where hot and chemically concentrated water merges. The autochthonous clay minerals in such environments are marked by various species depending on fluid temperature, oxidation-reduction processes, and fluid/rock ratio. For instance, drilling holes in Pacific hydrothermal systems show different mineral evolutions. In the hydrothermal mounds of the Galapagos spreading center, the fluids are rich in silicon and iron and of a low temperature (20°–30°C) throughout the 30 m-thick sedimentary column; this gives way in oxidized conditions to the precipitation of Fe-smectite as greenish layers interbedded in biogenic oozes that at depth evolve into glauconite by addition of potassium (**Figure 3A**). By contrast the detrital to authigenic deposits of the Middle Valley of Juan de Fuca ridge show on a 40 m-thick series the *in situ* formation from high temperature Mg-rich fluids (200°C) of a downwards sequence characterized by saponite (a Mg-smectite), corrensite (a regular chlorite-smectite mixed layer), swelling chlorite, and chlorite (**Figure 3B**). At this site geochemical and isotope investigations reflect a noticeable downhole increase of temperature and strong changes in the fluid composition.

A more widespread process consists of the **hydrogenous** formation of clay at the sediment–seawater interface, in deep-sea environments characterized by water depths >4000 m, insignificant terrigenous supply, and very low sedimentation rate (<1 mm/1000 years). This is particularly the case for some Central and South Pacific basins. The sediments mostly consist of reddish-brown oozes rich in Fe and Mn oxides (i.e., 'deep sea red clay'). There iron-rich smectites of the nontronite group may form in significant proportions, probably due to long-term low temperature interactions between (1) metal

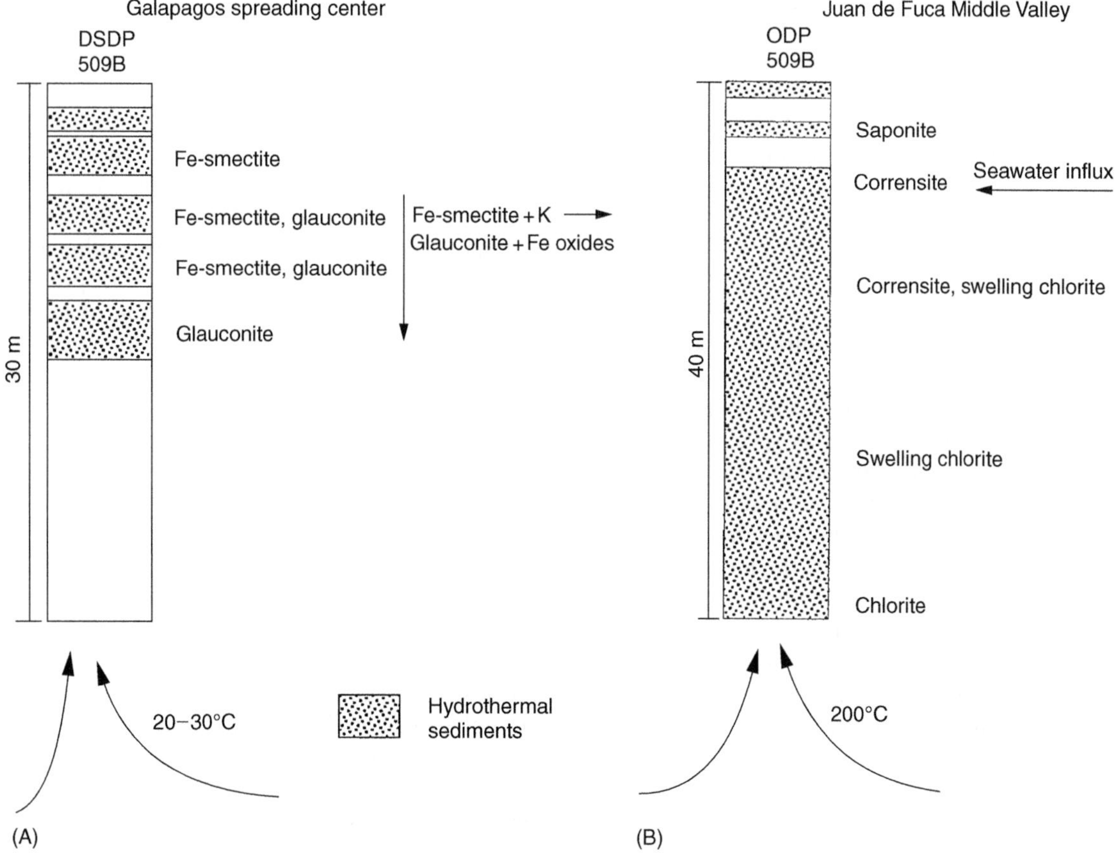

Figure 3 Schematic vertical distribution of typically hydrothermal clay minerals in the sedimentary systems of (A) the Galapagos spreading center, and (B) the Juan de Fuca Middle Valley. (Reproduced with permission from Buatier and Karpoff, 1995.)

Table 1 Examples of chemical composition of hydrothermal to hydrogenous smectites in Central and South Pacific sediments

Type of smectite	*Tetrahedra*			*Octahedra*			*Interlayers*		
	Si	*Al*	*Fe*	*Al*	*Fe*	*Mg*	*Ca*	NH_4	*K*
Pure hydrothermal (Galapagos mounds field)	3.94	0.06	–	0.03	1.59	0.38	0.03	–	0.36
Hydrothermal and hydrogenous (Bauer Deep)	3.97	0.03	–	0.44	1.07	0.54	0.05	–	0.06
Hydrogenous>hydrothermal (Galapagos spreading centre)	3.97	0.03	–	1.12	0.48	0.37	0.09	–	0.11
Pure hydrogenous (?) (North Marquises fracture zone)	3.37	0.63	–	0.39	1.12	0.46	0.46	–	0.17

(Reproduced with permission from Chamley, 1989.)

oxyhydroxides supplying the iron, (2) sea-water supplying the magnesium and other minor to trace elements, (3) biogenic silica supplying the silicon, and (4) allochthonous accessory particles (e.g., aeolian clay) supplying the other chemical elements (e.g., Al). Notice that the distinction between pure hydrothermal and pure hydrogenous clay minerals forming on the deep-sea floor necessitates detailed chemical analyses (**Table 1**) and often additional microprobe and isotope investigations.

To summarize, the distribution of clay minerals in deep sea deposits marked by active volcanic-hydrothermal activity and by very low sedimentation rates depends on various and complex *in situ* influences among which the hydrogenous processes quantitatively prevail. The distinction of these autochthonous influences is complicated both in the vicinity of land masses where terrigenous supply becomes active, and in shallower areas where biogenic influences may intervene more intensely (e.g., Nazca plate, southeast Pacific).

Ferriferous Clay Granules

Iron-rich clay granules are traditionally called glauconite, which is somewhat incorrect as glauconite is a specific clay mineral, whereas clay granules may include various iron-bearing clay species. Ferriferous clay granules form on continental margins at water depths not exceeding 1000 m, and comprise two major types characterized by specific colors, clay minerals, and habits. Glaucony, the most widespread type, constitutes dark green to brown clayey aggregates, and may comprise different varieties of iron-rich illite- and smectite-like minerals such as glauconite (Fe- and K-rich illitic clay), Fe-smectite, and Fe illite-smectite mixed layers. Glaucony may form at latitudes as high as 50° and in water depths as great as 1000 m, but usually occurs in 150–300 m water depths at the shelf-slope transition of temperate-warm to equatorial regions. Verdine, which is less ubiquitous and has been identified more recently, constitutes light green to light brown granules characterized by phyllite V or odinite, a ferriferous clay mineral of the kaolinite family (described by G.S. Odin, who has developed outstanding investigations on clay granules). Verdine forms in rather shallow water sediments (maximum 50–80 m) of intertropical regions, and depends on the supply of abundant dissolved iron by low latitude rivers.

Ferriferous clay granules form at the sediment–water interface and evolve at burial depths rarely exceeding a few decimeters. They develop in semi-confined environments at the expense of various substrates submitted to 'greening': chiefly fecal pellets and microfossil chambers (e.g., foraminifera), calcareous or siliceous bioclasts, minerals (especially micas), and rock debris. The formation of glaucony (which somewhat leads to diffuse habits), occurs in successive stages marked by a rapid and strong enrichment of iron and then potassium, a volume increase causing external cracks, and the obliteration of the initial shape (**Figure 4**). The formation of verdine still has to be documented, but both clay granule types correspond to true authigenic formation rather than to transformation of pre-existing clay minerals. The chemical evolution of ferriferous clay granules vanishes either after a long exposure at the sediment–water interface (10^5–10^6 years for glaucony), or after significant burying.

Organic Environments

The influence of living organisms on clay-rich sediments is mainly marked by physical processes referred to as bioturbation, and concerns various marine environments, especially on continental shelves. Chemical modifications of clay associations are only occasionally reported and seem to affect the crystalline status of chlorite and associated random mixed layer clays locally through ingestion and digestion processes of shallow water crustaceans, annelids or copepods. The chemical interactions developing in digestive tracts between clay minerals and organic acids appear to have small quantitative

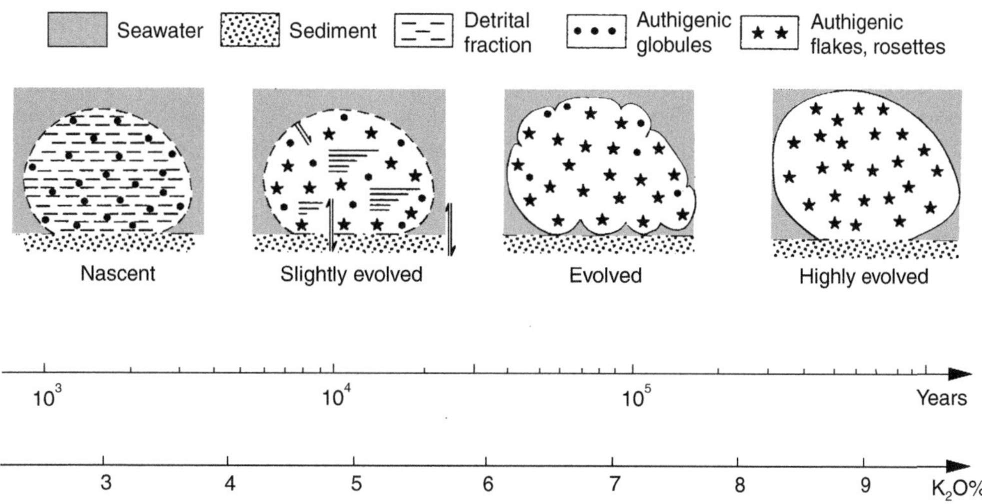

Figure 4 Successive stages of glaucony formation from a pre-existing substrate. (Reproduced with permission from Odin, 1998.)

effects, as the marine clay associations are roughly the same as the terrestrial associations.

The chemical impact on clay mineral stability of the organic matter incorporated in deep marine sediments is variable. Most sedimentary series containing significant amounts of dispersed organic matter (i.e., 1–3%) do not display any specific clay mineral composition. For example, this is the case for black shales deposited during the Cretaceous period in the Atlantic, where clay mineral associations may comprise vulnerable species such as smectite and palygorskite, the abundance and crystalline status of which vary independently of the content and distribution of the organic matter. In contrast, the sapropels developing in the eastern Mediterranean during the late Cenozoic era, especially in Quaternary high sea level stages, show some *in situ* degradation processes of the detrital clay minerals (**Figure 5**). Submarine alteration affects the mineral species in successive stages depending on

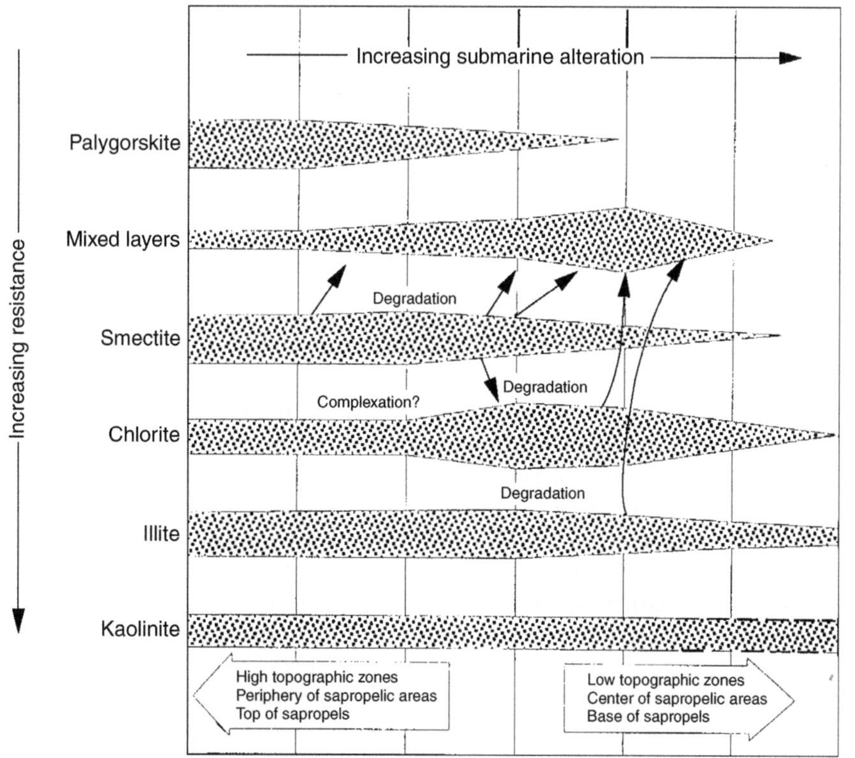

Figure 5 Characters of the clay mineral degradation in Quaternary sapropels of the eastern Mediterranean basins.

their ability to resist acid conditions. Palygorskite is the more vulnerable species and kaolinite the more resistant. The degradation of clay assemblages tends to increase toward the central and deepest parts of marine basins, in depressed morphological zones, and at the base of the decimeter-to-meter thick sapropels. The degradation of clay minerals under organic conditions has occurred close to the sediment–water interface and appears to depend on the chemical nature and evolution stage of the terrestrial and marine organic matter.

Paleoenvironmental Expression

Clay mineral assemblages of sediments successively deposited in marine basins express various environmental messages related to the geological history. A few examples from recent Quaternary to late Cenozoic series will be considered here. Similar messages may be preserved in much older series of Mesozoic and even Paleozoic ages, provided that the diagenetic imprint due to lithostatic overburden, geothermal gradient, and fluid circulation has remained moderate. Clay-rich, low permeability sedimentary formations 2–3 km thick and submitted to normal heat flow (c. 30°C/km) are usually prone to preserve such paleoenvironmental characteristics.

Climate

As clay minerals at the surface of the Earth are dominantly formed through pedogenic processes depending on climate and are particularly subjected to surficial erosion and reworking, their assemblages successively deposited in a given sedimentary basin are *a priori* able to reflect successive climatic conditions that prevailed on adjacent land masses. This implies that very little post-depositional, i.e., diagenetic changes have affected the clay assemblages after their storage in sediments. This is observed to be the case in many series drilled or cored in the oceans. The climatic message borne by clay has been documented by numerous investigations, and corroborated by the comparable range of variations recorded in the nature and proportions of clay minerals in both present-day soils outcropping at various latitudes and marine sedimentary columns. Marine clay mineral assemblages basically express the type and intensity of continental weathering, which depend predominantly on the ion leaching through the action of humidity and temperature, and secondarily on seasonal rainfall and drainage conditions.

Quaternary glacial–interglacial alternations caused terrestrial alternation of physical and chemical weathering processes, and this was reflected in the clay assemblages successively brought to marine sediments through soil erosion and river or wind transport. Sedimentary levels contemporary with cold periods are usually characterized by more abundant rock-derived minerals such as richly crystalline illite, chlorite, smectite and associated feldspars reworked from active physical weathering. Warm, humid periods generally correspond to increased supply of soil-derived kaolinite and metal oxides, poorly crystalline smectite and various random mixed layer clay minerals. For instance, the terrigenous fraction of hemipelagic sediments deposited from 500 000 to 100 000 years ago in the Northwestern Atlantic off New Jersey and dominantly derived from the erosion of Appalachian highlands shows increased proportions of chlorite in glacial isotopic stages, and of kaolinite in interglacial stages. This is clearly expressed by the kaolinite/chlorite ratio (**Figure 6**). Paleoclimatic reconstructions from clay mineral data are available for various geological periods, as for instance the passage since about 40 Ma from a non-glacial world dominated by chemical weathering (smectite, kaolinite) to a glacial world in which physical weathering was greater (chlorite, illite). The comparison of climatic curves provided by clay minerals and other indicators (oxygen isotopes, micro-faunas or -floras, magnetic susceptibility, etc.) allows a better understanding of the nature, intensity, and effect of the different factors characterizing the terrestrial and marine climate in given regions during given geological intervals.

High resolution studies show that clay assemblages may express terrestrial climatic variations at a centennial scale or even less, and that the influence of Earth's orbital parameters varies to different extents according to the latitude. For example, the clay minerals data of Quaternary North Atlantic deep sea sediments were submitted to cross-correlation spectral analyses on 5.5–14 m-long cores encompassing the last 300 000 years. The mineral composition displays a general 100 000-year cyclic signal (eccentricity) in the whole 45°–60°N range, a 41 000-year signal (obliquity) at highest latitudes related to dominant aeolian supply, and a 23 000-year signal (precession) at mid-latitudes related to dominant transport by marine currents (**Table 2**).

The paleoclimatic expression by clay mineral successions is direct or indirect, i.e., it either indicates the climate that actually prevailed at a given period, or reflects other events depending on climate: migration of lithospheric plates across successive climatic zones, varying extension of ice caps controlling the surficial erosion, variations in the marine circulation regime due to changing latitudinal and

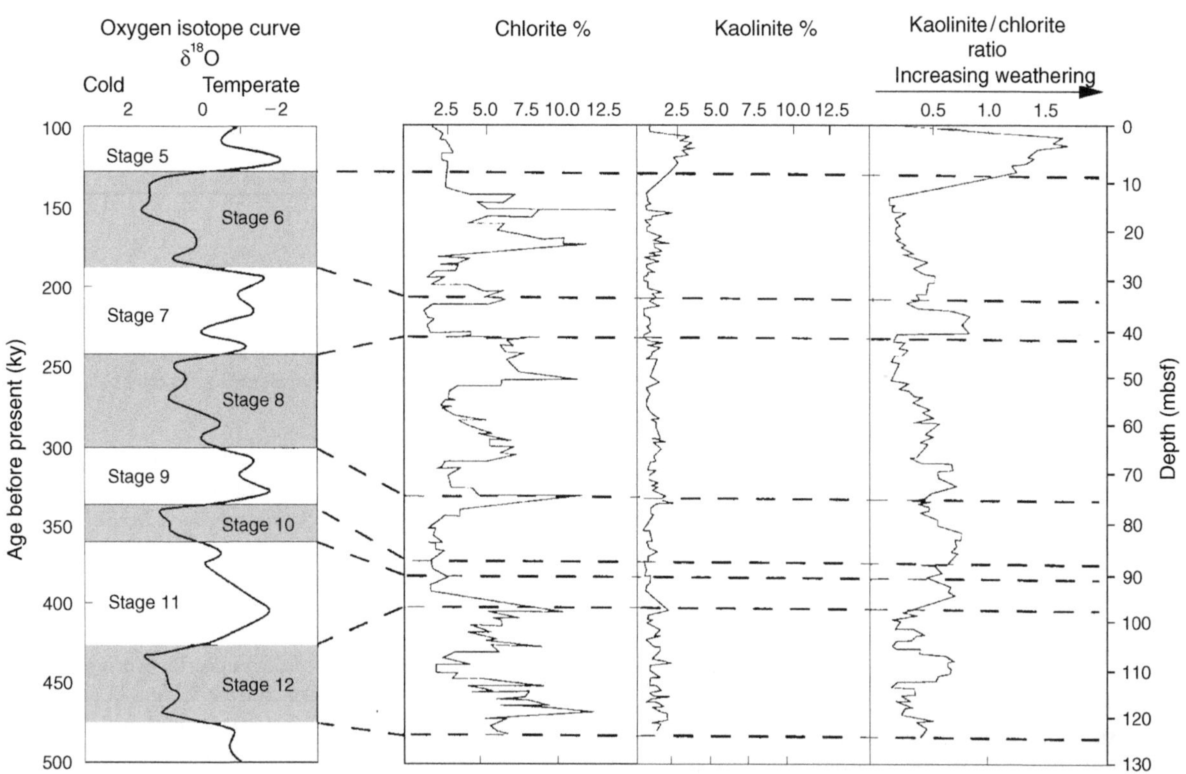

Figure 6 Comparison and climatic significance of clay mineral and oxygen isotope data from stages 12 to 5 at ODP Site 902, New Jersey continental margin. (Reproduced with permission from Vanderaveroet *et al.*, 1999.)

vertical heat transfers. The direct paleoclimatic reconstructions from clay mineral data are all the more reliable since the marine basins investigated are preserved from important erosion of paleosoils, changes in detrital sources, differential settling processes, longitudinal oceanic currents, and major geomorphological changes.

Marine Currents

The different marine water masses may carry the small and light clay mineral particles over long distances, and therefore leave an imprint within the sediments at the depth range they are moving. This has been demonstrated for late Quaternary sediments of the southwestern Atlantic, where the southward-flowing North Atlantic deep water is enriched in kaolinite supplied from rivers draining the intertropical South American continent, and the northward flowing Antarctic Atlantic bottom water supplies chlorite and smectite issuing from southernmost Argentina and Antarctica. Paleocurrent reconstructions from clay data exist mainly about Atlantic and Southern Oceans, which are marked by numerous and distinct terrigenous sources, vertical mixing and longitudinal heat transfers, and Tertiary to Quaternary changing conditions of the superimposed water masses volume and celerity.

Tectonic Activity

The tectonic instability determines some changes in the composition of clay mineral assemblages which are usually much more important than those due to climate or circulation. First, the subpermanent rejuvenation by neotectonics of continental relief increases the erosion potential and therefore impedes the development of pedogenic blankets where clay minerals tend to be in equilibrium with current climatic conditions. Such a chronic tectonic activity explains the abundance of rock-derived illite and chlorite in equatorial Indian Ocean basins depending on Himalayan output. Second, a continental tectonic uplift determines changes in the nature of clay minerals eroded from rocky substrates, while submarine uplift may determine morphological barriers to the clay transfer. This was the case for the Hellenic Trench in the eastern Mediterranean during late Pliocene to early Pleistocene periods, when the combined uplift of Peloponnese and of Mediterranean ridge both increased the terrigenous input of European illite and chlorite and blocked the supply of African palygorskite. Due to their sensitivity to

Table 2 General relationships between the clay mineral distribution and the three main Earth's orbital frequency bands according to latitude, from cross-correlation spectral analysis of X-ray diffraction data on North Atlantic cores

Core	SU 90-08			SU 90-12			SU 90-38			SU 90-33		
Latitude	44°N			51°N			54°N			60°N		
Orbital parameters	**E**	O	**P**	**E**	O	**P**	**E**	**O**	P	**E**	**O**	P
Illite	H	–	**V**	H	–	**V**	**V**	H	–	H	**V**	–
Chlorite	**V**	–	**V**	**V**	–	**V**	**V**	**V**	–	**V**	V	–
Kaolinite	H	–	**V**	H	–	–	**V**	**V**	–	H	**V**	–
Illite-vermiculite random mixed layer	–	–	–	–	–	**V**	–	–	–	–	–	–

E, eccentricity band, 100000 year; O, obliquity band, 41000 year; P, precession band, 23000 year; H, high variance power; V, very high variance power. Maximum correlations in bold characters. (Reproduced with permission from Bout-Roumazeilles *et al.*, 1997.)

geomorphological changes and their aptitude for long distance transportation, clay minerals are able to express slight and progressive epeirogenic changes as well as very remote tectonic events.

See also

Cenozoic Climate – Oxygen Isotope Evidence. Hydrothermal Vent Deposits. Rare Earth Elements and their Isotopes in the Ocean. River Inputs.

Further Reading

Bout-Roumazeilles V, Debrabant P, Labeyrie L, Chamley H, and Cortijo E (1997) Latitudinal control of astronomical forcing parameters on the high-resolution clay mineral distribution in the 45°–60° N range in the North Atlantic Ocean during the past 300,000 years. *Paleoceanography* 12: 671–686.

Buatier MD and Karpoff AM (1995) Authigenése et évolution d'argiles hydrothermales océaniques: exemples des monts des Galapagos et des sédiments de la ride de Juan de Fuca. *Bulletin de la Société Géologique de France* 166: 123–136.

Chamley H (1989) *Clay Sedimentology.* Berlin: Springer-Verlag.

Hoffert M (1980) Les 'argiles rouges des grands fonds' dans le Pacifique centre-est. *Sciences géologique.* Strasbourg, Mem 61: 257.

Millot G (1970) *Geology of Clays.* Berlin: Springer-Verlag.

Odin GS (ed.) (1988) *Green Marine Clays. Developments in Sedimentology*, 45, Amsterdam: Elsevier.

Robert C and Chamley H (1992) Late Eocene-early Oligocene evolution of climate and marine circulation: deep-sea clay mineral evidence. American Geophysical Union. *Antarctic Research Series* 56: 97–117.

Vanderaveroet P, Averbuch O, Deconinck JF, and Chamley H (1999) A record of glacial/interglacial alternations in Pleistocene sediments off New Jersey expressed by clay mineral, grain-size and magnetic susceptibility data. *Marine Geology* 159: 79–92.

Weaver CE (1999) *Clays, Muds, and Shales. Developments in Sedimentology*, 44. Amsterdam: Elsevier.

Windom HL (1976) Lithogenous material in marine sediments. *Chemical Oceanography* vol. 5, pp. 103–135. New York: Academic Press.

MID-OCEAN RIDGE GEOCHEMISTRY AND PETROLOGY

M. R. Perfit, Department of Geological Sciences, University of Florida, Gainsville, FL, USA

Introduction

The most volcanically active regions of our planet are concentrated along the axes of the globe, encircling mid-ocean ridges. These undersea mountain ranges, and most of the oceanic crust, result from the complex interplay between magmatic (i.e., eruptions of lavas on the surface and intrusion of magma at depth) and tectonic (i.e., faulting, thrusting, and rifting of the solid portions of the outer layer of the earth) processes. Magmatic and tectonic processes are directly related to the driving forces that cause plate tectonics and seafloor spreading. Exploration of mid-ocean ridges by submersible, remotely operated vehicles (ROV), deep-sea cameras, and other remote sensing devices has provided clear evidence of the effects of recent magmatic activity (e.g., young lavas, hot springs, hydrothermal vents and plumes) along these divergent plate boundaries. Eruptions are rarely observed because of their great depths and remote locations. However, over 60% of Earth's magma flux (approximately 21 km^3 $year^{-1}$) currently occurs along divergent plate margins. Geophysical imaging, detailed mapping, and sampling of mid-ocean ridges and fracture zones between ridge segments followed by laboratory petrologic and geochemical analyses of recovered rocks provide us with a great deal of information about the composition and evolution of the oceanic crust and the processes that generate mid-ocean ridge basalts (MORB).

Mid-ocean ridges are not continuous but rather broken up into various scale segments reflecting breaks in the volcanic plumbing systems that feed the axial zone of magmatism. Recent hypotheses suggest that the shallowest and widest portions of ridge segments correspond to robust areas of magmatism, whereas deep, narrow zones are relatively magma-starved. The unusually elevated segments of some ridges (e.g., south of Iceland, central portion of the Galapagos Rift, Mid-Atlantic Ridge near the Azores) are directly related to the influence of nearby mantle plumes or hot spots that result in voluminous magmatism.

Major differences in the morphology, structure, and scales of magmatism along mid-ocean ridges vary with the rate of spreading. Slowly diverging plate boundaries, which have low volcanic output, are dominated by faulting and tectonism whereas fast-spreading boundaries are controlled more by volcanism. The region along the plate boundary within which volcanic eruptions and high-temperature hydrothermal activity are concentrated is called the neovolcanic zone. The width of the neovolcanic zone, its structure, and the style of volcanism within it, vary considerably with spreading rate. In all cases, the neovolcanic zone on mid-ocean ridges is marked by a roughly linear depression or trough (axial summit collapse trough, ASCT), similar to rift zones in some subaerial volcanoes, but quite different from the circular craters and calderas associated with typical central-vent volcanoes. Not all mid-ocean ridge volcanism occurs along the neovolcanic zone. Relatively small (<1 km high), near-axis seamounts are common within a few tens of kilometers of fast and intermediate spreading ridges. Recent evidence also suggests that significant amounts of volcanism may occur up to 4 km from the axis as off-axis mounds and ridges, or associated with faulting and the formation of abyssal hills.

Lava morphology on slow spreading ridges is dominantly bulbous, pillow lava (**Figure 1A**), which tends to construct hummocks (<50 m high, <500 m diameter), hummocky ridges (1–2 km long), or small circular seamounts (10s–100s of meters high and 100s–1000s of meters in diameter) that commonly coalesce to form axial volcanic ridges (AVR) along the valley floor of the axial rift zone. On fast spreading ridges, lavas are dominantly oblong, lobate flows and fluid sheet flows that vary from remarkably flat and thin (<4 cm) to ropy and jumbled varieties (**Figure 1**). Although the data are somewhat limited, calculated volumes of individual flow units that have been documented on mid-ocean ridges show an inverse exponential relationship to spreading rate, contrary to what might be expected. The largest eruptive units are mounds and cones in the axis of the northern Mid-Atlantic Ridge whereas the smallest units are thin sheet/lobate flows on the East Pacific Rise. Morphologic, petrologic, and structural studies of many ridge segments suggest they evolve

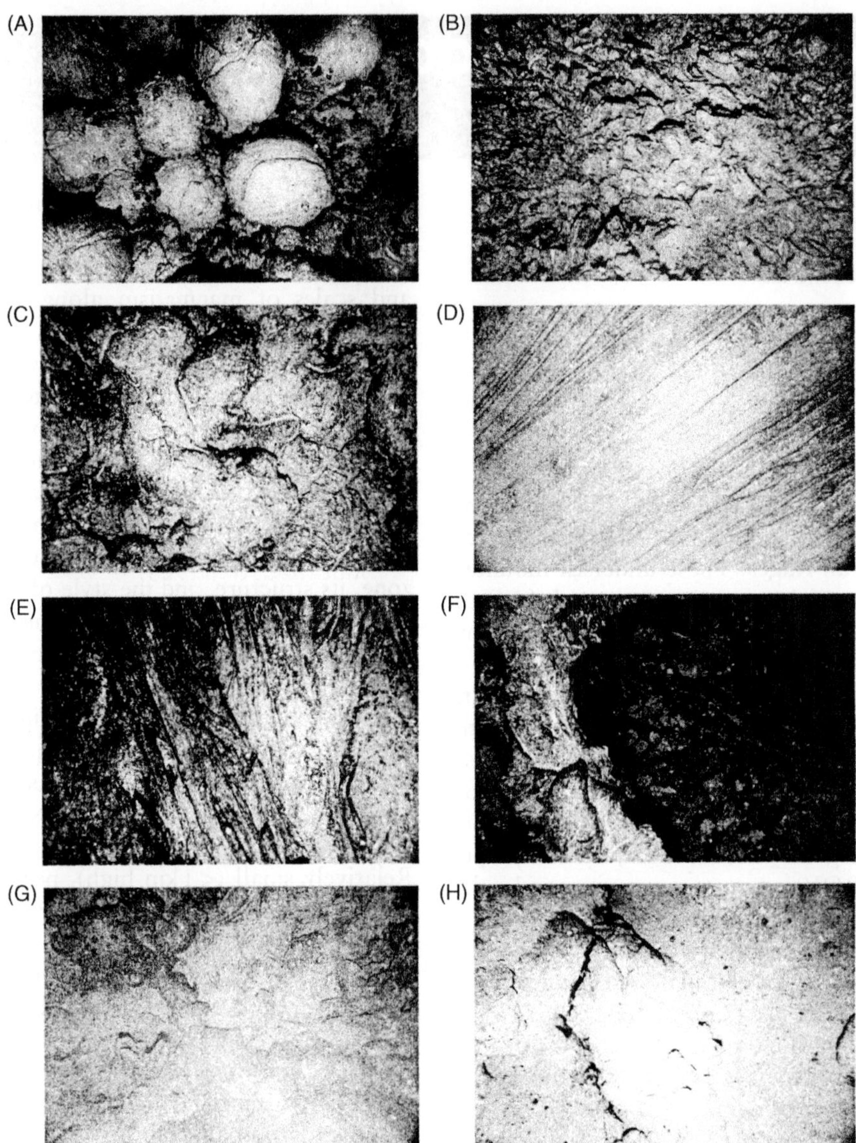

Figure 1 Examples of different morphologies, surface textures and sediment cover on lava flows on the northern East Pacific Rise. Digital images were taken from heights of 5–10 m above the seafloor using the Woods Hole Oceanographic Instution's camera system. The dimensions of the photographs are approximately 4.5 m × 3.0 m. (A) Pillow lava. (B) Hackly or scrambled flow. (C) Lobate lava. (D) Lineated sheet flow. (E) Ropy sheet flow. (F) Collapse structure in lobate flows. (G) A young flow contact on top of older flows. (H) Heavily sediment covered lobate flows with small fissure. Images from Kuras *et al.* 2000.

through cycles of accretion related to magmatic output followed by amagmatic periods dominated by faulting and extension.

Magma Generation

Primary MORB magmas are generated by partial melting of the upper mantle; believed to be composed of a rock type termed peridotite which is primarily composed of the minerals olivine, pyroxenes (enstatite and diopside), and minor spinel or garnet. Beneath ridges, mantle moves upward, in part, due to convection in the mantle but possibly more in response to the removal of the lithospheric lid above it, which is spreading laterally. Melting is affected by the decompression of hot, buoyant peridotite that crosses the melting point (solidus curve) for mantle material as it rises to shallow depths (<100 km), beneath the ridges. Melting continues as the mantle rises as long as the temperature of the peridotite remains above the solidus temperature at a given depth. As the seafloor spreads, basaltic melts formed in a broad region (10s to 100s of kilometers) beneath the ridge accumulate and focus so that they feed a relatively narrow region (a few kilometers) along the axis of the ridge (**Figure 1**).

During ascent from the mantle and cooling in the crust, primary mantle melts are subjected to a variety of physical and chemical processes such as fractional crystallization, magma mixing, crustal assimilation, and thermogravitational diffusion that modify and differentiate the original melt composition. Consequently, primary melts are unlikely to erupt on the seafloor without undergoing some modification. Picritic lavas and magnesian glasses thought to represent likely primary basalts have been recovered from a few ocean floor localities; commonly in transform faults (**Table 1**). MgO contents in these basalts range from ~10 wt% to over 15 wt% and the lavas typically contain significant amounts of olivine crystals. Based on comparisons with high-pressure melting experiments of likely mantle peridotites, the observed range of compositions may reflect variations in source composition and mineralogy (in part controlled by pressure), depth and percentage melting (largely due to temperature differences), and/or types of melting (e.g., batch vs. fractional).

Ocean Floor Volcanism and Construction of the Crust

Oceanic crust formed at spreading ridges is relatively homogeneous in thickness and composition compared to continental crust. On average, oceanic crust is 6–7 km thick and basaltic in composition as compared to the continental crust which averages 35–40 km thick and has a roughly andesitic composition. The entire thickness of the oceanic crust has not been sampled *in situ* and therefore the bulk composition has been estimated based on investigations of ophiolites (fragments of oceanic and back-arc crust that have been thrust up on to the continents), comparisons of the seismic structure of the oceanic crust with laboratory determinations of seismic velocities in known rock types, and samples recovered from the ocean floor by dredging, drilling, submersibles, and remotely operated vehicles.

Rapid cooling of MORB magmas when they come into contact with cold sea water results in the formation of glassy to finely crystalline pillows, lobate flows, or sheet flows (**Figure 1**). These lava flows typically have an ~0.5–1 cm-thick outer rind of glass and a fine-grained, crystalline interior containing only a few percent of millimeter-sized crystals of olivine, plagioclase, and more rarely clinopyroxene in a microscopic matrix of the same minerals. MORB lavas erupt, flow, and accumulate to form the uppermost volcanic layer (Seismic Layer 2A) of ocean crust (**Figure 2**). Magmas that do not reach the seafloor cool more slowly with increasing depth forming intrusive dikes at shallow levels (0.5–3 km) in the crust (layer 2B) and thick bodies of coarsely crystalline gabbros and cumulate ultramafic rocks at the lowest levels (3–7 km) of the crust (layer 3) (**Figure 2**).

Although most magma delivered to a MOR is focused within the neovolcanic zone, defined by the axial summit collapse trough or axial valley, off-axis volcanism and near-axis seamount formation appear to add significant volumes of material to the uppermost crust formed along ridge crests. In some portions of the fast spreading East Pacific Rise, off-axis eruptions appear to be related to syntectonic volcanism and the formation of abyssal hills. Near-axis seamount formation is common along both the East Pacific Rise and medium spreading rate Juan de Fuca Ridge. Even in areas where there are abundant off-axis seamounts they may add only a few percent to the volume of the extrusive crust. More detailed studies of off-axis sections of ridges are needed before accurate estimates of their contribution to the total volume of the oceanic crust can be made.

Oceanic transform faults are supposed to be plate boundaries where crust is neither created nor destroyed, but recent mapping and sampling indicate that magmatism occurs in some transform domains. Volcanism occurs in these locales either at short, intratransform spreading centers or at localized eruptive centers within shear zones or relay zones between the small spreading centers.

Mid-ocean Ridge Basalt Composition

Ocean floor lavas erupted along mid-ocean ridges are low-potassium tholeiites that can range in composition from picrites with high MgO contents to ferrobasalts and FeTi basalts containing lower MgO and high concentrations of FeO and TiO_2, and even to rare, silica-enriched lavas known as icelandites, ferroandesites and rhyodacites (**Table 1**). In most areas, the range of lava compositions, from MgO-rich basalt to FeTi basalt and ultimately to rhyodacite, is generally ascribed to the effects of shallow-level (low-pressure) fractional crystallization in a subaxial magma chamber or lens (**Figure 2**). A pronounced iron-enrichment trend with decreasing magnesium contents (related to decreasing temperature) in suites of genetically related lavas is, in part, what classifies MORB as tholeiitic or part of the tholeiitic magmatic suite (**Figure 3**).

Table 1 Average compositions of normal and enriched types of basalts from mid-ocean ridges and seamounts

	Normal							*Enriched*			
Oxide wt%	*Pacific*	*Atlantic*	*Galapagos*	*Seamounts*	*Pacific Picritic*	*Pacific Ferrobasalt*	*Pacific High-silica*	*Pacific*	*Atlantic*	*Galapagos*	*Seamounts*
SiO_2	50.49	50.64	50.41	50.03	48.80	50.61	55.37	50.10	51.02	49.17	50.19
TiO_2	1.78	1.43	1.54	1.28	0.97	2.36	2.10	1.86	1.46	1.94	1.74
Al_2O_3	14.55	15.17	14.75	15.97	17.12	13.30	12.92	15.69	15.36	16.86	16.71
FeO*	10.87	10.45	11.19	9.26	8.00	13.61	13.11	9.78	9.56	9.21	8.77
MnO	0.20	0.19	nd	0.15	0.14	0.23	0.21	0.19	0.18		0.15
MgO	7.22	7.53	7.49	8.06	10.28	5.92	3.64	7.00	7.31	6.93	6.80
CaO	11.58	11.62	11.69	12.21	11.93	10.43	8.05	11.17	11.54	10.90	10.67
Na_2O	2.74	2.51	2.28	2.68	2.32	2.74	3.33	3.04	2.52	3.16	3.38
K_2O	0.13	0.11	0.10	0.08	0.03	0.16	0.44	0.43	0.36	0.66	0.75
P_2O_5	0.17	0.14	0.14	0.13	0.07	0.22	0.40	0.24	0.19	0.32	0.33
Sum	99.62	99.61	99.60	99.73	100	99.39	99.40	99.37	99.31	99.14	99.34
K/Ti	7.49	7.70	6.24	6.10	3.0	7.04	13.80	22.26	23.67	32.77	39.05
$N=$	2303	2148	867	623	10	706	97	304	972	65	197

Analyses done by electron microprobe on natural glasses at the Smithsonian Institution in Washington, D.C. (by W. Melson and T.O'Hearn) except the picritic samples that were analyzed at the USGS in Denver, Co. Enriched MORB in this compilation are any that have K/Ti values greater than 13. High-silica lavas have SiO_2 values between 52 and 64.
$K/Ti = (K_2O/TiO_2) \times 100$. N = number of samples used in average. FeO* = total Fe as FeO.

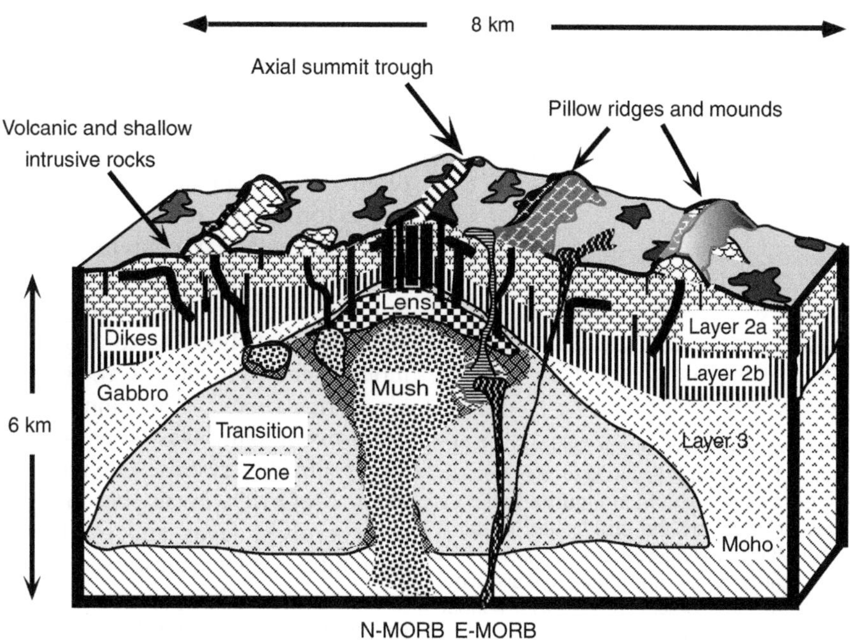

Figure 2 Diagrammatic three-dimensional representation of oceanic crust formed along a fast-spreading ridge showing the seismically determined layers and their known or inferred petrologic composition. Note that although most of the volcanism at mid-ocean ridges appears to be focused within the axial summit trough, a significant amount of off-axis volcanism (often forming pillow mounds or ridges) is believed to occur. Much of the geochemical variability that is observed in MORB probably occurs within the crystal–liquid mush zone and thin magma lens that underlie the ridge crest. The Moho marks the seismic boundary between plutonic rocks that are gabbroic in composition and those that are mostly ultramafic but may have formed by crystal accumulation in the crust.

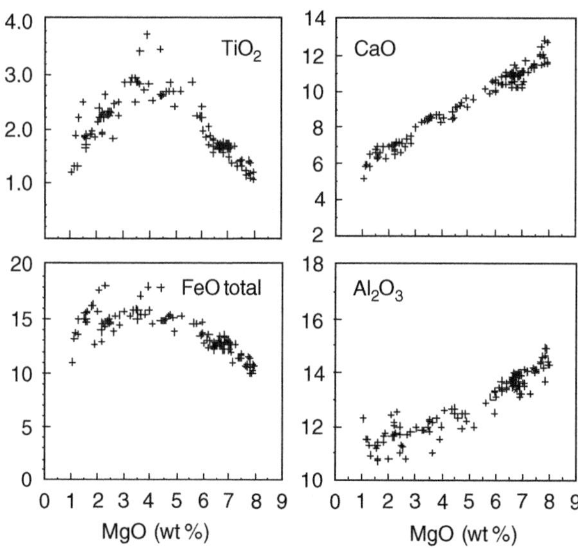

Figure 3 Major element variations in MORB from the Eastern Galapagos Spreading Center showing the chemical trends generated by shallow-level fractional crystallization in the oceanic crust. The rocks range in composition from basalt to ferrobasalt and FeTi basalt to andesite.

Although MORB are petrologically similar to tholeiitic basalts erupted on oceanic islands (OIB), MORB are readily distinguished from OIB based on their comparatively low concentrations of large ion lithophile elements (including K, Rb, Ba, Cs), light rare earth elements (LREE), volatile elements and other trace elements such as Th, U, Nb, Ta, and Pb that are considered highly incompatible during melting of mantle mineral assemblages. In other words, the most incompatible elements will be the most highly concentrated in partial melts from primitive mantle peridotite. On normalized elemental abundance diagrams and rare earth element plots (**Figure 4**), normal MORB (N-type or N-MORB) exhibit characteristic smooth concave-down patterns reflecting the fact that they were derived from incompatible element-depleted mantle. Isotopic investigations have conclusively shown that values of the radiogenic isotopes of Sr, Nd, Hf and Pb in N-MORB are consistent with their depleted characteristics and indicate incompatible element depletion via one or more episodes of partial melting of upper mantle sources beginning more than 1 billion years ago. Compared to ocean island basalts and lavas erupted in arc or continental settings, MORB comprise a relatively homogeneous and easily distiguishable rock association. Even so, MORB vary from very depleted varieties (D-MORB) to those containing moderately elevated incompatible element abundances and more radiogenic isotopes. These less-depleted MORB are called E-types (E-MORB)

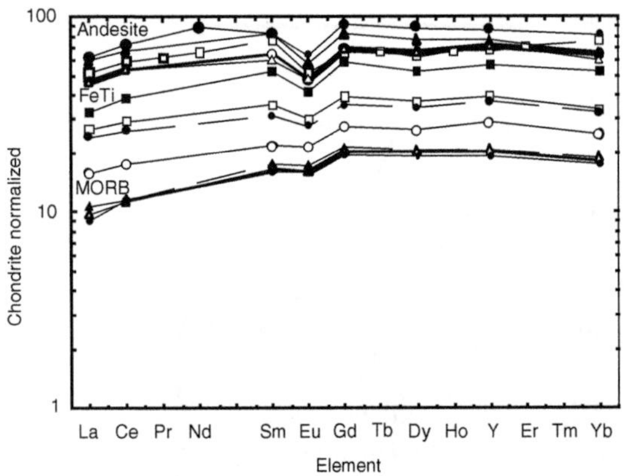

Figure 4 Chondrite-normalized rare earth element (REE) abundances in a suite of cogenetic lavas from the Eastern Galapagos Spreading Center (also shown in **Figures 3** and **6**). Increasing abundances of REE and the size of the negative europium anomaly from MORB to andesite are consistent with evolution of the suite primarily by fractional crystallization. Concave-down patterns are an indication of their 'normal' depleted chemical character (N-MORB).

or P-types, indicative of an 'enriched' or 'plume' component (**Table 1**) typically associated with intraplate 'hot spots'. Transitional varieties are classified as T-MORB. Enriched MORB are volumetrically minor on most normal ridge segments, but can comprise a significant proportion of the crust around regions influenced by plume magmatism such as the Galapagos Islands, the Azores, Tristan, Bouvet, and Iceland.

Mineralogy of Mid-ocean Ridge Basalts

The minerals that crystallize from MORB magmas are not only dependent on the composition of the melt, but also the temperature and pressure during crystallization. Because the majority of MORB magmas have relatively similar major element compositions and probably begin to crystallize within the uppermost mantle and oceanic crust (pressures less than 0.3 GPa), they have similar mineralogy. Textures (including grain size) vary depending on nucleation and crystallization rates. Hence lavas, that are quenched when erupted into sea water, have few phenocrysts in a glassy to cryptocrystalline matrix. Conversely, magmas that cool slowly in subaxial reservoirs or magma chambers form gabbros that are totally crystalline (holocrystalline) and composed of well-formed minerals that can be up to a few centimeters long. Many of the gabbros recovered from the ocean floor do not represent melt compositions but rather reflect the accumulation of crystals and percolation of melt that occurs during convection, deformation and fractional crystallization in the mush zone hypothesized to exist beneath some mid-ocean ridges (**Figure 2**). These cumulate gabbros are composed of minerals that have settled (or floated) out of cooling MORB magmas and their textures often reflect compaction, magmatic sedimentation, and deformation.

MOR lavas may contain millimeter-sized phenocrysts of the silicate minerals plagioclase (solid solution that ranges from $CaAl_2Si_2O_8$ to $NaAlSi_3O_8$) and olivine (Mg_2SiO_4 to Fe_2SiO_4) and less commonly, clinopyroxene ($Ca[Mg,Fe]Si_2O_6$). Spinel, a Cr-Al rich oxide, is a common accessory phase in more magnesian lavas where it is often enclosed in larger olivine crystals. Olivine is abundant in the most MgO-rich lavas, becomes less abundant in more evolved lavas and is ultimately replaced by pigeonite (a low-Ca pyroxene) in FeO-rich basalts and andesite. Clinopyroxene is only common as a phenocryst phase in relatively evolved lavas. Titanomagnetite, ilmenite and rare apatite are present as microphenocrysts, although not abundantly, in basaltic andesites and andesites.

Intrusive rocks, which cool slowly within the oceanic crust, have similar mineralogy but are holocrystalline and typically much coarser grained. Dikes form fine- to medium-grained diabase containing olivine, plagioclase and clinopyroxene as the major phases, with minor amounts of ilmenite and magnetite. Gabbros vary from medium-grained to very coarse-grained with crystals up to a few centimeters in length. Because of their cumulate nature and extended cooling histories, gabbros often exhibit layering of crystals and have the widest mineralogic variation. Similar to MORB, the least-evolved varieties (troctolites) consist almost entirely of plagioclase and olivine. Some gabbros can be nearly monomineralic such as anorthosites (plagioclase-rich) or contain monomineralic layers (such as olivine that forms layers or lenses of a rock called dunite). The most commonly recovered varieties of gabbro are composed of plagioclase, augite (a clinopyroxene) and hypersthene (orthopyroxene) with minor amounts of olivine, ilmenite and magnetite and, in some cases, hornblende (a hydrous Fe-Mg silicate that forms during the latest stages of crystallization). Highly evolved liquids cool to form ferrogabbros and even rarer silica-rich plutonics known as trondhjemites or plagiogranites.

The descriptions above pertain only to those portions of the oceanic crust that have not been tectonized or chemically altered. Because of the dynamic nature of oceanic ridges and the pervasive hydrothermal circulation related to magmatism, it is common for the basaltic rocks comprising the crust to be chemically altered and metamorphosed. When this occurs, the primary minerals are recrystallized or replaced by a variety of secondary minerals such as smectite, albite, chlorite, epidote, and amphibole that are more stable under lower temperature and more hydrous conditions. MOR basalts, diabases and gabbros are commonly metamorphosed to greenschists and amphibolites. Plutonic rocks and portions of the upper mantle rich in olivine and pyroxene are transformed into serpentinites. Oceanic metamorphic rocks are commonly recovered from transform faults, fracture zones and slowly spreading segments of the MOR where tectonism and faulting facilitate deep penetration of sea water into the crust and upper mantle.

Chemical Variability

Although MORB form a relatively homogeneous population of rock types when compared to lavas erupted at other tectonic localities, there are subtle, yet significant, chemical differences in their chemistry due to variability in source composition, depth and extent of melting, magma mixing, and processes that modify primary magmas in the shallow lithosphere. Chemical differences between MORB exist on all scales, from individual flows erupted along the same ridge segment (e.g., CoAxial Segment of the Juan de Fuca Ridge) to the average composition of basalts from the global ridge system (e.g. Mid-Atlantic Ridge vs. East Pacific Rise). High-density sampling along several MOR segments has shown that quite a diversity of lava compositions can be erupted over short time (10s–100 years) and length scales (100 m to a few kilometers). Slow spreading ridges, which do not have steady-state magma bodies, generally erupt more mafic lavas compared to fast spreading ridges where magmas are more heavily influenced by fractional crystallization in shallow magma bodies. Intermediate rate-spreading centers, where magma lenses may be small and intermittent, show characteristics of both slow- and-fast spreading centers. In environments where magma supply is low or mixing is inhibited, such as proximal to transform faults, propagating rift tips and overlapping spreading centers, compositionally diverse and highly differentiated lavas are commonly found (such as the Eastern Galapagos Spreading Center, **Figures 3, 4** and **6**). In these environments, extensive fractional crystallization is a consequence of relatively cooler thermal regimes and the magmatic processes associated with rift propagation.

Local variability in MORB can be divided into two categories: (1) those due to processes that affect an individual parental magma (e.g., fractional crystallization, assimilation) and (2) those created via partial melting and transport in a single melting regime (e.g., melting in a rising diapir). In contrast, global variations reflect regional variations in mantle source chemistry and temperature, as well as the averaging of melts derived from diverse melting regimes (e.g. accumulative polybaric fractional melting). At any given segment of MOR, variations may be due to various combinations of these processes.

Local Variability

Chemical trends defined by suites of related MOR lavas are primarily due to progressive fractional crystallization of variable combinations and proportions of olivine, plagioclase and clinopyroxene as a magma cools. The compositional 'path' that a magma takes is known as its liquid line of descent (LLD). Slightly different trajectories of LLDs (**Figure** 5) are a consequence of the order of crystallization and the different proportions of crystallizing phases that are controlled by initial (and subsequent changing) liquid composition, temperature, and pressure. In some MORB suites, linear elemental trends may be due to mixing of primitive magmas with more evolved magmas that have evolved along an LLD.

Suites of MORB glasses often define distinctive LLDs that match those determined by experimental crystallization of MORB at low to moderate

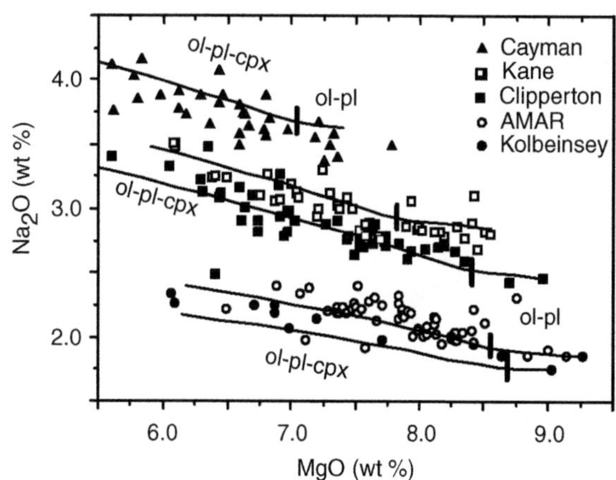

Figure 5 MgO vs. Na_2O in MORB from five different Ridge segments (Mid Cayman Rise in the Caribbean; near Kane Fracture Zone on the Mid-Atlantic Ridge, 23°N; AMAR on the Mid-Atlantic Ridge around 37°N; East Pacific Rise near the Clipperton Fracture Zone around 10°N; Kolbeinsey Ridge north of Iceland. Lines are calculated Liquid Lines of Descent (LLDs) from high MgO parents. Bar shows where clinopyroxene joins plagioclase and olivine as a fractionating phase. Na_8 is determined by the values of Na_2O when the LLD is at MgO of 8 wt%. (Adapted with permission from Langmuir *et al.*, 1992.)

pressures that correspond to depths of ~1 to 10 km within the oceanic crust and upper mantle. Much of the major element data from fast-spreading ridges like the East Pacific Rise are best explained by low-pressure (~0.1 GPa) fractional crystallization whereas at slow-spreading ridges like the Mid-Atlantic Ridge data require higher pressure crystallization (~0.5–1.0 GPa). This is consistent with other evidence suggesting that magmas at fast-spreading ridges evolve in a shallow magma lens or chambers and that magmas at slow-spreading ridges evolve at significantly greater depths; possibly in the mantle lithosphere or at the crust–mantle boundary. Estimated depths of crystallization correlate with increased depths of magma lens or fault rupture depth related to decreasing spreading rate.

Cogenetic lavas (those from the same or similar primary melts) generated by fractional crystallization exhibit up to 10-fold enrichments of incompatible trace elements (e.g., Zr, Nb, Y, Ba, Rb, REE) that covary with indices of fractionation such as decreasing MgO (**Figure 6**) and increasing K_2O concentrations and relatively constant incompatible trace element ratios irrespective of rock type. In general, the rare earth elements show systematic increases in abundance through the fractionation sequence from MORB to andesite (**Figure 4**) with a slight increase in light rare earth elements relative to the heavy-rare earth elements. The overall enrichments in the trivalent rare earth elements is a consequence of their incompatibility in the crystals separating from the cooling magma. Increasing negative Eu anomalies develop in more fractionated lavas due to the continued removal of plagioclase during crystallization because Eu partially substitutes for Ca in plagioclase which is removed during fractional crystallization.

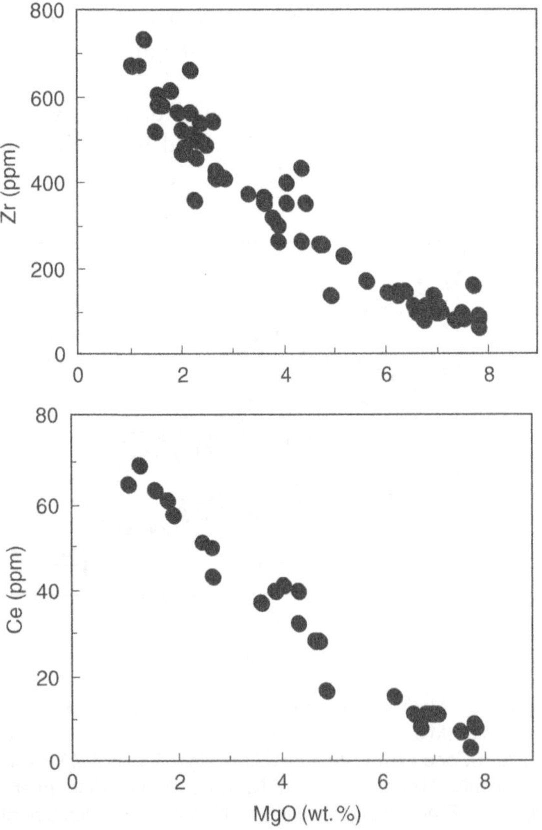

Figure 6 Trace element (Zr and Ce) versus MgO variation diagram showing the systematic enrichments of these highly incompatible elements with increasing fractionation in a suite of cogenetic lavas from the Eastern Galapagos Spreading Center.

Global Variability

MORB chemistry of individual ridge segments (local scale) is, in general, controlled by the relative balance between tectonic and magmatic activity, which in turn may determine whether a steady-state magma chamber exists, and for how long. Ultimately, the tectonomagmatic evolution is controlled by temporal variations in input of melt from the mantle. Global correlation of abyssal peridotite and MORB geochemical data suggest that the extent of mantle melting beneath normal ridge segments increases with increasing spreading rate and that both ridge morphology and lava composition are related to spreading rate.

The depths at which primary MORB melts form and equilibrate with surrounding mantle remain controversial (possibly 30 to 100 km), as does the mechanism(s) of flow of magma and solid mantle beneath divergent plate boundaries. The debate is critical for understanding the dynamics of plate spreading and is focused on whether flow is 'passive' plate driven flow or 'active' buoyantly driven solid convection. At present, geological and geophysical observations support passive flow which causes melts from a broad region of upwelling and melting to converge in a narrow zone at ridge crests.

It has also been hypothesized that melting beneath ridges is a dynamic, near-fractional process during which the pressure, temperature, and composition of the upper mantle change. Variations in these parameters as well as in the geometry of the melting region result in the generation of MORB with different chemical characteristics.

Differences in the major element compositions of MORB from different parts of the world's oceans (global scale) have been recognized for some time. In general, it has been shown that N-type MORB from slow-spreading ridges such as the Mid-Atlantic Ridge are more primitive (higher MgO) and have greater Na_2O, Al_2O_3 and lower FeO and CaO/Al_2O_3 contents at given MgO values than lavas from medium- and fast-spreading ridges (**Figure** 7). A comparison of ocean floor glass compositions (over 9000) analyzed by electron microprobe at the Smithsonian Institution from major spreading centers and seamounts is presented in **Table 1**. The analyses have been filtered into normal (N-MORB) and enriched (E-MORB) varieties based on their K/Ti ratios (E-MORB $[K_2O/TiO_2] \times 100 > 13$) which reflect enrichment in the highly incompatible elements. These data indicate that on average, MORB are relatively differentiated compared to magmas that might be generated directly from the mantle (compare averages with picritic basalts from the Pacific in **Table 1**). Furthermore, given the variability of glass compositions in each region, N-MORB have quite similar average major element compositions (most elemental concentrations overlap at the 1-sigma level). E-MORB, are more evolved than N-MORB

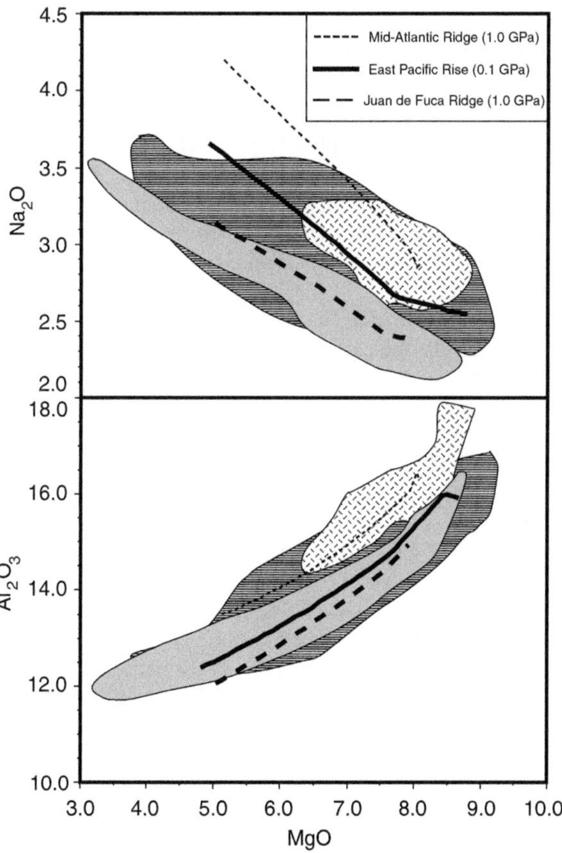

Figure 7 Major element variation diagrams showing compositional ranges from different spreading rate ridges. Generally higher Na_2O and Al_2O_3 concentrations in Mid-Atlantic Ridge (hatchured field) lavas in comparison to MORB from the Juan de Fuca (grey field) and East Pacific Rise (dark field) are shown. Lines show calculated liquid lines of descent at 0.1 and 1.0 GPa for parental magmas from each ridge.

from comparable regions of the ocean and there are a higher proportion of E-MORB in the Atlantic (31%) compared to the Pacific (12%) and Galapagos Spreading Center region (7%). Unlike the Atlantic where E-MORB are typically associated with inflated portions of the ridge due to the effects of plume–ridge interaction, East Pacific Rise E-MORB are randomly dispersed along-axis and more commonly recovered off-axis. As well as having higher K_2O contents than N-MORB, E-MORB have higher concentrations of P_2O_5, TiO_2, Al_2O_3 and Na_2O and lower concentrations of SiO_2, FeO and CaO. Positive correlations exist between these characteristics, incompatible element enrichments and more radiogenic Sr and Nd isotopes in progressively more enriched MORB.

Direct comparison of elemental abundances between individual MORB (or even groups) is difficult because of the effects of fractional crystallization. Consequently, fundamental differences in chemical characteristics are generally expressed as differences in parameters such as Na_8, Fe_8, Al_8, Si_8 etc. which are the values of these oxides calculated at an MgO content of 8.0 wt% (**Figure 5** and **8**). When using these normalized values, regionally averaged major element data show a strong correlation with ridge depth and possibly, crustal thickness. MORB with high FeO and low Na_2O are sampled from shallow ridge crests with thick crust whereas low FeO– high Na_2O MORB are typically recovered from deep ridges with thin crust (**Figure 8**). This chemical/tectonic correlation gives rise to the so-called 'global array'. Major element melting models indicate there is a strong correlation between the initial depth of melting and the total amount of melt formed. As a consequence, when temperatures are high enough to initiate melting at great depths, the primary MORB melts contain high FeO, low Na_2O and low SiO_2. Conversely, if the geothermal gradient is low, melting is restricted to the uppermost part of the upper mantle, and little melt is generated (hence thinner crust) and the basaltic melts contain low FeO, high Na_2O and relatively high SiO_2.

Although the global systematics appear robust, detailed sampling of individual ridge segments have shown MORB from limited areas commonly exhibit chemical correlations that form a 'local trend' opposite to the chemical correlations observed globally (e.g., FeO and Na_2O show a positive correlation). A local trend may reflect the spectrum of melts formed at different depths beneath one ridge crest rather than the aggregate of all the melt increments.

Although the original hypothesis that global variations in MORB major element chemistry are a consequence of total extents of mantle melting and

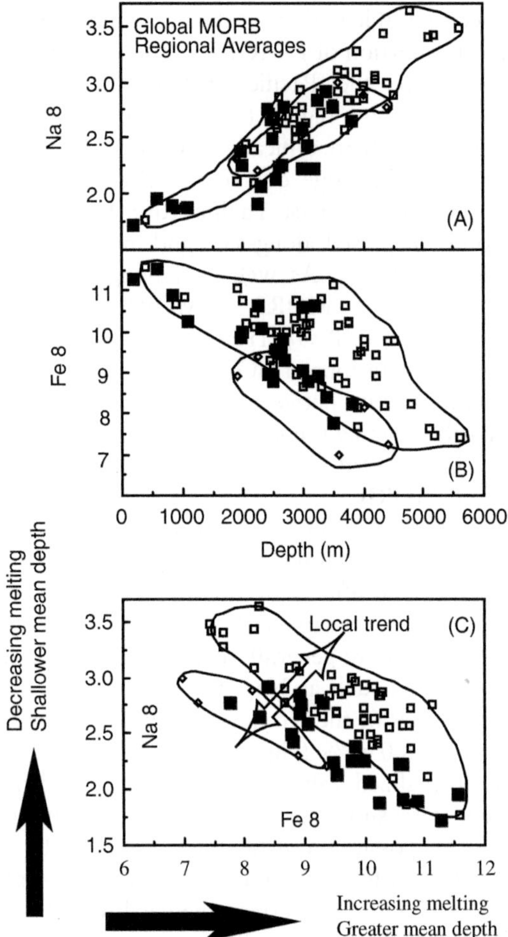

Figure 8 (A) and (B) Global correlations between regional averages of ridge axial depth and the Na_8 and Fe_8 of MORBs. Different groups of MORB are distinguished. □, Normal ridge segments; ◇, ridges behind island arcs; ■, ridges influenced by hot spots. (C) Global trend of Na_8 vs. Fe_8 due to differences in extents and depths of melting. Representative 'Local trend' is common along individual portions of some ridges. (Adapted with permission from Langmuir *et al.*, 1992.)

mean pressure of extraction due to variations in mantle temperature, more recent evidence suggests that heterogeneity in the mantle also plays an important role in defining both global and local chemical trends. In particular, U-series data suggest some MDRB melts equilibrate with highly depleted mantle at shallow depths whereas others equilibrate with less depleted garnet pesidatite at depths greater than ~80 km.

Conclusions

Passive rise of the mantle beneath oceanic spreading centers results in the decompression melting of upwelling peridotite which gives rise to a spectrum of MORB compositions varying from extremely depleted to moderately enriched varieties. The compositional variability in primary MORB result from combinations of differing source compositions, extents and styles of partial melting, and depths of melt formation. The moderately evolved composition of most MORB primarily reflects the effects of crystal fractionation that occurs as the primary melts ascend from the mantle into the cooler crust. Although MORB are relatively homogeneous compared to basalts from other tectonic environments, they exhibit a range of compositions that provide us with information about the composition of the mantle, the influence of plumes, and dynamic magmatic processes that occur to form the most voluminous part of the Earth's crust.

Further Reading

Batiza R and Niu Y (1992) Petrology and magma chamber processes at the East Pacific Rise – 9°30′N. *Journal of Geophysical Research* 97: 6779–6797.

Grove TL, Kinzler RJ, and Bryan WB (1992) Fractionation of Mid-Ocean Ridge Basalt (MORB). In: Phipps-Morgan J, Blackman DK, and Sinton J (eds.) *Mantle Flow and Melt Generation at Mid-ocean Ridges,* Geophys. Monograph 71, pp. 281–310. Washington, DC: American Geophysical Union.

Klein EM and Langmuir CH (1987) Global correlations of ocean ridge basalt chemistry with axial depth and crustal thickness. *Journal of Geophysical Research* 92: 8089–8115.

Kurras GJ, Fomari DJ, Edwards MH, Perfit MR, and Smith MC (2000) Volcanic morphology of the East Pacific Rise Crest 9° 49′–52′N:1. Implications for volcanic emplacement processes at fast-spreading mid-ocean ridges. *Marine Geophysical Research* 21: 23–41.

Langmuir CH, Klein EM, and Plank T (1992) Petrological systematics of mid-ocean ridge basalts: constraints on melt generation beneath ocean ridges. In: Phipps-Morgan J, Blackman DK and Sinton J (eds) *Mantle Flow and Melt Generation at Mid-ocean Ridges*, Geophysics Monograph 71, Washington, DC: American Geophysical Union, p. 183–280.

Lundstrom CC, Sampson DE, Perfit MR, Gill J, and Williams Q (1999) Insight, into mid-ocean ridge basalt petrogenesis: U-series disequilibrium from the Siqueiro, Transform, lamont seamounts, and East Pacific Rise *Journal of Geophysical Research* 104: 13035–13048.

Macdonald KC (1998) Linkages between faulting, volcanism, hydrothermal activity and segmentation on fast spreading centers. In: Buck WR, Delaney PT, Karson JA and Lagabrielle Y (eds) *Faulting and Magmatism at Mid-ocean ridges*, American Geophysics Monograph 106, Washington, DC: American Geophysical Union, p. 27–59.

Nicolas A (1990) *The Mid-Ocean Ridges*. Springer Verlag, Berlin.

Niu YL and Batiza R (1997) Trace element evidence from seamounts for recycled oceanic crust in the Eastern Pacific mantle. *Earth Planet. Sci. Lett* 148: 471–483.

Perfit MR and ChadwickWW (1998) Magmatism at mid-ocean ridges: constraints from volcanological and geochemical investigations. In: Buck WR, Delaney PT, Karson JA and Lagabrielle Y (eds), *Faulting and Magmatism at Mid-ocean Ridges*. American Geophysics Monograph 106, Washington, DC: American Geophysics Union, p. 59–115.

Perfit MR and Davidson JP (2000) Plate tectonics and volcanism. In: Sigurdsoon H (ed.) *Encyclopedia of Volcanoes*. San Diego: Academic Press.

Perfit MR, Ridley WI, and Jonasson I (1998) Geologic, petrologic and geochemical relationships between magmatism and massive sulfide mineralization along the eastern Galapagos Spreading Center. *Review in Economic Geology* 8(4): 75–99.

Shen Y and Forsyth DW (1995) Geochemical constraints on initial and final depths of melting beneath mid-ocean ridges. *Journal of Geophysical Research* 100: 2211–2237.

Sigurdsson H (ed.) (2000) *Encyclopedia of Volcanoes*. Academic Press

Sinton JM and Detrick RS (1992) Mid-ocean ridge magma chambers. *Journal of Geophysical Research* 97: 197–216.

Smithsonian Catalog of Basalt Glasses [http://www.nmnh.si.edu/minsci/research/glass/index.htm]

Thompson RN (1987) Phase-equilibria constraints on the genesis and magmatic evolution of oceanic basalts. *Earth Science Review* 24: 161–210.

MANGANESE NODULES

D. S. Cronan, Royal School of Mines, London, UK

Introduction

Manganese nodules, together with micronodules and encrustations, are ferromanganese oxide deposits which contain variable amounts of other elements (**Table 1**). They occur throughout the oceans, although the economically interesting varieties have a much more restricted distribution. Manganese nodules are spherical to oblate in shape and range in size from less than 1 cm in diameter up to 10 cm or more. Most accrete around a nucleus of some sort, usually a volcanic fragment but sometimes biological remains.

The deposits were first described in detail in the Challenger Reports. This work was co-authored by J. Murray and A. Renard, who between them initiated

Table 1 Average abundances of elements in ferromanganese oxide deposits

	Pacific Ocean	*Atlantic Ocean*	*Indian Ocean*	*Southern Ocean*	*World Ocean average*	*Crustal abundance*	*Enrichment factor*	*Shallow marine*	*Lakes*
B	0.0277	—	—	—	—	0.0010	27.7		
Na	2.054	1.88	—	—	1.9409	2.36	0.822	0.81	0.22
Mg	1.710	1.89	—	—	1.8234	2.33	0.782	0.55	0.26
Al	3.060	3.27	2.49	—	2.82	8.23	0.342	1.80	1.16
Si	8.320	9.58	11.40	—	8.624	28.15	0.306	8.76	5.38
P	0.235	0.098	—	—	0.2244	0.105	2.13	0.91	0.15
K	0.753	0.567	—	—	0.6427	2.09	0.307	1.30	0.40
Ca	1.960	2.96	2.37	—	2.47	4.15	0.595	2.40	1.14
Sc	0.00097	—	—	—	—	0.0022	0.441		
Ti	0.674	0.421	0.662	0.640	0.647	0.570	1.14	0.212	0.338
V	0.053	0.053	0.044	0.060	0.0558	0.0135	4.13	0.012	0.001
Cr	0.0013	0.007	0.0029	—	0.0035	0.01	0.35	0.002	0.006
Mn	19.78	15.78	15.10	11.69	16.02	0.095	168.6	11.88	12.61
Fe	11.96	20.78	14.74	15.78	15.55	5.63	2.76	21.67	21.59
Co	0.335	0.318	0.230	0.240	0.284	0.0025	113.6	0.008	0.013
Ni	0.634	0.328	0.464	0.450	0.480	0.0075	64.0	0.014	0.022
Cu	0.392	0.116	0.294	0.210	0.259	0.0055	47.01	0.002	0.003
Zn	0.068	0.084	0.069	0.060	0.078	0.007	11.15	0.011	0.051
Ga	0.001	—	—	—	—	0.0015	0.666		
Sr	0.085	0.093	0.086	0.080	0.0825	0.0375	2.20		
Y	0.031	—	—	—	—	0.0033	9.39	0.002	0.002
Zr	0.052	—	—	0.070	0.0648	0.0165	3.92	0.004	0.045
Mo	0.044	0.049	0.029	0.040	0.0412	0.00015	274.66	0.004	0.003
Pd	0.602^{-6}	0.574^{-6}	0.391^{-6}	—	0.553^{-6}	0.665^{-6}	0.832		
Ag	0.0006	—	—	—	—	0.000007	85.71		
Cd	0.0007	0.0011	—	—	0.00079	0.00002	39.50		
Sn	0.00027	—	—	—	—	0.00002	13.50		
Te	0.0050	—	—	—	—	—	—		
Ba	0.276	0.498	0.182	0.100	0.2012	0.0425	4.73	0.287	0.910
La	0.016	—	—	—	—	0.0030	5.33		0.027
Yb	0.0031	—	—	—	—	0.0003	10.33		
W	0.006	—	—	—	—	0.00015	40.00		
Ir	0.939^{-6}	0.932^{-6}	—	—	0.935^{-6}	0.132^{-7}	70.83		
Au	0.266^{-6}	0.302^{-6}	0.811^{-7}	—	0.248^{-6}	0.400^{-6}	0.62		
Hg	0.82^{-4}	0.16^{-4}	0.15^{-6}	—	0.50^{-4}	0.80^{-5}	6.25		
Tl	0.017	0.0077	0.010	—	0.0129	0.000045	286.66		
Pb	0.0846	0.127	0.093	—	0.090	0.00125	72.72	0.002	0.063
Bi	0.0006	0.0005	0.0014	—	0.0008	0.000017	47.05		

Note: Superscript numbers denote powers of ten, e.g. $^{-6} = \times 10^{-6}$.
Reproduced with permission from Cronan (1980).

the first great manganese nodule controversy. Murray believed the deposits to have been formed by submarine volcanic processes whereas Renard believed that they had precipitated from continental runoff products in sea water. This controversy remained unresolved until it was realized that nodules could obtain their metals from either or both sources. The evidence for this included the finding of abundant nodules in the Baltic Sea where there are no volcanic influences, and the finding of rapidly grown ferromanganese oxide crusts associated with submarine hydrothermal activity of volcanic origin on the Mid-Atlantic Ridge. Subsequently, a third source of metals to the deposits was discovered, diagenetic remobilization from underlying sediments. Thus marine ferromanganese oxides can be represented on a triangular diagram (**Figure 1**), the corners being occupied by hydrothermal (volcanically derived), hydrogenous (seawater derived) and diagenetic (sediment interstitial water derived) constituents.

There appears to be a continuous compositional transition between hydrogenous and diagenetic deposits, all of which are formed relatively slowly at normal deep seafloor temperatures. By contrast, although theoretically possible, no continuous compositional gradation has been reported between hydrogenous and hydrothermal deposits, although mixtures of the two do occur. This may be partly because (1) the growth rates of hydrogenous and hydrothermal deposits are very different with the latter accumulating much more rapidly than the former leading to the incorporation of only limited amounts of the more slowly accumulating hydrogenous material in them, and (2) the temperatures of formation of the deposits are different leading to mineralogical differences between them which can affect their chemical composition. Similarly, a continuous compositional gradation between hydrothermal and diagenetic ferromanganese oxide deposits has not been found, although again this is theoretically possible. However, the depositional conditions with which the respective deposits are associated i.e., high temperature hydrothermal activity in mainly sediment-free elevated volcanic areas on the one hand, and low-temperature accumulation of organic rich sediments in basin areas on the other, would preclude much mixing between the two. Possibly they may occur in sedimented active submarine volcanic areas.

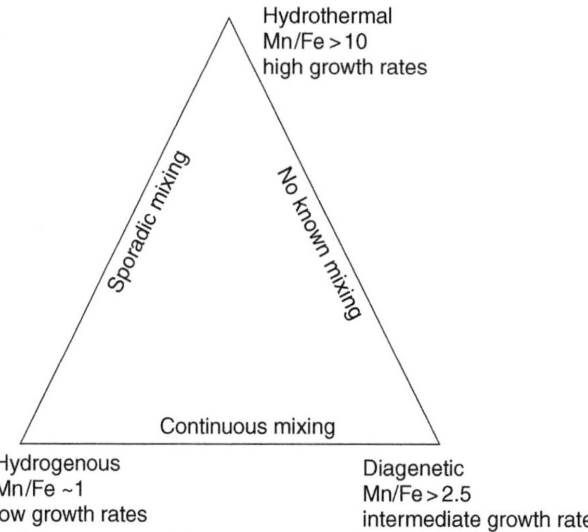

Figure 1 Triangular representation of marine ferromanganese oxide deposits.

Internal Structure

The main feature of the internal structure of nodules is concentric banding which is developed to a greater or lesser extent in most of them (**Figure 2**). The bands represent thin layers of varying reflectivity in polished section, the more highly reflective layers being generally richer in manganese than the more poorly reflective ones. They are thought to possibly represent varying growth conditions.

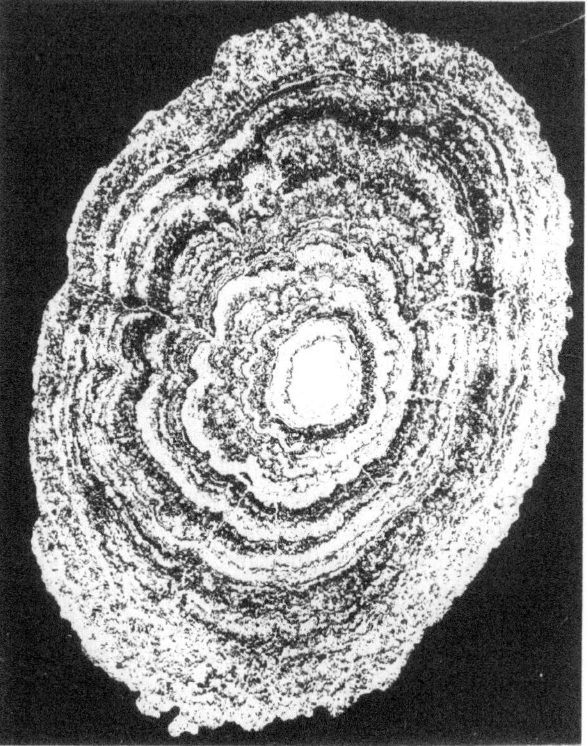

Figure 2 Concentric banding in a manganese nodule. (Reproduced by kind permission of CNEXO, France.)

On a microscopic scale, a great variety of structures and textures are apparent in nodules, some of them indicative of postdepositional alteration of nodule interiors. One of the most commonly observed and most easily recognizable is that of collomorphic globular segregations of ferromanganese oxides on a scale of tenths of a millimeter or less, which often persist throughout much of the nodule interior. Often the segregations become linked into polygons or cusps elongated radially in the direction of growth of the nodules. Several workers have also recognized organic structures within manganese nodules. Furthermore, cracks and fissures of various sorts are a common feature of nodule interiors. Fracturing of nodules is a process which can lead to their breakup on the seafloor, in some cases as a result of the activity of benthic organisms, or of bottom currents. Fracturing is an important process in limiting the overall size of nodules growing under any particular set of conditions.

Growth Rates

It is possible to assess the rate of growth of nodules either by dating their nuclei, which gives a minimum rate of growth, or by measuring age differences between their different layers. Most radiometric dating techniques indicate a slow growth rate for nodules, from a few to a few tens of millimeters per million years. Existing radiometric and other techniques for nodule dating include uranium series disequilibrium methods utilizing ^{230}Th ^{231}Pa, the ^{10}Be method, the K-Ar method, fission track dating of nodule nuclei, and hydration rind dating.

In spite of the overwhelming evidence for slow growth, data have been accumulating from a number of sources which indicate that the growth of nodules may be variable with periods of rapid accumulation being separated by periods of slower, or little or no growth. In general, the most important factor influencing nodule growth rate is likely to be the rate at which elements are supplied to the deposits, diagenetic sources generally supplying elements at a faster rate than hydrogenous sources (**Figure 1**). Further, the tops, bottoms and sides of nodules do not necessarily accumulate elements at the same rate, leading to the formation of asymmetric nodules in certain circumstances (**Figure 3**). Differences in the surface morphology between the tops, bottoms and sides of nodules *in situ* may also be partly related to growth rate differences. The tops receive slowly accumulating elements hydrogenously supplied from seawater and are smooth, whereas the bottoms receive more rapidly accumulating elements diagenetically supplied from the interstitial waters of the sediments and are rough (**Figure 3**). The 'equatorial bulges' at the sediment–water interface on some nodules have a greater abundance of organisms on them than elsewhere on the nodule surface, suggesting that the bulges may be due to rapid growth promoted by the organisms.

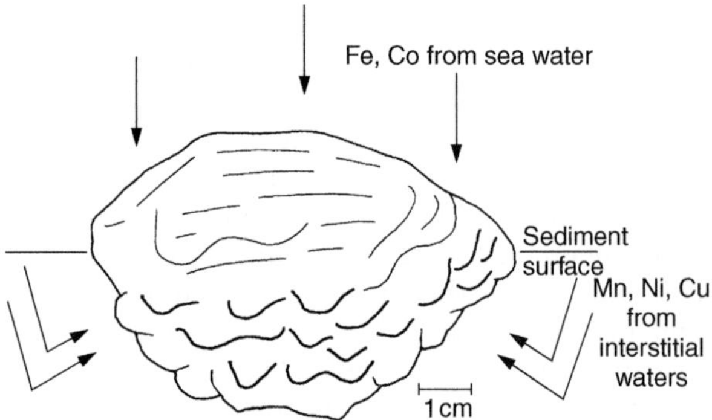

Figure 3 Morphological and compositional differences between the top and bottom of a Pacific nodule. (Reproduced with permission from Cronan, 1980.)

It is evident therefore that nodule growth cannot be regarded as being continuous or regular. Nodules may accrete material at different rates at different times and on different surfaces. They may also be completely buried for periods of time during which it is possible that they may grow from interstitial waters at rates different from those while on the surface, or possibly not grow at all for some periods. Some even undergo dissolution, as occurs in the Peru Basin where some nodules get buried in suboxic to reducing sediments.

Distribution of Manganese Nodules

The distribution and abundance of manganese nodules is very variable on an oceanwide basis, and can also be highly variable on a scale of a kilometer or less. Nevertheless, there are certain regional regularities in average nodule abundance that permit some broad areas of the oceans to be categorized as containing abundant nodules, and others containing few nodules (**Figure 4**), although it should always be borne in mind that within these regions local variations in nodule abundance do occur.

The distribution of nodules on the seafloor is a function of a variety of factors which include the presence of nucleating agents and/or the nature and age of the substrate, the proximity of sources of elements, sedimentation rates and the influence of organisms. The presence of potential nuclei on the seafloor is of prime importance in determining nodule distribution. As most nodule nuclei are volcanic in origin, patterns of volcanic activity and the

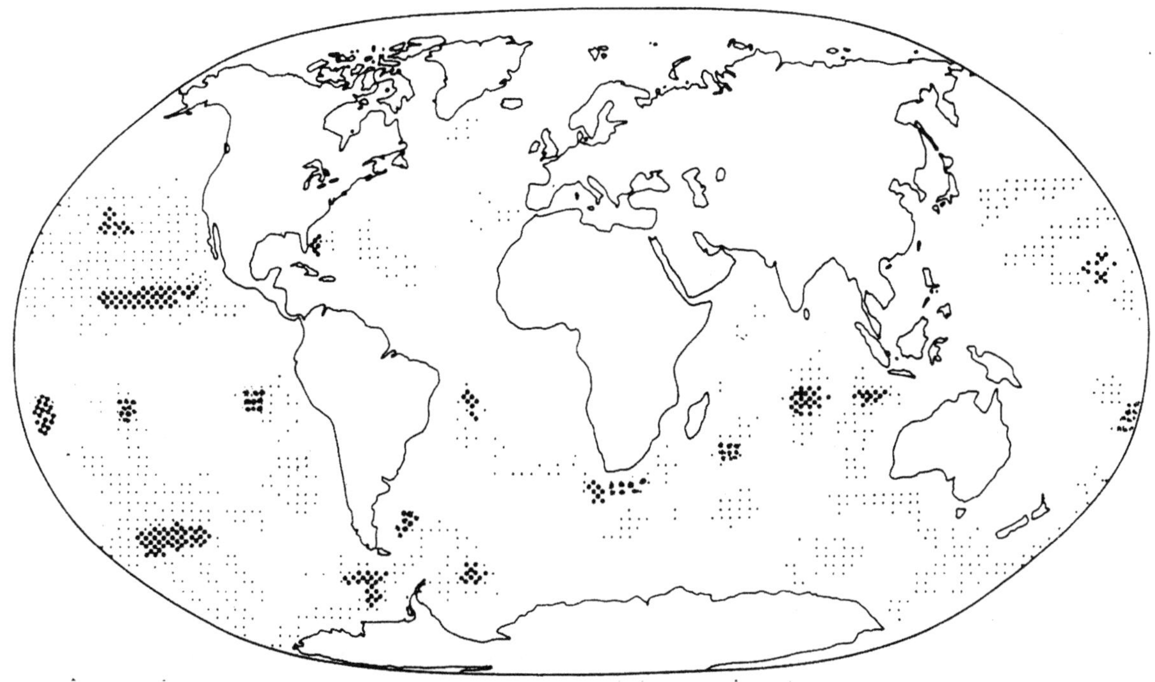

Figure 4 Distribution of mangancese nodules in the oceans (updated from Cronan, 1980 after various sources.) :::, Areas of nodule coverage; ⁝⁝⁝, areas where nodules are locally abundant.

subsequent dispersal of volcanic materials have an important influence on where and in what amounts nodules occur. Other materials can also be important as nodule nuclei. Biogenic debris such as sharks' teeth, can be locally abundant in areas of slow sedimentation and their distribution will in time influence the abundance of nodules in such areas.

As most nuclei are subject to replacement with time, old nodules have sometimes completely replaced their nuclei and have fractured, thus providing abundant nodule fragments to serve as fresh nuclei for ferromanganese oxide deposition. In this way, given sufficient time, areas which initially contained only limited nuclei may become covered with nodules.

One of the most important factors affecting nodule abundance on the seafloor is the rate of accumulation of their associated sediments, low sedimentation rates favoring high nodule abundances. Areas of the seafloor where sedimentation is rapid are generally only sparsely covered with nodules. For example, most continental margin areas have sedimentation rates that are too rapid for appreciable nodule development, as do turbidite-floored deep-sea abyssal plains. Low rates of sedimentation can result either from a minimal sediment supply to the seafloor or currents inhibiting its deposition. Large areas in the centers of ocean basins receive minimal sediment input. Under these conditions substantial accumulation of nodules at the sediment surface is favored.

Worldwide Nodule Distribution Patterns

Pacific Ocean As shown in, nodules are abundant in the Pacific Ocean in a broad area, called the Clarion–Clipperton Zone, between about 6°N and 20°N, extending from approximately 120°W to 160°W. The limits of the area are largely determined by sedimentation rates. Nodules are also locally abundant further west in the Central Pacific Basin. Sediments in the northern part of the areas of abundant nodules in the North Pacific are red clays with accumulation rates of around 1 mm per thousand years whereas in the south they are siliceous oozes with accumulation rates of 3 mm per thousand years, or more.

Nodule distribution appears to be more irregular in the South Pacific than in the North Pacific, possibly as a result of the greater topographic and sedimentological diversity of the South Pacific. The nodules are most abundant in basin environments such as those of the south-western Pacific Basin, Peru Basin, Tiki Basin, Penrhyn Basin, and the Circum-Antarctic area.

Indian Ocean In the Indian Ocean the most extensive areas of nodule coverage are to the south of the equator. Few nodules have been recorded in the Arabian Sea or the Bay of Bengal, most probably because of the high rates of terrigenous sediment input in these regions from the south Asian rivers. The equatorial zone is also largely devoid of nodules. High nodule concentrations have been recorded in parts of the Crozet Basin, in the Central Indian Ocean Basin and in the Wharton Basin.

Atlantic Ocean Nodule abundance in the Atlantic Ocean appears to be more limited than in the Pacific or Indian Oceans, probably as a result of its relatively high sedimentation rates. Another feature which inhibits nodule abundance in the Atlantic is that much of the seafloor is above the calcium carbonate compensation depth (CCD). The areas of the Atlantic where nodules do occur in appreciable amounts are those where sedimentation is inhibited. The deep water basins on either side of the Mid-Atlantic Ridge which are below the CCD and which accumulate only limited sediment contain nodules in reasonable abundance, particularly in the western Atlantic. Similarly, there is a widespread occurrence of nodules and encrustations in the Drake Passage–Scotia Sea area probably due to the strong bottom currents under the Circum-Antarctic current inhibiting sediment deposition in this region. Abundant nodule deposits on the Blake Plateau can also be related to high bottom currents.

Buried nodules Most workers on the subject agree that the preferential concentration of nodules at the sediment surface is due to the activity of benthic organisms which can slightly move the nodules. Buried nodules have, however, been found in all the oceans of the world. Their abundance is highly variable, but it is possible that it may not be entirely random. Buried nodules recovered in large diameter cores are sometimes concentrated in distinct layers. These layers may represent ancient erosion surfaces or surfaces of nondeposition on which manganese nodules were concentrated in the past. By contrast, in the Peru Basin large asymmetrical nodules get buried when their bottoms get stuck in tenacious suboxic sediment just below the surface layer.

Compositional Variability of Manganese Nodules

Manganese nodules exhibit a continuous mixing from diagenetic end members which contain the mineral 10Å manganite (todorokite) and are enriched in Mn, Ni and Cu, to hydrogenous end members which contain the mineral δ MnO_2 (vernadite) and are enriched in Fe and Co. The diagenetic deposits derive their metals at least in part from the recycling through the sediment interstitial waters of elements originally contained in organic phases on their decay and dissolution in the sediments, whereas the hydrogenous deposits receive their metals from normal sea water or diagenetically unenriched interstitial waters. Potentially ore-grade manganese nodules of resource interest fall near the diagenetic end member in composition. These are nodules that are variably enriched in Ni and Cu, up to a maximum of about 3.0% combined.

One of the most striking features shown by chemical data on nodules are enrichments of many elements over and above their normal crustal abundances (**Table 1**). Some elements such as Mn, Co, Mo and Tl are concentrated about 100-fold or more; Ni, Ag, Ir and Pb are concentrated from about 50- to 100-fold, B, Cu, Zn, Cd, Yb, W and Bi from about 10 to 50-fold and P, V, Fe, Sr, Y, Zr, Ba, La and Hg up to about 10-fold above crustal abundances.

Regional Compositional Variability

Pacific Ocean In the Pacific, potentially ore-grade nodules are generally confined to two zones running roughly east–west in the tropical regions, which are well separated in the eastern Pacific but which converge at about 170°–180°W (**Figure 5**). They follow the isolines of intermediate biological productivity, strongly suggestive of a biological control on their distribution. Within these zones, the nodules preferentially occupy basin areas near or below the CCD. Thus they are found in the Peru Basin, Tiki Basin, Penrhyn Basin, Nova Canton Trough area, Central Pacific Basin and Clarion–Clipperton Zone (**Figure 5**). Nodules in all these areas have features in common and are thought to have attained their distinctive composition by similar processes.

The potentially ore-grade manganese nodule field in the Peru Basin, centered at about 7°–8°S and 90°W (**Figure 5**), is situated under the southern flank of the equatorial zone of high biological productivity on a seafloor composed of pelagic brown mud with variable amounts of siliceous and calcareous remains. Nodules from near the CCD at around 4250 m are characterized by diagenetic growth and are enriched in Mn, Ni and Cu, whereas those from shallower depth are characterized mainly by hydrogenous growth. The Mn/Fe ratio increases from south to north as productivity increases, whereas the Ni and Cu contents reach maximum values in the middle of the area where Mn/Fe ratios are about 5.

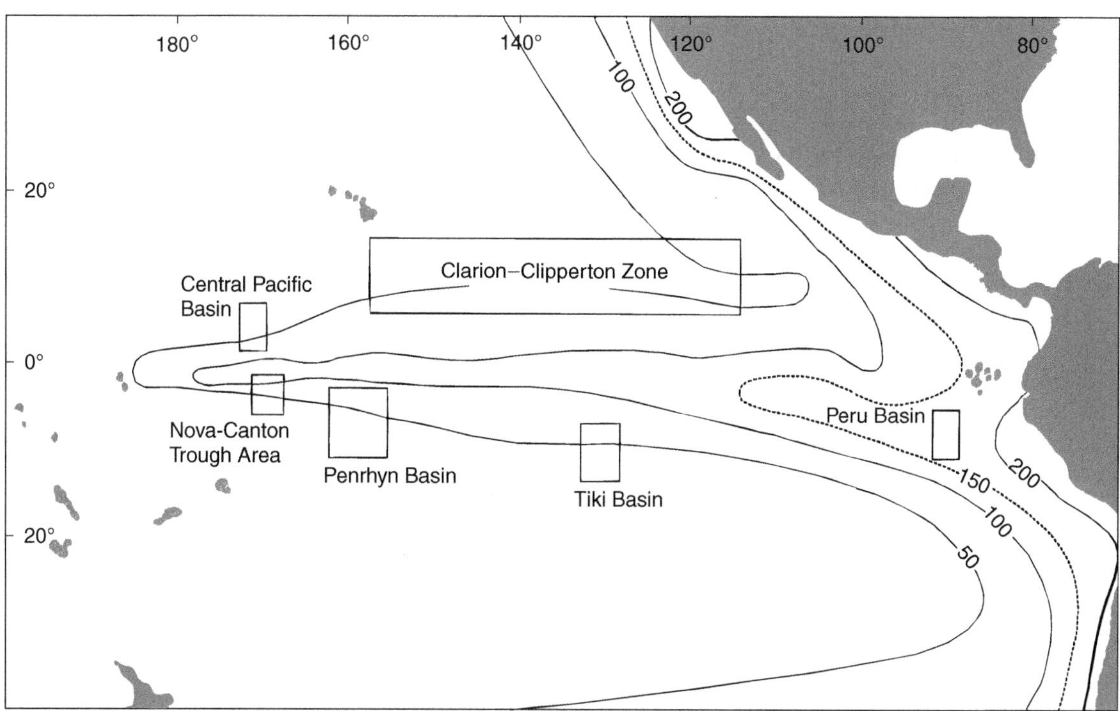

Figure 5 Approximate limits of areas of nickel- and copper-rich nodules in the subequatorial Pacific referred to in the text (productivity isolines if $g\,C\,m^{-2}\,y^{-1}$).

In the Tiki Basin there is also an increase in the Mn/Fe ratio of the nodules from south to north. All Ni + Cu values are above the lower limit expected in diagenetically supplied material.

The Penrhyn Basin nodules fall compositionally within the lower and middle parts of the Mn/Fe range for Pacific nodules as a whole. However, nodules from the northern part of the Basin have the highest Mn/Fe ratios and highest Mn, Ni and Cu concentrations reflecting diagenetic supply of metals to them, although Ni and Cu decrease slightly as the equator is approached. Superimposed on this trend are variations in nodule composition with their distance above or below the CCD. In the Mn-, Ni-, and Cu-rich nodule area, maximum values of these metals in nodules occur within about 200 m above and below the CCD. The latititudinal variation in Mn, Ni and Cu in Penrhyn Basin nodules may be due to there being a hydrogenous source of these metals throughout the Basin, superimposed on which is a diagenetic source of them between about 2° and 6°S at depths near the CCD, but less so in the very north of the Basin (0–2°S) where siliceous sedimentation prevails under highest productivity waters.

In the Nova Canton Trough area, manganese concentrations in the nodules are at a maximum between the equator and 2.5°S, where the Mn/Fe ratio is also highest. Manganese shows a tendency to decrease towards the south. Nickel and copper show similar trends to Mn, with maximum values of these elements being centered just south of the equator at depths of 5300–5500 m, just below the CCD.

In the central part of the Central Pacific Basin, between the Magellan Trough and the Nova Canton Trough, diagenetic nodules are found associated with siliceous ooze and clay sedimentation below the CCD. Their Ni and Cu contents increase southeastwards reaching a maximum at about 2.5°–3°N and then decrease again towards the equator where productivity is highest.

The Clarion–Clipperton Zone deposits rest largely on slowly accumulated siliceous ooze and pelagic clay below the CCD. The axis of highest average Mn/Fe ratio and Mn, Ni and Cu concentrations runs roughly southwest–northeast with values of these elements decreasing both to the north and south as productivity declines respectively to the north and increases towards the equatorial maximum in the south.

Indian Ocean In the Indian Ocean, Mn-, Ni-, and Cu-rich nodules are present in the Central Indian Ocean Basin between about 5° and 15°S. They are largely diagenetic in origin and rest on siliceous sediments below the CCD under high productivity waters. The deposits show north–south compositional variability with the highest grades occurring in the north.

Atlantic Ocean In the Atlantic Ocean, diagenetic Mn-, Ni-, and Cu-rich nodules occur most notably in the Angola Basin and to a lesser extent in the Cape/Agulhas Basin and the East Georgia Basin. These three areas have in common elevated biological productivity and elevated organic carbon contents in their sediments, which coupled with their depth near or below the CCD would help to explain the composition of their nodules. However, Ni and Cu contents are lower in them than in areas of diagenetic nodules in the Pacific and Indian Oceans.

Economic Potential

Interest in manganese nodules commenced around the mid-1960s and developed during the 1970s, at the same time as the Third United Nations Law of the Sea Conference. However, the outcome of that Conference, in 1982, was widely regarded as unfavorable for the mining industry. This, coupled with a general downturn in metal prices, resulted in a lessening of mining company interest in nodules. About this time, however, several government-backed consortia became interested in them and this work expanded as evaluation of the deposits by mining companies declined. Part 11 of the 1982 Law of the Sea Convention, that part dealing with deep-sea mining, was substantially amended in an agreement on 28 July 1994 which ameliorated some of the provisions relating to deep-sea mining. The Convention entered into force in November 1994.

During the 1980s interest in manganese nodules in exclusive economic zones (EEZs) started to increase. An important result of the Third Law of the Sea Conference, was the acceptance of a 200-nautical-mile EEZ in which the adjacent coastal state could claim any mineral deposits as their own. The nodules found in EEZs are similar to those found in adjacent parts of the International Seabed Area, and are of greatest economic potential in the EEZs of the South Pacific.

At the beginning of the twenty-first century, the out-look for manganese nodule mining remains rather unclear. It is likely to commence some time in this century, although it is not possible to give a precise estimate as to when. The year 2015 has been suggested as the earliest possible date for nodule mining outside of the EEZs. It is possible, however, that EEZ mining for nodules might commence earlier if conditions were favorable. It would depend upon many factors; economic, technological, and political.

Discussion

A model to explain the compositional variability of nodules in the Penrhyn Basin can be summarized as follows. Under the flanks of the high productivity area, reduced sedimentation rates near the CCD due to calcium carbonate dissolution enhance the content of metal-bearing organic carbon rich phases (fecal material, marine snow, etc.) in the sediments, the decay of which drives the diagenetic reactions that in turn promote the enrichment of Mn, Ni, and Cu in the nodules via the sediment interstitial waters. Away from the CCD, organic carbon concentrating processes are less effective. Further south as productivity declines, there is probably insufficient organic carbon supplied to the seafloor to promote the formation of diagenetic nodules at any depth. Under the equator, siliceous ooze replaces pelagic clay as the main sediment builder at and below the CCD, and when its rate of accumulation is high it dilutes the concentrations of organic carbon-bearing material at all depths to levels below that at which diagenetic Mn, Ni, and Cu rich nodules can form.

To a greater or lesser extent, this model can account for much of the variability in nodule composition found in the other South Pacific areas described, although local factors may also apply. In the Peru Basin, as in the Penrhyn Basin, diagenetic Mn-, Ni-, and Cu-rich nodules are concentrated near the CCD and their Ni and Cu contents reach a maximum south of the highest productivity waters. In the Tiki Basin, the greatest diagenetic influences are also found in the north of the Basin. As the South Pacific basins deepen to the west, the areas of diagenetic nodules tend to occur below the CCD as, for example, in the Nova Canton Trough area. This may be because the settling rates of large organic particles are quite fast in the deep ocean. Probably only limited decay of this material takes place between it settling through the CCD and reaching the seafloor, and enough probably gets sedimented to extend the depth of diagenetic nodule formation to well below the CCD under high productivity waters where there is limited siliceous sediment accumulation.

In the North Pacific, the trends in nodule composition in relation to the equatorial zone are the mirror image of those in the south. Thus in both the Central Pacific Basin and the Clarion–Clipperton Zone the highest nodule grades occur in diagenetic nodules on the northern flanks of the high productivity area and decline both to the north and south. The general model erected to explain the Penrhyn Basin nodule variability thus probably applies, at least in part, to these areas also.

The model also has some applicability in the Indian Ocean but less in the Atlantic. In the Indian Ocean, diagenetic nodules associated with sediments containing moderate amounts of organic carbon occur resting on siliceous ooze to the south of the equatorial zone in the Central Indian Ocean Basin. Farther to the south these nodules give way to hydrogenous varieties resting on pelagic clay. However, in the north the changes in nodule composition that might be expected under higher productivity waters do not occur, probably because terrigenous sedimentation becomes important in those areas which in turn reduces the Mn, Ni, and Cu content of nodules. In the Atlantic, the influence of equatorial high productivity on nodule composition that is evident in the Pacific is not seen, mainly because the seafloor in the equatorial area is largely above the CCD. Where diagenetic nodules do occur, as in the Angola, Cape and East Georgia Basins, productivity is also elevated, but the seafloor is near or below the CCD leading to reduced sedimentation rates.

Conclusions

Manganese nodules, although not being mined today, are a considerable resource for the future. They consist of ferromanganese oxides variably enriched in Ni, Cu, and other metals. They generally accumulate around a nucleus and exhibit internal layering on both a macro- and microscale. Growth rates are generally slow. The most potentially economic varieties of the deposits occur in the subequatorial Pacific under the flanks of the equatorial zone of high biological productivity, at depths near the CCD. Similar nodules occur in the Indian Ocean under similar conditions.

See also

Authigenic Deposits. Hydrothermal Vent Fluids, Chemistry of.

Further Reading

Cronan DS (1980) *Underwater Minerals*. London: Academic Press.

Cronan DS (1992) *Marine Minerals in Exclusive Economic Zones*. London: Chapman and Hall.

Cronan DS (ed.) (2000) *Handbook of Marine Mineral Deposits*. Boca Raton: CRC Press.

Cronan DS (2000) Origin of manganese nodule 'ore provinces'. *Proceedings of the 31st International Geological Congress*, Rio de Janero, Brazil, August 2000.

Earney FC (1990) *Marine Mineral Resources*. London: Routledge.

Glasby GP (ed.) (1977) *Marine Manganese Deposits*. Amsterdam: Elsevier.

Halbach P, Friedrich G, and von Stackelberg U (eds.) (1988) *The Manganese Nodule Belt of the Pacific Ocean*. Stuttgart: Enke.

Nicholson K. Hein J, Buhn B, Dasgupta S (eds.) (1997) *Manganese Mineralisation: Geochemistry and Mineralogy of Terrestrial and Marine Deposits*. Geological Society Special Publication 119, London.

Roy S (1981) *Manganese Deposits*. London: Academic Press.

Teleki PG, Dobson MR, Moore JR, and von Stackelberg U (eds.) (1987) *Marine Minerals: Advances in Research and Resource Assessment*. Dordrecht: D. Riedel.

HYDROTHERMAL VENT DEPOSITS

R. M. Haymon, University of California, CA, USA

Introduction

In April 1979, submersible divers exploring the mid-ocean ridge crest at latitude 21°N on the East Pacific Rise discovered superheated (380±30°C) fluids, blackened by tiny metal-sulfide mineral crystals, spewing from the seafloor through tall mineral conduits (*see* Hydrothermal Vent Fluids, Chemistry of). The crystalline conduits at these 'black smoker' hydrothermal vents were made of minerals rich in copper, iron, zinc, and other metals. Since 1979, hundreds of similar hydrothermal deposits have been located along the midocean ridge. It is now clear that deposition of hydrothermal mineral deposits is a common process, and is integrally linked to cracking, magmatism, and cooling of new seafloor as it accretes and spreads away from the ridge (*see* Mid-Ocean Ridge Geochemistry and Petrology).

For thousands of years before mid-ocean ridge hot springs were discovered in the oceans, people mined copper from mineral deposits that were originally formed on oceanic spreading ridges. These fossil deposits are embedded in old fragments of seafloor called 'ophiolites' that have been uplifted and emplaced onto land by fault movements. The copper-rich mineral deposits in the Troodos ophiolite of Cyprus are well-known examples of fossil ocean-ridge deposits that have been mined for at least 2500 years; in fact, the word 'copper' is derived from the Latin word 'cyprium' which means 'from Cyprus.'

The mineral deposits accumulating today at hot springs along the mid-ocean ridge are habitats for a variety of remarkable organisms ranging in size from tiny microbes to large worms. The properties of the mineral deposits are inextricably linked to the organisms that inhabit them. The mineral deposits contain important clues about the physical–chemical environments in which some of these organisms live, and also preserve fossils of some organisms, creating a geologic record of their existence.

Hydrothermal vent deposits are thus a renewable source of metals and a record of the physical, chemical, biological, and geological processes at modern and ancient submarine vents.

Where Deposits Form: Geologic Controls

Less than 2% of the total area of the mid-ocean ridge crest has been studied at a resolution sufficient to reveal the spatial distribution of hydrothermal vents, mineral deposits, and other significant small-scale geologic features. Nevertheless, because study areas have been carefully selected and strategically surveyed, much has been learned about where vents and deposits form, and about the geologic controls on their distribution. The basic requirements for hydrothermal systems include heat to drive fluid circulation, and high-permeability pathways to facilitate fluid flow through crustal rocks. On mid-ocean ridges, vents and deposits are forming at sites where ascending magma intrusions introduce heat into the permeable shallow crust, and at sites where deep cracks provide permeability and fluid access to heat sources at depth.

Fast-spreading Ridges

Near- and on-bottom studies along the fast-spreading East Pacific Rise suggest that most hydrothermal mineral deposits form along the summit of the ridge crest within a narrow 'axial zone' less than 500 m wide. Only a few active sites of mineral deposition have been located outside this zone; however, more exploration of the vast area outside the axial zone is needed to establish unequivocally whether or not mineral deposition is uncommon in this region. The overall spatial distribution of hydrothermal vents and mineral deposits along fast-spreading ridges traces the segmented configuration of cracks and magma sources along the ridge crest (*see* Mid-Ocean Ridge Geochemistry and Petrology).

Within the axial zone, mineral deposition is concentrated along the floors and walls of axial troughs created by volcanic collapse and/or faulting along the summit of the ridge crest. The majority of the deposits are located along fissures that have opened above magmatic dike intrusions, and along collapsed lava ponds formed above these fissures by pooling and drainage of erupted lava. Where fault-bounded troughs have formed along the summit of the ridge crest, mineral deposition is focused along the bounding faults and also along fissures and collapsed lava ponds in the trough floor. Hydrothermal vents appear to be most abundant along magmatically inflated segments of fast-spreading ridges; however, the mineral deposits precipitated on the seafloor on magmatically active segments are often buried beneath frequent eruptions of new lava flows. The greatest number of deposits, therefore, are observed on inflated ridge segments that are surfaced by somewhat older flows, i.e., along segments where: (1) much heat is available to power hydrothermal vents; and (2) mineral deposits have had time to develop but have not yet been buried by renewed eruptions.

Intermediate- and Slow-spreading Ridges

Most hydrothermal deposits that have been found on intermediate- and slow-spreading ridge crests are focused along faults, fissures, and volcanic structures within large rift valleys that are several kilometers wide. The fault scarps along the margins of rift valleys are common sites for hydrothermal venting and mineral deposition. Fault intersections are thought to be particularly favorable sites for hydrothermal mineral deposition because they are zones of high permeability that can focus fluid flow. Mineral deposition on rift valley floors is observed along fissures above dike intrusions, along eruptive fissures and volcanic collapse troughs, and on top of volcanic mounds, cones and other constructions. In general at slower-spreading ridges, faults appear to play a greater role in controlling the distribution of hydrothermal vents and mineral deposits than they do at fast-spreading ridges, where magmatic fissures are clearly a dominant geologic control on where vents and deposits are forming.

Structures, Morphologies, and Sizes of Deposits

A typical hydrothermal mineral deposit on an unsedimented mid-ocean ridge accumulates directly on top of the volcanic flows covering the ridge crest. On sedimented ridges, minerals are deposited within and on top of the sediments. Beneath seafloor mineral deposits are networks of feeder cracks through which fluids travel to the seafloor. Precipitation of hydrothermal minerals in these cracks and in the surrounding rocks or sediments creates a subseafloor zone of mineralization called a 'stockwork'. In hydrothermal systems where fluid flow is weak, unfocused, or where the fluids mix extensively with sea water beneath the seafloor, most of the minerals will precipitate in the stockwork rather than on the seafloor.

Hydrothermal deposits on mid-ocean ridges are composed of: (1) vertical structures, including individual conduits known as 'chimneys' (**Figure 1**) and larger structures of coalesced conduits that are often called 'edifices'; (2) horizontal 'flange' structures that extend outward from chimneys and edifices (**Figure 1**); (3) mounds of accumulated mineral precipitates (**Figure 1**); and (4) horizontal layers of hydrothermal sediments, debris, and encrustations. Chimneys are initially built directly on top of the seabed around focused jets of high-temperature effluents. Chimneys and edifices are physically unstable and often break or collapse into pieces that accumulate into piles of debris. The debris piles are cemented into consolidated mounds by precipitation of minerals from solutions percolating through the

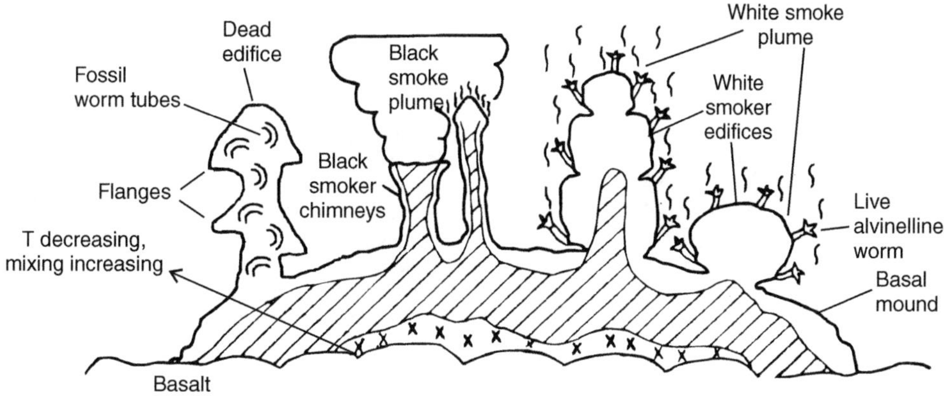

Figure 1 Composite sketch of the mineral structures and zones in hydrothermal mineral deposits on unsedimented ridge crests (modified after Haymon, 1989). Although mound interiors are seldom observed on the seafloor, the simplified sketch of mineral zoning within the mound is predicted by analogy with chimneys and massive sulfide deposits exposed in ophiolites. An outer peripheral zone (unshaded) of anhydrite + amorphous silica + Zn-rich sulfide, dominantly $ZnS + FeS_2$, is replaced in the interior by an inner zone (hatched) of Cu-rich sulfide ($CuFeS_2 + FeS_2$) + minor anhydrite and amorphous silica. The inner zone may be replaced by a basal zone (cross-pattern) of Cu-rich sulfide ($CuFeS_2 + FeS_2$) + quartz. Zones migrate as thermochemical conditions within the mound evolve. Although not shown here, it is expected that zoning around individual fractures cutting through the mound will be superimposed on the simplified zone structure in this sketch.

piles. New chimneys are constructed on top of the mounds as the mounds grow in size. Hydrothermal plume particles and particulate debris from chimneys settle around the periphery of the mounds to form layers of hydrothermal sediment. Diffuse seepage of fluids also precipitates mineral encrustations on mound surfaces, on volcanic flows and sediments, and on biological substrates, such as microbial mats or the shells and tubes of sessile macrofauna.

The morphologies of chimneys are highly variable and evolve as the chimneys grow, becoming more complex with time. Black smoker chimneys are often colonized by organisms and evolve into 'white smokers' that emit diffuse, diluted vent fluids through a porous carapace of worm tubes (**Figure 1**). Fluid compositions and temperatures, flow dynamics, and biota are all factors that influence the development of chimney morphology. The complexity of the interactions between these factors, and the high degree of spatial–temporal heterogeneity in the physical, chemical, biological and geological conditions influencing chimney growth, account for the diverse morphologies exhibited by chimneys, and present a challenge to researchers attempting to unravel the processes producing these morphologies.

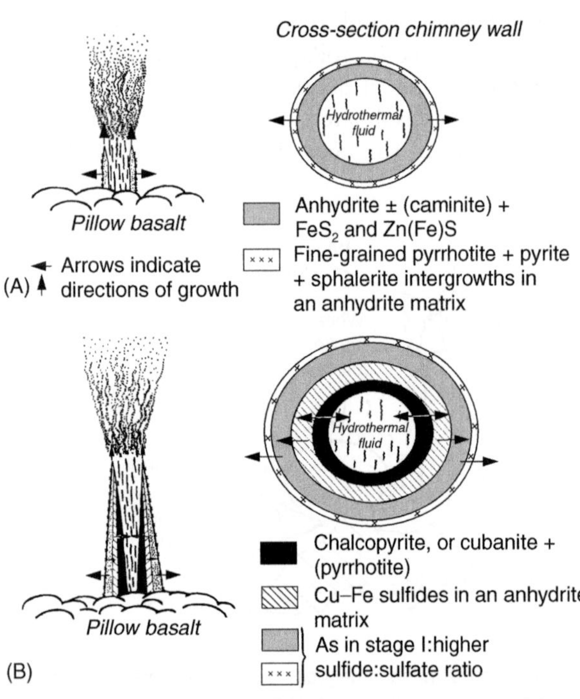

Figure 2 Two-stage model of black smoker chimney growth. (A) Stage I, sulfate-dominated stage; (B) stage II, sulfide replacement stage. During stage II, several different sulfide mineral zonation sequences develop, depending on permeability and thickness of chimney walls, hydrodynamic variables, and hydrothermal fluid composition.

The sizes of hydrothermal mineral deposits on ridges also vary widely. It has been suggested that the largest deposits are accumulating on sedimented ridges, where almost all of the metals in the fluids are

Table 1 Minerals occurring in ocean ridge hydrothermal mineral deposits

Mineral group/name	*Chemical formula*
Sulfides/Sulfosalts	
Most abundant	
Sphalerite	Zn(Fe)S
Wurtzite	Zn(Fe)S
Pyrite	FeS_2
Chalcopyrite	$CuFeS_2$
Less abundant	
Iss-Isocubanite	Variable $CuFe_2S_3$
Marcasite	FeS_2
Melnicovite	FeS_{2-x}
Pyrrhotite	$Fe_{1-x}S$
Bornite-Chalcocite	Cu_5FeS_4–Cu_2S
Covellite	CuS
Digenite	Cu_9S_5
Idaite	$Cu_{5.5}FeS_{6.5}$
Galena	PbS
Jordanite	$Pb_9As_4S_{15}$
Tennantite	$(Cu,Ag)_{10}(Fe,Zn,Cu)_2As_4S_{23}$
Valeriite	$2(Cu,Fe)_2S_23(Mg,Al)(OH)_2$
Sulfates	
Anhydrite	$CaSO_4$
Gypsum	$CaSO_4 \cdot H_2O$
Barite	$BaSO_4$
Caminite	$MgSO_4 \cdot xMg(OH)_2 \cdot (1-2x)H_2O$
Jarosite–Natrojarosite	$(K,Na)Fe_3(SO_4)_2(OH)_6$
Chalcanthite	$CuSO_4 \cdot 5H_2O$
Carbonate	
Magnesite	$MgCO_3$
Calcite	$CaCO_3$
Elements	
Sulfur	S
Oxides/Oxyhydroxides	
Goethite	FeO(OH)
Lepidocrocite	FeO(OH)
Hematite	Fe_2O_3
Magnetite	Fe_3O_4
Psilomelane	$(Ba,H_2O)_2Mn_5O_{10}$
'Amorphous' Fe-compounds	
'Amorphous' Mn-compounds	
Silicates	
Opaline silica	$SiO_2 \cdot nH_2O$
Quartz	SiO_2
Talc	$Mg_3Si_4O_{10}(OH)_2$
Nontronite	$(Fe,Al,Mg)_2(Si_{3.66}Al_{0.34})O_{10}(OH)_2$
Illite-smectite	
Aluminosilicate colloid	
Hydroxychlorides	
Atacamite	$Cu_2Cl(OH)_3$

deposited within the sediments rather than being dispersed into the oceans by hydrothermal plumes. On unsedimented ridges, the structures deposited on the seabed at fast spreading rates are usually relatively small in dimension (mounds are typically less than a few meters in thickness and less than tens of meters in length, and vertical structures are <15 m high). On intermediate- and slow-spreading ridges, mounds are sometimes much larger (up to tens of meters in thickness, and up to 300 m in length). On the Endeavour Segment of the Juan de Fuca Ridge, vertical structures reach heights of 45 m. The size of a deposit depends on many factors, including: magnitude of the heat source, which influences the duration of venting and mineral deposition; tendency of venting and mineral deposition to recur episodically at a particular site, which depends on the nature of the heat source and plumbing system, and the rate of seafloor spreading; frequency with which deposits are buried beneath lava flows; and the compositions of the vent fluids and minerals. The large deposits found on slower-spreading ridge crests are located on faults that have moved slowly away from the ridge axis and have experienced repeated episodes of venting and accumulated mineral deposition over thousands of years, without being buried by lava flows. The tall Endeavor Segment edifices are formed because ammonia-enriched fluid compositions favor precipitation of silica in the edifice walls. The silica is strong enough to stabilize these structures so that they do not collapse as they grow taller.

How Do Chimneys Grow?

A relatively simple two-stage inorganic growth model has been advanced to explain the basic characteristics of black smoker chimneys (**Figure 2**). In this model, a chimney wall composed largely of anhydrite (calcium sulfate) precipitates initially from sea water that is heated around discharging jets of hydrothermal fluid. The anhydrite-rich chimney wall precipitated during stage I contains only a small component of metal sulfide mineral particles that crystallize because of rapid chilling of the hydrothermal fluids. In stage II, the anhydrite-rich wall continues to grow upward and to thicken radially, protecting the fluid flowing through the chimney from very rapid chilling and dilution by sea water. This allows metal sulfide minerals to precipitate into the central conduit of the chimney from the hydrothermal fluid. The hydrothermal fluid percolates outward through the chimney wall, gradually replacing anhydrite and filling voids with metal sulfide minerals. During stage II, the chimney increases in height, girth and wall thickness, and both the calcium sulfate/metal sulfide ratio and permeability of the walls decrease. Equilibration of minerals with pore fluid in the walls occurs continuously along steep, time–variant temperature and chemical gradients between fluids in the central conduit and sea water surrounding the chimney. This equilibration produces sequences of concentric mineral zones across chimney walls that evolve with changes in thermal and chemical gradients and wall permeability.

The model of chimney growth described above is accurate but incomplete, as it does not include the effects on chimney development of fluid phase separation, biological activity, or variations in fluid composition. Augmented models that address these complexities are needed to fully characterize the processes governing chimney growth.

Table 2 Ranges of elemental compositions in bulk midocean ridge hydrothermal mineral deposits

Element	*Ranges*[a]
Cu	0.1–15.0 wt%
Fe	2.0–44.0 wt%
Zn	<0.1–48.7 wt%
Pb	0.003–0.6 wt%
S	13.0–52.2 wt%
SiO_2	<0.1–28.0 wt%
Ba	<0.01–32.5 wt%
Ca	<0.1–16.5 wt%
Au	<0.1–4.6 p.p.m.
Ag	3.0–303.0 p.p.m.
As	7.0–918.0 p.p.m.
Sb	2.0–375.0 p.p.m.
Co	<2.0–3500.0 p.p.m.
Se	<2.0–224.0 p.p.m.
Ni	<1.5–226.0 p.p.m.
Cd	<5–1448 p.p.m.
Mo	1.0–290.0 p.p.m.
Mn	36.0–1847.0 p.p.m.
Sr	2.0–4300.0 p.p.m.

[a]Data sources: Hannington *et al.* (1995) and Haymon (1989).

Elemental and Mineral Compositions of Deposits

Ridge crest hydrothermal deposits are composed predominantly of iron-, copper- and zinc sulfide minerals, calcium- and barium-sulfate minerals, iron oxide and iron oxyhydroxide minerals, and silicate

Morphological and Mineralogical Evolution of Chimneys
on the East Pacific Rise at 9°-10°N

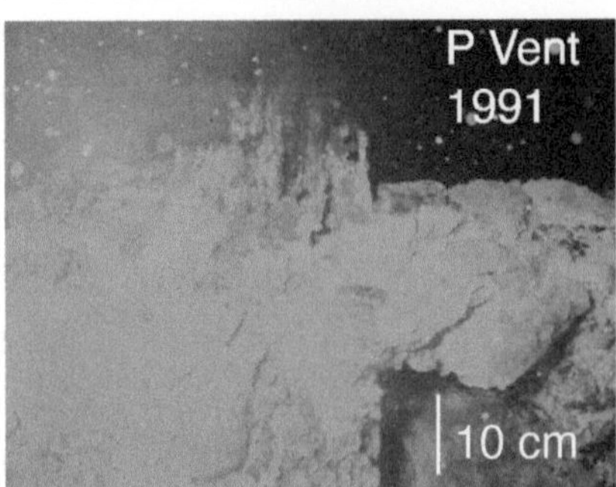

Stage 2

1991 "Proto-chimney"
Anhydrite-dominated
T = 389°-403°C

Stage 2

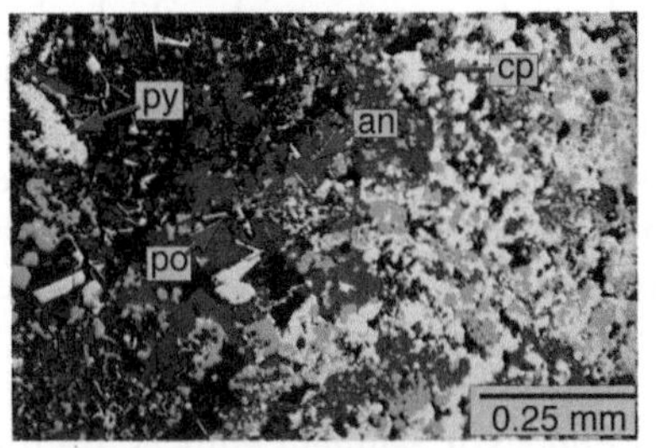

1992-1995
CuFe-sulfide-dominated
T = 340°-392°C

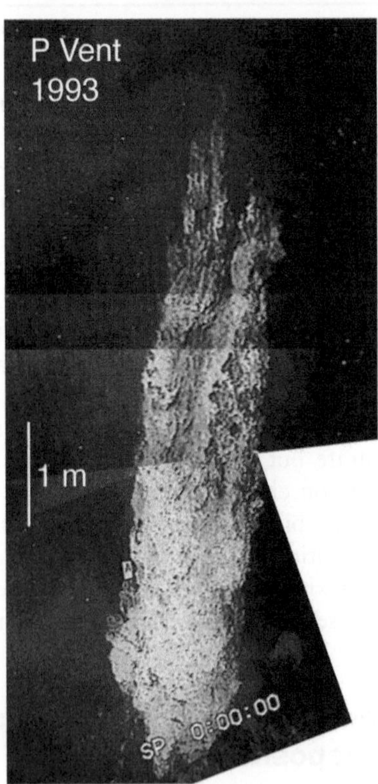

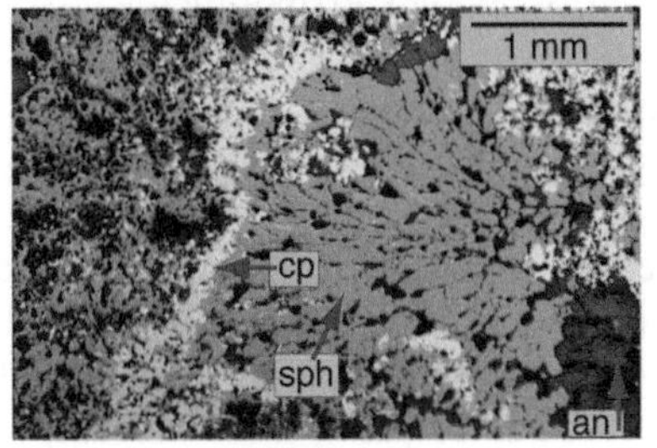

1992-1995
Zn-sulfide-dominated
T = 264°-340°C

KEY
an – anhydrite cp – chalcopyrite po – pyrrhotite py – pyrite

minerals (**Table 1**). These minerals precipitate from diverse processes, including: heating of sea water; cooling of hydrothermal fluid; mixing between sea water and hydrothermal fluid; reaction of hydrothermal minerals with fluid, sea water, or fluid–sea water mixtures; reaction between hydrothermal fluid and seafloor rocks and sediments; and reactions that are mediated or catalyzed biologically. This diversity in the processes and environments of mineral precipitation results in the deposition of many different minerals and elements (**Tables 1** and **2**). High concentrations of strategic and precious metals are found in some deposits (**Table 2**). The deposits are potentially valuable, if economic and environmentally safe methods of mining them can be developed.

Chimneys can be classified broadly by composition into four groups: sulfate-rich, copper-rich, zinc-rich and silica-rich structures. Copper-rich chimney compositions are indicative of formation at temperatures above 300°C. Sulfate-rich compositions are characteristic of active and immature chimneys. Many chimneys are mineralogically zoned, with hot interior regions enriched in copper, and cooler exterior zones enriched in iron, zinc, and sulfate (**Figures 1** and **2**). Mounds exhibit a similar gross mineral zoning, and those which are exposed by erosion in ophiolites often have silicified (quartz-rich) interiors (**Figure 2**). Seafloor weathering of deposits after active venting ceases results in dissolution of anhydrite, and oxidation and dissolution of metal-sulfide minerals. Small deposits that are not sealed by silicification or buried by lava flows will not be well preserved in the geologic record (**Figure 3**).

Chimneys as Habitats

Chimney and mound surfaces are substrates populated by microbial colonies and sessile organisms such as vestimentiferan and polychaete worms, limpets, mussels, and clams. It is likely that pore spaces in exterior regions of chimney walls are also inhabited by microbes. All of these organisms that are dependent on chemosynthesis benefit from the seepage of hydrothermal fluid through active mineral structures, and from the thermal and chemical gradients across mineral structures. The structures provide an interface between sea water and hydrothermal fluid that maintains tolerable temperatures for biota, and allows organisms simultaneous access to the chemical constituents in both sea water and hydrothermal fluid. However, organisms attached to active mineral structures must cope with changes in fluid flow across chimney walls (which sometimes occur rapidly), and with ongoing engulfment by mineral precipitation.

Some organisms actively participate in the precipitation of minerals; for example, sulfide-oxidizing microbes mediate the crystallization of native sulfur crystals, and microbes are also thought to participate in the precipitation of marcasite and iron oxide minerals. Additionally, the surfaces of organisms provide favorable sites for nucleation and growth of amorphous silica, metal sulfide and metal oxide crystals, and this facilitates mineral precipitation and fossilization of vent fauna (**Figure 3**).

Fossil Record of Hydrothermal Vent Organisms

Fossil molds and casts of worm tubes, mollusc shells, and microbial filaments have been identified in both modern ridge hydrothermal deposits and in Cretaceous, Jurassic, Devonian, and Silurian deposits. This fossil record establishes the antiquity of vent communities and the long evolutionary history of specific faunal groups. The singular Jurassic fossil assemblage preserved in a small ophiolite-hosted deposit in central California is particularly interesting because it contains fossils of vestimentiferan worms, gastropods and brachiopods, but no clam or mussel fossils. In contrast, modern and Paleozoic faunal assemblages described thus far include clams, mussels and gastropods, but no brachiopods. Does this mean that brachiopods have competed with molluscs for ecological niches at vents, and have

Figure 3 On left: a time series of seafloor photographs showing the morphological development of a chimney that grew on top of lava flows erupted in 1991 on the crest of the East Pacific Rise near 9°50.3′N (Haymon *et al.*, 1993). Within a few days-to-weeks after the eruption, anhydrite-rich 'Stage 1 Protochimneys' a few cm high had formed where hot fluids emerged from volcanic outcrops covered with white microbial mats (top left). Eleven months later, the chimney consisted of cylindrical 'Stage 2' anhydrite-sulfide mineral spires approximately one meter in height, and as-yet unpopulated by macrofauna (middle left). Three and a half years after the eruption, the cylindrical conduits had coalesced into a 7 m-high chimneys structure that was covered with inhabited Alvinelline worms tubes (bottom left). On right: photomicrographs of chimney samples from the eruption area that show how the chimneys evolved from Stage 1 (anhydrite-dominated; top right) to Stage 2 (metal-sulfide dominated) mineral compositions (see text). As the fluids passing through the chimneys cooled below ~330°C during Stage 2, the CuFe-sulfide minerals in the chimney walls (middle right) were replaced by Zn- and Fe-sulfide minerals (bottom right).

moved in and out of the hydrothermal vent environment over time? Fossilization of organisms is a selective process that does not preserve all the fauna that are present at vents. Identification of fossils at the species level is often difficult, especially where microbes are concerned. Notwithstanding, it is important to search for more examples of ancient fossil assemblages and to trace the fossil record of life at hydrothermal vents back as far as possible to shed light on how vent communities have evolved, and whether life on earth might have originated at submarine hydrothermal vents.

Summary

Formation of hydrothermal deposits is an integral aspect of seafloor accretion at mid-ocean ridges. These deposits are valuable for their metals, for the role that they play in fostering hydrothermal vent ecosystems, for the clues that they hold to understanding spatial–temporal variability in hydrothermal vent systems, and as geologic records of how life at hydrothermal vents has evolved. From these deposits we may gain insights about biogeochemical processes at high temperatures and pressures that can be applied to understanding life in inaccessible realms within the earth's crust or on other planetary bodies. We are only beginning to unravel the complexities of ridge hydrothermal vent deposits. Much exploration and interdisciplinary study remains to be done to obtain the valuable information that they contain.

See also

Hydrothermal Vent Fluids, Chemistry of. Mid-Ocean Ridge Geochemistry and Petrology.

Further Reading

Dilek Y, Moores E, Elthon D, and Nicolas A (eds.) (2000) *Ophiolites and Oceanic Crust: New Insights from Field Studies and the Ocean Drilling Program. Geological Society of America Memoir.* Boulder: Geological Society of America.

Haymon RM (1989) Hydrothermal processes and products on the Galapagos Rift and East Pacific Rise, 1989. In: Winterer EL, Hussong DM, and Decker RW (eds.) *The Geology of North America: The Eastern Pacific Ocean and Hawaii*, vol. N, pp. 125–144. Boulder: Geological Society of America.

Haymon RM (1996) The response of ridge crest hydrothermal systems to segmented, episodic magma supply. In: MacLeod CJ, Tyler P, and Walker CL (eds.) *Tectonic, Magmatic, Hydrothermal, and Biological Segmentation of Mid-Ocean Ridges*, vol. Special Publication 118, pp. 157–168. London: Geological Society.

Humphris SE, Zierenberg RA, Mullineaux LS, and Thomson RE (eds.) (1995) *Seafloor Hydrothermal Systems: Physical, Chemical, Biological, and Geological Interactions, Geophysical Monograph*, vol. 91. Washington, DC: American Geophysical Union.

Little CTS, Herrington RJ, Haymon RM, and Danelian T (1999) Early Jurassic hydrothermal vent community from the Franciscan Complex, San Rafael Mountains, California. *Geology* 27: 167–170.

Tivey MK, Stakes DS, Cook TL, Hannington MD, and Petersen S (1999) A model for growth of steep-sided vent structures on the Endeavour Segment of the Juan de Fuca Ridge: results of a petrological and geochemical study. *Journal of Geophysical Research* 104: 22859–22883.

PORE WATER CHEMISTRY

D. Hammond, University of Southern California, Los Angeles, CA, USA

Introduction

As marine sediments are deposited, they trap sea water in the pore space between grains. This pore water (sometimes called interstitial water) may represent more than 90% of the volume of the bulk sediment in fine-grained deposits near the sediment–water interface. The volume fraction of the bulk sediment that is water is called the porosity. Porosity usually decreases rapidly with increasing depth through the uppermost sediments, with the profile depending on lithology and accumulation rate. Ultimately, porosity may decrease to only 5–20% as the sediment becomes lithified. As sediments are compacted during burial, the pore fluid is squeezed upward, traveling through fractures, burrows, or perhaps the sediment pore space itself.

As sediment is buried, its composition may be modified by chemical reactions, a process called chemical diagenesis. Pore water provides a medium that permits a solute to migrate from a site where it is produced to another site where it may be removed. For example, in organic-rich sediments, pyrite (FeS_2) is a common end product of diagenesis. While the intermediate steps in pyrite formation are not fully understood, they involve sulfate reduction to sulfide at sites where reactive organic matter is found, and reduction of insoluble ferric oxides to form soluble ferrous iron at sites of iron-bearing minerals. Iron sulfides have very low solubility, and their deposition is usually localized at one of the two sites. If sulfide is released faster than ferrous iron, the sulfide diffuses from its site of production on an organic-rich particle to the mineral containing iron, where it forms insoluble iron sulfides. This can produce pyrite overgrowths on iron-bearing minerals. Alternatively, if the iron is more readily released, it diffuses to form iron sulfides near the sites of organic particles such as shells or localized pockets of organic substrate.

Because diagenesis may alter only a small fraction of the solid phases, its impact may be difficult to detect from studies of solid phases alone. Pore water chemistry is much more sensitive to such changes. For example, in a sediment of 80% porosity, dissolution of 0.1 weight percent $CaCO_3$ from the solid phase (near the detection limit measurable in solid phases) would make a change of 6 mmol kg^{-1} in the concentration of dissolved calcium. This change in pore water would be easily detectable because it results in a concentration 60% greater than that in the starting sea water. Of course, this calculation assumes that the pore water acts as a closed system, which is generally not the case as noted below. However, this example illustrates that pore water chemistry is more sensitive than solid phase chemistry to diagenesis.

Studies of pore waters have become a standard tool for understanding the biogeochemical processes that influence sediments, and considerable efforts have been invested during the past several decades to develop techniques to collect samples, evaluate whether vertical profiles exhibit artifacts introduced during collection and handling, and develop approaches to model the observed profiles and obtain quantitative estimates of reaction kinetics and stoichiometry. Usually, modeling approaches assume steady-state behavior, but when time-dependent constraints can be established, nonsteady-state approaches can be applied.

Reasons to Study Pore Water Composition

The study of pore waters can reveal many processes that are important in regulating the biogeochemical cycles of the ocean and in evaluating the impact of diagenesis on the sedimentary record recorded in solid phases. Some applications of pore water studies are given below.

Calculation of Mineral Stability

Thermodynamic calculations can be carried out to determine which solid phases should be dissolving, precipitating, or in equilibrium with the pore fluid chemistry. While these calculations do not guarantee the presence of minerals that are at or above saturation, or the absence of minerals that are undersaturated, they are a very useful indicator of whether it may be worthwhile to search for minerals that could be present in only trace abundance.

Identification of Sites of Reaction and Reaction Stoichiometry

Maxima in pore water profiles indicate localized inputs, and minima define sinks for solutes. However, defining the transition between source and sink regions requires location of inflection points in the pore water profile. If the pore water profile is in steady-state, material balance calculations can be carried out for solutes if the transport mechanisms are known. This usually involves fitting a reaction-transport model that defines the depth dependence of the net reaction-kinetics. Transport processes are discussed below, and may include molecular diffusion, advection, and macrofaunal irrigation. The impact of reactions occurring at depths beyond the range of sampling may also be evident in pore water profiles.

Interpretation of the Sedimentary Record

Are changes in solid phase profiles with depth due to diagenetic reactions during burial, or due to temporal variations in the composition or rain rate of solid phase inputs? Pore water profiles provide a way to evaluate the contemporary rates of reactions (assuming they are in steady-state) and predict the effect of diagenetic reactions on solid phase profiles. Solid phase changes that exceed the diagenetic effects derived from modeling pore water profiles must reflect nonsteady-state behavior in the input of solid phases to the sediment column.

Estimation of Benthic Exchange Rates

Sediments are a sink or source for many solutes in the water column. Thus, they can play an important role in regulating the composition of the overlying waters. This approach provides information that can be compared to direct measurements of benthic fluxes.

Recovery of Deep Pore Water that may be Fossil Water

The trapping of pore fluids as sediments are buried may potentially preserve fluid from a time when ocean composition differed from the present, such as the last glacial period when salinity should have been greater than at present and the isotopic composition of water should have been heavier. However, pore water is an open system, and diffusion facilitates the re-equilibration between fossil pore water and bottom waters. Consequently, relict signals may be difficult to detect, even in the absence of any influence of diagenetic reactions.

Sampling Techniques

Initial studies of pore waters utilized retrieval of cores, sectioning them into intervals, and centrifugation to separate pore waters from the associated solids. This approach works well for many solutes in sediments with high porosity, as long as appropriate precautions are taken to minimize artifacts (changes in composition attributable to recovering and processing samples). During sample processing, it is often critical to regulate temperature and eliminate contact between reducing sediments and oxygen, depending on the solute of interest. For studies related to near-surface diagenesis, it is essential to obtain cores that have undisturbed interfaces and with bottom water still in contact with the sediment. To extract water from low porosity sediments, or minimize contact with gas phases, squeezing techniques have been developed. Several kinds of squeezing devices have been utilized, but all rely on compressing sediments while permitting water to escape through a filtration assembly.

To avoid artifacts associated with retrieving cores from the deep sea, devices have been developed to carry out filtration *in situ*. These devices avoid the pressure- and temperature-dependent perturbations associated with core retrieval, but have their own logistical difficulties in deployment to minimize leakage and obtain accurate sampling resolution. One strategy drives a probe called a harpoon into sediments; openings at various distances along the harpoon shaft permit water to be drawn through filters into sample reservoirs. Another device collects cores, seals the bottom, and squeezes water by driving a piston and filter pack down onto the core; sequential aliquots of water are collected and assumed to represent water from progressively deeper intervals. Other strategies have relied on inserting probes that contain water that may communicate with pore water through a dialysis membrane. These devices, named ‘peepers’, require several days to equilibrate with pore waters and must be initially filled with a solution that will not significantly contaminate the surrounding sediment with exotic solutes.

Several of the artifacts noted above may be avoided through the use of *in situ* electrodes that can be inserted directly into sediment and measure activities of various solutes. Systems to measure oxygen and pH are often used. Very recently, new electrodes to measure pCO_2, sulfide, iron, and manganese have been developed. By using microelectrodes, gradients over short distances can be resolved.

Finally, some tools have been developed to retrieve pressurized cores and extract pore fluids onboard ships at *in situ* pressures. These tools are particularly important where high quantities of methane are found, either dissolved in pore fluid or as a gas hydrate.

Limitations of Pore Water Studies

Observation of the Net Process

Some solutes may be involved in more than one reaction. This limits the ability to uniquely define reaction stoichiometry.

Required Assumptions

Pore water profiles are usually assumed to be steady-state. If boundary conditions vary, pore waters respond, but there is a temporal lag in response that increases with depth. The profile should exhibit concentrations that are roughly averaged over this response time. A second problem may be the existence of unidentified transport processes, such as macrofaunal irrigation (see below). A third problem is that patchiness of organisms on the seafloor may lead to localized effects, such as caches of freshly deposited organic matter in burrows. It is not always possible to collect and process sufficient cores to evaluate the spatial heterogeneity introduced by these effects, that may occur on horizontal scales of centimeters to meters. In the Equatorial Pacific, for example, benthic fluxes of oxygen calculated from pore water profiles collected with replicate cores at the same site have been shown to vary by 30%, and inferences based on a single core have an inherent uncertainty.

Sampling Resolution

Gradients may exist over very short vertical or horizontal distances that cannot be easily resolved during sampling. In organic-rich slope sediments, for example, microelectrode measurements show that the thickness of oxygenated sediments may be only 1–2 mm. Furthermore, if micro-environments are present within burrows or inside shells, this can lead to localized sites of reaction, but the pore water measurement of sectioned cores defines an average for the zone sampled. Most studies are designed to evaluate vertical gradients, as it is usually difficult to evaluate any horizontal gradients, if they are present.

Sampling Artifacts

These may be created by changes in temperature, pressure, exposure to oxygen (or perhaps any gas phase), activities of stressed organisms, or deterioration of samples between collection and storage (see **Table 1**). Some of these changes appear to be reversible, while others are not. Some changes involve the direct reaction of the solute in question, while others are indirect due to the co-precipitation or adsorption of one solute with the solid formed by the direct reaction of another.

Diagenetic Reactions and Biogeochemical Zonation

The seafloor receives a rain of sediment that is a mixture of biogenic debris and detrital materials. Some of these components are rather reactive and undergo diagenesis at very shallow depth. Of paramount importance are reactions related to the oxidation of reduced carbon in organic material to form carbon dioxide. The details of organic carbon diagenesis are not well understood, and a detailed discussion of relevant reactions is beyond the scope of this article. However, organic carbon diagenesis involves microbial catalysis of reactions that result in decreasing the free energy of the system through the transfer of electrons from the organic material to terminal electron acceptors. Dissolved organic carbon is produced, and some escapes from sediments

Table 1 Known artifacts in pore water studies

1. **Changes in temperature.** Consistently observed for boron, potassium, and silicon; sometimes observed for acid, calcium, and magnesium.
2. **Changes in pressure.** Precipitation of carbonate during retrieval of cores from deep water is consistently observed in carbonate-rich sediments. As a consequence, phosphate and uranium are often lost.
3. **Exposure to oxygen.** Consistently observed for ferrous iron. As a consequence, phosphate, silicon, and perhaps other metals are affected by precipitating ferric oxyhydroxides.
4. **Gas exchange.** Exposure to a gas phase permits any gas dissolved in the sample to partition among the available phases present. In sediments under pressure, high concentrations of dissolved gases (e.g. methane,) can accumulate and form bubbles when retrieved to the surface. Also, some containers are permeable to certain types of gases.
5. **Stressed organisms.** Some animals may excrete large amounts of ammonia when they are stressed. Stress can occur due to temperature and pressure changes as cores are retrieved, or due to physical disturbance. Some bacteria may contain vacuoles rich in nitrate; these vacuoles may break when samples are centrifuged.
6. **Deterioration during sample storage.** Filtering samples can screen out most bacteria, but may not inhibit reactions that might occur inorganically. For example, oxygen will eventually diffuse into plastic sample bottles and oxidize ferrous iron. Precipitation of ferric iron can be inhibited by acidification of the sample.

into the overlying water, but most of these reactions result in production of carbon dioxide, which reacts with water to form carbonic acid. A consistent pattern has been observed in the distribution of the principal terminal electron acceptors with increasing distance from an oxic water column. Oxygen, nitrate, manganese dioxide, ferric oxides, sulfate, and finally carbon dioxide serve as the principal electron acceptors. Some representative reactions are illustrated in **Table 2**.

This sequence of reactions has led to the concept of biogeochemical zonation, with each zone named for the solute that serves as the principal electron acceptor or that is the principal product (**Figure 1**). The sequence of zones is determined by the chemical free energy yields released by possible reactants. The thickness of each zone is dependent on the rate of reaction consuming the electron acceptor, the rate of a reactant's transport through sediments, and the concentration of the acceptor in bottom waters or in solid phases. Zones may overlap, and tracer studies have shown that they need not be mutually exclusive. The existence of the deeper zones depends on the availability of sufficient reactive organic matter.

Boundaries between zones are often interesting sites, and may provide environments where specialized bacteria thrive. As soluble reduced reaction products form at depth, they may diffuse upward into the overlying zone, where they are oxidized. One example is illustrated in **Figure 2**. In iron-rich systems, ferrous iron produced at depth diffuses upward, until it encounters nitrate or oxygen diffusing downward. At this horizon, ferrous iron is oxidized to ferric iron that precipitates as an oxyhydroxide, often leaving a visible thin red band in the sediments that marks the ferrous/ferric transition. Continued accumulation of new sediments transports the ferric oxyhydroxide downward relative to the sediment–water interface, beyond the penetration depth of oxidants, where the ferric iron is again reduced to ferrous iron; the ferrous iron diffuses upward again and is re-oxidized. Studies of pore water have confirmed this redox shuttle system, which maintains a horizon of sediments rich in ferric iron at a consistent depth relative to the sediment–water interface. Manganese can undergo a similar cycle to produce a manganese-rich horizon. In sulfur-rich sediments that are overlain by oxygenated bottom waters, the

Table 2 Biogeochemical zonation and proton balance

(A) *Principal reactions involving electrons and protons*

Oxygen reduction zone

$O_2 + C_{org} + H_2O = HCO_3^- + H^+$

$NH_3)_{org} + 2O_2 = H^+ + NO_3^- + H_2O$

$2Mn^{2+} + O_2 + 2H_2O = 2MnO_2 + 4H^+$

Nitrate, manganese reduction zone

$2NO_3^- + 2.5C_{org} + 1.5H_2O = 2.5HCO_3^- + N_2 + 0.5H^+$

$5Fe^{2+} + NO_3^- + 12H_2O = 5Fe(OH)_3 + 0.5N_2 + 9H^+$

$2MnO_2 + C_{org} + 3H^+ = HCO_3^- + H_2O + 2Mn^{2+}$

$NH_3)org + H^+ = NH_4^+$ (also occurs in deeper zones)

Sulfate, iron reduction zone

$4Fe(OH)_3 + C_{org} + 7H^+ = HCO_3^- + 4Fe^{2+} + 9H_2O (Fe > S)$

$SO4^{-2} + 2C_{org} + 2H_2O = HS^- + 2HCO_3^- + H^+ (S > Fe)$

with Fe oxide source:

$2SO_4^{-2} + Fe(OH)_3 + 3.75C_{org} + 0.25H_2O + 0.25H^+ = FeS_2 + 3.75HCO_3^-$

with Fe silicate source:

$2SO_4^{2-} + (Fe^{3+})_{clay} + Mg^{2+} + 3.75C_{org} + 3.25H_2O = FeS_2 + (Mg^{2+}H^+)_{clay} + 3.75HCO_3^- + 1.75H^+$

Methane production zone

$2CH_2O + H_2O = CH_4 + HCO_3^- + H^+$ (fermentation)

$(HCO_3^- + H^+ + 4H_2)_{org} = CH_4 + 3H_2O$ (CO_2 reduction)

Carbonate dissolution (primarily in oxygen reduction zone)

$CaCO_3 + H^+ = Ca^{2+} + HCO_3^-$

(B) *Net production of protons from organic matter diagenesis*

Electron acceptor		Protons per mole C_{ox} (ignoring NH_3)
Oxygen		1.0
Nitrate		0.2
Manganese		−3.0
Iron		−7.0
Sulfate	(iron-free)	0.5
	(ferric oxide)	−0.07
	(Fe–Mg exchange)	0.47

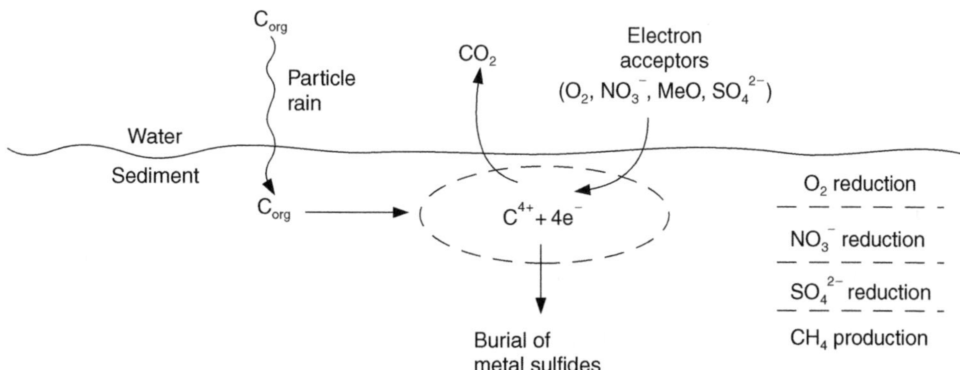

Figure 1 Biogeochemical zonation. The rain of organic carbon to the seafloor and its burial provide a substrate for metabolic activity, as microbial communities transfer electrons from the organic carbon to terminal electron acceptors. This results in the conversion of organic carbon into carbon dioxide, and may be accompanied by the conversion of oxidized forms of nitrogen, sulfur, and metals (MeO) into reduced forms: molecular nitrogen that escapes and metal sulfides that are buried. Sediments can be divided into zones, characterized by the principal acceptor that is present, or by the key product (in the case of methane). In some cases, distinct zones may be observed where manganese and iron are the principal acceptors, but these often overlap with the nitrate and sulfate zones. This schematic does not include the details of transport, but acceptors migrate downward from overlying waters, or are produced in an upper zone and diffuse downward. The drawing is not to scale. The relative thickness of each zone varies, depending on input of reactive organic material and the availability of different acceptors, and the deeper zones do not form where the rain of labile organic materials is too low. The oxygen reduction zone may be only a few millimeters thick in margin sediments, tens of centimeters thick under open-ocean equatorial sediments, and many meters thick in open-ocean sediments that underlie oligotrophic waters. The geometry of each zone may be convoluted due to the presence of macrofaunal burrows or other heterogeneities.

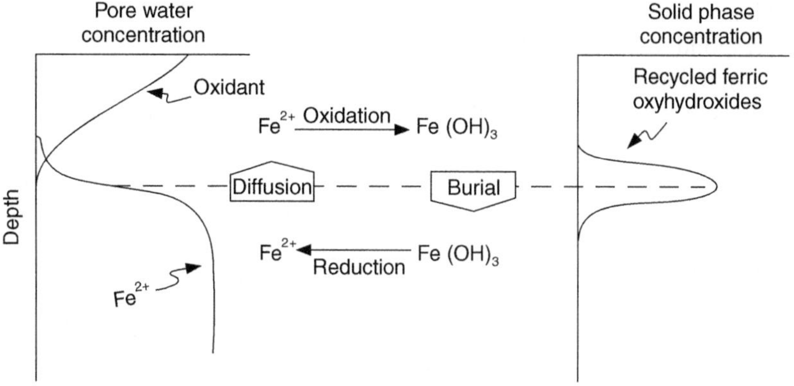

Figure 2 Schematic illustration of iron cycling to maintain a diagenetic front at a constant depth near the sediment–water interface, as explained in the text. Only the recycled component of iron is shown.

upward diffusion of sulfide through pore waters brings it into contact with dissolved oxygen. This provides a unique environment that may be exploited by the sulfur-oxidizing bacteria *Beggiatoa*, an organism that utilizes the energy released by sulfide oxidation and forms bacterial mats frequently found where low oxygen bottom waters may exclude predators.

Diagenetic Modeling

Substantial advances have been made in the development of mathematical models to quantify the effects of the diagenetic processes that influence pore water profiles. Important processes include those mentioned below, but a full discussion of appropriate formulations to describe these processes is beyond the scope of this article (see the Further Reading section for the mathematical development).

Chemical Reactions

Some solids dissolve and others precipitate. Many reactions are driven by redox processes associated with the oxidation of organic matter, largely catalyzed by microbial activities as they extract metabolic energy. Some of these are illustrated in

Table 2. These redox reactions can also result in the production or consumption of acid, depending on the suite of reactants available. The addition of acid should be buffered by the dissolution of carbonate solids, if they are present, and removal of acid may promote precipitation of carbonates. For example, as shown in **Table 2B**, utilization of oxygen as a terminal electron acceptor is quite effective in producing acid, and should favor the dissolution of any carbonate minerals in the oxygenated zone. If manganese or iron serves as the terminal electron acceptor, acid is consumed, and carbonate may precipitate from the pore waters. If sulfate serves as the terminal electron acceptor, and no iron oxides are available to permit iron sulfides to precipitate, pore waters are acidified and may dissolve carbonates. A similar behavior occurs if iron can be leached from silicates and exchanged for dissolved magnesium. However, if iron oxyhydroxides are present, they dissolve and favor carbonate precipitation. This example illustrates the importance of iron availability in helping regulate the pH of marine pore waters. The behavior of nitrogen (oxidized to nitric acid if oxygen is present, and consuming a proton if it is released as ammonia in the absence of oxygen) also contributes to pH buffering. Many other solid phases respond to these pH changes, changes that are largely influenced by the oxidation of organic matter and the behavior of carbonate and iron-bearing minerals.

Another interesting effect is the co-precipitation of some trace constituents with phases created by cycles of a more abundant substance. For example, as ferrous iron diffuses upward, other trace metals or phosphate may co-precipitate with the oxyhydroxides when oxygen or nitrate is encountered. These constituents may re-dissolve as the ferric oxyhydroxides are buried more deeply.

Reversible Adsorption and Desorption

Pore waters are in intimate contact with solid surfaces, and may exchange solutes reversibly on short timescales. Consequently, a change in pore water concentration is accompanied by an additional change in the adsorbed inventory. If a solute is strongly adsorbed, its transport will be dominated by movement of the particulate phase, rather than by migration through the dissolved phase. The ionic speciation of the solute is often critical in determining the degree of adsorption. Adsorption and desorption equilibria are temperature dependent, and are responsible for some of the artifacts noted earlier.

Diffusion

The random motion of solutes in pore waters results in a net transport from regions of high concentration to regions of low concentration, as described by Fick's laws for diffusion. This is the principal mechanism for transport over short distance scales, and the rate of diffusion depends on the solute in question, the porosity of the sediment, the temperature, and to some extent on the ensemble of other ions present and their concentration gradients. This last effect results from the speciation of the ion and cross-coupling among ion gradients that produces electrical potentials. A rough estimate of the relationship between length- and time-scales for which diffusion is effective can be obtained from using Fick's Second Law to evaluate the time for a diffusive front to migrate from a perturbation introduced in a one-dimensional system. The time is given by the relationship $t = x^2/(2D)$, where t is the time for response at a distance x from the perturbation in a sediment with a diffusivity of D. In deep-sea sediments, many solutes have diffusivities in the range of $5 \times 10^{-6}\ cm^2\,s^{-1}$, so the diffusive front migrates approximately 1 cm in 1 day, but requires 30 years to migrate 1 m and 0.3 million years to migrate 100 m.

Advection

It is most convenient to define pore water spatial coordinates relative to the sediment–water interface. If sediments are accumulating and contain some water, pore waters must be moving relative to this interface. In addition, flow may be driven by strong heating at depth, or other externally imposed forcing. These directed flows are considered advection and may be an important transport mechanism. The relative importance of advection in transporting a solute can be evaluated by considering the dimensionless Peclet number, D/UL, where D represents the diffusivity through sediments, U is the flow velocity, and L is the length scale over which transport must be accomplished. If this number is much larger than one, advection can be ignored, and transport is dominated by diffusion.

Irrigation

In coastal and slope sediments, organisms create burrows that act as conduits through which bottom waters may move. Flow through a burrow may be driven by active pumping by the organism, or by hydrodynamic effects created by the flow of bottom waters past the burrow orifice. Thus, the presence of burrows creates a complex geometry for the effective shape of the sediment–water interface and for the biogeochemical zonation that should parallel the burrow walls. Burrows may provide a pathway for communication between deep sediment and overlying water that does not depend on vertical diffusion. Because communication depends only on transport through the dissolved phase, it is called irrigation. This transport is distinct from bioturbation, a process that may also be important, but it is

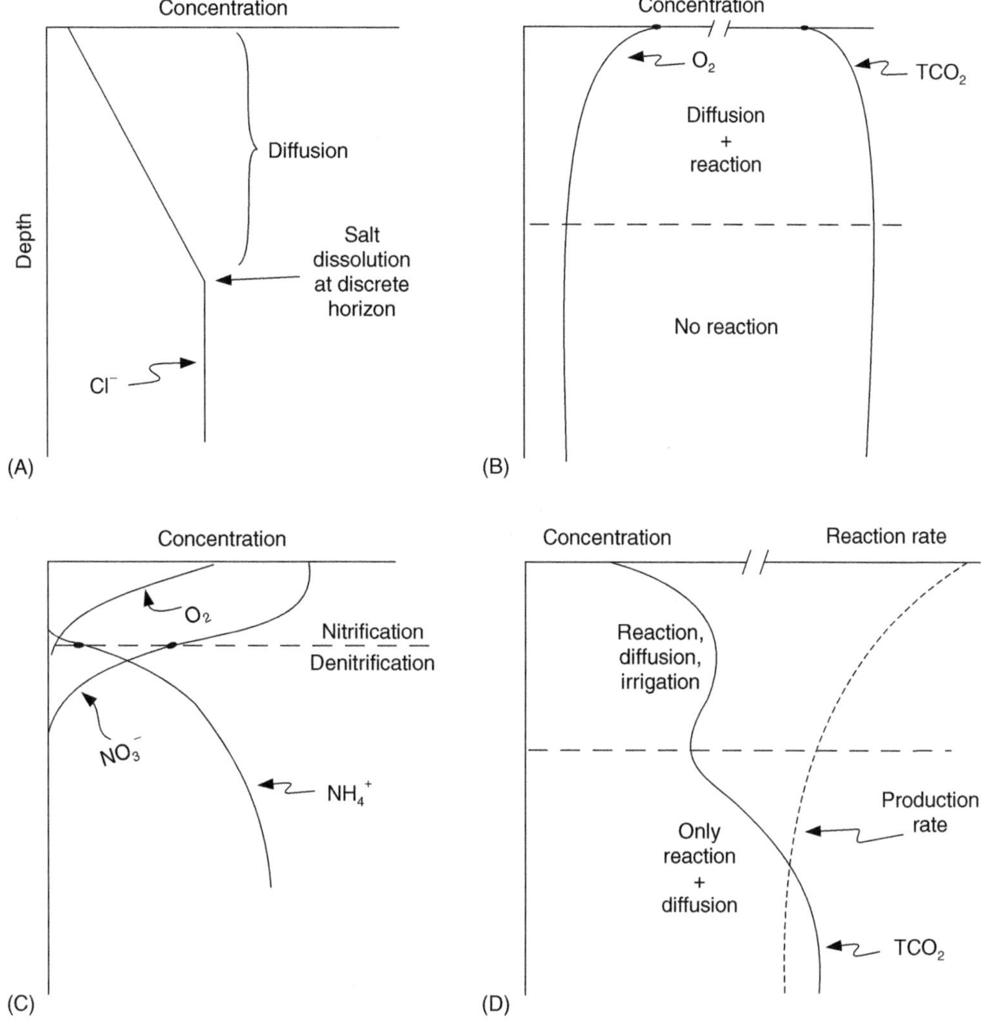

Figure 3 Schematic pore water profiles for various combinations of reaction and transport that conceptually illustrate observations made in field studies. Depth scales vary widely for these examples. (A) Reaction at a discrete horizon, with diffusive transport from the site of reaction to the overlying water. An example of this is the profile of chloride in sediments of the Mediterranean Sea created by dissolution of deeply buried (several hundred meters) salt deposits. Another example of this style of profile includes observations in ODP pore waters that reveal interactions of basaltic basement rocks where they contact pore waters in the overlying sediment, removing magnesium and releasing calcium. (B) Coupled reaction–diffusion profiles. In this example, organic carbon is oxidized to carbon dioxide (TCO_2) utilizing oxygen as the electron acceptor. The reaction is assumed to occur only above the dashed line, and the curvature defines whether the solute is consumed (O_2) or produced (TCO_2). (C) Coupled nitrification and denitrification. In this example, it is assumed that any ammonia released by degradation of organic matter is completely oxidized to nitrate (nitrification) if it enters the oxic zone (including ammonia diffusing from below). The nitrate produced diffuses downward and is converted to N_2 in the nitrate reduction zone (denitrification). Inflection points in the curves define the horizon separating these zones. Because of the competing production of nitrate in the oxic zone and removal in the nitrate reduction zone, sediments may be a net source or net sink for nitrate in the overlying water column, depending on the relative availabilities of reactive organic matter and oxygen. In this example, the competing reactions balance so there is no nitrate gradient at the sediment–water interface. Open-ocean sediments are relatively efficient at recycling their fixed nitrogen during diagenesis, while denitrification in margin sediments may lose 30–70% of the fixed nitrogen that rains to the seafloor in those locations. (D) Reaction, diffusion, and irrigation. In this example, TCO_2 is produced throughout the sediment column, but the rate decreases with increasing depth (dotted line). Reaction products are transported by diffusion and irrigation in the upper zone, and by diffusion only in the lower zone. If sampling defines the average concentration at each depth, the nonlocal irrigation effect provides an apparent sink for TCO_2.

slower and involves the physical mixing of both solids and water by organismal activities. Accurately quantifying the effect of irrigation has posed a significant modeling challenge, and for convenience it is usually introduced into equations as a parameter called non-local transport. Irrigation effect are commonly observed in estuarine, shelf, and slope locations overlain by oxygenated bottom water. The importance of irrigation in promoting benthic exchange depends on the density of burrows, the depth to which they penetrate, how frequently they are flushed, and the length scale over which diagenetic reactions for the solute of interest take place. Effect of Irrigation on pore water profiles have not been detected in sediments overlain by bottom waters low in oxygen, due to the exclusion of macrofauna, or in deep-sea settings where macrofauna are less abundant than in margin settings. Irrigation effects are greatest in the upper 10 cm, but pore water profiles in some deep basins of the California Borderland indicate observable effects to nearly 2 m.

Boundary Conditions

Pore water profiles are dependent on boundary conditions. An upper boundary condition is imposed by the solute concentration in the overlying water, although complications exist if the solute is very reactive. Above the sediment–water interface is a diffusive sublayer that may be several hundred micrometers thick in the deep sea, and thinner in more energetic environments. A significant concentration gradient may exist through this zone if the scale length characterizing the solute profile in the sediment is small enough to approach this distance. It can be more difficult to identify a lower boundary condition. Sometimes the solute will approach a constant value if its reactions cease at depth. In other cases, the solute may reach a constant value dictated by solubility constraints or by its disappearance.

Interpretation of Pore Water Profiles

Some examples of pore water profiles are illustrated in **Figure 3**, drawn schematically to illustrate the range of behavior that may be observed when different factors are important. Several assumptions have been made in drawing these profiles. One is that they represent steady-state, relative to the sediment–water interface. A second is that the diffusivity is not depth-dependent. A third is that any reactions go to zero at infinite depth. A fourth is that advection has been ignored. The shape of a profile is a clue to interpret what factors are important and their depth dependence, but the shape may not have a unique interpretation, particularly if these assumptions are not valid. It is also important to remember that concentration gradients of solutes adjust until transport is equal to the net reactions occurring. If a solute is involved in competing reactions, such as dissolution of one phase and precipitation of a less soluble phase, the net reaction could be zero and no concentration gradient would be created.

See also

Authigenic Deposits. Calcium Carbonates.

Further Reading

Aller RC (1988) Benthic fauna and biogeochemical processes in marine sediments: The role of burrow structure. In: Blackburn TH and Sorensen J (eds.) *Nitrogen Cycling in Coastal Marine Environments*, pp. 301–338. New York: John Wiley.

Berner RA (1974) Kinetic models for early diagenesis of nitrogen, sulfur, phosphorus, and silicon. In: Goldberg ED (ed.) *The Sea*, vol. 5, pp. 427–450. New York: John Wiley.

Berner RA (1980) *Early Diagenesis*. Princeton: Princeton Press.

Boudreau BP (1997) *Diagenetic Models and Their Implementation*. Berlin: Springer-Verlag.

Boudreau BP and Jorgensen BB (2000) *The Benthic Boundary Layer: Transport Processes and Biogeochemistry*. New York: Oxford University Press.

Boudreau BP (2000) The mathematics of early diagenesis: from worms to waves. *Reviews of Geophysics* 38: 389–416.

Burdige DJ (1993) The biogeochemistry of manganese and iron reduction in marine sediments. *Earth Science Reviews* 35: 249–284.

Fanning KA and Manheim FT (eds.) (1982) *The Dynamic Environment of the Ocean Floor*. Lexington, MA: DC Heath and Co.

Froelich PN, Klinkhammer GP, Bender ML, *et al.* (1979) Early oxidation of organic matter in pelagic sediments of the eastern equatorial Atlantic: Suboxic diagenesis. *Geochimica Cosmochimica Acta* 43: 1075–1090.

Lerman A (1977) Migrational processes and chemical reactions in interstitial waters. In: Goldberg E, McCave I, O'Brien J, and Steele J (eds.) *The Sea*, vol. 6, pp. 695–738. New York: John Wiley.

Li Y-H and Gregory S (1974) Diffusion of ions in sea water and deep-sea sediments. *Geochimica Cosmochimica Acta* 38: 703–714.

Luther GW III, Reimers CE, Nuzzio DB, and Lovalvo D (1999) In situ deployment of voltammetric, potentiometric, and amperometric microelectrodes from a ROV to determine dissolved O2, Mn, Fe, S(-2), and pH in porewaters. *Environmental Science and Technology* 33: 4352–4356.

Manheim FT and Sayles FL (1974) Composition and origin of interstitial waters of marine sediments, based on deep sea drill cores. In: Goldberg ED (ed.) *The Sea*, vol. 5, pp. 527–568. New York: John Wiley.

Nealson KH (1997) Sediment bacteria: Who's there, what are they doing and what's new? *Annual Reviews of Earth and Planetary Science* 25: 403–434.

Van Der Weijden C (1992) Early diagenesis and marine pore water. In: Wolf KH and Chilingarian GV (eds.) *Diagenesis III, Developments in Sedimentology*, pp. 13–134. New York: Elsevier.

BENTHIC FORAMINIFERA

A. J. Gooday, Southampton Oceanography Centre, Southampton, UK

Introduction

Foraminifera are enormously successful organisms and a dominant deep-sea life form. These amoeboid protists are characterized by a netlike (granuloreticulate) system of pseudopodia and a life cycle that is often complex but typically involves an alternation of sexual and asexual generations. The most obvious characteristic of foraminifera is the presence of a shell or 'test' that largely encloses the cytoplasmic body and is composed of one or more chambers. In some groups, the test is constructed from foreign particles (e.g., mineral grains, sponge spicules, shells of other foraminifera) stuck together ('agglutinated') by an organic or calcareous/organic cement. In others, it is composed of calcium carbonate (usually calcite, occasionally aragonite) or organic material secreted by the organism itself.

Although the test forms the basis of foraminiferal classification, and is the only structure to survive fossilization, the cell body is equally remarkable and important. It gives rise to the complex, highly mobile, and pervasive network of granuloreticulose pseudopodia. These versatile organelles perform a variety of functions (locomotion, food gathering, test construction, and respiration) that are probably fundamental to the ecological success of foraminifera in marine environments.

As well as being an important component of modern deep-sea communities, foraminifera have an outstandingly good fossil record and are studied intensively by geologists. Much of their research uses knowledge of modern faunas to interpret fossil assemblages. The study of deep-sea benthic foraminifera, therefore, lies at the interface between biology and geology. This articles addresses both these facets.

History of Study

Benthic foraminifera attracted the attention of some pioneer deep-sea biologists in the late 1860s. The monograph of H.B. Brady, published in 1884 and based on material collected in the *Challenger* round-the-world expedition of 1872–76, still underpins our knowledge of the group. Later biological expeditions added to this knowledge. For much of the 1900s, however, the study of deep-sea foraminifera was conducted largely by geologists, notably J.A. Cushman, F.B. Phleger, and their students, who amassed an extensive literature dealing with the taxonomy and distribution of calcareous and other hard-shelled taxa. In recent decades, the emphasis has shifted toward the use of benthic species in paleoceanographic reconstructions. Interest in deep-sea foraminifera has also increased among biologists since the 1970s, stimulated in part by the description of the Komokiacea, a superfamily of delicate, soft-shelled foraminifera, by O.S. Tendal and R.R. Hessler. This exclusively deep-sea taxon is a dominant component of the macrofauna in some abyssal regions.

Morphological and Taxonomic Diversity

Foraminifera are relatively large protists. Their tests range from simple agglutinated spheres a few tens of micrometers in diameter to those of giant tubular species that reach lengths of 10 cm or more. However, most are a few hundred micrometers in size. They exhibit an extraordinary range of morphologies (**Figures 1** and **2**), including spheres, flasks, various types of branched or unbranched tubes, and chambers arranged in linear, biserial, triserial, or coiled (spiral) patterns. In most species, the test has an aperture that assumes a variety of forms and is sometimes associated with a toothlike structure. The komokiaceans display morphologies not traditionally associated with the foraminifera. The test forms a treelike, bushlike, spherical, or lumpish body that consists of a complex system of fine, branching tubules (**Figure 2A–C**).

The foraminifera (variously regarded as a subphylum, class, or order) are highly diverse with around 900 living genera and an estimated 10 000 described living species, in addition to large numbers of fossil taxa. Foraminiferal taxonomy is based very largely on test characteristics. Organic, agglutinated, and different kinds of calcareous wall structure serve to distinguish the main groupings (orders or suborders). At lower taxonomic levels, the nature and position of the aperture and the number, shape, and arrangement of the chambers are important.

Figure 1 Scanning electron micrographs of selected deep-sea foraminifera (maximum dimensions are given in parentheses). (A) *Epistominella exigua*; 4850 m water depth, Porcupine Abyssal Plain, NE Atlantic (190 μm). (B) *Nonionella iridea*; 1345 m depth, Porcupine Seabight, NE Atlantic (110 μm). (C) *Nonionella stella*; 550 m depth, Santa Barbara Basin, California Borderland (220 μm). (D) *Brizalina tumida*; 550 m depth, Santa Barbara Basin, California Borderland (680 μm). (E) *Melonis barleaanum*; 1345 m depth, Porcupine Seabight, NE Atlantic (450 μm). (F) *Hormosina* sp., 4495 m depth, Porcupine Abyssal Plain (1.5 mm). (G) *Pyrgoella* sp.; 4550 m depth, foothills of Mid-Atlantic Ridge (620 μm). (H) *Technitella legumen*; 997–1037 m depth, NW African margin (8 mm). (A)–(E) and (G) have calcareous tests, (F) and (H) have agglutinated tests. (C) and (D), photographs courtesy of Joan Bernhard.

Methodology

Qualitative deep-sea samples for foraminiferal studies are collected using nets (e.g., trawls) that are dragged across the seafloor. Much of the *Challenger* material studied by Brady was collected in this way. Modern quantitative studies, however, require the use of coring devices. The two most popular corers used in the deep sea are the box corer, which obtains a large (e.g., 0.25 m^2) sample, and the multiple corer, which collects simultaneously a battery of up to 12 smaller cores. The main advantage of the multiple corer is that it obtains the sediment–water interface in a virtually undisturbed condition.

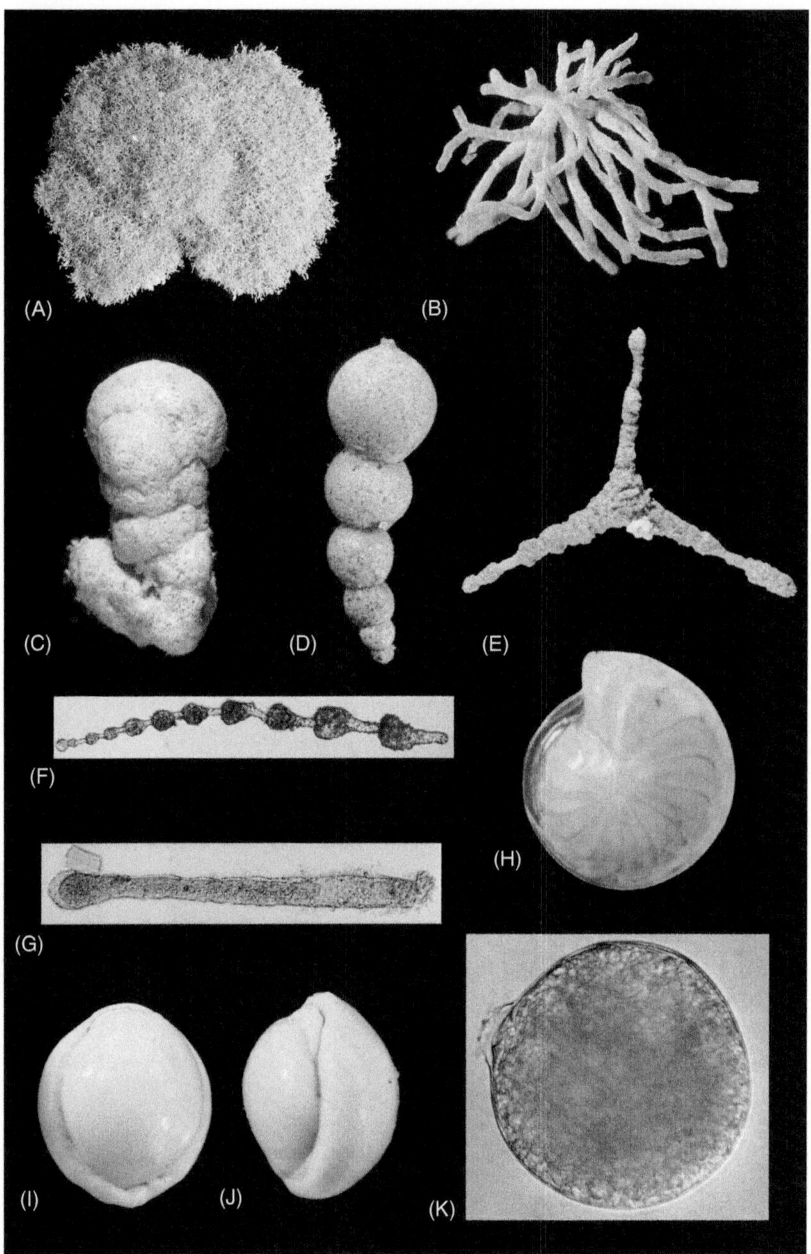

Figure 2 Light micrographs of deep-sea foraminifera (maximum dimensions are given in parentheses). (A) Species of *Lana* in which pad-like test consists of tightly meshed system of fine tubules; 5432 m water depth, Great Meteor East region, NE Atlantic (7.4 mm). (B) *Septuma* sp.; same locality (2 mm). (C) *Edgertonia* mudball; same locality (3.8 mm). (D) *Hormosina globulifera*; 4004 m depth, NW African margin (6.4 mm). (E) *Rhabdammina parabyssorum*; 3392 m depth, Oman margin, NW Arabian Sea (18 mm). (F) *Leptohalysis* sp.; 3400 m depth, Oman margin, NW Arabian Sea (520 μm). (G) Minute species of *Hyperammina*; 3400 m depth, Oman margin, NW Arabian Sea (400 μm). (H) *Lenticularia* sp.; 997–1037 m depth, NW African margin (2.5 mm). (I, J) *Biloculinella* sp.; 4004 m depth, NW African margin (3 mm). (K) Spherical allogromiid; 3400 m depth, Oman margin, NW Arabian Sea (105 μm). Specimens illustrated in (A)–(G) have agglutinated tests, in (H)–(J) calcareous tests and in (K) an organic test. (A)–(C) belong to the superfamily Komokiacea.

Foraminifera are extracted from sieved sediment residues. Studies are often based on dried residues and concern 'total' assemblages (i.e. including both live and dead individuals). To distinguish individuals that were living at the time of collection from dead tests, it is necessary to preserve sediment samples in either alcohol or formalin and then stain them with rose Bengal solution. This colors the cytoplasm red and is most obvious when residues are examined in water. Stained assemblages provide a snapshot of the foraminifera that were living when the samples were collected. Since the live assemblage varies in both time and space, it is also instructive to examine the dead assemblage that provides an averaged view of the foraminiferal fauna. Deep-sea foraminiferal assemblages are typically very diverse and therefore faunal data are often condensed mathematically by using multivariate approaches such as principal components or factor analysis.

The mesh size of the sieve strongly influences the species composition of the foraminiferal assemblage retained. Most deep-sea studies have been based on >63 μm, 125 μm, 150 μm, 250 μm, or even 500 μm meshes. In recent years, the use of a fine 63 μm mesh has become more prevalent with the realization that some small but important species are not adequately retained by coarser sieves. However, the additional information gained by examining fine fractions must be weighed against the considerable time and effort required to sort foraminifera from them.

Ecology

Abundance and Diversity

Foraminifera typically make up >50% of the soft-bottom, deep-sea meiofauna (**Table 1**). They are also often a major component of the macrofauna. In the central North Pacific, for example, foraminifera (mainly komokiaceans) outnumber all metazoans combined by at least an order of magnitude. A few species are large enough to be easily visible to the unaided eye and constitute part of the megafauna. These include the tubular species *Bathysiphon filiformis*, which is sometimes abundant on continental slopes (**Figure 3**). Some xenophyophores, agglutinated protists that are probably closely related to the foraminifera, are even larger (up to 24 cm maximum dimension!). These giant protists may dominate the megafauna in regions of sloped topography (e.g., seamounts) or high surface productivity. In well-oxygenated areas of the deep-seafloor, foraminiferal assemblages are very species rich, with well over 100 species occurring in relatively small volumes of surface sediment (**Figure 4**). Many are undescribed delicate, soft-shelled forms. There is an urgent need to describe at least some of these species as a step toward estimating global levels of deep-sea species diversity. The common species are often widely distributed, particularly at abyssal depths, although endemic species undoubtedly also occur.

Table 1 The percentage contribution of foraminifera to the deep-sea meiofauna at sites where bottom water is well oxygenated

Area	*Depth (m)*	*Percentage of foraminifera*	*Number of samples*
NW Atlantic			
Off North Carolina	500–2500	11.0–90.4	14
Off North Carolina	400–4000	7.6–85.9	28
Off Martha's Vineyard	146–567	3.4–10.6	4
NE Atlantic			
Porcupine Seabight	1345	47.0–59.2	8
Porcupine Abyssal Plain[a]	4850	61.8–76.3	3
Madeira Abyssal Plain[a]	4950	61.4–76.1	3
Cape Verde Abyssal Plain[a]	4550	70.2	1
Off Mauretania	250–4250	4–27	26
46°N, 16–17°W	4000–4800	0.5–8.3	9
Indian Ocean			
NW Arabian Sea[b]	3350	54.4	1
Pacific			
Western Pacific	2000–6000	36.0–69.3	11
Central North Pacific	5821–5874	49.5	2
Arctic	1000–2600	14.5–84.1	74
Southern Ocean	1661–1680	2.2–23.7	2

[a] Data from Gooday AJ (1996) Epifaunal and shallow infaunal foraminiferal communities at three abyssal NE Atlantic sites subject to differing phytodetritus input regimes. *Deep-Sea Research I* 43: 1395–1421.

[b] Data from Gooday AJ, Bernhard JM, Levin LA and Suhr SB (2000) Foraminifera in the Arabian Sea oxygen minimum zone and other oxygen-deficient settings: taxonomic composition, diversity, and relation to metazoan faunas. *Deep-Sea Research II* 47: 25–54.

Based on Gooday AJ (1986) Meiofaunal foraminiferans from the bathyal Porcupine Seabight (northeast Atlantic): size structure, standing stock, taxonomic composition, species diversity and vertical distribution in the sediment. *Deep-Sea Research* 35: 1345–1373; with permission from Elsevier Science.

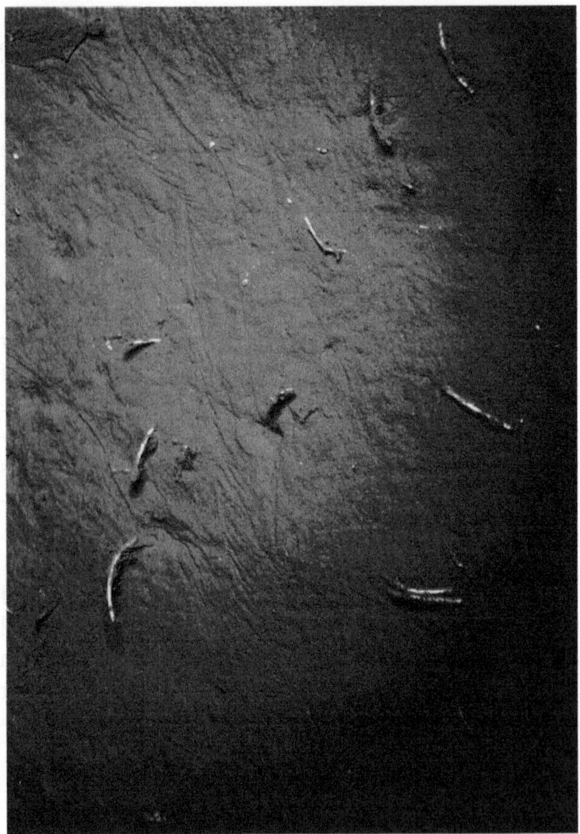

Figure 3 *Bathysiphon filiformis*, a large tubular agglutinated foraminifer, photographed from the Johnson Sealink submersible on the North Carolina continental slope (850 m water depth). The tubes reach a maximum length of about 10 cm. (Photograph courtesy of Lisa Levin.)

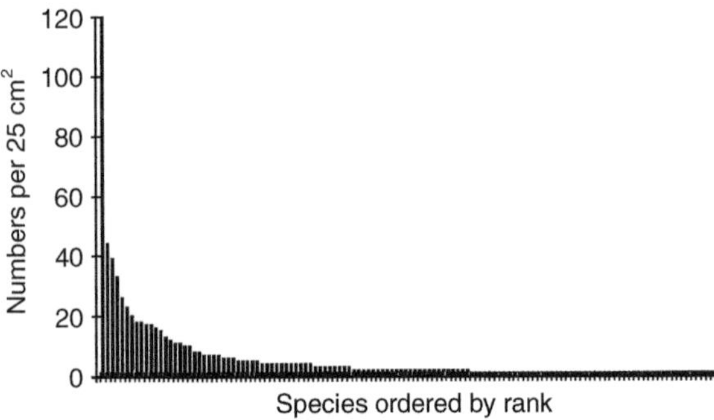

Figure 4 Deep-sea foraminiferal diversity: all species from a single multiple corer sample collected at the Porcupine Abyssal Plain, NE Atlantic (4850 m water depth), ranked by abundance. Each bar represents one 'live' (rose Bengal-stained) species. The sample was 25.5 cm^2 surface area, 0–1 cm depth, and sieved on a 63 μm mesh sieve. It contained 705 'live' specimens and 130 species.

Foraminifera are also a dominant constituent of deep-sea hard-substrate communities. Dense populations encrust the surfaces of manganese nodules as well as experimental settlement plates deployed on the sea floor for periods of months. They include various undescribed matlike taxa and branched tubular forms, as well as a variety of small coiled agglutinated species (many in the superfamily Trochamminacea), and calcareous forms.

Role in Benthic Communities

The abundance of foraminifera suggests that they play an important ecological role in deep-sea communities, although many aspects of this role remain poorly understood. One of the defining features of these protists, their highly mobile and pervasive pseudopodial net, enables them to gather food particles very efficiently. As a group, foraminifera exhibit a wide variety of trophic mechanisms (e.g., suspension feeding, deposit feeding, parasitism, symbiosis) and diets (herbivory, carnivory, detritus feeding, use of dissolved organic matter). Many deep-sea species appear to feed at a low trophic level on organic detritus, sediment particles, and bacteria. Foraminifera are prey, in turn, for specialist deep-sea predators (scaphopod mollusks and certain asellote isopods), and also ingested (probably incidentally) in large numbers by surface deposit feeders such as holothurians. They may therefore provide a link between lower and higher levels of deep-sea food webs.

Some deep-sea foraminifera exhibit opportunistic characteristics – rapid reproduction and population growth responses to episodic food inputs. Well-known examples are *Epistominella exigua, Alabaminella weddellensis* and *Eponides pusillus*. These small (generally $<200\,\mu m$), calcareous species feed on fresh algal detritus ('phytodetritus') that sinks through the water column to the deep-ocean floor after the spring bloom (a seasonal burst of phytoplankton primary production that occurs most strongly in temperate latitudes). Utilizing energy from this labile food source, they reproduce rapidly to build up large populations that then decline when their ephemeral food source has been consumed. Moreover, certain large foraminifera can reduce their metabolism or consume cytoplasmic reserves when food is scarce, and then rapidly increase their metabolic rate when food again becomes available. These characteristics, together with the sheer abundance of foraminifera, suggest that their role in the cycling of organic carbon on the deep-seafloor is very significant.

The tests of large foraminifera are an important source of environmental heterogeneity in the deep sea, providing habitats and attachment substrates for other foraminifera and metazoans. Mobile infaunal species bioturbate the sediment as they move through it. Conversely, the pseudopodial systems of foraminifera may help to bind together and stabilize deep-sea sediments, although this has not yet been clearly demonstrated.

Microhabitats and Temporal Variability

Like many smaller organisms, foraminifera reside above, on and within deep-sea sediments. Various factors influence their overall distribution pattern within the sediment profile, but food availability and geochemical (redox) gradients are probably the most important. In oligotrophic regions, the flux of organic matter (food) to the seafloor is low and most foraminifera live on or near the sediment surface where food is concentrated. At the other extreme, in eutrophic regions, the high organic-matter flux causes pore water oxygen concentrations to decrease rapidly with depth into the sediment, restricting access to the deeper layers to those species that can tolerate low oxygen levels. Foraminifera penetrate most deeply into the sediment where organic inputs are of intermediate intensity and the availability of food and oxygen within the sediment is well balanced.

Underlying these patterns are the distributions of individual species. Foraminifera occupy more or less distinct zones or microenvironments ('microhabitats'). For descriptive purposes, it is useful to recognize a number of different microhabitats: epifaunal and shallow infaunal for species living close to the sediment surface (upper 2 cm); intermediate infaunal for species living between about 1 cm and 4 cm (**Figure 5**); and deep infaunal for species that occur at depths down to 10 cm or more (**Figure 6**). A few deep-water foraminifera, including the well-known calcareous species *Cibicidoides wuellerstorfi*, occur on hard substrates (e.g., stones) that are raised above the sediment–water interface (elevated epifaunal microhabitat). There is a general relation between test morphotypes and microhabitat preferences. Epifaunal and shallow infaunal species are often trochospiral with large pores opening on the spiral side of the test; infaunal species tend to be planispiral, spherical, or ovate with small, evenly distributed pores. It is important to appreciate that foraminiferal microhabitats are by no means fixed. They may vary between sites and over time and are modified by the burrowing activities of macrofauna. Foraminiferal microhabitats should therefore be regarded as dynamic rather than static. This tendency is most pronounced in shallow-water settings where environmental conditions are more changeable and macrofaunal activity is more intense than in the deep sea.

The microhabitats occupied by species reflect the same factors that constrain the overall distribution patterns of foraminifera within the sediment. Epifaunal and shallow infaunal species cannot tolerate low oxygen concentrations and also require a diet of relatively fresh organic matter. Deep infaunal foraminifera are less opportunistic but are more tolerant of oxygen depletion than are species living close to the sediment–water interface (**Figure 6**). It has been suggested that species of genera such as *Globobulimina* may consume either sulfate-reducing bacteria or labile organic matter released by the metabolic

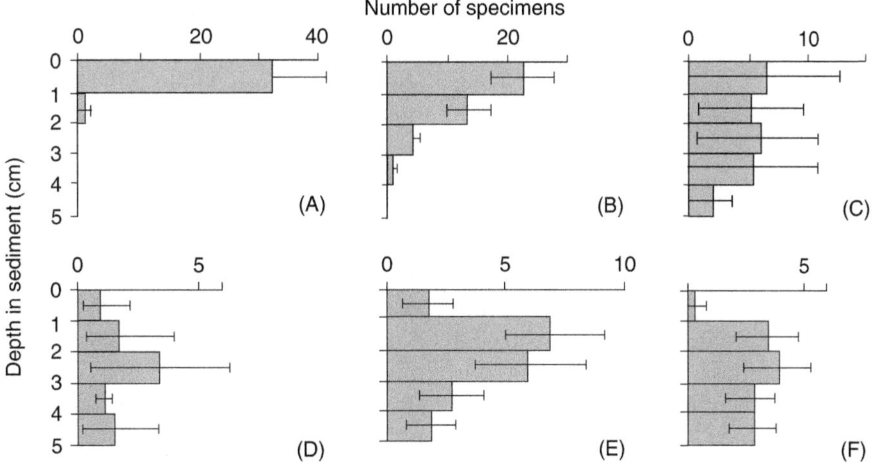

Figure 5 Vertical distribution patterns within the top 5 cm of sediment of common foraminiferal species ('live', rose Bengal-stained specimens) in the Porcupine Seabight, NW Atlantic (51°36′N, 13°00′W; 1345 m water depth). Based on >63 μm sieve fraction. (A) *Ovammina* sp. (mean of 20 samples). (B) *Nonionella iridea* (20 samples). (C) *Leptohalysis* aff. *catenata* (7 samples). (D) *Melonis barleeanum* (9 samples). (E) *Haplophragmoides bradyi* (19 samples). (F) *'Turritellella' laevigata* (21 samples). (Amended and reprinted from Gooday AJ (1986) Meiofaunal foraminiferans from the bathyal Porcupine Seabight (northeast Atlantic): size structure, standing stock, taxonomic composition, species diversity and vertical distribution in the sediment. *Deep-Sea Research* 35: 1345–1373; permission from Elsevier Science.)

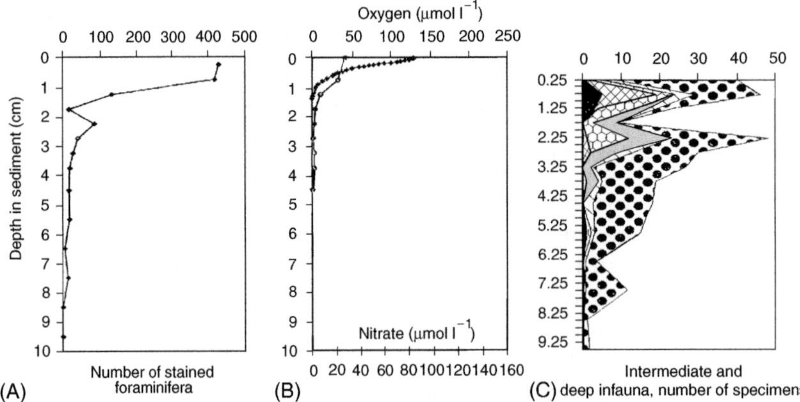

Figure 6 Vertical distribution of (A) total 'live' (rose Bengal-stained) foraminifera, (B) pore water oxygen and nitrate concentrations, and (C) intermediate and deep infaunal foraminiferal species within the top 10 cm of sediment on the north-west African margin (21°28.8′N, 17°57.2′W, 1195 m). All foraminiferal counts based on $>150\,\mu m$ sieve fraction, standardized to a $34\,cm^3$ volume. Species are indicated as follows: *Pullenia salisburyi* (black), *Melonis barleeanum* (crossed pattern), *Chilostomella oolina* (honeycomb pattern), *Fursenkoina mexicana* (grey), *Globobulimina pyrula* (diagonal lines), *Bulimina marginata* (large dotted pattern). (Adopted and reprinted from Jorissen FJ, Wittling I, Peypouquet JP, Rabouille C and Relexans JC (1998) Live benthic foraminiferal faunas off Cape Blanc, northwest Africa: community structure and microhabitats. *Deep-Sea Research I* 45: 2157–2158; with permission from Elsevier Science.)

activities of these bacteria. These species move closer to the sediment surface as redox zones shift upward in the sediment under conditions of extreme oxygen depletion. Although deep-infaunal foraminifera must endure a harsh microenvironment, they are exposed to less pressure from predators and competitors than those occupying the more densely populated surface sediments.

Deep-sea foraminifera may undergo temporal fluctuations that reflect cycles of food and oxygen availability. Changes over seasonal timescales in the abundance of species and entire assemblages have been described in continental slope settings (**Figure 7**). These changes are related to fluctuations in pore water oxygen concentrations resulting from episodic (seasonal) organic matter inputs to the seafloor. In some cases, the foraminifera migrate up and down in the sediment, tracking critical oxygen levels or redox fronts. Population fluctuations also occur in abyssal settings where food is a limiting ecological factor. In these cases, foraminiferal population dynamics reflect the seasonal availability of phytodetritus ('food'). As a result of these temporal processes, living foraminifera sampled during one season often provide an incomplete view of the live fauna as a whole.

Environmental Controls on Foraminiferal Distributions

Our understanding of the factors that control the distribution of foraminifera on the deep-ocean floor is very incomplete, yet lack of knowledge has not prevented the development of ideas. It is likely that foraminiferal distribution patterns reflect a combination of influences. The most important first-order factor is calcium carbonate dissolution. Above the carbonate compensation depth (CCD), faunas include calcareous, agglutinated, and allogromiid taxa. Below the CCD, calcareous species are almost entirely absent. At oceanwide or basinwide scales, the organic carbon flux to the seafloor (and its seasonality) and bottom-water hydrography appear to be particularly important, both above and below the CCD.

Studies conducted in the 1950s and 1960s emphasized bathymetry (water depth) as an important controlling factor. However, it soon became apparent that the bathymetric distribution of foraminiferal species beyond the shelf break is not consistent geographically. Analyses of modern assemblages in the North Atlantic, carried out in the 1970s, revealed a much closer correlation between the distribution of foraminiferal species and bottom-water masses. For example, *Cibicidoides wuellerstorfi* was linked to North Atlantic Deep Water (NADW) and *Nuttallides umbonifera* to Antarctic Bottom Water (AABW). At this time, it was difficult to explain how slight physical and chemical differences between water masses could influence foraminiferal distributions. However, recent work in the south-east Atlantic, where hydrographic contrasts are strongly developed, suggests that the distributions of certain foraminiferal species are controlled in part by the lateral advection of water masses. In the case of *N. umboniferus* there is good evidence that the main

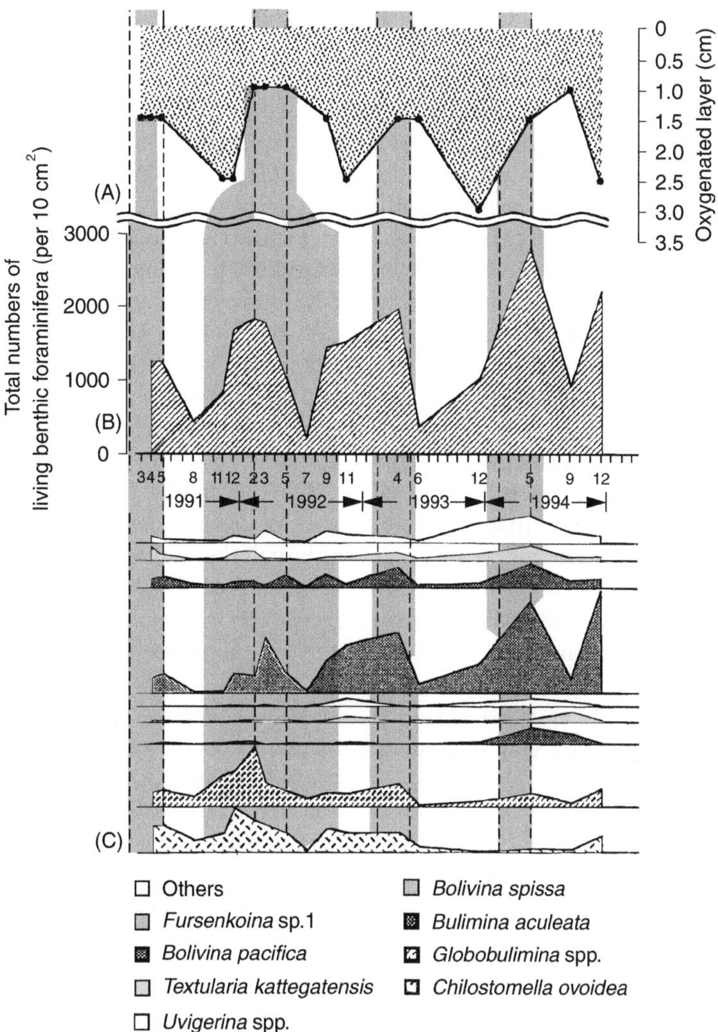

Figure 7 Seasonal changes over a 4-year period (March 1991 to December 1994) in (A) the thickness of the oxygenated layer, (B) the total population density of live benthic foraminifera, and (C) the abundances of the most common species at a 1450 m deep site in Sagami Bay, Japan. (Reprinted from Ohga T and Kitazato H (1997) Seasonal changes in bathyal foraminiferal populations in response to the flux of organic matter (Sagami Bay, Japan). *Terra Nova* 9: 33–37; with permission from Blackwell Science Ltd.)

factor is the degree of undersaturation of the bottom water in calcium carbonate. This abyssal species is found typically in the carbonate-corrosive (and highly oligotrophic) environment between the calcite lysocline and the CCD, a zone that may coincide approximately with AABW. Where water masses are more poorly delineated, as in the Indian and Pacific Oceans, links with faunal distributions are less clear.

During the past 15 years, attention has focused on the impact on foraminiferal ecology of organic matter fluxes to the seafloor. The abundance of dead foraminiferal shells $>150\,\mu m$ in size correlates well with flux values. There is also compelling evidence that the distributions of species and species associations are linked to flux intensity. Infaunal species, such as *Melonis barleeanum, Uvigerina peregrina, Chilostomella ovoidea* and *Globobulimina affinis*, predominate in organically enriched areas, e.g. beneath upwelling zones. Epifaunal species such as *Cibicidoides wuellerstorfi* and *Nuttallides umbonifera* are common in oligotrophic areas, e.g. the central oceanic abyss. In addition to flux intensity, the degree of seasonality of the food supply (i.e., whether it is pulsed or continuous) is a significant factor. *Epistominella exigua*, one of the opportunists that exploit phytodetritus, occurs in relatively oligotrophic areas where phytodetritus is deposited seasonally.

Recent analysis of a large dataset relating the relative abundance of 'live' (stained) foraminiferal assemblages in the north-east Atlantic and Arctic Oceans to flux rates to the seafloor has provided a quantitative framework for these observations. Although species are associated with a wide flux range, this range diminishes as a species become relatively more abundant and conditions become increasingly optimum for it. When dominant occurrences (i.e., where species represent a high percentages of the fauna) are plotted against flux and water depth, species fall into fields bounded by particular flux and depth values (**Figure 8**). Despite a good deal of overlap, it is possible to distinguish a series of dominant species that succeed each other bathymetrically on relatively eutrophic continental slopes and other species that dominate on the more oligotrophic abyssal plains.

Other environmental attributes undoubtedly modify the species composition of foraminiferal assemblages in the deep sea. Agglutinated species with tubular or spherical tests are found in areas where the seafloor is periodically disturbed by strong currents capable of eroding sediments. Forms projecting into the water column may be abundant where steady flow rates convey a continuous supply of suspended food particles. Other species associations may be linked to sedimentary characteristics.

Low-Oxygen Environments

Oxygen availability is a particularly important ecological parameter. Since oxygen is consumed during the degradation of organic matter, concentrations of oxygen in bottom water and sediment pore water are inversely related to the organic flux derived from surface production. In the deep sea, persistent oxygen depletion ($O_2 < 1\,ml\,l^{-1}$) occurs at bathyal depths ($<1000\,m$) in basins (e.g., on the California Borderland) where circulation is restricted by a sill and in areas where high primary productivity resulting from the upwelling of nutrient-rich water leads to the development of an oxygen minimum zone (OMZ; e.g., north-west Arabian Sea and the Peru margin). Subsurface sediments also represent an oxygen-limited setting, although oxygen penetration is generally greater in oligotrophic deep-sea sediments than in fine-grained sediments on continental shelves.

On the whole, foraminifera exhibit greater tolerance of oxygen deficiency than most metazoan taxa, although the degree of tolerance varies among species. Oxygen probably only becomes an important limiting factor for foraminifera at concentrations well below $1\,ml\,l^{-1}$. Some species are abundant at levels of $0.1\,ml\,l^{-1}$ or less. A few apparently live in permanently anoxic sediments, although anoxia sooner or later results in death when accompanied by high concentrations of hydrogen sulfide. Oxygen-deficient areas are characterized by high foraminiferal densities but low, sometimes very low (<10), species numbers. This assemblage structure (high dominance, low species richness) arises because (i) low oxygen

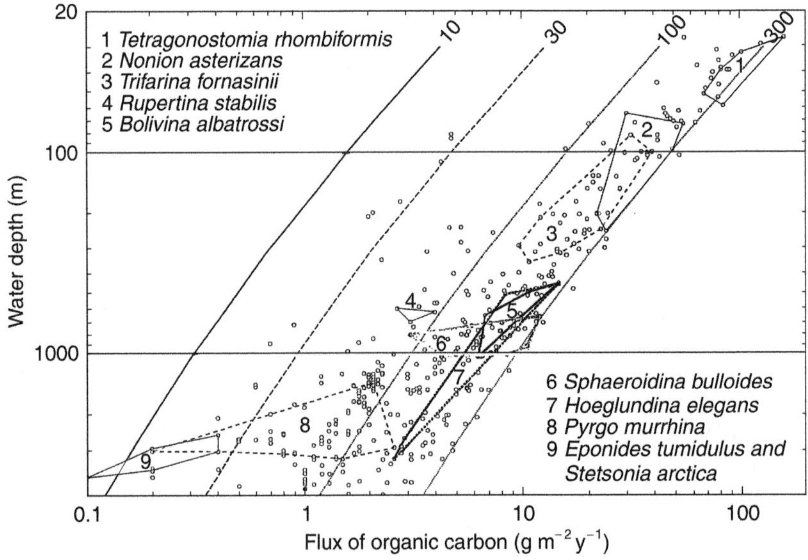

Figure 8 Dominant 'live' (rose Bengal-stained) occurrences of foraminiferal species in relation to water depth and flux or organic carbon to seafloor in the North Atlantic from the Guinea Basin to the Arctic Ocean. Each open circle corresponds to a data point. The polygonal areas indicate the combination of water depth and flux conditions under which nine different species are a dominant faunal component. The diagonal lines indicate levels of primary production (10, 30, 100, 300 g m^{-2} y^{-1}) that result in observed flux rates. Based on $>250\,\mu m$ sieve fraction plus 63–250 μm fraction from Guinea Basin and Arctic Ocean. (Reprinted from Altenbach AV, Pflaumann U, Schiebel R *et al.* (1999) Scaling percentages and distribution patterns of benthic foraminifera with flux rates of organic carbon. *Journal of Foraminiferal Research* 29: 173–185; with permission from The Cushman Foundation.)

concentration acts as a filter that excludes non-tolerant species and (ii) the tolerant species that do survive are able to flourish because food is abundant and predation is reduced. Utrastructural studies of some species have revealed features, e.g., bacterial symbionts and unusually high abundances of peroxisomes, that may be adaptations to extreme oxygen depletion. In addition, mitachondria-laden pseudopodia have the potential to extend into overlying sediment layers where some oxygen may be present.

Many low-oxygen-tolerant foraminifera belong to the Orders Rotaliida and Buliminida. They often have thin-walled, calcareous tests with either flattened, elongate biserial or triserial morphologies (e.g., *Bolivina, Bulimina, Globobulimina, Fursenkoina, Loxotomum, Uvigerina*) or planispiral/lenticular morphologies (e.g., *Cassidulina, Chilostomella, Epistominella, Loxotomum, Nonion, Nonionella*). Some agglutinated foraminifera, e.g., *Textularia, Trochammina* (both multilocular), *Bathysiphon*, and *Psammosphaera* (both unilocular), are also abundant. However, miliolids, allogromiids, and other soft-shelled foraminifera are generally rare in low-oxygen environments. It is important to note that no foraminiferal taxon is currently known to be confined entirely to oxygen-depleted environments.

Deep-Sea Foraminifera in Paleo-Oceanography

Geologists require proxy indicators of important environmental variables in order to reconstruct ancient oceans. Benthic foraminifera provide good proxies for seafloor parameters because they are widely distributed, highly sensitive to environmental conditions, and abundant in Cenozoic and Cretaceous deep-sea sediments (note that deep-sea

Table 2 Benthic foraminiferal proxies or indicators (both faunal and chemical) useful in paleo-oceanographic reconstruction

Environmental parameter/property	*Proxy or indicator*	*Remarks*
Water depth	Bathymetric ranges of abundant species in modern oceans	Depth zonation largely local although broad distinction between shelf, slope and abyssal depth zones possible
Distribution of bottom water masses	Characteristic associations of epifaunal species	Relations between species and water masses may reflect lateral advection
Carbonate corrosiveness of bottom water	Abundance of *Nuttallides umbonifera*	Corrosive bottom water often broadly corresponds to Antarctic Bottom Water
Deep-ocean thermohaline circulation	Cd/Ca ratios and $\delta^{13}C$ values for calcareous tests	Proxies reflect 'age' of bottom watermasses; i.e., period of time elapsed since formation at ocean surface
Oxygen-deficient bottom-water and pore water	Characteristic species associations; high-dominance, low-diversity assemblages	Species not consistently associated with particular range of oxygen concentrations and also found in high-productivity areas
Primary productivity	Abundance of foraminiferal tests $>150\,\mu m$	Transfer function links productivity to test abundance (corrected for differences in sedimentation rates between sites) in oxygenated sediments
Organic matter flux to seafloor	(i) Assemblages of high productivity taxa (e.g. *Globobulimina, Melonis barleeanum*)	Assemblages indicate high organic matter flux to seafloor, with or without corresponding decrease in oxygen concentrations
	(ii) Ratio between infaunal and epifaunal morphotypes	
	(iii) Ratio between planktonic and benthic tests	
Seasonality in organic matter flux	Relative abundance of 'phytodetritus species'	Reflects seasonally pulsed inputs of labile organic matter to seafloor
Methane release	Large decrease (2–3‰) in $\delta^{13}C$ values of benthic and planktonic tests	Inferred sudden release of ^{12}C enriched methane from clathrate deposits following temperature rise

sediments older than the middle Jurassic age have been destroyed by subduction, except where preserved in ophiolite complexes).

Foraminiferal faunas, and the chemical tracers preserved in the tests of calcitic species, can be used to reconstruct a variety of paleoenvironmental parameters and attributes. The main emphasis has been on organic matter fluxes and bottom-water/pore water oxygen concentrations (inversely related parameters), the distribution of bottom-water masses, and the development of thermohaline circulation (**Table 2**). Modern deep-sea faunas became established during the Middle Miocene (10–15 million years ago), and these assemblages can often be interpreted in terms of modern analogues. This approach is difficult or impossible to apply to sediments from the Cretaceous and earlier Cenozoic, which contain many foraminiferal species that are now extinct. In these cases, it can be useful to work with test morphotypes (e.g., trochospiral, cylindrical, biserial/triserial) rather than species. The relative abundance of infaunal morphotypes, for example, has been used as an index of bottom-water oxygenation or relative intensities of organic matter inputs. The trace element (e.g., cadmium) content and stable isotope (δ^{13}C; i.e., the deviation from a standard ^{12}C : ^{12}C ratio) chemistry of the calcium carbonate shells of benthic foraminifera provide powerful tools for making paleo-oceanographic reconstructions, particularly during the climatically unstable Quaternary period. The cadmium/calcium ratio is a proxy for the nutrient (phosphate) content of sea water that reflects abyssal circulation patterns. Carbon isotope ratios also reflect deep-ocean circulation and the strength of organic matter fluxes to the seafloor.

It is important to appreciate that the accuracy with which fossil foraminifera can be used to reconstruct ancient deep-sea environments is often limited. These limitations reflect the complexities of deep-sea foraminiferal biology, many aspects of which remain poorly understood. Moreover, simple relationships between the composition of foraminiferal assemblages and environmental variables are elusive, and it is often difficult to identify faunal characteristics that can be used as precise proxies for paleo-oceanographic parameters. For example, geologists often wish to establish paleobathymetry. However, the bathymetric distributions of foraminiferal species are inconsistent and depend largely on the organic flux to the seafloor, which decreases with increasing depth (**Figure 8**) and is strongly influenced by surface productivity. Thus, foraminifera can be used only to discriminate in a general way between shelf, slope, and abyssal faunas, but not to estimate precise paleodepths. Oxygen concentrations and organic matter inputs are particularly problematic. Certain species and morphotypes dominate in low-oxygen habitats that also are usually characterized by high organic loadings. However, the same foraminifera may occur in organically enriched settings where oxygen levels are

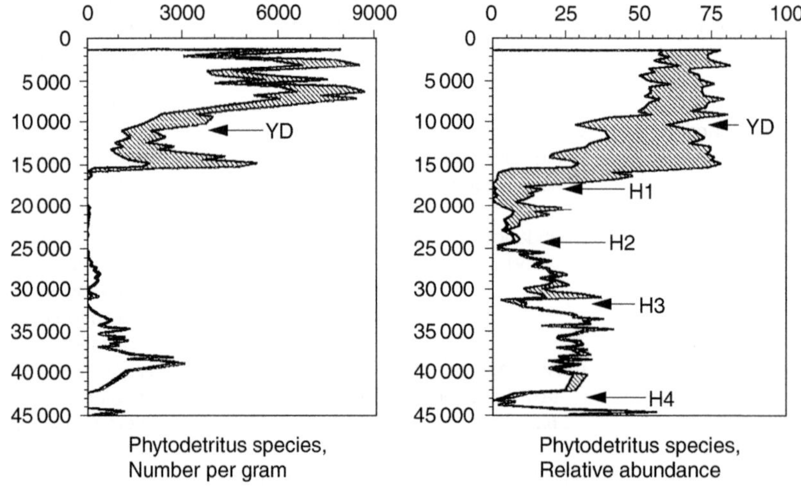

Figure 9 (A) Absolute (specimens per gram of dry sediment) and (B) relative (percentage) abundances of *Alabaminella weddellensis* and *Epistominella exigua* (>63 μm fraction) in a long-sediment core from the North Atlantic (50°41.3′N, 21°51.9′W, 3547 m water depth). In modern oceans, these two species respond to pulsed inputs of organic matter ('phytodetritus') derived from surface primary production. Note that they increased in abundance around 15 000 years ago, corresponding to the main Northern Hemisphere deglaciation and the retreat of the Polar Front. Short period climatic fluctuations (YD = Younger Dryas; H1–4 = Heinrich events, periods of very high meltwater production) are also evident in the record of these two species. (Reprinted from Thomas E, Booth L, Maslin M and Shackleton NJ (1995). Northeast Atlantic benthic foraminifera during the last 45000 years: change in productivity seen from the bottom up. *Paleoceanography*: 10: 545–562; with permission from the American Geophysical Union.)

not severely depressed, making it difficult for paleo-oceanographers to disentangle the influence of these two variables. Finally, biological factors such as microhabitat preferences and the exploitation of phytodetrital aggregates ('floc') influence the stable isotope chemistry of foraminiferal tests.

There are many examples of the use of benthic foraminiferal faunas to interpret the geological history of the oceans. Only one is given here. Cores collected at 50°41′N, 21°52′W (3547 m water depth) and 58°37′N, 19°26′W (1756 m water depth) were used by E. Thomas and colleagues to study changes in the North Atlantic over the past 45 000 years. The cores yielded fossil specimens of two foraminiferal species, *Epistominella exigua* and *Alabaminella weddellensis*, both of which are associated with seasonal inputs of organic matter (phytodetritus) in modern oceans. In the core from 51°N, these 'phytodetritus species' were uncommon during the last glacial maximum but increased sharply in absolute and relative abundance during the period of deglaciation 15 000–16 000 years ago (**Figure 9**). At the same time there was a decrease in the abundance of *Neogloboquadrina pachyderma*, a planktonic foraminifer found in polar regions, and an increase in the abundance of *Globigerina bulloides*, a planktonic species characteristic of warmer water. These changes were interpreted as follows. Surface primary productivity was low at high latitudes in the glacial North Atlantic, but was much higher to the south of the Polar Front. At the end of the glacial period, the ice sheet shrank and the Polar Front retreated northwards. The 51°N site was now overlain by more productive surface water characterized by a strong spring bloom and a seasonal flux of phytodetritus to the seafloor. This episodic food source favored opportunistic species, particularly *E. exigua* and *A. weddellensis*, which became much more abundant both in absolute terms and as a proportion of the entire foraminiferal assemblage.

Conclusions

Benthic foraminifera are a major component of deep-sea communities, play an important role in ecosystem functioning and biogeochemical cycling, and are enormously diverse in terms of species numbers and test morphology. These testate (shell-bearing) protists are also the most abundant benthic organisms preserved in the deep-sea fossil record and provide powerful tools for making paleo-oceanographic reconstructions. Our understanding of their biology has advanced considerably during the last two decades, although much remains to be learnt.

See also

Anthropogenic Trace Elements in the Ocean. Cenozoic Oceans - Carbon Cycle Models. Ocean Carbon System, Modeling of. Radiocarbon. Stable Carbon Isotope Variations in the Ocean. Tracers of Ocean Productivity.

Further Reading

Fischer G and Wefer G (1999) *Use of Proxies in Paleoceanography: Examples from the South Atlantic.* Berlin: Springer-Verlag.

Gooday AJ, Levin LA, Linke P, and Heeger T (1992) The role of benthic foraminifera in deep-sea food webs and carbon cycling. In: Rowe GT and Pariente V (eds.) *Deep-Sea Food Chains and the Global Carbon Cycle*, pp. 63–91. Dordrecht: Kluwer Academic.

Jones RW (1994) *The Challenger Foraminifera.* Oxford: Oxford University Press.

Loeblich AR and Tappan H (1987) *Foraminiferal Genera and their Classification*, vols 1, 2. New York: Van Nostrand Reinhold.

Murray JW (1991) *Ecology and Palaeoecology of Benthic Foraminifera.* New York: Wiley; Harlow: Longman Scientific and Technical.

SenGupta BK (ed.) (1999) *Modern Foraminifera.* Dordrecht: Kluwer Academic.

Tendal OS and Hessler RR (1977) An introduction to the biology and systematics of Komokiacea. *Galathea Report* 14: 165–194, plates 9–26.

Van der Zwan GJ, Duijnstee IAP, den Dulk M, *et al.* (1999) Benthic foraminifers:: proxies or problems? A review of paleoecological concepts. *Earth Sciences Reviews* 46: 213–236.

COCCOLITHOPHORES

T. Tyrrell, National Oceanography Centre, Southampton, UK
J. R. Young, The Natural History Museum, London, UK

Introduction

Coccolithophores (**Figure 1**) are a group of marine phytoplankton belonging to the division Haptophyta. Like the other free-floating marine plants (phytoplankton), the coccolithophores are microscopic (they range in size between about 0.003 and 0.040 mm diameter) single-celled organisms which obtain their energy from sunlight. They are typically spherical in shape. They are distinguished from other phytoplankton by their construction of calcium carbonate ($CaCO_3$) plates (called coccoliths) with which they surround their cells. While not quite the only phytoplankton to use $CaCO_3$ (there are also some calcareous dinoflagellates), they are by far the most numerous; indeed they are one of the most abundant of phytoplankton groups, comprising in the order of 10% of total global phytoplankton biomass.

The first recorded observations of coccoliths were made in 1836 by Christian Gottfried Ehrenberg, a founding figure in micropaleontology. The name 'coccoliths' (Greek for 'seed-stones') was coined by Thomas Henry Huxley (famous as 'Darwin's bulldog') in 1857 as he studied marine sediment samples. Both Ehrenberg and Huxley attributed coccoliths to an inorganic origin. This was soon challenged by Henry Clifton Sorby and George Charles Wallich who inferred from the complexity of coccoliths that they must be of biological origin, and supported this with observations of groups of coccoliths aggregated into empty spheres.

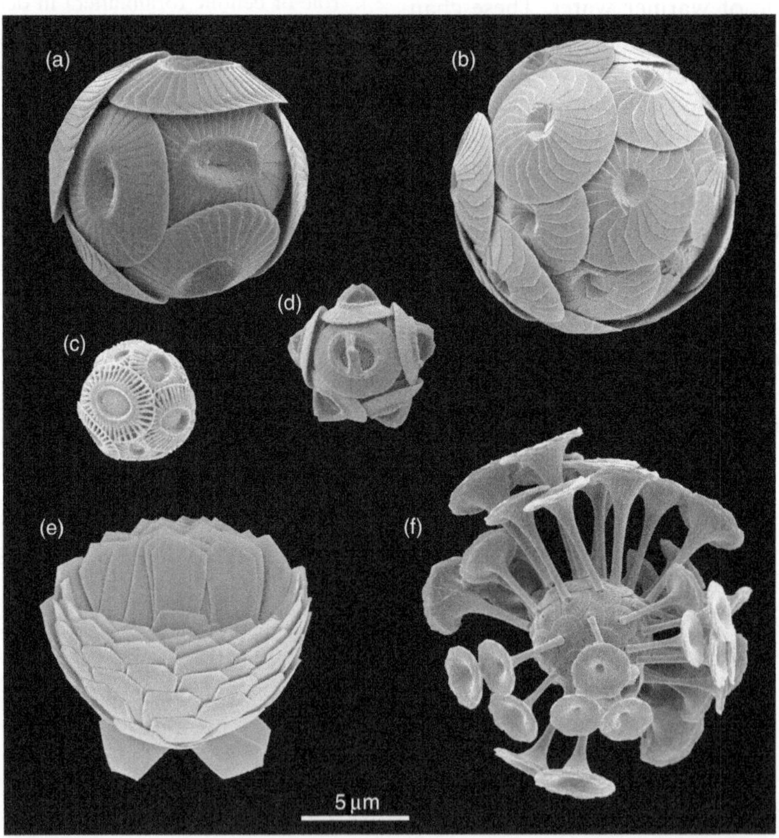

Figure 1 Electron microscope images of some major coccolithophore species: (a) *Coccolithus pelagicus*, (b) *Calcidiscus quadriperforatus*, (c) *Emiliania huxleyi*, (d) *Gephyrocapsa oceanica*, (e) *Florisphaera profunda*, (f) *Discosphaera tubifera*.

Species and Distribution

Approximately 200 species of coccolithophore have been formally described, separated into 65 genera. However, the true number of authentic modern coccolithophore species is rather unclear, for a couple of reasons. First, it is now realized that pairs of species, previously thought to be distinct and rather unrelated, are actually different life-cycle stages of the same species; coccolithophores typically have life cycles in which the haploid (single set of chromosomes, as in sex cells) and diploid (double set of chromosomes) phases can form different coccolith types: 'heterococcoliths' during the diploid life stage, and 'holococcoliths' during the haploid life stage. This type of life cycle has long been known from classic studies of laboratory cultures. It has only recently been appreciated that it is a very widespread pattern, as a result of observations of combination coccospheres representing the transition between the two life-cycle phases, that is to say possessing half a covering of heterococcoliths and half a covering of holococcoliths (**Figure 2**). Fifty of the 200 described coccolithophores are taxa known only from their holococcolith-producing phase and so may prove to be part of the life cycle of a heterococcolith-producing species. The second factor making diversity estimates difficult is that recent research combining studies of fine-scale morphology, biogeography, and molecular genetics has suggested that many described species are actually clusters of a few closely related, but genetically distinct, species. Indeed, as a result of such studies, numerous additional morphotypes have been recognized and await formal description.

Figure 2 A combination coccosphere, upper half (inner layer) heterococcoliths, lower half (outer layer) holococcoliths, of the species *Calcidiscus quadriperforatus* (the two stages were previously regarded as two separate species – *Calcidiscus leptoporus* and *Syracolithus quadriperforatus* – prior to discovery of this combination coccosphere). Scale = 1 μm. Electron microscope image provided by Markus Geisen (Alfred-Wegener-Institute, Germany).

Coccolithophores occur widely throughout the world's oceans, with the exception of the very-high-latitude polar oceans. Most individual species have more restricted biogeographical ranges than the range of the group as a whole, but still typically have interoceanic distributions. Unlike diatoms (the other major group of phytoplankton that make hard mineral shields), they are absent from almost all freshwater rivers and lakes. They occur in the brackish (more saline than freshwater but less saline than seawater) Black Sea, but not in the brackish Baltic Sea. In contrast once more to diatoms, coccolithophores are almost exclusively planktonic. There are very few bottom-dwelling species, even at shallow depths experiencing adequate light levels. Most species today live in warm, nutrient-poor conditions of the subtropical oceanic gyres, where they form a prominent component of the phytoplankton; there are fewer species that inhabit coastal and temperate or subpolar waters.

Emiliania huxleyi (**Figure 1(c)**) is the best-known species, primarily because it forms intense blooms which are clearly visible in satellite images, appearing as pale turquoise swirls in the ocean (**Figure 3**). While *E. huxleyi* frequently dominates phytoplankton counts in seawater samples, at least in terms of numbers of cells, their cells (and therefore also the coccoliths that surround them) are rather small, with the cells about 5 μm across and the coccoliths about 3 μm long. No other coccolithophore species regularly forms blooms, although occasional blooms of other species, for instance *Gephyrocapsa oceanica* and *Coccolithus pelagicus*, have been recorded. Many other species, for example *Calcidiscus quadriperforatus* and *Umbilicosphaera sibogae*, are most successful in low-productivity waters but do not bloom there. Although these species are almost always much less numerous than *E. huxleyi* in water samples, they are on the other hand also significantly larger, with typical cell diameters greater than 10 μm and correspondingly larger coccoliths. Most species of coccolithophores are adapted to life in the surface mixed layer, but some species, such as *Florisphaeara profunda*, are confined to the deep photic zone where they make an important contribution to the 'shade flora'.

Figure 3 *SeaWiFS* satellite image from June 1998 of *E. huxleyi* blooms (the turquoise patches) along the west coast of Norway and to the southwest of Iceland. The perspective is from a point over the Arctic Ocean, looking southward down the North Atlantic. The Greenland ice sheet is visible in the center foreground. Imagery provided with permission by GeoEye and NASA SeaWiFS Project.

In oligotrophic surface waters, typical concentrations of coccolithophores are in the range 5000–50 000 cells per liter. To put this in context, a teaspoonful (5 ml) of typical surface open ocean seawater will contain between 25 and 250 coccolithophore cells. Blooms of *E. huxleyi* have been defined as concentrations exceeding 1 million cells per liter; the densest bloom ever recorded, in a Norwegian fiord, had a concentration of 115 000 000 cells per liter. Blooms of *E. huxleyi* can cover large areas; the largest ever recorded bloom occurred in June 1998 (see **Figure 3**) in the North Atlantic south of Iceland and covered about 1 million km^2, 4 times the area of the United Kingdom.

Coccoliths

Coccolithophores, and the coccolith shields with which they surround themselves, are incredibly small. And yet, despite their small size, coccoliths are elegant and ornate structures, which, if the water chemistry is suitable, are produced reliably with few malformations. This efficient manufacture occurs at a miniature scale: the diameter of an *E. huxleyi* coccolith 'spoke' (**Figure 1(c)**) is of the order 50 nm, considerably smaller than the wavelength range of visible light (400–700 nm). Calcite is mostly transparent to visible light (unsurprisingly, given that coccolithophores are photosynthetic) and the small coccoliths are often at the limit of discrimination, even under high magnification. However, under cross-polarized light, coccoliths produce distinctive patterns which are closely related to their structure. As a result most coccoliths can be accurately identified by light microscopy. However, the details and beauty of coccoliths can only be properly appreciated using electron microscopy (**Figure 1**).

Coccolithophores synthesize different types of coccoliths during different life-cycle stages. Here we concentrate on the heterococcoliths associated with the diploid life stage. These heterococcoliths are formed from crystal units with complex shapes, in contrast to holococcoliths which are constructed out of smaller and simpler crystal constituents. Coccoliths are typically synthesized intracellularly (within a vesicle), probably one at a time, and subsequently extruded to the cell surface. The time taken to form a single coccolith can be less than 1 h for *E. huxleyi*. Coccoliths continue to be produced until a complete coccosphere covering (made up of maybe 20 coccoliths, depending on species) is produced.

Most coccolithophores construct only as many coccoliths as are required in order to provide a complete single layer around their cell. *Emiliania huxleyi* is unusual in that, under certain conditions, it overproduces coccoliths; many more coccoliths are built than are needed to cover the cell. In these conditions, multiple layers of coccoliths accumulate around the *E. huxleyi* cell until the excessive covering eventually becomes unstable and some of the coccoliths slough off to drift free in the water. The large number of unattached coccoliths accompanying an *E. huxleyi* bloom contributes to a great extent to the turbidity of the water and to the perturbations to optics that make the blooms so apparent from space.

Curiously, the functions of coccoliths are still uncertain. It is probable that a major function is to provide some protection from grazing by zooplankton, but many alternative hypotheses have also been advanced. For instance, the coccoliths may increase the rate of sinking of the cells through the water (and therefore also enhance the rate at which nutrient-containing water flows past the cell surface) or they may provide protection against the entry of viruses or bacteria to the cell. At one time it was thought that coccoliths might provide protection against very high light intensities, which could explain the resistance to photoinhibition apparent in *E. huxleyi*, but various experimental results make this explanation unlikely. One species, *F. profunda*, a member of the deeper 'shade flora', orients its coccoliths in such a way that they conceivably act as a light-focusing apparatus maximizing photon capture in the darker waters it inhabits (**Figure 1(e)**). Some species produce trumpet-like protrusions from each coccolith (**Figure 1(f)**), again for an unknown purpose. Currently there is a paucity of hard data with which to discriminate between the various hypotheses for coccolith function, and the diversity of coccolith morphology makes it likely that they have been adapted to perform a range of functions.

Life Cycle

Many details are still obscure, and data are only available from a limited number of species, but it appears that most coccolithophores alternate between fully armored (heterococcolith-covered) diploid life stages and less-well-armoured (either holococcolith-covered or else naked) haploid phases. Both phases are capable of indefinite asexual reproduction, which is rather unusual among protists. That sexual reproduction also occurs fairly frequently is evidenced by the observation of significant genetic diversity within coccolithophore blooms. Bloom populations do not consist of just one clone (just one genetic variant of the organism). Coccolithophore gametes (haploid stages) are radically different from those of larger (multicellular) organisms in the sense of being equipped for an independent existence: they can move about, acquire energy (photosynthesize), and divide asexually by binary fission. Naked diploid phases can be induced in cultures, but these may be mutations which are not viable in the wild. There are no confirmed identifications of resting spore or cyst stages in coccolithophores.

Coccolithophores, in common with other phytoplankton, experience only an ephemeral existence. Typical life spans of phytoplankton in nature are measured in days. Comparison of the rate at which $CaCO_3$ is being produced in open ocean waters (as measured by the rate of uptake of isotopically labeled carbon), to the amounts present (the 'standing-stock'), has led to the calculation that the average turnover (replacement) time for $CaCO_3$ averages about 3 days, ranging between a minimum of <1 and a maximum of 7 days at different locations in the Atlantic Ocean. This implies that if a surface-dwelling coccolithophore synthesizes coccoliths on a Monday, the coccoliths are fairly unlikely to still be there on the Friday, either because they have redissolved or else because they have sunk down to deeper waters.

The genome of one species, *E. huxleyi*, has recently been sequenced, but at the time of writing its analysis is at an early stage.

Calcification

Calcification is the synthesis of solid calcium carbonate from dissolved substances, whether passively by spontaneous formation of crystals in a supersaturated solution (inorganic calcification) or actively through the intervention of organisms (biocalcification). The building of coccoliths by coccolithophores is a major fraction of the total biocalcification taking place in seawater. Inorganic calcification is not commonplace or quantitatively significant in the global budget, with the exception of 'whitings' that occur in just a few unusual locations in the world's oceans, such as the Persian Gulf and the Bahamas Banks. The chemical equation for calcification is

$$Ca^{2+} + 2HCO_3^- \Rightarrow CaCO_3 + H_2O + CO_2$$

Heterococcoliths are constructed out of calcite (a form of calcium carbonate; corals by contrast synthesize aragonite, which has the same chemical composition but a different lattice structure). Heterococcolith calcite typically has a very low magnesium content, making coccoliths relatively dissolution-resistant (susceptibility to dissolution increases with increasing magnesium content).

Dissolved inorganic carbon in seawater is comprised of three different components: bicarbonate ions (HCO_3^-), carbonate ions (CO_3^{2-}), and dissolved CO_2 gas ($CO_2(aq)$), of which it appears that bicarbonate or carbonate ions are taken up to provide the carbon source for $CaCO_3$ (coccoliths have a $\delta^{13}C$ isotopic composition that is very different from dissolved CO_2 gas). The exact physiological mechanisms of calcium and carbon assimilation remain to be established. Calcification (coccolith genesis) is stimulated by light but inhibited in most cases by plentiful nutrients. Separate experiments have found that increased rates of calcification in cultures can be induced by starving the cultures of phosphorus, nitrogen, and zinc. Low levels of magnesium also enhance calcification, and high levels inhibit it, but in this case probably because Mg atoms can substitute for Ca atoms in the crystalline lattice and thereby 'poison' the lattice. Calcification shows the opposite response to levels of calcium, unsurprisingly. Progressive depletion of calcium in the growth medium induces progressively less normal (smaller and more malformed) coccoliths. The calcification to photosynthesis (C:P) ratio in nutrient-replete, Ca-replete cultures is often in the vicinity of 1:1 (i.e., more or less equivalent rates of carbon uptake into the two processes). Low levels of iron appear to depress calcification and photosynthesis equally.

Measurements at sea suggest that the total amount of carbon taken up by the whole phytoplankton community to form new $CaCO_3$ is rather small compared to the total amount of carbon taken up to form new organic matter. Both calcification carbon demand and photosynthetic carbon demand have recently been measured on a long transect in the Atlantic Ocean and the ratio of the two was found to average 0.05; or, in other words, for every 20 atoms of carbon taken up by phytoplankton, only one on average was taken up into solid $CaCO_3$.

Ecological Niche

In addition to our lack of knowledge about the exact benefit of a coccosphere, we also have rather little definite knowledge as to the ecological conditions that favor coccolithophore success. There is certainly variation between species, with some being adapted to relatively eutrophic conditions (although diatoms invariably dominate the main spring blooms in temperate waters, as well as the first blooms in nutrient-rich, recently upwelled waters) and some to oligotrophic conditions. Most species are best adapted to living near to the surface, but some others to the darker conditions prevailing in the thermocline. Most species today live in warm, nutrient-poor, open ocean conditions; the highest diversity occurs in subtropical oceanic gyres, whereas lower diversity occurs in coastal and temperate waters.

Much of our knowledge of coccolithophore physiology and ecology comes from studies of *E. huxleyi*, which has attracted more scientific interest than the other coccolithophore species because of its ease of culturing and the visibility of its blooms from space. The ability to map bloom distributions from space provides unique information on the ecology of this species. Blooms of the species *E. huxleyi* occur preferentially in strongly stratified waters experiencing high light levels. Coccolithophore success may be indirectly promoted by exhaustion of silicate, due to exclusion of the more competitive diatoms. By analogy with diatoms, whose success is contingent on silicate availability for their shell building, coccolithophores might be expected to be more successful at high $CaCO_3$ saturation state Ω ($= [CO_3{}^{2-}][Ca^{2+}]/K_{sp}$), because the value of Ω controls inorganic calcification and dissolution. Such a dependency would render coccolithophores vulnerable to ocean acidification, as discussed further below. It was formerly thought that *E. huxleyi* was particularly successful in phosphate-deficient waters, but a reassessment has suggested that this is not a critical factor. Many coccolithophores are restricted to the warmer parts of the oceans, although this may be coincidental rather than due to a direct temperature effect. *Emiliania huxleyi* is found to grow well at low iron concentrations, in culture experiments.

Biogeochemical Impacts

Coccolithophores assimilate carbon during photosynthesis, leading to similar biogeochemical impacts to other phytoplankton that do not possess mineral shells. They also, however, assimilate carbon into biomass.

Following death, some of the coccolith $CaCO_3$ dissolves in the surface waters inhabited by coccolithophores, with the rest of the coccolith $CaCO_3$ sinking out of the surface waters within zooplankton fecal pellets or marine snow aggregates. The exact means by which some coccoliths are dissolved in near-surface waters are unclear (dissolution within zooplankton guts may be important), but regardless of mechanisms several lines of evidence suggest that near-surface dissolution does occur. The size of coccoliths precludes the likelihood of single coccoliths sinking at all rapidly under gravity, because of the considerable viscosity of water with respect to such small particles (Stokes' law). Stokes' law can be overcome if coccoliths become part of larger aggregates, either marine snow or zooplankton fecal pellets. Another possible fate for coccoliths is to become incorporated into the shells of tintinnid microzooplankton, which when grazing on coccolithophores make use of the coccoliths in their own shells (**Figure 4**). Regardless of their immediate fate, the coccoliths must eventually either dissolve or else sink toward the seafloor.

The construction of $CaCO_3$ coccoliths (calcification) leads to additional impacts, over and above those associated with the photosynthesis carried out by all species. The first and perhaps the most important of these is that $CaCO_3$ contains carbon and the vertical downward flux of coccoliths thereby removes carbon from the surface oceans. It might be expected that this would lead to additional removal of CO_2 from the atmosphere to the oceans, to replace that taken up into coccoliths, but in fact, because of the complex effect of calcification ($CaCO_3$ synthesis) on seawater chemistry, the production of coccoliths actually increases the partial pressure of CO_2 in surface seawater and promotes outgassing rather than ingassing. Determining the exact nature and magnitude of the overall net effect is complicated by a possible additional role of coccoliths as 'ballast' (coccoliths are denser than water and hence when

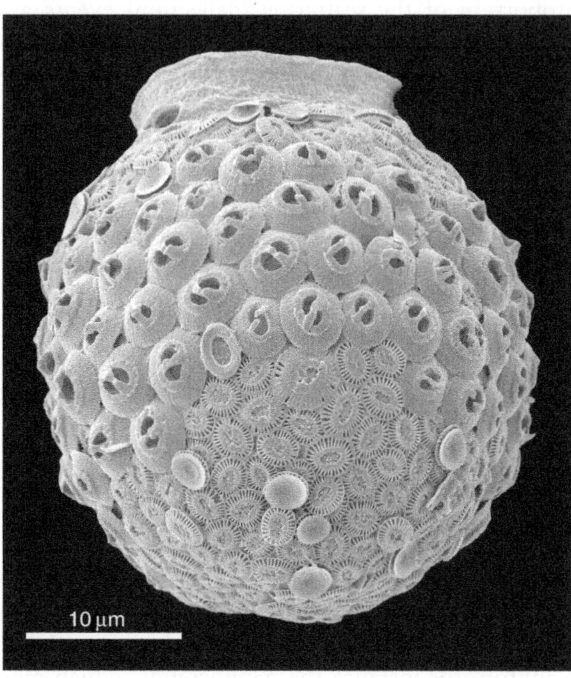

Figure 4 Tintinnid lorica (casing) with embedded coccoliths.

incorporated into aggregates of particulate fecal material may drag down extra organic carbon into the ocean interior).

Microscopic examination of seafloor sediments (if shallow enough that the $CaCO_3$ does not dissolve) and of material caught in sediment traps has revealed that much of the calcium carbonate in the samples consists of coccoliths. The flux of coccoliths probably accounts for *c.* 50% of the total vertical $CaCO_3$ flux in open ocean waters (in other words, about 50% of the inorganic carbon pump), with foraminifera shells responsible for most of the rest. It is usually not the most numerous species (*E. huxleyi*) but rather larger species (e.g., *Calcidiscus quadriperforatus* and *Coccolithus pelagicus*) that make the greatest contributions to the total coccolith flux.

Coccolithophores also impact on climate in other ways, ones that are unconnected with carbon. Coccolithophores are intense producers of a chemical called dimethylsulfoniopropionate (DMSP). The production of DMSP leads eventually (via several chemical transformations) to additional cloud condensation nuclei in the atmosphere and thereby to increased cloud cover.

Coccoliths also scatter light, polarizing it in the process. They do not reflect or block light (this would clearly be disadvantageous for the photosynthetic cell underneath), but the difference between the refractive indices of water and of calcium carbonate means that the trajectories of photons are deflected by encounters with coccoliths. A small proportion of the scattering (deflection) events are through angles greater than 90°, leading to photons being deflected into upward directions and eventually passing back out through the sea surface. Because of this light-scattering property of coccoliths, their bulk effect is to make the global oceans slightly brighter than they would otherwise be. It has been calculated that the Earth would become slightly dimmer (the albedo of the Earth would decrease by about 0.1% from its average global value of about 30%) were coccolithophores to disappear from the oceans. The effect of coccoliths in enhancing water brightness is seen in its most extreme form during coccolithophore blooms (**Figure 3**).

The Past

Coccolithophores are currently the dominant type of calcifying phytoplankton, but further back in time there were other abundant calcifiying phytoplankton, for instance the nannoconids, which may or may not have been coccolithophores. The fossil calcifying phytoplankton are referred to collectively as calcareous nannoplankton.

The first calcareous nannoplankton are seen in the fossil record *c.* 225 Ma, in the late Triassic period. Abundance and biodiversity increased slowly over time, although they were at first restricted to shallow seas. During the early Cretaceous (145–100 Ma), calcareous nannoplankton also colonized the open ocean. They reached their peak, both in terms of abundance and number of different species (different morphotypes) in the late Cretaceous (100–65 Mya). 'The Chalk' was formed at this time, consisting of thick beds of calcium carbonate, predominantly coccoliths. Thick deposits of chalk are most noticeable in various striking sea cliffs, including the white cliffs of Dover in the United Kingdom, and the Isle of Rugen in the Baltic Sea. The chalk deposits were laid down in the shallow seas that were widespread and extensive at that time, because of a high sea level.

Calcareous nannoplankton, along with other biological groups, underwent long intervals of slowly but gradually increasing species richness interspersed with occasional extinction events. Their heyday in the late Cretaceous was brought to an abrupt end by the largest extinction event of all at the K/T boundary (65 Ma), at which point ~93% of all species (~85% of genera) suddenly went extinct. Although biodiversity recovered rapidly in the early Cenozoic, calcareous nannoplankton have probably never since re-attained their late Cretaceous levels.

Because the chemical and isotopic composition of coccoliths is influenced by the chemistry of the seawater that they are synthesized from, coccoliths from ancient sediments have the potential to record details of past environments. Coccoliths are therefore a widely used tool by paleooceanographers attempting to reconstruct the nature of ancient oceans. Some of the various ways in which coccoliths are put to use in interpreting past conditions are as follows: (1) elemental ratios such as Sr/Ca and Mg/Ca are used to infer past seawater chemistry, ocean productivity, and temperatures; (2) the isotopic composition ($\delta^{13}C$, $\delta^{18}O$) of the calcium carbonate is used to infer past carbon cycling, temperatures, and ice volumes; (3) the species assemblage of coccoliths (some species assemblages are characteristic of eutrophic conditions, some of oligotrophic conditions) is used to infer trophic status and productivity. Some of the organic constituents of coccolithophores are also used for paleoenvironmental reconstructions. In particular, there is a distinctive group of ketones, termed long-chain alkenones, which are specific to one family of coccolithophores and closely related haptophytes. These alkenones tend to survive degradation in sediments, and the ratio of one type of alkenones to another (the U_K^{37} index) can be used to estimate past ocean temperatures. Calcareous nannofossils are also extremely useful in determining the age of different layers in cores of ocean sediments (biostratigraphy).

The Future

The pH of the oceans is falling (they are becoming increasingly acidic), because of the invasion of fossil fuel-derived CO_2 into the oceans. Surface ocean pH has already dropped by 0.1 units and may eventually drop by as much as 0.7 units, compared to preindustrial times, depending on future CO_2 emissions. The distribution of dissolved inorganic carbon (DIC) between bicarbonate, carbonate, and dissolved CO_2 gas changes with pH in such a way that carbonate ion concentration (and therefore saturation state, Ω) is decreasing even as DIC is increasing due to the invading anthropogenic CO_2. It is predicted that, by the end of this century, carbonate ion concentration and Ω may have fallen to as little as 50% of preindustrial values. If emissions continue for decades and centuries without regulation then the surface oceans will eventually become undersaturated with respect to calcium carbonate, first with respect to the more soluble aragonite used by corals, and some time later also with respect to the calcite formed by coccolithophores.

There has been an increasing appreciation over the last few years that declining saturation states may well have significant impacts on marine life, and, in particular, on marine organisms that synthesize $CaCO_3$. Experiments on different classes of marine calcifiers ($CaCO_3$ synthesizers) have demonstrated a reduction in calcification rate in high CO_2 seawater. One such experiment showed a strong decline in coccolithophore calcification rate (and a notable increase in the numbers of malformed coccoliths) at high CO_2 (low saturation state), although some other experiments have obtained different results. If coccolithophore biocalcification is controlled by Ω then the explanation could be linked to the importance of Ω in controlling inorganic calcification, although coccolithophores calcify intracellularly and so such a link is not guaranteed. At the time of writing, further research is being undertaken to determine whether, as the oceans become more acidic, coccolithophores will continue to be able to synthesize coccoliths and subsequently maintain them against dissolution.

Our ability to predict the consequences of ocean acidification on coccolithophores is hampered by our poor understanding of the function of coccoliths (what they are for, and therefore how the cells will be affected by their absence), and also by our poor understanding of the possibilities for evolutionary adaptation to a low-pH ocean. These constraints can be overcome to an extent by examining the geological history of coccolithophores, and their (in)ability to survive previous acid ocean events in Earth history. Although coccoliths (and other calcareous nannofossils) have been widely studied by geologists, it is only recently that there has been a concerted effort to study their species turnover through events in Earth history when the oceans were more acidic than now.

Although many authors have taken the success of coccolithophores during the high-CO_2 late Cretaceous as reassuring with respect to their future prospects, the reasoning is fallacious. Levels of calcium are thought to have been higher than now during the Cretaceous, and the CCD (the depth at which $CaCO_3$ disappears from sediments due to dissolution, which is a function of deep-water Ω) was only slightly shallower than today, indicating that Cretaceous seawater conditions were not analogous to those to be expected in a future high-CO_2 world.

It turns out that coccolithophores survived the Paleocene–Eocene Thermal Maximum event (thought to more closely resemble the predicted future) fairly well, with a modest increase in extinction rates matched by a similar increase in speciation rates. On the other hand, the environmental changes at the Cretaceous–Tertiary boundary (the K/T impact event), which also appears to have induced acidification, led to a mass extinction of 93% of all coccolithophore species, as well as to extinction of many other calcifying marine organisms including ammonites. It is necessary to more accurately characterize the environmental changes that took place across such events, in order to better determine how well they correspond to the ongoing and future ocean acidification.

See also

Calcium Carbonates. Benthic Foraminifera.

Further Reading

Gibbs SJ, Bown PR, Sessa JA, Bralower TJ, and Wilson PA (2007) Nannoplankton extinction and origination across the Paleocene–Eocene Thermal Maximum. *Science* 314: 1770–1773 (doi: 10.1126/science.1133902).

Holligan PM, Fernandez E, Aiken J, *et al.* (1993) A biogeochemical study of the coccolithophore *Emiliania huxleyi* in the North Atlantic. *Global Biogeochemical Cycles* 7: 879–900.

Paasche E (2002) A review of the coccolithophorid *Emiliania huxleyi* (Prymnesiophyceae), with particular reference to growth, coccolith formation, and calcification–photosynthesis interactions. *Phycologia* 40: 503–529.

Poulton AJ, Sanders R, Holligan PM, *et al.* (2006) Phytoplankton mineralization in the tropical and subtropical Atlantic Ocean. *Global Biogeochemical Cycles* 20: GB4002 (doi: 10.1029/2006GB002712).

Riebesell U, Zonderva I, Rost B, Tortell PD, Zeebe RE, and Morel FMM (2000) Reduced calcification in marine plankton in response to increased atmospheric CO_2. *Nature* 407: 634–637.

Thierstein HR and Young JR (eds.) (2004) *Coccolithophores: From Molecular Processes to Global Impact.* Berlin: Springer.

Tyrrell T, Holligan PM, and Mobley CD (1999) Optical impacts of oceanic coccolithophore blooms. *Journal of Geophysical Research, Oceans* 104: 3223–3241.

Winter A and Siesser WG (eds.) (1994) *Coccolithophores.* Cambridge, UK: Cambridge University Press.

Young JR, Geisen M, Cros L, *et al.* (2003) *Special Issue: A Guide to Extant Coccolithophore Taxonomy. Journal of Nannoplankton Research* 1–125.

Relevant Websites

http://cics.umd.edu
– Blooms of the Coccolithophorid *Emiliania huxleyi* in Global and US Coastal Waters, CICS.

http://www.ucl.ac.uk
– Calcareous Nannofossils, MIRACLE, UCL.

http://www.nanotax.org
– Calcareous Nannofossil Taxonomy.

http://www.emidas.org
– Electronic Microfossil Image Database System.

http://www.noc.soton.ac.uk
– *Emiliania huxleyi* Home Page, National Oceanography Centre, Southampton.

http://www.nhm.ac.uk
– International Nannoplankton Association page, hosted at Natural History Museum website.

CENOZOIC CLIMATE – OXYGEN ISOTOPE EVIDENCE

J. D. Wright, Rutgers University, Piscataway, NJ, USA

Discoveries of fossil remains of 50 million year old alligators on Ellesmere Island and 30–40 million year-old forests on Antarctica contrast sharply with our present vision of polar climates. These are not isolated discoveries or quirks of nature. An ever-growing body of faunal, floral, and geochemical evidence shows that the first half of the Cenozoic Era was much warmer than the present time. What maintained such a warm climate and could it be an analog for future global warming? To address these and other questions, one needs more than a qualitative estimate of planetary temperatures. Quantitative temperature estimates (both magnitudes and rates of change) are required to depict how the Earth's climate has changed through time. One of the most powerful tools used to reconstruct past climates during the Cenozoic (the last 65 million years of Earth's history) is the analysis of oxygen isotopes in the fossil shells of marine organisms. The calcium carbonate shells of the protist foraminifera are the most often analyzed organisms because the different species are distributed throughout surface (planktonic) and deep (benthic) marine environments.

Oxygen Isotope Systematics

The stable isotopes of oxygen used in paleooceanographic reconstructions are ^{16}O and ^{18}O. There are about 500 ^{16}O atoms for every ^{18}O atom in the ocean/atmosphere environment. During the 1940s, Harold Urey at the University of Chicago predicted that the $^{18}O/^{16}O$ ratio in calcite ($CaCO_3$) should vary as a function of the temperature at which the mineral precipitated. His prediction spurred on experiments by himself and others at the University of Chicago who measured $^{18}O/^{16}O$ ratios in $CaCO_3$ precipitated in a wide range of temperatures, leading to the use of stable oxygen isotope measurements as a paleothermometer.

To determine oxygen isotopic ratios, unknown $^{18}O/^{16}O$ ratios are compared to the known $^{18}O/^{16}O$ ratio of a standard. The resulting values are expressed in delta notation, $\delta^{18}O$, where:

$$\delta^{18}O = \frac{^{18}O/^{16}O_{sample} - ^{18}O/^{16}O_{standard}}{^{18}O/^{16}O_{standard}} \times 1000 \quad [1]$$

Carbonate samples are reacted in phosphoric acid to produce CO_2. To analyze water samples, CO_2 gas is equilibrated with water samples at a constant temperature. Given time, the CO_2 will isotopically equilibriate with the water. For both the carbonate and water samples, the isotopic composition of CO_2 gas is compared with CO_2 gas of known isotopic composition. There are two standards for reporting $\delta^{18}O$ values. For carbonate samples, the reference standard is PDB, which was a crushed belemnite shell (*Belemnitella americana*) from the Peedee formation of Cretaceous age in South Carolina. The original PDB material has been exhausted, but other standards have been calibrated to PDB and are used as an intermediate reference standard through which a PDB value can be calculated. For measuring the isotopic composition of water samples, Standard Mean Ocean Water (SMOW) is used as the reference. The SMOW reference was developed so that its $\delta^{18}O_{water}$ value is 0.0‰ (parts per thousand) and approximates the average oxygen isotopic composition of the whole ocean. Deep ocean $\delta^{18}O_{water}$ values are close to the SMOW value, ranging from -0.2 to 0.2‰. In contrast, surface ocean $\delta^{18}O_{water}$ values exhibit a much greater variability, varying between -0.5 and $+1.5$‰.

Oxygen Isotope Paleothermometry

Early studies into the natural variations in oxygen isotopes led to the development of a paleotemperature equation. The temperature during the precipitation of calcite can be estimated by measuring the $\delta^{18}O$ value in calcite-secreting organisms (foraminifera, corals, and mollusks) and the value of the water in which the organisms live. The various paleotemperature equations all follow the original proposed by Sam Epstein and his colleagues (University of Chicago):

$$T = 16.5 - 4.3 \times (\delta^{18}O_{calcite} - \delta^{18}O_{water}) + 0.14 \times (\delta^{18}O_{calcite} - \delta^{18}O_{water})^2 \quad [2]$$

where T and $\delta^{18}O_{water}$ are the temperature (°C) and oxygen isotope value of the water in which the organism lived[1] and $\delta^{18}O_{calcite}$ is the oxygen isotope value of calcite measured in the mass spectrometer.

Eqn [2] shows that the changes in $\delta^{18}O_{calcite}$ are a function of the water temperature and $\delta^{18}O_{water}$ value. A one-to-one relationship between $\delta^{18}O_{calcite}$ and $\delta^{18}O_{water}$ values dictates that a change in the $\delta^{18}O_{water}$ term will cause a similar change in the measured $\delta^{18}O_{calcite}$ value. However, an inverse relationship between $\delta^{18}O_{calcite}$ and T changes dictates that for every 1°C increase in temperature, there is a 0.23‰ decrease in the measured $\delta^{18}O_{calcite}$ value. These relationships enable us to interpret $\delta^{18}O_{calcite}$ changes generated from foraminifera, corals, and mollusks. For many years, the convention was to plot $\delta^{18}O_{calcite}$ values with the axis reversed (higher values to the left or bottom) so that $\delta^{18}O$ records reflect climate changes (e.g., colder to the left or bottom). More recently, there has been a trend among some scientists to plot $\delta^{18}O_{calcite}$ values without reversing the axis.

The paleotemperature equation contains two unknowns (temperature, $\delta^{18}O_{water}$). Although temperature is the main target in reconstructions, one cannot ignore the $\delta^{18}O_{water}$ term. In the modern ocean, the equator-to-pole gradient measured in planktonic foraminifera $\delta^{18}O_{calcite}$ values is ~5.0‰ and largely reflects the temperature gradient (~28°C). However, if temperature were the sole influence on modern $\delta^{18}O_{calcite}$ values, the equator-to-pole gradient would be ~6.5‰ (28°C × 0.23‰/°C). The attenuated $\delta^{18}O_{calcite}$ gradient measured in planktonic foraminifera reflects the surface ocean $\delta^{18}O_{water}$ variability. Therefore, a key to using $\delta^{18}O_{calcite}$ records as indicators of past climates is to understand the hydrographic parameters that produce the modern $\delta^{18}O_{calcite}$ gradient. For instance, ignoring the $\delta^{18}O_{water}$ term results in a 5–6°C underestimation compared to the observed temperature gradient. This occurs largely because tropical temperature estimates will be too cold (~4°C) whereas polar estimates will be warm (~1–2°C).

$\delta^{18}O$ Variation in the Natural Environment

$\delta^{18}O_{water}$ values in the ocean/atmosphere system vary both spatially and temporally because fractionation between the $H_2{}^{18}O$ and $H_2{}^{16}O$ molecules is temperature-dependent in the hydrologic cycle and follows the Rayleigh Distillation model (**Figure 1**). In general, water vapor evaporates at low latitudes and precipitates at higher latitudes. Fractionation during evaporation concentrates the lighter $H_2{}^{16}O$ molecule in the water vapor, leaving the water enriched in $H_2{}^{18}O$ and $H_2{}^{16}O$. On average, the $\delta^{18}O_{water}$ value of water vapor is 9‰ lower than its source water (**Figure 1**). Fractionation during condensation concentrates the $H_2{}^{18}O$ molecules in the precipitation (rain/snow) by ~9‰. Therefore, if all of the water evaporated in the tropics rained back into the tropical oceans, there would be no net change in the $\delta^{18}O_{water}$ term. However, some water vapor is transported to higher latitudes. If the clouds remain a closed system (i.e., mid-to-high latitude evaporation does not influence the $\delta^{18}O_{water}$ value in the clouds[2]), then precipitation will further deplete the clouds (water vapor) in $H_2{}^{18}O$ relative to $H_2{}^{16}O$. Consequently, the $\delta^{18}O$ value of water vapor decreases from the original value as water vapor condenses into precipitation (**Figure 1**) and the cloud that formed from the evaporation in the tropics will eventually lose moisture, fractionating the $\delta^{18}O_{water}$ value of the remaining water vapor (**Figure 1** and **2**).

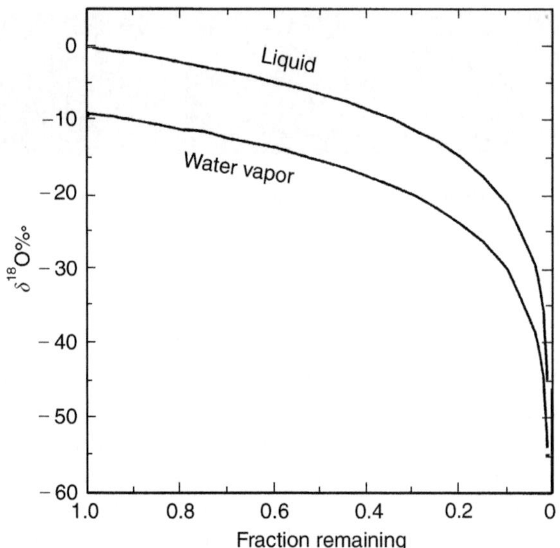

Figure 1 Rayleigh distillation model showing the effects of evaporation and precipitation on the $\delta^{18}O$ values in the vapor and liquid phases. The initial conditions are a temperature of 25°C and $\delta^{18}O_{water}$ value of 0‰. This model also assumes that it is a closed system, meaning that water vapor is not added once the cloud moves away from the source regions. As clouds lose moisture, fractionation during the condensation further lowers the $\delta^{18}O_{water}$ value in the water vapor.

[1] $\delta^{18}O$ calcareous deposits are commonly reported relative to a carbonate standard, PDB (Peedee belemnite), and not SMOW (Standard Mean Ocean Water). PDB is 22‰ relative to SMOW.

[2] Many island or coastal regions have significantly higher $\delta^{18}O_{water}$ values relative to continental locations at similar latitudes. This occurs because local evaporation increases the $\delta^{18}O_{water}$ values, thus resetting the initial conditions for Rayleigh distillation to occur.

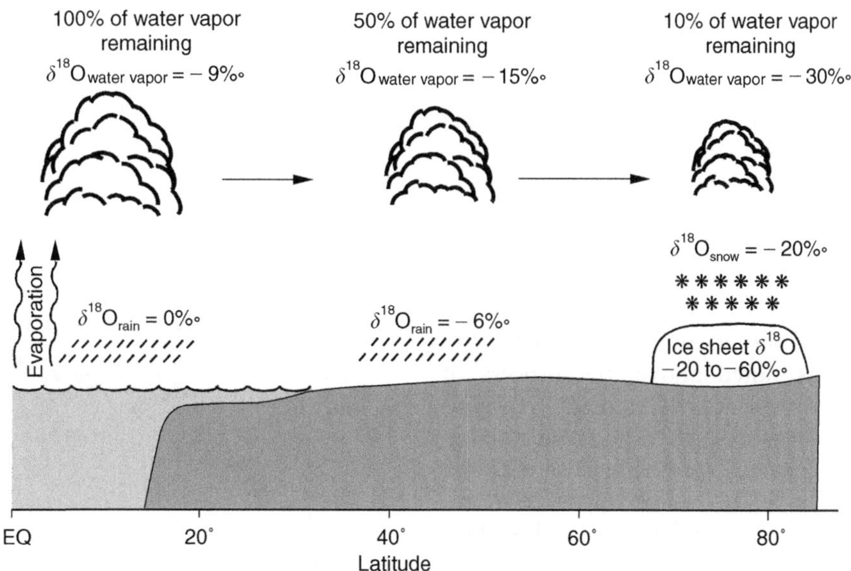

Figure 2 Illustration of the Rayleigh distillation process on $\delta^{18}O$ values as clouds move over land and into the polar regions. Decreasing air temperatures cause moisture to rain/snow out of the cloud. Fractionation of the oxygen isotopes during condensation further decreases values. By the point that a cloud reaches the high latitudes, less than 10% of the original water vapor remains. Snowfall on Antarctica has values between −20 and −60‰.The average $\delta^{18}O$ value for ice on Antarctica is ~ −40‰.

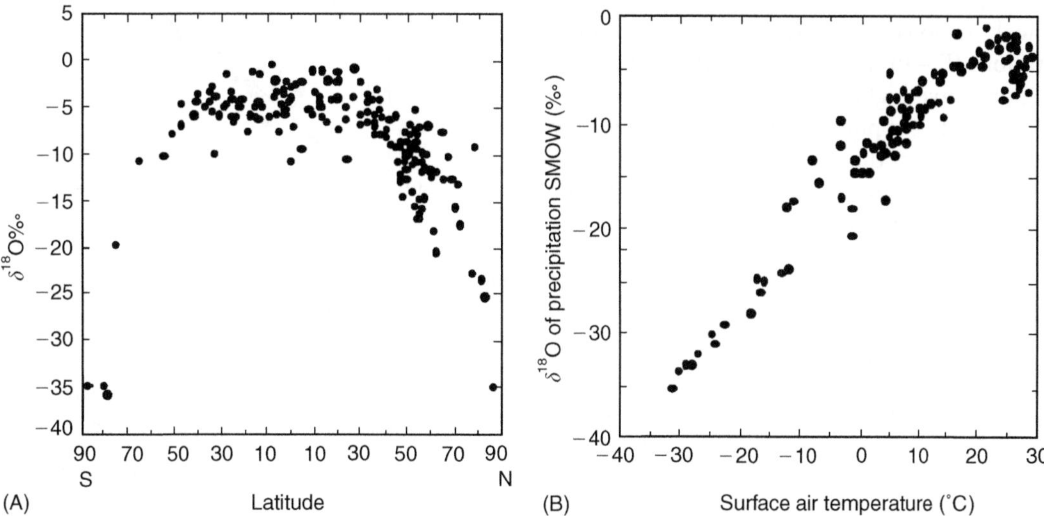

Figure 3 Mean annual $\delta^{18}O$ water of precipitation (rain/snow) versus mean annual temperatures. The correlation between $\delta^{18}O$ values and latitude (A) is a function of temperature (B). The rainout/fraction of water remaining, and hence the fraction of $\delta^{18}O$ values, is determined by the cloud temperatures. Latitude is the dominant effect shown here. The scatter among sites at similar latitude results from elevation differences as well as differences in the distance from the ocean.

By the time 50% of the initial moisture precipitates, the $\delta^{18}O$ value of the water vapor will be ~ −15%, while precipitation will be ~ −6‰. Once the cloud reaches the poles, over 90% of the initial water vapor will have been lost, producing $\delta^{18}O$ values of snow less than −20‰. Snow at the South Pole approaches −60‰. There is a strong relationship between $\delta^{18}O$ values in precipitation and air mass temperatures because air temperature dictates how much water vapor it can hold, and the $\delta^{18}O$ values of the precipitation is a function of the amount of water remaining in the clouds (**Figure 3**).

Spatial Variations in $\delta^{18}O_{water}$ of Modern Sea Water

The evaporation/precipitation process that determines the $\delta^{18}O_{water}$ values of precipitation (e.g., **Figure 1**) also controls the $\delta^{18}O_{water}$ values in regions in the ocean. At any one time, the volume of water being transported through the hydrologic cycle (e.g., atmosphere, lakes, rivers, and groundwater) is small compared to the volume of water in the oceans (1:130). Therefore, the hydrologic cycle can influence the whole ocean $\delta^{18}O_{water}$ value only by creating a new or enlarging an existing reservoir (e.g., glacier/ice sheets). In contrast, evaporation/precipitation processes will change the $\delta^{18}O_{water}$ and salinity values in the surface waters because only the thin surface layer of the ocean communicates with the atmosphere. As noted above, the process of evaporation enriches surface water in $H_2{}^{18}O$ molecules and salt because the water vapor is enriched in $H_2{}^{16}O$ molecules. For this reason, high salinity sea water has a high $\delta^{18}O_{water}$ value and vice versa. More specifically, tropical and subtropical surface water $\delta^{18}O_{water}$ values are ~1‰ higher than mean ocean water values (**Figure 4**). Interestingly, subtropical $\delta^{18}O_{water}$ values are higher than tropical values even though evaporation is higher in the tropics. Atmospheric circulation patterns produce intense rainfall in the tropics to offset some of the evaporation, whereas very little rain falls in the subtropical regions. Because evaporation minus precipitation (E − P) is greater in the subtropics, these regions have higher salinity and $\delta^{18}O_{water}$ values. In contrast, subpolar and polar regions have greater precipitation than evaporation; hence, high-latitude surface waters have low salinity and $\delta^{18}O_{water}$ values that approach − 0.5‰ (**Figure 4**).

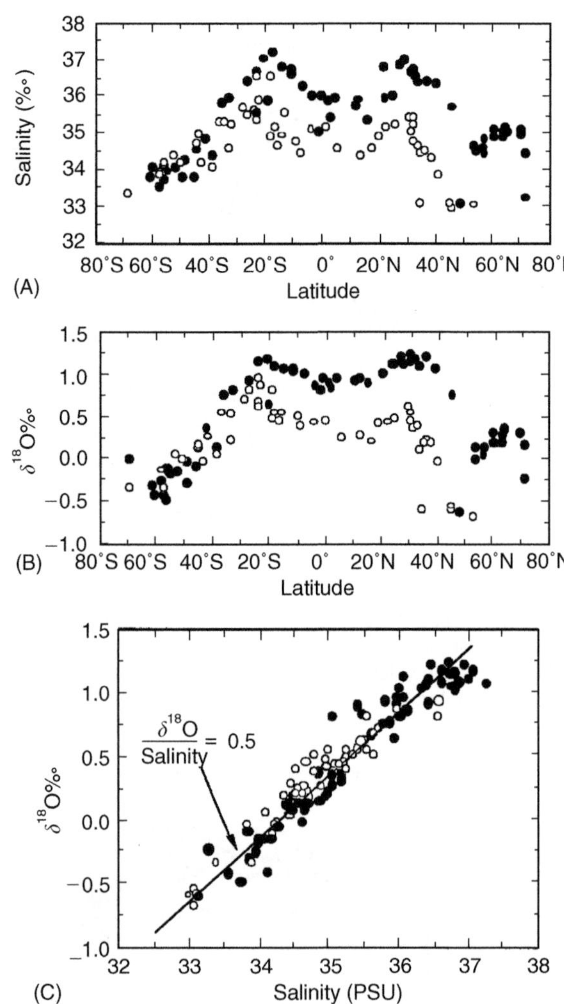

Figure 4 The salinity (A) and $\delta^{18}O_{water}$ values (B) measured in the open Atlantic (•) and Pacific (○) Oceans. Note the higher values in the tropical and subtropical region relative to the subpolar and polar regions. Evaporation and precipitation/runoff processes produce similar patterns in salinity and $\delta^{18}O_{water}$ values which is illustrated by the linearity in the $\delta^{18}O$ versus linity plot (C). The ocean-to-ocean difference between the Atlantic and Pacific results from a net transfer of fresh water from the Atlantic to the Pacific.

Temporal Variations

Variations in the amount of water stored on land through time, usually in the form of ice, can have a significant effect on the mean ocean $\delta^{18}O_{water}$ value, and hence, the marine $\delta^{18}O_{calcite}$ record. At present, high-latitude precipitation returns to the oceans through summer ice/snow melting. During glacial periods, snow and ice accumulate into large ice sheets. Because the difference in ice sheet and mean ocean values is large ($\delta^{18}O_{ice}$ = − 35 to − 40‰ vs. $\delta^{18}O_{water}$ mean ocean = ~0‰), ice sheet fluctuations are reflected in mean oceanic $\delta^{18}O_{water}$ values. This relationship can be illustrated by examining how the mean ocean $\delta^{18}O_{water}$ value increased during the last glacial maximum (LGM) relative to the present (**Figure 5**). During the LGM, water stored in continental ice lowered global sea level by 120 m, removing ~3% of the ocean's volume. Thus, the mean ocean $\delta^{18}O_{water}$ value increased by 1.2‰ during the LGM relative to the present (**Figure 5**).

Pleistocene Oxygen Isotope Variations

The first systematic downcore examination of the marine stable isotope record was made by Cesaré Emiliani during the 1950s on $\delta^{18}O_{calcite}$ records generated from planktonic foraminifera in Caribbean deep-sea cores. Emiliani recognized the cyclic pattern of low and high $\delta^{18}O_{calcite}$ values and concluded that these represented glacial–interglacial intervals. Emiliani identified the seven most recent climate cycles and estimated that they spanned the last 280 000 years. (Current age estimates indicate that the duration of the cycles is approximately

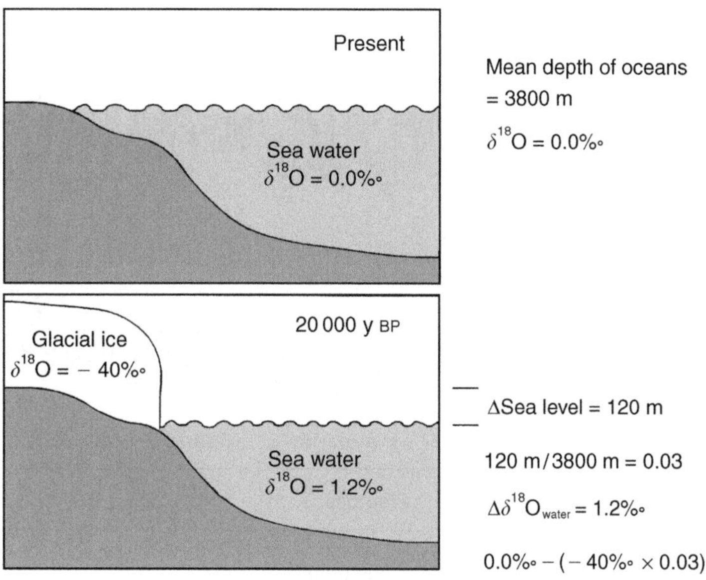

Figure 5 The effect of building or removing large ice sheets on the $\delta^{18}O$ composition of the ocean can be significant. The removal of 3% of the ocean's water during the last glacial maximum lowered sea level by 120 m. The $\delta^{18}O$ difference between the ocean and the ice is 40‰, causing a whole ocean $\delta^{18}O$ change of 1.2‰. The reverse process occurs during the melting of large ice sheets. If the Antarctic and Greenland ice were to melt, then sea level would rise ~70 m. The volume of water stored in these ice sheets is equivalent to ~2% of the water in the ocean. Therefore, the mean $\delta^{18}O$ value of the ocean would decrease by 0.7–0.8‰ (relative to PDB).

525 000 years.) To apply the paleotemperature equation to these records, Emiliani estimated that ice sheet-induced ocean $\delta^{18}O_{water}$ variability was relatively small, 0.3‰. (As shown above, the maximum glacial–interglacial ice sheet signal was closer to 1.2‰.) Therefore, most of the $\delta^{18}O_{calcite}$ variability between glacial and interglacial intervals represented temperature changes of 5–10°C. Emiliani divided the $\delta^{18}O_{calcite}$ record into warm stages (designated with odd numbers counting down from the Holocene) and cold stages (even numbers). Hence, 'Isotope Stage 1' refers to the present interglacial interval and 'Isotope Stage 2' refers to the LGM (**Figure 6**). During the 1960s and 1970s, many argued that most of the glacial to interglacial difference in $\delta^{18}O_{calcite}$ values resulted from ice volume changes. Nicholas Shackleton of Cambridge University made the key observation that benthic foraminiferal $\delta^{18}O$ values show a glacial to interglacial difference of ~1.8‰. If the ice volume contribution was only 0.3‰ as argued by Emiliani, then the deep ocean temperatures would have been 6–7°C colder than the present temperatures of 0–3°C. Sea water freezes at −1.8°C, precluding Emiliani's 'low' ice volume estimate. By the early 1970s, numerous $\delta^{18}O$ records had been generated and showed a cyclic variation through the Pleistocene and into the late Pliocene. One hundred oxygen isotope stages, representing 50 glacial–interglacial cycles, have been identified for the interval since 2.6 million years ago (Ma) (**Figure 6**).

Cenozoic $\delta^{18}O$ Records

The first Cenozoic $\delta^{18}O$ syntheses based on foraminiferal $\delta^{18}O$ records were produced during the mid-1970s. Nicholas Shackleton and James Kennett produced a composite benthic $\delta^{18}O$ record for the Cenozoic from cores to the south of Australia. A second group led by Samuel Savin generated low-latitude planktonic and benthic foraminiferal $\delta^{18}O$ syntheses. Both records are important to understanding Cenozoic climate changes. Benthic foraminiferal records best reflect global temperature and ice volume changes. Additional advantages of the benthic foraminiferal composite include: (1) deep-ocean temperatures are more uniform with respect to horizontal and vertical gradients; (2) deep-ocean $\delta^{18}O_{water}$ values are less variable compared to the large surface water changes; (3) the deep ocean approximates high-latitude surface water conditions where deep waters originated during the Cenozoic (i.e., Antarctica, northern North Atlantic); and (4) many benthic foraminifera taxa are long-lived so that one species can be used to construct records spanning several millions of years in contrast to planktonic taxa which have shorter durations and

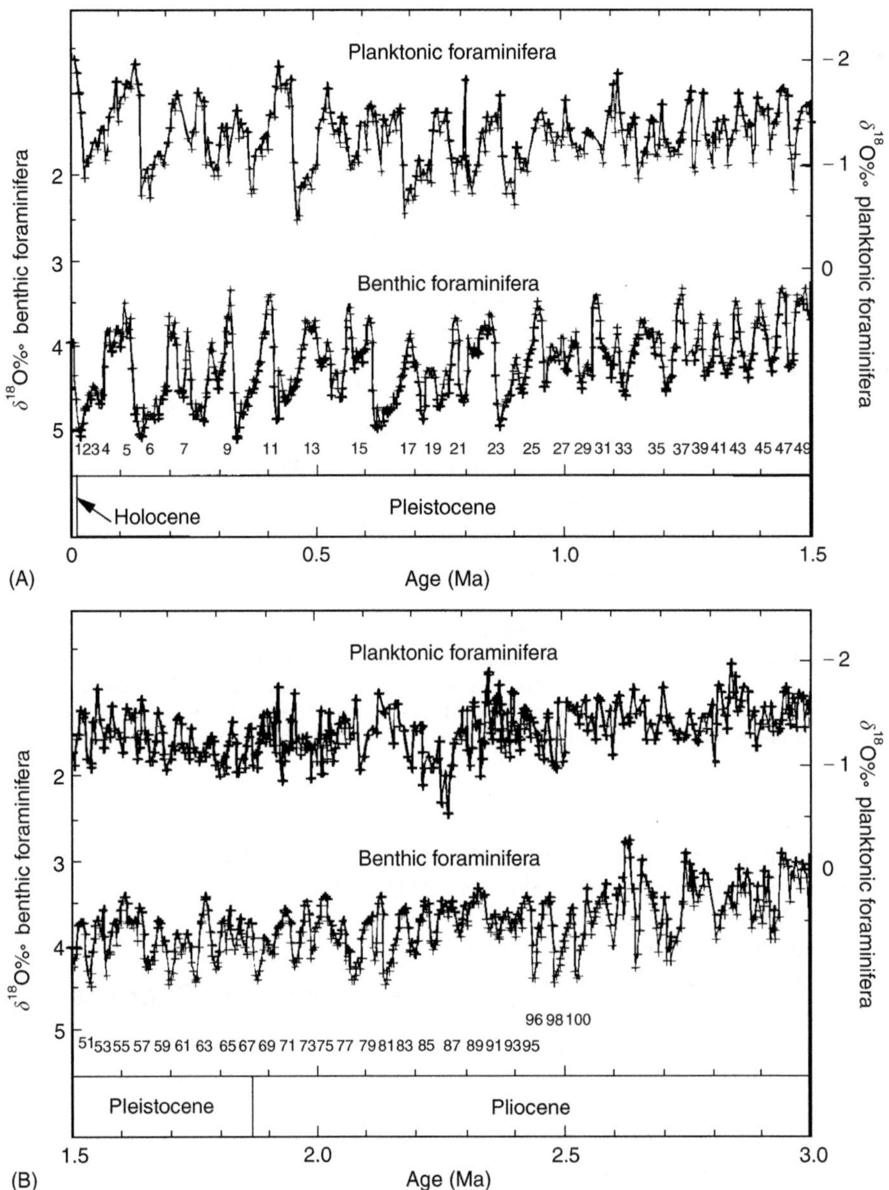

Figure 6 Planktonic and benthic foraminiferal $\delta^{18}O$ records for the last 3 million years. Note the high frequency signals in the records. For the interval between 3 and 1 Ma, a 40 000 year cycle dominates the records. After 1 Ma, the beat changes to a 100 000 year cycle and the amplitudes increase (relative to PDB).

require records to be spliced together from several species.

Low-latitude planktonic foraminiferal $\delta^{18}O$ records are good proxies for tropical sea surface temperatures. Tropical temperatures are an important component of the climate system because they influence evaporation, and hence, total moisture in the atmosphere. Planktonic and benthic foraminiferal $\delta^{18}O$ comparison allows one to assess equator-to-pole as well as vertical temperature gradients during the Cenozoic, and thus, to determine planetary temperature changes. Finally, much of the climatic change in the last 65 million years has been ascribed to poleward heat transport or greenhouse gas fluctuations. General circulation models indicate that each mechanism should produce different temperature patterns that can be approximated with the planktonic and benthic $\delta^{18}O$ records.

The first benthic $\delta^{18}O$ syntheses generated, as well as more recent compilations, show the same long-term patterns. After the Cretaceous–Tertiary (K/T) boundary events, deep-water $\delta^{18}O$ values remained relatively constant for the first 7 million years of the Paleocene (**Figure 7A**). At 58 Ma, benthic foraminiferal $\delta^{18}O$ values began a decrease over the next 6 My that culminated during the early Eocene

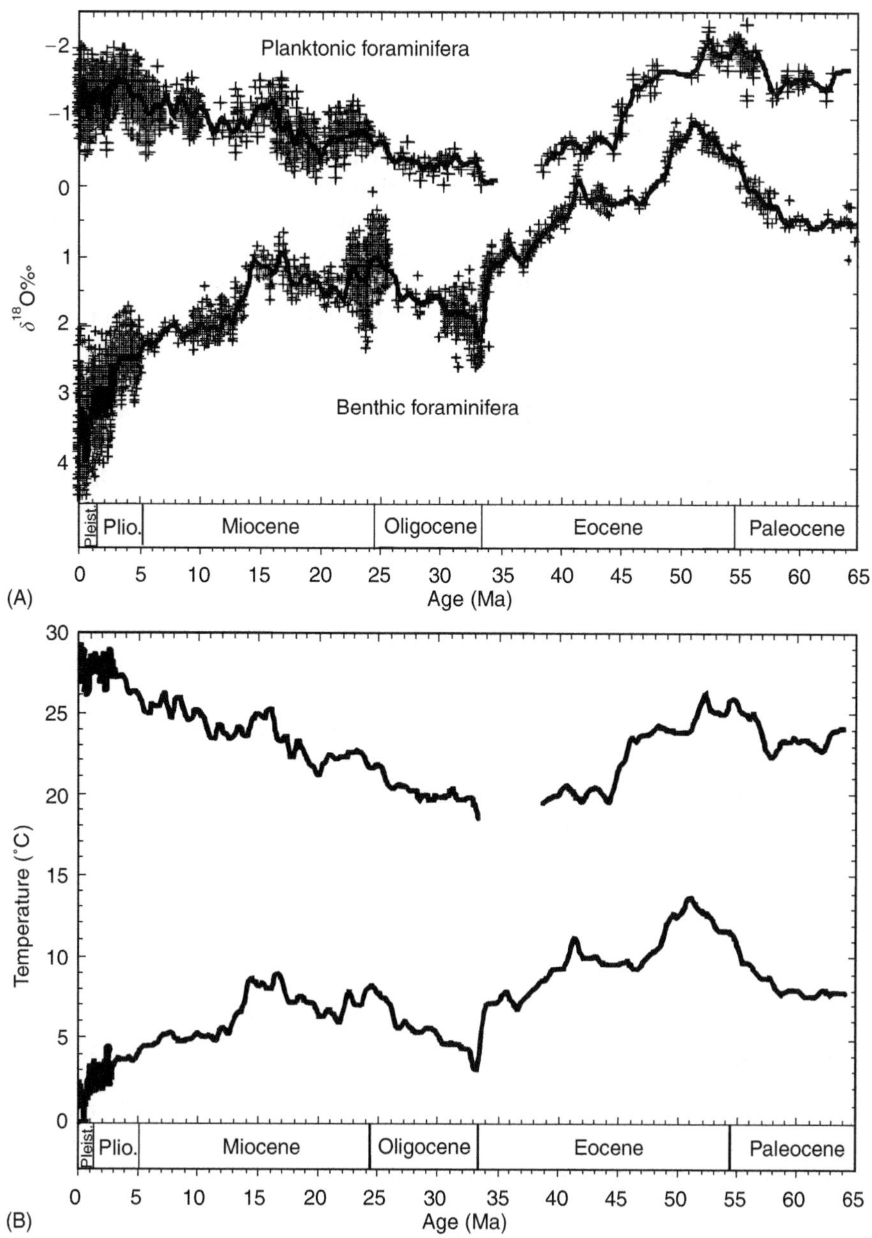

Figure 7 (A) Planktonic and benthic foraminiferal $\delta^{18}O$ composite records representing the tropical surface and deep ocean conditions (relative to PDB). The thick line through both records was generated using a 1 million year Gaussian filter. (B) Temperature estimates based on planktonic and benthic records and ice volume estimates discussed in the text.

with the lowest recorded value (− 0.5‰) of the Cenozoic. Following this minimum at 52 Ma, $\delta^{18}O$ values increased by 5.5‰, recording maximum values (~5‰) during the glacial intervals of Pleistocene (**Figure 6**). The first part of this long-term change was a gradual increase of 2‰ through the end of the Eocene (52–34 Ma). The remainder of the increase was accomplished through large steps at the Eocene/Oligocene boundary (~33.5 Ma), during the middle Miocene (ca. 15–13 Ma) and late Pliocene (ca. 3.2–2.6 Ma). After 2.6 Ma, the amplitude of the high-frequency signal increased to > 1‰, reaching 1.8‰ over the past 800 thousand years.

Planktonic and benthic foraminiferal $\delta^{18}O$ values co-varied in general during the early Cenozoic (6.5–34 Ma). Values averaged about − 1‰ between 65 and 58 Ma, before decreasing to − 2.5‰ during the early Eocene, recording the lowest values of the Cenozoic (**Figure 7A**). From 52 to 33 Ma, planktonic foraminiferal values increased by 2‰. In spite of a break in the latest Eocene record, it appears that the tropical ocean differed from the deep ocean across the Eocene/Oligocene boundary. For much of the Oligocene (~33–25 Ma), planktonic foraminiferal $\delta^{18}O$ values remained unusually high, averaging −0.5‰. Beginning around the Oligocene/Miocene boundary (~25 Ma), planktonic foraminiferal $\delta^{18}O$ values began a long-term decrease, culminating in the Pleistocene with average values of −1.5‰. In contrast, the benthic $\delta^{18}O$ record permanently changed during the middle Miocene $\delta^{18}O$ shift and late Pliocene increase.

Apportioning the $\delta^{18}O$ changes recorded by the benthic and planktonic foraminifera between temperature and ice volume changes requires knowledge of, or reasonable estimates for, one of these parameters. One promising tool that may help discriminate between each effect is the Mg/Ca ratio measured in benthic foraminifera. Initial studies using Mg/Ca ratios confirmed the long-term temperature changes during the Cenozoic calculated using the $\delta^{18}O$ record and other climate proxies. If verified, this record implies that small ice sheets grew during the middle and late Eocene and fluctuated in size throughout the Oligocene to Miocene. At present, the Mg/Ca record lacks the resolution for key intervals and still requires verification of interspecies offsets before it can be applied unequivocally to isolate the ice volume-induced $\delta^{18}O_{water}$ component in the foraminiferal $\delta^{18}O$ records. For the discussion that follows, glacialogical evidence is used to estimate the ice volume/$\delta^{18}O_{water}$ variations.

The Greenhouse World

The oldest unequivocal evidence for ice sheets on Antarctica, ice-rafted detritus (IRD) deposited by icebergs in the ocean, places the first large ice sheet in the earliest Oligocene. Thus, it is reasonable to assume that ice sheets were small to absent and that surface and deep-water temperature changes controlled much if not all the $\delta^{18}O$ change prior to 34 Ma. The modern Antarctic and Greenland ice sheets lock up ~2% of the total water in the world's ocean. If melted, these ice sheets would raise global sea level by ~70–75 m and mean ocean $\delta^{18}O_{water}$ value would decrease to −0.9‰ PDB (see above). One can then apply eqn [2] to the benthic and planktonic foraminiferal $\delta^{18}O$ records to estimate deep- and surface-ocean temperatures for the first half of the Cenozoic (*c.* 65–34 Ma).

During the early to middle Paleocene, deep-water temperatures remained close to 10°C (**Figure 7B**). The 1‰ decrease between 58 and 52 Ma translates into a deep-water warming of 4°C, reaching a high of 14°C. This is in sharp contrast to the modern deep-water temperatures, which range between 0 and 3°C. Following the peak warmth at 52 Ma, the 2‰ increase in benthic foraminiferal values indicates that the deep waters cooled by 7°C and were 7°C by the end of the Eocene. If small ice sheets existed during the Paleocene and Eocene, then temperature estimates would be on the order of 1°C warmer than those calculated for the ice-free assumption. (Some data indicate that smaller ice sheets may have existed on the inland parts of Antarctica during the late Eocene. However, these were not large enough to deposit IRD in the ocean. Therefore, their effect on the $\delta^{18}O$ values of the ocean was probably less than 0.3‰.)

Tropical surface water temperatures warmed from 22 to 24°C, based on eqn [2], at the beginning of the Cenozoic to 28°C during the early Eocene (52 Ma; **Figure 7B**). The higher estimate is similar to temperatures in the equatorial regions of the modern oceans. Planktonic foraminiferal $\delta^{18}O$ values recorded a long-term increase of by 2‰ (−2.5 to −0.5‰) through the remainder of the Eocene. Just prior to the Eocene/Oligocene boundary, tropical surface water temperatures were ~21°C, ending the long-term tropical cooling of 7°C from 52 to 34 Ma.

The Ice House World of the Last 33 Million Years

As mentioned above, southern ocean cores contain IRD at and above the Eocene/Oligocene boundary. Widely distributed IRD and glacial tills on parts of the Antarctic continental margin representing the Oligocene to Recent mark the onset of large ice sheets. Whether these sediments represent persistent or periodic ice cover is uncertain. At least some ice was present on Antarctica during the Oligocene to early Miocene. The Antarctic ice sheet has been a fixture since the middle Miocene (~15 Ma). Our record of Northern Hemisphere ice sheets suggests that they were small or nonexistent prior to the late Pliocene. For the purpose of estimating surface and deep temperatures, an ice volume estimate slightly lower than the modern will be applied for the interval that spans from the Oligocene into the middle Miocene (33–15 Ma). For the interval between 15 and 3 Ma, ice volumes were probably similar to those of today. From 3 Ma, ice volumes ranged between the modern and LGM. Using these broad estimates for ice volumes, mean ocean $\delta^{18}O_{water}$ values for those three intervals were −0.5, −0.22, and 0.4‰ PDB, respectively. The 0.4‰ estimate reflects the average between the maximum and minimum conditions during the Plio-Pleistocene. As noted above, the largest portion of the high frequency signal is controlled by ice volume changes.

The benthic foraminiferal $\delta^{18}O$ increase at the Eocene/Oligocene boundary occurred rapidly (~10 000 years; **Figure 8A**). At the peak of the Eocene/Oligocene boundary event, benthic foraminifera recorded $\delta^{18}O$ values similar to modern values. Using the ice volume assumption from above, deep-water temperatures approached modern deep-ocean temperatures (3°C). This marks an important transition from the relatively warm oceans of the Paleocene and Eocene to the cold deep waters of the Oligocene to present. This switch to a cold ocean where bottom waters formed at near-freezing temperatures heralded the development of the psychrosphere. Following the Eocene/Oligocene boundary, deep-water temperatures began a long-term warming over the next 18 million years (33–15 Ma). The coldest deep-water temperatures of 3°C were recorded at 33 Ma, while temperatures reached 9°C at ~25 and ~15 Ma (**Figure 7B**).

There is a gap in the planktonic foraminiferal $\delta^{18}O$ record for the latest Eocene that hampers our assessment of tropical response during Eocene/Oligocene climate event. However, it is clear from the data that do exist that the planktonic response across the Eocene/Oligocene boundary differed from the benthic response. The planktonic foraminiferal $\delta^{18}O$ values for the early Oligocene are similar to late Eocene values, whereas the benthic values recorded a 1.5‰ increase. Planktonic foraminiferal records from other regions that span the Eocene/Oligocene boundary indicate that the surface water $\delta^{18}O$ increase was on the order of 0.5‰. This change is approximately equal to the effect of the modern Antarctic ice sheet. Combined with the physical evidence, it seems probable that the planktonic foraminiferal $\delta^{18}O$ increase at the Eocene/Oligocene boundary recorded the ice volume influence with little temperature effect. Therefore, tropical surface temperatures remained around 22°C while the deep ocean cooled during

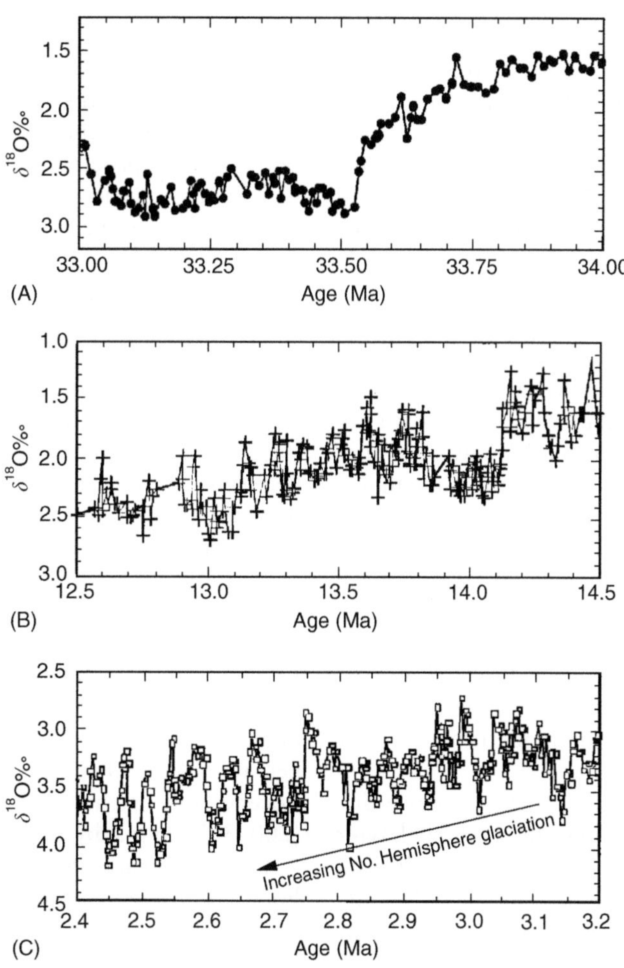

Figure 8 High-resolution $\delta^{18}O$ records representing the Eocene/Oligocene boundary (A), middle Miocene (B), and late Pliocene (C) $\delta^{18}O$ shifts (relative to PDB).

this $\delta^{18}O$ shift. Following the boundary event, planktonic foraminifera $\delta^{18}O$ record during the Oligocene and early Miocene mirrored the benthic record in many respects. For much of the Oligocene and early Miocene, the absolute values are close to $-0.5‰$, which translates into a temperature estimate of 21°C (**Figure 7B**). By 15 Ma, tropical surface waters had warmed to 26°C.

The middle Miocene $\delta^{18}O$ shift represents an increase of 1.5‰ in the benthic record between 15 and 13 Ma. This transition is composed of two sharp increases around 14 and 13 Ma (**Figure 8B**). These $\delta^{18}O$ steps occurred in less than 200 000 years with each recording an increase of $\sim 1‰$ followed by a small decrease. During these two shifts, deep waters cooled from 9 to 5°C. The planktonic foraminiferal $\delta^{18}O$ record from 15 to 13 Ma shows two increases as recorded in the benthic foraminiferal record (**Figure 7**). However, it does not show the large permanent shift recorded by the benthic foraminifera, indicating a small cooling from 26 to 24°C. From 13 to 3 Ma, the deep ocean cooled slightly from 5 to 3°C while the surface waters warmed from 24 to 26°C (**Figure 7B**).

The last of the large $\delta^{18}O$ steps in the Cenozoic was recorded during the late Pliocene from 3.2 to 2.6 Ma. This 'step' is better characterized as a series of $\delta^{18}O$ cycles with increasing amplitudes and values over this interval (**Figures 6** and **8C**). The cycles have been subsequently determined to be 40 000 year cycles related to variations in the solar radiation received in the high latitudes. This interval ushered in the large-scale Northern Hemisphere ice sheets that have since dominated Earth's climate. At 2.6 Ma, the first IRD was deposited in the open North Atlantic and was coeval with the $\delta^{18}O$ maximum. Prior to 2.6 Ma, IRD was confined to the marginal basins to the north, Greenland's and Iceland's continental margins. Subsequent $\delta^{18}O$ maxima were associated with IRD. Between 2.6 and 1 Ma, large Northern Hemisphere ice sheets waxed and waned on the 40 000 year beat. Beginning around 1 Ma, the ice sheets increased in size and switched to a 100 000 year beat (**Figure 6**). During this interval, deep-water temperatures remained similar to those in the modern ocean (0 to 3°C).

The planktonic foraminiferal $\delta^{18}O$ response during the late Pliocene event shows the cyclic behavior, but not the overall increase recorded by the benthic foraminifera. As with the middle Miocene $\delta^{18}O$ shift, the late Pliocene increase represents the cyclic build-up of ice sheets accompanied by deep water cooling. The tropical surface water temperatures, however, varied between 26 and 28°C.

Mechanisms for Climate Change

Most climate change hypotheses for the Cenozoic focus on either oceanic heat transport and/or greenhouse gas concentrations. Each mechanism produces different responses in the equatorial-to-pole and surface-to-deep temperature gradients. An increase in the meridional heat transport generally cools the tropics and warms the poles. If poleward transport of heat decreases, then the tropics will warm and the poles will cool. Variations in greenhouse gas concentrations should produce similar changes in both the tropical and polar regions.

Tropical surface water and deep-ocean records covaried for the first part of the Cenozoic. The warming and subsequent cooling between 65 and 34 Ma are most often ascribed to changing greenhouse gas concentrations. The interval of warming that began around 58 Ma and peaked at 52 Ma coincided with the release of large amounts of CO_2 into the atmosphere as a consequence of tectonic processes. The eruption of the Thulean basalts in the northeastern Atlantic Ocean began during the Paleocene and peaked around 54 Ma. It is also recognized that there was a large-scale reorganization of the mid-ocean ridge hydrothermal system which began during the late Paleocene and extended into the Eocene. Both tectonic processes accelerate mantle degassing which raises atmospheric levels of CO_2. Recently, evidence for another potentially large CO_2 reservoir was found along the eastern continental margin of North America. Methane hydrates frozen within the sediments appear to have released catastrophically at least once and possibly multiple times during the latest Paleocene and early Eocene ($\sim$58–52 Ma). One or all of these sources could have contributed to the build-up of greenhouse gases in the atmosphere between 58 and 52 Ma. Following the thermal maximum, the long-term cooling in both the surface and deep waters implies that greenhouse gas concentrations slowly decreased. Proxies for estimating pCO_2 concentrations ($\delta^{13}C$ fractionation within organic carbon and boron isotopes) are still being developed and refined. However, preliminary indications are that atmospheric pCO_2 levels were high (>1000 p.p.m.) during the early Eocene, dropped to $\sim$400–500 p.p.m. during the middle to late Eocene, and reached late Pleistocene concentrations (200–300 p.p.m.) by the early Oligocene.

The deep-water temperature cooling across the Eocene/Oligocene boundary (**Figure 7B**) was not accompanied by tropical cooling, and resulted from the first step in the thermal isolation of Antarctica. In modern ocean, the Antarctic Circumpolar Current is a vigorous surface-to-bottom current that provides an effective barrier to southward-flowing warm surface waters. The development of this current during the Cenozoic hinged on the deepening for the Tasman Rise and opening of the Drake Passage. Recent drilling indicates that marine connections developed across the Tasman Rise at or near the Eocene/Oligocene boundary (33.5 Ma). Tectonic constraints on the separation of the Drake Passage are less precise. Estimates range from 35 to 22 Ma for the opening of this gateway. The uncertainty lies in the tectonic complexity of the region and what constitutes an effective opening for water to flow through. The climatic consequence of creating a circumpolar flow was to thermally isolate Antarctica and promote the growth of the Antarctic ice sheet. As noted above, the first large ice sheet grew at the beginning at the Eocene/Oligocene boundary.

The most notable divergence in the δ^{18}O records occurred during the middle Miocene (~15 Ma). For the first time during the Cenozoic, the tropical surface and deep waters recorded a clear divergence in δ^{18}O values, a trend that increased in magnitude and reached a maximum in the modern ocean. Any poleward transport of heat appears to have been effectively severed from Antarctica by 15 Ma, promoting further cooling. On the other hand, the tropics have been warming over the past 15 My. A combination of different factors fueled this warming. First, less heat was being transported out of the low- and mid-latitude regions to the high southern latitudes. Second, the opening of the Southern Hemisphere gateways that promoted the formation of the circumpolar circulation led to the destruction of the circumequatorial circulation. The effects of the closure of the Tethys Ocean (predecessor to the Mediterranean), shoaling of the Panamanian Isthmus (4.5–2.6 Ma), and constriction in the Indonesian Passage (~3 Ma to present) allowed the east-to-west flowing surface waters in the tropics to 'pile' up and absorb more solar radiation. A consequence of the equatorial warming and high-latitude cooling was an increase in the equator-to-pole temperature gradient. As the gradient increased, winds increased, promoting the organization of the surface ocean circulation patterns that persist today.

Some Caveats

A concern in generating marine isotope records is that the isotopic analyses should be made on the same species. This is important because $\delta^{18}O_{calcite}$ values can vary among the different species of organisms. Coexisting taxa of benthic foraminifera record δ^{18}O values that can differ by as much as 1‰. In planktonic foraminifera, variations between species can be as great as 1.5‰. For both the planktonic and benthic foraminifera, interspecific differences are as large as the glacial–interglacial signal. These interspecific $\delta^{18}O_{calcite}$ variations are often ascribed to a vital effect or kinetic fractionation of the oxygen isotopes within the organism. However, some of the differences in the planktonic taxa results from different seasonal or depth habitats and therefore provides important information about properties in the upper part of the water column. It is noteworthy that the first δ^{18}O syntheses were based on mixed species analyses and yet basic features captured in these curves still persist today. This attests to the robustness of these records and method for reconstructing climate changes in the ocean.

The high-frequency signal that dominates the late Pliocene to Pleistocene records is also present in the Miocene and Oligocene intervals. The cloud of points about the mean shown in **Figure 7** reflects records that were sampled at a resolution sufficient to document the high frequency signal. For the interval between 35 and 1 Ma, the benthic foraminiferal δ^{18}O record has a 40 000 year frequency superimposed on the long-term means that are represented by the smoothed line. The origin of the 40 000 year cycles lies in variations in the tilt of the earth's axis that influences the amount of solar radiation received in the high latitudes. This insolation signal is transmitted to the deep ocean because the high latitudes were the source regions for deep waters during much, if not all, of the Cenozoic. The record prior to 35 Ma is unclear with regard to the presence or absence of 40 000 year cycles.

See also

Oxygen Isotopes in the Ocean.

Further Reading

Craig H (1957) Isotopic standards for carbon and oxygen and correction factors for mass spectrometric analysis of carbon dioxide. *Geochemica et Cosmochemica Acta* 12: 133–149.

Craig H (1965) The measurement of oxygen isotope paleotemperatures. In: Tongiorgi E (ed.) *Stable Isotopes in Oceanographic Studies and Paleotemperatures*, pp. 161–182. Spoleto: Consiglio Nazionale delle Ricerche, Laboratorio di Geologica Nucleare, Pisa.

Craig H and Gordon LI (1965) Deuterium and oxygen-18 variations in the oceans and marine atmosphere. In: Tongiorgi E (ed.) *Stable Isotopes in Oceanographic*

Studies and Paleotemperatures, pp. 1–122. Spoleto: Consiglio Nazionale delle Ricerche, Laboratorio di Geologica Nucleare, Pisa.

Emiliani C (1955) Pleistocene temperatures. *Journal of Geology* 63: 539–578.

Epstein S, Buchsbaum R, Lowenstam H, and Urey HC (1953) Revised carbonate-water temperature scale. *Bulletin of the Geological Society of America* 64: 1315–1326.

Fairbanks RG, Charles CD, and Wright JD (1992) Origin of Melt Water Pulses. In: Taylor RE, Long A, and Kra RS (eds.) *Radiocarbon After Four Decades*, pp. 473–500. New York: Springer-Verlag.

Imbrie J, Hays JD, Martinson DG, *et al.* (1984) The orbital theory of Pleistocene climate: support from a revised chronology of the marine $\delta^{18}O$ record. In: Berger AL, Imbrie J, Hays JD, Kukla G, and Saltzman B (eds.) *Milankovitch and Climate, part I*, pp. 269–305. Dordrecht: Reidel.

Lear CH, Elderfield H, and Wilson PA (1999) Cenozoic deep-sea temperatures and global ice volumes from Mg/Ca in benthic foraminiferal calcite. *Science* 287: 269–272.

Miller KG, Fairbanks RG, and Mountain GS (1987) Tertiary oxygen isotope synthesis, sea-level history, and continental margin erosion. *Paleoceanography* 2: 1–19.

Miller KG, Wright JD, and Fairbanks RG (1991) Unlocking the Ice House: Oligocene–Miocene oxygen isotopes, eustasy, and margin erosion. *Journal of Geophysical Research* 96: 6829–6848.

Pagani M, Arthur MA, and Freeman KH (1999) Miocene evolution of atmospheric carbon dioxide. *Paleoceanography* 14: 273–292.

Palmer MR, Pearson PN, and Cobb SJ (1998) Reconstructing past ocean pH-depth profiles. *Science* 282: 1468–1471.

Pearson PN and Palmer MR (1999) Middle Eocene seawater pH and atmospheric carbon dioxide concentrations. *Science* 284: 1824–1826.

Rozanski K, Araguas-Araguas L and Gonfiantini R (1993) Isotopic patterns in modern global precipitation. In: Swart PK, McKenzie J and Savin S (eds) *Climate Change in Continental Isotopic Records*. Geophysical Monograph 78, pp. 1–35. Washington, DC: American Geophysical Union.

Rye DM and Sommer MA (1980) Reconstructing paleotemperature and paleosalinity regimes with oxygen isotopes. In: Rhoads DC and Lutz RA (eds.) *Skeletal Growth of Aquatic Organisms*, pp. 162–202. New York: Plenum.

Savin SM, Douglas RG, and Stehli FG (1975) Tertiary marine paleotemperatures. *Geological Society of America Bulletin* 86: 1499–1510.

Shackleton NJ (1967) Oxygen isotope analyses and Pleistocene temperatures re-assessed. *Nature* 215: 115–117.

Shackleton NJ, Berger A, and Peltier WR (1990) An alternative astronomical calibration of the Lower Pleistocene time scale based on ODP Site 677. *Transactions of the Royal Society of Edinburgh, Earth Science* 81: 251–261.

Shackleton NJ and Kennett JP (1975) Paleotemperature history of the Cenozoic and initiation of Antarctic glaciation. Oxygen and carbon isotopic analysis in DSDP Sites 277, 279, and 281. *Initial Report Deep Sea Drilling Project* 29: 743–755.

Shackleton NJ and Opdyke ND (1973) Oxygen isotope and paleomagnetic stratigraphy of equatorial Pacific core V28-238. Oxygen isotope temperatures and ice volumes on a 10^5 year and 10^6 year scale. *Quaternary Research* 3: 39–55.

Tiedemann RM, Sarnthein M, and Shackleton NJ (1994) Astronomic calibration for the Pliocene Atlantic $\delta^{18}O$ and dust flux records of Ocean Drilling Program Site 659. *Paleoceanography* 9: 619–638.

Urey HC (1947) The thermodynamic properties of isotopic substances. *Journal of the Chemical Society* pp. 562–581.

PAST CLIMATE FROM CORALS

A. G. Grottoli, University of Pennsylvania, Philadelphia, PA, USA

Introduction

The influence of the tropics on global climate is well recognized. Our ability to understand, model and predict the interannual, decadal and long-term variability in tropical climate depends on our knowledge of past climate. However, our understanding of the natural variability in tropical climate is limited because long-term instrumental records prior to 1950 are sparse or nonexistent in many tropical regions. Continuous satellite monitoring did not begin until the 1970s and *in situ* equatorial Pacific Ocean monitoring has only existed since the 1980s. Therefore, we depend on proxy records to provide information about past climate. Proxy records are indirect measurements of the physical and chemical structure of past environmental conditions chronicled in natural archives such as ice cores, sediment cores, coral cores and tree rings.

In the tropical oceans, the isotopic, trace and minor elemental signatures of coral skeletons can vary as a result of environmental conditions such as temperature, salinity, cloud cover and upwelling. As such, coral cores offer a suite of proxy records with potential for reconstructing tropical paleoclimate on intraannual-to-centennial timescales. Massive, symbiotic stony corals are good tropical climate proxy recorders because: (1) they are widely distributed throughout the tropics; (2) their unperturbed annual skeletal banding pattern offers excellent chronological control; (3) they incorporate a variety of climate tracers from which paleo sea surface temperature (SST), sea surface salinity (SSS), cloud cover, upwelling, ocean circulation, ocean mixing patterns, and other climatic and oceanic features can be reconstructed; (4) their proxy records can be almost as good as instrumental records; (5) their records can span several centuries; and (6) their high skeletal growth rate (usually in the range of 5–25 mm y^{-1}) permits subseasonal sampling resolution. Thus proxy records in corals provide the best means of obtaining long seasonal-to-centennial timescale paleoclimate information in the tropics.

Coral-based paleoclimate research has grown tremendously in the last few decades. Most of the published records come from living corals and report the reconstructed SST, SSS, rainfall, water circulation pattern, or some combination of these variables. Records from fossil and deep-sea corals are also increasing. The goal of this chapter is to provide an overview of the current state of coral-based paleoclimate research and a list of further readings for those interested in more detailed information.

Coral Biology

General

Corals are animals in the cnidarian family, order Scleractinia (class Anthozoa). Their basic body plan consists of a polyp containing unicellular endosymbiotic algae known as zooxanthellae overlaying a calcium carbonate exoskeleton (**Figure 1**). Most coral species are colonial and include many polyps interconnected by a lateral layer of tissue. The polyp, consisting of tentacles (used to capture prey), an oral opening and a gastrovascular cavity, has three tissue layers: the epidermis, mesoglea, and gastrodermis. The symbiotic zooxanthellae are located in the gastrodermis. Corals deposit a calcium carbonate ($CaCO_3$) aragonite skeleton below the basal epidermis. Corals reproduce sexually during mass-spawning events by releasing egg and/or sperm, or egg–sperm bundles into the water column. Mass spawning events typically occur a few times a year for each species, and are triggered by the lunar cycle. Corals can also reproduce asexually by fragmentation.

Animal–Zooxanthellae Symbiosis

Corals acquire the greater part of their food energy by two mechanisms: photosynthesis and heterotrophy (direct ingestion of zooplankton and other

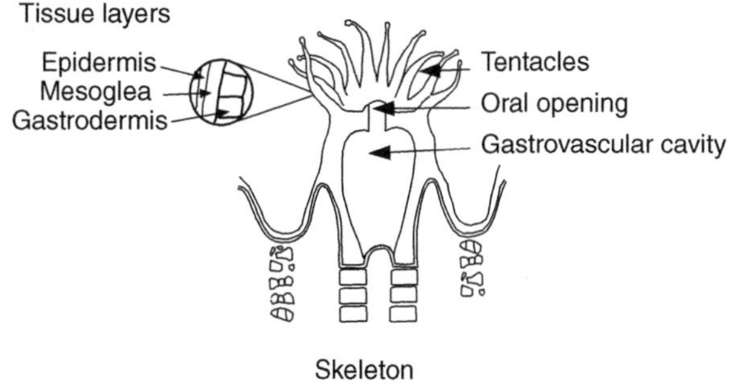

Figure 1 Cross-section of coral polyp and skeleton.

organic particles in the water column by the coral animal). Photosynthesis is carried out by the endosymbiotic zooxanthellae. The bulk of photosynthetically fixed carbon is translocated directly to the coral host. In some cases, the coral animal can obtain all of its daily energy requirements via photosynthesis alone. In general, as light intensity increases, photosynthesis increases.

Skeleton

Corals deposit skeleton below the basal epidermis. Typically, corals deposit one high-density and one low-density band of skeleton each year. The high-density band has thicker skeletal elements than the low-density band. Each band is often composed of several finer bands called dissepiments, deposited directly at the base of the coral tissue. At discrete intervals, the polyp presumably detaches from the dissepiment, and begins to lay down a new skeletal dissepiment. Some evidence suggests that dissepiments may form on a lunar cycle.

High- and low-density bands are deposited seasonally. Overall, high-density bands form during suboptimal temperature conditions and low-density bands form during optimal temperature conditions. At higher latitudes (i.e., Hawaii, Florida) optimal growth temperature occurs in summer. At lower latitudes (i.e., Galápagos, equatorial Pacific regions, Australian Great Barrier Reef), optimal growth temperatures occur in the cooler months. The width and density of growth bands also vary with environmental variables such as light, sedimentation, season length, and salinity. In general, as light levels decrease due to increased cloud cover, increased sedimentation or due to increasing depth, maximum linear skeletal extension decreases, calcification decreases, and skeletal density increases.

The Interpretation of Isotopes, Trace Elements, and Minor Elements in Corals

Several environmental variables can be reconstructed by measuring changes in the skeletal isotope ratios, trace and minor elemental composition, and growth rate records in coral cores (**Table 1**). The width, density, and chemical composition of each band are generally thought to reflect the average environmental conditions that prevailed during the time over which that portion of the skeleton was calcified. Reconstructions of seawater temperature, salinity, light levels (cloud cover), upwelling, nutrient composition and other environmental parameters have been obtained from coral records (**Table 1**).

Table 1 Environmental variable(s) that can be reconstructed from coral skeletal isotopes, trace and minor elements, and growth records

Proxy	*Environmental variable*
Isotopes	
$\delta^{13}C$	Light (seasonal cloud cover), nutrients/zooplankton levels
$\delta^{18}O$	Sea surface temperature, sea surface salinity
$\Delta^{14}C$	Ocean ventilation, water mass circulation
Trace and minor elements	
Sr/Ca	Sea surface temperature
Mg/Ca	Sea surface temperature
U/Ca	Sea surface temperature
Mn/Ca	Wind anomalies, upwelling
Cd/Ca	Upwelling
$\delta^{11}B$	pH
F	Sea surface temperature
Ba/Ca	Upwelling, river outflow, sea surface temperature
Skeleton	
Skeletal growth bands	Light (seasonal changes), stress, water motion, sedimentation, sea surface temperature
Fluorescence	River outflow

Method

Continuous records of past tropical climate conditions can be obtained by extracting a core from an individual massive coral head along its major axis of growth. Typically, this involves placing a coring device on the top and center of the coral head (**Figure 2A**). The extracted core is cut longitudinally into slabs ranging in thickness from 0.7 to 1 cm that are then X-rayed. X-ray-positive prints reveal the banding pattern of the slab and are used: (1) as a guide for sample drilling and (2) to establish a chronology for the entire coral record when the banding pattern is clear (**Figure 2B**). Samples are drilled out along the major axis of growth by grinding the skeletal material with a diamond-tipped dental drill. For high-resolution climate reconstructions, samples are extracted every millimeter or less down the entire length of the core. Since corals grow about 5–15 mm per year, this sampling method can yield approximately bimonthly-to-monthly resolution. Much higher resolution sampling is possible, yielding approximately weekly samples, but this is not commonly performed. In most cases, the $\delta^{13}C$ (the per mil deviation of the ratio of $^{13}C/^{12}C$ relative to the Peedee Belemnite (PDB) Limestone Standard) and $\delta^{18}O$ (ratio of $^{18}O/^{16}O$ relative to PDB) values of each sample are

(A)

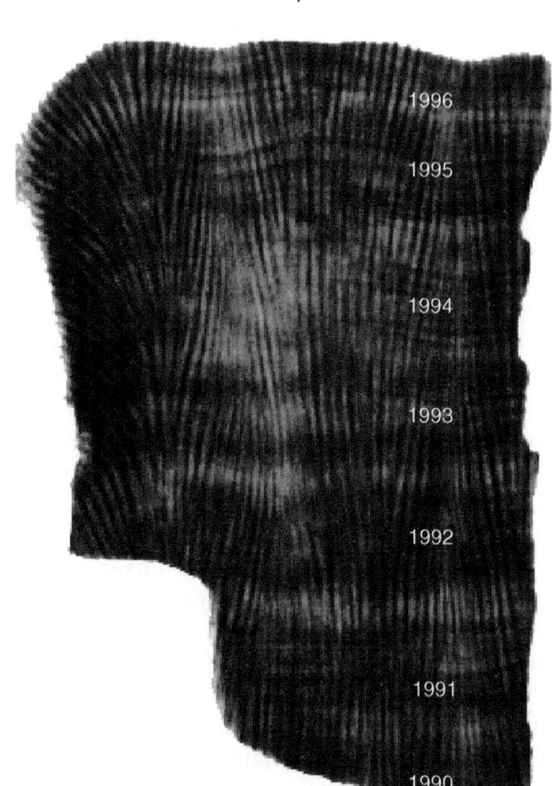

(B)

Figure 2 Collecting coral cores. (A) Coral core being extracted from top and center of an individual massive coral head using a pneumatic coring device. (Photo courtesy of M. Kazmers/Shark Song Tax ID #374-50-5314.) (B) X-ray positive print reveals the banding pattern of the slab and is used to help establish a chronology for the entire coral record.

measured. Since the δ^{13}C and/or δ^{18}O compositions of corals usually have a strong seasonal component, they are often used to establish an accurate chronology and/or to confirm the chronology established from the X-rays.

Temperature and Salinity Reconstructions

Coral skeletal δ^{18}O reflects a combination of the local SST and SSS. In the many regions of the tropical ocean where the natural variation in salinity is small, changes in coral skeletal δ^{18}O primarily reflect changes in SST. The δ^{18}O of coral skeleton responds to changes in temperature usually according to the standard paleotemperature relationship for carbonates. Based on empirical studies, a 1°C increase in water temperature corresponds to a decrease of about 0.22‰ (parts per thousand) in δ^{18}O. Precipitation has a low δ^{18}O value relative to that of sea water. Therefore, in regions with pronounced variability in rainfall and/or river runoff, coral δ^{18}O values reflect changes in SSS. Thus, depending on the nature of the coral collection site, the δ^{18}O record is used to reconstruct the SST and/or SSS. Additional studies show that other proxy indicators of temperature include the ratios of strontium/calcium (Sr/Ca), magnesium/calcium (Mg/Ca), and uranium/calcium (U/Ca), fluorine levels (F) and skeletal band thickness (**Table 1**). The ratios of Sr/Ca, Mg/Ca, and U/Ca incorporated into the skeleton is largely determined by the temperature-dependent distribution coefficient of Sr/Ca, Mg/Ca, and U/Ca between aragonite and sea water. As temperatures increase, the Sr/Ca and U/Ca ratios decrease and the Mg/Ca ratio increases.

Cloud Cover and Upwelling

δ^{13}C seems to indicate seasonal changes in cloud cover and upwelling. Thus far, only a small number of studies have used δ^{13}C records to confirm seasonal rainfall patterns established using the δ^{18}O signature. Only one study has directly linked a δ^{13}C record with seasonal upwelling. δ^{13}C in coral skeletons has been difficult to use as a paleoclimate tracer because it is heavily influenced by metabolic processes, namely photosynthesis and heterotrophy. Firstly, as light levels decrease due to cloud cover, the rate of photosynthesis by the coral's symbiotic zooxanthellae decreases, and skeletal $\delta^{13}C$ decreases. The reverse occurs when light levels increase. Secondly, zooplankton have a low $\delta^{13}C$ value relative to coral. During upwelling events in the Red Sea, nutrient and zooplankton level increases have been linked to decreases in coral skeletal $\delta^{13}C$ values. Other upwelling tracers include cadmium (Cd) and barium (Ba) concentrations, and $\Delta^{14}C$. Cadmium and barium are trace elements whose concentrations are greater in deep water than in surface water. During upwelling events, deep water is driven to the surface and cadmium/calcium (Cd/Ca) and barium/ calcium (Ba/Ca) ratios in the surface water, and consequently in the coral skeleton, increase (**Table 1**). Although SST also influences Ba/Ca ratios, most of the variation in Ba/ Ca ratios in corals is due to nutrient fluxes and upwelling. $\Delta^{14}C$ is also an excellent tracer for detecting upwelling and changes in seawater circulation ($\Delta^{14}C$ is the per mil deviation of the ratio of $^{14}C/^{12}C$ relative to a nineteenth century wood standard). For example, in the eastern equatorial Pacific Ocean, the $\Delta^{14}C$ value of deep water tends to be very low relative to the $\Delta^{14}C$ of surface water. Here, increased upwelling or increases in the proportion of deep water contributing to surface water results in a decrease in the $\Delta^{14}C$ of the coral skeleton. Manganese (Mn) is a trace element whose concentration is highest in surface waters and decreases with depth. Therefore, during upwelling events, Mn/Ca ratios decrease. The ratio of Mn/Ca can also record prolonged and sustained changes in winds. In at least one case, Mn/Ca ratios from a Tarawa Atoll coral increased during El Niño events as a result of strong and prolonged wind reversals that had remobilized manganese from the lagoon sediments.

Other Proxy Indicators

Other environmental parameters that can be inferred from coral skeleton structure and composition are river outflow (fluorescence bands) and pH (boron isotope levels) (**Table 1**). Large pulses in river outflow can result in an ultraviolet-sensitive fluorescent band in the coral record. New evidence strongly suggests that the fluorescent patterns in coral skeletal records are due to changes in skeletal density, not terrestrially derived humics as previously thought. Variations in salinity associated with fresh water discharge pulses from rivers appear to cause changes in coral skeletal growth density which can be observed in the skeletal fluorescence pattern. In the case of boron, $\delta^{11}B$ levels in sea water increase as pH increases. Changes in the pH at the site of coral calcification seem to reflect changes in productivity of the symbiotic zooxanthellae. As photosynthesis increases, pH increases, and $\delta^{11}B$ levels in the coral skeleton increase.

Coral Records: What has been Learned About Climate From Corals?

To date, there are over 100 sites where coral cores have been recovered and analyzed (**Figure 3**). Of these, at least 22 have records that exceed 120 years in length. In most cases, $\delta^{13}C$ and $\delta^{18}O$ have been measured at annual-to-subannual resolution. $\delta^{18}O$ as a SST and/or SSS proxy is the best understood and most widely reported of all the coral proxy measurements. In a few cores, other isotopic, trace, and minor elements, or skeletal density and growth measurements have also been made.

Typically, coral-derived paleoclimate records are studied on three timescales: seasonal, interannual-to-decadal, and long-term trends. The seasonal variation refers to one warm and one cool phase each year. An abrupt shift in the proxy's long-term mean often indicates a decadal modulation in the data. Long-term trends are usually associated with a gradual increase or decrease in the measured proxy over the course of several decades or centuries. The following sections explore some of the seasonal, decadal and long-term trends in coral-derived paleoclimate records and some of the limitations associated with interpreting coral proxy records.

Seasonal Variation in Coral Climate Records

Seasonal variation accounts for the single largest percentage of the variance in most coral isotope, trace, and minor element records. In regions such as Japan and the Galápagos with distinct SST seasonality, the annual periodicity in $\delta^{18}O$, Sr/Ca, Mg/Ca, and U/Ca is pronounced. In regions heavily affected by monsoonal rains such as Tarawa Atoll, the seasonal variation in $\delta^{13}C$ is regular and pronounced. $\Delta^{14}C$ also has an annual periodicity in coral from regions with a strong seasonal upwelling regime such as is seen in the Galápagos (see next section). The strength and duration of the upwelling season is reflected in the length and degree of $\Delta^{14}C$ decrease in the coral record.

Interannual-to-decadal Variation in Climate and El Niño Southern Oscillation (ENSO)

The second largest component of the variance in Pacific coral isotope records is associated with the interannual-to-decadal variation in the El Niño

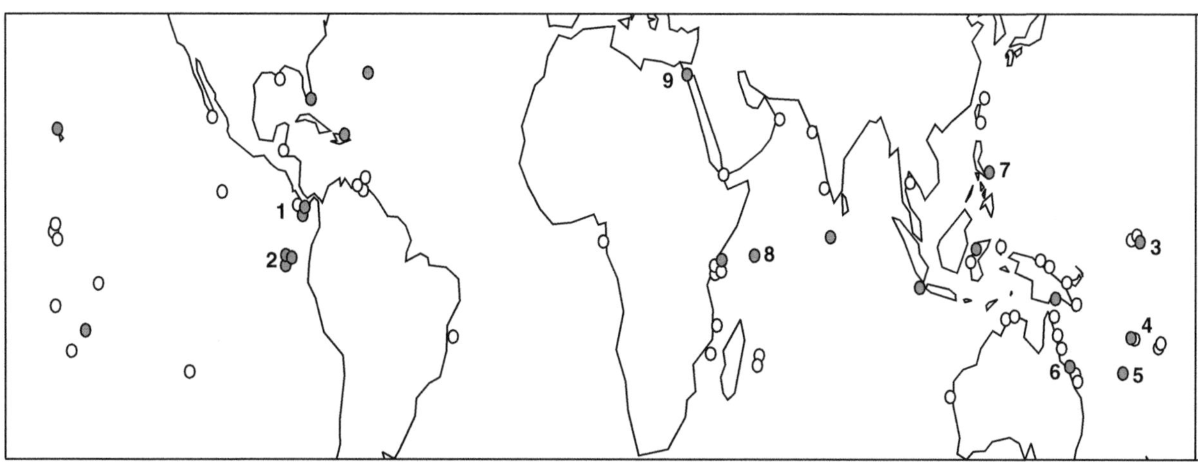

Figure 3 Map indicating the approximate locations of current paleoclimate research. The coral sites involve the work of many investigators and may be incomplete. The $\delta^{18}O$ records from the numbered sites are shown in **Figure 5** and are discussed in the text. , sites with records longer than 120 years (most are published); ○, sites where cores have been recovered and data collection is underway. (Reproduced from Gagan *et al.* (2000), with permission from Elsevier Science.)

Southern Oscillation (ENSO)[1]. Several of the longer Pacific $\delta^{18}O$ coral records reveal that the frequency of ENSO has changed on decadal timescales over the past few centuries. Over the last 300 years, the dominant mode of ENSO recorded by a Galápagos coral has been at 4.6 years. However, during that time period there have been shifts in that mode from 4.6 to 7 years during the 1600s, 3–4.6 years from 1700–1750, and 3.5 years from 1800–1850. These major shifts in ENSO frequency may indicate major reorganizations in Pacific climate at various intervals over time. A 101-year long $\delta^{18}O$ record from Clipperton Atoll reveals a pronounced period of reduced ENSO frequency from ~1925 to 1940 suggesting a reduced coupling between the eastern and western Pacific. At Clipperton, decadal timescale variability represents the largest percentage of the variance in $\delta^{18}O$ and appears to be related to the processes influencing the Pacific Decadal Oscillation phenomenon (PDO)[2].

Another component of ENSO variability recovered from coral $\delta^{18}O$ records is the shift in rainfall patterns during El Niños associated with: (1) the migration of the Indonesian Low pressure cell to the region of the date line and the equator in the western Pacific, and (2) the northern migration of the intertropical convergence zone (ITCZ) in the eastern Pacific. Eastward migration of the Indonesian Low results in decreased precipitation in the Indian Ocean and increased precipitation in the western and central Pacific. These phenomena are reflected in the $\delta^{18}O$ record of Seychelles and Tarawa Atoll corals, respectively. Decadal variability in the Seychelles

[1]ENSO refers to the full range of variability observed in the Southern Oscillation, including both El Niño and La Niña events in the Pacific. The Southern Oscillation Index (SOI) is a measure of the normalized difference in the surface air pressure between Tahiti, French Polynesia and Darwin, Australia. Most of the year, under normal seasonal Southern Oscillation cool phase conditions, easterly trade winds induce upwelling in the eastern equatorial Pacific and westward near-equatorial surface flow. The westward flowing water warms and piles up in the western Pacific creating a warm pool and elevating sea level. The wind-driven-transport of this water from the eastern Pacific leads to an upward tilt of the thermocline and increases the efficiency of the local trade-wind-driven equatorial upwelling to cool the surface resulting in an SST cold tongue that extends from the coast of South America to near the international date line. Normal seasonal Southern Oscillation warm-phase conditions are marked by a relaxation of the zonal component of trade winds, reduced upwelling, and a weakening or reversal of the westward flowing current coupled with a deepening of the thermocline in the eastern equatorial Pacific Ocean, and increased SST in the central and eastern equatorial Pacific. This oscillation between cool and warm phases normally occurs annually. Exaggerated and/or prolonged warm-phase conditions are called El Niño events. They usually last 6–18 months, occur irregularly at intervals of 2–7 years, and average about once every 3–4 years. The SOI is low during El Niñ events. Exaggerated and/ or prolonged ENSO cool phase conditions are called La El Niña events. They often follow El Niñ events (but not necessarily). La El Niña events are marked by unusually low surface temperatures in the eastern and central equatorial Pacific and a high SOI.

[2]The PDO appears to be a robust, recurring two-to-three decade pattern of ocean–atmosphere climate variability in the North Pacific. A positive PDO index is characterized by cooler than average SST in the central North Pacific and warmer than average SST in the Gulf of Alaska and along the Pacific Coast of North America and corresponds to warm phases of ENSO. The reverse is true with a negative PDO.

record suggests that regional rainfall variability may originate from the ocean. Decadal variability in a 280-year $\delta^{18}O$ record from a Panamanian coral indicates decadal periods in the strength and position of the ITCZ.

Changes in the decadal variability of coral skeletal $\Delta^{14}C$ reveals information about the natural variability in ocean circulation, water mass movement and ventilation rates in surface water. Biennial-to-decadal shifts in $\Delta^{14}C$ between 1880 and 1955 in a Bermuda coral indicates that rapid pulses of increased mixing between surface and subsurface waters occurred in the North Atlantic Ocean during the past century and that these pulses appeared to correlate with fluctuations in the North Atlantic Oscillation. In a post-bomb Galápagos coral record, abrupt increases in monthly $\Delta^{14}C$ values during the upwelling season after 1976 suggest a decadal timescale shift in the vertical thermal structure of the eastern tropical Pacific (**Figure 4**). The decadal variability in $\Delta^{14}C$ in the Bermuda and Galápagos records are testimony to the power of coral proxy records to provide information about ocean circulation patterns. Additional $\Delta^{14}C$ records from Nauru, Fanning Island, Great Barrier Reef, Florida, Belize, Guam, Brazil, Cape Verde, French Frigate Shoals, Tahiti, Fiji, Hawaii and a few other locations are either published or in progress. As the number of coral $\Delta^{14}C$ records increases, our understanding of the relationship between climate and ocean circulation patterns will also increase.

Some decadal-to-centennial trends in climate are consistent among many of the longer $\delta^{18}O$ coral records. For example in **Figure 5**, all six records longer than 200 years show a cooler/dryer period from AD 1800 to 1840. Cooling may be related to enhanced volcanism during this period. Following this cooler/dryer interval, four of the six records show shifts towards warmer/wetter conditions around 1840–1860 and five of the six show another warming around 1925–1940. These abrupt shifts towards warmer/wetter conditions detected in corals from a variety of tropical locations suggest that corals may be responding to global climate forcing.

Long-term Trends in Climate

There are three major long-term trends observed in several coral records: (1) a prolonged cool phase prior to 1900 generally consistent with the Little Ice Age; (2) a gradual warming/freshening trend over the past century; and (3) evidence of increased burning of fossil fuels. First, in three of the four longest $\delta^{18}O$ coral records the cool/dry period of the Little Ice Age is observed from the beginning of their respective records, up to the mid to late 1800s (**Figure 5**). However, the lack of this cool/dry period in the Galápagos coral indicates that the Little Ice Age

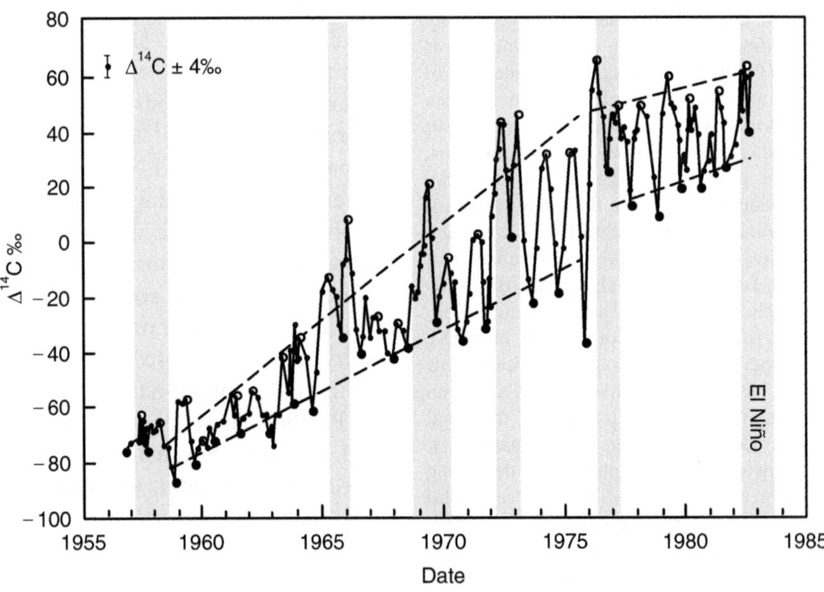

Figure 4 Galápagos coral $\Delta^{14}C\Delta$ record from 1957 to 1983. El Niños are indicated by the shaded bars. ●, upwelling maxima; ○, nonupwelling season. Dashed lines indicated linear trend in the upwelling and nonupwelling seasons. The seasonal variation in $\Delta^{14}C$ is pronounced with high $\Delta^{14}C$ during the nonupwelling season and low $\Delta^{14}C$ values during the upwelling season. A shift in $\Delta^{14}C$ baselines began in 1976. (Reproduced from Guilderson TP and Schrag DP (1998) Abrupt shift in subsurface temperatures in the tropical Pacific associated with changes in El Niño. *Science* 281: 240–243; with permission from the American Association for the Advancement of Science.)

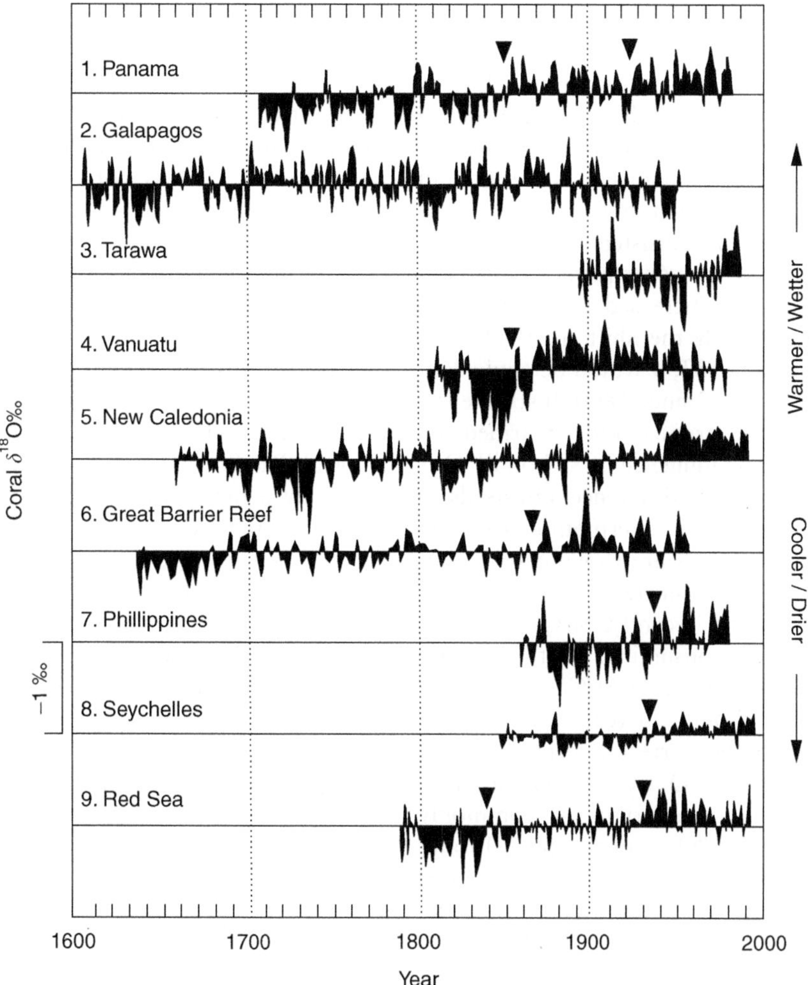

Figure 5 Annual mean coral δ^{18}O records in the Pacific and Indian Ocean region extending back at least for 100 years (locations of cores are indicated in **Figure 3**). Mean δ^{18}O values for each site indicated with a horizontal line. Abrupt shifts in δ^{18}O towards warmer/wetter conditions indicated by black triangles. Data are from the World Data Center-A for Paleoclimatology, NOAA/NGDC Paleoclimatology Program, Boulder, Colorado, USA (http://www.ngdc.noaa.gov/paleo/corals.html) and the original references. Core details list locality, species name, record length, and original reference: 1, Gulf of Chiriqui, Panama, *Porites lobata* 1708–1984 Linsley BK, Dunbar RB, Wellington GM, Mucciarone DA (1994) A coral-based reconstruction of intertropical convergence zone variability over Central America since 1707. *Journal of Geophysical Research* 99: 9977–9994); 2, Urvina Bay, Galápagos, *Pavona clavus* and *Pavona gigantea*, 1607–1981 Dunbar RG, Wellington GM, Colgan MW, Glynn PW (1994) Eastern Pacific sea surface temperature since 1600 A.D.: the δ^{18}O record of climate variability in Galápagos corals. *Paleoceanography* 9: 291–315; 3, Tarawa Atoll, Republic of Kiribati, *Porites* spp., 1893–1989 Cole JE, Fairbanks RG, Shen GT (1993) Recent variability in the Southern Oscillation: isotopic results from Tarawa Atoll coral. *Science* 260: 1790–1793; 4, Espiritu Santo, Vanuatu, *Platygyra lamellina*, 1806–1979 Quinn TM, Taylor FW, Crowley TJ (1993) A 173 year stable isotope record from a tropical south Pacific coral. *Quaternary Science Review* 12: 407–418; 5, Amedee Lighthouse, New Caledonia, *Porites lutea*, 1657–1992 Quinn TM, Crowley TJ, Taylor FW, Henin C, Joannot P, Join Y (1998) A multicentury stable isotope record from a New Caledonia coral: Interannual and decadal sea surface temperature variability in the southwest Pacific since 1657 A.D. *Paleoceanography* 13: 412–426; 6, Abraham Reef, Great Barrier Reef, Australia, *Porites australiensis*, 1635–1957 Druffel ERM, Griffin S (1993) Large variations of surface ocean radiocarbon: evidence of circulation changes in the southwestern Pacific. *Journal of Geophysical Research* 98: 20 249–22 259; 7, Cebu, Philippines, *Porites lobata*, 1859–1980 Pätzold J (1986) Temperature and CO_2 changes in the tropical surface waters of the Philippines during the past 120 years: record in the stable isotopes of hermatypic corals. Berichte Reports, Gol.-Paläont, Inst. Univ. Kiel, 12; 8, Mahe Island, Seychelles, *Porites lutea*, 1846–1995 Charles CD, Hunter ED, Fairbanks RG (1997) Interaction between the ENSO and the Asian monsoon in a coral record of tropical climate. *Science* 277: 925–928; 9, Aqaba, Red Sea, *Porites* sp., 1788–1992 Heiss GA (1996) Annual band width variation in *Porites* sp. from Aquaba, Gulf of Aquaba, Red Sea. *Bulletin of Marine Science* 59: 393–403; (Reproduced from Gagan (2000), with permission from Elsevier Science.)

effects may not have been uniform throughout the tropical oceans.

Second, this cool phase was followed by a general warming/freshening of the global tropical ocean beginning during the nineteenth century (**Figure 5**). This overall warming/freshening trend is observed in seven of the nine records. The timing of the onset of this warming/freshening is consistent with the onset of industrialization and the consequent increases in greenhouse gases due to increased emissions from fossil fuel consumption. If the shift in $\delta^{18}O$ were solely due to increases in SST, it would be equivalent to an increase of 0.3–2.0°C since 1800. Instrumental data indicate that the tropics only warmed by ~0.5°C since 1850. The influence of SSS on $\delta^{18}O$ is probably responsible for the difference and needs to be taken into account when interpreting $\delta^{18}O$ records. Although Sr/Ca ratios are thought to be unaffected by SSS, only a few shorter coral records are currently published. Until recently, Sr/Ca measurements were very time-consuming. With recently developed technology, the use of Sr/Ca as a paleothermometer proxy should increase. Two main limitations exist with the correct interpretation of decadal and long-term $\delta^{18}O$ trends: (1) an interdecadal cycle of unknown origin is commonly identified in long coral $\delta^{18}O$ records; and (2) long-term trends of increasing $\delta^{18}O$ are observed in some coral while other coral $\delta^{18}O$ records show a decreasing trend. Whether these trends are due to biological processes or are the result of gradual environmental changes (i.e., global warming) is unclear.

Finally, evidence of increased fossil fuel emission into the atmosphere can be seen in the general decrease in $\Delta^{14}C$ from 1850 to 1955 in shallow corals from the Atlantic and Pacific Oceans. This phenomenon, referred to as the Suess Effect, is mainly the result of ^{14}C-free CO_2 produced from combusted fossil fuel entering the atmosphere, the oceans and eventually, the coral skeleton (post-1950, coral $\Delta^{14}C$ values skyrocketed as a result of ^{14}C produced by thermonuclear bombs effectively swamping out the Suess effect).

Fossil Corals

Fossil corals provide windows into past climate. Records covering several decades to centuries offer the opportunity to compare the same three components (seasonal, interannual-to-decadal, and long term) in climate in the distant past to the present. A 3.0 million-year-old south-western Florida coral reveals a seasonal $\delta^{18}O$ derived temperature pattern similar to today but ~3.5°C cooler. A North Sulawesi, Indonesian coral indicates that 124 000 years BP the variability in ENSO was similar to modern ENSO frequency from 1856 to 1976. However, the shift in ENSO frequency observed in modern records after 1976 is not found in the fossil coral record nor in pre-1976 instrumental records. This suggests that the current state of ENSO frequency is outside of the natural range of ENSO variability. Perhaps anthropogenic effects are having an effect on ENSO frequency. Finally, long-term changes in climate can also be reconstructed from fossil coral records. A series of coral records from Vanuatu indicate that ~10 300 years BP the south-western tropical Pacific was 6.5°C cooler than today followed by a rapid rise in temperature over the subsequent 15 000 years. This rapid rise in temperature lags the post-Younger Dryas warming of the Atlantic by ~3000 years suggesting that the mechanism for deglacial climate change may not have been globally uniform. How seasonal, decadal (ENSO) and long-term climate changes varied in the distant past throughout the tropics can be addressed using fossil coral records and can offer us a better idea of the natural variability in tropical climate over geologic time.

Deep-sea Corals

Deep sea corals do deposit calcium carbonate exoskeleton but do not contain endosymbiotic zooxanthellae and are not colonial. Their isotopic and trace mineral composition reflects variation in ambient conditions on the seafloor. Although this does not directly reflect changes in climate on the surface, ocean circulation patterns are tightly coupled with atmospheric climatic conditions. Understanding the history of deep and intermediate water circulation lends itself to a better understanding of climate. For example, the origin of the Younger Dryas cooling event (13 000 to 11 700 years BP) has recently been attributed to a cessation or slowing of North Atlantic deep water formation and subsequent reduction in heat flux. Isotopic evidence from deep-sea corals suggests that profound changes in intermediate-water circulation also occurred during the Younger Dryas. Other studies of deep-sea corals show rapid changes in deep ocean circulation on decadal-to-centennial timescales at other intervals during the last deglaciation. Reconstructing intermediate and deep ocean circulation patterns and their relationship to climate using isotopic, trace element and minor element records in deep sea coral promises to be an expanding line of paleoclimate research.

Discussion

The geochemical composition of coral skeletons currently offers the only means of recovering multicentury records of seasonal-to-centennial timescale variation in tropical climate. $\delta^{18}O$-derived SST and SSS records are the workhorse of coral-based paleoclimate reconstructions to date. Improved methodologies are now making high resolution, multicentury Sr/Ca records feasible. Since Sr/Ca is potentially a less ambiguous SST recorder, coupling Sr/Ca with $\delta^{18}O$ records could yield more reliable SST and SSS reconstructions. $\delta^{13}C$ as a paleorecorder of seasonal variation in cloud cover and upwelling is also gaining credibility. However, more experimental research needs to be done before $\delta^{13}C$ records can be used more widely for paleoclimate reconstructions. Coral $\Delta^{14}C$ records are highly valued as an ocean circulation/ventilation proxy. Increasing numbers of high-resolution $\Delta^{14}C$ coral records are being published shedding invaluable new light on links between climate and ocean circulation processes. Coral trace and minor element records are also becoming more common and can add critical information about past upwelling regimes, wind patterns, pH, river discharge patterns, and SST.

The growing number of multicentury coral oxygen isotope records is yielding new information on the natural variability in tropical climate. Eastern equatorial Pacific corals track ENSO-related changes in SST and upwelling. Further west, coral records track ENSO-related changes in SST and SSS related to the displacement of rainfall associated with the Indonesian Low. Decadal timescale changes in ENSO frequency and in ocean circulation and water mass movement detected in $\delta^{18}O$ and $\Delta^{14}C$ records, respectively, indicate a major reorganization in Pacific climate at various intervals over time. Long-term trends in coral oxygen isotope records point to a gradual warming/freshening of the oceans over the past century suggesting that the tropics are responding to global forcings.

Although the coral-based paleoclimate records reconstructed to date are impressive, much work remains to be done. It is necessary to develop multiple tracer records from each coral record in order to establish a more comprehensive reconstruction of several concurrent climatic features. In addition, replication of long isotopic and elemental records from multiple sites is invaluable for establishing better signal precision and reproducibility. Coupled with fossil and deep-sea coral records, coral proxy records offer a comprehensive and effective means of reconstructing tropical paleoclimates.

Acknowledgments

I thank B Linsley, E Druffel, T Guilderson, J Adkins and an anonymous reviewer for their comments on the manuscript. I thank the Henry and Camille Dreyfus Foundation for financial support.

Further Reading

Beck JW, Recy J, Taylor F, Edwards RL, and Cabioch G (1997) Abrupt changes in early Holocene tropical sea surface temperature derived from coral records. *Nature* 385: 705–707.

Druffel ERM (1997) Geochemistry of corals: proxies of past ocean chemistry, ocean circulation, and climate. *Proceedings of the National Academy of Sciences of the USA* 94: 8354–8361.

Druffel ERM, Dunbar RB, Wellington GM, and Minnis SS (1990) Reef-building corals and identification of ENSO warming episodes. In: Glynn PW (ed.) *Global Ecological Consequences of the 1982–83 El Niño – Southern Oscillation pp. 233–253* Elsevier Oceanography, Series 52. New York: Elsevier.

Dunbar RB and Cole JE (1993) Coral records of ocean-atmosphere variability. *NOAA Climate and Global Change Program Special Report* No. 10, Boulder, CO: UCAR.

Dunbar RB and Cole JE (1999) *Annual Records of Tropical Systems (ARTS)*. Kauai ARTS Workshop, September 1996. Pages workshop report series 99–1.

Fairbanks RG, Evans MN, Rubenstone JL *et al.* (1997) Evaluating climate indices and their geochemical proxies measured in corals. *Coral Reefs* 16 suppl.: s93–s100.

Felis T, Pätzold J, Loya Y, and Wefer G (1998) Vertical water mass mixing and plankton blooms recorded in skeletal stable carbon isotopes of a Red Sea coral. *Journal of Geophysical Research* 103: 30731–30739.

Gagan MK, Ayliffe LK, Beck JW, *et al.* (2000) New views of tropical paleoclimates from corals. *Quaternary Science Reviews* 19: 45–64.

Grottoli AG (2000) Stable carbon isotopes ($\delta^{13}C$) in coral skeletons. *Oceanography* 13: 93–97.

Linsley BK, Ren L, Dunbar RB, and Howe SS (2000) El Niño Southern Oscillation (ENSO) and decadal-scale climate variability at 10°N in the eastern Pacific from 1893 to 1994: a coral-based reconstruction from Clipperton Atoll. *Paleoceanography* 15: 322–335. http:// pangea. stanford. edu/Oceans/ARTS/arts_report/arts_ report_ home. html

Hudson JH, Shinn EA, Halley RB, and Lidz B (1981) Sclerochronology: a tool for interpreting past environments. *Geology* 4: 361–364.

NOAA/NGCD Paleoclimatology Program http://www.ngdc. noaa.gov/paleo/corals.thml

Shen GT (1993) Reconstruction of El Niño history from reef corals. *Bull. Inst. fr. études andines* 22(1): 125–158.

Smith JE, Risk MJ, Schwarcz HP, and McConnaughey TA (1997) Rapid climate change in the North Atlantic during the Younger Dryas recorded by deep-sea corals. *Nature* 386: 818–820.

Swart PK (1983) Carbon and oxygen isotope fractionation in scleractinian corals: a review. *Earth-Science Reviews* 19: 51–80.

Weil SM, Buddemeier RW, Smith SV, and Kroopnick PM (1981) The stable isotopic composition of coral skeletons: control by environmental variables. *Geochimica et Cosmochimica Acta* 45: 1147–1153.

Wellington GM, Dunbar RB, and Merlen G (1996) Calibration of stable oxygen isotope signatures in Galapagos corals. *Paleoceanography* 11: 467–480.

PLIO-PLEISTOCENE GLACIAL CYCLES AND MILANKOVITCH VARIABILITY

K. H. Nisancioglu, Bjerknes Centre for Climate Research, University of Bergen, Bergen, Norway

Introduction

A tremendous amount of data on past climate has been collected from deep-sea sediment cores, ice cores, and terrestrial archives such as lake sediments. However, several of the most fundamental questions posed by this data remain unanswered. In particular, the Plio-Pleistocene glacial cycles which dominated climate during the past ~2.8 My have puzzled scientists. More often than not during this period large parts of North America and northern Europe were covered by massive ice sheets up to 3 km thick, which at regular intervals rapidly retreated, giving a sea level rise of as much as 120 m.

The prevalent theory is that these major fluctuations in global climate, associated with the glacial cycles, were caused by variations in insolation at critical latitudes and seasons. In particular, ice sheet growth and retreat is thought to be sensitive to high northern-latitude summer insolation as proposed by Milankovitch in his original astronomical theory.

Brief History of the Astronomical Theory

Long before the first astronomical theory of the ice ages, the people of northern Europe had been puzzled by the large erratic boulders scattered a long way from the Alpine mountains where they originated. Based on these observations the Swiss geologist and zoologist Louis Agassiz presented his ice age theory at a meeting of the Swiss Society of Natural Sciences in Neuchatel in 1837, where he claimed that the large boulders had been transported by Alpine glaciers covering most of Switzerland in a past ice age.

A few years later, the French mathematician Joseph Alphonse Adhemar was the first to suggest that the observed ice ages were controlled by variations in the Earth's orbit around the Sun. At this point it was known that there had been multiple glaciations, and Adhemar proposed that there had been alternating ice ages between the North and the South Pole following the precession of the equinoxes. Indeed, the winter is warmer when the Earth is at the point on its orbit closest to the Sun, and colder when the Earth is furthest from the Sun. Adhemar correctly deduced that the precession of the equinoxes had a period of approximately 21 000 years, giving alternating cold and warm winters in the two hemispheres every 10 500 years.

In 1864, James Croll expanded on the work by Adhemar and described the influence of changing eccentricity on the precession of the equinoxes. He assumed that winter insolation controlled glacial advances and retreats, and determined that the precession of the equinoxes played an important role in regulating the amount of insolation received during winter. Based on this, he estimated that the last ice age lasted from about 240 000 to 80 000 years ago. Croll was aware of the fact that the amplitude of the variations in insolation was relatively small, and introduced the concept of positive feedbacks due to changing surface snow and ice cover as well as changes in atmosphere and ocean circulation.

In parallel to the work of Croll, geologists in Europe and America found evidence of multiple glacial phases separated by interglacial periods with milder climate similar to that of the present day, or even warmer. These periodic glaciations were consistent with Croll's astronomical theory. However, most geologists abandoned his theory after mounting evidence from varved lake sediments in Scandinavia and North America showed that the last glacial period ended as late as 15 000 years ago, and not 80 000 years ago as suggested by Croll.

Milankovitch's Astronomical Theory of the Ice Ages

Following Croll's astronomical theory there was a period where scientists such as Chamberlin and Arrhenius tried to explain the ice ages by natural variations in the atmospheric content of carbon dioxide. The focus of the scientific community on an astronomical cause of the ice ages was not renewed until the publication of Milankovitch's theory in a textbook on climate by the well-known geologists Wladimir Köppen and Alfred Wegener in 1924. This was the first comprehensive astronomical theory of the Pleistocene glacial cycles, including detailed calculations of the orbitally induced changes in insolation.

Milutin Milankovitch was of Serbian origin, born in 1879. He obtained his PhD in Vienna in 1904 and was later appointed Professor of Applied Mathematics at the University of Belgrade. He was captured during World War I, but allowed to work at the Hungarian Academy of Sciences, where he completed his calculation of the variations of the orbital parameters of the Earth and their impact on insolation and climate. Milankovitch's basic idea was that at times of reduced summer insolation, snow and ice could persist at high latitudes through the summer melt season. At the same time, the cool summer seasons were accompanied by mild winter seasons leading to enhanced winter accumulation of snow. When combined, reduced summer melt and a slight increase in winter accumulation, enhanced by a positive snow albedo feedback, could eventually lead to full glacial conditions.

During World War II, Milankovitch worked on a complete revision of his astronomical theory which was published as the *Kanon der Erdbestrahlung* in 1941. However, the scientific establishment was critical of Milankovitch, and his theory was largely rejected until the early 1970s. By this time, great advances in sediment coring, deep-sea drilling, and dating techniques had made it possible to recover climate records covering the last 500 000 years. By studying the variations in oxygen isotopes of foraminifera in deep-sea sediment cores as well as by reconstructing past sea level from terraces of fossil coral reefs, new support was emerging for an astronomical phasing of the glacial cycles. The oxygen isotope data from the long deep-sea cores presented in a paper in 1976 by Hays *et al.* were considered as proof of the Milankovitch theory, as they showed cycles with lengths of roughly 20 000 and 40 000 years as well as 100 000 years in agreement with Milankovitch's original calculations.

Orbital Parameters and Insolation

The Earth's orbit around the Sun is an ellipse where the degree to which the orbit departs from a circle is measured by its eccentricity (e). The point on the orbit closest to the Sun is called the perihelion, and the point most distant from the Sun the aphelion (**Figure 1**). If the distance from the Earth to the Sun is r_p at perihelion, and r_a at aphelion, then the eccentricity is defined as $e = (r_a - r_p)/(r_a + r_p)$.

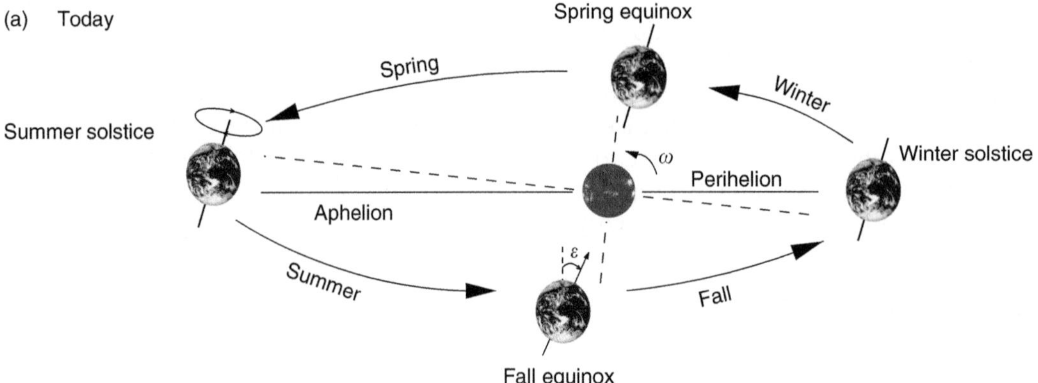

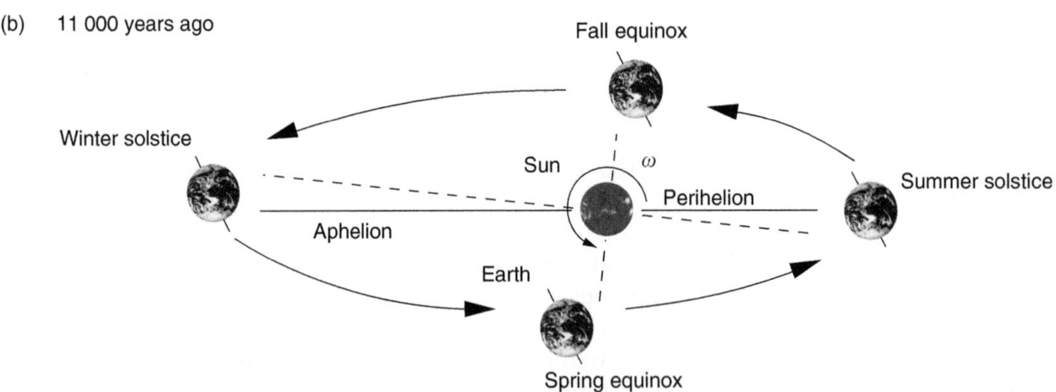

Figure 1 Sketch of the Earth's orbit around the Sun today and at the end of the last glacial cycle (11 000 years ago), showing the positions of the solstices and equinoxes relative to perihelion. The longitude of perihelion (ω) is measured as the angle between the line to the Earth from the Sun at spring equinox and the line to the Earth at perihelion.

Variations in the eccentricity of the Earth's orbit follow cycles of 100 000 and 400 000 years giving a change in annual mean insolation on the order of 0.2% or less. This change in insolation is believed to be too small to produce any notable effect on climate.

A more significant change in insolation is caused by variations in the seasonal and latitudinal distribution of insolation due to obliquity. Obliquity (ε) is the angle between Earth's axis of rotation and the normal to the Earth's plane around the Sun (**Figure 1**). This angle is 23.5° today, but varies between values of 22.1° and 24.5° with a period of 41 000 years. A decrease in obliquity decreases the seasonal insolation contrast, with the largest impact at high latitudes. At the same time, annual mean insolation at high latitudes is decreased compared to low latitudes. An example of the effect of obliquity variations on seasonal insolation is shown in **Figure 2(a)**. During times when obliquity is small, high-latitude summertime insolation decreases, whereas mid-latitude wintertime insolation increases. The magnitude of the change in high-latitude summer insolation due to obliquity variations can be as large as 10%.

The third and last variable affecting insolation is the longitude of perihelion (ω). This parameter is defined as the angle between the line to the Earth from the Sun at spring equinox and the line to the Earth at perihelion (**Figure 1**). It determines the direction of the Earth's rotational axis relative to the orientation of the Earth's orbit around the Sun, thereby giving the position of the seasons on the orbit relative to perihelion. Changes in the longitude of perihelion result in the Earth being closest to the Sun at different times of the year. Today, the Earth is closest to the Sun in early January, or very near winter solstice in the Northern Hemisphere. All other things being equal, this will result in relatively warm winter and cool summer seasons in the Northern Hemisphere, whereas the opposite is the case in the Southern Hemisphere. At the time of the last deglaciation, 11 000 years ago the Earth was closest to the Sun at summer solstice, resulting in extra warm summers and cool winters in the Northern Hemisphere. An example of the effect of changes in precession on seasonal insolation is shown in **Figure 2(b)**.

If the Earth's orbit were a circle, the distance to the Sun would remain constant at all times of the year and it would not make any difference where on the orbit the seasons were positioned. Therefore, the impact of variations in the longitude of perihelion depends on the eccentricity of the Earth's orbit and is described by

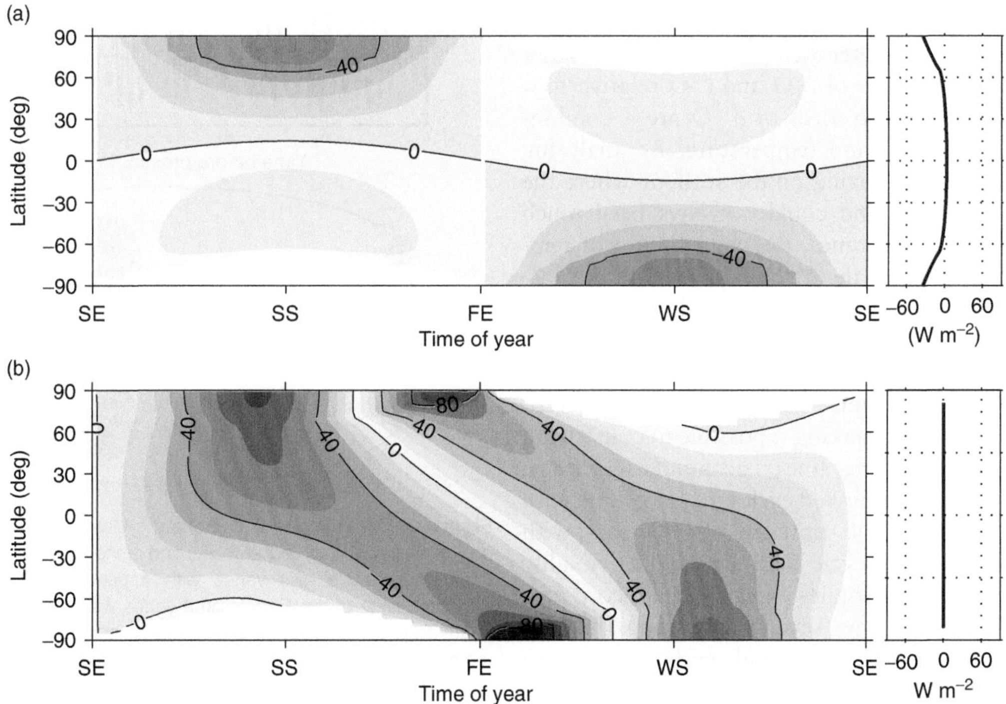

Figure 2 Insolation difference in units of W m^{-2} as a function of latitude and season: (a) when decreasing obliquity from 24.5° to 22° in the case of a perfectly circular orbit ($e=0$); and (b) for a change in precession going from summer solstice at perihelion to summer solstice at aphelion while keeping obliquity at today's value ($\varepsilon=23.5°$) and using a mean value for eccentricity ($e=0.03$). The annual mean insolation difference is shown to the right of each figure and the seasons are defined as follows: FE, fall equinox; SE, spring equinox; SS, summer solstice; WS, winter solstice.

the precession parameter ($e \sin \omega$). The combined effect of eccentricity and longitude of perihelion can give changes in high-latitude summer insolation on the order of 15% and varies with periods of 19 000 and 23 000 years, but is modulated by the longer-period variations in eccentricity. **Figure 3** shows the variations in obliquity (ε), eccentricity (e), and the precession parameter ($e \sin \omega$).

Plio-Pleistocene Glacial Cycles

Some of the longest continuous records of past climate come from deep-sea sediment cores. Ocean sediments are laid down over time, and by drilling into the seafloor, layered sediment cores can be extracted containing valuable information about the conditions at the time when the layers were formed. By studying the relative abundance of oxygen isotopes in shells of tiny marine organisms (foraminifera) found in the sediments, it is possible to estimate the amount of water tied up in the continental ice sheets and glaciers. This is because water molecules containing the lighter isotope of oxygen (^{16}O) are more readily evaporated and transported from the oceans to be deposited as ice on land. Thus, leaving the ocean water enriched with the heavy oxygen isotope (^{18}O) during glacial periods. However, the fractionation of the oxygen isotopes when forming the shells of the foraminifera also depends on the surrounding water temperature: low water temperature gives higher δ^{18}O values (the ratio of ^{18}O and ^{16}O relative to a standard). Therefore records of δ^{18}O are a combination of ice volume and temperature. By analyzing benthic foraminifera living on the seafloor where the ocean is very cold, and could not have been much colder during glacial times, the contribution of temperature variations to the δ^{18}O value is reduced.

The benthic δ^{18}O ice volume record of Hays *et al.* from 1976 was one of the very first continuous records of the late Pleistocene extending back to the Brunhes–Matuyama magnetic reversal event (780 000 years ago), making it possible to construct a timescale by assuming linear accumulation rates. Analysis of the data showed cycles in ice volume with periods of about 20 000 years and 40 000 years, with a particularly strong cycle with a period of roughly 100 000 years. Later studies extended the record past the Brunhes–Matuyama reversal, showing that the late Pliocene (3.6–1.8 Ma) and early Pleistocene records (1.8–0.8 Ma) were dominated by smaller-amplitude cycles with a period of 41 000 years, rather than the large 100 000 years cycles of the late Pleistocene (0.8–0 Ma).

Many records generated since this time have confirmed these early observations, namely:

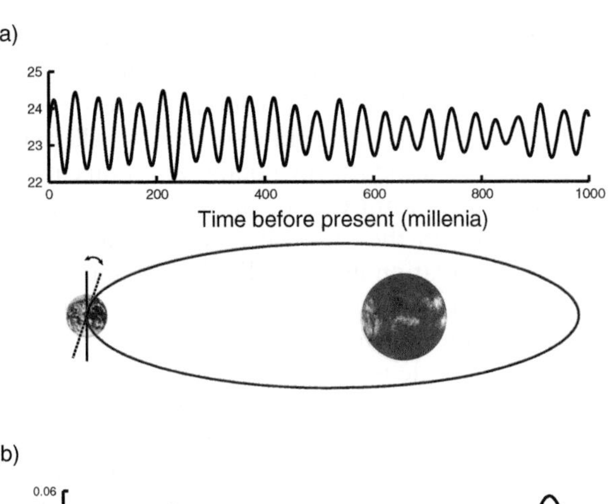

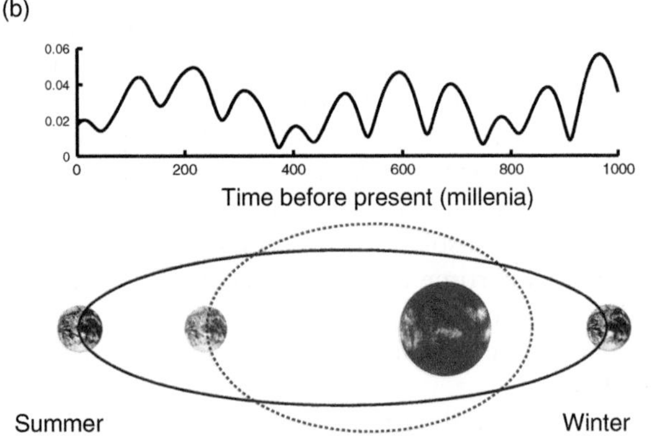

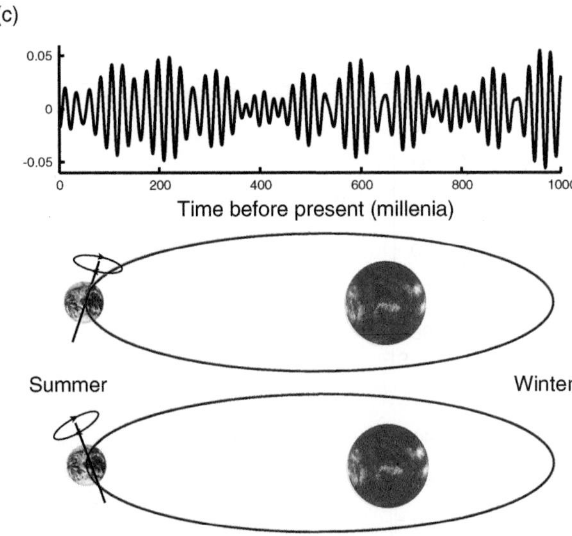

Figure 3 The three most important cycles regulating insolation on Earth are obliquity, eccentricity, and precession: (a) obliquity, or tilt of the Earth's axis varies with a period of 41 000 years; (b) eccentricity of the Earth's orbit varies with periods of 100 000 years and 400 000 years; and (c) precession of the equinoxes has a dominant period of 21 000 years and is modulated by eccentricity.

1. from about 3 to 0.8 Ma, the main period of ice volume change was 41 000 years, which is the dominant period of orbital obliquity;
2. after about 0.8 Ma, ice sheets varied with a period of roughly 100 000 years and the amplitude of

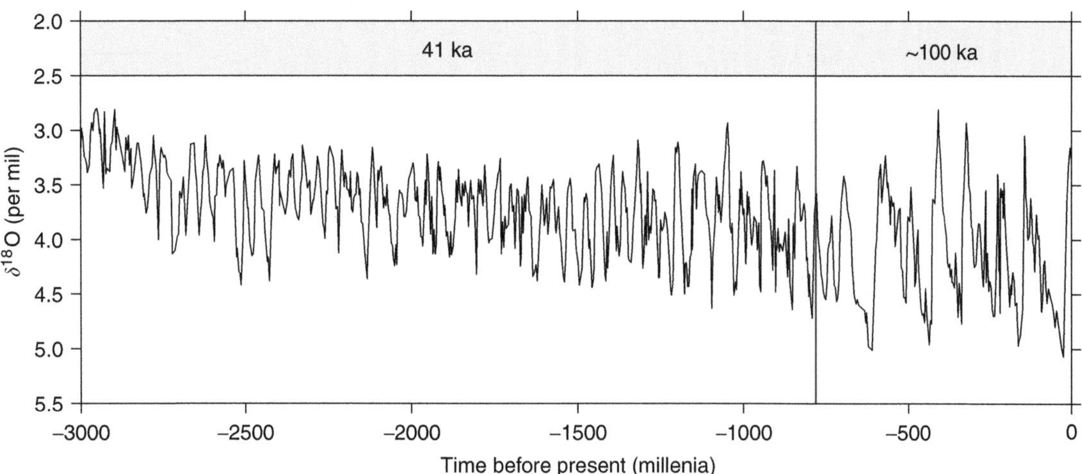

Figure 4 Benthic foraminiferal δ^{18}O ice volume record from the North Atlantic plotted to a paleomagnetic timescale covering the last 3 My. The transition from a dominant 41 000 to a 100 000-year periodicity in ice volume occurs close to the Brunhes–Matuyama magnetic reversal event ($\sim$780 000 years ago).

oscillations in δ^{18}O increases, implying growth of larger ice sheets.

The long benthic δ^{18}O record from a deep-sea sediment core extracted from the North Atlantic shown in **Figure 4** illustrates both of these points. This isotope record is plotted with a paleomagnetic timescale determined by the depth of magnetic field reversals recorded by ferromagnetic grains in the sediment core. Using this simple timescale, which is not biased by orbital tuning, one can clearly observe the 41 000-year periodicity of the late Pliocene and early Pleistocene (3.0–0.8 My), as well as the dominance of the stronger $\sim$100 000-year periodicity of the late Pleistocene (last 800 000 years).

Note that the main periods of orbital precession (19 000 and 23 000 years) are of less importance in the benthic ice volume record, whereas it is known that they increase in strength after about 800 000 years (the mid-Pleistocene transition). The lack of an imprint from orbital precession in the early part of the record and the reason for the dominance of roughly 100 000 years periodicity in the recent part of the record are some of the major unanswered questions in the field.

Only eccentricity varies with periods matching the roughly 100 000 years periods observed in the late Pleistocene. Although eccentricity is the only orbital parameter which changes the annual mean global insolation received on Earth, it has a very small impact. This was known to Croll and Milankovitch, who saw little direct importance in variations in eccentricity and assumed that changes in precession and obliquity would dominate climate by varying the amount of seasonal, rather than annual mean insolation received at high latitudes. Milankovitch postulated that the total amount of energy received from the Sun during the summer at high northern latitudes is most important for controlling the growth and melt of ice. To calculate this insolation energy, he divided the year into two time periods of equal duration, where each day of the summer season received more insolation than any day of the winter season. The seasons following these requirements were defined as the caloric summer and caloric winter half-years. These caloric half-years are of equal duration through time and the amount of insolation energy received in each can be compared from year to year.

For Milankovitch's caloric summer half-year insolation (**Figure 5**),obliquity (ε) dominates at high latitudes ($>65°$ N), whereas climatic precession ($e \sin \omega$) dominates at low latitudes ($<55°$ N). In the mid-latitudes ($\sim 55 - 65°$ N), the contribution by obliquity and climatic precession are of similar magnitude. In the Southern Hemisphere, variations in caloric half-year insolation due to obliquity are in phase with the Northern Hemisphere and could potentially amplify the global signal, whereas variations due to climatic precession are out of phase. By taking into account the positive snow albedo feedback, Milankovitch used his caloric insolation curves to reconstruct the maximum extent of the glacial ice sheets back in time (**Figure 6**).

Milankovitch's predicted cold periods occurred roughly every 40 000–80 000 years, which fit reasonably well with the glacial advances known to geologists at that time. However, as the marine sediment core data improved, it became clear that the last several glacial periods were longer and had a preferred period of roughly 100 000 years (**Figure 4**), which was not consistent with Milankovitch's original predictions. Based on these observations, and without knowledge of the 41 000 years cycles of the

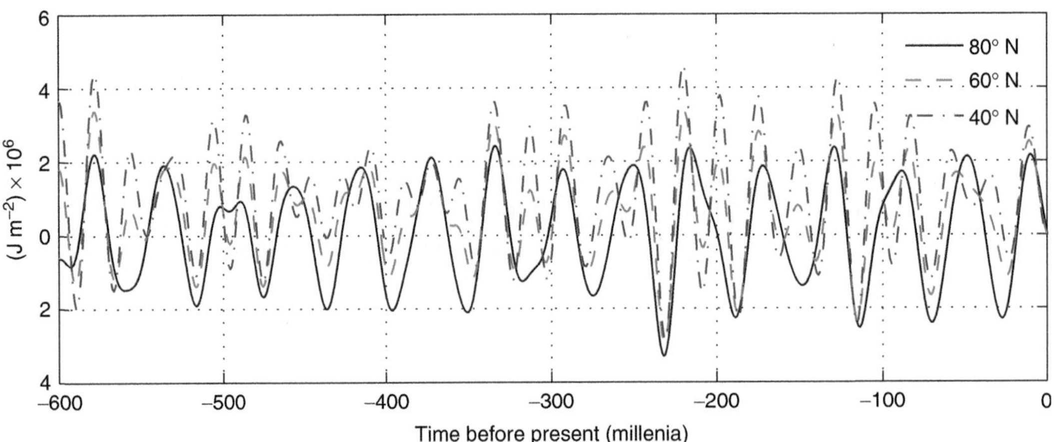

Figure 5 Caloric summer half-year insolation following Milankovitch's definition plotted for the latitudes of 40° N, 60° N, and 80° N. Caloric half-years are periods of equal duration where each day of the summer half-year receives more insolation than any day of the winter half-year.

early Pliocene and late Pleistocene, scientists reasoned that climatic precession and its modulation by eccentricity must play the leading role in past climates. Following Andre Berger and others, who recalculated and improved the records of orbital insolation, most researchers replaced the caloric half-year insolation as a driver of glacial climate by mid-month, or monthly mean insolation, for example, June or July at 65° N (**Figure** 7).

As can be seen from **Figure** 7, monthly mean insolation is dominated by precession. As insolation time series at a given time of the year (e.g., June or July) are in phase across all latitudes of the same hemisphere, the proxy records could be compared equally well with insolation from other latitudes than the typical choice of 65° N shown here. This means that any direct response of climate at high latitudes to monthly or daily insolation requires a strong presence of precession in the geologic record. Although both the frequencies of precession and obliquity are clearly found in the proxy records, a simple linear relationship between summer insolation and glacial cycles is not possible. This is particularly true for the main terminations spaced at roughly 100 000 years, which must involve strongly nonlinear mechanisms.

The strong positive feedback on global climate caused by greenhouse gases, such as CO_2, was pointed out as early as 1896 by the Swedish physical chemist Svante Arrhenius. Shortly thereafter, the American geologist Thomas Chamberlin suggested a possible link between changing levels of CO_2 and glacial cycles. From the long ice cores extracted from Antarctica, covering the last 740 000 years, it is now known that atmospheric levels of CO_2 closely follow the glacial temperature record (**Figure** 8). Although greenhouse gases, such as CO_2, cannot explain the timing and rapidity of glacial terminations, the changing levels of atmospheric greenhouse gases clearly contributed by amplifying the temperature changes observed during the glacial cycles.

Modeling the Glacial Cycles

Following the discovery of the orbital periods in the proxy records, a considerable effort has gone into modeling and understanding the physical mechanisms involved in the climate system's response to variations in insolation and changes in the orbital parameters. In this work, which requires modeling climate on orbital timescales ($>$10 000 years), the typical general circulation models (GCMs) used for studying modern climate and the impact of future changes in greenhouse gases require too much computing power. These GCMs can be used for simulations covering a few thousand years at most, but provide valuable equilibrium simulations of the past climates, such as the Last Glacial Maximum (LGM).

Instead of the GCMs, it has been common to use Energy Balance Models (EBMs) to study changes in climate on orbital timescales. These types of models can be grouped into four categories: (1) annual mean atmospheric models; (2) seasonal atmospheric models with a mixed layer ocean; (3) Northern Hemisphere ice sheet models; and (4) coupled climate–ice sheet models, which in some cases include a representation of the deep ocean.

Studies with the first type of simple climate models were pioneered by the early work of Budyko in the 1960s, who investigated the sensitivity of climate to changes in global annual mean insolation. However, changes in the Earth's orbital parameters result in a

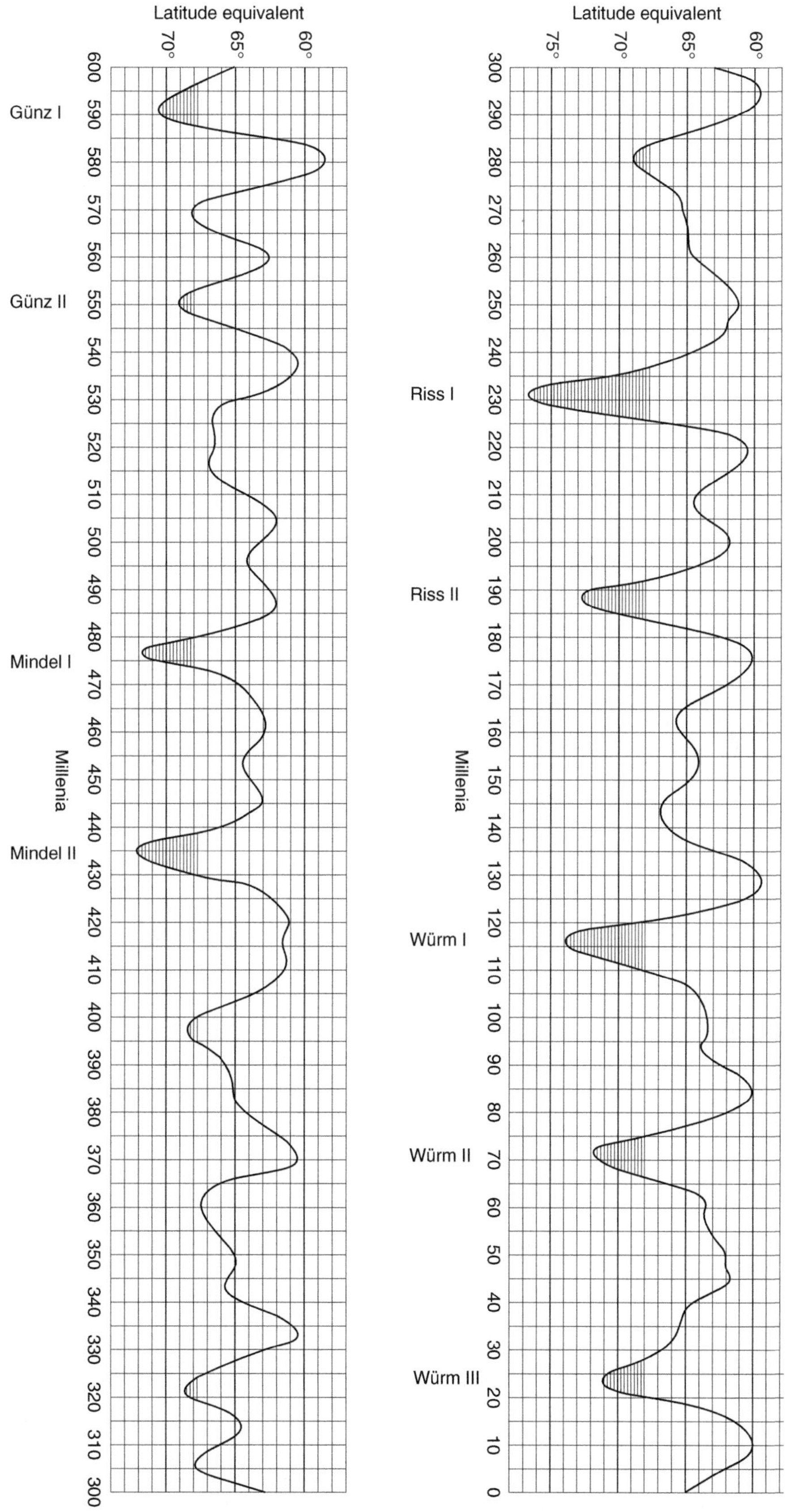

Figure 6 Milankovitch's reconstructed maximum glacial ice extent for the past 600 000 years. From Milankovitch M (1998) *Canon of Insolation and the Ice-Age Problem* (orig. publ. 1941). Belgrade: Zavod za Udzbenike I Nastavna Sredstva.

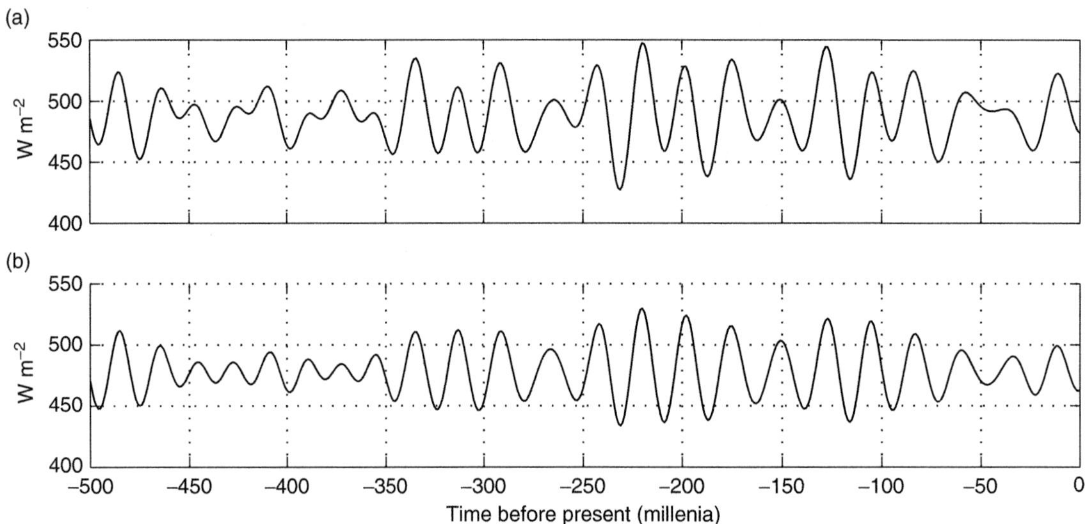

Figure 7 Summer solstice insolation at (a) 65° N and (b) 25° N for the past 500 000 years.

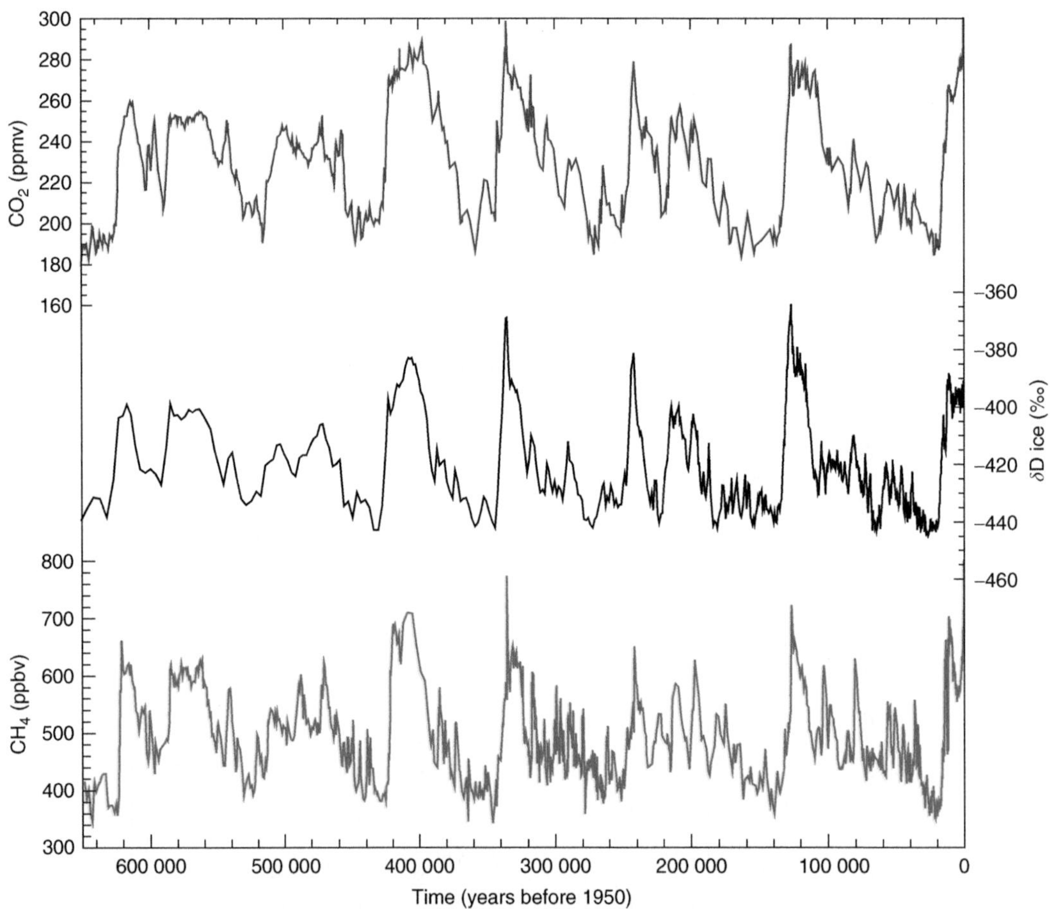

Figure 8 Variations in deuterium (δD; black), a proxy for local temperature, and atmospheric concentrations of the greenhouse gases carbon dioxide (CO_2; red), and methane (CH_4; green), from measurements of air trapped within Antarctic ice cores. Data from Spahni R, Chappellaz J, Stocker TF, *et al.* (2005) Atmospheric methane and nitrous oxide of the late Pleistocene from Antarctica ice cores. *Science* 310: 1317–1321 and Siegenthaler U, Stocker TF, Monnin E, *et al.* (2005) Stable carbon cycle-climate relationship during the late Pleistocene. *Science* 310: 1313–1317.

redistribution of insolation with latitude and time of the year, with a negligible impact on global annual mean insolation. Therefore, annual mean models are not adequate when investigating the impact of orbital insolation on climate, as they cannot capture the parts of the insolation variations which are seasonal and translate them into long-term climate change.

The second type of models includes a representation of the seasonal cycle, and has been used to investigate the orbital theory of Milankovitch. In this case, the seasonal variations in orbital insolation are resolved. However, as for the first type of models, past changes in ice cover are assumed to follow the simulated variations in the extent of perennial snow. This approach assumes that ice cover and the powerful ice albedo feedback are governed only by temperature, as the extent of snow in these models is fixed to the latitude with a temperature of 0 °C. In reality, the growth and decay of land-based ice sheets are governed by the balance of accumulation and ablation. Therefore, when investigating changes in ice cover, it is necessary to include an appropriate representation of the dynamics and mass balance of ice sheets in the model.

The third type of models improves upon this by focusing on modeling past changes in mass balance and size of typical Northern Hemisphere ice sheets, such as the Laurentide. This type of studies was initiated by Weertman in the 1960s who used simple ice sheet models, forced by a prescribed distribution of accumulation minus ablation, to predict ice thickness versus latitude. These models do not calculate the atmospheric energy balance in order to estimate snowfall and surface melt; instead, changes to the prescribed distribution of net accumulation follow variations in mean summer insolation.

The fourth type of models include zonal mean seasonal climate models coupled to the simple Weertman-type ice sheet model, as well as earth models of intermediate complexity (EMICs) coupled to a dynamic ice sheet. These models give a more realistic representation of the climate as compared with the simpler models.

Partly due to the lack of good data on variations in global ice volume older than about half a million years, most model studies have focused on understanding the more recent records dominated by the ~100 000 years glacial cycles. All of these models respond with periods close to the precession and obliquity periods of the insolation forcing. However, the amplitude of the response is in most cases significantly smaller than what is observed in the proxy records. At the same time, the dominant ~100 000 year cycles of the ice volume record, characterized by rapid deglaciations, are only found when including a time lag in the response of the model. Such an internal time lag can be produced by taking into account bedrock depression under the load of the ice, or by adding a parametrization of ice calving into proglacial lakes, or marine incursions at the margin of the ice sheet. Alternatively, the ~100 000-year cycles have been explained as free, self-sustained oscillations, which might be phase-locked to oscillations in orbital insolation.

One of the very few model studies that have investigated variations in ice volume before the late Pleistocene transition (~800 000 years ago) used a two-dimensional climate model developed at Louvain-la-Neuve in Belgium. It falls within the definition of an EMIC and includes a simple atmosphere coupled to a mixed layer ocean, sea ice and ice sheets. By forcing this model with insolation and steadily decreasing atmospheric CO_2 concentrations, the model reproduces some of the characteristics of the ice volume record. The 41 000-year periodicity is present in the simulated ice volume for most of the past 3 My and the strength of the 100 000-year signal increases after about 1 My. However, a longer 400 000-year year period is also present and often dominates the simulated Northern Hemisphere ice volume record.

This nicely illustrates the remaining questions in the field. It is expected that models responding to the 100 000-year period will also respond to the longer 400 000-year period of eccentricity. However, this later period is not present in the ice volume record. At the same time, the late Pleistocene transition from a dominance of 41 to ~100 000-year period oscillations in ice volume is not well understood. Explanations for the transition which have been tested in models are: a steady decrease in CO_2 forcing and its associated slow global cooling; or a shift from a soft to a hard sediment bed underlying the North American ice sheet through glacial erosion and exposure of unweathered bedrock. Neither of these changes are in themselves abrupt, but could cause a transition in the response of the ice sheets to insolation as the ice sheets grew to a sufficiently large size.

In addition to the challenge of modeling the mid-Pleistocene transition, no model has successfully reproduced the relatively clean 41 000-year cycles preceding the transition. Following the transition, the models only exhibit a good match with the observed glacial cycles when forced with reconstructed CO_2 from Antarctic ice cores together with orbital insolation.

Summary

The Plio-Pleistocene glacial cycles represent some of the largest and most significant changes in past climate, with a clear imprint in terrestrial and marine proxy records. Many of the physical mechanisms driving these large cycles in ice volume are not well understood. However, the pursuit to explain these climate changes has greatly advanced our understanding of the climate system and its future response to man-made forcing. New and better resolved proxy records will improve our spatial and temporal picture of the glacial cycles. Together with the advent of comprehensive climate models able to simulate longer periods of the glacial record, scientists will be able to better resolve the interaction of the atmosphere, ocean, biosphere, and ice sheets and the mechanisms linking them to the astronomical forcing.

Nomenclature

e	eccentricity
r_a	aphelion
r_p	perihelion
$\delta^{18}O$	oxygen isotope ratio (ratio of ^{18}O and ^{16}O relative to a standard)
ε	obliquity
ω	longitude of perihelion

See also

Oxygen Isotopes in the Ocean. Stable Carbon Isotope Variations in the Ocean.

Further Reading

Bard E (2004) Greenhouse effect and ice ages: Historical perspective. *Comptes Rendus Geoscience* 336: 603–638.

Berger A, Li XS, and Loutre MF (1999) Modelling Northern Hemisphere ice volume over the last 3 Ma. *Quaternary Science Reviews* 18: 1–11.

Budyko MI (1969) The effect of solar radiation variations on the climate of the Earth. *Tellus* 5: 611–619.

Crowley TJ and North GR (1991) *Paleoclimatology*. New York: Oxford University Press.

Hays JD, Imbrie J, and Shackleton NJ (1976) Variations in the Earth's orbit: Pacemakers of the ice ages. *Science* 194: 1121–1132.

Imbrie J and Imbrie KP (1979) *Ice Ages, Solving the Mystery*. Cambridge, MA: Harvard University Press.

Köppen W and Wegener A (1924) *Die Klimate Der Geologischen Vorzeit*. Berlin: Gebrüder Borntraeger.

Milankovitch M (1998) *Canon of Insolation and the Ice-Age Problem* (orig. publ. 1941). Belgrade: Zavod za Udzbenike I Nastavna Sredstva.

Paillard D (2001) Glacial cycles: Toward a new paradigm. *Reviews of Geophysics* 39: 325–346.

Saltzman B (2002) *Dynamical Paleoclimatology*. San Diego, CA: Academic Press.

Siegenthaler U, Stocker TF, Monnin E, *et al.* (2005) Stable carbon cycle–climate relationship during the late Pleistocene. *Science* 310: 1313–1317.

Spahni R, Chappellaz J, Stocker TF, *et al.* (2005) Atmospheric methane and nitrous oxide of the late Pleistocene from Antarctica ice cores. *Science* 310: 1317–1321.

Weertman J (1976) Milankovitch solar radiation variations and ice age ice sheet sizes. *Nature* 261: 17–20.

COASTAL OCEAN

CHEMICAL PROCESSES IN ESTUARINE SEDIMENTS

W. R. Martin, Woods Hole Oceanographic Institution, Woods Hole, MA, USA

Introduction

The physical and chemical environments of estuarine sediments vary over wide ranges. The sediments may lie beneath nearly fresh water or may have salinities near that of the open ocean. They may be intertidal or always under several meters of water. They may be covered by grasses, by algal or bacterial mats, or may be free of surface biological cover. They may be permeable sands, affected by groundwater flow, or may be impermeable muds. Nonetheless, common features determine their chemical environments. They exist in productive, coastal waters and therefore experience large fluxes of particulate organic matter. They experience large inputs of terrestrial material that supply Fe and Mn oxides. Most estuarine sediments lie underneath oxic water columns, and support abundant macrofauna that both mix sedimentary particles and irrigate the seabed. Common features such as oxic bottom water, abundant supplies of organic matter, Fe and Mn oxides, a supply of sulfur from seawater (except for freshwater end members of estuarine systems), and active macrofauna produce the set of biogeochemical processes that determines the chemical environment and its effects on carbon, nutrient, and contaminant cycling in the coastal ocean.

The upper sediment column in estuarine and coastal environments can be regarded as a slowly stirred reactor to which substrates are added at the top and mixed downward. Transport processes, both in the dissolved and solid phases, distribute reactants within the reactor, while at the same time biogeochemical processes alter the particles and the pore waters surrounding them. This combination of transport and reaction drives solute exchanges between the sediments and overlying waters that are important to coastal carbon, nutrient, and contaminant cycles. It transforms the 'reactive' component of sedimentary particles so that the accumulating sediments differ in important ways from the particles that fall through the water column to the seafloor. These processes are outlined in **Figure 1**.

Because the rates of chemical reactions in this system are faster than mixing rates, the reactor is not chemically uniform; rather, roughly speaking, it has a layered structure. In all but a few locations, there is a thin, oxic layer at the sediment surface. As particles pass through this layer and through the underlying anoxic sediments, they are altered by heterotrophic respiration: organic matter is gradually oxidized, and a series of oxidants – O_2, N(V) in NO_3^-, Mn(IV) and Fe(III) in particulate oxides, and S(VI) in SO_4^{2-} – are consumed. Some of the energy originally stored in organic matter remains in the reduced products of these reactions: Fe(II), and S(–II), and, to some extent, Mn(II). This energy is available to chemolithotrophic bacteria, and the result is rapid internal cycling between oxidized and reduced forms of Mn, Fe, and S. Reduced Mn, Fe, and S that escape bacterial oxidation may be oxidized abiotically by O_2 in the surface oxic layer. The result of these processes is sediments with an oxic 'cap', in which precipitation of newly formed Fe and Mn oxides may limit benthic/water column exchange of many solutes, and an underlying anoxic layer in which rapid cycling between oxidized and reduced forms limits the rate of burial of products of heterotrophic respiration. The chemical processes defining this system are oxidation–reduction reactions, including 'heterotrophic respiration', by which organic matter is oxidized, 'chemolithotrophic respiration', most importantly oxidizing Fe(II) and S(–II), and 'abiotic oxidation'; and 'authigenic mineral formation', which converts dissolved products of these reactions to solid phases. The stirring within the sedimentary reactor, which is vital to its operation, is accomplished by solute diffusion and particle mixing, accomplished predominantly by the feeding and burrowing activities of sedimentary fauna.

In the following paragraphs, oxidation/reduction reactions in sediments and authigenic mineral formation are discussed. The effects of these reactions on the sedimentary environment are illustrated using profiles of solutes in sedimentary pore waters at a near-shore location. Then, there is a brief discussion of the role that sedimentary chemical processes play in nutrient cycles and the cycling of anthropogenic contaminants.

It is important to note that this article is not a comprehensive review, but instead focuses on the range of chemical reactions occurring in coastal

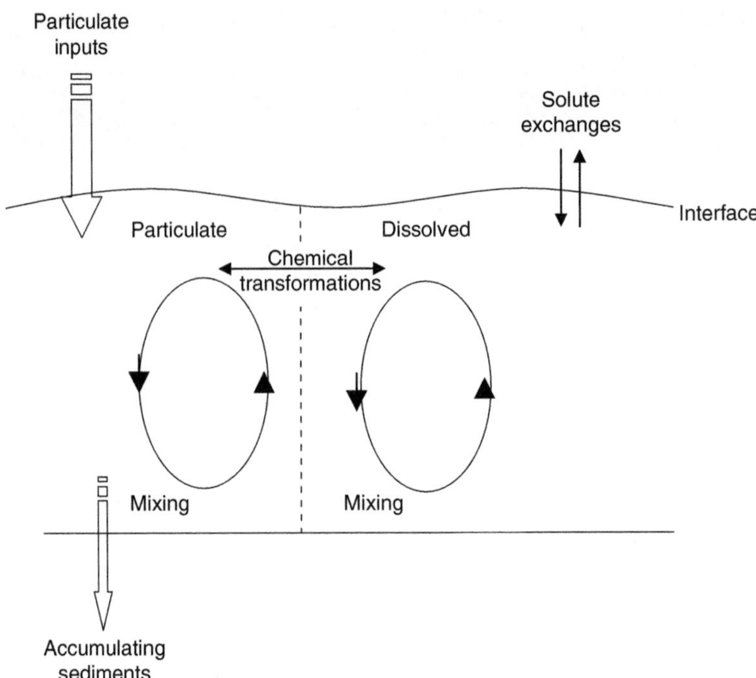

Figure 1 Biogeochemical cycles in estuarine and coastal sediments. Particles that fall to the sediment–water interface simultaneously undergo chemical reactions and are mixed through the upper few centimeters of the sediment column. Chemical transformations alter the particles and the sedimentary pore waters surrounding them, driving solute exchange between sediments and the water column and determining the composition of accumulating sediments.

sediments and their role in creating sedimentary environments that are important in coastal carbon, nutrient, and contaminant cycles. Therefore, a single type of coastal sediment – muddy, impermeable sediments underlying oxic bottom water – is discussed. Other sediment types are important: for example, permeable, sandy sediments are widespread, and sediments underlying permanently or seasonally anoxic bottom water can be important. In some high-energy areas, the effects of sediment resuspension can dominate sediment–water-column exchange. Although this article emphasizes sediments as sites of respiration, primary production at the surface of sediments in shallow waters can be important. There are important differences between sedimentary environments that determine their involvement in coastal biogeochemical processes. Nonetheless, similar chemical processes occur in all of these environments. The goal of this article is to illustrate these fundamental chemical processes, so that their role in a range of sedimentary environments can be examined.

Oxidation/Reduction Reactions

The distinctive chemical environment of estuarine and coastal sediments is determined to a large extent by electron transfer, or 'redox' reactions, in which a chemical species with a greater affinity for electrons accepts these carriers of negative charge from a species with a lesser affinity. In the process, energy is released. Although these reactions are often slow, they are catalyzed by organisms – in particular, microbes – that can harness the released energy. To understand the chemical environment of sediments, it is necessary to know which redox reactions will occur.

Oxidants and Reductants

In chemical terms, we can consider some constituents of sediments as oxidants, capable of accepting electrons under conditions extant in the sediments, and reductants, capable of donating them. The primary sources of reductants are the organic matter, formed by production in the water column or on the sediment surface or by anthropogenic activities, that reaches the sediments, and the reduced products of sedimentary reactions. Oxidants diffuse from bottom water (O_2, NO_3^-, and SO_4^{2-}), arrive to the sediments with the particulate flux (Fe and Mn oxides), or are formed by sedimentary processes (NO_3^- and authigenic Fe and Mn oxides). The principal electron donors and acceptors are listed in **Table 1**.

pe: A Measure of the Affinity to Accept Electrons

Free electrons do not exist in aqueous solution. Nonetheless, since each electron acceptor (or donor)

Table 1 Oxidants and reductants in shallow-water sediments

e^- *acceptor/donor*	*Formal oxidation state*	*Chemical form*	*Main sources*
Oxidants			
O	0	Dissolved O_2	Solute transport from bottom water
N	+V	NO_3^-	Sedimentary nitrification
Mn	+IV, +III	Mn oxyhydroxides	Terrigenous material Cycling within sediments
Fe	+III	Fe oxyhydroxides	Terrigenous material Cycling within sediments
S	+VI	SO_4^{2-}	Solute transport from bottom water Cycling within sediments
Reductants			
C	A range of oxidation states	Organic matter	Marine production and anthropogenic
N			
S			
Mn	+II	Dissolved Mn^{2+}	Product of microbial respiration
Fe	+II	Dissolved Fe^{2+} FeS	Product of microbial respiration
S	−II	Dissolved HS^- FeS	Product of microbial respiration

can, in principle, react with several different electron donors (or acceptors), it is useful to consider 'half-reactions' of the sort

$$\tfrac{1}{4}O_2 + H^+ + e^- \Leftrightarrow \tfrac{1}{2}H_2O \qquad [1]$$

In this reaction, the oxygen atom starts in the 0 oxidation state and ends in the –II state. To undergo this change, each O atom 'accepts' two electrons (from an unspecified source). Thus, 1/4 of a mole of O_2 molecules accepts 1 mol of electrons. In formal terms, we can write an 'equilibrium constant' for the half-reaction:

$$K = \frac{(H_2O)^{1/2}}{(O_2)^{1/4}(H^+)(e^-)} \qquad [2]$$

in which parentheses denote activities of the species. Using the definitions, $pH = -\log(H^+)$ and $pe = -\log(e^-)$, we can rewrite this equation:

$$pe = \log\{K\} - \log\left\{\frac{(H_2O)^{1/2}}{(O_2)^{1/4}}\right\} - pH \qquad [3]$$

The equilibrium constant, K, can – in principle – be determined from the free energies of formation of the reactants and products through the relationship between ΔG^0 and K. However, since free electrons do not exist in aqueous solution, chemists have devised a way of eliminating the free electron from the reaction. Half-reactions of the sort shown in eqn [1] are combined with with the oxidation of $H_2(g)$ to $H^+(aq)$. This reaction releases one electron and is 'defined' to have a free energy change of 0. Some algebraic manipulation will show that this combination of half-reactions allows the calculation of ΔG^0 for reaction [1] as

$$\Delta G^0 = \tfrac{1}{2}G^0_f(H_2O(l)) - \tfrac{1}{4}G^0_f(O_2(aq)) \qquad [4]$$

Then,

$$\log(K) = -\frac{\Delta G^0}{2.303RT} \equiv pe^0 \qquad [5]$$

can be calculated from tabulated thermodynamic data. In these equations, ΔG^0 is the Gibbs free energy yield of the reaction, when all reactants and products are in their standard states, R is the gas constant, and T is temperature (in K). Now, eqn [3] can be rewritten as

$$pe = pe^0 - \log\left\{\frac{(H_2O)^{1/2}}{(O_2)^{1/4}}\right\} - pH \qquad [6]$$

A similar equation can be derived for any half-reaction, written as a reduction of the reactant involving a single electron. Then, three quantities can be calculated:

1. pe^0, with all reactants and products having activity = 1;
2. pe(sw), which is applicable to reactions in seawater, and therefore is calculated with pH = 7.5 and $[HCO_3^-] = 2$ mM, but other species with activity = 1;
3. pe, with all species at concentrations that are observed in the environment.

pe^0 (calculation (1)) is not very useful for environmental calculations, as *in situ* activities are far from 1.

Table 2 Redox couples important to shallow water marine sediments.

Reaction	*pe*			
	pe⁰	*pe(sw)*	*pe*	
$\frac{1}{4}O_2 + H^+ + e^- \rightarrow \frac{1}{2}H_2O$	21.49	13.99	$O_2 = 350\ \mu M$	13.1
			$= 1\ \mu M$	12.5
$\frac{1}{5}NO_3^- + \frac{6}{5}H^+ + e^- \rightarrow \frac{1}{10}N_2 + \frac{3}{5}H_2O$	20.81	11.81	$N_2 = 500\ \mu M$	
			$NO_3^- = 1\ \mu M$	10.94
			$= 30\ \mu M$	11.24
$\frac{1}{2}MnO_2 + 2H^+ + e^- \rightarrow \frac{1}{2}Mn^{2+} + H_2O$	21.54	6.54	$Mn^{2+} = 1\ \mu M$	9.54
			$= 50\ \mu M$	8.69
$FeOOH + 3H^+ + e^- \rightarrow Fe^{2+} + 2H_2O$	15.98	−6.52	$Fe^{2+} = 1\ \mu M$	−0.52
			$= 400\ \mu M$	−3.12
$\frac{1}{8}SO_4^{2-} + \frac{9}{8}H^+ + e^- \rightarrow \frac{1}{8}HS^- + \frac{1}{2}H_2O$	4.21	−4.23	$HS^-/SO_4^{2-} = 0.1\ \mu M/28\ mM$	−4.12
			$= 1\ mM/1\ mM$	−4.23
$\frac{1}{4}HCO_3^- + \frac{5}{4}H^+ + e^- \rightarrow \frac{1}{4}CH_2O + \frac{1}{2}H_2O$	1.43	−7.95	$HCO_3^- = 2\ mM$	−8.62
$\frac{1}{8}NO_3^- + \frac{5}{4}H^+ + e^- \rightarrow \frac{1}{8}NH_4^+ + \frac{3}{8}H_2O$	14.88	5.41		

The reactions are written as one-electron half-reactions. pe values are calculated from thermodynamic data from Stumm and Morgan (1996). Note that the reduced form of carbon is shown as 'CH_2O'. This commonly used shorthand denotes a carbohydrate (not formaldehyde), and its standard state free energy is taken to be one-sixth that of glucose ($C_6H_{12}O_6$, or, in this notation, '$(CH_2O)_6$').

pe(sw) is often used because the difference between this calculation (2) and pe (3) is often small.

pe^0, pe(sw), and pe values are listed in **Table 2** for the electron acceptors and donors in **Table 1**, and they are shown graphically in **Figure 2**. The order of decreasing pe (**Figure 2**) is the order of decreasing affinity of the oxidant species for electrons. All reactions in **Table 2** and **Figure 2** are written as reductions (a species accepts an electron) involving the transfer of one electron. Thus, complete redox reactions, in which one chemical species is reduced while a second is oxidized, can be constructed by combining the reaction as written in **Table 2** for the oxidant (which is reduced) with the reverse of a reaction that is below it on the pe scale. The resulting reaction will have $\Delta G < 0$:

$$\Delta G_{\text{redox}} = -2.303RT(\text{pe}_{\text{reduction}} - \text{pe}_{\text{oxidation}}) \quad [7]$$

When the species that is oxidized lies below the species that is reduced on the pe scale, the reaction will release energy and can occur either abiotically or via microbial catalysis.

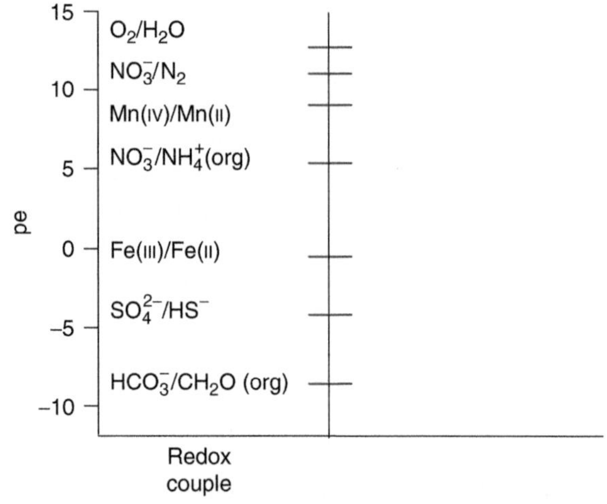

Figure 2 A graphical depiction of the pe of several environmentally important reduction reactions. A reduction as depicted in this figure, when combined with the oxidation of the reduced member of a couple listed lower on the figure, will yield a $\Delta G < 0$ under the conditions at which the pe values were calculated.

The Oxidation of Organic Matter in Estuarine Sediments

Most of the organic matter falling to the seafloor in estuarine and coastal environments is formed by local primary production. A useful (but not quite accurate) representation of marine organic matter is

$$(CH_2O)_{106}(NH_3)_{16}(H_3PO_4) \quad [8]$$

In this simple model, organic C is assumed to be in the form of a carbohydrate, and organic N in the form of ammonia (or a primary amine). In actuality, organic C has a lower average oxidation state than shown in this representation, but the model is still useful and widely used. We have used this organic matter stoichiometry, and neglected P since it is not oxidized during organic matter breakdown, to calculate the relative free energy yield for the oxidation of the C and N in organic matter by the different

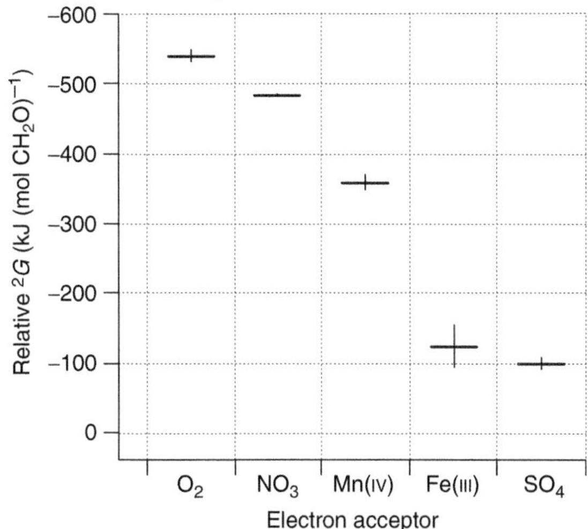

Figure 3 The relative free energy yield of the oxidation of a 'model' organic compound by the different electron acceptors. The model compound was $(CH_2O)(NH_3)_{16/106}$. ΔG values were calculated using the data in **Table 2**. NH_3 was assumed to be oxidized to NO_3^- by O_2 and to N_2 by NO_3^-.

electron acceptors. CH_2O is oxidized to CO_2 (converted to HCO_3^- in the marine environment) and NH_3 to NO_3^- (oxidation by O_2), N_2 (oxidation by NO_3^-), or NH_4^+ (Mn(IV), Fe(III), S(VI)). The result is shown in **Figure 3**. The vertical bars shown for each electron acceptor represent the range in ΔG expected over the ranges concentrations of reactants and products observed in sediments. There is a distinct order of ΔG. In order of free energy yield per mole of CH_2O, the order of yield by reaction with the different electron acceptors is

$$O_2 > NO_3^- > \mathrm{Mn(IV)} > \mathrm{Fe(III)} \geq SO_4^{2-} \quad [9]$$

Observations in a variety of sedimentary environments confirm that this order of electron acceptor use is followed. Apparently, organisms that can obtain a higher energy yield per mole of organic matter oxidized are favored over those obtaining less energy per mole. When an electron acceptor with a higher free energy yield is depleted, the next in the order is used.

Under the circumstances depicted in **Figure 1**, with reactants supplied at the sediment surface and mixed downward, the sequence of reactions by which organic matter is oxidized leads to a predictable sequence of solute concentrations. Starting from the sediment surface and moving downward: first, O_2 is removed from the pore waters; over the depth range where O_2 is removed, NO_3^- is added; it is then removed as NO_3^- is used as an electron acceptor; when NO_3^- is depleted, Mn^{2+} is added to the pore waters, followed by Fe^{2+}; then, SO_4^{2-} is removed and H_2S is added.

An example of this sequence of concentrations is shown in pore waters from a coastal site in **Figure 4**. The site from which the pore water profiles shown were taken lies under a 5-m water column with salinity ~31 psu and bottom water dissolved O_2 concentrations ranging from 240 to 390 μM during the year. The annually averaged organic carbon oxidation rate is high, ~800 μmol C cm^{-2} yr^{-1}. Clearly, under these conditions, O_2 is very rapidly consumed in the sediments. The absence of dissolved Fe^{2+}, which is very rapidly oxidized by O_2, in the upper 2 mm of the sediments indicates that the O_2 flux from overlying water is consumed within the upper 2 mm of the sediment column. There is little NO_3^- in the water overlying the sediments, but, as for O_2, its supply to the sediments is consumed in the upper 2 mm. Dissolved Mn^{2+} appears in the upper 2 mm, indicating that its production by oxidation of organic matter (and potentially other reduced phases) occurs very near the sediment/water interface. The shape of the dissolved Fe^{2+} profile shows that its release begins within 3–7 mm of the sediment–water interface. Finally, the SO_4^{2-} profile shows that sulfate reduction begins by ~20 mm below the interface. In these high-carbon-flux environments, the redox zones are compressed into a very small region near the sediment–water interface; nonetheless, the pore water data are consistent with the order of use suggested by ΔG calculations. This order has been observed at many locations. The site shown falls in the class of 'sulfide-dominated' sediments, in which the sulfate reduction rate is rapid enough to lead to a buildup of H_2S in the pore waters. Other sediments, either with lower salinity (hence a smaller source of SO_4^{2-}) or lower organic matter oxidation rates (hence a lower sulfate reduction rate), do not show this sulfide buildup. These sites are often considered 'iron-dominated' sediments because the pore water dissolved Fe^{2+} concentration never drops to near 0 and H_2S never increases to high levels.

Authigenic Mineral Formation

Reduced Fe and Mn Phases

The dissolved SO_4^{2-} profile in **Figure 4** clearly shows that sulfate reduction, which produces H_2S, occurs in the region of the sediments where there are very large concentrations of Fe^{2+} in the pore waters. However, no H_2S is present in the pore waters in this depth region, and H_2S does not appear until the dissolved Fe^{2+} concentration begins its rapid decline to values <1 μM. The explanation for both the absence of H_2S and the decline of Fe^{2+} is the formation of solid-phase FeS. It is believed that the solid phase

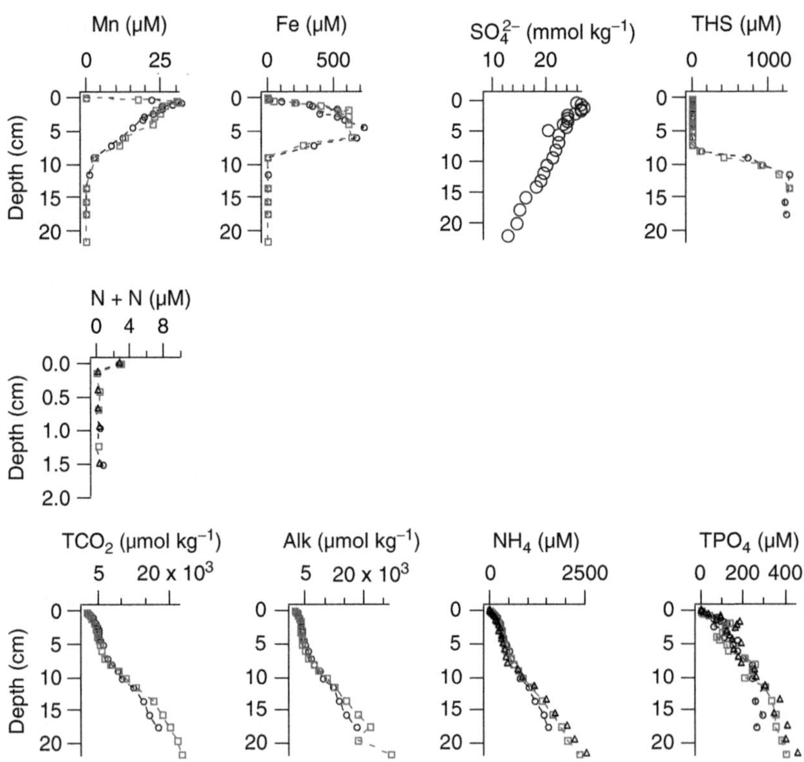

Figure 4 Pore water concentration vs. depth profiles, measured at a site in Hingham Bay in Boston Harbor, Massachusetts, in October 2001. Different Symbols represent replicate cores taken at the site, all at the same time. The depth ranges over which chemical reactions occur in the sediments and the rates of these reactions can be inferred from the solute profiles. The methods used for making these inferences are outlined in a classic paper by Froelich *et al.* Two important assumptions underlying the interpretation of the profiles are: (1) They are approximately in 'steady state'. That is, the rates of the transport and reaction processes whose balance sets the shape of the profiles are rapid relative to the rates at which environmental variability causes them to fluctuate over time. (2) The most important transport process in the sediments is vertically oriented molecular (or ionic) diffusion in the pore waters. Possible deviation from these simplifications can be seen in the pore water TCO_2, alkalinity ('Alk'), and NH_4^+ profiles, which all have inflections near the depth at which H_2S builds up in pore waters. These inflections could be caused either by transient changes in solute concentrations or by removal of the solutes from the sediments by 'sediment irrigation', a solute transport process due to the activities of animals living in the sediments. Data from Morford JL, Martin WR, Kalnejais LH, François R, Bothner M, and Karle I-M (2007) Insights on geochemical cycling of U, Re, and Mo from seasonal sampling in Boston Harbor, Massachusetts, USA. *Geochimica et Cosmochimica Acta* 71: 895–917.

formed is a fine-grained, poorly crystalline form of mackinawite (which we will call FeS_m). Studies of the dissolution of mackinawite have shown that, in alkaline solutions (such as seawater), FeS_m may be in equilibrium with either dissolved Fe^{2+} and HS^- or with the uncharged species, FeS^0_{aq}. In either case, the total dissolved Fe^{2+} is expected to be $\sim 1\,\mu M$. Thus, the relationship between $[Fe^{2+}]$, $[SO_4^{2-}]$, and total $[H_2S]$ is explained by precipitation of FeS_m until the supply of dissolved Fe^{2+} from dissimilatory Fe reduction is exhausted. Then, dissolved $[H_2S]$ begins to increase. In many cases, Fe(II) and S($-$II) do not remain sequestered in mackinawite. Mackinawite is metastable, and is subject to oxidation, either to Fe(III) and S(VI) or to the stable mineral, pyrite (FeS_2). In 'sulfide-dominated' systems such as that shown in **Figure 4**, pyrite (FeS_2) can form either through oxidation of S in FeS by H_2S:

$$FeS + H_2S \Rightarrow FeS_2 + H_2 \quad [10]$$

or by the reaction of aqueous FeS 'clusters' with polysulfide:

$$FeS_{aq} + S_n^{2-} \Rightarrow FeS_2 + S_{n-1}^{2-} \quad [11]$$

Thus, mackinawite is best viewed as an unstable intermediate product of iron and sulfate reduction. The fate of Fe and S present in mackinawite may be recycling to Fe(III) and S(VI), to be used again as electron acceptors for organic matter oxidation, or sequestration in pyrite.

As noted above, not all sediments – particularly those in low-salinity environments – have elevated

H_2S levels in pore waters. In these locations, Fe phosphate minerals may control Fe solubility. The combination of laboratory equilibration studies and pore water solute concentration measurements led Martens *et al.*, for example, to conclude that deep pore waters were in equilibrium with vivianite – $Fe_3PO_4 \cdot 8H_2O$ – in a coastal sediment. Hyacinthe *et al.* found that iron was sequestered as an Fe(III) phosphate in low-salinity, estuarine sediments. This ferric phosphate may have been formed in surface sediments (see below) or in the water column.

Figure 4 also shows the removal of Mn^{2+} from pore waters below the zone where it is added by reductive dissolution of Mn oxides. Equilibrium calculations have been used to infer that the mineral controlling Mn solubility in coastal and estuarine sediments is most likely to be a mixed Mn, Ca carbonate,

$$xMn^{2+} + (1-x)Ca^{2+} + CO_3{}^{2-} \Rightarrow Mn_xCa_{(1-x)}CO_3 \qquad [12]$$

with a solubility product between those of calcite ($CaCO_3$) and rhodochrosite ($MnCO_3$):

$$\log K_{Mn_xCa_{(1-x)}}CO_3 = x \log K_{MnCO_3} + (1-x)\log K_{CaCO_3} \qquad [13]$$

Oxidized Phases

The removal of reduced Fe, Mn, and S to solid phases is readily apparent from the pore water solute profiles in **Figure 4**. Close inspection shows that dissolved Fe^{2+} is also removed from solution above its pore water maximum, as dissolved $[Fe^{2+}]$ drops to near zero before Fe^{2+} reaches the sediment–water interface by diffusion. In addition, there is a small maximum in dissolved $[SO_4{}^{2-}]$ just below the sediment–water interface, indicating its addition by oxidation of S(–II). Finally, although the pore water Mn^{2+} profile does not clearly show its removal, direct measurement of the Mn^{2+} flux across the sediment–water interface at this site (using *in situ* benthic flux chambers) showed that the actual flux is significantly smaller than is calculated by diffusion driven by the pore water concentration gradient: Mn^{2+} must be removed in a thin layer at the sediment–water interface.

These Fe and Mn removal processes are shown more clearly by pore water profiles at a site with a lower organic matter oxidation rate than the **Figure 4** site. **Figure 5** shows the seasonality in oxygen penetration depth that is common in shallow-water sediments. In March, the depth of removal of Fe^{2+} from pore waters coincides closely with the O_2 penetration depth, suggesting that there is rapid oxidation of Fe^{2+} by O_2 to insoluble oxides. In August, O_2 penetration is much shallower and Fe^{2+} production during organic matter oxidation is much more rapid, but the dissolved Fe^{2+} profile still shows evidence of removal (the upward curvature of the profile is indicative of removal from solution). In this case, the removal coincides with both the zone of Mn^{2+} release into the pore waters and with O_2 penetration; oxidation by both O_2 and Mn oxides may be occurring. The figure shows removal of Fe from solution by upward diffusion of dissolved Fe^{2+} and its oxidation. As at the **Figure 4** site, directly measured fluxes of dissolved Mn^{2+} across the sediment–water interface show that removal of dissolved Mn in a thin layer at the sediment–water interface impedes its transport from sedimentary pore waters to bottom water.

The removal of Fe(II), S(–II), and Mn(II) when they are transported toward the sediment–water interface is predicted by the pe scale shown in **Figure 2**. The oxidation of Fe^{2+} (the reverse of the reaction in **Table 2**) could be coupled to the reduction of MnO_2,

$$Fe^{2+} + \tfrac{1}{2}MnO_2 + H_2O \Rightarrow FeOOH + \tfrac{1}{2}Mn^{2+} + H^+$$
$$\Delta G_{7.5} = -2.303RT(8.69 + 0.52) = -53\frac{\text{kJ}}{\text{mol Fe}} \qquad [14]$$

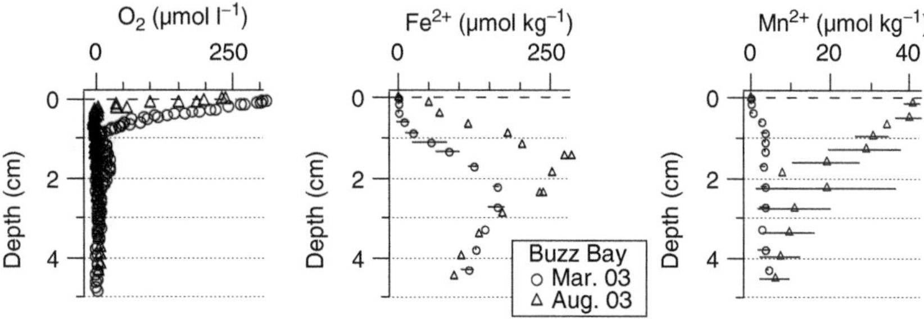

Figure 5 Pore water concentration vs. depth profiles, measured at a site in Buzzards Bay, Massachusetts, in Mar. and Aug., 2003.

or to the reduction of O_2,

$$Fe^{2+} + \frac{1}{4}O_2 + \frac{3}{2}H_2O \Rightarrow FeOOH + 2H^+$$
$$\Delta G = -2.303RT(13.1 + 0.52) = -78\frac{kJ}{mol\ Fe} \quad [15]$$

These reactions release energy, and can occur both abiotically or by microbial catalysis. Similar calculations using the data in **Table 2** and **Figure 2** show that S(− II) can be oxidized by Mn oxides or O_2, and Mn(II) can be oxidized by O_2. The rapid oxidation of Fe and S by O_2 is well known and occurs abiotically. Oxidation by O_2 was demonstrated by Aller using sediments from Long Island Sound. Aller and Rude used incubation experiments to demonstrate the occurrence of microbially catalyzed oxidation of solid-phase Fe(II) and S(− II) by Mn oxides.

Figure 5 illustrates these oxidative removal processes through loss of dissolved components of pore waters. Equally important is the upward mixing of solid phase Fe sulfides to zones of O_2 or Mn oxide reduction, followed by their oxidation. Particle mixing by bioturbation is an important mechanism for transport of reduced phases toward the sediment–water interface, particularly in warm-weather (and high productivity) months.

Measurement

The above discussion highlights the key role that authigenic solid phases play in sedimentary processes in estuarine sediments. Because rapid particle transport due to the activities of benthic fauna is present in virtually all of these sediments, solid phases are important participants in cycling between reduced and oxidized forms of Fe, S, and Mn. Unfortunately, the quantification of the solid phases in these sediments is difficult. Because the solids are predominantly made up of nonreactive, terrigenous minerals, and because the authigenic minerals tend to be very fine-grained and poorly crystalline, they cannot be measured directly. Instead, their presence is inferred by removal of dissolved constituents from pore waters and calculation of solubilities based on laboratory studies. The only means currently available to quantify their concentrations are selective chemical leaches of sediments. Several studies have evaluated the use of reducing agents to determine reactive Fe and Mn oxides in sediments. Hyacinthe *et al.* showed recently that ligand-enhanced reductive dissolution using a buffered ascorbate–citrate solution dissolves approximately the same amount of Fe as microbial processes, but the correspondence is still quite uncertain. The quantification of reduced phases is still more difficult. Cornwell and Morse compared several procedures for quantifying reduced sulfur in sediments, concluding that dissolution in 6 N or 12 N HCl released S associated with FeS (and other relatively reactive reduced S species), but that the procedures also release Fe associated with oxides and some silicates. The procedure achieved reasonable separation of FeS from pyrite, which was dissolved by a more intense chemical leach. Rickard and Morse have emphasized the difficulty in interpreting the results of these procedures. Clearly, given the importance of cycling of solid phases in estuarine and coastal sediments, the quantification of their concentrations remains an important problem.

Elemental Cycling within Estuarine and Coastal Sediments

It is clear that chemical processes do not involve just alteration of particles that fall to the sediment–water interface. Rather, these systems are dominated by cycling between oxidized and reduced chemical forms within the sediment column. Heterotrophic respiration leads to the oxidation of organic matter, releasing C, N, and P into solution, and the reduction of O_2, Fe, Mn, and S. The latter three elements are then subject to transport, both in dissolved and solid forms, back toward the sediment–water interface, where they may be reoxidized either abiotically or by lithotrophic bacteria. The burial of reduced Fe, Mn, and S is a slow leak from these rapid internal cycles (**Figure 6**).

The upward transport and oxidation of reduced phases introduces complications into the determination of the relative importance of different organic matter oxidation pathways. For instance, O_2 is consumed not only for direct organic matter oxidation, but also for oxidation of reduced products of organic matter oxidation by Mn and Fe oxides and SO_4^{2-} (Mn(II), Fe(II), S(− II)). Similarly, Mn oxides are consumed by oxidation of Fe(II) and S(− II) as well as by oxidation of organic matter. Thus, in considering the role played by each electron acceptor in organic matter decomposition, allowance must be made for oxidation of chemical constituents other than organic matter. Jorgensen used measurements of dissolved fluxes across the sediment–water interface to infer that oxidation of organic matter by O_2 and SO_4^{2-} were of roughly equal importance in shallow-water marine sediments. More recently, Thamdrup used sediment incubation techniques to show that those results overestimated the role of O_2 but underestimated that of Fe. The uncertainties in earlier estimates were both due to the failure to account for oxidation of species other than organic matter and due to the failure to account for the

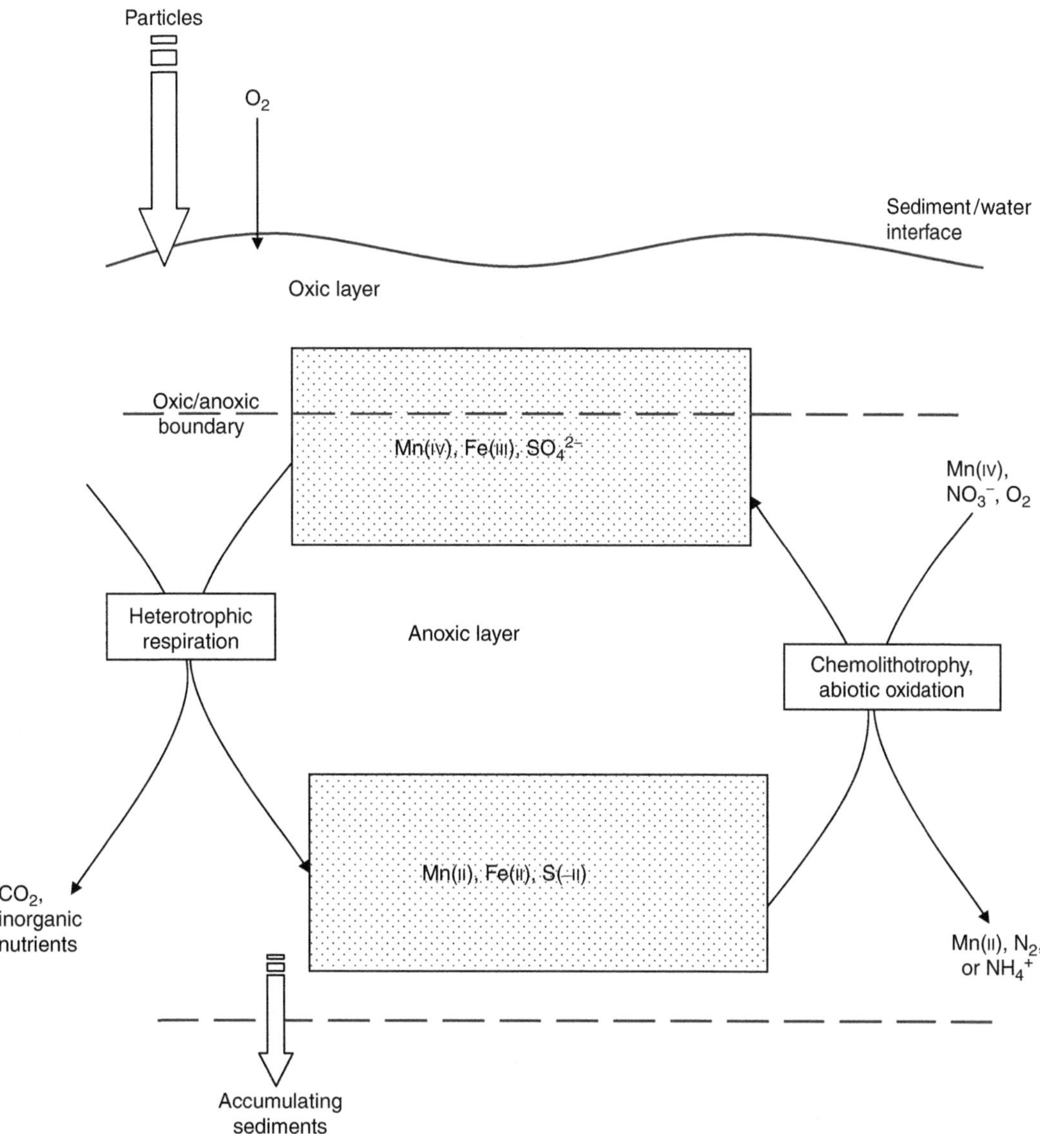

Figure 6 The cycling of Fe, Mn, and S within the sediment column. Transport of reduced Fe, Mn, and S occurs both in the dissolved and solid phases.

formation of solid-phase FeS. O_2 is still likely to be of great importance, however. For instance, while the Fe(III)/Fe(II) pathway may be used directly for organic matter oxidation, it is the internal cycling between Fe(III) and Fe(II) that allows this pathway to be important: the ultimate sink for electrons is likely to be O_2. This linking of oxidation/reduction reactions in series is illustrated in **Figure 7** using a diagram of the sort introduced by Aller.

The Effect of Sedimentary Redox Cycling on Nutrient Cycles

Sedimentary respiration results in the decomposition of the majority of the organic matter that falls from the water column to the sediments. However, sedimentary processes can result in less efficient cycling of P and N than C, with potentially significant implications for coastal productivity.

Phosphate has a strong tendency to adsorb to Fe oxides. Therefore, the precipitation of Fe oxides near the interface of coastal and estuarine sediments can impede the return of phosphate, released into solution by the decomposition of organic matter, to the water column. **Figure 4** shows that dissolved reactive phosphate ('TRP') can reach very high levels in sediments to which there is a large flux of organic matter. In fact, the TRP level in **Figure 4** is significantly greater than would be predicted by the decomposition of organic matter alone. This phenomenon is further illustrated in **Figure 8**.

Figure 8(a) shows that P is highly enriched in the solid phase in the upper centimeter of the sediment column. This enrichment cannot be explained simply by larger organic matter concentrations in surface seciments. **Figure 8**(b) illustrates that internal P cycling with Fe establishes the enrichment. The dashed line in the figure shows the slope of a pore water TPO_4 versus TCO_2 plot if TPO_4 is simply released by organic matter oxidation. If the only process affecting dissolved TPO_4 were decomposition of organic matter, then the pore water concentration data would plot along a line of that slope from the origin. The pore water data deviate from that simple relationship in two ways: TPO_4 is essentially completely removed from pore waters near the sediment–water interface by upward diffusion and removal with freshly formed Fe oxides. When Fe(III) is reduced to oxidize organic matter, the rate of P release is much greater than can be accounted for by the organic matter alone: the Fe oxide-associated P is released. The result is a strong internal cycle of P. A fraction of the P that arrives at the sediment surface with organic matter is sequestered within the sediments by precipitation with Fe oxides. **Figure 8** illustrates this process at a marine sediment site. Hyacinthe *et al.* showed that P can also be sequestered in estuarine sediments when P that is removed with Fe oxides remains in the solid phase after incomplete reduction of the oxide phase.

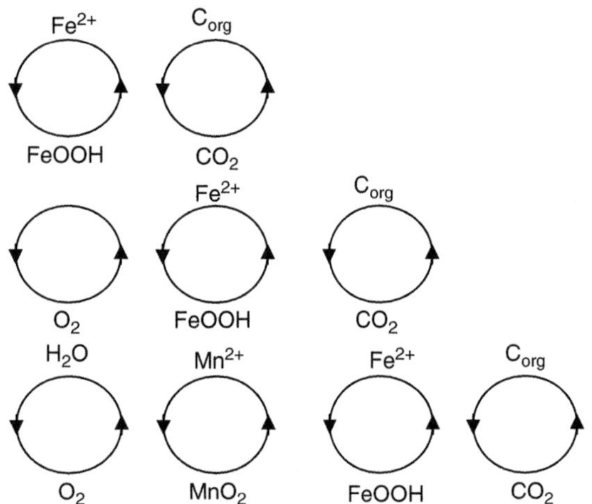

Figure 7 The oxidation of organic matter by FeOOH. In each panel, the oxidation of organic carbon ('C_{org}') to CO_2 is coupled to the reduction of FeOOH to Fe^{2+}. In the top panel, the source of the FeOOH is the particle flux to the sediments. The lower two panels show that internal cycling within the sediments can also supply Fe(III) for organic matter oxidation. In the middle panel, Fe(II) that was produced by organic matter oxidation is oxidized by O_2 to Fe(III), to be reused ot oxidize organic matter. In the lower panel, Fe(II) is oxidized to Fe(III) by MnO_2. The MnO_2 for this reaction, in turn, may be supplied either by the particle flux to the sediments or by oxidation of Mn(II) to Mn(IV) by O_2. The figure illustrates that the coupling of organic matter oxidation to Fe(III) reduction can be quantitatively important even if the supply of reactive Fe(III) through the fall or particles to the seafloor is small. When this occurs, the ultimate sink for the electrons for organic matter oxidation by the Fe(III)/Fe(II) pathway is O_2.

The sediments are also important sites for ther removal of fixed nitrogen from coastal waters. Because there is relatively little NO_3^- in the shallow waters overlying coastal and estuarine sediments, diffusion of NO_3^- from bottom water is not a major source of N for denitrification. However, rapid organic matter oxidation results in the release of NH_4^+ to the pore waters. NH_4^+ can be converted to N_2 by a nitrification/denitrification cycle or by NH_4^+ oxidation coupled to the reduction of NO_3^-:

$$5NH_4^+ + 3NO_3^- \Rightarrow 4N_2 + 9H_2O + 2H^+ \quad [16]$$

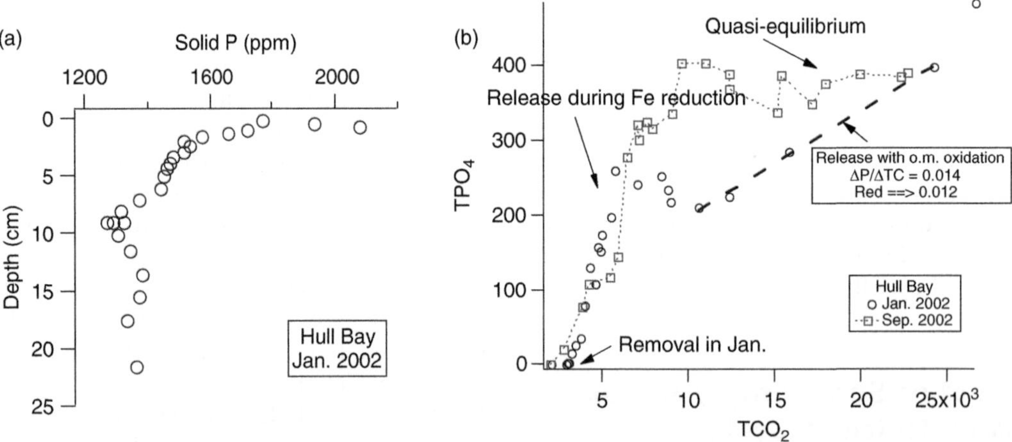

Figure 8 (a) Solid-phase P, measured at a site in Hingham Bay, Boston Harbor. (b) The relationship between the pore water concentrations of total reactive phosphate (TPO_4) and TCO_2 in pore waters at that site. Concentrations of both species are in $\mu mol\,l^{-1}$.

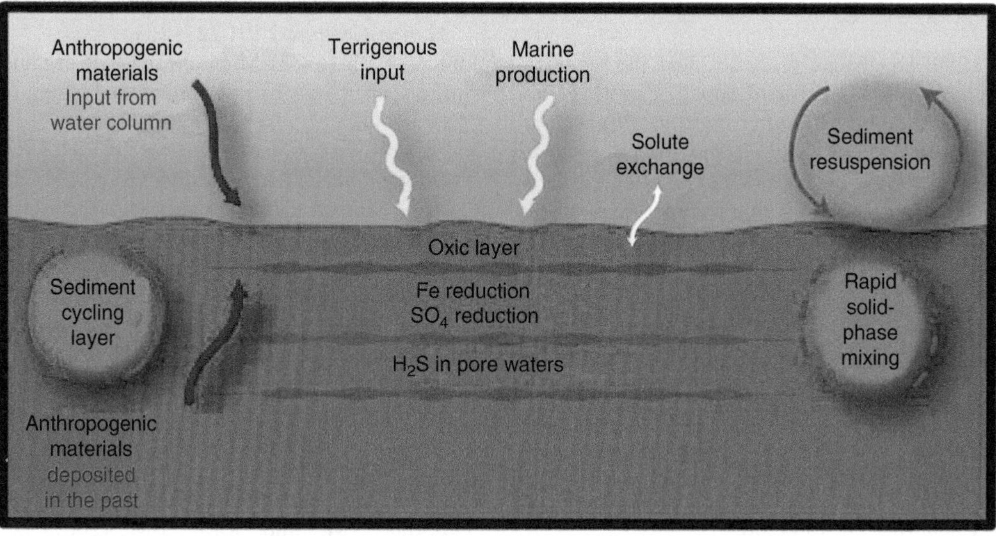

Figure 9 The structure of the chemical environment in shallow-water marine sediments.

Thamdrup and Dalsgaard illustrated the occurrence of this N_2 formation mechanism. The result of these processes is that 15–75% of the flux of N to the surface of coastal and estuarine sediments may be removed to N_2.

Contaminant Cycling: Anthropogenic Metals

The 'layered' structure of the chemical environment in estuarine and coastal sediments has important implications for the cycling of contaminants. Anthropogenic materials can both be present in the sediments from deposition in the past and can arrive with current deposits. They encounter an environment similar to that depicted in **Figure 9**. The upper centimeters of the sediment can be seen as an active, 'sediment-cycling layer', in which particles are rapidly mixed by bioturbation. This layer consists of an oxic 'cap' just below the sediment–water interface, where Fe and Mn oxides precipitate. Below that is a region of Fe and S reduction, but without buildup of dissolved H_2S. Below this layer – if the supply of organic matter supports extensive sulfate reduction – is a layer of elevated dissolved H_2S. An extensive discussion of contaminant cycling in estuarine and coastal sediments is beyond the scope of this article. Two examples are cited to show the importance of the redox structure of these environments to contaminant cycling.

Many contaminant metals – Cu, Ag, and Pb are examples – both form insoluble sulfides and tend to sorb onto precipitating Fe oxides. Thus, they are released to pore waters when Fe oxide reduction occurs, but tend to be immobilized by precipitation as sulfides. Dissolved fluxes to the overlying water are determined by a balance between their rates of diffusion from the site of Fe reduction toward the sediment/water interface and of removal by precipitation with Fe oxides. Because both the thickness of the oxic layer and the rate of Fe reduction vary seasonally and spatially, these metals have seasonally and spatially variable rates of release from sediments to the water column. Their precipitation with Fe oxides can lead to enrichments in fine-grained solids at the sediment–water interface. Thus, sediment resuspension and subsequent reaction and transport in the water column can be an important mechanism for redistribution of contaminant metals. Because the FeS that is formed by precipitation of the products of Fe(III) and SO_4^{2-} reduction is metastable, the formation of pyrite may be an important step in the immobilization of these metals. One example of a study of these processes in coastal systems is that of Kalnejais.

A different type of metal is Hg. The toxic, bioavailable form of Hg (methylmercury) may be formed in sediments when sulfate-reducing bacteria incorporate HgS^0 and methylate the Hg. This process appears to occur where sulfate reduction is important, but dissolved sulfide levels are less than $\sim 10\,\mu M$. Under these conditions, sediments can be an important source of methylmercury to the coastal water column.

See also

Nitrogen Isotopes in the Ocean. Phosphorus Cycle.

Further Reading

Aller RC (1980) Diagenetic processes near the sediment–water interface of Long Island Sound. Part II: Fe and Mn. In: Saltzman B (ed.) *Estuarine Physics and Chemistry: Studies in Long Island Sound*, vol. 22, pp. 351–415. New York: Academic Press.

Aller RC and Rude PD (1988) Complete oxidation of solid phase sulfides by manganese and bacteria in anoxic marine sediments. *Geochimica et Cosmochimica Acta* 52: 751–765.

Benoit JM, Gilmour CC, and Mason RP (2001) The influence of sulfide on solid-phase mercury bioavailability for methylation by pure cultures of *Desulfobulbus propionicus*. *Environmental Science and Technology* 35: 127–132.

Cornwell JC and Morse JW (1987) The characterization of iron sulfide minerals in anoxic marine sediments. *Marine Chemistry* 22: 193–206.

Elderfield H, Luedtke, McCaffrey RJ, and Bender M (1981) Benthic studies in Narragansett Bay. *American Journal of Science* 281: 768–787.

Froelich PN, Klinkhammer GP, Luedtke NA, *et al.* (1979) Early oxidation of organic matter in pelagic sediments of the eastern equatorial Atlantic: Suboxic diagenesis. *Geochimica et Cosmochimica Acta* 43: 1075–1090.

Hammerschmidt CR, Fitzgerald WC, Lamborg CH, Balcom PH, and Visscher PT (2004) Biogeochemistry of methylmercury in sediments of Long Island Sound. *Marine Chemistry* 90: 31–52.

Huerta-Diaz MA and Morse JW (1992) Pyritization of trace metals in anoxic marine sediments. *Geochimica et Cosmochimica Acta* 56: 2681–2702.

Hyacinthe C, Bonneville S, and Van Cappellen P (2006) Reactive iron(III) in sediments: Chemical versus microbial extractions. *Geochimica et Cosmochimica Acta* 70: 4166–4180.

Hyacinthe C and Van Cappellen P (2004) An authigenic iron phosphate phase in estuarine sediments: Composition, formation and chemical reactivity. *Marine Chemistry* 91: 227–251.

Jorgensen BB (1982) Mineralization of organic matter in the sea bed – the role of sulphate reduction. *Nature* 296: 643–645.

Kalnejais LH (2005) *Mechanisms of Metal Release from Contaminated Coastal Sediments*, p. 238. Woods Hole, MA: Massachusetts Institute of Technology, Woods Hole Oceanographic Institution.

Kostka JE and Luther GW (1994) Partitioning and speciation of solid phase iron in saltmarsh sediments. *Geochimica et Cosmochimica Acta* 58: 1701–1710.

Martens CS, Berner RA, and Rosenfeld JK (1978) Interstitial water chemistry of anoxic Long Island Sound sediments. Part 2: Nutrient regeneration and phosphate removal. *Limnology and Oceanography* 23(4): 605–617.

Morford JL, Martin WR, Kalnejais LH, François R, Bothner M, and Karle I-M (2007) Insights on geochemical cycling of U, Re, and Mo from seasonal sampling in Boston Harbor, Massachusetts, USA. *Geochimica et Cosmochimica Acta* 71: 895–917.

Rickard D (2006) The solubility of FeS. *Geochimica et Cosmochimica Acta* 70: 5779–5789.

Rickard D and Morse JW (2005) Acid volatile sulfide (AVS). *Marine Chemistry* 97: 141–197.

Seitzinger SP (1988) Denitrification in freshwater and coastal marine ecosystems: Ecological and geochemical significance. *Limnology and Oceanography* 33: 702–724.

Stumm W and Morgan JJ (1996) *Aquatic Chemistry: Chemical Equilibria and Rates in Natural Waters*, 3rd edn. New York: Wiley.

Thamdrup B (2000) Bacterial manganese and iron reduction in aquatic sediments. *Advances in Microbial Ecology* 16: 41–84.

Thamdrup B and Dalsgaard T (2002) Production of N_2 through anaerobic ammonium oxidation coouled to nitrate reduction in marine sediments. *Applied and Environmental Microbiology* 68: 1312–1318.

RIVER INPUTS

J. D. Milliman, College of William and Mary, Gloucester, VA, USA

Introduction

Rivers represent the major link between land and the ocean. Presently rivers annually discharge about 35 000 km^3 of freshwater and 20–22 × 10^9 tonnes of solid and dissolved sediment to the global ocean. The freshwater discharge compensates for most of the net evaporation loss over the ocean surface, groundwater and ice-melt discharge accounting for the remainder. As such, rivers can play a major role in defining the physical, chemical, and geological character of the estuaries and coastal areas into which they flow.

Historically many oceanographers have considered the land–ocean boundary to lie at the mouth of an estuary, and some view it as being at the head of an estuary (or, said another way, at the mouth of the river). A more holistic view of the land–ocean interface, however, might include the river basins that drain into the estuary. This rather unconventional view of the land–sea interface is particularly important when considering the impact of short- and medium-term changes in land use and climate, and how they may affect the coastal and global ocean.

Uneven Global Database

A major hurdle in assessing and quantifying fluvial discharge to the global ocean is the uneven quantity and quality of river data. Because they are more likely to be utilized for transportation, irrigation, and damming, large rivers are generally better documented than smaller rivers, even though, as will be seen below, the many thousands of small rivers collectively play an important role in the transfer of terrigenous sediment to the global ocean. Moreover, while the database for many North American and European rivers spans 50 years or longer, many rivers in Central and South America, Africa and Asia are poorly documented, despite the fact that many of these rivers have large water and sediment inputs and are particularly susceptible to natural and anthropogenic changes.

The problem of data quality is magnified by the fact that the available database spans the latter half of the twentieth century, some of the data having been collected >20–40 years ago, when flow patterns may have been considerably different to present-day patterns. In many cases, more recent data can reflect anthropogenically influenced conditions, often augmented by natural change. The Yellow River in northern China, for example, is considered to have one of the highest sediment loads in the world, 1.1 billion tonnes y^{-1}. In recent years, however, its load has averaged <100 million tons, in response to drought and increased human removal of river water. How, then, should the average sediment load of the Yellow River be reported – as a long-term average or in its presently reduced state?

Finally, mean discharge values for rivers cannot reflect short-term events, nor do they necessarily reflect the flux for a given year. Floods (often related to the El Ninõ/La Niña events) can have particularly large impacts on smaller and/or arid rivers, such that mean discharges or sediment loads may have little relevance to short-term values. Despite these caveats, mean values can offer sedimentologists and geochemists considerable insight into the fluxes (and fates) from land to the sea.

Because of the uneven database (and the often difficult access to many of the data), only in recent years has it been possible to gather a sufficient quantity and diversity of data to permit a quantitative understanding of the factors controlling fluvial fluxes to the ocean. Recent efforts by Meybeck and Ragu (1996) and Milliman and Farnsworth (2002) collectively have resulted in a database for about 1500 rivers, whose drainage basin areas collectively represent about 85% of the land area emptying into the global ocean. It is on this database that much of the following discussion is based.

Quantity and Quality of Fluvial Discharge

Freshwater Discharge

River discharge is a function of meteorological runoff (precipitation minus evaporation) and drainage basin area. River basins with high runoff but small drainage area (e.g., rivers draining Indonesia, the Philippines, and Taiwan) can have discharges as great as rivers with much larger basin areas but low runoff (**Table 1**). In contrast, some rivers with low

Table 1 Basin area, runoff, and discharge for various global rivers

River	*Basin area* ($\times 10^3 km^2$)	*Runoff* (mm y^{-1})	*Discharge* ($km^3 y^{-1}$)
Amazon	6300	1000	6300
Congo	3800	360	1350
Ganges/Bramaputra	1650	680	1120
Orinoco	1100	1000	1100
Yangtze (Changjiang)	1800	510	910
Parana/Uruguay	2800	240	670
Yenisei	2600	240	620
Mekong	800	690	550
Lena	2500	210	520
Mississippi	3300	150	490
Choshui (Taiwan)	3.1	1970	6.1
James (USA)	20	310	6.2
Cunene (Angola)	100	68	6.8
Limpopo (Mozambique)	380	14	5.3

The first 10 rivers represent the highest discharge of all world rivers, the Amazon having discharge equal to the combined discharge of the next seven largest rivers. Basin areas of these rivers vary from 6300 (Amazon) to 800 (Mekong) $\times 10^3$ km^2, and runoffs vary from 1000 (Amazon) to 150 mm y^{-1} (Mississippi). The great variation in runoff can be seen in the example of the last four rivers, each of which has roughly the same mean annual discharge (5.3–6.8 km^3 y^{-1}), but whose basin areas and runoffs vary by roughly two orders of magnitude.

runoff (such as the Lena and Yenisei) have high discharges by virtue of their large drainage basin areas. The Amazon River has both a large basin (comprising 35% of South America) and a high runoff; as such its freshwater discharge equals the combined discharge of the next seven largest rivers (**Table 1**). Not only are the coastal waters along north-eastern South America affected by this enormous discharge, but the Amazon influence can be seen as far north as the Caribbean >2000 km away. The influence of basin area and runoff in controlling discharge is particularly evident in the bottom four rivers listed in **Table 1**, all of which have similar discharges (5.3–6.2 km^3 y^{-1}) even though their drainage basin areas and discharges can vary by two orders of magnitude.

Sediment Discharge

A river's sediment load consists of both suspended sediment and bed load. The latter, which moves by traction or saltation along the river bed, is generally assumed to represent 10% (or less) of the total sediment load. Suspended sediment includes both wash load (mostly clay and silt that is more or less continually in suspension) and bed material load (which is suspended only during higher flow); bed material includes coarse silt and sand that may move as bed load during lower flow. A sediment-rating curve is used to calculate suspended sediment concentration (or load), which relates measured concentrations (or loads) with river flow. Sediment load generally increases exponentially with flow, so that a two-order of magnitude increase in flow may result in three to four orders of magnitude greater suspended sediment concentration.

In contrast to water discharge, which is mainly a function of runoff and basin area, the quantity and character of a river's sediment load also depend on the topography and geology of the drainage basin, land use, and climate (which influences vegetation as well as the impact of episodic floods). Mountainous rivers tend to have higher loads than rivers draining lower elevations, and sediment loads in rivers eroding young, soft rock (e.g., siltstone) are greater than rivers flowing over old, hard rock (e.g., granite) (**Figure 1**). Areas with high rainfall generally have higher rates of erosion, although heavy floods in arid climates periodically can carry huge amounts of sediment.

The size of the drainage basin also plays an important role. Small rivers can have one to two orders of magnitude greater sediment load per unit basin area (commonly termed sediment yield) (**Figure 1**) than larger rivers because they have less flood plain area on which sediment can be stored, which means a greater possibility of eroded sediment being discharged directly downstream. Stated another way, a considerable amount of the sediment eroded in the headwaters of a large river may be stored along the river course, whereas most of the sediment eroded along a small river can be discharged directly to the sea, with little or no storage.

Small rivers are also more susceptible to floods, during which large volumes of sediment can be transported. Large river basins, in contrast, tend to be self-modulating, peak floods in one part of the basin are often offset by normal or dry conditions in another part of the basin. The impact of a flood on a small, arid river can be illustrated by two 1-day floods on the Santa Clara River (north of Los Angeles) in January and February 1969, during which more sediment was transported than the river's cumulative sediment load for the preceding 25 years! The combined effect of high sediment yields from small rivers can be seen in New Guinea, whose more than 250 rivers collectively discharge more sediment to the ocean than the Amazon River, whose basin area is seven times larger than the entire island.

Our expanded database allows us to group river basins on the basis of geology, climate, and basin area, thereby providing a better understanding of how these factors individually and collectively influence sediment load. Rivers draining the young,

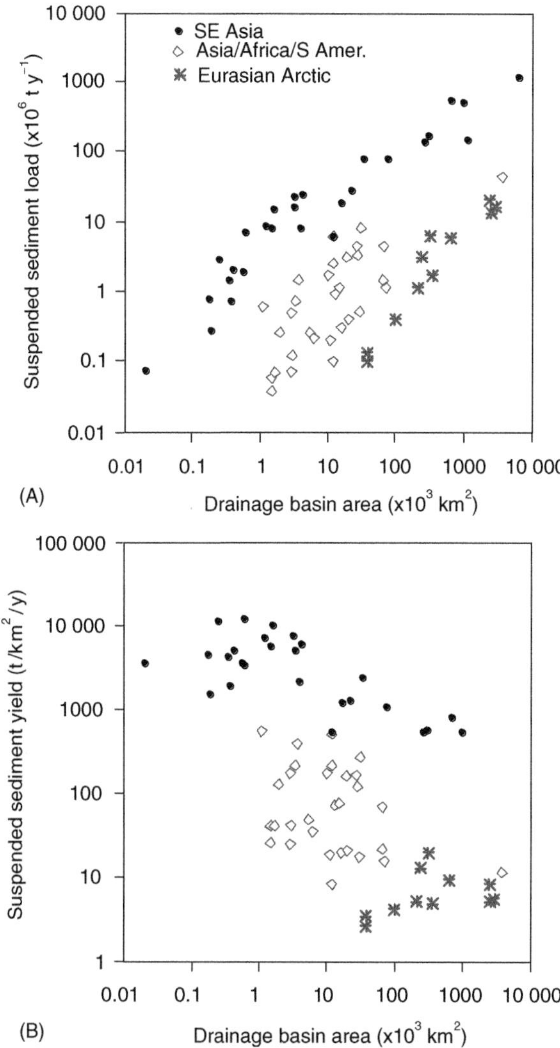

Figure 1 Suspended sediment discharge from rivers with various sized basins draining wet mountains in south-east Asia with young (assumed to be easily erodable) rocks (dots); wet mountains in south Asia, north-eastern South America, and west Africa with old (assumed to be less erodable) rocks (open diamonds); and semi-arid to humid mountainous and upland rivers in the Eurasian Arctic. Sediment loads generally increase with basin size, and are one to two orders of magnitude greater for south-east Asian rivers than for rivers with equal rainfall but old rocks, which in turn are greater than for rivers with old rocks but less precipitation (A). Sediment yields for smallest south-east Asian rivers are as much as two orders of magnitude greater than for the largest, whereas the yield for Eurasian Arctic rivers shows little change with basin size (B).

easily eroded rocks in the wet mountains of south-east Asia, for instance, have one to two orders of magnitude greater sediment loads (and sediment yields) than similar-sized rivers draining older rocks in wet Asian mountains (e.g., in India or Malaysia), which in turn have higher loads and yields than the rivers from the older, drier mountains in the Eurasian Arctic (**Figure 1**).

Because of the many variables that determine a river's sediment load, it is extremely difficult to calculate the cumulative sediment load discharged from a land mass without knowing the area, morphology, geology, and climate for every river draining that land mass. This problem can be seen in the last four rivers listed in **Table 2**. The Fitzroy-East drains a seasonally arid, low-lying, older terrain in north-eastern Australia, whereas the Mad River in northern California (with a much smaller drainage basin), drains rainy, young mountains. Although the mean annual loads of the two rivers are equal, the Mad has 112-fold greater sediment yield than the Fitzroy-East. The Santa Ynez, located just north of Santa Barbara, California, has a similar load and yield, but because it has much less runoff, its average suspended sediment concentration is 13 times greater than the Mad's.

Dissolved Solid Discharge

The amount of dissolved material discharged from a river depends on the concentration of dissolved ions and the quantity of water flow. Because dissolved concentrations often vary inversely with flow, a river's dissolved load often is more constant throughout the year than its suspended load.

Dissolved solid concentrations in river water reflect the nature of the rock over which the water flows, but the character of the dissolved ions is controlled by climate as well as lithology. Rivers with different climates but draining similar lithologies can have similar total dissolved loads compared with rivers with similar climates but draining different rock types. For example, high-latitude rivers, such as those draining the Eurasian Arctic, have similar dissolved solid concentrations to rivers draining older lithologies in southern Asia, north-eastern South America and west Africa, but lower concentrations than rivers draining young, wet mountains in south-east Asia (**Figure 2**).

The concentration of dissolved solids shows little variation with basin area, small rivers often having concentrations as high as large rivers (**Figure 2B**). At first this seems surprising, since it might be assumed that small rivers would have low dissolved concentrations, given the short residence time of flowing water. This suggests an important role of ground water in both the dissolution and storage of river water, allowing even rivers draining small basins to discharge relatively high concentrations of dissolved ions.

Four of the 10 rivers with highest dissolved solid discharge (Salween, MacKenzie, St Lawrence, and Rhine) in **Table 3** are not among the world leaders in

Table 2 Basin area, TSS (mg l^{-1}), annual sediment load, and sediment yield (t km^2 per year) for various global rivers

River	*Basin area* ($\times 10^3 km^2$)	*TSS* (gl^{-1})	*Sediment load* ($\times 10^6 t\ y^{-1}$)	*Sediment yield* ($t\ km^{-2} y^{-1}$)
Amazon	6300	0.19	1200	190
Yellow (Huanghe)	750	25	1100	1500
Ganges/Bramaputra	1650	0.95	1060	640
Yangtze (Changjiang)	1800	0.51	470	260
Mississippi	3300	0.82	400 (150)	120 (45)
Irrawaddy	430	0.6	260	600
Indus	980	2.8	250 (100)	250 (100)
Orinoco	1100	0.19	210	190
Copper	63	1.4	130(?)	2100
Magdalena	260	0.61	140	540
Fitzroy-East (Australia)	140	0.31	2.2	16
Arno (Italy)	8.2	0.69	2.2	270
Santa Ynez (USA)	2	23	2.3	1100
Mad (USA)	1.2	1.7	2.2	1800

Loads and yields in parentheses represent present-day values, the result of river damming and diversion. The first 10 rivers represent the highest sediment loads of all world rivers, the Amazon, Ganges/Brahmaputra and Yellow rivers all having approximately equal loads of about $1100 \times 10^6\ t\ y^{-1}$. No other river has an average sediment load greater than $470 \times 10^6\ t\ y^{-1}$). Discharges for these rivers vary from 6300 (Amazon) to $43 \times km^3\ y^{-1}$ (Yellow), and basin areas from 6300 (Amazon) to $63 \times 10^3\ km^2\ y^{-1}$ (Copper). Corresponding average suspended matter concentrations and yields vary from 0.19 to 25 and 190 to 2100, respectively. The great variation in values can be seen in the example of the last four rivers, which have similar annual sediment loads ($2.2\text{–}2.3 \times 10^6\ t\ y^{-1}$), but whose average sediment concentrations and yields vary by several orders of magnitude.

terms of either water or sediment discharge (**Tables 1** and 2). The prominence of the Rhine, which globally can be considered a second-order river in terms of basin area, discharge, and sediment load, stems from the very high dissolved ionic concentrations (19 times greater than the Amazon), largely the result of anthropogenic influence in its watershed (see below).

The last four rivers in **Table 3** reflect the diversity of rivers with similar dissolved solid discharges. The Cunene, in Angola, is an arid river that discharges about as much water as the Citandy in Indonesia, but drains more than 20 times the watershed area. As such, total dissolved solid (TDS) values for the two rivers are similar (51 vs 62 mg l^{-1}), but the dissolved yield (TDS divided by basin area) of the Cunene is <1% that of the Citandy. In contrast, the Ems River, in Germany, has a similar dissolved load to the Citandy, but concentrations are roughly three times greater.

Fluvial Discharge to the Global Ocean

Collectively, the world rivers annually discharge about 35 000 km^3 of fresh water to the ocean. More than half of this comes from the two areas with highest precipitation, south-east Asia/Oceania and north-eastern South America, even though these two areas collectively account for somewhat less than 20% of the total land area draining into the global ocean. Rivers from areas with little precipitation, such as the Canadian and Eurasian Arctic, have little discharge (4800 km^3) relative to the large land area that they drain ($>20 \times 10^6\ km^2$).

A total of about $20\text{–}22 \times 10^9\ t y^{-1}$ of solid and dissolved sediment is discharged annually. Our estimate for suspended sediment discharge (18×10^9 t y^{-1}) is less accurate than our estimate for dissolved sediment (3.9×10^9 t y^{-1}) because of the difficulty in factoring in both basin area (see above) and human impact (see below). Given the young, wet mountains and the large anthropogenic influence in south-east Asia, it is perhaps not surprising that this region accounts for 75% of the suspended sediment discharged to the global ocean. In fact, the six high-standing islands in the East Indies (Sumatra, Java, Celebes, Borneo, Timor, and New Guinea) collectively may discharge as much sediment ($4.1 \times 10^9\ t y^{-1}$) as all non-Asian rivers combined. Southeast Asia rivers are also the leading exporters of dissolved solids to the global ocean, $1.4 \times 10^9\ t y^{-1}$ (35% of the global total), but Europe and eastern North America are also important, accounting for another 25%.

Another way to view river fluxes is to consider the ocean basins into which they empty. While the Arctic Ocean occupies <5% of the global ocean basin area ($17 \times 10^6\ km^2$), the total watershed draining into the Arctic is $21 \times 10^6\ km^2$, meaning a land/ocean ratio of 1.2. In contrast, the South Pacific accounts for one-quarter of the global ocean area, but its land/ocean ratio is only 0.05 (**Table 4**). The greatest fresh water input occurs in the North Atlantic (largely

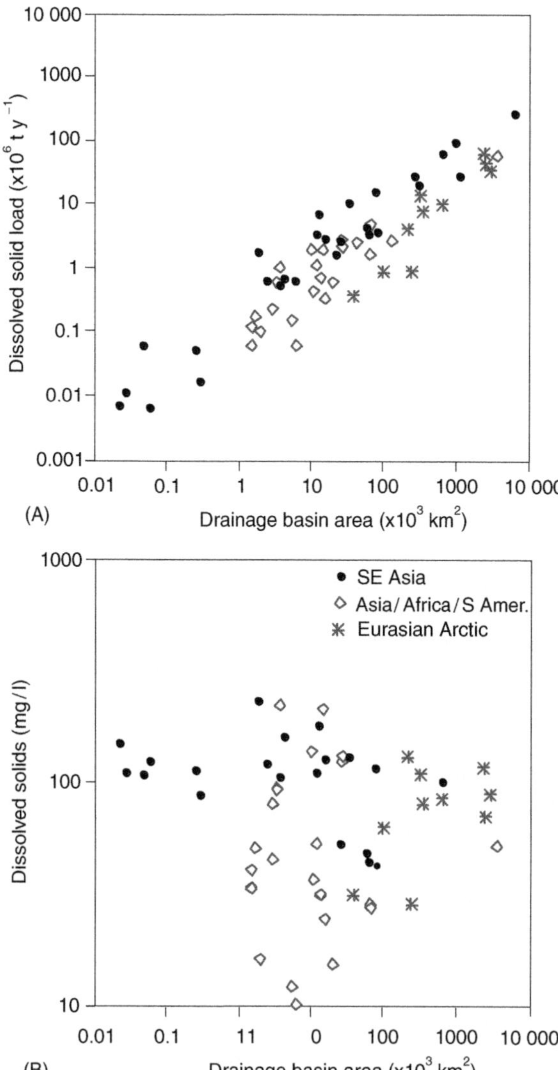

Figure 2 Dissolved solid load versus basin area for rivers draining south-east Asian rivers (solid dots), south Asia, west Africa, north-east South America (open diamonds – high rainfall, old rocks), and the Canadian-Eurasian Arctic (asterisks – low precipitation, old rocks). Note that while the dissolved loads relative to basin size are much closer than they are for suspended load, the difference seems to increase with decreasing basin area (A). The lack of correlation between dissolved concentrations and basin area (B) suggests that residence time of a river's surface water may play less of a role than ground water in determining the quantity and character of the dissolved solid fraction.

because of the Amazon and, to a lesser extent, the Orinoco and Mississippi), but the greatest input per unit basin area is in the Arctic (28 cm y^{-1} if evenly distributed over the entire basin). The least discharge per unit area of ocean basin is the South Pacific (4.5 cm y^{-1} distributed over the entire basin). In terms of suspended sediment load, the major sinks are the Pacific and Indian oceans (11.1 and 4×10^9 t y^{-1}, respectively), whereas the North Atlantic is the major sink for dissolved solids (1.35×10^9 t y^{-1}; **Table 4**), largely due to the high dissolved loads of European and eastern North American rivers.

This is not to say, of course, that the fresh water or its suspended and dissolved loads are evenly distributed throughout the ocean basins. In fact, most fluvial identity is lost soon after the river discharges into the ocean due to mixing, flocculation, and chemical uptake. In many rivers, much of the sediment is sequestered on deltas or in the coastal zone. In most broad, passive margins, in fact, little sediment presently escapes the inner shelf, and very little reaches the outer continental margin. During low stands of sea level, on the other hand, most fluvial sediment is discharged directly to the deep sea, where it can be redistributed far from its source(s) via mass wasting (e.g., slumping and turbidity currents). This contrast between sediment discharge from passive margins during high and low stands of sea level is the underlying basis for sequence stratigraphy. However, in narrow active margins, often bordered by young mountains, many rivers are relatively small (e.g., the western Americas, East Indies) and therefore often more responsive to episodic events. As such, sediment can escape the relatively narrow continental shelves, although the ultimate fate of this sediment is still not well documented.

Changes in Fluvial Processes and Fluxes – Natural and Anthropogenic

The preceding discussion is based mostly on data collected in the past 50 years. In most cases it reflects neither natural conditions nor long-term conditions representing the geological past. While the subject is still being actively debated, there seems little question that river discharge during the last glacial maximum (LGM), 15 000–20 000 years ago, differed greatly from present-day patterns. Northern rivers were either seasonal or did not flow except during periodic ice melts. Humid and sub-humid tropical watersheds, on the other hand, may have experienced far less precipitation than they do at present.

With the post-LGM climatic warming and ensuing ice melt, river flow increased. Scattered terrestrial and marine data suggest that during the latest Pleistocene and earliest Holocene, erosion rates and sediment delivery increased dramatically as glacially eroded debris was transported by increased river flow. As vegetation became more firmly established in the mid-Holocene, however, erosion rates apparently decreased, and perhaps would have remained relatively low except for human interference.

Table 3 Basin area, TDS (mg l^{-1}), annual dissolved load, and dissolved yield (t/km^2 per year) for various global rivers

River	*Basin area* ($\times 10^3 km^2$)	*TDS* ($mg\ l^{-1}$)	*Dissolved load* ($\times 10^6 t\ y^{-1}$)	*Dissolved yield* ($t\ km^{-2} y^{-1}$)
Amazon	6300	43	270	43
Yangtze (Changjiang)	1800	200	180	100
Ganges/Bramaputra	1650	130	150	91
Mississippi	3300	280	140	42
Irrawaddy	430	230	98	230
Salween	320	310	65	200
MacKenzie	1800	210	64	35
Parana/Uruguay	2800	92	62	22
St Lawrence	1200	180	62	52
Rhine	220	810	60	270
Cunene (Angola)	110	51	0.35	3
Torne (Norway)	39	30	0.37	9
Ems (Germany)	8	180	0.34	42
Citandy (Indonesia)	4.8	62	0.38	79

The first 10 rivers represent the highest dissolved loads of all world rivers, only the Amazon, Yangtze, Ganges/Brahmaputra, and Mississippi rivers having annual loads >100 $\times 10^6$ t y^{-1}. Discharges vary from 6300 (Amazon) to 74 $\times$ km^3 y^{-1} (Rhine), and basin areas from 6300 (Amazon) to 220 $\times 10^3$ km^2 (Rhine). Corresponding average dissolved concentrations and yields vary from 43 to 810 and 43 to 270, respectively. The bottom rivers have similar annual dissolved loads (0.34–0.38 $\times 10^6$ t y^{-1}), but their average sediment concentrations and yields vary by factors of 6–26 (respectively), reflecting both natural and anthropogenic influences.

Few terrestrial environments have been as affected by man's activities as have river basins, which is not surprising considering the variety of uses that humans have for rivers and their drainage basins: agriculture and irrigation, navigation, hydroelectric power, flood control, industry, etc. Few, if any, modern rivers have escaped human impact, and with exception of the Amazon and a few northern rivers, it is difficult to imagine a river whose flow has not been strongly affected by anthropogenic activities. Natural ground cover helps the landscape resist erosion, whereas deforestation, road construction, agriculture, and urbanization all can result in channelized flow and increased erosion. Erosion rates and corresponding sediment discharge of rivers draining much of southern Asia and Oceania have increased substantially because of human activities, locally as much as 10-fold. While land erosion in some areas of the world recently has been decreasing (e.g., Italy, France, and Spain) due to decreased farming and increased reforestation; deforestation and poor land conservation practices elsewhere, particularly in the developing world, have led to accelerated increases in both land erosion and fluvial sediment loads.

While increased river basin management has led to very low suspended loads for most northern European rivers, mining and industrial activities have resulted in greatly elevated dissolved concentrations.

Table 4 Cumulative oceanic areas, drainage basin areas, and discharge of water, suspended and dissolved solids of rivers draining into various areas of the global ocean

Oceanic area	*Basin area* ($\times 10^6 km^2$)	*Drainage basin* ($\times 10^6 km^2$)	*Water discharge* ($km^3 y^{-1}$)	*Sediment load* ($\times 10^6 t\ y^{-1}$)	*Dissolved load* ($\times 10^6 t\ y^{-1}$)
North Atlantic	44	30	12 800	2500	1350
South Atlantic	46	12	3300	400	240
North Pacific	83	15	6000	7200	660
South Pacific	94	5	4300	3900	650
Indian	74	14	4000	4000	520
Arctic	17	21	4800	350	480
Total	358	98	35 200	18 000	3900

For this compilation, it is assumed that Sumatra and Java empty into the Indian Ocean, and that the other high-standing islands in Indonesia discharge into the Pacific Ocean. Rivers discharging into the Black Sea and Mediterranean area are assumed to be part of the North Atlantic drainage system. (Data from Milliman and Farnsworth, 2001.)

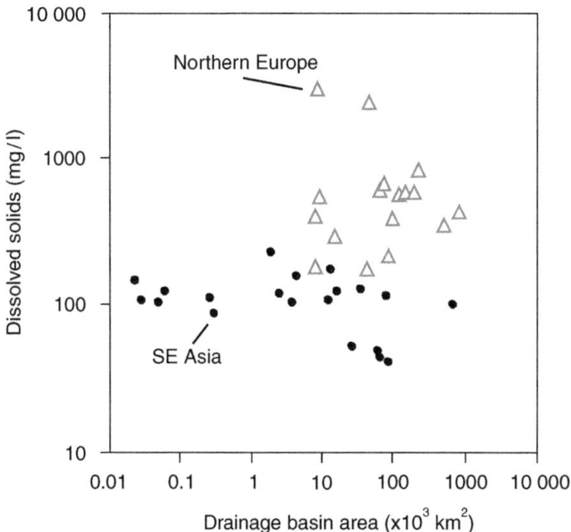

Figure 3 Dissolved solid concentration for south-east Asian rivers (also shown in **Figure 2**) compared with the much higher concentrations seen in northern European rivers (excluding those in Scandinavia). The markedly higher levels of dissolved solids in northern European rivers almost certainly reflects greater mining and industrial activity.

Compared to south-east Asian rivers, whose waters generally contain 100–200 mg l^{-1} dissolved solids, some European rivers, such as the Elbe, can have concentrations greater than 1000 mg l^{-1} (**Figure 3**), in large part due to the mining of salt deposits for potassium. Since the fall of the Soviet empire, attempts have been made to clean up many European rivers, but as of the late 1990s, they still had the highest dissolved concentrations in the world.

Interestingly, while sediment loads of rivers may be increasing, sediment flux to the ocean may be decreasing because of increased river diversion (e.g., irrigation and flood protection through levees) and stoppage (dams). As of the late 1990s there were >24 000 major dams in operation around the world. The nearly 200 dams along the Mississippi River, for example, have reduced sediment discharge to the Gulf of Mexico by >60%. In the Indus River, construction of irrigation barrages in the late 1940s led to an 80% reduction in the river's sediment load. Even more impressive is the almost complete cessation of sediment discharge from the Colorado and Nile rivers in response to dam construction. Not only have water discharge and sediment flux decreased, but high and low flows have been greatly modulated; the effect of this modulated flow, in contrast to strong seasonal signals, on the health of estuaries is still not clear.

Decreased freshwater discharge can affect the circulation of shelf waters. Many of the nutrients utilized in coastal primary production are derived from upwelled outer shelf and slope waters as they are advected landward to offset offshore flow of surface waters. Decreased river discharge from the Yangtze River resulting from construction of the Three Gorges Dam might decrease the shoreward flow of nutrient-rich bottom shelf waters in the East China Sea, which could decrease primary production in an area that is highly dependent on its rich fisheries.

Retaining river water within man-made reservoirs also can affect water quality. For example, reservoir retention of silicate-rich river water can lead to diatom blooms within the man-made lakes and thus depletion of silicate within the river water. One result is that increased ratios of dissolved nitrate and phosphate to dissolved silica may have helped change primary production in coastal areas from diatom-based to dinoflagellates and coccolithophorids. One result of this altered production may be increased hypoxia in coastal and shelf waters in the north-western Black Sea and other areas off large rivers.

Taken in total, negative human impact on river systems seems to be increasing, and these anthropogenic changes are occurring faster in the developing world than in Europe or North America. Increased land degradation (partly the result of increased population pressures) in northern Africa, for instance, contrasts strongly with decreased erosion in neighboring southern Europe. Dam construction in Europe and North America has slackened in recent years, both in response to environmental concerns and the lack of further sites of dam construction, but dams continue to be built in Africa and Asia.

Considering increased water management and usage, together with increased land degradation, the coastal ocean almost certainly will look different 100 years from now. One can only hope that the engineers and planners in the future have the foresight to understand potential impacts of drainage basin change and to minimize their effects.

Further Reading

Berner RA and Berner EK (1997) Silicate weathering and climate. In: Ruddiman RF (ed.) *Tectonic Uplift and Climate Change*, pp. 353–365. New York: Plenum Press.

Chen CTA (2000) The Three Gorges Dam: reducing the upwelling and thus productivity in the East China Sea. *Geophysics Research Letters* 27: 381–383.

Douglas I (1996) The impact of land-use changes, especially logging, shifting cultivation, mining and urbanization on sediment yields in tropical Southeast

Asia: a review with special reference to Borneo. *Int. Assoc. Hydrol. Sci. Publ.* 236: 463–472.

Edmond JM and Huh YS (1997) Chemical weathering yields from basement and orogenic terrains in hot and cold climates. In: Ruddiman RF (ed.) *Tectonic Uplift and Climate Change*, pp. 329–351. New York: Plenum Press.

Humborg C, Conley DJ, Rahm L, *et al.* (2000) Silicon retention in river basins: far-reaching effects on biogeochemistry and aquatic food webs in coastal marine environments. *Ambio*.

Lisitzin AP (1996) *Oceanic Sedimentation, Lithology and Geochemistry.* Washington, DC: American Geophysics Union.

Meybeck M (1994) Origin and variable composition of present day river-borne material. *Material Fluxes on the Surface of the Earth, National Research Council Studies in Geophysics*, pp. 61–73. Washington: National Academy Press.

Meybeck M and Ragu A (1996) *River Discharges to the Oceans. An Assessment of Suspended Solids, Major Ions and Nutrients.* GEMS-EAP Report.

Milliman JD (1995) Sediment discharge to the ocean from small mountainous rivers: the New Guinea example. *Geo-Mar Lett* 15: 127–133.

Milliman JD and Farnsworth KL (2002) *River Runoff, Erosion and Delivery to the Coastal Ocean: A Global Analysis.* Oxford University Press (in press).

Milliman JD and Meade RH (1983) World-wide delivery of river sediment to the oceans. *J Geol* 91: 1–21.

Milliman JD and Syvitski JPM (1992) Geomorphic/tectonic control of sediment discharge to the ocean: the importance of small mountainous rivers. *J Geol* 100: 525–544.

Milliman JD, Ren M-E, Qin YS, and Saito Y (1987) Man's influence on the erosion and transport of sediment by Asian rivers: the Yellow River (Huanghe) example. *J Geol* 95: 751–762.

Milliman JD, Farnworth KL, and Albertin CS (1999) Flux and fate of fluvial sediments leaving large islands in the East Indies. *Journal of Sea Research* 41: 97–107.

Thomas MF and Thorp MB (1995) Geomorphic response to rapid climatic and hydrologic change during the late Pleistocene and early Holocene in the humid and sub-humid tropics. *Quarterly Science Review* 14: 193–207.

Walling DE (1995) Suspended sediment yields in a changing environment. In: Gurnell A and Petts G (eds.) *Changing River Channels*, pp. 149–176. Chichester: John Wiley Sons.

GROUNDWATER FLOW TO THE COASTAL OCEAN

A. E. Mulligan and M. A. Charette, Woods Hole Oceanographic Institution, Woods Hole, MA, USA

Introduction

Water flowing through the terrestrial landscape ultimately delivers fresh water and dissolved solutes to the coastal ocean. Because surface water inputs (e.g., rivers and streams) are easily seen and are typically large point sources to the coast, they have been well studied and their contributions to ocean geochemical budgets are fairly well known. Similarly, the hydrodynamics and geochemical importance of surface estuaries are well known. Only recently has significant attention turned toward the role of groundwater inputs to the ocean. Historically, such inputs were considered insignificant because groundwater flow is so much slower than riverine flow. Recent work however has shown that groundwater flow through coastal sediments and subsequent discharge to the coastal ocean can have a significant impact on geochemical cycling and it is therefore a process that must be better understood.

Groundwater discharge into the coastal ocean generally occurs as a slow diffuse flow but can be found as large point sources in certain terrain, such as karst. In addition to typically low flow rates, groundwater discharge is temporally and spatially variable, complicating efforts to characterize site-specific flow regimes. Nonetheless, the importance of submarine groundwater discharge (SGD) as a source of dissolved solids to coastal waters has become increasingly recognized, with recent studies suggesting that SGD-derived chemical loading may rival surface water inputs in many coastal areas. So while the volume of water discharged as SGD may be small relative to surface discharge, the input of dissolved solids from SGD can surpass that of surface water inputs. For example, SGD often represents a major source of nutrients in estuaries and embayments. Excess nitrogen loading can result in eutrophication and its associated secondary effects including decreased oxygen content, fish kills, and shifts in the dominant flora.

First, let us define 'groundwater' in a coastal context. We use the term to refer to any water that resides in the pore spaces of sediments at the land–ocean boundary. Hence, such water can be fresh terrestrially derived water that originates as rainwater infiltrating through the subsurface or it can represent saline oceanic water that flows through the sediments (**Figure 1**). Therefore, groundwater discharging to coastal waters can have salinity that spans a large range, being some mixture of the two end members. We therefore use the terms fresh SGD and saline SGD to distinguish these sources of fluid and brackish SGD to mean a mixture of the fresh and saline end members.

Basics of Groundwater Flow

Groundwater flow in the subsurface is driven by differences in energy – water flows from high energy areas to low energy. The mechanical energy of a unit volume of water is determined by the sum of gravitational potential energy, pressure energy, and kinetic energy:

$$\text{Energy per unit volume} = \rho g z + P + \frac{\rho V^2}{2} \quad [1]$$

where ρ is fluid density, g is gravitational acceleration, z is elevation of the measuring point relative to a datum, P is fluid pressure at the measurement point, and V is fluid velocity. Because groundwater flows very slowly (on the order of 1 m day^{-1} or less), its kinetic energy is very small relative to its gravitational potential and pressure energies and the kinetic energy term is therefore ignored. By removing the kinetic energy term and rearranging eqn [1] to express energy in terms of mechanical energy per unit weight, the concept of hydraulic head is developed:

$$\text{Energy per unit weight} = \text{hydraulic head} = z + \frac{P}{\rho g} \quad [2]$$

Groundwater therefore flows from regions of high hydraulic head to areas of low hydraulic head.

Because groundwater flows through a porous media, the rate of flow depends on soil properties such as the degree to which pore spaces are interconnected. The property of interest in groundwater flow is the permeability, **k**, which is a measure of the ease with which a fluid flows through the soil matrix. Groundwater flow rate can then be calculated using Darcy's law, which says that the flow rate is linearly proportional to the hydraulic gradient:

$$\mathbf{q} = -\frac{\rho g \mathbf{k}}{\mu}(\nabla h) \quad [3]$$

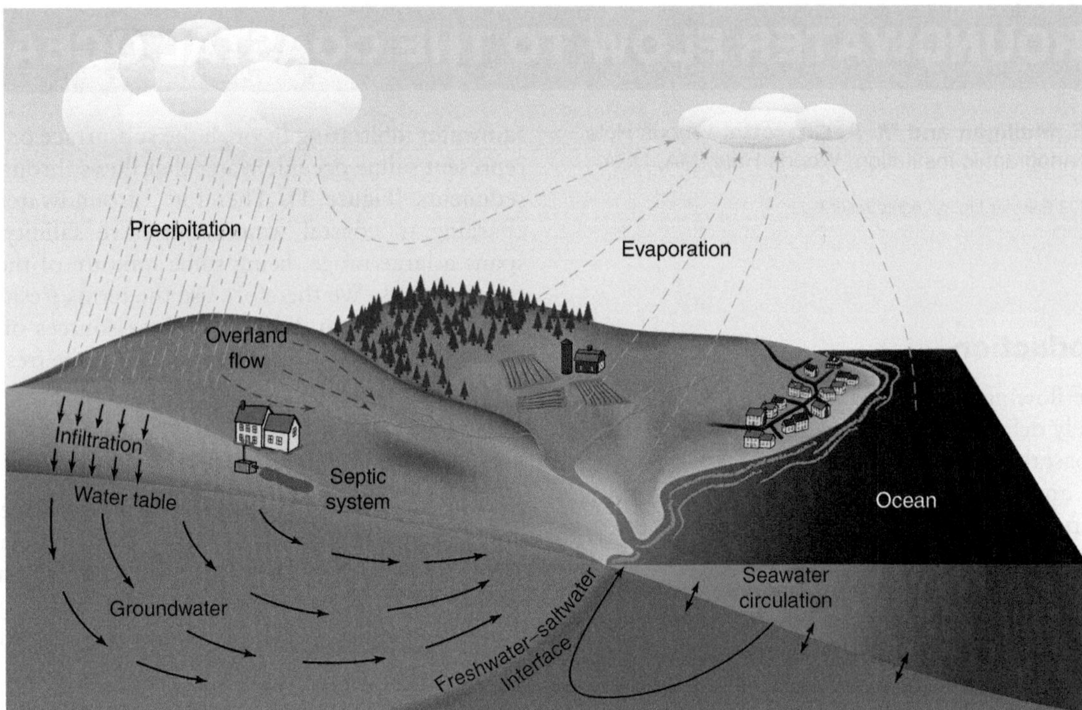

Figure 1 Simplified water cycle at the coastal margin. Modified from Heath R (1998) Basic ground-water hydrology. *US Geological Survey Water Supply Paper 2220*. Washington, DC: USGS.

where **q** is the Darcy flux, or flow rate per unit surface area, and μ is fluid viscosity. A more general expression of Darcy's law is:

$$\mathbf{q} = -\frac{\mathbf{k}}{\mu}(\nabla P + \rho \mathbf{g} \nabla z) \qquad [4]$$

In inland aquifers, the density of groundwater is constant and eqn [4] is reduced to the simpler form of Darcy's law (eqn [3]). In coastal aquifers, however, the presence of saline water along the coast means that the assumption of constant density is not valid and so the more inclusive form of Darcy's law, eqn [4], is required.

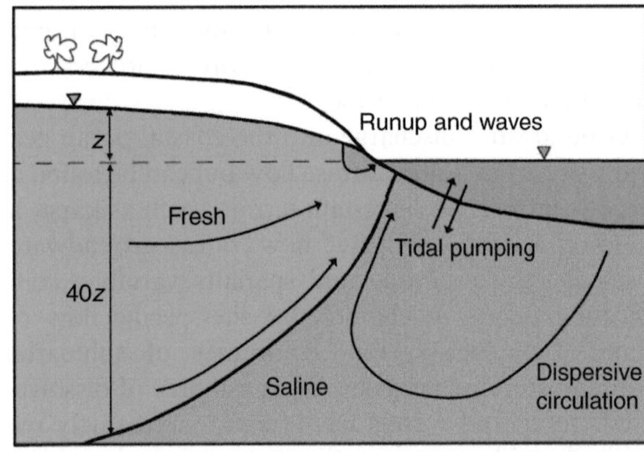

Figure 2 Simplification of an unconfined coastal groundwater system. Water flow is driven by the inland hydraulic gradient, tides, beach runup and waves, and dispersive circulation. Other forcing mechanisms can drive fluid through coastal sediments, including seasonal changes in recharge to the inland groundwater system and tidal differences across islands.

Groundwater Flow at the Coast

Several forces drive groundwater flow through coastal aquifers, leading to a complex flow regime with significant variability in space and time (**Figure 2**). The primary driving force of fresh SGD is the hydraulic gradient from the upland region of a watershed to the surface water discharge location at the coast. Freshwater flux is also influenced by several other forces at the coastal boundary that also drive seawater through the sediments. For example, seawater circulates through a coastal aquifer under the force of gravity, from oceanic forces such as waves and tides, as a result of dispersive circulation along the freshwater–saltwater boundary within the aquifer, and from changes in upland recharge. Several other forcing mechanisms exist, but they are generally only present in specific settings. For example, tidal height differences across many islands can drive flow through the subsurface. All of these forcing mechanisms affect the rate of fluid flow for both fresh and saline groundwater and are ultimately important in controlling the submarine discharge of both fluids.

The analysis of coastal groundwater flow must account for the presence of both fresh- and saline water components. When appropriate, such as in regional-scale analysis or for coarse estimation purposes, an assumption can be made that there is a sharp transition, or interface, between the fresh and saline groundwater. While this assumption is not strictly true, it is often appropriate, and invoking it results in simplifying the analysis. For example, we can estimate the position of the freshwater–saltwater interface by assuming a sharp interface, no flow within the saltwater region, and only horizontal flow within the fresh groundwater. Invoking these assumptions means that the pressures at adjacent points along the interface on both the freshwater and saltwater sides are equal. Equating these pressures and rearranging, the depth to the interface can be calculated as follows:

$$\text{Interface depth} = \frac{\rho_1}{\rho_2 - \rho_1} z = 40z \qquad [5]$$

where ρ_1 is the density of fresh water ($1000\,\mathrm{kg\,m^{-3}}$) and ρ_2 is the density of seawater ($1025\,\mathrm{kg\,m^{-3}}$). This equation states that the depth of the interface is 40 times the elevation of the water table relative to mean sea level. While eqn [5] is only an approximation of the interface location, it is very helpful in thinking about freshwater and saltwater movement in response to changes in fresh groundwater levels. As recharge to an aquifer increases, water levels increase and the interface is pushed downward. This is also equivalent to pushing the interface seaward and the net effect is to force saltwater out of the subsurface and to replace it with fresh water. The opposite flows occur during times of little to no recharge or if groundwater pumping becomes excessive.

While the sharp-interface approach is useful for conceptualizing flow at the coast, particularly in large-scale problems, the reality is more complex. Not only does the saline groundwater flow but also a zone of intermediate salinity extends between the fresh and saline end members, establishing what many refer to as a 'subterranean estuary'. Like their surface water counterparts, these zones are hotbeds of chemical reactions. Because the water in the interface zone ultimately discharges into coastal waters, the flow and chemical dynamics within the zone are critically important to understand. Research into these issues has only just begun.

Detecting and Quantifying Submarine Groundwater Discharge

As a first step in quantifying chemical loads to coastal waters, the amount of water flowing out of the subsurface must be determined. This is particularly challenging because groundwater flow is spatially and temporally variable. A number of qualitative and quantitative techniques have been developed to sample SGD, with each method sampling a particular spatial and temporal scale. Because of limitations with each sampling method, several techniques should be used at any particular site.

Physical Approaches

Infrared thermography Infrared imaging has been used to identify the location and spatial variability of SGD by exploiting the temperature difference between surface water and groundwater at certain times of the year. While this technique is quite useful for identifying spatial discharge patterns, it has not yet been applied to estimating flow rates.

An example of thermal infrared imagery is shown in **Figure 3**, an image of the head of Waquoit Bay, a small semi-enclosed estuary on Cape Cod, Massachusetts, USA. In late summer, the groundwater temperature is approximately 13 °C, whereas surface water is about 7–10 °C warmer. Locations of SGD can be seen in the infrared image as locations along the beach face with cooler temperature than the surrounding surface water. The image clearly shows spatial variability in SGD along the beach face, information that is extremely valuable in designing an appropriate field sampling campaign.

Hydrologic approaches There are two hydrologic approaches to estimating SGD: the mass balance method and Darcy's law calculation. Both methods are typically applied to estimating fresh groundwater discharge, although Darcy's law can also be used to estimate saline flow into and out of the seafloor.

To apply Darcy's law, one must measure the soil permeability and hydraulic head at several locations (at least two) at the field site. Data must also be gathered to determine the cross-sectional flow area. The field data are then used with Darcy's law to calculate a groundwater flow rate into the coastal ocean. The main disadvantages of this approach include the fact that permeability is highly heterogeneous, often ranging over several orders of magnitude, and so an 'average' value to use with Darcy's law is seldom, if ever, well known. Furthermore, hydraulic head measurements require invasive, typically

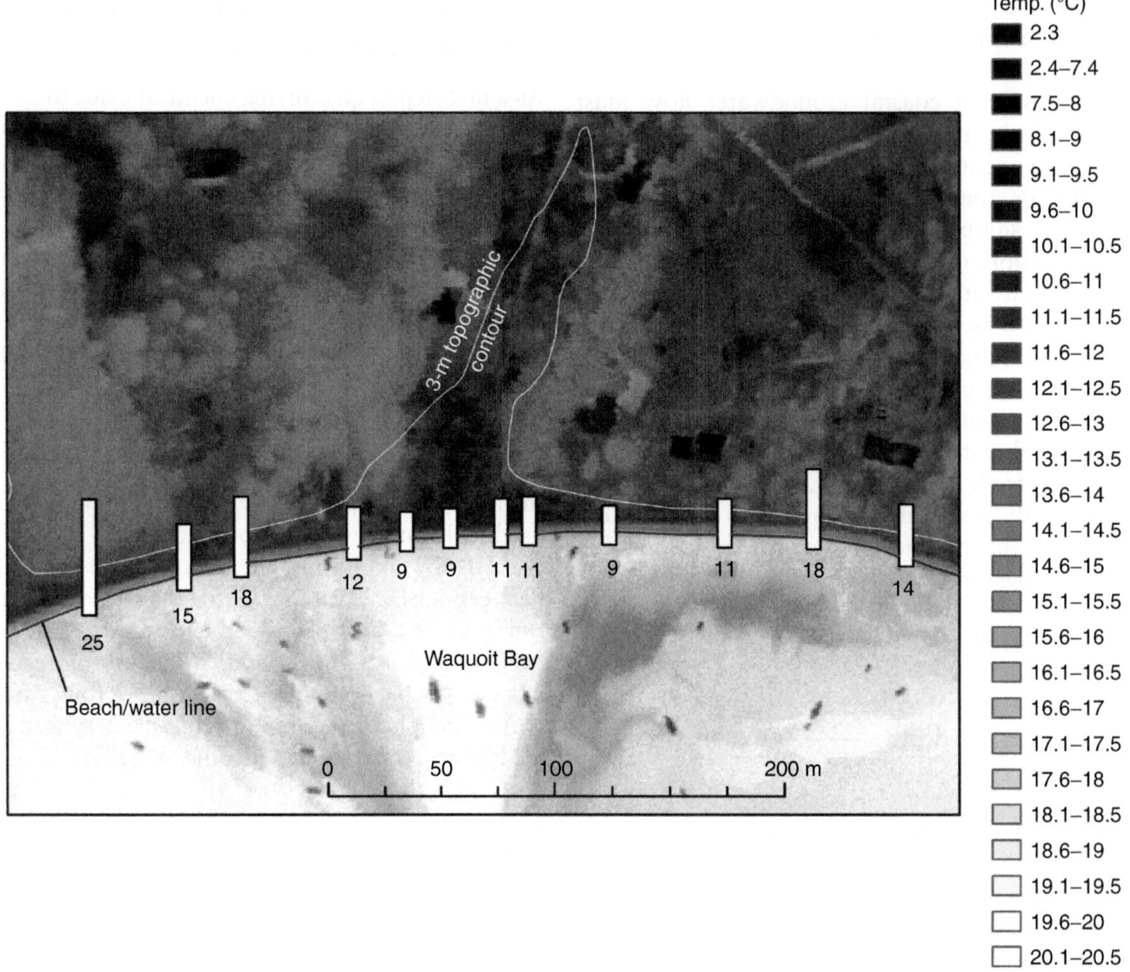

Figure 3 Thermal infrared image of the head of Waquoit Bay, Massachusetts. The beach/water line is shown as a red curve. Light grays imply higher temperatures and darker shades show lower temperatures. Lower temperatures within the bay indicate regions influenced by SGD. The bars show average groundwater seepage rates as measured by manual seep meters from high tide to low tide. Numbers below the bars are average rates in cm d^{-1}. Modified from Mulligan AE and Charette MA (2006) Intercomparison of submarine ground water discharge estimates from a sandy unconfined aquifer. *Journal of Hydrology* 327: 411–425.

expensive, well installations. Finally, hydraulic head is a point measurement and capturing the spatial variability therefore requires installing many wells. The primary advantage of this approach is that it is well established and easy to implement: head measurements are easy to collect once wells are installed and the flux calculations are simple.

The mass balance approach to estimating SGD requires ascertaining all inputs and outputs of water flow, except SGD, through the groundwatershed. Assuming a steady-state condition over a specified time frame, the groundwater discharge rate is calculated as the difference between all inputs and all outputs. Implementing this approach can be quite simple or can result in complex field campaigns, but the quality of the data obviously affect the level of uncertainty. Even with extensive field sampling, water budgets are seldom known with certainty and so should be used with that in mind. Furthermore, if the spatial and temporal variability of SGD is needed for a particular study, the mass balance approach is not appropriate.

Direct measurements: Seepage meters SGD can be measured directly with seepage meters. Manual seepage meters (**Figure 4**) are constructed using the tops of 55 gallon drums, where one end is open and placed into the sediment. The top of the seepage meter has a valve through which water can flow; a plastic bag prefilled with a known volume of water is attached to the valve, so that inflow to or outflow from the sediments can be determined. After a set length of time, the bag is removed and the volume of water in the bag is measured. The change in water volume over the sampling period is then used to determine the average flow rate of fluid across the water–sediment boundary over the length of the sampling period. These meters are very simple to operate, but they are manually intensive and are sensitive to wave disturbance and currents. Furthermore, they only sample a small flow area and so many meters are needed to characterize the spatial variability seen at most sites (**Figure 3**).

Recently, several other technologies have been applied toward developing automated seep meters. Technologies include the heat-pulse method, continuous heat, ultrasonic, and dye dilution. These meters can be left in place for days and often weeks and will measure seepage without the manual intervention needed using traditional seepage meters. The trade-off with these meters is that they are expensive and therefore only a limited number are typically employed at any given time. An example of seepage measurements made using the dye dilution meter is shown in **Figure 5**. Note that the seepage rates are inversely proportional to tidal height. As the tide rises, the hydraulic gradient from land to sea is reduced, seepage slows, and the flow reverses, indicating that seawater is flowing into the aquifer at this location during high tide. Conversely, at low tide, the hydraulic gradient is at its steepest and SGD increases.

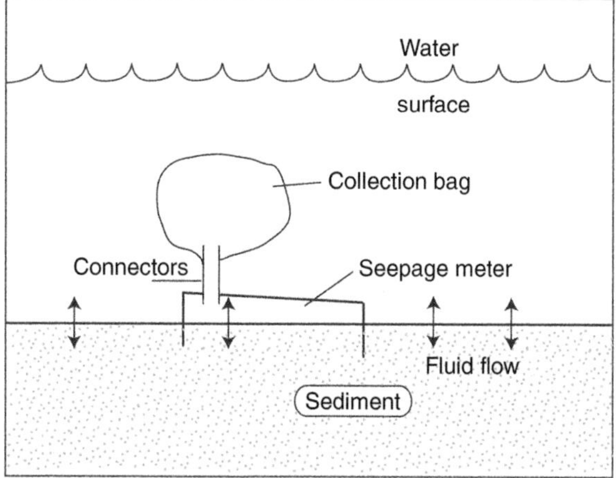

Figure 4 Graphic of a manual seepage meter deployed at the sediment–water interface. These meters can be constructed using the top of a 55 gallon drum. The collection bag serves as a fluid reservoir so that both inflow to and outflow from the sediments can be measured. Modified from Lee DR (1977) A device for measuring seepage in lakes and estuaries. *Limnology and Oceanography* 22: 140–147.

Chemical Tracers

The chemical tracer approach to quantifying SGD has an advantage over seepage meters in that it provides an integrated flux over a wide range of spatial scales from estuaries to continental shelves. The principle of using a chemical tracer is simple: find an element or isotope that is highly enriched (or depleted) in groundwater relative to other sources of water, like rivers or rainfall, to the system under study. If SGD is occurring, then the flux of this element via groundwater will lead to enrichment in the coastal zone that is well above background levels in the open ocean (**Figure 6**). A simple mass balance/box model for the system under study can be performed, where all sources of the tracer other than groundwater are subtracted from the total inventory of the chemical.

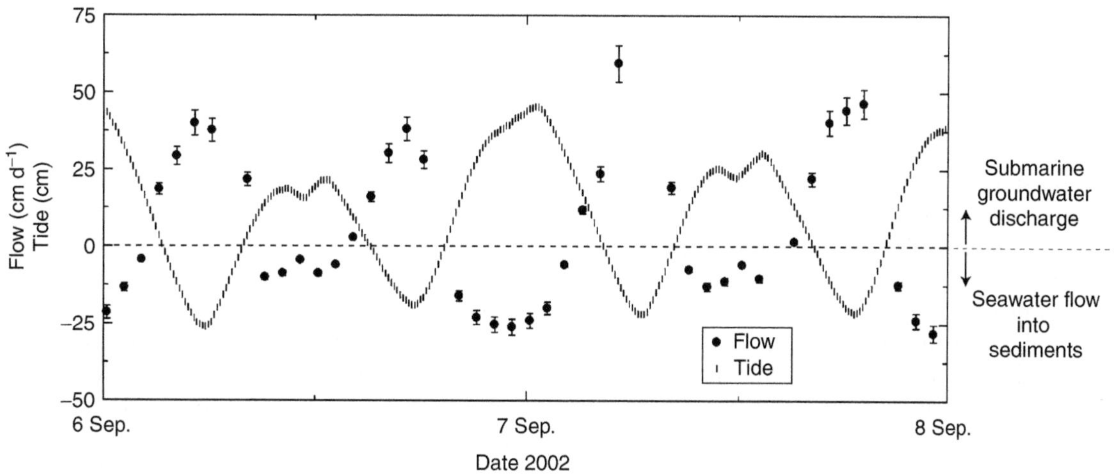

Figure 5 Seepage rates at Waquoit Bay, Massachusetts, USA, measured using a dye dilution automated seepage meter. Modified from Sholkovitz ER, Herbold C, and Charette MA (2003) An automated dye-dilution based seepage meter for the time-series measurement of submarine groundwater discharge. *Limnology and Oceanography: Methods* 1: 17–29.

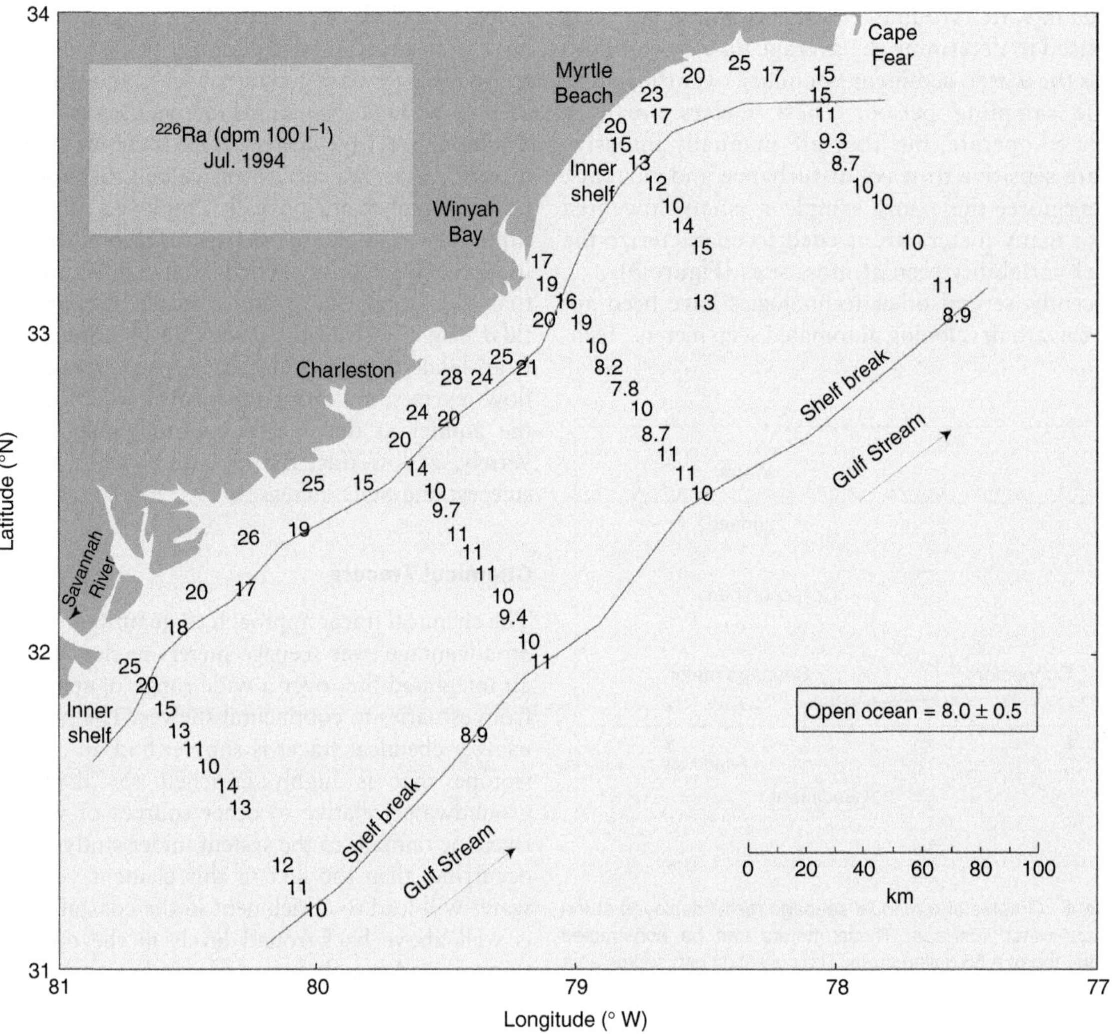

Figure 6 The distribution of ^{226}Ra offshore South Carolina reveals high activities on the inner shelf that decrease offshore. Moore (1996) used the excess ^{226}Ra to estimate regional SGD fluxes.

The residual inventory, or 'excess', is then divided by the concentration of the tracer in the discharging groundwater to calculate the groundwater flow rate.

Naturally occurring radionuclides such as radium isotopes and radon-222 have gained popularity as tracers of SGD due to their enrichment in groundwater relative to other sources and their built-in radioactive 'clocks'. The enrichment of these tracers is owed to the fact that the water–sediment ratio in aquifers is usually quite small and that aquifer sediments (and sediments in general) are enriched in many U and Th series isotopes; while many of these isotopes are particle reactive and remain bound to the sediments, some like Ra can easily partition into the aqueous phase. Radon-222 ($t_{1/2} = 3.82$ days) is the daughter product of ^{226}Ra ($t_{1/2} = 1600$ years) and a noble gas; therefore, it is even more enriched in groundwater than radium.

A key issue when comparing different techniques for measuring SGD is the need to define the fluid composition that each method is measuring (i.e., fresh, saline, or brackish SGD). For example, whereas hydrogeological techniques are estimates of fresh SGD, the radium and radon methods include a component of recirculated seawater. Therefore, it is often not possible to directly compare the utility of these techniques. Instead, they should be regarded as complementary.

One of the seminal studies that showed how SGD can impact chemical budgets on the scale of an entire coastline was conducted off North and South Carolina, USA. Literature estimates of water residence time, riverine discharge/suspended sediment load, and the activity of desorbable ^{226}Ra on riverine particles were used to determined that only ~50% of the ^{226}Ra inventory on the inner continental shelf

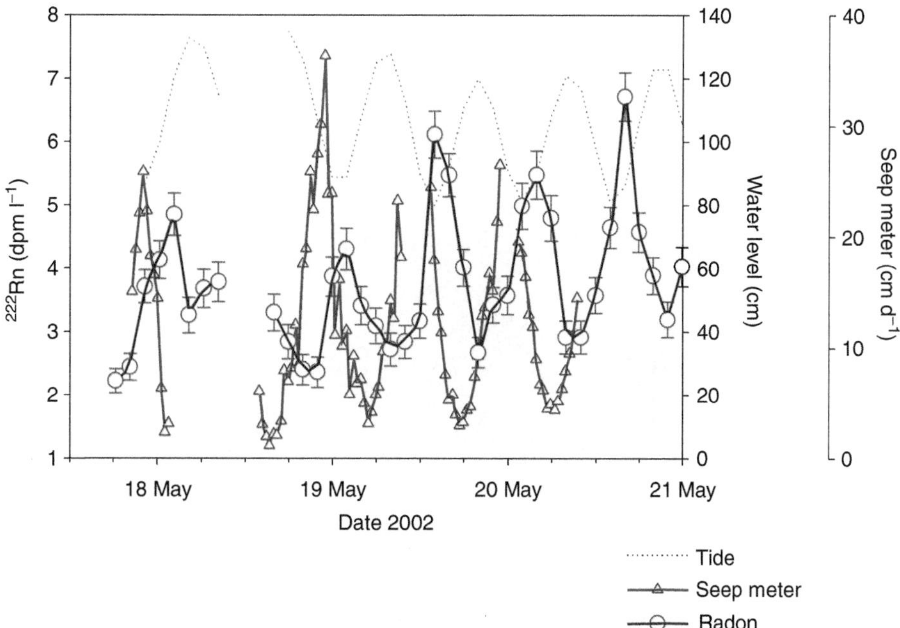

Figure 7 Plot comparing variations in SGD at Shelter Island, NY, between an automated seepage meter and the continuous radon method.

off North and South Carolina could be explained by surface inputs (**Figure 6**). The remaining inventory enters the system via SGD. Using an estimate of groundwater ^{226}Ra, it is calculated that the groundwater flux to this region of the coastline is on the order of 40% of the river water flux.

The approach for quantifying SGD using ^{222}Rn is similar to that for radium (^{226}Ra), except for a few key differences: (1) ^{222}Rn loss to the atmosphere must be accounted for in many situations, (2) there is no significant source from particles in rivers, and (3) decay must be accounted for owing to its relatively short half-life. The first example of ^{222}Rn use to quantify SGD to the coastal zone occurred in a study in the northeastern Gulf of Mexico. The strong pycnocline that develops in the summer time means that fluid that flows from the sediments into the bottom boundary layer is isolated from the atmosphere and therefore no correction for the air–sea loss of ^{222}Rn is needed. Using this approach, diffuse SGD in a 620 km^2 area of the inner shelf was estimated to be equivalent to ~20 first-magnitude springs (a first-magnitude spring has a flow equal to or greater than 245 000 m^3 water per day, which is equivalent to ~60% of the daily water supply for the entire state of Rhode Island, USA).

Since 2000, a number of SGD estimation technique intercomparison experiments have been conducted through a project sponsored by the Scientific Committee on Oceanic Research (SCOR) and the Land–Ocean Interaction in the Coastal Zone (LOICZ) Project. During these intercomparisons, several methods (chemical tracers, different types of seep meters, hydrogeologic approaches, etc.) were run side by side to evaluate their relative strengths and weaknesses. **Figure 7** displays a comparison from the Shelter Island, NY, experiment of calculated radon fluxes (based on measurements from a continuous radon monitor), with seepage rates measured directly via the dye dilution seepage meter. During the period (17–20 May) when both devices were operating at the same time, there is a clear and reproducible pattern of higher radon and water fluxes during the low tides. There is also a suggestion that the seepage spikes slightly led the radon fluxes, which is consistent with the notion that the groundwater seepage is the source of the radon (the radon monitor was located offshore of the seepage meter). The excellent agreement in patterns and overlapping calculated advection rates (seepage meter: 2–37 cm day^{-1}, average = 12 ± 8 cm day^{-1}; radon model: 0–34 cm day^{-1}, average = 11 ± 7 cm day^{-1}) by these two completely independent assessment tools is reassuring.

Geochemistry of the Subterranean Estuary

The magnitude of chemical fluxes carried by SGD is influenced by biogeochemical processes occurring in the subterranean estuary, defined as the mixing zone between groundwater and seawater in a coastal

(a)

(b)

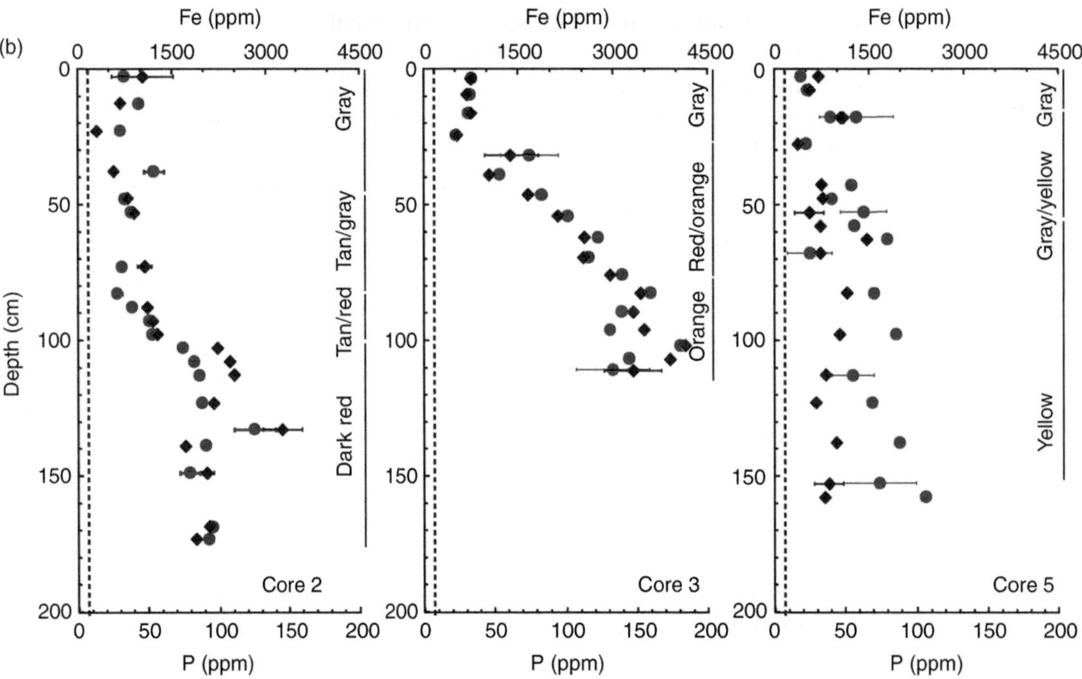

Figure 8 (a) Scientists extruding a sediment core taken through the subterranean estuary of Waquoit Bay, MA. Note the presence of iron oxides within the sediments at the bottom of the core (orange-stained sediments in foreground). (b) Changes in iron and phosphate concentration with depth in three sediment cores similar to the one shown in (a). The red circles indicate Fe concentration (ppm:μg Fe/g dry sediment) while the blue diamonds represent P (ppm:μg P/g dry sediment). Error bars indicate the standard deviation for triplicate leaches performed on a selected number of samples. The dashed lines represent the concentration of Fe and P in 'off-site' quartz sand. Also shown is the approximate color stratigraphy for each core. The R^2 value for Fe vs. P in cores 2, 3, and 5 is 0.80, 0.91, and 0.16, respectively.

aquifer. The biogeochemical reactions in such underground estuaries are presumed to be similar to the surface estuary (river water/seawater) counterpart, though few comprehensive studies of chemical cycling in subterranean estuaries have been undertaken. Drivers of biogeochemical reactions in these environments include oxidation–reduction gradients, desorption–sorption processes, and microbially driven diagenesis.

In a study of the Waquoit Bay subterranean estuary, a large accumulation of iron (hydr)oxide-coated sediments within the fresh–saline interface was encountered. These iron-oxide-rich sands could act as a geochemical barrier by retaining and accumulating certain dissolved chemical species carried to the subterranean estuary by groundwater and/or coastal seawater. Significant accumulation of phosphorus in the iron oxide zones of the Waquoit cores exemplifies this process (**Figure 8**).

Phosphorous is not the only nutrient that can be retained/removed via reactions in the subterranean estuary. The microbial reduction of nitrate to inert dinitrogen gas, a process known as denitrification, is known to occur in the redox gradients associated with fresh and saline groundwater mixing. Conversely, ammonium, which is more soluble in saline environments, may be released within the subterranean estuary's mixing zone. While the overall importance of SGD on the 'global cycle' of certain chemical species remains to be seen, there is little doubt that SGD is important at the local scale both within the United States and throughout the world.

Summary

Groundwater discharge to the coastal ocean can be an important source of fresh water and dissolved solutes. Although a significant amount of research into the role of SGD in solute budgets has only occurred in the past decade or so, there is increasing evidence that solute loading from groundwater can be significant enough to affect solute budgets and even ecosystem health. Proper estimation of solute loads from groundwater requires confident estimates of both the groundwater discharge rate and average solute concentrations in the discharging fluid, neither of which are easily determined.

Estimating groundwater discharge rates is complicated by the spatial and temporal variability of groundwater flow. A multitude of time-varying driving mechanisms complicate analysis, as does geologic heterogeneity. Nonetheless, a suite of tools from hydrogeology, geophysics, and geochemistry have been developed for sampling and measuring SGD.

The nature of geochemical reactions within nearshore sediments is not well understood, yet recent studies have shown that important transformations occur over small spatial scales. This is an exciting and critically important area of research that will reveal important process in the near future.

See also

Chemical Processes in Estuarine Sediments. Gas Exchange in Estuaries. Inverse Modeling of Tracers and Nutrients. Long-Term Tracer Changes. Tracer Release Experiments. Tracers of Ocean Productivity.

Further Reading

Burnett WC, Aggarwal PK, Bokuniewicz H, *et al.* (2006) Quantifying submarine groundwater discharge in the coastal zone via multiple methods. *Science of the Total Environment* 367: 498–543.

Burnett WC, Bokuniewicz H, Huettel M, Moore WS, and Taniguchi M (2003) Groundwater and pore water inputs to the coastal zone. *Biogeochemistry* 66: 3–33.

Cable JE, Burnett WC, Chanton JP, and Weatherly GL (1996) Estimating groundwater discharge into the northeastern Gulf of Mexico using radon-222. *Earth and Planetary Science Letters* 144: 591–604.

Charette MA and Sholkovitz ER (2002) Oxidative precipitation of groundwater-derived ferrous iron in the subterranean estuary of a coastal bay. *Geophysical Research Letters* 29: 1444.

Charette MA and Sholkovitz ER (2006) Trace element cycling in a subterranean estuary. Part 2: Geochemistry of the pore water. *Geochimica et Cosmochimica Acta* 70: 811–826.

Heath R (1998) Basic ground-water hydrology. *US Geological Survey Water Supply Paper 2220*. Washington, DC: USGS.

Kohout F (1960) Cyclic flow of salt water in the Biscayne aquifer of southeastern Florida. *Journal of Geophysical Research* 65: 2133–2141.

Lee DR (1977) A device for measuring seepage in lakes and estuaries. *Limnology and Oceanography* 22: 140–147.

Li L, Barry DA, Stagnitti F, and Parlange J-Y (1999) Submarine groundwater discharge and associated chemical input to a coastal sea. *Water Resources Research* 35: 3253–3259.

Michael HA, Mulligan AE, and Harvey CF (2005) Seasonal water exchange between aquifers and the coastal ocean. *Nature* 436: 1145–1148 (doi:10.1038/nature03935).

Miller DC and Ullman WJ (2004) Ecological consequences of ground water discharge to Delaware Bay, United States. *Ground Water* 42: 959–970.

Moore WS (1996) Large groundwater inputs to coastal waters revealed by Ra-226 enrichments. *Nature* 380: 612–614.

Moore WS (1999) The subterranean estuary: A reaction zone of ground water and sea water. *Marine Chemistry* 65: 111–125.

Mulligan AE and Charette MA (2006) Intercomparison of submarine ground water discharge estimates from a sandy unconfined aquifer. *Journal of Hydrology* 327: 411–425.

Paulsen RJ, Smith CF, O'Rourke D, and Wong T (2001) Development and evaluation of an ultrasonic ground water seepage meter. *Ground Water* 39: 904–911.

Portnoy JW, Nowicki BL, Roman CT, and Urish DW (1998) The discharge of nitrate-contaminated groundwater from developed shoreline to marsh-fringed estuary. *Water Resources Research* 34: 3095–3104.

Shinn EA, Reich CD, and Hickey TD (2002) Seepage meters and Bernoulli's revenge. *Estuaries* 25: 126–132.

Sholkovitz ER, Herbold C, and Charette MA (2003) An automated dye-dilution based seepage meter for the time-series measurement of submarine groundwater discharge. *Limnology and Oceanography: Methods* 1: 17–29.

Slomp CP and Van Cappellen P (2004) Nutrient inputs to the coastal ocean through submarine groundwater discharge: Controls and potential impact. *Journal of Hydrology* 295: 64–86.

Taniguchi M, Burnett WC, Cable JE, and Turner JV (2002) Investigation of submarine groundwater discharge. *Hydrological Processes* 16: 2115–2129.

Valiela I, Foreman K, LaMontagne M, *et al.* (1992) Couplings of watersheds and coastal waters: Sources and consequences of nutrient enrichment in Waquoit Bay, Massachusetts. *Estuaries* 15: 443–457.

GAS EXCHANGE IN ESTUARIES

M. I. Scranton, State University of New York, Stony Brook, NY, USA
M. A. de Angelis, Humboldt State University, Arcata, CA, USA

Introduction

Many atmospherically important gases are present in estuarine waters in excess over levels that would be predicted from simple equilibrium between the atmosphere and surface waters. Since estuaries are defined as semi-enclosed coastal bodies of water that have free connections with the open sea and within which sea water is measurably diluted with fresh water derived from land drainage, they tend to be supplied with much larger amounts of organic matter and other compounds than other coastal areas. Thus, production of many gases is enhanced in estuaries relative to the rest of the ocean. The geometry of estuaries, which typically have relatively large surface areas compared to their depths, is such that flux of material (including gases) from the sediments, and fluxes of gases across the air/water interface, can have a much greater impact on the water composition than would be the case in the open ocean. Riverine and tidal currents are often quite marked, which also can greatly affect concentrations of biogenic gases.

Gas Solubility

The direction and magnitude of the exchange of gases across an air/water interface are determined by the difference between the surface-water concentration of a given gas and its equilibrium concentration or gas solubility with respect to the atmosphere. The concentration of a specific gas in equilibrium with the atmosphere (C_{eq}) is given by Henry's Law:

$$C_{eq} = p_{gas}/K_H \qquad [1]$$

where p_{gas} is the partial pressure of the gas in the atmosphere, and K_H is the Henry's Law constant for the gas. Typically as temperature and salinity increase, gas solubility decreases. (Note that Henry's Law and the Henry's Law constant also may be commonly expressed in terms of the mole fraction of the gas in either the gas or liquid phase.) For gases that make up a large fraction of the atmosphere (O_2, N_2, Ar), p_{gas} does not vary temporally or spatially. For trace atmospheric gases (carbon dioxide (CO_2), methane (CH_4), hydrogen (H_2), nitrous oxide (N_2O), and others), p_{gas} may vary considerably geographically or seasonally, and may be affected by anthropogenic activity or local natural sources.

Gas Exchange (Flux) Across the Air/Water Interface

The rate of gas exchange across the air/water interface for a specific gas is determined by the degree of disequilibrium between the actual surface concentration of a gas (C_{surf}) and its equilibrium concentration (C_{eq}), commonly expressed as R:

$$R = C_{surf}/C_{eq} \qquad [2]$$

If $R = 1$, the dissolved gas is in equilibrium with the atmosphere and no net flux or exchange with the atmosphere occurs. For gases with $R < 1$, the dissolved gas is undersaturated with respect to the atmosphere and there is a net flux of the gas from the atmosphere to the water. For gases with $R > 1$, a net flux of the gas from the water to the atmosphere occurs.

Models of Gas Exchange

The magnitude of the flux (F) in units of mass of gas per unit area per unit time across the air/water interface is a function of the magnitude of the difference between the dissolved gas concentration and its equilibrium concentration as given by Fick's First Law of Diffusion:

$$F = k(C_{surf} - C_{eq}) \qquad [3]$$

where k is a first order rate constant, which is a function of the specific gas and surface water conditions. The rate constant, k, also known as the transfer coefficient, has units of velocity and is frequently given as

$$k = D/z \qquad [4]$$

where D is the molecular diffusivity (in units of cm^2 s^{-1}), and z is the thickness of the laminar layer at the air/water interface, which limits the diffusion of gas across the interface.

In aquatic systems, C_{surf} is easily measured by gas chromatographic analysis, and C_{eq} may be calculated readily if the temperature and salinity of the water are known. In order to determine the flux of gas (F) in or out of the water, the transfer coefficient (k) needs to be determined. The value of k is a function of the surface roughness of the water. In open bodies of water, wind speed is the main determinant of surface roughness. A number of studies have established a relationship between wind speed and either the transfer coefficient, k, or the liquid laminar layer thickness, z (**Figure 1**). The transfer coefficient is also related to the Schmidt number, Sc, defined as:

$$Sc = \nu/D \qquad [5]$$

where ν is the kinematic viscosity of the water. In calmer waters, corresponding to wind speeds of $<5\ \mathrm{m\,s^{-1}}$, k is proportional to $Sc^{-2/3}$. At higher wind speeds, but where breaking waves are rare, k is proportional to $S^{-1/2}$. Therefore, if the transfer coefficient of one gas is known, the k value for any

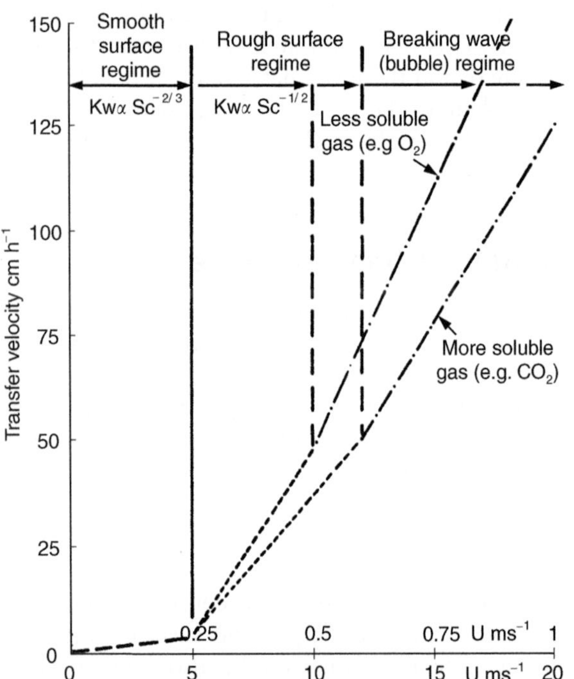

Figure 1 Idealized plot of transfer coefficient (Kw) as a function of wind speed (u) and friction velocity (u*). (Adapted with permission from Liss PS and Merlivat L (1985) Air–sea gas exchange rates. In: Buat-Menard P (ed.) The Role of Air–sea Exchange in Geochemical Cycling, p. 117. *NATO ASI Series C, vol. 185*. Dordrecht: Reidel.)

other gas can be determined as:

$$k_1/k_2 = Sc_1^n/Sc_2^n \qquad [6]$$

where n = the exponent. For short-term steady winds, the transfer coefficient for CO_2 has been derived as

$$k_{CO_2} = 0.31(U_{10})^2(Sc/600)^{-0.5} \qquad [7]$$

where U_{10} is the wind speed at a height of 10 m above the water surface. Eqns [6] and [7] can be used to estimate k for gases other than CO_2. In most estuarine studies wind speeds are measured closer to the water surface. In such cases, the wind speed measured at 2 cm above the water surface can be approximated as $0.5U_{10}$.

In restricted estuaries and tidally influenced rivers, wind speed may not be a good predictor of wind speed due to limited fetch or blockage of prevailing winds by shore vegetation. Instead, streambed-generated turbulence is likely to be more important than wind stress in determining water surface roughness. In such circumstances, the large eddy model may be used to approximate k as:

$$k = 1.46(D{*}u{*}l^{-1})^{1/2} \qquad [8]$$

where u is the current velocity, and l is equivalent to the mean depth in shallow turbulent systems. Much of the reported uncertainty (and study to study variability in fluxes) is caused by differences in assumptions related to the transfer coefficient rather than to large changes in concentration of the gas in the estuary.

Direct Gas Exchange Measurements

Gas exchange with the atmosphere for gases for which water column consumption and production processes are known can be estimated using a dissolved gas budget. Through time-series measurements of biological and chemical cycling, gas loss or gain across the air/water interface can be determined by difference. For example, in the case of dissolved O_2, the total change in dissolved O_2 concentration over time can be attributed to air/water exchange and biological processes. The contribution of biological processes to temporal changes in dissolved O_2 may be estimated from concurrent measurements of phosphate and an assumed Redfield stoichiometry, and subtracted from the total change to yield an estimate of air/water O_2 exchange.

Gas fluxes also may be measured directly using a flux chamber that floats on the water surface. The headspace of the chamber is collected and analyzed over several time points to obtain an estimate of the net amount of gas crossing the air/water interface enclosed by the chamber. If the surface area enclosed by the chamber is known, a net gas flux can be determined. Some flux chambers are equipped with small fans that simulate ambient wind conditions. However, most chambers do not use fans and so do not take account of the effects of wind-induced turbulence on gas exchange. Despite this limitation, flux chambers are important tools for measuring gas exchange in environments (such as estuaries or streams) where limited fetch or wind breaks produced by shoreline vegetation make wind less important than current-induced turbulence in shallow systems. While flux chambers may alter the surface roughness and, hence, gas/exchange rates via diffusion, flux chambers or other enclosed gas capturing devices also are the best method for determining loss of gases across the air/water interface due to ebullition of gas bubbles from the sediment.

Measurement of radon (Rn) deficiencies in the upper water column can be used to determine gas exchange coefficients and laminar layer thickness, which can then be applied to other gases using eqn [3]. In this method, gaseous ^{222}Rn, produced by radioactive decay of ^{226}Ra, is assumed to be in secular equilibrium within the water column. ^{222}Rn is relatively short-lived and has an atmospheric concentration of essentially zero. Therefore, in near-surface waters, a ^{222}Rn deficiency is observed, due to flux of ^{222}Rn across the air/water interface. The flux of ^{222}Rn across the air/water interface can be determined by the depth-integrated difference in measured ^{222}Rn and that which should occur based on the ^{226}Ra inventory. From this flux, the liquid laminar layer thickness, z, can be calculated.

Other volatile tracers have been used in estuaries to determine gas exchange coefficients. These tracers, such as chlorofluorocarbons (CFCs) or sulfur hexafluoride (SF_6), are synthetic compounds with no known natural source. Unlike ^{222}Rn, these gases are stable in solution. These tracers may be added to the aquatic system and the decrease of the gas due to flux across the air/water interface monitored over time. In some estuaries, point sources of these compounds may exist and the decrease of the tracer with distance downstream may be used to determine k or z values for the estuary.

Individual Gases

Methane (CH_4)

Atmospheric methane plays an important role in the Earth's radiative budget as a potent greenhouse gas, which is 3.7 times more effective than carbon dioxide in absorbing infrared radiation. Despite being present in trace quantities, atmospheric methane plays an important role controlling atmospheric chemistry, including serving as a regulator of tropospheric ozone concentrations and a major sink for hydroxyl radicals in the stratosphere. Methane concentrations have been increasing annually at the rate of approximately 1–2% over the last two centuries. While the contribution of estuaries to the global atmospheric methane budget is small because of the relatively small estuarine global surface area, estuaries have been identified as sources of methane to the atmosphere and coastal ocean and contribute a significant fraction of the marine methane emissions to the atmosphere.

Surface methane concentrations reported primarily from estuaries in North America and Europe range from 1 to >2000 nM throughout the tidal portion of the estuaries. Methane in estuarine surface waters is generally observed to be supersaturated (100% saturation ~2–3 nM CH_4) with R-values ranging from 0.7 to 1600 (**Table 1**). In general, estuarine methane concentrations are highest at the freshwater end of the estuary and decrease with salinity. This trend reflects riverine input as the major source of methane to most estuaries, with reported riverine methane concentrations ranging from 5 to 10 000 nM. Estuaries with large plumes have been

Table 1 Methane saturation values (R) and estimated fluxes to the atmosphere for US and European estuaries

Geographical region[a]	*R*[b]	*Flux CH_4 ($\mu mol\,m^{-2}\,h^{-1}$)*
North Pacific coast, USA	3–290	6.2–41.7[c]
North Pacific coast, USA	1–550	3.6–8.3[c]
Columbia River, USA	78	26.0[c]
Hudson River, USA	18–376	4.7–40.4[c]
Tomales Bay, CA, USA	2–37	17.4–26.3[c]
Baltic Sea, Germany	10.5–1550	9.4–15.6[c]
European Atlantic coast	0.7–1580	5.5[c]
Atlantic coast, USA	n.a.	102–1107[c]
Pettaquamscutt Estuary, USA	81–111	0.8–14.2[c] 541–3375[d]

[a]For studies that report values for a single estuary, the major river feeding the estuary is provided. For studies that report values for more than one estuary, the oceanic area being fed by the estuaries is given.

[b]R = degree of saturation = measured concentration/atmospheric equilibrium concentration. n.a. indicates values not available in reference.

[c]Diffusive flux.

[d]Ebullition (gas bubble) flux.

observed to cause elevated methane concentrations in adjacent coastal oceans. In addition to riverine input, sources of methane to estuaries include intertidal flats and marshes, ground-water input, runoff from agricultural and pasture land, petroleum pollution, lateral input from exposed bank soils, wastewater discharge and emission from organic-rich anaerobic sediments, either diffusively or via ebullition (transport of gas from sediments as bubbles) and subsequent dissolution within the water column. Anthropogenically impacted estuaries or estuaries supplied from impacted rivers tend to be characterized by higher water-column methane concentrations relative to pristine estuarine systems. Seasonally, methane levels in estuaries are higher in summer compared with winter, primarily due to increased bacterial methane production (methanogenesis) in estuarine and riverine sediments.

Methane can be removed from estuarine waters by microbial methane oxidation and emission to the atmosphere. Methane oxidation within estuaries can be quite rapid, with methane turnover times of <2 h to several days. Methane oxidation appears to be most rapid at salinities of less than about 6 (on the practical salinity units scale) and is strongly dependent on temperature, with highest oxidation rates occurring during the summer, when water temperatures are highest. Methane oxidation rates decrease rapidly with higher salinities.

Methane diffusive fluxes to the atmosphere reported for estuaries (**Table 1**) fall within a narrow range of 3.6–41.7 $\mu mol\ m^{-2}\ h^{-1}$ (2–16 mg $CH_4\ m^{-2}\ day^{-1}$). Using a global surface area for estuaries of $1.4 \times 10^6\ km^2$ yields an annual emission of methane to the atmosphere from estuaries of 1–8 $Tg\ y^{-1}$. Because the higher flux estimates given in **Table 1** generally were obtained close to the freshwater endmember of the estuary, the global methane estuarine emission is most likely within the range of 1–3 $Tg\ y^{-1}$, corresponding to approximately 10% of the total global oceanic methane flux to the atmosphere, despite the much smaller global surface area of estuaries relative to the open ocean.

Methane is also released to the atmosphere directly from anaerobic estuarine sediments via bubble formation and injection into the water column. Although small amounts of methane from bubbles may dissolve within the water column, the relatively shallow nature of the estuarine environment results in the majority of methane in bubbles reaching the atmosphere. The quantitative release of methane via this mechanism is difficult to evaluate due to the irregular and sporadic spatial and temporal extent of ebullition. Where ebullition occurs, the flux of methane to the atmosphere is considerably higher than diffusive flux (**Table 1**), but the areal extent of bubbling is relatively smaller than that of diffusive flux and, except in organic-rich stagnant areas such as tidal marshes, probably does not contribute significantly to estuarine methane emissions to the atmosphere. Methane emission via ebullition has been observed to be at least partially controlled by tidal changes in hydrostatic pressure, with release of methane occurring at or near low tide when hydrostatic pressure is at a minimum.

Nitrous Oxide (N_2O)

Nitrous oxide is another important greenhouse gas that is present in elevated concentrations in estuarine environments. At present, N_2O is responsible for about 5–6% of the anthropogenic greenhouse effect and is increasing in the atmosphere at a rate of about 0.25% per year. However, the role of estuaries in the global budget of the gas has only been addressed recently.

Nitrous oxide is produced primarily as an intermediate during both nitrification (the oxidation of ammonium to nitrate) and denitrification (the reduction of nitrate, via nitrite and N_2O, to nitrogen gas), although production by dissimilatory nitrate reduction to ammonium is also possible. In estuaries, nitrification and denitrification are both thought to be important sources. Factors such as the oxygen level in the estuary and the nitrate and ammonium concentrations of the water can influence which pathway is dominant, with denitrification dominating at very low, but non-zero, oxygen concentrations. Nitrous oxide concentrations are typically highest in the portions of the estuary closest to the rivers, and decrease with distance downstream. A number of workers have reported nitrous oxide maxima in estuarine waters at low salinities (<5–10 on the PSU scale), but this is not always the case. The turbidity maximum has been reported to be the site of maximum nitrification (presumably because of increased residence time for bacteria attached to suspended particulate matter, combined with elevated substrate (oxygen and ammonium)).

Table 2 presents a summary of the data published for degree of saturation and air–estuary flux of nitrous oxide from a variety of estuaries, all of which are located in Europe and North America. Concentrations are commonly above that predicted from air–sea equilibrium, and estimates of fluxes range from 0.01 $\mu mol\ m^{-2}\ h^{-1}$ to 5 $\mu mol\ m^{-2}\ h^{-1}$. Ebullition is not important for nitrous oxide because it is much more soluble than methane. Researchers have estimated the size of the global estuarine source for N_2O based on fluxes from individual estuaries

Table 2 Nitrous oxide saturation values (R) and estimated fluxes to the atmosphere for US and European estuaries

Estuary	*R*	*Flux*[a] N_2O ($\mu mol\, m^{-2} h^{-1}$)
Europe		
Gironde River	1.1–1.6	n.a.
Gironde River	≈1.0–3.2	n.a.
Oder River	0.9–3.1	0.014–0.165
Elbe	2.0–16	n.a.
Scheldt	≈1.0–31	1.27–4.77
Scheldt	≈1.2–30	3.56
UK		
Colne	0.9–13.6	1.3
Tamar	1–3.3	0.41
Humber	2–40	1.8
Tweed	0.96–1.1	≈0
Mediterranean		
Amvraikos Gulf	0.9–1.1	0.043 ± 0.0468
North-west USA		
Yaquina Bay	1.0–4.0	0.165–0.699
Alsea River	0.9–2.4	0.047–0.72
East coast USA		
Chesapeake Bay	0.9–1.4	n.a.
Merrimack	1.2–4.5	n.a.

[a]All fluxes given are for diffusive flux to the atmosphere. n.a. indicates that insufficient data were given to permit calculation of flux.

Table 3 Fluxes of carbon dioxide from estuaries in Europe and eastern USA

Estuary	*R*	*Flux*[a] CO_2 ($mmol\, m^{-2} h^{-1}$)
European rivers		
Northern Europe	0.7–61.1	1.0–31.7
Scheldt estuary	0.35–26.2	4.2–50
Portugal	1.6–15.8	10–31.7
UK	1.1–14.4	4.4–10.4
Clyde estuary	≈0.7–1.8	n.a.
East coast USA		
Hudson River (tidal freshwater)	1.2–5.4	0.67–1.54
Georgia rivers	Slight supersaturation to 22.9	1.7–23

[a]n.a., insufficient data were available to permit calculation of this value.

multiplied by the global area occupied by estuaries to range from $0.22\,Tg\,N_2O\,y^{-1}$ to $5.7\,Tg\,y^{-1}$ depending on the characteristics of the rivers studied. Independent estimates based on budgets of nitrogen input to rivers, assumptions about the fraction of inorganic nitrogen species removed by nitrification or denitrification, and the fractional 'yield' of nitrous oxide production during these processes indicate that nitrous oxide fluxes to the atmosphere from estuaries is about $0.06–0.34\,Tg\,N_2O\,y^{-1}$.

Carbon Dioxide (CO_2) and Oxygen (O_2)

Estuaries are typically heterotrophic systems, which means that the amount of organic matter respired within the estuary exceeds the amount of organic matter fixed by primary producers (phytoplankton and macrophytes). Since production of carbon dioxide then exceeds biological removal of carbon dioxide, it follows that estuaries are likely to be sources of the gas to the atmosphere. At the same time, since oxidation of organic matter to CO_2 requires oxygen, the heterotrophic nature of estuaries suggests that they represent sinks for atmospheric oxygen. In many estuaries, primary productivity is severely limited by the amount of light that penetrates into the water due to high particulate loadings in the water. In addition, large amounts of organic matter may be supplied to the estuary by runoff from agricultural and forested land, from ground water, from sewage effluent, and from organic matter in the river itself. There are many reports of estuarine systems with oxygen saturations below 1 (undersaturated with respect to the atmosphere), but few studies in which oxygen flux to the estuary has been reported. However, estuaries are often dramatically supersaturated with respect to saturation with CO_2, especially at low salinities, and a number of workers have reported estimates of carbon dioxide flux from these systems (**Table 3**).

Dimethylsulfide (DMS)

DMS is an atmospheric trace gas that plays important roles in tropospheric chemistry and climate regulation. In the estuarine environment, DMS is produced primarily from the breakdown of the phytoplankton osmoregulator 3-(dimethylsulfonium)-propionate (DMSP). DMS concentrations reported in estuaries are generally supersaturated, ranging from 0.5 to 22 nM, and increase with increasing salinity. DMS levels in the water column represent a balance between tightly coupled production from DMSP and microbial consumption. Only 10% of DMS produced from DMSP in the estuarine water column is believed to escape to the atmosphere from estuarine surface water, since the biological turnover of DMS (turnover time of 3–7 days) is approximately 10 times faster than DMS exchange across the air/water interface. A large part of the estuarine DMS flux to the atmosphere may occur over short time periods on the order of weeks, corresponding to phytoplankton blooms. The DMS flux for an estuary in Florida, USA, was estimated to be on the order of <1 nmoles m^{-2} h^{-1}. Insufficient data are available to determine reliable global DMS air/water exchanges from estuaries.

Hydrogen (H_2)

Hydrogen is an important intermediate in many microbial catabolic reactions, and the efficiency of hydrogen transfer among microbial organisms within an environment helps determine the pathways of organic matter decomposition. Hydrogen is generally supersaturated in the surface waters of the few estuaries that have been analyzed for dissolved hydrogen with R-values of 1.5–67. Hydrogen flux to the atmosphere from estuaries has been reported to be on the order of 0.06–0.27 nmol m^{-2} h^{-1}. The contribution of estuaries to the global atmospheric H_2 flux cannot be determined from the few available data.

Carbon Monoxide (CO)

Carbon monoxide in surface waters is produced primarily from the photo-oxidation of dissolved organic matter by UV radiation. Since estuarine waters are characterized by high dissolved organic carbon levels, the surface waters of estuaries are highly supersaturated and are a strong source of CO to the atmosphere. Reported R-values for CO range from approximately 10 to >10 000. Because of the highly variable distributions of dissolved CO within surface waters (primarily as the result of the highly variable production of CO), it is impossible to derive a meaningful value for CO emissions to the atmosphere from estuaries.

Carbonyl Sulfide (OCS)

Carbonyl sulfide makes up approximately 80% of the total sulfur content of the atmosphere and is the major source of stratospheric aerosols. Carbonyl sulfide is produced within surface waters by photolysis of dissolved organosulfur compounds. Therefore, surface water OCS levels within estuaries exhibit a strong diel trend. Carbonyl sulfide is also added to the water column by diffusion from anoxic sediments, where its production appears to be coupled to microbial sulfate reduction. Diffusion of OCS from the sediment to the water column accounts for ~75% of the OCS supplied to the water column and is responsible for the higher OCS concentrations in estuaries relative to the open ocean. While supersaturations of OCS are observed throughout estuarine surface waters, no trends with salinity have been observed. Atmospheric OCS fluxes to the atmosphere from Chesapeake Bay have been reported to range from 10.4 to 56.2 nmol m^{-2} h^{-1}. These areal fluxes are over 50 times greater than those determined for the open ocean.

Elemental Mercury (Hg^0)

Elemental mercury is produced in estuarine environments by biologically mediated processes. Both algae and bacteria are able to convert dissolved inorganic mercury to volatile forms, which include organic species (monomethyl- and dimethyl-mercury) and Hg^0. Under suboxic conditions, elemental mercury also may be the thermodynamically stable form of the metal. In the Scheldt River estuary, Hg^0 correlated well with phytoplankton pigments, suggesting that phytoplankton were the dominant factors, at least in that system. Factors that may affect elemental mercury concentrations include the type of phytoplankton present, photo-catalytic reduction of ionic Hg in surface waters, the extent of bacterial activity that removes oxygen from the estuary, and removal of mercury by particulate scavenging and sulfide precipitation. Fluxes of elemental mercury to the atmosphere have been estimated for the Pettaquamscutt estuary in Rhode Island, USA, and for the Scheldt, and range from 4.2–29 pmol m^{-2} h^{-1}, although the values are strongly dependent on the model used to estimate gas exchange coefficients.

Volatile Organic Compounds (VOCs)

In addition to gases produced naturally in the environment, estuaries tend to be enriched in by-products of industry and other human activity. A few studies have investigated volatile organic pollutants such as chlorinated hydrocarbons (chloroform, tetrachloromethane, 1,1-dichloroethane, 1,2-dichloroethane, 1,1,1-trichloroethane, trichloroethylene and tetrachloroethylene) and monocyclic aromatic hydrocarbons (benzene, toluene, ethylbenzene, o-xylene and m- and p-xylene). Concentrations of VOCs are controlled primarily by the location of the sources, dilution of river water with clean marine water within the estuary, gas exchange, and in some cases, adsorption onto suspended or settling solids. In some cases (for example, chloroform) there also may be natural biotic sources of the gas. Volatilization to the atmosphere can be an important 'cleansing' mechanism for the estuary system. Since the only estuaries studied to date are heavily impacted by human activity (the Elbe and the Scheldt), it is not possible to make generalizations about the importance of these systems on a global scale.

Conclusions

Many estuaries are supersaturated with a variety of gases, making them locally, and occasionally regionally, important sources to the atmosphere. However, estuarine systems are also highly variable in the amount of gases they contain. Since most estuaries studied to date are in Europe or the North American continent, more data are needed before global budgets can be reliably prepared.

Glossary

Air–water interface The boundary between the gaseous phase (the atmosphere) and the liquid phase (the water).

Catabolic Biochemical process resulting in breakdown of organic molecules into smaller molecules yielding energy.

Denitrification Reduction of nitrate via nitrite to gaseous endproducts (nitrous oxide and dinitrogen gas).

Ebullition Gas transport by bubbles, usually from sediments.

Estuary Semi-enclosed coastal body of water with free connection to the open sea and within which sea water is measurably diluted with fresh water derived from land drainage.

Gas solubility The amount of gas that will dissolve in a liquid when the liquid is in equilibrium with the overlying gas phase.

Henry's Law Constant Proportionality constant relating the vapor pressure of a solute to its mole fraction in solution.

Liquid laminar thickness The thickness of a layer at the air/water interface where transport of a dissolved species is controlled by molecular (rather than turbulent) diffusion.

Molecular diffusivity (D) The molecular diffusion coefficient.

Nitrification Oxidation of ammonium to nitrite and nitrate.

Practical salinity scale A dimensionless scale for salinity.

Redfield stoichiometry Redfield and colleagues noted that organisms in the sea consistently removed nutrient elements from the water in a fixed ratio (C : N : P=106 : 16 : 1). Subsequent workers have found that nutrient concentrations in the sea typically are present in those same ratios.

Transfer coefficient The rate constant which determines the rate of transfer of gas from liquid to gas phase.

See also

Air–Sea Transfer: Dimethyl Sulfide, COS, CS_2, NH_4, Non-Methane Hydrocarbons, Organo-Halogens. Air–Sea Transfer: N_2O, NO, CH_4, CO.

Further Reading

Bange HW, Rapsomanikis S, and Andreae MO (1996) Nitrous oxide in coastal waters. *Global Biogeochemical Cycles* 10: 197–207.

Bange HW, Dahlke S, Ramesh R, Meyer-Reil L-A, Rapsomanikis S, and Andreae MO (1998) Seasonal study of methane and nitrous oxide in the coastal waters of the southern Baltic Sea. *Estuarine, Coastal and Shelf Science* 47: 807–817.

Barnes J and Owens NJP (1998) Denitrification and nitrous oxide concentrations in the Humber estuary, UK and adjacent coastal zones. *Marine Pollution Bulletin* 37: 247–260.

Cai W-J and Wang Y (1998) The chemistry, fluxes and sources of carbon dioxide in the estuarine waters of the Satilla and Altamaha Rivers, Georgia. *Limnology and Oceanography* 43: 657–668.

de Angelis MA and Lilley MD (1987) Methane in surface waters of Oregon estuaries and rivers. *Limnology and Oceanography* 32: 716–722.

de Wilde HPJ and de Bie MJM (2000) Nitrous oxide in the Schelde estuary: production by nitrification and emission to the atmosphere. *Marine Chemistry* 69: 203–216.

Elkins JW, Wofsy SC, McElroy MB, Kolb CE, and Kaplan WA (1978) Aquatic sources and sinks for nitrous oxide. *Nature* 275: 602–606.

Frankignoulle M, Abril G, Borges A, *et al.* (1998) Carbon dioxide emission from European estuaries. *Science* 282: 434–436.

Frost T and Upstill-Goddard RC (1999) Air–sea exchange into the millenium: progress and uncertainties. *Oceanography and Marine Biology: An Annual Review* 37: 1–45.

Law CS, Rees AP, and Owens NJP (1992) Nitrous oxide: estuarine sources and atmospheric flux. *Estuarine, Coastal and Shelf Science* 35: 301–314.

Liss PS and Merlivat L (1986) Air–sea gas exchange rates: introduction and synthesis. In: Baut-Menard P (ed.) *The Role of Air–Sea Exchange in Geochemical Cycling*, pp. 113–127. Dordrecht: Riedel.

Muller FLL, Balls PW, and Tranter M (1995) Processes controlling chemical distributions in the Firth of Clyde (Scotland). *Oceanologica Acta* 18: 493–509.

Raymond PA, Caraco NF, and Cole JJ (1997) Carbon dioxide concentration and atmospheric flux in the Hudson River. *Estuaries* 20: 381–390.

Robinson AD, Nedwell DB, Harrison RM, and Ogilvie BG (1998) Hypernutrified estuaries as sources of N_2O emission to the atmosphere: the estuary of the River Colne, Essex, UK. *Marine Ecology Progress Series* 164: 59–71.

Sansone FJ, Rust TM, and Smith SV (1998) Methane distribution and cycling in Tomales Bay, California. *Estuaries* 21: 66–77.

Sansone FJ, Holmes ME, and Popp BN (1999) Methane stable isotopic ratios and concentrations as indicators of methane dynamics in estuaries. *Global Biogeochemical Cycles* 463–474.

Scranton MI, Crill P, de Angelis MA, Donaghay PL, and Sieburth JM (1993) The importance of episodic events in controlling the flux of methane from an anoxic basin. *Global Biogeochemical Cycles* 7: 491–507.

Seitzinger SP and Kroeze C (1998) Global distribution of nitrous oxide production and N input in freshwater and coastal marine ecosystems. *Global Biogeochemical Cycles* 12: 93–113.

CYCLES OF THE NUCLIDES

LAND–SEA GLOBAL TRANSFERS

F. T. Mackenzie and L. M. Ver, University of Hawaii, Honolulu, HI, USA

Introduction

The interface between the land and the sea is an important boundary connecting processes operating on land with those in the ocean. It is a site of rapid population growth, industrial and agricultural practices, and urban development. Large river drainage basins connect the vast interiors of continents with the coastal zone through river and groundwater discharges. The atmosphere is a medium of transport of substances from the land to the sea surface and from that surface back to the land. During the past several centuries, the activities of humankind have significantly modified the exchange of materials between the land and sea on a global scale – humans have become, along with natural processes, agents of environmental change. For example, because of the combustion of the fossil fuels coal, oil, and gas and changes in land-use practices including deforestation, shifting cultivation, and urbanization, the direction of net atmospheric transport of carbon (C) and sulfur (S) gases between the land and sea has been reversed. The global ocean prior to these human activities was a source of the gas carbon dioxide (CO_2) to the atmosphere and hence to the continents. The ocean now absorbs more CO_2 than it releases. In preindustrial time, the flux of reduced sulfur gases to the atmosphere and their subsequent oxidation and deposition on the continental surface exceeded the transport of oxidized sulfur to the ocean via the atmosphere. The situation is now reversed because of the emissions of sulfur to the atmosphere from the burning of fossil fuels and biomass on land. In addition, river and groundwater fluxes of the bioessential elements carbon, nitrogen (N), and phosphorus (P), and certain trace elements have increased because of human activities on land. For example, the increased global riverine (and atmospheric) transport of lead (Pb) corresponds to its increased industrial use. Also, recent changes in the concentration of lead in coastal sediments appear to be directly related to changes in the use of leaded gasoline in internal combustion engines. Synthetic substances manufactured by modern society, such as pesticides and pharmaceutical products, are now appearing in river and groundwater flows, thus moving toward the sea in a greater degree than before.

The Changing Picture of Land–Sea Global Transfers

Although the exchange through the atmosphere of certain trace metals and gases between the land and the sea surface is important, rivers are the main purveyors of materials to the ocean. The total water discharge of the major rivers of the world to the ocean is 36 000 $km^3\,yr^{-1}$. At any time, the world's rivers contain about 0.000 1% of the total water volume of $1459 \times 10^6\,km^3$ near the surface of the Earth and have a total dissolved and suspended solid concentration of *c*. 110 and 540 ppm l^{-1}, respectively. The residence time of water in the world's rivers calculated with respect to total net precipitation on land is only 18 days. Thus the water in the world's rivers is replaced every 18 days by precipitation. The global annual direct discharge to the ocean of groundwater is about 10% of the surface flow, with a recent estimate of 2400 $km^3\,yr^{-1}$. The dissolved constituent content of groundwater is poorly known, but one recent estimate of the dissolved salt groundwater flux is $1300 \times 10^6\,t\,yr^{-1}$.

The chemical composition of average river water is shown in **Table 1**. Note that the major anion in river water is bicarbonate, HCO_3^-; the major cation is calcium, Ca^{2+}, and that even the major constituent concentrations of river water on a global scale are influenced by human activities. The dissolved load of the major constituents of the world's rivers is derived from the following sources: about 7% from beds of salt and disseminated salt in sedimentary rocks, 10% from gypsum beds and sulfate salts disseminated in rocks, 38% from carbonates, and 45% from the weathering of silicate minerals. Two-thirds of the HCO_3^- in river waters are derived from atmospheric CO_2 via the respiration and decomposition of organic matter and subsequent conversion to HCO_3^- through the chemical weathering of silicate (~30% of total) or carbonate (~70% of total) minerals. The other third of the river HCO_3^- comes directly from carbonate minerals undergoing weathering.

It is estimated that only about 20% of the world's drainage basins have pristine water quality. The organic productivity of coastal aquatic environments has been heavily impacted by changes in the

Table 1 Chemical composition of average river water

By continent	*River water concentration*[a] *(mg l^{-1})*												*Water discharge (10^3 km^3 yr^{-1})*	*Runoff ratio*[c]
	Ca^{2+}	Mg^{2+}	Na^{+}	K^{+}	Cl^{-}	SO_4^{2-}	HCO_3^{-}	SO_2	*TDS*[b]	*TOC*[b]	*TDN*[b]	*TDP*[b]		
Africa														
Actual	5.7	2.2	4.4	1.4	4.1	4.2	36.9	12.0	60.5				3.41	0.28
Natural	5.3	2.2	3.8	1.4	3.4	3.2	26.7	12.0	57.8					
Asia														
Actual	17.8	4.6	8.7	1.7	10.0	13.3	67.1	11.0	134.6				12.47	0.54
Natural	16.6	4.3	6.6	1.6	7.6	9.7	66.2	11.0	123.5					
S. America														
Actual	6.3	1.4	3.3	1.0	4.1	3.8	24.4	10.3	54.6				11.04	0.41
Natural	6.3	1.4	3.3	1.0	4.1	3.5	24.4	10.3	54.3					
N. America														
Actual	21.2	4.9	8.4	1.5	9.2	18.0	72.3	7.2	142.6				5.53	0.38
Natural	20.1	4.9	6.5	1.5	7.0	14.9	71.4	7.2	133.5					
Europe														
Actual	31.7	6.7	16.5	1.8	20.0	35.5	86.0	6.8	212.8				2.56	0.42
Natural	24.2	5.2	3.2	1.1	4.7	15.1	80.1	6.8	140.3					
Oceania														
Actual	15.2	3.8	7.6	1.1	6.8	7.7	65.6	16.3	125.3				2.40	
Natural	15.0	3.8	7.0	1.1	5.9	6.5	65.1	16.3	120.3					
World average														
Actual	14.7	3.7	7.2	1.4	8.3	11.5	53.0	10.4	110.1	12.57	0.574	0.053	37.40	0.46
Natural (unpolluted)	13.4	3.4	5.2	1.3	5.8	5.3	52.0	10.4	99.6	9.89	0.386	0.027	37.40	0.46
Pollution	1.3	0.3	2.0	0.1	2.5	6.2	1.0	0.0	10.5	2.67	0.187	0.027		
World % pollutive	9%	8%	28%	7%	30%	54%	2%	0%	10%	21%	33%	50%		

[a]Actual concentrations include pollution. Natural concentrations are corrected for pollution.
[b]TDS, total dissolved solids; TOC, total organic carbon; TDN, total dissolved nitrogen; TDP, total dissolved phosphorus.
[c]Runoff ratio = average runoff per unit area/average rainfall.
Revised after Meybeck M (1979) Concentrations des eaux fluviales en elements majeurs et apports en solution aux oceans. *Revue de Geologie Dynamique et de Geographie Physique* 21: 215–246; Meybeck M (1982) Carbon, nitrogen, and phosphorus transport by world rivers. *American Journal of Science* 282: 401–450; Meybeck M (1983) C, N, P and S in rivers: From sources to global inputs. In: Wollast R, Mackenzie FT, and Chou L (eds.) *Interactions of C, N, P and S Biogeochemical Cycles and Global Change*, pp. 163–193. Berlin: Springer.

dissolved and particulate river fluxes of three of the major bioessential elements found in organic matter, C, N, and P (the other three are S, hydrogen (H), and oxygen (O)). Although these elements are considered minor constituents of river water, their fluxes may have doubled over their pristine values on a global scale because of human activities. Excessive river-borne nutrients and the cultural eutrophication of freshwater and coastal marine ecosystems go hand in hand. In turn, these fluxes have become sensitive indicators of the broader global change issues of population growth and land-use change (including water resources engineering works) in the coastal zone and upland drainage basins, climatic change, and sea level rise.

In contrast to the situation for the major elements, delivery of some trace elements from land to the oceans via the atmosphere can rival riverine inputs. The strength of the atmospheric sources strongly depends on geography and meteorology. Hence the North Atlantic, western North Pacific, and Indian Oceans, and their inland seas, are subjected to large atmospheric inputs because of their proximity to both deserts and industrial sources. Crustal dust is the primary terrestrial source of these atmospheric inputs to the ocean. Because of the low solubility of dust in both atmospheric precipitation and seawater and the overwhelming inputs from river sources, dissolved sources of the elements are generally less important. However, because the oceans contain only trace amounts of iron (Fe), aluminum (Al), and manganese (Mn) (concentrations are in the ppb level), even the small amount of dissolution in seawater (~10% of the element in the solid phase) results in eolian dust being the primary source for the dissolved transport of these elements toremote areas of the ocean. Atmospheric transport of the major nutrients N, silicon (Si), and Fe to the ocean has been hypothesized to affect and perhaps limit primary productivity in certain regions of the ocean at certain times. Modern processes of fossil fuel combustion and biomass burning have significantly modified the atmospheric transport from land to the ocean of trace metals like Pb, copper (Cu), and zinc (Zn), C in elemental and organic forms, and nutrient N.

As an example of global land–sea transfers involving gases and the effect of human activities on the exchange, consider the behavior of CO_2 gas. Prior to human influence on the system, there was a net flux of CO_2 out of the ocean owing to organic metabolism (net heterotrophy). This flux was mainly supported by the decay of organic matter produced by phytoplankton in the oceans and part of that transported by rivers to the oceans. An example overall reaction is:

$$C_{106}H_{263}O_{110}N_{16}S_2P + 141O_2 \Rightarrow 106CO_2 + 6HNO_3 + 2H_2SO_4 + H_3PO_4 + 120H_2O \quad [1]$$

Carbon dioxide was also released to the atmosphere due to the precipitation of carbonate minerals in the oceans. The reaction is:

$$Ca^{2+} + 2HCO_3^- \Rightarrow CaCO_3 + CO_2 + H_2O \quad [2]$$

The CO_2 in both reactions initially entered the dissolved inorganic carbon (DIC) pool of seawater and was subsequently released to the atmosphere at an annual rate of about 0.2×10^9 t of carbon as CO_2 gas. It should be recognized that this is a small number compared with the 200×10^9 t of carbon that exchanges between the ocean and atmosphere each year because of primary production of organic matter and its subsequent respiration.

Despite the maintenance of the net heterotrophic status of the ocean and the continued release of CO_2 to the ocean–atmosphere owing to the formation of calcium carbonate in the ocean, the modern ocean and the atmosphere have become net sinks of anthropogenic CO_2 from the burning of fossil fuels and the practice of deforestation. Over the past 200 years, as CO_2 has accumulated in the atmosphere, the gradient of CO_2 concentration across the atmosphere–ocean interface has changed, favoring uptake of anthropogenic CO_2 into the ocean. The average oceanic carbon uptake for the decade of the 1990s was *c.* 2×10^9 t annually. The waters of the ocean have accumulated about 130×10^9 t of anthropogenic CO_2 over the past 300 years.

The Coastal Zone and Land–Sea Exchange Fluxes

The global coastal zone environment is an important depositional and recycling site for terrigenous and oceanic biogenic materials. The past three centuries have been the time of well-documented human activities that have become an important geological factor affecting the continental and oceanic surface environment. In particular, historical increases in the global population in the areas of the major river drainage basins and close to oceanic coastlines have been responsible for increasing changes in land-use practices and discharges of various substances into oceanic coastal waters. As a consequence, the global C cycle and the cycles of N and P that closely interact with the carbon cycle have been greatly affected. Several major perturbations of the past three centuries of the industrial age have affected the processes of transport from land, deposition of terrigenous materials, and *in situ* production of organic matter in coastal zone environments. In addition, potential future changes in oceanic circulation may have significant effects on the biogeochemistry and CO_2 exchange in the coastal zone.

The coastal zone is that environment of continental shelves, including bays, lagoons, estuaries, and near-shore banks that occupy 7% of the surface area of the ocean ($36 \times 10^{12}\,m^2$) or *c.* 9% of the volume of the surface mixed layer of the ocean ($3 \times 10^{15}\,m^3$). The continental shelves average 75 km in width, with a bottom slope of $1.7\,m\,km^{-1}$. They are generally viewed as divisible into the interior or proximal shelf, and the exterior or distal shelf. The mean depth of the global continental shelf is usually taken as the depth of the break between the continental shelf and slope at *c.* 200 m, although this depth varies considerably throughout the world's oceans. In the Atlantic, the median depth of the shelf-slope break is at 120 m, with a range from 80 to 180 m. The depths of the continental shelf are near 200 m in the European section of the Atlantic, but they are close to 100 m on the African and North American coasts. Coastal zone environments that have high sedimentation rates, as great as $30–60\,cm\,ky^{-1}$ in active depositional areas, act as traps and filters of natural and human-generated materials transported from continents to the oceans via river and groundwater flows and through the atmosphere. At present a large fraction (~80%) of the land-derived organic and inorganic materials that are transported to the oceans is trapped on the proximal continental shelves. The coastal zone also accounts for 30–50% of total carbonate and 80% of organic carbon accumulation in the ocean. Coastal zone environments are also regions of higher biological production relative to that of average oceanic surface waters, making them an important factor in the global carbon cycle. The higher primary production is variably attributable to the nutrient inflows from land as well as from coastal upwelling of deeper ocean waters.

Fluvial and atmospheric transport links the coastal zone to the land; gas exchange and deposition are its links with the atmosphere; net advective transport of water, dissolved solids and particles, and coastal upwelling connect it with the open ocean. In addition, coastal marine sediments are repositories for much of the material delivered to the coastal zone. In the last several centuries, human activities on land have become a geologically important agent affecting the land–sea exchange of materials. In particular, river and groundwater flows and atmospheric transport of materials to the coastal zone have been substantially altered.

Bioessential Elements

Continuous increase in the global population and its industrial and agricultural activities have created four major perturbations on the coupled system of the biogeochemical cycles of the bioessential elements C, N, P, and S. These changes have led to major alterations in the exchanges of these elements between the land and sea. The perturbations are: (1) emissions of C, N, and S to the atmosphere from fossil-fuel burning and subsequent partitioning of anthropogenic C and deposition of N and S; (2) changes in land-use practices that affect the recycling of C, N, P, and S on land, their uptake by plants and release from decaying organic matter, and the rates of land surface denudation; (3) additions of N and P in chemical fertilizers to cultivated land area; and (4) releases of organic wastes containing highly reactive C, N, and P that ultimately enter the coastal zone. A fifth major perturbation is a climatic one: (5) the rise in mean global temperature of the lower atmosphere of about 1 °C in the past 300 years, with a projected increase of about 1.4–5.8 °C relative to 1990 by the year 2100. **Figure 1** shows how the fluxes associated with these activities have changed during the past three centuries with projections to the year 2040.

Partially as a result of these activities on land, the fluxes of materials to the coastal zone have changed historically. **Figure 2** shows the historical and projected future changes in the river fluxes of dissolved inorganic and organic carbon (DIC, DOC), nitrogen (DIN, DON), and phosphorus (DIP, DOP), and fluxes associated with the atmospheric deposition and denitrification of N, and accumulation of C in organic matter in coastal marine sediments. It can be seen in **Figure 2** that the riverine fluxes of C, N, and P all increase in the dissolved inorganic and organic phases from about 1850 projected to 2040. For example, for carbon, the total flux (organic + inorganic) increases by about 35% during this period. These increased fluxes are mainly due to changes in land-use practices, including deforestation, conversion of forest to grassland, pastureland, and urban centers, and regrowth of forests, and application of fertilizers to croplands and the subsequent leaching of N and P into aquatic systems.

Inputs of nutrient N and P to the coastal zone which support new primary production are from the land by riverine and groundwater flows, from the open ocean by coastal upwelling and onwelling, and to a lesser extent by atmospheric deposition of nitrogen. New primary production depends on the

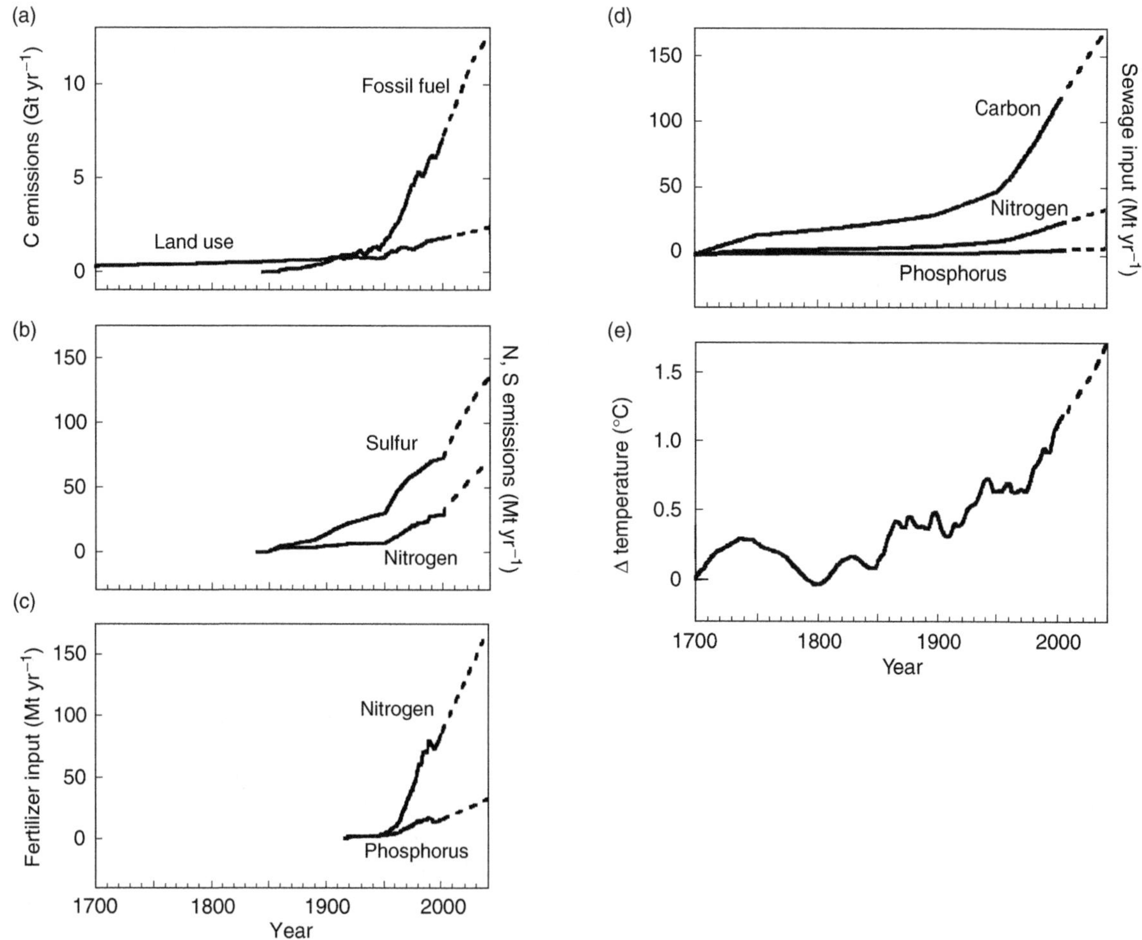

Figure 1 Major perturbations on the Earth system over the past 300 years and projections for the future: (a) emissions of CO_2 and (b) gaseous N and S from fossil-fuel burning and land-use activities; (c) application of inorganic N and P in chemical fertilizers to cultivated land; (d) loading of highly reactive C, N, and P into rivers and the coastal ocean from municipal sewage and wastewater disposal; and (e) rise in mean global temperature of the lower atmosphere relative to 1700. Revised after Ver LM, Mackenzie FT, and Lerman A (1999) Biogeochemical responses of the carbon cycle to natural and human perturbations: Past, present, and future. *American Journal of Science* 299: 762–801.

availability of nutrients from these external inputs, without consideration of internal recycling of nutrients. Thus any changes in the supply of nutrients to the coastal zone owing to changes in the magnitude of these source fluxes are likely to affect the cycling pathways and balances of the nutrient elements. In particular, input of nutrients from the open ocean by coastal upwelling is quantitatively greater than the combined inputs from land and the atmosphere. This makes it likely that there could be significant effects on coastal primary production because of changes in ocean circulation. For example, because of global warming, the oceans could become more strongly stratified owing to freshening of polar oceanic waters and warming of the ocean in the tropical zone. This could lead to a reduction in the intensity of the oceanic thermohaline circulation (oceanic circulation owing to differences in density of water masses, also popularly known as the 'conveyor belt') and hence the rate at which nutrient-rich waters upwell into coastal environments.

Another potential consequence of the reduction in the rate of nutrient inputs to the coastal zone by upwelling is the change in the CO_2 balance of coastal waters: reduction in the input of DIC to the coastal zone from the deeper ocean means less dissolved CO_2, HCO_3^-, and CO_3^{2-} coming from that source. With increasing accumulation of anthropogenic CO_2 in the atmosphere, the increased dissolution of atmospheric CO_2 in coastal water is favored. The combined result of a decrease in the upwelling flux of DIC and an enhancement in the transfer of atmospheric CO_2 across the air–sea interface of coastal waters is a lower saturation state for coastal waters with respect to the carbonate minerals calcite, aragonite, and a variety of magnesian calcites. The

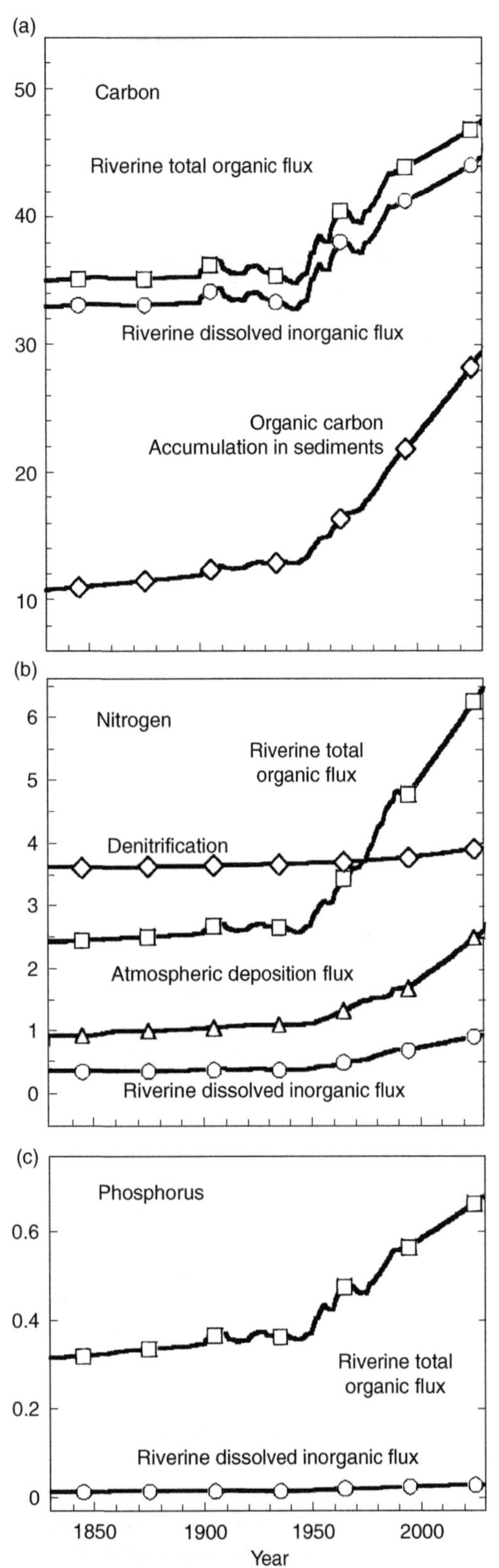

Figure 2 Past, present, and predicted fluxes of carbon, nitrogen, and phosphorus into or out of the global coastal margin, in 10^{12} mol yr^{-1}.

lower saturation state in turn leads to the likelihood of lower rates of inorganic and biological precipitation of carbonate and hence deposition and accumulation of sedimentary carbonate.

In addition, the present-day burial rate of organic carbon in the ocean may be about double that of the late Holocene flux, supported by increased fluxes of organic carbon to the ocean via rivers and groundwater flows and increased *in situ* new primary production supported by increased inputs of inorganic N and P from land and of N deposited from the atmosphere. The organic carbon flux into sediments may constitute a sink of anthropogenic CO_2 and a minor negative feedback on accumulation of CO_2 in the atmosphere.

The increased flux of land-derived organic carbon delivered to the ocean by rivers may accumulate there or be respired, with subsequent emission of CO_2 back to the atmosphere. This release flux of CO_2 may be great enough to offset the increased burial flux of organic carbon to the seafloor due to enhanced fertilization of the ocean by nutrients derived from human activities. The magnitude of the CO_2 exchange is a poorly constrained flux today. One area for which there is a substantial lack of knowledge is the Asian Pacific region. This is an area of several large seas, a region of important river inputs to the ocean of N, P, organic carbon, and sediments from land, and a region of important CO_2 exchange between the ocean and the atmosphere.

Anticipated Response to Global Warming

From 1850 to modern times, the direction of the net flux of CO_2 between coastal zone waters and the atmosphere due to organic metabolism and calcium carbonate accumulation in coastal marine sediments was from the coastal surface ocean to the atmosphere (negative flux, **Figure 3**).This flux in 1850 was on the order of $-0.2\times10^9\,\mathrm{t\,yr^{-1}}$. In a condition not disturbed by changes in the stratification and thermohaline circulation of the ocean brought about by a global warming of the Earth, the direction of this flux is projected to remain negative (flux out of the coastal ocean to the atmosphere) until early in the twenty-first century. The increasing partial pressure of CO_2 in the atmosphere because of emissions from anthropogenic sources leads to a reversal in the gradient of CO_2 across the air–sea interface of coastal zone waters and, hence, invasion of CO_2 into the coastal waters. From that time on the coastal ocean will begin to operate as a net sink (positive flux) of atmospheric CO_2 (**Figure 3**). The role of the open ocean as a sink for anthropogenic CO_2 is slightly reduced while that of the coastal oceans

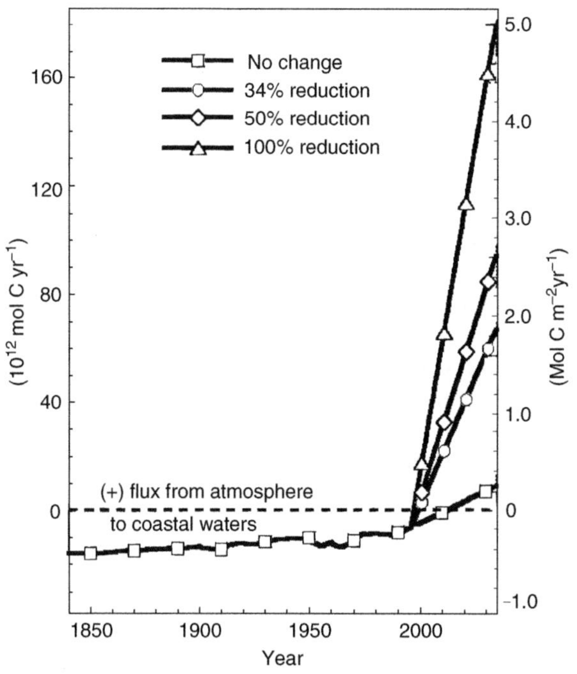

Figure 3 The net flux of CO_2 between coastal zone waters and the atmosphere due to organic metabolism and calcium carbonate accumulation in coastal marine sediments, under three scenarios of changing thermohaline circulation rate compared to a business-as-usual scenario, in units of 10^{12} mol C yr^{-1}.

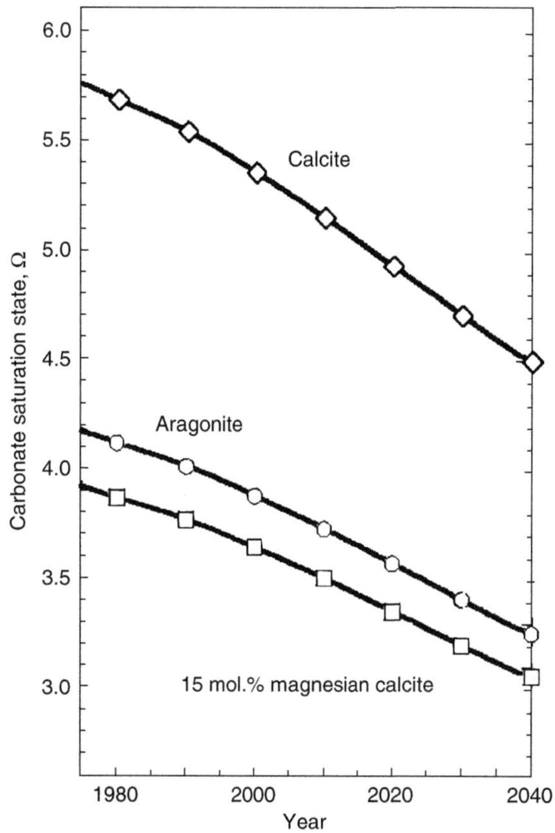

Figure 4 Changes in saturation state with respect to carbonate minerals of surface waters of the coastal ocean projected from 1999 to 2040. Calculations are for a temperature of 25 °C.

increases. The net result is the maintenance of the role of the global oceans as a net sink for anthropogenic CO_2.

The saturation index (Ω) for calcite or aragonite (both $CaCO_3$) is the ratio of the ion activity product IAP in coastal waters to the respective equilibrium constant K at the *in situ* temperature. For aqueous species, $IAP = aCa^{2+} \times aCO_3^{2-}$ (where a is the activity; note that for 15 mol.% magnesian calcite, the IAP also includes the activity of the magnesium cation, aMg^{2+}). Most coastal waters and open-ocean surface waters currently are supersaturated with respect to aragonite, calcite, and magnesian calcite containing 15 mol.% Mg, that is, $\Omega_{calcite}$, $\Omega_{aragonite}$, and $\Omega_{15\%magnesian\ calcite}$ are >1. Because of global warming and the increasing land-to-atmosphere-to-seawater transport of CO_2 (due to the continuing combustion of fossil fuels and burning of biomass), the concentration of the aqueous CO_2 species in seawater increases and the pH of the water decreases slightly. This results in a decrease in the concentration of the carbonate ion, CO_3^{2-}, resulting in a decrease in the degree of supersaturation of coastal zone waters. **Figure 4** shows how the degree of saturation might change into the next century because of rising atmospheric CO_2 concentrations. The overall reduction in the saturation state of coastal zone waters with respect to aragonite from 1997 projected to 2040 is about 16%, from 3.89 to 3.26.

Modern carbonate sediments deposited in shoal-water ('shallow-water') marine environments (including shelves, banks, lagoons, and coral reef tracts) are predominantly biogenic in origin derived from the skeletons and tests of benthic and pelagic organisms, such as corals, foraminifera, echinoids, mollusks, algae, and sponges. One exception to this statement is some aragonitic muds that may, at least in part, result from the abiotic precipitation of aragonite from seawater in the form of whitings. Another exception is the sand-sized, carbonate oöids composed of either aragonite with a laminated internal structure or magnesian calcite with a radial internal structure. In addition, early diagenetic carbonate cements found in shoal-water marine sediments and in reefs are principally aragonite or magnesian calcite. Thus carbonate production and accumulation in shoal-water environments are dominated by a range of metastable carbonate minerals associated with skeletogenesis and abiotic processes, including calcite, aragonite, and a variety of magnesian calcite compositions.

With little doubt, as has been documented in a number of observational and experimental studies, a reduction in the saturation state of ocean waters will lead to a reduction in the rate of precipitation of both inorganic and skeletal calcium carbonate. Conversely, increases in the degree of supersaturation and temperature will increase the precipitation rates of calcite and aragonite from seawater. During global warming, rising sea surface temperatures and declining carbonate saturation states due to the absorption of anthropogenic CO_2 by surface ocean waters result in opposing effects. However, experimental evidence suggests that within the range of temperature change predicted for the next century due to global warming, the effect of changes in saturation state will be the predominant factor affecting precipitation rate. Thus decreases in precipitation rates should lead to a decrease in the production and accumulation of shallow-water carbonate sediments and perhaps changes in the types and distribution of calcifying biotic species found in shallow-water environments.

Anticipated Response to Heightened Human Perturbation: The Asian Scenario

In the preceding sections it was shown that the fluxes of C, N, and P from land to ocean have increased because of human activities (refer to **Table 1** for data comparing the actual and natural concentrations of C, N, P, and other elements in average river water). During the industrial era, these fluxes mainly had their origin in the present industrialized and developed countries. This is changing as the industrializing and developing countries move into the twenty-first century. A case in point is the countries of Asia.

Asia is a continent of potentially increasing contributions to the loading of the environment owing to a combination of such factors as its increasing population, increasing industrialization dependent on fossil fuels, concentration of its population along the major river drainage basins and in coastal urban centers, and expansion of land-use practices. It is anticipated that Asia will experience similar, possibly even greater, loss of storage of C and nutrient N and P on land and increased storage in coastal marine sediments per unit area than was shown by the developed countries during their period of industrialization. The relatively rapid growth of Asia's population along the oceanic coastal zone indicates that higher inputs of both dissolved and particulate organic nutrients may be expected to enter coastal waters.

A similar trend of increasing population concentration in agricultural areas inland, within the drainage basins of the main rivers flowing into the ocean, is also expected to result in increased dissolved and particulate organic nutrient loads that may eventually reach the ocean. Inputs from inland regions to the ocean would be relatively more important if no entrapment or depositional storage occurred en route, such as in the dammed sections of rivers or in alluvial plains. In the case of many of China's rivers, the decline in sediment discharge from large rivers such as the Yangtze and the Yellow Rivers is expected to continue due to the increased construction of dams. The average decadal sediment discharge from the Yellow River, for example, has decreased by 50% from the 1950s to the 1980s. If the evidence proposed for the continental United States applies to Asia, the damming of major rivers would not effectively reduce the suspended material flow to the ocean because of the changes in the erosional patterns on land that accompany river damming and more intensive land-use practices. These flows on land and into coastal ocean waters are contributing factors to the relative importance of autotrophic and heterotrophic processes, competition between the two, and the consequences for carbon exchange between the atmosphere and land, and the atmosphere and ocean water. The change from the practices of land fertilization by manure to the more recent usage of chemical fertilizers in Asia suggests a shift away from solid organic nutrients and therefore a reduced flow of materials that might promote heterotrophy in coastal environments.

Sulfur is an excellent example of how parts of Asia can play an important role in changing land–sea transfers of materials. Prior to extensive human interference in the global cycle of sulfur, biogenically produced sulfur was emitted from the sea surface mainly in the form of the reduced gas dimethyl sulfide (DMS). DMS was the major global natural source of sulfur for the atmosphere, excluding sulfur in sea salt and soil dust. Some of this gas traveled far from its source of origin. During transport the reduced gas was oxidized to micrometer-size sulfate aerosol particles and rained out of the atmosphere onto the sea and continental surface. The global sulfur cycle has been dramatically perturbed by the industrial and biomass burning activities of human society. The flux of gaseous sulfur dioxide to the atmosphere from the combustion of fossil fuels in some regions of the world and its conversion to sulfate aerosol greatly exceeds natural fluxes of sulfur gases from the land surface. It is estimated that this flux for the year 1990 was equivalent to 73×10^6 $t\,yr^{-1}$, nearly 4 times the natural DMS flux from the ocean. This has led to a net transport of sulfur from the land to the ocean via the atmosphere, completely reversing the flow direction in preindustrial times. In addition, the sulfate aerosol content of the atmosphere derived from human activities has increased. Sulfate aerosols affect global climate directly as particles that scatter incoming solar radiation and indirectly as cloud condensation nuclei (CCNs), which lead to an increased number of cloud droplets and an increase in the solar reflectance of clouds. Both effects cause the cooling of the planetary surface. As can be seen in **Figure 5** the eastern Asian

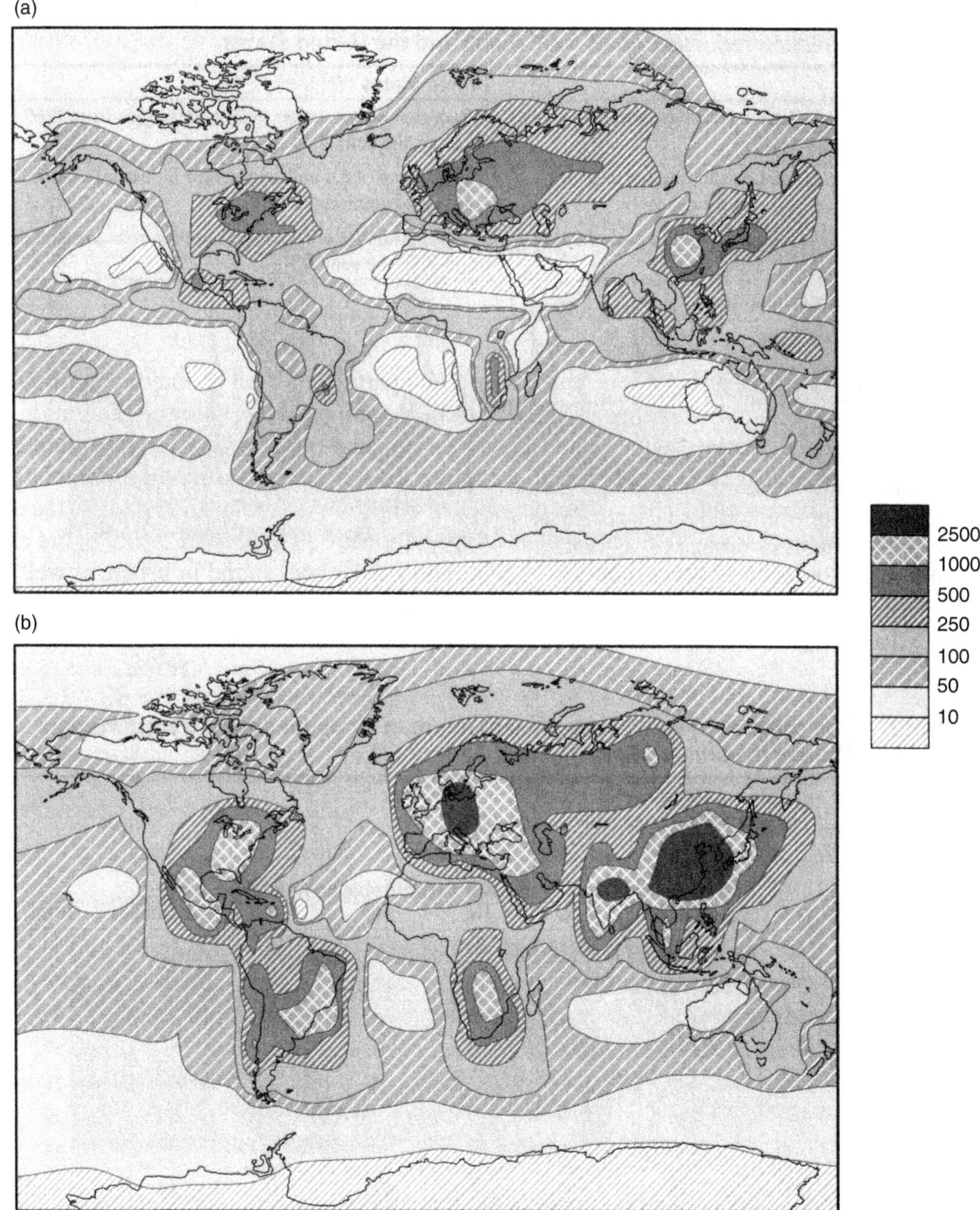

Figure 5 Comparison of the magnitude of atmospheric sulfur deposition for the years 1990 (a) and 2050 (b). Note the large increases in both spatial extent and intensity of sulfur deposition in both hemispheres and the increase in importance of Asia, Africa, and South America as sites of sulfur deposition between 1990 and 2050. The values on the diagrams are in units of kg S $m^{-2} yr^{-1}$. Revised after Mackenzie FT (1998) *Our Changing Planet: An Introduction to Earth System Science and Global Environmental Change*. Upper Saddle River, NJ: Prentice Hall; Rodhe H, Langner J, Gallardo L, and Kjellström E (1995) Global transport of acidifying pollutants. *Water, Air and Soil Pollution* 85: 37–50.

region is an important regional source of sulfate aerosol because of the combustion of fossil fuels, particularly coal. This source is predicted to grow in strength during the early- to mid-twenty-first century (**Figure 5**).

Conclusion

Land–sea exchange processes and fluxes of the bioessential elements are critical to life. In several cases documented above, these exchanges have been substantially modified by human activities. These modifications have led to a number of environmental issues including global warming, acid deposition, excess atmospheric nitrogen deposition, and production of photochemical smog. All these issues have consequences for the biosphere – some well known, others not so well known. It is likely that the developing world, with increasing population pressure and industrial development and with no major changes in agricultural technology and energy consumption rates, will become a more important source of airborne gases and aerosols and materials for river and groundwater systems in the future. This will lead to further modification of land–sea global transfers. The region of southern and eastern Asia is particularly well poised to influence significantly these global transfers.

See also

Air–Sea Transfer: Dimethyl Sulfide, COS, CS_2, NH_4, Non-Methane Hydrocarbons, Organo-Halogens. Carbon Cycle. Carbon Dioxide (CO_2) Cycle. Nitrogen Cycle. Ocean Carbon System, Modeling of. Past Climate From Corals. Phosphorus Cycle.

Further Reading

Berner EA and Berner RA (1996) *Global Environment: Water, Air and Geochemical Cycles*. Upper Saddle River, NJ: Prentice Hall.

Galloway JN and Melillo JM (eds.) (1998) *Asian Change in the Context of Global Change*. Cambridge, MA: Cambridge University Press.

Mackenzie FT (1998) *Our Changing Planet: An Introduction to Earth System Science and Global Environmental Change*. Upper Saddle River, NJ: Prentice Hall.

Mackenzie FT and Lerman A (2006) *Carbon in the Geobiosphere – Earth's Outer Shell*. Dordrecht: Springer.

Meybeck M (1979) Concentrations des eaux fluviales en elements majeurs et apports en solution aux oceans. *Revue de Geologie Dynamique et de Geographie Physique* 21: 215–246.

Meybeck M (1982) Carbon, nitrogen, and phosphorus transport by world rivers. *American Journal of Science* 282: 401–450.

Meybeck M (1983) C, N, P and S in rivers: From sources to global inputs. In: Wollast R, Mackenzie FT, and Chou L (eds.) *Interactions of C, N, P and S Biogeochemical Cycles and Global Change*, pp. 163–193. Berlin: Springer.

Rodhe H, Langner J, Gallardo L, and Kjellström E (1995) Global transport of acidifying pollutants. *Water, Air and Soil Pollution* 85: 37–50.

Schlesinger WH (1997) *Biogeochemistry: An Analysis of Global Change*. San Diego, CA: Academic Press.

Smith SV and Mackenzie FT (1987) The ocean as a net heterotrophic system: Implications from the carbon biogeochemical cycle. *Global Biogeochemical Cycles* 1: 187–198.

Ver LM, Mackenzie FT, and Lerman A (1999) Biogeochemical responses of the carbon cycle to natural and human perturbations: Past, present, and future. *American Journal of Science* 299: 762–801.

Vitousek PM, Aber JD, and Howarth RW (1997) Human alteration of the global nitrogen cycle: Sources and consequences. *Ecological Applications* 7(3): 737–750.

Wollast R and Mackenzie FT (1989) Global biogeochemical cycles and climate. In: Berger A, Schneider S, and Duplessy JC (eds.) *Climate and Geo-Sciences*, pp. 453–473. Dordrecht: Kluwer Academic Publishers.

CARBON CYCLE

C. A. Carlson, University of California, Santa Barbara, CA, USA
N. R. Bates, Bermuda Biological Station for Research, St George's, Bermuda, USA
D. A. Hansell, University of Miami, Miami FL, USA
D. K. Steinberg, College of William and Mary, Gloucester Pt, VA, USA

Introduction

Why is carbon an important element? Carbon has several unique properties that make it an important component of life, energy flow, and climate regulation. It is present on the Earth in many different inorganic and organic forms. Importantly, it has the ability to form complex, stable carbon compounds, such as proteins and carbohydrates, which are the fundamental building blocks of life. Photosynthesis provides marine plants (phytoplankton) with an ability to transform energy from sunlight, and inorganic carbon and nutrients dissolved in sea water, into complex organic carbon materials. All organisms, including autotrophs and heterotrophs, then catabolize these organic compounds to their inorganic constituents via respiration, yielding energy for their metabolic requirements. Production, consumption, and transformation of these organic materials provide the energy to be transferred between all the trophic states of the ocean ecosystem.

In its inorganic gaseous phases (carbon dioxide, CO_2; methane, CH_4; carbon monoxide, CO), carbon has important greenhouse properties that can influence climate. Greenhouse gases in the atmosphere act to trap long-wave radiation escaping from Earth to space. As a result, the Earth's surface warms, an effect necessary to maintain liquid water and life on Earth. Human activities have led to a rapid increase in greenhouse gas concentrations, potentially impacting the world's climate through the effects of global warming. Because of the importance of carbon for life and climate, much research effort has been focused on understanding the global carbon cycle and, in particular, the functioning of the ocean carbon cycle. Biological and chemical processes in the marine environment respond to and influence climate by helping to regulate the concentration of CO_2 in the atmosphere. We will discuss (1) the importance of the ocean to the global carbon cycle; (2) the mechanisms of carbon exchange between the ocean and atmosphere; (3) how carbon is redistributed throughout the ocean by ocean circulation; and (4) the roles of the 'solubility', 'biological,' and 'carbonate' pumps in the ocean carbon cycle.

Global Carbon Cycle

The global carbon cycle describes the complex transformations and fluxes of carbon between the major components of the Earth system. Carbon is stored in four major Earth reservoirs, including the atmosphere, lithosphere, biosphere, and hydrosphere. Each reservoir contains a variety of organic and inorganic carbon compounds ranging in amounts. In addition, the exchange and storage times for each carbon reservoir can vary from a few years to millions of years. For example, the lithosphere contains the largest amount of carbon (10^{23} g C), buried in sedimentary rocks in the form of carbonate minerals ($CaCO_3$, $CaMgCO_3$, and $FeCO_3$) and organic compounds such as oil, natural gas, and coal (fossil fuels). Carbon in the lithosphere is redistributed to other carbon reservoirs on timescales of millions of years by slow geological processes such as chemical weathering and sedimentation. Thus, the lithosphere is considered to be a relatively inactive component of the global carbon cycle (though the fossil fuels are now being added to the biologically active reservoirs at unnaturally high rates). The Earth's active carbon reservoirs contain approximately 43×10^{18} g of carbon, which is partitioned between the atmosphere (750×10^{15} g C), the terrestrial biosphere (2190×10^{15} g C), and the ocean ($39\,973 \times 10^{15}$ g C; **Figure 1**). While the absolute sum of carbon found in the active reservoirs is maintained in near steady state by slow geological processes, more rapid biogeochemical processes drive the redistribution of carbon among the active reservoirs.

Human activities, such as use of fossil fuels and deforestation, have significantly altered the amount of carbon stored in the atmosphere and perturbed the fluxes of carbon between the atmosphere, the terrestrial biosphere, and the ocean. Since the emergence of the industrial age 200 years ago, the release of CO_2 from fossil fuel use, cement manufacture, and deforestation has increased the partial pressure of atmospheric CO_2 from 280 ppm to present day values of 360 ppm; an increase of 25% in the last century (**Figure 2**). Currently, as a result of human activities, approximately 5.5×10^{15} g of

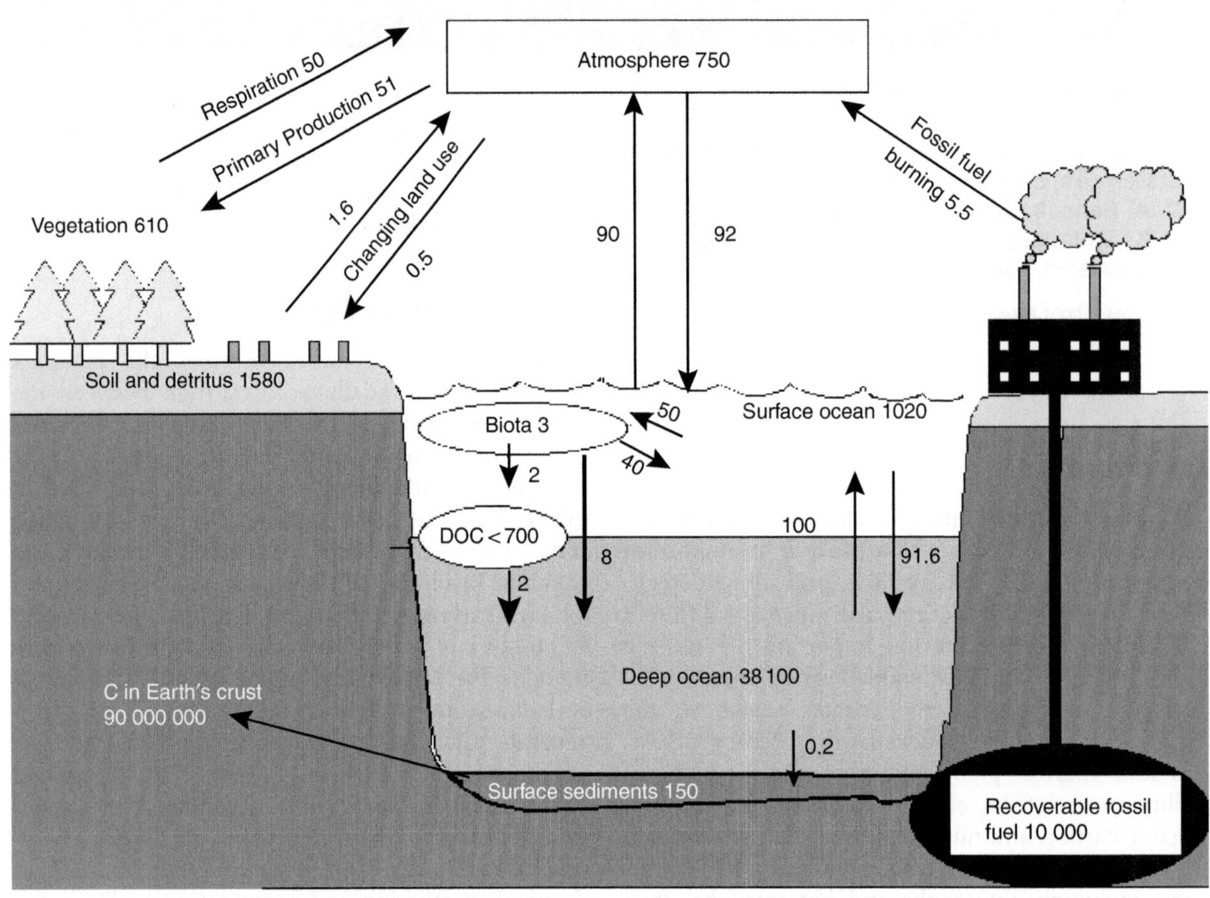

Figure 1 The global carbon cycle. Arrows indicate fluxes of carbon between the various reservoirs of the atmosphere, lithosphere, terrestrial biosphere, and the ocean. All stocks are expressed as 10^{15} g C. All fluxes are decadal means and expressed as 10^{15} g C y^{-1}. (Adapted with permission from Sigenthaler and Sarmiento, 1993), copyright 1993, Macmillan Magazines Ltd.). Data used to construct this figure came from Sigenthaler and Sarmiento (1993), Hansell and Carlson (1998), and Sarmiento and Wofsy (1999).

'anthropogenic' carbon is added to the atmosphere every year. About half of the anthropogenic CO_2 is retained in the atmosphere, while the remaining carbon is transferred to and stored in the ocean and the terrestrial biosphere. Carbon reservoirs that remove and sequester CO_2 from the atmosphere are referred to as carbon 'sinks'. The partitioning of anthropogenic carbon between oceanic and terrestrial sinks is not well known. Quantifying controls on the partitioning is necessary for understanding the dynamics of the global carbon cycle. The terrestrial biosphere may be a significant sink for anthropogenic carbon, but scientific understanding of the causative processes is hindered by the complexity of terrestrial ecosystems.

Global ocean research programs such as Geochemical Ocean Sections (GEOSEC), the Joint Global Ocean Flux Study (JGOFS), and the JGOFS/World Ocean Circulation Experiment (WOCE) Ocean CO_2 Survey have resulted in improvements in our understanding of physical circulation and biological processes of the ocean. These studies have also allowed oceanographers to better constrain the role of the ocean in CO_2 sequestration compared to terrestrial systems. Based on numerical models of ocean circulation and ecosystem processes, oceanographers estimate that 70% (2×10^{15} g C) of the anthropogenic CO_2 is absorbed by the ocean each year. The fate of the remaining 30% (0.75×10^{15} g) of anthropogenic CO_2 is unknown. Determining the magnitude of the oceanic sink of anthropogenic CO_2 is dependent on understanding the interplay of various chemical, physical, and biological factors.

Oceanic Carbon Cycle

The ocean is the largest reservoir of the Earth's active carbon, containing $39\ 973 \times 10^{15}$ g C. Oceanic carbon occurs as a variety of inorganic and organic

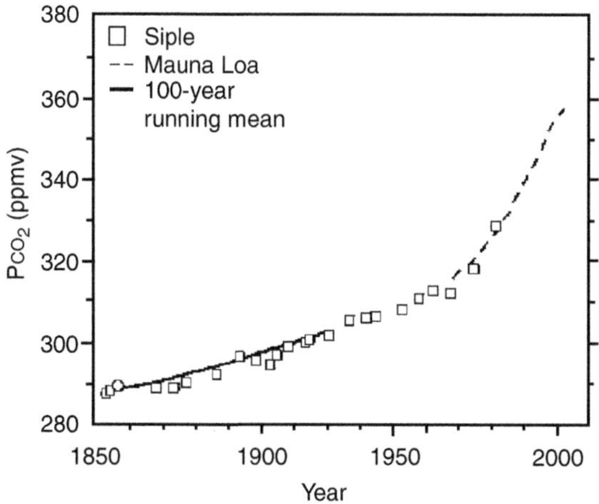

Figure 2 Atmospheric CO_2 concentrations from 1850 to 1996. These data illustrate an increase in atmospheric CO_2 concentration from pre-industrial concentration of 280 ppmv to present-day concentrations of 360 ppmv. Human activities of fossil fuel burning and deforestation have caused this observed increase in atmospheric CO_2. (Adapted from Houghton *et al.* (1996) with permission from Intergovernmental Panel on Climate Change (IPCC). The original figure was constructed from Siple ice core data and (from 1958) data collected at the Mauna Loa sampling site.)

forms, including dissolved CO_2, bicarbonate (HCO_3^-), carbonate (CO_3^{2-}) and organic compounds. CO_2 is one of the most soluble of the major gases in sea water and the ocean has an enormous capacity to buffer changes in the atmospheric CO_2 content.

The concentration of dissolved CO_2 in sea water is relatively small because CO_2 reacts with water to form a weak acid, carbonic acid (H_2CO_3), which rapidly dissociates (within milliseconds) to form HCO_3^- and CO_3^{2-}(eqn [I]).

$$CO_2(gas) + H_2O \rightleftharpoons H_2CO_3(aq) \rightleftharpoons H^+(aq) + HCO_3^-(aq) \rightleftharpoons 2H^+(aq) + CO_3^{2-}(aq) \quad [I]$$

For every 20 molecules of CO_2 absorbed by the ocean, 19 molecules are rapidly converted to HCO_3^- and CO_3^{2-}; at the typical range of pH in sea water (7.8–8.2; see below), most inorganic carbon is found in the form of HCO_3^-. These reactions (eqn [I]) provide a chemical buffer, maintain the pH of the ocean within a small range, and constrain the amount of atmospheric CO_2 that can be taken up by the ocean.

The amount of dissolved CO_2 in sea water cannot be determined analytically but can be calculated after measuring other inorganic carbon species. Dissolved inorganic carbon (DIC) refers to the total amount of CO_2, HCO_3^- plus CO_3^{2-} in sea water, while the partial pressure of CO_2 (P_{CO_2}) measures the contribution of CO_2 to total gas pressure. The alkalinity of sea water (A) is a measure of the bases present in sea water, consisting mainly of HCO_3^- and CO_3^{2-} ($A[HCO_3^-]+2[CO_3^{2-}]$) and minor constituents such as borate (BO_4) and hydrogen ions (H^+). Changes in DIC concentration and alkalinity affect the solubility of CO_2 in sea water (i.e., the ability of sea water to absorb CO_2) (see below).

The concentrations of inorganic carbon species in sea water are controlled not only by the chemical reactions outlined above (i.e., eqn [I]) but also by various physical and biological processes, including the exchange of CO_2 between ocean and atmosphere; the solubility of CO_2; photosynthesis and respiration; and the formation and dissolution of calcium carbonate ($CaCO_3$).

Typical surface sea water ranges from pH of 7.8 to 8.2. On addition of acid (i.e., H^+), the chemical reactions shift toward a higher concentration of CO_2 in sea water (eqn [IIa]) and pH decreases from 8.0 to 7.8 and then pH will rise from 8.0 to 8.2.

$$H^+ + HCO_3^- \rightarrow H_2CO_3 \rightarrow CO_2(aq) + H_2O \quad [IIa]$$

If base is added to sea water (eqn [IIb]), then pH will rise.

$$H_2CO_3 \rightarrow H^+ + HCO_3 \quad [IIb]$$

Solubility and Exchange of CO_2 between the Ocean and Atmosphere

The solubility of CO_2 in sea water is an important factor in controlling the exchange of carbon between the ocean and atmosphere. Henry's law (eqn [1]) describes the relationship between solubility and sea water properties, where S equals the solubility of gas in liquid, k is the solubility constant (k is a function mainly of temperature) and P is the overlying pressure of the gas in the atmosphere.

$$S = kP \quad [1]$$

Sea water properties such as temperature, salinity, and partial pressure of CO_2 determine the solubility of CO_2. For example, at 0°C in sea water, the solubility of CO_2 is double that in sea water at 20°C; thus colder water will tend to absorb more CO_2 than warmer water.

Henry's law also describes the relationship between the partial pressure of CO_2 in solution (P_{CO_2}) and its concentration (i.e., $[CO_2]$). Colder waters tend to have lower P_{CO_2} than warmer waters: for every 1°C temperature increase, sea water P_{CO_2} increases by ~4%. Sea water P_{CO_2} is also influenced by complicated thermodynamic relationships

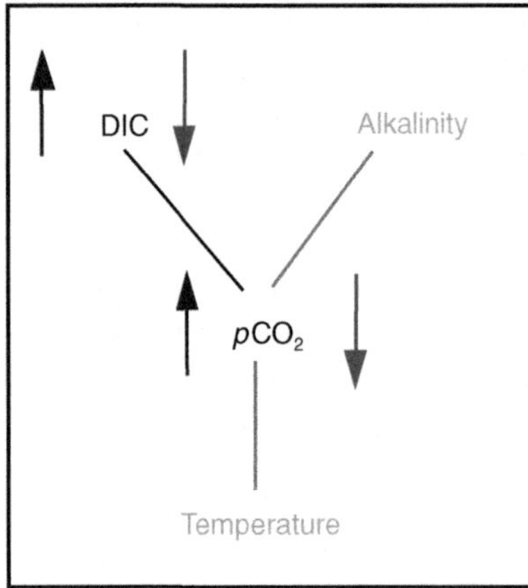

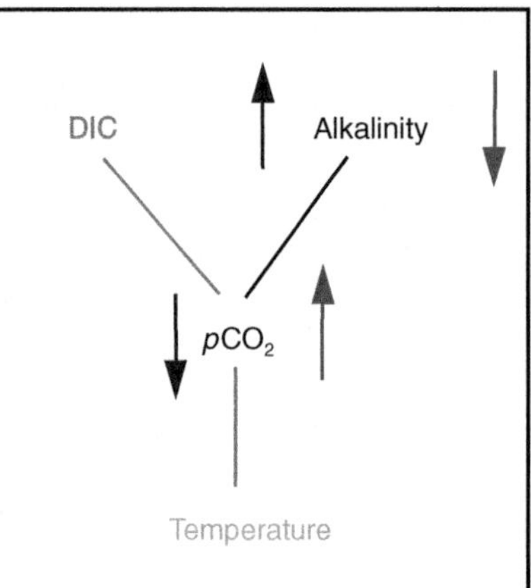

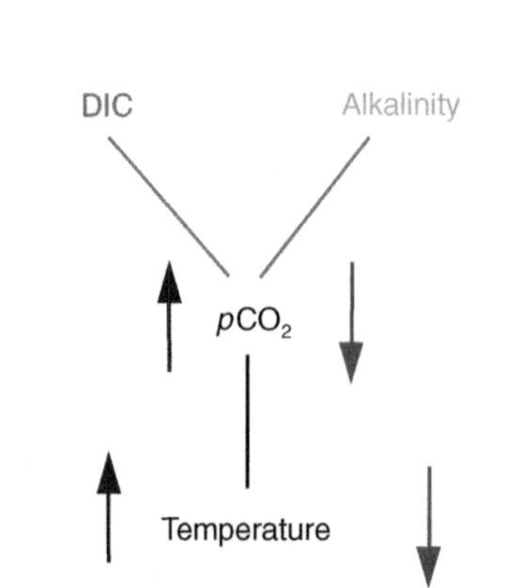

Figure 3 The response of $P\text{co}_2$ to changes in the sea water properties of (A) DIC concentration, (B) alkalinity and (C) temperature. Each panel describes how $P\text{co}_2$ will respond to the changes in the relevant sea water property. The blue arrows illustrate the response of $P\text{co}_2$ to an increase in the sea water property and the red arrows illustrate the response to a decrease in the sea water property. For example, as DIC or temperature increases, $P\text{co}_2$ increases; whereas an increase in alkalinity results in a decrease in $P\text{co}_2$.

between the different carbon species. For example, a decrease in sea water DIC or temperature acts to decrease $P\text{co}_2$, while a decrease in alkalinity acts to increase $P\text{co}_2$ (**Figure 3**).

Carbon dioxide is transferred across the air–sea interface by molecular diffusion and turbulence at the ocean surface. The flux (F) of CO_2 between the atmosphere and ocean is driven by the concentration difference between the reservoirs (eqn [2]).

$$F = \Delta P\text{co}_2 K_W \qquad [2]$$

In eqn [2] $\Delta P\text{co}_2$ is the difference in $P\text{co}_2$ between the ocean and atmosphere and K_w is the transfer coefficient across the air–sea interface, termed the piston velocity. In cold waters, sea water $P\text{co}_2$ tends to be lower than atmospheric $P\text{co}_2$, thus driving the direction of CO_2 gas exchange from atmosphere to ocean (**Figure 3**). In warmer waters, sea water $P\text{co}_2$ is greater than atmospheric $P\text{co}_2$, and CO_2 gas exchange occurs in the opposite direction, from the ocean to the atmosphere. The rate at which CO_2 is transferred between the ocean and the atmosphere depends not only on the $P\text{co}_2$ difference but on turbulence at the ocean surface. The piston velocity of CO_2 is related to solubility and the strength of the wind blowing on the sea surface. As wind speed increases, the rate of air–sea CO_2 exchange also increase. Turbulence caused by breaking waves also influences gas exchange because air bubbles may dissolve following entrainment into the ocean mixed layer.

Ocean Structure

Physically, the ocean can be thought of as two concentric spheres, the surface ocean and the deep ocean, separated by a density discontinuity called the pycnocline. The surface ocean occupies the upper few hundred meters of the water column and contains approximately 1020×10^{15} g C of DIC (**Figure 1**). The absorption of CO_2 by the ocean through gas exchange takes place in the mixed layer, the upper portion of the surface ocean that makes direct contact with the atmosphere. The surface ocean reaches equilibrium with the atmosphere within one year. The partial pressure of CO_2 in the surface ocean is slightly less than or greater than that of the atmosphere, depending on the controlling variables as described above, and varies temporally and spatially with changing environmental conditions. The deeper ocean represents the remainder of the ocean volume and is supersaturated with CO_2, with a DIC stock of $38\,100 \times 10^{15}$ g C (**Figure 1**), or 50 times the DIC contained in the atmosphere.

CO_2 absorbed by the ocean through gas exchange has a variety of fates. Physical and biological mechanisms can return the CO_2 back to the atmosphere or transfer carbon from the surface ocean to the deep ocean and ocean sediments through several transport processes termed the 'solubility', 'biological', and 'carbonate' pumps.

The Solubility Pump, Oceanic Circulation, and Carbon Redistribution

The 'solubility pump' is defined as the exchange of carbon between the atmosphere and the ocean as mediated by physical processes such as heat flux, advection and diffusion, and ocean circulation. It assists in the transfer of atmospheric CO_2 to the deep ocean. This transfer is controlled by circulation patterns of the surface ocean (wind-driven circulation) and the deep ocean (thermohaline circulation). These circulation patterns assist in the transfer of atmospheric CO_2 to the deep ocean and help to maintain the vertical gradient of DIC found in the ocean (**Figure 4**). The ability of the ocean to take up anthropogenic CO_2 via the solubility pump is limited by the physical structure of the ocean, the distribution of oceanic DIC, ocean circulation patterns, and the exchange between the surface and deep ocean layers. To be an effective sink for anthropogenic carbon, CO_2 must be transferred to the deep ocean by mixing and biological processes (see below).

Wind-driven circulation occurs as a consequence of friction and turbulence imparted by wind blowing over the sea surface. This circulation pattern is primarily horizontal in movement and is responsible for transporting warm water from lower latitudes (warm) to higher latitudes (cold). Surface currents move water and carbon great distances within ocean basins on timescales of months to years. As surface sea water moves from low latitudes to high latitudes, the increasing solubility of CO_2 in the sea water (due to sea surface cooling) allows atmospheric CO_2 to invade the surface mixed layer (**Figure 5** and **6A**).

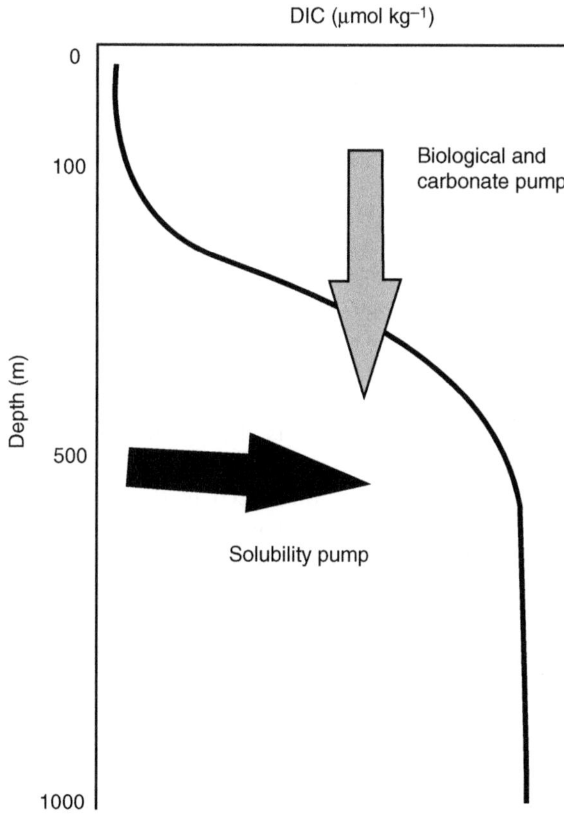

Figure 4 Illustration of the vertical gradient of DIC in the ocean. The uptake of DIC by phytoplankton and conversion into sinking organic matter ('biological pump'; gray arrow) and sinking calcium carbonate skeletal matter ('carbonate pump'; gray arrow) contributes to the maintenance of the vertical gradient. Introduction of DIC to the deep waters via the 'solubility pump' at high latitudes and subsequent deep water formation also helps maintain this vertical gradient (black arrow; see **Figure 5**).

Exchange of surface waters with the deep ocean through wind-driven mixing is limited because of strong density stratification of the water column over the majority of the world's oceans. However, thermohaline (overturning) circulation at high latitudes provides a mechanism for surface waters to exchange with the deep ocean. Passage of cold and dry air masses over high-latitude regions, such as the Greenland and Labrador Seas in the North Atlantic or the Weddell Sea in the Southern Ocean, forms cold and very dense sea water ('deep water' formation). Once formed, these dense water masses sink vertically until they reach a depth at which water is of similar density (i.e., 2000–4000 m deep). Following sinking, the dense waters are transported slowly throughout all of the deep ocean basins by advection and diffusion, displacing other deep water that eventually is brought back to the surface by upwelling (**Figure 5**).

Because of the smaller volume and faster circulation, the residence time of the surface ocean is only one decade compared to 600–1000 years for the deep ocean. The process of deep water formation transfers CO_2, absorbed from the atmosphere by the solubility pump, into the deep ocean. The effect is that DIC concentration increases with depth in all ocean basins (**Figure 4, 5** and **6A**,). As a result of the long residence time of the deep ocean, carbon, once removed from the surface ocean to the deep ocean through the effects of solubility and deep water formation, is stored without contact with the atmosphere for hundreds to thousands of years. At present, deep water formed at the surface that is in equilibrium with the atmosphere (sea water P_{CO_2} of $\sim$360 ppm), carries more CO_2 to depth than deep water formed prior to the industrial age (e.g., $\sim$280 ppm). Furthermore, P_{CO_2} of upwelled deep water is less than that in the recently formed deep water, indicating that the deep water formation and the 'solubility pump' allow the ocean to be a net sink for anthropogenic CO_2. The vertical gradient in DIC (**Figure 4**) and the ability of the ocean to take up atmospheric CO_2 is augmented by biological processes known as the 'biological pump'.

The Biological Pump

Although the standing stock of marine biota in the ocean is relatively small (3×10^{15} g C), the activity

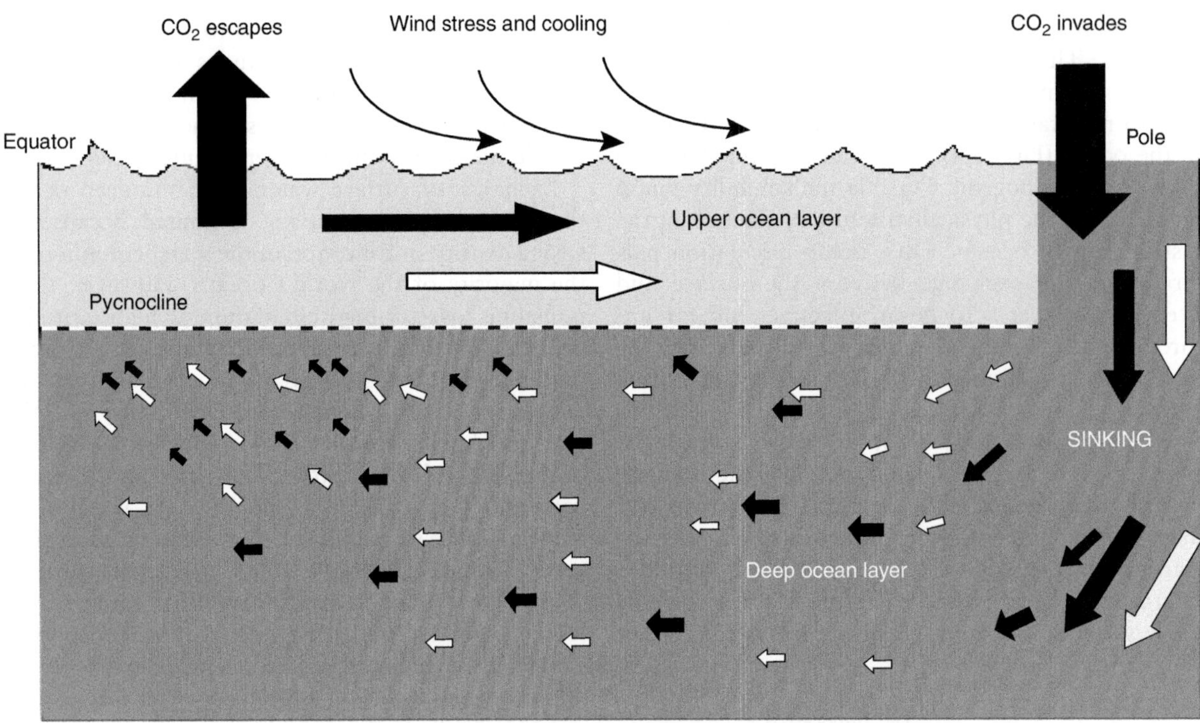

Figure 5 Conceptual model of the 'solubility pump'. White arrows represent movement of water; black arrows represent movement of CO_2 within, and into and out of, the ocean. Cooling increases the solubility of CO_2 and results in a flux of CO_2 from the atmosphere to the surface ocean. At subpolar latitudes the water density increases and the CO_2-enriched water sinks rapidly. At depth, the CO_2-enriched water moves slowly as is it is dispersed throughout the deep ocean. The sinking water displaces water that is returned to the surface ocean in upwelling regions. As the water warms, P_{CO_2} increases, resulting in escape of CO_2 from the surface water to the atmosphere.

associated with the biota is extremely important to the cycling of carbon between the atmosphere and the ocean. The largest and most rapid fluxes in the global carbon cycle are those that link atmospheric CO_2 to photosynthetic production (primary production) on the land and in the ocean. Globally, marine phytoplankton are responsible for more than one-third of the total gross photosynthetic production (50×10^{15} g C y^{-1}). In the sea, photosynthesis is limited to the euphotic zone, the upper 100–150 m of the water column where light can penetrate. Photosynthetic organisms use light energy to reduce CO_2 to high-energy organic compounds. In turn, a portion of these synthesized organic compounds are utilized by heterotrophic organisms as an energy source, being remineralized to CO_2 via respiration. Eqn [III] represents the overall reactions of photosynthesis and respiration.

$$CO_2(\text{gas}) + H_2O \underset{\text{Metabolic energy (respiration)}}{\overset{\text{Light energy (photosynthesis)}}{\rightleftharpoons}} (CH_2O)_n + O_2(\text{gas}) \quad [III]$$

In the sea, net primary production (primary production in excess of respiration) converts CO_2 to organic matter that is stored as particulate organic carbon (POC; in living and detrital particles) and as dissolved organic carbon (DOC). In stratified regions of the ocean (lower latitudes), net primary production results in a drawdown of DIC and an accumulation of organic matter as POC and DOC (**Figure 6B**). However, it is the portion of organic carbon production that can be exported from the surface ocean and remineralized in the deep ocean that is important in the exchange of CO_2 between the atmosphere and the ocean. The biological pump refers to the processes that convert CO_2 (thereby drawing down DIC) to organic matter by photosynthesis, and remove the organic carbon to depth (where it is respired) via sinking, mixing, and active transport mechanisms (**Figure** 7). Once at great depth, it is effectively removed from exchange with the atmosphere. As living biomass is produced, some particles becomes senescent and form sinking aggregates, while other particles are consumed by herbivores and sinking fecal pellets (POC) are formed. These sinking aggregates and pellets remove carbon from the surface to be remineralized at depth via decomposition by bacteria or consumption by zooplankton and fish (**Figure** 7). In addition, DOC produced by phytoplankton or by animal excretion

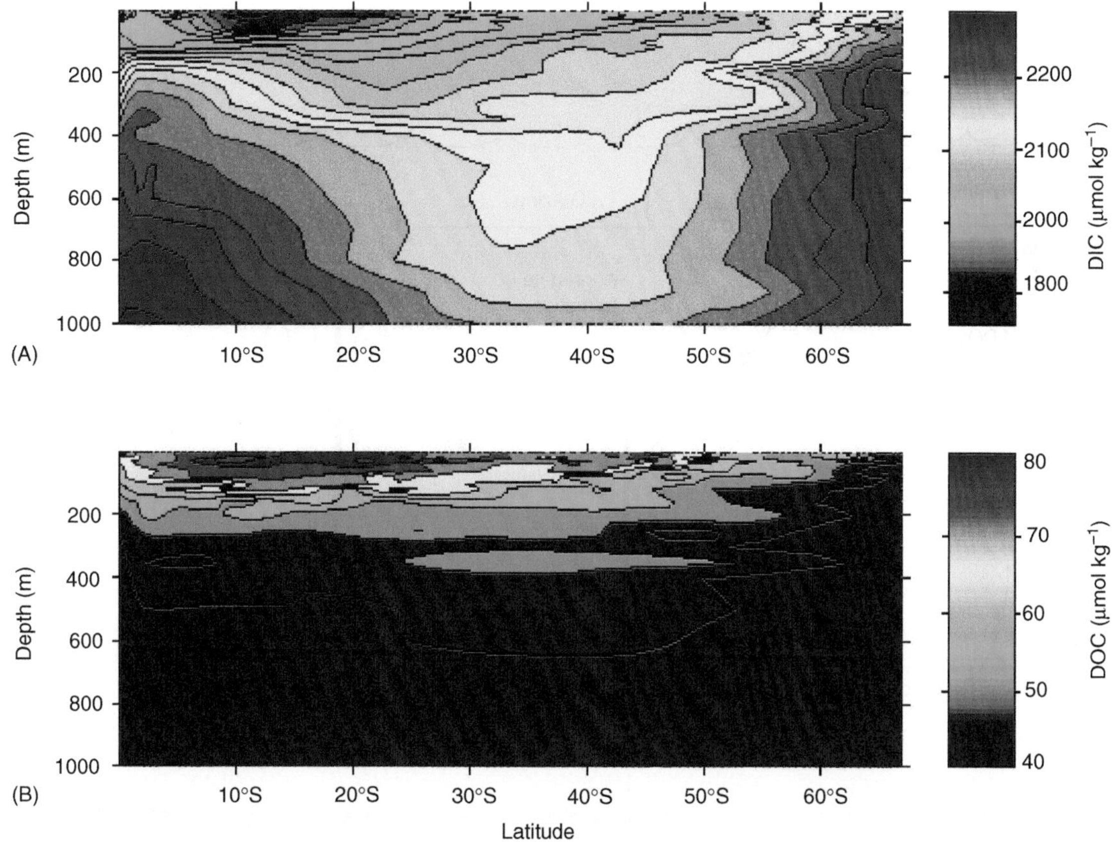

Figure 6 Contour plot of (A) DIC and (B) DOC along a transect line in the South Pacific between the equator (0°; 170° W) and the Antarctic Polar Front (66° S; 170° W). Note that in the low-latitude stratified waters DIC concentrations are depleted in surface water relative to deep water, as a result of net primary production and air–sea exchange. DOC concentrations are elevated relative to deep water. In high-latitude regions, DIC concentration are elevated in the surface water as a result of increased solubility of cooler surface waters.

in surface waters can also be transported downward by subduction or convective mixing of surface waters (**Figure** 7). Finally, vertically migrating zooplankton that feed in the surface waters at night and return to deep waters during the day actively transport dissolved and particulate material to depth, where a portion is metabolized (**Figure** 7).

Production via photosynthesis can occur only in the surface ocean, whereas remineralization can occur throughout the water column. The biological pump serves to spatially separate the net photosynthetic from net respiratory processes. Thus, the conversion of DIC to exportable organic matter acts to reduce the DIC concentration in the surface water and its subsequent remineralization increases DIC concentration in the deep ocean (**Figure** 6). The biological pump is important to the maintenance of a vertical DIC profile of undersaturation in the surface and supersaturation at depth (**Figures** 4 and **5A**). Undersaturation of DIC in the surface mixed layer, created by the biological pump, allows for the influx of CO_2 from the atmosphere (see Henry's law above; **Figure** 7).

Gross export of organic matter out of the surface waters is approximately 10×10^{15} g C y^{-1} (**Figure 1**). Less than 1% of the organic matter exported from the surface waters is stored in the abyssal sediment. In fact, most of the exported organic matter is remineralized to DIC in the upper 500 m of the water column. It is released back to the atmosphere on timescales of months to years via upwelling, mixing, or ventilation of high-density water at high latitudes. It is that fraction of exported organic matter that actually reaches the deep ocean (>1000 m) that is important for long-term atmospheric CO_2 regulation. Once in the deep ocean, the organic matter either remains as long-lived DOC or is remineralized to DIC and is removed from interaction with the atmosphere on timescales of centuries to millennia. Thus, even though less than 1% of the exported carbon is stored in marine sediments, the activities of the biological pump are very important in mediating

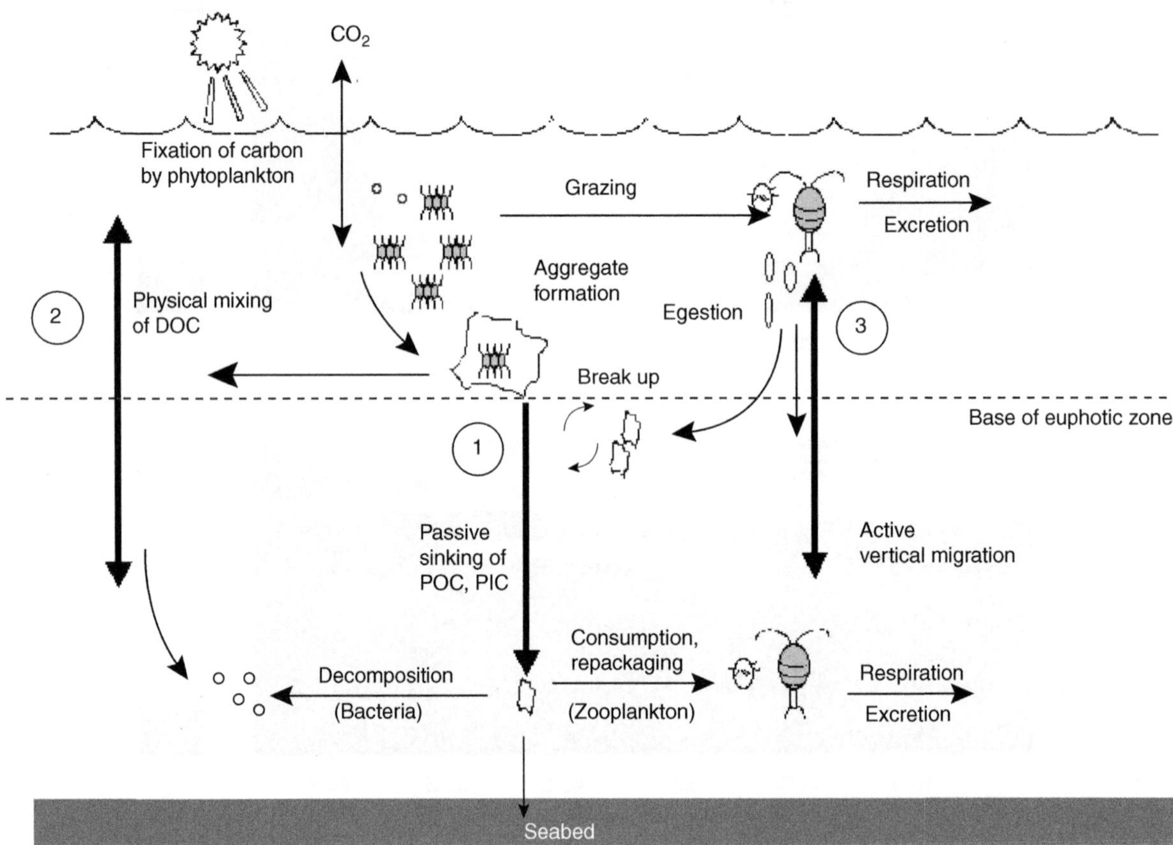

Figure 7 Conceptual diagram depicting components of the 'biological pump'. CO_2 is taken up by phytoplankton and organic matter is produced. As this organic matter is processed through the marine food web, fecal pellets or aggregates are produced, a portion of which sink from the surface waters to depth (1). As organic matter is processed through the food web, DOC is also produced. DOC is removed from the surface waters to depth via physical mixing of the water by convective overturn (2). DOC and DIC are also actively transported to depth by vertically migrating organisms such as copepods that feed in surface waters and excrete and respire the consumed organic carbon at depth (3).

the air–sea transfer of CO_2. Without this pump in action, atmospheric CO_2 concentration might be as high as 500 to 1000 ppm versus the 360 ppm observed today.

Contribution of POC Versus DOC in the Biological Pump

Historically, sinking particles were thought of as the dominant export mechanism of the biological pump and the primary driver of respiration in the ocean's interior. However, downward mixing of surface water can also transport large quantities of DOC trapped within the sinking water mass. In order for DOC to be an important contributor to the biological pump two sets of conditions must exist. First, the producer–consumer dynamics in the surface waters must yield DOC of a quality that is resistant to rapid remineralization by bacteria and lead to net DOC production. Second, the physical system must undergo periods of deep convective mixing or subduction in order to remove surface waters and DOC to depth. Although approximately 80% of the globally exported carbon is in the form of POC, DOC can represent 30–50% of carbon export in the upper 500 m of the water column at specific ocean sites. The biological/physical controls on DOC export are complex and are currently being assessed for various regions of the worlds' ocean.

Factors That Affect the Efficiency of the Biological Pump

An efficient biological pump means that a large fraction of the system's net production is removed from the surface waters via export mechanisms. Factors that affect the efficiency of the biological pump are numerous and include nutrient supply and plankton community structure.

Nutrient supply Does an increased partial pressure of atmospheric CO_2 lead to a more efficient biological pump? Not necessarily, since net primary production is limited by the availability of other inorganic nutrients such as nitrogen, phosphorus, silicon and iron. Because these inorganic nutrients are continuously being removed from the surface waters with vertical export of organic particles, their concentrations are often below detection limits in highly stratified water columns. As a result, primary production becomes limited by the rate at which these nutrients can be re-supplied to the surface ocean by mixing, by atmospheric deposition, or by heterotrophic recycling. Primary production supported by the recycling of nutrients in the surface ocean is referred to as 'regenerative' production and contributes little to the biological pump. Primary production supported by the introduction of new nutrients from outside the system, via mixing from below or by atmospheric deposition (e.g. dust), is referred to as 'new' production. New nutrients enhance the amount of net production that can be exported (new production). Because CO_2 is not considered to be a limiting nutrient in marine systems, the increase in atmospheric CO_2 is not likely to stimulate net production for most of the world's ocean unless it indirectly affects the introduction of new nutrients as well.

Community structure Food web structure also plays an important role in determining the size distribution of the organic particles produced and whether the organic carbon and associated nutrients are exported from or recycled within the surface waters. The production of large, rapidly settling cells will make a greater contribution to the biological pump than the production of small, suspended particles. Factors such as the number of trophic links and the size of the primary producers help determine the overall contribution of sinking particles. The number of trophic steps is inversely related to the magnitude of the export flux. For example, in systems where picoplankton are the dominant primary producers there may be 4–5 steps before reaching a trophic level capable of producing

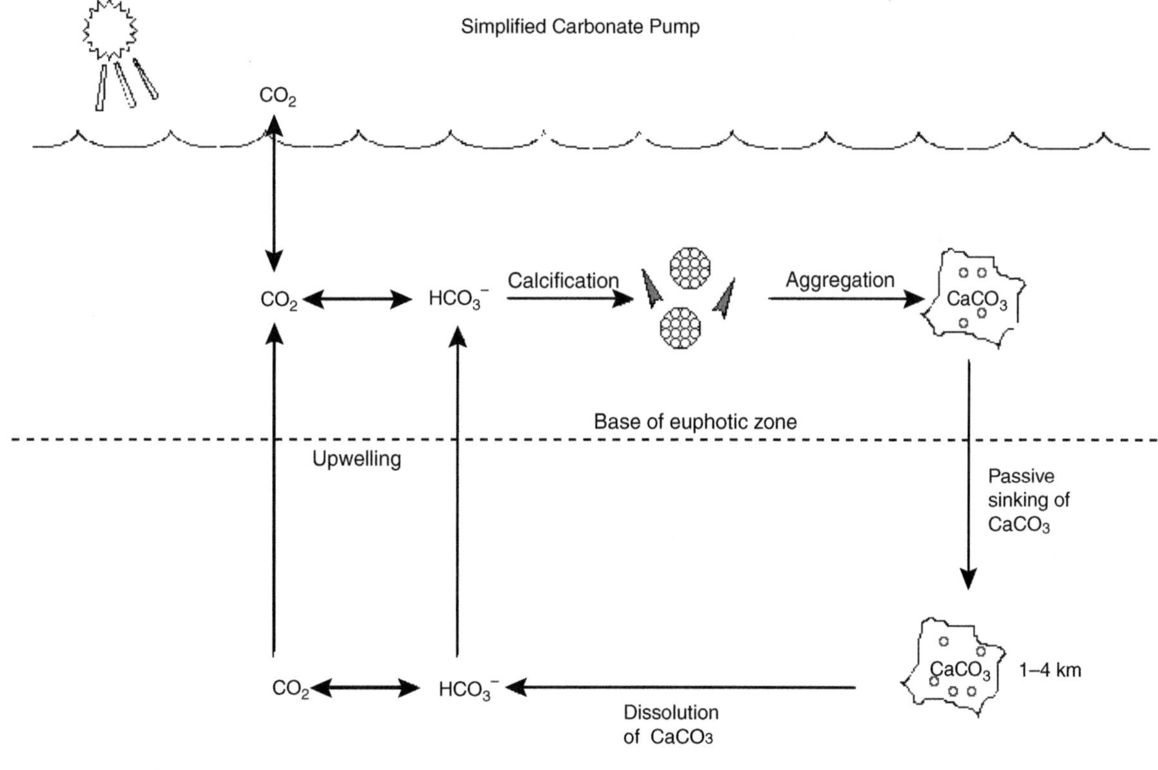

Figure 8 Conceptual diagram of a simplified 'carbonate pump'. Some marine organisms form calcareous skeletal material, a portion of which sinks as calcium carbonate aggregates. These aggregates are preserved in shallow ocean sediments or dissolve at greater depths (3000–5000 m), thus increasing DIC concentrations in the deep ocean. The calcium and bicarbonate are returned to the surface ocean through upwelling.

sinking particles. With each trophic transfer, a percentage (50–70%) of the organic carbon is respired, so only a small fraction of the original primary production forms sinking particles. Although picoplankton may dominate primary production in oceanic systems, their production is considered 'regenerative' and contributes little to the production of sinking material. Alternatively, production by larger phytoplankton such as diatoms (>20 μm in size) may represent a smaller fraction of primary production, but their contribution to the biological pump is larger because fewer trophic steps are taken to produce sinking particles.

The Carbonate Pump

A process considered part of the biological pump (depending how it is defined) is the formation and sinking of calcareous skeletal material by some marine phytoplankton (e.g., coccolithophores) and animals (e.g., pteropods and foraminifera). Calcification is the process by which marine organisms combine calcium with carbonate ions to form hard body parts. The resulting calcium carbonate ($CaCO_3$) is dense and sinks out of the surface water with export production (**Figure 8**). The global mean ratio for carbon sinking from the surface ocean as $CaCO_3$ or organic carbon is 1:4. However, unlike organic matter, $CaCO_3$ is not remineralized as it sinks; it only begins to dissolve in intermediate and deep waters, waters undersaturated with respect to $CaCO_3$. Complete dissolution of $CaCO_3$ skeletons typically occurs at depths of 1–4 km (in the north Pacific Ocean) to 5 km (in the North Atlantic). This depth zone is known as the carbonate compensation depth. $CaCO_3$ is only found in sediments shallower than the carbonate compensation depth. Globally, the CO_2 sink in sedimentary rock is four times greater than the sink in organic sediments.

Summary

In summary, the biological and physical processes of the oceanic carbon cycle play an important role in the regulation of atmospheric CO_2. However, the intricacies of the oceanic carbon cycle are vast and continued ocean research is essential to better understand the controls of the Earth's climate.

See also

Atmospheric Input of Pollutants. Carbon Dioxide (CO_2) Cycle.

Further Reading

Bates NR, Michaels AF, and Knap AH (1996) Seasonal and interannual variability of oceanic carbon dioxide species at the U.S. JGOFS Bermuda Atlantic Time-series Study (BATS) site. *Deep-Sea Research II* 43: 347–383.

Bolin B (ed.) (1983) *The Major Biogeochemical Cycles and Their Interactions: SCOPE 21*. New York: Wiley.

Carlson CA, Ducklow HW, and Michaels AF (1994) Annual flux of dissolved organic carbon from the euphotic zone in the northwestern Sargasso Sea. *Nature* 371: 405–408.

Denman K, Hofman H, and Marchant H (1996) Marine biotic responses to environmental change and feedbacks to climate. In: Houghton JT, Meira Filho LG, and Callander BA, *et al.* (eds.) *Climate Change 1995: The Science of Climate Change*. New York: Cambridge University Press.

Follows MJ, Williams RG, and Marshall JC (1996) The solubility pump of carbon in the subtropical gyre of the North Atlantic. *Journal of Marine Research* 54: 605–630.

Hansell DA and Carlson CA (1998) Net community production of dissolved organic carbon. *Global Biogeochemical Cycles* 12: 443–453.

Holmén K (1992) The global carbon cycle. In: Butcher SS, Charlson RJ, Orians GH, and Wolfe GV (eds.) *Global Biogeochemical Cycles*, pp. 239–262. New York: Academic Press.

Houghton JT, Meira Filho LG, and Callander BA, *et al.* (eds.) (1996) *Climate Change 1995: The Science of Climate Change*. New York: Cambridge University Press.

Michaels AF and Silver MW (1988) Primary producers, sinking fluxes and the microbial food web. *Deep-Sea Research* 35: 473–490.

Sarmiento JL and Wofsy (eds.) (1999) *A U.S. Carbon Cycle Science Plan*. Washington, DC: U.S. Global Change Research Program.

Sarmiento JL, Hughes TMC, Stouffer RJ, and Manabe S (1998) Simulated response of the ocean carbon cycle to anthropogenic climate warming. *Nature* 393: 245–249.

Schlesinger WH (1997) *Biogeochemistry: An Analysis of Global Change*. New York: Academic Press.

Siegenthaler U and Sarmiento JL (1993) Atmospheric carbon dioxide and the ocean. *Nature* 365: 119–125.

Steinberg DK, Carlson CA, Bates NR, Goldthwait SA, Madin LP, and Michaels AF (2000) Zooplankton vertical migration and the active transport of dissolved organic and inorganic carbon in the Sargasso Sea. *Deep-Sea Research I* 47: 137–158.

Takahashi T, Tans PP, and Fung I (1992) Balancing the budget: carbon dioxide sources and sinks, and the effect of industry. *Oceanus* 35: 18–28.

Varney M (1996) The marine carbonate system. In: Summerhayes CP and Thorpe SA (eds.) *Oceanography an Illustrated Guide*, pp. 182–194. London: Manson Publishing.

CARBON DIOXIDE (CO_2) CYCLE

T. Takahashi, Lamont Doherty Earth Observatory, Columbia University, Palisades, NY, USA

Introduction

The oceans, the terrestrial biosphere, and the atmosphere are the three major dynamic reservoirs for carbon on the earth. Through the exchange of CO_2 between them, the atmospheric concentration of CO_2 that affects the heat balance of the earth, and hence the climate, is regulated. Since carbon is one of the fundamental constituents of living matter, how it cycles through these natural reservoirs has been one of the fundamental questions in environmental sciences. The oceans contain about 50 times as much carbon (about 40 000 Pg-C or10^{15} g as carbon) as the atmosphere (about 750Pg-C). The terrestrial biosphere contains about three times as muchcarbon (610 Pg-C in living vegetation and 1580 Pg-C in soil organic matter) as the atmosphere. The air–sea exchange of CO_2 occurs via gas exchange processes across the sea surface; the natural air-to-sea and sea-to-air fluxes have been estimated to be about 90 Pg-C y^{-1} each. The unperturbed uptake flux of CO_2 by global terrestrial photosynthesis is roughly balanced with the release flux by respiration, and both have been estimated to be about 60 Pg-C y^{-1}. Accordingly, atmospheric CO_2 is cycled through the ocean and terrestrial biosphere with a time scale of about 7 years.

The lithosphere contains a huge amount of carbon (about 100 000 000 Pg-C) in the form of limestones ((Ca, Mg) CO_3), coal, petroleum, and other formsof organic matter, and exchanges carbon slowly with the other carbon reservoirs via such natural processes as chemical weathering and burial of carbonate andorganic carbon. The rate of removal of atmospheric CO_2 by chemical weathering has been estimated to be of the order of 1 Pg-Cy^{-1}. Since the industrial revolution in the nineteenth century, the combustion of fossil fuels and the manufacturing of cement have transferred the lithospheric carbon into the atmosphere at rates comparable to the natural CO_2 exchange fluxes between the major carbon reservoirs, and thus have perturbed the natural balance significantly (6 Pg-Cy^{-1} is about an order of magnitude less than the natural exchanges with the oceans (90 Pg-C y^{-1} and land (60 Pg-C y^{-1})). The industrial carbon emission rate has been about 6 Pg-C y^{-1} for the 1990s, and the cumulative industrial emissions since the nineteenth century to the end of the twentieth century have been estimated to be about 250 Pg-C. Presently, the atmospheric CO_2 content is increasing at a rate ofabout 3.5 Pg-C y^{-1} (equivalent to about 50% of theannual emission) and the remainder of the CO_2 emitted into the atmosphere is absorbed by the oceans and terrestrial biosphere in approximately equal proportions. These industrial CO_2 emissions have caused the atmospheric CO_2 concentration to increase by as much as 30% from about 280 ppm (parts per million mole fraction in dry air) in the pre-industrial year 1850 to about 362 ppm in the year 2000. Theatmospheric CO_2 concentration may reach 580 ppm, double thepre-industrial value, by the mid-twenty first century. This represents a significant change that is wholly attributable to human activities on the Earth.

It is well known that the oceans play an important role in regulating our living environment by providing water vapor into the atmosphere and transporting heat from the tropics to high latitude areas. In addition to these physical influences, the oceans partially ameliorate the potential CO_2-induced climate changes by absorbing industrial CO_2 in the atmosphere.

Therefore, it is important to understand howthe oceans take up CO_2 from the atmosphere and how they store CO_2 in circulating ocean water. Furthermore, in order to predict the future course of the atmospheric CO_2 changes, we need to understand how the capacity of the ocean carbon reservoir might be changed in response to the Earth's climate changes, that may, in turn, alter the circulation of ocean water. Since the capacity of the ocean carbon reservoir is governed by complex interactions of physical, biological, and chemical processes, it is presently not possible to identify and predict reliably various climate feedback mechanisms that affect the ocean CO_2 storage capacity.

Units

In scientific and technical literature, the amount of carbon has often been expressed in three different units: giga tons of carbon (Gt-C), petagrams of carbon (Pg-C) and moles of carbon or CO_2. Their relationships are: 1Gt-C = 1 Pg-C = 1×10^{15} g of carbon = 1000 million metric tonnes of carbon = $(1/12) \times 10^{15}$ moles of carbon. The equivalent quantity as CO_2 may be obtained by multiplying the above numbers by 3.67 (= 44/12 = the molecular weight of CO_2 divided by the atomic weight of carbon).

The magnitude of CO_2 disequilibrium between the atmosphere and ocean water is expressed by the difference between the partial pressure of CO_2 of ocean water, $(pCO_2)sw$, and that in the overlying air, $(pCO_2)air$. This difference represents the thermodynamic driving potential for CO_2 gas transfer across the sea surface. The pCO_2 in the air may be estimated using the concentration of CO_2 in air, that is commonly expressed in terms of ppm (parts per million) in mole fraction of CO_2 in dry air, in the relationship:

$$p(CO_2)air = (CO_2\ conc.)air \times (Pb - pH_2O) \quad [1]$$

where Pb is the barometric pressure and pH_2O is the vapor pressure of water at the sea water temperature. The partial pressure of CO_2 in sea water, $(pCO_2)sw$, may be measured by equilibration methods or computed using thermodynamic relationships. The unit of microatmospheres (μatm) or 10^{-6} atm is commonly used in the oceanographic literature.

History

The air–sea exchange of $CO_2$was first investigated in the 1910s through the 1930s by a group of scientists including K. Buch, H. Wattenberg, and G.E.R. Deacon. Buch and his collaborators determined in land-based laboratories CO_2 solubility, the dissociation constants for carbonic and boric acids in sea water, and their dependence on temperature and chlorinity (the chloride ion concentrationin sea water). Based upon these dissociation constants along with the shipboard measurements of pH and titration alkalinity, they computed the partial pressure of CO_2 in surface ocean waters. The Atlantic Ocean was investigated from the Arctic to Antarctic regions during the period 1917–1935, especially during the METEOR Expedition 1925–27, in the North and South Atlantic. They discovered that temperate and cold oceans had lower pCO_2 than air (hence the sea water was a sink for atmospheric CO_2), especially during spring and summer seasons, due to the assimilation of CO_2 by plants. They also observed that the upwelling areas of deep water (such as African coastal areas) had greater pCO_2 than the air (hence the sea water was a CO_2 source) due to the presence of respired CO_2 in deep waters.

With the advent of the high-precision infrared CO_2 gas analyzer, a new method for shipboard measurements of pCO_2 in sea water and in air was introduced during the International Geophysical Year, 1956–59. The precision of measurements was improved by more than an order of magnitude. The global oceans were investigated by this new method, which rapidly yielded high precision data. The equatorial Pacific was identified as a major CO_2 source area. The GEOSECS Program of the International Decade of Ocean Exploration, 1970–80, produced a global data set that began to show systematic patterns for the distribution of CO_2 sink and source areas over the global oceans.

Methods

The net flux of CO_2 across these a surface, Fs-a, may be estimated by:

$$\begin{aligned} \text{Fs-a} &= \boldsymbol{E} \times [(pCO_2)sw - (pCO_2)air] \\ &= \boldsymbol{k} \times \alpha \times [(pCO_2)sw - (pCO_2)air] \end{aligned} \quad [2]$$

where E is the CO_2 gas transfer coefficient expressed commonly in (moles $CO_2/m^2/y/uatm$); $\boldsymbol{k}$ is the gas transfer piston velocity (e.g. in (cmh^{-1})) and α is the solubility of CO_2 in sea water at a given temperature and salinity (e.g. (moles $CO_2\,kg\text{-}sw^{-1}\ atm^{-1}$)). If $(pCO_2)sw < (pCO_2)$ air, the net flux of CO_2 is from the sea to the air and the ocean is a source of CO_2; if $(pCO_2)sw < (pCO_2)$ air, the ocean water is a sink for atmospheric CO_2. The sea–air pCO_2 difference may be measured at sea and α has been determined experimentally as a function of temperature and salinity. However, the values of E and $\boldsymbol{k}$ that depend on the magnitude of turbulence near the air–water interface cannot be simply characterized over complex ocean surface conditions. Nevertheless, these two variables have been commonly parameterized in terms of wind speed over the ocean. A number of experiments have been performed to determine the wind speed dependence under various wind tunnel conditions as well as ocean and lake environments using different nonreactive tracer gases such as SF_6 and ^{222}Rn. However, the published results differ by as much as 50% over the wind speed range of oceanographic interests.

Since ^{14}C is in the form of CO_2 in the atmosphere and enters into the surface ocean water as CO_2 in a timescale of decades, its partition between the atmosphere and the oceans yields a reliable estimate for the mean CO_2 gas transfer rate over the global oceans. This yields a CO_2 gas exchange rate of 20 ± 3 mol $CO_2\,m^{-2}y^{-1}$ that corresponds to a sea–air CO_2 transfer coefficient of 0.067 mol $CO_2\ m^{-2}y^{-1}\ uatm^{-1}$. Wanninkhof in 1992 presented an expression that satisfies the mean global CO_2 transfer coefficient based on ^{14}C and takes other field and wind tunnel results into consideration. His equation for variable wind speed conditions is:

$$\boldsymbol{k}\left(cm\,h^{-1}\right) = 0.39 \times (u_{av})^2 \times (Sc/660)^{-0.5} \quad [3]$$

where u_{av} is the average wind speed in ms^{-1} corrected to 10 m above sea surface; Sc(dimensionless) is the Schmidt number (kinematic viscosity ofwater)/(diffusion coefficient of CO_2 gas inwater); and 660 represents the Schmidt number for CO_2 in seawater at 20°C.

In view of the difficulties in determining gas transfer coefficients accurately, direct methods for CO_2 flux measurements aboard the ship are desirable. Sea–air CO_2 flux was measured directly by means of the shipboard eddy-covariance method over the North Atlantic Ocean by Wanninkhof and McGillis in 1999. The net flux of CO_2 across the sea surface was determined by a covariance analysis of the tri-axial motion of air with CO_2 concentrations in the moving air measured in short time intervals ($\sim$ms) as a ship moved over the ocean. The results obtained over awind speed range of 2–13.5 m s^{-1} are consistent with eqn [3] within about $\pm 20\%$. If the data obtainedin wind speeds up to 15 m s^{-1} are taken into consideration, they indicate that the gas transfer piston velocity tends to increase as a cubeof wind speed. However, because of a large scatter ($\pm 35\%$) ofthe flux values at high wind speeds, further work is needed to confirm the cubic dependence.

In addition to the uncertainties in the gas transfer coefficient (or piston velocity), the CO_2 fluxestimated with eqn [2] is subject to errors in $(pCO_2)sw$ caused by the difference between the bulk water temperature and the temperature of the thin skin of ocean water at the sea–air interface. Ordinarily the $(pCO_2)sw$ is obtained at the bulk seawater temperature, whereas the relevant value for the flux calculation is $(pCO_2)sw$ at the 'skin'temperature, that depends on the rate of evaporation, the incoming solar radiation, the wind speed, and the degree of turbulence near the interface. The 'skin' temperature is often cooler than the bulk water temperatureby as much as 0.5°C if the water evaporates rapidly to a dry air mass, but is not always so if a warm humid air mass covers over the ocean. Presently, the time–space distribution of the 'skin' temperature is not well known. This, therefore, could introduce errors in $(pCO_2)sw$ up to about 6 µatm or 2%.

CO_2 Sink/Source Areas of the Global Ocean

The oceanic sink and source areas for atmospheric CO_2 and the magnitude of the sea–air CO_2 flux over the global ocean vary seasonally and annually as well as geographically. These changes are the manifestation of changes in the partial pressure of sea water, $(pCO_2)sw$, which are caused primarily by changes in the water temperature, in the biological utilization of CO_2, and in the lateral/vertical circulation of ocean waters including the upwelling of deep water rich in CO_2. Over the global oceans, sea water temperatures change from the pole to the equator by about 32°C. Since the pCO_2 in sea water doubles with each 16°C of warming, temperature changes should cause a factor of 4 change in pCO_2. Biological utilization of CO_2 over the global oceans is about 200 µmol $CO_2\,kg^{-1}$, which should reduce pCO_2 in sea water by a factor of 3. If this is accompanied with growths of $CaCO_3$-secreting organisms, the reduction of pCO_2 could be somewhat smaller. While these effects are similar in magnitude, they tend to counteract each other seasonally, since the biological utilization tends to be large when waters are warm. In subpolar and polar areas, winter cooling of surface waters induces deep convective mixing that brings high pCO_2 deep waters to the surface. The lowering effect on CO_2 by winter cooling is often compensated for or some times over compensated for by the increasing effect of the upwelling of high CO_2 deep waters. Thus, in high latitude oceans, surface waters may become a source for atmospheric CO_2 during the winter time when the water is coldest.

In **Figure 1**, the global distribution map of the sea–air pCO_2 differences for February and August 1995, are shown. These maps were constructed on the basis of about a half million pairs of atmospheric and seawater pCO_2 measurements made at sea over the 40-year period, 1958–98, by many investigators. Since the measurements were made in different years, during which the atmospheric pCO_2 was increasing, they were corrected to a single reference year (arbitrarily chosen to be 1995) on the basis of the following observations. Warm surface waters in subtropical gyres communicate slowly with the underlying subsurface waters due to the presence of a strong stratification at the base of the mixed layer. This allows a long time for the surface mixed-layer-waters ($\sim$75 m thick) to exchange CO_2 with the atmosphere. Therefore, their CO_2 chemistry tends to follow the atmospheric CO_2 increase. Accordingly, the pCO_2 in the warm water follows the increasing trend of atmospheric CO_2, and the sea–air pCO_2 difference tends to be independent of the year of measurements. On the other hand, since surface waters in high latitude regions are replaced partially with subsurface waters by deep convection during the winter, the effect of increased atmospheric CO_2 is diluted to undetectable levels and their CO_2 properties tend to remain unchanged from year to year. Accordingly, the sea–air pCO_2 difference measured in a given year increases as the atmospheric CO_2

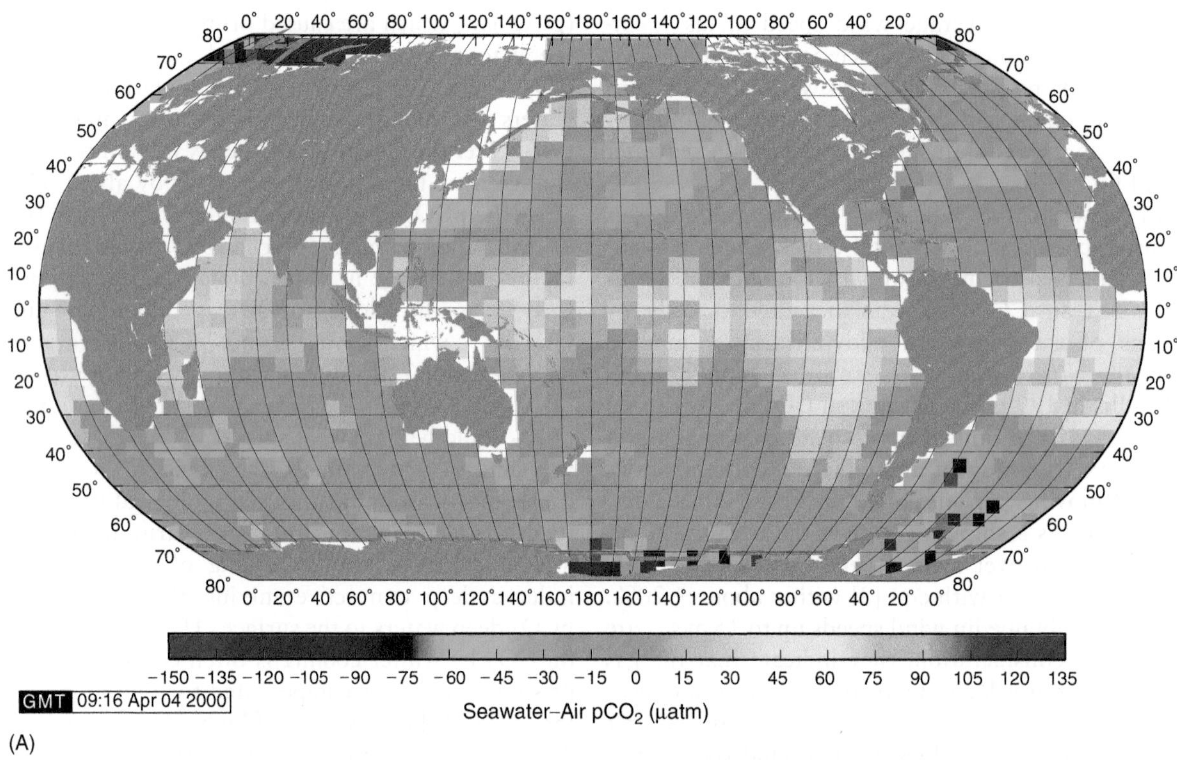

(A)

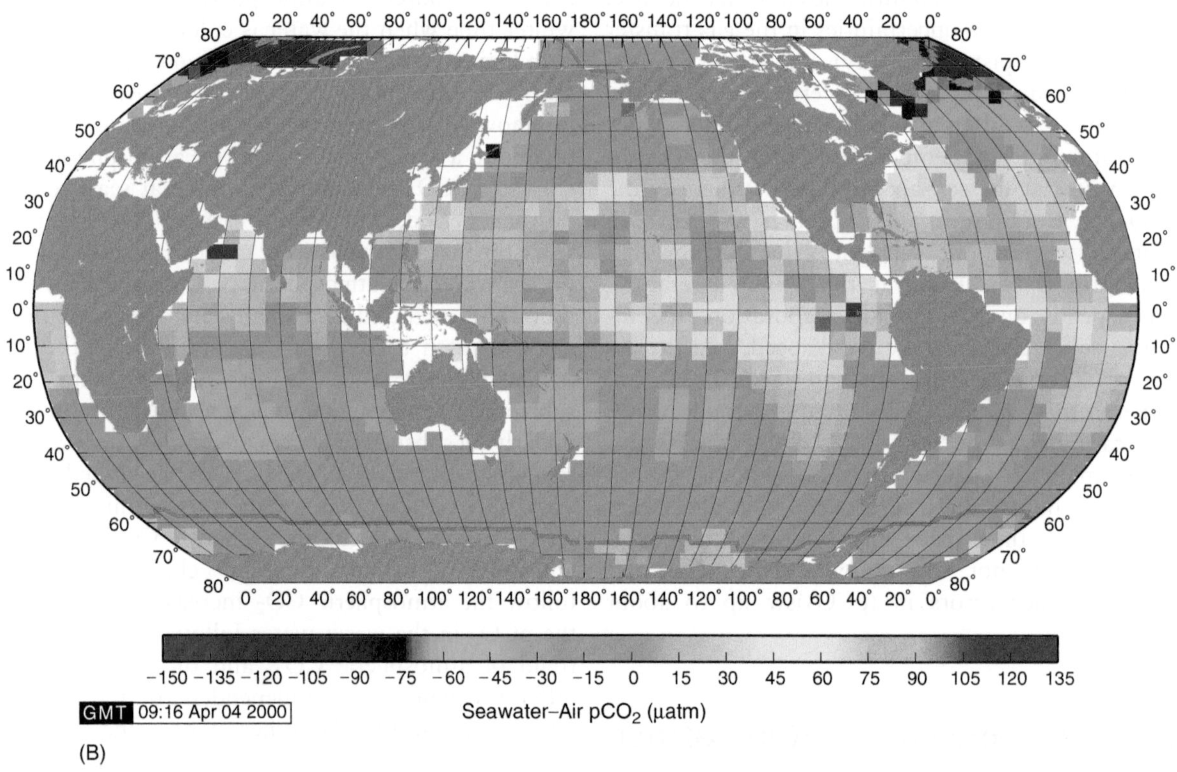

(B)

Figure 1 The sea–air pCO_2 difference in μatm (ΔpCO_2) for (A) February and (B) August for the reference year 1995. The purple-blue areas indicate that the ocean is a sink for atmospheric CO_2, and the red-yellow areas indicate that the ocean is source. The pink lines in the polar regions indicate the edges of ice fields.

concentration increases with time. This effect was corrected to the reference year using the observed increase in the atmospheric CO_2 concentration. During El Niño periods, sea–air pCO_2 differences over the equatorial belt of the Pacific Ocean, which are large in normal years, are reduced significantly and observations are scarce. Therefore, observations made between 10°N and 10°S in the equatorial Pacific for these periods were excluded from the maps. Accordingly, these maps represent the climatological means for non-El Niño period oceans for the past 40 years. The purple-blue areas indicate that the ocean is a sink for atmospheric CO_2, and the red-yellow areas indicate that the ocean is a source.

Strong CO_2 sinks (blue and purple areas) are present during the winter months in the Northern (**Figure 1A**) and Southern (**Figure 1B**) Hemispheres along the poleward edges of the subtropical gyres, where major warm currents are located. The Gulf Stream in the North Atlantic and the Kuroshio Current in the North Pacific are both major CO_2 sinks (**Figure 1A**) due primarily to cooling as they flow from warm tropical oceans to subpolar zones.

Similarly, in the Southern Hemisphere, CO_2 sink areas are formed by the cooling of poleward-flowing currents such as the Brazil Current located along eastern South America, the Agulhus Current located south of South Africa, and the East Australian Current located along south-eastern Australia. These warm water currents meet with cold currents flowing equator ward from the Antarctic zone along the northern border of the Southern (or Antarctic) Ocean. As the sub Antarctic waters rich in nutrients flow northward to more sunlit regions, CO_2 is drawn down by photosynthesis, thus creating strong CO_2 sink conditions, as exemplified by the Falkland Current in the western South Atlantic (**Figure 1A**). Confluence of subtropical waters with polar waters forms broad and strong CO_2 sink zones as a result of the juxta position of the lowering effects on pCO_2 of the cooling of warm waters and the photosynthetic drawdown of CO_2 in nutrient-rich subpolar waters. This feature is clearly depicted in azone between 40°S and 60°S in **Figure 1A** representing the austral summer, and between 20°S and 40°S in **Figure 1B** representing the austral winter.

During the summer months, the high latitude areas of the North Atlantic Ocean (**Figure 1A**) and the Weddell and Ross Seas, Antarctica(**Figure 1B**), are intense sink areas for CO_2. This is attributed to the intense biological utilization of CO_2 within the strongly stratified surface layer caused by solar warming and ice melting during the summer. The winter convective mixing of deep waters rich in CO_2 and nutrient seliminates the strong CO_2 sink and replenishes the depleted nutrients in the surface waters.

The Pacific equatorial belt is a strong CO_2 source which is caused by the warming of upwelled deep waters along the coast of South America as well as by the upward entrainment of the equatorial under current water. The source strengths are most intense in the eastern equatorial Pacific due to the strong upwelling, and decrease to the west as a result of the biological utilization of CO_2 and nutrients during the westward flow of the surface water.

Small but strong source areas in the north-western subArctic Pacific Ocean are due to the winter convective mixing of deep waters (**Figure 1A**). The lowering effect on pCO_2 of cooling in the winter is surpassed by the increasing effect of highCO_2 concentration in the upwelled deep waters. During the summer (**Figure 1B**), however, these source areas become a sink for atmospheric CO_2 due to the intense biological utilization that overwhelms the increasing effect on pCO_2 of warming. A similar area is found in the Arabian Sea, where upwelling of deep-waters is induced by the south-west monsoon during July–August(**Figure 1B**), causing the area tobecome a strong CO_2 source. This source area is eliminated by the photosynthetic utilization of CO_2 following the end of the upwelling period (**Figure 1A**).

As illustrated in **Figure 1A** and **B**, the distribution of oceanic sink and source areas for atmospheric CO_2 varies over a wide range in space and time. Surface ocean waters are out of equilibrium with respect to atmospheric CO_2 by as much as $\pm 200\,\mu atm$ (or by$\pm 60\%$).The large magnitudes of CO_2 disequilibrium between the sea and theair is in contrast with the behavior of oxygen, another biologically mediated gas, that shows only up to $\pm 10\%$ sea–air disequilibrium. The large CO_2 disequilibrium may be attributed to the fact that the internal ocean processes that control pCO_2 in sea water, such as the temperature of water, the photosynthesis, and the upwelling of deep waters,occur at much faster rates than the sea–air CO_2 transfer rates. The slow rate of CO_2 transfer across the sea surface is due to the slow hydration rates of CO_2 as well as to the large solubility of CO_2 in sea water attributable to the formation of bicarbonate and carbonate ions. The latter effect does not exist at all for oxygen.

Net CO_2 Flux Across the Sea Surface

The net sea–air CO_2 flux over the global oceans may be computed using eqns [2] and [3]. **Figure 2** shows the climatological mean distribution of the annual sea–air CO_2 flux for the reference year 1995 using the following set of information. (1) The monthly mean ΔpCO_2 values in $4^\circ \times 5^\circ$ pixel areas for the reference year 1995 (**Figure 1A** and **B** for all other months); (2) the Wanninkhof formulation, eqn [3], for the effect of wind speed on the CO_2 gas transfer coefficient; and (3) the climatological mean wind speeds for each month compiled by Esbensen and Kushnir in 1981. This set yields a mean global gas transfer rate of 0.063 mole $CO_2\,m^{-2}\,\mu atm^{-1}\,y^{-1}$, that is consistent with 20 moles $CO_2\ m^{-2}y^{-1}$ estimated on the basis of carbon-14 distribution in the atmosphere and the oceans.

Figure 2 shows that the equatorial Pacific is a strong CO_2 source. On the other hand, the areas along the poleward edges of the temperate gyres in both hemispheres are strong sinks for atmospheric CO_2. This feature is particularly prominent in the southern Indian and Atlantic Oceans between 40°S and 60°S, and is attributable to the combined effects of negative sea–air pCO_2 differences with strong winds ('the roaring 40 s') that accelerate sea–air gas transfer rates. Similarly strong sink zones are formed in the North Pacific and North Atlantic between 45°N and 60°N. In the high latitude Atlantic, strong sink areas extend into the Norwegian and Greenland Seas. Over the high latitude Southern Ocean areas, the sea–air gas transfer is impeded by the field of ice that covers the sea surface for ⩾ 6 months in a year.

The net sea–air CO_2 fluxes computed for each ocean basin for the reference year of 1995, representing non-El Niño conditions, are summarized in **Table 1**. The annual net CO_2 uptake by the global ocean is estimated to be about 2.0 Pg-C y^{-1}. This is consistent with estimates obtained on the basis of a number of different ocean–atmosphere models including multi-box diffusion advection models and three-dimensional general circulation models.

The uptake flux for the Northern Hemisphere ocean (north of 14°N) is 1.2 Pg-C y^{-1}, whereas that for the Southern Hemisphere ocean (south of 14°S) is 1.7 Pg-C y^{-1}. Thus, the Southern Hemisphere ocean is astronger CO_2 sink by about 0.5 Pg-C y^{-1}. This is due partially to the much greater oceanic areas in the Southern Hemisphere. In addition, the Southern Ocean south of 50°S is an efficient CO_2 sink, for it takes up about 26% of the global ocean CO_2 uptake, while it has only 10% of the global ocean area. Cold temperature and moderate photosynthesis are both

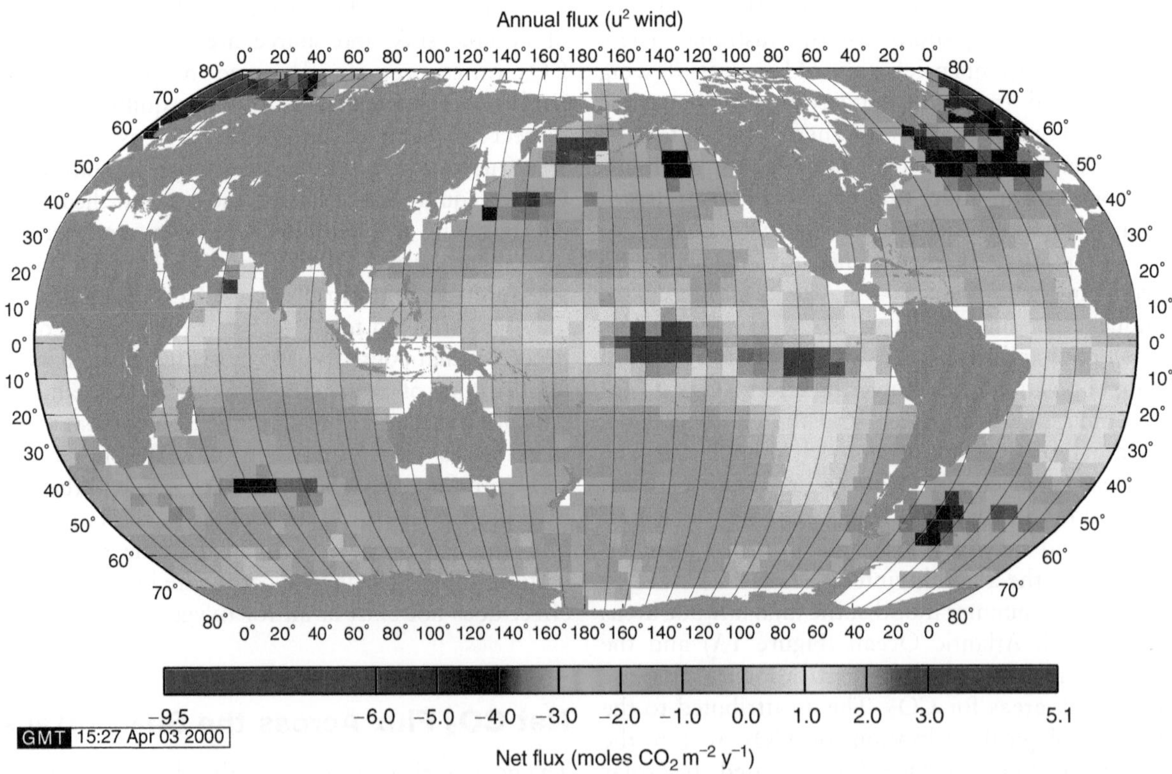

Figure 2 The mean annual sea–air flux of CO_2 in the reference year 1995. The red-yellow areas indicate that the flux is from sea to air, whereas blue-purple areas indicate that the flux is from air to sea. The flux is given in moles of $CO_2\ m^{-2}y^{-1}$. The map gives a total annual air-to-sea flux of 2.0 Pg-C y^{-1}.

Table 1 The net sea–air flux of CO_2 estimated for a reference year of 1995 using the effect of wind speed on the CO_2 gas transfer coefficient, eqn [3], of Wanninkhof and the monthly wind field of Esbensen and Kushnir

Latitudes	*Pacific Ocean*	*Atlantic Ocean*	*Indian Ocean*	*Southern Ocean*	*Global Oceans*
	Sea–air flux in 10^{15}g Carbon y^{-1}				
North of 50°N	−0.02	−0.44	—	—	−0.47
50°N–14°N	−0.47	−0.27	+0.03	—	−0.73
14°N–14°S	+0.64	+0.13	+0.09	—	+0.86
14°S–50°S	−0.37	−0.20	−0.60	—	−1.17
South of 50°S	—	—	—	−0.52	−0.52
Total	−0.23	−0.78	−0.47	−0.52	−2.00
%Uptake	11%	39%	24%	26%	100%
Area (10^6 km^2)	151.6	72.7	53.2	31.7	309.1
Area (%)	49.0%	23.5%	17.2%	10.2%	100%

Positive values indicate sea-to-air fluxes, and negative values indicate air-to-sea fluxes.

responsible for the large uptake by the Southern Ocean.

The Atlantic Ocean is the largest net sink for atmospheric CO_2 (39%); the Southern Ocean (26%) and the Indian Ocean (24%) are next; and the Pacific Ocean (11%) is the smallest. The intense biological drawdown of CO_2 in the high latitude areas of the North Atlantic and Arctic seasduring the summer months is responsible for the Atlantic being a major sink. This is also due to the fact that the upwelling deep waters in the North Atlantic contain low CO_2 concentrations, which are in turn caused primarily by the short residence time ($\sim$80y) of the North Atlantic Deep Waters. The small uptake flux of the Pacific can be attributed to the fact that the combined sink flux of the northern and southern subtropical gyres is roughly balanced by the source flux from the equatorial Pacific during non-El Niño periods. On the other hand, the equatorial Pacific CO_2 source flux is significantly reduced or eliminated during El Niño events. As a result the equatorial zone is covered with the eastward spreading of the warm, low pCO_2 western Pacific waters in response to the relaxation of the trade wind. Although the effects of El Niño and Southern Ocean Oscillation may be far reaching beyond the equatorial zone as far as to the polar areas, the El Niño effects on the equatorial Pacific alone could reduce the equatorial CO_2 source. Hence, this could increase the global ocean uptake flux by up to 0.6 Pg-C y^{-1} during an El Niño year.

The sea–air CO_2 flux estimated above is subject to three sources of error: (1) biases in sea–air ΔpCO_2 values interpolated from relatively sparse observations, (2) the 'skin' temperature effect, and(3) uncertainties in the gas transfer coefficient estimated on the basis of the wind speed dependence. Possible biases in ΔpCO_2 differences have been tested using sea surface temperatures (SST) as a proxy. The systematic error in the global sea–air CO_2 flux resulting from sampling and interpolation has been estimated to be about $\pm$30% or $\pm$0.6 Pg-C y^{-1}. The'skin' temperature of ocean water may affect ΔpCO_2 by as much as ± 6 μatm depending upon time and place, as discussed earlier.

Although the distribution of the 'skin' temperature over the global ocean is not known, it may be cooler than the bulk water temperature by a few tenths of a degree on the global average. This may result in an under estimation of the ocean uptake by 0.4 Pg-C y^{-1}. The estimated global sea–air flux depends on the wind speed data used. Since the gas transfer rate increases nonlinearly with wind speed, the estimated CO_2 fluxes tend to be smaller when mean monthly wind speeds are used instead of high frequency wind data.

Furthermore, the wind speed dependence on the CO_2 gas transfer coefficient in high wind speed regimes is still questionable. If the gas transfer rate is taken to be a cubic function of wind speed instead of the square dependence as shown above, the global ocean uptake would be increased by about 1 Pg-C y^{-1}. The effect is particularly significant over the high latitude oceans where the winds are strong. Considering various uncertainties discussed above, the global ocean CO_2 uptake presented in **Table 1** is uncertain by about 1 Pg-C y^{-1}.

See also

Carbon Cycle. Ocean Carbon System, Modeling of. Radiocarbon. Stable Carbon Isotope Variations in the Ocean.

Further Reading

Broecker WS and Peng TH (1982) *Tracers in the Sea.* Palisades, NY: Eldigio Press.

Broecker WS, Ledwell JR, Takahashi, *et al.* (1986) Isotopic versus micrometeorologic ocean CO_2 fluxes a: serious conflict. *Journal of Geophysical Research* 91: 10517–10527.

Keeling R, Piper SC, and Heinmann M (1996) Global and hemispheric CO_2 sinks deduced from changes in atmospheric O_2 concentration. *Nature* 381: 218–221.

Sarmiento JL, Murnane R, and Le Quere C (1995) Air–sea CO_2 transfer and the carbon budget of the North Atlantic. *Philosophical Transactions of the Royal Society of London, series B* 343: 211–219.

Sundquist ET (1985) Geological perspectives on carbon dioxide and carbon cycle. In: Sundquist ET and Broecker WS (eds.) *The Carbon Cycle and Atmospheric CO_2 N:atural Variations, Archean to Present, Geophysical Monograph* 32, pp. 5–59. Washington, DC: American Geophysical Union.

Takashahi T, Olafsson J, Goddard J, Chipman DW, and Sutherland SC (1993) Seasonal variation of CO_2 and nutrients in the high-latitude surface oceans a: comparative study. *Global Biogeochemical Cycles* 7: 843–878.

Takahashi T, Feely RA, Weiss R, *et al.* (1997) Global air–sea flux of CO_2 a:n estimate based on measurements of sea–air pCO_2 difference. *Proceedings of the National Academy of science USA* 94: 8292–8299.

Tans PP, Fung IY, and Takahashi T (1990) Observational constraints on the global atmospheric CO_2 budget. *Sciece* 247: 1431–1438.

Wanninkhof R (1992) Relationship between wind speed and gas exchange. *Journal of Geophysical Research* 97: 7373–7382.

Wanninkhof R and McGillis WM (1999) A cubic relationship between gas transfer and wind speed. *Geophysical Research Letters* 26: 1889–1893.

OCEAN CARBON SYSTEM, MODELING OF

S. C. Doney and D. M. Glover, Woods Hole Oceanographic Institution, Woods Hole, MA, USA

Introduction

Chemical species such as radiocarbon, chlorofluorocarbons, and tritium–^{3}He are important tools for ocean carbon cycle research because they can be used to trace circulation pathways, estimate timescales, and determine absolute rates. Such species, often termed chemical tracers, typically have rather simple water-column geochemistry (e.g., conservative or exponential radioactive decay), and reasonably well-known time histories in the atmosphere or surface ocean. Large-scale ocean gradients of nutrients, oxygen, and dissolved inorganic carbon reflect a combination of circulation, mixing, and the production, transport, and oxidation (or remineralization) of organic matter. Tracers provide additional, often independent, information useful in separating these biogeochemical and physical processes. Biogeochemical and tracer observations are often framed in terms of ocean circulation models, ranging from simple, idealized models to full three-dimensional (3-D) simulations. Model advection and diffusion rates are typically calibrated or evaluated against transient tracer data. Idealized models are straightforward to construct and computationally inexpensive and are thus conducive to hypothesis testing and extensive exploration of parameter space. More complete and sophisticated dynamics can be incorporated into three-dimensional models, which are also more amenable for direct comparisons with field data. Models of both classes are used commonly to examine specific biogeochemical process, quantify the uptake of anthropogenic carbon, and study the carbon cycle responses to climate change. All models have potential drawbacks, however, and part of the art of numerical modeling is deciding on the appropriate model(s) for the particular question at hand.

Carbon plays a unique role in the Earth's environment, bridging the physical and biogeochemical systems. Carbon dioxide (CO_2), a minor constituent in the atmosphere, is a so-called greenhouse gas that helps to modulate the planet's climate and temperature. Given sunlight and nutrients, plants and some microorganisms convert CO_2 via photosynthesis into organic carbon, serving as the building blocks and energy source for most of the world's biota. The concentration of CO_2 in the atmosphere is affected by the net balance of photosynthesis and the reverse reaction respiration on land and in the ocean. Changes in ocean circulation and temperature can also change CO_2 levels because carbon dioxide is quite soluble in sea water. In fact, the total amount of dissolved inorganic carbon (DIC) in the ocean is about 50 times larger than the atmospheric inventory. The air–sea exchange of carbon is governed by the gas transfer velocity and the surface water partial pressure of CO_2 (pCO_2), which increases with warmer temperatures, higher DIC, and lower alkalinity levels. The natural carbon cycle has undergone large fluctuations in the past, the most striking during glacial periods when atmospheric CO_2 levels were about 30% lower than preindustrial values. The ocean must have been involved in such a large redistribution of carbon, but the exact mechanism is still not agreed upon.

Human activities, including fossil-fuel burning and land-use practices such as deforestation and biomass burning, are altering the natural carbon cycle. Currently about 7.5 Pg C yr^{-1} ($1 \text{ Pg} = 10^{15}\text{g}$) are emitted into the atmosphere, and direct measurements show that the atmospheric CO_2 concentration is indeed growing rapidly with time. Elevated atmospheric CO_2 levels are projected to heat the Earth's surface, and the evidence for climate warming is mounting. Only about 40% of the released anthropogenic carbon remains in the atmosphere, the remainder is taken up in about equal portions (or 2 Pg C yr^{-1}) by land and ocean sinks (**Figure 1**). The future magnitude of these sinks is not well known, however, and is one of the major uncertainties in climate simulations.

Solving this problem is complicated because human impacts appear as relatively small perturbations on a large natural background. In the ocean, the reservoir of organic carbon locked up as living organisms, mostly plankton, is only about 3 Pg C. The marine biota in the sunlit surface ocean are quite productive though, producing roughly 50 Pg of new organic carbon per year. Most of this material is recycled near the ocean surface by zooplankton grazing or microbial consumption. A small fraction, something like 10–20% on average, is exported to the deep ocean as sinking particles or as dissolved organic matter moving with the ocean circulation. Bacteria and other organisms in the deep ocean feed on this source of organic matter from above, releasing DIC and associated nutrients back into the

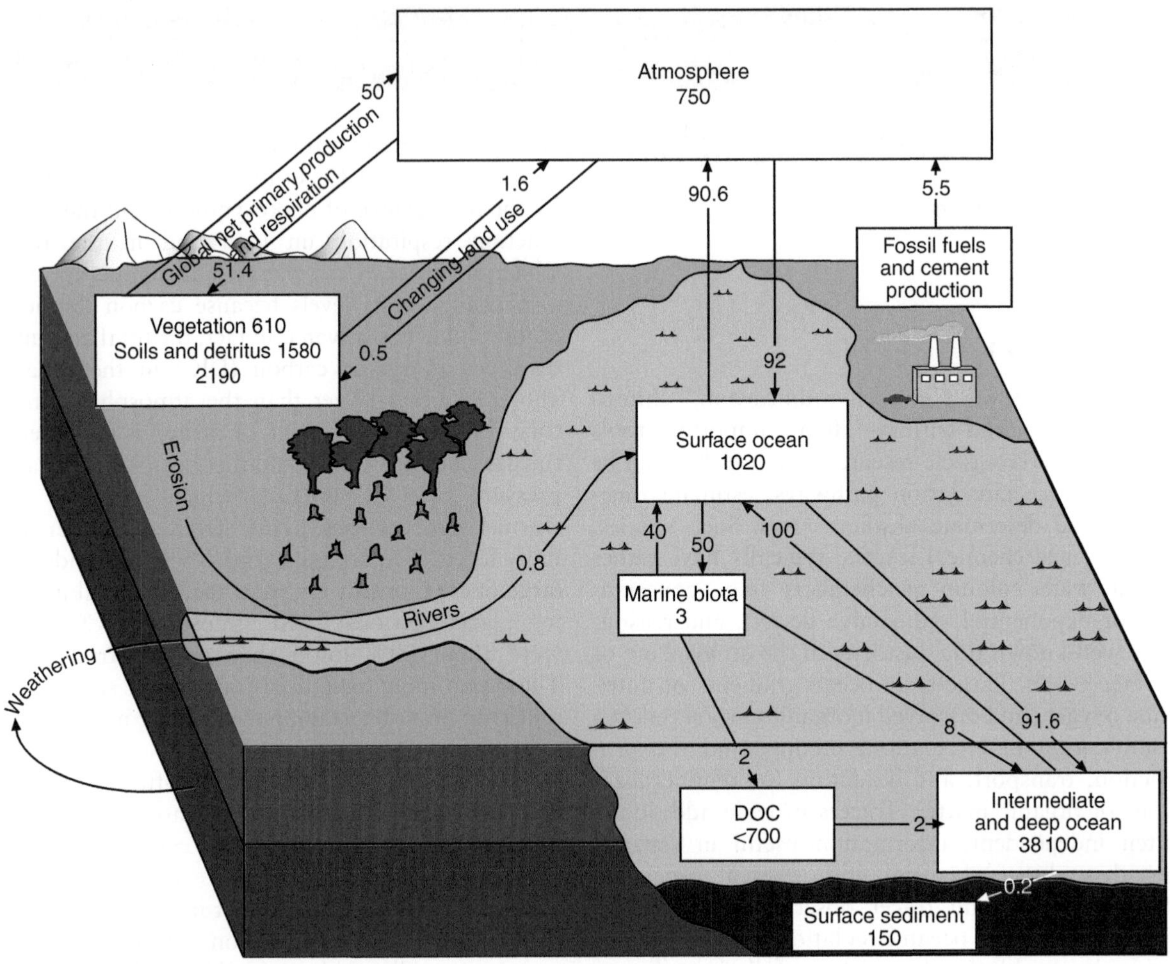

Figure 1 Schematic of global carbon cycle for the 1980s including natural background and human perturbations. Carbon inventories are in Pg C (1 Pg $= 10^{15}$ g) and fluxes are in Pg C yr^{-1}. DOC, dissolved organic carbon. Adapted with permission from Schimel D, Enting IG, Heimann M, *et al.* (1995) CO_2 and the carbon cycle. In: Houghton JT, Meira Filho LG, Bruce J, *et al.* (eds.) *Climate Change 1994, Intergovernmental Panel on Climate Change*, pp. 39–71. Cambridge, UK: Cambridge University Press.

water, a process termed respiration or remineralization. The export flux from the surface ocean is a key factor driving the marine biogeochemical cycles of carbon, oxygen, nitrogen, phosphorus, silicon, and trace metals such as iron.

The surface export and subsurface remineralization of organic matter are difficult to measure directly. Biogeochemical rates, therefore, are often inferred based on the large-scale distributions of DIC, alkalinity, inorganic nutrients (nitrate, phosphate, and silicate), and dissolved oxygen. The elemental stoichiometry of marine organic matter, referred to as the Redfield ratio, is with some interesting exceptions relatively constant in the ocean, simplifying the problem of interrelating the various biogeochemical fields. Geochemical distributions have the advantage that they integrate over much of the localized time/space variability in the ocean and can be used to extrapolate to region and basin scales. Property fields, though, reflect a combination of the net biogeochemical uptake and release as well as physical circulation and turbulent mixing. Additional information is required to separate these signals and can come from a mix of dynamical constraints, numerical models, and ocean process tracers.

The latter two approaches are related because natural and artificial tracers are used to calibrate or evaluate ocean models. A key aspect of these tracers is that they provide independent information on timescale, either because they decay or are produced at some known rate, for example, due to radioactivity, or because they are released into the ocean with a known time history. The different chemical tracers can be roughly divided into two classes. Circulation tracers such as radiocarbon, tritium–^{3}He, and the chlorofluorocarbons are not strongly impacted by biogeochemical cycling and are used primarily to quantify physical advection and mixing rates. These tracers are the major focus here. The distribution of other tracer species is more closely governed by biology and chemistry, for example, the thorium isotope series, which is used to study export production, particle scavenging, vertical transport, and remineralization rates.

Ocean Tracers and Dynamics: A One-dimensional (1-D) Example

Natural radiocarbon (^{14}C), a radioactive isotope of carbon, is a prototypical example of a (mostly) passive ocean circulation tracer. Radiocarbon is produced by cosmic rays in the upper atmosphere and enters the surface ocean as radiolabeled carbon dioxide ($^{14}CO_2$) via air–sea gas exchange. The ^{14}C DIC concentrations in the ocean decrease away from the surface, reflecting the passage of time since the water was last exposed to the atmosphere. Some radiolabeled carbon is transported to the deep ocean in sinking particulate organic matter, which can be largely corrected for in the analysis. The ^{14}C deficits relative to the surface water can be converted into age estimates for ocean deep waters using the radioactive decay half-life (5730 years). Natural radiocarbon is most effective for describing the slow thermohaline overturning circulation of the deep ocean, which has timescales of roughly a few hundred to a thousand years.

The main thermocline of the ocean, from the surface down to about 1 km or so, has more rapid ventilation timescales, from a few years to a few decades. Tracers useful in this regard are chlorofluorocarbons, tritium and its decay product ^{3}He, and bomb radiocarbon, which along with tritium was released into the atmosphere in large quantities in the 1950s and 1960s by atmospheric nuclear weapons testing.

When properly formulated, the combination of ocean process tracers and numerical models provides powerful tools for studying ocean biogeochemistry. At their most basic level, models are simply a mathematical statement quantifying the rates of the essential physical and biogeochemical processes. For example, advection–diffusion models are structured around coupled sets of differential equations:

$$\frac{\partial C}{\partial t} = -\nabla \cdot (\vec{u}C) + \nabla \cdot (K\nabla C) + J \qquad [1]$$

describing the time rate of change of a generic species C. The first and second terms on the right-hand side of eqn [1] stand for the local divergence due to physical advection and turbulent mixing, respectively. All of the details of the biogeochemistry are hidden in the net source/sink term J, which for radiocarbon would include net input from particle remineralization (R) and radioactive loss ($-\lambda^{14}C$).

One of the first applications of ocean radiocarbon data was as a constraint on the vertical diffusivity, upwelling, and oxygen consumption rates in the deep waters below the main thermocline. As illustrated in **Figure 2**, the oxygen and radiocarbon concentrations in the North Pacific show a minimum at mid-depth and then increase toward the ocean seabed. This reflects particle remineralization in the water column and the inflow and gradual upwelling of more recently ventilated bottom waters from the Southern Ocean. Mathematically, the vertical profiles for radiocarbon, oxygen (O_2), and a conservative tracer salinity (S) can be posed as steady-state, 1-D balances:

$$0 = K_z\frac{d^2S}{dz^2} - w\frac{dS}{dz} \qquad [2]$$

$$0 = K_z\frac{d^2O_2}{dz^2} - w\frac{dO_2}{dz} + R_{O_2} \qquad [3]$$

$$0 = K_z\frac{d^2\,{}^{14}C}{dz^2} - w\frac{d^{14}C}{dz} + (^{14}C:O_2)R_{O_2} - \lambda^{14}C \qquad [4]$$

K_z and w are the vertical diffusivity and upwelling rates, and $^{14}C:O_2$ is a conversion factor.

Looking carefully at eqn [2], one sees that the solution depends on the ratio K_z/w but not K_z or w separately. Similarly the equation for oxygen gives us information on the relative rates of upwelling and remineralization. It is only by the inclusion of radiocarbon, with its independent clock due to radioactive decay, that we can solve for the absolute physical and biological rates. The solutions to eqns [2]–[4] can be derived analytically, and as shown in **Figure 2** parameter values of $w = 2.3 \times 10^{-5}\,cm\,s^{-1}$, $K_z = 1.3\,cm^2\,s^{-2}$, and $R_{O_2} = 0.13 \times 10^{-6}\,mol\,kg^{-1}\,yr^{-1}$ fit the data reasonably well. The 1-D model-derived vertical diffusivity is about an order of magnitude larger than estimates from deliberate tracer release experiments and microscale turbulence measurements in the upper thermocline. However, they may be consistent with recent observations of enhanced deep-water vertical mixing over regions of rough bottom topography.

Ocean Circulation and Biogeochemical Models

The 1-D example shows the basic principles behind the application of tracer data to the ocean carbon

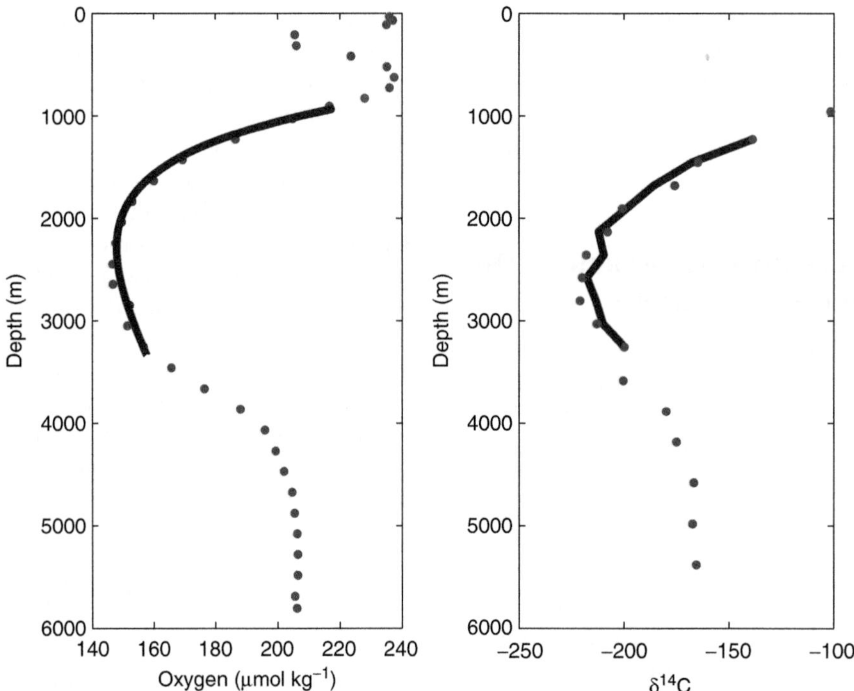

Figure 2 Observed vertical profiles of oxygen (O_2) and radiocarbon ($\delta^{14}C$) in the North Pacific. The solid curves are the model solution of a 1-D advection–diffusion equation.

cycle, but the complexity (if not always the sophistication) of the models and analysis has grown with time. Ocean carbon models can be roughly divided into idealized models (multi-box, 1-D and 2-D advection–diffusion models) and 3-D general circulation models (GCMs). Although the distinction can be blurry at times, idealized models are characterized typically by reduced dimensionality and/or kinematic rather than dynamic physics. That is, the circulation and mixing are specified rather than computed by the model and are often adjusted to best match the transient tracer data.

Global Box Models

An example of a simple, high-latitude outcrop box model is shown in **Figures 3(a)** and **3(b)**. The boxes represent the atmosphere, the high- and low-latitude surface ocean, and the deep interior. In the model, the ocean thermohaline circulation is represented by one-way flow with high-latitude sinking, low-latitude upwelling, and poleward surface return flow. Horizontal and vertical mixing is included by two-way exchange of water between each pair of boxes. The physical parameters are constrained so that model natural radiocarbon values roughly match observations. Note that the ^{14}C concentration in the deep water is significantly depleted relative to the surface boxes and amounts to a mean deep-water ventilation age of about 1150 years.

The model circulation also transports phosphate, inorganic carbon, and alkalinity. Biological production, particle export, and remineralization are simulated by the uptake of these species in the surface boxes and release in the deep box. The model allows for air–sea fluxes of CO_2 between the surface boxes and the atmosphere. The low-latitude nutrient concentrations are set near zero as observed. The surface nutrients in the high-latitude box are allowed to vary and are never completely depleted in the simulation of modern conditions. Similar regions of 'high-nutrient, low-chlorophyll' concentrations are observed in the subpolar North Pacific and Southern Oceans and are thought to be maintained by a combination of light and iron limitation as well as zooplankton grazing. The nutrient and DIC concentrations in the deep box are higher than either of the surface boxes, reflecting the remineralization of sinking organic particles.

The three-box ocean outcrop model predicts that atmospheric CO_2 is controlled primarily by the degree of nutrient utilization in high-latitude surface regions. Marine production and remineralization occur with approximately fixed carbon-to-nutrient ratios, the elevated nutrients in the deep box are associated with an equivalent increase in DIC and pCO_2. Large adjustments in the partitioning of

(a)

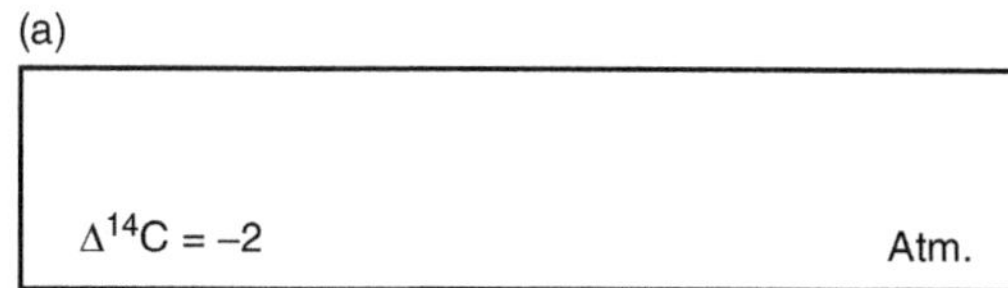

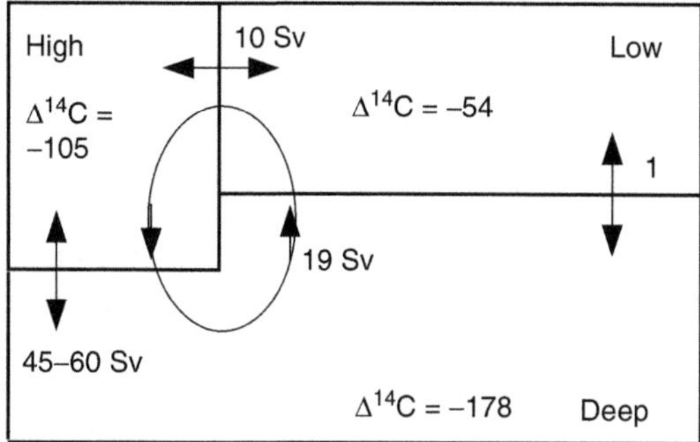

(b)

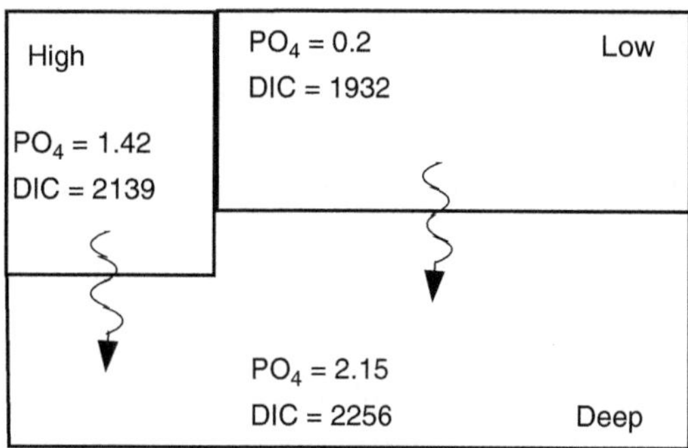

Figure 3 Results from a simple three-box ocean carbon cycle model. (a) The physical circulation and modeled radiocarbon ($\Delta^{14}C$) values. (b) The model biogeochemical fields, ocean DIC, and phosphate (PO_4) and atmospheric pCO_2. From Toggweiler JR and Sarmiento JL, Glacial to inter-glacial changes in atmospheric carbon dioxide: The critical role of ocean surface waters in high latitudes, *The Carbon Cycle and Atmospheric CO_2: Natural Variations Archean to Present*, Sundquist ET and Broecker WS (eds.), pp. 163–184, 1985, Copyright [1985]. American Geophysical Union. Adapted by permission of American Geophysical Union.

carbon between the ocean and atmosphere can occur only where this close coupling of the carbon and nutrient cycles breaks down. When subsurface water is brought to the surface at low latitude, production draws the nutrients down to near zero and removes to first order all of the excess seawater DIC and pCO_2. Modifications in the upward nutrient flux to the low latitudes have relatively little impact on the model atmospheric CO_2 as long as the surface nutrient concentrations stay near zero.

At high latitudes, however, the nutrients and excess DIC are only partially utilized, resulting in higher surface water pCO_2 and, over decades to centuries, higher atmospheric CO_2 concentrations. Depending on the polar biological efficiency, the model atmosphere effectively sees more or less of the high DIC concentrations (and pCO_2 levels) of the deep ocean. Thus changes in ocean biology and physics can have a correspondingly large impact on atmospheric CO_2. On longer timescales (approaching a few millennia), these variations are damped to some extent by adjustments of the marine calcium carbonate cycle and ocean alkalinity.

The three-box outcrop model is a rather crude representation of the ocean, and a series of geographical refinements have been pursued. Additional boxes can be added to differentiate the individual ocean basins (e.g., Atlantic, Pacific, and Indian), regions (e.g., Tropics and subtropics), and depths (e.g., thermocline, intermediate, deep, and bottom waters), leading to a class of models with a half-dozen to a few dozen boxes. The larger number of unknown advective flows and turbulent exchange parameters, however, complicates the tuning procedure. Other model designs take advantage of the vertical structure in the tracer and biogeochemical profile data. The deep box (es) is discretized in the vertical, essentially creating a continuous interior akin to a 1-D advection–diffusion model. This type of model was often used in the 1970s and 1980s for the initial anthropogenic CO_2 uptake calculations, where it is important to differentiate between the decadal ventilation timescales of the thermocline and the centennial timescales of the deep water.

Intermediate Complexity and Inverse Models

In terms of global models, the next step up in sophistication from box models is intermediate complexity models. These models typically have higher resolution and/or include more physical dynamics but fall well short of being full GCMs. Perhaps the most common examples for ocean carbon cycle research are zonal average basin models. The dynamical equations are similar to a GCM but are integrated in 2-D rather than in 3-D, the third east–west dimension removed by averaging zonally across the basin. In some versions, multiple basins are connected by an east–west Southern Ocean channel. The zonal average models often have a fair representation of the shallow wind-driven Ekman and deep thermohaline overturning circulations but obviously lack western boundary currents and gyre circulations. Tracer data remain an important element in tuning some of the mixing coefficients and surface boundary conditions and in evaluating the model solutions.

Based on resolution, many inverse models can also be categorized as intermediate complexity, but their mode of operation differs considerably from the models considered so far. In an inverse model, the circulation field and biogeochemical net source/sink (the J terms in the notation above) are solved for using the observed large-scale hydrographic and tracer distributions as constraints. Additional dynamic information may also be incorporated such as the geostrophic velocity field, general water mass properties, or float and mooring velocities.

Inverse calculations are typically posed as a large set of simultaneous linear equations, which are then solved using standard linear algebra methods. The inverse techniques are most commonly applied to steady-state tracers, though some exploration of transient tracers has been carried out. The beauty of the inverse approach is that it tries to produce dynamically consistent physical and/or biogeochemical solutions that match the data within some assigned error. The solutions are often underdetermined in practice, however, which indicates the existence of a range of possible solutions. From a biogeochemical perspective, the inverse circulation models provide estimates of the net source/sink patterns, which can then be related to potential mechanisms.

Thermocline Models

Ocean process tracers and idealized models have also been used extensively to study the ventilation of the main thermocline. The main thermocline includes the upper 1 km of the tropical to subpolar ocean where the temperature and potential density vertical gradients are particularly steep. Thermocline ventilation refers to the downward transport of surface water recently exposed to the atmosphere, replenishing the oxygen and other properties of the subsurface interior. Based on the vertical profiles of tritium and ^{3}He as well as simple 1-D and box models, researchers in the early 1980s showed that ventilation of the main thermocline in the subtropical gyres occurs predominately as a horizontal process along surfaces of constant density rather than by local vertical mixing. Later work on basin-scale bomb-tritium distributions confirmed this result and suggested the total magnitude of subtropical ventilation is large, comparable to the total wind-driven gyre circulation.

Two dimensional gyre-scale tracer models have been fruitfully applied to observed isopycnal tritium–^{3}He and chlorofluorocarbon patterns. As shown in **Figure 4**, recently ventilated water (near-zero tracer age) enters the thermocline on the poleward side of the gyre and is swept around the clockwise circulation of the gyre (for a Northern Hemisphere case). Comparisons of model tracer patterns and property–property relationships constrain the absolute ventilation rate and the relative effects of isopycnal advection versus turbulent mixing by depth and region.

Thermocline tracer observations are also used to estimate water parcel age, from which biogeochemical rate can be derived. For example, remineralization produces an apparent oxygen deficit relative to atmospheric solubility. Combining the oxygen deficit with an age estimate, one can compute the rate of oxygen utilization. Similar geochemical approaches have been or can be applied to a host of problems: nitrogen fixation, denitrification, dissolved organic matter remineralization, and nutrient resupply to the upper ocean. The biogeochemical application of 2-D gyre models has not been pursued in as much detail.

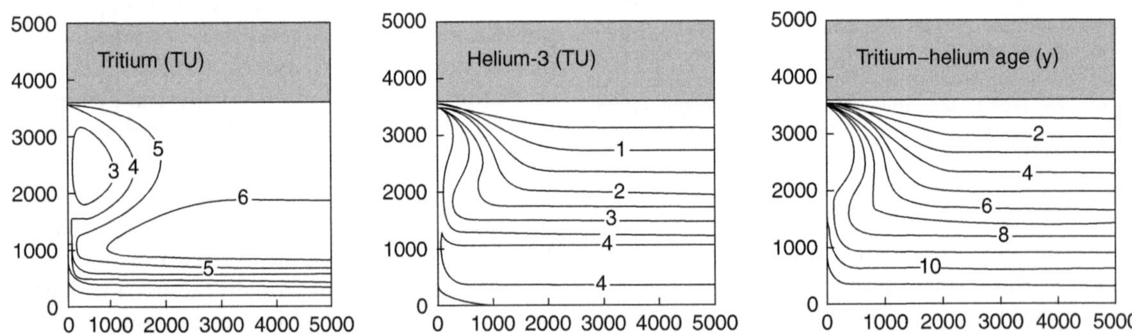

Figure 4 Tracer results from a two-dimensional gyre model. The model represents the circulation on a constant density surface (isopycnal) in the main thermocline that outcrops along the northern boundary (shaded region). Thermocline ventilation is indicated by the gradual increase of tritium–^{3}He ages around the clockwise flowing gyre circulation. TU, tritium unit (1 TU = 1 ^{3}H atom/10^{18} H atoms). Reproduced with permission from Musgrave DL (1990) Numerical studies of tritium and helium-3 in the thermocline. *Journal of Physical Oceanography* 20: 344–373.

This form of modeling, however, is particularly useful for describing regional patterns and in the areas where simple tracer age approaches breakdown.

Three-Dimensional (3-D) Biogeochemical Simulations

By their very nature, idealized models neglect many important aspects of ocean dynamics. The alternative is full 3-D ocean GCMs, which incorporate more realistic spatial and temporal geometry and a much fuller suite of physics. There are several different families of ocean physical models characterized by the underlying governing equations and the vertical discretization schemes. Within model families, individual simulations will differ in important factors of surface forcing and choice of physical parametrizations. Often these parametrizations account for complex processes, such as turbulent mixing, that occur on small space scales that are not directly resolved or computed by the model but which can have important impacts on the larger-scale ocean circulation; often an exact description of specific events is not required and a statistical representation of these subgrid scale processes is sufficient. For example, turbulent mixing is commonly treated using equations analogous to those for Fickian molecular diffusion. Ocean GCMs, particularly coarse-resolution global versions, are sensitive to the subgrid scale parametrizations used to account for unresolved processes such as mesoscale eddy mixing, surface and bottom boundary layer dynamics, and air–sea and ocean–ice interactions. Different models will be better or worse for different biogeochemical problems.

Considerable progress has been achieved over the last two decades on the incorporation of chemical tracers, biogeochemistry, and ecosystem dynamics into both regional and global 3-D models. Modeling groups now routinely simulate the more commonly measured tracers (e.g., radiocarbon, chlorofluorocarbons, and tritium–^{3}He). Most of these tracers have surface sources and, therefore, provide clear indications of the pathways and timescales of subsurface ventilation in the model simulations. The bottom panel of **Figure 5** shows from a simulation of natural radiocarbon the layered structure of the intermediate and deep water flows in the Atlantic basin, with Antarctic Intermediate and Bottom Waters penetrating from the south and southward-flowing North Atlantic Deep Water sandwiched in between.

In a similar fashion, biogeochemical modules have been incorporated to simulate the 3-D fields of nutrients, oxygen, DIC, etc. Just as with the physics, the degree of biological complexity can vary considerably across ocean carbon models. At one end are diagnostic models where the production of organic matter in the surface ocean is prescribed based on satellite ocean color data or computed implicitly by forcing the simulated surface nutrient field to match observations (assuming that any net reduction in nutrients is driven by organic matter production). At the other extreme are simulations that couple prognostic

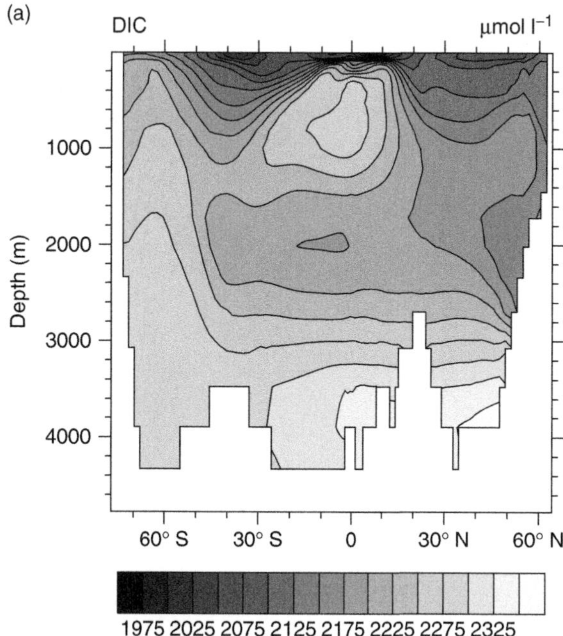

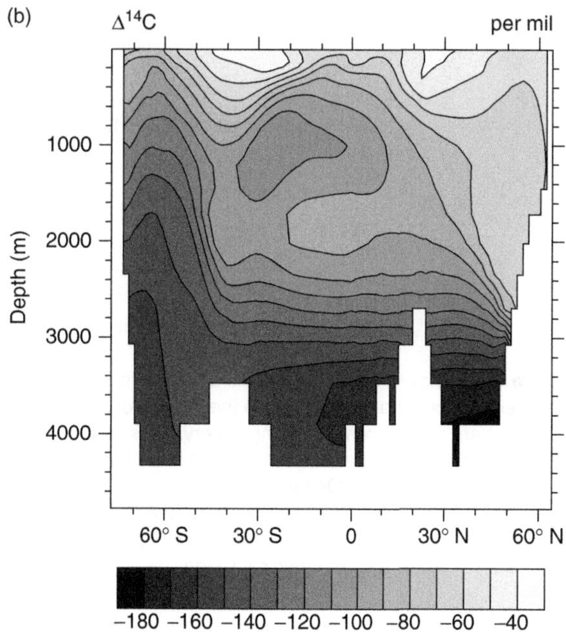

Figure 5 Simulated radiocarbon and carbon cycle results from a three-dimensional global ocean biogeochemical model. Depth–latitude sections are shown for (a) DIC concentration (μmol l^{-1}) and (b) natural (preindustrial) radiocarbon (Δ^{14}C, per mil) along the prime meridian in the Atlantic Ocean.

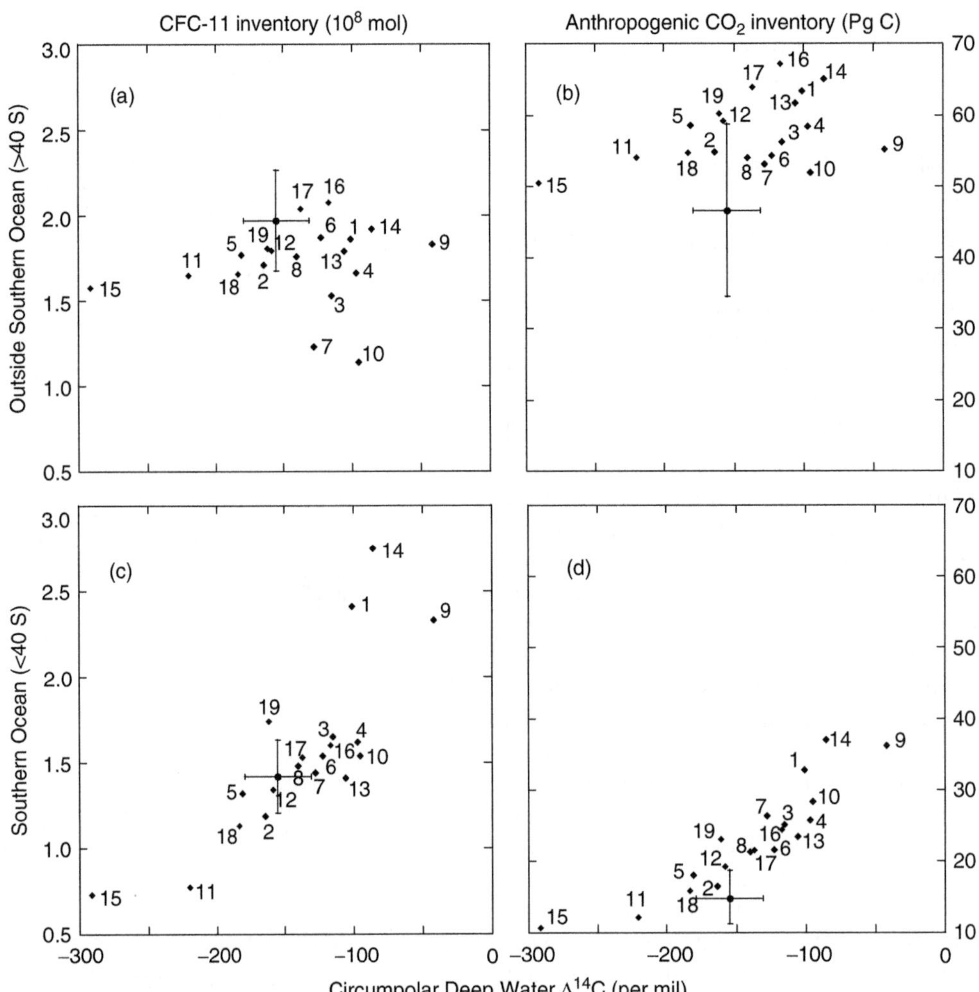

Figure 6 Transient tracer constraints on numerical model ocean carbon dynamics. Each panel shows results for the suite of global ocean simulations that participated in the international Ocean Carbon-Cycle Model Intercomparison Project (OCMIP). The points with the error bars are the corresponding observations. The *x*-axis in all panels is the average Circumpolar Deep Water radiocarbon level ($\Delta^{14}C$, per mil). The *y*-axis in the left column is the chlorofluorocarbon (CFC-11) inventory (10^8 mol) outside of the Southern Ocean (top row) and in the Southern Ocean. The *y*-axis in the right column is the corresponding anthropogenic carbon inventory (Pg C). From Matsumoto K, Sarmiento JL, Key RM, *et al.*, Evaluation of ocean carbon cycle models with data-based metrics, *Geophysical Research Letters*, Vol. 31, L07303 (doi:10.1029/2003GL018970), 2004. Copyright [2004] American Geophysical Union. Reproduced by permission of American Geophysical Union.

treatments of multiple phytoplankton, zooplankton, and bacteria species to the biogeochemical tracers. Transient tracers enter the picture because the biogeochemical distributions are also strongly influenced circulation. This is nicely illustrated by comparing the simulated radiocarbon and DIC sections in **Figure 5**. To first order, older waters marked by low radiocarbon exhibit high DIC levels due to the accumulation of carbon released from particle remineralization. Tracers can be used to guide the improvement of the physical circulation and highlight areas where the circulation is poor and thus the biogeochemical tracers may be suspect. They can also be used more quantitatively to create model skill metrics; this approach can be applied across multiple model simulations to provide the best overall constraint on quantities like the ocean uptake of anthropogenic CO_2 or export production (**Figure 6**).

Models as Research Tools

Designed around a particular question or hypothesis, conceptual models attempt to capture the basic elements of the problem while remaining amenable to straightforward analysis and interpretation. They are easy to construct and computationally inexpensive, requiring only a desktop PC rather than a supercomputer. When well-formulated, idealized models provide a practical method to analyze ocean physical and biogeochemical dynamics and in some cases to quantitatively constrain specific rates. Their application is closely tied to ocean tracer observations, which are generally required for physical calibration and evaluation. Idealized models remain a valuable tool for estimating the oceanic uptake of anthropogenic carbon and the long time-scale responses (centuries to millennia) of the natural carbon cycle. Also, some of the more memorable and lasting advances in tracer oceanography are directly linked to simple conceptual models. Examples include constraints on the deep-water large-scale vertical diffusivity and demonstration of the dominance of lateral over vertical ventilation of the subtropical main thermocline.

Three-dimensional models offer more realism, at least apparently, but with the cost of greater complexity, a more limited number of simulations, and a higher probability of crucial regional errors in the base solutions, which may compromise direct, quantitative model-data comparisons. Ocean GCM solutions, however, should be exploited to address exactly those problems that are intractable for simpler conceptual and reduced dimensional models. For example, two key assumptions of the 1-D advection–diffusion model presented in **Figure 2** are that the upwelling occurs uniformly in the horizontal and vertical and that mid-depth horizontal advection is not significant. Ocean GCMs and tracer data, by contrast, show a rich three-dimensional circulation pattern in the deep Pacific.

The behavior of idealized models and GCMs can diverge, and it is not always clear that complexity necessarily leads to more accurate results. In the end, the choice of which model to use depends on the scientific problem and the judgment of the researcher. Probably the best advice is to explore solutions from a hierarchy of models and to thoroughly evaluate the skill of the models against a range of tracers and other dynamical measures. Just because a model does a good job reproducing the distribution of one tracer field does not mean that it can be applied indiscriminately to another variable, especially if the underlying dynamics or timescales differ.

Models can be quite alluring in the sense that they provide concrete answers to questions that are often difficult or nearly impossible to address from sparsely sampled field data. However, one should not forget that numerical models are simply a set of tools for doing science. They are no better than the foundations upon which they are built and should not be carried out in isolation from observations of the real ocean. For ocean carbon cycle models, the two key elements are the ocean physical circulation and the biogeochemical processes. Even the best biogeochemical model will perform poorly in an ill-constructed physical model. Conversely, if the underlying biogeochemical mechanisms are poorly known, a model may be able to correctly reproduce the distributions of biogeochemical tracers but for the wrong reasons. Mechanistic-based models are critical in order to understand and predict natural variability and the response of ocean biogeochemistry to perturbations such as climate change.

Glossary

Anthropogenic carbon The additional carbon that has been released to the environment over the last several centuries by human activities including fossil-fuel combustion, agriculture, forestry, and biomass burning.

Excess ^{3}He Computed as the ^{3}He in excess of gas solubility equilibrium with the atmosphere.

Export production That part of the organic matter formed in the surface layer by photosynthesis that is transported out of the surface layer and into the interior of the ocean by particle sinking, mixing, and circulation, or active transport by organisms.

Radiocarbon ^{14}C Either δ^{14}C or Δ^{14}C where δ^{14}C $= [(^{14}C/^{12}C)_{sample}/(^{14}C/^{12}C)_{standard} - 1] \times 1000$ (in parts per thousand or 'per mil'); Δ^{14}C is similar but corrects the sample ^{14}C for biological fractionation using the sample $^{13}C/^{12}C$ ratio.

Transient tracers Chemical tracers that contain time information either because they are radioactive or because their source, usually anthropogenic, has evolved with time.

Tritium–^{3}He age An age is computed assuming that all of the excess ^{3}He in a sample is due to the radioactive decay of tritium, age $= \ln[(^3H + {}^3He)/{}^3H]/\lambda$, where λ is the decay constant for tritium.

Ventilation The physical process by which surface properties are transported into the ocean interior.

See also

Carbon Cycle. Carbon Dioxide (CO_2) Cycle. CFCs in the Ocean. Nitrogen Cycle. Radiocarbon. Tritium–Helium Dating.

Further Reading

Broecker WS and Peng T-H (1982) *Tracers in the Sea*. Palisades, NY: Lamont-Doherty Geological Observatory, Columbia University.

Charnock H, Lovelock JE, Liss P, and Whitfield M (eds.) (1988) *Tracers in the Ocean*. London: The Royal Society.

Doney SC, Lindsay K, Caldeira K, *et al.* (2004) Evaluating global ocean carbon models: The importance of realistic physics. *Global Biogeochemical Cycles* 18: GB3017 (doi:10.1029/2003GB002150).

England MH and Maier-Reimer E (2001) Using chemical tracers in ocean models. *Reviews in Geophysics* 39: 29–70.

Fasham M (ed.) (2003) *Ocean Biogeochemistry*. New York: Springer.

Jenkins WJ (1980) Tritium and ^{3}He in the Sargasso Sea. *Journal of Marine Research* 38: 533–569.

Kasibhatla P, Heimann M, and Rayner P (eds.) (2000) *Inverse Methods in Global Biogeochemical Cycles*. Washington, DC: American Geophysical Union.

Matsumoto K, Sarmiento JL, Key RM, *et al.* (2004) Evaluation of ocean carbon cycle models with data-based metrics. *Geophysical Research Letters* 31: L07303 (doi:10.1029/2003GL018970).

Munk WH (1966) Abyssal recipes. *Deep-Sea Research* 13: 707–730.

Musgrave DL (1990) Numerical studies of tritium and helium-3 in the thermocline. *Journal of Physical Oceanography* 20: 344–373.

Sarmiento JL and Gruber N (2006) *Ocean Biogeochemical Dynamics*. Princeton, NJ: Princeton University Press.

Schimel D, Enting IG, Heimann M, *et al.* (1995) CO_2 and the carbon cycle. In: Houghton JT, Meira Filho LG, and Bruce J, *et al.* (eds.) *Climate Change 1994, Intergovernmental Panel on Climate Change*, pp. 39–71. Cambridge, UK: Cambridge University Press.

Siedler G, Church J, and Gould J (eds.) (2001) *Ocean Circulation and Climate*. New York: Academic Press.

Siegenthaler U and Oeschger H (1978) Predicting future atmospheric carbon dioxide levels. *Science* 199: 388–395.

Toggweiler JR and Sarmiento JL (1985) Glacial to interglacial changes in atmospheric carbon dioxide: The critical role of ocean surface waters in high latitudes. In: Sundquist ET and Broecker WS (eds.) *The Carbon Cycle and Atmospheric CO_2: Natural Variations Archean to Present*, pp. 163–184. Washington, DC: American Geophysical Union.

Relevant Websites

http://cdiac.ornl.gov
– Global Ocean Data Analysis Project, Carbon Dioxide Information Analysis Center.

http://w3eos.whoi.edu
– Modeling, Data Analysis, and Numerical Techniques for Geochemistry (resource page for MIT/WHOI course number 12.747), at Woods Hole Oceanographic Institution.

http://us-osb.org
– Ocean Carbon and Biogeochemistry (OCB) Program

http://www.ipsl.jussieu.fr
– Ocean Carbon-Cycle Model Intercomparison Project (OCMIP), Institut Pierre Simon Laplace.

CENOZOIC OCEANS – CARBON CYCLE MODELS

L. François and Y. Goddéris, University of Liège, Liège, Belgium

Introduction

The story of the Cenozoic is essentially a story of global cooling. The last 65 million years of the Earth's history mark the transition from the Cretaceous 'greenhouse' climate toward the present-day 'icehouse' conditions. Particularly, the cooling by about 8–10°C of deep ocean waters since the Cretaceous was linked to a reorganization of the oceanic circulation triggered by tectonic plate movements. These oceanic circulation changes were coeval with continental climatic change, as demonstrated by abundant evidence for global cooling (pollen, faunal assemblages, development of glaciers, etc.). For instance, most of western Europe and the western United States had a subtropical climate during the Eocene, despite the fact that they were located at the same latitude as today. Another striking feature of the changes that have occurred during Cenozoic times is the decrease of the partial pressure of atmospheric CO_2 (P_{CO_2}). Since CO_2 is a greenhouse gas, there might be a causal relationship between the decrease in P_{CO_2} and the general cooling trend of Cenozoic climate. The global cooling might be the result of the changes in oceanic circulation and atmospheric CO_2, both probably influencing each other and possibly initiated by tectonic processes.

Indicators of Atmospheric CO_2 Change

It should be kept in mind that there are no direct proxies of ancient levels of CO_2 in the atmosphere. Methods rely on three indirect indicators.

1. The $\delta^{13}C$ measured in ancient soil carbonates can be directly linked to the atmospheric P_{CO_2}. This method reveals declining atmospheric P_{CO_2} over the last 65 million years, from about 650 ppm by volume (ppmv) in the Paleocene (**Figure 1**).
2. The biological isotopic fractionation occurring during assimilation of carbon by the marine biosphere (ε_p) depends on the partial pressure of CO_2 dissolved in sea water, itself directly related to the atmospheric P_{CO_2}. The estimation of ε_p for ancient organic sediments indicates high P_{CO_2} values in the Eocene (620 ppmv), followed by a constant decline toward the present-day pressure through the Cenozoic (**Figure 1**).
3. The measured boron isotopic composition of marine carbonates gives insight into the pH of ancient sea water. Assuming a plausible history for the ocean alkalinity, lower or higher pH values can be respectively related to higher or lower P_{CO_2}. This method has been applied to the last 60 million years, showing values as high as 3500 ppmv CO_2 during the Paleocene. The decline in P_{CO_2} was then roughly linear through time until the late Eocene. During the last 25 million years of the Earth's history, P_{CO_2} was relatively constant, possibly displaying lower values than present-day ones during Miocene. No data are available for the Oligocene (**Figure 1**).

Despite some disagreements between the three reconstructions, they all indicate a major reduction of the atmospheric CO_2 partial pressure during the Cenozoic, which might potentially play an important role in the coeval global cooling. Any exploration of the cause of the decline in P_{CO_2} requires the identification of the sources and sinks of oceanic and atmospheric carbon, and some knowledge of their relative changes during the Cenozoic.

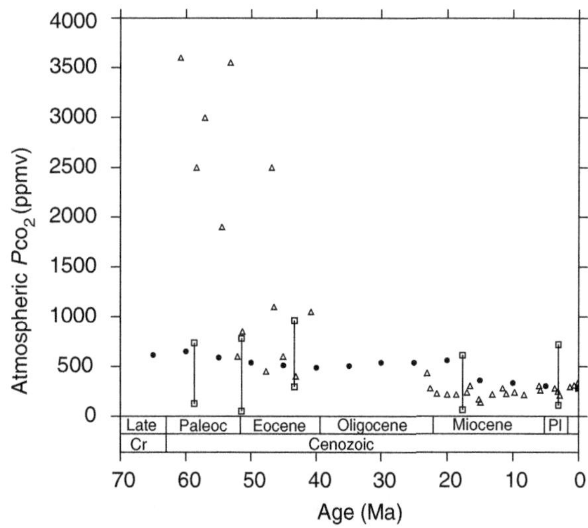

Figure 1 The reconstructed partial pressure of atmospheric CO_2. Δ, from boron isotopes (Pearson and Palmer (2000)) ; •, from ε_p (Kump and Arthur, 1997); □, from paleosoils analysis (compilation by Berner (1998)). Timescale according to Harland *et al.* (1990) (CR = Cretaceous; PALEOC = Paleocene; PL = Pleistocene.)

Carbon Cycle Changes and Processes Involved

Long-term Regulation of Atmospheric CO_2

On the geological timescale, and neglecting at this point the possible impact of sedimentary organic carbon cycling, the sources of carbon for the ocean–atmosphere system are the degassing of the mantle and metamorphic processes. Carbon is injected as CO_2 into the ocean–atmosphere system today through the degassing of fresh basalts along mid-oceanic ridges (MOR) ($1.5–2.2 \times 10^{12}$ mol y^{-1}), and through plume events and arc volcanism ($1.5–5.5 \times 10^{12}$ mol y^{-1}). These various sources account for the total degassing flux F_{VOL}. Once released, this carbon is rapidly (within 10^3 years) redistributed between the atmosphere and the ocean, reaching a steady-state repartition after a negligible time compared to the geological timescale. Carbon can leave the system mainly through the deposition of carbonate minerals on the seafloor. The rate of carbon removal through carbonate deposition is controlled by the saturation state of the ocean, and thus by the rate of supply of alkalinity by the chemical weathering of continental minerals. Carbonate and silicate minerals exposed at the continental surface weather under the corrosive action of atmospheric CO_2 dissolved in rain water as carbonic acid, and eventually concentrated by the microbial respiration in soils. Regarding the chemical weathering of carbonate minerals, the net budget of the dissolution reaction can be written as follows eqn [I].

$$\begin{aligned} &CaCO_3(\text{rock}) + H_2CO_3(\text{atmosphere}) \\ &\rightarrow Ca^{2+}(\text{rivers}) + 2HCO_3^-(\text{rivers}) \end{aligned} \qquad [I]$$

In this reaction, there is a net transfer of Ca^{2+} (or Mg^{2+} in the case of magnesium carbonates) from the continental crust to the ocean, a creation of two equivalents of alkalinity and a transfer to the river system of two moles of carbon per mole of Ca^{2+}, one coming from the crust, the other one from the atmosphere. Once the weathering products reach the ocean, they will increase the saturation state of surface waters with respect to calcite and induce rapidly (within 10^3 years) the biologically driven precipitation of one mole of $CaCO_3$ followed by its deposition on the seafloor. The precipitation–deposition reaction is the reverse of reaction [1]. The net carbon budget of the weathering of carbonate minerals followed by deposition of sedimentary carbonate is thus equal to zero.

The chemical weathering of continental silicate rocks is fundamentally different, since silicate rocks do not contain carbon. The budget can be written as

$$\begin{aligned} &CaSiO_3(\text{rock}) + 2H_2CO_3(\text{atmosphere}) + H_2O \\ &\rightarrow Ca^{2+}(\text{rivers}) + 2HCO_3^-(\text{rivers}) + H_4SiO_4(\text{rivers}) \end{aligned} \qquad [II]$$

Here $CaSiO_3$ stands for a 'generic' silicate mineral. The weathering reaction of more realistic Ca- (or Mg-) silicate minerals, if more complex, displays the same budget in terms of alkalinity versus carbon fluxes. Again, once reaching the ocean, the excess Ca^{2+} will precipitate as $CaCO_3$, thus removing one mole of carbon from the ocean per mole of weathered silicates. The net budget of this reaction, after sedimentary carbonate precipitation, is the transfer of exospheric carbon to the crust. Chemical weathering of continental silicate minerals thus acts as the main sink of carbon on the geological timescale. Today, about 6×10^{12} mol y^{-1} of Ca^{2+} and Mg^{2+} are released from silicate weathering.

The size of the exospheric carbon pool (ocean + atmosphere) is about 3.2×10^{18} mol today. As mentioned above, the fluxes entering and leaving this reservoir are of the order of $10^{12}–10^{13}$ mol y^{-1}. A relatively small imbalance between the input and output of carbon of 10^{12} mol y^{-1}, the output being higher, but persisting for several million years, will result in a drastic reduction in the exospheric content. Three million years will be sufficient to remove all the carbon from the exospheric system, thus forcing the atmospheric P_{CO_2} to zero. There is no lithological, fossil, or geochemical record of such a dramatic event during the Cenozoic, or event during the complete Phanerozoic. To avoid the occurrence of such events for which there is no evidence, the perturbations of the carbon cycle had to be limited in time and amplitude, and thus the past exospheric carbon cycle was not strictly at, but always close to, steady state. The same considerations apply to the alkalinity budget. These steady-state conditions require that the amount of carbon removed from the atmosphere–ocean system by continental silicate weathering must always closely track the amount of carbon released by degassing. Mathematically, these conditions translate into eqn [1].

$$F_{SW} = F_{VOL} \qquad [1]$$

The question is now how to physically force F_{SW} to follow the degassing. The answer lies in the fact that the chemical weathering of continental silicates appears to be dependent on air temperature, the dissolution being enhanced during warmer climates. This dependence provides a negative feedback that not only allows equilibration of the carbon and alkalinity budgets on the geological timescale, but also stabilizes the Earth's climate. When the degassing increases suddenly, for instance as a result of a higher spreading rate of the oceanic floor, the amount of carbon in the ocean and atmosphere will first increase, increasing the atmospheric $P\text{CO}_2$. Because CO_2 is a greenhouse gas, the climate will become warmer, and this will enhance the weathering of silicate rocks. As a result, any increase in the input of carbon will be counterbalanced by an increase of the output through silicate weathering, thus stabilizing the system through a negative feedback loop. The $P\text{CO}_2$ will stabilize at a somewhat higher level than before the perturbation. Similarly, the decline in $P\text{CO}_2$ through the Cenozoic could be due to a decreasing degassing rate, which acts as the driving force of changes. This simple process is the basis of all existing long-term geochemical cycle models. It was first identified in 1981 by Walker, Hays, and Kasting. Breaking this feedback loop would result in fluctuations in calculated $P\text{CO}_2$ and thus presumably in climate, that are not reflected in the geological record.

Himalayan Uplift, $^{87}Sr/^{86}Sr$ Record, and Possible Implications for Weathering History

M.E. Raymo in 1991 put forward another explanation of the global Cenozoic $P\text{CO}_2$ decline. Instead of a decreased degassing rate, she suggested that continental silicate weathering rates increased drastically over the last 40 million years, although degassing conditions remained more or less constant. This assertion was originally based on the Cenozoic carbonate record of $^{87}Sr/^{86}Sr$. The isotope ^{86}Sr is stable, whereas ^{87}Sr is produced by the radioactive β-decay of ^{87}Rb. Strontium ions easily replace calcium ions in mineral lattices, since their ionic diameters are comparable. The present-day sea water $^{87}Sr/^{86}Sr$ equals 0.709. Two main processes impinge on this ratio: the chemical weathering of continents, delivering strontium with a mean $^{87}Sr/^{86}Sr$ of 0.712, and the exchanges between seafloor basalts and sea water, resulting in the release of mantle strontium into the ocean ($^{87}Sr/^{86}Sr = 0.703$). In other words, chemical weathering of continental rocks tends to increase the strontium isotopic ratio of sea water, while exchanges with seafloor basalts at low or high temperature tend to decrease it.

The sea water $^{87}Sr/^{86}Sr$ recorded over the last 65 million years displays a major increase, starting about 37–38 million years ago (**Figure 2**). An event approximately coeval with the sea water $^{87}Sr/^{86}Sr$ upward shift is the Himalayan uplift, which was

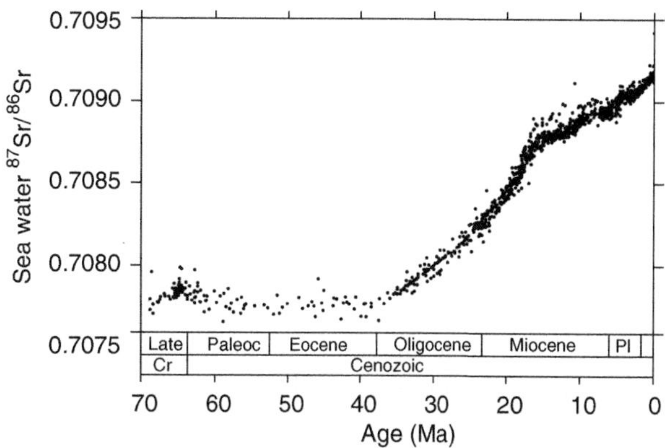

Figure 2 Sea water $^{87}Sr/^{86}Sr$ recorded in ancient carbonate sediments (Ottawa-Bochum database: http://www.science.uottawa.ca/geology/isotope_data/). Timescale abbreviations as **Figure 1**.

initiated by the India–Asia collision some 50 million years ago. Raymo has proposed that, in an uplifted area, the mechanical breakdown of rocks increases owing to the cooling and development of glaciers, to the development of steep slopes, and to temperatures oscillating below and above the freezing point at high altitudes. Furthermore, the development of the monsoon regime, about 10 million years ago, resulted in increased runoff, and thus an enhanced water availability for weathering, over at least the southern side of the Himalayan range. All these uplift-related changes might result in an enhanced chemical dissolution of minerals, since the surface in contact with the corrosive solutions increases when rocks fragment. The consequence of the Himalayan uplift might thus be an increase in the consumption of exospheric carbon by enhanced weathering on the continents, a process recorded in the sea water $^{87}Sr/^{86}Sr$ rise. The system depicted in this hypothesis is new, compared to the hypothesis described in the previous section. Here, tectonic processes result in uplift, followed by enhanced weathering, itself consuming atmospheric CO_2, thus cooling the climate. This cooling favors the development of glaciers not only in the uplifted area, but also globally, producing a global increase in mechanical and subsequent chemical weathering, a positive feedback that further cools the Earth. In Raymo's hypothesis, the negative feedback proposed by Walker *et al.*, stabilizing $P\text{CO}_2$ no longer exists. Chemical weathering is mainly controlled by tectonic processes with high rates in a cool world (Raymo's world), while it was controlled by climate and $P\text{CO}_2$ with high rates in a warm world in the Walker hypothesis (Walker's world). However, as mentioned above, negative feedbacks are needed to stabilize $P\text{CO}_2$, especially since the degassing remained more or less constant over the period of interest. Raymo's world has the ability to exhaust atmospheric CO_2 within a few million years.

In an attempt to reconcile the two approaches, François and Walker proposed in 1992 the addition of a new CO_2 consumption flux to the carbon cycle, identified as the precipitation of abiotic carbonates within the oceanic crust, subsequent to its alteration at low temperature. This flux is directly dependent on deep water temperature, which has decreased by ~8°C over the Cenozoic. An increase in the continental weathering rate might be compatible with a constant degassing rate, since the sink of carbon through low-temperature alteration of the oceanic crust is decreasing. The balance between input and output is thus still in place. However, this additional sink of carbon is poorly constrained. The present-day consumption of carbon is estimated to be about 1.4×10^{12} mol y^{-1}, but the kinetics of the process is essentially unknown. This attractive hypothesis still needs experimental verification.

Finally, it should be noted that Raymo's hypothesis interprets the increase in the sea water $^{87}Sr/^{86}Sr$ in terms of an increase in the weathering fluxes. However, silicate minerals exposed in the Himalayan area, particularly in the High Himalayan Crystalline Series, display unusually high isotopic ratios (reaching 0.740). Sediments of Proterozoic age with a $^{87}Sr/^{86}Sr$ reaching 0.8 are also exposed in the Lesser Himalaya area. For this reason, rivers draining the Himalayan area (Ganges, Brahmapoutra, etc.) display an isotopic ratio (0.725 for the Ganges) higher than the mean global value (0.712). At least part of the Cenozoic increase in the sea water $^{87}Sr/^{86}Sr$ might thus be due to changes in the isotopic composition of source rocks.

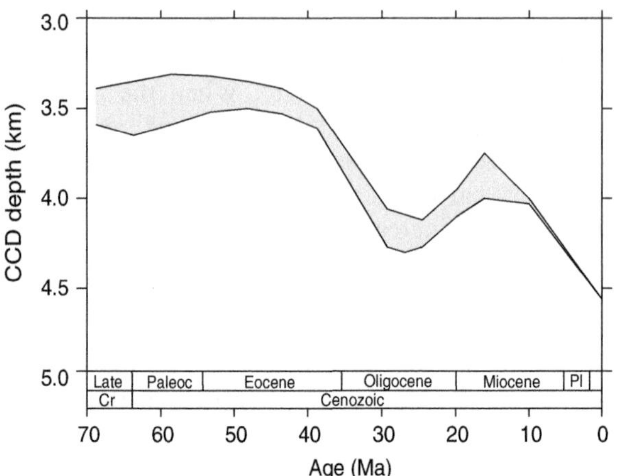

Figure 3 Reconstructed Carbonate Compensation Depth (CCD) through the Cenozoic (Van Andel, 1975; Broecker and Peng, 1982). Timescale as **Figure 1**.

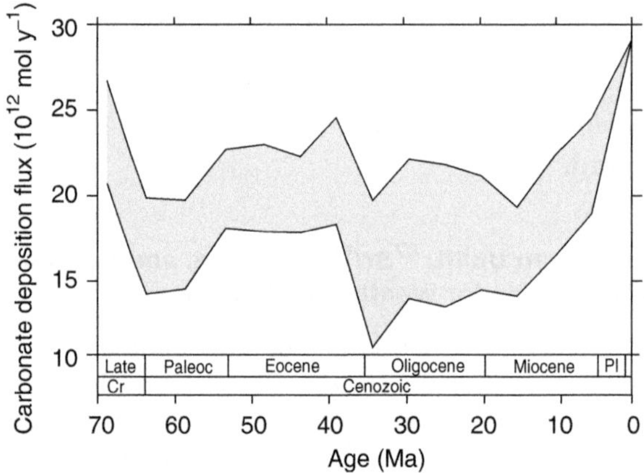

Figure 4 Total carbonate accumulation flux reconstructed from paleodata (Opdyke and Wilkinson, 1988). Timescale as **Figure 1**.

Lysocline and Carbonate Accumulation Changes

Other indicators of a possible increase in the continental weathering rate over the course of Cenozoic exist. For instance, the global mean Calcite Compensation Depth (CCD) sank by about 1 km over the last 40 million years (**Figure 3**), a change possibly linked to an increased supply of alkalinity from rivers caused by the Himalayan uplift. Paradoxically, there is no evidence of major changes in the carbonate accumulation flux during the Cenozoic (**Figure 4**). The deepening of the CCD might thus be linked, at least partially, to the global Cenozoic marine regression, reducing the area of shallow epicontinental seas and thus the area available for the accumulation of coral reefs. In that case, carbon and alkalinity will be preferentially removed from the ocean through enhanced formation of calcitic shells in open waters, leading to the deepening of the CCD. This process might have been favored by the coeval development of new foraminiferal species. The cause of the CCD deepening thus remains unresolved.

Organic Carbon Subcycle

The Cenozoic history of sea water $\delta^{13}C$ recorded in marine limestones (**Figure 5**) is marked by an ample fluctuation in the Paleocene and early Eocene, a roughly constant background value with superimposed high-frequency variations from the middle Eocene to the middle Miocene, and a sharp decrease from the middle Miocene to the present. Since organic matter is enriched in the lighter ^{12}C isotope with respect to sea water (owing to photosynthetic fractionation), this $\delta^{13}C$ record can be used to constrain the temporal changes in the organic fluxes of the carbon cycle. The burial of organic matter on the sea floor preferentially removes ^{12}C from the ocean and hence tends to increase seawater $\delta^{13}C$. Conversely, the oxidation of old organic carbon

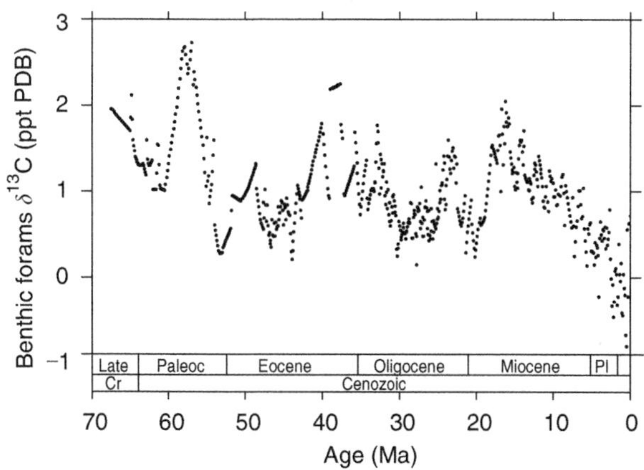

Figure 5 Sea water δ^{13}C recorded in ancient carbonate sediments (Ottawa-Bochum database: http://www.science.uottawa.ca/geology/isotope_data/). Timescale as **Figure 1**. ppt PDB = parts per thousand PeeDee Belemnite.

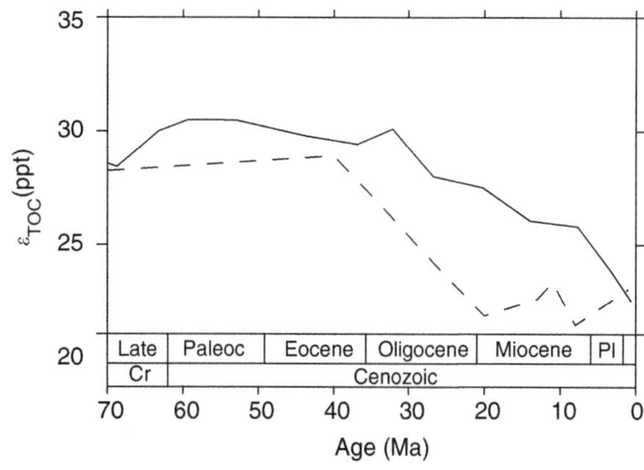

Figure 6 Average carbon isotopic fractionation (ε_{TOC}) between total organic carbon and sedimentary carbonate (solid line, Hayes *et al.* (1999); dashed line, Freeman and Hayes (1992)). Timescale as **Figure 1**.

(kerogen) contained in weathered sedimentary rocks is a source of isotopically light carbon for the ocean, tending to decrease its δ^{13}C. The recent decrease in δ^{13}C since mid-Miocene times might thus be interpreted as the result of kerogen carbon oxidation being larger than organic carbon burial during that period. Similarly, the overall constancy of sea water δ^{13}C from the middle Eocene to the middle Miocene may suggest that the organic subcycle was essentially balanced at that time. However, the average carbon isotopic fractionation (ε_{TOC}) between total organic carbon and sedimentary carbonate (which is close to coeval sea water) has decreased from the Eocene to the present (**Figure 6**). With a balanced organic subcycle, this change in $_{TOC}$ would imply a decrease of the sea water δ^{13}C over time, as it forces the δ^{13}C of organic deposits to become closer to the sea water value than it is for kerogen carbon. For the isotopic composition of the ocean to remain constant from the middle Eocene to the middle Miocene, the trend associated with ε_{TOC} variations must be compensated for by an imbalance in the organic subcycle in which the burial of organic carbon exceeds kerogen oxidation. This imbalance was progressively reduced after mid-Miocene times, but may have persisted until very recently.

The late Cenozoic was therefore a time of unusually high organic carbon deposition rates, leading to an increase in the size of the sedimentary organic carbon reservoir. The organic subcycle thus acted as a carbon sink over the course of the Himalayan uplift. There are two possible causes of this evolution.

1. The increase in chemical weathering rates in the Himalayan region during the uplift (Raymo's world) leads to enhanced delivery of nutrients to the ocean, forcing the oceanic primary productivity to increase. This might result in an increased burial of organic matter.
2. Enhanced mechanical weathering in the Himalayan region increased the sedimentation rate on the ocean floor, so that organic carbon was more easily preserved. This hypothesis does not require any increase in the chemical weathering rate in the Himalayan region. This facilitated burial might have significantly contributed to the Cenozoic $P\text{CO}_2$ decrease, since carbon is stored in a sedimentary reservoir. C. France-Lanord and L. Derry argued in 1997 that the consumption of CO_2 through organic carbon burial might be three times more important today than the amount of CO_2 consumed by silicate weathering within the orogen. Even if this hypothesis still links the climatic cooling with the Himalayan uplift, the origin of the CO_2 sink is quite different from that hypothesized in Raymo's world.

Observational data argue toward the second hypothesis, indicating that the Cenozoic increase in the sea water ^{87}Sr/^{86}Sr might be of isotopic origin. Calcium silicates are indeed not the most common mineral exposed in the Himalayan orogen, and thus cannot contribute widely to the CO_2 consumption. Furthermore, reverse weathering reactions take place in the Bengal Fan, releasing CO_2 and thus reducing the impact of the Himalayan silicate weathering on $P\text{CO}_2$.

It has been suggested that the emission at some time in the past of large amounts of methane from gas hydrates may have influenced the δ^{13}C of the ocean. This may invalidate the interpretation of the carbon isotopic record if the gas hydrate reservoir has had long-term as well as shorter-term effects.

Organic carbon deposition on the seafloor is linked to ocean biological productivity, itself

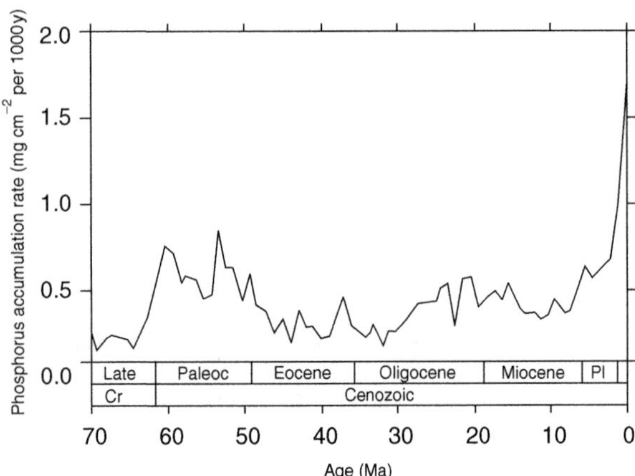

Figure 7 Average accumulation of phosphorus in sediments through the Cenozoic (Föllmi, 1995). Timescale as **Figure 1**.

depending on the availability of nutrients, among which phosphorus is thought to play a key role. The global phosphorus accumulation into sediments increased by a factor of about 4 over the last 10 million years (**Figure** 7), interpreted as the record of a global continental weathering enhanced by the onset of large ice sheets and glaciers at the end of Cenozoic. However, the question whether this increase is related to an increase in continental chemical weathering, or in mechanical weathering alone, is not clear.

Modelling: An Attempt to Integrate the Records into a Unified Framework

The Concept of Box Models

Biogeochemical cycles are usually described with box models. Such models provide a simple mathematical framework appropriate for calculating the geochemical evolution of the Earth through geological times. The Earth system is split into a relatively small number of components or reservoirs assumed to be homogeneous, such as the atmosphere, the ocean, the biosphere, the continental or oceanic crust, and the (upper) mantle. These reservoirs are connected by a series of 'arrows' representing the flows of material between them. The biogeochemical cycle of each element is thus represented as a set of interconnected reservoirs and, at any time, its state is characterized by the reservoir sizes or contents q_i (amount of the element in reservoir '*I*', units: mol or kg) and the fluxes F_{ij} (amount of the element transferred per unit time from reservoir '*I*' to reservoir '*j*', units: mol y^{-1} or kg y^{-1}).

The temporal evolution of the system can be calculated by making a budget of input and output fluxes for each reservoir (eqn [2]).

$$\frac{dq_i}{dt} = \sum_{j=1j\neq i}^{N} F_{ji} - \sum_{j=1j\neq i}^{N} F_{ij}(i = 1, \ldots, N) \qquad [2]$$

To solve this system of differential equations, the values of the fluxes must be provided at each time step. Kinetic rate laws describing the dependence of the fluxes F_{ij} on the reservoir contents q_i, time t, or some external forcing are thus needed. Defining such kinetic rate laws is the most critical task of modeling. The reliability of the solution and hence the usefulness of the results depend strongly on the adopted rate laws. The challenge is clearly to get at least a first-order estimate of the fluxes from a very broad knowledge of the system, i.e., from the values of its state variables $q_1,\ldots, q_N$.

A useful concept in box modeling is that of turnover time. The turnover time of an element in a given reservoir is defined as the ratio between its reservoir content and its total output flux (eqn [3]).

$$\tau_i = \frac{q_i}{\sum_{j=1j\neq i}^{N} F_{ij}} \qquad [3]$$

The turnover time can be seen as the time needed to empty the reservoir if the input happened to stop suddenly and the current output flux were held constant. It provides a first-order idea of the evolution timescale of a reservoir. At steady state (i.e., when input and output fluxes balance each other), the turnover time is equal to the residence time, which is the average time spent by individual atoms of the element in the reservoir. Finally, the response time of a reservoir characterizes the time needed for the reservoir to adjust to a new equilibrium after a perturbation.

Models of the Carbon Cycle

Figure 8 illustrates the present state of the long-term carbon cycle from a recent (unpublished) box model simulation of the authors. The reservoirs and fluxes that have been included in this figure are those that are important to describe the evolution of atmospheric CO_2 at the geological timescale. The values of reservoir sizes and fluxes are consistent with current knowledge of the system. Crustal reservoirs include continental (5000×10^{18} mol C) and pelagic (150×10^{18} mol C) carbonates, as well organic carbon (1250×10^{18} mol C) from the sedimentary cover. The atmosphere and ocean have been lumped into one single reservoir containing 3.2×10^{18} mol C, since the time necessary for the atmosphere to reach equilibrium with the ocean is much shorter than ~ 1 My, the timescale of geological processes. Indeed,

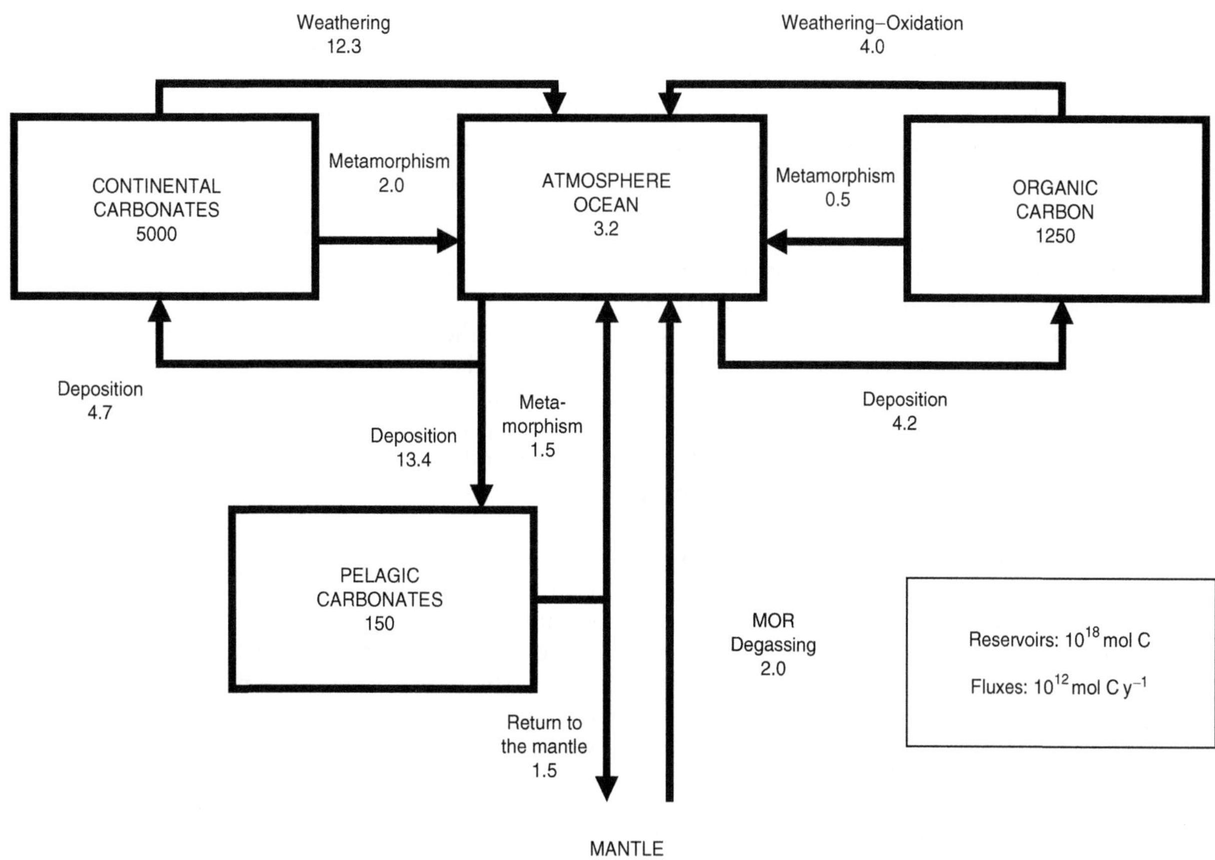

Figure 8 Present-day state of the long-term carbon cycle. Numbers represent 10^{18} mol C for reservoirs (boxes) and 10^{12} mol C y^{-1} for fluxes (arrows).

with a modern atmosphere-ocean exchange flux of 7.5×10^{15} mol C y^{-1} (i.e., 90 Gt C y^{-1}) and an atmospheric content of 62.5×10^{15} mol C (i.e., 750 Gt C), the turnover time of carbon in the atmosphere can be calculated to be only 8.33 years. Similarly, the terrestrial biosphere has not been included, since its size is small compared to other reservoirs and it can be assumed in equilibrium with the atmosphere–ocean system. The fluxes involved in the cycle are MOR/metamorphic CO_2 release ('volcanism'), weathering fluxes, and deposition of carbonates or organic carbon on the seafloor. The reported values of these fluxes are long-term averages, that is, they should be thought of as averages over several glacial–interglacial oscillations of the Pleistocene, although such averages cannot always be estimated from presently available data. The turnover time of carbon in the atmosphere–ocean reservoir in this 'geological' system can be calculated to be 143 000 y. Owing to this relatively short turnover time with respect to the timescale of long-term geological changes, the atmosphere–ocean system is essentially at equilibrium. By contrast, crustal reservoirs that exhibit much larger turnover times are not at equilibrium. This is clearly the case of continental ($\tau_i = 350$ My) and pelagic ($\tau_i = 50$ My) carbonate reservoirs, as a result of the Cenozoic deepening of the ocean lysocline and the associated transfer of carbonate deposition from the shelf to the pelagic environment.

To distribute the carbon content of the atmosphere–ocean system among its two components, and hence to derive the atmospheric P_{CO_2} value (and its effect on the climate), it is necessary to know the alkalinity content of the ocean. For this reason, the evolution of the ocean alkalinity (A_T) and ocean–atmosphere carbon (C_T) content are always calculated in parallel. Writing eqn [2] for these two variables yields eqns [4a] and [4b].

$$\frac{dA_T}{dt} = 2F_{SW} + 2F_{CW} - 2F_{CD} \quad [4a]$$

$$\frac{dC_T}{dt} = F_{VOL} + F_{CW} - F_{CD} + F_{OW} - F_{OD} \quad [4b]$$

F_{VOL} represents the total CO_2 release flux from volcanic origin (i.e., the sum of all metamorphic and MOR fluxes in **Figure 8**), F_{CW} and F_{SW} are the weathering fluxes from respectively carbonate and silicate rocks expressed in moles of divalent ions (Ca^{2+} or Mg^{2+}) per unit of time (i.e., the rates of reactions [I] and [II]), F_{CD} is the carbonate deposition flux, F_{OW} is the carbon input flux from weathering-oxidation of crustal organic carbon, and F_{OD} is the organic carbon deposition flux. Note that the silicate weathering flux does not appear in the carbon budget, eqn [4b], since silicate weathering (reaction [II]) transfers carbon from the atmosphere to the ocean but does not remove it from the atmosphere–ocean system. The factor of 2 in eqn [4a] results from the fact that two equivalents of alkalinity are transferred to the ocean when one Ca^{2+} or Mg^{2+} ion is delivered to the ocean by rivers (reactions [I] and [II]). The same factor of 2 holds for carbonate deposition, which is the reverse of reaction [I]. As already mentioned, the atmosphere–ocean system must be close to equilibrium, so that the derivatives on the left-hand side of eqns [4a] and [4b] can be set to zero. This assumption transforms the differential equation system into a set of two algebraic equations, which can be solved to yield eqn [5].

$$F_{VOL} - F_{SW} = F_{OD} - F_{OW} \qquad [5]$$

This equation leads to eqn [1] if the effect of the organic subcycle is neglected (i.e., when this subcycle is set to equilibrium). Hence, eqn [5] is a generalization of the Walker, Hays, and Kasting budget. It states that the disequilibrium of the inorganic part of the carbon cycle must be compensated for by a disequilibrium of opposite sign in the organic subcycle.

Use of Isotopic Data (Inverse Modeling)

To solve eqns [4] or [5], some kinetic laws must be provided for the fluxes, that is, the relations between these fluxes, time t, and the reservoir contents, or atmospheric P_{CO_2}, must be known. Such kinetic laws are, however, poorly known, so it may be preferable, at least for some fluxes, to use forcing functions in the calculation of these fluxes. For example, volcanic fluxes are often made proportional to the seafloor spreading rate and weathering fluxes to land area, for which past reconstructions are available. Ocean isotopic records, such as those presented earlier, can also be used to force the model. Budget equations similar to [2] are then written for the relevant isotopes and transformed into equations containing isotopic ratios r (or δ, the relative departure of the isotopic ratio from a standard). The sea water $^{87}Sr/^{86}Sr$ ratio has been used in this way to estimate the silicate weathering flux F_{SW}, but as discussed earlier the results are strongly dependent on the hypothesis of constancy for the isotopic ratios of weathered products. The ^{13}C isotopic history of the ocean has been used in many models, since the beginning of the 1980s, to constrain the organic carbon subcycle. The ^{13}C isotopic budget for the ocean can be written as eqn [6].

$$C_T\frac{d\delta_{OC}}{dt} = (\delta_{VOL} - \delta_{OC})F_{VOL} + (\delta_{CW} - \delta_{OC})F_{CW} + (\delta_{OW} - \delta_{OC})F_{OW} - (\delta_{OD} - \delta_{OC})F_{OD} \qquad [6]$$

δ_{oc} here is the $\delta^{13}C$ of the ocean (more precisely, this should be the $\delta^{13}C$ of the atmosphere–ocean system); δ_{VOL}, δ_{CW}, and δ_{OW} are the $\delta^{13}C$ of the carbon inputs from respectively volcanic, carbonate weathering, and crustal organic carbon weathering–oxidation fluxes. It is assumed that no fractionation occurs with respect to average oceanic carbon during carbonate precipitation, so that this flux does not appear in the equation. $\delta_{OD} = \delta_{oc} - \Delta$ is the $\delta^{13}C$ of the organic carbon deposited on the seafloor, with θ being the average fractionation of photosynthesis with respect to oceanic carbon (this includes both terrestrial and marine photosynthesis). The past values of δ_{oc} are known from the ^{13}C isotopic history of sea water (**Figure 5**). Equation [6] can then be solved with respect to F_{OD} and the resulting expression for F_{OD} is then used in eqn [4b] or [5]. The isotopic composition of the input fluxes must, however, be known or derived from similar isotopic budgets for the crustal reservoirs. This procedure is actually an inverse method, since it derives model parameters (fluxes) from an observed signal (ocean isotopic composition) linked to the model parameters through a mathematical operator (the isotopic budget equation). Y. Goddéris and L.M. François in 1996, and L.R. Kump and M.A. Arthur in 1997, published two separate models inverting the oceanic $\delta^{13}C$ signal over the Cenozoic, making use of an isotopic fractionation Δ variable with age and derived from paleodata. The Cenozoic histories of silicate weathering from these models are compared in **Figure 9**. The predicted trend of the carbonate deposition flux is broadly consistent with an available reconstruction based on carbonate accumulation data (**Figure 10**).

A classical example of a box model using ^{13}C isotopic data to constrain the organic carbon subcycle is the BLAG model of Lasaga, Berner, and Garrels published in 1985. R.A. Berner in 1990 also used such an isotopic budget in GEOCARB to calculate the history of atmospheric CO_2 over the Phanerozoic. The results show a decreasing trend of atmospheric CO_2 over the Cenozoic. The trend is consistent with the overall trend reconstructed with other models (e.g., François and Walker, 1992) or

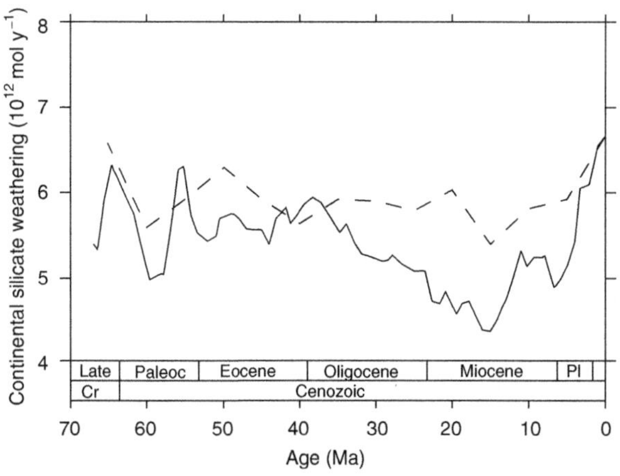

Figure 9 Model silicate weathering flux (solid line, Goddéris and François (1996); dashed line, Kump and Arthur (1997)). Timescale as **Figure 1**.

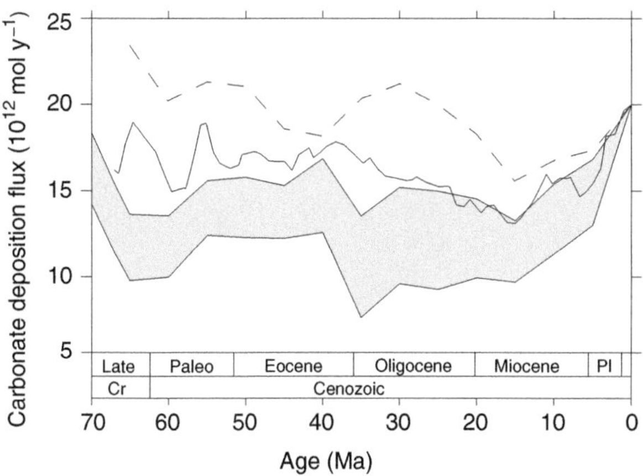

Figure 10 Model carbonate accumulation flux compared to the reconstruction of **Figure 4** normalized to the present day value of 20×10^{12} mol y^{-1} (solid line: Goddéris and François (1996); dashed line, Kump and Arthur (1997); shaded area, reconstruction from **Figure 4**).

from various paleoindicators (**Figure 1**). This does not mean, however, that we understand the carbon cycle (and climate) trends of the Cenozoic, since different models can produce similar trends from completely different underlying mechanisms. To be reliable, models should not rest only on a limited set of data but should be able to explain a wide range of geochemical records.

Conclusions

Proxy records indicate that the Earth's climate cooled gradually over the Cenozoic. This cooling trend was accompanied by a decrease of atmospheric P_{CO_2}. Other striking features of the Cenozoic are the sharp increase of the $^{87}Sr/^{86}Sr$ ratio of sea water and the overall deepening of the lysocline from late Eocene time to the present, and after the mid-Miocene a marked decrease of ocean $\delta^{13}C$ together with an increase in total carbonate accumulation and possibly phosphorus deposition. Are these environmental changes related? The role of models is to synthesize and provide a coherent explanation of such records, and then reconstruct the history of other key variables not directly accessible from paleodata. Today, we are still far from this goal. It is fundamental that models use multiple proxy data both as forcings and for validation, implying that other biogeochemical cycles for which proxies are available are modeled together with the carbon cycle. A coupling to other major biogeochemical cycles is also essential because of the interactions with the carbon cycle and the feedbacks involved.

See also

Calcium Carbonates. Carbon Cycle. Cenozoic Climate - Oxygen Isotope Evidence. Ocean Carbon System, Modeling of. Past Climate From Corals. River Inputs. Stable Carbon Isotope Variations in the Ocean.

Further Reading

Berner RA (1990) Atmospheric carbon dioxide levels over Phanerozoic time. *Science* 249: 1382–1386.

Berner RA (1998) The carbon cycle and CO_2 over Phanerozoic time: the role of land plants. *Philosophical Transactions of the Royal Society of London B* 353: 75–82.

Broecker WS and Peng TH (1982) *Tracers in the Sea.* Palisades: Eldigio Press.

Butcher SS, Charlson RJ, Orians GH, and Wolfe GV (eds.) (1992) *Global Biogeochemical Cycles.* London: Academic Press.

Chameides WL and Perdue EM (1997) *Biogeochemical Cycles: A Computer-Interactive Study of Earth System Science and Global Change.* Oxford: Oxford University Press.

Föllmi KB (1995) 160 My record of marine sedimentary phosphorus burial: coupling of climate and continental weathering under greenhouse and icehouse conditions. *Geology* 23: 859–862.

France-Lanord C and Derry LA (1997) Organic carbon burial forcing of the carbon cycle from Himalayan erosion. *Nature* 390: 65–67.

François LM and Walker JCG (1992) Modelling the Phanerozoic carbon cycle and climate: constraints from the $^{87}Sr/^{86}Sr$ isotopic ratio of sea water. *American Journal of Science* 292: 81–135.

Freeman KH and Hayes JM (1992) Fractionation of carbon isotopes by phytoplankton and estimates of ancient CO_2 levels. *Global Biogeochemical Cycles* 6: 185–198.

Goddéris Y and François LM (1996) Balancing the Cenozoic carbon and alkalinity cycles: constraints from isotopic records. *Geophysical Research Letters* 23: 3743–3746.

Harland WB, Armstrong RL, Cox AV, *et al.* (1990) *A Geologic Time Scale 1989*. Cambridge: Cambridge University Press.

Hayes JM, Strauss H, and Kaufman AJ (1999) The abundance of ^{13}C in marine organic matter and isotopic fractionation in the global biogeochemical cycle of carbon during the past 800 Ma. *Chemical Geology* 161: 103–125.

Kump LR and Arthur MA (1997) Global chemical erosion during the Cenozoic: weatherability balances the budgets. In: Ruddiman WF (ed.) *Tectonic Uplift and Climate Change*. New York: Plenum Press.

Kump LR, Kasting JF, and Crane RG (1999) *The Earth System*. New Jersey: Prentice Hall.

Lasaga AC, Berner RA, and Garrels RM (1985) An improved geochemical model of atmospheric CO_2 fluctuations over the past 100 million years. In: Sundquist E and Broecker WS (eds.) *The Carbon Cycle and Atmospheric CO_2: Natural Variations Archean to Present Geophysical Monograph*, vol. 32, pp. 397–411. Washington, DC: American Geophysical Union.

Opdyke BN and Wilkinson BH (1988) Sea surface area control of shallow cratonic to deep marine carbonate accumulation. *Paleoceanography* 3: 685–703.

Pearson PN and Palmer MR (2000) Atmospheric carbon dioxide concentrations over the past 60 million years. *Nature* 406: 695–699.

Raymo ME (1991) Geochemical evidence supporting T.C. Chamberlin's theory of glaciation. *Geology* 19: 344–347.

Ruddiman WF (ed.) (1997) *Tectonic Uplift and Climate Change*. New York: Plenum Press.

Van Andel TH (1975) Mesozoic-Cenozoic calcite compensation depth and the global distribution of calcareous sediments. *Earth and Planetary Science Letters* 26: 187–194.

Van Andel TH (1994) *New Views on an Old Planet: a History of Global Change*, 2nd edn. Cambridge: Cambridge Universitys Press.

Walker JCG, Hays PB, and Kasting JF (1981) A negative feedback mechanism for the long-term stabilization of Earth's surface temperature. *Journal of Geophysical Research* 86: 9776–9782.

MARINE SILICA CYCLE

D. J. DeMaster, North Carolina State University, Raleigh, NC, USA

Introduction

Silicate, or silicic acid (H_4SiO_4), is a very important nutrient in the ocean. Unlike the other major nutrients such as phosphate, nitrate, or ammonium, which are needed by almost all marine plankton, silicate is an essential chemical requirement only for certain biota such as diatoms, radiolaria, silicoflagellates, and siliceous sponges. The dissolved silicate in the ocean is converted by these various plants and animals into particulate silica (SiO_2), which serves primarily as structural material (i.e., the biota's hard parts). The reason silicate cycling has received significant scientific attention is that some researchers believe that diatoms (one of the silica-secreting biota) are one of the dominant phytoplankton responsible for export production from the surface ocean (Dugdale *et al.*, 1995). Export production (sometimes called new production) is the transport of particulate material from the euphotic zone (where photosynthesis occurs) down into the deep ocean. The relevance of this process can be appreciated because it takes dissolved inorganic carbon from surface ocean waters, where it is exchanging with carbon dioxide in the atmosphere, turns it into particulate organic matter, and then transports it to depth, where most of it is regenerated back into the dissolved form. This process, known as the 'biological pump', along with deep-ocean circulation is responsible for the transfer of inorganic carbon into the deep ocean, where it is unable to exchange with the atmosphere for hundreds or even thousands of years. Consequently, silicate and silica play an important role in the global carbon cycle, which affects the world's climate through greenhouse feedback mechanisms. In addition, the accumulation of biogenic silica on the ocean floor can tell us where in the ocean export production has occurred on timescales ranging from hundreds to millions of years, which in turn reveals important information concerning ocean circulation and nutrient distributions.

Basic Concepts

In understanding the cycling of silicate in the oceans, the concept of mean oceanic residence time is commonly used. Mean oceanic residence time is defined as in eqn. [1].

$$\frac{\text{(amount of dissolved material in a reservoir)}}{\text{(steady-state flux into or out of the reservoir)}} \qquad [1]$$

For silicate there are approximately 7×10^{16} moles of dissolved silicon in ocean water. (One mole is equal to 6×10^{23} molecules of a substance, which in the case of silicic acid has a mass of approximately 96 g.) As described later, the various sources of silicate to the ocean supply approximately 7×10^{12} mol y^{-1}, which is approximately equal to our best estimates of the removal rate. Most scientists believe that there has been a reasonably good balance between supply and removal of silicate from the oceans on thousand-year timescales because there is little evidence in the oceanic sedimentary record of massive abiological precipitation of silica (indicating enhanced silicate concentrations relative to today), nor is there any evidence in the fossil record over the past several hundred million years that siliceous biota have been absent for any extended period (indicating extremely low silicate levels). Dividing the amount of dissolved silicate in the ocean by the supply/removal rate yields a mean oceanic residence time of approximately 10 000 years. Basically, what this means is that an atom of dissolved silicon supplied to the ocean will remain on average in the water column or surface seabed (being transformed between dissolved and particulate material as part of the silicate cycle) for approximately 10 000 years before it is permanently removed from the oceanic system via long-term burial in the seabed.

Distribution of Silicate in the Marine Environment

Because of biological activity, surface waters throughout most of the marine realm are depleted in dissolved silicate, reaching values as low as a few micromoles per liter (μmol l^{-1}). When the siliceous biota die, their skeletons settle through the water column, where more than 90% of the silica is regenerated via inorganic dissolution. This process enriches the deep water in silicate, causing oceanic bottom waters to have as much as 10–100 times more silicate than surface waters in tropical and temperate regions. The magnitude of the deep-ocean silicate concentration depends on the location within the deep thermohaline circulation system. In general, deep water originates in North Atlantic and Antarctic surface waters. The deep water forming in the North Atlantic moves southward, where it joins with Antarctic water, on its way to feeding the deep Indian Ocean basin and then flowing from south to north in the Pacific basin. All along this 'conveyor belt' of deep-ocean water, siliceous biota are continually settling out from surface waters and dissolving at depth, which further increases the silicate concentration of the deep water downstream. Consequently, deep-ocean water in the Atlantic (fairly near the surface ocean source) is not very enriched in silicate (only 60 μmol l^{-1}), whereas the Indian Ocean deep water exhibits moderate enrichment ($\sim$100 μmol l^{-1}), and the north Pacific deep water is the most enriched ($\sim$180 μmol l^{-1}). This trend of increasing concentration is observed as well in the other nutrients such as nitrate and phosphate. Generalized vertical profiles of silicate are shown for the Atlantic and Pacific basins in **Figure 1**. The depth of the silicate maximum in these basins (typically 2000–3000 m depth) is deeper than the nutrient maxima for phosphate or nitrate, primarily because organic matter (the source of the phosphate and nitrate) is generally regenerated at shallower depths in the ocean than is silica. The nutrient concentrations in oceanic deep waters can affect the chemical composition of particles settling through the water column because the vertical transport of nutrients from depth via upwelling and turbulence drives the biological production in surface waters. For example, the ratio of biogenic silica to organic carbon in particles settling between 1000 and 4000 m depth in the North Pacific Ocean (typically about 2–3) is substantially higher than that observed in the Arabian Sea ($\sim$0.7) and much higher than typical values in the Atlantic Ocean (<0.3). This chemical trend in particle flux, which is caused in part by changes in planktonic species assemblage, is consistent with the systematic increase in silicate and other nutrients along the thermohaline-driven conveyor belt of deep-ocean circulation. The change in the biogenic silica to organic carbon ratio throughout the ocean basins of the world turns out to be one of the most important parameters controlling the nature of biogenic sedimentation in the world (see the global ocean sediment model of Heinze and colleagues, listed in Further Reading).

Silicate concentrations also can be used to distinguish different water masses. The most obvious example is at the Southern Ocean Polar Front (see **Figure 2**), which separates Antarctic Surface Water from the Subantarctic system. The silicate and nitrate concentration gradients across these Southern Ocean waters occur in different locations (in a manner similar to the distinct maxima in their vertical profiles). The high concentrations of silicate (50–100 μmol l^{-1}) south of the Polar Front result from wind-induced upwelling bringing silicate to the surface faster than the local biota can turn it into particulate silica. Turnover times between surface waters and

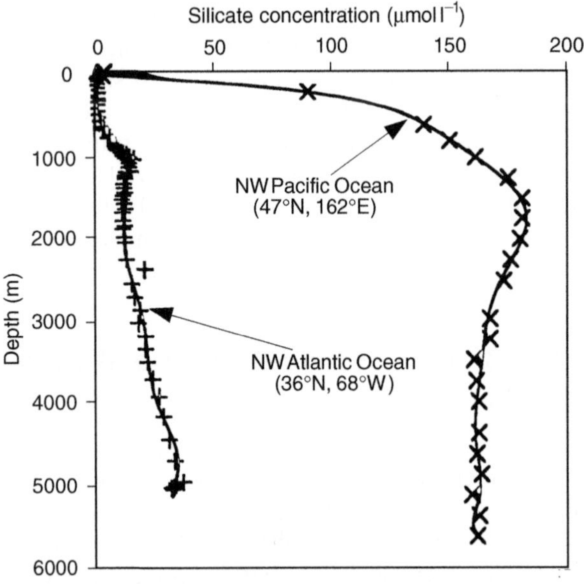

Figure 1 Vertical distribution of dissolved silicate in the Atlantic Ocean and the Pacific Ocean. The Atlantic data come from Spencer (1972), whereas the Pacific data are from Nozaki *et al.* (1997).

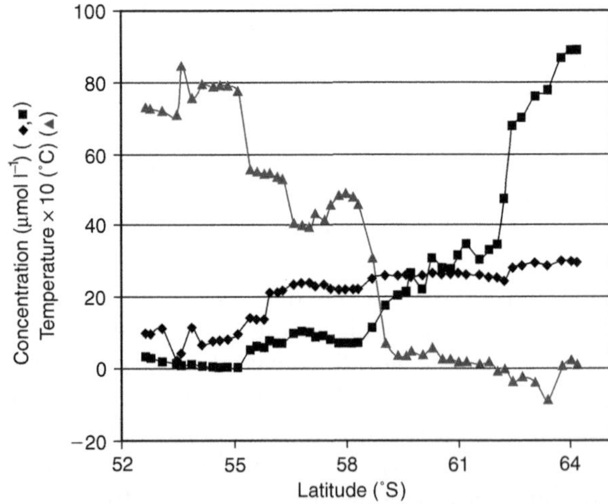

Figure 2 Distribution of silicate (■), nitrate (♦), and temperature (▲) across the Drake Passage illustrating different water masses and frontal features during November, 1999.

deep waters in the Southern Ocean are on the order of 100 years as compared to values of 500–1000 years in the Atlantic, Indian, and Pacific Oceans.

In terms of chemical equilibrium, biogenic silica is undersaturated by a factor of 10–1000-fold in surface waters and by at least a factor of 5 in deep waters. Therefore, siliceous biota must expend a good deal of energy concentrating silicate in their cells and bodies before precipitation can take place. This is quite different from the case of calcium carbonate (a material used by another type of plankton to form hard parts), which is supersaturated severalfold in most tropical and temperate surface waters. Deep waters everywhere are undersaturated with respect to biogenic silica (although to different extents). Therefore, inorganic dissolution of silica takes place in the water column as soon as the plankton's protective organic matter is removed from the biota (typically by microbial or grazing activities). It is not until the siliceous skeletal material is buried in the seabed that the water surrounding the silica even approaches saturation levels (see later discussion on sedimentary recycling), which diminishes the rate of dissolution and enhances preservation and burial.

The Marine Silica Cycle

Sources of Dissolved Silicate to the Ocean

Figure 3 shows the main features of the marine silica cycle as portrayed in a STELLA^{TM} model. The main source of silicate to the oceans as a whole is rivers, which commonly contain $\sim$150 μmol l^{-1} silicate but depending on location, climate, and local rock type, can range from 30 to 250 μmol l^{-1}. The silicate in rivers results directly from chemical weathering of rocks on land, which is most intense in areas that are warm and wet and exhibit major changes in relief (i.e., elevation). The best estimate of the riverine flux of dissolved silicate to the oceans is $\sim 6 \times 10^{12}$ mol Si y^{-1}. Other sources include hydrothermal fluxes ($\sim 0.3 \times 10^{12}$ mol Si y^{-1}), dissolution of eolian particles (0.5×10^{12} mol Si y^{-1}) submarine volcanic activity (negligible), and submarine weathering of volcanic rocks ($\sim 0.4 \times 10^{12}$ mol Si y^{-1}). Ground waters may contribute additional silicate to the marine realm, but the magnitude of this flux is difficult to quantify and it is believed to be small relative to the riverine flux. A more detailed discussion of the various sources of silicate to the marine environment can be

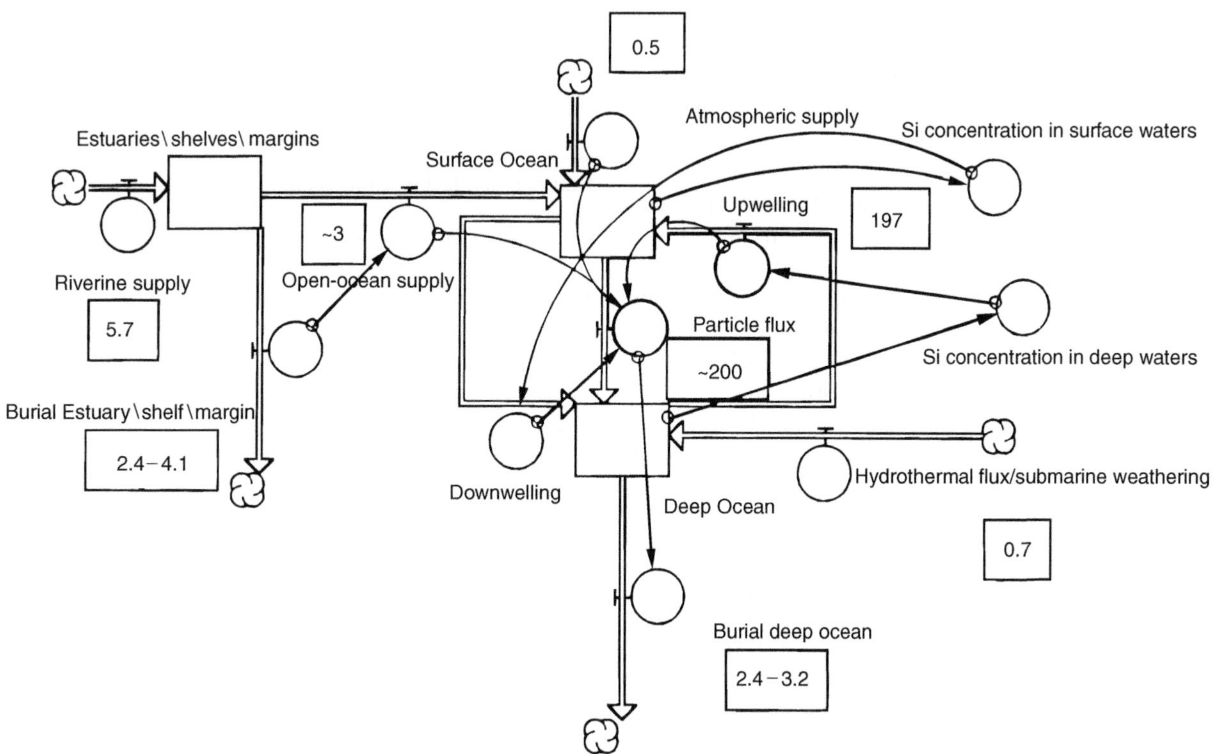

Figure 3 STELLA^{TM} model of the global marine silica cycle showing internal and external sources of silicate to the system, internal recycling, and burial of biogenic silica in the seabed. The various reservoirs are shown as rectangles, whereas the fluxes in and out of the reservoirs are shown as arrows with regulating valves (indicating relationships and functional equations). The flux values (indicated by numbers inside the boxes) have units of 10^{12} mol y^{-1}.

found in the works by Treguer *et al.* and DeMaster (listed in Further Reading).

All of these sources of silicate to the oceanic water column are considered to be external. As shown in **Figure 3**, there are internal sources supplying silicate to oceanic surface waters and they are oceanic upwelling and turbulence. Because of the strong gradient in silicate with depth, upwelling of subsurface water (100–200 m depth) by wind-induced processes and turbulence can bring substantial amounts of this nutrient (and others) to surface waters that typically would be depleted in these valuable chemical resources as a result of biological activity. This upwelling flux ($\sim 100 - 300 \times 10^{12}$ mol Si y^{-1}), in fact, is 20–50 times greater than the riverine flux. The extraction of silicate from surface waters by siliceous biota is so efficient that nearly 100% of the nutrient reaching the surface is converted into biogenic silica. Therefore, the production of biogenic silica in oceanic surface waters is comparable to the flux from upwelling and turbulent transport. Riverine sources may be the dominant external source of silicate to the oceans, but they sustain only a negligible amount (only a few percent) of the overall marine silica production. Internal recycling, upwelling, and turbulence provide nearly all of the silicate necessary to sustain the gross silica production in marine surface waters. Therefore, changes in oceanic stratification and wind intensity may significantly affect the flux of nutrients to the surface and the overall efficiency of the biological pump. Silicate dynamics in the water column have been simulated using a general circulation model. The results of this study by Gnanadesikan suggest that the model distributions of silicate in the ocean are very sensitive to the parametrization of the turbulent flux. In addition, according to the model, the Southern Ocean and the North Pacific were the two major open-ocean sites where net silica production occurred, accounting for nearly 80% of the biogenic silica leaving the photic zone.

If the entire ocean (surface waters, deep waters, and near-surface sediments) is considered as a single box, the external fluxes of silicate to the ocean (mentioned above) must be balanced by removal terms in order to maintain the silicate levels in the ocean at more or less a constant value over geological time. From this point of view the flux of silicate from oceanic upwelling and turbulence can be treated as part of an internal cycle. The dominant mechanism removing silica from this system is burial of biogenic silica. There is some controversy about where some of this burial takes place, but most scientists believe that burial of biogenic silica (or some chemically altered by-product thereof) is the primary way that silicate is removed from the ocean.

Removal of Silica from the Ocean

The sediments with the highest rate of silica accumulation (on an areal basis) occur beneath the coastal upwelling zones, where strong winds bring extensive amounts of nutrients to the surface. Examples of these upwelling areas include the west coast of Peru, the Gulf of California, and Walvis Bay (off the west coast of South Africa). Diatom skeletons are the dominant form of biogenic silica in these deposits. The sediments in these upwelling areas accumulate at rates of 0.1–1.0 cm y^{-1} and they contain as much as 40% biogenic silica by weight. The burial rate for silica in these areas can be as high as 1.7 mol cm^{-2} y^{-1}. In calculating the total contribution of these areas to the overall marine silica budget, the accumulation rates (calculated on an areal basis) must be multiplied by the area covered by the particular sedimentary regime. Because these upwelling regimes are confined to such small areas, the overall contribution of coastal upwelling sites to the marine silica budget is quite small (<10%, see **Table 1**), despite the fact that (for a given area) they bury biogenic silica more rapidly than anywhere else in the marine realm.

The sediments containing the highest fraction of biogenic silica in the world occur in a 1000 km-wide

Table 1 The marine silica budget

Source/Sink	*Flux (10^{12}mol Si y^{-1})*
Sources of silicate to the ocean	
Rivers	5.7
Hydrothermal emanations	0.3
Eolian flux (soluble fraction)	0.5
Submarine weathering	0.4
Total supply of silicate	$\sim 7 \times 10^{12}$
Sites of biogenic silica burial	
Deep-Sea Sediments	2.4–3.2
Antarctic	
Polar Front	0.7–0.9
Non Polar Front	0.7–1.1
Bering Sea	0.5
North Pacific	0.3
Sea of Okhotsk	0.2
Equatorial Pacific	0.02
Poorly siliceous sediments	<0.2
Continental margin sediments	2.4–4.1
Estuaries	<0.6
Coastal upwelling areas (e.g., Peru, Walvis Bay, Gulf of California)	0.4–0.5
Antarctic margin	0.2
Other continental margins	1.8–2.8
Total burial of biogenic silica	$5–7 \times 10^{12}$

belt surrounding Antarctica. These sediments typically contain ~60% biogenic silica by weight (the majority of which are the skeletons of diatoms). Most of these sediments, however, accumulate at rates of only a few centimeters per thousand years, so their silica burial rate is quite small (<0.008 mol Si cm^{-2} y^{-1}), accounting for 1×10^{12} mol y^{-1} of silica burial or less than 20% of the total burial in the marine environment. Beneath the Polar Front (corresponding to the northern 200–300 km of the Southern Ocean siliceous belt), however, the sediment accumulation rates increase dramatically to values as high as 50×10^{-3} cm y^{-1}. Regional averages can be as high as 19×10^{-3} cm y^{-1}, yielding a silica burial rate of 0.08 mol Si cm^{-2} y^{-1}. Unfortunately, many of the siliceous deposits beneath the Polar Front occur in areas of very rugged bottom topography (because of the submarine Antarctic Ridge), where sediments are focused into the deeper basins from the flanks of the oceanic ridge crests. This distribution of accumulation rates would not create a bias if all of the sedimentary environments are sampled equally. However, it is more likely to collect sediment cores in the deep basins, where the deposits are thicker and accumulating more rapidly, than it is on the flanks where the sediment coverage is thinner. The effects of this sediment focusing can be assessed by measuring the amount in the seabed of a naturally occurring, particle-reactive radioisotope, thorium-230 (^{230}Th). If there were no sediment focusing, the amount of excess ^{230}Th in the sediments would equal the production from its parent, uranium-234, in the overlying water column. In some Polar Front Antarctic cores there is 12 times more excess ^{230}Th in the sediment column than produced in the waters above, indicating that sediment focusing is active. Initial estimates of the biogenic silica accumulation beneath the Polar Front were as high as 3×10^{12} mol Si y^{-1}, but tracer-corrected values are on the order of 1×10^{12} mol Si y^{-1}.

There are other high-latitude areas accumulating substantial amounts of biogenic silica, including the Bering Sea, the Sea of Okhotsk, and much of the North Pacific Ocean; however, the accumulation rates are not as high as in the Southern Ocean (see **Table 1**). The high rate of silica burial in the high-latitude sediments may be attributed in part to the facts that cold waters occurring at the surface and at depth retard the rate of silica dissolution and that many of the diatom species in high latitudes have more robust skeletons than do their counterparts in lower latitudes. Moderately high silica production rates and elevated silica preservation efficiencies (approximately double the world average) combine to yield high-latitude siliceous deposits accounting for approximately one-third of the world's biogenic silica burial.

If the focusing-corrected biogenic silica accumulation rates are correct for the Polar Front, then a large sink for biogenic silica (~$1–2 \times 10^{12}$ mol Si y^{-1}) needs to be identified in order to maintain agreement between the sources and sinks in the marine silica budget. Continental margin sediments are a likely regime because these environments have fairly high surface productivity (much of which is diatomaceous), a relatively shallow water column (resulting in reduced water column regeneration as compared to the deep sea), rapid sediment accumulation rates ($10–100 \times 10^{-3}$ cm y^{-1}) and abundant aluminosilicate minerals (see Biogenic Silica Preservation below). The amount of marine organic matter buried in shelf and upper slope deposits is on the order of 3×10^{12} mol C y^{-1}. When this flux is multiplied by the silica/organic carbon mole ratio (Si/C_{org}) of sediments in productive continental margin settings ($Si/C_{org} = 0.6$), the result suggests that these nearshore depositional environments can account for sufficient biogenic silica burial ($1.8–2.8 \times 10^{12}$ mol Si y^{-1}) to bring the silica budget into near balance (i.e., within the errors of calculation).

Biogenic Silica Preservation

As mentioned earlier, all ocean waters are undersaturated with respect to biogenic silica. Surface waters may be more than two orders of magnitude undersaturated, whereas bottom waters are 5–15-fold undersaturated. The solubility of biogenic silica is greater in warm surface waters than in colder deep waters, which, coupled with the increasing silicate concentration with depth in most ocean basins, diminishes the silicate/silica disequilibrium (or corrosiveness of the water) as particles sink into the deep sea. This disequilibrium drives silica regeneration in oceanic waters along with other factors and processes such as particle residence time in the water column, organic and inorganic surface coatings, particle chemistry, particle aggregation, fecal pellet formation, as well as particle surface area. Recycling of biogenic silica occurs via inorganic dissolution; however, the organic coating that siliceous biota use to cover their skeletons (inhibiting dissolution) must be removed by microbial or zooplankton grazing prior to dissolution. This association is highlighted by the fact that bacterial assemblages can accelerate the dissolution of biogenic silica in the water column.

An important aspect of biogenic silica dissolution pertains to surface chemistry and clay-mineral formation on the surface of siliceous tests. Incorporation of aluminum in the initial skeleton as well as aluminosilicate formation on skeletal surfaces during settling and burial greatly decrease the solubility of biogenic silica, in some cases by as much as a factor of 5–10. It appears that some of this 'armoring' of siliceous skeletons occurs up in the water column (possibly in aggregates or fecal pellets), although some aluminosilicate formation may occur in flocs just above the seabed as well as deeper in the sediment column. The occurrence of clay minerals on skeletal surfaces has been documented using a variety of instruments (e.g., the scanning electron microscope). The nature of the settling particles also affects dissolution rates in the water column. If siliceous skeletons settle individually, they settle so slowly (a timescale of years to decades) that most particles dissolve before reaching the seabed. However, if the siliceous skeletons aggregate or are packaged into a fecal pellet by zooplankton, sinking velocities can be enhanced by several orders of magnitude, favoring preservation during passage through the water column. Siliceous tests that have high surface areas (lots of protruding spines and ornate surface structures) also are prone to high dissolution rates and low preservation in the water column relative to species that have more robust skeletons and more compact structures.

Very few studies have documented silica production rates in surface waters, established the vertical fluxes of silica in the water column, and then also examined regeneration and burial rates in the seabed. One place that all of these measurements have been made is in the Ross Sea, Antarctica. In this high-latitude environment, approximately one-third of the biogenic silica produced in surface waters is exported from the euphotic zone, with most of this material (27% of production) making it to the seabed some 500–900 m below. Seabed preservation efficiencies (silica burial rate divided by silica rain rate to the seafloor) vary from 1% to 86%, depending primarily on sediment accumulation rate, but average 22% for the shelf as a whole. Consequently, the overall preservation rate (water column and seabed) is estimated to be $\sim$6% in the Ross Sea. On a global basis, approximately 3% of the biogenic silica produced in surface waters is buried in the seabed. The total preservation efficiencies for different ocean basins vary, with the Atlantic and Indian Oceans having values on the order of 0.4–0.8% and the Pacific and Southern Oceans having values of approximately 5–10%.

Sediment accumulation rate can make a large difference in seabed preservation efficiency. In the Ross Sea, for example, increasing the sediment accumulation rate from 1–2 to 16×10^{-3} cm y^{-1}, increases the seabed preservation efficiency from 1–5% up to 50–60%. In most slowly accumulating deep-sea sediments (rates of 2×10^{-3} cm y^{-1} or less), nearly all of the biogenic silica deposited on the seafloor dissolves prior to long-term burial. Increasing the sedimentation rate decreases the time that siliceous particles are exposed to the corrosive oceanic bottom waters, by burying them in the seabed where silicate, aluminum, and cation concentrations are high, favoring aluminosilicate formation and preservation. Consequently, continental margin sediments with accumulation rates of $10\text{–}100 \times 10^{-3}$ cm y^{-1} are deposits expected to have high preservation efficiencies for biogenic silica and are believed to be an important burial site for this biogenic phase.

Estuaries extend across the river–ocean boundary and are generally regions of high nutrient flux and rapid sediment burial (0.1–10 cm y^{-1}). They commonly exhibit extensive diatom production in surface waters, but may not account for substantial biogenic silica burial because of extensive dissolution in the water column. For example, on the Amazon shelf approximately 20% of the world's river water mixes with ocean water and silicate dynamics have been studied in detail. Although nutrient concentrations are highest in the low-salinity regions of the Amazon mixing zone, biological nutrient uptake is limited because light cannot penetrate more than a few centimeters into the water column as a result of the high turbidity in the river (primarily from natural weathering of the Andes Mountains). After the terrigenous particles have flocculated in the river–ocean mixing zone, light is able to penetrate the warm surface waters, leading to some of the highest biogenic silica production rates in the world. However, resuspension on the shelf, zooplankton grazing, and high water temperatures lead to fairly efficient recycling in the water column and nearly all of the dissolved silicate coming down the river makes it out to the open ocean. The Amazon shelf seabed does appear to exhibit clay-mineral formation (primarily through replacement of dissolving diatoms), but the burial fluxes are expected to be small relative to the offshore transport of silicate and biogenic silica.

Biogenic Silica in Marine Sediments

As mentioned above, the primary biota that construct siliceous skeletons are diatoms, radiolaria, silicoflagellates, and siliceous sponges. Diatoms are marine algae. These phytoplankton account for 20–40% of the primary production in the ocean and an even greater percentage of the export production from the photic zone. Diatom skeletons are the primary form of biogenic silica in deposits associated with coastal upwelling areas, high-latitude oceans (predominantly in the Pacific and the Southern Oceans), and the continental margins (**Figure 4**). In equatorial upwelling areas radiolarian skeletons commonly occur in marine sediments along with the diatom frustules. Radiolaria are zooplankton that live in the upper few hundred meters of the water column. Their skeletons are larger and more robust than many diatoms; consequently their preservation in marine sediments is greater than that of most diatoms. Silicoflagellates account for a very small fraction of the biogenic silica in marine sediments because most of them dissolve up in the water column or in surface sediments. They have been used in some continental margin sediments as a paleo-indicator of upwelling intensity. Siliceous sponge spicules can make up a significant fraction of the near-interface sediments in areas in which the sediment accumulation rate is low ($<5\times10^{-3}$ cm y^{-1}). For example, on the Ross Sea continental shelf, fine sediments accumulate in the basins, whereas the topographic highs (<400 m water depth) have minimal fine-grained material (because of strong currents and turbulence). As a result, mats of siliceous sponge spicules occur in high abundance on some of these banks.

To measure the biogenic silica content of marine sediments, hot (85°C) alkaline solutions are used to dissolve biogenic silica over a period of 5–6 hours. The silicate concentration in the leaching solution is measured colorimetrically on a spectrophotometer and related to the dry weight of the original sedimentary material. In many sediments, coexisting clay minerals also may yield silicate during this leaching process; however, this contribution to the leaching solution can be assessed by measuring the silicate concentration in the leaching solution hourly over the course of the dissolution. Most biogenic silica dissolves within 2 hours, whereas clay minerals release silicate at a fairly constant rate over the entire leaching period. Consequently, the contributions of biogenic silica and clay-mineral silica can be resolved using a graphical approach (see **Figure 5**).

Measuring Rates of Processes in the Marine Silica Cycle

There are several useful chemical tracers for assessing rates of silicate uptake, silica dissolution in the water column, and particle transport in the seabed. Most of these techniques are based on various isotopes of silicon, some of which are stable and some of which are radioactive. Most of the stable silicon occurring naturally in the ocean and crust is ^{28}Si (92.2%) with minor amounts of ^{29}Si (4.7%) and ^{30}Si (3.1%). By adding known quantities of dissolved ^{29}Si or ^{30}Si to surface ocean waters, the natural abundance ratios of Si can be altered, allowing resolution of existing biogenic silica from silica produced after spiking in incubation studies. Similarly, if the silicate content of ocean water is spiked with either dissolved ^{29}Si or ^{30}Si, then, as biogenic silica dissolves, the ratio of the silicon isotopes will change in proportion

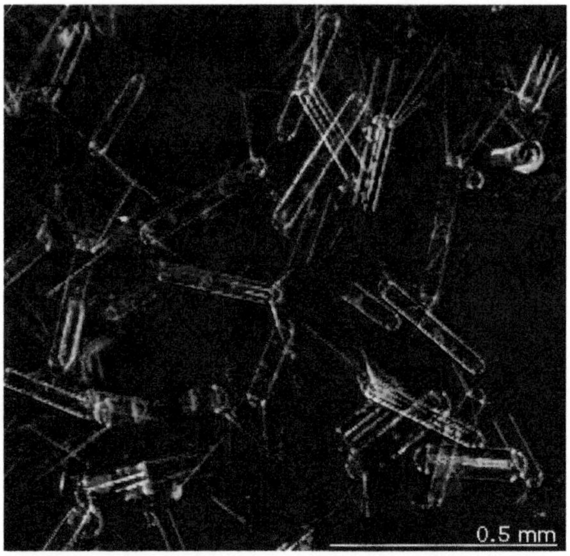

Figure 4 Micrograph of diatoms (genus *Corethron*) collected from an Antarctic plankton tow near Palmer Station.

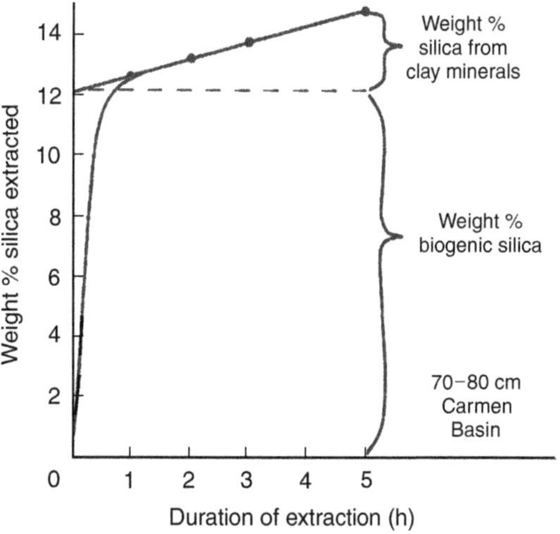

Figure 5 Graphical approach to resolving silicate originating via biogenic silica dissolution from that generated via clay-mineral dissolution during the alkaline leach technique used to quantify biogenic silica abundance. This sample was from the Gulf of California, Carmen Basin.

to the amount of silica dissolved (enabling characterization of dissolution rates). In addition, the measurement of natural silicon isotopes in sea water and in siliceous sediments has been suggested as a means of assessing the extent of silicate utilization in surface waters on timescales ranging from years to millennia. Addition of radioactive ^{32}Si (half-life 160 y) to incubation solutions recently has been used to simplify the measurement of silica production rates in surface ocean waters. In the past, ^{32}Si has been difficult to obtain, but recent advances in production and isolation protocols have made it possible to produce this radioisotope for oceanographic studies. Distributions of naturally occurring ^{32}Si in the water column and seabed can be used to determine deep-ocean upwelling rates as well as the intensity of eddy diffusion (or turbulence) in the deep ocean. This same radioactive isotope can be used to evaluate rates of bioturbation (biological particle mixing) in the seabed on timescales of hundreds of years.

See also

Carbon Cycle.

Further Reading

Craig H, Somayajulu BLK, and Turekian KK (2000) Paradox lost, silicon-32 and the global ocean silica cycle. *Earth and Planetary Science Letters* 175: 297–308.

DeMaster DJ (1981) The supply and removal of silica from the marine environment. *Geochimica et Cosmochimica Acta* 45: 1715–1732.

Dugdale RC, Wilkerson FP, and Minas HJ (1995) The role of a silicate pump in driving new production. *Deep-Sea Research* 42: 697–719.

Gnanadesikan A (1999) A global model of silicon cycling: sensitivity to eddy parameterization and dissolution. *Global Biogeochemical Cycles* 13: 199–220.

Heinze C, Maier-Reimer E, Winguth AME, and Archer D (1999) A global oceanic sediment model for long-term climate studies. *Global Biogeochemical Cycles* 13: 221–250.

Nelson DM, DeMaster DJ, Dunbar RB, and Smith WO Jr (1996) Cycling of organic carbon and biogenic silica in the southern Ocean: estimates of water-column and sedimentary fluxes on the Ross Sea continental shelf. *Journal of Geophysical Research* 101: 18519–18532.

Nelson DM, Treguer P, Brzezinski MA, Leynaert A, and Queguiner B (1995) Production and dissolution of biogenic silica in the ocean: revised global estimates, comparison with regional data and relationship to biogenic sedimentation. *Global Biogeochemical Cycles* 9: 359–372.

Nozaki Y, Zhang J, and Takeda A (1997) ^{210}Pb and ^{210}Po in the equatorial Pacific and the Bering Sea: the effects of biological productivity and boundary scavenging. *Deep-Sea Research II* 44: 2203–2220.

Ragueneau O, Treguer P, Leynaert A, *et al.* (2000) A review of the Si cycle in the modern ocean: recent progress and missing gaps in the application of biogenic opal as a paleoproductivity proxy. *Global and Planetary Change* 26: 317–365.

Spencer D (1972) GEOSECS II. The 1970 North Atlantic Station: Hydrographic features, oxygen, and nutrients. *Earth and Planetary Science Letters* 16: 91–102.

Treguer P, Nelson DM, Van Bennekom AJ, *et al.* (1995) The silica balance in the world ocean: A re-estimate. *Science* 268: 375–379.

NITROGEN CYCLE

D. M. Karl, University of Hawaii at Manoa, Honolulu, HI, USA
A. F. Michaels, University of Southern California, Los Angeles, CA, USA

Introduction

The continued production of organic matter in the sea requires the availability of the many building blocks of life, including essential major elements such as carbon (C), nitrogen (N), and phosphorus (P); essential minor elements such as iron, zinc, and cobalt; and, for many marine organisms, essential trace organic nutrients that they cannot manufacture themselves (e.g., amino acids and vitamins). These required nutrients have diverse structural and metabolic function and, by definition, marine organisms cannot survive in their absence.

The marine nitrogen cycle is part of the much larger and interconnected hydrosphere–lithosphere–atmosphere–biosphere nitrogen cycle of the Earth. Furthermore, the oceanic cycles of carbon, nitrogen, and phosphorus are inextricably linked together through the production and remineralization of organic matter, especially near surface ocean phytoplankton production. This coordinated web of major bioelements can be viewed as the nutrient 'super-cycle.'

The dominant form of nitrogen in the sea is dissolved gaseous dinitrogen (N_2) which accounts for more than 95% of the total nitrogen inventory. However, the relative stability of the triple bond of N_2 renders this form nearly inert. Although N_2 can serve as a biologically-available nitrogen source for specialized N_2 – fixing microorganisms, these organisms are relatively rare in most marine ecosystems. Consequently, chemically 'fixed' or 'reactive' nitrogen compounds such as nitrate (NO_3^-), nitrate (NO_2^-), ammonium (NH_4^+), and dissolved and particulate organic nitrogen (DON/PON) serve as the principal sources of nitrogen to sustain biological processes.

For more than a century, oceanographers have been concerned with the identification of growth-and production-rate limiting factors. This has stimulated investigations of the marine nitrogen cycle including both inventory determinations and pathways and controls of nitrogen transformations from one form to another. Contemporaneous ocean investigations have documented an inextricable link between nitrogen and phosphorus cycles, as well as the importance of trace inorganic nutrients. It now appears almost certain that nitrogen is only one of several key elements for life in the sea, neither more nor less important than the others. Although the basic features of the marine nitrogen cycle were established nearly 50 years ago, new pathways and novel microorganisms continue to be discovered. Consequently, our conceptual view of the nitrogen cycle is a flexible framework, always poised for readjustment.

Methods and Units

The analytical determinations of the various dissolved and particulate forms of nitrogen in the sea rely largely on methods that have been in routine use for several decades. Determinations of NO_3^-, NO_2^-, and NH_4^+ generally employ automated shipboard, colorimetric assays, although surface waters of open ocean ecosystems demand the use of modern high-sensitivity chemiluminescence and fluorometric detection systems. PON is measured by high-temperature combustion followed by chromatographic detection of the by-product (N_2), usually with a commercial C–N analyzer. Total dissolved nitrogen (TDN) determination employs sample oxidation, by chemical or photolytic means, followed by measurement of NO_3^-. DON is calculated as the difference between TDN and the measured dissolved, reactive inorganic forms of N (NO_3^-, NO_2^-, NH_4^+) present in the original sample. Gaseous forms of nitrogen, including N_2, nitrous oxide (N_2O), and nitric oxide (NO) are generally measured by gas chromatography.

Nitrogen exists naturally as two stable isotopes, ^{14}N (99.6% by atoms) and ^{15}N (0.4% by atoms). These isotopes can be used to study the marine nitrogen cycle by examination of natural variations in the $^{14}N/^{15}N$ ratio, or by the addition of specific tracers that are artificially enriched in ^{15}N.

Most studies of oceanic nitrogen inventories or transformations use either molar or mass units; conversion between the two is straightforward (1 mole N = 14 g N, keeping in mind that the molecular weight of N_2 gas is 28).

Components of the Marine Nitrogen Cycle

The systematic transformation of one form of nitrogen to another is referred to as the nitrogen cycle (**Figure 1**). In the sea, the nitrogen cycle revolves around the metabolic activities of selected microorganisms and it is reasonable to refer to it as the microbial nitrogen cycle because it depends on bacteria (**Table 1**). During most of these nitrogen transformations there is a gain or loss of electrons and, therefore, a change in the oxidation state of nitrogen from the most oxidized form, NO_3^- (+5), to the most reduced form, NH_4^+ (−3). Transformations in the nitrogen cycle are generally either energy-requiring (reductions) or energy-yielding (oxidations). The gaseous forms of nitrogen in the surface ocean can freely exchange with the atmosphere, so there is a constant flux of nitrogen between these two pools.

The natural, stepwise process for the regeneration of NO_3^- from PON can be reproduced in a simple 'decomposition experiment' in an enclosed bottle of sea water (**Figure 2**). During a 3-month incubation period, the nitrogen contained in particulate matter is first released as NH_4^+ (the process of ammonification), then transformed to NO_2^- (first step of nitrification), and finally, and quantitatively, to NO_3^- (the second step of nitrification). These transformations are almost exclusively a result of the metabolic

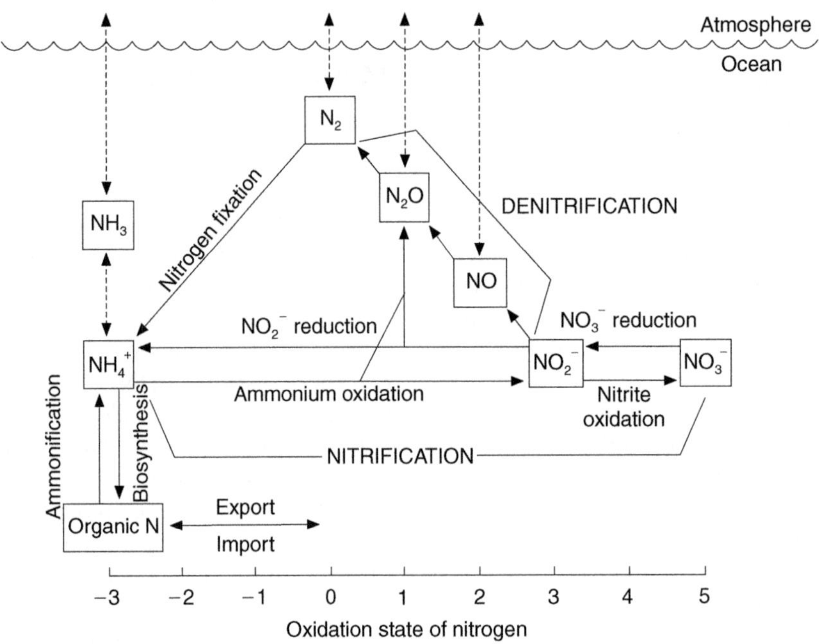

Figure 1 Schematic representation of the various transformations from one form of nitrogen to another that compose the marine nitrogen cycle. Shown at the bottom is the oxidation state of nitrogen for each of the components. Most transformations are microbiological and most involve nitrogen reduction or oxidation. (Adapted from Capone, ch. 14 of Rogers and Whitman (1991).)

Table 1 Marine nitrogen cycle

	Credits		
Process	*Bacteria*	*Phytoplankton*[a]	*Zooplankton/Fish*[b]
NH_4^+ production from DON/PON (ammonification)	+	+	+
$NH_4^+/NO_2^-/NO_3^-$/DON assimilation	+	+	−
PON ingestion	−	+	+
$NH_4^+ \rightarrow NO_2^-$ (nitrification, step 1)	+	−	−
$NO_2^- \rightarrow NO_3^-/N_2O$ (nitrification, step 2)	+	−	−
$NO_3^-/NO_2^- \rightarrow N_2/N_2O$ (denitrification)	+	+	−
$N_2 \rightarrow NH_4^+$/organic N (N_2 fixation)	+	−	−

[a] Phytoplankton – eukaryotic phytoplankton.
[b] Zooplankton – including protozoans and metazoans.
Abbreviations: NH_4^+, ammonium; NO_2^-, nitrite; NO_3^-, nitrate; N_2O, nitrous oxide; N_2, dinitrogen; DON, dissolved organic N; PON, particulate organic N; organic N, DON and PON.

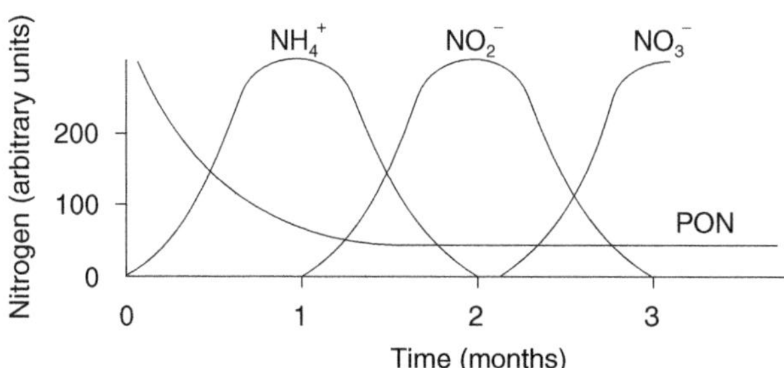

Figure 2 Stepwise decomposition of particulate organic nitrogen (PON) in arbitrary units versus time during a dark incubation. Nitrogen is transformed, first to NH_4^+ by bacterial ammonification and finally to NO_2^- and NO_3^- by the two-step process of bacterial nitrification. These same processes are responsible for the global ocean formation of NO_3^- in the deep sea. These data are the idealized results of pioneering nitrogen cycle investigators, T. von Brand and N. Rakestraw, who unraveled these processes more than 50 years ago.

activities of bacteria. This set of regeneration reactions is vital to the nitrogen cycle, and since most deep water nitrogen (excluding N_2) is in the form of NO_3^-, bacterial nitrification must be a very important process (see Nitrogen Distributions in the Sea, below).

Nitrogen Assimilation

Several forms of nitrogen can be directly transported across cell membranes and assimilated into new cellular materials as required for biosynthesis and growth. Most microorganisms readily transport NH_4^+, NO_2^-, NO_3^-, and selected DON compounds such as amino acids, urea, and nucleic acid bases. By comparison, the ability to utilize N_2 as a nitrogen source for biosynthesis is restricted to a very few species of specialized microbes. Many protozoans, including both photosynthetic and heterotrophic species, and all metazoans obtain nitrogen primarily by ingestion of PON.

Once inside the cell or organism, nitrogen is digested and, if necessary, reduced to NH_4^+. If oxidized compounds such as NO_3^- or NO_2^- are utilized, cellular energy must be invested to reduce these substrates to ammonium for incorporation into organic matter. The process of reduction of NO_3^- (or NO_2^-) for the purpose of cell growth is referred to as assimilatory nitrogen (NO_3^-/NO_2^-) reduction and most microorganisms, both bacteria and phytoplankton, possess this metabolic capability (**Table 1**). In theory, there should be a metabolic preference for NH_4^+ over either NO_3^- or NO_2^-, based strictly on energetic considerations. However, it should be emphasized that preferential utilization of NH_4^+ does not always occur. For example, two closely related and abundant planktonic cyanobacteria that coexist in tropical and subtropical marine habitats have devised alternate metabolic strategies: *Synechococcus* prefers NO_3^- and *Prochlorococcus* prefers NH_4^+. In fact, *Prochlorococcus* cannot reduce NO_3^- to NH_4^+, presumably because the critical enzyme systems are absent.

Nitrification

As nitrogen is oxidized from NH_4^+ through NO_2^- to NO_3^-, energy is released (**Figure 1**), a portion of which can be coupled to the reduction of carbon dioxide (CO_2) to organic matter (CH_2O) by nitrifying bacteria. These specialized bacteria, one group capable only of the oxidation of NH_4^+ to NO_2^- and the second capable only of the oxidation of NO_2^- to NO_3^-, are termed 'chemolithoautotrophic' because they can fix CO_2 in the dark at the expense of chemical energy. Other related chemolithoautotrophs can oxidize reduced sulfur compounds, and this pathway of organic matter production has been hypothesized as the basis for life at deep-sea hydrothermal vents.

It is essential to emphasize an important ecological aspect of NH_4^+/NO_2^- chemolithoautotrophy. First, the oxidation of NH_4^+ to NO_2^- and of NO_2^- to NO_3^- usually requires oxygen and these processes are ultimately coupled to the photosynthetic production of oxygen in the surface water. Second, the continued formation of reduced nitrogen, in the form of NH_4^+ or organic nitrogen, is also dependent, ultimately, on photosynthesis. In this regard the CO_2 reduced via this 'autotrophic' pathway must be considered secondary, not primary, production from an ecological energetics perspective.

Marine nitrifying bacteria, especially the NO_2^- oxidizers are ubiquitous in the world ocean and key to the regeneration of NO_3^-, which dominates waters below the well-illuminated, euphotic zone. However they are never very abundant and, at least for those species in culture, grow very slowly. Certain heterotrophic bacteria can also oxidize NH_4^+ to both NO_2^- and NO_3^- during metabolism of preformed organic matter. However, very little is known about the potential for 'heterotrophic nitrification' in the sea.

Denitrification

Under conditions of reduced oxygen (O_2) availability, selected species of marine bacteria can use NO_3^- as a terminal acceptor for electrons during metabolism, a process termed NO_3^- respiration or dissimilatory NO_3^- reduction. This process allows microorganisms to utilize organic matter in low-O_2

or anoxic habitats with only a slight loss of efficiency relative to O_2-based metabolism. A majority of marine bacteria have the ability for NO_3^- respiration under the appropriate environmental conditions (**Table 1**). Potential by-products of NO_3^- respiration are NO_2^-, N_2, and N_2O; if a gas is formed (N_2/N_2O) then the process is termed denitrification because the net effect is to remove bioavailable nitrogen from the local environment. The total rate of denitrification is generally limited by the availability of NO_3^-, and a continued supply of NO_3^- via nitrification is dependent upon the availability of NH_4^+ and free O_2. Consequently denitrification typically occurs at boundaries between low-O_2 and anoxic conditions where the supply of NH_4^+ from the anoxic zone sustains a high rate of NO_3^- production via nitrification to fuel- sustained NO_3^- respiration and denitrification. Recently a new group of microorganisms has been isolated that are capable of simultaneously using both O_2 and NO_3^-/NO_2^- as terminal electron acceptors. This process is termed 'aerobic denitrification.' Likewise, there are exceptional microorganisms that are able to carry out anaerobic nitrification (oxidation of NH_4^+ in the absence of O_2). It appears difficult to establish any hard-and-fast rules regarding marine nitrogen cycle processes.

N_2 Fixation

The ability to use N_2 as a growth substrate is restricted to a relatively small group of microorganisms. Open ocean ecosystems that are chronically depleted in fixed nitrogen would appear to be ideal habitats for the proliferation of N_2-fixing microorganisms. However, the enzyme that is required for reduction of N_2 to NH_4^+ is also inhibited by O_2, so specialized structural, molecular, and behavioral adaptations have evolved to promote oceanic N_2 fixation.

Fixation of molecular nitrogen in the open ocean may also be limited by the availability of iron, which is an essential cofactor for the N_2 reduction enzyme system. Changes in iron loading are caused by climate variations, in particular the areal extent of global deserts, by the intensity of atmospheric circulation, and more recently by changes in land use practices. Conversion of deserts into irrigated croplands may cause a change in the pattern and intensity of dust production and, therefore, of iron transport to the sea. Humanity is also altering the global nitrogen cycle by enhancing the fixation of N_2 by the manufacturing of fertilizer. At the present time, the industrial fixation of N_2 is approximately equivalent to the pre-industrial, natural N_2 fixation rate. Eventually some of this artificially fixed N_2 will make its way to the sea, and this may lead to a perturbation in the natural nitrogen cycle.

On a global scale and over relatively long timescales, the total rate of N_2 fixation is more or less in balance with total denitrification, so that the nitrogen cycle is mass-balanced. However, significant net deficits or excesses can be observed locally or even on ocean basin space scales and on decade to century timescales. These nitrogen imbalances may impact the global carbon and phosphorus cycles as well, including the net balance of CO_2 between the ocean and the atmosphere.

Nitrogen Distributions in the Sea

Required growth nutrients, like nitrogen, typically have uneven distributions in the open sea, with deficits in areas where net organic matter is produced and exported, and excesses in areas where organic matter is decomposed. For example, surface ocean NO_3^- distributions in the Pacific basin reveal a coherent pattern with excess NO_3^- in high latitudes, especially in the Southern Ocean (south of 60° S), and along the Equator (especially east of the dateline), and generally depleted NO_3^- concentrations in the middle latitudes of both hemispheres (**Figure 3**). These distributions are a result of the balance between NO_3^- supply mostly by ocean mixing and NO_3^- demand or net photosynthesis. The very large NO_3^- inventory in the surface waters of the Southern Ocean implies that factors other than fixed nitrogen availability control photosynthesis in these regions. It has been hypothesized that the availability of iron is key in this and perhaps other regions of the open ocean. The much smaller but very distinctive band of elevated NO_3^- along the Equator is the result of upwelling of NO_3^--enriched waters from depth to the surface. This process has a large seasonal and, especially, interannual variability, and it is almost absent during El Niño conditions.

Excluding these high-latitude and equatorial regions, the remainder of the surface waters of the North and South Pacific Oceans from about 40°N to 40°S are relatively depleted in NO_3^-. In fact surface (0–50 m) NO_3^- concentrations in the North Pacific subtropical gyre near Hawaii are typically below 0.01 $\mu mol\,l^{-1}$ (**Figure 4**). Within the upper 200 m, the major pools of fixed nitrogen (e.g., NO_3^-, DON, and PON) have different depth distributions. In the sunlit surface zone, NO_3^- is removed to sustain organic matter production and export. Beneath 100 m, there is a steep concentration versus depth gradient (referred to as the nutricline), which reaches a

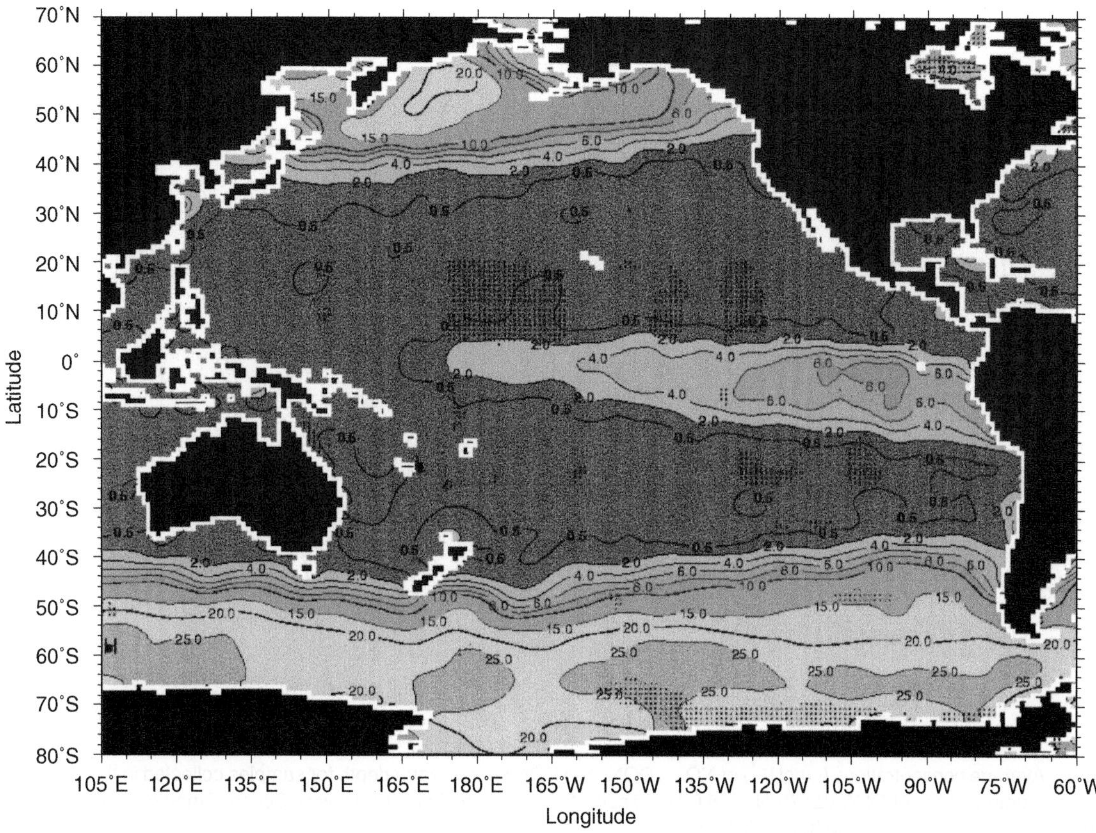

Figure 3 Mean annual NO_3^- concentration ($\mu mol\, l^{-1}$) at the sea surface for samples collected in the Pacific Ocean basin and Pacific sector of the Southern Ocean. (From Conkright *et al.* 1998.)

maximum of about 40–45 it $\mu mol\, l^{-1}$ at about 1000 m in the North Pacific Ocean. PON concentration is greatest in the near-surface waters where the production of organic matter via photosynthesis is highest (**Figure 4**). PON includes both living (biomass) and nonliving (detrital) components; usually biomass nitrogen is less than 50% of the total PON in near-surface waters, and less than 10% beneath the euphotic zone (>150 m). DON concentration is also highest in the euphotic zone (~5–6 $\mu mol\, l^{-1}$) and decreases systematically with depth to a minimum of 2–3 $\mu mol\, l^{-1}$ at 800–1000 m. The main sources for DON in the surface ocean are the combined processes of excretion, grazing, death, and cell lysis. Consequently, DON is a complex mixture of cell-derived biochemicals; at present, less than 20% of the total DON has been chemically characterized. Dissolved N_2 (not shown) is always high (~800 $\mu mol\, l^{-1}$) and increases systematically with depth. The major controls of N_2 concentration are temperature and salinity, which together determine gas solubility. Marine life has little impact on N_2 distributions in the open sea even though some microorganisms can utilize N_2 as a growth substrate and others can produce N_2 as a metabolic by-product. These transformations are simply too small to significantly impact the large N_2 inventories in most regions of the world ocean.

Another important feature of the global distribution of NO_3^- is the regional variability in the deep water inventory (**Figure 5**). Deep ocean circulation can be viewed as a conveyor-belt-like flow, with the youngest waters in the North Atlantic and the oldest in the North Pacific. The transit time is in excess of 1000 y, during which time NO_3^- is continuously regenerated from exported particulate and dissolved organic matter via coupled ammonification and nitrification (**Figure 2**). Consequently, the deep Pacific Ocean has nearly twice as much NO_3^- as comparable depths in the North Atlantic (**Figure 5**).

Nitrous Oxide Production

Nitrous oxide (N_2O) is a potent greenhouse gas that has also been implicated in stratospheric ozone depletion. The atmospheric inventory of N_2O is presently increasing, so there is a renewed interest in the marine ecosystem as a potential source of N_2O. Nitrous oxide is a trace gas in sea water, with typical concentrations ranging from 5 to 50 $nmol\, l^{-1}$.

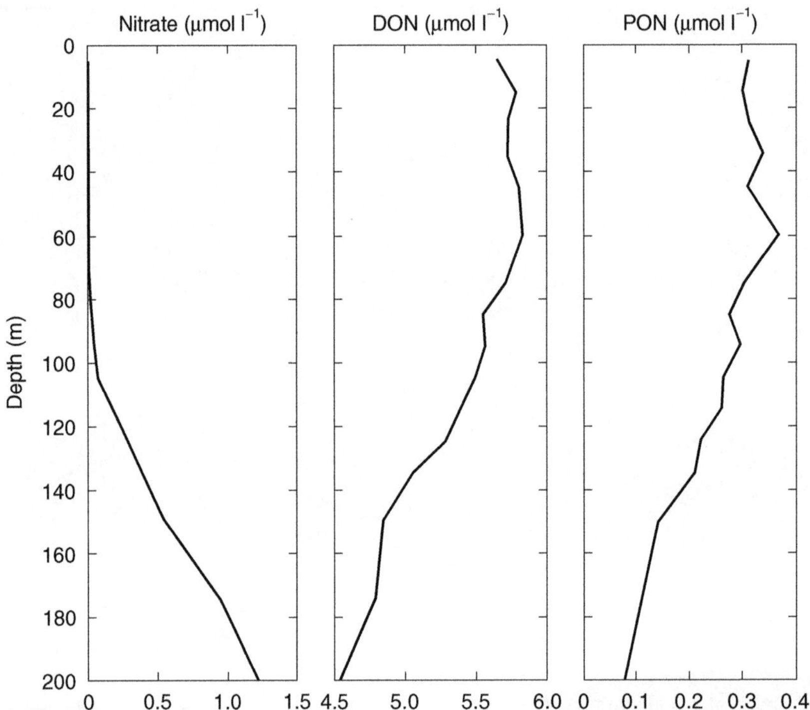

Figure 4 Average concentrations (μmol l^{-1}) of NO_3^-, DON, and PON versus water depth for samples collected in the upper 200 m of the water column at Sta. ALOHA (22.75°N, 158.0°W). These field data are from the Hawaii Ocean Time-series program and are available at http://hahana.soest.edu/hot_jgofs.html).

Concentrations of N_2O in oceanic surface waters are generally in slight excess of air saturation, implying both a local source and a sustained ocean-to-atmosphere flux. Typically there is a mid-water (500–1000 m) peak in N_2O concentration that coincides with the dissolved oxygen minimum. At these intermediate water depths, N_2O can exceed 300% saturation relative to atmospheric equilibrium. The two most probable sources of N_2O in the ocean are bacterial nitrification and bacterial denitrification, although to date it has been difficult to quantify the relative contribution of each pathway for a given habitat. Isotopic measurements of nitrogen and oxygen could prove invaluable in this regard. Because the various nitrogen cycle reactions are interconnected, changes in the rate of any one process will likely have an impact on the others. For example, selection for N_2-fixing organisms as a consequence of dust deposition or deliberate iron fertilization would increase the local NH_4^+ inventory and lead to accelerated rates of nitrification and, hence, enhanced N_2O production in the surface ocean and flux to the atmosphere.

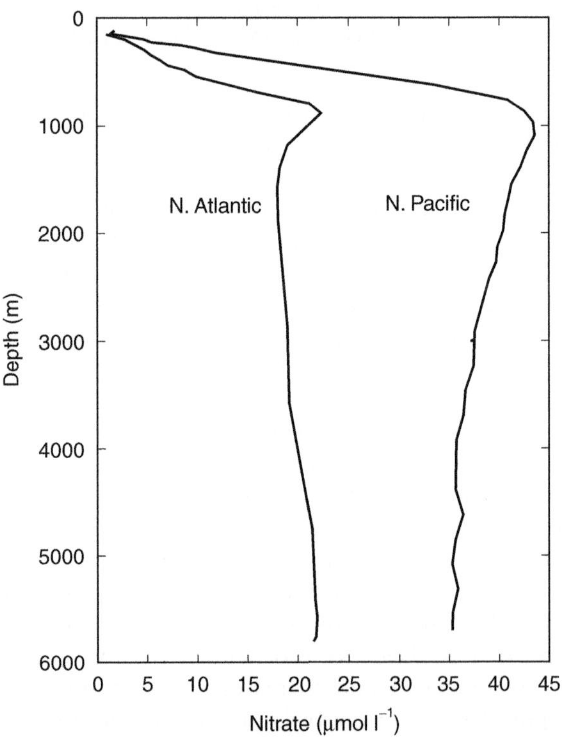

Figure 5 Nitrate concentrations (μmol l^{-1}) versus water depth at two contrasting stations located in the North Atlantic (31.8°N, 50.9°W) and North Pacific (30°N, 160.3°W) Oceans. These data were collected in the 1970s during the worldwide GEOSECS expedition, stations #119 and #212, respectively.

Primary Nitrite Maximum

An interesting, almost cosmopolitan feature of the world ocean is the existence of a primary NO_2^-

maximum (PNM) near the base of the euphotic zone ($\sim$100–150 m; **Figure 6**). Nitrite is a key intermediate between NO_3^- and NH_4^+, so there are several potential pathways, both oxidative and reductive, that might lead to its accumulation in sea water. First, phototrophic organisms growing on NO_3^- may partially reduce the substrate to NO_2^- as the first, and least energy-consuming, step in the assimilatory NO_3^- reduction pathway. However, the next step, reduction of NO_2^- to NH_4^+, requires a substantial amount of energy, so when energy is scarce (e.g., light limitation) NO_2^- accumulates inside the cells. Because NO_2^- is the salt of a weak acid, nitrous acid (HNO_2) forms in the slightly acidic intracellular environment, diffuses out of the cell and ionizes to form NO_2^- in the alkaline sea water. This $NO_3^- \rightarrow NO_2^-$ phytoplankton pump, under the control of light intensity, could provide a source of NO_2^- necessary to create and maintain the PNM. Alternatively, local regeneration of dissolved and particulate organic matter could produce NH_4^+ (via ammonification) that is partially oxidized in place to produce a relative excess of NO_2^- (the first step of nitrification). Kinetic controls on this process would be rates of NH_4^+ production and NO_2^- oxidation to NO_3^- (the second and final step in nitrification). Sunlight, even at very low levels, appears to disrupt the normal coupling between NO_2^- production and NO_2^- oxidation, in favor of NO_2^- accumulation. Finally, it is possible, though perhaps less likely, that NO_3^- respiration (terminating at NO_2^-), followed by excretion of NO_2^- (into the surrounding sea water might also contribute to the accumulation of NO_2^-) near the base of the euphotic zone. Because the global ocean at the depth of the PNM is characteristically well-oxygenated, one would need to invoke microenvironments like animal guts or large particles as the habitats for this nitrogen cycle pathway. The use of ^{15}N-labeled substrates, selective metabolic inhibitors, and other experimental manipulations provides an opportunity for direct assessment of the role of each of these potential processes. In all likelihood, more than one of these processes contributes to the observed PNM. Whatever the cause, light appears to be an important determinant that might explain the relative position, with regard to depth, of this global feature.

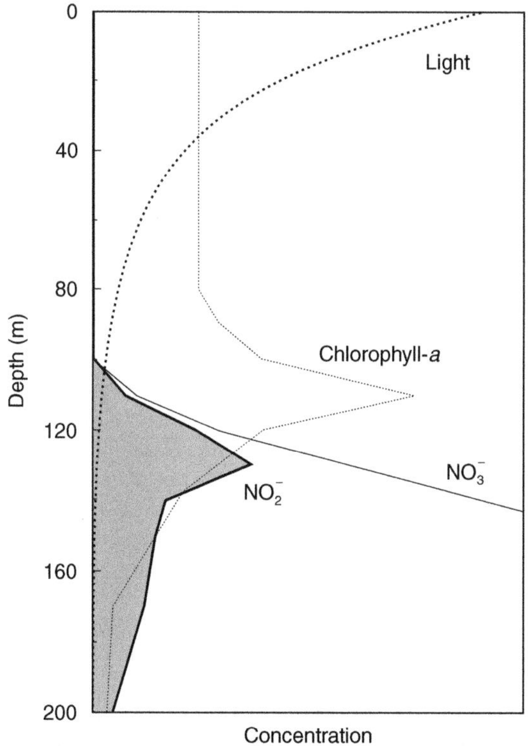

Figure 6 Schematic representation of the depth distributions of sunlight, chlorophyll-*a*, NO_2^-, and NO_3^- for a representative station in the subtropical North Pacific Ocean showing the relationship of the primary NO_2^- maximum (PNM) zone (shaded) to the other environmental variables. (Modified from J. E. Dore, Microbial nitrification in the marine euphotic zone, Ph.D. Dissertation, University of Hawaii, redrawn with permission of the author.)

Nitrogen Cycle and Ocean Productivity

Because nitrogen transformations include both the formation and decomposition of organic matter, much of the nitrogen used in photosynthesis is locally recycled back to NH_4^+ or NO_3^- to support another pass through the cycle. The net removal of nitrogen in particulate, dissolved, or gaseous form can cause the cycle to slow down or even terminate unless new nitrogen is imported from an external source. A unifying concept in the study of nutrient dynamics in the sea is the 'new' versus 'regenerated' nitrogen dichotomy (**Figure 7**). New nitrogen is imported from surrounding regions (e.g., NO_3^- injection from below) or locally created (e.g., NH_4^+/organic N from N_2 fixation). Regenerated nitrogen is locally recycled (e.g., NH_4^+ from ammonification, NO_2^-/ NO_3^- from nitrification, or DON from grazing or cell lysis). Under steady-state conditions, the amount of new nitrogen entering an ecosystem will determine the total amount that can be exported without the system running down.

In shallow, coastal regions runoff from land or movement upward from the sediments are potentially major sources of NH_4^+, NO_3^- and DON for water column processes. In certain regions, atmospheric deposition (both wet and dry) may also supply bioavailable nitrogen to the system. However, in

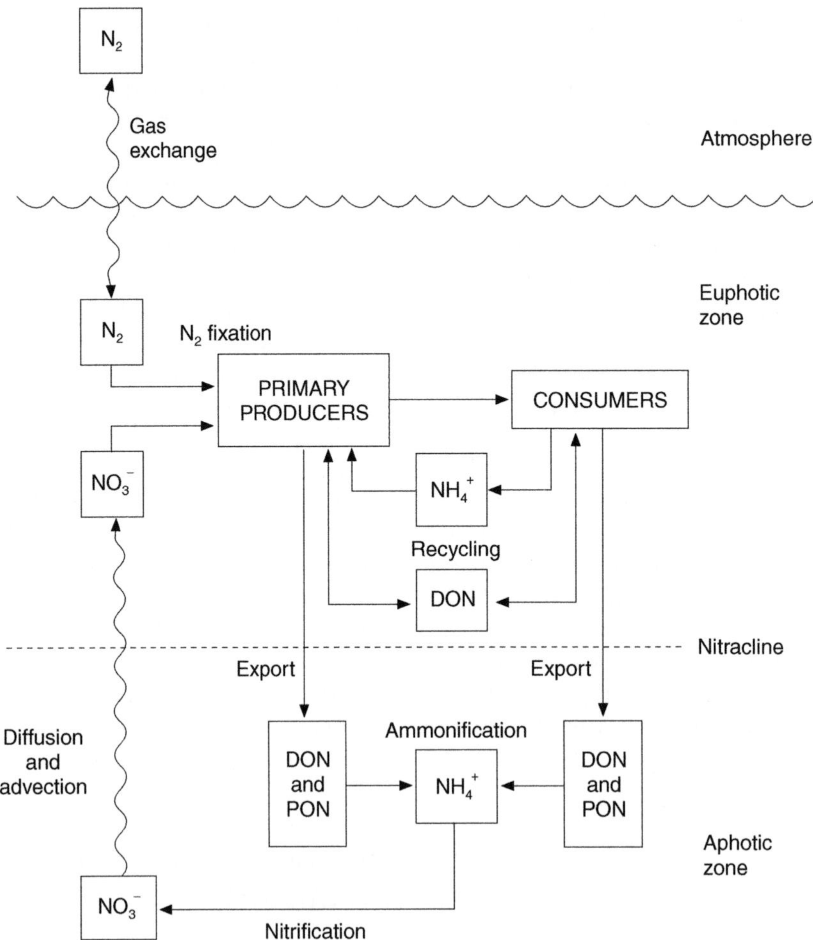

Figure 7 Schematic representation of the major pools and transformations/fluxes of nitrogen in a typical open ocean ecosystem. New sources of bioavailable N (NO_3^- and N_2 in this presentation) continuously resupply nitrogen that is lost via DON and PON export. These interactions and ocean processes form the conceptual framework for the 'new' versus 'regenerated' paradigm of nitrogen dynamics in the sea that was originally proposed by R. Dugdale and J. Goering.

most open ocean environments, new sources of nitrogen required to balance the net losses from the euphotic zone are restricted to upward diffusion or mixing of NO_3^- from deep water and to local fixation of N_2 gas. In a balanced steady state, the importation rate of new sources of bioavailable nitrogen will constrain the export of nitrogen (including fisheries production and harvesting). If all other required nutrients are available, export-rich ecosystems are those characterized by high bioavailable nitrogen loading such as coastal and open ocean upwelling regions. These are also the major regions of fish production in the sea.

See also

Atmospheric Input of Pollutants. Nitrogen Isotopes in the Ocean. Phosphorus Cycle.

Further Reading

Carpenter EJ and Capone DG (eds.) (1983) *Nitrogen in the Marine Environment*. New York: Academic Press.

Conkright ME, O'Brien TD, Levitus S, *et al.* (1998) *NOAA Atlas NESDIS 37, WORLD OCEAN ATLAS 1998*, vol. II: Nutrients and Chlorophyll of the Pacific Ocean. Washington, DC: US Department of Commerce.

Harvey HW (1966) *The Chemistry and Fertility of Sea Waters*. London: Cambridge University Press.

Kirchman DL (ed.) (2000) *Microbial Ecology of the Oceans*. New York: Wiley-Liss.

Rogers JE and Whitman WB (eds.) (1991) *Microbial Production and Consumption of Greenhouse Gases: Methane, Nitrogen Oxides, and Halomethanes*. Washington, DC: American Society for Microbiology.

Schlesinger WH (1997) *Biogeochemistry: An Analysis of Global Change*. San Diego: Academic Press.

Wada E and Hattori A (1991) *Nitrogen in the Sea: Forms, Abundances, and Rate Processes*. Boca Raton, FL: CRC Press.

NITROGEN ISOTOPES IN THE OCEAN

D. M. Sigman and K. L. Karsh, Princeton University, Princeton, NJ, USA
K. L. Casciotti, Woods Hole Oceanographic Institution, Woods Hole, MA, USA

Introduction

Nitrogen has two stable isotopes, ^{14}N and ^{15}N (atomic masses of 14 and 15, respectively). ^{14}N is the more abundant of the two, comprising 99.63% of the nitrogen found in nature. Physical, chemical, and biological processes discriminate between the two isotopes, leading to subtle but measurable differences in the ratio of ^{15}N to ^{14}N among different forms of nitrogen found in the marine environment.

Nitrogen is a central component of marine biomass and one of the major nutrients required by all phytoplankton. In this sense, biologically available (or 'fixed', i.e., non-N_2) N is representative of the fundamental patterns of biogeochemical cycling in the ocean. However, N differs from other nutrients in that its oceanic sources and sinks are dominantly internal and biological, with marine N_2 fixation supplying much of the fixed N in the ocean and marine denitrification removing it. The N isotopes provide a means of studying both the input/output budget of oceanic fixed N and its cycling within the ocean. In this overview, we outline the isotope systematics of N cycle processes and their impacts on the isotopic composition of the major N reservoirs in the ocean. This information provides a starting point for considering the wide range of questions in ocean sciences to which the N isotopes can be applied.

Terms and Units

Mass spectrometry can measure precisely the ratio of the N isotopes relative to a N reference containing a constant isotopic ratio. The universal reference for N isotopes is atmospheric N_2, with an $^{15}N/^{14}N$ ratio of $0.367\,65\% \pm 0.000\,81\%$. Natural samples exhibit small deviations from the standard ratio, which are expressed in δ notation (in units of per mil, ‰):

$$\delta^{15}N(‰) = \left(\frac{(^{15}N/^{14}N)_{sample}}{(^{15}N/^{14}N)_{standard}} - 1\right) \times 1000 \quad [1]$$

In this notation, the $\delta^{15}N$ of atmospheric N_2 is 0‰.

Special terms are also used to characterize the amplitude of isotopic fractionation caused by a given process. Isotope fractionation results from both equilibrium processes ('equilibrium fractionation') and unidirectional reactions ('kinetic fractionation'). Nitrogen isotope variations in the ocean are typically dominated by kinetic fractionation associated with the conversions of N from one form to another. The kinetic isotope effect, ε, of a given reaction is defined by the ratio of rates with which the two N isotopes are converted from reactant to product:

$$\varepsilon(‰) = (^{14}k/^{15}k - 1) \times 1000 \quad [2]$$

where ^{14}k and ^{15}k are the rate coefficients of the reaction for ^{14}N- and ^{15}N-containing reactant, respectively. For $\varepsilon \ll 1000‰$, ε is approximated by the difference in $\delta^{15}N$ between the reactant and its instantaneous product. That is, if a reaction has an ε of 5‰, then the $\delta^{15}N$ of the product N generated at any given time will be ~5‰ lower than the $\delta^{15}N$ of the reactant N at that time.

Measurements

The isotopic analysis of N relies on the generation of a stable gas as the analyte for isotope ratio mass spectrometry. Online combustion to N_2 is currently the standard method for the preparation of a N sample for isotopic analysis. With 'off-the-shelf' technology, a typical sample size requirement is 1–2 μmol N per analysis. Gas chromatography followed by combustion to N_2 is improving as a technique for specific organic compounds, amino acids in particular, although the polarity of many N compounds remains a challenge. Liquid chromatography is also being explored. There are standard methods of collection for most bulk forms of particulate N (PN) in the ocean. Shallow and deep samples of suspended PN are filtered onto glass fiber filters. Sinking PN is collected by sediment traps. Zooplankton can be picked from filtered samples or net tows, and particulates can be separated into size classes. In the case of dissolved forms of N, the species of interest must be converted selectively to a gas or other extractable form for collection. Since the 1970s, the $\delta^{15}N$ values of marine nitrate (NO_3^-), nitrite (NO_2^-), and ammonium (NH_4^+) have been analyzed by conversion to ammonia gas and collection of the cationic ammonium form for subsequent conversion to N_2 (often referred to as the ammonia 'distillation' and 'diffusion' methods). Recently, more sensitive isotope analysis methods (requiring only 5–10 nmol of N per analysis) have been developed for nitrate and nitrite in which these species are converted to nitrous oxide (N_2O), followed by isotopic analysis of this gas (the 'bacterial' or 'denitrifier' method and the 'chemical' or 'azide' method). The N_2O produced by these methods (or naturally occuring N_2O) is analyzed by a purge and trap system, followed by gas chromatography and isotope ratio mass spectrometry. The N_2O-based methods also allow for oxygen isotope analysis of nitrate and nitrite, a measurement not previously possible in seawater. In addition, they provide a cornerstone for isotopic analysis of other dissolved forms of N, such as dissolved organic N (DON) and ammonium (NH_4^+), which can be converted to nitrate and/or nitrite. With respect to dissolved gases, methods of collection and isotopic analysis have been developed for N_2 and N_2O, with recent progress on isotopomer analysis of N_2O (i.e., distinguishing $^{15}N^{14}N^{16}O$ from $^{14}N^{15}N^{16}O$).

Models

Two simple models, the 'Rayleigh' model and the 'steady-state' model, are frequently used to interpret N isotope data from the ocean. In both of these models, the degree of consumption of the reactant N pool (f) is a key parameter, and the $\delta^{15}N$ of the initial reactant N pool ($\delta^{15}N_{initial}$) and kinetic isotope effect (ε) are the two central isotopic parameters. If a transformation proceeds with a constant isotope effect and if the reactant and product N pools are neither replenished nor lost from the system during the progress of the transformation, then the process can be described in terms of Rayleigh fractionation kinetics, which define the isotopic variation of the reactant N pool ($\delta^{15}N_{reactant}$; eqn [3]), the instantaneously generated product N ($\delta^{15}N_{instantaneous}$; eqn [4]), and the integrated product N pool ($\delta^{15}N_{integrated}$; eqn [5]) as a given reservoir of reactant N is consumed (**Figure 1**):

$$\delta^{15}N_{reactant} = \delta^{15}N_{initial} - \varepsilon\{\ln(f)\} \qquad [3]$$

$$\delta^{15}N_{instantaneous} = \delta^{15}N_{reactant} - \varepsilon \qquad [4]$$

$$\delta^{15}N_{integrated} = \delta^{15}N_{initial} + \varepsilon\{f/(1-f)\}\ln(f) \qquad [5]$$

where f is the fraction of the reactant remaining, $\delta^{15}N_{initial}$ is the $\delta^{15}N$ of the initial reactant N pool, and ε is the kinetic isotope effect of the transformation. These equations are simplified, approximate forms of the full expressions. They are typically adequate, but their error is greater for higher consumption (lower f) and higher ε. The Rayleigh model is often used to describe events in the ocean, such as the uptake of nitrate by phytoplankton during a bloom in a stratified surface layer.

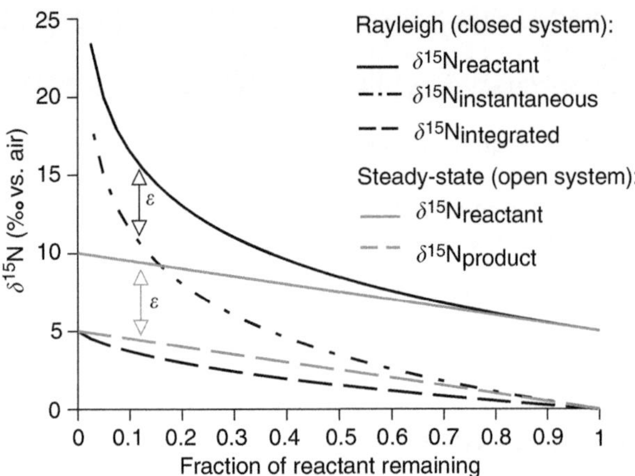

Figure 1 The $\delta^{15}N$ of reactant and product N pools of a single unidirectional reaction as a function of the fraction of the initial reactant supply that is left unconsumed, for two different models of reactant supply and consumption, following the approximate equations given in the text. The Rayleigh model (black lines) applies when a closed pool of reactant N is consumed. The steady-state model (gray lines) applies when reactant N is supplied continuously. The same isotopic parameters, an isotope effect (ε) of 5‰ and a $\delta^{15}N$ of 5‰ for the initial reactant supply, are used for both the Rayleigh and steady-state models. ε is approximately equal to the isotopic difference between reactant N and its product (the instantaneous product in the case of the Rayleigh model).

The end-member alternative to the Rayleigh model is the steady-state model, in which reactant N is continuously supplied and partially consumed, with residual reactant N being exported at a steady-state rate such that the gross supply of reactant N equals the sum of the product N and the residual reactant N exported. In this case, the following approximate expessions apply to the reactant N pool ($\delta^{15}N_{reactant}$; eqn [6]) and the product N pool ($\delta^{15}N_{product}$; eqn [7]) (**Figure 1**):

$$\delta^{15}N_{reactant} = \delta^{15}N_{initial} + \varepsilon(1-f) \qquad [6]$$

$$\delta^{15}N_{product} = \delta^{15}N_{initial} - \varepsilon(f) \qquad [7]$$

The steady-state model and modified forms of it, such as the more spatially complex 'reaction–diffusion' model, are used to quantify uptake processes where supply and uptake are simultaneous and relatively time-invariant, such as in the consumption of nitrate by denitrification in the ocean interior or in sediments.

Processes

Inputs

N_2 fixation is the major input of fixed N to the ocean (**Figure 2**). N_2 fixation is carried out by N_2 fixers, cyanobacteria and other microorganisms able to convert N_2 into biomass N. Subsequent remineralization of this biomass supplies new N to the dissolved fixed N pools in the surface and subsurface ocean. Field collections of *Trichodesmium* colonies, the best-known genus of open ocean N_2 fixer, have yielded a $\delta^{15}N$ of *c.* −2‰ to +0.5‰. Taking into account the $\delta^{15}N$ of dissolved N_2 (0.6‰ in the surface mixed layer), this range in $\delta^{15}N$ is consistent with, but perhaps less variable than, the range in isotope effects estimated from culture studies of marine and terrestrial N_2 fixers, ~0–4‰ (**Table 1**). An average $\delta^{15}N$ of −1‰ has been suggested for the fixed N input to the ocean from N_2 fixation.

Other inputs of fixed N to the marine environment include terrestrial runoff and atmospheric precipitation, the N isotopic compositions of which are poorly constrained (**Figure 2**). Dissolved and particulate $\delta^{15}N$ in pristine river systems ranges mostly from 0‰ to 5‰. However, biological processes along the flow path and in estuaries (in particular, by denitrification; see below) can alter the $\delta^{15}N$ of the final inputs from terrestrial runoff in complex ways. Anthropogenic inputs often increase the $\delta^{15}N$ of a system because they encourage denitrification. In atmospheric inputs, a wide range in the $\delta^{15}N$ of inorganic (*c.* −16‰ to 10‰) and organic (*c.* −8‰ to 1‰) N has been observed, with increasing evidence that at least some of this variability can provide insight into sources and processes. In the face of large uncertainties, a preindustrial mean $\delta^{15}N$ of 4‰ for terrestrial runoff and −2‰ for atmospheric precipitation has been suggested by some workers.

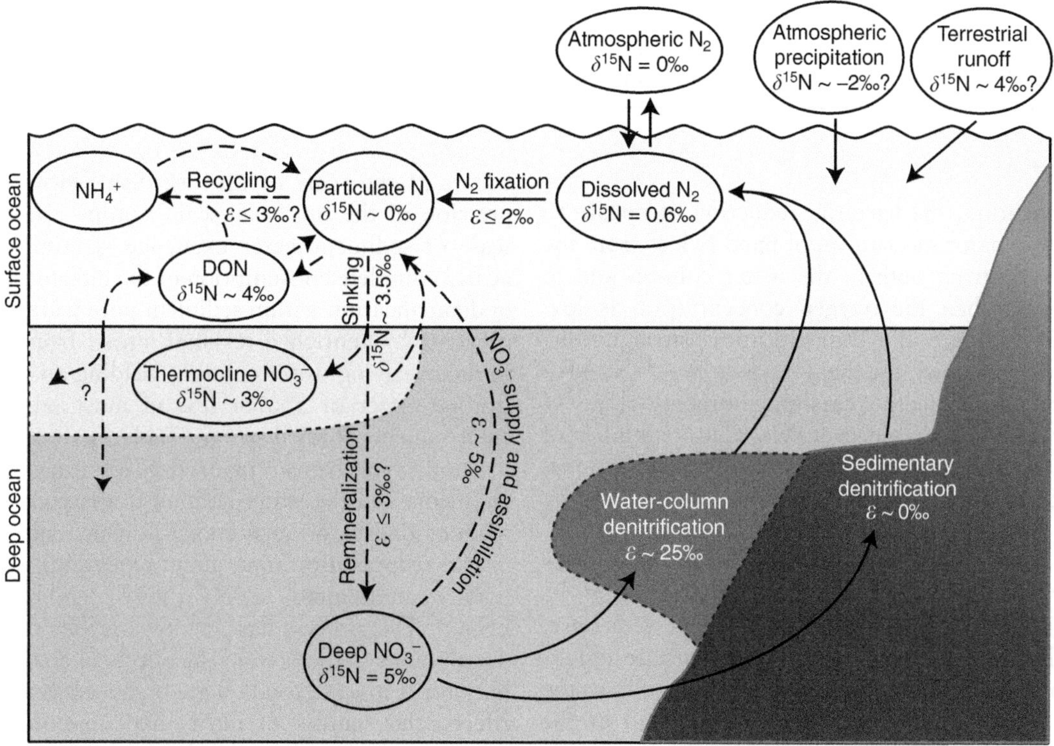

Figure 2 Processes affecting the distribution of nitrogen isotopes in the sea. The inputs and outputs (solid arrows) control the ocean's inventory of fixed N, the majority of which is in the form of nitrate (NO_3^-). Inputs are marine N_2 fixation in the surface ocean, terrestrial runoff, and atmospheric precipitation. Outputs indicated are sedimentary and water column denitrification. As discussed in the text, water-column denitrification in low-oxygen regions of the ocean interior leads to elevated $\delta^{15}N$ of nitrate (dark area). Remineralization of newly fixed N can explain the low $\delta^{15}N$ of nitrate in the shallow subsurface, or thermocline, of low-latitude regions (light area). Internal cycling is represented with dashed arrows. Nitrate supplied from the deep ocean and thermocline is assimilated in the surface ocean. PN is recycled in the surface ocean, degraded to ammonium (NH_4^+) that is subsequently assimilated; the role of DON in this recycling is not yet clear. Sinking and remineralization (via degradation and nitrification, see text) returns N from the particulate pool to nitrate. For simplicity, possible nitrification in the surface ocean is not shown. The isotope fractionation associated with nitrification in the ocean interior is also excluded because this process generally goes to completion (see text). The $\delta^{15}N$ of particulate suspended and sinking N, DON, and thermocline nitrate is taken from the subtropical North Atlantic. Question marks indicate the greatest uncertainties, due to variation in the available data and/or insufficient data.

Table 1 Representative estimates of isotope effects for N cycles processes

Process	*Isotope effect* (ε)	*Details*
N_2 fixation ($N_2 \rightarrow PN$)	1.8–3.0‰	*Trichodesmium spp.*
	−0.2‰	Western Tropical North Pacific
	−0.4–0.2‰	Western Tropical Atlantic
Denitrification ($NO_3^- \rightarrow N_2$)	5–30‰	*Pseudomonas stutzeri* (marine)
	18–28‰	*Paracoccus denitrificans* (terrestrial)
Water column	25 − 30‰	Eastern Tropical North Pacific and Arabian Sea
Sedimentary	≤3‰	Coastal sites, eastern N. Pacific margin, and deep Bering Sea
NO_3^- assimilation ($NO_3^- \rightarrow PN$)	5–17‰	*Thalassiosira weissflogii*
	5–6‰	Coastal Antarctic
	5–9‰	Open Antarctic and Subantarctic
	5‰	Subarctic Pacific
	5‰	Equatorial Pacific
NH_4^+ assimilation ($NH_4^+ \rightarrow PN$)	20‰	*Thalassiosira pseudonana*
	8–27‰	*Skeletonema costatum*
	4–27‰	Bacterial assemblage
	6.5–8‰	Chesapeake Bay
	9‰	Delaware Estuary
	18.5‰	Scheldt Esturary
Nitrification		
Ammonia oxidation ($NH_3 \rightarrow NO_2^-$)	14‰	*Nitrosomonas marina* (marine)
	19‰	*Nitrosomonas* C-113a (marine)
	35–38‰	*Nitrosomonas europaea* (terrestrial)
	12–16‰	Chesapeake Bay
Nitrite oxidation ($NO_2^- \rightarrow NO_3^-$)	Unknown	

Outputs

Denitrification, the bacterial reduction of nitrate to N_2, is the major mechanism of fixed N loss from the ocean, occurring both in the water column and in sediments when the oxygen concentration is low ($<5\,\mu M$) (**Figure 2**). Denitrification strongly discriminates against the heavier isotope, ^{15}N, progressively enriching the remaining nitrate pool in ^{15}N as nitrate consumption proceeds. Culture studies of denitrifying bacteria suggest (with some exceptions) an isotope effect of ~20–30‰, a range supported by water column estimates (**Table 1**). The isotopic discrimination during denitrification likely takes place as nitrate is reduced intracellularly to nitrite by the dissimilatory form of the enzyme nitrate reductase, such that unconsumed, ^{15}N-enriched nitrate effluxing from the cell back into ambient waters allows the enzyme-level isotope effect to be expressed at the environmental scale. Where it occurs in low-oxygen regions of the mid-depth ocean, water column denitrification causes a clear elevation in the $\delta^{15}N$ of nitrate, and it is the reason that global ocean nitrate $\delta^{15}N$ is higher than that of the N source from N_2 fixation, the dominant input.

In contrast to water-column denitrification, denitrification in sediments leads to little increase in the $\delta^{15}N$ of water-column nitrate. The high $\delta^{15}N$ of nitrate within the pore waters of actively denitrifying sediments demonstrates that isotopic discrimination occurs at the scale of the organism. However, expression of the organism-scale isotope effect at the scale of sediment/water exchange is minimized by nearly complete consumption of the nitrate at the site of denitrification within sediment pore waters, which prevents ^{15}N-enriched residual nitrate from evading to the overlying water column, yielding an 'effective' isotope effect of 3‰ or less in most sedimentary environments studied so far (**Table 1**).

Another mechanism of fixed N loss that occurs in sediments and the water column is anaerobic ammonium oxidation, or 'anammox', in which nitrite (from nitrate reduction or ammonium oxidation) is used to oxidize ammonium to N_2 ($NO_2^- + NH_4^+ \rightarrow N_2 + 2H_2O$). This process has unknown effects on isotope distributions in the ocean. The effects of anammox on N isotopes must depend on the organism-scale isotope effects, the sources of nitrite and ammonium substrates for the reaction, and the degree to which these substrates are consumed. For instance, if nitrate reduction by denitrifiers is the source of the nitrite, remineralization processes are the source of the ammonium, and both the nitrite and ammonium are completely consumed in the environment where anammox occurs, then the isotope discrimination would simplify to that of the nitrate reduction by denitrifiers averaged with any isotope discrimination during the remineralization that produces the needed ammonium. It should be noted that many water-column-derived isotope effect estimates for 'denitrification' have inherently included the effect of anammox, in that they regress the nitrate $\delta^{15}N$ increase against the total nitrate deficit relative to phosphate, and ammonium is not observed to accumulate.

Internal Cycling

The fluxes associated with internal cycling are neither sources nor sinks of fixed N but affect the distributions of N species and isotopes in the ocean.

N assimilation In the surface ocean, phytoplankton assimilate fixed N (nitrate and ammonium, as well as nitrite, urea, and other organic N compounds) (**Figure 3**). Culture studies indicate that different forms of fixed N are assimilated with distinct isotope effects, although these isotope effects may vary with physiological conditions. For all studied forms, phytoplankton preferentially consume ^{14}N relative to ^{15}N (**Figures 3** and **4**). Nitrate is the deep-water source of fixed N for phytoplankton growth, and the degree of its consumption varies across the surface ocean. The isotope effect of nitrate assimilation therefore has a major impact on the isotopic distributions of all N forms in the ocean. Field-based estimates of the isotope effect of nitrate assimilation range from 4‰ to 10‰, with most estimates closer to 5–8‰ (**Table 1**). Culture-based estimates are more variable. Physiological studies suggest that isotopic fractionation associated with nitrate assimilation is imparted by the intracellular assimilatory nitrate reductase enzyme, which has an estimated intrinsic isotope effect of 15–30‰. The enzyme-level isotope effect is expressed by efflux of unconsumed nitrate out of the cell, as also appears to be the case with denitrifiers. The lower range of isotope effect estimates associated with algal nitrate assimilation than with denitrification suggests proportionally less nitrate efflux by algae, perhaps related to the fact that fixed N is often scarce in the surface ocean. Some studies suggest the degree of efflux and isotope effect of nitrate assimilation may vary with growth conditions; one set of studies of the diatom *Thalassiosira weissflogii* showed a higher isotope effect under light-limited growth than under growth limited by iron or temperature. The isotope effect of nitrate assimilation is an integrative characteristic of the upper ocean biota that can be measured without perturbing the system, and its magnitude is of broad significance in the application of N isotopes to various questions in the modern and past oceans. Thus, it is a priority to develop a predictive understanding of its controls.

Other forms of fixed N assimilated by phytoplankton (ammonium, nitrite, and urea) are

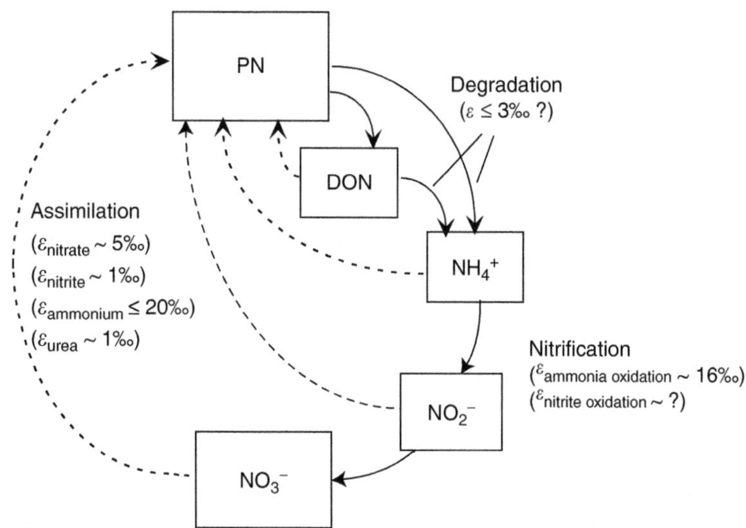

Figure 3 Schematic diagram of the processes and pools central to the internal cycling of N in the ocean. The isotope effects shown here are based on laboratory studies. Dashed arrows represent assimilation of dissolved species into particulate matter, and solid arrows represent remineralization. Complete consumption of the ammonium pool by assimilation in the surface ocean or by nitrification in the ocean interior causes the relatively high isotope effects associated with these processes to have little effect on N isotope dynamics. However, in regions where ammonium assimilation and nitrification co-occur, their isotope effects will impact the $\delta^{15}N$ of their respective products, PN and nitrate. In nitrification, ammonia (NH_3), rather than the protonated form ammonium (NH_4^+), is oxidized. However, ammonium is the dominant species in seawater, and there is isotope discrimination in the ammonium–ammonia interconversion. Thus, the isotope effects for 'ammonia oxidation' given here and elsewhere in the text refer specifically to consumption of ammonium. The processes surrounding DON production and utilization are not well understood from an isotopic perspective but are thought to play an important role in N cycling.

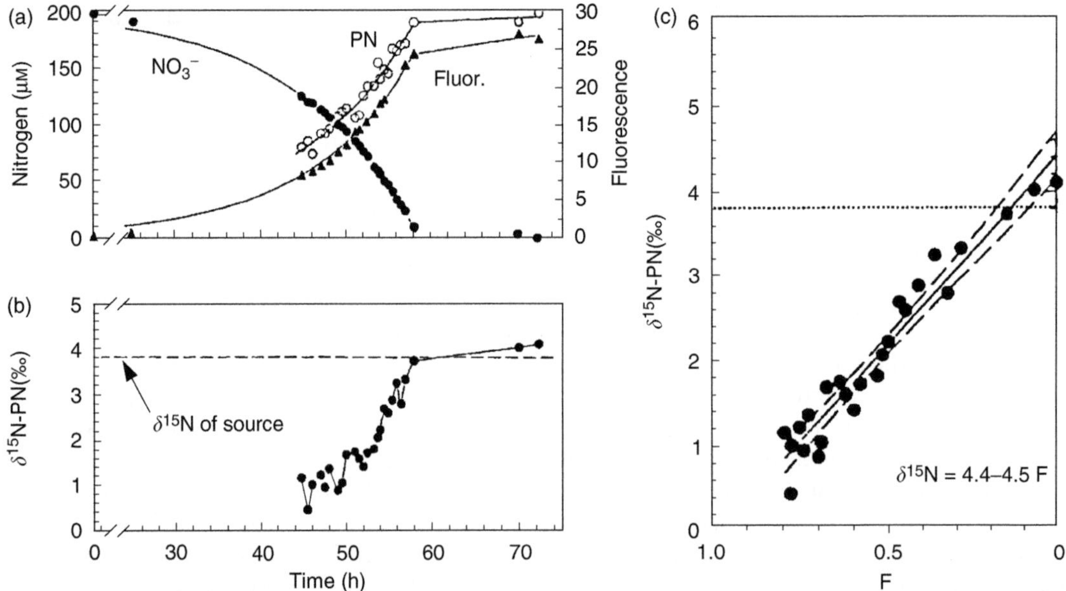

Figure 4 Isotopic fractionation during nitrate (NO_3^-) assimilation by a culture of the marine diatom *Thalassiosira pseudonana*. (a) Time series of NO_3^- concentration in the culture medium (filled circles), PN (open circles), and fluorescence, a measure of phytoplankton biomass (filled triangles); (b) $\delta^{15}N$ of PN accumulated over the same time period; (c) $\delta^{15}N$ of PN accumulated during log phase of growth plotted vs. F, a measure of nitrate utilization. F is $[-f/(1-f)]\ln f$, where f is the fraction of initial NO_3^- remaining at the time the culture is sampled. ε is calculated to be 4.5‰ from the slope of the regression in (c), according to the Rayleigh integrated product equation (see text, eqn [5]). Dashed horizontal lines in (b) and (c) represent the initial $\delta^{15}N$ of the source NO_3^- (3.8‰). Reprinted from Waser NAD, Harrison PJ, Nielsen B, Calvert SE, and Turpin DH (1998) Nitrogen isotope fractionation during the uptake and assimilation of nitrate, nitrite, ammonium, and urea by a marine diatom. *Limnology and Oceanography* 43: 215–224.

produced and nearly completely consumed within the open ocean surface mixed layer (**Figure 3**). Culture studies suggest an isotope effect for ammonium assimilation of up to ~20‰, decreasing as ammonium concentration decreases, and minimal isotope effects (<1‰) for assimilation of nitrite and urea. In estuaries, where ammonium can accumulate in the shallow subsurface and be entrained into the surface layer, ammonium assimilation causes a clear increase in the $\delta^{15}N$ of the remaining ammonium pool. Isotope effect estimates based on ammonium concentrations and $\delta^{15}N$ in these environments range from 6.5‰ to 18.5‰ (**Table 1**).

Remineralization The return of organic N to nitrate occurs in two steps, the degradation of organic N to ammonium and the bacterial oxidation of ammonium to nitrate, or 'nitrification' (**Figure 3**). Nitrification itself occurs in two steps, the oxidation of ammonium to nitrite and the oxidation of nitrite to nitrate, mediated by distinct groups of microorganisms. Isotopic discrimination may occur at all steps involved in remineralization. Field studies generally suggest that both bacteria and zooplankton preferentially degrade low-$\delta^{15}N$ PN to ammonium, yielding residual organic matter relatively high in ^{15}N. The wide spectrum of reactions involved in organic N degradation and the heterogeneous nature of organic matter (comprised of compounds with distinct $\delta^{15}N$ that degrade at various rates) make quantifying the isotope effect associated with degradation difficult. A few laboratory studies have quantified the isotope effects of individual processes such as thermal peptide bond cleavage, bacterial amino acid uptake and transamination, and zooplankton ammonium release. Laboratory studies attempting to mimic degradation as a whole suggest a net isotope effect of ≤3‰.

Culture studies indicate a large isotope effect for the conversion of ammonium to nitrite, the first step in nitrification (**Table 1**). Estimates of the isotope effect for marine ammonia-oxidizing bacteria (~14–19‰) are lower than those for terrestrial ammonia-oxidizing bacteria (as high as ~38‰), possibly due to phylogenetic differences. The isotope effect of nitrification estimated from ammonium concentration and $\delta^{15}N$ measurements in the Chesapeake Bay is 12–16‰, similar to the culture results for marine ammonia-oxidizing bacteria (**Table 1**). Thus far, no culture-based information is available regarding isotope effects for ammonia-oxidizing crenarchaea or nitrite-oxidizing bacteria.

Nitrogen Reservoirs

Dissolved Nitrogen

Nitrate Nitrate accounts for most of the fixed N in the ocean. The $\delta^{15}N$ of deep ocean nitrate is typically ~5‰. Regionally, the $\delta^{15}N$ of nitrate varies between 2‰ and 20‰ due to the effects of N_2 fixation, nitrate assimilation, and denitrification (**Figure 5**). Nitrate $\delta^{15}N$ significantly lower than deep-ocean nitrate has been observed in the upper thermocline of the low-latitude oligotrophic ocean (**Figure 6**). This ^{15}N depletion is most likely due to the oxidation of newly fixed N, which, as described above, has a $\delta^{15}N$ of *c.* −1‰. Values higher than 5‰ result from discrimination associated with nitrate assimilation by phytoplankton at the ocean surface (**Figure 7**) or denitrification in oxygen-deficient zones of the ocean interior (**Figure 8**).

Nitrate assimilation by phytoplankton leads to elevated $\delta^{15}N$ of nitrate in regions of the ocean where nitrate is incompletely consumed in surface waters, such as the high-latitude, nutrient-rich regions of the Southern Ocean and the subarctic Pacific, and the low-latitude upwelling regions of the California Current and the Equatorial Pacific. In the surface waters of these regions, there is a strong correlation between the degree of nitrate consumption by phytoplankton and the $\delta^{15}N$ of the nitrate remaining in the water (**Figure 7**). However, while nitrate assimilation elevates the $\delta^{15}N$ of nitrate in the surface ocean and causes modest ^{15}N enrichment in some newly formed thermocline waters, it does not appear to affect greatly the $\delta^{15}N$ of nitrate in the deep ocean. Below 2.5–3.0-km depth in the ocean, nitrate $\delta^{15}N$ is relatively constant at ~5‰, despite large inter-basin differences in nitrate concentration. The minimal degree of isotopic variation in the nitrate of the deep ocean is due to the fact that, in most surface waters, the nitrate supply from below is almost completely consumed by phytoplankton, such that the organic N exported from the surface ocean converges on the $\delta^{15}N$ of the nitrate supply. Because the sinking flux $\delta^{15}N$ is close to that of the nitrate supplied from the ocean interior, remineralization of the sinking flux in the ocean interior does not alter greatly the $\delta^{15}N$ of deep nitrate. In this respect, the oceanic cycling of N isotopes differs markedly from that of the carbon isotopes.

Because water-column denitrification occurs in the subsurface and because it consumes only a fraction of the nitrate plus nitrite available, its isotope effect is more completely expressed in the $\delta^{15}N$ of subsurface nitrate. In denitrifying regions of the water column, the $\delta^{15}N$ of nitrate in the subsurface is commonly elevated to 15‰ or higher (**Figure 8**). The subsurface $\delta^{15}N$ maximum occurs in the core of the oxygen minimum and is correlated with the estimated degree of nitrate consumption by water-column denitrification.

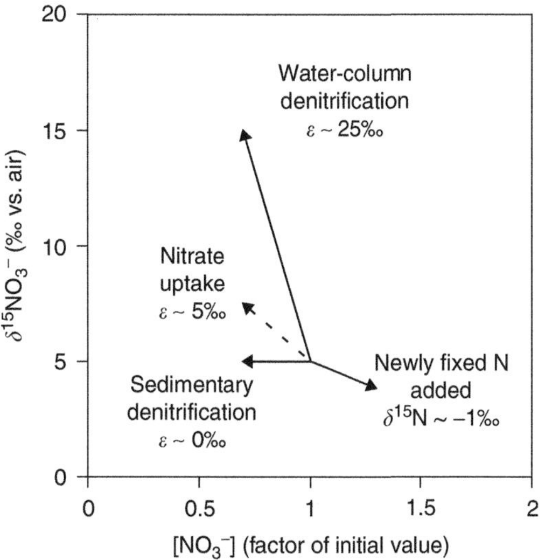

Figure 5 The effect of different marine N cycle processes on nitrate $\delta^{15}N$ and concentration, assuming an initial nitrate $\delta^{15}N$ of 5‰. The trajectories are for reasonable estimates of the isotope effects, and they depend on the initial nitrate $\delta^{15}N$ as well as the relative amplitude of the changes in nitrate concentration (30% for each process in this figure). A solid arrow denotes a process that adds or removes fixed N from the ocean, while a dashed line denotes a component of the internal cycling of oceanic fixed N. The effects of these two types of processes can be distinguished in many cases by their effect on the concentration ratio of nitrate to phosphate in seawater. The actual impact of the different processes on the N isotopes varies with environment. For instance, if phytoplankton completely consume the available nitrate in a given environment, the isotope effect of nitrate uptake plays no major role in the $\delta^{15}N$ of the various N pools and fluxes; the effect of nitrate generation by organic matter degradation and nitrification, not shown here, will depend on this dynamic. Similarly, the lack of a large isotope effect for sedimentary denitrification is due to the fact that nitrate consumption by this process can approach completion within sedimentary pore waters.

Denitrification, both in the water column and sediments, exerts a direct control on the $\delta^{15}N$ of mean deep ocean nitrate. When the ocean N budget is at steady state, the $\delta^{15}N$ of the fixed N removed (through water-column and sedimentary denitrification) will equal the $\delta^{15}N$ of the fixed N added (*c.* −1‰, approximating N_2 fixation as the sole source) (**Figure 9**). If denitrification with an isotope effect of 20–30‰ were occurring homogenously in the ocean water column and responsible for all fixed N loss, the $\delta^{15}N$ of mean oceanic nitrate would be 19–29‰ to achieve a $\delta^{15}N$ of −1‰ for N loss. That the modern mean oceanic nitrate $\delta^{15}N$ is ~5‰, much lower than 19–29‰, reflects at least two factors:

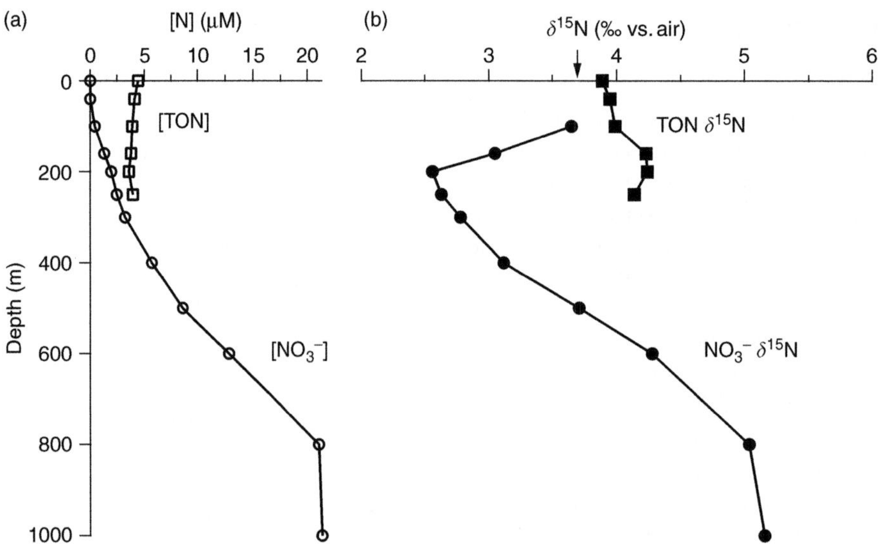

Figure 6 Depth profiles of (a) [NO_3^-] (open circles) and [TON] (total organic N, or DON plus the small pool of PN) (open squares) and (b) nitrate $\delta^{15}N$ (filled circles) and TON $\delta^{15}N$ (filled squares) at the Bermuda Atlantic Time-Series Study site in the oligotrophic Sargasso Sea. Low nitrate $\delta^{15}N$ in the thermocline has been proposed to reflect N_2 fixation at the surface, the N from which sinks and is remineralized at depth. The increase in nitrate $\delta^{15}N$ above 200 m reflects fractionation associated with nitrate assimilation. [TON] increases slightly into the surface layer while TON $\delta^{15}N$ decreases, suggesting a possible source of low-^{15}N DON in the surface. The $\delta^{15}N$ of the TON pool is lower than that of mean ocean nitrate and deep nitrate at Bermuda (~5‰) but higher than that of thermocline nitrate at Bermuda (~2.5‰). High nitrate concentrations below 250 m prevent accurate assessment of TON $\delta^{15}N$ using published methods (see text). The $\delta^{15}N$ of sinking particles collected at 100 m (3.7‰) is indicated by the arrow at top of (b); surface suspended PN $\delta^{15}N$ is *c.* −0.2‰ (not shown). Nitrate and TON data are the means of monthly measurements between June 2000 and May 2001. Modified from Knapp AH, Sigman DM, and Lipschultz F (2005) N isotopic composition of dissolved organic nitrogen and nitrate at the Bermuda Atlantic time-series study site. *Global Biogeochemical Cycles* 19: GB1018 (doi:10.1029/2004GB002320). Sinking and suspended PN $\delta^{15}N$ data from Altabet MA (1988) Variations in nitrogen isotopic composition between sinking and suspended particles: Implications for nitrogen cycling and particle transformation in the open ocean. *Deep Sea Research* 35: 535–554.

(1) the importance of sedimentary denitrification, which appears to express a minimal isotope effect; and (2) the localized nature of water-column denitrification. With regard to the second, because denitrification consumes a significant fraction of the ambient nitrate in the ocean's suboxic zones and elevates its $\delta^{15}N$ above that of the mean ocean (**Figures 5** and 8), the $\delta^{15}N$ of nitrate being removed by water column denitrification is higher than if the substrate for denitrification had the mean ocean $\delta^{15}N$. Much as with sedimentary denitrification, this reduces the expression of the organism-level isotope effect of water column denitrification and thus lowers the mean $\delta^{15}N$ of nitrate required to achieve an isotope balance between inputs and outputs. With these considerations, one study estimates that water-column denitrification is responsible for 30% of fixed N loss from the modern ocean, with sedimentary denitrification responsible for the remainder. Still, this isotope-based budget for marine fixed N remains uncertain.

One limitation of using the N isotopes to investigate N cycling in the ocean is their inability to separate co-occurring processes with competing N isotopic signatures, such as denitrification/N_2 fixation and nitrate assimilation/nitrification. Coupled analysis of N and O isotopes in nitrate promises to disentangle such otherwise overprinting processes. Culture studies have demonstrated that the two most important nitrate-consuming processes, nitrate assimilation and denitrification, fractionate the N and O in nitrate with a ratio close to 1:1 (**Figure 10**). Deviations in the ratio of $\delta^{18}O$ and $\delta^{15}N$ in nitrate from 1:1 can therefore provide information about the nitrate being added by nitrification, such as if it derives from newly fixed N or if there is otherwise undetected nitrate cycling within a water parcel.

Nitrite Nitrite is an intermediate in oxidative and reductive processes such as nitrification and denitrification. It also serves as a substrate for anammox and can be assimilated by phytoplankton. Only when these processes become uncoupled, such as at the base of the euphotic zone and in oxygen minimum zones, does nitrite accumulate to significant levels (0.1–10 μM). In some oxygen-deficient regions of the water column in the eastern tropical North Pacific, eastern tropical South Pacific, and the Arabian Sea, nitrite can represent as much as 25% of the pool of nitrogen oxides ($NO_3^- + NO_2^-$).

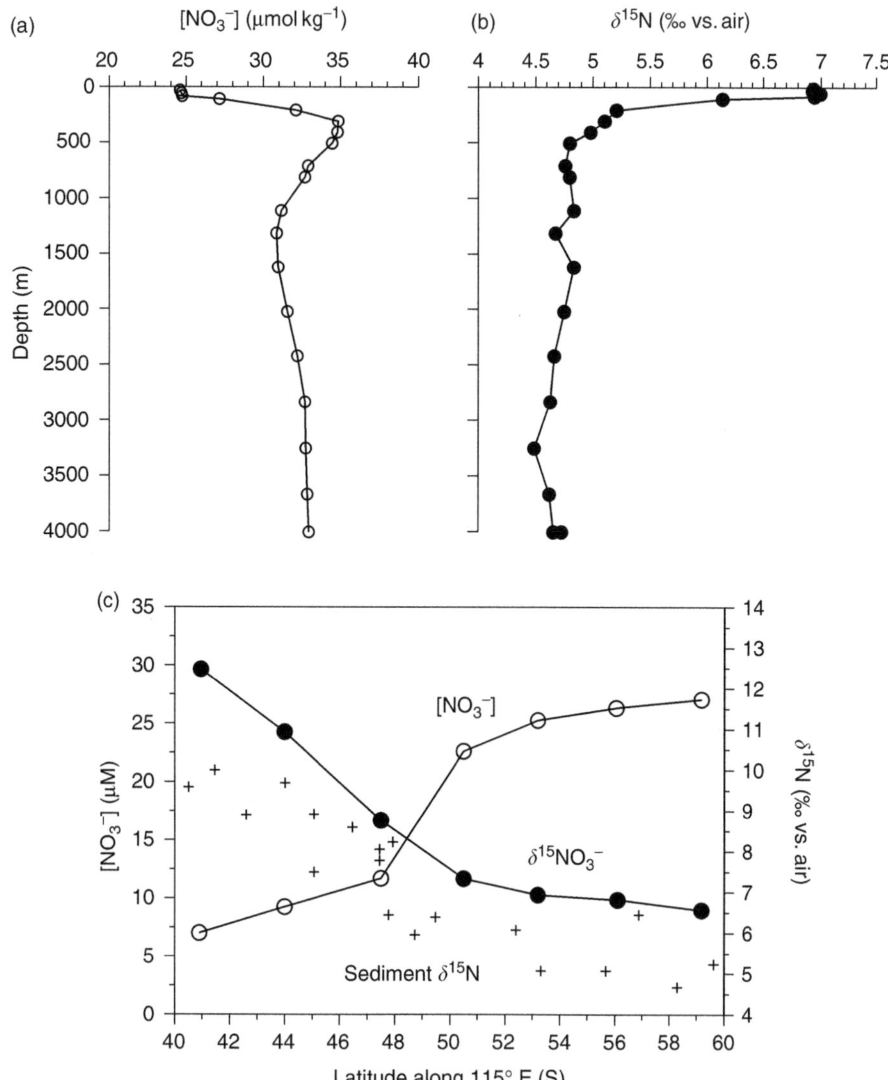

Figure 7 Nitrate δ^{15}N data from the Southern Ocean show the effect of uptake by phytoplankton. A depth profile of nitrate concentration (a) and δ^{15}N (b) from the Antarctic region of the east Indian Ocean (53.2° S, 115° E) shows a decrease in nitrate concentration into the surface layer and an associated increase in nitrate δ^{15}N, both resulting from nitrate uptake by phytoplankton. A meridional transect (c) of nitrate concentration (open circles) and δ^{15}N (filled circles) in the surface mixed layer of the Southern Ocean along 115° E shows that nitrate δ^{15}N is spatially correlated with variations in the utilization of nitrate by phytoplankton, as reflected by the equatorward decrease in nitrate concentration. Nitrate δ^{15}N is lowest where the nitrate concentration is highest (and nitrate utilization is lowest), in Antarctic waters, south of ~51° S in this region. The meridional gradient in nitrate δ^{15}N is recorded in bulk sediment δ^{15}N along 115° E (c, crosses) through the sinking of PN out of the surface ocean. Across the transect, sediment δ^{15}N is ~5‰ higher than expectations for and measurements of sinking N δ^{15}N, probably mainly due to isotopic alteration of this N at the seafloor. Nevertheless, the link between nitrate consumption and sediment δ^{15}N provides a possible avenue for paleoceanographic reconstruction of nitrate utilization by phytoplankton. The left axis in (c) is scaled to 35 μM, the approximate nitrate concentration of Upper Circumpolar Deep Water that upwells in the Antarctic (see (a)). (a, b) Reprinted from Sigman DM, Altabet MA, Michener RH, McCorkle DC, Fry B, and Holmes RM (1997) Natural abundance-level measurement of the nitrogen isotopic composition of oceanic nitrate: An adaptation of the ammonia diffusion method. *Marine Chemistry* 57: 227–242. (c) Modified from Sigman DM, Altabet MA, McCorkle DC, François R, and Fischer G (2000) The δ^{15}N of nitrate in the Southern Ocean: Nitrate consumption in surface waters. *Global Biogeochemical Cycles* 13: 1149–1166, with the sediment data taken from Altabet MA and François R (1994) Sedimentary nitrogen isotopic ratio records surface ocean nitrate utilization. *Global Biogeochemical Cycles* 8: 103–116.

The δ^{15}N of nitrite is expected to reflect the balance of isotopic fractionation during nitrite production and consumption processes. In the eastern tropical North Pacific, the δ^{15}N of nitrite is very low (−18‰ to −7‰). This δ^{15}N is ~30‰ lower than the nitrate in the same environment and is lower than expected from denitrification alone, give the known isotope effects for nitrate and nitrite reduction. The

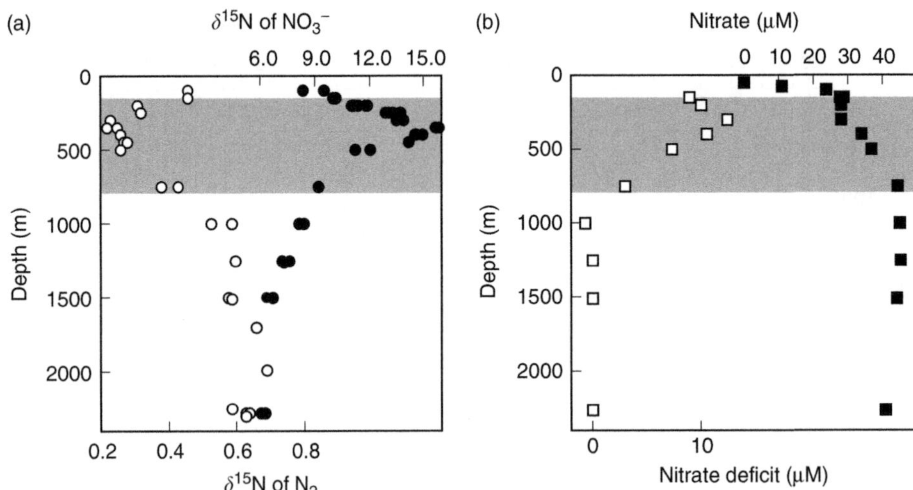

Figure 8 (a) The $\delta^{15}N$ of NO_3^- (filled circles) and N_2 (open circles) in water column profiles through an intense denitrification zone in the eastern tropical North Pacific (22° N, 107° W). The shaded interval indicates the depth range with dissolved O_2 concentration $<10\,\mu M$ where denitrification is encouraged (b, concentration shown in filled squares) and leads to a characteristic NO_3^- deficit (b, open squares) relative to phosphate. Measurements indicate enrichment of NO_3^- in ^{15}N and concurrent depletion of N_2 in ^{15}N, arising from isotope discrimination during denitrification, with the conversion of NO_3^- to N_2. ε for denitrification in this environment was estimated to be ~25‰. Modified from Brandes JA, Devol AH, Yoshinari T, Jayakumar DA, and Naqvi SWA (1998) Isotopic composition of nitrate in the central Arabian Sea and eastern tropical North Pacific: A tracer for mixing and nitrogen cycles. *Limnology and Oceanography* 43: 1680–1689.

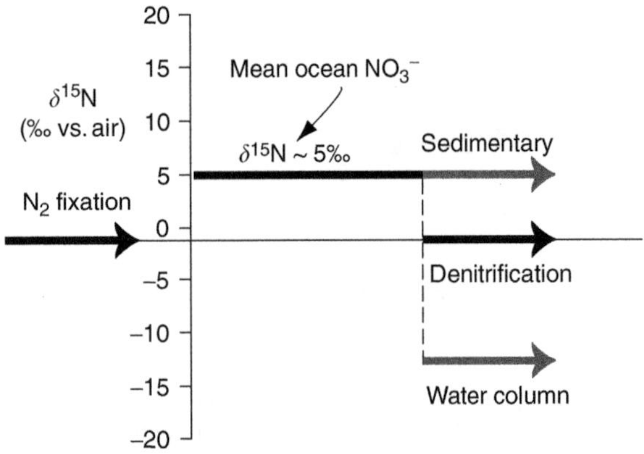

Figure 9 Simplified global ocean N isotope budget. The *y*-axis indicates the $\delta^{15}N$ of a given flux or pool. The $\delta^{15}N$ of N from oceanic N_2 fixation, the dominant N input to the ocean, is *c.* −1‰ ('N_2 fixation' on the left). At steady state, the total denitrification loss ('denitrification' on the right) must have the same $\delta^{15}N$ as the input. The $\delta^{15}N$ of mean ocean nitrate is ~5‰. Water column denitrification removes nitrate with a low $\delta^{15}N$ ('water column' at lower right), while sedimentary denitrification removes nitrate with a $\delta^{15}N$ similar to that of mean ocean nitrate ('sedimentary' at upper right). The need for the flux-weighted $\delta^{15}N$ of the denitrification loss to be *c.* −1‰ leads to estimates of partitioning between water-column and sedimentary denitrification in which sedimentary denitrification is found to drive the greater part of the total N loss.

meaning of this low $\delta^{15}N$ is still not well understood but likely points to other processes acting on the NO_2^- pool. In these regions, nitrite can represent a significant reservoir of N that is depleted in ^{15}N.

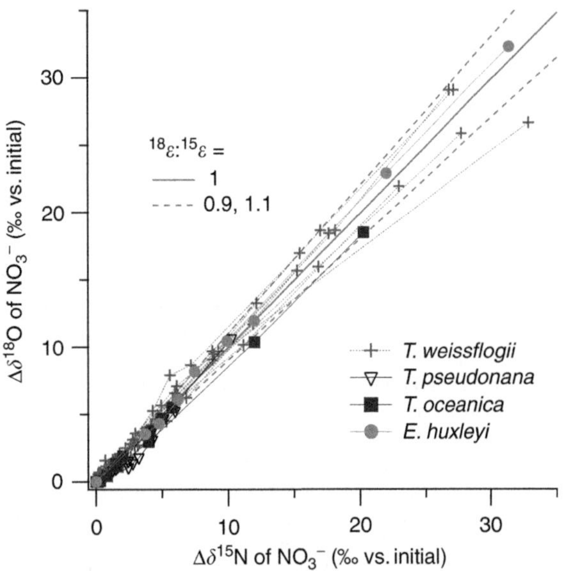

Figure 10 The $\delta^{18}O$ vs. $\delta^{15}N$ in nitrate as it is progressively assimilated by four eukaryotic species of marine phytoplankton. Both $\delta^{15}N$ and $\delta^{18}O$ in nitrate increase as nitrate is consumed, and they do so with an O:N ratio for isotopic discrimination ($^{18}\varepsilon$:$^{15}\varepsilon$) of ~1. Dashed lines show slopes of 1.1 and 0.9 for comparison. Modified from Granger J, Sigman DM, Needoba JA, and Harrison PJ (2004) Coupled nitrogen and oxygen isotope fractionation of nitrate during assimilation by cultures of marine phytoplankton. *Limnology and Oceanography* 49: 1763–1773.

Until recently, the two species nitrate and nitrite have typically been combined in isotopic analysis, such that the presence of low $\delta^{15}N$ nitrite may have masked some of the ^{15}N enrichment of nitrate in the oxygen-deficient zone.

Ammonium The $\delta^{15}N$ of ammonium reflects the production of ammonium by the degradation of organic N and its consumption by nitrification, ammonium assimilation, and perhaps anammox (**Figure 3**). Analytical constraints have limited isotopic studies of ammonium to environments with ammonium concentrations greater than 1 μM, excluding studies in the open ocean. In estuarine systems, where ammonium can be abundant, its $\delta^{15}N$ is often high (commonly higher than +10‰, with one observation of +70‰) and it increases as the ammonium concentration decreases along transects from riverine to marine waters, due to discrimination associated with ammonium consumption by nitrification and/or ammonium assimilation.

In the open ocean interior, below the depth of algal assimilation, essentially all ammonium generated from particles is oxidized to nitrite and then nitrate before it can be transported into or out of a given region. Thus, nitrification should be of limited importance for the isotope dynamics of both particulate and dissolved N once the former has sunk out of the upper ocean. In the open ocean surface mixed layer, it is generally assumed that ammonium generated by remineralization is quickly and entirely assimilated by plankton, in which case the isotope effect associated with its consumption would not play an important role in N isotope dynamics of the open ocean. However, in at least some regions of the upper ocean, ammonium oxidation and assimilation are likely to co-occur. If the isotope effect of ammonium oxidation is greater than that of ammonium assimilation, low-$\delta^{15}N$ N will preferentially be routed to the nitrate pool by oxidation and high-$\delta^{15}N$ N will be routed back to the PN pool by assimilation. If the isotope effect of oxidation is less than that of assimilation, the opposite will occur. Thus, with better constaints on the isotope effects of ammonium-consuming processes, the isotopes of upper ocean N pools promise to provide an integrative constraint on the relative rate of nitrification in the upper ocean, especially when paired with the O isotopes of nitrate (see above).

Dissolved organic nitrogen DON concentrations are significant in the open ocean, typically ≥4 μM in surface waters, decreasing to ~2 μM in deep water. Fluxes associated with the DON pool are among the least constrained terms in the modern marine N budget and may be important. Studies to date of bulk DON have been in the subtropical ocean, where DON is by far the dominant N pool in the surface ocean. In the surface mixed layer at the Bermuda Atlantic Time-series site in the Sargasso Sea, the concentration and $\delta^{15}N$ of TON are ~4 μM and ~4‰ (TON being total organic N, or DON plus the small pool of PN) (**Figure 6**). This $\delta^{15}N$ is similar to or slightly higher than the shallow subsurface nitrate that is entrained into the euphotic zone during wintertime vertical mixing (**Figure 6(b)**). Minimal gradients in the concentration and $\delta^{15}N$ of DON in this region of the upper ocean hinder reconstruction of fluxes of DON or the $\delta^{15}N$ of those fluxes. There is a weak increase in the concentration of TON into the surface layer and an accompanying decrease in its $\delta^{15}N$ (**Figures 6(a)** and **6(b)**). Thus, there may be an input of low-$\delta^{15}N$ N into the surface DON pool, which is remineralized at depth, but this requires further validation. Progress on DON $\delta^{15}N$ dynamics would be aided by a method to remove nitrate from samples without compromising the DON pool, which would make subsurface waters and high-nitrate surface waters more accessible to study. Ongoing work on separable fractions of the DON pool (e.g., the high-molecular-weight fraction and its components) is also promising.

Dissolved gases Dissolved N_2 in equilibrium with atmospheric N_2 at the surface has a $\delta^{15}N$ of 0.6‰. The isotopic composition of dissolved N_2 does not vary greatly in ocean profiles, except in zones of denitrification. Production of low-$\delta^{15}N$ N_2 in denitrification zones results in measured N_2 $\delta^{15}N$ as low as 0.2‰ (**Figure 8(a)**). Since N_2 is the main product of denitrification, its $\delta^{15}N$ provides a test of the nitrate-based estimates of the isotope effect for this process.

Dissolved N_2O is produced by nitrification and is both produced and consumed by denitrification. The marine flux of N_2O is perhaps one-third of the global flux of this greenhouse gas to the atmosphere; therefore, an understanding of the mechanisms of N_2O production and their regulation in the ocean is an important goal. Culture studies indicate that bacterial production of N_2O by nitrification and denitrification produces gas depleted in ^{15}N and ^{18}O relative to the source material. Consumption of N_2O by denitrification leaves the residual gas enriched in ^{15}N and ^{18}O, with $\delta^{15}N$ of N_2O as high as 40‰ measured in the Arabian Sea. In oxygenated waters of the open ocean, nitrification likely dominates N_2O production and its isotopic profile. A depth profile in the subtropical North Pacific shows three main features (**Figure 11**): (1) isotopic equilibrium with atmospheric N_2O at the surface; (2) a subsurface $\delta^{15}N$ minimum attributed to nitrification; and (3) a broad $\delta^{15}N$ maximum in deeper waters probably due to N_2O consumption, perhaps in the denitrifying waters of the eastern Pacific margin. In and near denitrification zones, a strong maximum in the $\delta^{15}N$ of N_2O is observed, presumably due to isotope fractionation associated with N_2O consumption (via reduction to N_2).

Particulate Nitrogen

Suspended particles A typical profile of suspended particles has its lowest $\delta^{15}N$ in the surface layer, increasing below the euphotic zone (**Figure 12**). The $\delta^{15}N$ of suspended particles reflects in part the $\delta^{15}N$ of nitrate supplied to the surface ocean and, in nutrient-rich surface waters, isotope discrimination associated with its incomplete assimilation. However, the low $\delta^{15}N$ in the surface layer is typically lower than what would be expected solely from nitrate assimilation. This low $\delta^{15}N$ has two competing explanations: N_2 fixation and N recycling. As described earlier, N_2 fixation is expected to add fixed N with a $\delta^{15}N$ of *c.* −1‰ to surface waters. The isotopic effect of N recycling originates from heterotrophic processes. Zooplankton appear to release ammonium which has a lower $\delta^{15}N$ than their food source, making their tissues and solid wastes ~3‰ higher in $\delta^{15}N$ than their food source. The low-$\delta^{15}N$ ammonium is consumed by phytoplankton and thus retained in the surface ocean N pool, while the ^{15}N-enriched PN is preferentially exported as sinking particles, leading to a lower $\delta^{15}N$ of surface PN in regions where recycled N is an important component of the gross N supply to phytoplankton. Low $\delta^{15}N$ observed in suspended PN from the Antarctic and other high-latitude regions, beyond what is expected from isotope discrimination during nitrate assimilation, is unlikely to be due to N_2 fixation and thus likely reflects N recycling. In the low-latitude, low-nutrient ocean surface, such as the Sargasso Sea and western tropical Pacific, the relative importance of N_2 fixation and N recycling in producing low-$\delta^{15}N$ surface particles is uncertain. Because of its implications for the rates of N_2 fixation and N recycling, this question deserves further study.

The $\delta^{15}N$ of suspended particles in the subsurface is typically ~6‰ higher than suspended particles in the surface ocean and ~3‰ higher than the sinking flux (**Figure 12**). The $\delta^{15}N$ of deep

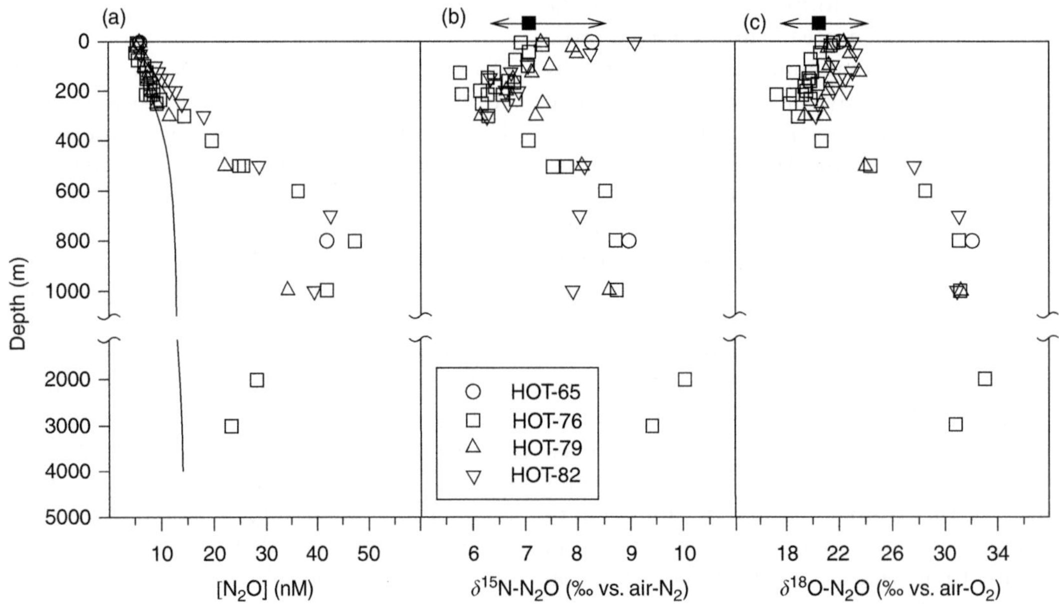

Figure 11 Depth profiles of (a) N_2O concentration and (b) $\delta^{15}N$ and (c) $\delta^{18}O$ of N_2O at station ALOHA in the subtropical North Pacific (22° 45′ N, 158° W) during four separate cruises. The solid line in (a) indicates theoretical saturation with atmospheric N_2O at *in situ* temperatures and salinities. The minima in $\delta^{15}N$ and $\delta^{18}O$ around 200 m are thought to be due to significant *in situ* production of N_2O from nitrification. The broad isotopic maxima at depth are likely due to N_2O consumption, perhaps in the denitrifying waters along the eastern Pacific margin. The filled squares at the top of (b) and (c) represent measurements of $\delta^{15}N$ and $\delta^{18}O$ of atmospheric N_2O during the Hawaii Ocean Time-series 76 cruise, and arrows indicate the range of historical measurements as of the late 1980s. Reprinted from Dore JE, Popp BN, Karl DM, and Sansone FJ (1998) A large source of atmospheric nitrous oxide from subtropical North Pacific surface waters. *Nature* 396: 63–66.

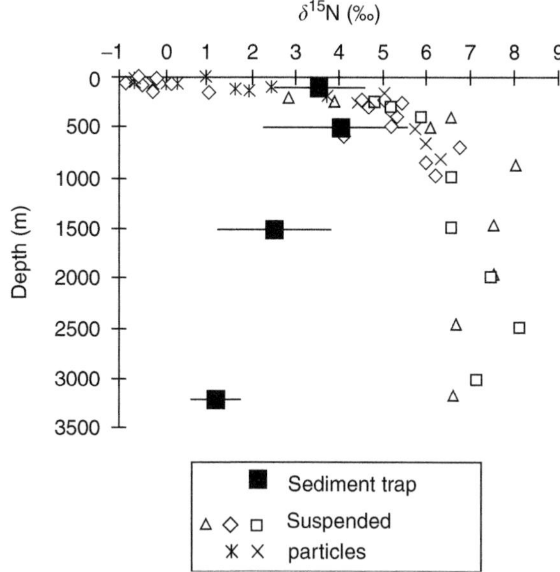

Figure 12 Nitrogen isotopic values of suspended particulate matter and sinking particles (as collected by sediment traps) in the subtropical North Atlantic Ocean near BATS (31° 50′ N, 64° 10′ W; see **Figure 6**). The profiles of suspended PN show the representative depth gradient in δ^{15}N, with lower δ^{15}N in the surface ocean than at depth. The δ^{15}N of the sinking flux shows a decrease with depth, which has also been observed in other regions. Reprinted from Altabet MA, Deuser WG, Honjo S, and Stienen C (1991) Seasonal and depth-related changes in the source of sinking particles in the North Atlantic. *Nature* 354: 136–139.

particles is consistent with the inference that deep particles are the breakdown products of material exported from the surface, and that bacteria preferentially remineralize low-δ^{15}N PN.

Isotopic analysis of zooplankton and organisms at higher trophic levels can provide insights into the marine N cycle. The 'trophic effect', an observed ~3‰ increase in δ^{15}N per trophic level that presumably results from isotopic discrimination during metabolism of N-bearing organic matter, is used widely in food-web studies. N isotopic analysis of specific amino acids within organisms and organic matter promises new insights, as some amino acids increase in δ^{15}N with trophic level while others preserve the δ^{15}N of the food source.

Sinking particulate Nitrogen and sedimentary Nitrogen Because vertical sinking is an important mode of N export from the surface ocean, the δ^{15}N of the sinking flux is one of the most valuable N isotopic constraints on modern ocean processes. Combined with other isotopic data, sinking flux δ^{15}N data can provide information on the routes and mechanisms of nitrate supply and can be used to constrain other sources of N to the surface. The sinking flux also transfers the isotopic signal from the surface ocean to the seafloor, providing the link through which the sediment column records the history of surface ocean processes (**Figure** 7(c)). Sinking particles collected in depth arrays of sediment traps often show a modest decrease in δ^{15}N with depth (**Figure 12**). This trend runs contrary to our expectations for the isotopic change of particulate matter as it degrades, and it currently lacks a compelling explanation.

There is generally a good correlation between the δ^{15}N of surface sediments and sinking particulate δ^{15}N from the overlying water column. In regions of the ocean where a relatively large fraction of the organic rain is preserved in the sediment column, as occurs along continental margins, this correlation is excellent. In open ocean sediments where only a very small fraction of N is preserved, spatial patterns in the δ^{15}N of sediment core tops mirror those in the water column above (**Figure** 7(c)), but a significant ^{15}N enrichment (of ~2–5‰) is observed in the sediment N relative to sinking particles. Upon burial, reactions in the shallow sediment column known collectively as 'diagenesis' can cause a clear increase in the δ^{15}N of PN as it is incorporated into the sediment mixed layer. While some studies have found that sedimentary diagenesis has not greatly affected the paleoceanographic information provided by specific sedimentary records, it cannot be assumed that changes in the 'diagenetic offset' are unimportant. To address concerns regarding alteration of both sinking and sedimentary bulk δ^{15}N, studies are increasingly focusing on isolating specific N components, the δ^{15}N of which is insensitive to diagenesis, such as N bound within the mineral matrix of microfossils, or that does not change in δ^{15}N as it is degraded, such as chlorophyll degradation products.

Nitrogen isotopes in the sedimentary record The isotopes of sedimentary N are used to investigate past changes in the marine N budget and the internal cycling of N within the ocean. The processes and parameters reflected by the δ^{15}N of sedimentary N include (1) mean ocean nitrate δ^{15}N, (2) regional subsurface nitrate ^{15}N depletion or enrichment relative to the global ocean owing to N_2 fixation or denitrification, (3) regional isotope dynamics associated with partial nitrate consumption in surface waters, and (4) possible direct contribution of newly fixed N to sinking PN. Paleoceanographers have focused on sediment δ^{15}N records underlying three environments where a single process or parameter is thought to dominate changes in sinking δ^{15}N. In oligotrophic regions, sediment δ^{15}N is assumed to reflect the δ^{15}N of mean ocean nitrate and therefore the global ocean balance of

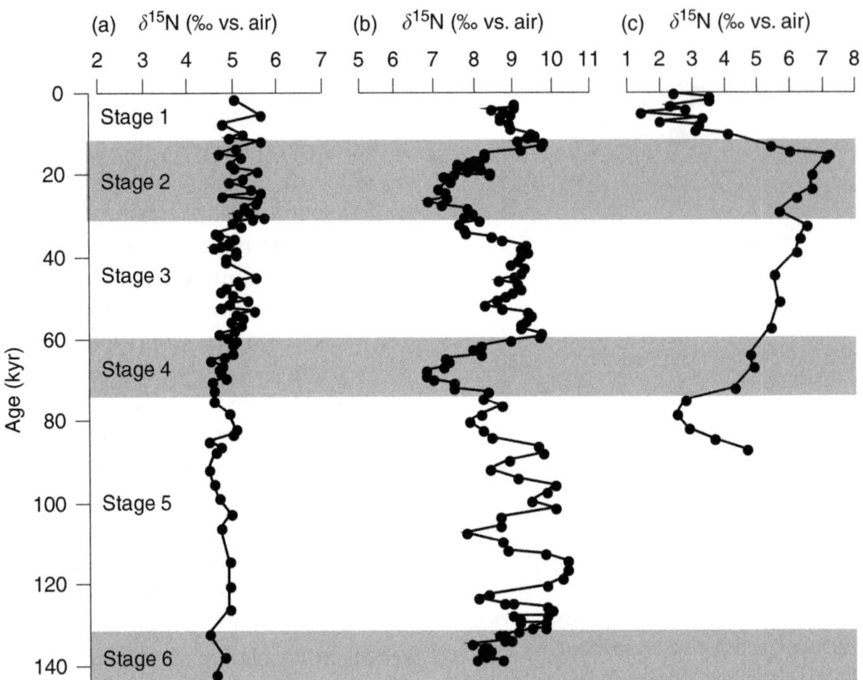

Figure 13 Sedimentary δ^{15}N records spanning the past 140 ky, which encompass recent ice ages (shaded, marine oxygen isotope stages 2, 4, and 6, with 2 and 6 being the most extreme) and interglacials (stages 1, 3, and 5, with 1 and 5 being the most extreme). (a) Sediment record underlying the oligotrophic South China Sea (8 °30.4 N, 112 °19.9 E), where sedimentary δ^{15}N is expected to track, with some offset, the δ^{15}N of nitrate in the western Pacific thermocline. The western Pacific thermocline nitrate, in turn, is assumed to have maintained a constant isotopic relationship with deep ocean nitrate. The small magnitude of variation in δ^{15}N (<1.5‰) and lack of correlation with glacial/interglacial transitions suggest that mean ocean nitrate δ^{15}N remained unchanged through shifts in Earth's climate. (b) Sedimentary δ^{15}N record underlying the eastern tropical North Pacific (22 °23.3 N, 107 °04.5 W), a major region of denitrification. Interglacials are characterized by high δ^{15}N (8–9‰), with δ^{15}N 2–3‰ lower during glacials. Low δ^{15}N, along with coincident evidence for decreased productivity and higher oxygen content in the mid-depth water-column, indicates decreased water column denitrification during glacial periods. (c) Sedimentary δ^{15}N record from the high-nitrate Antarctic Zone of the Southern Ocean (54 °55 S, 73 °50 E) shows higher δ^{15}N during the period spanning glacial stages 2–4, suggesting greater algal utilization of nitrate in the surface ocean. Coupled with evidence of lower glacial productivity, the glacial ^{15}N enrichment suggests reduced nutrient supply from below. All data shown are of bulk sediment. The sedimentary δ^{15}N record shown in (b) is from a region of high organic matter preservation in the sediments, where bulk sedimentary δ^{15}N correlates well with sinking δ^{15}N. The records in (a) and (c) are from regions where a diagenetically driven difference is observed between sinking and sedimentary N, which introduces uncertainties in interpretation. (a) Core 17961 from Kienast M (2000) Unchanged nitrogen isotopic composition of organic matter in the South China Sea during the last climatic cycle: Global implications. *Paleoceanography* 15: 244–253. (b) Core NH8P from Ganeshram RS, Pederson TF, Calvert SE, and Murray JW (1995) Evidence from nitrogen isotopes for large changes in glacial–interglacial oceanic nutrient inventories. *Nature* 376: 755–758. (c) Core MD84-552 from François R, Altabet MA, Yu E-F, *et al.* (1997) Contribution of Southern Ocean surface-water stratification to low atmospheric CO_2 concentrations during the last glacial period. *Nature* 389: 929–935.

inputs and outputs of fixed N (**Figure 13(a)**). In denitrifying regions, sediment δ^{15}N has been taken to largely reflect changes in regional ^{15}N enrichment due to water column denitrification (**Figure 13(b)**). In high-nutrient regions, sediment δ^{15}N primarily records the degree of nitrate consumption by algal assimilation, providing insight into changes in balance between gross nitrate supply to surface waters and export of organic N from the surface (**Figure 13(c)**). However, it must be kept in mind that multiple processes may affect the δ^{15}N of sediments in any given region. For example, even in low-nutrient regions far from denitrification zones, the sediment δ^{15}N record may record changes not only in mean ocean nitrate δ^{15}N but also in the δ^{15}N of nitrate imported from subpolar regions and in the isotopic imprint of N_2 fixation.

Concluding Remarks

The study of the N isotopes in the ocean is young relative to that of the other light isotopes (e.g., carbon, oxygen, and sulfur), with much of the work to date developing the methods needed to measure different forms of oceanic N and establishing the isotope systematics of N cycle processes that are necessary to interpret observed patterns. Over the previous decades, the N isotopes have had perhaps their greatest impact on foodweb studies and in paleoceanographic work. In the case of the latter, this reflects the ability of the N isotopes to provide basic constraints on environmental conditions when there are few other indicators available. Recent and ongoing method development is greatly improving our ability to measure diverse N pools in the ocean. This is yielding a new generation of N isotope studies that promise to provide geochemical estimates for the rates and distributions of N fluxes in the modern ocean, complementing instantaneous 'bottle' measurements of these fluxes as well as other geochemical approaches. Fundamental aspects of the oceanic N cycle are still poorly understood, and the N isotopes provide an important tool for their study.

See also

Nitrogen Cycle.

Further Reading

Altabet MA (2005) Isotopic tracers of the marine nitrogen cycle: Present and past. In: Hutzinger O (ed.) *The Handbook of Environmental Chemistry, Vol. 2: Marine Organic Matter: Chemical and Biological Markers*, pp. 251–294. Berlin: Springer.

Altabet MA and Francois R (1994) The use of nitrogen isotopic ratio for reconstruction of past changes in surface ocean nutrient utilization. In: Zahn R, Kaminski M, Labeyrie L, and Pederson TF (eds.) *Carbon Cycling in the Glacial Ocean: Constraints on the Ocean's Role in Global Change*, pp. 281–306. Berlin: Springer.

Brandes JA and Devol AH (2002) A global marine fixed nitrogen isotopic budget: Implications for Holoene nitrogen cycling. *Global Biogeochemical Cycles* 16: 1120 (doi:10.1029/2001GB001856).

Casciotti KL and McIlvin MR (2007) Isotopic analyses of nitrate and nitrite from references mixtures and application to Eastern Tropical North Pacific waters. *Marine Chemistry* 107: 184–201.

Chang CCY, Silva SR, Kendall C, Michalski G, Casciotti KL, and Wankel S (2004) Preparation and Analysis of Nitrogen-bearing Compounds in Water for Stable Isotope Ratio Measurement. In: deGroot PA (ed.) *Handbook of Stable Isotope Analytical Techniques*, vol. 1, pp. 305–354. Amsterdam: Elsevier.

Fogel ML and Cifuentes LA (1993) Isotope fractionation during primary production. In: Engel MH and Macko SA (eds.) *Organic Geochemistry*, pp. 73–98. New York: Plenum.

Fry B (2006) *Stable Isotope Ecology*. New York: Springer.

Galbraith ED, Sigman DM, Robinson RS, and Pedersen TF (in press) Nitrogen in past marine environments. In: Bronk DA, Mulholland MR, and Capone DG (eds.) Nitrogen in the Marine Environment.

Goericke R, Montoya JP, and Fry B (1994) Physiology of isotope fractionation in algae and cyanobacteria, In: Lajtha K and Michener R (eds.) *Stable Isotopes in Ecology and Environmental Science,* 1st edn., pp. 187–221. Oxford: Blackwell Scientific Publications.

Macko SA, Engel MH, and Parker PL (1993) Early diagenesis of organic matter in sediments: Assessment of mechanisms and preservation by the use of isotopic molecular approaches. In: Engel MH and Macko SA (eds.) *Organic Geochemistry*, pp. 211–224. New York: Plenum.

Michener R and Lajtha K (2007) *Stable Isotopes in Ecology and Environmental Science*, 2nd edn., New York: Wiley-Blackwell.

Montoya JP (1994) Nitrogen isotope fractionation in the modern ocean: Implications for the sedimentary record. In: Zahn R, Kaminski M, Labeyrie L, and Pederson TF (eds.) *Carbon Cycling in the Glacial Ocean: Constraints on the Ocean's Role in Global Change*, pp. 259–279. Berlin: Springer.

Needoba JA, Sigman DM, and Harrison PJ (2004) The mechanism of isotope fractionation during algal nitrate assimilation as illuminated by the $^{15}N/^{14}N$ of intracellular nitrate. *Journal of Phycology* 40: 517–522.

Owens NJP (1987) Natural variations in ^{15}N in the marine environment. *Advances in Marine Biology* 24: 390–451.

Peterson BJ and Fry B (1987) Stable isotopes in ecosystem studies. *Annual Review of Ecology and Systematics* 18: 293–320.

Sigman DM, Granger J, DiFiore PJ, Lehmann MF, Ho R, Cane G, and van Geen A (2005) Coupled nitrogen and oxygen isotope measurements of nitrate along the eastern North Pacific margin. *Global Biogeochemical Cycles* 19: GB4022 (doi:10.1029/2005GB002458).

Wada E (1980) Nitrogen isotope fractionation and its significance in biogeochemical processes occurring in marine environments. In: Goldberg ED, Horibe Y, and Saruhashi K (eds.) *Isotope Marine Chemistry*, pp. 375–398. Tokyo: Uchida Rokakudo.

OXYGEN ISOTOPES IN THE OCEAN

K. K. Turekian, Yale University, New Haven, CT, USA

The oxygen isotope signature of sea water varies as a function of the processing of water in the oceanic cycle. The two chemical parameters, salinity and oxygen isotope ratio, are distinctive for various water types. The oxygen-18 to oxygen-16 ratio is represented in comparison to a standard. The notation is δO^{18} which is defined as follows:

$$\left[\frac{^{18}O/^{16}O_{\text{sample}}}{^{18}O/^{16}O_{\text{standard}}} - 1\right] \times 1000$$

Figure 1 shows the results for the world oceans from Craig and Gordon (1965). During the GEOSECS program the oxygen isotope ratios of seawater samples were also determined. The features resemble those in **Figure 1.** The GEOSECS data are available in the shore-based measurements volume of the GEOSECS Atlas (1987) published by the US National Science Foundation.

The tracking of fresh water from streams draining into the ocean at different latitudes has been used to

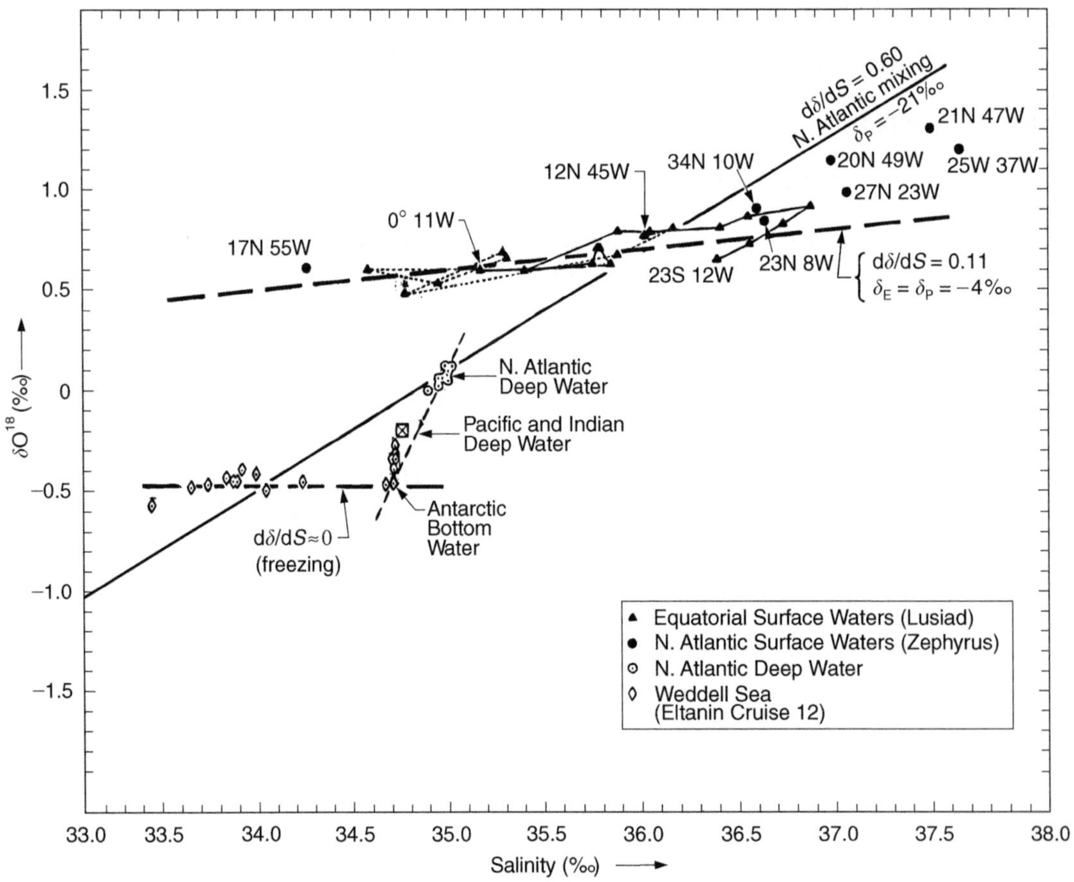

Figure 1 Oxygen-18–salinity relationships in Atlantic surface and deep waters. δ_E and δ_P refer to the isotopic composition of evaporating vapor and precipitation, respectively. (From Craig and Gordon, 1965.)

study several coastal oceanic regimes. The work of Fairbanks (1982) is one of the earliest of these efforts.

See also

Cenozoic Climate - Oxygen Isotope Evidence. River Inputs.

Further Reading

Craig H and Gordon LI (1965) Deuterium and oxygen-18 variations in the ocean and marine atmosphere. *Stable Isotopes in Oceanographic Studies and Paleotemperatures, Consiglio Nazionale Delle Ricerche, Laboratorio di Geologia Nucleare-Pisa, 122 pp. (Also in Symposium on Marine Chemistry, Publ. 3., Kingston, Graduate School of Oceanography, University of Rhode Island,* 277–374).

Fairbanks RG (1982) The origin of continental shelf and slope water in the New York Bight and Gulf of Maine: evidence from $H_2^{18}O/H_2^{16}O$ ratio measurements. *Journal of Geophysical Research* 87: 5796–5808.

GEOSECS Atlantic, Pacific, and Indian Ocean Expeditions, Volume 7, Shorebased Data and Graphics (1987) National Science Foundation. 200pp.

STABLE CARBON ISOTOPE VARIATIONS IN THE OCEAN

K. K. Turekian, Yale University, New Haven, CT, USA

The two stable isotopes of carbon, ^{12}C and ^{13}C, vary in proportions in different reservoirs on earth. The ratio of ^{13}C to ^{12}C is commonly given relative to a standard (a belemnite from the Peedee formation in South Carolina and therefore called PDB). On the basis of this standard $\delta^{13}C$ is defined as:

$$\left[\frac{^{13}C/^{12}C_{\text{sample}}}{^{13}C/^{12}C_{\text{standard}}} - 1\right] \times 1000$$

The values for some major carbon reservoirs are: marine limestones, $\delta^{13}C = 0$; C-3 plants, $\delta^{13}C = -25$; air CO_2, $\delta^{13}C = -7$. The inorganic carbon in the surface ocean is in isotopic equilibrium with atmospheric CO_2 and has a value of about 2. Organic matter in the shallow ocean ranges from -19 at high latitudes to -28 at low latitudes. The midlatitude value is around -21. The transport of organic matter to depth and subsequent metabolism adds inorganic carbon to the water. The isotopic composition of dissolved inorganic carbon then reflects the amount of addition of this metabolic carbon. **Figure 1** is a profile of $\delta^{13}C$ for the North Pacific. It is typical of other profiles in the oceans.

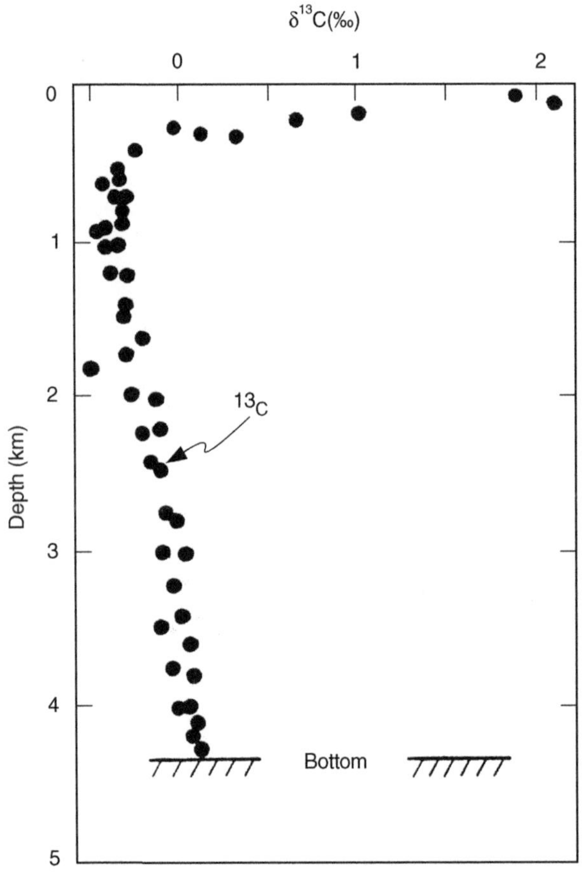

Carbon isotope measurements in all the oceans were made on the GEOSECS expedition. These values are given in the *GEOSECS Atlas* (1987).

Figure 1 Variation of $\delta^{13}C$ in dissolved inorganic carbon with depth in the Pacific Ocean at GEOSECS Station 346 (28°N, 121°W) (Kroopnick, Deuser and Craig, 1970).

See also

Carbon Cycle. Carbon Dioxide (CO_2) Cycle.

Further Reading

Chesselet R, Fontagne M, Buat-Menard P, Ezat U, and Lambert CE (1981) The origin of particulate organic matter in the marine atmosphere as indicated by its stable carbon isotopic composition. *Geophysical Research Letters* 8: 345–348.

GEOSECS Atlantic, Pacific, and Indian Ocean Expeditions, vol 7: *Shorebased Data and Graphics*. National Science Foundation. 200 pp. (1987).

Kroopnick P, Deuser WG, and Craig H (1970) Carbon-13 measurements on dissolved inorganic carbon at the North Pacific (1969) GEOSECS station. *Journal of Geophysical Research* 75: 7668–7671.

Sackett WM (1964) The depositional history and isotopic organic composition of marine sediments. *Marine Geology* 2: 173–185.

PHOSPHORUS CYCLE

K. C. Ruttenberg, Woods Hole Oceanographic Institution, Woods Hole, MA, USA

Introduction

The global phosphorus cycle has four major components: (i) tectonic uplift and exposure of phosphorus-bearing rocks to the forces of weathering; (ii) physical erosion and chemical weathering of rocks producing soils and providing dissolved and particulate phosphorus to rivers; (iii) riverine transport of phosphorus to lakes and the ocean; and (iv) sedimentation of phosphorus associated with organic and mineral matter and burial in sediments (**Figure 1**). The cycle begins anew with uplift of sediments into the weathering regime.

Phosphorus is an essential nutrient for all life forms. It is a key player in fundamental biochemical reactions involving genetic material (DNA, RNA) and energy transfer (adenosine triphosphate, ATP), and in structural support of organisms provided by membranes (phospholipids) and bone (the biomineral hydroxyapatite). Photosynthetic organisms utilize dissolved phosphorus, carbon, and other essential nutrients to build their tissues using energy from the sun. Biological productivity is contingent upon the availability of phosphorus to these organisms, which constitute the base of the food chain in both terrestrial and aquatic systems.

Phosphorus locked up in bedrock, soils, and sediments is not directly available to organisms. Conversion of unavailable forms to dissolved orthophosphate, which can be directly assimilated, occurs through geochemical and biochemical reactions at various stages in the global phosphorus cycle. Production of biomass fueled by phosphorus bioavailability results in the deposition of organic matter in soil and sediments, where it acts as a source of fuel and nutrients to microbial communities. Microbial activity in soils and sediments, in turn, strongly influences the concentration and chemical form of phosphorus incorporated into the geological record.

This article begins with a brief overview of the various components of the global phosphorus cycle. Estimates of the mass of important phosphorus reservoirs, transport rates (fluxes) between reservoirs, and residence times are given in **Tables 1** and **2**. As is clear from the large uncertainties associated with these estimates of reservoir size and flux, there remain many aspects of the global phosphorus cycle that are poorly understood. The second half of the article describes current efforts underway to advance our understanding of the global phosphorus cycle. These include (i) the use of phosphate oxygen isotopes ($\delta^{18}O\text{-}PO_4$) as a tool for identifying the role of microbes in the transformer of phosphate from one reservoir to another; (ii) the use of naturally occurring cosmogenic isotopes of phosphorus (^{32}P and ^{33}P) to provide insight into phosphorus-cycling pathways in the surface ocean; (iii) critical evaluation of the potential role of phosphate limitation in coastal and open ocean ecosystems; (iv) reevaluation of the oceanic residence time of phosphorus; and (v) rethinking the global phosphorus-cycle on geological timescales, with implications for atmospheric oxygen and phosphorus limitation primary productivity in the ocean.

The Global Phosphorus Cycle: Overview

The Terrestrial Phosphorus Cycle

In terrestrial systems, phosphorus resides in three pools: bedrock, soil, and living organisms (biomass) (**Table 1**). Weathering of continental bedrock is the principal source of phosphorus to the soils that support continental vegetation (F_{12}); atmospheric deposition is relatively unimportant (F_{82}). Phosphorus is weathered from bedrock by dissolution of phosphorus-bearing minerals such as apatite ($Ca_{10}(PO_4)_6(OH, F, Cl)_2$), the most abundant primary phosphorus mineral in crustal rocks. Weathering reactions are driven by exposure of minerals to naturally occurring acids derived mainly from microbial activity. Phosphate solubilized during weathering is available for uptake by terrestrial plants, and is returned to the soil by decay of litterfall (**Figure 1**).

Soil solution phosphate concentrations are maintained at low levels as a result of absorption of phosphorus by various soil constituents, particularly ferric iron and aluminum oxyhydroxides. Sorption is considered the most important process controlling terrestrial phosphorus bioavailability. Plants have different physiological strategies for obtaining phosphorus despite low soil solution concentrations.

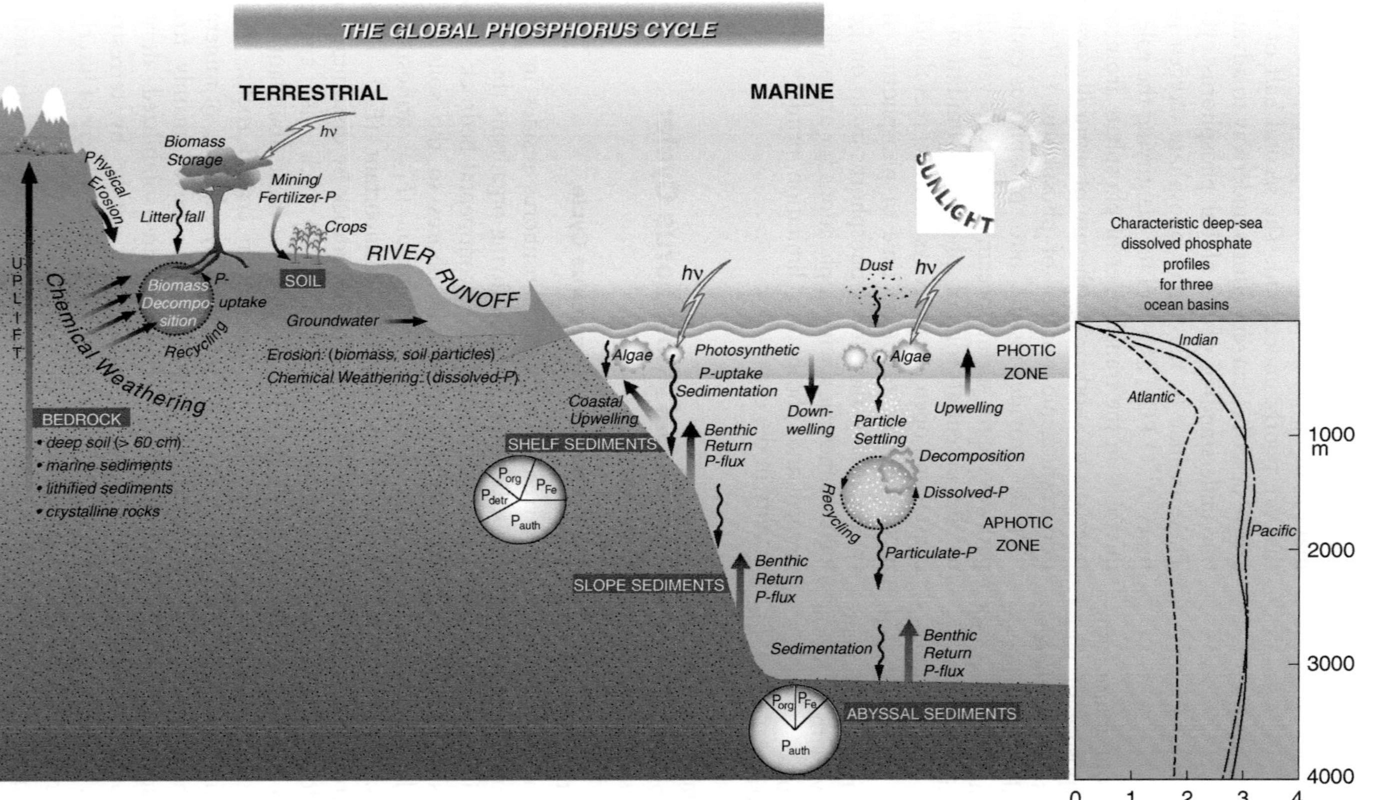

Figure 1 Cartoon illustrating the major reservoirs and fluxes of phosphorus described in the text and summarized in Tables **Table 1** and **2**. The oceanic photic zone, idealized in the cartoon, is typically thinner in coastal environments owing to turbidity from continental terrigenous input, and deepens as the water column clarifies with distance away from the continental margins. The distribution of phosphorus among different chemical/mineral forms in marine sediments is given in the pie diagrams, where the abbreviations used are: P_{org}, organic phosphorus; P_{Fe}, iron-bound phosphorus; P_{detr}, detrital apatite; P_{auth}, authigenic/biogenic apatite. The P_{org}, P_{Fe}, and P_{auth} reservoirs represent potentially reactive phosphorus pools (see text and **Tables 2** and **5** for discussion), whereas the P_{detr} pool reflects mainly detrital apatite weathered off the continents and passively deposited in marine sediments (note that P_{detr} is not an important sedimentary phosphorus component in abyssal sediments, far from continents).Continental margin phosphorus speciation data were compiled from Louchouarn P, Lucotte M, Duchemin E and de Vernal A (1997) Early diagenetic processes in recent sediments of the Gulf of St-Lawrence: Phosphorus, carbon and iron burial rates. *Marine Geology* 139(1/4): 181–200, and Ruttenberg KC and Berner RA (1993) Authigenic apatite formation and burial in sediments from non-upwelling continental margin environments. *Geochimica et Cosmochimica Acta* 57: 991–1007. Abyssal sediment phosphorus speciation data were compiled from Filippelli GM and Delaney ML (1996) Phosphorus geochemistry of equatorial Pacific sediments. *Geochimica et Cosmochimica Acta* 60: 1479}1495, and Ruttenberg KC (1990) *Diagenesis and burial of phosphorus in marine sediments: implications for the marine phosphorus budget.* PhD thesis, Yale University. The global phosphorus cycle cartoon is from Ruttenberg (2000). The global phosphorus cycle. In: *The Encyclopedia of Global Change*, Oxford University Press. (in Press), with permission. The vertical water column distributions of phosphate typically observed in the three ocean basins are shown in the panel to the right of the global phosphorus cycle cartoon, and are from Sverdrup HV, Johnson MW and Fleming RH (1942) *The Oceans, Their Physics, Chemistry and General Biology*. New York: Prentice Hall, © 1942 Prentice Hall; used with permission.

Table 1 Major reservoirs active in the global phosphorus cycle and associated residence times

Reservoir no.	*Reservoir description*	*Reservoir size (10^{12} mol P)*	*Reference*	*Residence time τ (years)*
R1	Sediments (crustal rocks and soil > 60 cm deep and marine sediments)	0.27×10^{8}–1.3×10^{8}	b, a = c = d	$42\text{–}201\times10^{6}$
R2	Land (≈ total soil < 60 cm deep: organic + inorganic)	3100–6450	b, a = c = d	425–2311
R3	Land biota	83.9–96.8	b, a = c = d	13–48
R4	Surface ocean, 0–300 m (total dissolved P)	87.4	a = c	2.46–4.39
R5	Deep sea, 300–3300 m (total dissolved P)	2810	a = c & d	1502
R6	Oceanic biota	1.61–4.45	b & d, a = c & d	0.044–0.217 (16–78 d)
R7	Minable P	323–645	a = c, b & d	718–1654
R8	Atmospheric P	0.0009	b = c = d	0.009 (80 h)

[a]*Notes*

(1) Ranges are reported for those reservoirs for which a consensus on a single best estimated reservoir size does not exist. Maximum and minimum estimates found in a survey of the literature are reported. References cited before the comma refer to the first (lowest) estimate, those after the comma refer to the second (higher) estimate. References that give identical values are designated by an equality sign, references giving similar values are indicated by an ampersand. As indicated by the wide ranges reported for some reservoirs, all calculations of reservoir size have associated with them a large degree of uncertainty. Methods of calculation, underlying assumptions, and sources of error are given in the references cited.

(2) Residence times are calculated by dividing the concentration of phosphorus contained in a given reservoir by the sum of fluxes out of the reservoir. Where ranges are reported for reservoir size and flux, maximum and minimum residence time values are given; these ranges reflect the uncertainties inherent in reservoir size and flux estimates. Fluxes used to calculate residence times for each reservoir are as follows: R1 (F_{12}), R2 ($F_{23} + F_{28} + F_{24(d)} + F_{24(p)}$), R3 ($F_{32}$), R4 ($F_{45} + F_{46}$), R5 ($F_{54}$), R6 ($F_{64} + F_{65}$), R7 ($F_{72}$), R8 ($F_{82} + F_{84}$). Flux estimates are given in **Table 2**. The residence time of R5 is decreased to 1492 y by inclusion of the scavenged flux of deep-sea phosphate at hydrothermal mid-ocean ridge systems, mostly onto ferric oxide and oxyhydroxide phases (Wheat CG, Feely RA and Mottl MJ (1996). Phosphate removal by oceanic hydrothermal processes: an update of the phosphorus budget in the oceans. *Geochimica et Cosmochimica Acta* 60(19): 3593–3608).

(3) Estimates for the partitioning of the oceanic reservoir between dissolved inorganic phosphorus and particulate phosphorus are given in references b and d as follows: $2581\text{–}2600\times10^{12}$ mol dissolved inorganic phosphorus (b, d) and $20\text{–}21\times10^{12}$ mol particulate phosphorus (d, b).

(4) The residence times estimated for the minable phosphorus reservoir reflect estimates of current mining rates; if mining activity increases or diminishes the residence time will change accordingly.

References

(a) Lerman A, Mackenzie FT and Garrels RM (1975) *Geological Society of America Memoir* 142: 205.

(b) Richey JE (1983). In: Bolin B and Cook RB (eds) *The Major Biogeochemical Cycles and Their Interactions*, SCOPE 21, pp. 51–56. Chichester: Wiley.

(c) Jahnke RA (1992) In: Butcher SS *et al.* (eds) *Global Geochemical Cycles*, pp. 301–315. San Diego: Academic Press. (Values identical to Lerman *et al.*, except the inclusion of the atmospheric reservoir estimate taken from Graham WF and Duce RA (1979) *Geochimica et Cosmochimica Acta* 43: 1195.)

(d) Mackenzie FT, Ver LM, Sabine C, Lane M and Lerman A (1993) In: Wollast R, Mackenzie FT and Chou L (eds) *Interactions of C, N, P and S Biogeochemical Cycles and Global Change.* NATO ASI Series 1, vol. 4, pp. 1–61. Berlin: Springer-Verlag.

For example, some plants can increase root volume and surface area to optimize uptake potential. Alternatively, plant roots and/or associated fungi can produce chelating compounds that solubilize ferric iron and calcium-bound phosphorus, enzymes and/or acids that solubilize phosphate in the root vicinity. Plants also minimize phosphorus loss by resorbing much of their phosphorus prior to litterfall, and by efficient recycling from fallen litter. In extremely unfertile soils (e.g., in tropical rain forests) phosphorus recycling is so efficient that topsoil contains virtually no phosphorus; it is all tied up in biomass.

Systematic changes in the total amount and chemical form of phosphorus occur during soil development. In initial stages, phosphorus is present mainly as primary minerals such as apatite. In mid-stage soils, the reservoir of primary apatite is diminished; less-soluble secondary minerals and organic phosphorus make up an increasing fraction of soil phosphorus. Late in soil development, phosphorus is partitioned mainly between refractory minerals and organic phosphorus (**Figure 2**).

Transport of Phosphorus from Continents to the Ocean

Phosphorus is transferred from the continental to the oceanic reservoir primarily by rivers (F_{24}). Deposition

Table 2 Fluxes between the major phosphorus reservoirs[a]

Flux no.	*Description of flux*	*Flux (10^{12} mol P y^{-1})*	*References and comments*
Reservoir fluxes			
F_{12}	Rocks/sediments → soils (erosion/weathering, soil accumulation)	0.645	a = c & d
F_{21}	Soils → rocks/sediments (deep burial, lithification)	0.301–0.603	d, a = c
F_{23}	Soils → land biota	2.03–6.45	a = c, b & d
F_{32}	Land biota → soils	2.03–6.45	a = c, b & d
$F_{24(d)}$	Soil → surface ocean (river total dissolved P flux)	0.032–0.058	e, a = c; ~ > 50% of TDP is DOP (e)
$F_{24(p)}$	Soil → surface ocean (river particulate P flux)	0.59–0.65	d, e; ~40% of RSPM-P (Riverine Suspended Particulate Matter-Phosphorus) is organic P (e); it is estimated that between 25–45% is reactive once it enters the ocean (f).
F_{46}	Surface ocean → oceanic biota	19.35–35	b, d; a = c = 33.5, b reports upper limit of 32.3; d reports lower limit of 28.2
F_{64}	Oceanic biota → surface ocean	19.35–35	b, d; a & c = 32.2, b reports upper limit of 32.3, d reports lower limit of 28.2
F_{65}	Oceanic biota → deep sea (particulate rain)	1.13–1.35	d, a = c
F_{45}	Surface ocean → deep sea (downwelling)	0.581	a = c
F_{54}	Deep sea → surface ocean (upwelling)	1.87	a = c
F_{42}	Surface ocean → land (fisheries)	0.01	d
F_{72}	Minable P → land (soil)	0.39–0.45	a = c = d, b
F_{28}	Land (soil) → atmosphere	0.14	b = c = d
F_{82}	Atmosphere → land (soil)	0.1	b = c = d
F_{48}	Surface ocean → atmosphere	0.01	b = c = d
F_{84}	Atmosphere → surface ocean	0.02–0.05	c, b; d gives 0.04; ~30% of atmospheric aerosol P is soluble (g)
Subreservoir fluxes: marine sediments			
sF_{ms}	Marine sediment accumulation (total)	0.265–0.280	i, j; for higher estimate (j), use of sediment P concentration below the diagenesis zone implicitly accounts for P loss via benthic remineralization flux and yields pre-anthropogenic net burial flux. For estimates of reactive P burial see note (j).
sF_{cs}	Continental margin ocean sediments → burial	0.150–0.223	j, i; values reported reflect total P, reactive P burial constitutes from 40–75% of total P (h). These values reflect pre-agricultural fluxes, modern value estimated as 0.33 (d).
sF_{as}	Abyssal (deep sea) sediments → burial	0.042–0.130	i, j; a = c gives a value of 0.055. It is estimated that 90–100% of this flux is reactive P (h). These values reflect pre-agricultural fluxes, modern value estimates range from 0.32 (d) to 0.419 (b).

(*Continued*)

Table 2 *Continued*

Flux no.	*Description of flux*	*Flux (10^{12} mol P y^{-1})*	*References and comments*
sF_{cbf}	Coastal sediments → coastal waters (remineralization, benthic flux)	0.51–0.84	d, k; these values reflect pre-agricultural fluxes, modern value estimated as 1.21 with uncertainties ±40% (k)
sF_{abf}	Abyssal sediments → deep sea (remineralization, benthic flux)	0.41	k; this value reflects pre-agricultural fluxes, modern value estimated as 0.52, uncertainty ±30% (k)

[a]*Notes*

(1) Reservoir fluxes (*F*) represent the P-flux between reservoirs #R1–R8 defined in **Table 1**. The subreservoir fluxes (*sF*) refer to the flux of phosphorus into the marine sediment portion of reservoir #1 via sediment burial, and the flux of diagenetically mobilized phosphorus out of marine sediments via benthic return flux. These subfluxes have been calculated as described in references h–k. Note that the large magnitude of these sub-fluxes relative to those into and out of reservoir #1 as a whole, and the short oceanic-phosphorus residence time they imply (**Tables 1** and **5**), highlight the dynamic nature of the marine phosphorus cycle.

(2) Ranges are reported where consensus on a single best estimate does not exist. References cited before the comma refer to the first (lowest) estimate, those after the comma refer to the second (higher) estimate. References that give identical values are designated by an equality sign, references giving similar values are indicated by an ampersand. Maximum and minimum estimates found in a survey of the literature are reported. In some cases this range subsumes ranges reported in the primary references. As indicated by the wide ranges reported, all flux calculations have associated with them a large degree of uncertainty. Methods of calculation, underlying assumptions, and sources of error are given in the references cited.

References

(a) Lerman A, Mackenzie FT and Garrels RM (1975) *Geological Society of America Memoir* 142: 205.

(b) Richey JE (1983). In: Bolin B and Cook RB (eds), *The Major Biogeochemical Cycles and Their Interactions*, SCOPE 21, pp. 51–56. Chichester: Wiley.

(c) Jahnke RA (1992) In: Butcher SS *et al.* (eds), *Global Geochemical Cycles*, pp. 301–315. San Diego: Academic Press. (Values are identical to those found in Lerman *et al.* (1975) except for atmospheric phosphorus fluxes taken from Graham WF and Duce RA (1979) *Geochimica et Cosmochimica Acta* 43: 1195.)

(d) Mackenzie FT, Ver LM, Sabine C, Lane M and Lerman A (1993) In: Wollast R, Mackenzie FT and Chou L (eds), *Interactions of C, N, P and S Biogeochemical Cycles and Global Change*, NATO ASI Series 1, vol. 4, pp. 1–61. Berlin: Springer-Verlag.

(e) Meybeck M. (1982). *American Journal of Science* 282(4): 401.

(f) The range of riverine suspended particulate matter that may be solubilized once it enters the marine realm (e.g. so-called 'reactive phosphorus') is derived from three sources. Colman AS and Holland HD ((2000). In: Glenn, C.R., Prévôt-Lucas L and Lucas J (eds) *Marine Authigenesis: From Global to Microbial*, SEPM Special Publication No. 66. pp. 53–75) estimate that 45% may be reactive, based on RSPM-P compositional data from a number of rivers and estimated burial efficiency of this material in marine sediments. Berner RA and Rao J-L ((1994) *Geochimica et Cosmochimica Acta* 58: 2333) and Ruttenberg KC and Canfield DE ((1994) *EOS, Transactions of the American Geophysical Union* 75: 110) estimate that 35% and 31% of RSPM-P is released upon entering the ocean, based on comparison of RSPM-P and adjacent deltaic surface sediment phosphorus in the Amazon and Mississippi systems, respectively. Lower estimates have been published: 8% (Ramirez AJ and Rose AW (1992) *American Journal of Science* 292: 421); 18% (Froelich PN (1988) *Limnology and Oceanography* 33: 649); 18%: (Compton J, Mallinson D, Glenn CR *et al.* (2000) In: Glenn CR, Prévôt-Lucas L and Lucas J (eds) *Marine Authigenesis: From Global to Microbial*, SEPM Special Publication No. 66, pp. 21–33). Higher estimates have also been published: 69% (Howarth RW, Jensen HS, Marine R and Postma H (1995) In: Tiessen H (ed) *Phosphorus in the Global Environment*, SCOPE 54, pp. 323–345. Chichester: Wiley). Howarth *et al.* (1995) also estimate the total flux of riverine particulate P to the oceans at 0.23×10^{12} moles P y^{-1}, an estimate likely too low because it uses the suspended sediment flux from Milliman JD and Meade RH ((1983) Journal of Geology 91: 1), which does not include the high sediment flux rivers from tropical mountainous terranes (Milliman JD and Syvitski JPM (1992) *Journal of Geology* 100: 525).

(g) Duce RA, Liss PS, Merrill JT *et al.* (1991) *Global Biogeochemical Cycles* 5: 193.

(h) Ruttenberg KC (1993) *Chemical Geology* 107: 405.

(i) Howarth RW, Jensen HS, Marino R and Postma H (1995) In: Tiessen H (ed.) *Phosphorus in the Global Environment*, SCOPE 54, pp. 323–345, Chichester: Wiley.

(j) Phosphorus-burial flux estimates as reported in Ruttenberg ((1993) *Chemical Geology* 107: 405) modified using pre-agricultural sediment fluxes updated by Colman AS and Holland HD ((2000) In: Glenn CR, Prévôt-Lucas L and Lucas J (eds) *Marine Authigenesis: From Global to Microbial*, SEPM Special Publication No. 66, pp. 53–75). Using these total phosphorus burial fluxes and the ranges of likely reactive phosphorus given in the table, the best estimate for reactive phosphorus burial flux in the oceans lies between 0.177 and 0.242×10^{12} moles P y^{-1}. Other estimates of whole-ocean reactive phosphorus burial fluxes range from, at the low end, $0.032–0.081 \times 10^{12}$ moles P y^{-1} (Compton J, Mallinson D, Glenn CR *et al.* (2000) In: Glenn CR, Prévôt-Lucas L and Lucas J (eds) *Marine Authigenesis: From Global to Microbial*, SEPM Special Publication No. 66, pp. 21–33), and 0.09×10^{12} moles P y^{-1} (Wheat CG, Feely RA and Mottl MJ (1996) *Geochimica et Cosmochimica Acta* 60: 3593); to values more comparable to those derived from the table above (0.21×10^{12} moles P y^{-1}: Filippelli GM and Delaney ML (1996) *Geochimica Cosmochimica Acta* 60: 1479)

(k) Colman AS and Holland HD (2000) In: Glenn CR, Prévôt-Lucas L and Lucas J (eds) *Marine Authigenesis: From Global to Microbial*, SEPM Special Publication No. 66, pp. 53–75.

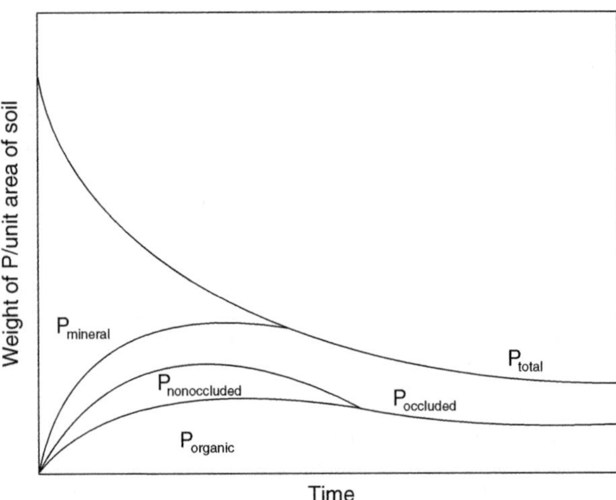

Figure 2 The fate of phosphorus during soil formation can be viewed as the progressive dissolution of primary mineral P (dominantly apatite), some of which is lost from the system by leaching (decrease in P_{total}), and some of which is reincorporated into nonoccluded, occluded and organic fractions within the soil. Nonoccluded P is defined as phosphate sorbed to surface of hydrous oxides of iron and aluminum, and calcium carbonate. Occluded P refers to P present within the mineral matrix of discrete mineral phases. The initial build-up in organic P results from organic matter return to soil from vegetation supported by the soil. The subsequent decline in $P_{organic}$ is due to progressive mineralization and soil leaching. The time-scale over which these transformations occur depends upon the initial soil composition, topographic, and climatic factors. Figure is after Walker and Syers (1976). The fate of phosphorus during pedogenesis. *Geoderma* 15: 1–19, with permission.

of atmospheric aerosols (F_{84}) is a minor flux. Groundwater seepage to the coastal ocean is a potentially important but undocumented flux.

Riverine phosphorus derives from weathered continental rocks and soils. Because phosphorus is particle-reactive, most riverine phosphorus is associated with particulate matter. By most estimates, over 90% of the phosphorus delivered by rivers to the ocean is as particulate phosphorus ($F_{24(p)}$). Dissolved phosphorus in rivers occurs in both inorganic and organic forms. The scant data on dissolved organic phosphorus suggest that it may account for 50% or more of dissolved riverine phosphorus. The chemical form of phosphorus associated with riverine particles is variable and depends upon the drainage basin geology, on the extent of weathering of the substrate, and on the nature of the river itself. Available data suggest that approximately 20–40% of phosphorus in suspended particulate matter is organic. Inorganic forms are partitioned mainly between ferric oxyhydroxides and apatite. Aluminum oxyhydroxides and clays may also be significant carriers of phosphorus.

The fate of phosphorus entering the ocean via rivers is variable. Dissolved phosphorus in estuaries at the continent–ocean interface typically displays nonconservative behavior. Both negative and positive deviations from conservative mixing can occur, sometimes changing seasonally within the same estuary. Net removal of phosphorus in estuaries is typically driven by flocculation of humic-iron complexes and biological uptake. Net phosphorus release is due to a combination of desorption from fresh water particles entering high-ionic-strength marine waters, and flux of diagenetically mobilized phosphorus from benthic sediments. Accurate estimates of bioavailable riverine phosphorus flux to the ocean must take into account, in addition to dissolved forms, the fraction of riverine particulate phosphorus released to solution upon entering the ocean.

Human impacts on the global phosphorus cycle The mining of phosphate rock (mostly from marine phosphorite deposits) for use as agricultural fertilizer (F_{72}) increased dramatically in the latter half of the twentieth century. In addition to fertilizer use, deforestation, increased cultivation, and urban and industrial waste disposal all have enhanced phosphorus transport from terrestrial to aquatic systems, often with deleterious results. For example, elevated phosphorus concentrations in rivers resulting from these activities have resulted in eutrophication in some lakes and coastal areas, stimulating nuisance algal blooms and promoting hypoxic or anoxic conditions that are harmful or lethal to natural populations.

Increased erosion due to forest clear-cutting and widespread cultivation has increased riverine suspended matter concentrations, and thus increased the riverine particulate phosphorus flux. Dams, in contrast, decrease sediment loads in rivers and therefore diminish phosphorus-flux to the oceans. However, increased erosion below dams and diagenetic mobilization of phosphorus in sediments trapped behind dams moderates this effect. The overall effect has been a 50–300% increase in riverine phosphorus flux to the oceans above pre-agricultural levels.

The Marine Phosphorus Cycle

Phosphorus in its simplest form, dissolved orthophosphate, is taken up by photosynthetic organisms at the base of the marine food web. When phosphate is exhausted, organisms may utilize more complex forms by converting them to orthophosphate via enzymatic and microbiological reactions. In the open ocean most phosphorus associated with biogenic particles is recycled within the upper water column.

Efficient stripping of phosphate from surface waters by photosynthesis combined with build-up at depth due to respiration of biogenic particles results in the classic oceanic dissolved nutrient profile. The progressive accumulation of respiration-derived phosphate at depth along the deep-water circulation trajectory results in higher phosphate concentrations in Pacific Ocean deep waters at the end of the trajectory than in the North Atlantic where deep water originates (**Figure 1**).

The sole means of phosphorus removal from the oceans is burial with marine sediments. The phosphorus flux to shelf and slope sediments is larger than the phosphorus flux to the deep sea (**Table 2**) for several reasons. Coastal waters receive continentally derived nutrients via rivers (including phosphorus, nitrogen, silicon, and iron), which stimulate high rates of primary productivity relative to the deep sea and result in a higher flux of organic matter to continental margin sediments. Organic matter is an important, perhaps primary, carrier of phosphorus to marine sediments. Owing to the shorter water column in coastal waters, less respiration occurs prior to deposition. The larger flux of marine organic phosphorus to margin sediments is accompanied by a larger direct terrigenous flux of particulate phosphorus (organic and inorganic), and higher sedimentation rates overall. These factors combine to enhance retention of sedimentary phosphorus. During high sea level stands, the sedimentary phosphorus reservoir on continental margins expands, increasing the phosphorus removal flux and therefore shortening the oceanic phosphorus residence time.

Terrigenous-dominated shelf and slope (hemipelagic) sediments and abyssal (pelagic) sediments have distinct phosphorus distributions. While both are dominated by authigenic Ca-P (mostly carbonate fluorapatite), this reservoir is more important in pelagic sediments. The remaining phosphorus in hemipelagic sediments is partitioned between ferric iron-bound phosphorus (mostly oxyhydroxides), detrital apatite, and organic phosphorus; in pelagic sediments detrital apatite is unimportant. Certain coastal environments characterized by extremely high, upwelling-driven biological productivity and low terrigenous input are enriched in authigenic apatite; these are proto-phosphorite deposits. A unique process contributing to the pelagic sedimentary Fe-P reservoir is sorptive removal of phosphate onto ferric oxyhydroxides in mid-ocean ridge hydrothermal systems.

Mobilization of sedimentary phosphorus by microbial activity during diagenesis causes dissolved phosphate build-up in sediment pore waters, promoting benthic efflux of phosphate to bottom waters or incorporation in secondary authigenic minerals. The combined benthic flux from coastal (sF_{cbf}) and abyssal (sF_{abf}) sediments is estimated to exceed the total riverine phosphorus flux ($F_{24(d+p)}$) to the ocean. Reprecipitation of diagenetically mobilized phosphorus in secondary phases significantly enhances phosphorus burial efficiency, impeding return of phosphate to the water column. Both processes impact the marine phosphorus cycle by affecting the primary productivity potential of surface waters.

Topics of Special Interest in Marine Phosphorus Research

Phosphate Oxygen Isotopes ($\delta^{18}O$-PO_4)

Use of the oxygen isotopic composition of phosphate in biogenic hydroxyapatite (bones, teeth) as a paleotemperature and climate indicator was pioneered by Longinelli in the late 1960s–early 1970s, and has since been fairly widely and successfully applied. A novel application of the oxygen isotope system in phosphates is its use as a tracer of biological turnover of phosphorus during metabolic processes. Phosphorus has only one stable isotope (^{31}P) and occurs almost exclusively as orthophosphate (PO_4^{3-}) under Earth surface conditions. The phosphorus–oxygen bond in phosphate is highly resistant to nonenzymatic oxygen isotope exchange reactions, but when phosphate is metabolized by living organisms, oxygen isotopic exchange is rapid and extensive. Such exchange results in temperature-dependent fractionations between phosphate and ambient water. This property renders phosphate oxygen isotopes useful as indicators of present or past metabolic activity of organisms, and allows distinction of biotic from abiotic processes operating in the cycling of phosphorus through the environment.

Currently, the $\delta^{18}O$-PO_4 system is being applied in a number of studies of marine phosphorus cycling, including (i) application to dissolved sea water inorganic phosphate as a tracer of phosphate source, water mass mixing, and biological productivity; (ii) use in phosphates associated with ferric iron oxyhydroxide precipitates in submarine ocean ridge sediments, where the $\delta^{18}O$-PO_4 indicates microbial phosphate turnover at elevated temperatures. This latter observation suggests that phosphate oxygen isotopes may be useful biomarkers for fossil hydrothermal vent systems.

Reevaluating the Role of Phosphorus as a Limiting Nutrient in the Ocean

In terrestrial soils and in the euphotic zone of lakes and the ocean, the concentration of dissolved orthophosphate is typically low. When bioavailable phosphorus is exhausted prior to more abundant nutrients, it limits the amount of sustainable biological productivity. Phosphorus limitation in lakes is widely accepted, and terrestrial soils are often phosphorus-limited. In the oceans, however, phosphorus limitation is the subject of controversy and debate.

The prevailing wisdom has favored nitrogen as the limiting macronutrient in the oceans. However, a growing body of literature convincingly demonstrates that phosphate limitation of marine primary productivity can and does occur in some marine systems. In the oligotrophic gyres of both the western North Atlantic and subtropical North Pacific, evidence in the form of dissolved nitrogen : phosphorous (N:P) ratios has been used to argue convincingly that these systems are currently phosphate-limited. The N(:)P ratio of phytoplankton during nutrient-sufficient conditions is 16N:1P (the Redfield ratio) (*see*). A positive deviation from this ratio indicates probable phosphate limitation, while a negative deviation indicates probable nitrogen limitation. In the North Pacific at the Hawaiian Ocean Time Series (HOT) site, there has been a shift since the 1988 inception of the time series to N:P ratios exceeding the Redfield ratio in both particulate and surface ocean dissolved nitrogen and phosphorus (**Figure 3**). Coincident with this shift has been an increase in the prevalence of the nitrogen-fixing cyanobacterium *Trichodesmium* (**Table 3**). Currently, it appears as though the supply of new nitrogen has shifted from a limiting flux of upwelled nitrate from below the euphotic zone to an unlimited pool of atmospheric N_2 rendered bioavailable by the action of nitrogen fixers. This shift is believed to result from climatic changes that promote water column stratification, a condition that selects for N_2-fixing microorganisms, thus driving the system to phosphate limitation. A similar situation exists in the subtropical Sargasso Sea at the Bermuda Ocean Time Series (BATS) site, where currently the dissolved phosphorus concentrations (especially dissolved inorganic phosphorus (DIP)) are significantly lower than at the HOT site, indicating even more severe phosphate limitation (**Table 3**).

A number of coastal systems also display evidence of phosphate limitation, sometimes shifting seasonally from nitrogen to phosphate limitation in concert with changes in environmental features such as upwelling and river runoff. On the Louisiana Shelf in the Gulf of Mexico, the Eel River Shelf of northern California (USA), the upper Chesapeake Bay (USA), and portions of the Baltic Sea, surface water column dissolved inorganic N:P ratios indicate seasonal phosphate limitation. The suggestion of phosphate limitation is reinforced in the Louisiana Shelf and Eel River Shelf studies by the occurrence or presence of alkaline phosphatase activity, an enzyme induced only under phosphate-limiting conditions. Alkaline phosphatase has also been observed seasonally in Narragansett Bay. Although these coastal sites are recipients of anthropogenically derived nutrients (nitrogen and phosphate) that stimulate primary productivity above 'natural' levels, the processes that result in shifts in the limiting nutrient are not

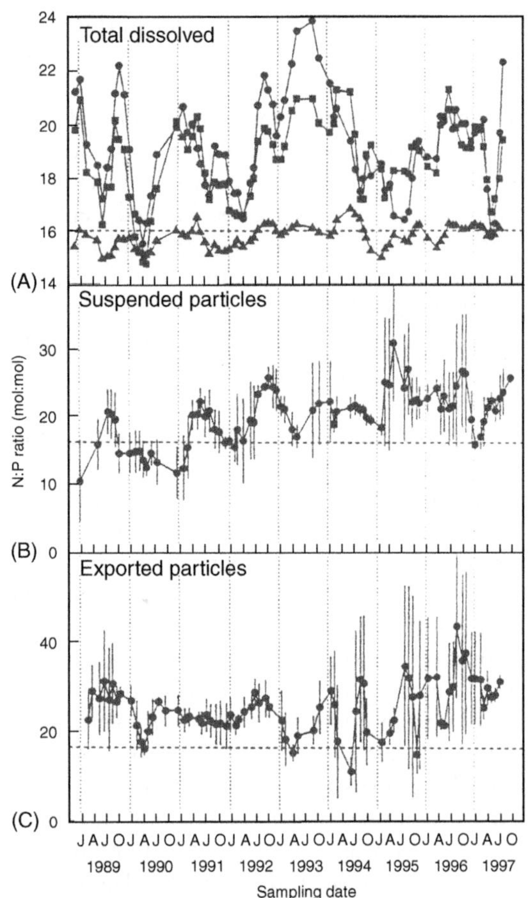

Figure 3 Time-series molar N:P ratios in (A) the dissolved pool, (B) suspended particulate matter, and (C) exported particulate matter from the HOT time series site at station ALOHA in the subtropical North Pacific near Hawaii. (A) The 3-point running mean N:P ratios for 0–100 m (circles), 100–200 m (squares), and 200–500 m (triangles). (B) The 3-point running mean (±1 SD) for the average suspended particulate matter in the upper water column (0–100 m). (C) The 3-point running mean (±1 SD) for the average N:P ratio of sediment trap-collected particulate matter at 150 m. The Redfield ratio (N:P16) is represented by a dashed line in all three panels. Particulate and upper water column dissolved pools show an increasing N:P ratio throughout the time-series, with a preponderance of values in excess of the Redfield ratio. (After Karl DM, Letelier R, Tupas L *et al.* (1997) The role of nitrogen fixation in the biogeochemical cycling in the subtropical North Pacific Ocean. *Nature* 388: 533–538, with permission.)

Table 3 Parameters affecting nutrient limitation: comparison between North Atlantic and North Pacific Gyres

Parameter	*Sargasso Sea*	*Pacific HOT site*
DIP	0.48 ± 0.27[a]	9–40[b]
TDN ($n mol l^{-1}$)	4512 ± 430	5680 ± 620[b]
TDP ($n mol l^{-1}$)	75 ± 42	222 ± 14[b]
TDN:TDP	60 ± 7	26 ± 3[b]
N_2-fixation rate ($mmol N m^{-2} y^{-1}$)	72[c]	31–51[c]

After Wu J, Sunda W, Boyle EA and Karl DM (2000) *Science* 289: 759, with permission.

[a]Average DIP between 26° and 31°N in Sargasso Sea surface waters in March 1998.

[b]North Pacific near Hawaii at station ALOHA (the HOT site) during 1991–1997.

[c]See Wu *et al.* (2000) for method of calculation or measurement.

necessarily related to anthropogenic effects. Other oceanic sites where phosphorus limitation of primary productivity has been documented include the Mediterranean Sea and Florida Bay (USA).

One key question that studies of nitrogen and phosphorus limitation must address before meaningful conclusions may be drawn about phosphorus versus nitrogen limitation of marine primary productivity is the extent to which the dissolved organic nutrient pools are accessible to phytoplankton. In brief, this is the question of bioavailability. Many studies of nutrient cycling and nutrient limitation do not include measurement of these quantitatively important nutrient pools, even though there is indisputable evidence that at least some portion of the dissolved organic nitrogen (DON) and dissolved organic phosphorus (DOP) pools is bioavailable. An important direction for future research is to characterize the DON and DOP pools at the molecular level, and to evaluate what fraction of these are bioavailable. The analytical challenge of identifying the molecular composition of the DOP pool is significant. Recent advances in ^{31}P-nuclear magnetic resonance (NMR) spectroscopy have permitted a look at the high molecular weight (>1 nm) fraction of the DOP pool. However, this fraction represents only one-third of the total DOP pool; the other two-thirds of this pool, made up of smaller-molecular-weight DOP compounds, remains outside our current window of analytical accessibility.

Debate continues among oceanographers about the most probable limiting nutrient on recent and on long, geological timescales. While there is an abundant reservoir of nitrogen (gaseous N_2) in the atmosphere that can be rendered bioavailable by nitrogen-fixing photosynthetic organisms, phosphorus supply to the ocean is limited to that weathered off the continents and delivered by rivers, with some minor atmospheric input. As a consequence of continental weathering control on phosphorus supply to the oceans, phosphorus limitation has been considered more likely than nitrogen limitation on geological timescales. As the recent studies reviewed in this section suggest, phosphorus limitation may be an important phenomenon in the modern ocean as well. Overall, current literature indicates that there is a growing appreciation of the complexities of nutrient limitation in general, and the role of phosphorus limitation in particular, in both fresh water and marine systems.

Cosmogenic ^{32}P and ^{33}P as Tracers of Phosphorus Cycling in Surface Waters

There are two radioactive isotopes of phosphorus, ^{32}P (half-life = 14.3 days) and ^{33}P (half-life = 25.3 days). Both have been widely used in the study of biologically mediated phosphorus cycling in aquatic systems. Until very recently, these experiments have been conducted by artificially introducing radiophosphorus into laboratory incubations or, far more rarely, by direct introduction into natural waters under controlled circumstances. Such experiments necessarily involve significant perturbation of the system, which can complicate interpretation of results. Recent advances in phosphorus sampling and radioisotope measurement have made it possible to use naturally produced ^{32}P and ^{33}P as *in situ* tracers of phosphorus recycling in surface waters. This advance has permitted studies of net phosphorus recycling in the absence of experimental perturbation caused by addition of artificially introduced radiophosphorus.

^{32}P and ^{33}P are produced naturally in the atmosphere by interaction of cosmic rays with atmospheric argon nuclei. They are then quickly scavenged onto aerosol particles and delivered to the ocean surface predominantly in rain. The ratio of $^{33}P/^{32}P$ introduced to the oceans by rainfall remains relatively constant, despite the fact that absolute concentrations can vary from one precipitation event to another. Once the dissolved phosphorus is incorporated into a given surface water phosphorus pool (e.g., by uptake by phytoplankton or bacteria, grazing of phytoplankton or bacteria by zooplankton, or abiotic sorption), the $^{33}P/^{32}P$ ratio will increase in a systematic way as a given pool ages. This increase in the $^{33}P/^{32}P$ ratio with time results from the different half-lives of the two phosphorus radioisotopes. By measuring the $^{33}P/^{32}P$ ratio in rain and in different marine phosphorus pools – e.g., DIP, DOP (sometimes called soluble nonreactive phosphorus, or SNP), and particulate phosphorus of various size classes corresponding to different levels in the food chain – the net age of phosphorus in any of these reservoirs can be determined (**Table 4**). New insights into phosphorus cycling in oceanic surface waters derived from recent work using the cosmogenically produced $^{33}P/^{32}P$ ratio include the following. (1) Turnover rates of dissolved inorganic phosphorus in coastal and oligotrophic oceanic surface waters ranges from 1 to 20 days. (2) Variable turnover rates in the DOP pool range from <1 week to >100 days, suggesting differences in either the demand for DOP, or the lability of DOP toward enzymatic breakdown. (3) In the Gulf of Maine, DOP turnover times vary seasonally, increasing from 28 days in July to >100 days in August, suggesting that the DOP pool may evolve compositionally during the growing seasons. (4) Comparison of the $^{33}P/^{32}P$ ratio in different particulate size classes indicates that the age of phosphorus generally increases at successive levels in the food chain. (5) Under some circumstances, the $^{33}P/^{32}P$ ratio can reveal which dissolved pool is being ingested by a particular size class of organisms. Utilization of this new tool highlights the dynamic nature of phosphorus cycling in surface waters by revealing the rapid rates and temporal variability of phosphorus turnover. It further stands to provide new insights into ecosystem nutrient dynamics by revealing, for example, that low phosphorus concentrations can support high primary productivity through rapid turnover rates, and that there is preferential utilization of particular dissolved phosphorus pools by certain classes of organisms.

The Oceanic Residence Time of Phosphorus

As phosphorus is the most likely limiting nutrient on geological timescales, an accurate determination of its oceanic residence time is crucial to understanding how levels of primary productivity may have varied in the Earth's past. Residence time provides a means of evaluating how rapidly the oceanic phosphorus inventory may have changed in response to variations in either input (e.g., continental weathering, dust flux) or output (e.g., burial with sediments). In its role as limiting nutrient, phosphorus will dictate the amount of surface-ocean net primary productivity, and hence atmospheric CO_2 drawdown that will occur by photosynthetic biomass production. It has been suggested that this so-called 'nutrient–CO_2' connection might link the oceanic phosphorus cycle to climate change due to reductions or increases in the atmospheric greenhouse gas inventory. Oceanic phosphorus residence time and response time (the inverse of residence time) will dictate the timescales over which such a phosphorus-induced climate effect may operate.

Over the past decade there have been several reevaluations of the marine phosphorus cycle in the literature, reflecting changes in our understanding of the identities and magnitudes of important phosphorus sources and sinks. One quantitatively important newly identified marine phosphorus sink is precipitation of disseminated authigenic carbonate fluoroapatite (CFA) in sediments in nonupwelling environments. CFA is the dominant phosphorus mineral in economic phosphorite deposits. Its presence in terrigenous-dominated continental margin environments is detectable only by indirect methods (coupled pore water/solid-phase chemical analyses), because dilution by terrigenous debris renders it below detection limits of direct methods (e.g. X-ray diffraction). Disseminated CFA has now been found in numerous continental margin environments, bearing out early proposals that this is an important marine phosphorus sink. A second class of authigenic phosphorus minerals identified as a quantitatively significant phosphorus sink in sandy continental margin sediments are aluminophosphates. Continental margins in general are quantitatively important sinks for organic and ferric iron-bound phosphorus, as well. When newly calculated phosphorus burial fluxes in continental margins, including the newly identified CFA and aluminophosphate (dominantly aluminum rare-earth phosphate) sinks are combined with older estimates of phosphorus burial fluxes in the deep sea, the overall burial flux results in a much shorter residence time than the canonical value of 100 000 years found in most text books (**Table 5**). This reduced residence time suggests that the oceanic phosphorus-cycle is subject to perturbations on shorter timescales than has previously been believed.

The revised, larger burial flux cannot be balanced by the dissolved riverine input alone. However, when the fraction of riverine particulate phosphorus that is believed to be released upon entering the marine realm is taken into account, the possibility of a balance between inputs and outputs becomes more feasible. Residence times estimated on the basis of phosphorus inputs that include this 'releasable' riverine particulate phosphorus fall within the range of residence time estimates derived from phosphorus burial fluxes (**Table 5**). Despite the large uncertainties associated with these numbers, as evidenced by the maximum and minimum values derived from both input and removal fluxes, these updated residence times are all significantly shorter than the canonical value of 100 000 years. Revised residence times on the order of 10 000–17 000 y make phosphorus-perturbations of the ocean–atmosphere CO_2

Table 4 Turnover rates of dissolved inorganic phosphorus (DIP) [a] and dissolved organic phosphorus (DOP) [b] in surface sea water

Phosphorus pool	*Phosphorus turnover rate*		
	Coastal	*Open Ocean*	*References*
DIP	<1 h to 10 d (>1000 d in Bedford Basin)	Weeks to months	c, d, e, f, g, h, i, j, k, l, m
Total DOP	3 to >90 d	50 to 300 d	l, m, n, o, p, q, r
Bioavailable DOP (model compounds)	2 to 30 d	1 to 4 d	k, s, t, u
Microplankton ($<1\,\mu m$)	>1 to 3 d	NA	m
Phytoplankton ($>1\,\mu m$)	<1 to 8 d	<1 week	m, v
Macrozooplankton ($>280\,\mu m$)	14 to 40 d	30 to 80 d	m, p, q, v, w

After Benitez-Nelson CR (2000) *Earth-Science Reviews* 51: 109, with permission.
[a] DIP is equivalent to the soluble reactive phosphorus (SRP) pool, which may include some phosphate derived from hydrolysis of DOP (e.g., see Monaghan EJ and Ruttenberg KC (1998) *Limnology and Oceanography* 44(7): 1702).
[b] DOP is equivalent to the soluble nonreactive phosphorus (SNP) pool which may include dissolved inorganic polyphosphates (e.g., see Karl DM and Yanagi K (1997) *Limnology and Oceanography* 42: 1398).
[c] Pomeroy LR (1960) *Science* 131: 1731.
[d] Duerden CF (1973) PhD thesis, Dalhousie University, Halifax.
[e] Taft JL, Taylor WR and McCarthy JJ (1975) *Marine Biology* 33: 21.
[f] Harrison WG, Azam F, Renger EH and Eppley RW (1977) *Marine Biology* 40: 9.
[g] Perry MJ and Eppley RW (1981) *Deep-Sea Research* 28: 39.
[h] Smith RE, Harrison WG and Harris L (1985) *Marine Biology* 86: 75.
[i] Sorokin YI (1985) *Marine Ecology Progress Series* 27: 87.
[j] Harrison WG and Harris LR (1986) *Marine Ecology Progress Series* 27: 253.
[k] Björkman K and Karl DM (1994) *Marine Ecology Progress Series* 111: 265.
[l] Björkman K, Thomson-Bulldis AL and Karl DM (2000) *Aquatic and Microbial Ecology* 22: 185.
[m] Benitez-Nelson CR and Buesseler KO (1999) *Nature* 398: 502.
[n] Jackson GA and Williams PM (1985) *Deep-Sea Research* 32: 223.
[o] Orrett K and Karl DM (1987) *Limnology and Oceanography* 32: 383.
[p] Lal D and Lee T (1988) *Nature* 333: 752.
[q] Lee T, Barg E and Lal D (1992) *Analytica Chimica Acta* 260: 113.
[r] Karl DM and Yanagi K (1997) *Limnology and Oceanography* 42: 1398.
[s] Ammerman JW and Azam F (1985) *Limnology and Oceanography* 36: 1437.
[t] Nawrocki MP and Karl DM (1989) *Marine Ecology Progress Series* 57: 35.
[u] Björkman K and Karl DM (1999) *Marine Ecology Progress Series* (submitted).
[v] Waser NAD, Bacon MP and Michaels AF (1996) *Deep-Sea Research* 43(2–3): 421.
[w] Lee T, Barg E and Lal D (1991) *Limnology and Oceanography* 36: 1044.

reservoir on the timescale of glacial–interglacial climate change feasible.

Long Time-scale Phosphorus Cycling, and Links to Other Biogeochemical Cycles

The biogeochemical cycles of phosphorus and carbon are linked through photosynthetic uptake and release during respiration. During times of elevated marine biological productivity, enhanced uptake of surface water CO_2 by photosynthetic organisms results in increased CO_2 evasion from the atmosphere, which persists until the supply of the least abundant nutrient is exhausted. On geological timescales, phosphorus is likely to function as the limiting nutrient and thus play a role in atmospheric CO_2 regulation by limiting CO_2 drawdown by oceanic photosynthetic activity. This connection between nutrients and atmospheric CO_2 could have played a role in triggering or enhancing the global cooling that resulted in glacial episodes in the geological past. It has recently been proposed that tectonics may play the ultimate role in controlling the exogenic phosphorus mass, resulting in long-term phosphorus-limited productivity in the ocean. In this formulation, the balance between subduction of phosphorus bound up in marine sediments and underlying crust and creation of new crystalline rock sets the mass of exogenic phosphorus.

Phosphorus and oxygen cycles are linked through the redox chemistry of iron. Ferrous iron is unstable at the Earth's surface in the presence of oxygen, and oxidizes to form ferric iron oxyhydroxide precipitates, which are extremely efficient scavengers of dissolved phosphate. Resupply of phosphate to surface waters where it can fertilize biological productivity is reduced when oceanic bottom waters are well oxygenated owing to scavenging of phosphate by ferric oxyhydroxides. In contrast, during times in Earth's history when oxygen was not abundant in the atmosphere (Precambrian), and when expanses of the deep ocean were anoxic (e.g., Cretaceous), the

Table 5 Revised oceanic phosphorus input fluxes, removal fluxes, and estimated oceanic residence time

	Flux description[a]	*Flux (10^{12} mol P y^{-1})*	*Residence time (y)*[e]
Input fluxes			
F_{84}	atmosphere → surface ocean	0.02–0.05	
$F_{24(d)}$	soil → surface ocean (river dissolved P flux)[b]	0.032–0.058	
$F_{24(p)}$	soil → surface ocean (river particulate P flux)[b]	0.59–0.65	
	Minimum reactive-P input flux	0.245	12 000
	Maximum reactive-P input flux	0.301	10 000
Removal fluxes			
sF_{cs}	Best estimate of total-P burial in continental margin marine sediments (**Table 2**, note j)[c]	0.150	
sF_{as}	Best estimate of total-P burial in abyssal marine sediments (**Table 2**, note j)[c]	0.130	
	Minimum estimate of reactive-P burial in marine sediments[d]	0.177	17 000
	Maximum estimate of reactive-P burial in marine sediments[d]	0.242	12 000

[a]All fluxes are from **Table 2**.

[b]As noted in **Table 2**, 30% of atmospheric aerosol phosphorus (Duce *et al.* (1991) *Global Biogeochemical Cycles* 5: 193) and 25–45% of the river particulate flux (see note (f) in **Table 2**) is believed to be mobilized upon entering the ocean. The reactive phosphorus input flux was calculated as the sum of $0.3(F_{84}) + F_{24(d)} + 0.35(F_{24(p)})$, where the mean value of the fraction of riverine particulate phosphorus flux estimated as reactive phosphorus (35%) was used. Reactive phosphorus is defined as that which passes through the dissolved oceanic phosphorus reservoir, and thus is available for biological uptake.

[c]These estimates are favored by the author, and reflect the minimum sF_{cs} and maximum sF_{as} fluxes given in **Table 2**. Because the reactive phosphorus contents of continental margin and abyssal sediments differ (see **Table 2** and note d, below), these fluxes must be listed separately in order to calculate the whole-ocean reactive phosphorus burial flux. See note (j) in **Table 2** for other published estimates of reactive-phosphorus burial flux.

[d]As noted in **Table 2**, between 40% and 75% of phosphorus buried in continental margin sediments is potentially reactive, and 90% to 100% of phosphorus buried in abyssal sediments is potentially reactive. The reactive phosphorus fraction of the total sedimentary phosphorus reservoir represents that which may have passed through the dissolved state in oceanic waters, and thus represents a true phosphorus sink from the ocean. The minimum reactive phosphorus burial flux was calculated as the sum of $0.4(sF_{cs}) + 0.9\ (sF_{as})$; the maximum reactive phosphorus burial flux was calculated as the sum of $0.75(sF_{cs}) + 1(sF_{as})$. Both the flux estimates and the % reactive phosphorus estimates have large uncertainties associated with them.

[e]Residence time estimates are calculated as the oceanic phosphorus inventory (reservoirs #4 and 5 (**Table 1**) = 3×10^{15} mol P) divided by the minimum and maximum input and removal fluxes.

potential for a larger oceanic dissolved phosphate inventory could have been realized due to the reduced importance of sequestering with ferric oxyhydroxides. This iron–phosphorus–oxygen coupling produces a negative feedback, which may have kept atmospheric O_2 within equable levels throughout the Phanerozoic. Thus, it is in the oceans that the role of phosphorus as limiting nutrient has the greatest repercussions for the global carbon and oxygen cycles.

Summary

The global cycle of phosphorus is truly a biogeochemical cycle, owing to the involvement of phosphorus in both biochemical and geochemical reactions and pathways. There have been marked advances in the last decade on numerous fronts of phosphorus research, resulting from application of new methods as well as rethinking of old assumptions and paradigms. An oceanic phosphorus residence time on the order of 10 000–20 000 y, a factor of 5–10 shorter than previously cited values, casts phosphorus in the role of a potential player in climate change through the nutrient–CO_2 connection. This possibility is bolstered by findings in a number of recent studies that phosphorus does function as the limiting nutrient in some modern oceanic settings. Both oxygen isotopes in phosphate ($\delta^{18}O\text{-}PO_4$) and *in situ*-produced radiophosphorus isotopes (^{33}P and ^{32}P) are providing new insights into how phosphorus is cycled through metabolic pathways in the marine environment. Finally, new ideas about global phosphorus cycling on long, geological timescales include a possible role for phosphorus in regulating atmospheric oxygen levels via the coupled iron–phosphorus–oxygen cycles, and the potential role of tectonics in setting the exogenic mass of phosphorus. The interplay of new findings in each of these areas is providing us with a fresh look at the marine phosphorus cycle, one that is sure to evolve further as these new areas are explored in more depth by future studies.

See also

Carbon Cycle. Nitrogen Cycle.

Further Reading

Blake RE, Alt JC, and Martini AM (2001) Oxygen isotope ratios of PO_4: an inorganic indicator of enzymatic activity and P metabolism and a new biomarker in the search for life. *Proceedings of the National Academy of Sciences of the USA* 98: 2148–2153.

Benitez-Nelson CR (2001) The biogeochemical cycling of phosphorus in marine systems. *Earth-Science Reviews* 51: 109–135.

Clark LL, Ingall ED, and Benner R (1999) Marine organic phosphorus cycling: novel insights from nuclear magnetic resonance. *American Journal of Science* 299: 724–737.

Colman AS, Holland HD, and Mackenzie FT (1996) Redox stabilization of the atmosphere and oceans by phosphorus limited marine productivity: discussion and reply. *Science* 276: 406–408.

Colman AS, Karl DM, Fogel ML, and Blake RE (2000) A new technique for the measurement of phosphate oxygen isotopes of dissolved inorganic phosphate in natural waters. *EOS, Transactions of the American Geophysical Union F176.*

Delaney ML (1998) Phosphorus accumulation in marine sediments and the oceanic phosphorus cycle. *Global Biogeochemical Cycles* 12(4): 563–572.

Falkowski PG (1997) Evolution of the nitrogen cycle and its influence on the biological sequestration of CO_2 in the ocean. *Nature* 387: 272–275.

Föllmi KB (1996) The phosphorus cycle, phosphogenesis and marine phosphate-rich deposits. *Earth-Science Reviews* 40: 55–124.

Guidry MW, Mackenzie FT and Arvidson RS (2000) Role of tectonics in phosphorus distribution and cycling. In: Glenn CR, Prévt-Lucas L and Lucas J (eds) *Marine Authigenesis: From Global to Microbial.* SEPM Special Publication No. 66, pp. 35–51. (See also in this volume Compton *et al.* (pp. 21–33), Colman and Holland (pp. 53–75), Rasmussen (pp. 89–101), and others.)

Howarth RW, Jensen HS, Marino R and Postma H (1995) Transport to and processing of P in near-shore and oceanic waters. In: Tiessen H (ed.) *Phosphorus in the Global Environment.* SCOPE 54, pp. 232–345. Chichester: Wiley. (See other chapters in this volume for additional information on P-cycling.)

Karl DM (1999) A sea of change: biogeochemical variability in the North Pacific Subtropical Gyre. *Ecosystems* 2: 181–214.

Longinelli A and Nuti S (1973) Revised phosphate–water isotopic temperature scale. *Earth and Planetary Science Letters* 19: 373–376.

Palenik B and Dyhrman ST (1998) Recent progress in understanding the regulation of marine primary productivity by phosphorus. In: Lynch JP and Deikman J (eds.) *Phosphorus in Plant Biology: Regulatory Roles in Molecular, Cellular, Organismic, and Ecosystem Processes*, pp. 26–38. American Society of Plant Physiologists

Ruttenberg KC (1993) Reassessment of the oceanic residence time of phosphorus. *Chemical Geology* 107: 405–409.

Tyrell T (1999) The relative influences of nitrogen and phosphorus on oceanic productivity. *Nature* 400: 525–531.

AIR SEA EXCHANGES

AIR–SEA TRANSFER: DIMETHYL SULFIDE, COS, CS_2, NH_4, NON-METHANE HYDROCARBONS, ORGANO-HALOGENS

J. W. Dacey, Woods Hole Oceanographic Institution, Woods Hole, MA, USA
H. J. Zemmelink, University of Groningen, Haren, The Netherlands

The oceans, which cover 70% of Earth's surface to an average depth of 4000 m, have an immense impact on the atmosphere's dynamics. Exchanges of heat and momentum, water and gases across the sea surface play major roles in global climate and biogeochemical cycling. The ocean can be thought of as a vast biological soup with myriad processes influencing the concentrations of gases dissolved in the surface waters. The quantities of mass flux across the surface interface, though perhaps small on a unit area basis, can be very important because of the extent of the ocean surface and the properties of the gases or their decomposition products in the atmosphere.

Gas exchange across the sea–air surface depends, in part, on differences in partial pressures of the gases between the ocean surface and the atmosphere. The partial pressure of a gas in the gas phase can be understood in terms of its contribution to the pressure in the gas mixture. So the partial pressure of O_2, for example, at 0.21 atm means that at 1 atmosphere total pressure, O_2 is present as 21% of the gas, or mixing, volume. Trace gases are present in the atmosphere at much lower levels, usually expressed as parts per million (10^{-6} atm), parts per billion (10^{-9} atm) or parts per trillion (10^{-12} atm, pptv). Dimethylsulfide (DMS), when present at 100 pptv, accounts for about 100 molecules per 10^{12} molecules of mixed gas phase, or about 10^{-10} of the gas volume.

In solution, a dissolved trace gas in equilibrium with the atmosphere would have the same partial pressure as the gas in the air. Its absolute concentration in terms of molecules or mass per unit volume of water depends on its solubility. Gas solubility varies over many orders of magnitude depending on the affinity of water for the gas molecules and the volatility of the gas. Gases range widely in their solubility in sea water, from the permanent gases like nitrogen (N_2), oxygen (O_2), nitrous oxide (N_2O) and methane (CH_4) that have a low solubility in sea water to the moderately soluble carbon dioxide (CO_2) and dimethylsulfide $(CH_3)_2S$, to highly soluble ammonia (NH_3 and its ionized form NH_4^+) and sulfur dioxide (SO_2). Sulfur dioxide is more than 10^6 times more soluble than O_2 or CH_4. Using the example above of an atmospheric DMS concentration of 100 pptv, the equilibrium concentration of DMS in surface water would be about $0.07\,nmol\,l^{-1}$. Generally the solubility of any individual gas increases at cooler water temperatures, and solubility of gases in sea water is somewhat less than for fresh water because of the so-called 'salting out' effect of dissolved species in sea water.

At any moment the partial pressure difference between surface water and the atmosphere depends on an array of variables. The gases in this article are biogenic, meaning that their mode of formation is the result of one or more immediate or proximate biological processes. These dissolved gases may also be consumed biologically, or removed by chemical processes in sea water, or they may flux across the sea surface to the atmosphere. The rates at which the source and sink processes occur determines the concentration of the dissolved gas in solution as well as the turnover, or residence time, of each compound. Similarly, there can be several source and sink processes for the gases in the atmosphere. Long-lived compounds in the atmosphere will tend to integrate more global processes, whereas short-lived compounds are concentrated near their source and reflect relatively short-term influences of source and sink. In this sense, carbonyl sulfide is a global gas. At it has a residence time of several years in the atmosphere, its concentration does not vary in the troposphere to any appreciable degree. On the other hand, the concentration of DMS varies on a diel basis and with elevation, with higher concentrations at night when atmospheric oxidants (most notable hydroxyl) are relatively depleted.

The extent of disequilibrium between the partial pressures of a gas in the surface water and in the atmosphere determines the thermodynamic gradient which drives gas flux. The kinetics of flux ultimately depend on molecular diffusion and larger-scale mixing processes. Molecular diffusivity is generally captured in a dimensionless parameter, the Schmidt number (ratio of viscosity of water to molecular diffusivity of gas in water), and varies widely between gases depending primarily on the molecular cross-section. From moment to moment, the flux of any particular gas is dependent on interfacial turbulence which is generated by shear between the wind and the sea surface whereby higher wind speed causes increasing turbulence and thus stimulating the onset of waves and eventually the production of bubbles and sea spray. There are considerable uncertainties relating gas exchange to wind speed. These arise due to the various sea-state factors (wave height, swell, breaking waves, bubble entrainment, surfactants, and others) whose individual dependencies on actual wind speed and wind history are not well quantified. The fluxes of gases across the air–sea interface are usually calculated using a wind-speed parameterization. These estimates are considered to be accurate to within a factor of 2 or so.

This article summarizes the characteristics of several important trace gases – dimethylsulfide, carbonyl sulfide, carbon disulfide, nonmethane hydrocarbons, ammonia and methylhalides – focusing on their production and fate as it is determined by biological and chemical processes.

Dimethylsulfide

Natural and anthropogenic sulfur aerosols play a major role in atmospheric chemistry and potentially in modulating global climate. One theory holds that a negative feedback links the emission of volatile organic sulfur (mostly as DMS) from the ocean with the formation of cloud condensation nuclei, thereby regulating, in a sense, the albedo and radiation balance of the earth. The direct (backscattering and reflection of solar radiation by sulfate aerosols) and indirect (cloud albedo) effects of sulfate aerosols may reduce the climatic forcing of trace greenhouse gases like CO_2, N_2O and CH_4. The oxidation products of DMS which also contribute to the acidity of rain, particularly in marine areas, result from industrialized and/or well-populated land.

Dimethylsulfide (DMS) is the most abundant volatile sulfur compound in sea water and constitutes about half of the global biogenic sulfur flux to the atmosphere. Studies of the concentration of DMS in the ocean have shown that average surface water concentrations may vary by up to a factor of 50 between summer and winter in mid and high latitudes. Furthermore, there are large-scale variations in DMS concentration associated with phytoplankton biomass, although there are generally poor correlations between local oceanic DMS concentrations and the biomass and productivity of phytoplankton (due to differences between plankton species in ability to produce DMS).

The nature and rates of the processes involved in the production and consumption of DMS in sea water are important in determining the surface concentrations and the concomitant flux to the atmosphere. The biogeochemical cycle of DMS (**Figure 1**) begins with its precursor, β-dimethylsulfoniopropionate (DMSP). DMSP is a cellular component in certain species of phytoplankton, notably some prymnesiophytes and dinoflagellates. The function of DMSP is unclear, although there is evidence for an

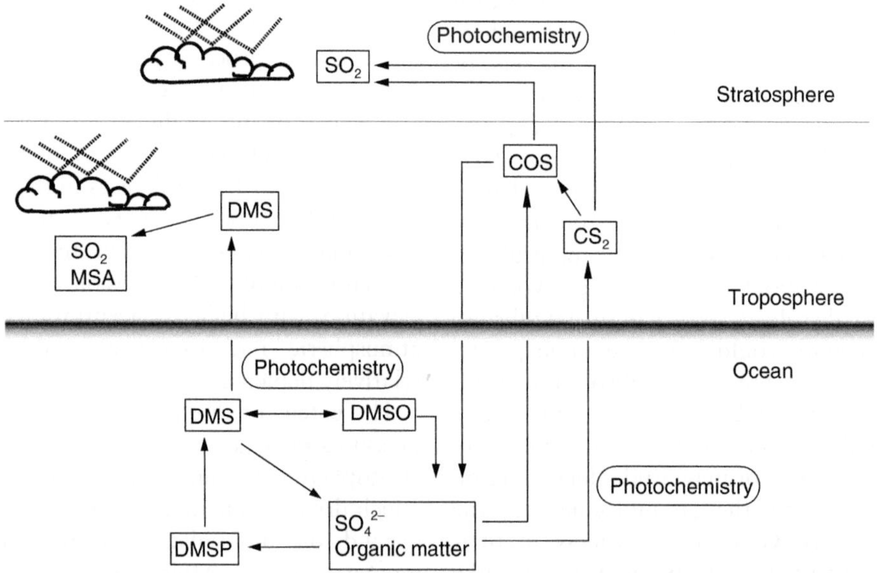

Figure 1 Fate and production of dimethylsulfide (DMS), carbonylsulfide (COS) and carbon disulfide. DMSO, dimethylsulfoxide; DMSP, dimethylsulfoniopropionate; MSA, methane sulfonic acid.

osmoregulatory role as its cellular concentrations have been found to vary with salinity. It is generally thought that healthy algal cells do not leak either DMSP or DMS, although mechanical release into the surrounding sea water can lead to DMS production during cell senescence and grazing by zooplankton or as a consequence of viral attack on phytoplankton cells. Oceanic regions dominated by prolific DMSP-producing phytoplankton tend to have high DMS and DMSP concentrations.

Breakdown of DMSP, presumably after transfer from the particulate algal (pDMSP) form to a dissolved (dDMSP) form in sea water, can proceed in different ways, mostly depending on microbiological conditions. One major pathway involves cleavage of DMSP to DMS and acrylic acid. Bacterial metabolism of dDMSP may be a major mechanism for DMS production in sea water, with acrylic acid residue acting as a carbon source for heterotrophic growth. Sulfonium compounds are vulnerable to attack by hydroxide ion; the resulting chemical elimination reaction occurs rapidly and quantitatively in strong base but only slowly at the pH of sea water.

DMS in sea water has many potential fates. The volatility of DMS and the concentration gradient across the sea–air interface lead to the ocean being the major source of DMS to the atmosphere. Estimates of the annual sulfur release (as DMS) vary from 13–37 Tg S y^{-1} (Kettle and Andreae, 1999). However, whereas the absolute flux of DMS from sea to air may be large on a global scale, sea–air exchange may represent only a minor sink for seawater DMS. It has been estimated that DMS loss to the atmosphere is only a very small percentage of the DMS sink, but this undoubtedly depends on the biogeochemical conditions in the water column at the time. Photochemical oxidation of DMS, either to dimethylsulfoxide (DMSO) or to other products, occurs via photosensitized reactions. The amount of photochemical decomposition depends on the amount of light of appropriate wavelengths and the concentration of colored organic compounds in solution to convert light energy into reactive radicals. Light declines exponentially with depth; the distribution of colored dissolved organic materials exhibits depth and seasonal variability. Microbial consumption of DMS, although extremely variable in both time and space in the ocean, appears to be a significant sink for oceanic DMS. The residence time of DMS is probably of the order of a day or two in most seawater systems.

Since the atmospheric residence time of DMS is about a day or two, the atmospheric consequences of DMS flux are mostly confined to the troposphere. In the troposphere, DMS is oxidized primarily by hydroxyl radical. The main atmospheric oxidation products are methane sulphonic acid, SO_2 and DMSO.

Carbonyl Sulfide

Carbonyl sulfide (COS, OCS) is the major sulfur gas in the atmosphere, present throughout the troposphere at 500 pptv. COS has a long atmospheric residence time ($\sim$4 years). Because of its relative inertness COS diffuses into the stratosphere where it oxidizes to sulfate particles and contributes in reactions involving stratospheric ozone chemistry. Unlike DMS which is photochemically oxidized in the troposphere, the major sink for COS is terrestrial vegetation and soils. COS is taken up by plants by passing through the stomata and subsequently hydrolyzing to CO_2 and H_2S through the action of carbonic anhydrase inside plant cells. There is no apparent physiological significance to the process; it appears to just occur accidentally to the normal physiology of plants.

COS is produced in the ocean by photochemical oxidation of organic sulfur compounds whereby dissolved organic matter acts as a photosensitizer. The aqueous concentration of COS manifests a strong diel cycle, with the highest concentrations in daytime (concentration range on the order of 0.03–0.1 nmol l^{-1}). COS hydrolyzes in water to H_2S at rates dependent on water temperature and pH. The flux of oceanic COS to the atmosphere may represent about one-third of the global COS flux.

Carbon Disulfide

Concentrations in surface water are around 10^{-11} mol l^{-1}. Although a number of studies have indicated that the ocean forms an important source for atmospheric CS_2, the underlying biochemical cycles still remain poorly understood. CS_2 is formed by photochemical reactions (possibly involving precursors such as DMS, DMSP and isothiocyanates). CS_2 formation has been observed to occur in bacteria in anoxic aquatic environments and in cultures of some marine algae species.

The residence time of CS_2 in the atmosphere is relatively short (about one week). Although CS_2 might contribute directly to SO_2 in the troposphere, its main significance is in the formation of COS via photochemical oxidation which results in the production of one molecule each of SO_2 and COS per molecule of CS_2 oxidized. The resulting COS may contribute to the stratospheric aerosol formation.

Concentrations around 14 pmol l^{-1} of carbon disulfide in the mid-Atlantic Ocean were first observed

(1974); higher concentrations have been found in coastal waters. More than a decade later CS_2 concentrations in the North Atlantic were found to be comparable to the earlier observations. However, in coastal waters CS_2 concentrations were found to be a factor 10 lower, respectively 33 and 300 pmol l^{-1}. The global CS_2 flux has been estimated on 6.7 Gmol S y^{-1}, and it has been concluded that the marine emission of CS_2 provides a significant indirect source of COS, but it forms an insignificant source of tropospheric SO_2.

Nonmethane Hydrocarbons

Nonmethane hydrocarbons (NMHCs) are important reactive gases in the atmosphere since they provide a sink for hydroxyl radicals and play key roles in the production and destruction of ozone in the troposphere. NMHCs generally refer to the C_2–C_4 series, notably ethane, ethene, acetylene, propane, propene, and n-butane, but also the five-carbon compound isoprene. Of these, ethene is generally the most abundant contributing 40% to the total NMHC pool in sea water. Published data of concentrations of NMHCs in sea water vary widely sometimes exceeding a factor 100. For example, in one extensive study, ethene and propane were found to be the most abundant species in the intertropical South Pacific, with mixing ratios of 2.7 to 58 and 6 to 75 pptv, respectively; whereas in the equatorial Atlantic these species showed mixing ratios of 20 pptv and 10 pptv, respectively.

The water-column dynamics of NMHCs are poorly understood. NMHCs have been detected in the surface sea and with maxima in the euphotic zone and tend to be present at concentrations in sea water at around 10^{-10} mol l^{-1}. Evidence suggests that photochemical oxidation of dissolved organic matter results in the formation of NMHCs. There can be very little doubt that the physiology of planktonic organisms is also involved in NMHC formation. Ethene and isoprene are freely produced by terrestrial plants where the former is a powerful plant hormone but the function of the latter less well understood. It is likely that similar processes occur in planktonic algae. NMHC production tends to correlate with light intensity, dissolved organic carbon and biological production. A simplified scheme of marine NMHC production is shown in **Figure 2**.

The flux of NMHCs to the atmosphere (with estimates ranging from $<10\,Mt\,y^{-1}$ to $50\,Mt\,y^{-1}$) is minor on a global scale, but has a potential significance in local atmospheric chemistry. Although oceans are known to act as sources of NMHCs, the sources of individual NMHCs in the marine boundary layer are not always clear. Those NMHCs with a life time of more than a week (e.g., ethane, ethyne, propane, cyclopropane) show latitudinal gradients consistent with a continental source, whereas variations of NMHCs with life times shorter than a week (all alkenes and pentane) are more consistent with a marine source.

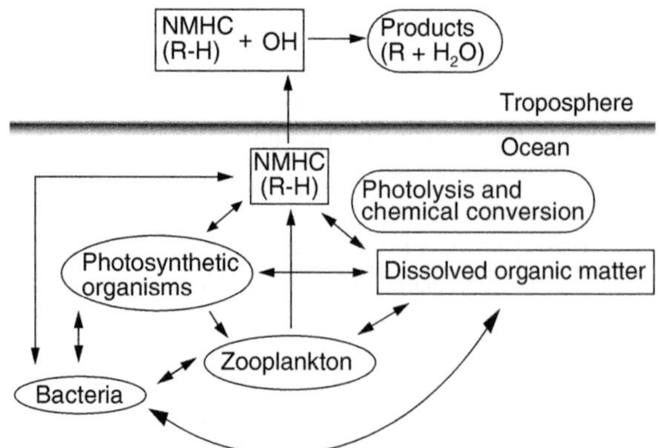

Figure 2 Simplified scheme of marine nonmethane hydrocarbon (NMHC) production. In the marine troposphere NMHC acts as a sink for hydroxyl (OH) radicals and thereby plays a key role in ozone chemistry.

Ammonia

Ammonia is an extremely soluble gas, reacting with water and dissociating into an ammonium ion at ambient pH. At pH 8.2, about one-tenth of dissolved ammonia is present as NH_3. Ammonium is also a rapidly cycling biological nutrient; it is taken up by bacteria and phytoplankton as a source of fixed nitrogen, and released by sundry physiological and decompositional processes in the food web. Anthropogenic loading of ammonium (and other nutrients) into the coastal marine environment results in increased phytoplankton growth in a phenomenon called eutrophication. Ammonium is oxidized to nitrate by bacteria in a process known as nitrification (**Figure 3**). Conversely, in anoxic environments, ammonium can be formed by nitrate-reducing bacteria.

Ammonia plays an important role in the acid–base chemistry in the troposphere where the unionized ammonia (NH_3) is converted into ionized ammonia (NH_4^+) via a reaction that neutralizes atmospheric acids as HNO_3 and H_2SO_4. This leads to the formation of ammonium aerosols such as the stable ammonium sulfate. Eventually the ammonia returns to the surface by dry or wet deposition.

Few data exist on the fluxes of NH_3 over marine environments. Evidence suggests that most of the ocean surface serves as a source of NH_3 to the

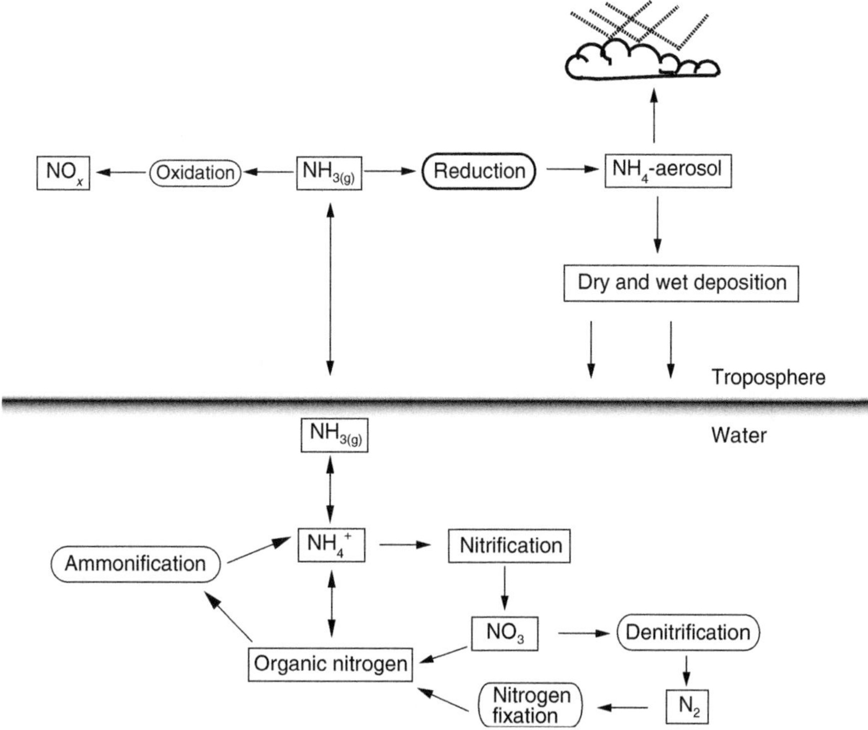

Figure 3 Simplified scheme of marine NH_x chemistry. In the marine boundary layer NO_3 acts as an initiator for the degradation of many organic compounds, in particular dimethylsulfide (DMS).

atmosphere, even in regions of very low nutrients. In the North Sea, an area situated in the middle of densely populated and industrialized countries of Western Europe, air from nearby terrestrial sources may act as a source of NH_3 into surface waters. It has been estimated that the annual biogenic emission of ammonia from European seas is around $30\,\mathrm{kt\,N\,y^{-1}}$, which is comparable to the emissions of smaller North European countries, leading to the conclusion (amongst others) that seas are among the largest sources of imported ammonium for maritime countries. The net emission of ammonia from coastal waters of the north-east Pacific Ocean to the atmosphere has been shown to be in the order of $10\,\mu\mathrm{mol\,m^{-2}\,d^{-1}}$.

Organohalogens

Halogenated compounds, such as methyl chloride (CH_3Cl), methyl bromide (CH_3Br) and methyl iodide (CH_3I) are a major source of halogens in the atmosphere, and subsequently form sources of reactive species capable of catalytically destroying ozone. Among these CH_3I is likely to play an important role in the budget of tropospheric ozone, through production of iodine atoms by photolysis. Due to their higher photochemical stability methyl chloride and methyl bromide are more important in stratospheric chemistry; it has been suggested that BrO species are responsible for losses of tropospheric ozone in the Arctic (**Figure 4**).

Atmospheric methyl halides, measured over the ocean by several cruise surveys, have been shown to have average atmospheric mixing ratios of: CH_3Cl, 550–600 pptv; CH_3Br, 10–12 pptv; CH_3I, 0.5–1 pptv. Their temporal and spatial variations are not well understood, neither is their production mechanism in the ocean known. Measurements of atmospheric and seawater concentrations of CH_3Cl and CH_3I have indicated that the oceans form natural sources of these methyl halides. In contrast, CH_3Br appears to be undersaturated in the open ocean and exhibits moderate to 100% supersaturation in coastal and upwelling regions, leading to a global atmosphere to ocean flux of $13\,\mathrm{Gg\,y^{-1}}$. Coastal salt marshes, although they constitute a minor area of the global marine environment, may produce roughly 10% of the total fluxes of atmospheric CH_3Br and CH_3Cl and thus contribute significantly to the global budgets.

Macrophytic and phytoplanktonic algae produce a wide range of volatile organohalogens including di- and tri-halomethanes and mixed organohalogens. There is evidence for the involvement of enzymatic synthesis of methyl halides, but the metabolic

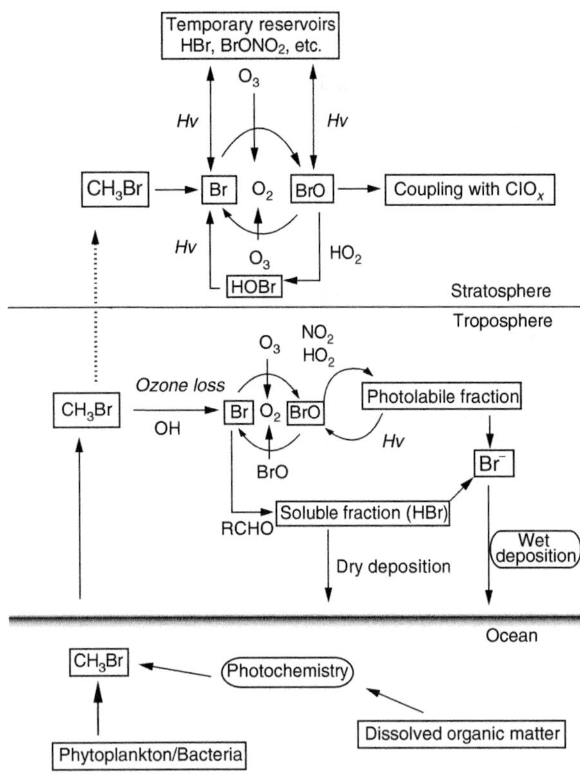

Figure 4 Schematic illustration of the circulation of methyl bromide.

production pathways are not well known. In free sea water, photochemical processes, ion substitution and, possibly the alkylation of halide ions (during the oxidation of organic matter by an electron acceptor such as Fe(III)) are also potential formation mechanisms. Sunlight or microbial mediation are not required for these reactions. In the ocean, chemical degradation of CH_3Br occurs by nucleophilic substitution by chloride and hydrolysis. Microbial consumption is also a likely sink for halogenated compounds.

Conclusions

The biogenic trace gases are influenced by the complete range of biological processes – from the biochemical and physiological to the ecological level of food web dynamics. The gases that are influenced directly by plant physiology (probably the light NMHCs and isoprene, for example) tend to be most closely related to phytoplankton biomass or primary productivity. Other gases produced during grazing and decomposition (e.g., DMS, NH_3), or gases formed by photochemical reactions in dissolved organic material show differing temporal dynamics and different dynamics spatially and with depth in the ocean. As conditions change in an apparently warming world, changes in the dynamics of surface ocean gases can be expected. The behavior of these trace gases or even the dynamics of the planktonic community are not understood sufficiently to allow good quantitative predictions about changes in trace gas flux to be made. Changes in flux of some gases could lead to an acceleration of warming, while changes in others could lead to cooling. It is, thus, important to understand the factors controlling trace gas dynamics in the surface ocean.

See also

Air–Sea Transfer: Dimethyl Sulfide, COS, CS_2, NH_4, Non-Methane Hydrocarbons, Organo-Halogens. Carbon Cycle. Chlorinated Hydrocarbons. Nitrogen Cycle.

Further Reading

Andreae MO (1990) Ocean–atmosphere interactions in the global biogeochemical sulfur cycle. *Marine Chemistry* 10: 1–29.

Andreae MO and Crutzen PJ (1997) Atmospheric aerosols: biogeochemical sources and role in atmospheric chemistry. *Science* 276: 1052–1058.

Barrett K (1998) Oceanic ammonia emissions in Europe and their transboundary fluxes. *Atmospheric Environment* 32(3): 381–391.

Chin M and Davis DD (1993) Global sources and sinks of OCS and CS_2 and their distributions. *Global Biogeochemical Cycles* 7: 321–337.

Cox RA, Rattigana OV, and Jones RL (1995) Laboratory studies of BrO reactions of interest for the atmospheric ozone balance. In: Bandy RA (ed.) *The Chemistry of the Atmosphere; Oxidants and Oxidation in the Earth's Atmosphere.* Cambridge: The Royal Society of Chemistry.

Crutzen PJ (1976) The possible importance of COS for the sulfate layer of the stratosphere. *Geosphysical Research Letters* 3: 73–76.

Graedel TE (1995) Tropospheric budget of reactive chlorine. *Global Biogeochemical Cycles* 9: 47–77.

Kettle AJ and Andreae MO (1999) Flux of dimethylsulfide from the oceans: a comparison of updated datasets and flux models. *Journal of Geophysical Research* 105: 26793–26808.

Lovelock JE (1974) CS_2 and the natural sulfur cycle. *Nature* 248: 625–626.

Turner SM and Liss PS (1985) Measurements of various sulfur gases in a coastal marine environment. *Journal of Atmospheric Chemistry* 2(3): 223–232.

AIR–SEA TRANSFER: N_2O, NO, CH_4, CO

C. S. Law, Plymouth Marine Laboratory, The Hoe, Plymouth, UK

Introduction

The atmospheric composition is maintained by abiotic and biotic processes in the terrestrial and marine ecosystems. The biogenic trace gases nitrous oxide (N_2O), nitric oxide (NO), methane (CH_4) and carbon monoxide (CO) are present in the surface mixed layer over most of the ocean, at concentrations which exceed those expected from equilibration with the atmosphere. As the oceans occupy 70% of the global surface area, exchange of these trace gases across the air–sea interface represents a source/sink for global atmospheric budgets and oceanic biogeochemical budgets, although marine emissions of NO are poorly characterized. These trace gases contribute to global change directly and indirectly, by influencing the atmospheric oxidation and radiative capacity (the 'greenhouse effect') and, together with their reaction products, impact stratospheric ozone chemistry (**Table 1**). The resultant changes in atmospheric forcing subsequently influence ocean circulation and biogeochemistry via feedback processes on a range of timescales. This article describes the marine sources, sinks, and spatial distribution of each trace gas and identifies the marine contribution to total atmosphere budgets. There is also a brief examination of the approaches used for determination of marine trace gas fluxes and the variability in current estimates.

Nitrous Oxide (N_2O)

The N_2O molecule is effective at retaining long-wave radiation with a relative radiative forcing 280 times that of a CO_2 molecule. Despite this the relatively low atmospheric N_2O concentration results in a contribution of only 5–6% of the present day 'greenhouse effect' with a direct radiative forcing of about 0.1 Wm^{-2}. In the stratosphere N_2O reacts with oxygen to produce NO radicals, which contribute to ozone depletion.

N_2O is a reduced gas which is produced in the ocean primarily by microbial nitrification and denitrification. N_2O is released during ammonium (NH_4^+) oxidation to nitrite (NO_2^-) (**Figure 1**), although the exact mechanism has yet to be confirmed. N_2O may be an intermediate of nitrification, or a by-product of the decomposition of other intermediates, such as nitrite or hydroxylamine. Nitrification is an aerobic process, and the N_2O yield under oxic conditions is low. However, as the nitrification rate decreases under low oxygen, the relative yield of N_2O to nitrate production increases and reaches a maximum at 10–20 μmol dm^3 oxygen (μmol $= 1 \times 10^{-6}$ mol). Conversely, denitrification is an anaerobic process in which soluble oxidized nitrogen

Table 1 The oceanic contribution and atmospheric increase and impact for methane, nitrous oxide, nitric oxide, and carbon monoxide[a]

Trace gas	*Atmospheric concentration (ppbv)*	*Atmospheric lifetime (years)*	*Major impact in atmosphere*	*Increase in atmosphere (1980–90)*	*Oceanic emission as % of total global emissions*
Nitrous oxide (N_2O)	315	110–180	Infrared active Ozone sink/source	0.25% (0.8 ppbv y^{-1})	7–34%
Nitric oxide (NO)	0.01	<0.2	Ozone sink/source OH sink/oxidation capacity	Not known	Not known
Methane (CH_4)	1760	10	Infrared active OH sink/oxidation capacity Ozone sink/source	0.8% (0.6 ppbv y^{-1})	1–10%
Carbon monoxide (CO)	120	0.2–0.8	OH sink/oxidation capacity Ozone sink/source Infrared active	− 13 to 0.6%	0.9–9%

[a] ppbv, parts per billion by volume. (Adapted from Houghton *et al.*, 1995.)

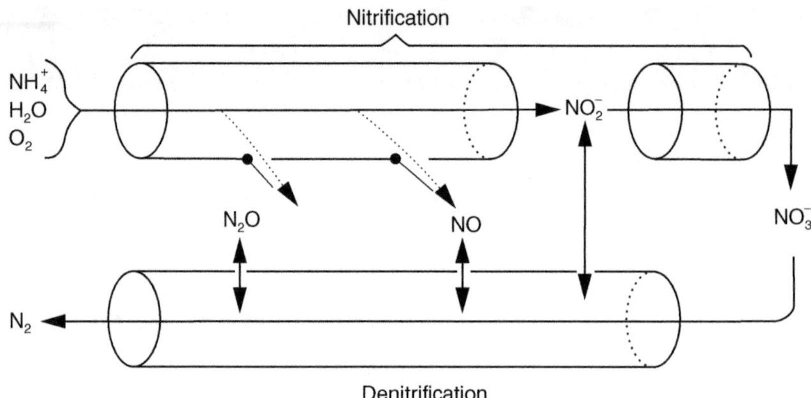

Figure 1 'Leaky Pipe' flow diagram of nitrification and denitrification indicating the potential exchange and intermediate role of NO and N_2O (Reprinted by permission from *Nature* copyright (1990), Macmillan Magazines Ltd.)

compounds, such as nitrate and nitrite, are converted to volatile reduced compounds (N_2O and N_2) in the absence of oxygen. Oxygen availability inhibits denitrification at ambient levels, and also determines the products of denitrification. An enzymatic gradient of sensitivity to oxygen results in the accumulation of N_2O under sub-oxia (3–10 μmol dm^3) due to the inhibition of the enzyme nitrous oxide reductase. At lower oxygen (<3 μmol dm^3) the reaction continues through to N_2 and so anoxic environments are sinks for N_2O. N_2O yields from nitrification are 0.2–0.5%, whereas denitrification yields may be as high as 5% at optimal levels of sub-oxia.

An inverse correlation between N_2O and oxygen, and associated linear relationship between nitrate and N_2O, suggest that N_2O in the ocean originates primarily from nitrification. This may not be the case for sediments, in which denitrification is the dominant source of N_2O under variable oxygen tension, with nitrification only contributing in a narrow suboxic band. Attribution of source is difficult as nitrification and denitrification may occur simultaneously and interact, with exchange of products and intermediates (**Figure 1**). This is further complicated, as denitrification will be limited to some extent by nitrate supply from nitrification. Isotopic data from the surface ocean in oligotrophic regions imply that N_2O originates primarily from nitrification. However, recent evidence from waters overlying oxygen-deficient intermediate layers suggests that the elevated surface mixed-layer N_2O arises from coupling between the two processes, as the observed isotope signatures cannot be explained by nitrification or denitrification alone. An additional N_2O source from the dissimilatory reduction of nitrate to ammonium is restricted to highly anoxic environments such as sediments.

The oceanic N_2O distribution is determined primarily by the oxygen and nutrient status of the water column. Estuaries and coastal waters show elevated supersaturation in response to high carbon and nitrogen loading, and the proximity of sub-oxic zones in sediment and the water column. As a result the total marine N_2O source tends to be dominated by the coastal region. The N_2O flux from shelf sea sediments is generally an order of magnitude lower than estuarine sediments, although the former have a greater spatial extent. A N_2O maximum at the base of the euphotic zone is apparent in shelf seas and the open ocean, and is attributed to production in suboxic microzones within detrital material. Oceanic surface waters generally exhibit low supersaturations (<105%), although N_2O supersaturations may exceed 300% in surface waters overlying low oxygen intermediate waters and upwelling regions, such as the Arabian Sea and eastern tropical North Pacific. These 'natural chimney' regions dominate the open ocean N_2O source, despite their limited surface area (**Table 2**). The surface N_2O in upwelling regions such as the Arabian Sea originates in part from the underlying low-oxygen water column at 100–1000 m, where favorable conditions result in the accumulation of N_2O to supersaturations exceeding 1200%. N_2O transfer into the surface mixed layer will be limited by vertical transport processes and a significant proportion of N_2O produced at these depths will be further reduced to N_2.

The oceans account for 1–5 Tg N-N_2O per annum ($Tg = 1 \times 10^{12}$ g) or 6–30% of total global N_2O emissions, although there is considerable uncertainty attached to this estimate (**Figure 2**). A recent estimate with greater representation of coastal sources has resulted in upward revision of the marine N_2O source to 7–10.8 Tg N-N_2O per annum; although this may represent an upper limit due to some bias from inclusion of estuaries with high N_2O supersaturation. However, this estimate is in agreement with a total

Table 2 N_2O and CH_4 regional surface water supersaturations (from *Bange et al.*, 1996; 1998) (supersaturation is >100%, undersaturation is <100% with equilibrium between atmosphere and water at 100%)

	Surface % N_2O saturation mean (range)	*Surface % CH_4 saturation mean (range)*
Estuaries	607 (101–2500)	1230 (146–29 000)
Coastal/shelf	109 (102–118)	395 (85–42 000)
Oligotrophic/transitional ocean	102.5 (102–104)	120 (80–200)
Upwelling ocean	176 (108–442)	200 (86–440)

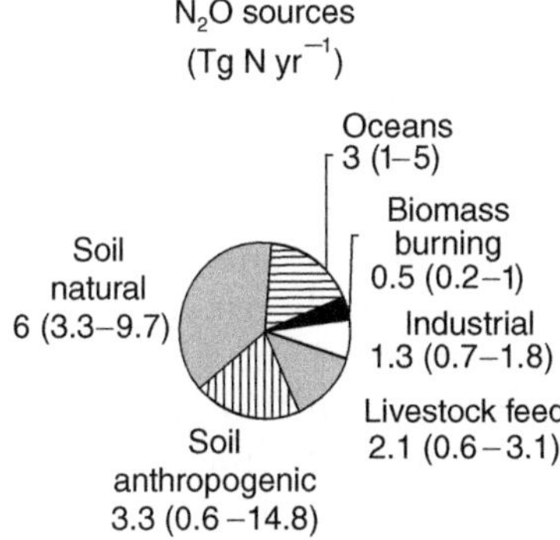

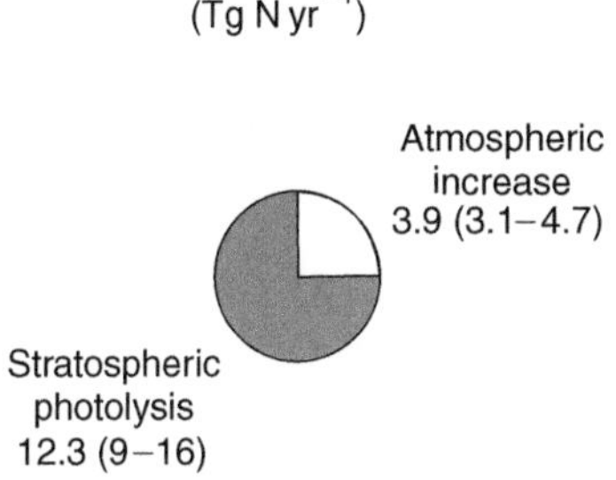

Figure 2 Atmospheric nitrous oxide sources and sinks (adapted from Houghton *et al.*, 1995). Units: $Tg = 1 \times 10^{12}$ g.

oceanic production rate of 11 Tg $N\text{-}N_2O$ per annum calculated from new production and nitrification.

Nitric Oxide (NO)

Nitric oxide (NO) plays a central role in atmospheric chemistry, influencing both ozone cycling and the tropospheric oxidation capacity through reactions with hydroperoxy- and organic peroxy-radicals. When the NO concentration exceeds ~40 pptv (pptv = parts per trillion by volume) it catalyzes the production of ozone (O_3):

$$CO + OH^* + O_2 \rightarrow HO_2^* + CO_2 \quad [1]$$

$$HO_2^* + NO \rightarrow OH^* + NO_2 \quad [2]$$

$$NO_2 + h\nu \rightarrow NO + O(^3P) \quad [3]$$

$$O + O_2 + M \rightarrow O_3 + M \quad [4]$$

At high concentrations (>50 ppbv; ppbv = parts per billion by volume), O_3 in the atmospheric boundary layer becomes a toxic pollutant that also has important radiative transfer properties. The production of nitric acid from NO influences atmospheric pH, and contributes to acid rain formation. In addition, the oxidation of NO to the nitrate (NO_3) radical at night influences the oxidizing capacity of the lower troposphere. Determination of the magnitude and location of NO sources is critical to modeling boundary layer and free tropospheric chemistry.

NO cycling in the ocean has received limited attention, as a result of its thermodynamic instability and high reactivity. Photolysis of nitrite in surface waters occurs via the formation of a nitrite radical with the production of NO:

$$NO_2^- + h\nu \rightarrow NO_2^{-*} + HOH \rightarrow NO + OH^- + OH$$

This reaction may account for 10% of nitrite loss in surface waters of the Central Equatorial Pacific, resulting in a 1000-fold increase in dissolved NO at a steady-state surface concentration of 5 pmol dm^{-3} during light periods ($pmol = 1 \times 10^{-12}$ mol). This photolytic production is balanced by a sink reaction with the superoxide radical (O_2^-) to produce peroxynitrite:

$$O_2^- + NO \rightarrow {}^-OONO$$

This reaction will be dependent upon steady-state concentration of the superoxide radical; however, as the reaction has a high rate constant, NO is rapidly turned over with a half-life on the order of 10–100 seconds.

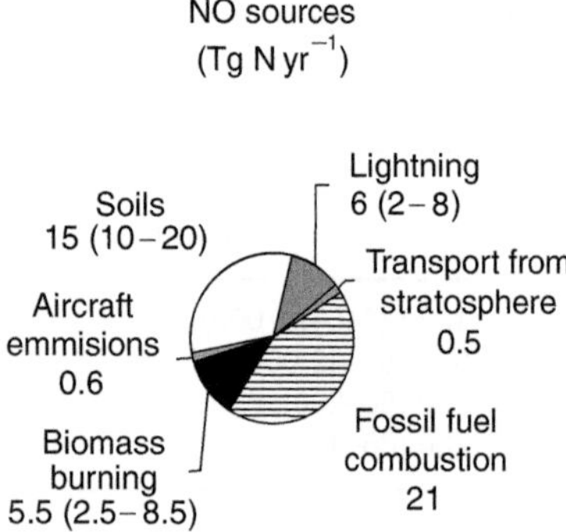

Figure 3 Atmospheric nitric oxide sources (from Graedel and Crutzen, 1992). Units: Tg = 1×10^{12} g.

As with N_2O, NO may also be produced as a by-product or intermediate of denitrification and nitrification (**Figure 1**). NO production by soils is better characterized than in marine systems, and is significant both in terms of nitrogen loss and the global NO budget (**Figure 3**). The greater oxygen availability in soils limits reduction of NO via denitrification and so enhances NO efflux. Sediment pore water NO maxima have been attributed to denitrification, although, as this process also represents a sink for NO (**Figure 1**), this may reflect poising at an optimal redox potential for NO production. Conversely, the NO maximum in low oxygen intermediate waters in the east tropical North Pacific derives from nitrification.

Current understanding of the oceanic NO distribution is that it is limited to the surface ocean and intermediate low oxygen water column. There is potential for higher NO concentrations in coastal and estuarine waters from sediment and photolytic sources, and nitrite photolysis to NO may also be significant in upwelling regions. Despite the short half-life of NO in surface waters, the maintenance of steady-state NO concentration suggests that photolytic production may support an, as yet unquantified, source of atmospheric NO. Surface concentrations in the Central Equatorial Pacific suggest that the oceanic NO source would not exceed 0.5 Tg N per annum, which is relatively insignificant when compared with other sources (**Figure 3**).

Methane (CH_4)

CH_4 is the most abundant organic volatile in the atmosphere and, next to CO_2, is responsible for 15% of the current greenhouse radiative forcing, with a direct radiative forcing of 0. 5 Wm^{-2}. CH_4 reacts with OH and so limits the tropospheric oxidation capacity and influences ozone and other greenhouse gases. The reaction with OH generates a feedback that leads to a reduction in the rate of CH_4 removal.

CH_4 is a reduced gas which, paradoxically, is supersaturated in the oxidized surface waters of the ocean (see **Table 2**). CH_4 is produced biotically and abiotically, although its oceanic distribution is controlled primarily by biological processes. Methanogenesis is classically defined as the formation of CH_4 from the fermentation and remineralization of organic carbon under anoxic conditions. Methanogens require a very low reducing potential and are generally obligate anaerobes, although there is evidence that they can tolerate some exposure to oxygen. However, methanogens cannot utilize complex organic molecules and often coexist with aerobic consortia to ensure a supply of simple C_1 substrates. Methanogens utilize formate, acetic acid, CO_2, and hydrogen in sulfate-rich anoxic environments, although they are generally out-competed by sulfate-reducing bacteria which have a greater substrate affinity. However, the methanogens can also utilize other noncompetitive substrates such as methanol, methylamines, and reduced methylated compounds, when out-competed for the C_1 compounds.

A significant fraction of CH_4 is oxidized before exiting the marine system and so the oxidation rate is critical in determining the air–sea flux. This is accomplished by methanotrophs that obtain their carbon and energy requirements from CH_4 oxidation under aerobic conditions via the following reactions:

$$\underset{\text{methane}}{CH_4} \rightarrow \underset{\text{methanol}}{CH_3OH} \rightarrow \underset{\text{formaldehyde}}{HCHO} \rightarrow \underset{\text{formate}}{HCOOC} \rightarrow \underset{\text{carbon dioxide}}{CO_2}$$

Methanotrophs are found in greater numbers in sediments than in oxic sea water, and consequently the oceanic water column CH_4 oxidation is an order of magnitude lower than in sediments. Methanotrophs have a high inorganic nitrogen requirement and so methanotrophy is highest at the oxic–anoxic interface where ammonium is available. Anaerobic CH_4 oxidation also occurs but is less well characterized. It is generally restricted to anaerobic marine sediments, utilizing sulfate as the only oxidant available, and is absent from anaerobic freshwater sediments which lack sulfate. A significant proportion of CH_4 produced in anaerobic subsurface layers in sediments is oxidized during diffusive transport through the sulfate-CH_4 transition zone by anaerobic oxidation and subsequently by aerobic oxidation in the overlying oxic layers. Anaerobic oxidation represents the main sink for CH_4 in marine sediments, where it may account for 97% of CH_4 production.

CH_4 production is characteristic of regions with high input of labile organic carbon such as wetlands and sediments, but is usually restricted to below the zone of sulfate depletion. The oceanic CH_4 source is dominated by coastal regions, which exhibit high CH_4 fluxes as a result of bubble ebullition from anoxic carbon-rich sediments, and also riverine and estuarine input. Some seasonality may result in temperate regions due to increased methanogenesis at higher temperatures. The predominant water column source in shelf seas and the open ocean is CH_4 production at the base of the euphotic zone. This may arise from lateral advection from sedimentary sources, and *in situ* CH_4 production. The latter is accomplished by oxygen-tolerant methanogens that utilize methylamines or methylated sulfur compounds in anoxic microsites within detrital particles and the guts of zooplankton and fish. Lateral advection and *in situ* production may be greater in upwelling regions, as suggested by the increased CH_4 supersaturation in surface waters in these regions. Oceanic CH_4 concentration profiles generally exhibit a decrease below 250 m due to oxidation. Methanogenesis is elevated in anoxic water columns, although these are not significant sources of atmospheric CH_4 due to limited ventilation and high oxidation rates. Other sources include CH_4 seeps in shelf regions from which CH_4 is transferred directly to the atmosphere by bubble ebullition, although their contribution is difficult to quantify. Abiotic CH_4 originating from high-temperature fluids at hydrothermal vents also elevates CH_4 in the deep and intermediate waters in the locality of oceanic ridges. A significant proportion is oxidized and although the contribution to the atmospheric CH_4 pool may be significant in localized regions this has yet to be constrained. Hydrates are crystalline solids in which methane gas is trapped within a cage of water molecules. These form at high pressures and low temperatures in seafloor sediments generally at depths below 500 m. Although CH_4 release from hydrates is only considered from anthropogenic activities in current budgets, there is evidence of catastrophic releases in the geological past due to temperature-induced hydrate dissociation. Although oceanic hydrate reservoirs contain 14 000 Gt CH_4, there is currently no evidence of significant warming of deep waters which would preempt release.

Other aquatic systems such as rivers and wetlands are more important sources than the marine environment. Shelf regions are the dominant source of CH_4 from the ocean (14(11–18) Tg CH_4 per annum), accounting for 75% of the ocean flux (**Table 2**). The ocean is not a major contributor to the atmospheric CH_4 budget, as confirmed by estimates of the oceanic CH_4 source (**Figure 4**).

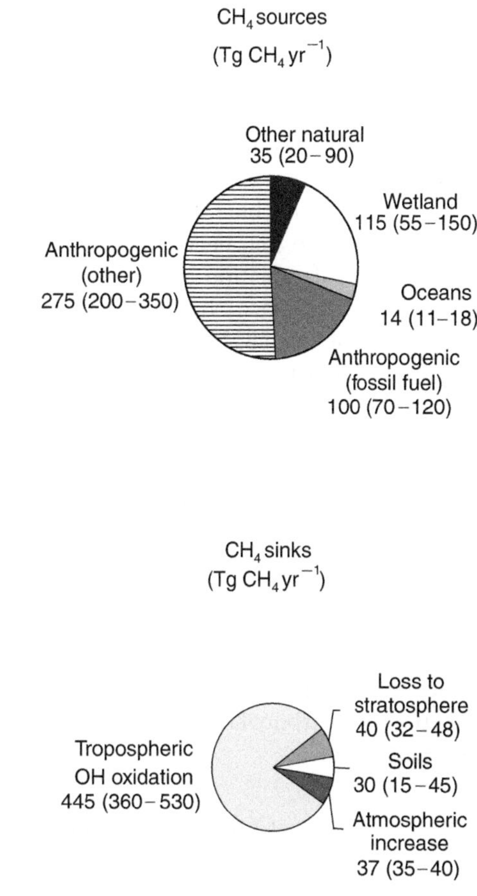

Figure 4 Atmospheric methane sources and sinks (adapted from Houghton *et al.* (1995)). Units: Tg $= 1 \times 10^{12}$ g.

Carbon Monoxide (CO)

The oxidation of CO provides the major control of hydroxyl radical content in the troposphere and limits the atmospheric oxidation capacity. This results in an increase in the atmospheric lifetime of species such as CH_4, N_2O, and halocarbons, and enhances their transfer to the stratosphere and the potential for subsequent ozone destruction. It has been suggested that decreasing stratospheric ozone and the resultant increase in incident ultraviolet (UV) radiation may increase marine production and efflux of CO, thereby generating a positive feedback loop. However, this may be compensated by a negative feedback in which increased UV reduces biological production and dissolved organic matter, so reducing the CO source. CO also influences tropospheric ozone by its interaction with NOx, and is a minor greenhouse gas with a radiative forcing of 0.06 Wm^{-2} at current atmospheric concentrations.

The principal source of dissolved CO is the abiotic photodegradation of dissolved organic matter (DOM) by UV-R, and CO represents one of the major photoproducts of DOM in the ocean. Quantum yields for CO are highest in the UV-B range (280–315 nm) and decrease with increasing wavelengths. However, the UV-A (315–390 nm) and blue portion of the visible spectrum contribute to marine CO production as a greater proportion of radiation at these wavelengths reaches the Earth's surface. Humics represent approximately half of the DOM and account for the majority of the chromophoric dissolved organic matter (CDOM), the colored portion of dissolved organic matter that absorbs light energy. The CO photoproduction potential of humics is dependent upon the degree of aromaticity. Terrestrial humics are characterized by an increased prevalence of phenolic groups, and addition of precursor compounds containing phenolic moieties to natural samples stimulates CO production. Direct photo-oxidation of humics and compounds containing carbonyl groups, such as aldehydes, ketones, and quinones, occurs via the production of a carbonyl radical during α-cleavage of an adjacent bond:-

$$RCOR' + h\nu \rightarrow \begin{cases} R + (COR')^* \rightarrow R' + CO \\ (RCO)^* + R' \rightarrow R' + CO \end{cases}$$

CO production may also occur indirectly by a photosensitized reaction in which light energy is transferred via an excited oxygen atom to a carbonyl compound. This may occur with ketonic groups via the photosensitized production of an acetyl radical. Whereas light and CDOM are the primary factors controlling CO production there may also be additional influence from secondary factors. For example, organo-metal complexes have increased light absorption coefficients and their photo-decomposition will enhance radical formation and CO production at higher levels of dissolved metals such as iron. There is also minor biotic production of CO by methanogens, but this does not appear to be significant.

CO can be oxidized to carbon dioxide by selected microbial groups including ammonia oxidizers and methylotrophs that have a broad substrate specificity and high affinity for CO. However, only the carboxidotrophs obtain energy from this reaction, and these may be unable to assimilate CO efficiently at *in situ* concentrations. CO turnover times of 4 hours are typical for coastal waters, whereas this varies between 1 and 17 days in the open ocean. The lower oxidation rate in the open ocean may be due to light inhibition of CO oxidation. Extrapolation from laboratory measurements suggests that only 10% of photochemically produced CO is microbially oxidized.

Dissolved CO exhibits diurnal variability in the surface ocean in response to its photolytic source, although this is also indicative of a strong sink term. The decline in the surface mixed-layer CO concentration in the dark results from a combination of CO oxidation, vertical mixing, and air–sea exchange. As the equilibration time between atmosphere and oceanic surface mixed layer is on the order of a month, this suggests that the former two processes dominate. Superimposed upon the diurnal cycle of CO in the surface ocean are spatial and seasonal gradients that result from the interaction of photoproduction and the sink processes. Below the euphotic zone CO is uniformly low throughout the intermediate water column.

CO production potential is highest in wetland regions, which are characterized by high CDOM and enhanced light attenuation. Photochemical production of CO represents a potential sink for terrestrial dissolved organic carbon (DOC) in estuaries and coastal waters. This pathway may account for some of the discrepancy between the total terrestrial DOC exported and the low proportion of terrestrial DOC observed in the marine pool. Although a strong lateral gradient in CDOM exists between rivers and the open ocean, estuarine CO production may be limited by reduced UV light penetration. CO photoproduction may occur down to 80 m in the open ocean, and 20 m in the coastal zone, but is restricted to the upper 1 m in wetlands and estuaries. In addition, estuarine and coastal CO flux may also be restricted by the higher CO oxidation rates. There is evidence that upwelling regions may support enhanced CO production, in response to upwelled CDOM that is biologically refractory but photolabile.

The presence of a CO gradient in the 10 m overlying the surface ocean suggests that the photolytic source of CO may influence the marine boundary layer. The marine source of CO is poorly constrained, with estimates varying from 10 to 220 Tg CO per annum. A flux of 1200 Tg CO per annum was estimated on the assumption that low rates of oceanic CO oxidation would only remove a small proportion of photoproduced CO, and that the residual would be ventilated to the atmosphere. The discrepancy between this and other flux estimates implies that a significant CO sink has been overlooked, although this may reflect shortcomings of different techniques. The oceanic contribution to the global source is between 1 and 20%, although Extrapolation of photochemical production rates from wetlands, estuaries, and coasts suggests that these

Table 3 Atmospheric CO sources and sinks[a] (adapted from *Zuo et al.*, 1998)

CO sources (Tg CO y^{-1})		*CO sinks (Tg CO y^{-1})*	
Industrial/fuel combustion	400–1000	Tropospheric hydroxyl oxidation	1400–2600
Biomass burning	300–2200	Soils	250–530
Vegetation and soils	50–200	Flux to stratosphere	80–140
Methane oxidation	300–1300		
NMHC oxidation	200–1800		
Ocean	10–220		
(Coast/Shelf	300–400)		
Total sources	1260–6720	Total sinks	1730–3270

[a]Note that a separate estimate of the coastal/shelf CO source is shown for comparison, but does not contribute to the total source. $Tg = 1 \times 10^{12}$ g.

alone may account for 20% of the total global CO flux. Although the marine source is responsible for <10% of the total global flux (**Table 3**), it may still dominate atmospheric oxidation conditions in remote regions at distance from land.

Air–Sea Exchange of Trace Gases

The flux of these trace gases across the air–sea interface is driven by physical transfer processes and the surface concentration anomaly, which represents the difference between the partial pressure observed in surface water and that expected from equilibrium with the atmosphere. Direct determination of the oceanic emission of a trace gas is difficult under field conditions. Atmospheric gradient measurements above the ocean surface require enhanced analytical resolution, whereas more advanced micrometeorological techniques have yet to be applied to these trace gases. Determination of the accumulation rate in a floating surface flux chamber is a simpler approach, but may generate artefactual results from the damping of wave- and wind-driven exchange, and enhanced transfer on the inner chamber surfaces. Consequently the majority of flux estimates are calculated indirectly rather than measured. The surface anomaly is derived from the difference between the measured surface concentration (Cw), and an equilibrium concentration calculated from the measured atmospheric concentration (Cg) and solubility coefficient ($\propto$) at ambient temperature and salinity. This is then converted to a flux by the application of a dynamic term, the gas transfer velocity, k:

$$F = k(Cw - \alpha Cg)$$

The transfer velocity k is the net result of a variety of molecular and turbulent processes that operate at different time and space scales. Wind is the primary driving force for most of these turbulent processes, and it is also relatively straightforward to obtain accurate measurements of wind speed. Consequently, k is generally parameterized in terms of wind speed, with the favored approaches assuming tri-linear and quadratic relationships between the two. These relationships are defined for CO_2 at 20°C in fresh water and sea water and referenced to other gases by a Schmidt number (Sc) relationship:

$$k\,gas = k\,ref(Sc\,gas/Sc\,ref)^n$$

where n is considered to be $-1/2$ at most wind speeds. This dependency of k is a function of the molecular diffusivity (D) of the gas and the kinematic viscosity of the water (μ), and is expressed in terms of the Schmidt number ($Sc = \mu/D$).

Determination of marine trace gas fluxes using different wind speed–transfer velocity relationships introduces uncertainty, which increases at medium-high wind speeds to a factor of two. Furthermore, additional uncertainty is introduced by the extrapolation of surface concentration gradient measurements to long-term climatological wind speeds. Current estimates of oceanic fluxes are also subject to significant spatial and temporal bias resulting from the fact that most studies focus on more productive regions and seasons. This uncertainty is compounded by the extrapolation of observational data sets to unchartered regions. With the exception of N_2O, the ocean does not represent a major source for these atmospheric trace gases, although spatial variability in oceanic source strength may result in localized impact, particularly in remote regions. In the near future, advances in micrometeorological techniques, improved transfer velocity parameterizations and the development of algorithms for prediction of surface ocean concentrations by remote sensing should provide further constraint in determination of the oceanic source of N_2O, NO, CH_4, and CO.

See also

Air–Sea Transfer: Dimethyl Sulfide, COS, CS_2, NH_4, Non-Methane Hydrocarbons, Organo-Halogens. Carbon Dioxide (CO_2) Cycle. Gas Exchange in Estuaries. Photochemical Processes.

Further Reading

Bange HW, Bartel UH, *et al.* (1994) Methane in the Baltic and North Seas and a reassessment of the marine emissions of methane. *Global Biogeochemical Cycles* 8: 465–480.

Bange HW, Rapsomanikis S, and Andreae MO (1996) Nitrous oxide in coastal waters. *Global Biogeochemical Cycles* 10: 197–207.

Carpenter EJ and Capone DG (eds.) (1983) *Nitrogen in the Marine Environment*. London: Academic Press.

Graedel TE and Crutzen PJ (eds.) (1992) *Atmospheric Change: An Earth System Perspective*. London: W. H. Freeman and Co.

Houghton JT, Meira Filho M, Bruce J, *et al.* (1995) *Climate Change 1994. Radiative Forcing of Climate Change and an Evaluation of the IPCC IS92 Emission Scenarios*, Intergovernmental Panel on Climate Change. Cambridge: Cambridge University Press.

Liss PS and Duce RA (eds.) (1997) *The Sea Surface and Global Change*. Cambridge: Cambridge University Press.

Zuo Y, Guerrero MA, and Jones RD (1998) Reassessment of the ocean to atmosphere flux of carbon monoxide. *Chemistry and Ecology* 14: 241–257.

APPENDICES

APPENDIX 1. PERIODIC TABLE OF THE ELEMENTS

Atomic number
Element symbol
Atomic mass

1 H 1.00794																	2 He 4.00260
3 Li 6.941	4 Be 9.01218											5 B 10.811	6 C 12.011	7 N 14.0067	8 O 15.9994	9 F 18.9984	10 Ne 20.1797
11 Na 22.9898	12 Mg 24.3050											13 Al 26.9815	14 Si 28.0855	15 P 30.9738	16 S 32.066	17 Cl 35.4527	18 Ar 39.948
19 K 39.0983	20 Ca 40.078	21 Sc 44.9559	22 Ti 47.88	23 V 50.9415	24 Cr 51.9961	25 Mn 54.9380	26 Fe 55.847	27 Co 58.9332	28 Ni 58.69	29 Cu 63.546	30 Zn 65.39	31 Ga 69.723	32 Ge 72.61	33 As 74.9216	34 Se 78.96	35 Br 79.904	36 Kr 83.80
37 Rb 85.4678	38 Sr 87.62	39 Y 88.9059	40 Zr 91.224	41 Nb 92.9064	42 Mo 95.94	43 Tc (98)	44 Ru 101.07	45 Rh 102.906	46 Pd 106.42	47 Ag 107.868	48 Cd 112.411	49 In 114.82	50 Sn 118.710	51 Sb 121.75	52 Te 127.60	53 I 126.905	54 Xe 131.29
55 Cs 132.905	56 Ba 137.327	57 La 138.906 ★	72 Hf 178.49	73 Ta 180.948	74 W 183.85	75 Re 186.207	76 Os 190.2	77 Ir 192.22	78 Pt 195.08	79 Au 196.967	80 Hg 200.59	81 Tl 204.383	82 Pb 207.2	83 Bi 208.980	84 Po (209)	85 At (210)	86 Rn (222)
87 Fr (223)	88 Ra 226.025	89 Ac 227.028 ▲	104 (261)	105 (262)	106 (263)	107 (262)	108 (265)	109 (267)									

★ Lanthanides	58 Ce 140.115	59 Pr 140.908	60 Nd 144.24	61 Pm (145)	62 Sm 150.36	63 Eu 151.965	64 Gd 157.25	65 Tb 158.925	66 Dy 162.50	67 Ho 164.930	68 Er 167.26	69 Tm 168.934	70 Yb 173.04	71 Lu 174.967
▲ Actinides	90 Th 232.038	91 Pa 231.036	92 U 238.029	93 Np 237.048	94 Pu (244)	95 Am (243)	96 Cm (247)	97 Bk (247)	98 Cf (251)	99 Es (252)	100 Fm (257)	101 Md (258)	102 No (259)	103 Lr (260)

APPENDIX 2. THE GEOLOGIC TIME SCALE

Eon	Era	Period	Epoch	Millions of Years Ago
Phanerozoic	Cenozoic	(Quaternary)	Holocene	0.011
			Pleistocene	1.82
		(Tertiary)	Pliocene	5.32
			Miocene	23
			Oligocene	33.7
			Eocene	55
			Paleocene	65
	Mesozoic	Cretaceous		144
		Jurassic		200
		Triassic		250
	Paleozoic	Permian		295
		Carboniferous Pennsylvanian		320
		Mississippian		355
		Devonian		410
		Silurian		440
		Ordovician		500
		Cambrian		540
Proterozoic				2500
Archean		Oldest Rock		4400
		Age of the Solar System		4550

APPENDIX 3. ESTIMATED MEAN OCEANIC CONCENTRATIONS OF THE ELEMENTS

Atomic number	Element	Species	Type of distribution[a]	Oceanic mean concentration (ng kg^{-1})
1	Hydrogen	H_2O	—	—
2	Helium	Dissolved gas	c	7.6
3	Lithium	Li^+	c	180×10^3
4	Beryllium	$BeOH^+$	s + n	0.21
5	Boron	$B(OH)_3$	c	4.5×10^6
6	Carbon	Inorganic ΣCO_2	n	27.0×10^6
7	Nitrogen	Dissolved N_2	c	8.3×10^6
		NO_3^-	n	0.42×10^6
8	Oxygen	Dissolved O_2	inverse n	2.8×10^6
9	Fluorine	F^-	c	1.3×10^6
10	Neon	Dissolved gas	c	160
11	Sodium	Na^+	c	10.78×10^9
12	Magnesium	Mg^{2+}	c	1.28×10^9
13	Aluminum	$Al(OH)_3^0$	s	30
14	Silicon	$H_4SiO_4^0$	n	2.8×10^6
15	Phosphorus	$NaHPO_4^-$	n	62×10^3
16	Sulfur	SO_4^{2-}	c	898×10^6
17	Chlorine	Cl^-	c	19.35×10^9
18	Argon	Dissolved gas	c	0.62×10^6
19	Potassium	K^+	c	399×10^6
20	Calcium	Ca^{2+}	almost c	412×10^6
21	Scandium	$Sc(OH)_3^0$	(s + n)	0.7
22	Titanium	$Ti(OH)_4^0$	s + n	6.5
23	Vanadium	$NaHVO_4^-$	almost c	2.0×10^3
24	Chromium	CrO_4^{2-}(VI)	r + n	210
		$Cr(OH)_3^0$ (III)	r + s	2
25	Manganese	Mn^{2+}	s	20
26	Iron	$Fe(OH)_3^0$	s + n	30
27	Cobalt	$Co(OH)_2^0$?	s	1.2
28	Nickel	Ni^{2+}	n	480
29	Copper	$CuCO_3^0$	s + n	150
30	Zinc	Zn^{2+}	n	350
31	Callium	$Ga(OH)_4^-$	s + n	1.2
32	Germanium	$H_4GeO_4^0$	n	5.5
33	Arsenic	$HAsO_4^{2-}$(V)	r + n	1.2×10^3
		$As(OH)_3^0$(III)	r + s	5.2
34	Selenium	SeO_4^{2-}(VI)	r + n	100
		SeO_3^{2-}(IV)	r + n	55
35	Bromine	Br^-	c	67×10^6
36	Krypton	Dissolved gas	c	310
37	Rubidium	Rb^+	c	0.12×10^6
38	Strontium	Sr^{2+}	almost c	7.8×10^6
39	Yttrium	YCO_3^+	n	17
40	Zirconium	$Zr(OH)_5^-$	s + n	15
41	Niobium	$Nb(OH)_6^-$	almost c	0.37
42	Molybdenum	MoO_4^{2-}	c	10×10^3
43	Technetium	TcO_4^-	—	—
44	Ruthenium	RuO_4^-	?	< 0.005
45	Rhodium	$Rh(OH)_3^0$?	n	0.08
46	Palladium	$PdCl_4^{2-}$?	n	0.06
47	Silver	$AgCl_2^-$	n	2
48	Cadmium	$CdCl_2^0$	n	70
49	Indium	$In(OH)_3^0$	s	0.01
50	Tin	$SnO(OH)_3^-$	s	0.5

Appendix 7 *Continued*

Atomic number	*Element*	*Species*	*Type of distribution*[a]	*Oceanic mean concentration (ng kg^{-1})*
51	Antimony	$Sb(OH)_6^-$	s?	200
52	Tellurium	$Te(OH)_6^0$	r + s	0.05
		$TeO(OH)_3^-$	r + s	0.02
53	Iodine	IO_3^-	almost c	58×10^3
		I^-	(r + s)	4.4
54	Xenon	Dissolved gas	c	66
55	Cesium	Cs^+	c	306
56	Barium	Ba^{2+}	n	15×10^3
57	Lanthanum	$LaCO_3^+$	n	5.6
58	Cerium	$Ce(OH)_4^0$	s	0.7
59	Praseodymium	$PrCO_3^+$	n	0.7
60	Neodymium	$NdCO_3^+$	n	3.3
61	Promethium	—	—	—
62	Samarium	$SmCO_3^+$	n	0.57
63	Europium	$EuCO_3^+$	n	0.17
64	Gadolinium	$GdCO_3^+$	n	0.9
65	Terbium	$TbCO_3^+$	n	0.17
66	Dysprosium	$DyCO_3^+$	n	1.1
67	Holmium	$HoCO_3^+$	n	0.36
68	Erbium	$ErCO_3^+$	n	1.2
69	Thulium	$TmCO_3^+$	n	0.2
70	Ytterbium	$YbCO_3^+$	n	1.2
71	Lutetium	$LuCO_3^+$	n	0.23
72	Hafnium	$Hf(OH)_5^-$	s + n	0.07
73	Tantalum	$Ta(OH)_5^0$	s + n	0.03
74	Tungsten	WO_4^{2-}	c	10
75	Rhenium	ReO_4^-	c	7.8
76	Osmium	OsO_4^0	almost c	0.009
77	Iridium	$Ir(OH)_3^0$	s?	0.00013
78	Platinum	$PtCl_4^{2-}$	c	0.05
79	Gold	$AuOH(H_2O)^0$	c	0.02
80	Mercury	$HgCl_4^{2-}$	(s + n)	0.14
81	Thallium	Tl^+	c	13
82	Lead	$PbCO_3^0$	anth. + s	2.7
83	Bismuth	$Bi(OH)_3^0$	s	0.03
84	Polonium	$PoO(OH)_3^-$	s	—
85	Astatine	—	—	—
86	Radon	Dissolved gas	c	—
87	Francium	Fr^+	—	—
88	Radium	Ra^{2+}	n	0.00013
89	Actinium	$AcCO_3^+$	n	—
90	Thorium	$Th(OH)_4^0$	s	0.02
91	Protactinium	$PaO_2(OH)^0$	s	—
92	Uranium	$UO_2(CO_3)_3^{4-}$	c	3.2×10^3
93	Neptunium	NpO_2^+	—	—
94	Plutonium	$PuO_2(CO_3)(OH)^-$	(r + s)	—
95	Americium	$AmCO_3^+$	(s + n)	—

[a]c, conservative; n, nutrient-like; s, scavenged; r, redox sensitive; anth., anthropogenic.

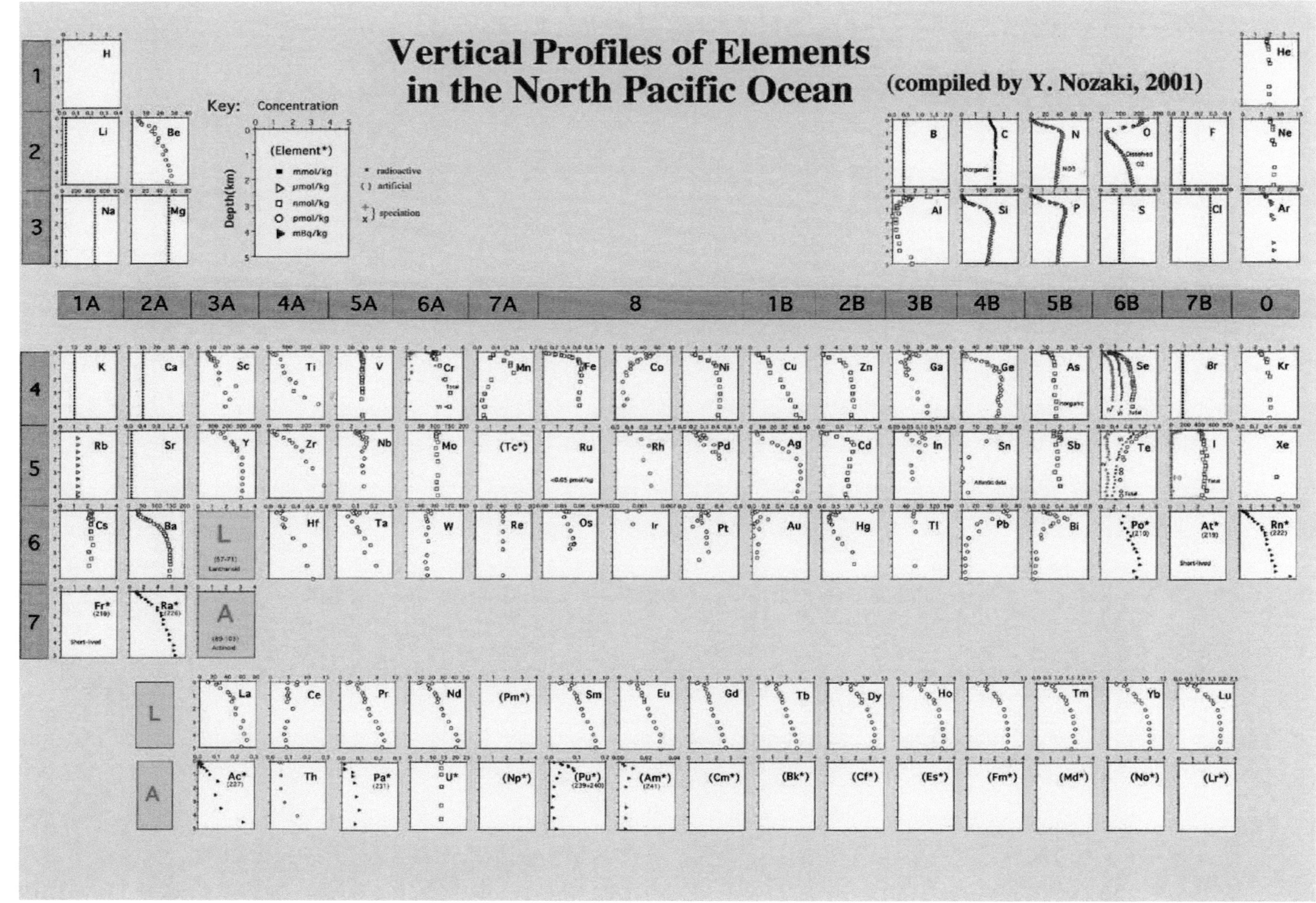
Vertical Profiles of Elements
in the North Pacific Ocean
(compiled by Y. Nozaki, 2001)
Key:
Concentration
Depth(km)
(Element*)
mmol/kg
µmol/kg
nmol/kg
pmol/kg
mBq/kg
* radioactive
( ) artificial
speciation

INDEX

Notes

Cross-reference terms in italics are general cross-references, or refer to subentry terms within the main entry (the main entry is not repeated to save space). Readers are also advised to refer to the end of each article for additional cross-references - not all of these cross-references have been included in the index cross-references.

The index is arranged in set-out style with a maximum of three levels of heading. Major discussion of a subject is indicated by bold page numbers. Page numbers suffixed by T and F refer to Tables and Figures respectively. vs. indicates a comparison.

This index is in letter-by-letter order, whereby hyphens and spaces within index headings are ignored in the alphabetization. For example, 'oceanography' is alphabetized before 'ocean optics.' Prefixes and terms in parentheses are excluded from the initial alphabetization.

Where index subentries and sub-subentries pertaining to a subject have the same page number, they have been listed to indicate the comprehensiveness of the text.

Abbreviations used in subentries

$\delta^{18}O$ - oxygen isotope ratio
DIC - dissolved inorganic carbon
DOC - dissolved organic carbon
ENSO - El Niño Southern Oscillation
MOR - mid-ocean ridge
NADW - North Atlantic Deep Water
POC - particulate organic carbon
SST - sea surface temperature

Additional abbreviations are to be found within the index.

A

B

E

F

G

H

I

M

N

O

P

S

U

Printed and bound by CPI Group (UK) Ltd, Croydon, CR0 4YY
03/10/2024
01040311-0013